GW01218193

Petroleum Systems of South Atlantic Margins

An outgrowth of the AAPG/ABGP Hedberg Research Symposium
Rio de Janeiro, Brazil, November 16–19, 1997

Edited by

Marcio Rocha Mello
and
Barry J. Katz

AAPG Memoir 73

Published jointly by

The American Association of Petroleum Geologists
and
PETROBRAS

Printed in the U.S.A.

Published 2000

ISBN: 0-89181-354-3

AAPG Association Editor: Neil F. Hurley
AAPG Geoscience Director: Robert C. Millspaugh
AAPG Publications Manager: Kenneth M. Wolgemuth
AAPG Bulletin and Book Editor: Carol E. Christopher
Copyediting and Production: Kathy A. Walker, EdTech, Renton, Washington
Printed by The Covington Group

This book and other AAPG publications are available from

The AAPG Bookstore
P.O. Box 979
Tulsa, OK 74101-0979
U.S.A.
Tel 1-918-584-2555 or 1-800-364-AAPG (U.S.A.)
Fax 1-918-584-0469 or 1-800-898-2274 (U.S.A.)
http://bookstore.aapg.org

Austrialian Mineral Foundation
AMF Bookshop
63 Conyngham Street
Glenside, South Australia 5065
Australia
Tel +61-8-8379-0444
Fax +61-8-8379-4634
www.amf.com.au/amf

Geological Society Publishing House
Unit 7, Brassmill Enterprise Centre
Brassmill Lane, Bath BA1 3JN
U.K.
Tel + 44-1225-445046
Fax +44-1225-442836
www.geolsoc.org.uk

Affiliated East-West Press Private Ltd.
G-1/16 Ansari Road Darya Ganj
New Dehli 110 002
India
Tel +91-11-3279113
Fax +91-11-3260538
e-mail: *affiliat@nda.vsnl.n*

Contents

AAPG and Petrobrás
wish to thank the following
for their generous contributions to

Petroleum Systems of South Atlantic Margins

BHP Petroleum (Americas) Inc.

Congo Basin Consortium, Department of Geosciences,
The Pennsylvania State University

ExxonMobil Upstream Research Company

Nigerian Association of Petroleum Explorationists

NNPC/Chevron Joint Venture

Repsol YPF

Texaco Inc.

Contributions are applied against the production costs of the publication, thus directly reducing the book's purchase price and making the volume available to a greater audience.

Preface

In 1995, the Research Committee of the American Association of Petroleum Geologists discussed and approved a Hedberg Research Symposium entitled "Petroleum Systems of the South Atlantic." This conference was held in November 1997 in Rio de Janeiro, Brazil. The convenors of this conference were the two editors of this volume, Marcio Rocha Mello and Barry Katz.

As a combined result of the opening of the oil industry in Brazil and a growing interest in deep-water regions along the west African continental margin, the meeting was highly successful. The symposium was attended by more than 200 geologists and geophysicists from four continents. A total of 48 oral and poster presentations were made covering nearly all of the sedimentary basins of the South Atlantic margins. Much of the information presented was described for the first time at this forum.

At the conclusion of the conference, a decision was made by many of the presenters and the two convenors to go forward with formal publication of a proceedings volume. A proposal was submitted to AAPG and approved in 1998. This volume represents the fruits of that process. It is believed by both volume editors that publication of many of the conference's presentations will have a significant impact on the region's future exploration through the release of much data that had previously been confidential.

The compilation and editing of this volume required the assistance and cooperation of many people. In addition to the numerous authors who contributed their time to prepare and revise each manuscript and their various organizations that released previously proprietary information, there was an army of reviewers who contributed their time to significantly improve the technical merit of each contribution. To the following reviewers we offer our thanks and that of the authors.

Andrew Bishop
George Claypool
Ed Colling
Brian Frost
Jim Granath
Randy Hunt
M. J. Kisucky
Louis Liro
Myron P. Maslanyj
Webster Mohriak
Thomas Oglesby
Alain Prinzhofer
David Reynolds
Roger Sassen
Jack Stonebraker
Carolyn L. Thompson-Rizer
Carlos Maria Urien
Lesli Wood
Ralph Burwood
Andrew Cohen
William C. Dawson
Wayne Gardiner
J. M. Guthrie
Garry Karner
Renato Kowsman
Anthony Lomando
Leslie B. Magoon III
J. M. Moldowan
Henrique Penteado
Michael Quearry
Vaughn D. Robison
Craig Schiefelbein
Lori L. Summa
Luis Antonio Trindade
Fred Weaver
Pinar Yilmaz
Richard Chuchla
Gary Cole
J. R. English
J. M. Gaulier
Michael Hoffman
Martin Keeley
Michael Kowalski
Susan Longacre
Steve May
P. K. Mukhopadhyay
Paul E. Potter
Rick Requejo
J. L. Rudkiewicz
Zvi Sofer
Peter Szatmari
Gregory Ulmishek
Edith Newton Wilson

J. B. L. Francolin of Petrobrás designed the dust cover.

Petrobrás, Texaco Inc., Chevron, and Exxon provided financial support for the Hedberg Research Symposium. The volume editors would also like to acknowledge the financial and logistical support of Petrobrás and Texaco Inc. during the compilation and editing of this volume.

Finally, the authors would like to acknowledge the support and patience of their wives, Lesley Rocha Mello and Terry Katz, without whom this volume would never have been completed.

Marcio Rocha Mello, Rio de Janeiro, Brazil

Barry J. Katz, Houston, Texas

About the Editors

Marcio Rocha Mello received his degree in geology from Brasilia University. In 1976, he joined Petrobrás as a well-site geologist working in the Recôncavo Basin located in northeastern Brazil. After working as a petroleum explorationist in the Ceará and Potiguar Basins in northeastern Brazil, in 1982 he became head of the Petrobrás Geochemistry Laboratory. In 1985, he entered Bristol University in England, where he received his Ph.D. in petroleum geochemistry. After returning to Petrobrás, he was promoted to head of the Geochemistry Section in 1988. Presently, he is the head of the Center of Excellence in Geochemistry at Petrobrás. During the last 10 years, Marcio has been applying the petroleum system approach to many Latin American sedimentary basins. In addition, Dr. Mello is president of the Brazilian Association of Petroleum Geologists, a recipient of AAPG's Distinguished Achievement Award, and an associate professor at several Brazilian universities.

Barry Katz received his Bachelor of Science degree in geology from Brooklyn College and his Ph.D. in marine geology and geophysics from the Rosenstiel School of Marine and Atmospheric Sciences, University of Miami, in 1979. He joined Texaco in 1979 and has held various positions within Texaco's research organization. Texaco's chairman named him a Texaco Fellow in May 1998 for his achievements in the applications of geochemistry to petroleum exploration.

Dr. Katz has served as an Associate Editor of the *AAPG Bulletin* and is chairman of the AAPG Committee on Research and a member of the AAPG Publication and Standing Technical Program Committees. He has also served as a member of the AAPG Marine Geology Committee. Barry is also a member of the JOIDES Pollution Prevention and Safety Panel and a member-at-large of the Organic Geochemistry Division of the Geochemical Society.

The major themes of his research have been related to the processes controlling the deposition of sedimentary organic matter and the characterization of organic facies and how this information can be applied to petroleum exploration. Over the past decade, much of his work has focused on lacustrine systems. He is currently a research consultant in Texaco's Upstream Technology Department in Houston, Texas.

Katz, B. J., and M. R. Mello, 2000, Petroleum systems of South Atlantic Marginal basins—an overview, *in* M. R. Mello and B. J. Katz, eds., Petroleum systems of South Atlantic margins: AAPG Memoir 73, p. 1–13.

Chapter 1

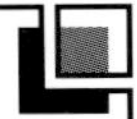

Petroleum Systems of South Atlantic Marginal Basins—An Overview

B. J. Katz

Texaco Group Inc.
Houston, Texas, U.S.A.

M. R. Mello

Petrobrás Research Center
Rio de Janeiro, Brazil

Abstract

The marginal basins along the South Atlantic have developed into one of the most active regions for petroleum exploration. The increase in the level of industry interest has resulted from numerous recent successes along both the eastern and western continental margins of the South Atlantic, the evolution of the region's political character, and an increase in the rate of permitting in deep and ultradeep waters. This heightened industry interest provided the rationale for a Hedberg Research Symposium on the petroleum systems of South Atlantic marginal basins.

Use of the petroleum system concept in South Atlantic marginal basins provides an effective means of classifying and characterizing the diversity of the systems and a way to aid in the selection of appropriate exploration analogs. South Atlantic marginal basins provide some of the best examples of how petroleum systems evolve through time with respect to both their levels of certainty and their areal and stratigraphic limits. A comparison of three basins from the South Atlantic—the Niger Delta, Lower Congo, and Campos Basins—provide examples of both the common traits that exist throughout the region as well as the differences among the individual basins. Differences are clear when the source and reservoir couplets are examined. In the Niger Delta, shallow water sands are charged from a Tertiary source with an important higher plant contribution. In the Lower Congo Basin, the lacustrine Bucomazi Formation (Neocomian–Barremian) charges primarily shelfal carbonates and sandstones. In the Campos Basin, the lacustrine Lagoa Feia (Barremian) Formation charges primarily Upper Cretaceous–Tertiary deep-water turbidite sandstones. A common trait appears to be the nature of the migration network which typically incorporates both normal faults and regional unconformities. The relative importance of vertical and lateral migration does differ among the basins, with vertical migration and short-distance lateral migration being dominant in the Campos and Gabon Basins and longer lateral distance migration being more important in the Niger Delta.

The basins of the South Atlantic also provide an excellent opportunity to examine the variety of lacustrine source rock settings. The depositional settings of these lakes range from freshwater to hypersaline. Source quality within these units also varies in response to their different depositional conditions and other factors that control or influence organic productivity and preservation.

INTRODUCTION

Over the past few years, the basins along the South Atlantic continental margins (Figures 1 and 2) have undergone a surge of expanded exploratory interest. This has in part been fueled by recent giant field discoveries such as the Bonga, Zafiro, N'Kossa, Girassol, Dalia, Rosa, and Kuito fields along the west African margin and the Roncador, Marlim, South Marlim, Barracuda, and Albacora Leste fields in the Campos Basin along the South American margin. Furthering this growth in activity has been an increase in political stabilization and other political changes along coastal Africa, denationalization and an end to the governmental exploration monopoly in the petroleum industry in Brazil and Argentina, and an overall increase in the rate of permitting activity in deep and ultradeep waters throughout the region. These factors have culminated in providing the industry with a unique opportunity to expand exploration into one of the last remaining high-potential frontier regions on earth.

Figure 1—Distribution of basins along the western South Atlantic margin.

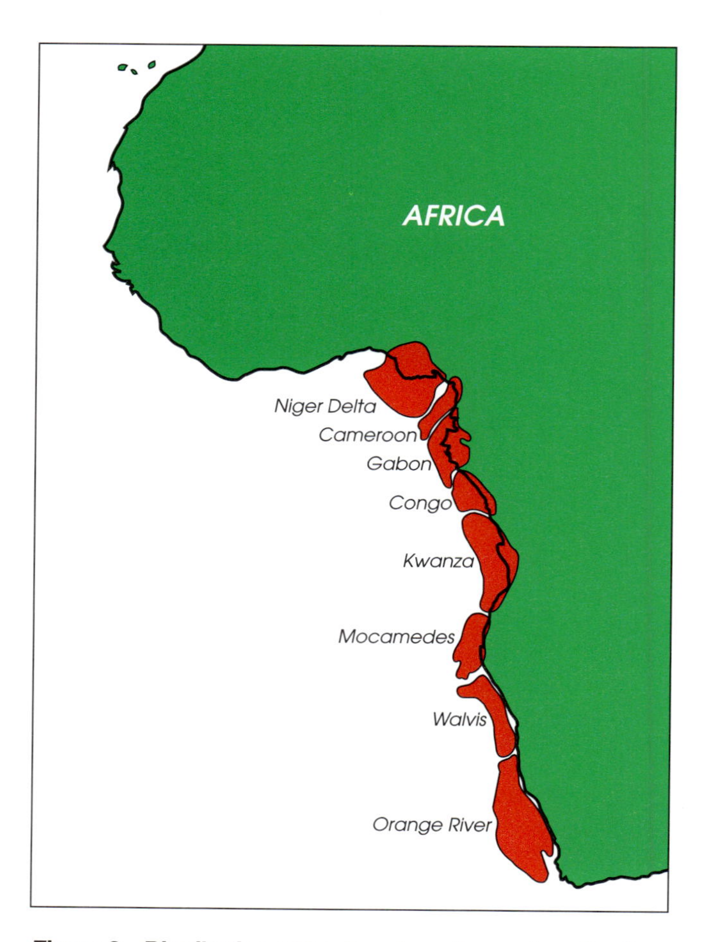

Figure 2—Distribution of basins along the eastern South Atlantic margin.

As activity has increased, it became clear through the numerous partnerships and working groups that there would be universal benefit to review and examine the current state of understanding of the petroleum systems present along the South Atlantic margins. Such an exercise would potentially provide the following:

1. A mechanism to relate the nature and effectiveness of the petroleum systems along the two conjugate margins;
2. A series of basin and/or system analogs as exploration extends into deeper waters and into basins where drilling density has been low;
3. Information on the general character and diversity of largely lacustrine-sourced petroleum systems that dominate much of the region's reserve base; and
4. A better understanding of the presence of marine petroleum systems in ultradeep waters of west Africa and possibly the southern Brazilian basins.

To accomplish these goals, a joint AAPG-ABGP Research Conference on the Petroleum Systems of the South Atlantic Margins was organized and held in Rio de Janiero, Brazil, on November 16–19, 1997. This volume represents a collection of papers presented at this meeting, with this chapter providing a general overview of the meeting's themes and conclusions. Most of the details are presented in the individual contributions.

PETROLEUM SYSTEM CONCEPT

The petroleum system is viewed in various ways by different organizations. Also, like many geoscience terms, although a rigorous definition has been proposed (Magoon, 1988), this definition has not been universally accepted or it has been modified to better fit individual organizational needs and circumstances. However, there is a common theme among the concept's numerous proponents. The *petroleum system* is viewed as a means of formalizing the relationship between the geologic elements in time and space that are required for the development of a commercial petroleum accumulation. The key elements necessary for the presence of a viable petroleum system are the source rock, reservoir, seal, trap, and necessary overburden for hydrocarbon generation to proceed. Furthermore, these individual elements must share the appropriate temporal and spatial relationships to permit hydrocarbons to accumulate and ultimately be preserved. The temporal, structural, and areal relationships among these different elements are often presented in a series of diagrams displaying the relative timing of depositional and other critical events such as trap development, along with cross-sections establishing stratigraphic and structural relationships among the different elements and maps defining the geographic limits of the system and its components.

The study of petroleum systems requires a paradigm shift from an emphasis on a basin's sedimentary fill to the fluids that it contains. The examination of a petroleum system differs further from a basin study in that it focuses on the hydrocarbons generated by a single source rock. Thus, by definition, a petroleum system is limited to a single geochemical oil family, although oil properties may vary as a function of alteration and thermal maturity. In fact, it has been proposed that a petroleum system be named for the source and dominant reservoir, that is, the reservoir unit in which the majority of the hydrocarbons are contained. In a basin, multiple oil families or petroleum systems may be present. One such basin is the Ogooué delta, where two distinct petroleum systems are present, each with its own source and primary reservoir units (Katz et al., Chapter 18, this volume). Within a basin, these different systems may overlap both stratigraphically and geographically, but often they are discrete and the presence of multiple petroleum systems extends both the stratigraphic and areal extent of a basin's exploratory opportunities.

In part, the interest in the use of the petroleum system concept as an exploration tool has developed as a direct consequence of the refinement of the geochemical and statistical tools necessary to establish the genetic relationships among oils (Mello et al., Chapter 4, this volume). For example, the combined use of stable isotopes and biomarker geochemistry has now become common. Geochemical data serve several roles in petroleum system assessment. First, these indices establish both similarities and differences among oils that define individual oil families or systems. Second, when samples of effective source rocks are present, they can establish genetic linkages with oil families. (*Source rocks* are defined as rocks that are thermally mature and contain sufficient quantities of the appropriate type of organic matter for petroleum generation to occur.) Third, the knowledge base associated with key geochemical marker (biomarker) compounds has expanded so that the chemical composition of an oil can be used to infer the nature of a source rock's depositional setting and in some cases chronostratigraphic position.

Geochemical information also establishes the level of certainty that exists for each system. Magoon (1988) suggested three levels of geochemical certainty: known, hypothetical, and speculative. In a *known system*, a definitive geochemical correlation can be established between source rock and oil family. In a *hypothetical system*, the presence of a source rock system can be established through its organic richness, generation potential, and kerogen character, but a definitive correlation with oils has yet to be established. Alternatively, the geochemical characteristics of an oil family can be used to establish the nature and stratigraphic position of a source even when no correlation has been established. In general, the lack of definitive correlation results either from a lack of sample availability or sample quality. In a *speculative system*, the necessary supporting data to establish the presence and character of a source are lacking and its presence is inferred through either geologic or geophysical data. The level of certainty of a system can and will be upgraded as access to new samples and data are made available.

The defining of a petroleum system along with its level of certainty can therefore be used to establish the exploratory risks within a region more effectively. Exploration risks increase with increasing geographic and stratigraphic displacement from a known petroleum system. This has been clearly shown by Demaison (1984) for the largely vertically migrating petroleum system present in the North Sea where exploration success is closely related to the areal limits of the generative (mature) Kimmeridge Clay.

A caveat, however, does exist in that the limits of a petroleum system are effectively controlled by drilling density. Numerous examples can be cited in which the limits of a petroleum system have been expanded significantly both stratigraphically and areally by additional drilling, as has been the case in the Campos Basin (Guardado et al., 1989). In the Campos Basin, initial exploration objectives were Albian carbonates limited mainly to the outer continental shelf and to water depths of less than 200 m. Today, these reservoirs account for less than 10% of the oil in place, with the largest discoveries having been made in Tertiary sandstone reservoirs in water depths typically greater than 500 m (Figure 3). In part, this expansion and redefining of the petroleum system has come about as a result of technological developments that have permitted exploration to be extended into deeper water. Note that, although the source for hydrocarbons has remained unchanged in the Campos Basin, the primary reservoir and hence petroleum system name has changed with time. Clearly, this is an example where the petroleum system has grown and evolved in a positive fashion.

At the same time, the lack of a key element or elements in a speculative petroleum system in either a basin or part of a basin results in exploratory failure. Such a critical flaw occurred in the Rawson Basin, Argentina (Otis and Schneiderman, Chapter 28, this volume) where drilling failed to prove the existence of two critical elements in the speculated petroleum system. Drilling revealed that both discrete reservoir sandstones and organic-rich shales were absent. Thus as known and hypothetical systems may grow and expand as information becomes available, speculative systems either will evolve into a hypothetical or known system depending on the nature of data that becomes available or will disappear if critical flaws associated with the absence of key elements are discovered through drilling.

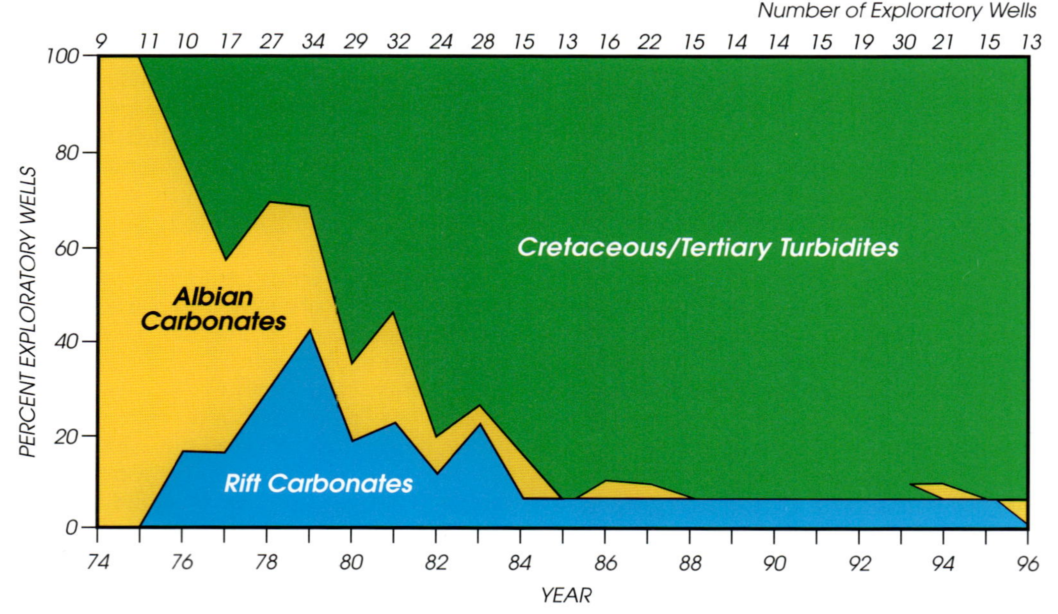

Figure 3—Evolution of primary drilling objectives in the Campos Basin. After Guardado et al. (1997).

SOUTH ATLANTIC CONJUGATE MARGINS

The sedimentary basins along the South American and African margins are traditionally considered to be independent basins. In part, this view has been fostered as a result of the lack of a common stratigraphic nomenclature and regional integration. An expanding knowledge base, however, allows both margins and their associated basins to be viewed as part of a larger single regional, structural, stratigraphic, and geochemical entity, upon which local characteristics can be overlain. This commonality among basins can be seen best in the geochemical characteristics of many of the oils. Available data show that although several oil types exist, there are common themes with respect to source rock depositional setting and effectiveness (Figure 4).

Such a view leads to a better understanding of the relationships among the different petroleum systems and the appropriateness of a basin or a specific petroleum system to be used as an analog for the less explored portions of the margins, including deep and ultradeep waters. For example, this approach demonstrates how the success in the Lagoa Feia–Carapebus system of the Campos Basin (which includes the Marlim field with 8.2 Bbbl of oil in place) (Guardado et al., 1989; Mello et al., 1994) can be used to explore more effectively in the more outer portions of the Kwanza and Lower Congo Basins where exploratory drilling is less mature. However, as Szatmari (Chapter 6, this volume) has pointed out, differences in basin development are such that the distribution of oil fields along the two margins should not be viewed as simple mirror images. It appears that the differences in basin evolution have created an asymmetric distribution of hydrocarbon resulting in an apparent alternation between productive and nonproductive portions in the two margins.

Although the presence of diapiric salt along much of the South Atlantic margin has complicated the imaging and analysis of synrift structures, merged geophysical data, including both seismic and satellite-derived gravity data, indicate a common tectonostratigraphic evolution for the two conjugate margins. Karner et al. (1997) noted that the South Atlantic margins formed as a result of rift propagation across the region during three rifting episodes: Berriasian–Hauterivian, Hauterivian–middle Barremian, and late Barremian–early Aptian. Each rifting event resulted in the formation of a series of basins. Fresh to brackish to saline lacustrine water conditions developed within these basins. Often these conditions led to both high levels of organic productivity and preservation. Each of these lacustrine basins was filled with an overall regressive package as a consequence of basin shallowing through in-filling.

The last rifting event resulted in the emplacement of oceanic crust and is consistent with the M0 magnetic anomaly along the Brazilian and Angolan margins. Two tectonic hinge zones have been defined along each of the margins. The inner hinge marks the limit of continental extension. The outer hinge consists of a series of segmented en echelon blocks. The en echelon character of these blocks plays a key role in controlling river drainage and hence the character of each rift segment's fill (Karner et al., 1997).

As noted above, the petroleum potential of the different marginal basins appears to be asymmetric, reflecting the nature of the South Atlantic's original rifting (Mello et

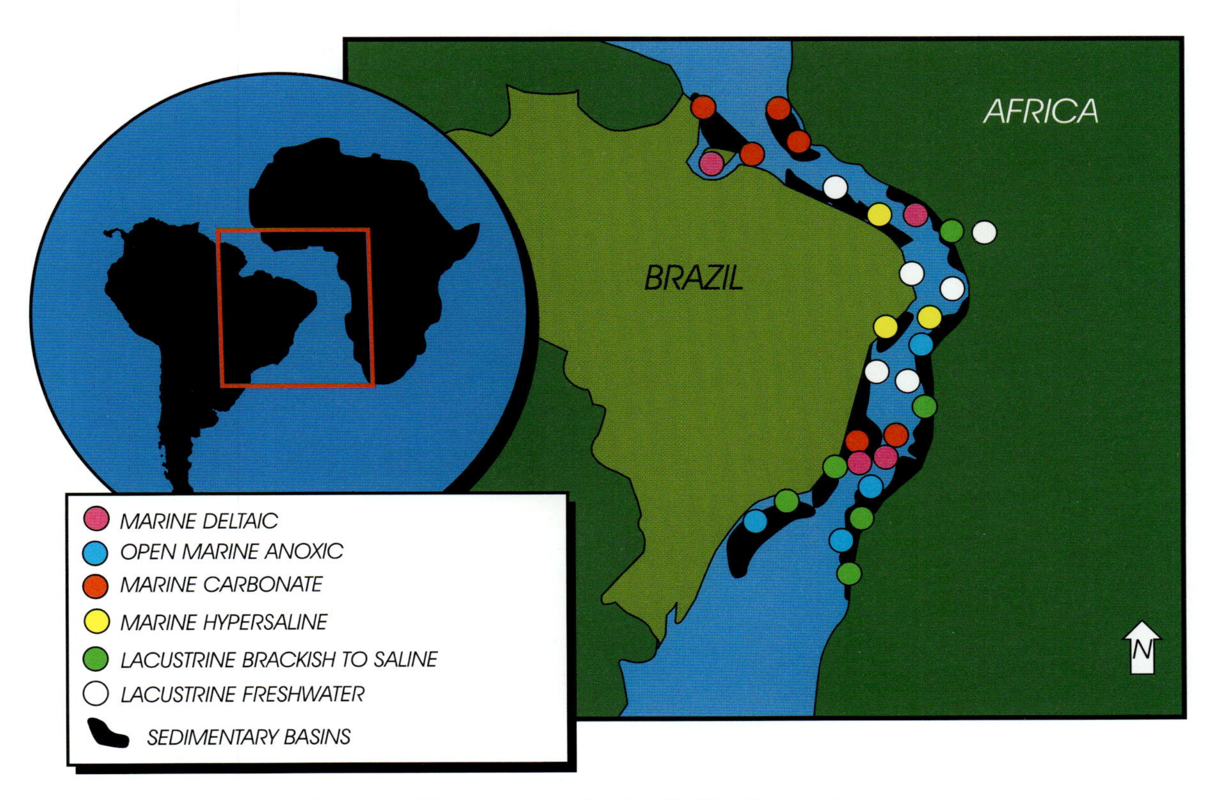

Figure 4—Oil types along the South Atlantic margins.

al., 1991). The asymmetric character of rifting along the margin is suggested by differences in the width of rifted continental crust. For example, the Campos and Santos Basins both display a wide zone of extended continental crust, while in the Bahia Sul and Sergipe–Alagoas Basins, the extended crust is restricted to a narrow band. The converse appears to be present along the African margin, where the wide zone of extended crust occurs in the north and the narrow band is present in the south. These differences in extent of the extended crust ultimately influence basin subsidence, basin fill, and thermal history, thus directly impacting the distribution and effectiveness of the margin's petroleum systems.

The onset of sea floor spreading was coincident with an apparent increase in salinity. Open marine conditions were established across the entire region by early Albian time. Initially, intermittent marine incursions from the south resulted in development of hypersaline conditions and deposition of thick evaporite sequences. These evaporitic conditions evolved into more normal marine conditions with time. During this initial open marine episode, the widespread deposition of carbonates occurred on both sides of the South Atlantic. From the middle Cretaceous onward, the deepening of the South Atlantic resulted in the accumulation of sediments over a wide bathymetric range (neritic to abyssal).

When viewed in broad terms, the stratigraphic columns of the South Atlantic marginal salt basins appear similar and can be divided into five *megasequences* (Figure 5): prerift, rift, transitional (or evaporitic), transgressive marine, and regressive marine. These megasequences are represented by time-equivalent successions along the two conjugate margins representing similar depositional environments. These five megasequences are separated by four major regional unconformities: a regional unconformity separating Paleozoic from Mesozoic (mainly Upper Jurassic) prerift sedimentary beds; a Lower Cretaceous unconformity between the prerift and rift sequences; a pre-Aptian unconformity that corresponds to the "breakup" unconformity; and a lower Tertiary unconformity associated with a Paleocene sea level drop (Henry et al., 1996). The Niger Delta, as a consequence of its different origin, does not fully share this common tectonostratigraphic succession (Doust and Omastsola, 1989).

Although the megasequence concept is useful, it is important to note that the distribution and geometry of the sedimentary fill differs from basin to basin. For example, along the Brazilian margin, the thickness of the drift sequence varies tremendously. The thickness of this sequence is much greater in the southern basins than in the northern basins. These differences can also be extended to the west African margin, where rift and prerift sedimentary rocks are exposed in outcrops close to Precambrian rocks in the Kwanza, Cabinda, and Gabon Basins. Such exposures do not exist within the conjugates along the southern Brazilian margin. Only in the

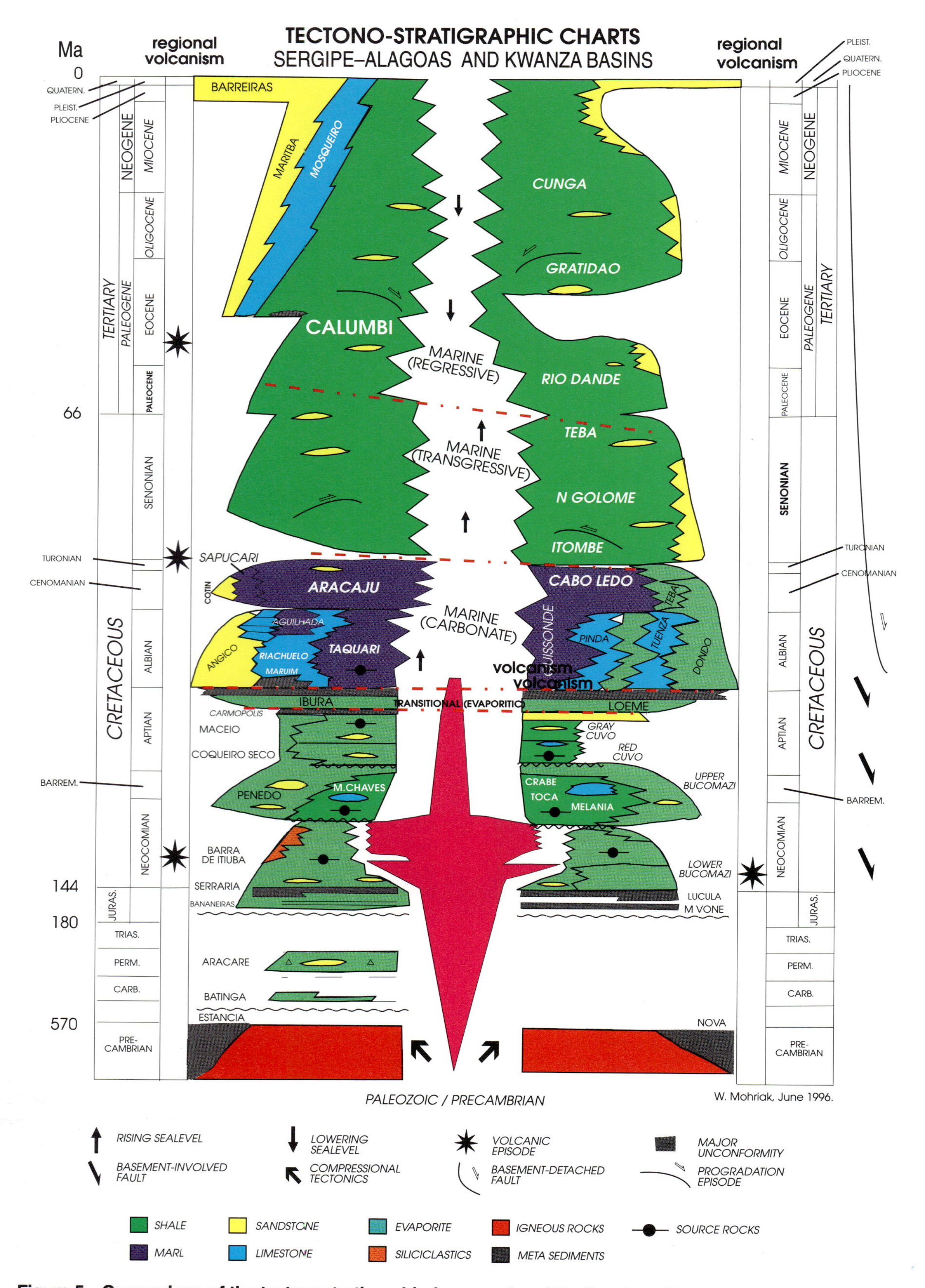

Figure 5—Comparison of the tectonostratigraphic frameworks of the Sergipe–Alagoas and Kwanza Basins.

Sergipe–Algoas and Recôncavo Basins can the prerift and rift sequence be mapped. These differences are partially a reflection of the previously noted asymmetric rifting along the margins (Mello et al., 1991).

SYSTEM DIVERSITY AND EFFECTIVENESS

The petroleum systems in the South Atlantic are highly diverse in character. This diversity is manifested in structural style, source rock depositional setting, reservoir lithology, and timing of fundamental events. The six distinct oil types present along the margin partially reflect the diversity of the petroleum systems. Each oil type reflects a unique source rock depositional environment that developed as the margins evolved. It is important to note, however, that individual oil accumulations may form through the mixing of various oil types. In stratigraphic succession (oldest first), the six oil types are briefly described here, following Mello et al. (1996).

Oil Types

The first oil type observed in the Ceará, Potiguar, Sergipe–Alagoas, Bahia Sul, Douala, Lower Congo, and Gabon Basins was derived from Neocomian source rocks deposited under lacustrine freshwater to brackish conditions (e.g., Pendência, Candeias, Bucomazi, and Melania Formations in the Potiguar, Bahia Sul, Lower Congo, and Gabon Basins, respectively). Much of this oil appears to be contained within Neocomian sandstones. Lateral and vertical migration distances appear limited. These oils are paraffinic (saturates >60%, paraffins >40%) and have low sulfur (<0.1%) and low naphthene contents and low gas–oil ratios (<60 m^3/m^3).

The second oil type, which is observed in the Santos, Campos, Espirito Santo, and Sergipe–Alagoas Basins and along much of west Africa from Angola to Cameroon, was derived from Neocomian–Barremian calcareous black shales deposited under brackish to saline conditions (e.g., Lagoa Feia, Bucomazi, and Melania Formations in the Campos, Lower Congo, and Gabon Basins, respectively). These oils are often reservoired in turbiditic sandstones ranging from Upper Cretaceous to Miocene and to a lesser degree in Cretaceous carbonates. Lateral migration distances away from the generative "kitchen" appear limited. Vertical migration associated with normal fault systems and through windows in the overlying salt may, however, be significant. These oils are naphthenic (saturates >60%, naphthenes >40%) and have low to medium sulfur contents (0.1–0.5%) and low to medium gas–oil ratios (<200 m^3/m^3).

The third distinct oil type identified in the Ceará, Potiguar, Sergipe–Alagoas, Bahia Sul, and Gabon Basins is derived from source rocks deposited under marine evaporitic conditions (e.g., the Paracuru, Upanema, Muribeca, and Gamba Formations in the Ceará, Potiguar, Sergipe–Alagoas, and Gabon Basins, respectively). The limited occurrence of this oil type along the west African margin is probably due to the margin's more limited volume of the evaporitic oil-prone source facies. Along the west African margin, this stratigraphic interval is dominated by shallow-water carbonates. Reservoirs for these oils range in age from late Aptian to early Eocene and are predominantly delta-front sandstones. These oils migrated vertically within the normal fault system and laterally along regional unconformities. They are naphthenic (saturates >60%, naphthenes >40%), with medium sulfur contents (0.1–0.5%) and low to medium gas–oil ratios (<200 m^3/m^3).

The fourth oil type is associated with accumulations in the Sergipe–Alagoas, Bahia Sul, Espirito Santo, Lower Congo, Kwanza, and Gabon Basins. These oils are derived from Albian–Aptian marls and calcareous black shales (e.g., Regencia and Iabe Formations of the Espirito Santo and Lower Congo Basins, respectively). The migration pathways of these oils appear torturous due to combined components of both vertical and lateral migration associated with transform fault systems. These oils are largely pooled in Upper Cretaceous shallow-water carbonates and sandstones. They are classified as naphthenic-aromatic (saturates <50%, naphthenes >25%), and they contain moderate quantities of sulfur (0.1–0.5 %) and display low gas–oil ratios (<60 m^3/m^3).

The fifth oil type is present in the Santos, Espirito Santo, and Sergipe–Alagoas Basins along the Brazilian margin and in Gabon, Lower Congo, and Kwanza Basins along the west African margin. These oils can be correlated to Cenomanian–Turonian marine black shales (e.g., Itajai Formation of the Santos Basin and the Azile and Anguille Formations of the Gabon Basin). They are found in reservoirs ranging from Albian carbonates to Eocene–Oligocene deep-water turbidites. They migrate principally along regional unconformities. These oils are characterized by their moderate sulfur (0.1–0.5%) and naphthene contents and by high gas–oil ratios (>200 m^3/m^3). In some basins along the Brazilian margin, the lack of significant accumulations of these oils appears to result from a lack of necessary overburden for generation to have occurred. Data indicate, however, that in some basins the viability of this potential petroleum system is enhanced and has not yet been tested by the limited drilling.

The sixth oil type has been recovered from the Espirito Santo and Lower Congo Basins and the Niger Delta. It can be correlated with rocks deposited within lower Tertiary marine deltaic deposits (e.g., Urucutuca Formation of the Espirito Santo Basin and the Akata Formation of the Niger Delta). These oils are pooled in Eocene and younger sandstone reservoirs. Hydrocarbon migration involves both vertical and lateral components that incorporate both growth fault systems and regional unconformities. These oils are paraffinic (saturates >60%, paraffins >40%) and have low sulfur contents (<0.1%) and high gas–oil ratios (>200 m^3/m^3).

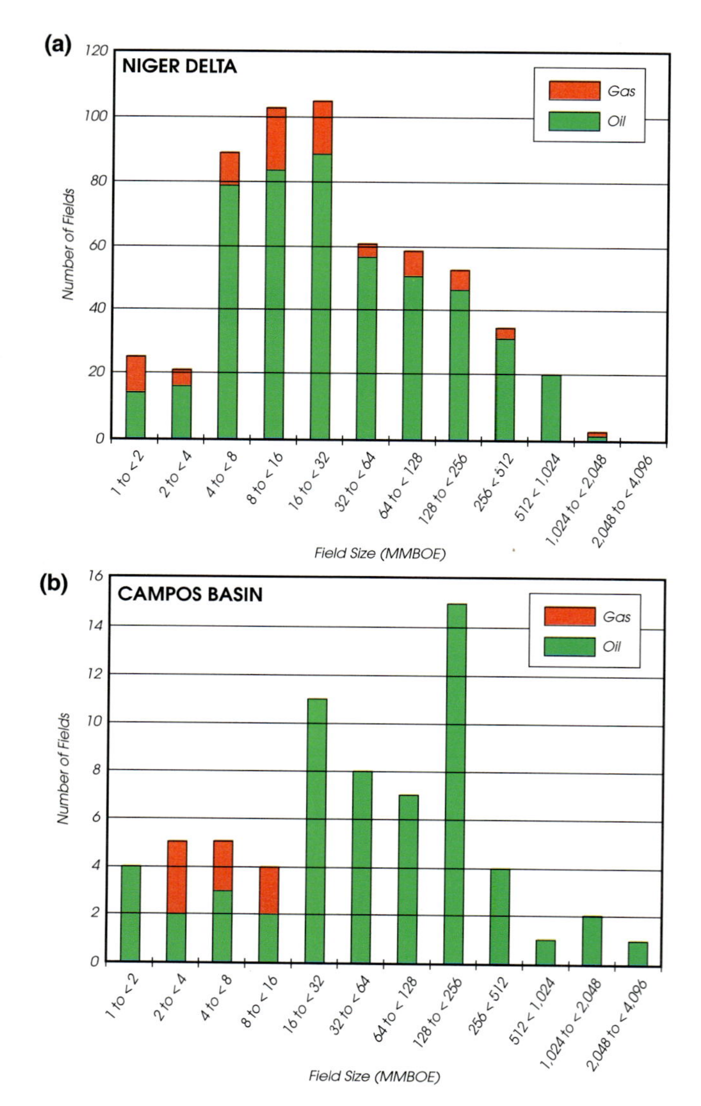

Figure 6—Field size distributions of (a) the Niger Delta and (b) the Campos Basin.

The nature and magnitude of South Atlantic petroleum system diversity can also be observed through a simple and brief comparison of the petroleum systems present in three of the South Atlantic basins: the Niger Delta, Lower Congo, and Campos Basins. Even before the details of the geology are examined, differences in these systems become apparent when field size distributions are examined (Figure 6a). For example, within the Niger Delta, the USGS (Klett et al., 1997) has estimated oil reserves of approximately 35 billion barrels contained in 491 oil fields, with a modal field size between 16 and 32 million barrels of oil equivalent. In contrast, they estimated current reserves in the Campos Basin at approximately 10 billion barrels of oil in 60 oil fields, with a modal field size between 128 and 256 million barrels of oil equivalent (Figure 6b).

Niger Delta

The Niger Delta began its development during the late Paleocene–Eocene as sediments progaded beyond the horst and graben associated with the breakup between Africa and South America (Doust and Omatsola, 1989). On the basis of newly generated geochemical data obtained on both oils and rocks, Haack et al. (Chapter 16, this volume) report the presence of three petroleum systems within the Niger Delta: a Lower Cretaceous lacustrine derived system, an Upper Cretaceous–lower Paleocene derived system, and a Tertiary (late Eocene–Pliocene) derived system. It is the Tertiary-sourced petroleum system that dominates the delta's reserve base. The Niger Delta oils display affinities typical of a mixed marine–terrestrial source rock system consistent with the recently identified source rocks. It is becoming apparent that this dominance by a Tertiary source rock is atypical among the South Atlantic basins, where Cretaceous lacustrine and marine source rocks are the general rule (Schiefelbein et al., Chapter 2, this volume).

Within the delta, all current hydrocarbon production is from the growth section above the regional décollement. The primary reservoirs are from the Oligocene–Miocene portions of the Agbada Formation, which is composed of a series of interbedded shallow marine and fluvial sandstones, siltstones, and claystones typical of most paralic settings. Barrier bar sandstones are typically more laterally continuous than those deposited in distributary channels. In general, the section is so sand rich that the reservoir is often considered a secondary risk to the seal (Doust and Omastsola, 1989).

The depositional setting of the Niger Delta reservoir is also different from the other two basins to be examined, which favor either deeper water clastics (Campos Basin) or carbonate reservoirs (Lower Congo Basin). Most of the hydrocarbon traps are structural and developed as a result of synsedimentary deformation, such as rollover anticlines associated with growth faults (Doust and Omatsola, 1989). Hydrocarbon generation is time transgressive across the delta as a result of the delta's progradation and appears to postdate the synsedimentary faulting. It also appears that hydrocarbon generation has been episodic, possibly as a result of depositional lobe switching. Petroleum generation is thought to have begun as early as late Eocene–early Oligocene in the most proximal parts of the delta (Ekweozor and Daukoru, 1994) and is still actively proceeding today. Hydrocarbon migration within the basin appears to be lateral along regional unconformities. Growth faults in the basin result in the redistribution of hydrocarbons, creating the common occurrence of multipay fields (Demaison and Huizinga, 1994).

Lower Congo Basin

The Lower Congo Basin began with Early Cretaceous rifting, and unlike the Niger Delta, its early rifting history has played a major role in controlling the distribution of hydrocarbons in the basin. The initial sediments were alluvial, fluvial, and lacustrine. Postrifting subsidence resulted in a marine incursion and deposition of the Loeme Evaporites. Following evaporite deposition, open marine conditions developed, leading first to the deposition of extensive carbonates that eventually evolved into a clastic-dominated sequence. Postrift structuring has been controlled by both salt tectonics and faulting within the Tertiary sequence.

As in the Niger Delta, three potential petroleum systems have been identified in the Lower Congo Basin: a Lower Cretaceous Bucomazi Formation sourced system, an Upper Cretaceous Iabe Formation sourced system, and Tertiary Malembo Formation derived system (Cole et al., Chapter 23, this volume; Schoellkopf and Patterson, Chapter 25, this volume). The system associated with the lacustrine Bucomazi Formation dominates in the basin. Bucomazi-derived oils are present in both presalt reservoirs (e.g., sandstones of the Lucala and Erva Formations and the Toca carbonates) and postsalt reservoirs (e.g., Pinda carbonates, shelf and beach sandstones of the Vermelha and Iabe Formations, and deep-water sandstones of the Malembo Formation). The Bucomazi Formation developed in synrift lakes. Burwood et al. (1995) noted that the source rock attributes of the Bucomazi are highly varied with respect to both quality and character. For example, Burwood noted that the basin fill portion of the Bucomazi Formation displays generally higher hydrocarbon generation potentials and different generation kinetics than the overlying sheet drape portion. In part, these differences appear to reflect changes in lake salinity and oxygen content through time. Hydrocarbon generation in the Bucomazi began during the Late Cretaceous and continues to the present day.

Unlike the Niger Delta, two other identified petroleum systems significantly contribute to the Lower Congo Basin's reserves. The outer shelf and slope facies of the Iabe Formation have also fed multiple reservoirs, including both the Pinda and Malembo Formations. The organic carbon content of the Iabe Formation averages 2–3%, ranging upward to ~5%. In the Iabe Formation, there is a general upward increase in source quality reflecting the transgressive character of the unit. Often, fields that have been charged from the Iabe have also received some contribution from the deeper lacustrine Bucomazi Formation. Generation from the Iabe began during the middle Miocene and continues to the present.

The source potential of the Malembo Formation is limited, with maximum organic carbon contents approaching about 4%. The limited source quality in the unit decreases up-section as a result of the regressive character of the unit. Generation from the Malembo Formation also appears to be restricted areally to areas where a thick Tertiary depocenter exists. As with the Iabe-derived oils, commercial accumulations that include Malembo-derived oils have also received a contribution from the Bucomazi Formation.

Campos Basin

As for the Lower Congo Basin, the Campos Basin also began with Early Cretaceous rifting and has a stratigraphic succession very similar to that of the Lower Congo, including an evaporitic facies in the upper part of the Lagoa Feia Formation, a carbonate sequence, followed by an open marine clastic sequence. Similarly, the basins share a similar postrifting structural history incorporating both salt tectonics and faulting. The amount of postrifting subsidence and sedimentation is less significant in the Campos than in the Lower Congo Basin.

The Campos Basin contains only a single active petroleum system. Although organic-rich Cenomanian–Turonian blacks shales are present, all of the known oils are derived from the synrift, brackish to saline, lacustrine black shales of the Barremian Lagoa Feia Formation. The organic carbon content of the Lagoa Feia Formation ranges typically from 2 to 6%, with values as high as 9%. Hydrocarbon generation potential typically exceeds 25 kg/ton rock and often exceeds 40 kg/ton. The geochemical attributes of the Lagoa Feia are also quite diverse, reflecting changes in depositional conditions such as an upward increase in lake salinity caused by intermittent marine incursions (Mello and Hessel, 1998).

Hydrocarbon generation from the Lagoa Feia began during the Late Cretaceous, but its apex was during the late Miocene and it continues to the present. This relatively late onset of hydrocarbon generation in comparison to the Lower Congo Basin is largely the result of differences in the amount of postrifting overburden that the two margins have received. The more limited amount of postrifting sediment also appears to explain the lack of effectiveness of the Cenomanian–Turonian source rocks. The Campos Basin oil reservoirs range in age from Neocomian to Miocene. Unlike the Lower Congo Basin, most of the Campos Basin reserves are located higher in the stratigraphic sequence in the Upper Cretaceous and Oligocene–Miocene deep-water, turbiditic sandstones. Migration is largely vertical through salt windows and along unconformities and faults.

Systems Summary

These three petroleum systems clearly reflect the different tectonostratigraphic positions of the effective source and primary reservoirs within the three basins. These three basins also show variations in system effectiveness that may exist mainly because of differences in thermal maturity resulting from variations in the amount of overburden. From this assessment, it is clear that, although the basins share many characteristics such as general tectonic and sedimentologic histories, local differences dominate the geologic details and control hydrocarbon generation and migration mechanisms. These differences must be clearly understood prior to attempting to extend an identified petroleum system or using one as an exploration analog.

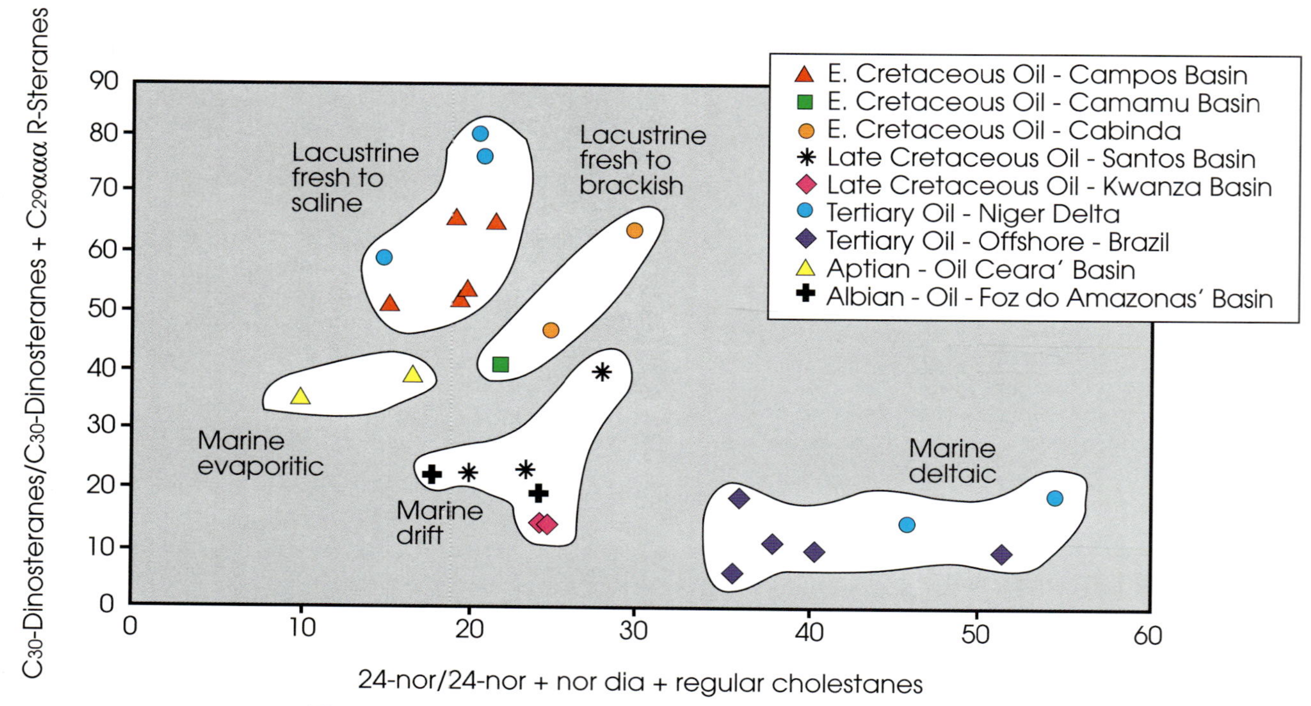

Figure 7—Geochemical differentiation of South Atlantic oil types.

SOUTH ATLANTIC LACUSTRINE SYSTEMS

Much of the available geologic and geochemical data indicate that a significant portion of the region's crude oil has been derived from lacustrine source rocks. Current estimates suggest that as much as 95% of discovered Brazilian oil and at least 15% of west African oil can be related to lacustrine source rocks. It should be noted that the west African reserves are heavily dominated by the Tertiary marine and deltaic sequence of the Niger Delta.

Although much of the known and potential future South Atlantic oils are derived from lacustrine source rocks, they display significant variability. These differences are largely the result of differences in lake water salinity (Figure 7) caused by a combination of factors such as climate changes and the magnitude and extent of marine incursions. Lake water salinity impacts the level and type of organic productivity, as well as its preservation and early diagenesis. These differences are reflected in the molecular chemistry of the source rock systems.

In addition to differences in oil chemistry and associated molecular chemistry in the source rocks, their bulk chemistry varies, reflecting their organic richness, oil proneness, and hydrocarbon generation kinetics. Available data indicate a long-term trend and oscillatory cycling of facies (Figures 8 and 9). The long-term trend is toward improved quality up-section, which maximizes prior to the development of lake-fill or drape facies. This trend in clearly shown by the Lagoa Feia Formation in the Campos Basin (Figure 8). The oscillatory changes are well represented in the Kwanza Basin (Figure 9). These cycles, which represent expansions and contractions of the lake body, most likely reflect climatic variations directly impacting water level and chemistry, as well as the amount of both organic and inorganic terrestrial input. Lake level low stands reflect intervals of poorer source quality largely due to poorer organic preservation.

Further examination of available data suggests that the areal distribution of oil-prone source material from saline versus freshwater lakes differs. The freshwater lakes appear to have been mainly restricted geographically to basin deeps. In contrast, the saline lakes tended to be shallower and areally more extensive, their distribution being strongly influenced by intermittent transgressions of marine waters from the south. These transgressions also introduced fresh nutrients into the systems, resulting is blooms of cyanobacteria. Such controls on lake character explain why the saline water oil type is volumetrically more important than the freshwater oil type and thus must be considered the dominant petroleum system in ultradeep water.

The complex anatomy of these synrift lacustrine systems has become apparent as additional geochemical and stratigraphic data become available. These data, however, reveal some consistent trends and patterns that can be effectively translated to less explored parts of the region as well as to other similar settings from elsewhere around the world.

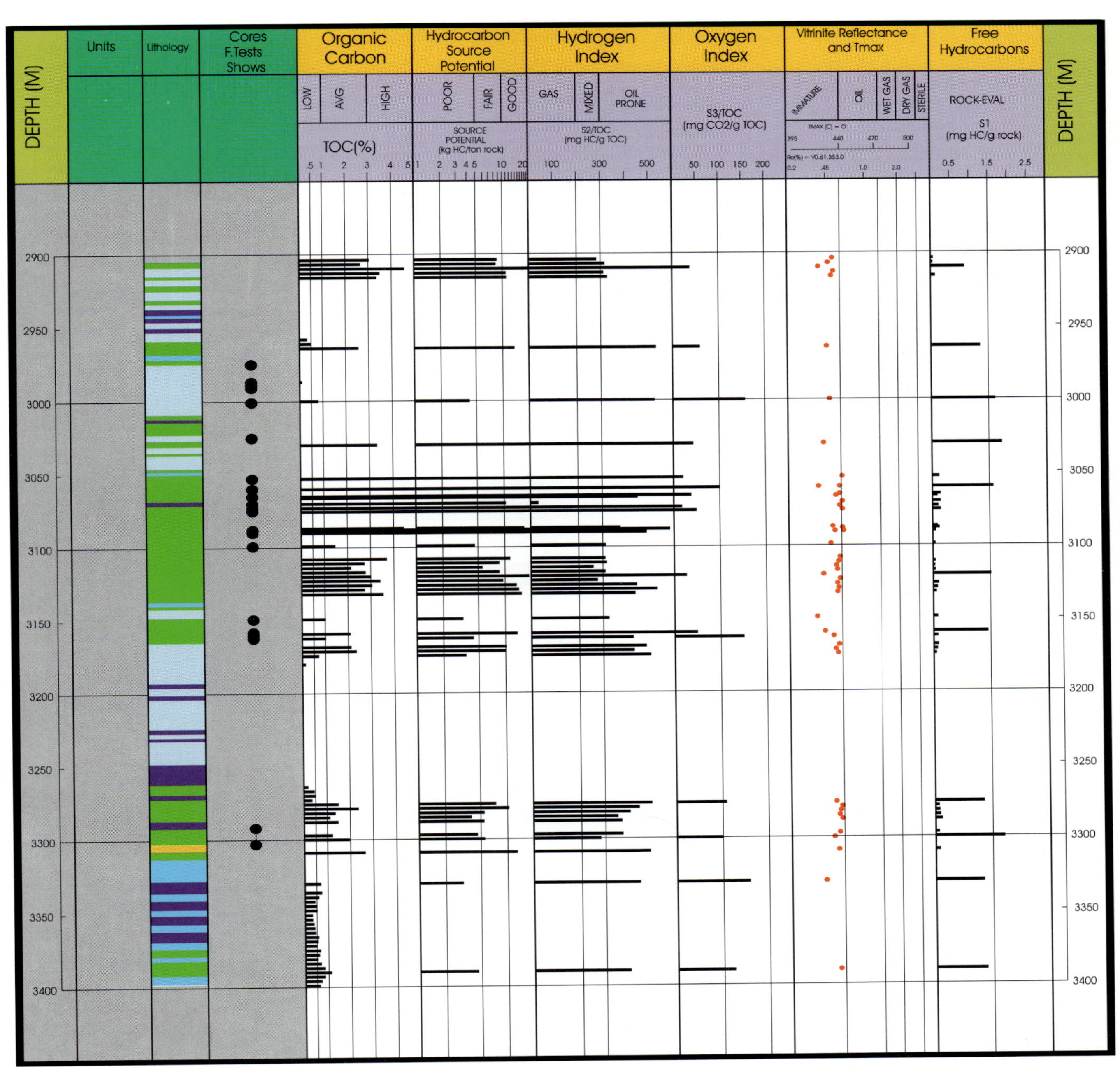

Figure 8—Geochemical log of the lacustrine sequence of the Campos Basin.

CONCLUSIONS

The petroleum system concept appears to be an effective tool in evaluating the remaining exploration potential of the South Atlantic. Such a multidisciplinary approach, focusing on the nature and distribution of hydrocarbon fluids, places the discovered hydrocarbons in a clear framework and provides a potential road map for future exploration. Available data reveal a general similarity between the South American and west African marginal basins with respect to their depositional sequences, including source rock facies and consequently oil types. Asymmetric rifting, however, has resulted in different sedimentary and subsidence histories, which has, in turn, created major differences in the distribution of oil types along the margins. As a result, the Brazilian marginal basins are dominated by lacustrine oils, whereas the west African margin is dominated by marine oils. Although marine source rocks are present in Brazilian marginal basins, their overburden is insufficient for generation to proceed. In constrast, sufficient overburden is present, at least locally, along the west African margin for generation to take place.

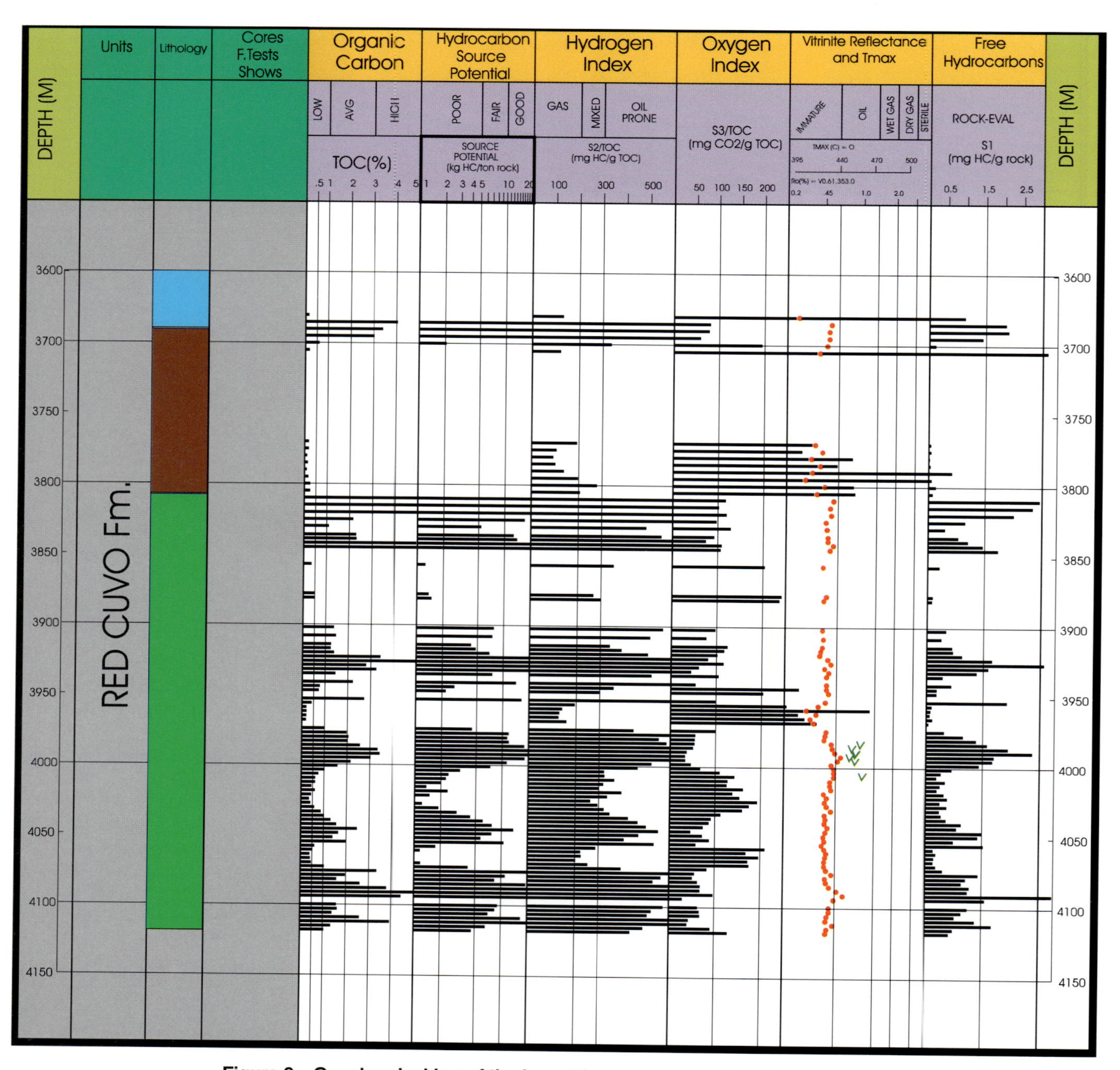

Figure 9—Geochemical log of the lacustrine sequence of the Kwanza Basin.

Consequently, when viewing the petroleum systems of the South Atlantic as possible exploration analogs, care must be taken. For example, this overview suggests that, although the deep-water success of the Brazilian Campos Basin associated with the Lagoa Feia–Carapebus system may be satisfactorily transferred to the west African margin, it does not appear that the Iabe–Pinda system of the Lower Congo will translate to the South American basins. This study also indicates that the synrift lacustrine facies is heterogeneous, reflecting differences in climate, water chemistry, and lake basin maturity. Yet, there do appear to be regular patterns that may be useful in the less well explored and documented parts of the South Atlantic as well as in other tropical to subtropical lacustrine settings.

Acknowledgments—*The authors wish to thank their respective managements for their permission to publish this introductory overview. Financial support for the joint AAPG/ABGP Research Conference on the Petroleum Systems of the South Atlantic Margins was provided by Petrobras, Texaco, Chevron, and Exxon.*

REFERENCES CITED

Burwood, R., S. M. De Witte, B. Mycke, and J. Paulet, 1995, Petroleum geochemical characterisation of the Lower Congo coastal basin Bucomazi Formation, *in* B. J. Katz, ed., Petroleum source rocks: Berlin, Springer-Verlag, p. 235–263.

Demaison, G., 1984, The generative basin concept, *in* G. Demaison and R. J. Murris, eds., Petroleum geochemistry and basin evaluation: AAPG Memoir 35, p. 1–14.

Demaison, G., and B. J. Huizinga, 1994. Genetic classification of petroleum systems using three factors: charge, migration, and entrapment, *in* L. B. Magoon and W. G. Dow, eds., The petroleum system—from source to trap: AAPG Memoir 60, p. 73–89.

Doust, H., and E. Omatsola, 1989, Niger Delta, *in* J. D. Edwards and P. A. Santogrossi, eds., Divergent/passive margin basins: AAPG Memoir 48, p. 201–238.

Ekweozor, C. M., and E. M. Daukoru, 1994, Northern delta depobelt portion of the Akata–Agbada(!) petroleum system, Niger Delta, Nigeria, *in* L. B. Magoon and W. G. Dow, eds., The petroleum system—from source to trap: AAPG Memoir 60, p. 599–613.

Guardado, L. R., L. A. P. Gamboa, and C. F. Lucchesi, 1989, Petroleum geology of the Campos basin, Brazil, a model for a producing Atlantic type basin, *in* J. D. Edwards and P. A. Santogrossi, eds., Divergent/passive margin basins: AAPG Memoir 48, p. 3–79.

Guardado, L. R., B. Wolff, and J. A. S. L. Brandao, 1997, Campos basin, Brazil, a model for producing Atlantic type basins: Proceedings of the 29th Annual Offshore Technology Conference, v. 3, p. 457–462.

Henry, S. G., W. U. Mohriak, and M. R. Mello, 1996, South Atlantic sag basins: new petroleum system components (abs.): AAPG Bulletin, v. 80, p. 1300.

Karner, G. D., N. W. Driscoll, J. P. McGinnis, W. D. Brumbaugh, and N. R. Cameron, 1997, Tectonic significance of syn-rift sediment packages across the Gabon–Cabinda continental margin: Marine and Petroleum Geology, v. 14, p. 973–1000.

Klett, T. R., T. S. Ahlbrandt, J. W. Schmoker, and G. L. Dolton, 1997, Ranking of the world's oil and gas provinces by known petroleum volumes: USGS Open-File Report 97-463, 207p.

Magoon, L. B., 1988, The petroleum system—a classification scheme for research, exploration, and resource assessment, *in* L. B. Magoon, ed., Petroleum systems of the United States: USGS Bulletin, v. 1870, p. 2–15.

Mello, M. R., W. U. Mohriak, E. A. M. Koutsoukos, and J. C. A. Figueira, 1991, Brazilian and west African oils: generation, migration, accumulation, and correlation: Proceedings of the 13th World Petroleum Congress, v. 2, 153–164.

Mello, M. R., E. A. M. Koutsoukos, W. U. Mohriak, and G. Bacoccoli, 1994, Selected petroleum systems of Brazil, *in* L. B. Magoon and W. G. Dow, eds., The Petroleum System—from Source to Trap: AAPG Memoir 60, p. 499–512.

Mello, M. R., E. A. M. Koutsoukos, W. U. Mohriak, and G. Bacoccoli, 1996. Petroleum systems of the Brazilian South Atlantic Margin (abs.): AAPG Bulletin, v. 80, p. 1314.

Mello, M. R., and M. H. Hessel, 1998, Biological marker and paleozoological characterization of the early marine incursion in the lacustrine sequences of the Campos Basin, Brazil (abs.): AAPG Annual Convention, Extended Abstracts, v. 2, p. A455.

Schiefelbein, C. F., J. E. Zumberge, N. C. Cameron, and S. W. Brown, 2000, Geochemical comparison of crude oil along the South Atlantic margins, *in* M. R. Mello and B. J. Katz, eds., Petroleum systems of South Atlantic margins: AAPG Memoir 73, p. 15–26.

Chapter 2

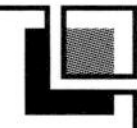

Geochemical Comparison of Crude Oil Along the South Atlantic Margins

C. F. Schiefelbein
J. E. Zumberge
N. C. Cameron
S. W. Brown

GeoMark Research, Inc.
Houston, Texas, U.S.A.

Abstract

The continental margin basins of Brazil and west Africa share very similar tectonostratigraphic units because of their proximity in Late Jurassic–Early Cretaceous time. As a result of the paleogeographic ties between the South American and the African plates, the oil habitat of the marginal basins of both continents can often be correlated. To better understand the petroleum systems along the South Atlantic margins, geochemical results obtained from the analysis of 290 oils from the major Brazilian and west African coastal basins were statistically evaluated to establish genetic relationships, distinguish source paleoenvironments and age, and identify different petroleum systems. A number of general oil families composed of genetically related oils were identified, several of which contain oils from both sides of the margin. Areas where oils of mixed provenance occur are also identified.

Presalt (Neocomian–Aptian) lacustrine oils can be separated into at least three distinct families and are present in the Congo, Lower Congo, and Benguela Basins of west Africa and the Recôncavo, Campos, Potiguar, Ceará, and Bahia Sul Basins of Brazil. Genetically related oils may be present in the Campos Basin and offshore central Angola and in the Recôncavo Basin and central Gabon and northern Angola.

Postsalt (Upper Cretaceous–Tertiary) marine oils are present in the Ivory Coast and the northern Gabon, Lower Congo, Kwanza, and Benguela Basins of west Africa and the Santos, Espírito Santo, and Sergipe–Alagoas Basins of Brazil. Oil mixing may have occurred in the Ivory Coast, Lower Congo, and Kwanza. Genetically related oils may occur in the Sergipe–Alagoas Basin and from northern Gabon. Oils from Nigeria, Benin, Equatorial Guinea, Cameroon, offshore northern Angola, and the Foz do Amazonas Basin of northern Brazil originated from Tertiary source rocks primarily composed of terrigenous organic matter deposited in fluvial deltaic or nearshore marine environments.

INTRODUCTION

The continental margin basins of Brazil and west Africa share very similar tectonostratigraphic units resulting from their proximity in Late Jurassic–Early Cretaceous time. As a result of the paleogeographic ties between the South American and African plates, the oil habitat of the marginal basins of both continents can often be correlated. The tectonic evolution and possible mechanistic causes have been discussed elsewhere (e.g., Lehner and De Riuter, 1977; Rabinowitz and La Brecque, 1979; Torquato and Cordani, 1981; Karner and Driscoll, 1997). In general, five stages of continental margin basin development can be described (Horn, 1980): the prerift intracratonic stage, the continental rift stage, the evaporite stage, the postevaporite transgressive stage, and the postevaporite regressive stage.

The Brazilian continental margin extends over 8000 km from 5° N latitude to 35° S latitude. More than a dozen Mesozoic–Cenozoic basins have been described, including (from north to south) Foz do Amazonas, Barreirinhas, Ceará, Potiguar, Sergipe–Alagoas, Recôncavo, Alameda, Bahia Sul, Espírito Santo, Campos, Santos, and Pelotas Basins. The west African marginal basins are located between the Walvis Ridge and the Guinea Rise.

Substantial oil production has been established along both sides of the South Atlantic margin. Most of the oil from both sides of the margin is believed to have been generated from lacustrine sediments deposited during Neocomian rifting of Africa and South America (e.g., Brice et al., 1980; Mello et al., 1988a, b; Burwood, 1997). Subsequent opening and invasion of marine seas in a restricted basinal setting allowed for deposition of a thick salt layer during Aptian time. Postsalt sediments with liquid hydrocarbon source potential were laid down in shallow marine and fluvial-deltaic environments as sea floor spreading continued into the Late Cretaceous (e.g., Mello et al., 1988a, b; Teisserenc and Villemin, 1990; Sofer, 1993; Burwood, 1997; Katz, et al., 1997).

The purpose of this investigation is to geochemically compare and contrast west African oils from presalt and postsalt strata with oils produced from offshore basins of Brazil. This geochemical comparison is an excellent way of identifying, evaluating, and comparing the various petroleum systems present in the region. It is accomplished by first establishing the number of compositionally distinct oil types or families. Geochemical data from oils are interpreted in such a way that oil/oil correlations are made and source rock inferences are proposed. The source inferences are possible because the geochemical characteristics of an oil reveal information about the source age and the paleoenvironmental conditions of its deposition (Moldowan et al., 1985; Peters et al., 1986; Powell, 1986; Zumberge, 1987; ten Haven et al., 1988; Mello et al., 1988a, b; Volkman, 1988; Pu et al., 1991). Also, a regional oil study approach is particularly useful because the generated and expelled oil can be considered to be representative of the "average" of a given source rock section, which can exhibit extreme variations in its geochemical characteristics (Moldowan et al., 1992; Katz et al., 1997).

Similarities and differences in the geochemistry of these oils relate to common or dissimilar source strata and have implications as to the tectonic control of source rock depositional environments and the timing of rifting and sea floor spreading. This investigation is possible because the suspected source sedimentary section for the Brazil and west African crude oils are stratigraphically and depositionally similar. This is particularly true for the oils derived from presalt lacustrine source rocks deposited during the Early Cretaceous (Neocomian–Barremian), a time when the South American and African plates were in close proximity. However, it is more difficult to establish correlation between oils derived from postsalt marine source rocks deposited after the two continents drifted apart and the depositional regimes were subjected to a higher degree of organic diversity.

Information obtained from the geochemical characterization of representative crude oils can be used to help define different petroleum systems active in a region. A thorough understanding of oil geochemistry is an integral component of the petroleum system concept. However, a complete definition of petroleum system implies the knowledge of many other varied factors controlling the occurrence of oil in nature, such as maturity of the source rock, expulsion, migration, accumulation, and retention (Demaison and Huizinga, 1991; Magoon and Dow, 1994). Oil geochemistry is a fundamental component of regional exploration and production programs when integrated within a meaningful geologic framework, including an understanding of the regional stratigraphic relationships, regional structural fabric, and evolution of structural elements.

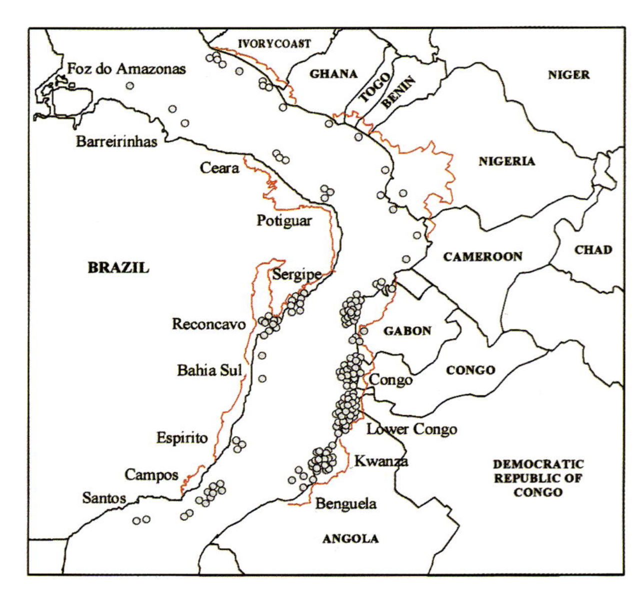

Figure 1—Map showing locations of various crude oils from Brazilian and west Africa basins included in study.

APPROACH

To better understand the petroleum systems along the South Atlantic margins, 290 oils from the major Brazilian and west African coastal basins have been geochemically characterized. Crude oil locations are shown in Figure 1. The detailed analytical program included measurements of bulk parameters, whole oil gas chromatography, stable carbon isotope composition of C_{15+} hydrocarbon fractions, and biomarker distributions using gas chromatography–mass spectrometry (GC-MS). Figure 2 shows terpane (m/z 191), sterane (m/z 259), and triaromatic dinosterane (m/z 245) mass chromatograms representing marine and lacustrine oil from along the South Atlantic margin.

Previous investigators have demonstrated the value of certain geochemical information on source inferences, particularly for differentiation between lacustrine- and marine-derived oils (Moldowan et al., 1985; Zumberge, 1987; Mello et al., 1988a, b; ten Haven et al., 1988; Schiefelbein et al., 1997). It is generally accepted that lacustrine-derived oils can be distinguished from marine-derived oils based on several different biomarker parameters, including higher proportions of C_{26} relative to C_{25} tricyclic terpanes, reduced abundances of extended pentacyclic terpanes, and low proportions of steranes relative to hopanes. It has also been observed that lacustrine-derived oils contain an abundance of an unknown

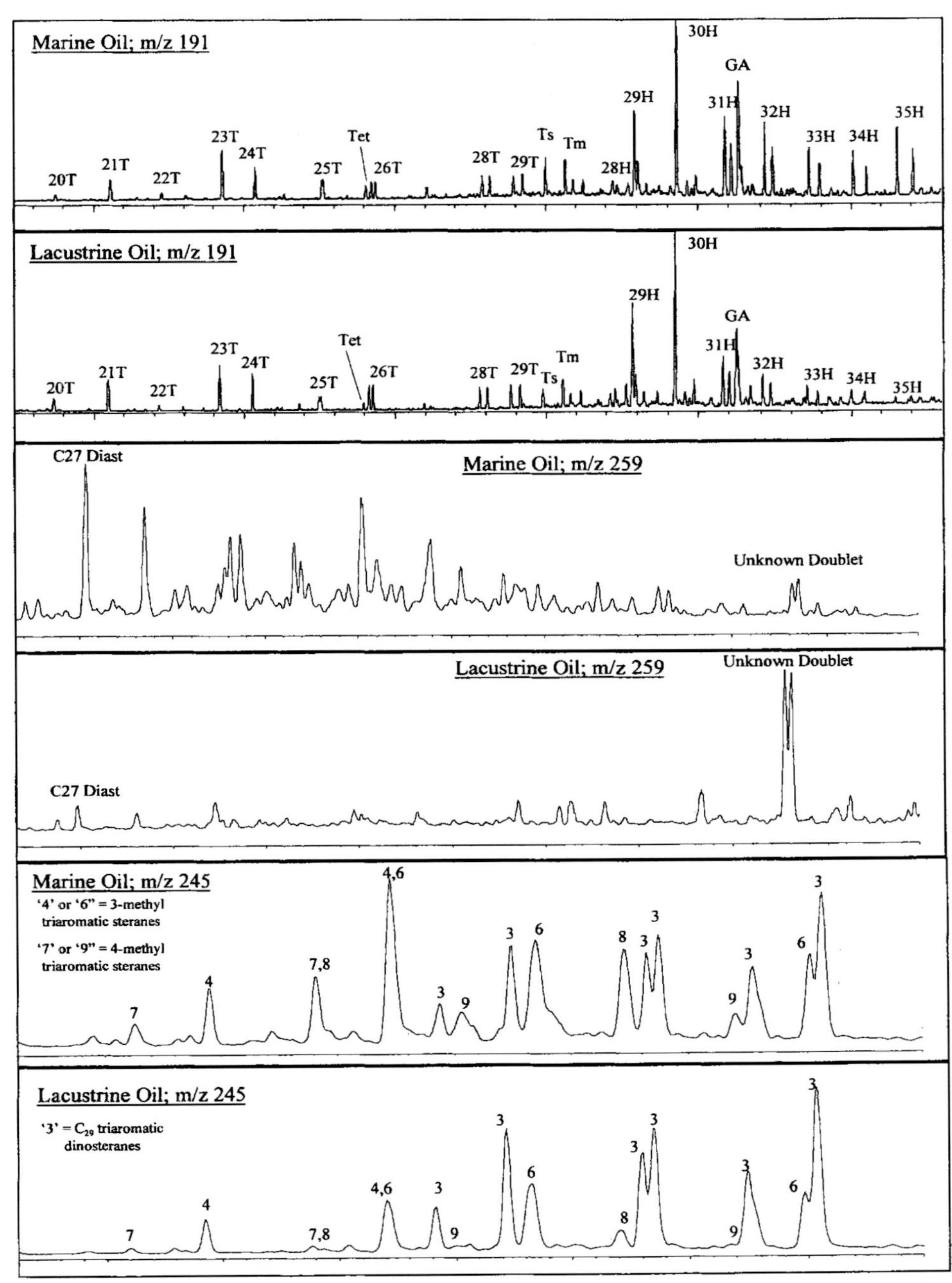

Figure 2—Representative terpane (*m/z* 191), rearranged sterane (*m/z* 259), and triaromatic dinosterane (*m/z* 245) mass chromatograms for marine and lacustrine oils from selected Brazilian and west Africa basins.

doublet (*m/z* 259) (Schiefelbein et al., 1997) relative to C_{27} diasteranes and that they exhibit different distributions of triaromatic dinosteranes (*m/z* 245) (Moldowan, et al., 1995). Several investigators (Didyk et al., 1978; ten Haven et al., 1985, 1988; Mello et al., 1988a, b) have shown that the pristane/phytane isoprenoid ratio can be used as a source indicator, but it probably also reflects the relationship between isoprenoid precursors and the chemistry of the environment. Sofer (1984) and Katz and Mertani (1989) have demonstrated that stable carbon isotopic compositions of lacustrine-derived oils are often quite variable because of multiple possible variations in the carbon budget and carbon cycle in lacustrine settings.

The analytical data were compared using multivariate statistical techniques such as hierarchical cluster analysis (HCA) (Figure 3) and principal component analysis (PCA) (Figure 4) using the commercial software program *Pirouette*. The primary objective of the statistical analyses was to reduce the dimensionality of the information to a few important components that best explain the variation

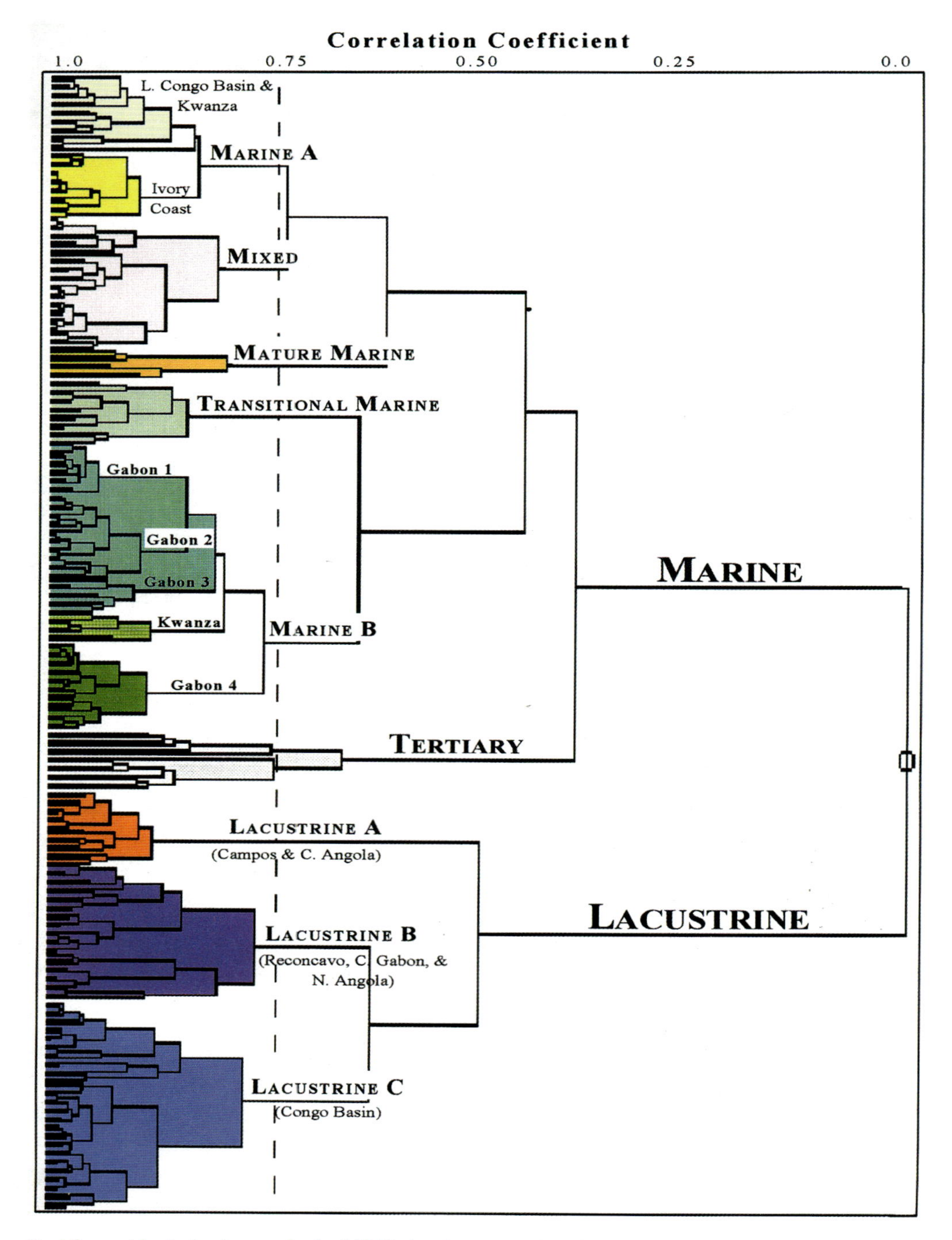

Figure 3—Hierarchical cluster analysis (HCA) dendrogram showing differentiation of identified oil families.

in the data. The geochemical data used in the statistical analyses are primarily source dependent. The 20 variables used in the present study include the pristane/phytane ratio, the stable carbon isotopic compositions of the C_{15+} saturate and aromatic hydrocarbon fractions, and 15 biomarker ratios. These ratios include the distribution of the 14β-, 17β-C27, -C28 and -C29 steranes (based on the abundance of the 20*R* stereoisomers measured from *m/z* 218), C_{31} hopane/C_{30} hopane, total steranes/total hopanes, C_{26}-tricyclic/C_{27}-pentacyclic terpanes, gammacerane/C_{30} hopane, C_{29} hopane/C_{30} hopane, and oleanane/C_{30} hopane. Ratios based on the distribution of tricyclic and tetracyclic terpanes were also used: C_{19}-tri/C_{23}-tri, C_{23}-tri/C_{24}-tri, C_{26}-tri/C_{25}-tri, and C_{24}-tetra/C_{26}-tri. Finally, the distribution of the C_{29}-triaromatic dinosteroids and triaromatic 3- and 4-methyl steroids measured from *m/z* 245 (Moldowan et al., 1995) were used, and the abundance of an unknown doublet relative to a C_{27} diasterane (measured from *m/z* 259) was also included. In this manner, 72% of the data variance was described by five principal components.

Using this approach, we identified a number of general oil families, differing from each other in terms of source paleoenvironment, lithology, and age. Figure 5 shows the distribution of these oil families. Gas chromatographic, stable carbon isotopic, and biomarker data are compared in Figures 6 through 9. Note that in Figures 3 through 9, oils from each of the different basins are represented by symbols of different colors and shapes.

RESULTS AND DISCUSSION

By combining several different source-dependent geochemical criteria and applying statistical analyses (HCA in Figure 3 and PCA in Figure 4), a number of different oil types (or families) with similar geochemical characteristics were identified. Clustering of the oil types (based on HCA) is shown in the form of a dendrogram in Figure 3, and their geographic distributions are shown in Figure 5. The South Atlantic margin oils can be separated into two broad groups: "lacustrine" and "marine" (Figure 3). Further detailed examination of each broad group of oils typically resulted in the establishment of subgroups of oils presumably representing specific source environments, such as fresh versus saline lacustrine depositional conditions, as well as oils of mixed provenance (Figure 4b).

The broad group of oils positioned at the top of the cluster analysis dendrogram (labeled as marine in Figure 3) typically has positive principal component 1 (PC1) values and variable PC2 values (Figure 4b), reflecting a diversity of oil types. For example, oils of mixed lacustrine–marine origin and oils derived from source rocks primarily composed of terrigenous organic matter deposited in fluvial-deltaic or nearshore marine environments also belong to this group. The broad group of oils in the lower cluster of the dendrogram is classified as lacustrine. These oils have negative PC1 values and variable PC2 values (Figure 4b), and based on HCA, can be further separated into three subgroups (lacustrine A, B, and C).

It is interesting to note that in several instances, lacustrine oils from both sides of the Atlantic margin cluster together in one family (Figure 5). This may suggest that these oils are genetically related, having originated from the same source bed or from different source rocks composed of similar organic matter.

Lacustrine Oils

In general, South Atlantic margin oils designated as "lacustrine" originated from source rocks composed mainly of algal kerogen (Powell, 1986). These oils exhibit the following general characteristics: low-sulfur content (<0.3 wt. %), enrichment in saturated hydrocarbons, pristane/phytane ratios between 1.5 and 3.0 (Figure 6), variable stable carbon isotopic compositions, $C_{26} > C_{25}$ tricyclic terpanes (Figure 7), low proportions of extended pentacyclic terpanes relative to C_{30} hopanes, low sterane/hopane ratios (Figure 8), and enrichment in C_{27}-$\alpha\beta\beta$ steranes (*m/z* 218). Lacustrine oils are often enriched in C_{29} triaromatic dinosteranes (Figure 9) and display a "simple" distribution of triaromatic dinosteranes (C_{29}-triaromatic dinosteranes are labeled as 3 in the *m/z* 245 chromatogram in Figure 2). Lacustrine oils also contain variable amounts of 4-methyl steranes (*m/z* 231) and have high concentrations of an unknown doublet relative to a C_{27} diasterane (*m/z* 259) (Figure 2). This relationship has also been observed for lacustrine-derived oils from the Far East (Schiefelbein et al., 1997).

The cluster labeled lacustrine A (Figure 5) includes oils from the Campos, Lower Congo, and Benguela Basins. These oils are typically from postsalt reservoirs and are characterized by intermediate to heavy isotopic compositions, variable amounts of gammacerane, and abundant 4-methylsteranes (*m/z* 231). According to Mello et al. (1988a, 1988b) and Guardado et al. (1997), calcareous black shales of the Aptian–Barremian Lagoa Feia Formation deposited in a brackish–saline lacustrine environment are responsible for the Campos Basin oils. For this type of oil from Angola, Burwood (1997) assigns the source as the Infra-Cuvo Formation (Neocomian–Barremian), a calcareous shale deposited in a brackish–saline lacustrine environment.

Another possible source for some of the lacustrine A oils has been suggested by several investigators (Brumbaugh et al., 1994; Henry et al., 1995; Abreu et al., 1997; Wilson et al., 1997). The lacustrine A oils that have the heaviest isotopic compositions and the greatest abundances of 4-methylsteranes may have originated from presalt transitional sediments with increasing marine influence in their upper section. This type of source rock may have been deposited in early thermal subsidence sag basins offshore from Angola, and it may have geochemical affinities with the upper section of the Lagoa Feia Formation in the Campos Basin (Mello and Hessel, 1998). These source rocks may have evolved from isolated deep rift to shallow saline lakes, culminating with the infill of the sag basin by large saline lakes to a marginally marine-restricted gulf.

The cluster labeled as lacustrine B (Figure 5) includes oils from the Recôncavo, Potiguar, Bahia Sul, and Ceará Basins from offshore Brazil and central Gabon and northern Angola. These oils are characterized by low-sulfur content, intermediate to light stable carbon isotopic compositions, low sterane/hopane ratios, and variable amounts of gammacerane and 4-methyl steranes. Lacustrine B oils from Brazil probably originated from algal source rocks deposited in similar fresh to saline lacustrine environments (Mello et al., 1988a, 1988b). Lacustrine B oils from northern Angola may have originated from late Barremian–Aptian synrift II shales deposited in brackish–saline ephemeral lacustrine environments (Burwood, 1997).

(a)

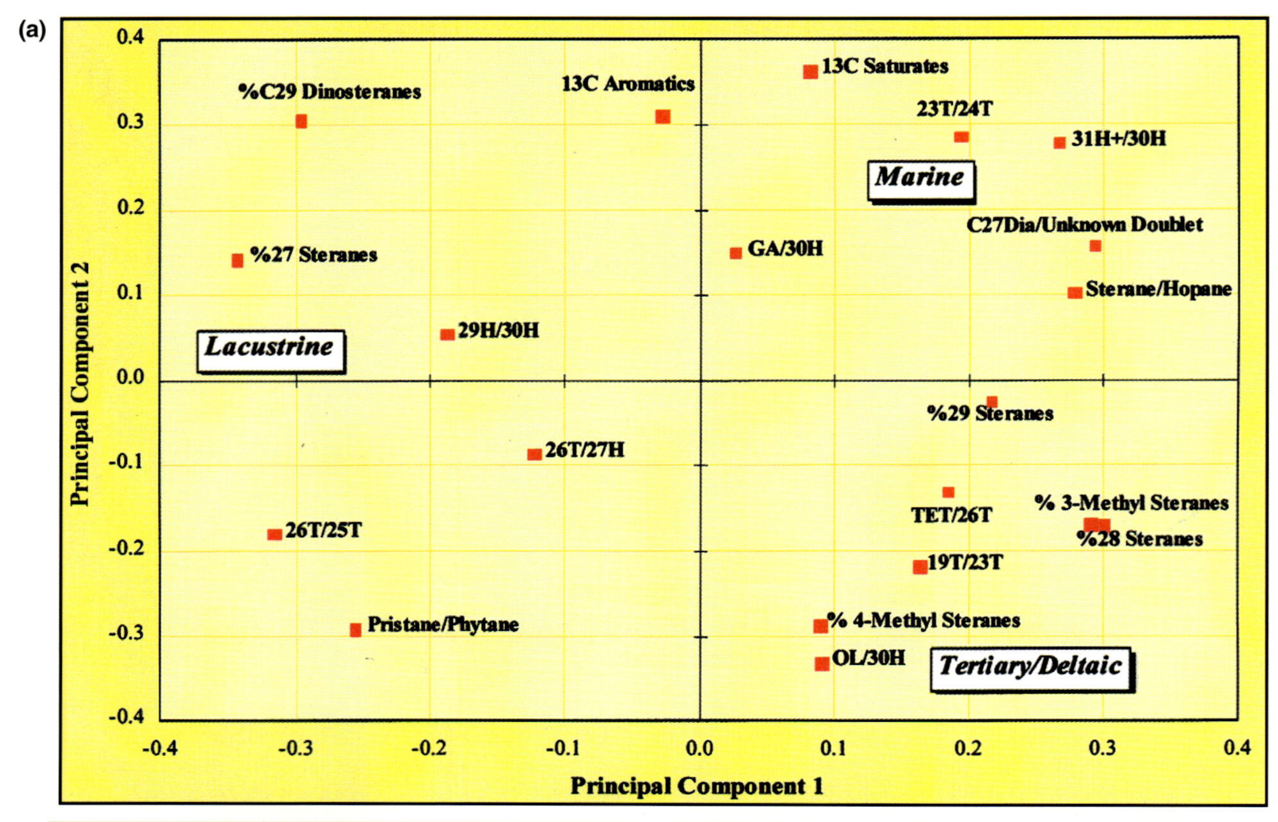

(b)

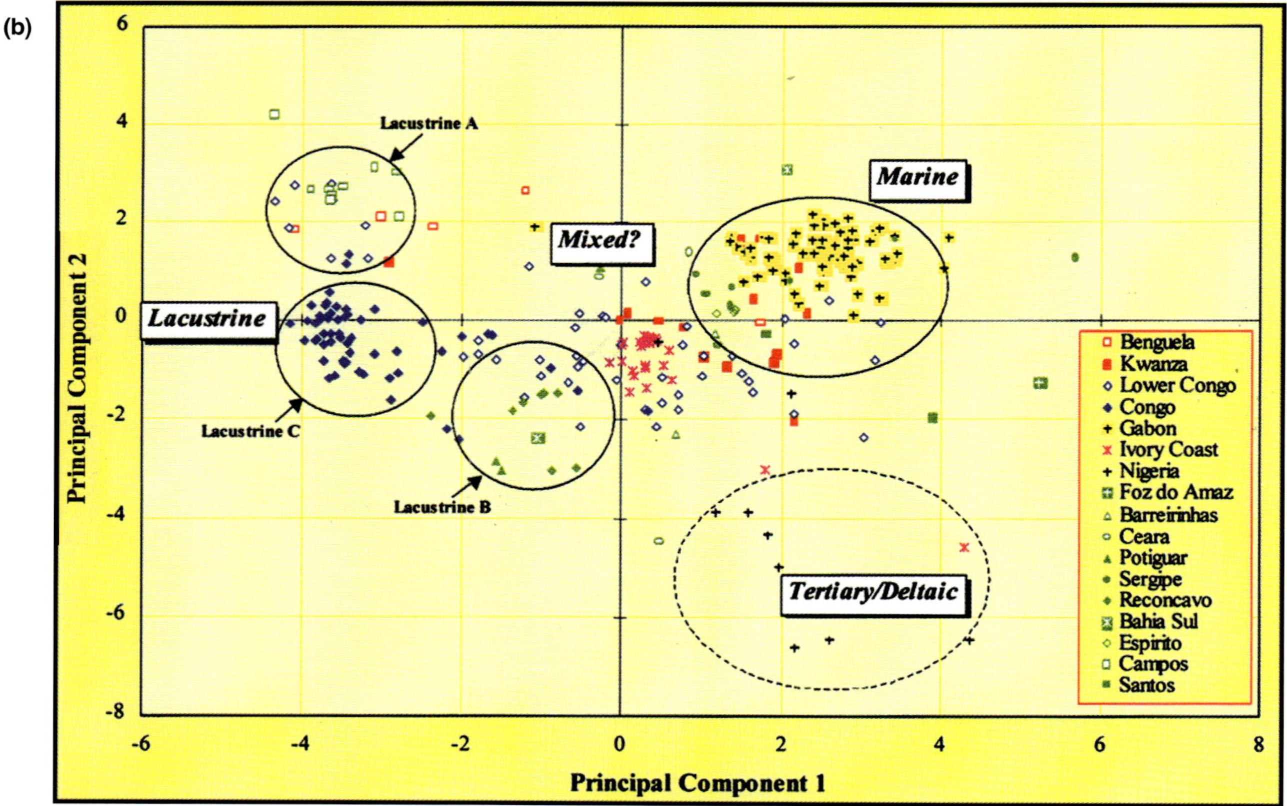

Figure 4—(a) Principal component analysis (PCA) loadings plot showing source-dependent geochemical parameters. (b) PCA scores plot showing differentiation of identified oil families.

The cluster labeled as lacustrine C (Figure 5) includes oils that are restricted to the Congo Basin (southern Gabon, Congo, Cabinda, and Zaire). None of the Brazilian oils in this study clustered with this type of oil from west Africa, perhaps suggesting a source depositional environment unique to west Africa. The oils are characterized by elevated isoprenoid ratios (pristine/phytane > 2.0), intermediate stable carbon isotopic compositions, negative canonical variables (Sofer, 1984), abundant extended tricyclic terpanes, enrichment in C_{27}-$\alpha\beta\beta$-steranes, and variable amounts of gammacerane and 4-methyl steranes.

The lacustrine C oils are produced from both presalt and postsalt reservoirs. Some oils from postsalt reservoirs have experienced varying degrees of bacterial alteration. For the oils from Cabinda and Zaire, Burwood (1997) showed an origin from algal source rocks of the Neocomian–Barremian (synrift I) "organic" Bucomazi Formation deposited in fresh, deep-water lacustrine environments. Gabon and Congo oils presumably originated from time-equivalent source rocks of the Melania and Marnes Noires Formations, respectively.

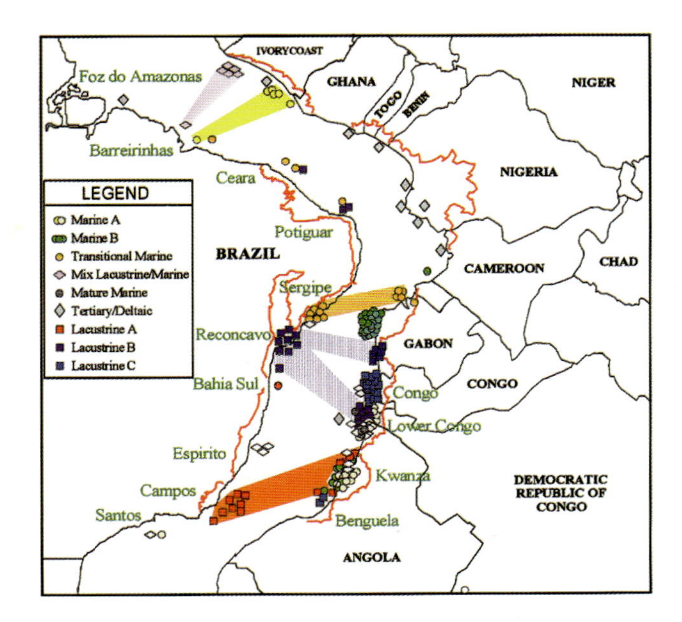

Figure 5—Map showing the distribution of identified oil families and their correlation from Brazilian to west African margins.

Marine Oils

South Atlantic margin oils designated as "marine" originated from source rocks of several different depositional environments. These oils typically have higher concentrations of sulfur (>0.5 wt. %), pristine/phytane ratios less than 1.5 (Figure 6), variable stable carbon isotopic compositions, $C_{25} > C_{26}$ tricyclic terpanes (Figure 7), elevated proportions of extended pentacyclic terpanes relative to C_{30} hopanes, and high sterane/hopane ratios (Figure 8). Marine oils have diverse triaromatic dinosterane compositions (Figure 9) that often contain abundant 3-methyl triaromatic steranes (labeled as 4 and 6 in Figure 2) and 4-methyl triaromatic steranes (labeled as 7 and 9 in Figure 2). According to Moldowan et al. (1990), marine-derived oils can also be identified on the basis of the presence of 24-*n*-propylsteranes, which are highly diagnostic biomarkers for marine algae as detected by metastable reaction monitoring (MRM) analyses.

A subset of the marine group includes oils derived from source rocks composed mainly of terrigenous organic matter deposited in nearshore (deltaic) marine paleoenvironments. These oils are characterized by elevated isoprenoid ratios (pristine/phytane > 2.0), positive canonical variables (Sofer, 1984), high abundances of low molecular weight C_{19}- and C_{20}-tricyclic terpanes and C_{24}-tetracyclic terpanes, and a predominance of C_{29} steranes. Oleanane, a specific biomarker derived from angio-sperms and associated with a Tertiary source due to the evolution of higher land plants (Ekweozor and Udo, 1987) is abundant in these oils (oleanane/C_{30} hopane > 0.2).

Another subset of the marine group includes oils that exhibit intermediate geochemical characteristics. These oils may represent lacustrine and marine oils that mixed in the reservoir. Alternatively, they may have originated from source rocks deposited in a sag basin setting (Brumbaugh, et al., 1994) that received input from both lacustrine and marine organic matter (i.e., mixed kerogen assemblage).

In Figure 3, the group labeled as marine A includes oils from the Ivory Coast, northern Angola, and the Kwanza, Santos, and Barreirinhas Basins (Figure 5). Marine source rocks responsible for the Ivory Coast oils were probably deposited during the Late Cretaceous synrift phase of deposition (Harms et al., 1996). Northern Angola oils are derived from Upper Cretaceous Iabe Formation source rocks, whereas the Kwanza Basin oils originated from either Cretaceous (Teba or Itombe Formation) or Eocene (Margas Negras Formation) source rocks (Burwood, 1997). Marine oils from the Santos and Barreirinhas Basins originated from Upper Cretaceous shales deposited in anoxic marine environments (Mello et al., 1988a, b).

The group labeled as marine B (Figures 3 and 5) includes oils from the northern Gabon and Kwanza Basins. The Gabon oils can be separated into at least four subgroups according to source lithofacies and relative anoxia in the depositional environment. Biomarkers most useful in differentiating oil types in northern Gabon include the relative abundance of 17α-, 18α-, 21β-25,28,30-trisnorhopane, 17α-, 18α-, 21β-28,30-bisnorhopane, gammacerane, diasteranes, and the proportion of C_{35} to C_{34} pentacyclic terpanes.

The cluster labeled as Gabon 1 in Figure 3 includes moderately mature to mature oils that are primarily characterized by an abundance of 17α-, 18α-, 21β-25,28,30-tris-

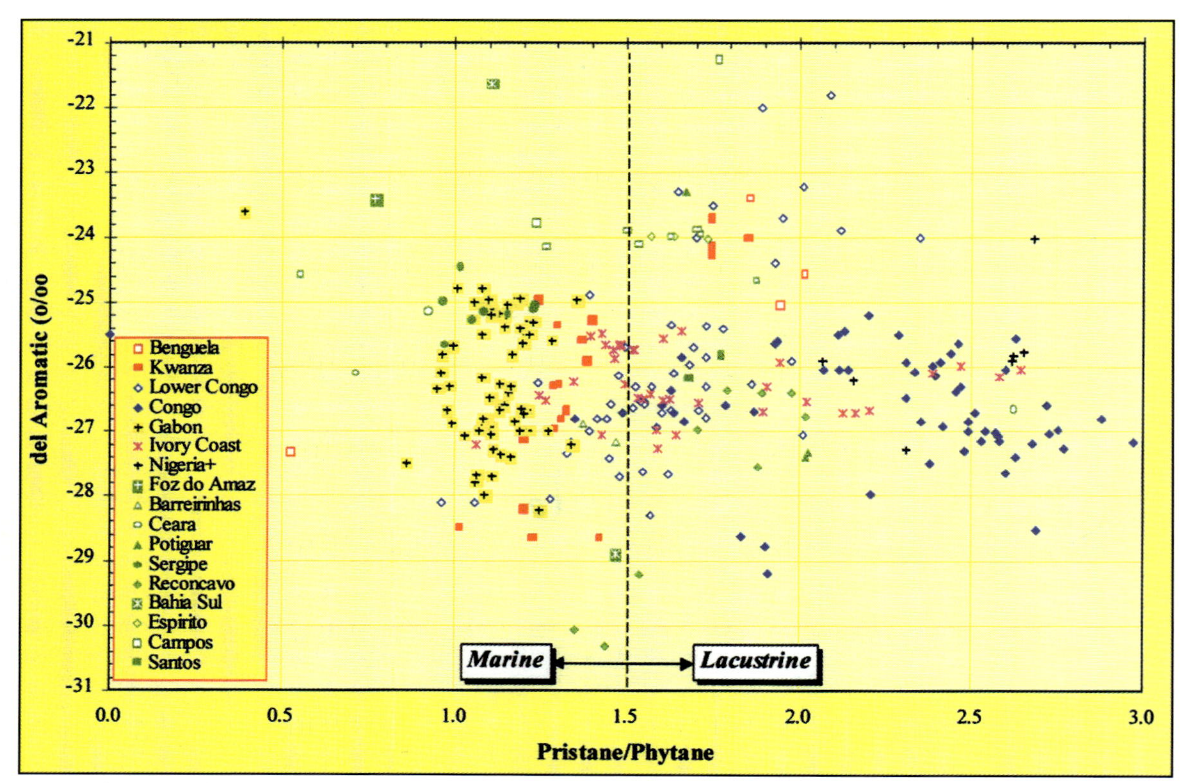

Figure 6—Plot of pristane/phytane versus the carbon isotopic composition of the aromatic hydrocarbons showing the difference between marine and lacustrine oils.

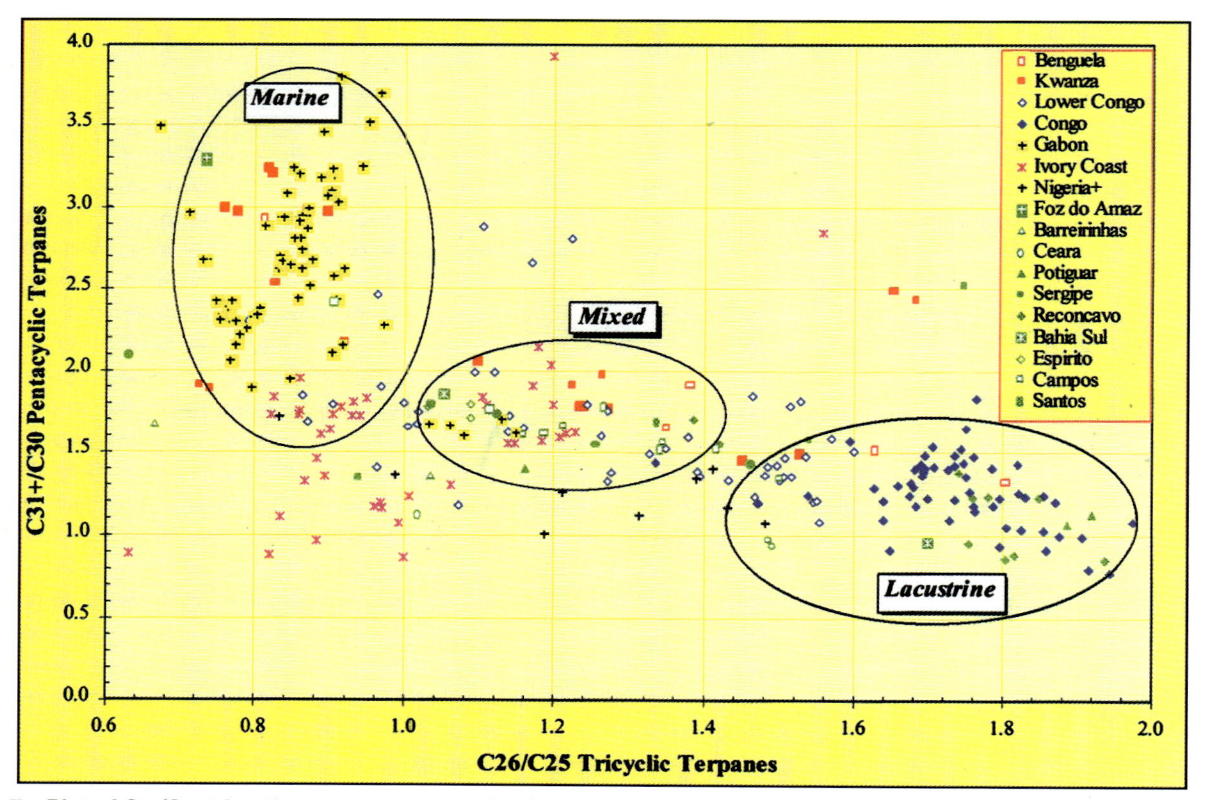

Figure 7—Plot of C_{26}/C_{25} tricyclic terpanes versus C_{31+}/C_{30} pentacyclic terpanes showing the difference between marine and lacustrine oils.

norhopane and gammacerane and moderate to low amounts of 17α-, 18α-, 21β-28,30-bisnorhopane. The cluster labeled as Gabon 2 contains oils from Batanga reservoirs that are isotopically heavier than the Gabon 1 oils. These oils are further characterized by their low to moderate maturity and the presence of abundant 17α-, 18α-, 21β-28,30-bisnorhopane and low amounts of 17α-, 18α-, 21β-25,28,30-trisnorhopane. The cluster labeled as Gabon 3 contains mature oils that are characterized by low to moderate amounts of 17α-, 18α-, 21β-25,28,30-trisnorhopane and 17α-, 18α-, 21β-28,30-bisnorhopane, abundant gammacerane and diasteranes, and elevated proportions of C_{35} to C_{34} hopanes.

Gabon 1, 2, and 3 oils are geographically situated within the same region described by Sofer (1993) as containing marine oils derived from source rocks in the Eocene Ozouri Formation. However, Katz et al. (1997) have assigned Batanga reservoired oils from this area (Gabon 2) as belonging to the Ezanga/Madiela–Batanga petroleum system. A pre-Cenomanian source deposited in a restricted marine environment is suggested. Katz et al. (1997) also suggested that it is unlikely that an Eocene source, if present, would be thermally mature. The differences in oil chemistries observed for the Gabon 1, 2, and 3 oils suggest that additional sources are present, significant source facies variations exist, and/or oil mixing may have occurred.

The cluster labeled as Gabon 4 (Figure 3) contains oils that are characterized by low abundances of 17α-, 18α-, 21β-25,28,30-trisnorhopane, 17α-, 18α-, 21β-28,30-bisnorhopane, gammacerane, and diasteranes. These oils are geographically situated in the same area defined by Katz et al. (1997) to contain oils belonging to the Azile/Anguille–Anguille petroleum system. Sofer (1993) suggested an Early–middle Cretaceous source for oils in this area.

The marine B group also contains a small cluster of marine-derived oils from the southern part of the onshore Kwanza Basin (Figure 5). These oils are from Aptian Binga reservoirs, have intermediate isotopic compositions, and contain appreciable gammacerane. Burwood (1997) suggested that oils from this area originated from source rocks within the Aptian–Albian Middle Binga Formation deposited in a transgressive restricted marine environment.

The broad marine cluster shown in Figure 3 also contains oils that display intermediate geochemical characteristics. These oils may represent a commingling of marine-derived and lacustrine-derived oils. Alternatively, these oils may have originated from source rocks composed of a mixed kerogen assemblage. This "mixed oil" family includes oils from the Ivory Coast, several oils from the Lower Congo Basin from postsalt reservoirs, plus oils from the Espírito Santo, Sergipe–Alagoas, Barreirinhas, and Kwanza Basins. Mixing of presalt (Organic Bucomazi Formation) and postsalt (Iabe Formation) oils in the Lower Congo Basin has been previously reported by Connan et al. (1987) and ten Haven (1996). An Aptian Cuvo Formation source deposited during a shallow to ephemeral, brackish–saline, lacustrine, initial marine transgressive phase may be responsible for the so-called mixed oils from the Kwanza Basin (Burwood, 1997).

The marine group in Figure 3 also includes a family of oils that display geochemical characteristics suggestive of an origin from marine marls and calcareous black shales. This "transitional oil" family includes oils from the Sergipe–Alagoas, Potiguar, and Ceará Basins and some oils from offshore northern Gabon. These low-sulfur oils have intermediate to heavy stable carbon isotopic compositions, contain abundant gammacerane, and have elevated proportions of steranes relative to hopanes. The Sergipe–Alagoas oils probably originated from Aptian–Albian source rocks within the Ibura or Maceio Formations. The northern Gabon oils may be derived from similar source rocks deposited within the Aptian Namina Formation (Kuo, 1998).

The last cluster of oils within the marine group (labeled as Tertiary in Figure 3) originated from moderately mature source rocks composed of mainly terrigenous organic matter deposited in fluvial deltaic or paralic marine environments. These oils, which contain abundant oleanane, have positive PC1 and negative PC2 values (Figure 4b). The South Atlantic margin oils belonging to this group are Tertiary oils from Nigeria, Benin, equatorial Guinea, Cameroon, offshore northern Angola, and the Foz do Amazonas Basin of northern Brazil (Figure 5).

CONCLUSIONS

Geochemical information obtained from the detailed analysis of representative oils from the South Atlantic margin can be used to establish genetic relationships and identify a number of oil families, differing in source age and depositional paleoenvironment. Areas where oil mixing has occurred can also be identified.

Several different geochemical criteria are useful in differentiating lacustrine-derived oils from marine-derived oils. These include the pristine/phytane ratio, the proportion of C_{26}- to C_{25}-tricyclic terpanes (Zumberge, 1987) and steranes to hopanes (Moldowan, et al., 1985), the relative abundance of extended pentacyclic terpanes and an unknown doublet (*m/z* 259) (Schiefelbein et al., 1997), and the distribution of triaromatic dinosteranes (*m/z* 245).

Lacustrine oils derived from different presalt (Neocomian–Aptian) source rocks are widespread in west Africa and offshore Brazil. These oils can be separated into at least three distinct families. Lacustrine oils from west Africa are present in the Congo, Lower Congo, and Benguela Basins. Brazilian oils with lacustrine affinities are present in the Recôncavo, Campos, Potiguar, Ceará ,and Bahia Sul Basins. Based on HCA, oils from the Campos Basin and offshore central Angola have similar chemical attributes and therefore may share a genetic origin. Similarly, oils from the Recôncavo Basin in

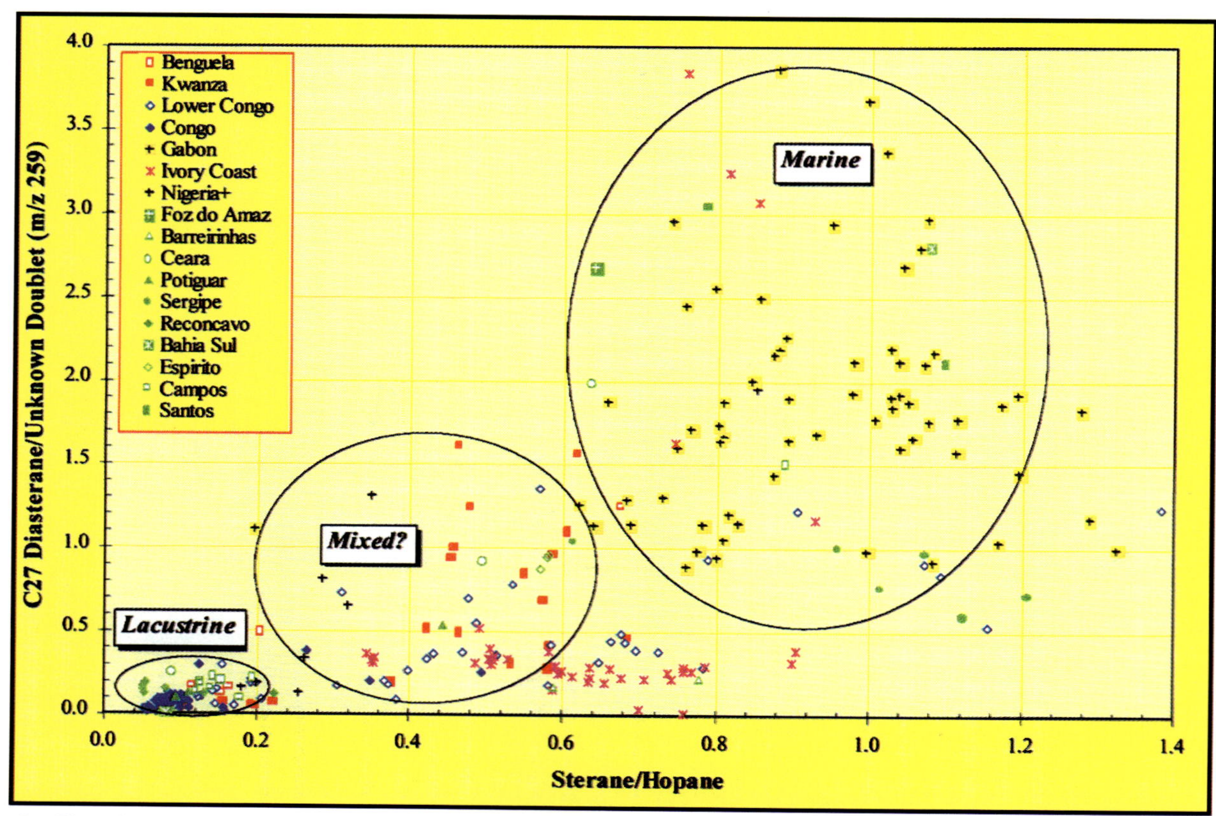

Figure 8—Plot of sterane/hopane versus C_{27} diasterane/unknown doublet (*m/z* 259) showing the difference between marine and lacustrine oils.

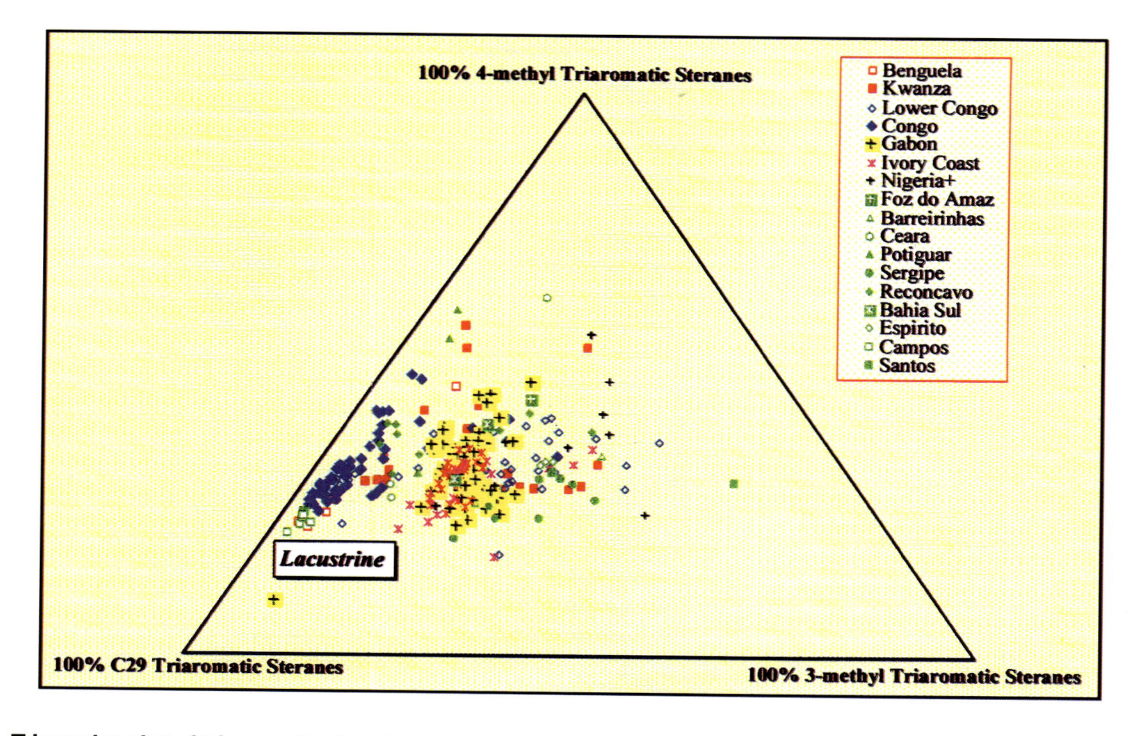

Figure 9—Triangular plot of triaromatic dinosterane compositions (*m/z* 245) showing an enrichment of C_{29} triaromatic steranes for lacustrine oils.

offshore Brazil and oils from central Gabon and northern Angola may also be genetically related.

Marine oils derived from different postsalt (Upper Cretaceous–Tertiary) source rocks are present on both sides of the South Atlantic. West African marine oils are primarily from the Ivory Coast and the Gabon (Port Gentil area), Lower Congo, Kwanza, and Benguela Basins. Marine oils from offshore Brazil are present in the Santos, Espírito Santo, and Sergipe–Alagoas Basins. Several oils from the Ivory Coast, Lower Congo Basin, and Kwanza Basin display intermediate geochemical characteristics, possibly suggesting commingling of marine-derived and lacustrine-derived oils and/or the presence of source rocks composed of a mixed kerogen assemblage. Brazilian oils from Sergipe–Alagoas Basin and oils from northern Gabon may be genetically related.

Several South Atlantic margin oils display geochemical characteristics suggestive of an origin from Tertiary source rocks primarily composed of terrigenous organic matter deposited in fluvial-deltaic or nearshore marine environments. These oils are from Nigeria, Benin, equatorial Guinea, Cameroon, offshore northern Angola, and the Foz do Amazonas Basin of northern Brazil.

REFERENCES CITED

Abreu, V., P. R. Vail, A. Bally, and E. Wilson, 1997, Geologic evolution of conjugate volcanic passive margins: influence on the petroleum systems of the South Atlantic: Joint Houston Geological Society and International Explorationists Meeting, December 15, Houston, Texas.

Brice , S., K. Kelts, and M. Arthur, 1980, Lower Cretaceous lacustrine source beds from early rifting phases of the South Atlantic (abs.): Program with Abstracts, AAPG Anuual Convention, Denver, Colorado, p. 30.

Brumbaugh, B., N. R. Cameron, and S. G. Henry, 1994, A new pre-salt source rock for central west African deep water? (abs.): Program with Abstracts, AAPG Annual Convention, Denver, Colorado, p. 109.

Burwood, R., 1997, Source rock control for Angolan Atlantic margin petroleum systems (abs.): AAPG/ABGP Joint Research Symposium, Petroleum Systems of the South Atlantic Margin, Rio de Janeiro, Brazil.

Connan, J., D. Levache, T. Salvatori, A. Riva, R. Burwood, and P. Leplat, 1987, Petroleum geochemistry of Angola: an Agelfi study (abs.): 13th International European Association of Organic Geochemists Meeting, Venice, Italy, p. 89–90.

Demaison, G., and B. Huizinga, 1991, Genetic classification of petroleum systems: AAPG Bulletin, v. 75, p. 1626–1643.

Didyk, B. M., B. R. T. Simoneit, S. C. Brassell, and G. Eglinton, 1978, Organic geochemical indicators of paleoenvironmental conditions of sedimentation: Nature, v. 272, p. 216–222.

Ekweozor, C. M., and O. T. Udo, 1987, The oleananes: origin, maturation, and limits of occurrence in southern Nigeria sedimentary basins: Organic Geochemistry, v. 13, p. 131–140.

Guardado, L. R., A. R. Spadini, M. R. Mello, and J. S. L. Brandao, 1997, Petroleum systems of the Campos Basin, Brazil (abs.): AAPG/ABGP Joint Research Symposium, Petroleum Systems of the South Atlantic Margin, Rio de Janeiro, Brazil.

Harms, J. C., J. M Bruso, Jr., R. L. Wallace, J. A. Canales, and N. Koffi, 1996, A petroliferous transform-margin basin, Cote d'Ivorie, west Africa (abs.): AAPG Annual Convention, Program with Abstracts, San Diego, California, v. 5, p. A61.

Henry, S. G, W. Brumbaugh, and N. R. Cameron, 1995, Pre-salt source rock development on Brazil's conjugate margin: west African examples (abs.): First Latin American Geophysical Conference, Rio de Janeiro, Brazil.

Horn, M. K., 1980, The habitat of oil and gas on continental margins: SEAPEX Proceedings, v. 1, p. 23–62.

Karner, G. D., and N. W. Driscoll, 1997, Timing, distribution, and structure of the rift-basins comprising the west African and Brazilian margins: AAPG/ABGP Joint Research Symposium, Petroleum Systems of the South Atlantic Margin, Rio de Janeiro, Brazil.

Katz, B. J., and B. Mertani, 1989, Central Sumatra: a geochemical paradox: Proceedings, 18th Annual Convention, Indonesian Petroleum Association, p. 403–435.

Katz, B., W. C. Dawson, L. M. Liro, V. D. Robison, and J. D. Stonebraker, 1997, The petroleum systems of the Ogooué Delta, offshore Gabon (abs.): AAPG/ABGP Joint Research Symposium, Petroleum Systems of the South Atlantic Margin, Rio de Janeiro, Brazil.

Kuo, L-C., 1998, Organic facies, richness, and quality of potential source rocks in the middle Cretaceous transition-drift sequence in northern Gabon (abs.): 1998 International AAPG Meeting, Rio de Janeiro, Brazil, AAPG Bulletin, v. 82, p. 1930–1931.

Lehner, P., and P. A. C. De Riuter, 1977, Structural history of Atlantic margin of Africa: AAPG Bulletin, v. 61, p. 961–981.

Magoon, L. B., and W. G. Dow, 1994, The petroleum system—from source to trap: AAPG Memoir 60, 655 p.

Mello, M. R., and M. H. Hessel, 1998, Biological marker and paleozoological evidence of marine incursion in the South Atlantic: example from the Campos Basin, Brazil (abs.): Latin American Congress on Organic Geochemistry, Margarita Island, Venezuela.

Mello, M. R., P. C. Gaglianone, S. C. Brassell, and J. R. Maxwell, 1988a, Geochemical and biological marker assessment of depositional environments using Brazilian offshore oils: Marine and Petroleum Geology, v. 5, p. 205–223.

Mello, M. R., N. Telnaes, P. C. Gaglianone, M. I. Chicarelli, S. C. Brassell, and J. R. Maxwell, 1988b, Organic geochemical characterization of depositional paleoenvironments of source rocks and oils in Brazilian marginal basins: Organic Geochemistry, v. 13, p. 31–45.

Moldowan, J. M., W. K. Seifert, and E. J. Gallegos, 1985, Relationship between petroleum composition and depositional environment of petroleum source rocks: AAPG Bulletin, v. 69, p. 1173–1180.

Moldowan, J. M., F. J. Fago, C. Y. Lee, S. R. Jacobson, D. S. Watt, N. C. Slougui, A. Jeganathan, and D. C. Young, 1990, Sedimentary 24-*n*-propylcholestanes, molecular fossils diagnostic of marine algae: Science, v. 247, p. 309–312.

Moldowan, J. W., C. Y. Lee, P. Sundararaman, T. Salvatori, A. Alajbeg, B. Gjukic, G. Demaison, N. E. Slougui, and D. S. Watt, 1992, Source correlation and maturity assessment of select oils and rocks from the Central Adriatic Basin, *in* J. M. Moldowan, P. Albrecht, and R. P. Philp, eds., Biological

markers in sediments and petroleum: New York, Prentice Hall, p. 370–401.
Moldowan, J. M., J. Dahl, F. J. Fago, R. Shetty, D. S. Watt, S. R. Jacobson, B. J. Huizinga, M. A. McCaffrey, and R. E. Summons, 1995, Correlation of biomarkers with geologic age (abs.): 17th International European Association of Organic Geochemists Meeting, San Sebastian, Spain, p. 418–420.
Peters, K. E., J. M. Moldowan, M. Schoell, and W. B. Hempkins, 1986, Petroleum isotopic and biomarker composition related to source rock organic matter and depositional environment: Organic Geochemistry, v. 10, p. 17–27.
Powell, T., 1986, Petroleum geochemistry and depositional setting of lacustrine source rocks: Marine and Petroleum Geology, v. 3, p. 200–219.
Pu, F., R. P. Philp, L. Zhenxi, Y. Xinke, and Y. Guangguo, 1991, Biomarker distributions in crude oils and source rocks from different sedimentary environments: Chemical Geology, v. 93, p. 61–78.
Rabinowitz, P. D., and J. La Brecque, 1979, The Mesozoic South Atlantic ocean and evolution of its continental margins: Journal of Geophysical Research, v. 84, p. 5973–6002.
Schiefelbein, C. F., J. E. Zumberge, and S. W. Brown, 1997, Petroleum systems in the Far East: Proceedings of the Petroleum Systems of SE Asia and Australasia Conference, Indonesian Petroleum Association, Jakarta, Indonesia, p. 101–113.
Sofer, Z., 1984, Stable carbon isotopic compositions of crude oils: application to source depositional environments and petroleum alteration: AAPG Bulletin, v. 68, p. 31–49.
Sofer, Z., 1993, Distribution of genetic oil families in west Africa based on biomarker ratios: Third Latin American Congress on Organic Geochemistry, Extended Abstracts, Manaus, Brazil, p. 134–137.
Teisserenc, P., and J. Villemin, 1990, Sedimentary basin of Gabon—geology and oil systems: AAPG Bulletin, v. 74, p. 117–199.
ten Haven, H., 1996, Applications and limitations of Mango's light hydrocarbon parameters in petroleum correlation studies: Organic Geochemistry, v. 24, n. 10/11, p. 957–976.
ten Haven, H., J. de Leeuw, and P. A. Schenck, 1985, Organic geochemical studies of a Messinian evaporitic basin, northern Apennines (Italy), I: hydrocarbon biological markers for a hypersaline environment: Geochimica et Cosmochimica Acta, v. 49, p. 2181–2191.
ten Haven, H., J. de Leeuw, J. Sinninghe Damste, P. Schenck, S. Palmer, and J. Zumberge, 1988, Application of biological markers in the recognition of paleo hypersaline environments, *in* K. Kelts, A. Fleet, and M. Talbot, eds., Lacustrine petroleum source rocks: London, Blackwell, p. 123–130.
Torquato, J. R., and U. G. Cordani, 1981, Brazil–Africa geologic links: Earth Science Reviews, v. 17, p. 155–176.
Volkman, J. K., 1988, Biological marker compounds as indicators of the depositional environment of petroleum source rocks, *in* K. Kelts, A. Fleet, and M. Talbot, eds., Lacustrine petroleum source rocks: London, Blackwell, p. 103–122.
Wilson, E., V. S. Abreu, M. P. Asley, M. Brandao, and A. S. Telles, 1997, Lower Cretaceous stratigraphy and source rock distribution in the pre-salt basins of the South Atlantic: comparison of Angola and Southern Brazil (abs.): AAPG/ABGP Joint Research Symposium, Petroleum Systems of the South Atlantic Margin, Rio de Janeiro, Brazil.
Zumberge J., 1987, Prediction of source rock characteristics based on terpane biomarkers in crude oils: a multivariate statistical approach: Geochimica et Cosmochimica Acta, v. 51, p. 1625–1637.

Rudkiewicz, J-L., H. L. B. Penteado, A. Vear, M. Vandenbroucke, F. Brigaud, J. Wendebourg, and S. Düppenbecker, 2000, Integrated basin modeling helps to decipher petroleum systems, *in* M. R. Mello and B. J. Katz, eds., Petroleum systems of South Atlantic margins: AAPG Memoir 73, p. 27–40.

Chapter 3

Integrated Basin Modeling Helps to Decipher Petroleum Systems

J-L. Rudkiewicz

Institut Français du Pétrole
Rueil-Malmaison, France

H. L. de B. Penteado

Petrobrás/Cenpes/CEGEQ
Rio de Janeiro, Brazil

A. Vear

BP Exploration
Houston, Texas, U.S.A.

Present address: *Woodside Energy*
Perth, Australia

M. Vandenbroucke

Institut Français du Pétrole
Rueil-Malmaison, France

F. Brigaud

Elf Aquitaine
Pau, France

J. Wendebourg

Institut Français du Pétrole
Rueil-Malmaison, France

S. Düppenbecker

BP Exploration
Houston, Texas, U.S.A.

Abstract

The behavior of a petroleum system often results from coupled physical interactions, whose effects are difficult to predict without adequate user experience and numerical tools. 2-D and 3-D basin simulators are such tools that are used to describe the generation and migration of hydrocarbons in sedimentary basins. We illustrate with two actual examples how these tools can be used to understand some key points of petroleum systems in an efficient way.

In the Brazilian Recôncavo basin, the possible migration patterns from the deep part of the basin toward the prerift and synrift reservoirs are examined, including the role of faults and the effect of long-range migration in the infilling of reservoirs. In the North Sea, the Jurassic petroleum system in the so-called high-pressure/high-temperature (HPHT) domain can be deciphered. There, oils survive temperatures as high as 190°C. Modeling tools that reconstruct the petroleum system in a geologic time frame show that this unusual feature has a dynamic behavior, linked to recent subsidence and sedimentation and to secondary cracking of the generated oils. Understanding petroleum migration and cracking processes in such settings demonstrates the capabilities and economic value of modeling tools.

INTRODUCTION

While exploring for oil and gas, many questions related to the petroleum system of a basin are often handled qualitatively to understand and then, if possible, to predict how traps are being charged. Nevertheless, when numerous physical and chemical interactions occur during geologic time, as is often the case in nature (Figure 1), considering only the major or the most obvious geologic processes can lead to severe bias in the exploration strategies. Basin simulators have been developed to overcome this drawback (Lerche, 1990; Ungerer et al., 1990). This chapter aims at showing how recently developed basin simulators can be used to describe the generation and migration of hydrocarbons in sedimentary basins in an efficient way.

Two- and three-dimensional simulators such as *Temispack* have been developed by the Institut Français du Pétrole (IFP) over the past few years, often in cooperation with partners from the oil industry (Rudkiewicz

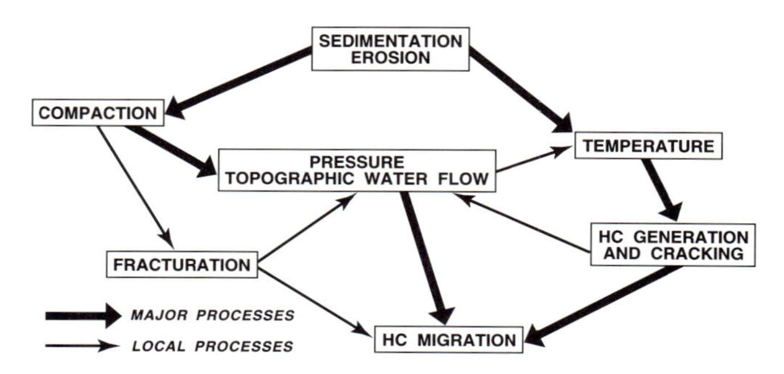

Figure 1—Schematic diagram showing active processes in a basin over geologic time. A basin simulator takes all these interactions into account. Heavy arrows represent dominant coupling, light arrows less effective interactions. The magnitude of these interactions may often be case-dependent based on the geologic context of a study.

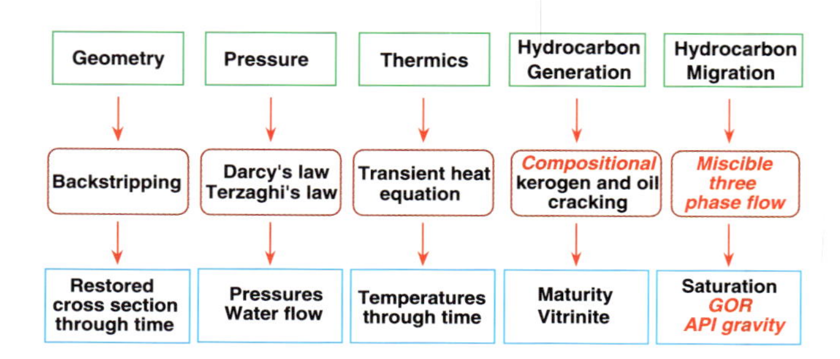

Figure 2—Schematic diagram showing stepwise solution of compositional generation and migration in 2-D and 3-D basin simulators. The processes to be solved are shown in the top boxes, and the physical laws used to model them in the middle boxes. Results are indicated in the lower boxes. Efficient modeling requires that successive steps from left to right must be solved sequentially.

et al., 1996). IFP's 2-D simulator currently aims at describing the nature of the hydrocarbon charge, whereas in 3-D models, migration volumes are being handled to predict the volume of the hydrocarbon charge to a given reservoir. The same physical laws are used to describe the evolution of the basin, its thermal history, and the generation and migration of hydrocarbons (Figure 2). The 2-D tool is able to describe the detailed composition of the generated fluids and to take into account two different phases for the hydrocarbon fluids. The 3-D tool allows for a simple fluid description with flow properties that do not change during generation and migration.

The Recôncavo Basin, located in NE Brazil is an aborted rift basin that contains source rocks with type I organic matter and numerous oil fields (Penteado, 1999). Accumulations occur both in prerift and synrift sequences. With basin simulation, the migration routes and the role of faults can be investigated, showing long distance migration and early entrapment of hydrocarbons.

We have also investigated a high-pressure/high-temperature (HPHT) petroleum system in the North Sea. This study was undertaken with the aim of testing several basin simulators, showing that they could be used to understand and then predict such HPHT systems. Another objective of the study was to define what could be the best approach in the industry toward such geologic settings.

One of the critical factors in the discovery of HPHT reservoirs was the fact that hydrocarbons in the oil range were discovered at temperatures as high as 190°C. At such high temperatures, only gas or gas condensate was expected to be seen. Being able to understand and predict the nature of hydrocarbons in such settings proves important for economic reasons. It might shift a field from a noncommercial to a commercial category. The practical implication of such a project is thus great, as it will shift the oil window toward greater depths.

RECÔNCAVO BASIN CASE

Recôncavo Basin Questions

The Recôncavo Basin (Figure 3) is one of the most prolific onshore Brazilian basins (Ghignone and Andrade, 1970; Figueiredo et al., 1994). The basin is an aborted rift basin of Late Jurassic–Early Cretaceous age, and its structural geology has been studied extensively (e.g., Magnavita and Cupertino, 1987; Milani et al., 1987; Milani and Davison, 1988; Magnavita, 1992). The lithostratigraphic units (from bottom to top) are as follows: the Afligidos Formation (Permian) and Água Grande Formation (early Berriasian), which belong to the prerift sequence; the Candeias Formation (Berriasian) to São Sebastião Formation (Barremian), which record the rifting event (Figure 3); and the sedimentary sequence of the Aptian Marizal Formation, a few tens of meters thick, which is the only indication of postrift subsidence and deposition.

Organic geochemistry has also been used extensively to recognize the main type I source rock units and to correlate reservoir oils and source rocks (Gaglianone and Trindade, 1988; Daniel et al., 1989). However, some questions remain about the petroleum systems of the basin and about the relationships between the composition of the source rock extracts and the oils in the reservoirs. According to Figueiredo et al. (1994), three petroleum subsystems exist in this basin:

1. **The prerift system**—The source rock is located in the synrift Candeias Formation, and the reservoirs are the prerift sandstones of the Água Grande, Sergi and Aliança Formations. The largest pools belong to this petroleum system, with 60% of known reserves (Ghignone and Andrade, 1970; Santana, 1989).

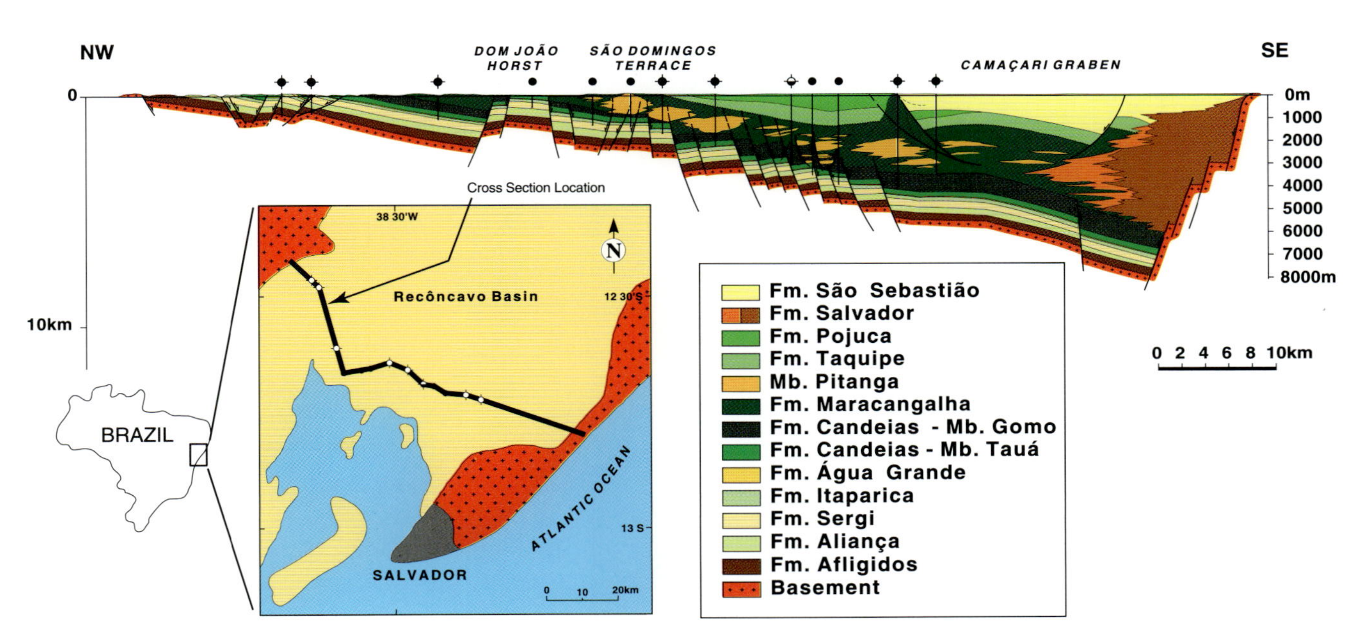

Figure 3—Selected Recôncavo Basin cross section and its location. The Tauá and Gomo Members are source rock intervals. The Pitanga, Água Grande, and Sergi Formations are the main reservoirs.

2. **The first synrift system**—The source rock is the Gomo Member of the Candeias Formation, and the reservoirs are intercalated turbidites in the Candeias and Maracangalha Formations. This system is responsible for 15% of the known reserves.
3. **The second synrift system**—The reservoirs are located in the Marfim and Pojuca Formations, sealed against growth faults and sourced from the underlying Gomo Member (Candeias Formation). It bears around 25% of known reserves.

We concentrate here on the long-distance migration from the deep kitchen in the Camaçari Low toward the reservoirs in the Dom João High and on the migration history in the first two petroleum systems.

Data Set

Figure 3 shows the location of the cross section that has been used to study the petroleum system in the southern part of the Recôncavo Basin. It connects the horst and the deepest parts of the graben. The rift basin is almost cylindrical, hence a 2-D cross section can be used to qualitatively analyze the influence of faults on migration. From the geologic cross section shown in Figure 3, a meshed cross section has been constructed (Figure 4). In the prerift part, lithofacies do not vary significantly along the cross section. However, in the synrift formations, the sandy turbidites of the Candeias Formation have been taken into account. Porosity and permeability in the reservoirs have been calibrated with well data. The sandstones of the Aliança, Sergi, and Água Grande Formations act as regional drains.

The geologic history of the basin has been taken into account and was based mainly on the assumptions of Magnavita et al. (1994). The following events reconstruct the basin's history up to the present day:

1. Rifting between Berriasian and Barremian time,
2. Erosion of the rift shoulders and of the basin during Aptian time,
3. Postrift subsidence between Aptian and Cenomanian time,
4. Almost no deposition from Cenomanian to Oligocene time,
5. Uplift and erosion during Oligocene time, and
6. Small subsidence between Miocene and the present.

The amount of erosion and post-Aptian subsidence, for which only scarce direct indications are available, had to be inferred from the regional data on adjacent basins and was included in the geologic history of the basin (Penteado, 1999). For instance, Magnavita et al. (1994) estimated that up to 700 m of postrift sedimentation was followed by erosion in the Tertiary.

Normal faults due to rifting have been taken into account as thin grid columns that connect or disconnect the faulted formations (Figure 4). These faults are currently seen as sealing, based on production data from the field. However, on a geologic time scale, the flow behavior of migrating fluids through the faults might have been different.

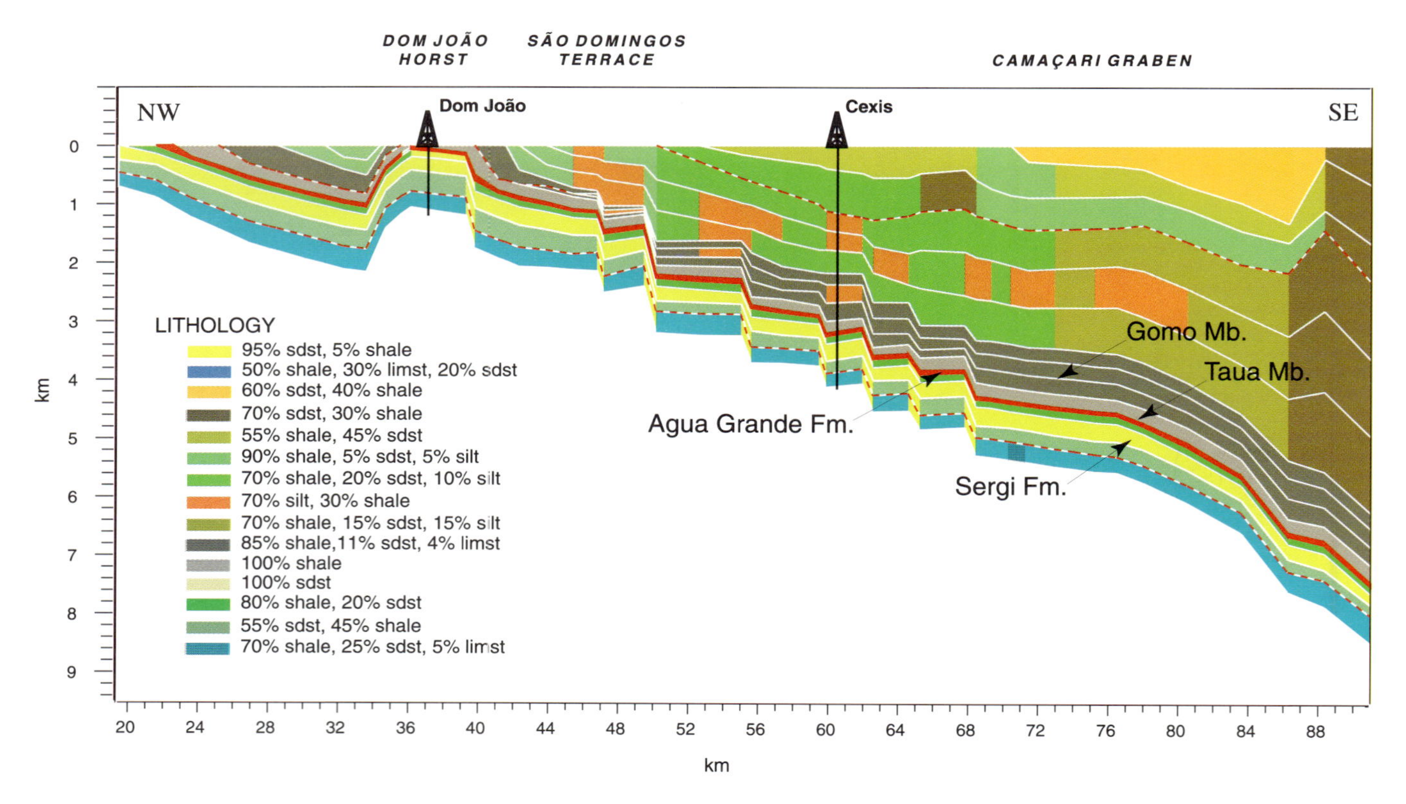

Figure 4—Cross section used to model generation and migration. The faults are seen as thin vertical cells each with a specific facies, allowing or preventing flow in both vertical and lateral directions depending on the given permeability.

A major constraint on the migration history of the oils is given by K/Ar dating of illites from the Dom João field. These ages (123–106 Ma) date the first arrival of oil into those reservoirs because the presence of oil stops illite precipitation.

Migration History

The temperature history of the basin through time has been calibrated based on present-day temperatures and recorded vitrinite reflectance. The initial source rock potential of the type I source rock was determined from carefully selected immature samples throughout the basin (Penteado, 1999). Several geologic scenarios with various estimates of amounts of postrift erosion (300, 700, 1200, and 1500 m) have been tested. We show only the results based on the estimate that allowed for the best calibration of thermal data, which was a maximum of 1200 m of postrift sedimentation.

The source rock began to mature and expel oil at 120 Ma in the deepest parts of the Camaçari half-graben. An interesting aspect of the migration pattern in the basin is the downward expulsion of petroleum from the Gomo Member source rock into the prerift Água Grande Formation. This formation acts as a regional carrier for the oil, and through connecting normal faults, oil can reach the Dom João High.

The major unknown parameter is fault permeability. With 2-D simulation using *Temispack*, the fault permeability decreases with porosity (hence depth) through a Kozeny–Carman relationship (Ungerer et al., 1990). However, this variation covers only one order of magnitude and is not large enough to cover all possible ranges from impermeable to permeable. In our study, three assumptions were tested:

1. Impermeable faults, with permeabilities ranging from 10^{-8} to 10^{-6} darcys (d) (Figure 5),
2. Low-permeability faults, with permeabilities ranging from 10^{-5} to 10^{-4} d (Figure 6), and
3. Permeable faults, with permeabilities ranging from 10^{-3} to 10^{-1} d (Figure 7).

Results show that low-permeability or permeable faults are required to account for the arrival of oil in the Dom João High. If faults are impermeable, hydrocarbons do not reach reservoirs in the high. Furthermore, relatively high permeabilities in the faults zones are needed for the early arrival of petroleum in the reservoirs.

Due to the large throw of the normal fault bounding the Dom João High to the east, both the Água Grande and the underlying Sergi reservoirs can be filled by petroleum ascending the fault zone without a direct lateral connection of reservoirs in the juxtaposed faulted blocks.

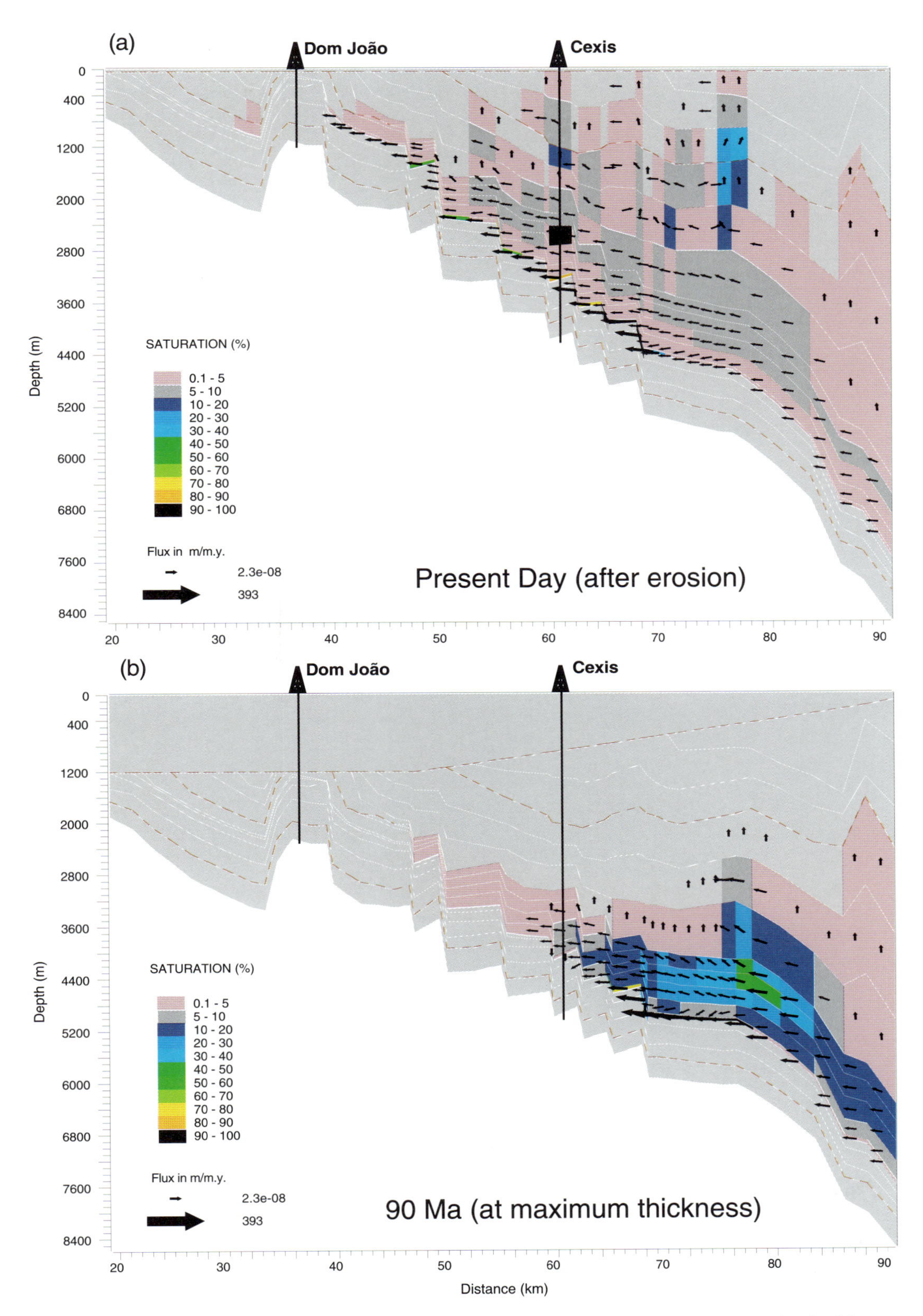

Figure 5—Recôncavo Basin cross section showing migration directions and hydrocarbon saturation in (a) present-day and (b) 90 Ma assuming the faults are impermeable (10^{-8}–10^{-6} d). Hydrocarbons do not reach the Dom João High early in the basin's history because the migration pathway through the Água Grande carrier is closed by faults. However, the Pitanga turbidites above the source rock are filled, even if when shale permeability is low.

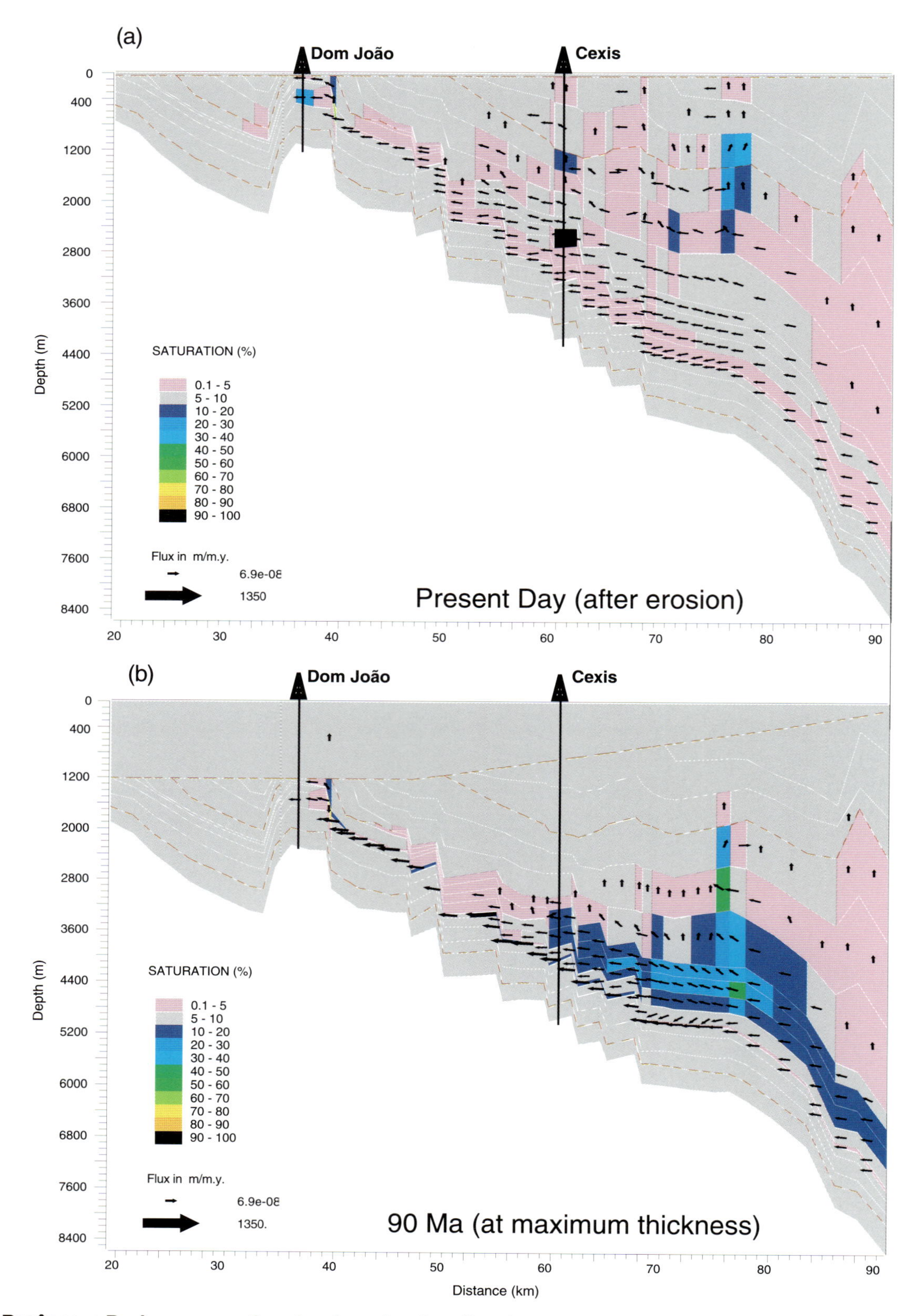

Figure 6—Recôncavo Basin cross section showing migration directions and hydrocarbon saturation in (a) present-day and (b) 90 Ma assuming the faults have low permeability (10^{-5}–10^{-4} d). Here the hydrocarbons reach the Dom João High early in the basin's history. The Pitanga turbidites are filled later than the Dom João High reservoirs because of the delay caused by crossing the less permeables shales.

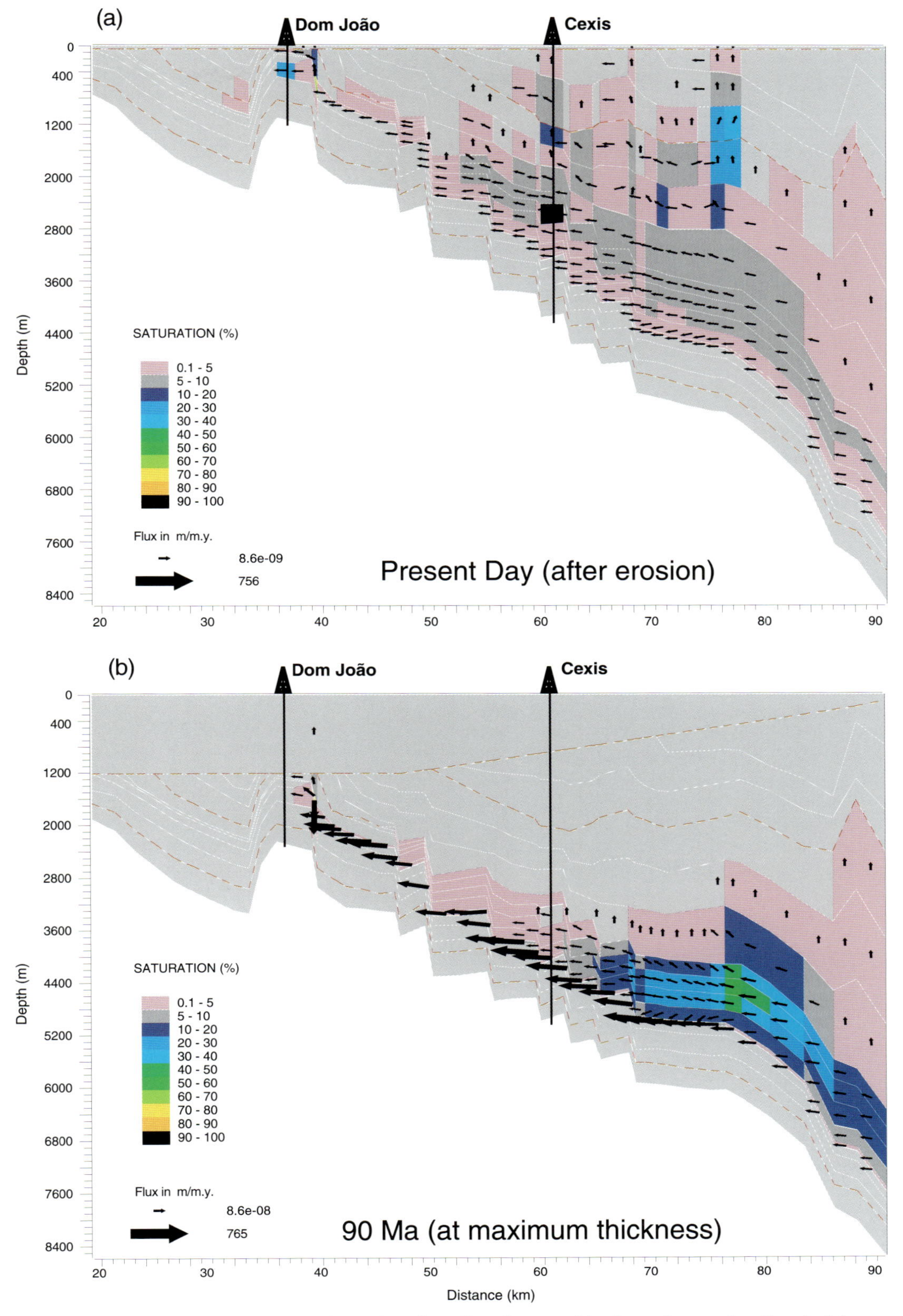

Figure 7—Recôncavo Basin cross section showing migration directions and hydrocarbon saturation in (a) present-day and (b) 90 Ma assuming the faults are permeable (10^{-1}–10^{-3} d). This picture is very similar to that in Figure 6 and shows the same features. Earlier arrival of petroleum in the Dom João High is in better agreement with dating of clay minerals.

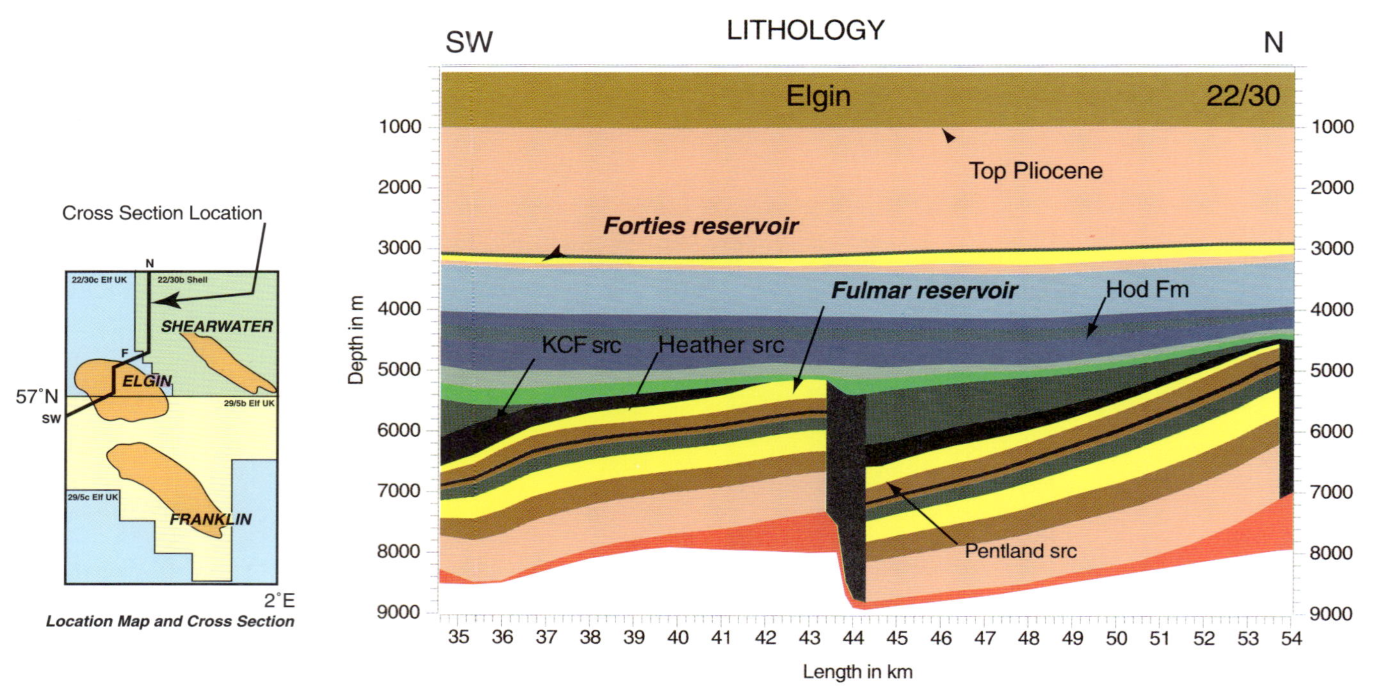

Figure 8—Two-dimensional cross section across the Elgin field in the North Sea and location map of the studied HPHT area. Source rocks are the Pentland, Heather, and Kimmeridge Clay. The main reservoir is the Brent sandstones. The Hod Formation is the impermeable layer that provides the seal to the reservoir and holds the high overpressures. The top of the Pliocene section shows the rapid burial during recent time.

The behavior of the first synrift petroleum system was also confirmed. Lateral migration occurs within the Gomo Member source rock and oil fills the interbedded turbidites. Because of overall low permeability, the filling up of reservoirs in this petroleum system was slower and occurred later (in Turonian time). Thus, the maturity of oils might be different in each separate turbidite along the cross section.

NORTH SEA HIGH-PRESSURE/ HIGH-TEMPERATURE CASE

North Sea Questions

The study area is the so-called high-pressure/high-temperature (HPHT) zone in the U.K. North Sea, 240 km off Aberdeen, Scotland (Figure 8). Pressures in the basin range up to 120 mega-Pascals (MPa) and temperatures up to 190°C. In this area, several fields were discovered with abnormal conditions, leading to severe constraints on exploration and drilling (Gaarenstroom et al., 1992; Grist et al., 1997). In this part of the North Sea, the petroleum system consists of Jurassic tilted blocks with three possible source rocks: the Pentland shales, the Heather shales, or the Kimmeridge Clay. Reservoirs are the Jurassic Fulmar sandstones and sometimes the Tertiary Forties sandstones (Figure 8).

The questions that arise are as follows:

1. What is the contribution of each source to the Jurassic reservoirs?
2. Why are some structures dry and others filled?
3. How can reservoirs at high temperatures (up to 190°C) and high pressures (up to 120 MPa) still bear high amounts of C_{15+} hydrocarbons?

We looked at the problem with a combined geochemical and modeling approach. Kerogens assumed to be representative for the source rocks were experimentally studied to determine cracking parameters. 2-D and 3-D data sets were built to model the generation and migration history. Details about the kinetics that have been used for both primary and secondary cracking can be found in Vandenbroucke et al. (1999).

Data Set

During the complete HPHT study, two major cross sections were constructed. The first section centers around the Franklin oilfield, and the second extends from southwest of Elgin northward to the northern fields. Here, we concentrate on the cross section going through Elgin. The section shows a double horst and graben system separated by an impermeable fault, with Elgin and the dry well 22/30 (Figure 8).

At present, all source rocks in the HPHT area are completely burned out. Hence, to derive adequate kinetics and richness, we had to rely on equivalent immature samples from adjacent areas. The Kimmeridge Clay and the Heather Formation are assumed to be type II source-bearing rocks, with initial richness of 8 and 2% total organic carbon (TOC), respectively. The Pentland Formation contains either type I or type III kerogen and has an initial richness of about 4%.

The compositional description of the hydrocarbon system relies on several specific classes or chemical components. This description goes back to the work of Behar et al. (1992), and has since been modified several times (Rudkiewicz et al., 1993; Behar et al., 1997; Vandenbroucke et al., 1999). Generated petroleum was described as a mixture of C_1, C_2–C_5, C_6–C_{14} and C_{15+} hydrocarbons with some additional subdivisions. The C_6–C_{14} and C_{15+} classes contain products with very different thermal stabilities, thus the carbon classes are again split into chemical classes accounting for the different stabilities of the aromatic and saturated fractions (Behar and Vandenbroucke, 1996). However, to display the results, the chemical classes are lumped together as C_{15+} saturates, C_{15+} aromatics, C_{15+} NSOs (nitrogen-, sulfur-, and oxygen-bearing compounds), C_6–C_{14} condensates, and C_1–C_5 gases.

Eight oil samples (C_{6+} fraction) from Elgin, Franklin, Erskine, and adjacent Kessog wells were analyzed in detail to describe their composition using the same chemical classes as those used in the kinetic scheme (Vandenbroucke et al., 1999). We should mention that such a detailed composition is often not available in a straightforward way. Reservoir engineers usually detail the composition up to C_{11} carbon numbers and group together all remaining products as C_{12+}. In contrast, organic geochemists typically provide a detailed description of C_{15+} products. Sometimes only specific products in the C_{15+} range are available, such as biomarkers, and no bulk description is provided. Here, those compositions were needed because they are being compared with the computed results.

Temperature and Pressure History

Pressures and overpressures were calibrated based on well data. The key parameters proved to be the permeability in the Hod Formation and the most recent sedimentary load. During Pleistocene time, almost 1000 m of sediments were deposited in less than 1.5 m.y (Figure 8). This rapid sedimentation and the very low permeability of the hardground in the Hod Formation made it difficult for fluids to migrate out of the reservoirs and kitchens. Permeabilities as low as 10^{-7} d were assigned to the Hod Formation to account for observed overpressures. The combination of fast loading and low permeability induced the observed high overpressures, as already observed in other basins with fast sedimentation rates (Burrus, 1998). Our calibration resulted in the overpressure profile shown in Figure 9. One can see from this

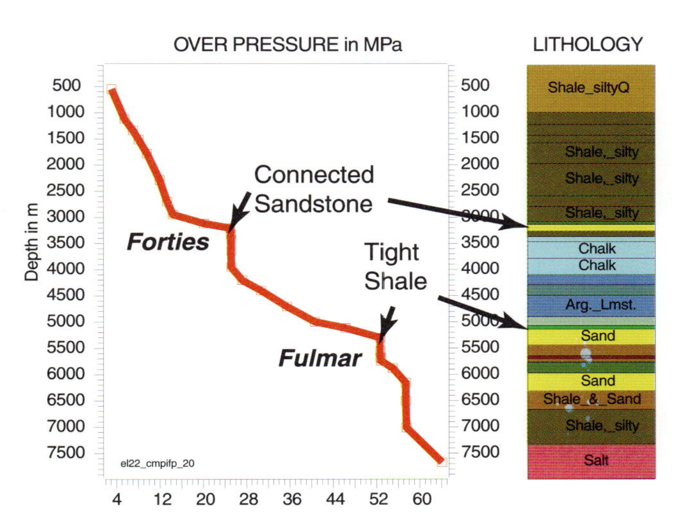

Figure 9—Vertical profile at Elgin field in the North Sea (at 43-km mark in Figure 8) showing the overpressure as modeled with Temispack. The continuous increase is related to the low permeability of the shaly layers. In formations with good permeabilities, such as the contiguous sandstones, the overpressure profile remains constant, indicating good connectivity.

profile that the overpressure increases with depth. It stays constant in areas with good vertical connectivity, such as in the chalk below the connected sandstones of the Forties and Fulmar reservoirs, and it increases rapidly in low-permeability formations such as the Hod shales.

Temperatures were calibrated on the basis of present-day temperatures and on vitrinite as a paleothermometer. A constant basal heat flow throughout the basin lifetime was assumed. Because of the rapid sedimentation during Pleistocene, there was a large thermal blanketing effect. Whereas basal heat flow at the bottom of 20-km-thick crust is 65 mW/m^2, the current heat flow at the base of the Triassic salt reaches only 52 mW/m^2. This shows that temperature, as well as pressure, is currently in a transient state and that the geologic history of the basin has to be taken into account to reconstruct the temperature history.

Figure 10 shows the temperature and pressure history for the Elgin reservoir based on our calibrations. The resulting temperature history shows a continuous increase related to ongoing subsidence. The sharp increase over the past 1.5 million years, when temperature jumped from 165° to 185°C and pressure from 80 to 110 MPa, is linked to the rapid Pleistocene sedimentation.

Hydrocarbon System

The hydrocarbon system was investigated using standard assumptions about source, TOC, and thermal regime. Then, the sensitivity of different parameters was tested. We concentrated on different types of kerogens for the various source rocks to see how this parameter could

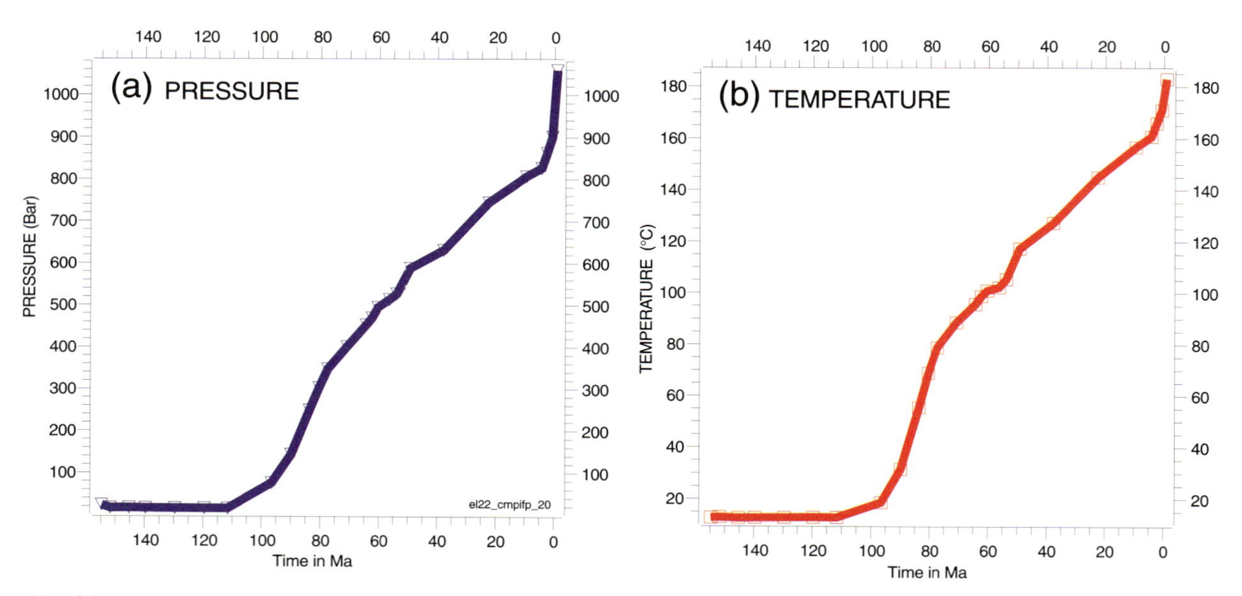

Figure 10—(a) Pressure and (b) temperature changes through time in the Elgin reservoir. Permeability is calibrated to observed pressures; basal heat flow is kept constant through time and is calibrated to present-day temperature and vitrinite reflectance. During the last 1.5 m.y., the temperature increased from 165 °C to 185°C and the pressure from 80 to 110 MPa because of recent sedimentation, showing the transient state of the basin.

influence the composition of predicted fluids in the reservoirs. The work flow that enabled an optimal use of computing time was as follows:

1. Calibrate pressure and temperature based on well data,
2. Calibrate capillary pressures and relative permeabilities using noncompositional runs, making simple assumptions about the nature of the hydrocarbons, and
3. Test the compositional description of fluids and compositional kinetics for primary and secondary cracking with full compositional runs.

Figure 11 shows the saturation and hydrocarbon composition in the Elgin reservoir through geologic time. Figure 11a shows the relative filling of the reservoir, and Figure 11b displays the "lumped" composition (in wt. %). It can be seen that the reservoir fills up continuously, though with a rapid rise in late Tertiary time. The first hydrocarbons reaching the reservoir were generated by primary cracking of the source from 40 to 16 Ma. The sharp rise from 4 Ma to present time is linked to secondary cracking in source rocks and carrier beds.

A noteworthy trend in the petroleum composition in the reservoir is the decrease in NSOs (geochemically, the unstable aromatic components) due to the increasing thermal stress. This shows that secondary cracking is also ongoing in the reservoir and not only in the source rock. The decrease in NSOs is correlated to the increase in solid organic components (coke) that precipitate and to the increase in lighter components, such as C_{15+} saturates, C_6–C_{14} condensates, or gases. Solid bitumens have been observed in cores from the Franklin field. This onset of secondary cracking in the reservoir, also shown by the decrease in C_{15+} saturates since 2 Ma, is linked to the global increase in temperature.

Generation first started in the Kimmeridge Clay kitchen at 50 and 38 Ma. Because of a delay in expulsion, the first hydrocarbons to reach the Elgin reservoir came from the deep kitchen to the east of the structure at 38 Ma (Figure 12). Later, they were followed by hydrocarbons originating in the Heather source from the inclined ramp to the southwest. Figure 12 shows that the Heather source was the most important feeder to the Elgin reservoir. In contrast, the Pentland source had only a limited contribution to reservoir feeding, probably because it is separated from the Fulmar sandstones by a low-permeability shale layer.

The contribution from the northern deep kitchen has continued through time since 38 Ma, with lighter hydrocarbons (condensates and gases) accounting for the composition of the reservoir. An attempt to model only the single tilted Elgin block without the deep kitchen (at 44–46 km in Figure 12) could not reproduce the observed accumulation and its composition. Also, the Kimmeridge Clay in the tilted block of the Elgin field did not contribute significantly to the feeding of the reservoir. Most of the flow dismigrated into the Cretaceous layers above. The main feeding from the source rock into the reservoir is exhausted today, as shown by the difference between the configurations at 38.6 Ma and at present day (almost no arrows from the deepest part of the southern kitchen). Scanning through time, it can be seen that the hydrocarbon flow into the Jurassic sandstone was at a maximum at 10.4 Ma.

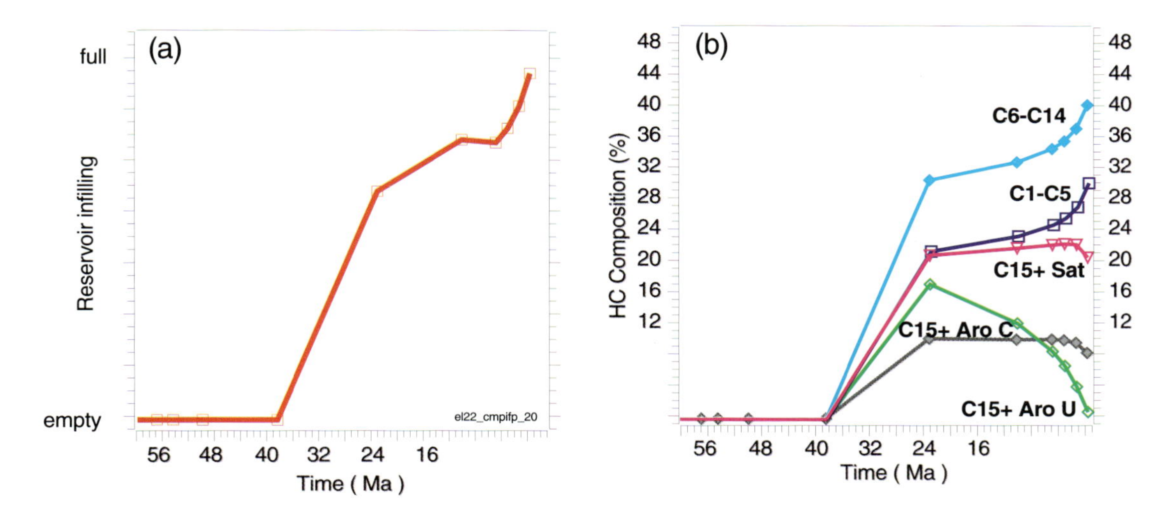

Figure 11—Modeled changes in (a) reservoir filling and (b) composition through time for the youngest Elgin reservoir. Abbreviations: C_{15+} Aro U, unstable aromatics; C_{15+} Aro C, condensed aromatics; C_{15+} Sat, saturates. These compounds have different thermal cracking rates. The decrease in C_{15+} shows the onset of secondary cracking in the reservoir and the transient state of the composition in the reservoir. It is probable that all C_{15+} saturates would be cracked down if the reservoir had been at 200°C for 10–15 Ma.

Modeling the generation and migration also gives information about the phase state through geologic history. An important feature is that all hydrocarbon fluids form a single phase. A free gas phase has never existed since generation and expulsion started. The hydrocarbon phase continuously changed in composition from a fluid rich in C_{5+} to one rich in C_{5-}. This means that a "gas feeding" of the reservoir by a gas charge sweeping the oil is not physically realistic.

Another distinct feature is the dismigration into the Forties sandstones that is observed in the northern faulted block. This results from a combination of two factors: (1) the greater charge coming from the deep and large kitchen areas, and (2) the higher structural position, which leads to a seal of poorer compaction and greater permeability. Seal failure probably explains why this structure is dry.

Finally, we devoted a large part of our study to checking the sensitivity of the hydrocarbon system to the different source rock types. We have studied several cases to test the sensitivity to (1) the presence or absence of each source rock and (2) the difference between a type III and a type I source rock in the Pentland Formation.

Generation and migration modeling allowed us to check these assumptions efficently. The implications of the presence or absence of source rock potential in one or more of the horizons is not easy to anticipate without such tools. From the runs, it can be seen that the final gas–oil ratio (GOR) in the Elgin reservoir does not depend, at a first approximation, on the source rock type. All GOR pictures look very much alike. The modeled GORs in the Elgin reservoir are also similar and range between 1500 and 2000 scf/bbl, which is consistent with observed values. In contrast, the saturation of the reservoirs depends on both the source rock types and on the geometry of the feeding system.

The homogeneity of GOR and the few differences observed between type I and type III sources can now be seen as the effect of high temperatures. In this area, the source rocks are all completely cooked. Hence, the chemical differences in the sources, due to differences in the generated products, is erased because of the ongoing secondary cracking of the heaviest generated products. This secondary cracking depends only on the chemical nature of the class and not on the kerogen. Hence, the resulting composition trapped in the reservoir mimics the secondary cracking rather than the primary cracking only. This is an unexpected result and is clearly linked to the chemical separation of the C_{15+} into thermally resistant and labile products.

CONCLUSIONS: WHAT HAVE WE LEARNED?

From these studies, we have learned several lessons that are worth disseminating into the industry. They are related to the following:

1. Set up and manipulation of 2-D and 3-D data sets,
2. Behavior of faults as migration pathways,
3. Behavior of shales and their capacity to hold high pressures, and
4. Increase in the oil window with depth in a geologic dynamic system.

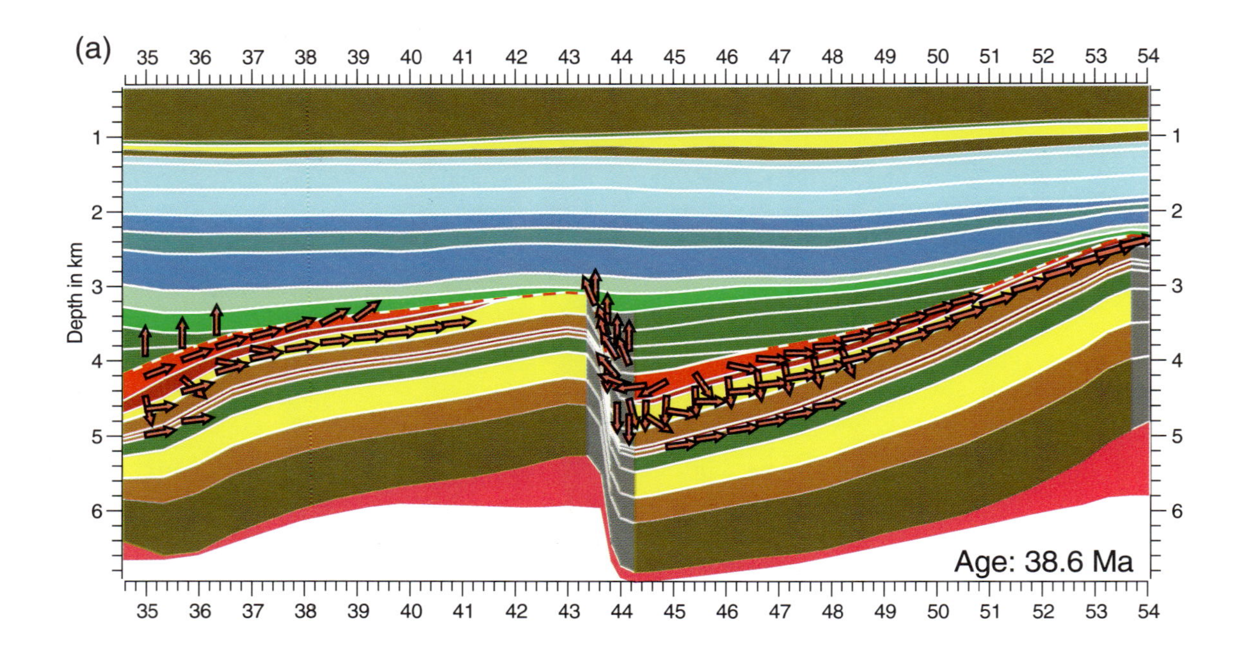

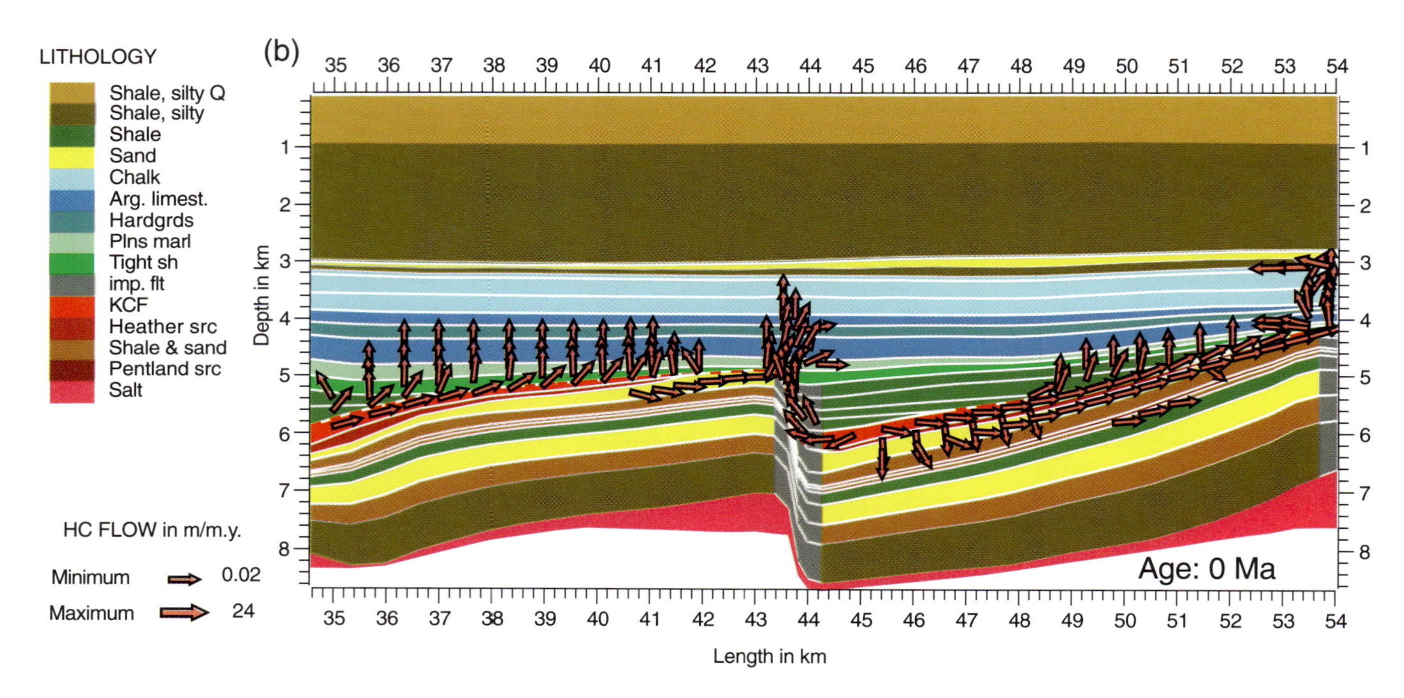

Figure 12—Cross sections of the Elgin reservoir showing computed migration arrows (a) at the onset of migration (38.6 Ma) and (b) at present time. Parts without arrows have no hydrocarbons (less than 3 km depth) or flow rates lower than 0.02 m/m.y. (more than 3 km depth). Note the early onset of Kimmeridge Clay feeding from the kitchen at 45 km, and the present-day dismigration form the northernmost end at 54 km, reaching the Forties sandstone at 3 km depth.

Why 2-D and 3-D?

One of the questions is related to the results from 2-D cross sections. Why should some work be performed with 2-D tools when nature is actually 3-D? Some answers are linked to technical reasons and some are operational. Technical reasons are linked to the features that can be currently simulated only with 2-D cross sections and are not available with 3-D tools. Operational reasons are linked to the availability of input data. Regional data were available as 2-D cross sections because a large number of seismic sections existed. They could be depth converted with a minimum amount of work. Their lithologies could also be interpreted consistently.

In contrast, a semi-regional 3-D data set was not available because such a data set has never been collected. The only available 3-D data set centered around the Franklin structure because that structure was on the point of being developed. Based on 3-D seismic, depth maps of all interesting horizons had been set up. Also, a correct picture of the facies distributions in the main reservoir and source layers existed.

A technical reason why 2-D cross sections were set up is that the 2-D tool enables a discussion of the composition of the generated and migrated hydrocarbons. At the time of the HPHT study in 1997, this was not possible with the 3-D tool. Computation time also plays a crucial role. All calibrations regarding pressure, temperature, permeability, and capillary pressures were obtained with the 2-D tool. In mid-1997, computing times in 2-D were on the order of 4–10 hours on a Sun Ultra 1, whereas they reached 4–10 days for a 3-D study. Now, in mid-2000, 12–24 hours of computing time is acceptable for large 3-D cases with several tens of thousands of cells on the same type of machine.

Faults as Migration Pathways

The Recôncavo Basin test case showed that faults can act as efficient migration pathways, even if they appear to be sealing at the present time. There have been numerous studies of fault behavior (e.g., Hooper, 1991; Grauls and Baleix, 1994; Moretti, 1998). Few, however, have concentrated on migration over geologic time. In our study, a clear difference between the two short-term properties of faults—sealing from the reservoir point of view and openess from the migration point of view—could be demonstrated.

Behavior of Shales and Capacity to Hold High Pressures

In the HPHT case from the North Sea, we have shown that overpressure can be modeled as a dynamic response to the sedimentary load deposited during the Pleistocene. Because of the impermeable nature of the Hod Formation, pressures did rise during this late stage of basin history. However, because the Hod Formation was already compacted at that time, overpressure development was not linked to significant undercompaction. This is a new feature, which has not been seen in other young overpressured basins such as the Gulf of Mexico or the Mahakam Delta, where overpressure and undercompaction are linked (Burrus, 1998).

Increase in Oil Window with Depth in a Dynamic System

Finally, the most interesting outcome of the HPHT project that will be of the greatest interest for oil exploration is the fact that some hydrocarbon species can survive temperatures as high as 190°C. However, this has to be in a geologic situation in which the temperature increase is recent. Because of the kinetics of secondary cracking, a mixture comprising C_{14} *n*-alkanes can resist thermal stress for several million years but not for several tens of million years.

A detailed sensitivity study of the type of organic matter showed that the important factor in this type of geologic environment is the secondary cracking scheme of the hydrocarbons. In a petroleum system where high temperatures cause a large amount of secondary cracking both in source rocks and in reservoirs, it seems difficult to predict the occurrence of oil or gas without taking into account the detailed chemical composition of products linked to their thermal stability.

Acknowledgments—*The HPHT part of this work was performed under the scope of the Thermie Project, no. OG/211/94-FR-UK: HPHT Petroleum System Prediction from Basin to Prospect Scale. We are grateful to IFP, Petrobrás, Elf, and BP management for permission to publish this paper.*

REFERENCES CITED

Behar, F., and M. Vandenbroucke, 1996, Experimental determination of the rate constants of the *n*-C_{25} thermal cracking at 120, 400, and 800 bar: implications for high-pressure/high-temperature prospects: Energy and Fuels, v. 10, p. 932–940.

Behar, F., S. Kressmann, J-L. Rudkiewicz, and M. Vandenbroucke, 1992, Experimental simulation in a confined system and kinetic modeling of kerogen and oil cracking: Organic Geochemistry, v. 19, p. 173–789.

Behar, F., M. Vandenbroucke, Y. Tang, F. Marquis, and J. Espitalié, 1997, Thermal cracking of kerogen in open and closed systems: determination of kinetic parameters and stoichiometric coefficients for oil and gas generation: Organic Geochemistry, v. 26, p. 321–339.

Burrus, J., 1998, Overpressure and undercompaction in shaly series: a critical reappraisal, *in* B. E. Law, G. F. Ulmishek, and V. I. Slavin, eds., Abnormal pressures in hydrocarbon environments: AAPG Memoir 70, p. 35–63.

Daniel, L. M. F., E. M. Souza, and L. F. Mato, 1989, Geoquímica e modelos de migração de hidrocarbonetos no Campo de Rio do Bu. Integração com o compartimento nordeste da Bacia do Recôncavo: Boletim de Geociências da Petrobrás, v. 3, p. 201–214.

Figueiredo, A. M. F., J. A. E. Braga, H. M. C. Zabalaga, J. J. Oliveira, G. A. Aguiar, O. B. Silva, L. F. Mato, L. M. F. Daniel, L. P. Magnavita, and C. H. L. Bruhn, 1994, Recôncavo Basin, Brazil: a prolific intracontinental rift basin, *in* S. M. Landon, ed., Interior rift basins: AAPG Memoir 59, p. 157–203.

Gaarenstroom, L., R. A. J. Tromp, M. C. De Jong, and A. M. Brandenburg, 1992, Overpressures in the Central North Sea: implications for trap integrity and drilling safety: Petroleum Geology of NW Europe Conference, Fourth Geological Society Meeting, v. 2, p. 1305–1313.

Gaglianone, P. C., and L. A. F. Trindade, 1988, Caracterização geoquímica dos óleos da Bacia do Recôncavo: Geochimica Brasiliensis, v. 2, p. 15–39.

Ghignone, J. A., and G. Andrade,1970, General geology and major oil fields of Recôncavo Basin, Brazil, *in* M. T. Halbouty, ed., Geology of giant petroleum fields: AAPG Memoir 14, p. 337–358.

Grauls, D. L., and J. M. Baleix, 1994, Role of overpressures and in situ stresses in faults controlled hydrocarbon migration: a case study: Marine and Petroleum Geology, v. 11, p. 734–742.

Grist, M., J. Meudell, J. Brubaker, and G. King, 1997, HPHT (high pressure, high temperature) reviewed: Offshore Engineering, January, p. 25–37.

Hooper, E. C., 1991, Fluid migration along growth faults in compacting sediments: Journal of Petroleum Geology, v. 14, p. 161–180.

Lerche, I., 1990, Basin analysis: quantitative methods: Academic Press Geology Series XIII, v. 1, 562 p., and v. II, 570 p.

Magnavita, L. P., 1992, Geometry and kinematics of the Recôncavo-Tucano-Jatobá rift, NE Brazil: Ph.D. dissertation, University of Oxford, Oxford, U.K., 492 p.

Magnavita, L. P., and J. A. Cupertino, 1987, Concepção atual sobre as bacias do Tucano e Jatobá, Nordeste do Brasil: Boletim de Geociências da Petrobrás, v. 1, p. 119–134.

Magnavita, L. P., I. Davison, and N. J. Kusznir, 1994, Rifting, erosion and uplift history of the Recôncavo-Tucano-Jatobá rift, northeast Brazil: Tectonics, v. 13, p. 367–388.

Milani, E. J., and I. Davison, 1988, Basement control and transfer tectonics in the Recôncavo-Tucano-Jatobá rift, NE Brazil: Tectonophysics, v. 154, p. 41–70.

Milani, E. J., M. C. Lana, and P. Szatmari, 1987, Mesozoic rift basins around the NE Brazilian microplate, *in* W. Manspeizer, ed., Triassic–Jurassic rifting and the opening of the Atlantic Ocean—continental breakup and the origin of the Atlantic Ocean and passive margins: Developments in Geotectonics, v. 27, p. 833–858.

Moretti, I., 1998, The role of faults in hydrocarbon migration: Petroleum Geoscience, v. 4, p. 81–94.

Penteado, H. L. B., 1999, Modélisation compositionnelle 2-D de la genèse, expulsion et migration du pétrole dans le Compartiment Sud du Bassin de Recôncavo, Brésil: Ph.D. dissertation, Université de Paris, Paris, 233 p.

Rudkiewicz, J-L., O. Brevart, J. Connan, and F. Montel, 1993, Primary migration behavior of hydrocarbons: from laboratory experiments to geologic situations through fluid flow models: Organic Geochemistry, v. 22, p. 631–639.

Rudkiewicz, J-L., S. Wolf, F. Behar, M. Vandenbroucke, and J. Wendebourg, 1996, Gases in reservoirs: a balance between generation and migration, part I: compositional modeling (abs.): AAPG Annual Convention, Program with Abstracts, San Diego, California, p. 33.

Santana, A. C., 1989, Continuidade dos reservatórios da Formação Sergi no Campo de Dom João—Bacia do Recôncavo: Boletim de Geociências da Petrobrás, v. 3, p. 337–345.

Ungerer, P., J. Burrus, B. Doligez, P. Y. Chenet, and F. Bessis, 1990, Basin evaluation by integrated 2-D modeling of heat transfer, fluid flow, hydrocarbon generation, and migration: AAPG Bulletin, v. 74, p. 309–335.

Vandenbroucke, M., F. Behar, and J-L. Rudkiewicz, 1999, Kinetic modelling of petroleum formation and cracking: implications from high-pressure/high-temperature Elgin field (U.K. North Sea): Organic Geochemistry, v. 30, p. 1105–1125.

Mello, M. R., J. M. Moldowan, J. Dahl, and A. G. Requejo, 2000, Petroleum geochemistry applied to petroleum system investigation, *in* M. R. Mello and B. J. Katz, eds., Petroleum systems of South Atlantic margins: AAPG Memoir 73, p. 41–52.

Chapter 4

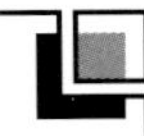

Petroleum Geochemistry Applied to Petroleum System Investigation

M. R. Mello

Petrobrás Research Center
Rio de Janeiro, Brazil

J. M. Moldowan

Stanford University
Stanford, California, U.S.A.

J. Dahl

Biomarker Technology, Stanford University
Stanford, California, U.S.A.

A. G. Requejo

Geochemical Solutions International
The Woodlands, Texas, U.S.A.

Abstract

Over the past 25 years, an evolution has taken place in the understanding of the distribution and habitat of petroleum source rocks, allowing their occurrence to be accurately predicted, their effectiveness to be modeled, and the resulting petroleum accumulations to be explained. Simultaneous developments in analytical methodology have established gas and molecular organic geochemistry as some of the most effective tools to characterize petroleum in terms of source rock characteristics, thermal evolution, migration pathways, and overall geologic history. The petroleum system concept requires a progressive shift in emphasis from rock to the fluid system to petroleum discoveries. It also defines the relationship between geologic elements and processes in time and space in order to understand the driving forces behind the distribution of hydrocarbon provinces. Several case histories from Brazilian sedimentary basins demonstrate the central role of new and emerging geochemical techniques in defining petroleum systems and identifying future exploration opportunities. Geochemistry is a powerful tool that enables the exploration geologist to target the most prospective oil and gas plays.

INTRODUCTION

Since the beginning of the century, geologists have used geochemical methods to identify hydrocarbon seeps (Tissot and Welte, 1984; Mello et al., 1996). With the acceptance of the organic origin for petroleum, one of the critical questions in unexplored basins became the location of organic-rich sediments mature enough to generate and expel hydrocarbons. These were initially addressed by the analysis of subsurface samples using geochemical measurements such as total organic carbon, Rock-Eval pyrolysis, organic petrography, and elemental analyses (described in Tissot and Welte, 1984; Peters and Cassa, 1994). More recently, the emergence of sophisticated analytical techniques such as high-resolution gas chromatography (GC) and mass spectrometry (MS) has established geochemistry as a key element of the petroleum exploration process, whereas earlier approaches emphasized geophysical information (e.g., Peters and Moldowan, 1993; Mello et al., 1994, 1996).

With the increased need to improve petroleum exploration success and reduce risk, the characterization of gases and oils using molecular and isotope geochemistry has provided a new impetus to the evaluation of petroleum systems, plays, and prospects. Using only samples of oils or gases, these techniques allow an assessment of source rock paleodepositional environment, age, and level of thermal evolution in the absence of actual source rock samples (e.g., Mello, 1988; Mello et al., 1988; Peters and Moldowan, 1993; Mello et al., 1998; Moldowan et al., 1998). Furthermore, molecular and isotope geochemistry allow determination of the degree of biodegradation and alteration, recognition of oil and gas mixtures, and assessment of migration distances in a sedimentary basin (e.g., Trindade, 1992; Peters and Moldowan, 1993; Moldowan et al., 1998; Mello et al., 1998; Prinzhofer et al., 2000).

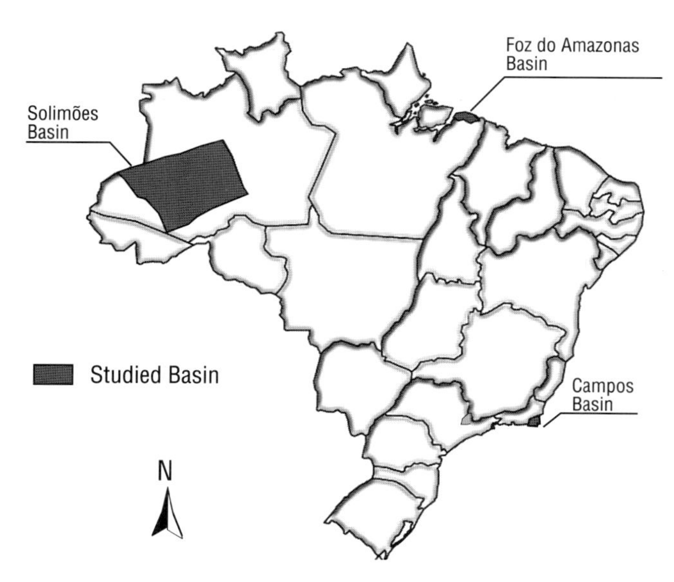

Figure 1—Location map of the sedimentary basins in which the samples were taken.

The integration of geochemical data with geologic information is crucial for proper understanding of a petroleum system and can identify new exploration plays in both frontier and mature basins. This chapter demonstrates the central role of molecular organic geochemistry in the petroleum system approach using oil and gas samples collected in Brazilian sedimentary basins.

RESULTS AND DISCUSSION

Petroleum systems investigations require initial identification of the hydrocarbon fluid system, followed by determination of the age and origin of the sedimentary rocks that are related to this fluid system (Magoon and Dow, 1994; Mello et al., 1994). Recently, a multidisciplinary approach involving surface and gas geochemistry, chemostratigraphy and diamondoid research, has greatly enhanced the level of understanding of some of the most representative petroleum systems in Brazilian sedimentary basins (Mello et al., 1994, 1998; Brooks et al., 1998; Dahl et al., 1999; Prinzhofer et al., 2000). Results of successful case studies in three Brazilian basins are presented (Figure 1). In each case, detailed information has been obtained concerning oil and gas types, the deadline (depth) for liquid hydrocarbons, and the degree of oil cracking. This information has proved integral to the identification of oil and gas prone regions and in identifying the most promising areas for future drilling.

Offshore Geochemical Prospecting

The basic assumption underlying the use of oil and gas seeps in offshore petroleum prospecting is that hydrocarbons from deep sedimentary sequences can migrate, either directly from source rocks or from reservoirs, to the sea floor. Geochemical analysis of surface and near-surface sediments in a generative basin should therefore detect a surface expression of the underlying hydrocarbons. Surface geochemical techniques that involve direct detection of hydrocarbons (seeps) have gained increasing acceptance as a risk-assessment tool in offshore exploration. The emergence of sophisticated analytical techniques using fluorescence, microbiology, GC, and MS now allows detection of low levels of migrated hydrocarbons.

Oil and gas seeps are of great significance to modern offshore petroleum exploration. Foremost, they indicate the presence of generative hydrocarbon source rocks, without which there can be no accumulations. Sediment cores retrieved using a ship-based piston core device can be analyzed in the laboratory to detect hydrocarbons derived from seeping oil and gas. This approach has been employed successfully in various offshore petroliferous basins, including the northern Gulf of Mexico, west Africa, and parts of Latin America such as Trinidad and Colombia (Sassen et al., 1993; Mello et al., 1998). In deep water where exploration costs are high and few wells have been drilled, the occurrence and nature of hydrocarbon seeps is one of the few available means for assessing prospective areas.

Piston coring has several advantages over other surface geochemical techniques. In addition to being one of the most cost-effective approaches, sediments record an integrated seep signal over time and typically yield sufficient hydrocarbon material to perform conventional geochemical analyses used to identify oil properties, thermal maturity, and source rock type. This information is an important element in proper play concepts which are necessary for a successful exploration program.

With an increased need to improve exploration success and reduce risks in frontier and unexplored deepwater areas of Campos Basin, Brazil, over 100 seabed piston cores were collected and analyzed (Figure 1). The Campos Basin is the most productive and prolific offshore Brazilian hydrocarbon province. It is located offshore Rio de Janeiro in southeastern Brazil and covers approximately 100,000 km^2. The giant deep-water Marlim Field is the largest oil field in the basin, with an approximate volume of 14 billion bbl of oil in place (Guardado et al., 1990). The broad objective of this study was to evaluate the presence of thermogenic hydrocarbon seepage over a deep-water turbidite associated with a salt diapir structure in order to evaluate the presence and nature of the hydrocarbon system.

Based on published literature (e.g., Guardado et al., 1990; Mohriak et al., 1990a, 1990b; Mello et al., 1994), the offshore Campos Basin shares key geologic attributes with other areas where sea floor seeps have been successfully detected, such as in the Gulf of Mexico (Sassen et al., 1993). The gross architecture of this basin has been profoundly influenced by salt tectonics as a result of rapid Tertiary sediment loading. Salt and faults serve as con-

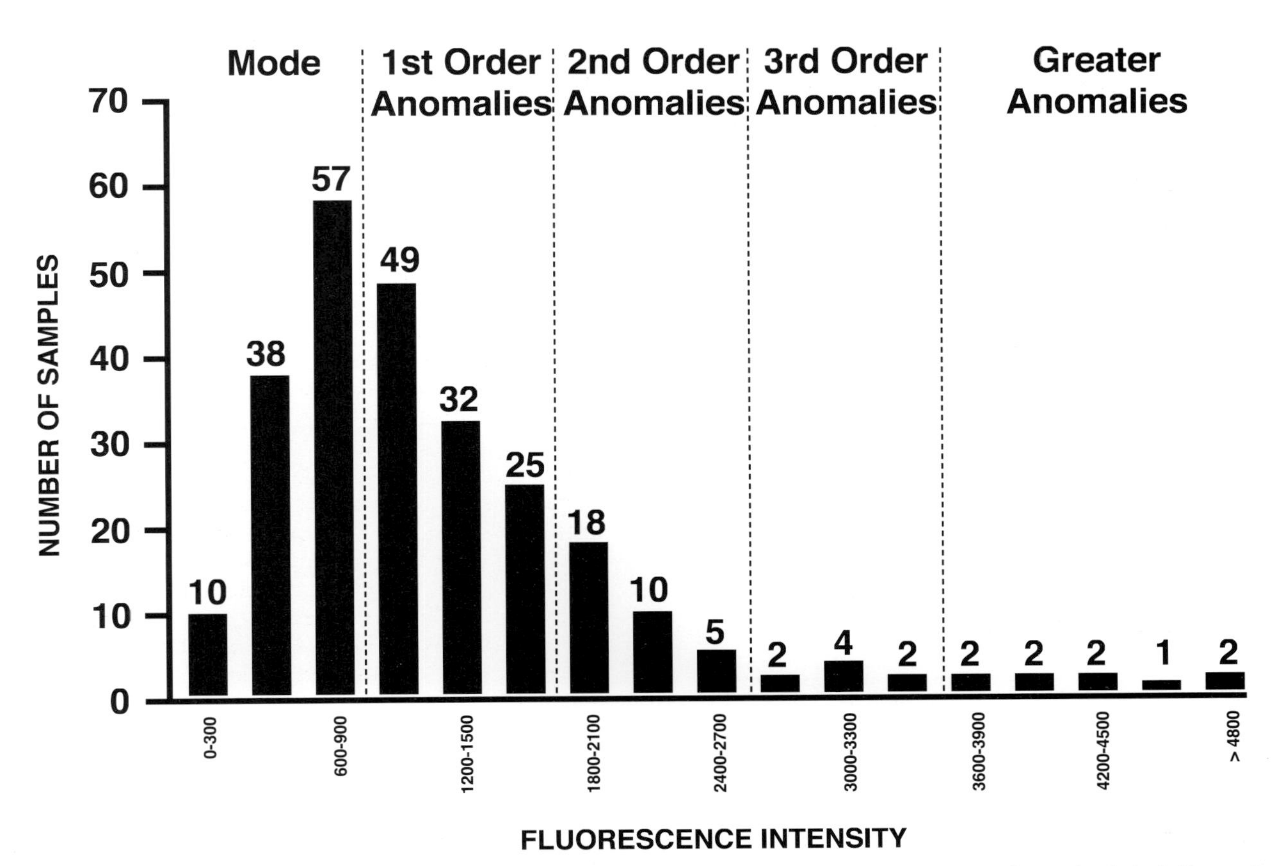

Figure 2—Fluorescence intensity distribution in Campos Basin sediments, presented in terms of standard deviations above a mode value.

duits for vertical migration, and many of these conduits intersect the sea floor. Also, the timing of hydrocarbon generation and migration is thought to be from geologically recent to the present. These features are conducive to the active migration of hydrocarbons to the sea floor.

Piston cores were sited using regional geophysical profiles to maximize the probability of sampling sea floor seeps connected to the subsurface hydrocarbon system. Core locations were selected to satisfy several objectives: (1) to provide a regional overview of the main structural elements so as to map the hydrocarbon fairway, (2) to provide calibration over known discoveries, (3) to provide prospect-focused locations for evaluating undrilled structures, and (4) to provide a limited number of "blank" locations intended to define the regional background.

Cores were analyzed by total scanning fluorescence (TSF), a sensitive method for the detection of petroleum in seabed cores. The majority of cores analyzed have relatively low fluorescence intensity (INT) values (Figure 2). In general, INT values less than 4000 units in sea floor sediments represent either the presence of microseepage (very low level hydrocarbon seepage) or regional background in the absence of seepage (Brooks et al., 1986). However, some of the highest fluorescence INT values (greater than 80,000 units) were detected in cores located over vertical conduits such as salt diapirs associated with faults that approach the sea floor. Such features have played an important role in migration pathways for Late Cretaceous turbidite accumulations. Salt diapirs at close proximity to the sea floor (Figure 3) can focus hydrocarbon migration from source and trap to the sea floor.

Figure 3 shows the distribution of several drop cores at a site where oil was recovered in a seabed core. The GC of the oil-bearing core (not shown) exhibited a broad unresolved complex mixture (UCM) "hump" characteristic of biodegraded crude oil. The detection of oil at this location is direct evidence of hydrocarbon migration from the deep subsurface to the sea floor, suggesting the presence of generative hydrocarbon source rocks. More detailed analyses of this seep by high-resolution gas chromatography–mass spectrometry–mass spectrometry (GCMS-MS) and carbon isotope composition of saturated and aromatic hydrocarbons (Figure 4) suggest great similarities between this oil seep and typical lacustrine saline oils from Campos Basin (Mello, 1988; Mello et al., 1988, 1995). The high abundance of 4-methyl steranes and the presence of 24-*n*-propylcholestanes in the sample (Figure 4) suggests that the oil was generated from upper Barremian calcareous mudstone facies interbedded in the upper Lagoa Feia Formation in which marine transgressions had occurred (Mello et al., 1998).

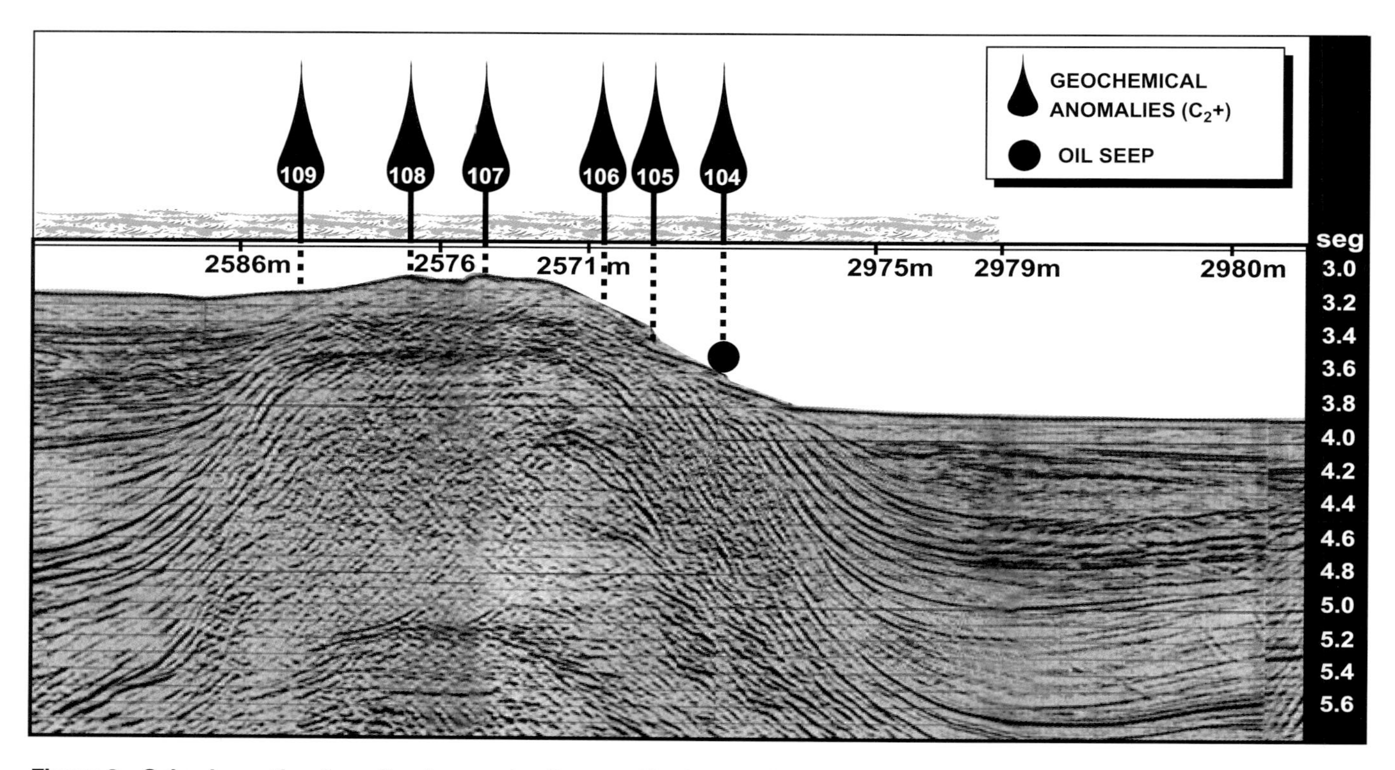

Figure 3—Seismic section from the deep-water Campos Basin showing salt diapir subcropping near the sea floor, with core location and samples showing oil and gas seeps.

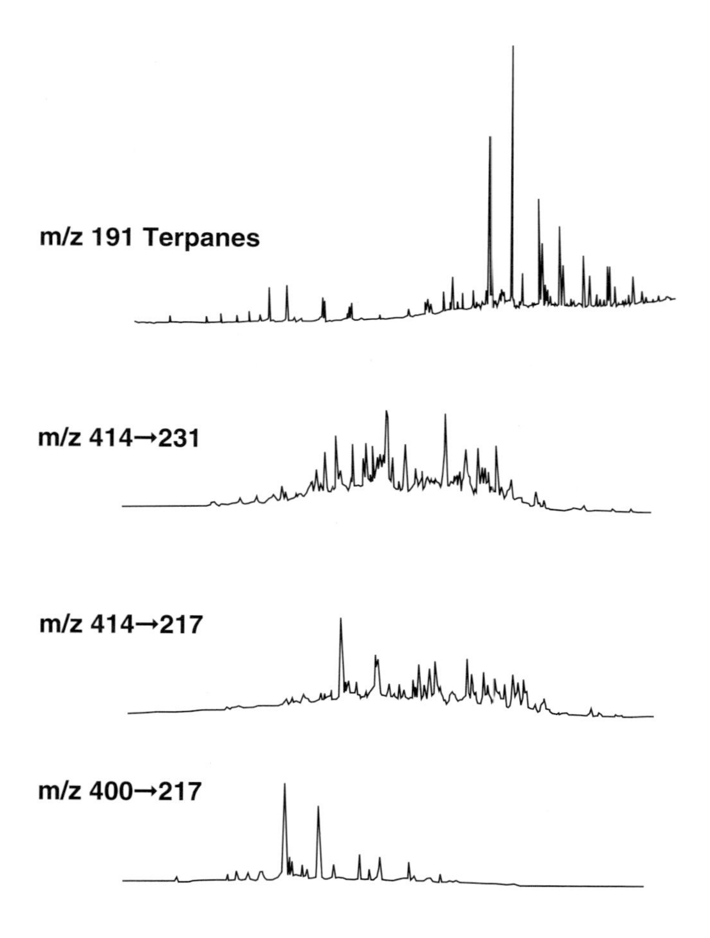

Also, gas macro- and microseeps were detected in the area using methods for detection of free gaseous hydrocarbons in the methane to pentane range (data not presented here). Using all sample results, a statistical treatment was performed with the concentrations of each gas from methane to pentane and with the sum of concentrations in the ethane to pentane ranges. The microseep anomalies for each gas were mapped and correlated with existing hydrocarbon-bearing structures. The results showed good correlation between gas geochemical anomalies and the structural traps mapped in the study area. Also, heavier gas anomalies (butane to pentane) were associated with oil field reservoirs containing heavier oils (lower API gravities), suggesting that such methods could also predict oil quality (Mello et al., 1998).

The clear association of the seepage detected with subsurface traps and salt tectonic-derived conduits, as shown in Figure 3, indicates the presence of subsurface hydrocarbon accumulations with active migration of hydrocarbons to the sea floor of Campos Basin. As observed in major offshore discoveries throughout the world, large discoveries in deepwater Brazil will almost certainly be marked by meaningful hydrocarbon seeps.

← Figure 4—Mass chromatogram (*m/z* 191) and GCMS-MS sterane data from an oil sample collected in Core 104 (shown in Figure 3).

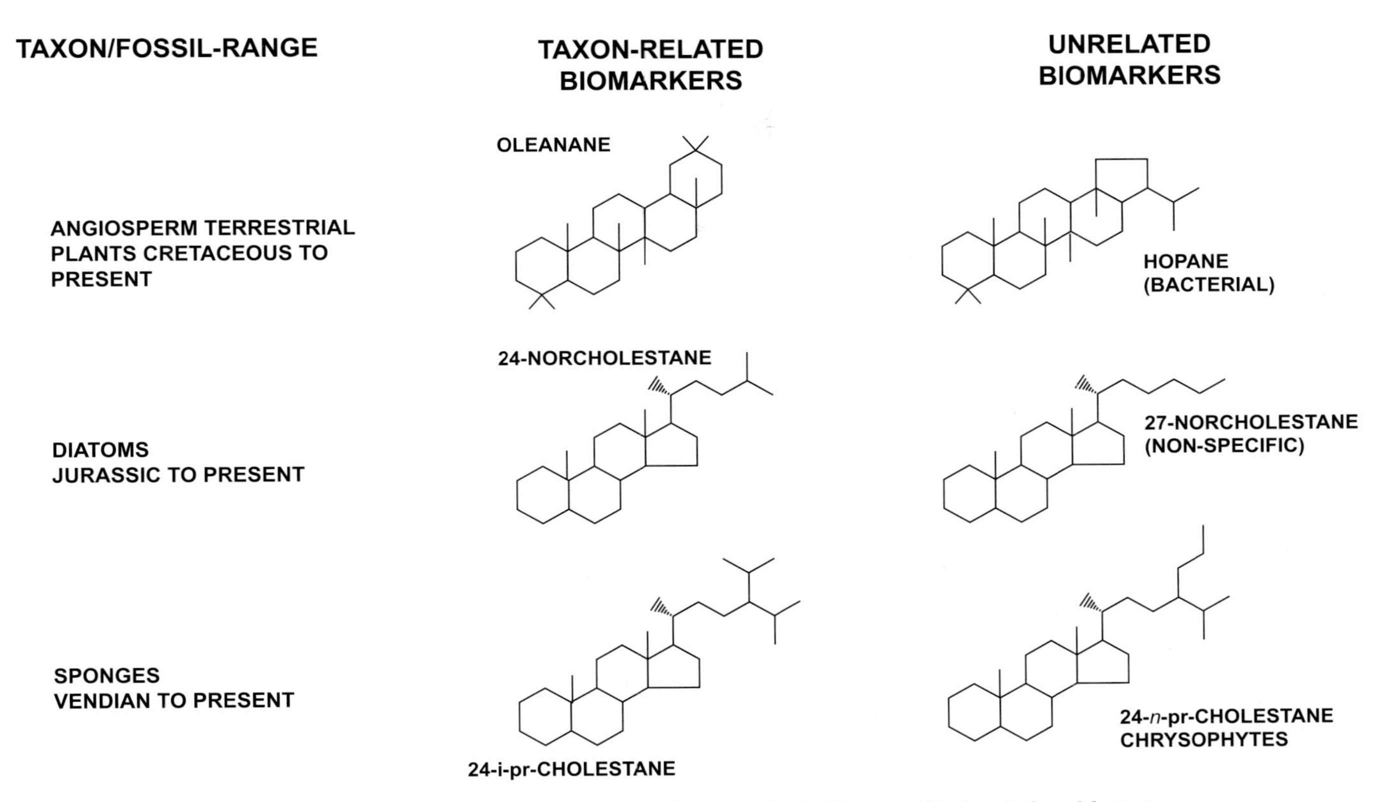

Figure 5—Occurrence of selected biological markers throughout geologic time and their relationship to taxons.

Chemostratigraphy

The method used to correlate oils and tie them to source rocks based on their chemical composition can be called *chemostratigraphy.* Just as micro- and macrofossils form the basis for biostratigraphy, biological markers (also known as biomarkers or molecular fossils) form the basis for chemostratigraphy. Although the study of chemostratigraphy is still in its infancy and therefore relatively crude in comparison to biostratigraphy, chemostratigraphy alone offers the possibility of age-dating oils. This is because molecular fossils, unlike macro- and microfossils, are oil soluble and migrate from the petroleum source rock along with generated oil into reservoirs. As a result, the oil maintains the age-diagnostic signature, providing a means to determine the age of an oil source rock without analyzing the rock itself. This can be very important because the source rock is usually not drilled. Similarly, some features of the biomarker distributions can be used to determine the depositional environment and lithology of the source, while others indicate the relative thermal maturity of the source at the time of oil expulsion. Recent research on age-related biomarkers has dramatically enhanced the resolution of chemostratigraphy. These biomarkers include the following (Figure 5):

1. **Oleanane**—This biomarker is derived from angiosperms. It distinguishes oils derived from source rocks of Cretaceous and younger age (Moldowan et al., 1994, 1998; Mello et al., 1997). This biomarker has proven to be extremely useful in recognizing Tertiary sources throughout the world, including Brazil, Colombia, Venezuela, west Africa, South Caspian, and the U.S. Gulf Coast.
2. **24-Isopropylcholestanes**—The ratio of these biomarkers (which are probably derived from sponges) to 24-*n*-propylcholestanes (derived from marine algae) can be used to indicate if an oil is derived from Cambrian or older source rocks (McCaffrey et al., 1994). This ratio was used to determine the source of the Benghewala-1 oil in India (Peters et al., 1995).
3. **Dinosteranes and triaromatic dinosteroids**—These biomarkers are derived from functionalized natural products originally synthesized by dinoflagellates. However, their recognition in source rocks older than Triassic (the period in which dinoflagellates evolved), especially those containing abundant achritarchs, led Moldowan and Talyzina (1998) to suggest a genetic dinoflagellate–achritarch link. The absence of triaromatic dinosteroids in marine-derived oils indicates a Triassic or older source (Moldowan et al., 1996).
4. **24-Norcholestanes and 24-nordiacholestanes**—The precursors to these molecular fossils are derived from diatoms (Holba et al., 1998). The

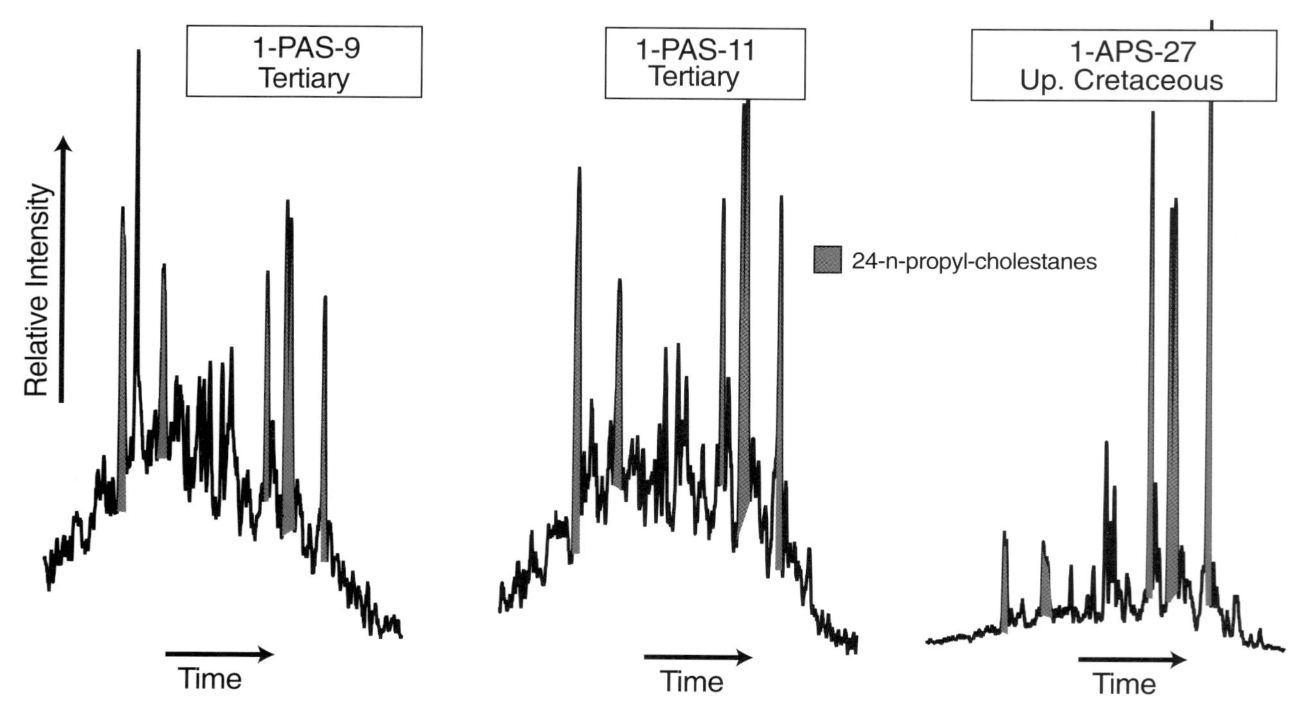

Figure 6—Metastable reaction monitoring–GCMS (MRM-GCMS) is used to provide key information about the paleoenvironment of oil source rocks from analysis of oils from Foz do Amazonas Basin. The presence of C_{30} steranes (24-*n*-propylcholestanes) is diagnostic for the marine algae contribution.

> ratio of these compounds to other C_{26} steranes, specifically 27-norcholestanes and 27-nordiacholestanes, can be used to recognize oils derived from Cretaceous and younger source rocks.

This chapter provides an example of the utility of these biomarkers for understanding the generative petroleum systems of Foz do Amazonas Basin, offshore Brazil.

Characteristic steroids show a high specificity in modern marine taxa, and their derived steroids reflect the fossil record of a given taxonomic group. The analysis of such steranes in oils provides information about the age of the source rock, even when the appropriate source rock is not available for analysis and correlation. These new parameters generally require the analytical specificity of GCMS-MS for reliable application. For example, certain steranes with 26 carbon atoms (24-norcholestanes) have a possible affinity to diatoms and their ratio to other 26-carbon steranes (27-norcholestanes) follows the record of diatom evolution in oils (Holba et al., 1998). This can be used to determine the age of oil within the diatom radiation envelope (Jurassic–Tertiary).

The geologic evolution and abundance of the biomarker oleanane has been shown to parallel the evolution of the flowering plants (Moldowan et al., 1994). According to the published data set, Tertiary oils with oleanane/(hopane + oleanane) ratios of >0.2 can generally be distinguished from older oils, which have ratios of <0.2. The oleanane analysis benefits from the higher specificity of metastable reaction monitoring–GCMS (MRM-GCMS) and GCMS-MS compared to GCMS by permitting oleanane amounts to be distinguished from other triterpanes, particularly when oleanane occurs in very low amounts relative to hopane.

The application of these and other molecular parameters are demonstrated by examples from the area of Foz do Amazonas Basin, Brazil (Figure 1). Oil, condensate, and potential source rock samples from this basin have been evaluated to determine their possible affinities to Tertiary or Upper Cretaceous source rocks believed to occur in the area (Mello, 1988; Mello et al., 1995).

In this study, selected oil and condensate samples were analyzed by carbon isotope, GC, and GCMS-MS techniques. All samples show significant amounts of C_{30} steranes and diasteranes (24-*n*-propylcholestanes and 24-*n*-propyldiacholestanes) (Figure 6). The occurrence of these compounds in oils indicates a contribution from marine algae, typically from a source rock deposited in an open marine to marine-influenced depositional environment (Mello et al., 1995). Thus, C_{30} steranes and diasteranes can show relative amounts of marine versus nonmarine input. Figure 7 shows how the ratio of C_{30} diasteranes is distributed in the analyzed samples. The samples of Cretaceous and Tertiary origin fall into two groups, mainly higher and lower C_{30} diasteranes (higher and lower marine input, respectively) based on this analysis. These data suggest a more marine character for the source rock that gave rise to the Cretaceous oils.

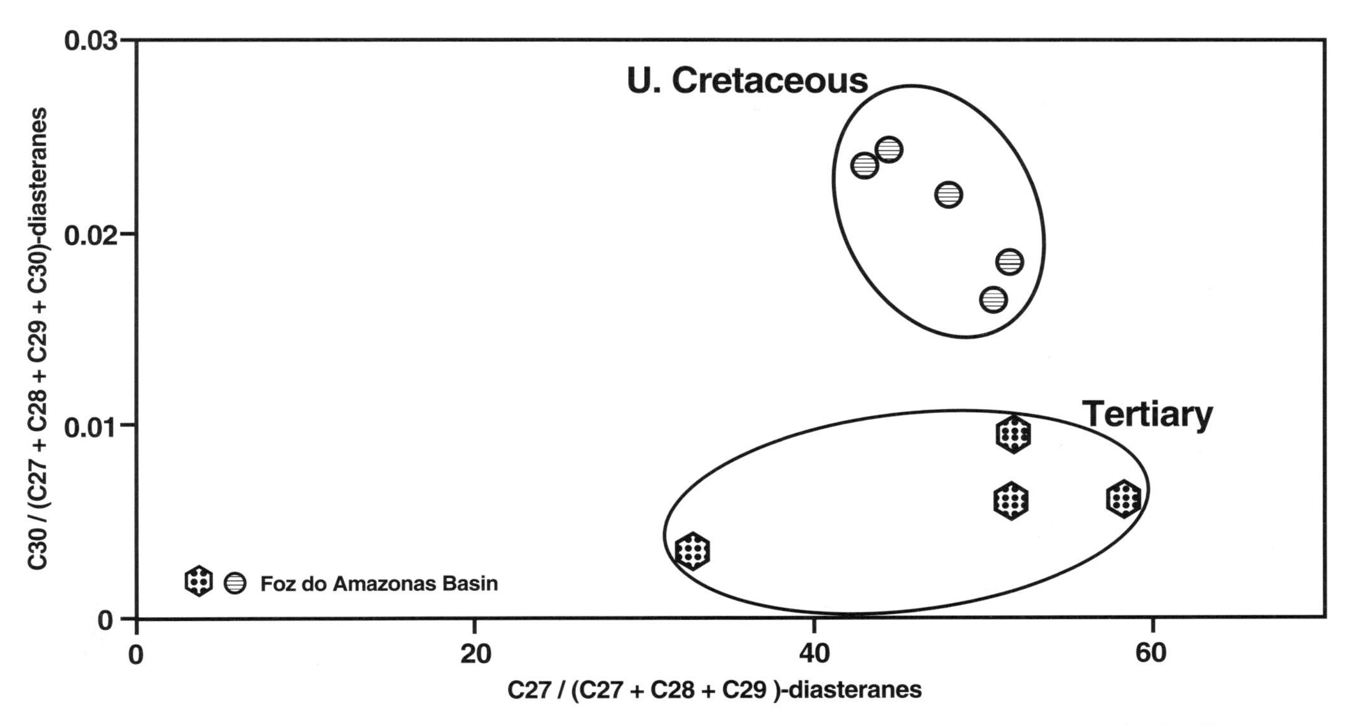

Figure 7—Plot based on biomarker ratios using an MRM-GCMS data of oils from Foz do Amazonas Basin, Brazil.

Oil sample APS-27 shows an absence of oleanane while oleanane/(oleanane + hopane) ratios for samples PAS-9 and PAS-11 are higher (Figures 8 and 9). All samples except APS-27 were found not to be biodegraded (Peters and Moldowan, 1993). Oil sample APS-27 presents a medium rank of biodegradation, in which *n*-alkanes and the isoprenoids pristane and phytane were degraded. In contrast, the biomarkers appear to have been unaffected. The partial MRM-GCMS trace shows a high oleanane/(oleanane + hopane) ratio for the PAS-9 and PAS-11 samples, which generally indicates a Tertiary source (Figure 8). It is interesting to note that, while all samples show the presence of C_{30} steranes and diasteranes (Figures 6 and 7), suggesting a marine origin, the APS-27 oil does not present oleanane, suggesting either an older source or much less terrestrial input (Figure 8).

A second age-related biomarker parameter, the nordiacholestane (NDC) ratio, does not prove a Tertiary age for samples PAS-9 and PAS-11. According to Holba et al. (1998), values for the NDC ratio of >0.5 indicate a Tertiary oil source. However, at low paleolatitudes, these ratios are lower in accordance with relatively lower diatom abundance near the equator (Holba et al., 1998). High oleanane values in these samples indicate a Tertiary marine deltaic source for the PAS oils and provide a way to calibrate the NDC ratio for the Tertiary samples in this area. In contrast, a low abundance of C_{30} diasteranes and an absence of oleanane in the APS sample suggest a marine carbonate source rock for this oil. As a final note, the NDC ratios for this study suggest that these ratios must be considered with care when determining age due to the importance of paleolatitude assessment, as they can show large differences when compared with data from higher latitudes (e.g., Holba et al., 1998).

The identification of two oil groups in the Foz do Amazonas Basin indicates the presence of more than one petroleum system in the area, thus opening new oil and gas frontiers for exploration in the entire basin.

Gas Geochemistry

For several decades, gas geochemistry has been used mainly to characterize the bacterial or thermogenic origins of a gas and, with respect to thermogenic origin, to estimate qualitatively the maturity of its source using essentially the carbon isotopic signature of methane and the wetness of the gas (Schoell, 1983). However, with the advent of new analytical techniques and a new breed of GC-MS equipment, it is now possible to characterize the physical and chemical processes affecting gases throughout geologic time, from generation to present accumulations in reservoirs (Mello et al., 1998; Prinzhofer et al., 2000, and Chapter 9, this volume). The relatively small number of different hydrocarbon species in gases creates a challenge in interpreting a maximum of information from only a dozen or so molecular percentages and carbon isotopic ratios. Furthermore, because all these gas molecules are small and have small molecular weights, they are highly affected and greatly fractionated by such processes as thermal and bacterial generation and degradation, migration, and partial loss.

In the past several years, a series of new experiments

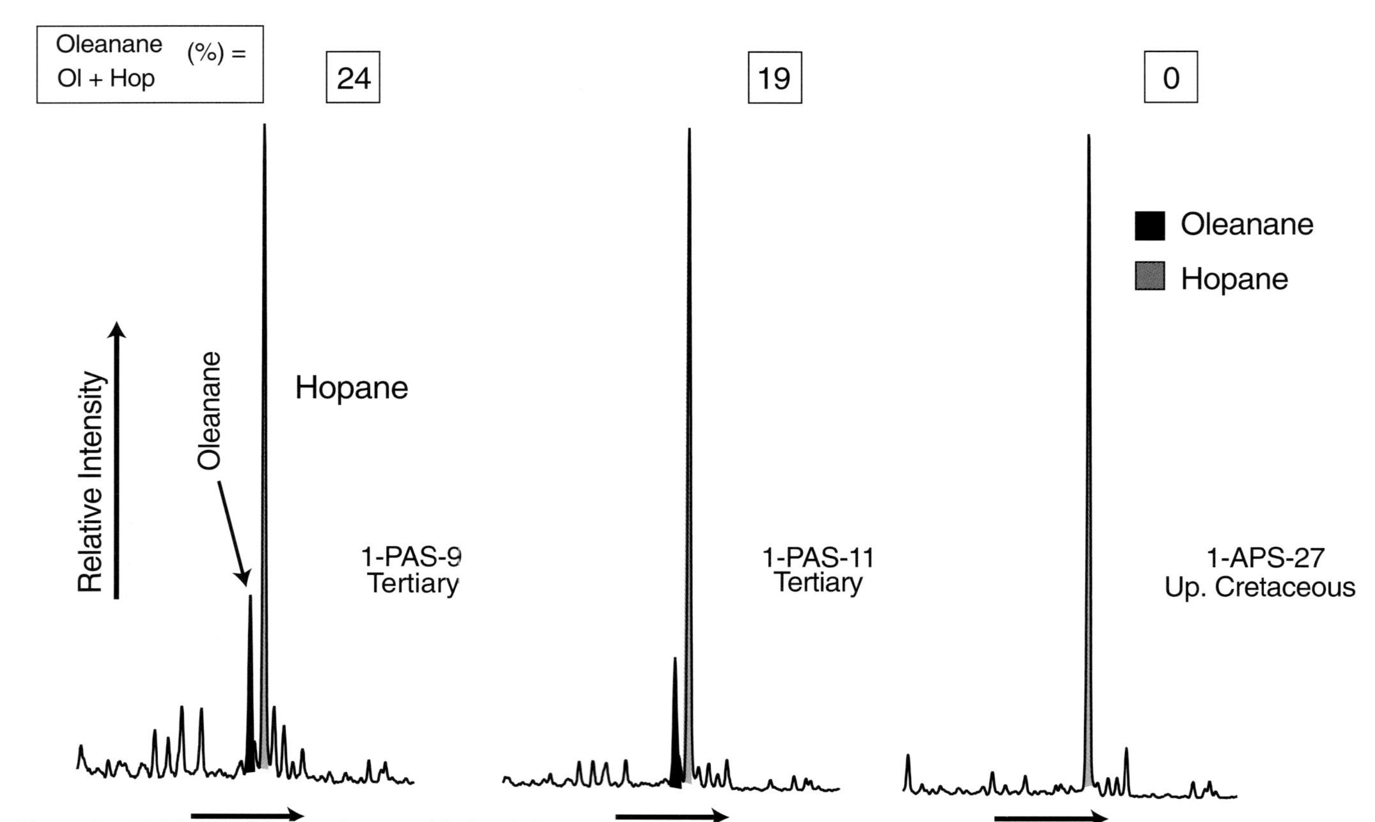

Figure 8—MRM-GCMS is used to provide key information about the age of oil source rocks from analysis of oils from Foz do Amazonas Basin. The presence of oleanane (C_{30}) is diagnostic for a Tertiary terrestrial plant contribution.

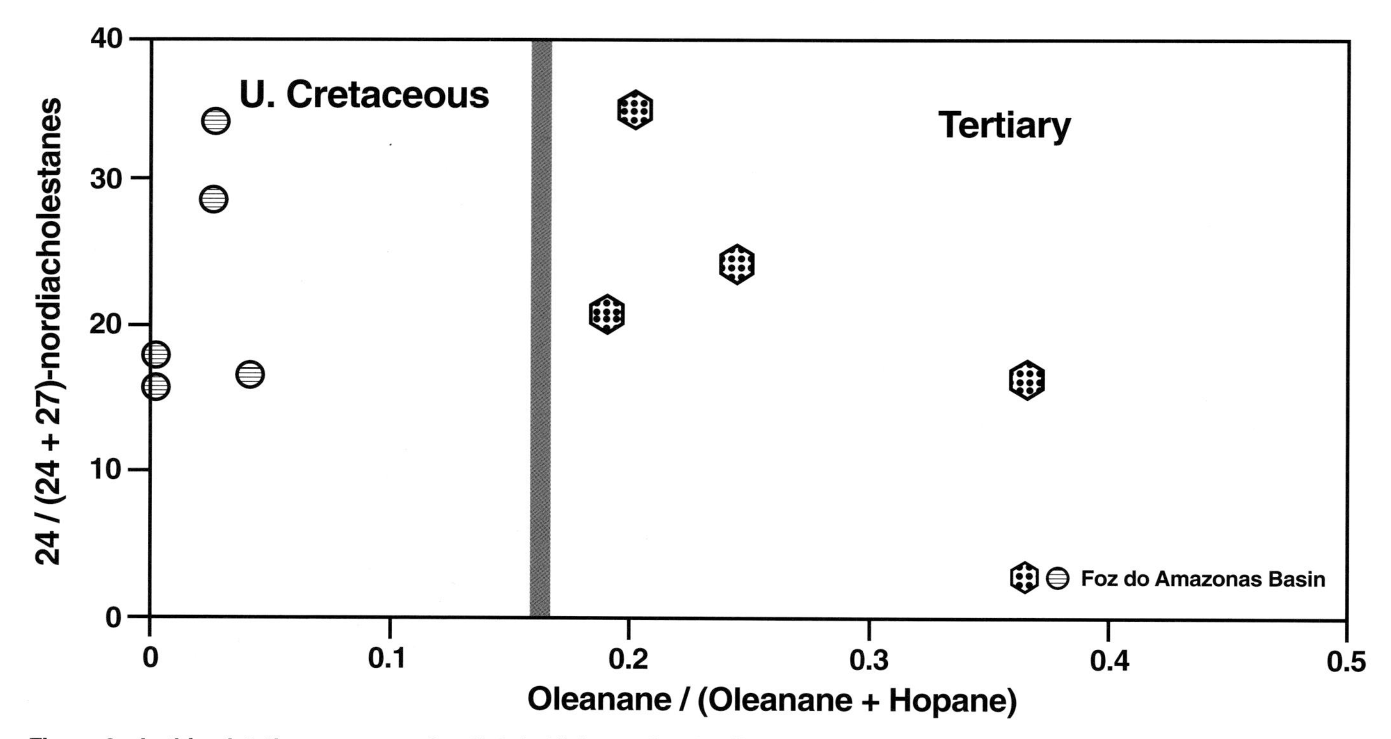

Figure 9—In this plot, the presence of a slightly higher ratio of 24/(24 + 27)-norcholestanes (C_{26} steranes, *m/z* 358, →217) supports a Tertiary age for the PAS oils, and high diasterane indices suggest that active clays are present in the source rock.

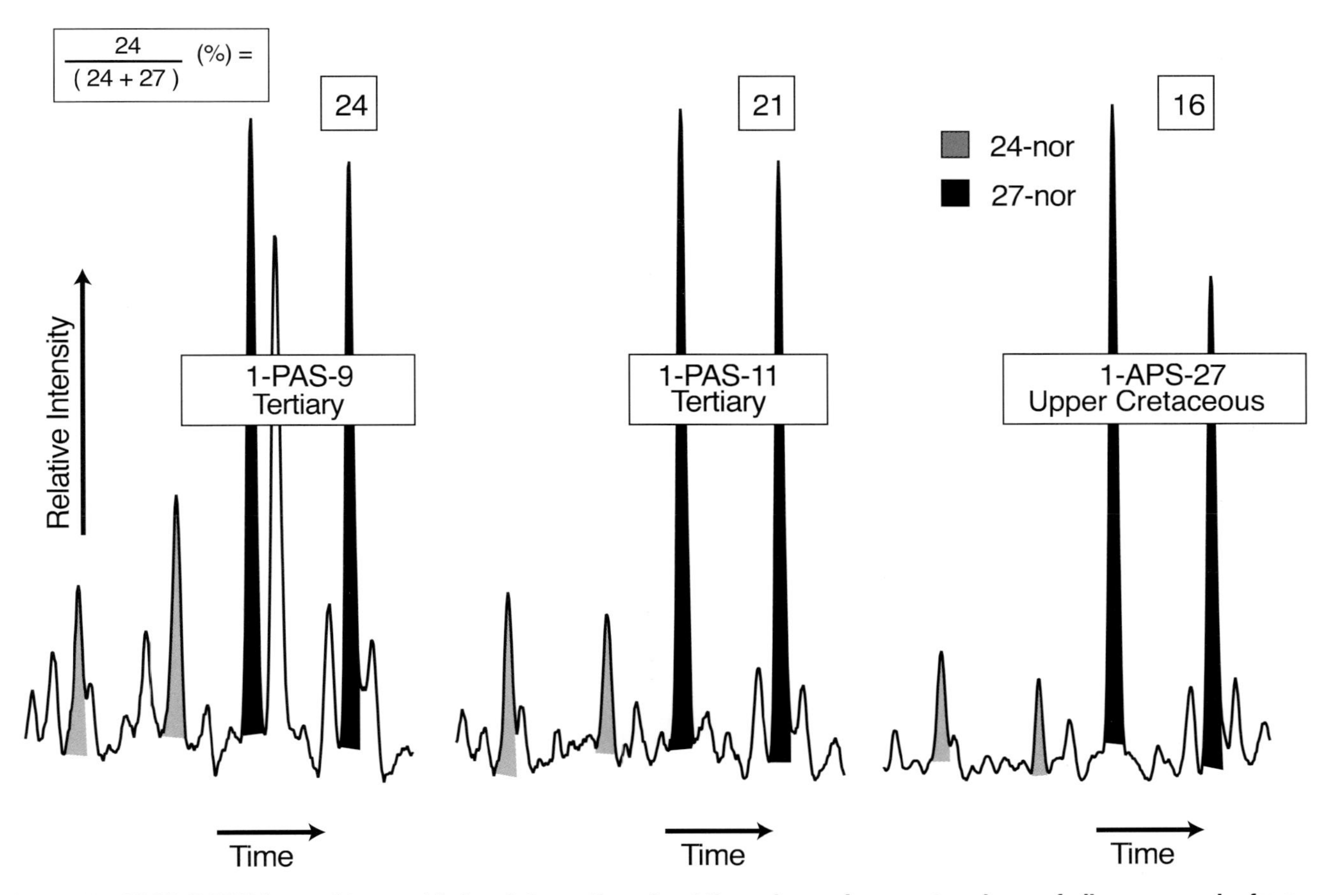

Figure 10—MRM-GCMS is used to provide key information about the paleoenvironment and age of oil source rocks from analysis of oils. The similar ratio of 24/(24 + 27)-norcholestanes (C_{26} steranes, *m/z* 358, →217) in PAS oils compared to APS oils may be due to the low paleolatitude during source rock deposition.

and models have been developed to monitor some of these processes (Prinzhofer and Pernaton, 1997). Laboratory simulation has provided the ability to characterize the source of a thermogenic gas (primary cracking of kerogen, secondary cracking of oil, and even tertiary cracking of the heavier part of gaseous hydrocarbons) and the efficiency of expulsion from the source rocks (Lorant et al., 1998). Also, experiments on gas migration have shown that, in some cases, methane may be affected by isotopic fractionation due to its transport by diffusive mechanisms (Prinzhofer and Pernaton, 1997). Cohesive integration of these various fractionations have been attempted with statistical analyses (principal component analysis, or PCA), yielding two parameters, V1 and V2, as a function of the chemical and isotopic signatures (Prinzhofer et al., 2000). These two parameters give a relative indication of maturity and a qualitative distance of migration, respectively. From these new experimental constraints, important implications for oil and gas evaluation have already been described (Prinzhofer et al., 2000, and Chapter 9, this volume).

As an example of the implications of this kind of study in a petroleum system investigation, a series of gas samples from deep-water wells from Marlim and Albacora fields in the Campos Basin were analyzed (Figure 11). The aim was to understand the gas distribution not only as fingerprints but also as processes affecting the geologic history of the petroleum habitat in the basin.

When plotting the Campos gas data (including samples from Marlim and Albacora gas fields) in the Schoell diagram (Figure 11), it appears that almost all the gases plot in the oil window without showing any consistent trend. Also, some very isotopically light data suggest a mixture between thermogenic and biogenic gases. In the C_2/C_3 diagram of Lorant et al. (1998) (Figure 12), however, the gases from Marlim and Albacora fields define a maturity trend going from the oil window to extremely mature fluids corresponding to the secondary alteration of oil and gas. This plot suggests the presence in the area of hydrocarbon kitchens buried at a large range of depths.

A principal component analysis (PCA) performed on a large database allowed the definition of two main parameters as the first two eigenvectors of the calculation (Prinzhofer et al., 2000). The first eigenvector is linked positively with the maturity of the studied sample, whereas the second eigenvector shows good correlation

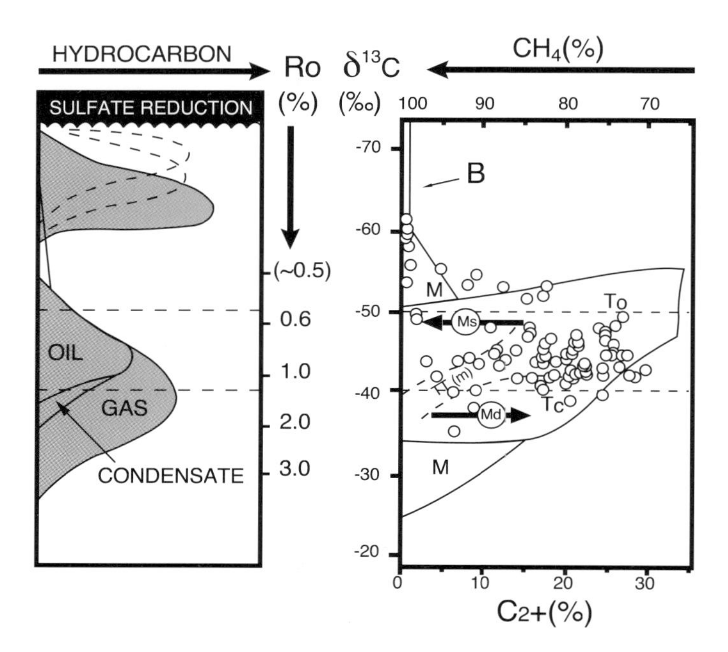

Figure 11—Genetic characterization of gases from Campos Basin, Brazil (adapted from Schoell, 1983).

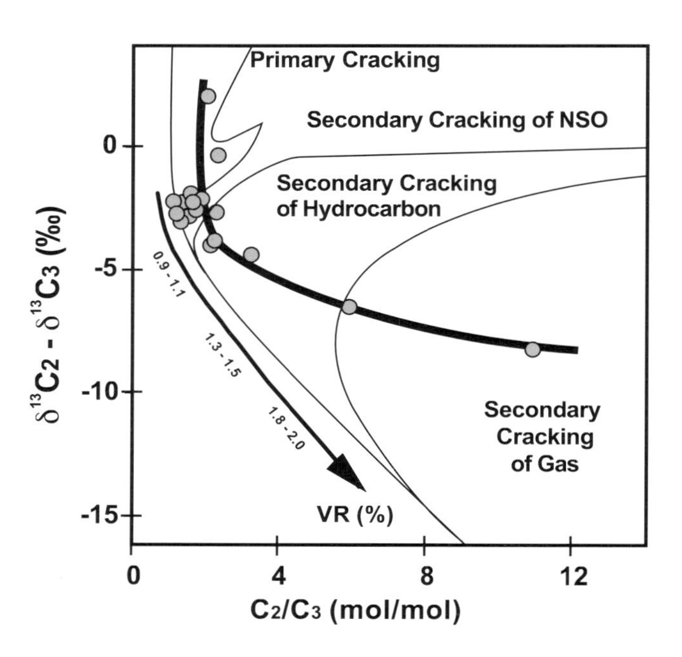

Figure 12—Simplified scheme of genetic events occurring during thermal degradation of organic matter, oil, and gas using C_2/C_3 composition and isotopic data from gases from Campos Basin, Brazil.

with migration distance. As can be observed on the V1 (maturity) versus V2 (migration) plot in Figure 13, the gases from Albacora Field wells, which have a small range of maturity, do not show evidence of long-distance secondary migration. In contrast, gases from Marlim Field, which exhibit a greater range of maturity from oil to gas window, show a clear trend of segregation due to long-distance secondary migration. Also, the Marlim gas samples suggest that two possible distinct trends of migration have occurred, corresponding to two ranges of maturity. Such data indicate the presence of different kitchens, one associated with long-distance migration processes. The gases from the Albacora Field, in contrast, show that they correspond to the same maturity range, with smaller fractionation, and therefore suggest a shallower and closer kitchen. Such results have had a direct impact on exploration because they not only predict the place and depth of the respective petroleum kitchens but also provide a favorable outlook for the ultra deep-water areas of Campos Basin for oil and gas exploration.

Diamondoids

Biomarkers give crucial source and thermal maturity information for oils generated in the main oil window (Mello, 1988). However, for oils generated by highly mature source rocks and those that have undergone secondary cracking, biomarker concentrations may be below conventional detection limits (Dahl et al., 1998). To compound the problem, low concentrations of biomarkers present in these oils may be picked up during oil migration and are thus not always indicative of the source. To circumvent these problems, recent workers have begun to analyze the more stable compounds in oil to determine source and thermal maturity. These include the light C_7 hydrocarbons and diamondoids (Mango, 1990; Dahl et al., 1998). The C_7 work was pioneered by Mango (1990), who analyzed hundreds of oils and found that the ratio of certain C_7 hydrocarbons was invariant. He attributed this invariance to a catalytic generation of oil. Subsequently, other authors have demonstrated that some C_7 ratios can be used to determine source and thermal maturity.

Diamondoids are hydrocarbon compounds that have the molecular structure of diamonds and, as such, have a relatively high degree of thermal stability. The smallest diamondoid is adamantane, consisting of one diamond structural subunit. Two subunits constitute diamantane, three triamantane, and so on. The presence of diamondoids in petroleum has been known since the 1930s, when adamantane was isolated from a Czech petroleum. A variety of substituted diamondoids are present in petroleum, having methyl, ethyl, propyl, and higher alkyl substitutions at a variety of positions. The ratio of some of these substituted diamondoids has been used to measure thermal maturity, in particular, the ratios of 1- to 2-methyladamantane and 1-, 3-, and 4-methyldiamantane (Chen et al., 1996). These ratios were used in a study of oils from the Tarim Basin in China and correlated very well to calculated vitrinite reflectance values.

Dahl et al. (1998, 1999) have recently developed a diamondoid cracking parameter to measure the amount of secondary cracking a particular oil has undergone. This parameter was applied to a series of oils from the

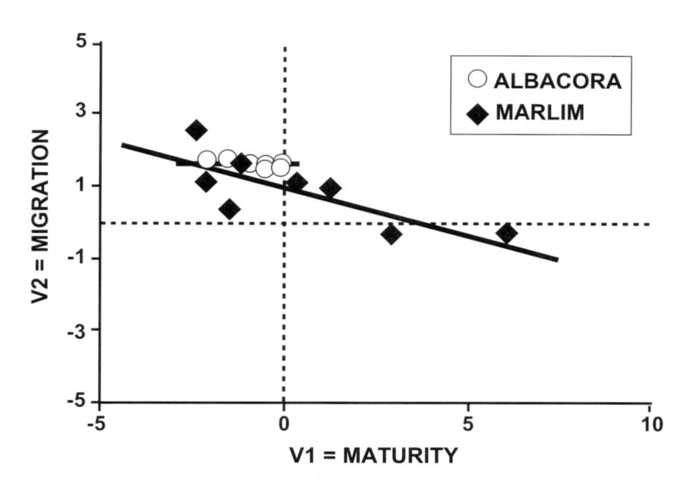

Figure 13—Principal component analysis (PCA) representation of gases from Marlim and Albacora fields, Campos Basin. The gases from Albacora exhibit a pure maturity trend, while the gases from Marlim show strong fractionation during migration.

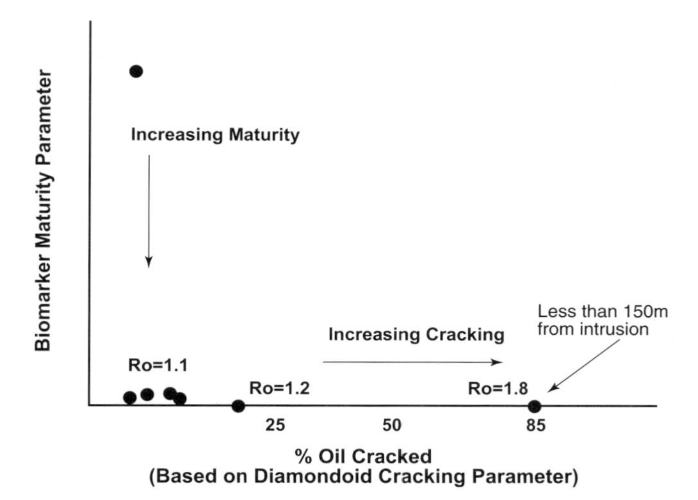

Figure 14—Plot of diamondoid and sterane data from condensate samples from Solimões Basin, Brazil. A newly developed method for the determination of oil cracking using diamondoids and biomarkers can be used to (1) determine the percentage of oil that has been converted to gas and coke, (2) determine the floor of the economic oil window, (3) calibrate oil-to-gas conversion models, (4) recognize mixes of highly mature oil with less mature oil, and (5) estimate expulsion efficiencies of "poor" source rocks. The ratio of biomarker to diamondoid thermal maturity exhibits a characteristic asymptotic curve in which the biomarker maturity measurement loses effectiveness just as the diamondoid maturity measurement becomes effective. Oils that are mixtures of highly mature (cracked) and mature oils plot away from the curve. The curve for each petroleum system is unique.

Paleozoic Solimões Basin in Brazil (Figure 14). These oils are in reservoirs located at varying distances from igneous intrusions (Caputo and Silva, 1990). The oil showing by far the highest estimated amount of secondary cracking based on diamondoids is located nearest to the intrusion (~150 m). Here, vitrinite reflectance values in surrounding rocks are about 1.8% R_o. The only two oils in the study, which appear to have undergone a slight amount of cracking, are at an intermediate distance from any intrusion. The other oils are farther from intrusions, and vitrinite reflectance values in rocks surrounding this reservoir are about 1.1% R_o. In conclusion, the application of diamondoids in the Solimões Basin helped to characterize not only the degree of oil cracking in the area but also to differentiate two distinct petroleum provinces: one gas-prone province (Jurua area) that shows about a 90% rate of liquid oil cracking and a condensate/light oil–prone province (Urucu area) that shows a very low level of oil cracking to gas (Figure 14).

CONCLUSIONS

In summary, advances in geochemical methods were discussed as they apply to fluid investigations in petroleum system case studies. Such studies can add significant value in the evaluation, assessment, and determination of exploration risk.

When plays or prospects are presented in the context of a petroleum system, it allows a fresh look at previously drilled areas. The geochemistry of fluids yields a different perspective from which to evaluate exploration risk. It provides an excellent mechanism to look at oil and gas accumulations in a basin and to see all the elements and processes that have occurred during and subsequent to their emplacement. Such data, when integrated with geologic and geophysical information, allows explorationists to understand petroleum systems comprehensively, thereby reducing the exploration risk.

REFERENCES CITED

Brooks, J. M., M. C. Kennicutt II, and B. D. Carey, 1986, Offshore surface geochemical exploration: Oil & Gas Journal, October 20.

Brooks, P. W., M. R. Mello, S. M. Barbanti, and J. M. Moldowan, 1998, The use of age related biomarkers in oils from southern Brazilian and west African marginal basins (abs.), *in* M. R. Mello and P. O. Yilmaz, eds., Extended Abstracts Volume, AAPG International Conference and Exhibition, Rio de Janeiro, p. 612–613.

Caputo, M. V., and O. B. Silva, 1990, Sedimentação e tectônica da bacia de Solimões, *in* G. P. Raja Gabaglia and E. J. Milani, eds., Em Origem e evolução de bacias sedimentares, Petroleo Brasileiro S.A., p. 169–191.

Chen, J., F. Jiamo, G. Sheng, D. Liu, and J. Zhang, 1996, Diamondoid hydrocarbon ratios: novel maturity indices for highly mature crude oils: Organic Geochemistry, v. 25, p. 179–190.

Dahl, J. E., J. M. Moldowan, K. E. Peters, and M. R. Mello, 1998, Diamondoid hydrocarbons as indicators of thermal maturity and oil cracking (abs.), *in* M. R. Mello and P. O. Yilmaz, eds, Extended Abstracts Volume, AAPG International Conference and Exhibition, Rio de Janeiro, p. 610–611.

Dahl, J. E., J. M. Moldowan, K. E. Peters, G. E. Claypool, M. A. Rooney, G. E. Michael, M. R. Mello, and M. L. Kohnen, 1999, Diamondoid hydrocarbons as indicators of natural oil cracking: Nature, v. 399, p. 54–57.

Guardado, L. R., L. A. P. Gamboa, and C. F. Lucchesi, 1990, Petroleum geology of the Campos Basin, Brazil, a model for a producing Atlantic type basin, *in* J. D. Edwards, ed., Divergent/passive margin basin: AAPG Memoir 48, p. 3–80.

Holba, A. G., E. W. Tegelaar, B. J. Huizinga, J. M. Moldowan, M. S. Singletary, M. A. McCaffrey, and L. I. P. Dzou, 1998, 24-Norcholestanes as age-sensitive molecular fossils: Geology, v. 26, p. 783–786.

Lorant, F., A. Prinzhofer, F. Behar, and A. Y. Huc, 1998, Carbon isotopic and molecular constraints on the formation and the expulsion of thermogenic hydrocarbon gases: Chemical Geology, v. 147, p. 249–264.

Magoon, L. B., and W. G. Dow, 1994, The petroleum system, in L. B. Magoon and W.G. Dow, eds., The petroleum system—from source to trap: AAPG Memoir 60, p. 3–24.

Mango, F. D., 1990, The origin of light cycloalkanes in petroleum: Geochimica et Cosmochimica Acta, v. 54, p. 23–27.

McCaffrey, M. A., J. M. Moldowan, P. A. Lipton, R. E. Summons, K. E. Peters, A. Jeganathan, and D. S. Watt, 1994, Paleoenvironmental implications of novel C_{30} steranes in Precambrian to Cenozoic age petroleum and bitumen: Geochimica et Cosmochimica Acta, v. 58, p. 529–532.

Mello, M. R., 1988, Geochemical and molecular studies of the deposition environments of source rocks and their derived oils from the Brazilian marginal basins: Ph.D. thesis, University of Bristol, Bristol, U.K., 240 p.

Mello, M. R., P. C. Gaglianone, S. C. Brassell, and J. R. Maxwell, 1988, Geochemical and biological environment using brasilian offshore oils: Marine and Petroleum Geology, v. 5, p. 205–223.

Mello, M. R., E. A. M. Koutsoukos, W. U. Mohriak, and G. Bacoccoli, 1994, Selected petroleum system in Brazil, *in* L. B. Magoon and W. G. Dow, eds., The petroleum system—from source to trap: AAPG Memoir 60, p. 499–512.

Mello, M. R., N. Telnaes, and J. R. Maxwell, 1995, The hydrocarbon source potential in the Brazilian marginal basins: a geochemical and paleoenvironmental assessment, *in* A. Y. Huc, ed., Paleogeography, paleoclimate, and source rocks: AAPG Studies in Geology, v. 40, p. 233–272.

Mello, M. R., F. T. T. Gonçalves, N. A. Babinski, and F. P. Miranda, 1996, Hydrocarbon prospecting in the Amazon rain forest: application of surface geochemical, microbiological, and remote sensing methods: AAPG Memoir 66, p. 401–411.

Mello, M. R., L. A. F. Trindade, E. Gil, E. Stoffer, N. Chigne, O. Luna, and J. Velandia, 1997, Geochemical characterization of the South American sub-Andean petroleum systems: Sixth Simpósio Bolivariano, p. 324–336.

Mello, M. R., T. Takaki, and A. Prinzhofer, A., 1998, Gas behavior in the Campos Basin: a new approach in exploration (abs.), *in* M. R. Mello and P. O. Yilmaz, eds., Extended Abstracts Volume, AAPG International Conference and Exhibition, Rio de Janeiro, p. 992–993.

Mohriak, W. U., M. R. Mello, J. F. Dewey, and J. R. Maxwell, 1990a, Petroleum geology of the Campos Basin, offshore Brazil, *in* J. Brooks, ed., Classic petroleum provinces: The Geologic Society, London, p. 119–142.

Mohriak, W. U., M. R. Mello, G. D. Karner, J. F. Dewey, and J. R. Maxwell, 1990b, Structural and stratigraphic evolution of Campos Basin, offshore Brazil: AAPG Memoir 46, p. 577–598.

Moldowan, J. M., and N. M. Talyzina, 1998, Biogeochemical evidence for dinoflagellate ancestors in the Early Cambrian: Science, v. 281, p. 1168–1170.

Moldowan, J. M., J. Dahl, B. J. Huizinga, F. J. Fago, L. J. Hickey, T. M. Peakman, and D. W. Taylor, 1994, The molecular fossil record of oleanane and its relation to angiosperms: Science, v. 265, p. 768–771.

Moldowan, J. M., J. Dahl, S. R. Jacobson, B. J. Huizinga, F. J. Fago, R. Shetty, D. S. Watt and K. E. Peters, 1996, Chemostratigraphic reconstruction of biofacies: molecular evidence linking cyst-forming dinoflagellates with pre-Triassic ancestors: Geology, v. 24, p. 159–162.

Moldowan, J. M., J. Dahl, and M. R. Mello, 1998, Characterization of South American oils to determine source-rock age and provenance using new high-resolution biomarker methods (abs.), *in* M. R. Mello and P. O. Yilmaz, eds, Extended Abstracts Volume, AAPG International Conference and Exhibition, Rio de Janeiro, p. 662–663.

Peters, K. E., and M. R. Cassa, 1994, Applied source rock geochemistry, *in* L. B. Magoon, and W.G. Dow, eds., The petroleum system—from source to trap: AAPG Memoir 60, p. 93–120.

Peters, K. E., and J. M. Moldowan, 1993, The Biomarker Guide: Interpreting Molecular Fossils in Petroleum and Ancient Sediments: Englewood Cliffs, N.J., Prentice Hall, 363 p.

Prinzhofer, A., and E. Pernaton, 1997, Isotopically light methane in natural gas: bacterial imprint or segregative migration?: Chemical Geology, v. 142, p. 193–200.

Prinzhofer, A., M. R. Mello, and T. Takaki, 2000, Geochemical characterization of natural gas: a physical multivariable approach and its applications in maturity and migration estimates: AAPG Bulletin, v. 84, n. 8, p. 1152–1172.

Sassen, R., J. M. Brooks, I. R. MacDonald, M. C. Kennicutt II, N. L. Guinasso, Jr., and A. G. Requejo, 1993, Association of oil seeps and chemosynthetic communities with oil discoveries, upper continental slope, Gulf of Mexico: Transactions of Gulf Coast Association of Geological Societies, v. 43, p. 349–355.

Schoell, M., 1983, Genetic characterization of natural gases: AAPG Bulletin, v. 67, n. 12, p. 2225–2238.

Tissot, B. P., and D. H. Welte, 1984, Petroleum Formation and Occurrence, 2nd ed.: Berlin, Springer, 699 p.

Trindade, L. F., 1992, Geochemical assessment of petroleum migration and mixing in the Potiguar and Sergipe–Alagoas Basins, Brazil: Ph.D. thesis, Stanford University, California, 305 p.

Magoon, L. B., and W. G. Dow, 2000, Mapping the petroleum system—an investigative technique to explore the hydrocarbon fluid system, *in* M. R. Mello and B. J. Katz, eds., Petroleum systems of South Atlantic margins: AAPG Memoir 73, p. 53–68.

Chapter 5

Mapping the Petroleum System— An Investigative Technique to Explore the Hydrocarbon Fluid System

L. B. Magoon
U.S. Geological Survey
Menlo Park, California, U.S.A.

W. G. Dow
DGSI
The Woodlands, Texas, U.S.A.

Abstract

Creating a petroleum system map includes a series of logical steps that require specific information to explain the origin in time and space of discovered hydrocarbon occurrences. If used creatively, this map provides a basis on which to develop complementary plays and prospects. The logical steps include the characterization of a petroleum system (that is, to identify, map, and name the hydrocarbon fluid system) and the summary of these results on a folio sheet. A petroleum system map is based on the understanding that there are several levels of certainty from "guessing" to "knowing" that specific oil and gas accumulations emanated from a particular pod of active source rock. Levels of certainty start with the close geographic proximity of two or more accumulations, continues with the close stratigraphic proximity, followed by the similarities in bulk properties, and then detailed geochemical properties. The highest level of certainty includes the positive geochemical correlation of the hydrocarbon fluid in the accumulations to the extract of the active source rock.

A petroleum system map is created when the following logic is implemented. Implementation starts when the oil and gas accumulations of a petroleum province are grouped stratigraphically and geographically. Bulk and geochemical properties are used to further refine the groups through the determination of genetically related oil and gas types. To this basic map, surface seeps and well shows are added. Similarly, the active source rock responsible for these hydrocarbon occurrences are mapped to further define the extent of the system. A folio sheet constructed for a hypothetical case study of the Deer-Boar(.) petroleum system illustrates this methodology.

INTRODUCTION

The reasons for carrying out oil and gas related investigations in a petroleum province are to reduce exploration risk in the search for undiscovered commercial quantities of petroleum. Published papers and meeting titles indicate that the petroleum system methodology is being widely used for this purpose (Howes, 1997a,b; Price and McNeil, 1997; Al Arouri et al., 1998; Chen et al., 1998; Hurley and Weimer, 1998). This chapter describes in detail how the petroleum system is mapped so that it can be explored more effectively and efficiently. Although this topic is discussed in a previous publication by the authors (Magoon and Dow, 1994b), in this chapter, we discuss certain aspects of mapping previously omitted. A properly mapped petroleum system provides a framework within which to collect additional information so the working hypothesis developed for that particular hydrocarbon fluid system can be continually tested as the exploration program progresses.

Petroleum System Concept

The term *oil system* was first introduced by Dow (1974) and is based on the concept of oil–source rock correlation. The term *petroleum system* was first used by Perrodon (1980). Independently, Demaison (1984) devised the term *generative basin*, Meissner et al. (1984) described their *hydrocarbon machine,* and Ulmishek (1986) identified an *independent petroliferous system.* All of these concepts are very similar to the oil system (Dow, 1974). Expanding upon previous work, Magoon (1987, 1988, 1989a,b, 1992a,b), Magoon and Beaumont (1999), and Magoon and Dow (1994a) formalized the criteria for identifying, mapping, and naming the petroleum system. Magoon (1995) added the complementary play and prospect to better explain the relationship between discovered and undiscovered oil and gas accumulations. Magoon (1997) has discussed the ability to determine risk objectively, that is, what is known versus what is unknown.

The *petroleum system* is defined as a naturally occurring hydrocarbon fluid system that encompasses a pod of active source rock (provenance), all related oil and gas, and the essential elements and processes needed for oil and gas accumulations to exist (Magoon and Dow, 1994b). The petroleum system concept infers that migration pathways must exist, either now or in the past, connecting the provenance of the petroleum with the accumulations. Using the principles of petroleum geology and geochemistry, this fluid system can be mapped to better understand how it evolved over time.

An important goal of the petroleum geologist is to map the evolution of the petroleum system to locate undiscovered hydrocarbons. A petroleum system case study provides an objective basis on which to determine exploration risk for a related or complementary prospect and play. As our ability to characterize and map a petroleum system improves, exploration risk decreases and the probability of success increases. The processes and their temporal relationships—trap, petroleum charge, and timing—provide the link between the complementary play and prospect and the petroleum system (Magoon, 1997; Magoon and Beaumont, 1999). Mapping the petroleum system in four dimensions, space and time, begins with a working hypothesis.

Working Hypothesis

A petroleum system investigation starts with a working hypothesis for generation, migration, and entrapment of petroleum in a province, based on available geologic and geochemical data, which evolves as more data become available (Figure 1A, B). The investigator starts with an oil and gas field map and related field data for the petroleum province of interest. The geographic location of the accumulations is important because accumulations located close together are more likely to have originated from the same pod of active source rock (Figure 2). Accumulations that occur in the same or nearly the same stratigraphic interval are also likely to be from the same active source rock. In contrast, accumulations separated by barren rock sections are presumed to have originated from different pods of active source rock. Accumulations of widely differing bulk properties, such as gas versus oil, API oil gravity, gas/oil ratios, and sulfur contents, can also be presumed to originate from different pods of active source rock. Detailed geochemical data on oil and gas samples provide the next level of evidence for determining whether a series of hydrocarbon accumulations originated from one or more pods of active source rock. Last, comparing the geochemistry of oils and gases to possible source rocks provides the highest level of certainty as to which active source rock generated which oil or gas type.

By acquiring and organizing information that addresses these issues of location, stratigraphic position, and geochemistry, an investigator can take a working hypothesis of how a particular petroleum system formed to increasing levels of certainty (Figure 2). The investigator organizes the information about the oil and gas accumulations into groups of like petroleum types on the oil and gas field map, cross section, and table (Figure 1A, B, C).

With this step completed, the investigator then locates all surface seeps on the oil and gas field map, which now becomes the petroleum system map. The seeps with the same geochemical composition as the subsurface accumulation provide geographic evidence for the end-point of a migration path. The stratigraphic unit from which the fluid emanates can be compared to the stratigraphic unit in which oil and gas accumulations are found to determine the complexity of their migration paths. If the stratigraphic units are the same, then the migration paths are simple. If they are different, migration may be more complex. Geochemical information from seeps can be compared with that of discovered accumulations to link the seep fluid to the proper petroleum system.

Oil and gas shows in exploratory wells are added to the petroleum system map and cross section to better define the migration paths. As this map and cross section evolve, the investigator is encouraged to anticipate how the final map will look based on the framework geology and petroleum fluid information. Intuitively, exploration risk is high if the petroleum system is complicated and hence less predictable; risk is lower if the petroleum system is simple and thus more predictable.

After similar hydrocarbon fluids in the petroleum system have been mapped, individual oil and gas accumulations are tabulated to better understand the size (by volume) and complexity of the petroleum system. The petroleum system table is organized by stratigraphic interval in each field (Figure 1C). These stratigraphic intervals are zones, members, and formations that produce or contain measurable amounts of oil and gas. The table should include the age of the stratigraphic interval, API gravity and sulfur content of the oil, gas/oil ratio (GOR), cumulative amount of oil and gas produced, and the remaining amount of oil and gas that can be produced. Other information the investigator may choose to include are lithology; gross and net thickness; porosity and permeability of the reservoir rock; geometry, closure, and area of the trap; and detailed geochemistry of the oil and gas. The information included in the table will depend on what is available and the objectives of the investigation. The required information is used to determine the size (by volume) of the petroleum system, to select a reservoir rock name to use for the name of the petroleum system, and to evaluate the complexity of the migration path.

Now the provenance or origin of the petroleum is mapped as the pod of active source rock. Only one pod of active source rock occurs in each petroleum system. A *pod* is a contiguous body of source rock that has or is expelling oil and gas. Because this pod has thickness and

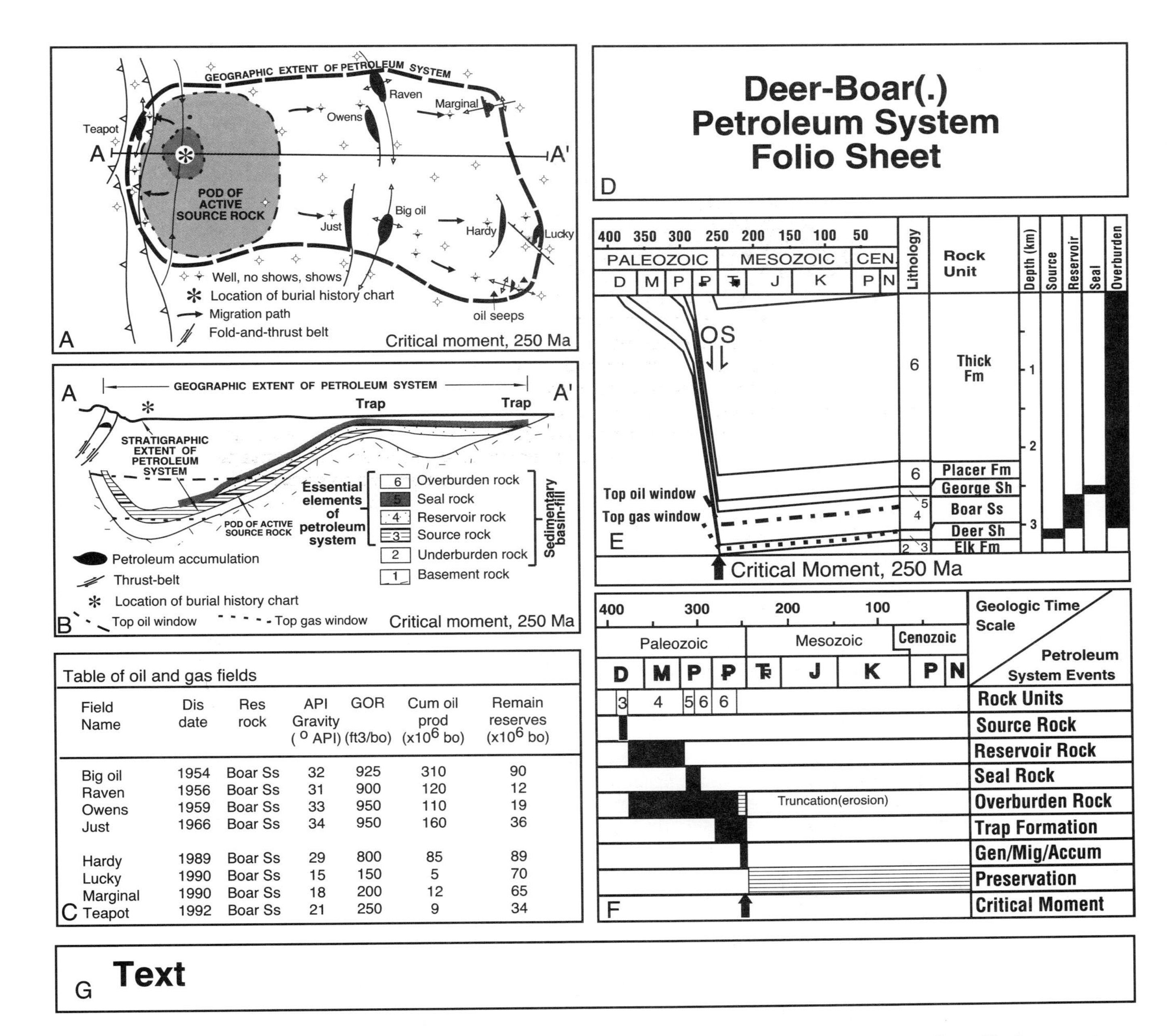

Field Name	Dis date	Res rock	API Gravity (° API)	GOR (ft3/bo)	Cum oil prod ($x10^6$ bo)	Remain reserves ($x10^6$ bo)
Big oil	1954	Boar Ss	32	925	310	90
Raven	1956	Boar Ss	31	900	120	12
Owens	1959	Boar Ss	33	950	110	19
Just	1966	Boar Ss	34	950	160	36
Hardy	1989	Boar Ss	29	800	85	89
Lucky	1990	Boar Ss	15	150	5	70
Marginal	1990	Boar Ss	18	200	12	65
Teapot	1992	Boar Ss	21	250	9	34

Figure 1—(A) Map showing the geographic extent of the so-called Deer-Boar(.) petroleum system at the critical moment (CM, 250 Ma). Thermally immature source rock is outside the oil window. The pod of active source rock lies within the oil and gas windows. (B) Cross section showing the stratigraphic extent of the Deer-Boar(.) petroleum system at the CM (250 Ma). Thermally immature source rock lies updip of the oil window. The pod of active source rock is downdip of the oil window (see Figure 5 for present-day cross section). (C) Table of oil and gas fields in the petroleum system, or the accumulations related to one pod of active source rock. (D) Name of the petroleum fluid system. (E) Burial history chart showing the CM and time of oil generation (260–240 Ma). This information is used on the events chart. All rock unit names used here are fictitious; location of burial history chart is shown in (A) and (B). (F) Events chart showing the relationship between the essential elements and processes as well as the preservation time and critical moment (CM). (G) Text and other figures needed to describe the petroleum system. Modified from Magoon and Dow (1994b) and Magoon (1997). Geologic time scale from Palmer (1983).

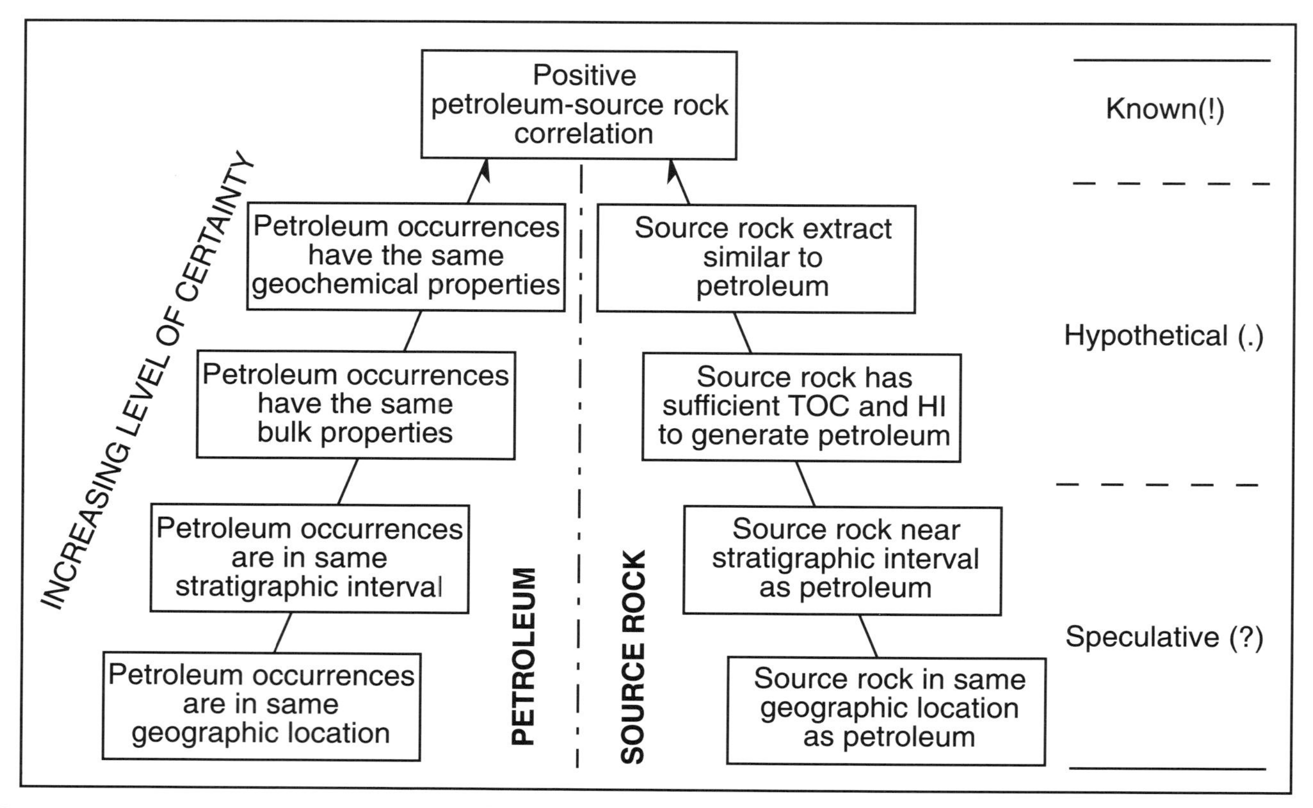

Figure 2—The logical sequence used for the levels of certainty from "guessing" to "knowing."

area, it can be mapped. When an organic-rich rock is in close, or reasonably close, proximity both stratigraphically and geographically to oil and gas accumulations, shows, or seeps, it is tentatively correlated with those fluids. Based on seismic, well, or outcrop data, the likelihood of correlation increases when the source rock's burial depth is known to reach 3 km. In the experience of the authors, this is a reasonable minimum burial depth for thermal maturity or when burial depth modeling (e.g., *Basin2,* Bethke et al., 1999) indicates that a source rock is in or below the oil window.

The correlation gains certainty if the source rock is established as being thermally mature by vitrinite reflectance or some other analytical technique. If the kerogen type of the source rock is consistent with that of the oil and gas, then confidence increases that the source rock is correctly correlated. If the geochemical composition of the organic matter in the source rock compares favorably with the migrated petroleum, the oil–source rock correlation is considered a match. Using seismic, well, and outcrop data, the suspected or confirmed active source rock is mapped as a contiguous, three-dimensional body or "pod" on the petroleum system map and cross section.

In this manner, the petroleum system map and cross section evolve, as the working hypothesis is taken to successive levels of certainty. To further refine this work, a burial history chart and an events chart are constructed and the petroleum system is named. Although this chapter discusses each of these petroleum system illustrations in sequence, they are usually drawn at the same time in an iterative manner so their relationships to one another evolve correctly. To organize these illustrations, the petroleum system folio sheet is used.

PETROLEUM SYSTEM FOLIO SHEET

The petroleum system folio sheet is a graphical way to depict the geographic, stratigraphic, and temporal evolution of the petroleum system (Figure 1). The folio sheet includes the following: (1) the petroleum system map (Figure 1A), (2) the petroleum system cross section (Figure 1B), (3) a table of genetically related accumulations (Figure 1C), (4) the petroleum system name (Figure 1D), (5) a burial history chart located over the pod of active source rock (Figure 1E), and (6) an events chart to summarize the history of the petroleum system (Figure 1F). In the ideal case, this folio sheet summarizes the detailed work of many specialists and provides the supportive information for the petroleum system map.

Petroleum System Map

The petroleum system map (Figure 1A) shows (1) the geographic extent of the petroleum system, (2) the pod of active source rock, (3) the genetically related oil and gas

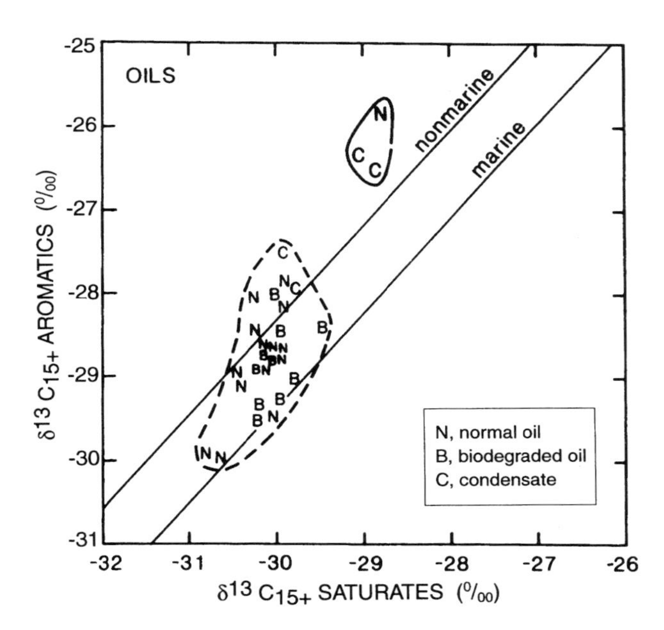

Figure 3—On the basis of carbon isotopes of saturated and aromatic hydrocarbons, normal oil (N), biodegraded oil (B), and condensate (C) samples from the Alaska Peninsula and Cook Inlet, Alaska, show two groups of oils. From Magoon and Anders (1992).

accumulations, shows, and seeps, (4) the location of the burial history chart, and (5) the location of the petroleum system cross section. Usually this map is drawn for the present day, but it can be refined later to depict the critical moment (CM) or the time when most of the hydrocarbons were generated and accumulating, especially if the petroleum system is old (see petroleum system age, Magoon and Dow, 1994a, p. 644). This map is in contrast to a geologic map, which depicts rock units and geometry at the surface, and an oil and gas field map previously mentioned.

Accumulations, Shows, and Seeps

By inference, the genetically related oil and gas fields or petroleum accumulations originated from the pod of active source rock. Thus, one should start with an oil and gas field map of the petroleum province. The accumulations shown on an oil and gas field map must be grouped into one or more possible petroleum systems based on their geographic and stratigraphic locations and on the bulk and geochemical properties of the fluids in each accumulation (Figure 1A). For example, the accumulations in the fictitious Deer-Boar(.) petroleum system include Teapot, in the thrust belt; Raven, Owens, Just, and Big Oil, in the anticlines adjacent to the foreland basin; and Marginal, Hardy, and Lucky, the most distal accumulations (see Table 2 for explanation of symbols in parentheses). These widely spaced accumulations are in the same stratigraphic interval, the Boar Sandstone, and all have similar geochemical properties as known from wells that sample these accumulations.

The exploratory wells indicate that some of the wells have oil shows in the Boar Sandstone. Surface oil seeps occur in the southeastern part of the map where the Boar Sandstone crops out. These oil occurrences have geochemical similarities with the oil accumulations. Based on this information, these oil occurrences are judged to belong to the same petroleum system.

As an example, on the basis of carbon isotopes of saturated and aromatic hydrocarbons, the normal oil (N), biodegraded oil (B), and condensate (C) samples from the Alaska Peninsula and Cook Inlet, Alaska, show two groups of geochemically related fluids (Magoon and Anders, 1992). These two geochemical groupings indicate two petroleum systems if each oil group came from its own pod of active source rock (Figure 3). There are more than two petroleum systems if either oil group came from two or more pods of active source rock.

Pod of Active Source Rock

The pod of active source rock is a contiguous volume of source rock that generated and expelled petroleum at the CM and is the provenance for a series of genetically related petroleum shows, seeps, and accumulations in a petroleum system. A pod of thermally mature source rock may be active, inactive, or spent (Magoon and Dow, 1994b). There is only one pod of active source rock for each petroleum system.

The kerogen type for the thermally mature source rock has been shown by numerous investigators to control the type and volume of petroleum expelled (e.g., Tissot and Welte, 1978, 1984; Hunt, 1979, 1995). Other investigators have provided explanations of the tools and techniques to characterize and map the pod of active source rock (e.g., Waples, 1981, 1985; Demaison and Murris, 1984; Bordenave, 1992; Peters and Moldowan, 1993). The pod of active source rock (also referred to as the "kitchen" or "oil and gas window") is a required feature of the petroleum system map because of its genetic control on petroleum accumulations.

For the fictitious Deer-Boar(.) system, the pod of active source rock is in the western part of the map area and is mapped using a dashed line (long and short dashes, Figure 1A,B). The Deer Shale is considered the most likely source rock because it is geographically near and stratigraphically below the Boar Sandstone, the reservoir rock for the accumulations, shows, and seeps. Thermal maturity data for the Deer Shale indicate that it is thermally mature in the foreland basin but immature in the thrust belt toward the west and on the craton toward the east. In Figures 1A and B, the long and short dashed lines correspond to the thermal maturity contours for the top of the oil window and the short dashes to the top of the gas window.

Geographic Extent

The geographic extent of the petroleum system at the CM is described by a line that circumscribes the pod of

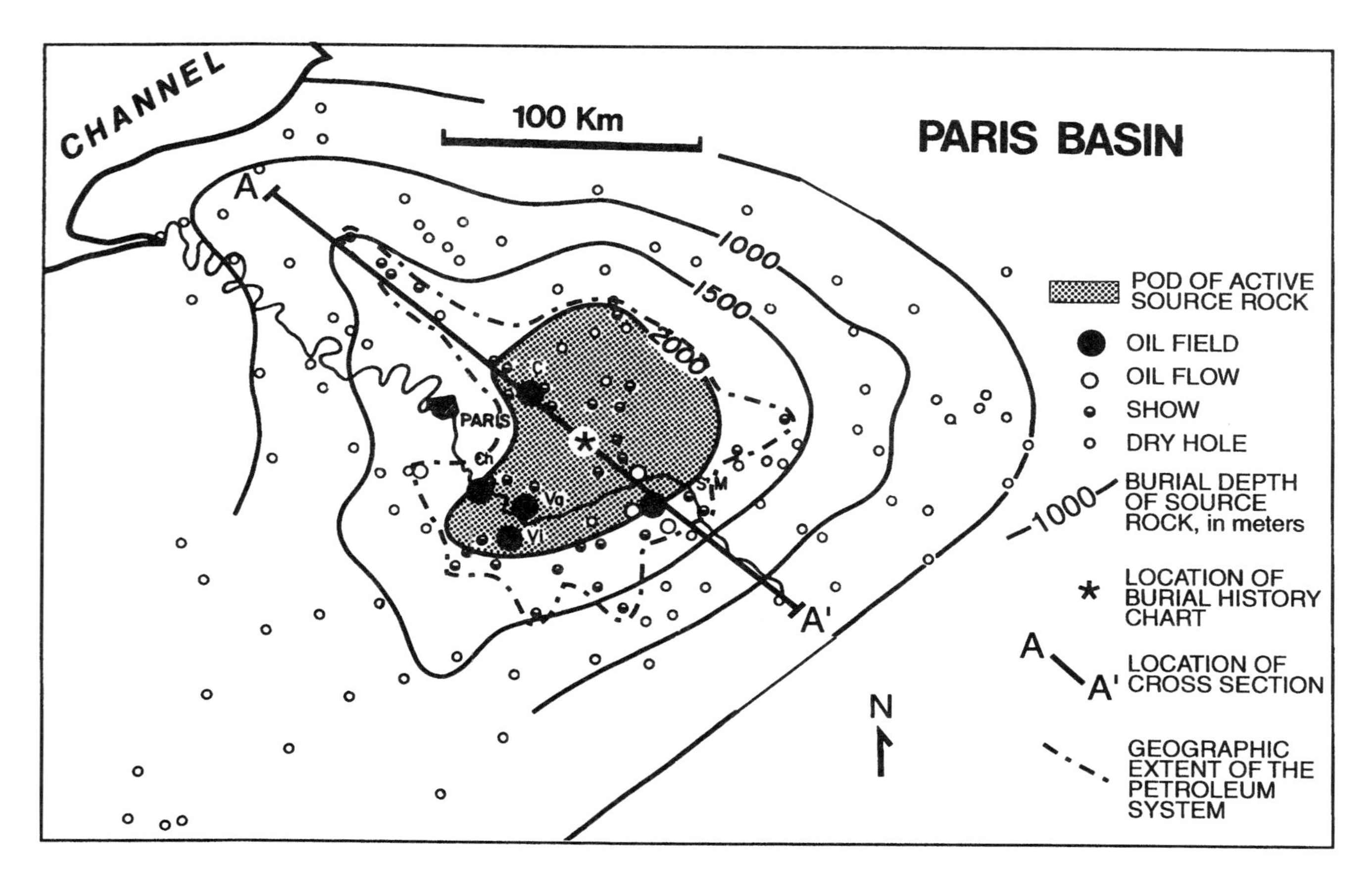

Figure 4—Map of a petroleum system in the Paris Basin showing petroleum occurrences, pod of active source rock, geographic extent, and locations of the cross section and burial history chart. Modified from Demaison (1984).

active source rock and includes all the discovered petroleum shows, seeps, and accumulations that originated from that pod. The CM is a snapshot of the petroleum system at the time when most of the hydrocarbons were generated and migrating. A map of the Deer-Boar(.) petroleum system, drawn at the end of Paleozoic time (250 Ma), includes a line that circumscribes the pod of active source rock and all related discovered hydrocarbons. This area represents the geographic extent or known extent of the petroleum system (Figure 1A).

As an example, a map of petroleum occurrences in the Paris Basin shown in Figure 4 (from Demaison, 1984) has been modified into a petroleum system map. To draw this petroleum system map, we assume that all petroleum occurrences emanate from the same pod of active source rock. The pod of active source rock is shown as a shaded area, and the depth to the source rock is assumed to be close to the thickness of overburden rock, which is contoured in thousands of meters. Based on the locations of the accumulations, drill-stem tests, and exploratory wells with and without oil shows (we assume "dry hole" lacks oil and gas shows), the geographic extent of the petroleum system is drawn. Using the guidelines discussed above, the location of the cross section and the burial chart are indicated. This petroleum system map emphasizes the short migration path of the discovered oil fields and shows that the long migration path to the northwest lacks an oil field. This map strongly suggests that the most likely place to find undiscovered oil is above or close to the pod of active source rock and along the preferential migration paths, as expressed by the geographic extent of the petroleum system.

Location of Cross Section

The cross section (Figure 1B) is placed on the map so that it passes through the largest accumulations and thickest overburden rock and extends over the entire geographic extent of the petroleum system. If possible, start with an available present-day cross section unless the petroleum system is so old (see Magoon and Dow, 1994a, p. 644) or structurally altered that a restored cross section representing a previous time is required to depict the time when most of the hydrocarbons migrated and accumulated. In Figure 1B, a cross section at the end of the Paleozoic (250 Ma) was used because the present-day cross section included a Tertiary rift graben (Figure 5). The largest accumulations are included because they are usually located on the simplest, most efficient migration path from the pod of active source rock. A transect

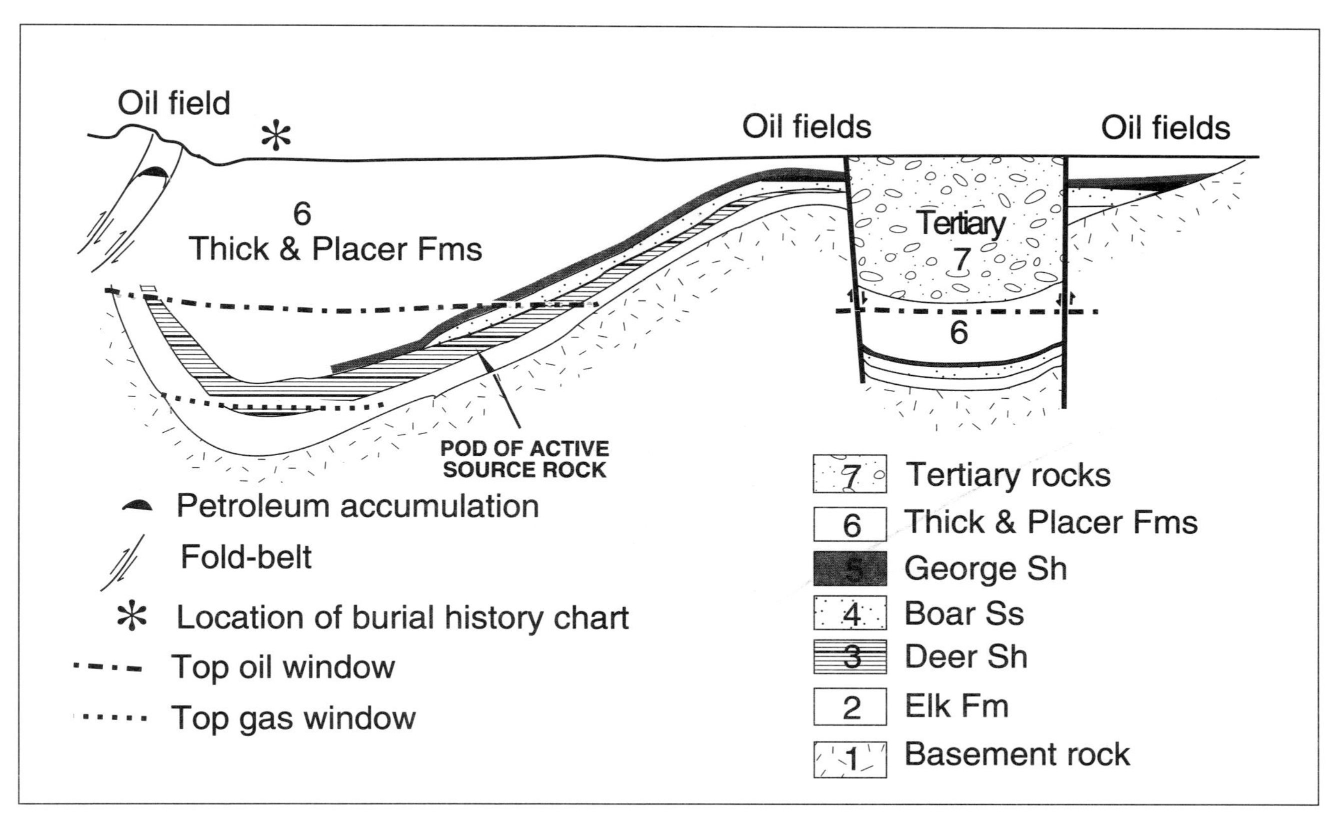

Figure 5—Present-day cross section showing the geographic extent of the Deer-Boar(.) petroleum system. From Peters and Cassa (1994).

through the thickest overburden rock shows the most likely area where the source rock would be thermally mature and, therefore, the provenance of the hydrocarbons. The cross section should transect the entire petroleum system so that the basis for the geographic extent can be demonstrated.

Location of Burial History Chart

The location of the burial history chart (Figure 1E) is along the cross section line within the pod of active source rock. At this location, the source rock must be thermally mature (active), otherwise petroleum would be absent in the conduits or migration paths. The reconstruction of the burial history provides the basis for the times of the onset (O) of generation, migration, and accumulation; the depletion (S, spent) of the source rock; and the critical moment (CM).

Table of Fields

The table (Figure 1C) showing all the oil and gas accumulations included in the folio sheet provides important information about the petroleum system. First, the discovery dates and sizes of the fields are useful for field size distributions and discovery rate modeling. Second, the complexity of the hydrocarbon plumbing system is suggested by the number of reservoir rocks. One reservoir rock for all fields indicates a simple plumbing system, whereas many reservoir rocks indicates a more complicated system. Third, the size of the petroleum system and the generation and expulsion efficiency can be determined by using the total volume of recoverable oil and gas for all fields. Finally, the reservoir rock with the highest percentage of oil or gas reserves is to be used in the petroleum system name. For example, if all the oil is in the Boar Sandstone, it is included in the name (Figure 2D).

For example, the Deer-Boar(.) is a 1.2-billion bbl petroleum system with a simple plumbing system. This size designation using recoverable petroleum is most useful to the explorationist who is interested in (1) comparing the sizes of different petroleum systems to rank or plan an exploration program, and (2) comparing the field sizes in a petroleum system to determine the most likely prospect size and what volumes can be produced. However, the size (volume of recoverable petroleum) of the petroleum system needed for material balance equations is quite different. Here, three additional types of information are estimated: (1) the in-place petroleum for each field, (2) what is left behind along the migration path, and (3) what was lost in surface seeps and exhumed accumulations. These estimates are made by the investigator. This volume of petroleum can then be compared to the estimated volume of petroleum generated in the source rock.

Table 1—Oil Fields that Produce Oil, Natural Gas Liquids (NGL), and Associated Gas, Green River(!) Petroleum System, Uinta Basin[a]

					Reservoir Rock						Cum Prod as of 12/89			Reserves as of 12/89		
	Year Disc	Trap Type[b]	Oil (°API)	GOR (ft^3/bbl)	Formation[c]	Lithology	Depth (ft)	Net Pay (ft)	Por (%)	Perm (md)	Oil ($\times 10^3$ bbl)	NGL ($\times 10^3$ bbl)	Gas ($\times 10^6$ ft^3)	Oil ($\times 10^3$ bbl)	NGL ($\times 10^3$ bbl)	Gas ($\times 10^6$ ft^3)
Altamont-Bluebell	1971	comb	42	1170	GR,W,C	ss, ls	12,400	575	5	—[d]	188,536	18,604	241,079	61,464	8746	112,921
Altamont-Bluebell	1949	comb	32	852	GR	ss, ls	9351	69	8	—	21,446	1519	19,519	3554	831	10,481
Antelope Creek	1983	—	40	3754	GR	ss	5634	269	—	—	962	0	3612	638	0	3888
Brennan Bottom	1953	strat	32	1075	GR	ls	6870	54	5	—	1221	0	1312	199	0	308
Brundage Canyon	1983	—	34	1662	GR	ss	4907	759	—	—	906	0	1506	394	0	744
Castle Peak	1962	comb	34	2728	GR	ss	3568	7245	—	—	489	0	1334	261	0	1216
Coyote Basin	1964	—	31	387	GR	ss	4202	2	—	—	1056	0	409	164	0	149
Duchesne	1958	comb	22	1696	GR	ss	2330	300	—	—	477	0	809	223	0	3181
Duchesne	1951	comb	20	647	C	ss	7486	100	—	—	402	0	260	198	0	206
Eight Mile Flat, North	1982	—	34	1900	GR	ss	3738	2326	—	—	1191	0	2263	509	0	1037
Horseshoe Bend	1964	comb	27	1346	GR	ss	6488	28	—	—	848	0	1141	544	0	945
Monument Butte	1964	comb	32	2555	GR	ss	4735	20	—	—	4461	0	11,396	2039	0	7504
Natural Buttes	1956	comb	29	7984	GR	ss, ls	3384	25	—	—	1343	0	10,723	257	0	8177
Pariette Bench	1962	comb	32	756	GR	ss	4862	1041	—	—	873	0	660	127	0	150
Peters Point	1954	comb	28	>20,000	GR	ss	2800	29	—	—	143	0	2753	0	0	1057
Red Wash	1951	comb	30	2396	GR	ss, ls	5500	90	—	95	137,399	1690	333,529	37,326	170	33,671
								Totals:			**361,610**	**21,813**	**629,552**	**107,897**	**9747**	**184,578**

[a]Production data through 1990 are from Utah Division of Oil, Gas and Mining (1991); reserve data through 1989 are from NRG Associates (1990).

[b]comb, combination; strat, stratigraphic.

[c]GR, Green River; W, Wasatch; C, Colton.

[d]—, no information.

As an example, Table 1 (from Fouch et al., 1994) shows the accumulations for the Green River(!) petroleum system and includes information about the reservoir rock and trap type as well as the hydrocarbon fluid in the trap. If the cumulative production and reserves are added together and gas is converted to oil equivalent (1 bbl = 6000 ft^3), the amount of recoverable petroleum is 636 billion bbl of oil equivalent. However, this recoverable volume omits the hydrocarbons bound up in the Sunnyside, Hill Creek, and other bituminous sandstone deposits estimated to be about 12.5 billion bbl (Fouch et al., 1994; Magoon and Valin, 1994). This amounts to billions of barrels of nonrecoverable asphaltic material that at the time of accumulation most likely ranged from 32° to 42° API oil like that being produced at the Altamont–Bluebell field. Due to uplift and erosion, the lighter portions of this oil were lost to evaporation (inspissation) and biodegradation to produce these bituminous sandstone deposits. If this bituminous material is added back into this petroleum system for material balance calculations, its size increases substantially. This increase improves the perceived effectiveness of the pod of active source rock for expelling oil for this petroleum system.

Cross Section

The petroleum system cross section (Figure 1B), drawn at the CM or at the time when most of the hydrocarbons were generated, shows the geographic and stratigraphic extent of the petroleum system and how each rock unit functions within the system to distribute the oil and gas. Stratigraphically, the petroleum system includes a petroleum source rock, reservoir rock, seal rock, and overburden rock. This cross section is in contrast to structural or stratigraphic cross sections.

The presence of adequate overburden rock in the correct geometry provides (1) the burial needed to thermally mature the source rock, (2) the updip vector needed for oil and gas to migrate to shallower depths, (3) the burial depth variations needed to form traps for petroleum accumulations, and (4) the burial depth of accumulations that allow for the preservation or biodegradation of oil. If the history of the petroleum system is to be modeled correctly, the age, thickness, and erosional history of the overburden rock is required. The cross section in Figure 1B, drawn to represent the end of the Paleozoic (250 Ma), shows the geometry or structural style of the essential elements at the time of hydrocarbon accumulation, or the CM, and best depicts the stratigraphic extent of the system.

As an example, the present-day cross section of the Mandel-Ekofisk(!) petroleum system in Figure 6 (from Cornford, 1994) shows the essential elements that function together to generate and distribute the expelled oil and gas to the various accumulations. The stratigraphic and geographic extents of the system are shown along with the top of the petroleum zone and the location of the burial history chart. Arrows indicate the migration direction for the petroleum and the most likely avenues it took to charge the producing interval.

Petroleum System Name

The name of a petroleum system (Figure 1D) labels the hydrocarbon fluid system or distribution network in much the same way that the name Colorado River designates an aqueous distribution system that includes the river and its tributaries. The name of the petroleum system includes the geologic formation names of the source rock, followed by the major reservoir rock (Figure 1C), and then the symbol expressing the level of certainty. For example, *Deer-Boar(.)* is the name of a hydrocarbon fluid system whose source rock, the Deer Shale, most likely generated the petroleum that charged one or more reservoir rocks, which in this case is the Boar Sandstone. It is the major reservoir rock because it contains the highest percentage by volume of hydrocarbons in the petroleum system.

A petroleum system can be identified at three levels of certainty: known, hypothetical, and speculative (Table 2) (Magoon and Dow, 1994b). At the end of the system's name, the level of certainty is indicated by (!) for known, (.) for hypothetical, or (?) for speculative. The symbol indicates the level of certainty that a particular pod of active source rock has generated the hydrocarbons on the table of accumulations (see Figure 1). In a known (!) petroleum system, a well-defined geochemical correlation exists between the source rock and the oil accumulations. In a hypothetical (.) petroleum system, geochemical information identifies a source rock, but no geochemical match exists between the source rock and the petroleum accumulation. In a speculative (?) petroleum system, the links between a source rock and the petroleum accumulations are postulated entirely on the basis of geologic or geophysical evidence.

In certain frontier areas of the world, especially offshore, stratigraphic units are often poorly understood and frequently undesignated. Here, the judgment of the investigator is required. The geologist should avoid using ages, such as Jurassic, in the name because it fails to identify the petroleum system uniquely. In other situations, it may be difficult or confusing to follow the naming rules. For example, when a rock unit that includes both the source and reservoir forms more than one petroleum system, the same name might be used more than once, such as the Monterey(!) petroleum systems. Here, a geographic modifier can be used to differentiate the systems. Another naming problem arises in frontier areas where formal rock units are lacking, so only ages or geophysically mappable units are used. A geographic name or the name of an accumulation in the petroleum system can be used. If it is impossible to follow the formal designation of a petroleum system, the investigator should select a unique name that identifies the fluid system, not the rock units.

Burial History Chart

The purpose of the burial history chart (Figure 1E) is to show the essential elements and three important hydrocarbon events, which are (1) the onset (O) of generation,

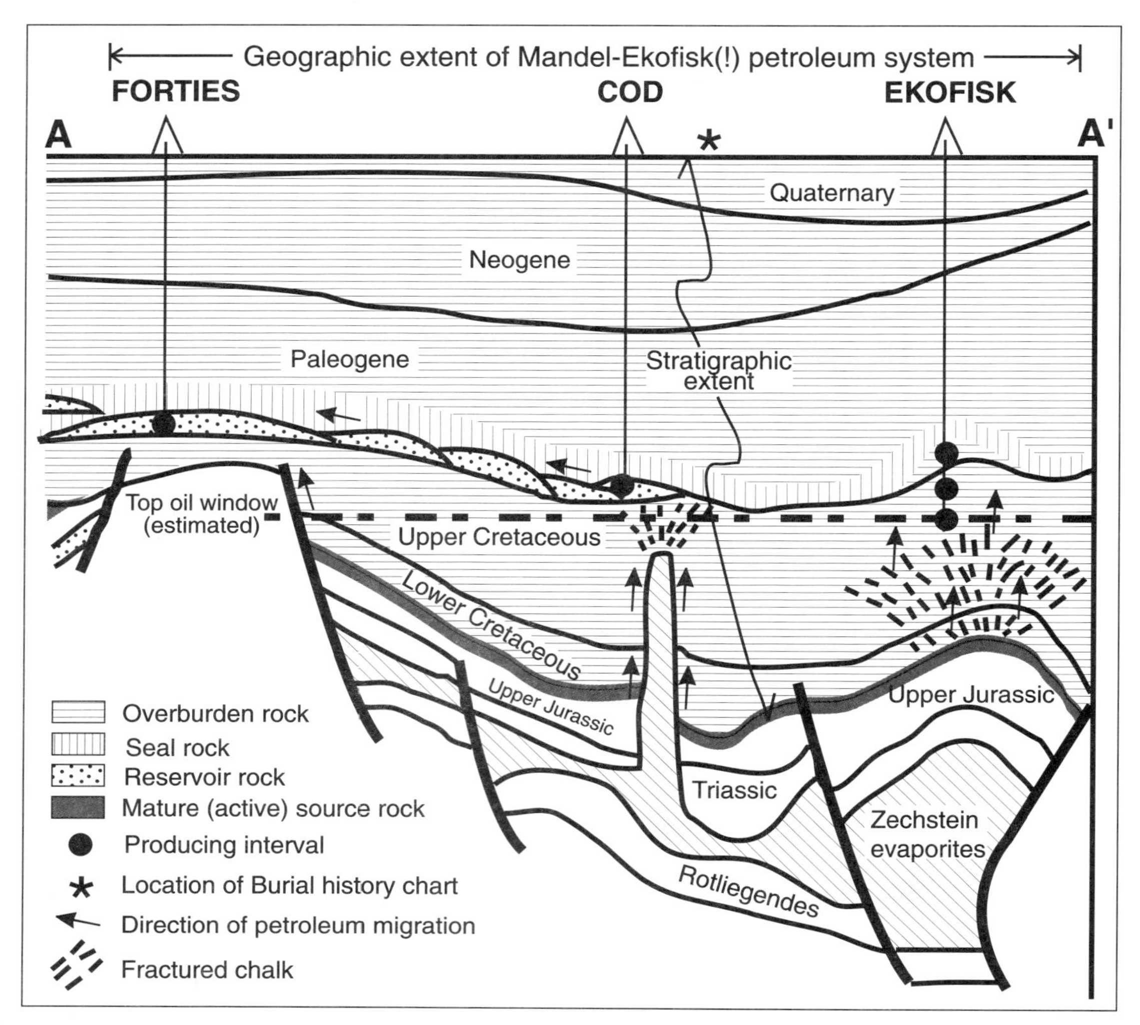

Figure 6—Petroleum system cross section showing the geographic extent of the Mandel-Ekofisk(!) system, essential elements, stratigraphic extent, vectors of petroleum migration (arrows), and location of burial history chart. Modified from Cornford (1994).

migration, and accumulation; (2) the partially spent (S) or depleted active source rock; and (3) the CM of the petroleum system. The tops of the oil and gas windows and the lithology and name of the rock units involved should also be shown. This chart uses sedimentologic and paleontologic evidence in the overburden rock to reconstruct the burial or thermal history of the source rock. The onset of generation, migration, and accumulation usually occurs when the source rock reaches thermal maturity at a vitrinite reflectance equivalence of 0.7 ± 0.1% R_0 and ends when the source rock is either uplifted before all the hydrocarbons can be expelled or depleted as the source rock is more deeply buried. The location of the burial history chart is shown on the petroleum system map and cross section.

In the example of Figure 1E, the Deer Shale (rock unit 3) is the source rock, the Boar Sandstone (unit 4) is the reservoir rock, the George Shale (unit 5) is the seal rock, and all the rock units above the Deer Shale (units 4, 5, and 6) comprise the overburden rock. The burial history chart is developed for the location where the overburden rock is thickest. In this case, it indicates that the onset of generation, migration, and accumulation started 260 Ma in Permian time (time scale of Palmer, 1983) and was at maximum burial depth 255 Ma. Oil generation ended about 240 Ma because the source rock was depleted at that time. The CM, as judged by the investigator, is 250 Ma because modeling indicates most (>50%) of the hydrocarbons have been expelled and are accumulating in their primary traps. Although the investigator would have been correct to choose anytime between 250 and 240 Ma, 250 Ma was chosen because this is the moment in time for which the best geologic information was available to reconstruct the map and cross section. The time of onset of generation-migration-accumulation ranges from 260 to 240 Ma and is used as the age of the petroleum system.

As previously mentioned in the cross section discussion, knowing the age and thickness for each increment

Table 2—Definitions of Levels of Certainty in Petroleum Systems

Level of Certainty	Criteria	Symbol
Known	Oil–source rock or gas–source rock correlation	(!)
Hypothetical	In absence of petroleum–source rock correlation, geochemical evidence indicates origin of oil and gas	(.)
Speculative	Geologic or geophysical evidence	(?)

of overburden rock is crucial for any modeling exercise. Each increment must be bounded by time lines whose ages are supported by paleontologic dates, isotopic age dates, or other means of dating strata. As the number of increments in the overburden rock increases, details of the burial history of the source rock become better understood. For example, in Figure 1E, the overburden rock is undifferentiated Permian. Suppose, however, that paleontologic dates indicated that 95% of the overburden rock was Early Permian and that the rest was Late Permian. This increase in increments in the overburden rocks would change the time when the underlying source rock became thermally mature. In our example, the change in time that the source rock became mature might be considered small, but in other examples, the difference could be large.

Location of Burial History Chart

The burial history chart chosen to show the three hydrocarbon events for a petroleum system should be developed for the location in the pod of active source rock where, in the judgment of the investigator, much of the oil and gas originated. Usually this location is downdip from a major migration path to the largest fields.

Petroleum systems are seldom so simple that only one burial history chart adequately describes the same three hydrocarbon events for every location in the pod of active source rock. The investigator chooses the burial history curve that best suits the purpose. If the investigator is presenting (oral or written) a petroleum system investigation, (s)he would develop the burial history curve for a location downdip from a major migration path to the largest fields. However, if the subject is a play or prospect, the burial history curve would be constructed for a location that is downdip on a suspected migration path to the area of the play or prospect.

Critical Moment

The generation-migration-accumulation of oil and gas in a petroleum system never starts when the source rock is being deposited and seldom extends into the present day. If a source rock was deposited in Paleozoic time, it could be Mesozoic time before it becomes thermally mature and charges adjacent traps. By the Cenozoic, this source rock would probably be depleted. The time over which the process of generation-migration-accumulation takes place could be tens of millions of years. This is a long period of time to chose from if an investigator must select the most appropriate moment during this process for which to make a map and cross section showing the petroleum system when most (>50%) of the hydrocarbons were migrating and accumulating. To help the investigator with this important exercise, the critical moment (CM) was introduced and incorporated into the petroleum system folio sheet.

Geologists use the concept of the CM for other exercises. Whenever a map such as an isopach map is constructed, it is frequently reconstructed to its original thickness at the moment of deposition. Although the kinematic development of a fold and thrust belt may occur over many millions of years, it is often represented by one cross section, which represents a snapshot in time. A structural cross section of a fold and thrust belt reconstructed at the end of the Cretaceous, for example, uses the CM concept.

The CM is the time that best depicts the generation-migration-accumulation of hydrocarbons in a petroleum system (Magoon and Dow, 1994a, p. 643). This definition requires an explanation and examples to be easily understood. A *moment* is a brief, indefinite interval of time that is of particular importance. When one takes a photo with a camera, the moment recorded is less than a second. In geology, the farther one goes back in time, the more the interval increases, becoming thousands or even millions of years. For the petroleum system, a moment relates to the shortest measurable time. The term *critical* refers to the moment that best represents, in the judgment of the investigator, the process of generation-migration-accumulation.

Best is a keyword in this definition. It represents the criteria that the investigator should use to select the appropriate moment. The best time must fulfill several criteria: (1) it must be within the age of the petroleum system (Magoon and Dow, 1994a, p. 644); (2) it must be when most, or more than half, of the hydrocarbons are migrating and accumulating; and (3) it must be shown as an arrow, not an interval, on the burial history and events charts.

The CM of a petroleum system can vary depending on its purpose. If the purpose is a petroleum system case study, then the CM should be representative of the entire system. However, if the purpose is related to an exploration play or prospect, then the CM should be chosen for that part of the pod of active source rock most likely to charge the traps in the play or prospect. Depending on the size, thickness, and variation of the thermal maturity of the pod of active source rock and the objective of the investigator, these could be different best times, neither of which is incorrect. In fact, the investigator may need to make numerous burial history charts of a large, thick pod of active source rock that has a wide range of thermal maturities to determine which best moment properly depicts generation-migration-accumulation for the intended audience.

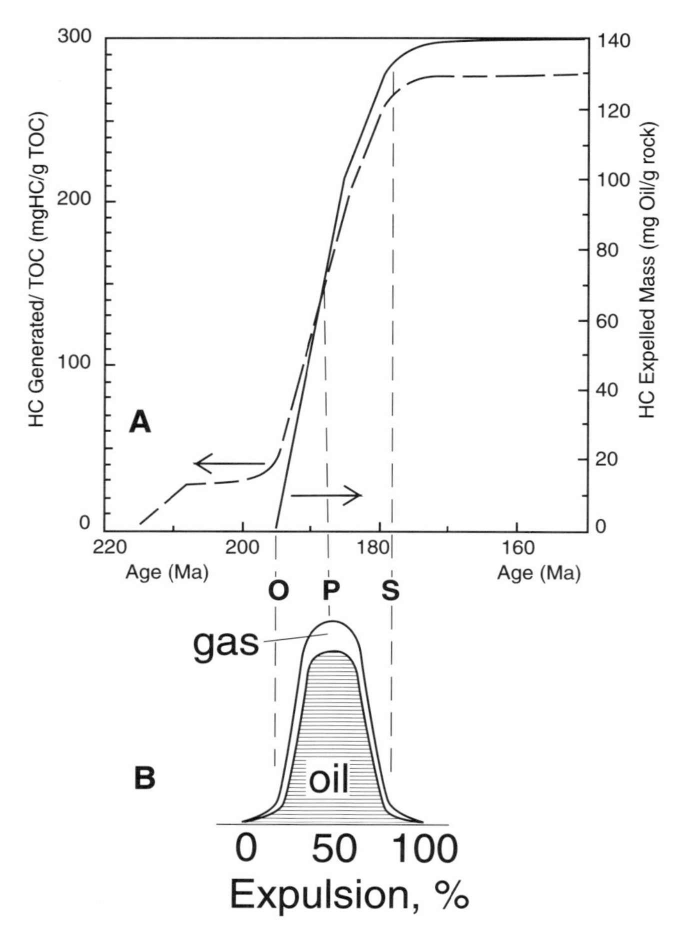

Figure 7—(A) Cumulative curves showing the time over which hydrocarbons are generated and expelled. (B) Distribution curve for oil and gas expulsion using the same information and showing the onset (O) of expulsion, peak (P) expulsion, and depletion (S, spent) of the source rock. The CM is selected to be any time between peak expulsion and depletion.

The burial history chart omits important information available in most modeling packages that explains the CM from a different perspective. The graph in Figure 7 shows cumulative volumes of generated or expelled oil and gas. Wherever the curves are horizontal, no volume was being added. This graph shows that the onset of generation (dashed curve) preceded expulsion (solid curve) by almost 20 m.y. According to this graph, most of the expulsion occurred over a period of 16 m.y. from 195 to 179 Ma. When this cumulative expulsion curve is transformed into a curve showing the distribution of expulsion, it shows peak expulsion at 188 Ma (Figure 7). At this time, at least half of the petroleum was migrating, so the CM could be chosen at any time between 188 and 179 Ma.

Now we give an example that takes into account the pod's area. The pod of active source rock expels oil and gas over the time the pod grows (Figure 8). This phenomenon can be explained in three stages using the distribution of the expulsion curve shown in Figure 7. In the first stage (Figure 8A), a small volume of chemically active source rock expels oil and gas into a carrier bed such that it migrates updip and is trapped at a shallower depth at a lower temperature than the thermally mature source rock. The thermal maturity of the source rock ranges from about 0.7% to 1.3% R_o and is represented by a single expulsion curve.

The second burial stage (Figure 8B) shows that the pod of active source rock has grown larger from deeper burial. The source rock involved in stage one has increased in thermal maturity to about 1.3–2.0% R_o and is now spent. The active source rock updip from this spent source rock is now expelling oil and gas into the same carrier bed and is migrating updip to the same traps available to stage one hydrocarbons. Because this hydrocarbon was expelled from the same source rock at the same thermal maturity, it is the same composition as the hydrocarbons expelled in the first stage. The two expulsion curves for the second stage occur where the thermal maturity of the source rock ranges from about 0.7% to 1.3% R_o (Figure 8B).

The last burial stage (Figure 8C) shows that the pod of active source rock has grown even larger from greater burial. The thermal maturity of the source rock involved in stage one now exceeds 2.0% R_o, and the source rock in stage two ranges from 1.3% to 2.0% R_o and is depleted. The source rock that is presently expelling oil and gas of the same composition as stage one and two is migrating updip to the same traps. Together, these three stages (Figure 8D) show the kinematics of the pod of active source rock over 40 m.y., or the age of the petroleum system, the time over which generation, migration, and accumulation took place. The CM is selected within the time when half or all of the active source rock volume exceeded 1.3% R_o, which was from 180 to 160 Ma. The burial history chart is constructed for the area where the source rock is buried deepest, but the CM is selected by the investigator using the volume of active source rock expelling hydrocarbons.

These maturity stages showing the evolution of the pod of active source rock provide guidance on how to select the CM and locate the burial history chart.

Events Chart

The petroleum system events chart (Figure 1F) shows the temporal relationship of the rock units, essential elements, processes, preservation, and critical moment for each petroleum system in bar graph form. The events chart concept is flexible and can be used in several ways (Figure 9).

The events chart (Figure 1F) shows the following items. The rock units include those within the stratigraphic extent of the petroleum system. On a certainty scale, the ages of the four essential elements (source, reservoir, seal, and overburden rocks) are usually better

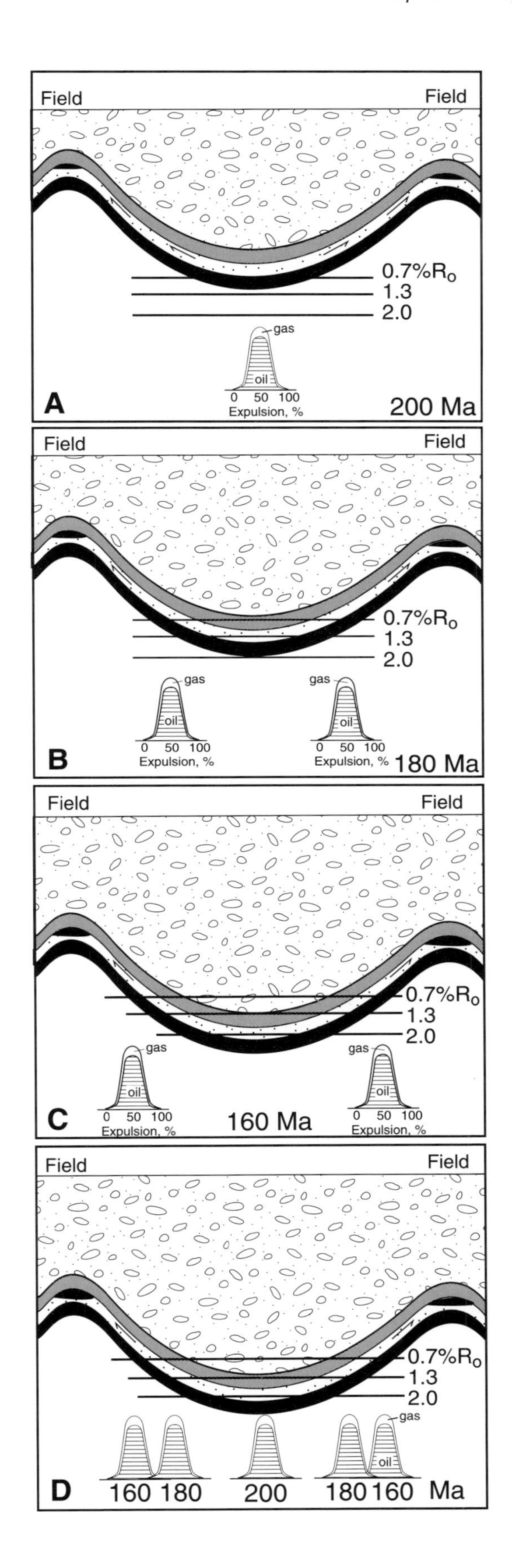

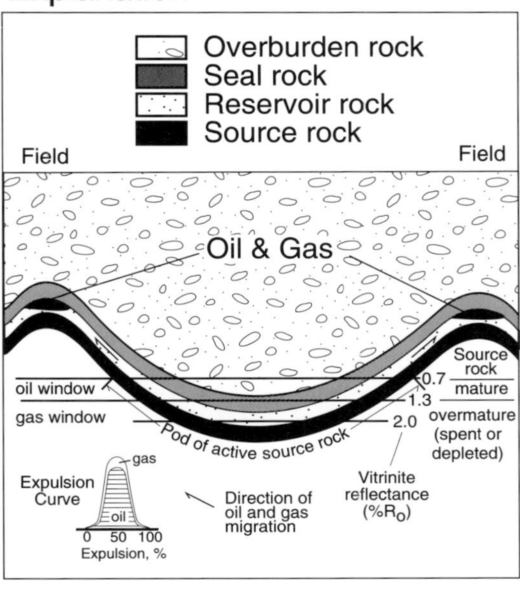

Figure 8—Schematic cross sections showing growth of the pod of active source rock with a corresponding increase in burial and thermal maturity. (A) When the source rock is only in the oil window (0.7–1.3% R_o), the pod is small. (B) When the source rock passes into the gas window (1.3–2.0% R_o), the pod grows larger. (C) Pod of active source rock continues to grow with increasing burial. (D) Summarizing these three burial histories provides a basis on which to better determine the critical moment. (E) Explanation of symbols.

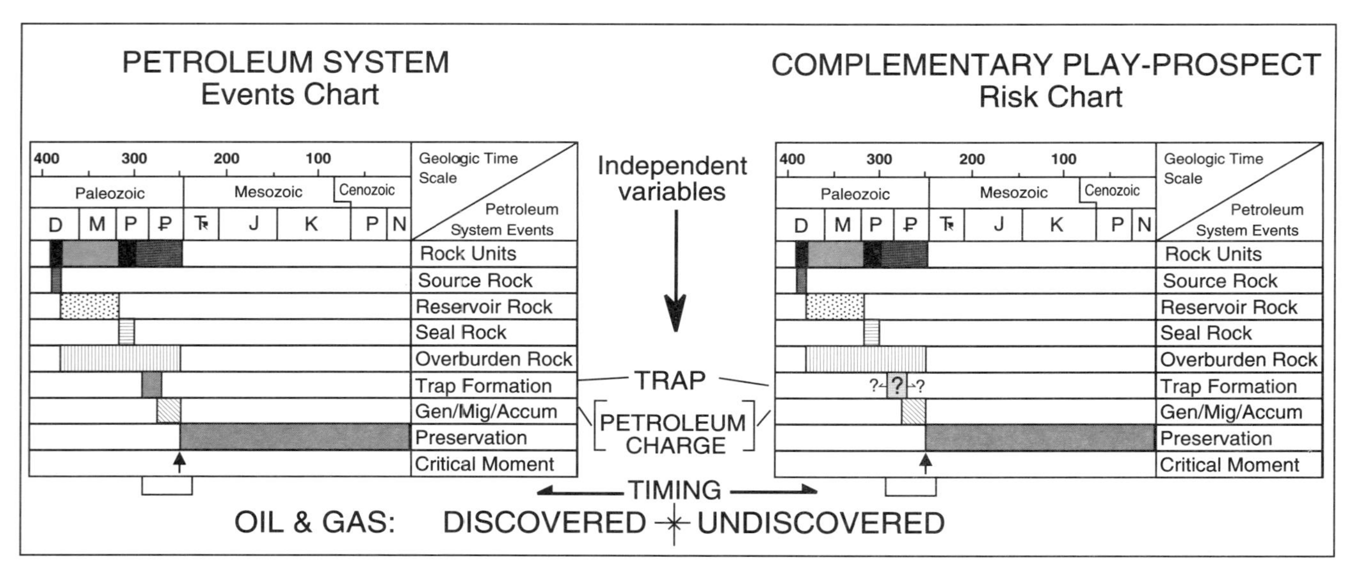

Figure 9—Events chart for a petroleum system compared to a risk chart. The risk chart is used to better evaluate uncertainty for a play or prospect relative to a system.

established from paleontology or radiometric age dates than those associated with the two processes (trap formation and generation-migration-accumulation). Determining the times over which the traps form and hydrocarbons accumulate (generation-migration-accumulation) is more interpretive because there is less precise temporal information about these processes. Therefore, risk or uncertainty with regard to the times over which the two processes takes place is higher than for the better established times of development of the four essential elements. This certainty relationship is important if a similar chart is constructed for a complementary prospect.

When an events chart is constructed for a complementary prospect, it becomes a "risk chart," as shown in Figure 9 (Magoon, 1997; Magoon and Beaumont, 1999). The risk chart is derived from the petroleum system events chart, which in turn is derived from the summation of the events chart for each oil and gas field in the petroleum system. These oil or gas fields are successful prospects. Unsuccessful prospects are dry holes. For example, if a risk chart for a prospect is similar to the petroleum system events chart, then it can be concluded that this prospect is more likely to contain petroleum than one that has a dissimilar risk chart. Conversely, if an events chart is constructed for each dry hole within a petroleum system, they should be dissimilar from producing fields. This dissimilarity indicates where the greater uncertainty lies. Used in this way, the events chart is a useful analytical tool to deal with uncertainty or risk.

For example, this simple bar graph clearly shows the sequence of events relative to petroleum entrapment. For an evolving petroleum system to trap migrating hydrocarbon fluids effectively, the trap-forming process must occur before or during the generation-migration-accumulation process in order for petroleum to accumulate.

When constructing an events chart, the following rules should be followed. First, there is only one pod of active source rock for each petroleum system. Second, every effective reservoir rock needs a seal, no matter how thin. Third, only reservoir rocks that contain petroleum accumulations, shows, or seeps should be shown. Fourth, eroded overburden rock should be shown with hatchures so that it can be incorporated into the modeling exercise. Fifth, the best information for timing of trap formation comes from oil and gas fields. Sixth, the best information for generation-migration-accumulation comes from geologic and geochemical information about the source rock that is then incorporated into the burial modeling and kinetics. This information indicates the onset, peak, and end of generation-migration-accumulation or when the active source rock is depleted (spent) or uplifted (inactive source rock). This process takes place over a relatively short period. Seventh, preservation time (see Magoon and Dow, 1994a, p. 644), by definition, starts when generation-migration-accumulation ends and continues to the present. Some petroleum systems have no preservation time. Finally, the time of the critical moment should be based on the judgment of the investigator, but modeling software packages are useful tools that show the time over which expulsion occurs for the pod of active source rock.

SUMMARY

The petroleum system folio sheet summarizes the interdependence of the essential elements and processes by showing how they function together in time and space to distribute oil and gas from its origin or provenance to its final destination. Everything about a petroleum system need not be known before this folio sheet is constructed. In fact, the earlier a folio sheet is constructed in the exploration cycle of a system, the better. This is because the working hypotheses about migration and accumulation of hydrocarbons can be better tested as plays that are used to define drillable prospects. Successful prospects or commercial discoveries increase one's confidence that this petroleum system is properly mapped, suggesting that the working hypothesis is correct. Used in this manner, every petroleum system folio sheet is a progress report. A preliminary progress report would be done for a lightly explored petroleum system, while a final progress report would cover a well-explored petroleum system. A final progress report can also be used as an analog for a preliminary report in another area, provided the petroleum systems are similar. For example, a well-explored petroleum system in a cratonic sag could be an analog for a lightly explored system in a similar geologic setting.

The juxtaposition of the figures and table on the petroleum system folio sheet points to their interdependence and the need for the petroleum system terms discussed in this chapter. For example, the term *critical moment* was needed to succinctly communicate to the user when the map and cross section was constructed and why. The geographic and stratigraphic extent of the system were needed to show which rock units are involved in the origin and distribution of hydrocarbons within that petroleum system (see Magoon and Dow, 1994a, for more definitions of terms).

A petroleum system study includes logical steps requiring specific information that explains the origin in time and space of discovered hydrocarbon accumulations, shows, and seeps. If used creatively, the map provides a basis from which to develop complementary prospects and plays. The logical steps include the characterization of a petroleum system (that is, to identify, map, and name the hydrocarbon fluid system) and the summary of the results on a folio sheet. The folio sheet depicts the geographic, stratigraphic, and temporal evolution of the petroleum system. The folio sheet includes the (1) petroleum system map, (2) petroleum system cross section, (3) table of genetically related accumulations, (4) petroleum system name, (5) burial history chart, and (5) events chart.

A petroleum system map is based on the understanding that there are several levels of certainty from "guessing" to "knowing" that certain oil and gas accumulations emanated from a particular pod of active source rock. The lowest level of certainty is based on the close geographic proximity of two or more accumulations. The level of certainty increases when there is close stratigraphic proximity of the accumulations and continues to increase with similarities in bulk petroleum properties and detailed geochemical properties. The highest level of proof is shown by the positive geochemical correlation of the hydrocarbon fluid in the accumulations to an extract from the active source rock. Parallel levels of proof can be applied to identifying the active source rock responsible for the genetically related hydrocarbon fluids.

***Acknowledgments**—The authors thank Marcio Mello and Barry Katz for the opportunity to present this paper as a keynote talk at the 1997 AAPG Hedberg Research Conference on "Petroleum Systems of the South Atlantic Margins" held in Rio de Janeiro, Brazil. Special thanks to Elisabeth Rowan, Francis Cole, Barry Katz, Gregory Ulmishek, and Louis Liro for their thoughtful reviews and helpful comments.*

REFERENCES CITED

Al Arouri, K., C. J. Boreham, D. M. McKirdy, and N. M. Lemon, 1998, Modeling of thermal maturation and hydrocarbon generation in two petroleum systems of the Taroom Trough, Australia: AAPG Bulletin, v. 82, p. 1504–1527.

Bethke, C. M., M. K. Lee, and J. Park, 1999, Basin modeling with *Basin2*, A guide to using the *Basin2* software package, Release 4: Hydrogeology Program, University of Illinois, Urbana, Illinois, 205 p.

Bordenave, M. L., 1992, Applied petroleum geochemistry: Éditions Technip, Paris, 524 p.

Chen, Honghan, Sitian Li, Sun Yongchuan, and Qiming Zhang, 1998, Two petroleum systems charge the YA13-1 gas field in Yinggehai and Qiongdongnan Basins, South China Sea: AAPG Bulletin, v. 82, p. 757–772.

Cornford, C., 1994, Mandal-Ekofisk(!) petroleum system in the Central Graben of the North Sea, *in* L. B. Magoon and W. G. Dow, eds., The petroleum system—from source to trap: AAPG Memoir 60, p. 537–571.

Demaison, G., 1984, The generative basin concept, *in* G. Demaison and R. J. Murris, eds., Petroleum geochemistry and basin evaluation: AAPG Memoir 35, p. 1–14.

Demaison, G., and R. J. Murris, eds., 1984, Petroleum geochemistry and basin evaluation: AAPG Memoir 35, 426 p.

Dow, W. G., 1974, Application of oil-correlation and source-rock data to exploration in Williston basin: AAPG Bulletin, v. 58, no. 7, p. 1253–1262.

Fouch, T. D., V. F. Nuccio, D. E. Anders, D. D. Rice, J. K. Pitman, and R. F. Mast, 1994, Green River(!) petroleum system, Uinta Basin, Utah, U.S.A., *in* L. B. Magoon and W. G. Dow, eds., The petroleum system—from source to trap: AAPG Memoir 60, p. 399–421.

Howes, J. V. C., 1997a, Petroleum resources and petroleum systems of SE Asia, Australia, Papua New Guinea, and New Zealand, *in* J. V. C. Howes and R. A. Noble, eds., Petroleum systems of SE Asia and Australasia: Indonesian Petroleum Association, Jakarta, Indonesia, p. 81–100.

Howes, J. V. C., 1997b, Petroleum systems, resources of southeast Asia, Australasia: Oil and Gas Journal, v. 95, n. 50, p. 52–56.

Hunt, J. M., 1979, Petroleum geochemistry and geology: San Francisco, W. H. Freeman, 617 p.

Hunt, J. M., 1995, Petroleum geochemistry and geology, 2nd ed.: San Francisco, W. H. Freeman, 743 p.

Hurley, N., and P. Weimer, eds., 1998, Gulf of Mexico Petroleum Systems: AAPG Bulletin, v. 82, n. 5B, p. 865–1112.

Magoon, L. B., 1987, The petroleum system—a classification scheme for research, resource assessment, and exploration (abs.): AAPG Bulletin, v. 71, no. 5, p. 587.

Magoon, L. B., 1988, The petroleum system—a classification scheme for research, exploration, and resource assessment,

in L. B. Magoon, Petroleum systems of the United States: USGS Bulletin 1870, p. 2–15.
Magoon, L. B., ed., 1989a, The petroleum system—status of research and methods, 1990: USGS Bulletin 1912, 88 p.
Magoon, L. B., 1989b, Identified petroleum systems within the United States—1990, *in* L. B. Magoon, ed., The petroleum system—status of research and methods, 1990: USGS Bulletin 1912, p. 2–9.
Magoon, L. B., ed., 1992a, The petroleum system—status of research and methods, 1992: USGS Bulletin 2007, 98 p.
Magoon, L. B., 1992b, Identified petroleum systems within the United States—1992, *in* L. B. Magoon, ed., The petroleum system—status of research and methods, 1992: USGS Bulletin 2007, p. 2–11.
Magoon, L. B., 1995, The play that complements the petroleum system—a new exploration equation: Oil and Gas Journal, v. 93, October 2, p. 85–87.
Magoon, L. B., 1997, The petroleum system—an exploratory tool to find oil and gas and to assist in risk management, *in* J. V. C. Howes and R. A. Noble, eds., Petroleum systems of SE Asia and Australasia: Indonesian Petroleum Association, Jakarta, Indonesia, p. 25-36.
Magoon, L. B., and D. E. Anders, 1992, Oil-to-source-rock correlation using carbon-isotopic data and biological marker compounds, Cook Inlet–Alaska Peninsula, Alaska, *in* J. M. Moldowan, P. Albrecht, and R. P. Philp, eds., Biological markers in sediments and petroleum: Herts, U.K., Prentice Hall, p. 241–274.
Magoon, L. B., and E. A. Beaumont, 1999, Petroleum systems, *in* E. A. Beaumont and N. H. Foster, eds., Exploring for oil and gas traps: AAPG Treatise of Petroleum Geology, p. 3.1–3.34.
Magoon, L. B., and W. G. Dow, eds., 1994a, The petroleum system—from source to trap: AAPG Memoir 60, 655 p.
Magoon, L. B., and W. G. Dow, 1994b, The petroleum system, *in* L. B. Magoon and W. G. Dow, eds., The petroleum system—from source to trap: AAPG Memoir 60, p. 3–24.
Magoon, L. B., and Z. C. Valin, 1994, Overview of petroleum system case studies, *in* L. B. Magoon and W. G. Dow, eds., The petroleum system—from source to trap: AAPG Memoir 60, p. 329–338.
Meissner, F. F., J. Woodward, and J. L. Clayton, 1984, Stratigraphic relationships and distribution of source rocks in the greater Rocky Mountain region, *in* J. Woodward, F. F. Meissner, and J. L. Clayton, eds., Hydrocarbon source rocks of the greater Rocky Mountain region: Denver, Rocky Mountain Association of Geologists, p. 1–34.
Palmer, A. R., 1983, The decade of North American geology, 1983 geologic time scale: Geology, v. 11, p. 503–504.
Perrodon, A., 1980, Géodynamique pétrolière. Genèse et répartition des gisements d'hydrocarbures: Paris, Masson–Elf-Aquitaine, 381 p.
Peters, K. E., and M. R. Cassa, 1994, Applied source rock geochemistry, *in* L. B. Magoon and W. G. Dow, eds., The petroleum system—from source to trap: AAPG Memoir 60, p. 93–117.
Peters, K. E., and J. M. Moldowan, 1993, The biomarker guide—interpreting molecular fossils in petroleum and ancient sediments: Englewood Cliffs, NJ, Prentice Hall, 363 p.
Price, L. C., and R. McNeil, 1997, Thoughts on the birth, evolution, and current state of petroleum geochemistry: Journal of Petroleum Geology, v. 20, p. 118–123.
Tissot, B. P., and D. H. Welte, 1978, Petroleum formation and occurrence—a new approach to oil and gas exploration: Berlin, Springer-Verlag, 538 p.
Tissot, B. P., and D. H. Welte, 1984, Petroleum formation and occurrence—second revised and enlarged edition: Berlin, Springer-Verlag, 699 p.
Ulmishek, G., 1986, Stratigraphic aspects of petroleum resource assessment, *in* D. D. Rice, ed., Oil and gas assessment—methods and applications: AAPG Studies in Geology #21, p. 59–68.
Waples, D. W., 1981, Organic geochemistry for exploration geologists: Minneapolis, MN, Burgess Publishing, 151 p.
Waples, D. W., 1985, Geochemistry in petroleum exploration: Boston, International Human Resources Development Corporation, 232 p.

Szatmari, P., 2000, Habitat of petroleum along the South Atlantic margins, *in* M. R. Mello and B. J. Katz, eds., Petroleum systems of South Atlantic margins: AAPG Memoir 73, p. 69–75.

Chapter 6

Habitat of Petroleum Along the South Atlantic Margins

Peter Szatmari

Petrobrás Research Center
Rio de Janeiro, Brazil

Abstract

On a predrift reconstruction of their facing margins, almost all of Brazil's and southwest Africa's petroleum reserves occur within a nearly continuous north-south-trending strip, the South Atlantic Megatrend, whose formation was controlled by continental breakup. It took almost 40 m.y., from earliest Valanginian to late Albian time, for Africa and South America to separate, as South America rotated clockwise relative to Africa about a pole located in northeast Brazil. Rifting mechanism and continental deformation were controlled by the distance from the shifting pole of differential rotation. The rift was cut into two approximately equal portions by the Ponta Grossa dike swarm and associated volcanics that formed over the Paraná–Tristan da Cunha hot spot. South of the dike swarm continental, breakup was fast and characterized by intense volcanic activity (seaward dipping reflectors). In this area, a euxinic sea formed by Aptian time over the oceanic crust. To the north of the dike swarm, continental breakup was slower, and a northward-tapering rift valley lying deep below sea level formed over extended continental crust. This depression was flooded catastrophically by sea water in late Aptian time, depositing a 2000-m-thick salt body.

The oil fields constituting the South Atlantic Megatrend occur either within the salt basin, both above and below the salt, or near the pole of differential rotation in transtensional inland rifts that contain no salt. The opening of the Atlantic divided this megatrend into segments that now alternate between the Brazilian and African margins: an oil-producing segment on one margin is faced by a relatively sterile segment on the other. This distribution may result from alternating polarity of rifting by simple shear.

Oil richness, defined here as the volume of original reserves per unit length of a segment, increases along the megatrend with distance from the pole of relative rotation. Far from the pole, as in the Lower Congo and especially in the Campos Basin where the rift is wide, the rifted continental crust extends far offshore, oil richness is high, and the largest petroleum accumulations occur offshore.

INTRODUCTION

On a prerift reconstruction of their facing margins, almost all of Brazil's and sub-Saharan western Africa's petroleum reserves occur within a narrow, straight, and (relative to South America) north-south-trending strip which is named here the South Atlantic Megatrend. This trend extends for more than 2000 km from southern Nigeria to northern Angola on the African side and from the Potiguar Basin to the northern part of the Santos Basin on the Brazilian side. Along this trend, oil is contained in basins and structures that owe their existence, directly or indirectly, to continental breakup. The mechanism of breakup thus exerts a controlling influence on the distribution of hydrocarbon deposits.

CONTINENTAL BREAKUP BY ROTATION ABOUT A POLE IN NORTHEASTERN BRAZIL

Our model is based on the separation of Africa and South America by differential rotation about a migrating pole that was located in northeastern Brazil (Rabinowitz and LaBrecque, 1979; Szatmari et al., 1985a) during most of the continental breakup (Figure 1). As a result, along Brazil's eastern South Atlantic margin the rift propagated to the north. Although such propagation locally resulted in distinct rifting episodes, such as the one observed at the end of the Early Cretaceous Rio da Serra stage, the crucial point of the model is that rift propagation by

Figure 1—Schematic view of compression and extension caused by separation of South America from Africa by clockwise rotation about a pole located along Brazil's equatorial margin.

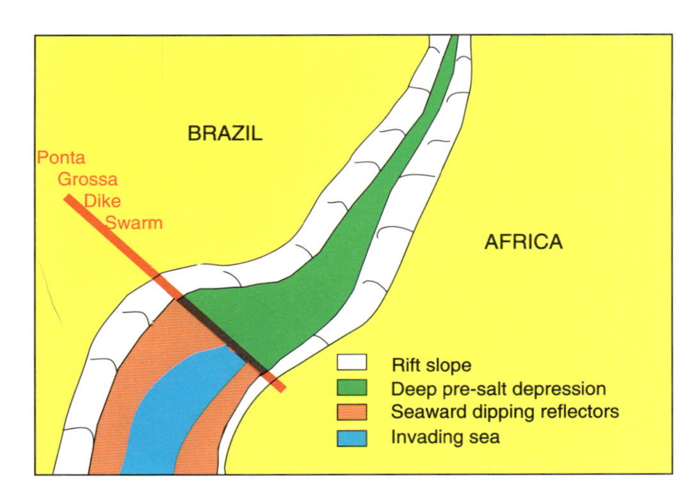

Figure 2—Paleogeographic pattern of the South Atlantic rift in middle Aptian time.

differential rotation of the separating continents was a lengthy process. The angular momentum for such large masses of thick continental lithosphere is very high, so that it takes a long time both for initial acceleration to occur and for differential rotation to grind to a halt. In the case of the South Atlantic, it took almost the entire Early Cretaceous, from earliest Valanginian to late Albian time, a period of nearly 40 m.y., for the two continents to separate.

Differential rotation of the two continents about a pole located in northeastern Brazil created the South Atlantic rift, which tapers northward toward the pole of rotation. The rift was divided into two structurally different portions of about equal lengths by the Ponta Grossa dike swarm (Figure 2) which formed over the Paraná–Tristan da Cunha hot spot at about 130 Ma. To the south of the dike swarm, in the southern Santos Basin, in the Pelotas Basin, and between Argentina and southern Africa, sublithospheric temperatures were high and continental breakup was rapid. It was accompanied by intense, mostly subaerial, volcanic activity that formed a volcanic margin marked by seaward-dipping reflectors. To the north of the dike swarm, from the northern Santos Basin to Recife, temperatures were lower, producing a rift floored by continental lithosphere that was wide near the Ponta Grossa dike swarm but narrowed to the northeast toward the pole of rotation (Figure 2).

Near the hot spot, in the Campos and Santos Basins, where the rift was wide, the pre-Aptian sequence is largely composed of volcanics (basalt flows and tuffs). These are slightly younger than the continental flood basalts of the Paraná Basin, reflecting the motion of the hot spot to the southeast relative to the South American plate. The thickness of pre-Aptian volcanics and the width of the rift decrease to the north, toward the pole of differential rotation, so that to the north of the Espírito Santo Basin, instead of volcanics, a thick sedimentary section occupies the pre-Aptian sequence. A similar decrease in volcanics also accompanies decreasing extension in the East African rift, away from the hot spot located in Ethiopia and the Afar triangle.

As a result of the differential rotation of the continents about a pole located in northeastern Brazil, rifting near the pole lasted longer and was completed later than in the south. Rifting in northeastern Brazil took place in two stages. During the first, pre-Aptian stage, faults propagated to the north, opening the Recôncavo–Tucano Basin, now located inland, while en-echelon north–south faults formed in the Sergipe–Alagoas Basin. During the second, Aptian–Albian stage, northeast-trending transtensional faults in the Sergipe–Alagoas Basin marked the transfer of the differential rotation between the two continents to the present continental margin (Szatmari et al., 1985a, 1987; Milani et al., 1987).

Both the inland Recôncavo–Tucano rift and the Sergipe–Alagoas Basin along the continental margin terminate to the north at the latitude of Recife, reflecting the strong obstacle to rift propagation presented by the transverse east–west trending Precambrian Pernambuco and Patos fault zones (Szatmari et al., 1985a, 1987). Therefore, north of Recife (between João Pessoa and Natal), the clockwise rotation of South America relative to Africa caused compression and left-lateral transpression instead of extension, with no major rifting taking place until Albian time (Françolin and Szatmari, 1987; Szatmari et al., 1985b, 1987). Thus, the northeastern tip of South America was turned counterclockwise relative to the rest of the continent as it was pulled away from the facing margin of Nigeria in the north while still being pressed against Africa in the east (Figure 1). This created a stress field that was inverted relative to most of Brazil's eastern margin, being characterized by east–west compression and north–south extension. The northeast-trending right-lateral transtensional Pendência rift in the Potiguar Basin formed in this stress regime (Szatmari et al., 1987; Françolin and Szatmari, 1987; Françolin et al., 1994). Along the rest of the equatorial margin, the differential rotation of the two continents created north–south compression (Figure 1) until Aptian time when the rift propagated westward along the entire length of the equatorial margin (Szatmari et al., 1985b, 1987; Zanotto and Szatmari, 1987).

The basic pattern of deformation can be interpreted in terms of simple mechanical principles (see Figure 1):

1. The rift along the eastern (South Atlantic) margin formed as South America moved away from the asthenospheric high in the east, which increasingly separated it from Africa, and onto the subduction zone in the west, at the open margin of Panthalassa (the Proto-Pacific).
2. This movement away from Africa was fast in the south but slight in the north because of friction along the equatorial margin. This difference induced the clockwise rotation of South America.
3. All observed effects resulted from this rotation.
4. The eastern margin, characterized by nearly east–west extension, tapered toward the pole located in the northeast.
5. Instead of extension, the rotation resulted in east–west to WNW-ESE compression at the northern end of the eastern margin, north of the east–west trending Pernambuco and Patos ductile shear zones.
6. East–west extension on the eastern margin took place at the expense of north–south compression on the equatorial margin as the continent rotated clockwise.
7. This compression decreased toward the pole of rotation in northeastern Brazil so that the easternmost segment of the equatorial margin was characterized by north–south extension and east–west compression.
8. North–south compression along the equatorial margin was in part alleviated by the breakaway and lesser clockwise rotation of northwestern Africa relative to the bulk of Africa.
9. Because the rotation of South America exceeded the rotation of northwestern Africa, compression along the equatorial margin increased westward.
10. Compression on the equatorial margin became compensated by left-lateral movement of continental slices along southeast–northwest trending faults toward the open Pacific subduction zone in the southwest.

ALLEVIATION OF PRESSURE BY NORTHEAST-TRENDING FAULTS ACROSS SOUTH AMERICA

Compression on the equatorial margin was alleviated by left-lateral strike slip movement along northeast-trending faults that cut through South America and may have formed as shear faults simultaneously with and parallel to Central and South Atlantic rifting (Figure 3).

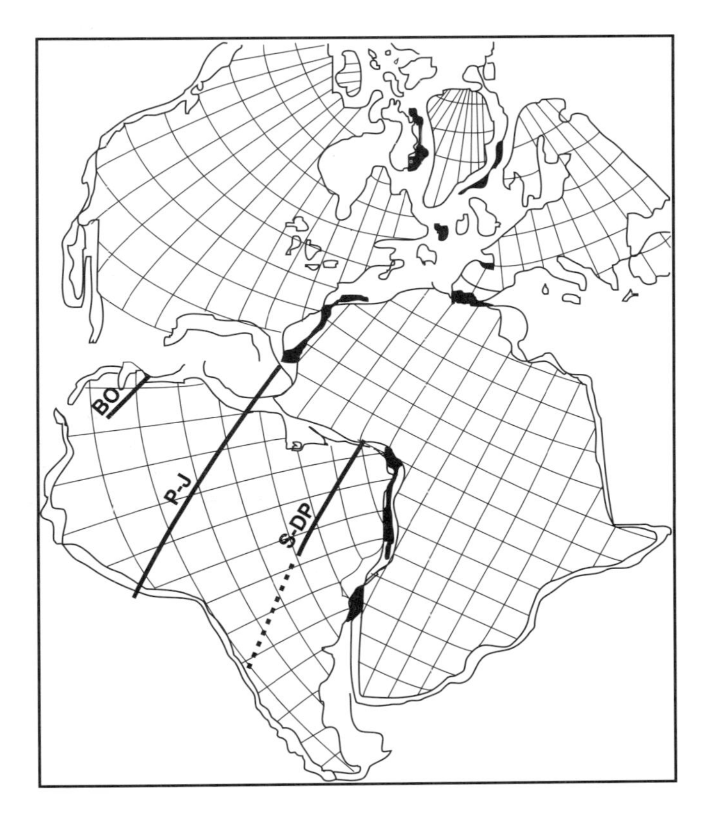

Figure 3—Position of the Bocono (B), Pisco–Jurua (P-J), and Sobral–Dom Pedro II (S-DP) faults on Bullard et al.'s (1965) reconstruction of Pangea. During South America's clockwise rotation, left-lateral strike-slip movement along these faults alleviated pressure on the continent's northern margin. (Modified after Szatmari, 1983.)

The westernmost of these faults is the *Bocono fault*, which trends northeastward in Venezuela. Present motion along this fault is right lateral due to east–west compression on the hot, hard-to-subduct Nazca and Cocos plates. However, the earlier stress field was characterized by north–south compression and east–west extension, resulting in left-lateral motion during Early Cretaceous time.

Farther to the east is the northeast-trending *Pisco–Jurua fault system* that cuts through South America from Georgetown on the Atlantic to Pisco on the Pacific margin (Szatmari, 1983). Left-lateral strike slip motion along this fault system was accompanied by increasing compression to the southwest. Wrench tectonics is evident throughout. In the northeast, the fault separates the Guayana shield into two structurally distinct provinces. In the exposed Precambrian basement, brecciation occurs along the fault. The Tacutu graben, located along the fault, is floored by Lower Jurassic basalt that is overlain by Jurassic and possibly Lower Cretaceous rift sedimentary beds containing a thick evaporite horizon. The east–northeast elongated graben presumably formed as the Central Atlantic rift opened during Early Jurassic time

(Szatmari, 1983). It was segmented by Early Cretaceous strike-slip faults created by the differential rotation of South America and Africa during Early Cretaceous continental breakup.

The Pisco–Jurua fault cuts through the east–west trending Upper Amazon or Solimões Basin. The preserved sedimentary sequence of the Upper Amazon Basin is mostly Paleozoic, ranging from Ordovician to Permian; it was intruded by Early Jurassic diabase sills and dikes simultaneously with Central Atlantic rifting. The sills are widespread and have more than 500 m of cumulative thickness. Both the sills and the Permian–Carboniferous evaporite-bearing sequence which they intruded were severely deformed in Early Cretaceous time (Szatmari, 1983; Mosmann et al., 1984). The exact time of deformation has not been determined; it is younger than the deformed Early Jurassic sills but older than the Aptian–Albian beds that discordantly overlie the deformed structures. Deformation is characterized mostly by reverse faulting and pop-up structures in the lower Paleozoic and Lower Carboniferous beds and also by folds in the Permian sequence. The axial plane of the folds is often inclined and lies along the upward continuation of the reverse faults. The structures are mostly elongated in an east–west to northeast–southwest direction, frequently forming arches concave to the northwest (possibly over reactivated ductile shear zones that originally formed along a Precambrian wrench fault system along the axis of the Amazon Basin). The heights of the folds and the vertical offsets along the faults vary but increase to the southwest; the structures hold significant commercial gas and light oil reserves.

To the southwest of the Amazon–Solimões Basin, the Pisco–Jurua fault passes through Precambrian basement covered by Cretaceous–Tertiary sedimentary beds until reaching the Eastern Cordillera of the Andes. At the Abancay deflection, the Pisco–Jurua fault intersects, left-laterally deflects, and offsets the Eastern Cordillera (Szatmari, 1983), which is composed of lower Paleozoic sedimentary rocks intruded by Permian granites. Northeast-trending fault planes, some with distinct horizontal striations, can be observed in both the lower Paleozoic sedimentary sequence and the Permian granites. They reach several hundred meters in height in the Urubamba Valley near Machu Picchu. The fault system reaches the Pacific coast near Pisco.

The next northeast-trending wrench fault system to the east is the *Sobral–Dom Pedro II fault zone.* It reaches the Atlantic coast near Fortaleza in Ceará State, Brazil. It is well expressed offshore, where it disturbs Cretaceous sedimentary beds at Icarai. On land, near Sobral, the fault is well exposed as mylonite zones cutting through the Precambrian basement, reactivating a fault formed during late Proterozoic continental collision (Braziliano cycle). A few kilometers to the southwest, at Santana de Acarau, the fault borders a sliver of Lower Cretaceous sedimentary beds that are nearly vertical along the fault but dip gently only a few hundred meters away from it (Destro et al., 1994). Partially sheared pebbles near the fault provide evidence for high temperatures. Farther to the southwest, the fault runs close to the southeastern border of the Paleozoic Parnaiba or Maranhão Basin; silicified fault crests with horizontal striations rise high above the surface along the fault.

Particularly interesting is the complex pattern of faults in northeastern Brazil, in the vicinity of the pole of rotation, created by Early Cretaceous reactivation of late Proterozoic ductile shear zones. The reactivated fault zones are approximately radial, pointing toward the center of the northeastern tip of South America. There is a set of concentric strike-slip faults about this pole, bordering the arcuate Brazilian coastline. Three major areas can be distinguished: (1) the Potiguar Basin, where the dominant movement was north–south extension (away from the equatorial margin of Nigeria); (2) the northeast coastline, where the motion was predominantly transcurrent; and (3) the southeast, where the predominant extension was east–west to southeast–northwest.

TECTONIC CONTROL OF THE SOUTH ATLANTIC SALT BASIN

The northward-tapering shape of the South Atlantic rift basin, created by the differential rotation of the two continents, is reflected by the wedge-like shape of the upper Aptian salt (Figure 2) that overlies the pre-Aptian and lower Aptian rift sequence. Geophysical data along the Brazilian margin indicate that the area of salt was not significantly increased by postdepositional oceanward salt flow during Late Cretaceous and Cenozoic time. Instead, extension of the salt nearshore was compensated by compression farther offshore (Demercian et al., 1993; Cobbold et al., 1996). Thrusting and folding of strata occurred over the salt near the limit of the oceanic crust which divided the salt body that was originally continuous from Brazil to Africa. Thus the present distribution of the salt permits reconstruction of the depositional limits of the Aptian salt basin.

The South Atlantic salt basin is about 400 km wide at its southern end in the Santos Basin, but it tapers toward the pole in northeastern Brazil to less than 100 km in the Sergipe–Alagoas Basin. The time of salt deposition is bracketed by the Aptian ages of two marine faunas located above and below the salt, respectively (Silva-Telles, 1996; E. A. M. Koutsoukos, personal communication, 1997). The subsalt fauna, in the Campos Basin, marks a short marine incursion before salt deposition and is of early Aptian age; the fauna above the salt, in the Sergipe–Alagoas Basin, is late Aptian. The salt interval itself is presumably short, on the order of a few hundred thousand years, reflecting very high rates of salt deposition (2–4 cm per year).

Subsalt relief was shaped in a southward-widening rift valley that was occupied, prior to its invasion by sea water, by partially sediment-filled lakes separated by

Figure 4—Distribution of hydrocarbon-bearing sedimentary basins on the facing margins of the South Atlantic.

intralake sedimentary, volcanic, and basement highs. Subsalt relief is more irregular near the pole in the Sergipe–Alagoas Basin, where transtensional rifting at the time of salt deposition was still active, than far from the pole in the Campos–Santos Basin, where the depression was wide and most of the faulting had ceased by the time salt deposition started.

Mass balance calculations indicate original salt thicknesses of about 2000 m. We believe that most of this salt was deposited in a preexisting rift depression dotted with lakes that were about 2000 m below sea level (Figure 2). This is because the rates of salt deposition were far too high for tectonic subsidence to create space for syntectonic salt deposition. Such a deep presalt depression is comparable to the present sub-sea-level depth of the surface of the Dead Sea (–397 m) which lies in a much smaller and less extended transtensional rift basin. Deposition in the South Atlantic depression may have taken place in cycles of recurrent catastrophic floods and evaporation, as the sea located between southern South America and southern Africa overflowed the Ponta Grossa dike swarm (Conceição et al., 1988) and associated volcanics of the Paraná–Tristan da Cunha hot spot and filled the presalt rift basin.

The flooding of the South Atlantic depression by the sea in late Aptian time is reflected by a major, short-lasting drop of sea level at 112 Ma (Haq et al., 1987; Silva-Telles, 1996). Assuming that the wedge-shaped depression was 400 km wide at its mouth, 2500 km long to Recife (where its width tapered to nearly zero), and 2000 m deep, its flooding required nearly 1 million cubic kilometers of sea water, causing about a 3-m drop in global sea level.

INCREASE IN OIL RICHNESS AWAY FROM THE POLE OF ROTATION

The oil-producing segments of the South Atlantic Megatrend now alternate between the Brazilian and the African margins; a productive segment on one margin is faced by a relatively sterile segment on the other. This distribution may result from the alternating polarity of master faults during rifting. The major productive segments are shown in the map of Figure 4.

The Gabon–Lower Congo Basin can be divided by political boundaries into the Gabon, Congo, and northern Angola segments, the latter including Cabinda. The productive segment of the Gabon Basin is duplicated by the parallel Recôncavo Basin in Brazil, which is separated from the Gabon Basin by the Salvador–Jacuipe horst.

In the productive segments, both the oil richness and the tectonic habitat of the oil are defined by the distance from the pole of differential rotation, situated in northeastern Brazil near Brazil´s equatorial margin. Near the

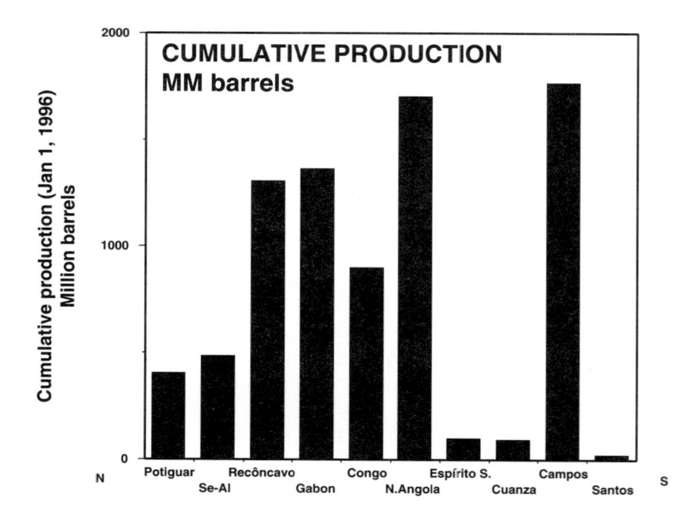

Figure 5—Variation in cumulative oil production from north to south along the South Atlantic Megatrend.

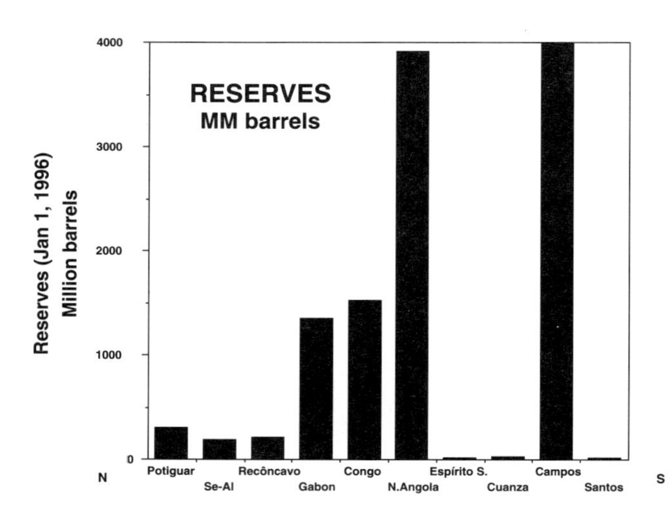

Figure 6—Variation in oil reserves from north to south along the South Atlantic Megatrend.

pole, in the Potiguar, Recôncavo, and Sergipe–Alagoas Basins, where the rift was relatively narrow and strike-slip motion was intense, petroleum occurs mostly on land, along largely transtensional rift border and cross-rift faults. Far from the pole, in the Lower Congo Basin and especially the Campos Basin, where the rift was wide and where rifted continental crust extends far offshore, oil fields occur mostly offshore. In the Campos Basin, part of the petroleum accumulated in lower Aptian subsalt sedimentary beds overlying thick synrift volcanics, but most of the large reserves are in Upper Cretaceous to lower Tertiary turbidites structured by salt flow.

The tectonic framework of the oil-producing basins created during rifting was subsequently modified by two interacting processes: salt flow and compressional reactivation of earlier faults during the Andean orogenic cycle (Szatmari and Mohriak, 1995). Compression started in the middle Cretaceous and culminated in early Tertiary time, contemporaneously with major global plate reorganization. Strike-slip movement and related mountain building were accompanied by hot spot activity in southeastern Brazil from Late Cretaceous to early Tertiary time, with several foci of igneous activity now marked by alkali intrusions onshore and volcanic piles offshore. The largest volcanic piles offshore formed in the area of the present Abrolhos Bank and controlled the deposition of extensive shallow-water carbonates. At the latitude of Rio de Janeiro, a platelet composed mostly of oceanic lithosphere was dislocated to the south–southwest along a reactivated late Precambrian ductile shear zone, creating transtensional grabens and transpressional mountain ranges (Serra do Mar and Serra da Mantiqueira). The sediments derived from these mountains were deposited in part as offshore turbidites; loading by these sediments controlled salt tectonics, which in turn controlled the distribution of turbidite channels.

The oil richness of a given segment along the South Atlantic Megatrend is indicated by the cumulative production (Figure 5) and the remaining reserves (Figure 6) of each basin or basin segment. These parameters, however, are strongly dependent on the rather arbitrary size of each segment, especially its length along the South Atlantic Megatrend. To obtain a value that is independent of basin size, we propose here to define *oil richness* as original reserves divided by the length of each basin or basin segment. Oil richness so defined is shown for the segments of the South Atlantic Megatrend along the facing margins of Brazil and Africa (Figures 7 and 8); the Niger Delta is omitted because of its different origin. Figure 8 shows a sharp rise in oil richness along the megatrend from north to south as the rift widens with increasing distance from the pole of differential rotation.

***Acknowledgments**—The author wishes to thank reviewers Juliano Macedo and Jan Golonka for helpful comments and Petrobrás for support and permission to publish. This is a contribution to IGCP Project No. 381.*

REFERENCES CITED

Bullard, E. C., J. E. Everett, and A. G. Smith, 1965, The fit of continents around the Atlantic, *in* A symposium on continental drift: Royal Society of London Philosophical Transactions, v. 258, p. 41–51.

Cobbold, P. R., P. Szatmari, L. S. Demercian, D. Coelho, and E. A. Rossello, 1996, Seismic and experimental evidence for thin-skinned horizontal shortening by convergent radial gliding on evaporites, deep-water Santos Basin, Brazil, *in* M. P. A. Jackson, D. G. Roberts, and S. Snelson,

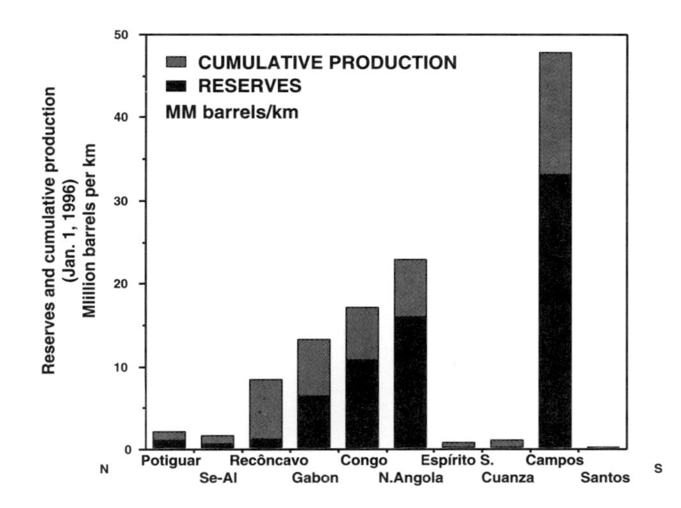

Figure 7—Variation in original oil reserves (reserves plus cumulative production) per unit length from north to south along the South Atlantic Megatrend.

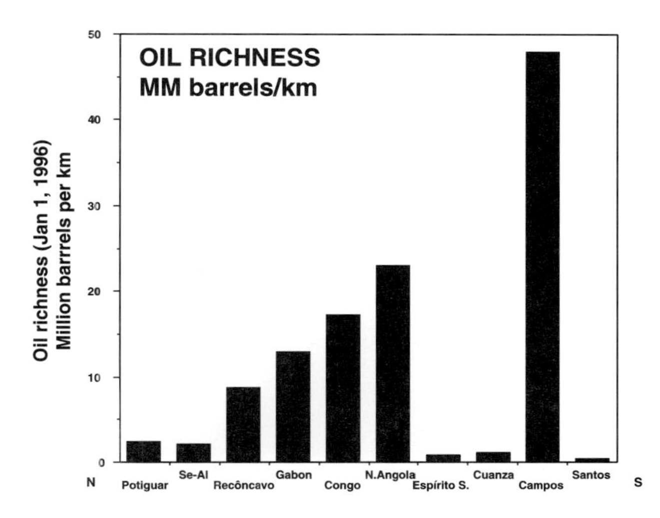

Figure 8—Sharp rise in oil richness, defined as original oil reserves per unit length, from north to south along the South Atlantic Megatrend. Gaps in southward-increasing trend may disappear with advancing exploration.

eds., Salt tectonics: a global perspective: AAPG Memoir 65, p. 305–321.

Conceição, J. C. J., P. V. Zalán, and S. Wolff, 1988, Mecanismo, evolução e cronologia do rifte Sul Atlântico: Boletim de Geociências da Petrobras, v. 2, n. 2/4, p. 255–265.

Demercian, S., P. Szatmari, and P. R. Cobbold, 1993, Style and pattern of salt diapirs due to thin-skinned gravitational gliding, Campos and Santos Basins, offshore Brazil: Tectonophysics, v. 228, p. 393–433.

Destro, N., P. Szatmari, and E. A. Ladeira, 1994, Post-Devonian transpressional reactivation of a Proterozoic ductile shear zone in Ceará, NE Brazil: Journal of Structural Geology, v. 16, p. 35–45.

Françolin, J. B. L., and P. Szatmari, 1987, Mecanismo de rifteamento da porção oriental da margem norte-brasileira: Revista Brasileiro de Geociências, v. 17, n. 2, p. 195-207.

Françolin, J. B. L., P. R. Cobbold, and P. Szatmari, 1994, Faulting in the Early Cretaceous Rio do Peixe basin (NE Brazil) and its significance for the opening of the Atlantic: Journal of Structural Geology, v. 16(5), p. 647–661.

Haq, B. U., J. Hardenbol, and P. R. Vail, 1987, The new chronostratigraphic basis of Cenozoic and Mesozoic sea level cycles: Cushman Foundation Foramifera Research Special Publication, v. 24, p. 7–13.

Milani, E. J., M. C. Lana, and P. Szatmari, 1987, Mesozoic rift basins around the NE-Brazilian microplate, *in* W. Manspeizer, ed., Triassic–Jurassic Rifting and the Opening of the Atlantic. New York, Elsevier, v. 2, p. 833–857.

Mosmann, R., F. U. H. Falkenheim, A. Gonçalves, and F. Nepomuceno-Filho, 1984, Oil and gas potential of the Amazon Paleozoic basins, *in* M. T. Halbouty, ed., Future petroleum provinces of the world: AAPG Memoir 40, p. 207–241.

Rabinowitz, P. D., and V. LaBrecque, 1979, The Mesozoic South Atlantic Ocean and evolution of its continental margin: Journal of Geophysical Research, v. 84(B11), p. 5973–6002.

Silva-Telles, Jr., A. C., 1996, Estratigrafia de sequências de alta resolução do Membro Coqueiros da Fm. Lagoa Feia (Barremiano?/Aptiano da Bacia de Campos–Brasil): M.S. Thesis, University of Federal Rio Grande do Sul, Porto Alegre.

Szatmari, P., 1983, Amazon rift and Pisco-Juruá fault; their relation to the separation of North America from Gondwana: Geology, v. 11, p. 300–304.

Szatmari, P., and W. U. Mohriak, 1995, Control of salt tectonics by young basement tectonics in Brazil's offshore basins: 1995 AAPG International Conference and Exhibition, September 8–11.

Szatmari, P., E. Milani, M. Lana, J. Conceição, and A. Lobo, 1985a, How South Atlantic rifting affects Brazilian oil reserves distribution: Oil & Gas Journal, January 14, p. 107–113.

Szatmari, P., O. Zanotto, J. Françolin, and S. Wolff, 1985b, Rifting and early tectonic evolution of the equatorial Atlantic: Abstracts with Programs, GSA 98th Annual Meeting, p. 731.

Szatmari, P., O. Zanotto, J. Françolin, and S. Wolff, 1987, Evolução tectônica da Margem Equatorial brasileira: Revista Brasileiro de Geociências, v. 17, n. 2, p. 180–188.

Zanotto, O., and P. Szatmari, 1987, Mecanismo de rifteamento da porção ocidental da Margem Equatorial: Revista Brasileiro de Geociências, v. 17, n. 2, p. 189–195.

Liro, L. M., and W. C. Dawson, 2000, Reservoir systems of selected basins of the South Atlantic, *in* M. R. Mello and B. J. Katz, eds., Petroleum systems of South Atlantic Margins: AAPG Memoir 73, p. 77–92.

Chapter 7

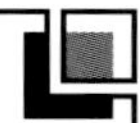

Reservoir Systems of Selected Basins of the South Atlantic

Louis M. Liro

Texaco Exploration
Houston, Texas, U.S.A.

Currently: Veritas Exploration Services
Houston, Texas, U.S.A.

William C. Dawson

Texaco Inc.
Houston, Texas, U.S.A.

Abstract

Early Cretaceous rifting of the South Atlantic basin resulted in the development of three unconformity-bound tectono-stratigraphic megasequences that are recognizable in petroliferous basins along the present-day margins of both Brazil and west Africa. These megasequences have been termed *nonmarine/synrift, transitional marine,* and *marine.* These megasequences provide a framework for understanding the character and distribution of hydrocarbon reservoirs within the South Atlantic petroleum systems. Hydrocarbons occur in nonmarine and marine stratal packages in both carbonates and siliciclastics. Reservoirs in the marine megasequences contain an estimated 70% of the region's known oil reserves, most of which have been discovered in the last two decades in deep-water fields.

Each reservoir system requires a comprehensive evaluation of depositional systems and diagenetic modification. Deep-water siliciclastic reservoirs are controlled by sediment provenance and transport mechanisms to the deep-water setting and less so by diagenesis. Synrift reservoirs are more affected by diagenesis, requiring detailed petrographic analysis. Carbonate reservoirs of the South Atlantic display the greatest degree of variability, thus requiring detailed petrographic work. Our results suggest that the carbonate reservoirs have the most complex diagenetic profiles and must be related to the timing of hydrocarbon migration to better understand charge risk.

INTRODUCTION

Over the course of the past decade, there has been a considerable increase in exploration interest along the west coast of Africa and the Brazilian margin. Although exploration in these basins has developed over the past 50 years (Brognon and Verrier, 1966; Reyre, 1984; Edwards and Bignell, 1988a, 1988b; Guardado et al., 1989; Teisserenc and Villemin, 1989; Dale et al., 1992; Horschutz et al., 1992), it has been exploration, particularly in deep-water regions, in only the last 15 years (Candido and Cora, 1992; Rangel et al., 1998; Barrett et al., 1998) that has resulted in significant discoveries (Table 1). In the Campos Basin of Brazil, Petrobras discovered the large Albacora and Marlim accumulations in the mid-1980s, although low oil prices and technical limitations prevented significant production of these reserves until the 1990s. More recent exploration has resulted in the discovery of the giant Roncador field in 1996.

A flurry of exploration activity along the west African margin, particularly the deep-water offshore Angola, has resulted in a notable number of significant discoveries (Table 1). What is particularly exciting about this activity is that discoveries are being made in Tertiary siliciclastic turbidites, which were largely untested by previous drilling campaigns in shallow water.

The basins of the South Atlantic originated as a result of the opening of the southern Atlantic Ocean beginning in the Early Cretaceous (Figure 1). Although these basins exhibit similar tectono-stratigraphy, particularly in their synrift strata, localized geologic factors require that each basin be explored independently (Mohriak et al., 1998).

In this chapter, we discuss petroliferous basins that are, at least in part, underlain by middle Cretaceous evaporites: the Gabon and Congo Basins on the west African margin, and the Campos and Espírito Santo Basins on the Brazilian Atlantic margin. The presence of these evaporites has had a profound effect on the nature of hydrocar-

Table 1—Recent Significant Discoveries (>100 MMBOE) Along the West African and Brazilian Margins[a]

Field Name	Year of Discovery	Basin and Country
Marlim	1985	Campos, Brazil
Albacora	1986	Campos, Brazil
Kuito	1996	Block 14, Congo, Angola
Roncador	1996	Campos, Brazil
Girassol	1996	Block 17, Congo, Angola
Dalia	1997	Block 17, Congo, Angola
Rosa	1997	Block 17, Congo, Angola
Kissanje	1998	Block 15, Congo, Angola
Marimba	1998	Block 15, Congo, Angola
Hungo	1998	Block 15, Congo, Angola
Dikanza	1998	Block 15, Congo, Angola

[a]Data from Guardado et al. (1989), Bray and Lawrence (1999), and Knight and Westwood (1999).

bon traps in these basins. In addition, we discuss the Recôncavo Basin of the Brazilian margin for additional description of the basin's synrift reservoir strata.

TECTONO-STRATIGRAPHY AND ITS IMPACT ON RESERVOIR DISTRIBUTION AND QUALITY

The basins of the South Atlantic margin (Figure 1) have their origin in a Jurassic–Early Cretaceous rifting episode (Lehner and de Ruiter, 1977; Rabinowitz and LaBrecque, 1979; Brice et al., 1982; Ojeda, 1982; Mohriak et al., 1998). This event created the initial opening of the South Atlantic ocean, separating the South American and African cratons. The basins that developed along each margin show rift–drift tectonic elements consistent with typical passive margin development.

Active rifting began in the latest Jurassic and Early Cretaceous. Synrift megasequence deposits accumulated from about the Neocomian through the Barremian (Ojeda, 1982; Guardado et al., 1989; Teisserenc and Villemin, 1989) (Figures 2 and 3). Initial deposits were Neocomian fluvial and fan delta coarse-grained siliciclastics. A major rifting event in the late Neocomian–Barremian created a series of tilted fault blocks in each basin (Brice et al., 1982). Ensuing subsidence, coupled with isolation from marine waters of the opening South Atlantic, resulted in a lacustrine depositional setting. During this episode, the major source rock interval of many of these basins was deposited (Brice et al., 1980; Abrahao and Warme, 1990; McHargue, 1990; Mello and Maxwell, 1990). The first marine incursions (transitional marine megasequence) into these basins resulted in the accumulation of evaporites. The interplay between source maturation and halokinesis allowed for a variety of source–reservoir pairs in these basins.

Continuing subsidence of the rift basins in the late synrift episode was coupled with arid climate and

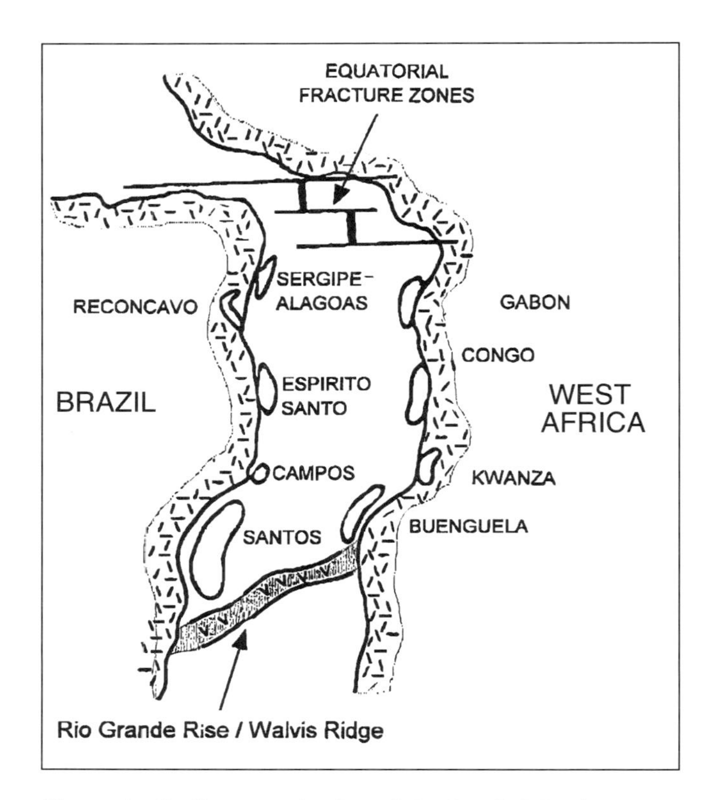

Figure 1—Sedimentary basins of the South Atlantic region. Paleogeographic reconstruction approximates Late Cretaceous position.

sporadic breaching of the Walvis Ridge to the south. The primary depositional result of this tectonic phase was the formation of a succession of evaporite strata conformably above the lacustrine sequence, including the Ezanga Formation in Gabon, the Loeme Formation in Angola, and Aptian salt of the Lagoa Feia Formation in the Campos Basin in Brazil. The extent of this Aptian salt defines the Aptian salt basin of west Africa. The extent of the salt deposits closely matches the maximum extent of the presalt lacustrine section along the west African margin. We do not have sufficient data at this time to infer this same relationship for the Brazilian margin.

With the onset of sea floor spreading in the South Atlantic during Albian time, the rift phase of deformation ended and the drift–subsidence phase began. Initial deposition of the drift phase (marine megasequence) resulted in development of a broad, open-marine carbonate platform extending virtually uninterrupted from Gabon to Angola in Africa and along the entire Brazilian margin, resulting in the Madiela Formation in Gabon, the Pinda Formation in Angola, and the Macaé Formation in Brazil.

With continued subsidence, the carbonate platform was drowned and deposition was replaced by clastics during rapid transgression in the latest Cretaceous (marine megasequence). Depending on provenance and maturity of the sediment delivery system, clastic deep-

	Northern Gabon	Southern Gabon	Congo	Cabinda	Zaire N. Angola	Tectono-stratigraphy
Lower Tertiary	Ozouri Ikando	Post — Madiéla Units	Madingo	Landana	Iabe	subsiding margin
Senonian	Ewongue Pointe Clairette Anguille		Emeraude	Iabe		
Turonian	Azile		Loango	Pinda / Vermelha	Iabe	late Cret. transgression
Cenoman.	Cap Lopez		Tchala / Likouala			
Albian	Madiéla	Madiéla	Sendji		Pinda / Vermelha	Carbonate sequence; Post-rift
Aptian	Ezanga Gamba	Loeme Gamba	Loeme Chela	Loeme Chela	Loeme Cuvo	Evaporite sequence; Transitional
Neocomian - Barremian	Upper Cocobeach Middle Cocobeach Lower Cocobeach pre-Cocobeach	Dentale Crabe Melania Lucina Kissenda Basal sand	Tchibota Toca Pointe Noire Djeno Sialivakou Vandji	Toca Bucomazi Lucula Nacanga	Bucomazi Lucula Zenze	Fluvial Lacustrine seq. sequence; Synrift
pre-Jurassic	*basement*	*basement*	*basement*	*basement*	*basement*	

Figure 2—Tectono-stratigraphic compilation of the west African margin basins.

water sandstones were deposited in these basins from the Late Cretaceous throughout much of the Tertiary. Neogene progradational episodes in most of the basins of the South Atlantic have provided few exploration targets, but they did play a significant role in the maturation and expulsion of hydrocarbons into Tertiary deepwater reservoirs. Progradation also promoted halokinetic structuring, which provides most of the hydrocarbon trapping mechanism in the Tertiary deepwater reservoirs.

Strata of the nonmarine–synrift megasequence developed a wide variety of lithostratigraphic and reservoir characteristics. Neocomian–Aptian nonmarine–synrift strata overlie extensive volcanics (120–140 Ma). Fractured vesicular basalts are locally oil productive (Badejo field, Campos Basin). Synrift strata, including the Lagoa Feia Formation (Campos Basin), the Kissenda Formation (Gabon Basin), and the Bucomazi Formation (Congo Basin) (Figures 2 and 3), record the development of lacustrine paleoenvironments within deep grabens containing as much as 2500 m of synrift strata. Interstratified lacustrine coquinas of ostracods, gastropods, and pelecypods (Lagoa Feia Formation of Campos Basin, and Viodo and Toca carbonates of Congo Basin) and siliciclastic lacustrine turbidites (Lucina Formation of Gabon Basin) are the principal reservoir types within the nonmarine megasequence (Bertani and Carozzi, 1984, 1985; Carvalho et al., 1998; Harris, Chapter 24, this volume). The distribution of coquinas is influenced by basement structure. Coquina reservoirs are extremely heterogeneous and discontinuous and have a mean porosity of 15% and permeabilities ranging from 50 to 500 md (Horschutz et al., 1992; Carvalho et al., 1998). Lacustrine carbonates have been subjected to a complex diagenetic history that includes partial to complete dolomitization, stylolitization, and chertification.

Fine-grained lacustrine turbidite reservoirs of the nonmarine–synrift megasequence have a mean porosity of 22% and permeabilities that range from 45 to 145 md. Comparable Lower Cretaceous lacustrine turbidite lithofacies exist in basins along western Africa and eastern Brazil (Smith, 1995). Characteristically, synrift reservoirs

	Reconcavo	Espirito Santo	Campos	Tectono-stratigraphy
Lower Tertiary		Urucutuca	Campos (Carapebus Mbr.)	subsiding margin
Senonian				
Turonian			(unconformity)	
Cenoman.		Barra Nova / Regencia Mbr.	Macae	
Albian				Carbonate sequence — Post-rift
Aptian	Marizal cgls.	Itaunas	(Alagoas evap.)	Evaporite sequence — Transitional
Neocomian - Barremian	Sao Sebastiao Pojuca Taquipe Marfim Candeias Itaparica	Mucuri	Lagoa Feia	Fluvial seq. / Lacustrine sequence — Synrift
pre-Jurassic	*basement*	*basement*	*basement*	

Figure 3—Tectono-stratigraphic compilation of the Brazilian margin basins.

are texturally and mineralogically immature and are thus highly susceptible to diagenetic alteration during progressive burial. Local differences in provenance influence relative diagenetic susceptibility. For example, the dissolution of volcanic lithoclasts commonly yields chlorite and zeolite cements that occlude pore systems (e.g., Moraes and Surdam, 1993). Secondary (dissolution) porosity is predominant in both coquina (Carvalho et al., 1998) and silicilastic turbidite reservoirs; lacustrine dolomites have intercrystalline and vuggy porosity. Porous horizons in synrift strata are commonly associated with paleosols (Abrahao and Warme, 1990). Microfracturing has enhanced permeability in presalt carbonate reservoirs. The relative timing of secondary porosity development appears to differ in each subbasin along the west African margin. Petrographic analyses in some presalt sandstones suggest that the formation of secondary porosity postdates oil migration.

Thick lacustrine shale sequences grade vertically and laterally into oil-saturated fluvial–deltaic lithofacies (Lucula Sandstone, offshore Cabinda). The fine-grained Lucula Sandstone is one of the more productive prerift reservoirs of west Africa.

The synrift megasequence is overlain unconformably by the transitional marine megasequence (Aptian) initially consisting of basal transgressive siliciclastic strata (Gamba and Chela sandstones). These "postrift" deposits consist of coarse-grained sandstones and conglomerates representing fluvial and marginal marine (estuarine) paleoenvironments. The Gamba Formation and its tectono-stratigraphic equivalent, the Chela Formation, are major reservoirs in the Gabon, Congo, and Kwanza area. Transitional marine strata are overlain by a thick sequence of evaporitic lithofacies, such as the Alagoas Formation (Brazil), the Ezanga Formation (Gabon Basin), and the Loeme Formation (Congo Basin). These evaporite units record an extensive restricted basin between Africa and South America. Subsequent salt tectonics resulted in the development of hydrocarbon traps in inversion structures in overlying formations.

Progressive northward opening of the South Atlantic Ocean resulted in the disruption of Walvis Ridge and São

Paulo Ridge topographic barriers and the deposition of an open-marine megasequence represented by the extensive Albian carbonate platform, including the Pinda, Madiela, and Sendji Formations (west Africa) and the Macaé Formation (Brazil) (Figures 2 and 3). Pinda carbonates of the Congo Basin have undergone pervasive dolomitization (Spaw and Koehler, 1981; Dawson et al., 1993), whereas the Brazilian Macaé carbonate platform is composed predominantly of undolomitized limestone (Falkenhein et al., 1981; Carozzi et al., 1983). Pinda carbonates contain interstratified lenses and wedges of fan delta coarse-grained arkosic sandstones. Ooid, oncoid, and skeletal grainstones and packstones, arranged into stacked, shoaling-upward subtidal to intertidal cycles (Tillement, 1987), are the main Pinda reservoirs, averaging 10% porosity and 75 md permeability. The secondary pore networks in Pinda and Macaé carbonates resulted from multistage dissolution during a prolonged burial history (Carozzi et al., 1979; Walgenwitz et al., 1990). Anhydrite cements have locally degraded pore systems in Pinda reservoirs. Core analyses reveal that dolomites and arkosic sandstones have greater average porosities than the interstratified limestones. Faulting appears to have strongly influenced the distribution, thickness, and dolomitization patterns of Pinda lithofacies. Shallow-marine Pinda and Macaé carbonates grade upward into Upper Cretaceous marls and shales, recording a progressive deepening of the basin, the Albian–Cenomanian eustatic drowning event, and the establishment of anoxic conditions during the late Cenomanian–early Turonian.

The Cenomanian–Turonian portion of the marine megasequence contains stacked sandstone-rich submarine fans. These turbidites exhibit abrupt lateral changes in thickness, and the sandstone distribution patterns have been influenced by syndepositional salt movement and faulting. Locally, these siliciclastic turbidites have excellent reservoir characteristics, such as in the Namorado and Albacora fields of Campos Basin. The overlying Senonian–Paleocene interval also contains sandstone-rich submarine fan deposits laid down during lowstand episodes. These turbidite sandstones (porosity 30% and permeability 800 md) have considerable economic importance; they include the Point Clairette and Batanga sandstones (Gabon Basin) and the Maastrichtian Carapebus and Ubatuba sandstones (Campos Basin). Sandstone-rich fairways appear to occur in paleotopographic lows between salt structures. The post-Paleogene interval includes turbidite complexes associated with Eocene, Oligocene, and Miocene lowstand events. These turbidites are the reservoir for the giant Marlim and Albacora oil fields in the Campos Basin. Petroliferous Cretaceous and Tertiary strata are overlain by a thick Miocene progradational sequence. Basal Miocene sandstones (Mandrove Formation, Gabon Basin) have undergone minor diagenesis and offer excellent reservoir characteristics. Regionally, the thick, rapidly deposited Miocene section provided the overburden needed to generate hydrocarbons in the South Atlantic margin.

PETROLEUM SYSTEMS OF THE SOUTH ATLANTIC

A petroleum system is defined by a mature source rock supplying hydrocarbons into a migration network and, ultimately, into a hydrocarbon trap (Magoon, 1988). Along the Brazilian margin, virtually all of the hydrocarbons discovered originated from Lower Cretaceous lacustrine facies, including the Lagoa Feia Formation (Mello and Maxwell, 1990) in the Campos Basin and an age-equivalent section in the Recôncavo and Espírito Santo Basins. Along the west African margin, most fields discovered early in the exploration history of the basins appear to have been sourced by lacustrine facies of the Lower Cretaceous Bucomazi Formation or its equivalents (Brice et al., 1980; McHargue, 1990; Levache and Dessort, 1998). Recent drilling, however, suggests to us that younger Cretaceous and even Tertiary source intervals with marine affinity may be supplying significant hydrocarbon volumes to the petroleum systems.

For the Lower Cretaceous source intervals, an extended period of generation appears to have first charged presalt reservoirs and subsequently migrated through salt "windows" into postsalt reservoirs. Through initial expulsion migration or structurally driven remigration, these hydrocarbons have accumulated in successively shallower reservoirs. Because of the inherent "leakiness" of these systems, the largest reserves are often in the youngest reservoir intervals (e.g., Dale et al., 1992), as exemplified by the Oligocene and Miocene turbidites of offshore Brazil and the Tertiary turbidites of offshore Angola. Only where early salt activity ceased, such as in the northern Gabon Basin, are the majority of reserves in older intervals less removed from the source rocks.

BASINS

Gabon Basin

The Gabon Basin (Figure 4) generally follows the basin development described earlier in this chapter. We have divided the Gabon Basin into the northern and southern subbasins, separated by the N'Komi fault zone (Teisserenc and Villemin, 1989). Although the Gabon Basin has relatively smaller discovered reserves than other South Atlantic basins discussed in this chapter (about 4 billion bbl based on data from Teisserenc and Villemin, 1989), the basin is noteworthy for displaying a variety of petroleum and reservoir systems. Unlike other basins of the South Atlantic, the presalt lacustrine source interval is not widespread, particularly offshore, and most of the presalt lacustrine-sourced oils are trapped in presalt reservoirs in the onshore part of the basin. However, there are fields of significant size with presalt reservoirs, such as the giant Rabi Konga field.

We believe that the lack of significant offshore presalt-sourced hydrocarbons in the Gabon Basin is due largely

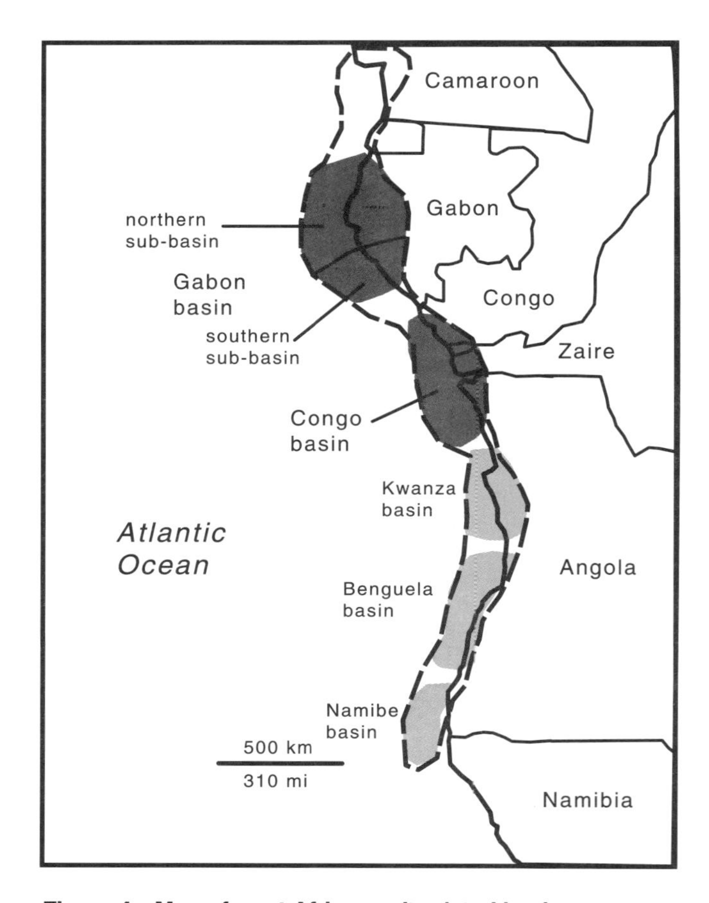

Figure 4—Map of west African salt-related basins.

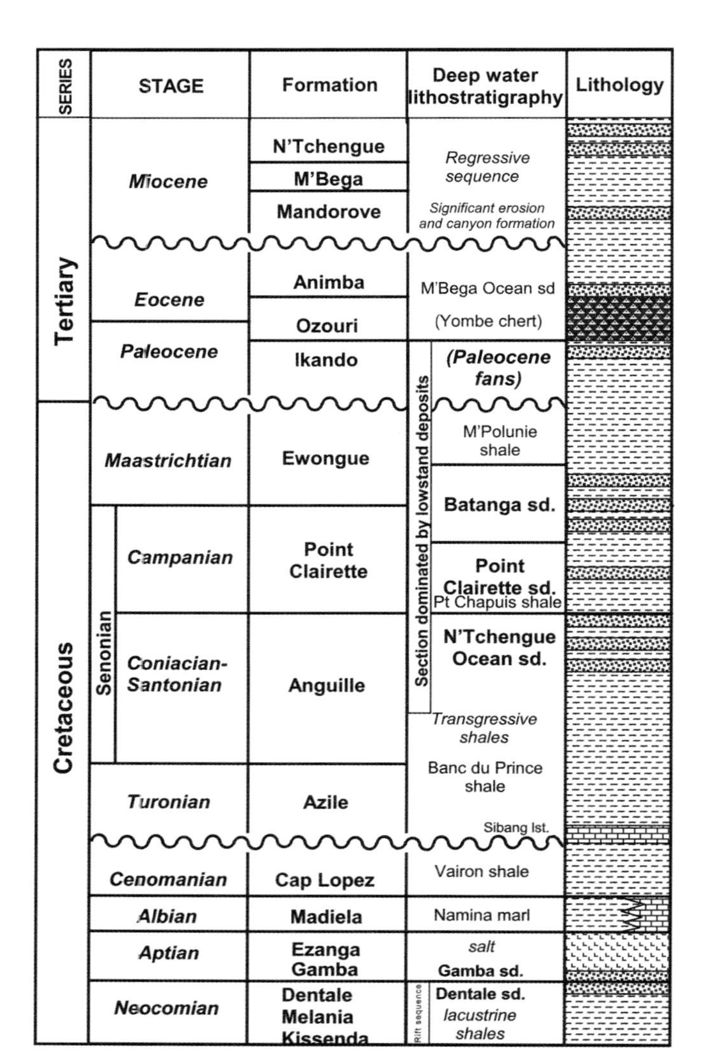

Figure 5—Composite Gabon Basin stratigraphic column.

to limited source distribution. Also limiting the viability of the presalt system is a lack of breaching of the salt layer in the northern subbasin to allow matured hydrocarbons into younger reservoirs, the lack of widespread thermal maturity of the presalt source interval in the southern subbasin, and a lack of a areally widespread presalt migration pathway system to efficiently distribute hydrocarbons to potential reservoirs. The major petroleum system of the Gabon Basin is in the northern subbasin and is related to mature Upper Cretaceous marine source intervals feeding Upper Cretaceous and Tertiary turbidite reservoirs. No petroleum system has been documented in the southern subbasin.

For purposes of describing the reservoir systems in this basin, we have subdivided the basin into the following reservoir intervals: Presalt, Madiela platform, and postAlbian–Tertiary clastic turbidites (Figure 5). These correspond to the nonmarine–synrift, early marine, and later marine megasequences.

Presalt Reservoirs

Lower Cretaceous presalt sandstones of the Melania, Lucina, and Dentale Formations (Figure 5) are important oil reservoirs in several fields in the Gabon Basin. These reservoirs represent nonmarine fluvial–deltaic, lacustrine, and lacustrine turbidite (Smith, 1995) paleoenvironments. A marine origin for the overlying Gamba Formation is supported by its fossils, glauconite, and textural maturity. The Gamba Formation represents an important transgressive interval; it has a distinct basal unconformity that truncates faulted and eroded subjacent formations (Teisserenc and Villemin, 1989).

Because of their restricted areal distribution, Melania and Lucina sandstones are prospective only in the onshore and shallow offshore regions of the Gabon Basin. According to Teisserenc and Villemin (1989), Melania and Lucina sandstones are restricted to grabens within the eastern part of the Dentale Trough. Although Gamba sandstones are more widely distributed than Melania and Lucina sandstones, the Gamba Formation shales out in the offshore basins, limiting its prospectivity in these areas (Edwards and Bignell, 1988b). The Dentale sandstone is more extensive than the other presalt sandstones in the Gabon Basin, attaining a gross thickness of 2–3 km within the Dentale Trough. Dentale sandstones also occur in offshore regions down to at least 200 m of present-day water depth. The distribution of encountered Dentale sandstones suggests linearities that mimic the underlying rift basin orientations, although this observation is based on limited well control (Teisserenc and Villemin, 1989).

Porosities in the Gamba and Dentale sandstones range from 15 to 30%, with permeabilities often more than 1 darcy (Teisserenc and Villemin, 1989). The principal control on presalt reservoir quality is burial diagenesis. The presence of authigenic clay minerals (illite and chlorite), compaction, and cementation by quartz, carbonates, and clay minerals has severely degraded presalt reservoir quality in wells examined in offshore areas, suggesting significant risk for these reservoirs as offshore targets (unpublished Texaco data, 1991).

Madiela Reservoirs

Madiela platform carbonates of Albian age (Figure 5) are stratigraphically equivalent to the Pinda Formation of offshore Angola (Reyre, 1984; Tillement, 1987) and to Macaé limestones in the Campos Basin of Brazil (Falkenhein et al., 1981). Both Pinda and Macaé produce significant quantities of oil, the Pinda principally from dolomite reservoirs and the Macaé entirely from limestone reservoirs. To date, no significant production has been obtained from the Gabon Basin's Madiela Formation. Examination of samples suggests that carbonate microfacies observed in the Madiela are consistent with a restricted marine shelf–lagoonal paleoenvironment.

Examined Madiela samples have very low porosity. Primary porosity has been occluded by early calcite cementation. Teisserenc and Villemin (1989) report stacked oolitic and bioclastic grainstone–packstone intervals with 20–30% porosity, suggesting that the Madiela may be an economic reservoir locally. We believe that more documentation of Madiela reservoir quality is needed to establish regional trends.

Post-Albian–Tertiary Clastic Turbidites

Upper Cretaceous (supra-Albian) siliciclastic sandstones (Figure 6) occur in a thick depocenter in the northern Gabon subbasin, immediately updip of the thermally mature part of the marine source rock facies. They account for about 70% of Gabon Basin's oil reserves (Teisserenc and Villemin, 1989). All of the hydrocarbon traps associated with the Upper Cretaceous sandstones are related to mobilization of the underlying salt.

Upper Anguille (Coniacian–Santonian) and Batanga (Maastrichtian) sandstones (Figure 5) are the principal reservoirs in this region. Anguille–Pointe Clairette sandstones have been interpreted as deep-marine turbidite-fan complexes; Batanga sandstones have been alternatively interpreted as turbiditic fan deposits (Teisserenc and Villemin, 1989) and as deltaic deposits (Edwards and Bignell, 1988a). A well-developed Batanga sandstone interval examined appears to be similar to inner to mid-fan channelized lobe complexes described by Moraes and Bruhn (1988) and Moraes (1989) for Campos Basin turbidites in Brazil.

Figure 6—Thin-section photomicrograph of postsalt (marine) sandstone core sample (from 12,018 ft depth), offshore Gabon. Fine- to medium-grained, moderately sorted, argillaceous, quartzose arenite with a framework of detrital quartz, feldspars, and micas. Clay minerals occlude much of the intergranular porosity (8%). Because of pervasive compaction and interstitial clay minerals, the pore system lacks well-developed interconnections. Width of image is 6 mm; plane polarized light.

Anguille–Pointe Clairette reservoirs are described as fine- to very fine-grained arkosic sandstones having a maximum porosity of 24% and permeabilities up to 700 md. Batanga sandstones are fine- to coarse-grained sandstones with porosities as high as 30% and permeabilities approaching 800 md. Batanga sandstones typically have high water saturations in shallow-water offshore wells.

Log studies reveal that producing fields located in shelfal waters typically contain 30–50% gross sandstone in the post-Albian Cretaceous interval, with this percentage decreasing to about 15% farther offshore. Individual sandstone packages within the post-Albian Cretaceous depopod typically range in thickness from 10 to 30 m.

In the southern Gabon subbasin, the post-Albian Cretaceous series is sand-poor, with less than 10% gross sandstone. The distribution of both Tertiary and Upper Cretaceous sandstones in the southern subbasin appears to be controlled by significant salt activity (Liro and Coen, 1995), and hence these sandstones are less areally extensive. We believe that the post-Albian Cretaceous turbidites of the northern subbasin formed from reworked and redeposited sediment originating in the area of the present-day Ogooué Delta.

The major clastic depocenter in the northern subbasin persisted from Late Cretaceous through the early Tertiary. Log studies indicate the presence of 15–35% gross sandstone locally in Paleocene and Miocene sequences. Sample descriptions reveal that these Tertiary sandstones are uncemented and have excellent reservoir characteristics.

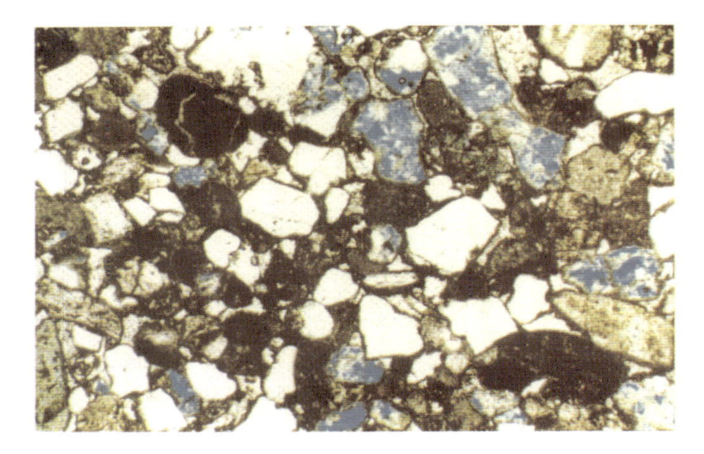

Figure 7—Thin-section photomicrograph of presalt (nonmarine) lithic arenite core sample (from 9360 ft depth), offshore Angola. Poorly sorted, medium-grained framework is composed of lithic grains (volcanic and sedimentary lithoclasts), detrital quartz, and feldspars. Isolated (grain moldic) pores resulted from preferential dissolution of feldspars (8% porosity). Note presence of authigenic clay minerals within secondary moldic pores. Framework has undergone pervasive compaction, and pores lack well-developed interconnections. Width of image is 6 mm; plane polarized light.

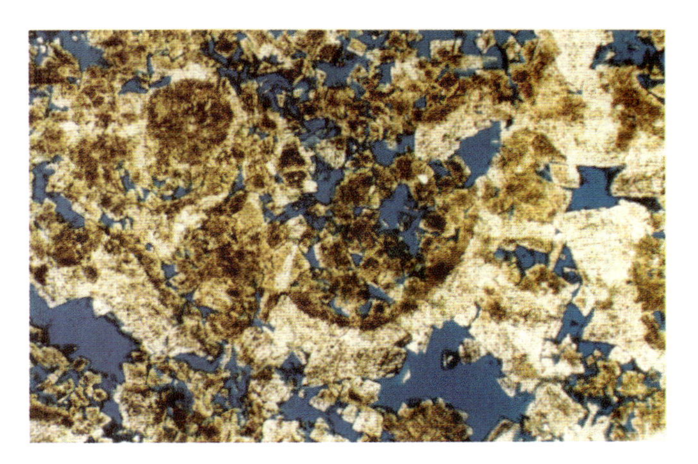

Figure 8—Thin-section photomicrograph of Pinda Formation reservoir core sample (from 9685 ft depth), offshore Angola. This porous, dolomitized, oncoid and ooid grainstone sample has been replaced by multiple episodes of dolomitization. Relicts of original spheroidal grains are evident (center). Abundant secondary porosity (23%) consists of intercrystalline and grain moldic pores. The pore system appears to be well interconnected. Width of image is 6 mm; plane polarized light.

Congo Basin

The Congo Basin of west Africa encompasses parts of offshore Congo, Cabinda, Zaire, and northern Angola (Figure 4). Sharing similar genetic traits with the other west African basins discussed here, the Congo Basin is composed of a nonmarine synrift interval and a normal marine carbonate and siliciclastic postrift section. The synrift and postrift sections are separated by Aptian salt, typically referred to as the Loeme salt (Figure 2). The major presalt hydrocarbon source is the lacustrine Point Noire marl; mature postsalt sources in the Late Cretaceous and Tertiary are inferred but undocumented in recent deep-water field discoveries (discussed below).

As with most of the circum–South Atlantic salt-related basins, virtually any interval with economic reservoir quality may be charged with hydrocarbons. Reservoirs in the Congo Basin range from the Neocomian Vandji sandstones that overlie the basement, other siliciclastics and nonmarine carbonates of the synrift interval (Figure 7), and salt-structured middle Cretaceous carbonates to Miocene deep-water siliciclastics of the postrift interval (Barrett et al., 1998).

Most of our experience has been with the Albian carbonate interval, which includes the Sendji Formation in Congo and the Pinda Formation in Angola (Figures 8 and 9). These formations are genetically related and essentially contemporaneous. The Albian sequence on the west African coast was deposited in an overall carbonate shelf environment and is generally devoid of reefal material. It does, however, contain reservoir rocks

Figure 9—Thin-section photomicrograph of Pinda Formation reservoir core sample (from 10,009 ft depth), offshore Angola. This is a porous, arenaceous coarsely crystalline dolomite. The groundmass of coarse, subhedral, dolomite crystals contains "floating" grains of detrital quartz and feldspars. Abundant secondary porosity (25%) consists of intercrystalline and microvuggy pores. Pore system has well-developed interconnections. Width of image is 6 mm; plane polarized light.

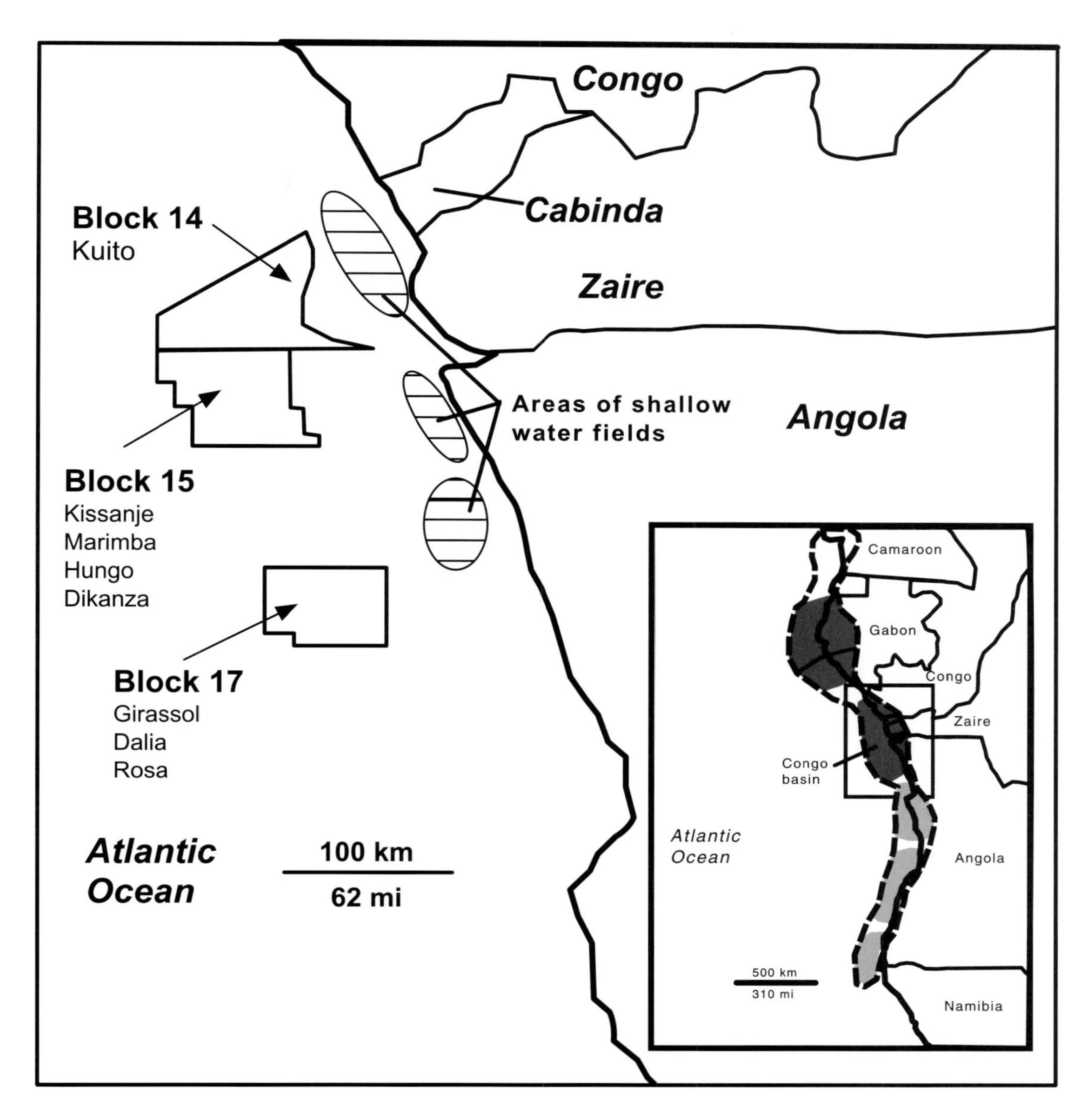

Figure 10—Map of recent significant discoveries in the deep-water Congo Basin.

that were deposited in high-energy oolitic shoals, siliciclastic and carbonate clastic tidal channel and offshore bar complexes, and generally fine-grained siliciclastics of the nearshore environment. Most of the Albian carbonates have been dolomitized to varying degrees and thus have diagenetic (secondary) porosity and permeability.

The Sendji field in offshore Zaire (Baudouy and Legorjus, 1991) is a structurally simple four-way anticlinal closure, formed as a result of underlying salt mobilization. The Albian reservoir, however, is a complex interval with both dolomitic and quartz sandstone and siltstone reservoirs, representing carbonates and siliciclastics deposited in a high-energy intertidal to subtidal setting. Hydrocarbon reserves are nearly equally distributed between carbonate and siliciclastic reservoirs. As for the Pinda Formation reservoirs, the Sendji carbonates are dolomitized. In the most productive zone at Sendji field, porosities range from 25 to 30% and permeabilities range from 300 to 1400 md (Baudouy and Legorjus, 1991). The thickness of this reservoir unit is controlled by paleostructure.

Based on our experience, we have found the following critical exploration parameters for Bucomazi-sourced hydrocarbons in Pinda and equivalent reservoirs:

1. The timing of diagenesis is difficult to determine. Detailed petrographic examination of the Pinda reveals that reservoir development is a complex

interplay of depositional and diagenetic processes and that there were possibly multiple episodes of reservoir creation.
2. The timing of trap development for Pinda reservoirs is partially structural, due to halokinesis, and partially stratigraphic, due to secondary porosity development.
3. Because the Bucomazi source and Pinda reservoirs are stratigraphically separated by the impermeable Loeme salt layer, the introduction of hydrocarbons into Pinda reservoirs was dependent either on the presence of salt evacuations to allow vertical migration from mature Bucomazi troughs or on the presence of growth faults that facilitated vertical migration from mature Bucomazi troughs, late synrift carrier beds, or preexisting traps in presalt reservoirs.

The Congo Basin of offshore deep-water Angola is currently the site of the most intense leasing and drilling activity along the west African margin. Since 1995, a series of discoveries (Dalia, Girassol, Kuito, Landana, Kissanje, Rosa, Marimba, and Hungo fields) has added significant reserves (Table 1 and Figure 10). Recent significant discoveries (>500 MMBOE) occur in the Turonian–Oligocene/Miocene petroleum system that was first suggested in the early part of this decade by the presence of marine-sourced geochemical signatures in encountered oils and verified through seismic interpretation and extensive deep-water drilling in the past few years (Amaral et al., 1998; Brock et al., 1998). Like the shelfal Bucomazi–Pinda petroleum system, the Turonian–Oligocene/Miocene system is controlled by Miocene wedge-driven maturation of the source interval. Vertical migration occurred through regional and semiregional growth fault systems. Equivalent reservoirs are typically unconsolidated and hence largely retain primary reservoir properties. The abundance of shallow geochemical anomalies and naturally occurring sea surface oil slicks suggests that the migration network is inherently "leaky." Because of the highly competitive nature of this current exploration activity, as well as the recent nature of the discoveries, specific information on these hydrocarbon accumulations is tightly held by individual companies and partnerships.

Campos Basin

Located along the southeastern coast of Brazil, the Campos Basin (Figures 1, 3, and 11) is to date the most hydrocarbon prolific basin along the Brazilian margin (Guardado et al., 1989; Jahnert et al., 1998). The basin is separated from the Espírito Santo Basin on the north by the Victoria High and from the Santos Basin to the south by the Cabo Frio Ridge. The importance of the Campos Basin to Brazilian exploration is relatively recent due to the discovery of significant reserves in deep-water fields in the past 15 years. Prior to this time, the Recôncavo Basin was the major exploration basin in Brazil.

Exploration in the Campos Basin began in the early 1970s with a series of discoveries starting in 1974 (Horschutz et al., 1992). More significantly, the discovery of the Marimba field in 1984 ushered in several significant deep-water discoveries: most importantly, Tertiary reservoirs of the Marlim and Albacora fields in the mid-1980s, and the Maastrichtian reservoir of the Roncador field in 1996 (Rangel et al., 1998) (Figure 11).

In the mid-1970s to early 1980s, Petrobras discovered Cretaceous (Namorado) and a number of Eocene (Enchova, Bonito) turbidite reservoir fields along the present-day shelf edge. Although the Cretaceous turbidite reservoir fields (Bruhn and Walker, 1995) tend to be structurally controlled due to late halokinesis, the Tertiary fields demonstrate significant off-structure stratigraphic accumulations as well.

Extension of the concept of off-structure accumulations to deep-water exploration resulted in two significant discoveries: Marlim (Oligocene) and Albacora (Oligocene–Miocene) reservoirs. The Marlim and Albacora fields were among the first significant turbidite reservoir oil fields discovered (Candido and Cora, 1992; Bruhn et al., 1998b). According to Caddah et al. (1998), Miocene sand-rich turbidite reservoirs account for about 85% of the total volume of oil in the Campos Basin. Both fields have associated seismic amplitude anomalies, due both to hydrocarbon content and lithologic contrasts.

Moraes and Bruhn (1988) and Bruhn et al. (1998a) have differentiated the turbidites of the Campos Basin into three major groupings based on their depositional geometries: (1) channel lobe complexes, which contain the majority of the Campos Basin reserves; (2) channel levee complexes; and (3) nonchannelized lobe complexes. Bruhn et al. (1998a) reported average reservoir porosities of 25% and permeabilities of 800 md. Arienti et al. (1998) have contrasted middle Eocene and Oligocene–Miocene deep-water reservoirs in the Campos Basin based on core data. They have demonstrated that the Eocene deep-water sandstones represent slope and valley fill, whereas the Oligocene and Miocene sandstones represent mainly high-density turbidite deposits (70% and 95%, respectively). De Ros (1998) has defined five petrofacies in the Brazilian margin turbidites based on composition, diagenesis, and depositional patterns.

In the Marlim field (Carminatti and Scarton, 1991) (Figure 11), the main Oligocene reservoir is characterized by three turbidite depositional episodes, with the upper episode comprising 70% of the sandstone reservoirs in the field. Five sandstone packages with distinct shale breaks suggest the presence of a higher order cyclicity. Sandstone depositional trends suggest unconfined turbidite deposition. Stratigraphic trapping is due to reservoir pinchout. Fault and sandstone isopach mapping suggests eastern fault control on sand distribution and accumulation. The source of the coarse sediment was the Brazilian platform, with relatively short (<200 km) transport distances (Peres, 1993). Sediment recycling via shelf margin incision and canyon formation was likely. The reservoir is composed of fine-grained, struc-

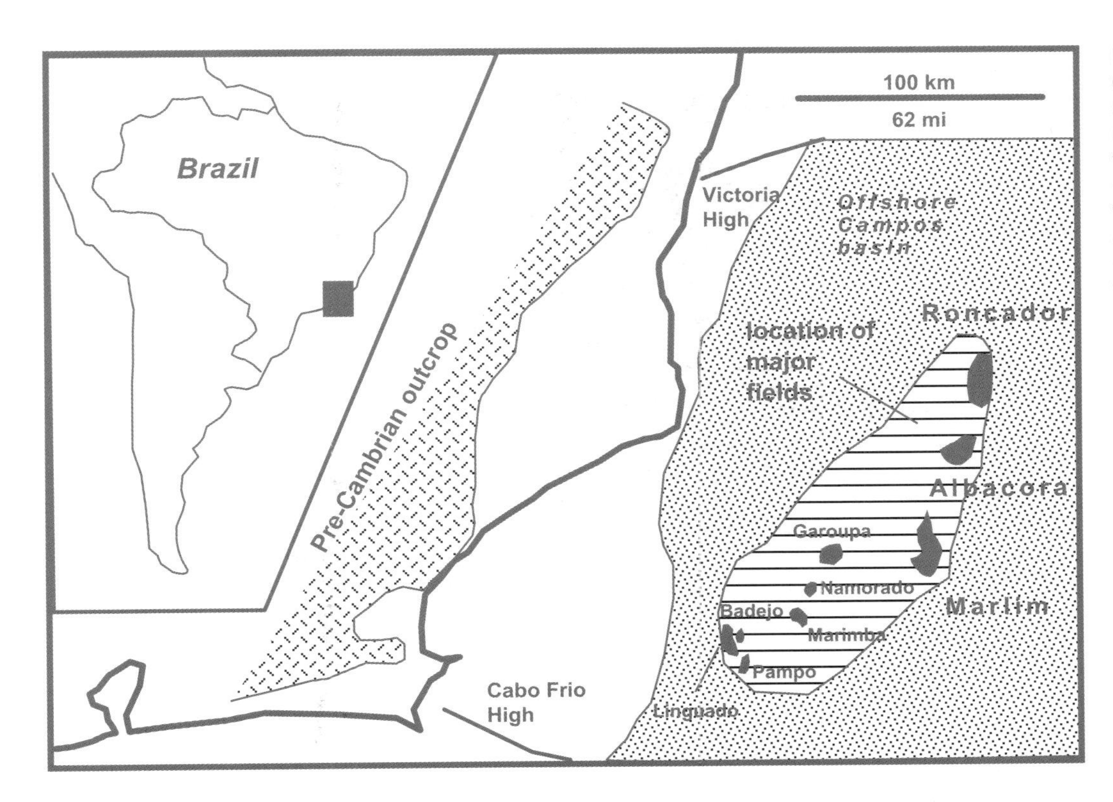

Figure 11—Map of the Campos Basin, offshore Brazil, with key fields noted. Most of the reservoir material in the Campos Basin turbidites was sourced from the onshore Precambrian outcrop belt. Adapted from Guardado et al. (1989).

tureless, well-sorted sandstones, with some facies ripple cross-laminated. This suggests massive sand deposition (Bouma Ta) reworked by bottom currents. Reservoir character is excellent, with 25–30% porosity and 2000–3000 md permeability typical.

Drilling history in the Campos Basin has generally favored discoveries of hydrocarbons in successively younger intervals as one moves basinward (Guardado et al., 1989). Discovery in 1996 of the giant Roncador field in the northern part of the Campos Basin (Figure 11), with significant reserves (estimated at 2–3 billion bbl) in Maastrichtian turbidites, suggests that considerable hydrocarbon potential exists in older intervals even in the deep-water.

Secondary production in the Campos Basin comes from a number of fields located on the present-day shelf (Figure 11). Reservoir intervals are oolitic and oncolitic limestone reservoirs of the Albian carbonate platform Macaé Formation (Bonito, Garoupa, and Linguado fields), Neocomian lacustrine coquinas (Pampo, Linguado, and Trilha fields) (Carvalho et al., 1998), and fractured basement subjacent to mature source rock intervals (Badejo and Linguado fields).

Recôncavo Basin

The Recôncavo Basin (Figures 1, 3, 12, and 13) was the first basin in Brazil to undergo extensive petroleum exploration, dating back to the 1930s, and in 1939 was the site of the first commercial oil production in the country (de Figueiredo, 1994). The Recôncavo Basin (Cohen, 1985; de Figueiredo, 1994) and its structural extension northward into the Tucano and Jatoba Basins (Figure 12) shared geologic affinities with other South Atlantic basins during the Early Cretaceous breakup of South America and Africa. The Recôncavo Basin, however, ceased rifting (and basin development) in the Aptian, and only a thin package of Upper Cretaceous and Tertiary covers the synrift succession in this basin. In this sense, the Recôncavo Basin is similar to the interior subbasin of the Gabon Basin.

Two strike-slip fault systems, the left-lateral Itanagra–Aracas and the right-lateral Mata–Catu, compartmentalize the Recôncavo Basin into three distinct structural blocks (Figures 12 and 13), and are responsible for creating most of the structural traps in the basin. The organic source rock that generated most of the reserves in the Recôncavo are the lacustrine shales of the Neocomian Candeias Formation (Figure 3).

Where structural movement has juxtaposed prerift strata of the Jurassic Sergi Formation against mature Candeias Formation shales, hydrocarbon accumulations occur, such as in the Agua Grande, Dom Joao, and Buracica fields (de Figueiredo, 1994). The Sergi Formation represents volumetrically the most significant hydrocarbon reservoir in the Recôncavo Basin. Reservoirs within the Sergi Formation were deposited in arid, nonmarine environments, with depositional settings ranging from braided fluvial to eolian. Porosities range from 10 to 25% and permeabilities from 20 to 1200 md (de Figueiredo, 1994). Grain size variations are significant, ranging from fine-grained to conglomeratic, and the reservoirs are compositionally heterogeneous. Early diagenesis consisted of paleosol formation due to subaerial exposure.

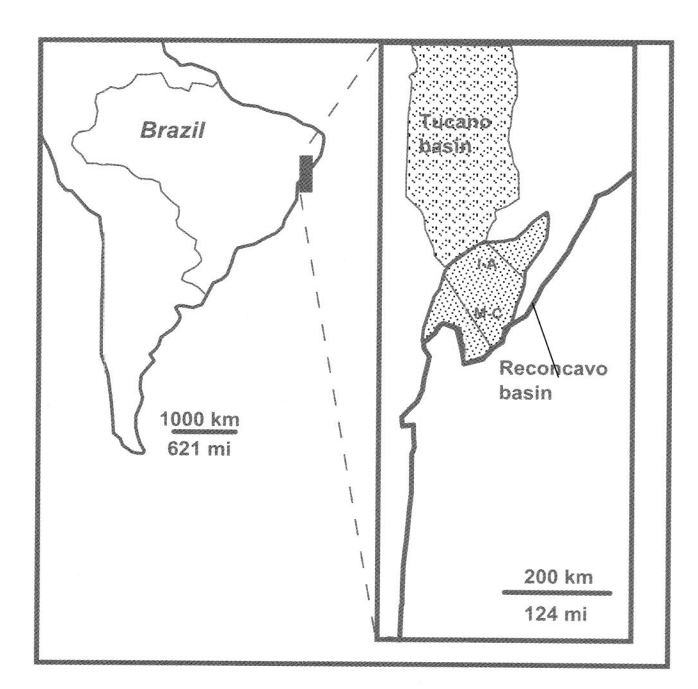

Figure 12—Map of the Recôncavo Basin, offshore Brazil. I-A indicates the left-lateral Itanagra–Aracas fault system; M-C indicates right-lateral Mata–Catu fault system. Adapted from de Figueiredo (1994).

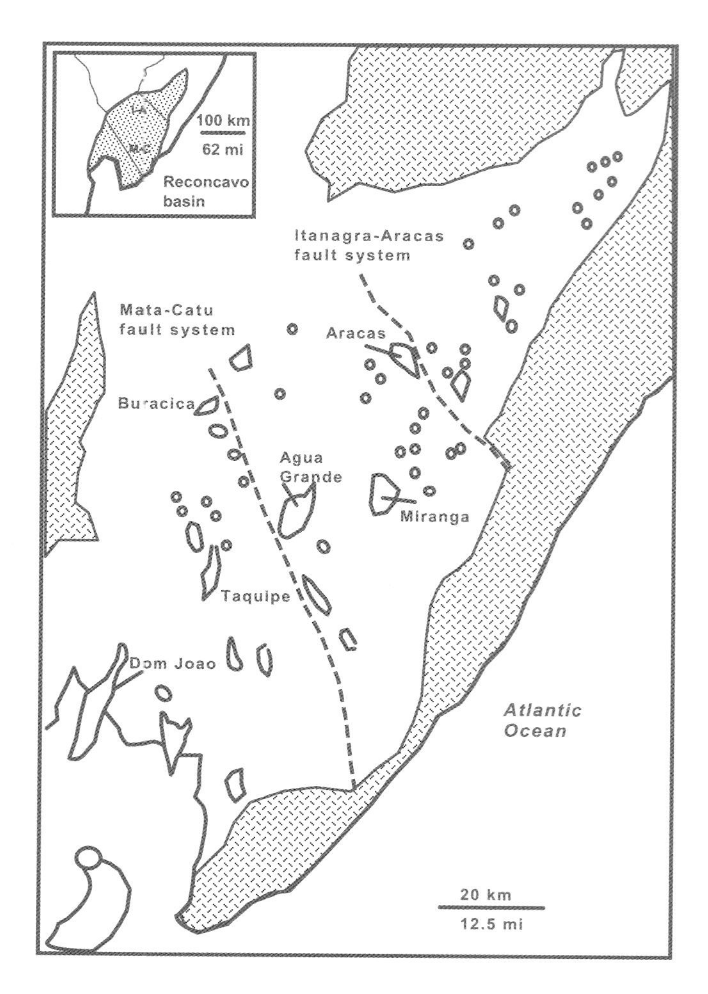

Figure 13—Oil and gas fields of the Recôncavo Basin. Adapted from de Figueiredo (1994).

Silcretes, caliches, and dolomites occur as nodules within paleosol horizons; combined with detrital clays, these materials represent early degradation of reservoir quality. Later cementation by calcite and quartz further degraded reservoir quality within the Sergi Formation. De Figueiredo (1994) lists late dissolution of calcite cements as the main agent for creation of secondary porosity in the Sergi.

Renewed tectonic activity in the Aptian facilitated mainly vertical migration of hydrocarbons into the synrift Ilhas and Sao Sebastiao Formations (Cohen, 1985), volumetrically the second most important reservoir interval in the Recôncavo Basin. Major fields with synrift reservoirs include the Miranga, Taquipe, and Aracas fields (de Figueiredo, 1994). These reservoirs are commonly lake margin to deltaic siliciclastic sandstones; the Taquipe Member of the Ilhas Formation is a lacustrine turbidite. Primary reduction in reservoir quality in these intervals is related to calcite and quartz cementation; dissolution of calcite and feldspar grains is the primary process in the creation of secondary porosity. The presence of clays and their diagenesis is of secondary importance to reservoir quality in the synrift succession. Moraes and Surdam (1993) present an excellent summation of the diagenetic history of reservoirs in the Recôncavo Basin.

Espírito Santo Basin

The Espírito Santo Basin (Figures 1, 3, and 14) has undergone a complete basin development, from Early Cretaceous synrift formation of structural grabens to Late Cretaceous and Tertiary marine successions (Palhares et al., 1992; D'avila et al., 1998). Although the basin has been extensively explored since the 1950s, only a few relatively small hydrocarbon accumulations have been discovered. Based on the work of Palhares et al. (1992), Estrella et al. (1984), and D'avila et al. (1998), the occurrence of trapped hydrocarbons in the Espírito Santo Basin appears to be controlled by a limited region of organically mature source rock and complex migration pathways outside of the Neocomian–Aptian source to reservoir succession. D'avila et al. (1998) have postulated the existence of a previously undocumented organic source interval within the Urucutuca Formation, suggesting local sourcing for Urucutuca deep-water sandstones.

On the basis of tectono-stratigraphy, we have determined three distinct reservoir systems present in the Espírito Santo Basin (Figure 3):

1. A Neocomian–Aptian synrift succession (Mucuri Member of the Mariricu Formation) of transitional

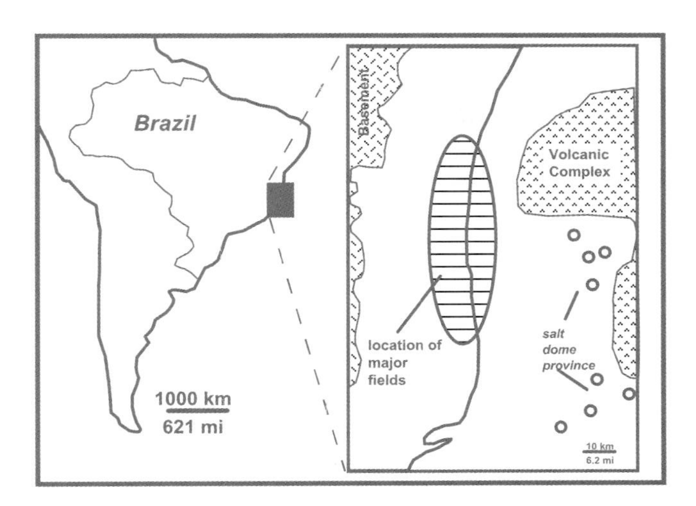

Figure 14—Map of the Espírito Santo Basin, offshore Brazil. Adapted from Estrella et al. (1984).

fluvial to shallow-marine sandstones and conglomerates, which are overlain by the Itaunas evaporites.

2. An Aptian–Cenomanian marine carbonate platform succession (oncolites and oolites of the Regência Member of the Barra Nova Formation) with age-equivalent sandstones (São Mateus Member of the Barra Nova Formation).
3. Open-marine Upper Cretaceous–lower Tertiary deep-water sandstones of the Urucutuca Formation (Estrella et al., 1984; Palhares et al., 1992; D'avila et al., 1998).

As with most of the synrift reservoirs of the South Atlantic, the Mucuri Member reservoir displays variable and diagenetically degraded reservoir quality. The clastic reservoir shows episodes of compaction, quartz cementation, and calcite cementation. The reservoir averages 15–20% porosity, with permeabilities averaging 10–400 md. Reservoir type is secondary intergranular porosity created by calcite dissolution (Palhares et al., 1992). The reservoir is not naturally fractured.

D'avila et al. (1998) have reported that the Urucutuca Formation turbidites are coarse- to fine-grained, massive and commonly amalgamated sandstones with good vertical hydraulic continuity.

EXPLORATION STRATEGIES FOR SOUTH ATLANTIC BASIN RESERVOIRS

Nothing is more basic in the development of an exploration strategy than the thorough understanding of the relevant petroleum systems in a basin. Historically, exploratory areas were examined for individual elements of a petroleum system, such as structural traps based on seismic interpretation, reservoirs extrapolated from well control into seismically defined traps, and organic maturation studies based on well-calibrated 1-D maturity modeling. However, it is the concept of the petroleum system that allows truly integrated study of the key elements controlling the occurrence and accumulation of hydrocarbons in a basin. Understanding of the workings of a petroleum system allows for the rational understanding of both dry holes and discoveries, and facilitates the exploration team in extracting the often subtle geologic characteristics and processes that distinguish the two.

Given the variety of reservoir ages, depositional settings, and diagenetic histories, it is imperative when exploring in South Atlantic basins to have a framework understanding of the history of development of specific reservoir objectives. While we have established in this chapter reservoir generalizations that link each of the reservoir types discussed, each exploration opportunity will be impacted by localized geologic factors and processes.

For any given exploration evaluation, we find it imperative to consider the following:

1. For deep-water siliciclastic reservoirs, sediment provenance, transport mechanism, and delivery process to deep water is vital information. These considerations alone have been the subject of voluminous recent literature (Barrett et al., 1998; Coterill et al., 1998; Dominey and Wiffe, 1998; Garfield et al., 1998; Marotta et al., 1998; Marton et al., 1998). Although it appears that many Tertiary deep-water reservoirs of the South Atlantic have undergone little diagenesis other than mechanical compaction, Mesozoic deep-water reservoirs display a variety of diagenetic episodes. In these cases, favorable linkage and timing of diagenetic episodes must be established and integrated into the exploration evaluation. Rarely can these factors be established from seismic (Ferraris et al., 1998) or well log data alone, emphasizing the need for core and outcrop study of analogous reservoirs.
2. Because the synrift reservoirs are highly impacted by sediment provenance and diagenesis, diagenetic profiles must be established for each potential reservoir. It has been our experience that these reservoirs require the highest degree of core study and thin section petrography. Because most of the South Atlantic basins synrift strata have no outcrop equivalent (the Brazilian Recôncavo Basin being an exception), it is often imperative to establish analog outcrop and field production histories.
3. All the carbonate reservoirs studied have undergone varying degrees of degradation and enhancement due to diagenesis. As with any carbonate reservoir, it is imperative to identify the diagenetic episodes that have affected the reservoir and to compose a diagenetic profile to understand the order of these episodes and the magnitude of

reservoir creation or destruction in each. We have found that integrating the diagenetic profile with the relevant organic source maturation and migration history often identifies specific geologic time intervals when the trapping of economic volumes of hydrocarbons was possible. Our experience has been that this integration allows far more detailed evaluation of prospect opportunities and can identify prospective zones independent of structural closure.

Acknowledgments—*We thank the organizers of the South Atlantic Petroleum Systems Hedberg Conference for inviting our participation in this program and this paper. We thank Texaco, Inc., for permission to present and publish this paper. We thank the Texaco geoscientists exploring for hydrocarbons in the west African and Brazilian basins for their insight and input. We thank our co-workers William Almon, Steve Johansen, and Vaughn Robison, who provided valuable technical input and support. Margy Walsh and George Griffith provided logistical support. Insightful and beneficial reviews of the original manuscript were given by Barry Katz, Susan Longacre, and Edith Wilson.*

REFERENCES CITED

Abrahao, D., and J. E. Warme, 1990, Lacustrine and associated deposits in a rifted continental margin—Lower Cretaceous Lagoa Feia Formation, Campos Basin, offshore Brazil, *in* B. J. Katz, ed., Lacustrine basin exploration—case studies and modern Analogs: AAPG Memoir 50, p. 287–305.

Amaral, J., J. J. Biteau, P. Zaroslinska, and L. DeCosta, 1998, Angola—the Lower Congo Basin Tertiary petroleum systems hydrocarbon distribution in relation with the structural and sedimentary evolution (abs.): AAPG International Conference and Exhibition, Rio de Janeiro, AAPG Bulletin, v. 82, n. 10, p. 1885.

Arienti, L. M., L. F. G. Caddah, E. B. Rodrigues, M. R. Becker, C. J. Abreu, and C. H. L. Bruhn, 1998, Contrasting middle Eocene and upper Oligocene/lower Miocene deep-water facies and processes from Campos Basin, Brazil (abs.): AAPG International Conference and Exhibition, Rio de Janeiro, AAPG Bulletin, v. 82, n. 10, p. 1887.

Barrett, M., A. Ruiter, T. Schirmer, and P. Kapela, 1998, Deep-water Tertiary turbidite channel exploration plays, Block 14, offshore Cabinda, Angola (abs.): AAPG International Conference and Exhibition, Rio de Janeiro, AAPG Bulletin, v. 82, n. 10, p. 1889.

Baudouy, S., and C. Legorjus, 1991, Sendji field—People's Republic of Congo, *in* N. H. Foster and E. A. Beaumont, eds., Structural traps V: AAPG Treatise of Petroleum Geology series, Atlas of Oil and Gas Fields, p. 121–149.

Bertani, R. T., and A. V. Carozzi, 1984, Microfacies, depositional models and diagenesis of Lagoa Feia Formation (Lower Cretaceous) Campos Basin, offshore Brazil: Petrobras–CENPES, Ciencia Tecnicia Petroleo, v. 14, 104 p.

Bertani, R. T., and A. V. Carozzi, 1985, Lagoa Feia Formation (Lower Cretaceous) Campos Basin, offshore Brazil: rift-valley type lacustrine carbonate reservoirs: Journal of Petroleum Geology, v. 8. p. 37–58.

Bray, R., and S. Lawrence, 1999, Nearby finds brighten outlook for equitorial Guinea and Namibia: Oil & Gas Journal v. 97, n. 5, 8 p.

Brice, S. E., K. R. Kelts, and M. D. Arthur, 1980, Lower Cretaceous lacustrine source beds from early rifting phases of South Atlantic (abs.): AAPG Bulletin, v. 64, p. 680–681.

Brice, S. E., M. D. Cochran, G. Pardo, and A. D. Edwards, 1982, Tectonics and sedimentation of the South Atlantic rift sequence: Cabinda, Angola, *in* J. S. Watkins and C. L. Drake, eds., Studies in continental margin geology: AAPG Memoir 34, p. 5–18.

Brock, D., L. Smith, and R. Sarg, 1998, Sequence stratigraphic prediction of reservoir and source deep-water Angola Block 24 (abs.): AAPG International Conference and Exhibition, Rio de Janeiro, AAPG Bulletin, v. 82, n. 10, p. 1896.

Brognon, G. P., and G. R. Verrier, 1966, Oil and geology in Cuanza Basin of Angola: AAPG Bulletin, v. 50, p. 108–158.

Bruhn, C. H. L., and R. G. Walker, 1995, High-resolution stratigraphy and sedimentary evolution of coarse-grained canyon-filling turbidites from the Upper Cretaceous transgressive megasequence, Campos Basin, offshore Brazil: Journal of Sedimentary Research, v. B65, n. 4, p. 426–442.

Bruhn, C. H. L., M. R. Becker, L. M. Arienti, E. B. Rodrigues, C. E. B. S. Abreu, R. R. P. Alves, D. D. Castro, R. A. Santos, L. C. S. Freitas, A. P. Baros, and D. J. Sarzenski, 1998a, Contrasting styles of Oligocene/Miocene turbidite reservoir from deep-water Campos Basin, Brazil: AAPG International Conference and Exhibition, Rio de Janeiro.

Bruhn, C. H. L., A. S. Barroso, M. R. F. Lopes, D. J. Sarzenski, C. J. Abreu, and C. M. A. Silva, 1998b, High-resolution stratigraphy and reservoir heterogeneities of upper Albian turbidite reservoirs of Albacora field, Campos Basin, offshore Brazil (abs.): AAPG International Conference and Exhibition, Rio de Janeiro.

Caddah, L. F. G. D. B. Alves, and A. M. P. Mizusaki, 1998, Turbidites associated with bentonites in the Upper Cretaceous of the Campos Basin, offshore Brazil: Sedimentary Geology, v. 115, p. 175–184.

Candido, A., and C. A. G. Cora, 1992, The Marlim and Albacora giant fields, Campos Basin, offshore Brazil, *in* M. T. Halbouty, ed., Giant oil and gas fields of the decade 1978–1988: AAPG Memoir 54, p. 123–135.

Carminatti, M., and J. C. Scarton, 1991, Sequence stratigraphy of the Oligocene turbidite complex of the Campos basin, offshore Brazil: an overview, *in* P. Weimer and M. H. Link, eds., Seismic facies and sedimentary processes of submarine fans and turbidite systems: New York, Springer-Verlag, p. 241–246.

Carozzi, A. V., F. U. H. Falkenhein, C. F. Lucchesi, and M. R. Franke, 1979, Depositional–diagenetic history of Macaé carbonate reservoirs (Albian–Cenomanian) Campos Basin, offshore Rio de Janeiro, Brazil (abs.): AAPG Bulletin, v. 63, p. 429.

Carozzi, A. V., F. U. H. Falkenhein, and M. R. Franke, 1983, Depositional environment, diagenesis, and reservoir properties of oncolitic packstones, Macaé Formation (Albian–Cenomanian) evolution of Cretaceous oncolitic packstone reservoirs, Campos basin, offshore Rio de Janeiro, Brazil, *in* T. Peryt, ed., Coated grains: New York, Springer-Verlag, p. 330–343.

Carvalho, M. D., R. J. Jahnert, and U. M. Praça, 1998, The coquinas sequence: lacustrine carbonate reservoirs in Campos Basin (abs.): AAPG International Conference and Exhibition, Rio de Janeiro, AAPG Bulletin, v. 82, n. 10, p. 1899.

Cohen, C. R., 1985, Role of fault rejuvenation in hydrocarbon accumulation and structural evolution of Recôncavo basin, northeastern Brazil: AAPG Bulletin, v. 69, p. 65–76.

Coterill, K., A. Champagne, J. Coleman, D. Marotta, M. Pasley, G. Tari, L. Binga, and H. Van Dierendonck, 1998, Sinuous morphologies in submarine channels—scale and geometries in seismic and outcrop indicating possible mechanisms for deposition (abs.): AAPG International Conference and Exhibition, Rio de Janeiro, AAPG Bulletin, v. 82, n. 10, p. 1904.

Dale, C. T., J. R. Lopes, and S. Abilio, 1992, Takula oil field and the greater Takula area, Cabinda, Angola, *in* M. T. Halbouty, ed., Giant oil and gas fields of the decade 1978–1988: AAPG Memoir 54, p. 197–215.

D'avila, R. S. F., A. S. Biassusi, A. C. Guirro, J. R. Brandão, and E. S. T. Frota, 1998, Urucutuca–Urucutuca(?): a new petroleum system in Espírito Santo Basin, Brazil (abs.): AAPG International Conference and Exhibition, Rio de Janeiro.

Dawson, W. C., T. C. O'Hearn, and E. E. Hiatt, 1993, Diagenetic and sedimentologic aspects of Pinda (Albian) dolomites, offshore west Africa (abs.): AAPG Annual Convention, Program with Abstracts, New Orleans, p. 90.

de Figueiredo, A. M. F., ed., 1994, Recôncavo basin, Brazil: a prolific intracontinental rift basin, *in* S. Landon, ed., Interior rift basins: AAPG Memoir 59, p. 157–203.

De Ros, L. F., 1998, Compositional genetic potential, diagenetic evolution, and reservoir quality of eastern Brazilian margin turbidites (abs.): AAPG International Conference and Exhibition, Rio de Janeiro, AAPG Bulletin, v. 82, n. 10, p. 1909.

Dominey, J. R., and S. Wiffe, 1998, Salt tectonics and sedimentation: an integrated interpretation ultra deep water area, Lower Congo Basin, offshore Angola (abs.): AAPG International Conference and Exhibition, Rio de Janeiro.

Edwards, A., and R. Bignell, 1988a, Hydrocarbon potential of west African salt basin: Oil & Gas Journal, December 12, p. 71–74.

Edwards, A., and R. Bignell, 1988b, Nine major play types recognized in west African salt basin: Oil & Gas Journal, December 19, p. 55–58.

Estrella, G., M. Rocha Mello, P. C. Gaglianone, R. L. M. Azevedo, D. Tsubone, E. Rossetti, J. Concha, and I. M. R. A. Bruning, 1984, The Espírito Santo basin (Brazil) source rock characterization and petroleum habitat, *in* G. Demaison and R. Murris, eds., Petroleum geochemistry and basin evaluation: AAPG Memoir 35, p. 253–271.

Falkenhein, F. U. H., M. R. Frank, and A. V. Carozzi, 1981, Petroleum geology of the Macaé Formation (Albian–Cenomanian), Campos basin, Brazil (carbonate microfacies–depositional and diagenetic models–natural and experimental porosity): Petrobras-CENPES, Ciencia Tecnica Petroleo Petrobras, v. 11, 140 p.

Ferraris, O., L. Pianelli, O. Acevedo, and D. Lorenzo, 1998, Avoiding failures in reservoir prediction from seismic: AAPG International Conference and Exhibition, Rio de Janeiro, AAPG Bulletin, v. 82, n. 10, p. 1913.

Garfield. T. R., D. C. Jennette, F. J. Goulding, and D. K. Sickafoose, 1998, An integrated approach to deep-water reservoir prediction: AAPG International Conference and Exhibition, Rio de Janeiro, AAPG Bulletin, v. 82, n. 10, p. 1916–1917.

Gilbert, D., 1984, Organic facies variations in the Mesozoic South Atlantic, *in* R. Amidei, ed., Initial Reports of the Deep Sea Drilling Project, v. 75, pt. 2, p. 1035–1049.

Guardado, L. R., L. A. P. Gamboa, and C. F. Lucchesi, 1989, Petroleum geology of the Campos basin, Brazil, a model for a producing Atlantic type basin, *in* J. D. Edwards and P. A. Santogrossi, eds., Divergent/passive margin basins: AAPG Memoir 48, p. 3–79.

Horschutz, P. M. C., L. C. S. de Freitas, C. V. Stank, A. da Silva Barroso, and W. M. Cruz, 1992, The Linguado, Carapeba, Vermelho, and Marimba giant oil fields, Campos basin, offshore Brazil, *in* M. T. Halbouty, ed., Giant oil and gas fields of the decade 1978–1988: AAPG Memoir 54, p. 137–153.

Jahnert, R., A. França, L. Trindade, C. Quintaes, P. Santos, J. Pessoa, and R. Bedregal, 1998, The petroleum system of Campos Basin: AAPG International Conference and Exhibition, Rio de Janeiro, AAPG Bulletin, v. 82, n. 10, p. 1926.

Knight, R., and J. Westwood, 1999, Long-term prospects very bright for deep-waters off west Africa: Oil & Gas Journal, v. 97, n. 3, 5 p.

Lehner, P., and P. A. C. de Ruiter, 1977, Structural history of the Atlantic margin of Africa: AAPG Bulletin, v. 61, p. 961–981.

Levache, D., and D. Dessort, 1998, Origin and maturity of oils from the Congo coastal basin (abs.): AAPG International Conference and Exhibition, Rio de Janeiro, AAPG Bulletin, v. 82, n. 10, p. 1933.

Liro, L. M., and R. M. Coen, 1995, Salt deformation history and postsalt structural trends, offshore southern Gabon, west Africa, *in* M. P. A. Jackson, D. G. Roberts, and S. Snelson, eds., Salt tectonics: a global perspective: AAPG Memoir 65, p. 323–331.

Magoon, L. B., 1988, The petroleum system—a classification scheme for research, exploration, and resource assessment, *in* L. B. Magoon, ed., Petroleum systems of the United States: USGS Bulletin 1870, p. 2–15.

Marotta, D., C. S. Alexander, K. Coterill, K. Hartman, M. Pasley, T. C. Stitelar, G. Tari, L Binga, and B. Lehner, 1998, The use of 3D visualization for understanding Tertiary deep-water clastic systems: a west Africa example (abs.): AAPG International Conference and Exhibition, Rio de Janeiro.

Marton, G., G. Tari, and C. Lehmann, 1998, Evolution of salt-related structures and their impact on the postsalt petroleum systems of the Lower Congo Basin, offshore Angola (abs.): AAPG International Conference and Exhibition, Rio de Janeiro.

McHargue, T. R., 1990, Stratigraphic development of proto-South Atlantic rifting in Cabinda, Angola—a petroliferous lake basin, *in* B. J. Katz, ed., Lacustrine basin exploration—case studies and modern analogs: AAPG Memoir 50, p. 307–326.

Mello, M. R., and J. R. Maxwell, 1990, Organic geochemical and biomarker characterization of source rocks and oils derived from lacustrine environments in the Brazilian continental margin, *in* B. J. Katz, ed., Lacustrine basin exploration—case studies and modern analogs: AAPG Memoir 50, p. 77–95.

Mohriak, W. U., P. R. Palagi, and M. R. Mello, 1998, Tectonic evolution of South Atlantic salt basins (abs.): AAPG

International Conference and Exhibition, Rio de Janeiro, AAPG Bulletin, v. 82, n. 10, p. 1945.

Moraes, M. A. S., 1989, Diagenetic evolution of Cretaceous–Tertiary turbidite reservoirs, Campos basin, Brazil: AAPG Bulletin, v. 73, p. 598–612.

Moraes, M. A. S., and C. H. L. Bruhn, 1988, Brazilian turbidite reservoirs: heterogeneity study from outcrops to subsurface (abs.): AAPG Bulletin, v. 72, p. 225.

Moraes, M. A. S., and R. C. Surdam, 1993, Diagenetic heterogeneity and reservoir quality: fluvial, deltaic, and turbiditic sandstone reservoirs, Potigar and Recôncavo rift basins: AAPG Bulletin, v. 77, p. 1142–1158.

Ojeda, H. A. O., 1982, Structural framework, stratigraphy and evolution of Brazilian marginal basins: AAPG Bulletin, v. 66, p. 732–749.

Palhares, A., H. D. Rangel, and A. M. F. de Figueiredo, 1992, Rio Itaunas field–Brazil–Espírito Santo basin, southeastern Brazil, *in* N. H. Foster and E. A. Beaumont, eds., Structural traps VI: AAPG Treatise of Petroleum Geology Series, Atlas of Oil and Gas Fields, p. 141–152.

Peres, W. E., 1993, Shelf-fed turbidite system model and its application to the Oligocene deposits of the Campos basin, Brazil: AAPG Bulletin, v. 77, p. 81–101.

Rabinowitz, P. D., and J. LaBrecque, 1979, The Mesozoic South Atlantic Ocean and evolution of its continental margin: Journal of Geophysical Research, v. 84, p. 5793–6002.

Rangel, H. D., P. R. Santos, C. M., and C. M. S. P. Quinpaes, 1998, Roncador field, a new giant in Campos basin, Brazil, *in* Proceedings, 30th Offshore Technology Conference, Houston, Paper OTC-8876, v. 1, p. 575–587.

Reyre, D., 1984, Petroleum characteristics and geological evolution of a passive margin—example of the Lower Congo–Gabon basin: Bulletin des Centres de Recherche Exploration Production ELF Aquitaine, v. 8, p. 303–332.

Smith, R. D., 1995, Reservoir architecture of synrift lacustrine turbidite systems, Early Cretaceous, offshore South Gabon, *in* J. J. Lambiase, ed., Hydrocarbon habitat in rift basins: Geological Society of London Special Publication 80, p. 197–202.

Spaw, R. H., and R. H. Koehler, 1981, Geology and reservoir distribution, Pinda Formation, offshore Zaire and southern offshore Cabinda (abs.): AAPG Bulletin, v. 65, p. 996.

Teisserenc, P., and J. Villemin, 1989, Sedimentary basin of Gabon—geology and oil systems, *in* J. D. Edwards and P. A. Santogrossi, eds., Divergent/passive margin basins: AAPG Memoir 48, p. 117–199.

Tillement, B. 1987, Insight into Albian carbonate geology in Angola: Bulletin of Canadian Petroleum Geology, v. 35, p. 65–74.

Walgenwitz, F., M. Pagel, A. Meyer, H. Maluski, and P. P. Monie, 1990, Thermo-chronological approach to reservoir diagenesis in the offshore Angola basin: a fluid inclusion, ^{40}Ar-^{39}Ar and K-Ar investigation: AAPG Bulletin, v. 74, p. 547–563.

Cunningham, R., and R. M. Lindholm, 2000, Seismic evidence for widespread gas hydrate formation, offshore west Africa, *in* M. R. Mello and B. J. Katz, eds., Petroleum systems of South Atlantic margins: AAPG Memoir 73, p. 93–105.

Chapter 8

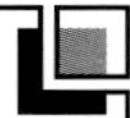

Seismic Evidence for Widespread Gas Hydrate Formation, Offshore West Africa

Robert Cunningham

Rosanne M. Lindholm

Exxon Exploration Company
Houston, Texas, U.S.A.

Abstract

Seismic reflection data recorded offshore west Africa reveal the presence of extensive bottom simulating reflectors (BSRs) on the continental slope off the Niger and Congo River Deltas. Cumulative surface areas encompassing the Nigeria and Congo BSR occurrences are approximately 11,000 km^2 and 4000 km^2, respectively. West African BSRs display many of the characteristics that have commonly been ascribed to submarine gas hydrates elsewhere in the world, including reflection polarity reversal relative to the sea floor reflection, consistency between depth of the BSR and base of the gas hydrate stability field, geometric relationships between BSR and bedding plane reflections, and a positive correlation between the subwater bottom depth of the BSR and the water depth. Widespread BSRs along the Nigeria and Congo continental slopes generally occur in areas of complex structure and where water depths are greater than 1200 m. The relationship between gas hydrate occurrence and structure exists, in part, because BSRs are more easily recognized on seismic sections where the BSRs cut across dipping stratigraphic reflections. However, faults and dipping stratigraphy also facilitate migration of biogenic or thermogenic gas to the base of the gas hydrate stability zone, enabling the formation of BSRs.

INTRODUCTION

Natural gas hydrates are solid, ice-like substances composed of water and gas, mainly methane, which may occur in marine continental margin sediments and in permafrost regions. Pressure–temperature conditions that characterize the marine settings allow gas hydrates to form within the upper several hundred meters of the sediment column provided that sufficient methane is supplied to the system so that concentrations in the pore fluids exceed methane solubility (Rempel and Buffet, 1997; Xu and Ruppel, 1999). The appearance on deep-water seismic records of a shallow bottom simulating reflector (BSR) that displays a negative impedance contrast, an often high amplitude, continuous character, and conformance to the base of the theoretical gas hydrate stability field has long been considered evidence for natural gas hydrates.

Observations of gas hydrates along South Atlantic continental margins have lagged behind those made in the Pacific, North Atlantic, and Arctic Oceans. The early global compilations of gas hydrate occurrences made by Kvenvolden and McMenamin (1980) and Kvenvolden and Barnard (1983) listed a few locations along the west African margin where seismic reflection geometries (Emery, 1974) or evidence for submarine slumps and slides (Summerhayes et al., 1979) suggested that gas hydrates may exist. These sites were deleted from a later compilation (Kvenvolden et al., 1993) due to lack of stronger corroborative evidence. The authors found the lack of compelling evidence incongruous with the widespread occurrences of gas hydrates along outer continental margins elsewhere in the world.

Now, with the explosive growth in deep-water oil exploration and the acquisition of high-quality seismic data off the west African countries and Brazil, it is becoming apparent that the South Atlantic may actually be a rich gas hydrate province (Figure 1). Regionally extensive BSRs, suggestive of gas hydrates, have been reported from seismic data off Brazil on the Amazon fan (Manley and Flood, 1988) and in the Pelotas Basin (Fontana, 1989; Fontana and Mussumeci, 1994), off Argentina in the Central Argentine Basin (Manley and Flood, 1988), and off Nigeria on the Niger fan (Hovland et al., 1997). In

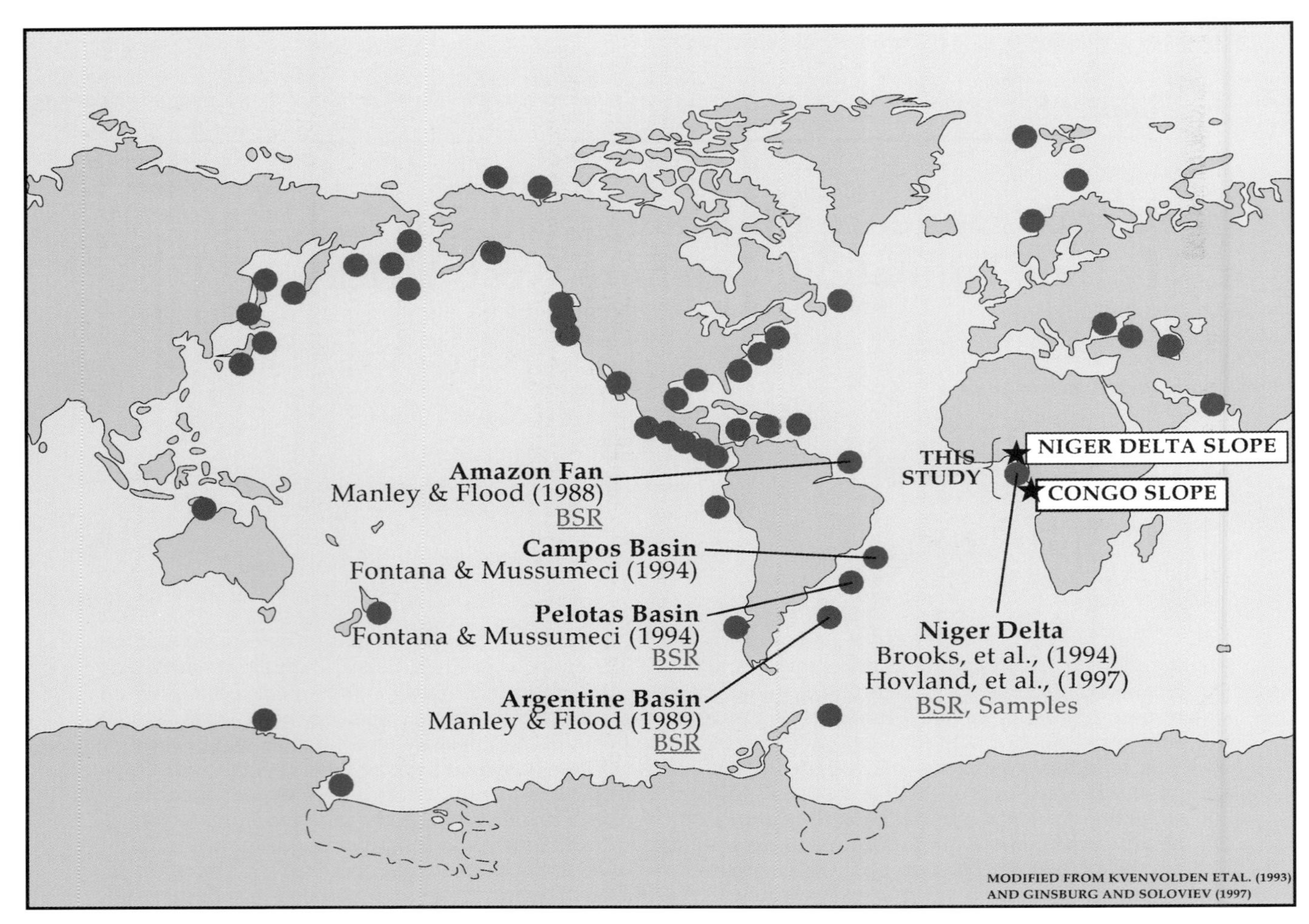

Figure 1—Global distribution of known and inferred deep water gas hydrates. Modified from Kvenvolden et al. (1993) and Ginsburg and Soloviev (1997).

addition to seismically observable bottom stimulating reflectors (BSRs), geologic evidence for gas hydrates includes sea floor mounds, pockmarks, mud volcanoes, submarine slumps and slides, and direct sampling of gas hydrates. This evidence undoubtedly continues to grow with the numerous seismic, geohazard, and surface geochemical surveys being conducted in exploration areas. For example, gas hydrates have been recovered in piston cores at several locations on the continental slope of the Niger Delta (Brooks et al., 1994).

In this chapter, we report on the regional distribution of gas hydrates as evidenced by BSRs off the west African margin from the Niger Delta to southern Angola and we discuss the geologic factors controlling their distribution. The study represents the results of a preliminary screening of seismic data available to us and is by no means an exhaustive survey of gas hydrate distribution. Three models are currently proposed for the formation of gas hydrates and associated BSRs in marine sediments. One model requires that the methane be derived locally from in situ microbial activity and supplied to the hydrated sediment presumably over relatively short distances (Claypool and Kaplan, 1974); this is called the *passive filtration model* by Kvenvolden et al. (1993). An important aspect to this model is that free gas builds up beneath the BSR due to the accumulation of indigenous biogenic methane and the continuous dissociation of hydrate at the base of the hydrate stability zone with continued sedimentation. The other model involves the movement of rising pore fluids carrying methane derived from some distance beneath the hydrate stability zone (Hyndman and Davis, 1992); this is called the *active filtration model* by Kvenvolden et al. (1993). For thick hydrates to form via this mechanism, longer distance vertical migration pathways are required. A third model has also been proposed for local hydrate occurrences that involves mass transport of gas-charged sediment into water depth ranges where gas hydrate is stable; this is called the *sedimentational model* by Kvenvolden et al. (1993). By analyzing BSR distribution patterns within the context of the regional structural and stratigraphic regimes off the west African margin, we evaluate the first two models in terms of origin and mechanism of gas flux to the hydrate stability zone.

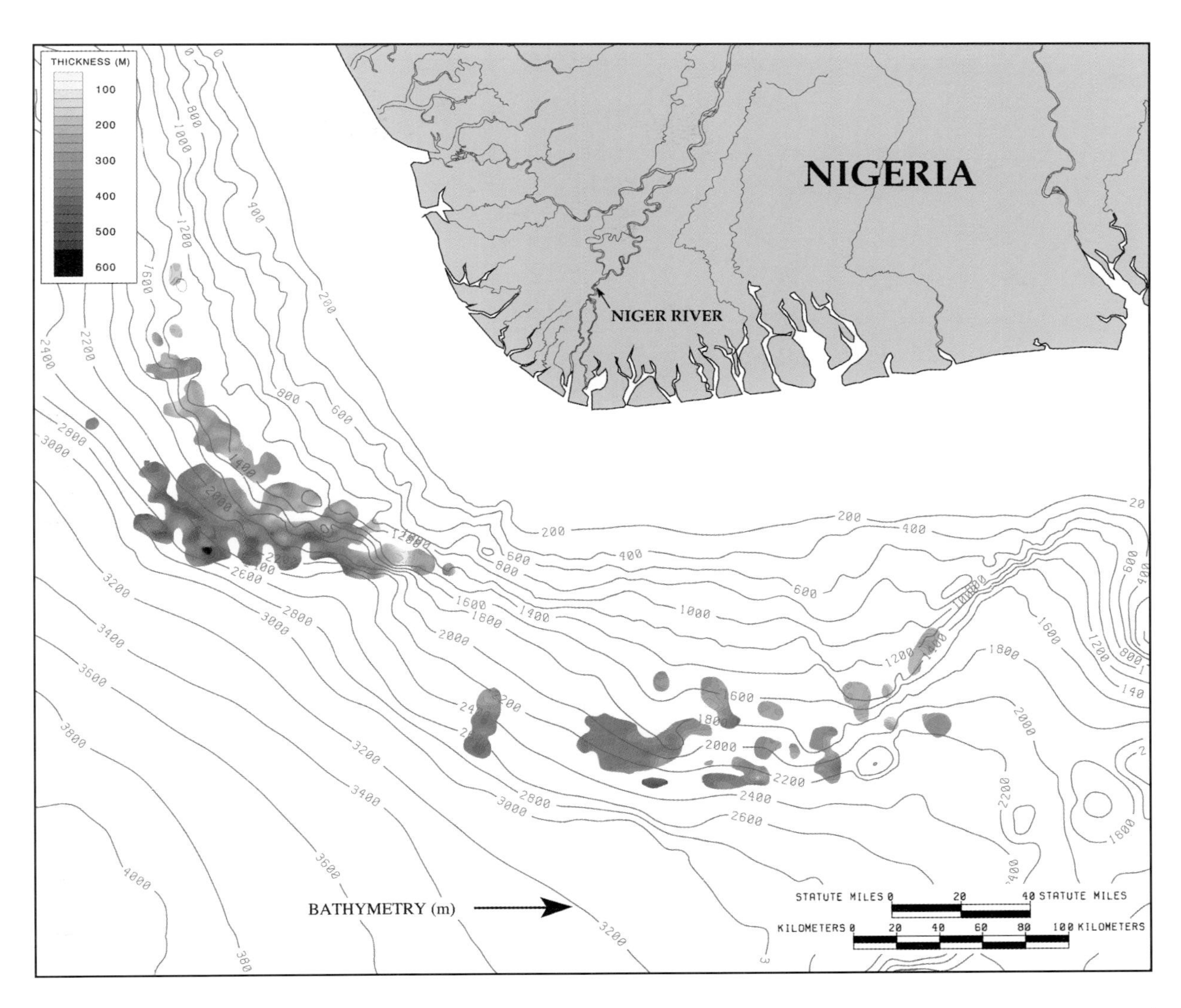

Figure 2—Isopach of interval from sea floor to bottom simulating reflector (BSR) off the Niger Delta (estimated assuming an average interval velocity from sea floor to BSR of 1800 m/s).

BOTTOM SIMULATING REFLECTOR DISTRIBUTION OFF THE NIGER DELTA

Along the Nigerian slope, a broad region of BSR occurrence is found off the Niger Delta front at water depths of 1200–2800 m (Figure 2). The isopach of the interval from the sea floor to the BSR shows a variation in thickness from 200 to 600 m and generally increases with increasing water depth. BSRs are typically associated with contractional structures, including complex shale-cored anticlines and imbricated fault-related folds (Figure 3a). These structures are associated with toe thrusts in the distal prodelta. A detailed view of the seismic character of the BSR is shown in Figure 3B. The BSR reflection generally displays reversed polarity relative to the sea floor reflection, indicating a negative impedance contrast downward through the BSR. High-amplitude, generally high-impedance reflections commonly occur above the BSR, indicating alternating permeable interbeds and/or variable gas hydrate concentration. Dimming or blanking of seismic reflections also occurs above the BSR in some places. The dimming can be either pervasive or restricted to certain intervals. Interpretations of the origin of acoustic blanking above BSRs include hydrate cementation (Shipley et al., 1979; Dillon et al., 1991; Lee et al., 1994), lithologic homogeneity (Holbrook et al., 1996), and overpressuring (Hovland et al., 1997). Since interbedding of sand and mud in these dim zones over the Nigerian BSRs is interpreted to be regionally consistent, the muting of lithologic impedance contrasts in these areas appears to be attributable to varying levels of hydrate saturation in the porosity of sand interbeds. Below the BSR, the presence of high-amplitude, low-impedance reflections and positive amplitude versus offset (AVO) response indicate the presence of free gas (Figure 4).

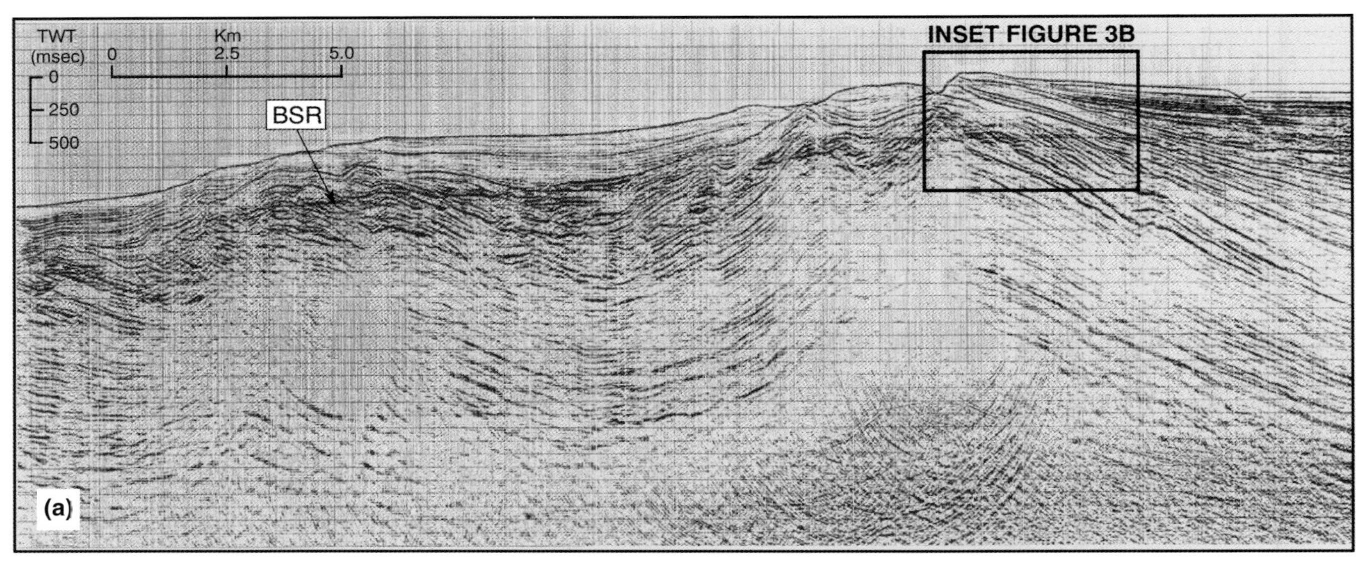

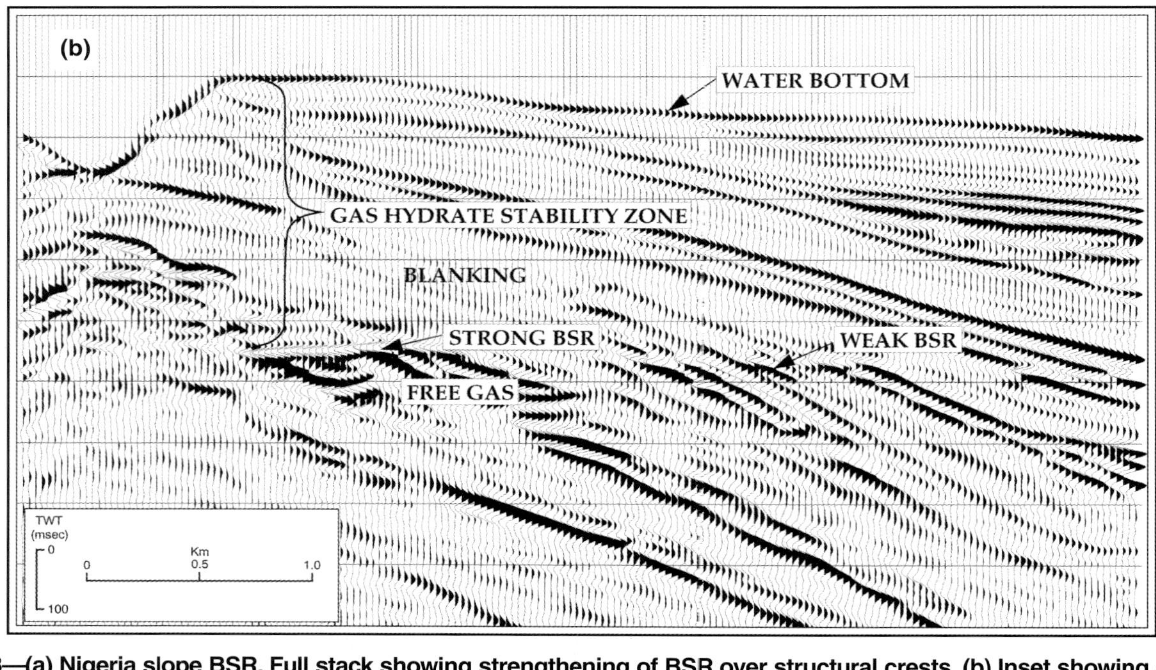

Figure 3—(a) Nigeria slope BSR. Full stack showing strengthening of BSR over structural crests. (b) Inset showing seismic character of the gas hydrate stability zone.

The coincidence of BSR development with structural crests along contractional trends on the Niger Delta front has previously been noted by Hovland et al. (1997). They observed that BSR strength typically declines moving off structure and is not observable where bedding becomes horizontal. We have made similar observations and further note that positive AVO beneath the BSR disappears as the BSR strength declines (Figure 4). Hovland et al. (1997) conclude that BSRs are a consequence of focused and diffusive vertical fluid escape that is facilitated by dipping or chaotic strata. Structural dip influences the gas flux to the hydrate stability zone by focusing biogenic gas that forms in the shallow section or migrates from older, more deeply buried biogenic gas accumulations and thermogenic gas generated deeper in the sedimentary section.

Besides gas hydrates that exist at depth in marine sediments, gas hydrates can also occur at the sea floor or in near-surface sediments. Shallow hydrates may occur with or without evidence of deeper BSRs. We have surveyed dip maps derived from 3-D seismic data collected on the Niger Delta slope and have observed positive-relief features associated with deep faults that we interpret to be hydrate mounds (Figure 5). We had piston cores taken on one of these mounds and recovered hydrate ice. Melting of the ice released gas that was 99%

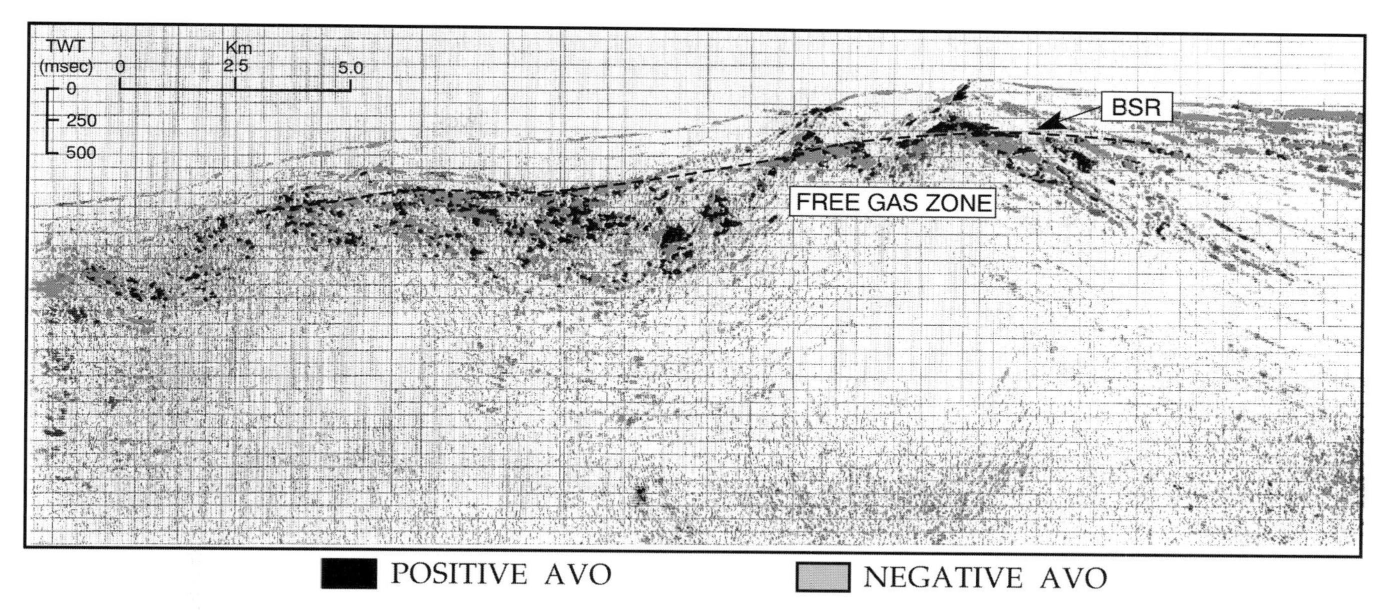

Figure 4—Nigeria slope BSR (angle stack envelope difference display) showing positive amplitude versus offset (AVO) response beneath the BSR. See Figure 3a for a full stack display of this seismic line.

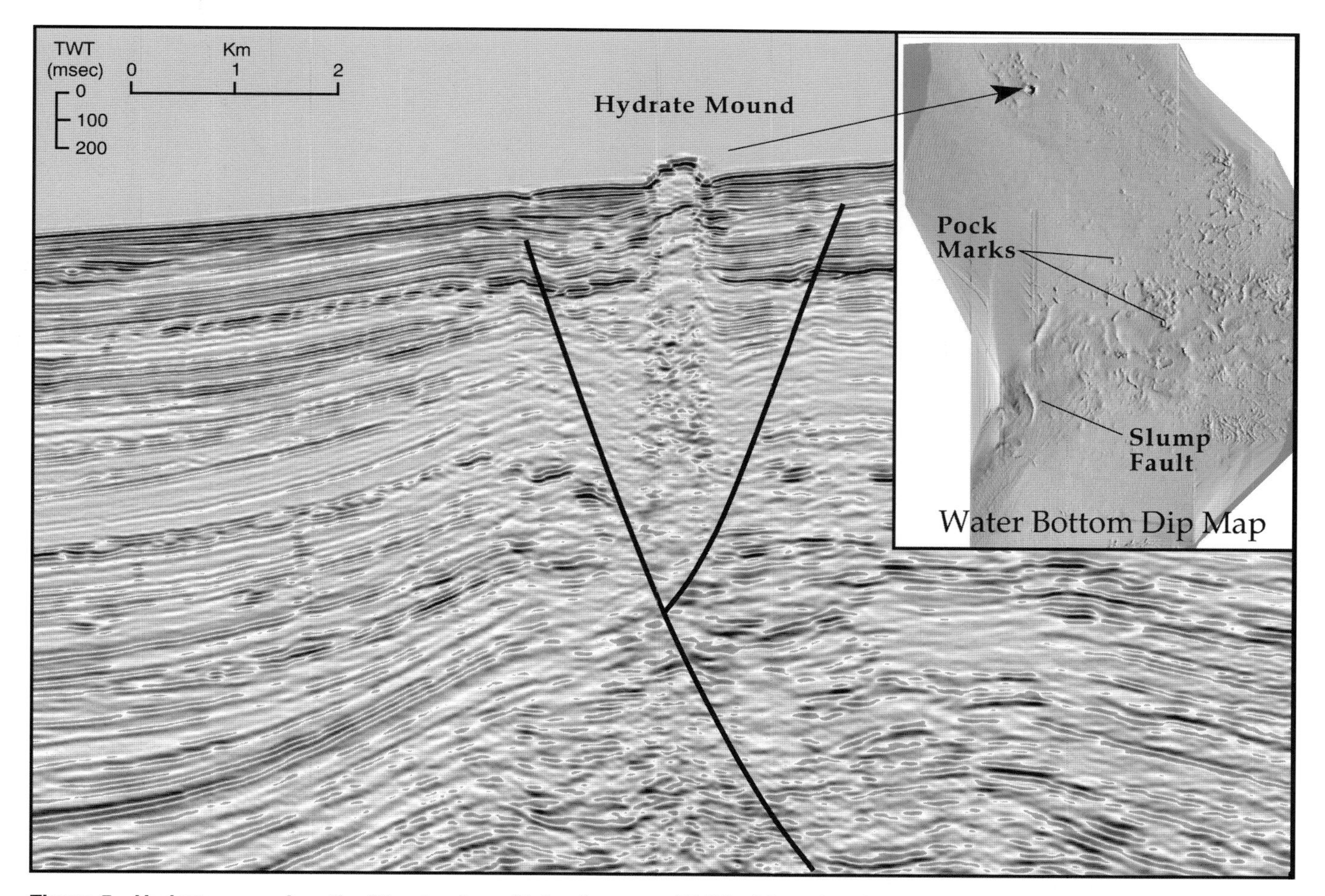

Figure 5—Hydrate mound on the Nigeria slope. Note absence of BSR, although seismic chimney and "flags" on faults indicate significant gas flux to surface.

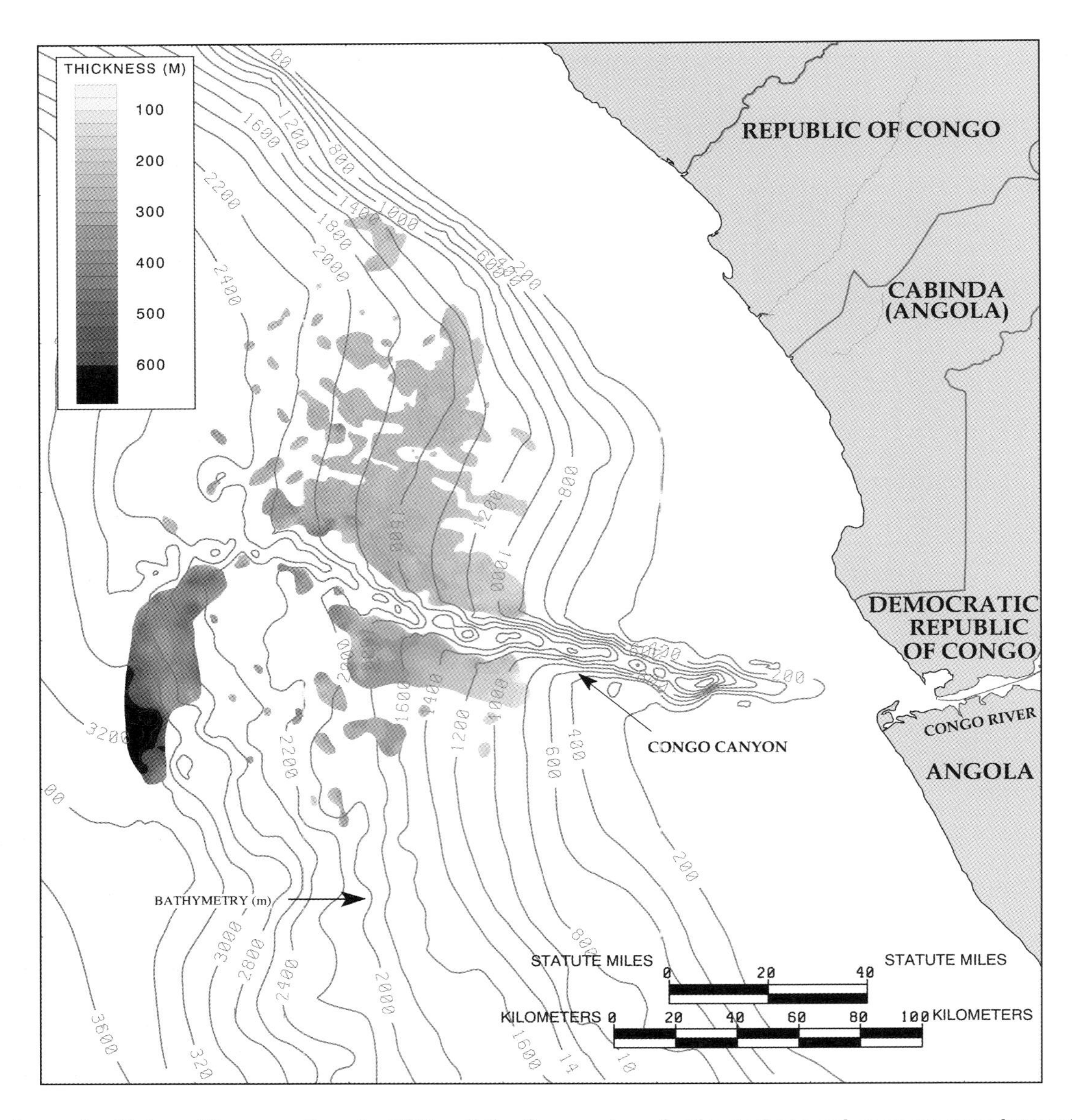

Figure 6—Isopach of interval from sea floor to BSRs off the Congo slope (estimated assuming an average interval velocity from sea floor to BSR of 1800 m/s).

methane, with concentrations of wet gas components that were significantly above the sedimentary background. The methane had a carbon isotopic ratio of –54‰. These data suggest that the hydrate may consist of a mixture of microbial methane with some thermogenic gas. Thermogenic methane contributions to gas hydrates in near-surface sediments have been detected by others on the Niger Delta slope (Brooks et al., 1994) and at several locations worldwide (e.g., Caspian Sea and Gulf of Mexico) where actual gas hydrate samples have been recovered (Brooks et al., 1984, 1986; Kvenvolden, 1993). The observation that shallow gas hydrates may not coincide with deeper BSRs may be due to efficient fault or stratigraphic migration pathways that allow gas to bypass the base of the hydrate stability zone enroute to the sea floor.

BOTTOM SIMULATING REFLECTOR DISTRIBUTION OVER THE CONGO FAN

Along the continental slope off the Republic of Congo, Democratic Republic of Congo, and Cabinda, a broad region of BSR occurrence is found on the Congo River fan at water depths of 800–3200 m (Figure 6). The isopach of the interval from the sea floor to the BSR shows a variation in thickness from about 200 to 600 m and generally increases with increasing water depth, as off the Niger Delta front. Structural features associated with BSRs on the Congo slope are generally more subdued than those offshore Nigeria. Here, the structural styles are dominated by deep-seated extension and salt withdrawal structures updip, to predominantly contractional features

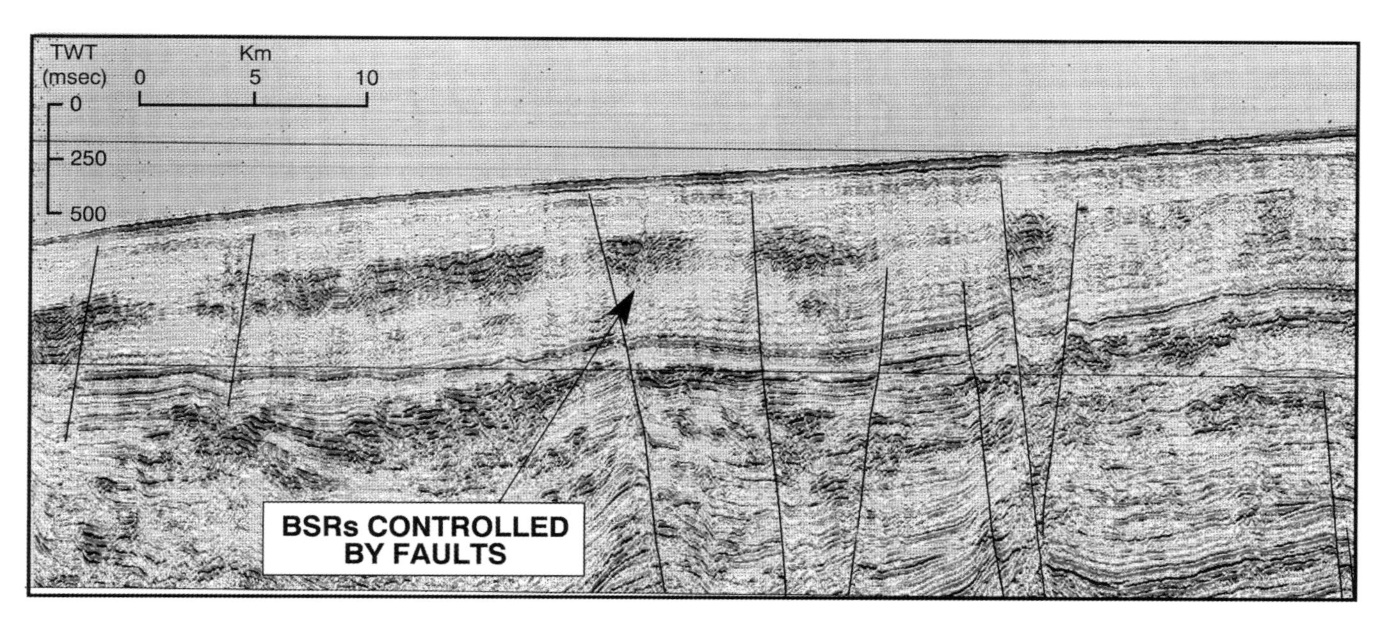

Figure 7—Congo slope BSR showing relationship to faults and higher amplitude character beneath the BSR.

downdip. In the updip area, the faults associated with structures commonly terminate before reaching the sea floor and have uniform dip in the shallow section.

Throughout the area of hydrate occurrence along the Congo fan, BSRs are best developed over the crests of underlying structures and above or adjacent to fault terminations (Figure 7). They commonly occur as "flags" on deep-seated faults associated with salt-cored anticlines and extensional grabens. The BSR generally occurs on the updip side of the fault regardless of the sense of throw. This indicates that gas migrates up the fault to the hydrate stability zone and then laterally beneath the BSR away from the fault. In some areas of the Congo fan, BSRs are controlled primarily by the presence of structure. Although BSRs typically conform uniformly to sea floor topography, they are occasionally seen to rise relative to the sea floor over underlying structures, especially if the structure is associated with salt and if the crests are faulted (Figure 8). This suggests that an increase in thermal gradient may exist due to the enhanced thermal conductivity of salt or to the vertical migration of fluids up faults. As with the Nigeria slope hydrates, the relationships between structuring and BSR occurrences on the Congo fan imply that gas flux from deeper in the sedimentary section may be involved in BSR development.

Stratigraphy and facies patterns play an important role in BSR distribution on the Congo slope. For example, BSRs occur in extensive sheets 20 km wide or greater, along both margins of the present Congo Canyon (Figures 6 and 8). The association of BSRs with the canyon margins may result from an increased potential for sourcing biogenic gas from the mud-rich levee–overbank facies that predominate in the canyon margin sedimentary section. BSRs also tend to overlie the large upper Miocene confined channel complexes interpreted to represent Congo Canyon predecessors. The higher net sand in the fill of these channels may lead to the temporary storage of biogenic and possibly thermogenic gas that ultimately migrates upward via extensional salt-related faults and smaller dewatering faults to the hydrate stability zone (Figure 9). In areas away from the canyon and its upper Miocene predecessors, structural elements appear to control the occurrence of BSRs, as previously discussed.

Seismic character associated with the BSRs on the Congo slope suggests that a significant free gas zone may exist beneath the BSR. Well-developed BSRs are typically underlain by 200–400 ms of high-amplitude reflections, suggesting elevated free gas concentrations (Figures 7, 8, and 9). Free gas zones have also been observed beneath BSRs in several other studies on active and passive margins (Miller et al., 1991; Rowe and Gettrust, 1993; MacKay et al., 1994; Minshull et al., 1994; Singh and Minshull, 1994; Andreassen et al., 1995; Holbrook et al., 1996; Yuan et al., 1996; Andreassen et al., 1997). These studies generally interpret the free gas zone to be thin, only tens of meters thick; however, a surprisingly thick (250-m) free gas zone has been interpreted on the Blake Ridge (Holbrook et al., 1996). This thickness is consistent with the possible free gas zone that we have observed on the Congo fan. Hedberg (1980) was one of the first to suggest that gas hydrates may act as a top seal for free gas accumulations. In addition to hydrate saturations in sediments, MacKay et al. (1994) suggest that the free gas that collects below the BSR, even in low concentrations of a few percent, can substantially lower permeability below that of water-saturated sediments, thus impeding vertical gas migration.

Seismic blanking commonly occurs above the BSRs on the Congo slope. This contrasts with BSRs off Nigeria, where high-amplitude reflections occur both above and

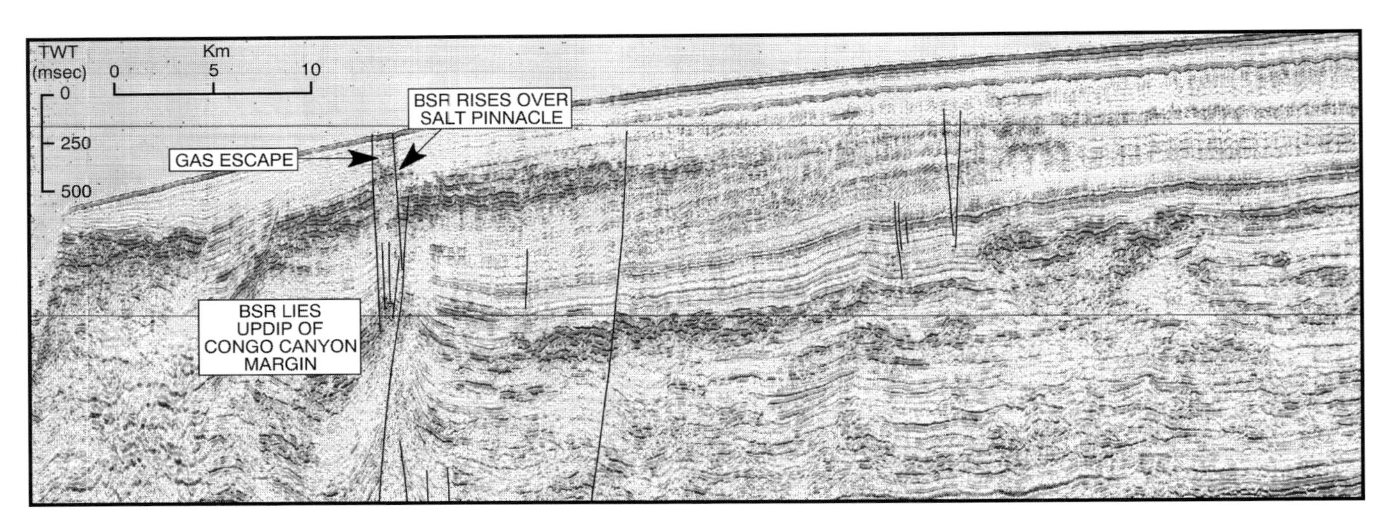

Figure 8—Congo slope BSR on the flank of the Congo canyon margin. Note that the BSR rises over a salt pinnacle.

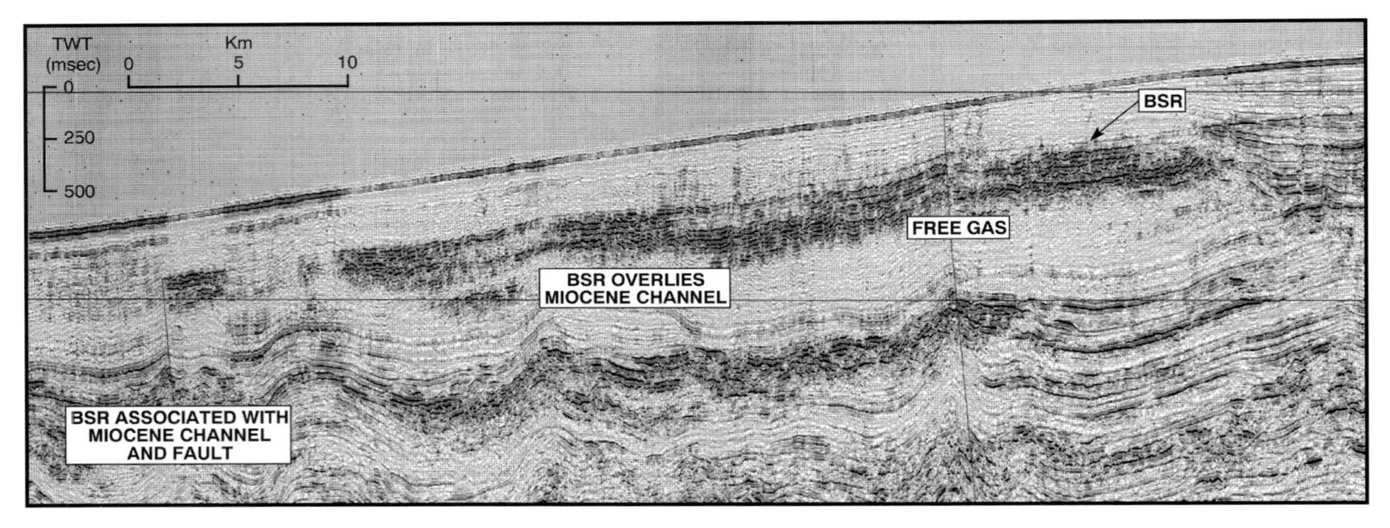

Figure 9—Congo slope BSR overlying a Miocene channel. A significant interval of higher amplitude character occurs beneath the BSR.

below the BSR. The seismic character of the gas hydrate stability zone on the Congo fan may be more similar to that on the Blake Ridge off South Carolina where seismic blanking is also common (Lee et al., 1994). There, the seismic character is attributed mainly to lithologic homogeneity rather than the relatively low hydrate saturation (5–7% of porosity) above the BSR (Holbrook et al., 1996). In support of this analogy, upper Neogene sediments are consistently mud-rich where they have been drilled and piston cored in the region of the BSRs on the Congo slope. Perhaps this muddy character has also enhanced the sealing quality of the sediments within the gas hydrate stability zone and contributed to the development of the thick free gas zone.

REGIONAL GEOLOGIC CONTROLS ON BOTTOM SIMULATING REFLECTOR DEVELOPMENT

The first-order control on gas hydrate development along continental margins is sufficient methane. If sediments are gas-saturated, then the next requirement is the specific pressure–temperature conditions allowing gas hydrates to form in the subsurface and at the sea floor. Given that high pressures and low temperatures are required for gas hydrate formation, their occurrence off west Africa is confined to subshelfal water depths. The known offshore Nigeria and Congo BSRs generally occur

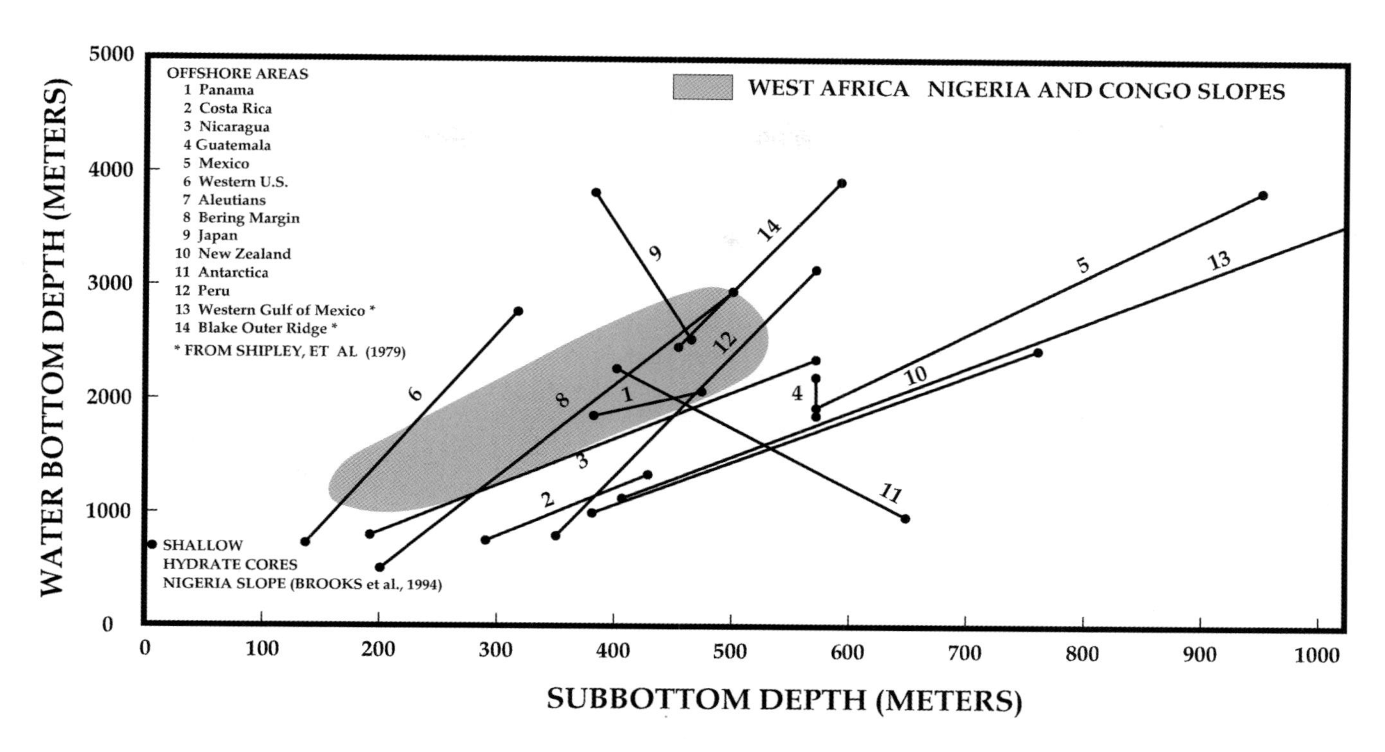

Figure 10—Relationship between water bottom depth and subbottom depth to BSRs on the Pacific and Atlantic margins. Modified from Kvenvolden and Claypool (1988).

at water depths of 800–3000 m and at subbottom depths of 200–600 m. The subbottom depths to BSRs generally increase with water depth for the entire west Africa BSR dataset (Figure 10). When projected to the sea floor, the trend suggests that hydrates may occur as shallow as about 500 m water depth. Surficial gas hydrates have been recovered off Nigeria in piston cores taken at water depths from 560 to 770 m (Brooks et al., 1994). A positive correlation has also been noted between water depth and subbottom depth to BSR for many occurrences worldwide (Shipley et al., 1979; Kvenvolden and Claypool, 1988). This occurs, in part, because an increase in water depth typically leads to an increase in pressure and to a decrease in water bottom temperature which enhances hydrate stability.

Although BSRs have been observed at many locations along the west African margin (Emery, 1974; Kvenvolden and McMenamin, 1980; Kvenvolden and Barnard, 1983; Hovland et al., 1997), we have found that the most numerous and extensive BSRs are associated with the Niger and Congo River fans (Figures 11a and b). The areas of BSR occurrence encompass approximately 11,000 km^2 and 4000 km^2 off the Niger and Congo Delta fronts, respectively. The reasons for this localization involve the favorable alignment of several factors that enhance the flux of methane to the hydrate stability zone.

Elevated organic carbon contents (>0.5 wt. %) in surficial sediments are required to yield an adequate supply of microbial methane to form and stabilize gas hydrates at the appropriate pressures and temperatures for hydrate stability (Kvenvolden and Claypool, 1988). Organic carbon contents in excess of this threshold may be expected off the west African margin since the region is well known for its strong oceanic upwelling and high biological productivity. Satellite-derived marine productivity observations for the globe indicate particularly high average biomass productivity in the regions of the Niger and Congo river fans (Moore and Bolin, 1986). River-induced phytoplankton activity has been found to contribute to the productivity caused by seasonal coastal upwelling off the Congo River (Jansen et al., 1984). The tropical catchment areas of the Niger and Congo Rivers would also be expected to provide significant amounts of terrigenous organic matter to the margins. At Ocean Drilling Program (ODP) Leg 175, Sites 1075–1077, located in deep water on the northern Congo River fan (Figure 11b), the upper 200 m of the sediment column contained high levels of microbial gas and a mixture of marine and terrestrial plant organic matter with total organic carbon contents averaging 2.5 wt. % (Wefer et al., 1998). However, there was no direct observational or geochemical evidence for gas hydrate layers, possibly because the sites were chosen to avoid obvious BSRs and gas chimneys (Meyers et al., 1998).

In addition to biogenic gas, the potential for thermogenic gas to be supplied to the hydrate stability zone along the west African margin is high off the Niger and Congo River Deltas. This is because the large depocenters (Figures 11a and b) fed by these river systems have experienced significant sediment loading, structural deforma-

(a)

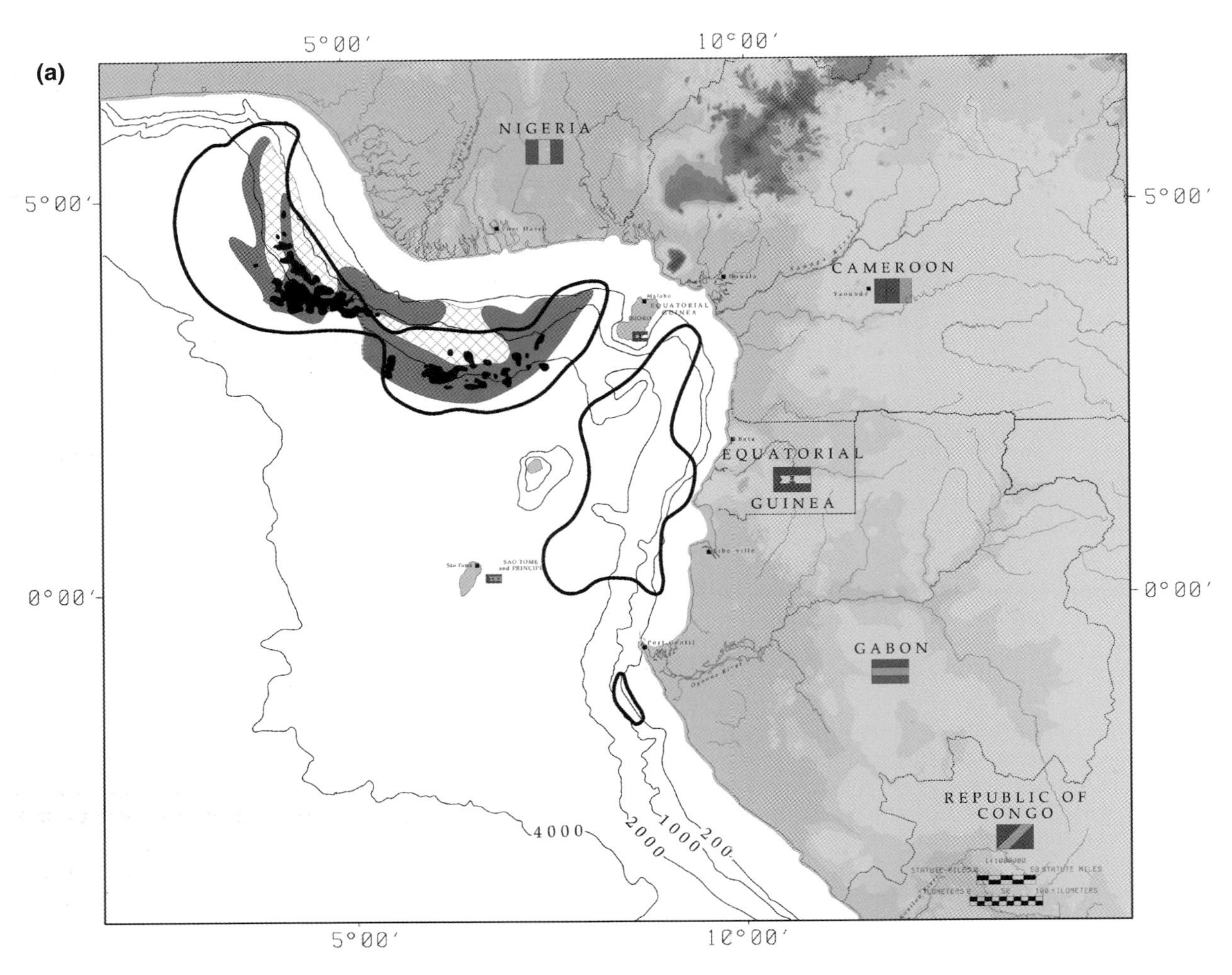

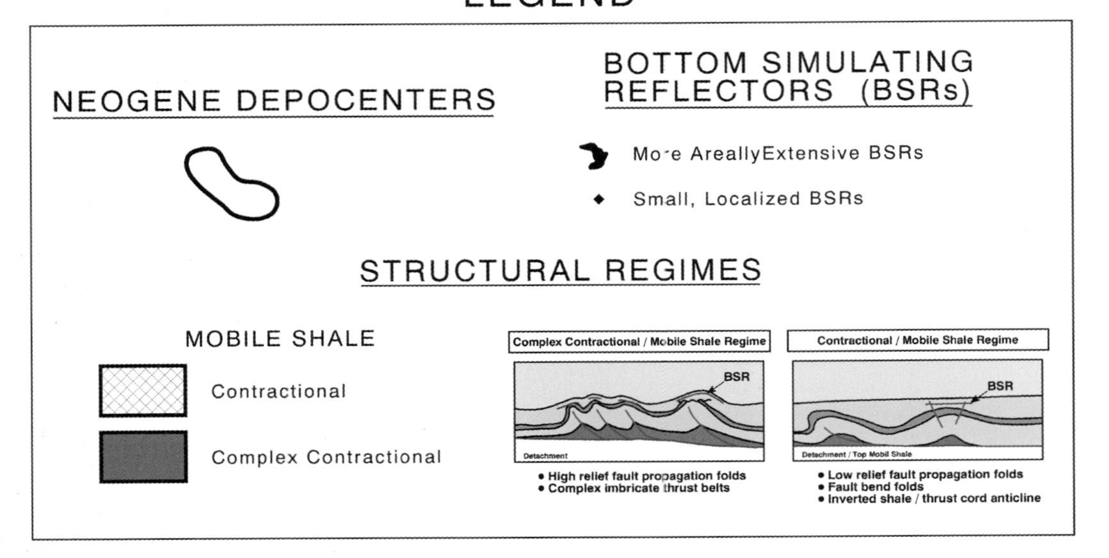

Figure 11—(a) Relationship of gas hydrate distribution to structural regimes and Neogene depocenters off west Africa (Nigeria to Congo margin). ***(Facing page)*** **(b) Relationship of gas hydrate distribution to structural regimes and Neogene depocenters off west Africa (Congo to Angola margin).**

(b)

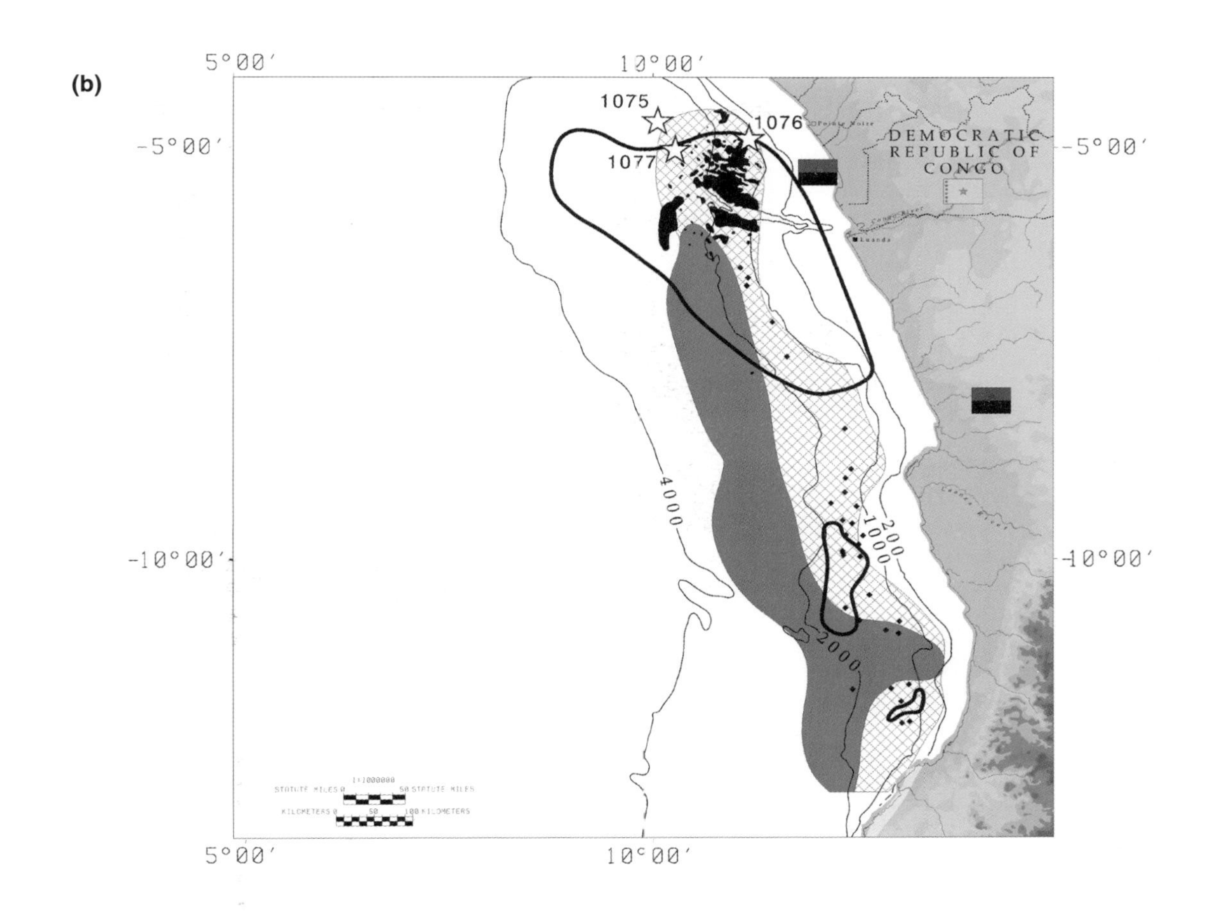

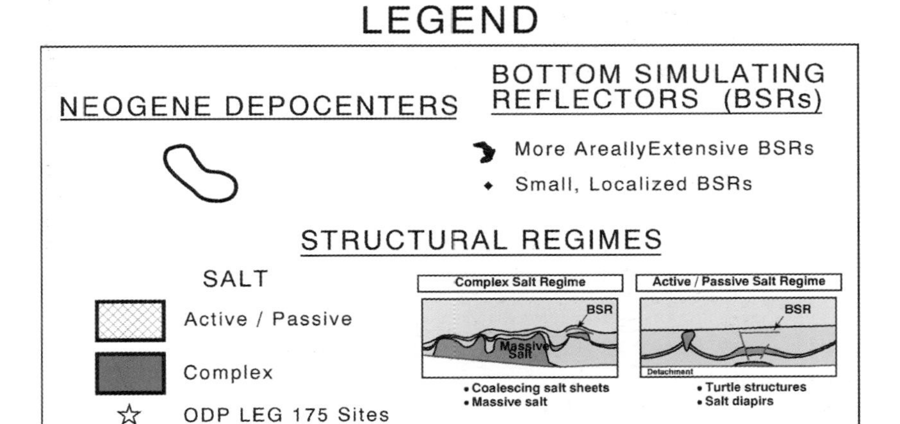

Figure 11(b)

tion, and hydrocarbon systems activity during the Neogene (Haack et al., 1998; Amaral et al., 1998). Oil and thermogenic gas seeps are common in this region of intense petroleum exploration activity (Brooks et al., 1997).

Favorable structural and stratigraphic features off west Africa appear to facilitate gas hydrate formation by improving gas migration efficiency and providing porosity trends in which gas hydrates may form. Although the areally extensive BSRs are generally associated with Neogene depocenters off the Niger and Congo Delta fronts, they are further confined to structural regimes in these regions with accentuated vertical deformation. These include the contractional mobile shale belt off the Niger Delta (Figure 11a) and the belt of active salt structuring along the Congo Delta front (Figure 11b). Both areas offer an abundance of structural culminations with steeply dipping strata that focus gas migration through lithologic or associated fault pathways updip to the hydrate stability zone. In addition to enhanced migration focus, these structural regimes also provide large drainage areas for the collection of biogenic and thermogenic gas. The biogenic gas may be formed through bacterial methanogenesis within and below the zone of hydrate stability while the thermogenic gas is generated at much deeper depths.

SUMMARY AND CONCLUSIONS

From the screening of 2-D seismic data off west Africa from western Nigeria to southern Angola and from shelf to lower slope water depths, we have observed numerous examples of BSRs. Many of these occur as isolated patches or groupings of smaller BSR occurrences, each up to a few tens of square kilometers in area and generally associated with shallow or more deeply buried structural culminations. Areally extensive occurrences, however, are found only off the Niger and Congo Delta fronts at water depths from 800 to 3000 m. In these settings, individual BSR occurrences reach hundreds of square kilometers in area and collectively may cover thousands of square kilometers. We have observed that the geographic distribution of the extensive BSRs appears to be most closely related to structural regimes with significant vertical deformation. On this basis, we conclude that to form, gas hydrates require focused migration of free gas or pore fluids containing dissolved gas to the base of the gas hydrate stability zone. The BSRs and associated gas hydrates may form from biogenic methane indigenous to the hydrate stability zone or from gases that have migrated over longer distances, such as more deeply buried biogenic methane and thermogenic gas formed at greater depths.

The gas hydrate occurrences that we have observed have implications for hydrocarbon exploration and production. A significant zone of high-amplitude reflections and positive AVO suggest that accumulations of free gas may be trapped below gas hydrates offshore west Africa. If the low gas saturations that have been estimated to occur in the free gas zone beneath the Blake Ridge BSR by Holbrook et al. (1996) (on the order of 1% of porosity) apply to west Africa, this gas does not represent a conventional exploration target. However, the free gas zones beneath BSRs may represent a shallow drilling hazard (Miller et al., 1991). It is possible that oil could also be trapped below hydrate layers, although the shallow burial depths and relatively low temperatures may lead to significant biodegradation of the oil. Other problems associated with hydrates include sediment instability and seabed erosion, corrosion and dissolution of materials, and promotion of biological communities (Borowski and Paull, 1997).

BSR distribution and character within the major Neogene depocenters off west Africa may also aid in defining elements of the hydrocarbon system such as thermal regime and migration. For example, subsurface temperatures can be estimated from a gas hydrate pressure–temperature phase diagram (e.g., Dickens and Quinby-Hunt, 1994) using an estimate of pressure at a BSR based on its depth below the sea floor. If sea floor temperature and sedimentary thermal conductivity data are available, thermal gradients and heat flow can be calculated. This approach has been found to yield results consistent with those derived using conventional means in the Nankai Trough off Japan, the Middle America Trench off Central America, the Peru margin, the Gulf of Mexico, and the Blake Outer Ridge (Yamano et al., 1982; Kvenvolden, 1993). Similarly, we have found consistency between conventional well-based estimates of heat flow and BSR-based estimates off west Africa.

Our study has shown that BSRs are clearly observed on seismic data offshore west Africa regardless of seismic data type (high-resolution single channel and 2-D and 3-D multichannel), processing, or vintage. Future ultra-deep-water seismic data sets, infill seismic data, and perhaps direct sampling through piston cores and drilling will further constrain gas hydrate distribution and origin.

REFERENCES CITED

Amaral, J., J. J. Biteau, L. DeCosta, and P. Zarolinska, 1998, Angola—the Lower Congo Basin Tertiary petroleum systems, hydrocarbon distribution in relation with the structural and sedimentary evolution (abs.), *in* M. R. Mello and P. O. Yilmaz, eds., Extended Abstracts Volume, AAPG International Conference, Rio de Janeiro, p. 924–925.

Andreassen, K, P. E. Hart, and A. Grantz, 1995, Seismic studies of a bottom simulating reflection related to gas hydrate beneath the continental margin of the Beaufort Sea: Journal of Geophysical Research, v. 100, p. 12,659–12,673.

Andreassen, K., P. E. Hart, and M. MacKay, 1997, Amplitude versus offset modeling of the bottom simulating reflection associated with submarine gas hydrates: Marine Geology, v. 17, p. 25–40.

Borowski, W. S., and C. K. Paull, 1997, The gas hydrate detection problem: recognition of shallow-subbottom gas hazards in deep-water areas: Offshore Technology Conference, OTC 8297, 6 p.

Brooks, J. M., M. C. Kennicutt II, R. R. Fay, T. L. McDonald, and R. Sassen, 1984, Thermogenic gas hydrates in the Gulf of Mexico: Science, v. 223, p. 696–698.

Brooks, J. M., H. B. Cox, W. R. Bryant, M. C. Kennicutt II, R. G. Mann, and T. J. McDonald, 1986, Association of gas hydrates and oil seepage in the Gulf of Mexico: Advances in Organic Geochemistry, v. 10, p. 221–234.

Brooks, J. M., A. L. Anderson, R. Sassen, I. R. MacDonald, M. C. Kennicutt II, and L. Guinasso, Jr., 1994, Hydrate occurrences in shallow subsurface cores by continental slope sediments, *in* E. D. Sloan, J. Happel, and M. A. Hnatow, eds., International Conference on Natural Gas Hydrates: Annals New York Academy of Sciences, v. 715, p. 381–391.

Claypool, G. E., and I. R. Kaplan, 1974, The origin and distribution of methane in marine sediments, *in* I. R. Kaplan, ed., Natural Gases in Marine Sediments: New York, Plenum, p. 99–139.

Dickens, G. R., and M. S. Quinby-Hunt, 1994, Methane hydrate stability in seawater: Geophysical Research Letters, v. 21, p. 2115–2118.

Dillon, W. P., M. W. Lee, K. Fehlhaber, and D. R. Hutchinson, 1991, Estimation of amounts of gas hydrate in marine sediments using amplitude reduction of seismic reflections (abs.): Journal of the Acoustic Society of America, v. 89, p. 1853.

Emery, K. O., 1974, Pagoda structures in marine sediment, *in* I. R. Kaplan, ed., Natural Gases in Marine Sediments: New York, Plenum, p. 309–317.

Fontana, R. L., 1989, Evidencias Geofisicas da presenca de hidratos de gas na Bacia de Pelotas-Brazil: Primeiro Congresso Brasileiro de Geofisica, Rio de Janeiro, v. 1, p. 234–248.

Fontana, R. L., and A. Mussumeci, 1994, Hydrates offshore Brazil, *in* E. D. Sloan, J. Happel, and M. A. Hnatow, eds., International Conference on Natural Gas Hydrates: Annals New York Academy of Sciences, v. 715, p. 106–113.

Haack, R. C., P. Sundararaman, J. O. Diedjomahor, N. J. Gant, and J. Dahl, 1998, Niger Delta petroleum systems (abs.), *in* M. R. Mello and P. O. Yilmaz, eds., Extended Abstracts Volume, AAPG International Conference, Rio de Janeiro, p. 936–937.

Hedberg, H. D., 1980, Methane generation and petroleum migration, *in* W. H. Roberts III and R. J. Cordell, eds., Problems of Petroleum Migration: AAPG Studies in Geology, v. 10, p. 179–206.

Holbrook, W. S., H. Hoskins, W. T. Wood, R. A. Stephan, and D. Lizarralde, 1996, Methane hydrate and free gas on the Blake Ridge from vertical seismic profiling: Science, v. 273, p. 1840–1843.

Hovland, M., J. W. Gallagher, M. B. Clennell, and K. Lekvam, 1997, Gas hydrate and free gas volumes in marine sediments: example from the Niger Delta front: Marine and Petroleum Geology, v. 14(3), p. 245–255.

Hyndman, R. D., and E. E. Davis, 1992, A mechanism for the formation of methane hydrate and bottom-simulating reflectors by vertical fluid expulsion: Journal of Geophysical Research, v. 97, p. 7025–7041.

Jansen, J. H. F., T. G. E. van Weering, R. Gieles, and J. van Iperen, 1984, Middle and late Quaternary oceanography and climatology of the Zaire-Congo fan and the adjacent eastern Angola Basin: Netherlands Journal of Sea Research, v. 17, p. 201–241.

Kvenvolden, K. A., 1993, Gas hydrates—geological perspective and global change: Reviews of Geophysics, v. 31, p. 173–187.

Kvenvolden, K. A., and M. A. McMenamin, 1980, Hydrates of natural gas—a review of their geologic occurrence: USGS Circular 825, 11 p.

Kvenvolden, K. A., and L. A. Barnard, 1983, Hydrates of natural gas in continental margins, *in* J. S. Watkins and C. L. Drake, eds., Studies in Continental Margin Geology: AAPG Memoir 34, p. 631–640.

Kvenvolden, K. A., and G. E. Claypool, 1988, Gas hydrates in oceanic sediment: USGS Open-File Report 88-216, 50 p.

Kvenvolden, K. A., G. D. Ginsburg, and V. A. Soloviev, 1993, Worldwide distribution of subaquatic gas hydrates: Geo-Marine Letters, v. 13, p. 32–40.

Lee, M. W., D. R. Hutchinson, W. F. Agena, W. P. Dillon, J. J. Miller, and B. A. Swift, 1994, Seismic character of gas hydrates on the southeastern U.S. continental margin: Marine Geophysical Research, v. 16, p. 163–184.

MacKay, M. E., R. D. Jarrad, G. K. Westbrook, R. D. Hyndman, and the Shipboard Scientific Party of ODP Leg 146, 1994, Origin of bottom-simulating reflectors: geophysical evidence from the Cascadia accretionary prism: Geology, v. 22, p. 459–462.

Manley, P. L., and R. D. Flood, 1988, Cyclic sediment deposition within Amazon deep-sea fan: AAPG Bulletin, v. 72, p. 912–925.

Meyers, P.A., and the Shipboard Scientific Party, 1998, Microbial gases in sediments from the southwest African margin, *in* G. Wefer, W. H. Berger, and C. Richter, et al., Proceedings of the Ocean Drilling Program, Initial Reports, v. 175, p. 555–560.

Miller, J. J., M. W. Lee, and R. von Huene, 1991, An analysis of a seismic reflection from the base of the gas hydrate zone, offshore Peru: AAPG Bulletin, v. 75, p. 910–924.

Minshull, T. A., S. C. Singh, and G. K. Westbrook, 1994, Seismic velocity structure at a gas hydrate reflector, offshore western Colombia, from full waveform inversion: Journal of Geophysical Research, v. 99, p. 4715–4734.

Moore, B., and B. Bolin, 1986, The oceans, carbon dioxide, and global climate change: Oceanus, v, 29, p. 9–15.

Rempel, A. W., and B. A. Buffett, 1997, Formation and accumulation of gas hydrate in porous media: Journal of Geophysical Research, v. 102, p. 10,151–10164.

Rowe, M. M., and J. F. Gettrust, 1993, Fine structure of methane hydrate-bearing sediments on the Blake Outer Ridge as determined from deep-tow multichannel seismic data: Journal of Geophysical Research, v. 98, p. 463–473.

Shipley, T. H., M. H. Houston, R. T. Buffler, F. J. Shaub, K. J. McMillen, J. W. Ladd, and J. L. Worzel, 1979, Seismic reflection evidence for the widespread occurrence of possible gas-hydrate horizons on continental slopes and rises: AAPG Bulletin, v. 63, p. 2204–2213.

Singh, S. C., and T. A. Minshull, 1994, Velocity structure of a gas hydrate reflector at Ocean Drilling Program Site 889 from a global seismic waveform inversion: Journal of Geophysical Research, v. 99, p. 24,221–24,223.

Summerhayes, C. P., B. D. Bornhold, and R. W. Embly, 1979, Surficial slides and slumps on the continental slopes and rises of Southwest Africa: a reconnaissance study: Marine Geology, v. 31, p. 265–277.

Wefer, G., W. H. Berger, C. Richter, and the Shipboard Scientific Party, 1998, Introduction: background, scientific objectives, and principal results for Leg 175 (Benguela Current and Angola-Benguela upwelling systems), *in* G. Wefer, W. H. Berger, and C. Richter, et al., Proceedings of the Ocean Drilling Program, Initial Reports, v. 175, p. 7–23.

Xu, W., and C. Ruppel, 1999, Predicting the occurrence, distribution, and evolution of methane gas in porous marine sediments: Journal of Geophysical Research, v. 104, p. 5081–5095.

Yamano, M., S. Uyeda, A. Y. Akoi, and T. H. Shipley, 1982, Estimates of heat flow derived from gas hydrates: Geology, v. 10, p. 339–343.

Yuan, T., R. D. Hyndman, G. D. Spence, and B. Desmons, 1996, Seismic velocity increase and deep sea gas hydrate concentration above a bottom-simulating reflector on the northern Cascadia continental slope: Journal of Geophysical Research, v. 101, p. 13,655–13,671.

Prinzhofer, A., M. R. Mello, L. C. da Silva Freitas, and T. Takaki, 2000, New geochemical characterization of natural gas and its use in oil and gas evaluation, *in* M. R. Mello and B. J. Katz, eds., Petroleum systems of South Atlantic margins: AAPG Memoir 73, p. 107–119.

Chapter 9

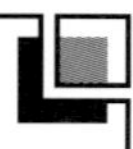

New Geochemical Characterization of Natural Gas and Its Use in Oil and Gas Evaluation

A. Prinzhofer

Division of Geology–Geochemistry
Institut Français du Pétrole
Rueil Malmaison, France

M. R. Mello

L. C. da Silva Freitas

T. Takaki

Petrobrás R & D Center/Cenpes
Rio de Janeiro, Brazil

Abstract

Gas signatures are used to characterize the processes occurring from generation of hydrocarbons in source rocks to their accumulation in reservoirs. Two main geologic parameters to consider are the maturity of the hydrocarbons (primary cracking of kerogen or secondary cracking of oils) and their degree of segregation during transport, linked to a migration distance. Several new diagrams and geochemical parameters are now suggested to characterize the directions and distances of migration of gases more quantitatively. Two examples from Brazil include the Ceará Basin and the Recôncavo Basin. In the Ceará Basin, a gas geochemical study shows that normal faulting associated with rifting of the passive margin has locally altered the global direction of hydrocarbon migration. Such estimations of global and local directions of hydrocarbon migration allow a better reconstruction of the filling of geologic traps with hydrocarbons. The Recôncavo Basin case study shows our enhanced ability to distinguish bacterial activity, mainly in the central part of the basin, from segregative transport, more visible in the southern part. We demonstrate the efficiency of a major fault (the Mata–Catu fault) as a conduit along which hydrocarbons migrate to distances of 40 km away from the oil kitchen. These geologic examples using gas geochemical and isotopic signatures provide a platform for further applications of these tools in modeling of hydrocarbon migration and accumulation for both basin exploration and reservoir estimation.

INTRODUCTION

Natural gas has garnered less theoretical and practical interest than liquid hydrocarbons. This is due to the lower economic value of gas until quite recently, and because of the relatively modest number of measurable parameters easily obtainable from a gas sample. Indeed, for a long time in the industry, only the wetness of the gas (C_2–C_5/C_1–C_5) and the carbon isotope composition of methane were typically used to interpret a gas series (Sackett et al., 1966; Stahl, 1977; Chung and Sackett, 1979). In some cases, hydrogen isotopic ratios of methane (Schoell, 1980, 1983; Faber et al., 1992) or carbon isotopic ratios of ethane and butane (Stahl and Carey, 1975; Faber, 1987; James, 1983, 1990; Clayton, 1991) were included. Thanks to the rapid development of routine analyses of carbon isotopes through continuous flow with a gas chromatographic–combustion–isotope ratio mass spectrometer (GC-C-IRMS), processes affecting natural gas from its generation (Berner et al., 1995; Lorant et al., 1998; Cramer et al., 1998) to its accumulation in a reservoir (Rooney et al., 1995; Galimov, 1988; Pernaton et al., 1996) are now better understood. Gas geochemistry is at a stage where it can now help reconstruct gas history and, in some cases, help determine the processes affecting associated oil, because some parameters are more sensitive and better constrained using gas than oil. The interpretation of both chemical and isotopic fractionation due to geochemical and geophysical processes adds a dimension to the geochemical database, leading to more adequate reconstruction of the gas history. This chapter summarizes several new ways of interpreting gas data and presents several examples of geologic applications in Brazilian sedimentary basins.

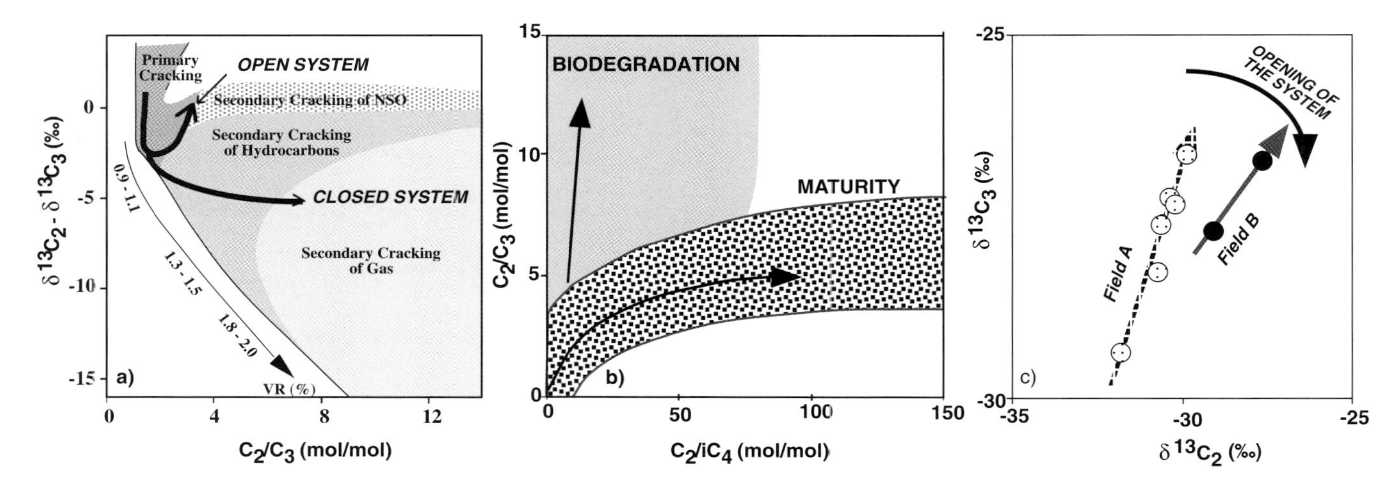

Figure 1—Three examples of geochemical diagrams characterizing the genesis of hydrocarbon gases. (a) C_2/C_3 diagram showing $\delta^{13}C_2/\delta^{13}C_3$ versus C_2/C_3 distinguishes gases generated from the primary cracking of kerogen, the secondary cracking of oil, and the tertiary cracking of the wet portion of the gases (Lorant et al., 1998). (b) C_2/C_3 versus $C_2/i\text{-}C_4$ diagram showing the possible biodegration of hydrocarbon gases (Lorant, personal communication, 1999). (c) $\delta^{13}C_3$ versus $\delta^{13}C_2$ diagram on which the relative position of several gas measurements indicates the relative efficiency of gas accumulation.

PHYSICO-CHEMICAL PROCESSES AFFECTING HYDROCARBON AND GAS GEOCHEMICAL SIGNATURES

Methane is the major hydrocarbon component of natural gas, but it is also the most difficult to interpret. Its genesis is connected to thermal maturation of organic matter and/or bacterial methanogenesis (this is also true for ethane but to a lesser degree because the proportion of bacterial ethane is always small). Methane can also be affected by isotopic fractionation due to various processes of migration, such as solubilization and diffusion in water, effusion as a gas phase in a porous medium, and adsorption onto the solid organic or mineral network. Because of this complexity, we generally begin a gas study using the chemical proportions and isotopic ratios of the other hydrocarbon molecules, such as ethane, propane, and the two isomers of butane. Pentane is generally too fractionated during the sampling between gas and liquid phases to provide measurements of any geochemical significance. Combining gas chemistry and gas isotopes allows the characterization of such parameters as indices of maturity, type of source (primary versus secondary cracking), efficiency of hydrocarbon accumulation, and traces of bacterial degradation. Figure 1 shows examples of geochemical diagrams used to characterize the genesis of gases.

For several decades, gas signatures were thought by some to be independent of any fractionation due to migration (Schoell, 1980, 1983; Fuex, 1980), whereas some work seemed to demonstrate a more complex behavior (Hoering and Moore, 1958; Colombo et al., 1965, 1966, 1970; Galimov, 1975; Bondar, 1987). Consistent new experimental work (Pernaton et al., 1996; Krooss et al., 1998) associated with evidence from natural case studies (Prinzhofer and Pernaton, 1997) presents evidence of significant isotopic shifts in methane during migration through porous media. To distinguish isotopically light methane coming from bacterial activity and methane segregated during migration, Prinzhofer and Pernaton (1997) suggested a diagram using C_2/C_1 versus $\delta^{13}C$ of C_1 (Figure 2).

To integrate these new constraints, more synthetic representations of gas data have been suggested (Prinzhofer et al., 2000). The GASTAR diagram (Figure 3) gives an overview of relative maturities of hydrocarbon gases, of segregation due to migration, and of efficiency of accumulation. Principal component analysis (PCA) (Figure 4), based on a large data set of gases (200 samples) from all over the world, has defined two main parameters, V_1 and V_2, which are important eigenvectors representing the maturity (V_1) and migration (V_2) of gases.

Using the various diagrams presented above, we now discuss two geologic applications from Brazilian basins in which we show that gas geochemistry may help the understanding of petroleum systems and reduce the risk during hydrocarbon exploration.

CEARÁ CASE STUDY

Four offshore fields have been sampled in detail in the Brazilian Ceará Basin on the continental margin of the Atlantic Ocean. The main points examined were (1) possible secondary alteration, (2) the maturity range of the gases, and (3) migration pathways of the hydrocarbons.

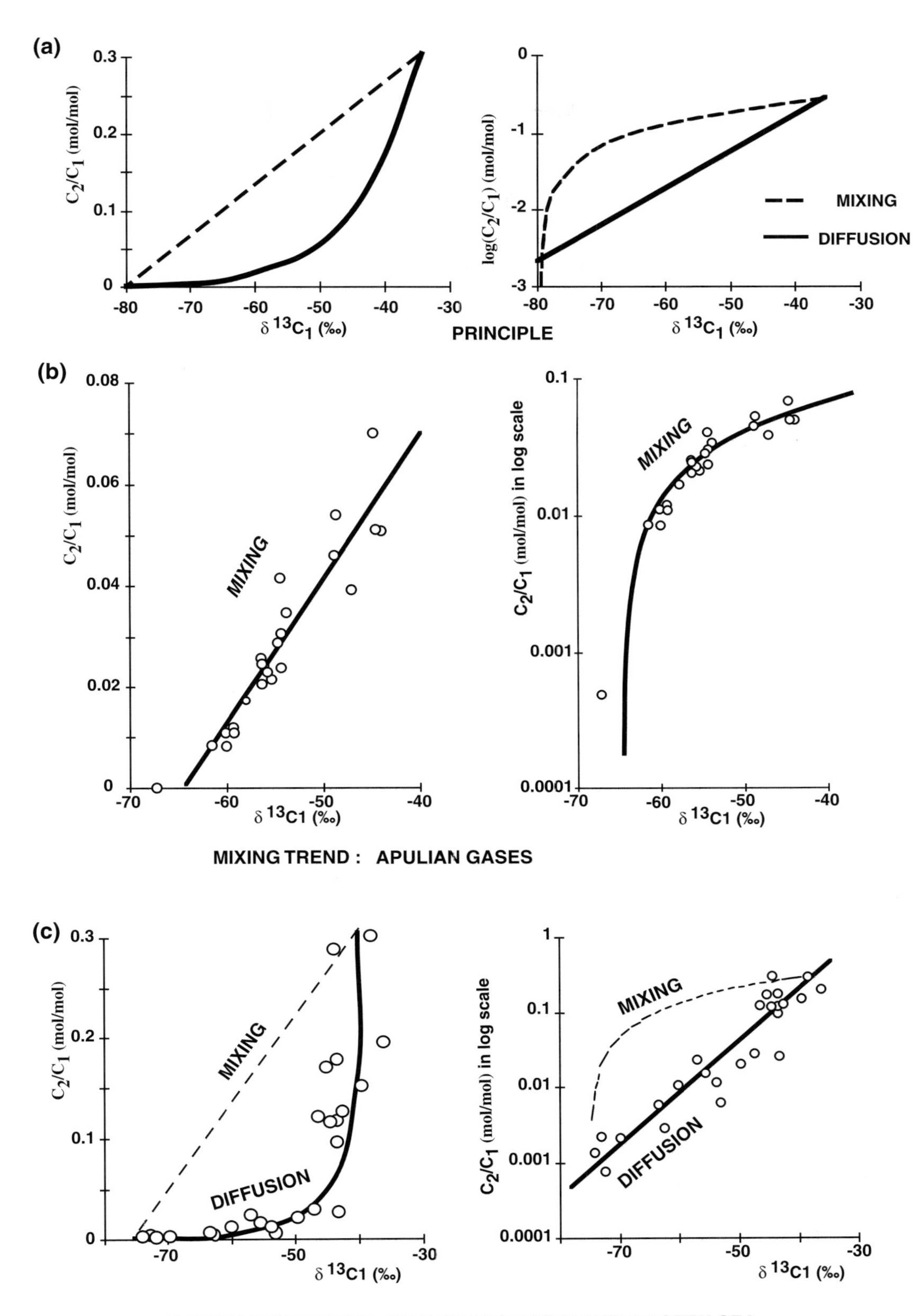

Figure 2—(a) C_1-C_2 diagrams showing the basic principle of their use. (b) C_1-C_2 diagrams showing a mixing trend with bacterial contamination in Apulian gases (from Ricchiuto and Schoell, 1988). (c) C_1-C_2 diagrams showing segregative migration in headspace gases from the North Sea. Modified from Prinzhofer and Pernaton (1997).

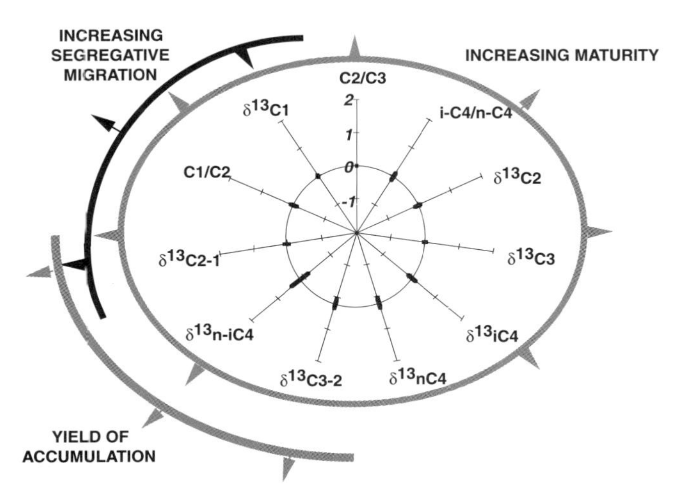

Figure 3—Principle of the GASTAR diagram. The average accuracy obtained is indicated on each axis. Three main parameters (maturity, segregative migration, and yield of accumulation) and their direction of variation, affecting the values of the 11 normalized parameters, are also represented. From Prinzhofer et al. (2000).

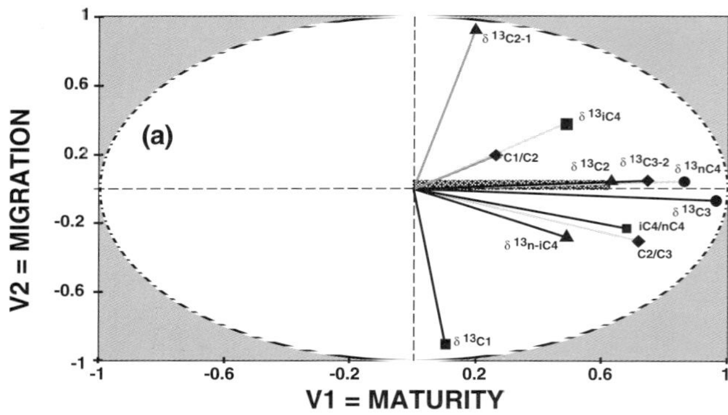

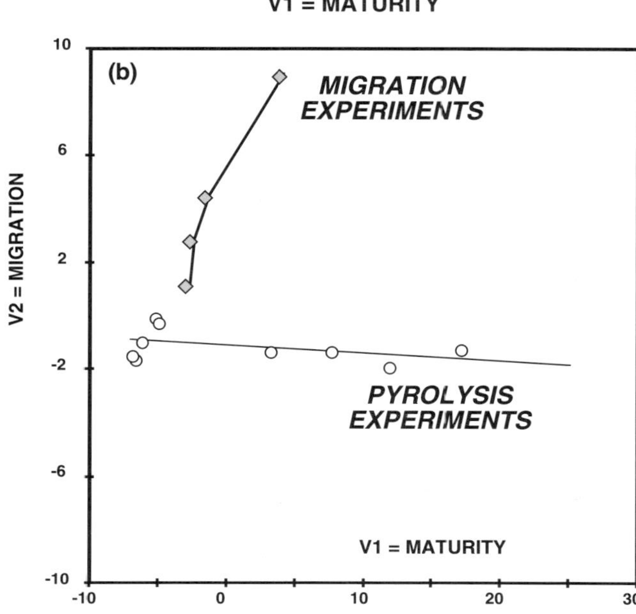

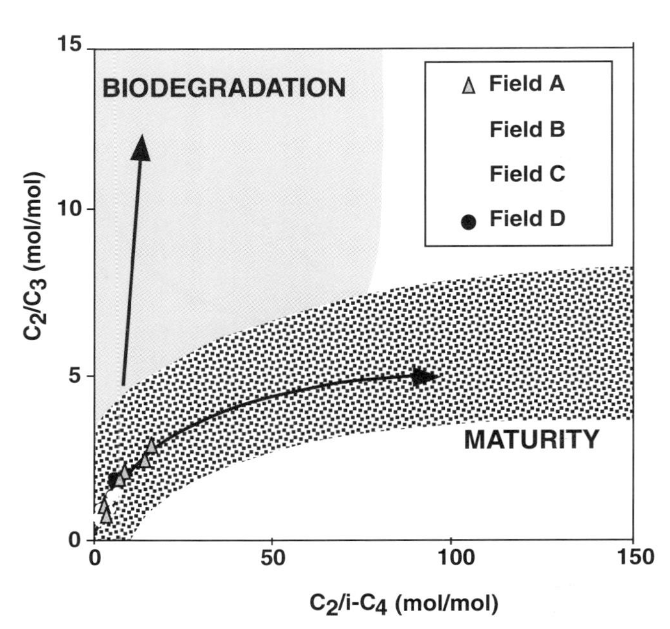

Figure 5—C_2/C_3 versus C_2/i-C_4 diagram showing the chemical signatures of gases from four selected fields in the Ceará Basin. No evidence of biodegradation can be seen from this study.

Bacterial Alteration

Figure 5 shows a plot of C_2/C_3 versus C_2/i-C_4 which enables gases fractionated only through thermal cracking to be distinguished from gases having suffered bacterial alteration. All the analyzed gases in the Ceará Basin are free of any trace of bacterial alteration.

Maturity

All the analyzed gases fit in the primary cracking area, with no evidence of any gas coming from thermal alteration of heavier hydrocarbons (Figures 6a and c). A close look at the data shows that the gases from three fields (A,

←Figure 4—Principle of the principal component analysis (PCA) diagram. (a) Projection of the 11 geochemical parameters on the two first eigenvectors in the V_1/V_2 plane, corresponding to the most discriminant dimensions of the database. The first eigenvector V_1 is interpreted as a good representation of maturity, the second one V_2 may be an indicator of segregative migration. (b) Experimental data represented in a V_1/V_2 diagram. Open circles represent dry confined system pyrolyses of the type II Menil sur Vair kerogen (from Paris Basin), with various times and temperature ranges. Diamonds correspond to gases that have migrated in saturated shales (Pernaton et al., 1996), with constant relative proportions and isotopic ratios of propane and the two butane isomers.

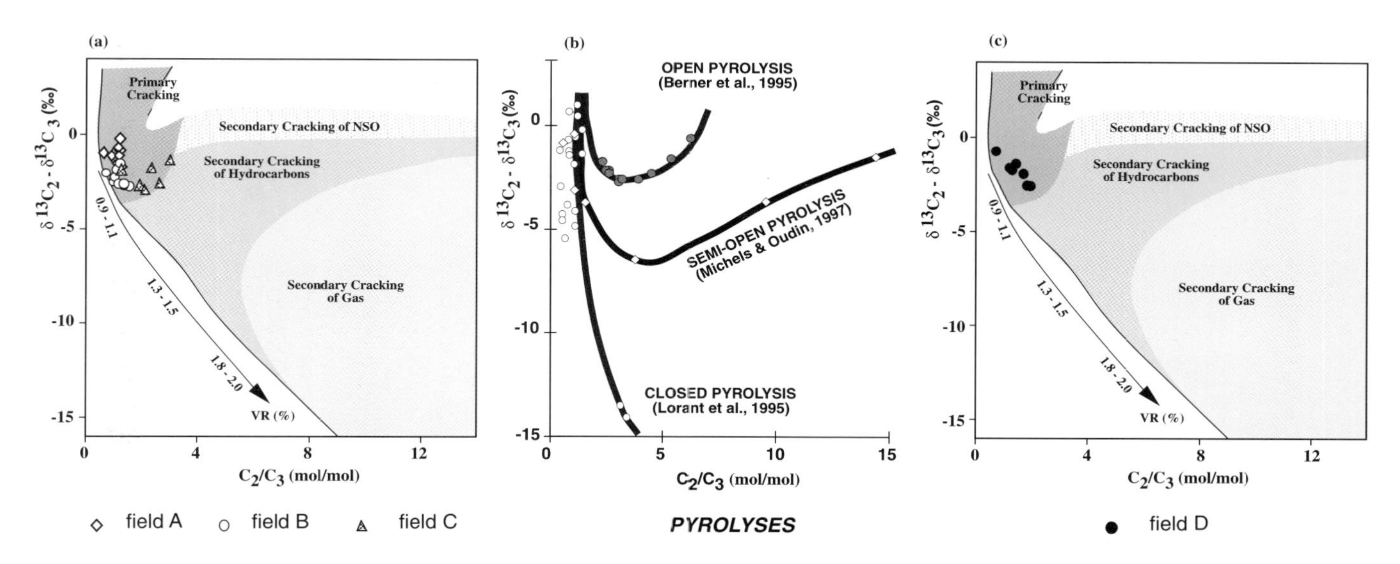

Figure 6—C_2/C_3 diagrams for gases from four selected fields in the Ceará Basin. (a) Fields A, B, and C show a trend of maturity corresponding to an open system. (b) Experimental evidence of maturity trends on the C_2/C_3 diagram for open pyrolysis (Berner et al., 1995), closed pyrolysis (Lorant et al., 1998), and semi-open pyrolysis (Michels and Oudin, 1997). (c) Field D shows a trend of maturity shifted from that of fields A, B, and C.

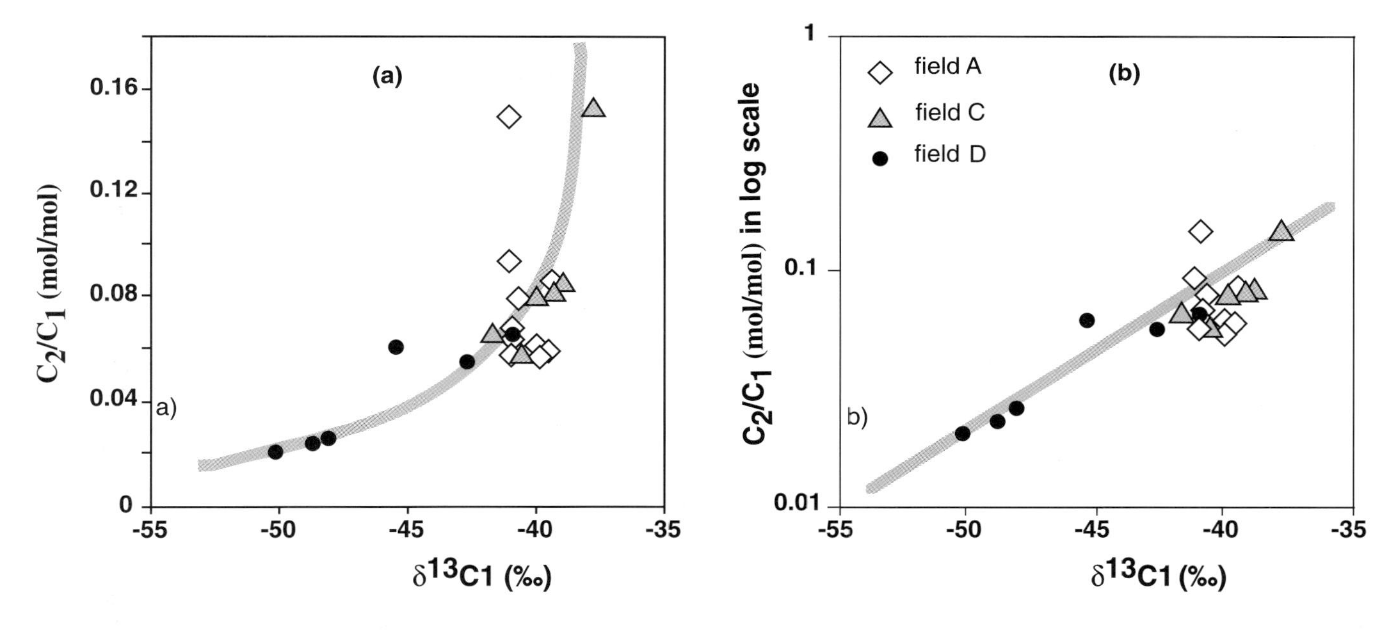

Figure 7—Gases from fields A, C, and D in the Ceará Basin plotted on a C_1-C_2 diagram both in (a) linear and (b) logarithmic scales, showing a trend of segregative migration.

B, and C) have a regular trend of maturity (with maturities ranked as A < B < C) (Figure 6a). According to experimental studies, this represents an "open system" pyrolysis trend (Figure 6b). The trend of the gases from the fourth field (D) is transverse to this trend, with a wide spread of maturity (Figure 6c).

Migration

Apart from field B, which has scattered data, other data plotted on the C_1-C_2 diagram show a clear trend of chemical and isotopic segregation during migration (Figure 7). This is shown by the shape of the trend, which

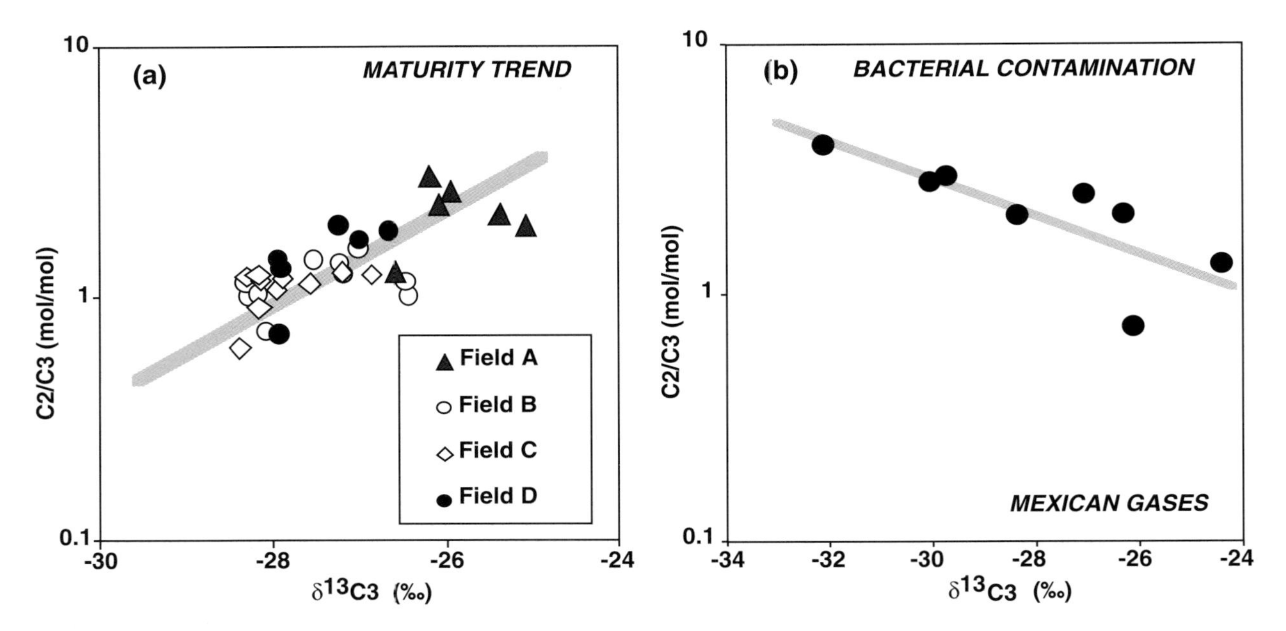

Figure 8—C_2/C_3 versus $\delta^{13}C_3$ diagrams showing a comparison of gases from (a) the Ceará Basin and (b) a basin in Mexico. Those from the Ceará Basin exhibit a trend of maturity, whereas the gases from Mexico are interpreted as the result of mixing of bacterial and thermogenic gases.

is curved on a linear scale (Figure 7a) and straight on a logarithmic scale (Figure 7b). The lighter isotopic ratios of methane cannot be explained by lower maturity because the gases would be wetter in that case, whereas they are dryer. Also, the same light values for the carbon isotopic ratios of methane cannot be explained by mixing with bacterial gas (contamination) because the trend in the C_2/C_1 versus $\delta^{13}C_1$ diagram would show a straight line on the linear scale and a curve on the logarithmic scale, whereas we observe the opposite. To prove that segregation is the process induced here, with no evidence of bacterial contamination, a C_2/C_3 versus $\delta^{13}C_3$ diagram shows a clear maturity trend (Figure 8a), while bacterial contamination gives an opposite correlation using the same diagram, as shown in Figure 8b by a documented example from Mexico. Field B also seems to be highly segregated, even though the data are scattered (Figure 9).

Values for both the $\delta^{13}C$ of methane and the V_2 parameter (representing migration) are shown in Figure 10 and the following table:

	$\delta^{13}C_1$ (‰)	V_2
Field A	−39.7	0.78
Field C	−40.4	0.82
Field B	−45.9	2.4
Field D	−46.3	3.0

These values demonstrate that the two distal offshore fields (A and C) are much less affected by segregation than the two more proximal offshore fields (B and D). This is consistent with a supply of hydrocarbons to those fields from an oil kitchen located farther offshore.

Geographic and Geologic Mapping at the Field Scale

Maps of the four studied fields have been drawn to characterize the history and direction of filling at the reservoir scale. Migration maps can be represented by isolines of $\delta^{13}C$ of methane (Figure 11) or by values of the V_2 parameter. Carbon isotopic ratios of methane have been shown to be related to migration in this basin, with little effect from maturity levels or bacterial contamination. The $\delta^{13}C_1$ maps in Figure 11 show consistent trends for fields B, C, and D, with the same direction of hydrocarbon supply from the northwest. For field A, in which gases are only slightly fractionated by migration, it seems that the direction of filling is more from the north.

When information about migration at the basin scale and the field scale are compared, it appears that the general trend of migration is from the northeast to the southwest (on the basis of the mean V_2 values in the four fields). However, it seems that at the field scale, local disturbances may be due to preferential pathways created by local geology, causing a minor change in the migration direction more toward the northwest–southeast. The one exception is field A, for which the local migrating direction seems to be north–south, consistent with the regional trend. Intensive normal faulting oriented northwest–southeast close to the shoreline may be the explanation for the direction of filling of hydrocarbon gases at the field scale.

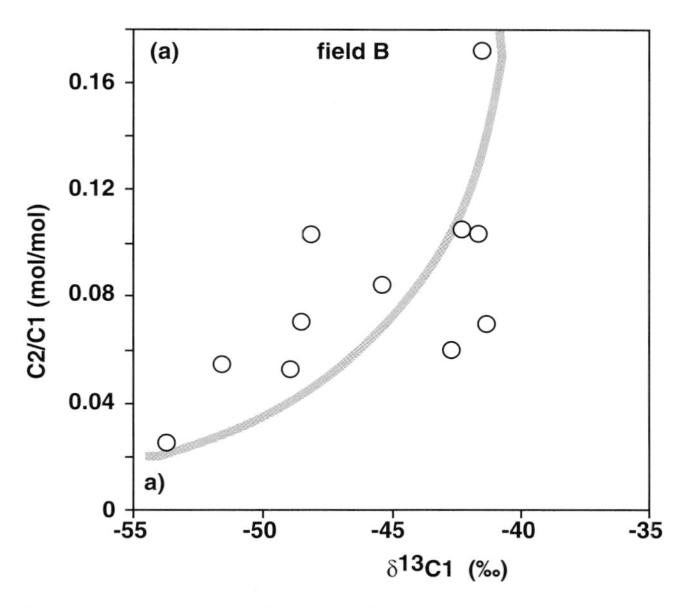

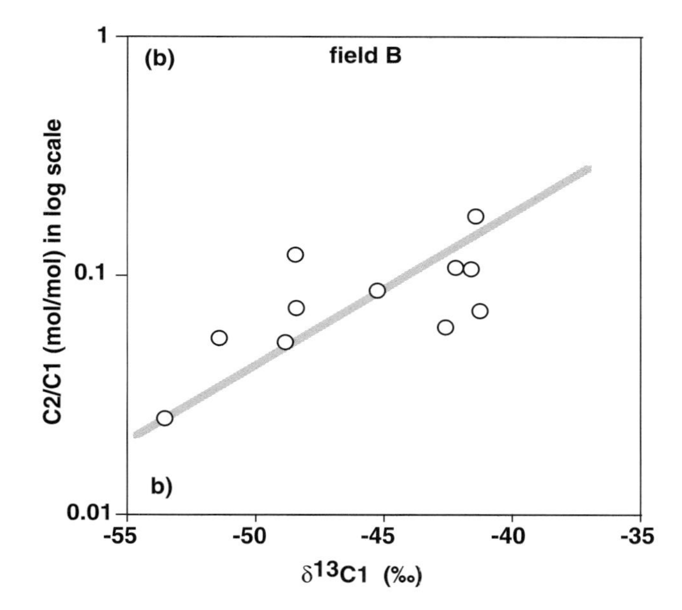

Figure 9—Gases from field B in the Ceará Basin plotted on a C_1-C_2 diagram both with (a) linear and (b) logarithmic scales, showing a larger scatter than those from fields A, C, and D (see Figure 7) but compatible with a trend of segregative migration.

RECÔNCAVO BASIN CASE STUDY

Gas Geochemistry at the Basin Scale

The Recôncavo Basin is an intracontinental north–south aborted rift basin (Figure 12) that formed immediately before the opening of the Atlantic Ocean (Figueiredo et al., 1994). Its source rocks are mainly lacustrine and have thermal maturities between the oil window and the beginning of the gas window. Chemical and carbon isotope analyses using C_1-C_2 diagrams have shown that, at the basin scale, major vertical gas segregation is apparent (Figure 13). Indeed, all the gas accumulations located between 1000 and 2300 m depth show a nonaltered maturity trend, whereas the deeper gas pools exhibit a residual character in terms of segregation. The shallower gas accumulations indicate a small fractionation due to segregative migration.

Comparison of Zones A and B

Looking specifically at some areas of the basin, it is possible to define areas where segregation can be distinguished from bacterial contamination. We decided to compare the oil and gas fields in selected zone A with zone B (Figure 12). Figure 14 presents a simplified block diagram of the hydrocarbon fields in zone A. Gas samples from four reservoirs were analyzed: two gas fields (fields 1 and 2), an oil field (field 3), and a condensate field (field 4). The $\delta^{13}C$ patterns of the hydrocarbons from methane to butane show a very consistent shift between the gas associated with the oil and the gas asso-

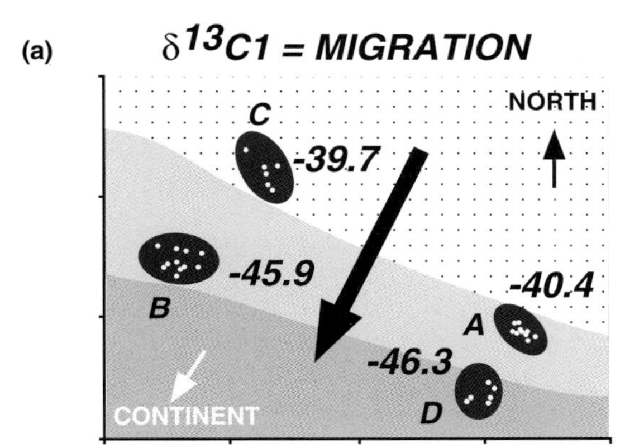

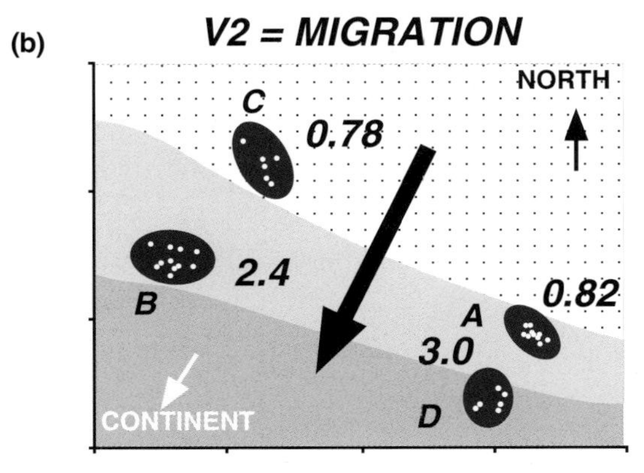

Figure 10—Basin-scale maps of two indicators of migration: (a) the carbon isotopic values of methane ($\delta^{13}C_1$) and (b) the V_2 parameter. Values given are average values for each field. A northeast–southwest direction of migration can be observed from the two maps.

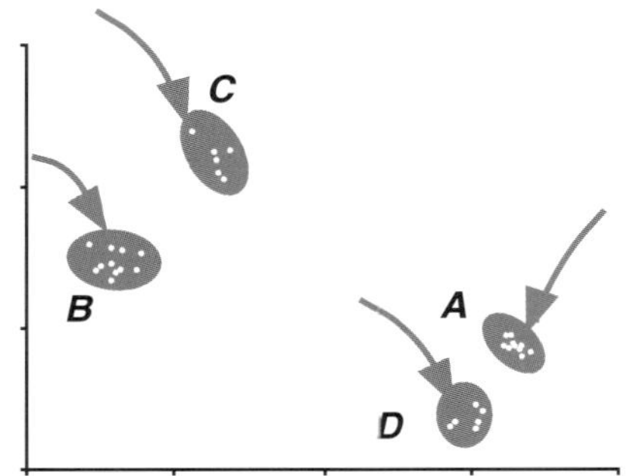

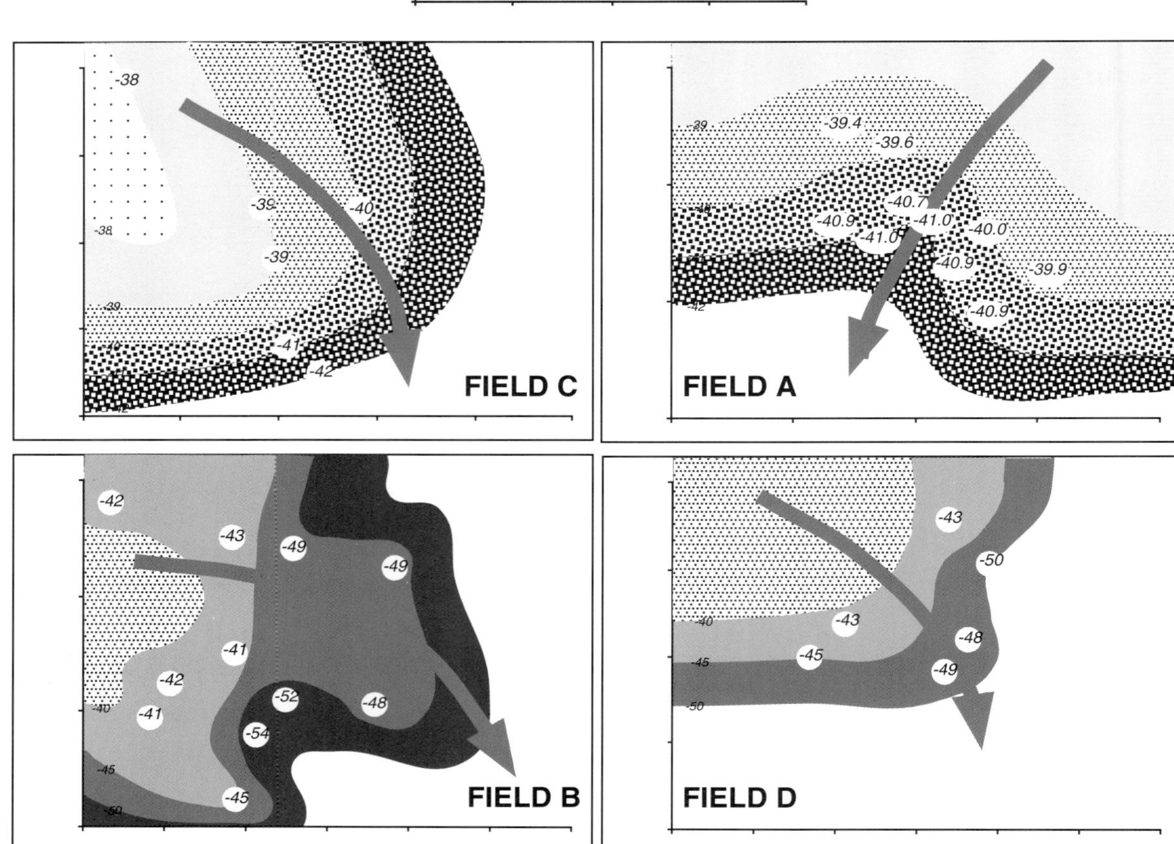

Figure 11—Reservoir-scale maps of one of the migration parameters, the carbon isotopic values of methane ($\delta^{13}C_1$). Fields B, C, and D are locally filled from the northwest, whereas field A is filled from the northeast.

ciated with the condensates. The $\delta^{13}C$ values of the oil-associated gas are all lighter, which is consistent with a genesis at lower maturity levels. However, the C_{2+} molecules and lighter carbon isotopic ratios for methane show that the two gas fields have gas signatures intermediate between the oils and the condensates. The C_1-C_2 diagram in Figure 15a shows that these four gases exhibit a concave curve, compatible with a trend of segregation. We interpret the origins of the two gas fields (fields 1 and 2) as resulting from segregative migration of hydrocarbon fluids, possibly a mixing of the two liquid fluids (oil and condensate), in which only the gas fraction would have migrated enough to fill the structures in these two fields.

Figure 16 shows another example of heterogeneous gas and oil fields from zone B in the Recôncavo Basin. We have superimposed layers of oil and gas, and the $\delta^{13}C$ patterns from shallower layers are clearly scattered compared to those from the deeper layers. This can be interpreted as evidence of bacterial alteration. When plotting the gas data from zone B on the C_1-C_2 diagram (Figure 15b), the straight-line data plot appears to indicate that all the gases may correspond to a mixing trend between a thermogenic end-member and a bacterial end-member. In this case, we do not see any evidence of migration, but contamination affecting mainly the shallower reservoirs.

Mata–Catu Fault

A final study made in the Recôncavo Basin concerns the vicinity of the Mata–Catu fault, a major northwest–southeast structural trend cross cutting the entire basin

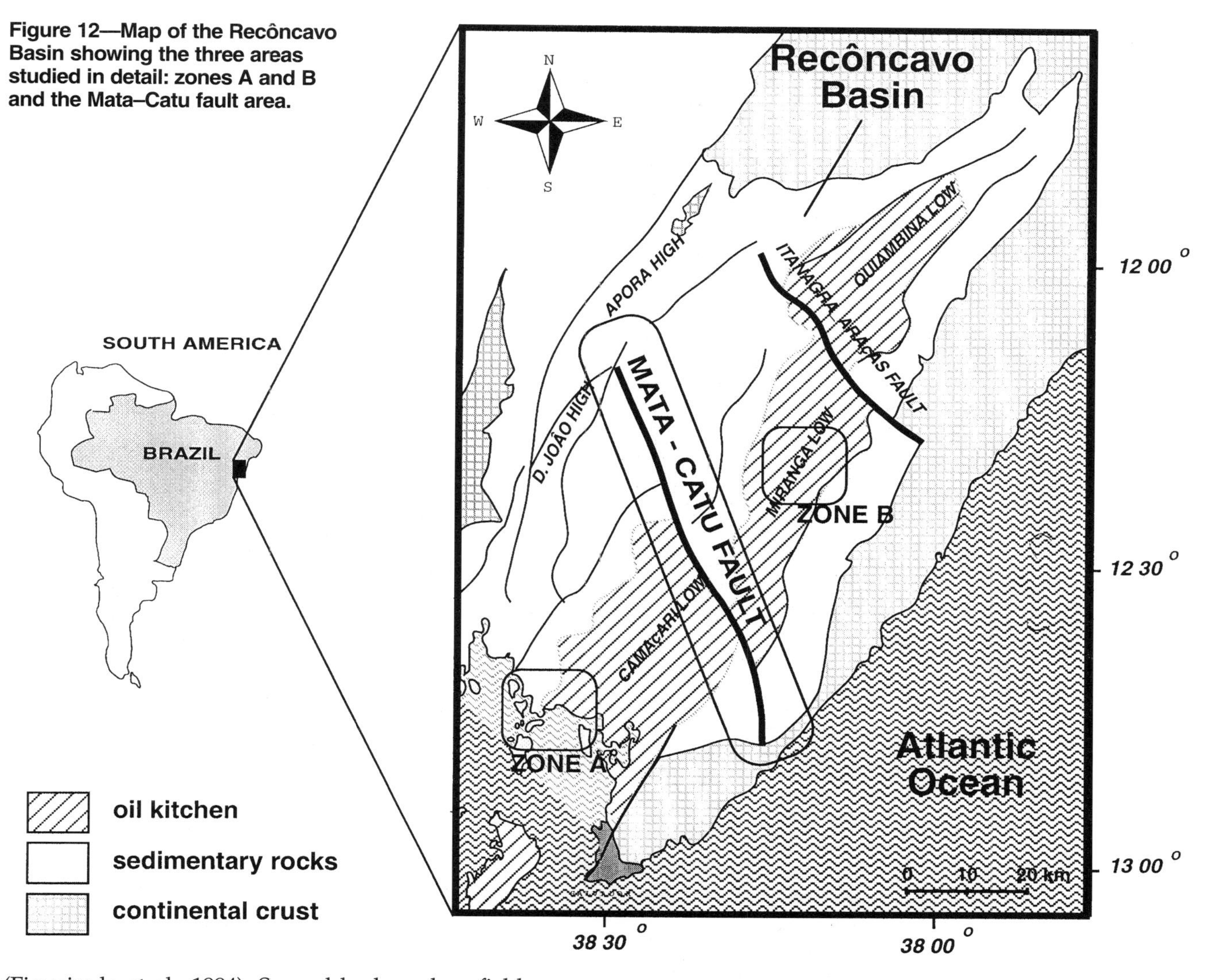

Figure 12—Map of the Recôncavo Basin showing the three areas studied in detail: zones A and B and the Mata–Catu fault area.

(Figueiredo et al., 1994). Several hydrocarbon fields are located along the fault, and others are found along a perpendicular line centered in the deepest part of the basin where the source rocks are the deepest, inside the oil or gas kitchens. Figure 17 presents the values of the synthetic parameters V_1 (maturity) and V_2 (migration) plotted against the distance along the fault from the center of the kitchen. It appears that the maturity of the gas does not show any systematic variation along the fault, but rather presents two distinct values of maturity that are independent of the distance from the kitchen. In contrast, the migration parameter clearly indicates a trend of segregation due to gas transport from the kitchen to the most remote reservoirs. This implies the absence of a second kitchen in the northwestern part of the basin.

If we compare the gases studied along the fault with those collected along a transect perpendicular to the fault using a V_1/V_2 diagram (Figure 18), we see that the data from gas at peak maturity in the source rock shows no evidence of migration and segregation, in contrast to the gas along the fault. This is the first demonstration using gas geochemistry that a fault may act as a segregative

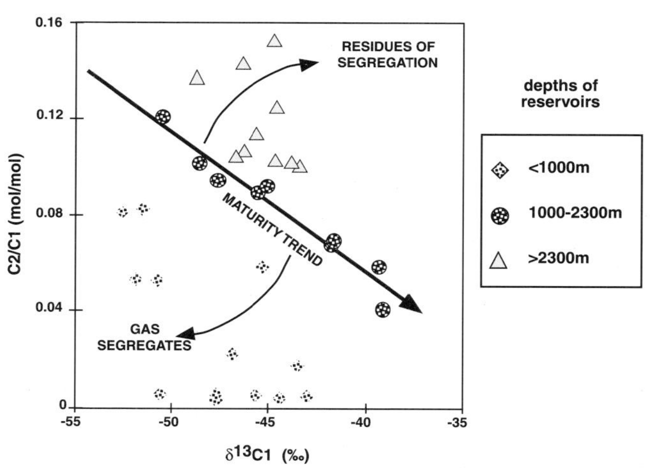

Figure 13— Recôncavo Basin gas signatures on the C_1-C_2 diagram, showing a normal maturity trend for the gases sampled in the between 1000 and 2300 m depth, a residual signature for the deeper gases, and a segregated signature for the shallower ones.

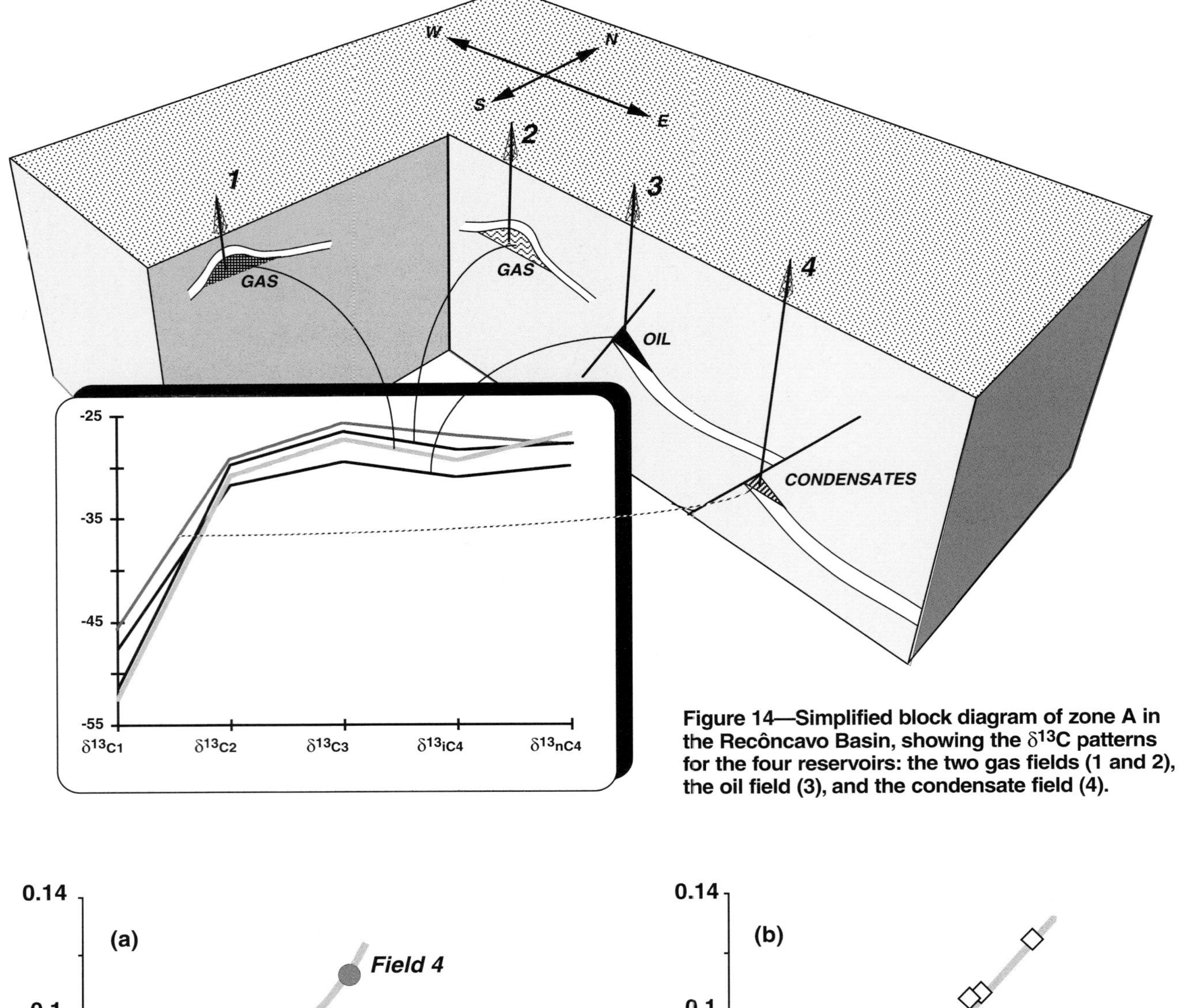

Figure 14—Simplified block diagram of zone A in the Recôncavo Basin, showing the $\delta^{13}C$ patterns for the four reservoirs: the two gas fields (1 and 2), the oil field (3), and the condensate field (4).

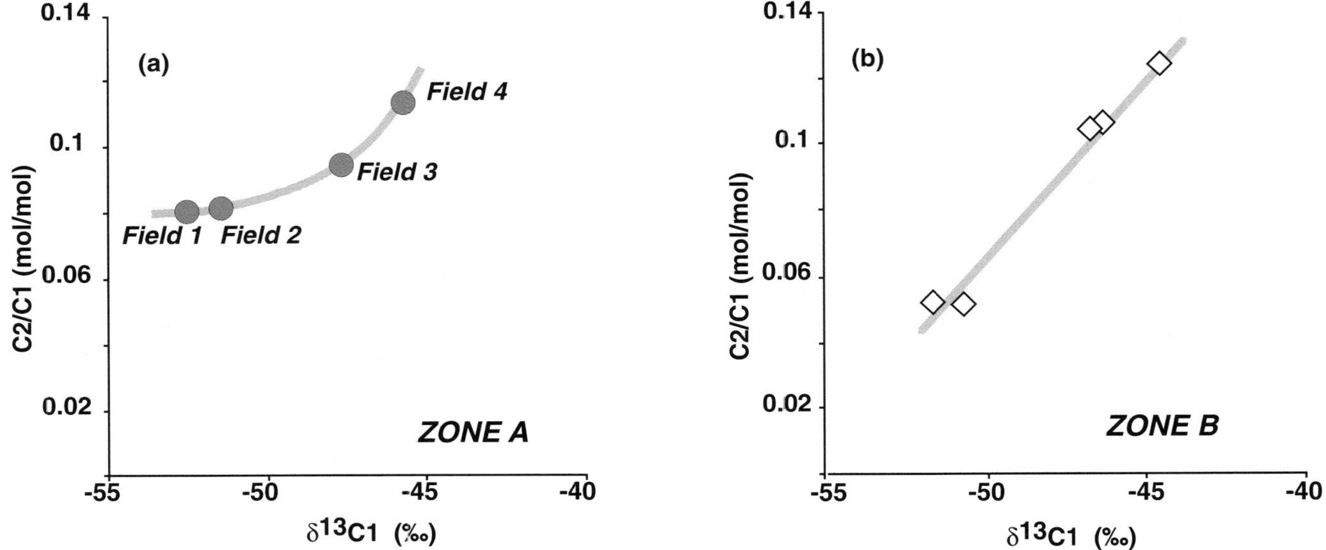

Figure 15—C_1-C_2 diagrams showing signatures of gases from zones A and B in the Recôncavo Basin. (a) The trend in zone A suggests segregation, while (b) the trend in zone B suggests bacterial contamination.

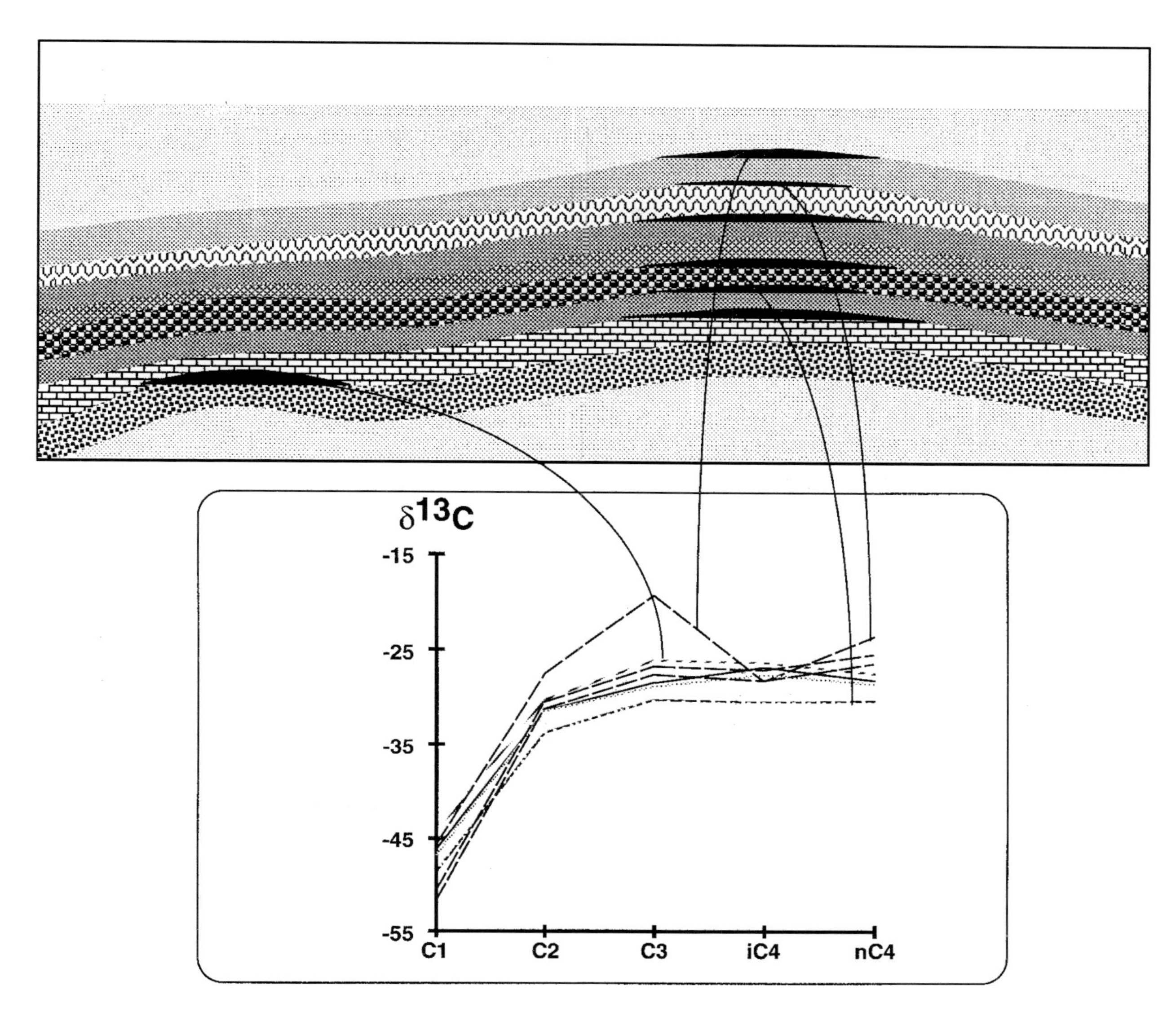

Figure 16—Simplified cross section of zone B in the Recôncavo Basin showing the $\delta^{13}C$ patterns for each reservoir.

drain. From this observation, directions and relative distances of migration can be assessed from the geochemical parameter V_2.

CONCLUSIONS

The recognition of gas behavior from its geochemical signature brings forth more and more complexity. Experimental controls and geologic case studies based on integrated statistical treatment of gas geochemistry enables us to reconstruct the hydrocarbon history in sedimentary basins. This results in constraints on maturity, migration pathways and directions, and segregation in fluids, providing an important method for evaluating petroleum systems.

REFERENCES CITED

Berner U., E. Faber, G. Scheeder, and D. Panten, 1995, Primary cracking of algal and landplant kerogens: kinetic models of isotope variations in methane, ethane, and propane: Chemical Geology, v. 126, n. 3–4, p. 233–246.

Bondar A. D., 1987, Role of diffusion in differentiation of carbon isotopes of methane of the earth's crust (in Russian): Geokhimiya, v. 9, p. 1274–1283.

Chung H. M., and W. M. Sackett, 1979, Use of stable isotope compositions of pyrolytically derived methane as maturity indices for carbonaceous materials: Geochimica et Cosmochimica Acta, v. 43, p. 1979–1988.

Clayton C., 1991, Carbon isotope fractionation during natural gas generation from kerogen: Marine and Petroleum Geology, v. 8, p. 232–240.

Colombo U., F. Gazzarrini, G. Sironi, R. Gonfiantini, and E. Tongiorgi, 1965, Carbon isotope composition of individual

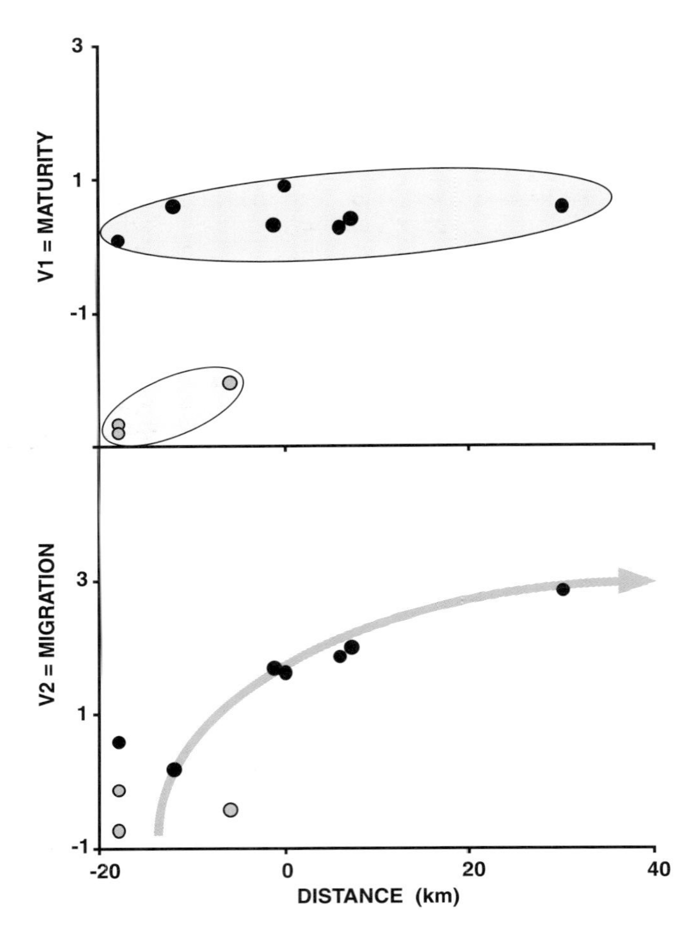

Figure 17—Values for the two synthetic parameters V_1 (maturity) and V_2 (migration), according to a PCA study (Prinzhofer et al., 2000), versus distance from the deepest part of the kitchen in the Mata–Catu fault area.

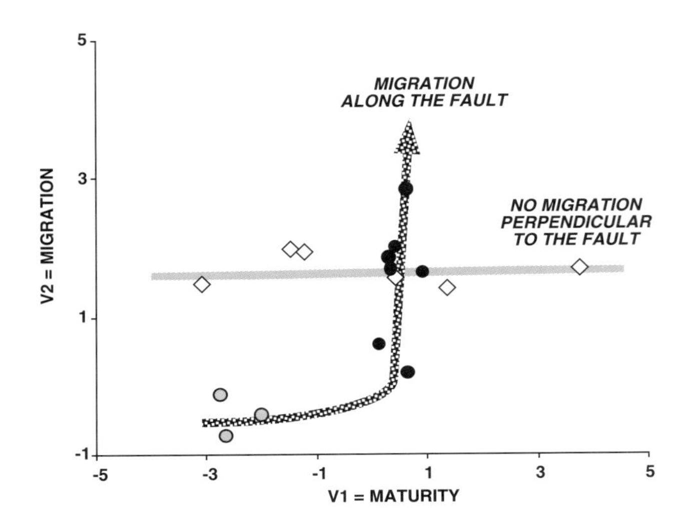

Figure 18—Plot of V_2 versus V_1 (migration versus maturity) for gases located along the Mata–Catu fault and for gases located in the vicinity of the kitchen on a trend perpendicular to the fault.

hydrocarbons from Italian natural gases: Nature, v. 205, p. 1303–1304.

Colombo U., F. Gazzarrini, R. Gonfiantini, G. Sironi, and E. Tongiorgi, 1966, Measurement of C_{13}/C_{12} isotope ratios on Italian natural gases and their geochemical interpretation, *in* P. A. Schenck and I. Havenaar, eds., Advances in organic geochemistry, 1964: Oxford, Pergamon Press, p. 279–292.

Colombo U., F. Gazzarrini, R. Gonfiantini, G. Kneuper, M. Teichmüller, and R. Teichmüller, 1970, Carbon isotope study on methane from German coal deposits, *in* G. D. Hobson and G. C. Speers, eds., Advances in organic geochemistry, 1966: Oxford, Pergamon Press, p. 1–26.

Cramer B., B. M. Krooss, and R. Littke, 1998, Modeling isotope fractionation during primary cracking of natural gas: a reaction kinetic approach: Chemical Geology, v. 149, p. 235–250.

Faber, E., 1987, Zur Isotopengeochemie gasförmiger Kohlenwasserstoffe: Erdöl, Erdgas, Kohle, v. 103, n. 5, p. 210–218.

Faber, E., W. J. Stahl, and M. J. Whiticar, 1992, Distinction of bacterial and thermogenic hydrocarbon gases, *in* R. Vially, ed., Bacterial gas: Paris, Editions Technip, p. 63–74.

Figueiredo, A. M. F., J. A. E. Braga, H. M. C. Zabalaga, J. J. Oliveira, G. A. Aguiar, O. B. Silva, L. F. Mato, L. M. F. Daniel, L. P. Magnavita, and C. H. L. Bruhn, 1994, Recôncavo basin, Brazil: a prolific intracontinental rift basin, *in* S. M. Landon, ed., Interior rift basins: AAPG Memoir 59, p. 157–203.

Fuex, A. N, 1980, Experimental evidence against an appreciable isotopic fractionation of methane during migration, *in* A. G. Douglas and J. R. Maxwell, eds., Advances in organic geochemistry, 1979: London, Pergamon Press, p. 725–732.

Galimov, E. M., 1975, Carbon isotopes in oil–gas geology, *in* NASA technical translation, NASA TT F-682: Washington, D.C., U.S. Government Printing Office.

Galimov, E. M., 1988, Sources and mechanisms of formation of gaseous hydrocarbons in sedimentary rocks, *in* M. Schoell, ed., Origins of methane in the earth: Chemical Geology Special Issue, v. 71, p. 77–95.

Hoering, T. C., and H. E. Moore, 1958, The isotopic composition of nitrogen in natural gases and associated crude oils: Geochimica et Cosmochimica Acta, v. 13, p. 225–232.

James, A. T., 1983, Correlation of natural gas by use of carbon isotopic distribution between hydrocarbon components: AAPG Bulletin, v. 67, n. 7, p. 1176–1191.

James, A. T., 1990, Correlation of reservoired gases using the carbon isotopic compositions of wet gas components: AAPG Bulletin, v. 74, n. 9, p. 1441–1458.

Krooss, B. M., S. Schlomer, R. Gaschnitz, and R. Littke, 1998, Aspects of natural gas generation and migration in sedimentary systems: Mineralogical Magazine, v. 62A, p. 818–819.

Lorant, F., A. Prinzhofer, F. Behar, and A. Y. Huc, 1998, Carbon isotopic and molecular constraints on the formation and the expulsion of thermogenic hydrocarbon gases: Chemical Geology, v. 147, p. 249–264.

Michels, R., and J. L. Oudin, 1997, Understanding of reservoir gas compositions in a natural case using semi-open artificial maturation (abs.): 18th International Meeting on

Organic Geochemistry, Maastricht, The Netherlands, p. 555–556.

Pernaton, E., A. Prinzhofer, and F. Schneider, 1996, Reconsideration of methane signature as a criterion for the genesis of natural gas: influence of migration on isotopic signature: Revue de l'IFP, v. 51, n. 5, p. 635–651.

Prinzhofer, A., and E. Pernaton, 1997, Isotopically light methane in natural gas: bacterial imprint or segregative migration?: Chemical Geology, v. 3–4, p. 193–200.

Prinzhofer, A., M. R. Mello, and T. Takaki, 2000, Geochemical characterization of natural gas: a physical multivariable approach and its applications in basin evolution: AAPG Bulletin, v. 84, n. 8, p. 1152–1172.

Ricchiuto, T., and M. Schoell, 1988, Origin of natural gases in the Apulian basin in south Italy: a case history of mixing of gases of deep and shallow origin, *in* L. Mattavelli and L. Novelli, eds., Advances in Organic Gechemistry, 1987: London, Pergamon Press, p. 311–318.

Rooney, M. A., G. E. Claypool, and G. E. Chung, 1995, Modeling thermogenic gas generation using carbon isotope ratios of natural gas hydrocarbons: Chemical Geology, v. 126, n. 3–4, p. 219–232.

Sackett, W. M., S. Nakaparksin, and D. Darlymple, 1970, Carbon isotope effects in methane production by thermal cracking, *in* G. D. Hobson and G. G. Speers, eds., Advances in organic geochemistry, 1966: Oxford, Pergamon Press, v. 3, p. 37–53.

Schoell, M., 1980, The hydrogen and carbon isotopic composition of methane from natural gases of various origins: Geochimica et Cosmochimica Acta, v. 44, p. 649–661.

Schoell, M., 1983, Genetic characterization of natural gases: AAPG Bulletin, v. 67, n. 12, p. 2225–2238.

Stahl, W. J., 1977, Carbon and nitrogen isotopes in hydrocarbon research and exploration: Chemical Geology, v. 20, p. 121–149.

Stahl, W., and B. D. Carey, 1975, Source rock identification by isotope analyses of natural gases from fields in the Val Verde and Delaware Basins, west Texas: Chemical Geology, v. 16, p. 257–267.

Wehr, F. L., L. H. Fairchild, M. R. Hudec, R. K. Shafto, W. T. Shea, and J. P. White, 2000, Fault seal: contrasts between the exploration and production problem, *in* M. R. Mello and B. J. Katz, eds., Petroleum systems of South Atlantic margins: AAPG Memoir 73, p. 121–132.

Chapter 10

Fault Seal: Contrasts Between the Exploration and Production Problem

F. L. Wehr

L. H. Fairchild

M. R. Hudec

R. K. Shafto

W. T. Shea

J. P. White

Exxon Production Research Company
Houston, Texas U.S.A.

Abstract

Faults can affect both volumetric estimates and reservoir management of a hydrocarbon-bearing structure. To the explorationist, the essential problem is to predict the potential for a fault-dependent trap to hold a commercial volume of hydrocarbons. To the production geologist or reservoir engineer, the question may be how intrareservoir faults influence fluid flow patterns and recovery efficiency during reservoir depletion. While similar overall, the two problems differ in detail, and miscommunication (particularly between geologists and reservoir engineers) is common when discussing whether or not a fault will "seal."

The capacity for a fault to hold hydrocarbons over geologic time depends primarily on the relationship between the minimum capillary entry pressure across the fault surface and the pressure exerted on the fault surface due to the buoyancy of the hydrocarbon column. In an exploration setting, the geologist typically uses structure maps from seismic data and stratigraphy from nearby wells to develop a model of reservoir juxtaposition and gouge distribution across a fault surface. Leakpoints are identified and a prediction of whether the fault will seal is made. Pressure data and hydrocarbon contact depths are used to calibrate fault seal predictions. Because fault zones are chaotic and exploration data density is low, irreducible uncertainty in exploration fault seal prediction is high. Fault sealing behavior can be sensitive to details of the fault surface stratigraphy and distribution of gouge. Thin thief sands and small gaps in an impermeable gouge can turn a sealing fault into a leaking fault over geologic time. Predicting exploration fault seals becomes complex and nonintuitive in traps with multiple faults and stacked reservoirs.

In the production setting, fluid flow across faults is a response to large pressure gradients over relatively short time periods. Under these circumstances, capillary effects are less important, and the flow behavior is governed by fault zone *transmissibility*, a function of width and permeability of the fault zone. A spectrum of fault sealing behaviors can occur: faults may be transparent to fluid flow, act as baffles, or complete seals. Because more data are available to the reservoir geologist and engineer than to the explorationist, the effect of faults on fluid flow in producing fields can be measured more precisely.

A recent study of a faulted North Sea oil field illustrates the effect of fault zone transmissibility on fluid flow patterns. The goal was to predict fault zone transmissibility from fault zone composition, calibrated to 13 years of field production history. Four models for fault transmissibility were compared in the simulation: no reduction (i.e., juxtaposition effects only), two different fault gouge models, and constant reduction irrespective of fault zone lithology. Model results showed that explicit modeling of fault transmissibility improved the water-cut match by 20–25% over the juxtaposition-only case. Subseismic faults were not required to match field performance, nor was it necessary to make nongeologically based changes to fault properties. Based on our experience in this field and elsewhere, careful attention to fault mapping combined with continual iteration between the geologist and engineer eliminates the need for most so-called anomalous fault seal behavior.

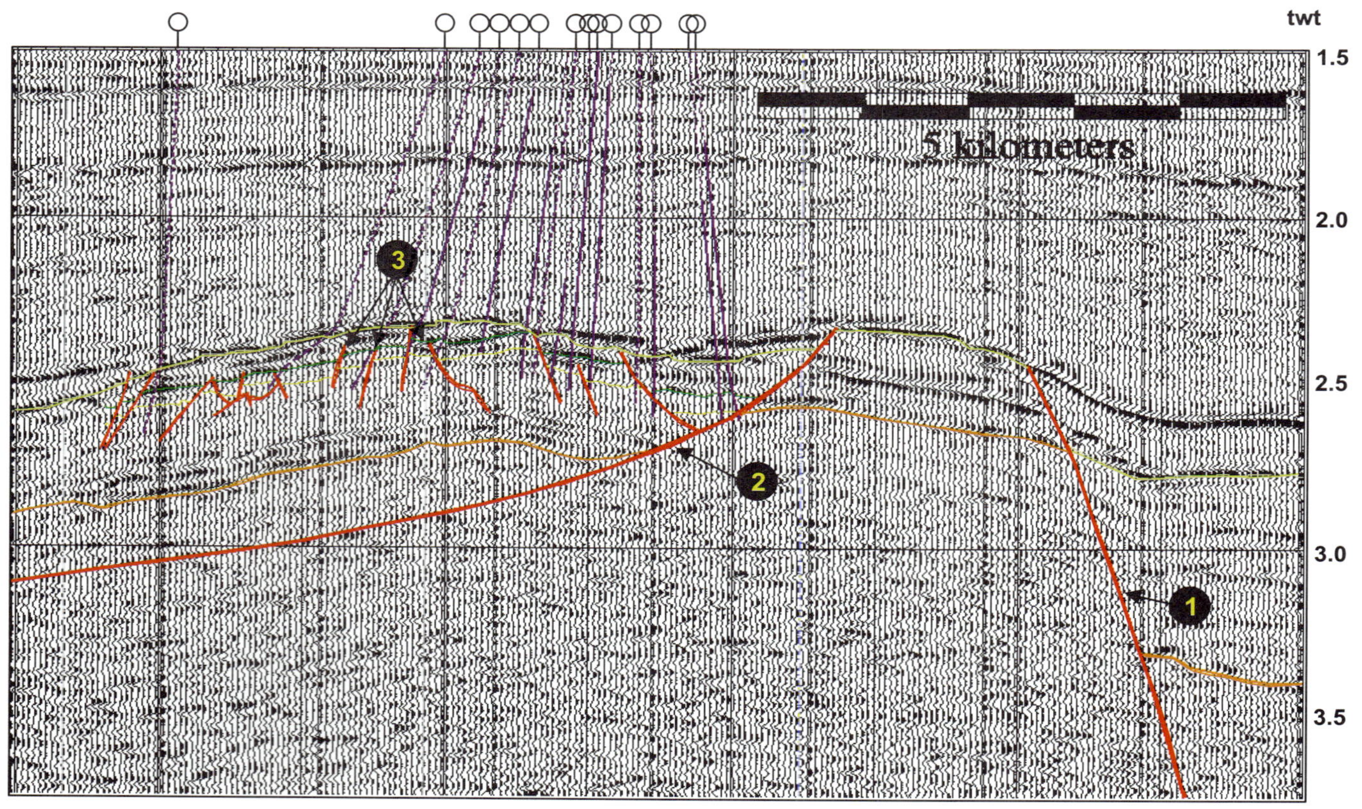

Figure 1—Seismic line from the East Shetland Basin, U.K. North Sea, showing an example of various scales of faults affecting a producing field. Fault 1 provides a migration conduit for hydrocarbons from mature source rocks to the trap; fault 2 sets up the closure, which is in part fault-dependent; and the numerous faults indicated by 3 segment the field and significantly reduce the total recovery factor.

INTRODUCTION

Faults are an integral part of hydrocarbon systems at all scales. At the basin scale, faults define regions of similar subsidence history, may control patterns of sedimentation, and may provide the critical migration pathways from source beds to reservoir. Most hydrocarbon traps are created at least to some degree by faulting, and faults at the reservoir scale can have a major effect on field producibility (Figure 1). Fault seal prediction can affect both volumetric estimates for undrilled prospects as well as reservoir management strategies for a producing field.

This chapter provides a brief review of our current understanding of the fault seal problem, from both the exploration and production perspectives. To the explorationist, the central problem is to predict the potential for a fault-dependent trap to hold a commercial volume of hydrocarbons over geologic time. To the production geologist or reservoir engineer, the question may be how intrareservoir faults influence fluid flow patterns and recovery efficiency during reservoir depletion. While similar overall, the two problems differ in detail, and miscommunication (particularly between geologists and reservoir engineers) is common when discussing whether or not a fault will "seal."

EXPLORATION FAULT SEAL

The central aim of fault seal analysis to the explorationist is typically a volumetric prediction of hydrocarbons in place. For a single reservoir trap with a fault-dependent closure, this may be as simple as predicting the sealing properties of a single fault. For a complexly faulted structure with stacked reservoirs, fault seal prediction becomes a complex 3-D problem of predicting leakpoint distribution on a network of faults. Nevertheless, the standard goal in the exploration setting remains to predict what portion of a fault-dependent closure will contain hydrocarbons.

Processes

The capacity of a single fault to seal is controlled by the relationship among the capillary entry pressure of the fault zone, the buoyant forces exerted upon the fault by the hydrocarbon column, and the fluid properties of the hydrocarbons (Smith, 1966; Schowalter, 1979; Downey, 1984; Watts, 1987). Capillary forces are considered to dominate in most exploration settings because of the low pressure gradients imposed by buoyant forces (typically tens of psi) combined with the long periods of time avail-

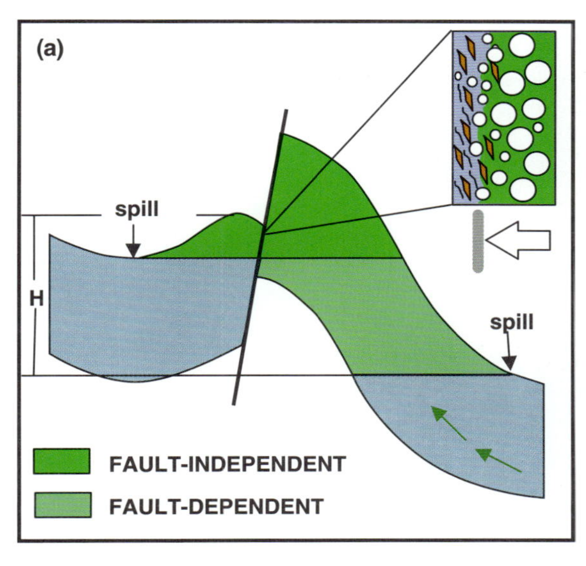

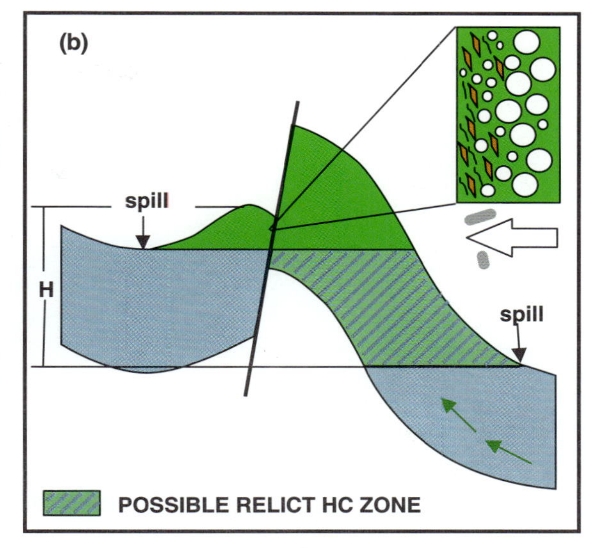

- ρ_c (min) of fault $> (\rho_{water} - \rho_{oil})gH$
- HC contact and reservoir pressure may differ across fault

- ρ_c (min) of fault $\leq (\rho_{water} - \rho_{oil})gH$
- HC contact and reservoir pressure similar across fault

Figure 2—Conceptual model for capillary fault seal, assuming hydrocarbon migration into the trap from the right. (a) For a "sealing" fault, the minimum capillary entry pressure (ρ_c) at every point along the fault surface is greater than the buoyant force exerted by the hydrocarbon column, $(\rho_{water} - \rho_{oil})gH$, where *g* is the gravitational constant and *H* is the column height. For a "leaking" fault, the capillary entry pressure has been overcome at least at one point along the fault surface, and the hydrocarbon columns and fluid pressures equilibrate across the fault. (b) A deeper, relict hydrocarbon column may occur in this case depending on the maximum ρ_c of the fault zone.

able for fluid flow. In this setting (assuming slow rates of hydrocarbon migration into a trap), a small capillary leak on a fault zone will exhibit the same result as a completely leaky fault. For a fault to seal hydrocarbons, the maximum capillary entry pressure at every point on the fault zone above the hydrocarbon contact must exceed the buoyant force exerted by the hydrocarbon column (Figure 2a). If the entry pressure is exceeded at any point, then the capillary seal will be broken and the hydrocarbon column will equilibrate across the fault up to the level of the leakpoint (Figure 2b).

As a working definition, a *sealing fault* to the explorationist is one that is observed to maintain separate contacts and/or fluid pressures. Conversely, a *leaking fault* is one that exhibits no change in contacts or pressures across it. In reality, most faults both leak and seal at various positions along the fault plane, and it is the distribution of leakpoints on the fault plane relative to closure that controls hydrocarbon contact positions.

Tools

A complete fault seal analysis may require detailed predictions of (1) fault architecture (position, throw distribution, and linkage among a set of faults), (2) fault plane stratigraphy (distribution of reservoir-to-reservoir juxtaposition points across a fault plane or planes), and (3) fault zone characterization (distribution of impermeable gouge either from clay content or fault zone diagenesis). In addition, external factors such as fault orientation relative to ancient or present-day stress fields may play a role in certain basins (e.g., Knott, 1993).

Fault seal prediction in an exploration setting uses a standard set of tools developed by structural geologists from a combination of subsurface studies, field observations, and kinematic analysis. The input data for fault seal analysis consists of two key elements: (1) one or more structure maps showing fault position and throw, and (2) a stratigraphic model, generally based on one or more nearby wells (Figure 3a). The first and most critical step in the analysis is to ensure the quality of the structural interpretation. Are the faults positioned correctly? Does the displacement of the fault vary along strike in a consistent manner? Does the combination of fault map pattern and displacement profile make kinematic sense? When these criteria are met, the next step is an analysis of reservoir juxtaposition using fault plane profiles (Figure 3b). A fault plane profile is a map of hanging-wall and footwall stratigraphy superimposed across the fault plane. Areas of reservoir juxtaposition can be mapped onto structure, and critical leakpoints identified and tested against hydrocarbon contact information. Finally, a prediction of impermeable gouge distribution can be superimposed on the fault plane (Figure 3c) based on empirically derived estimates of fault zone composition such as clay smear potential (Weber et al., 1978; Gibson, 1994) or gouge ratio

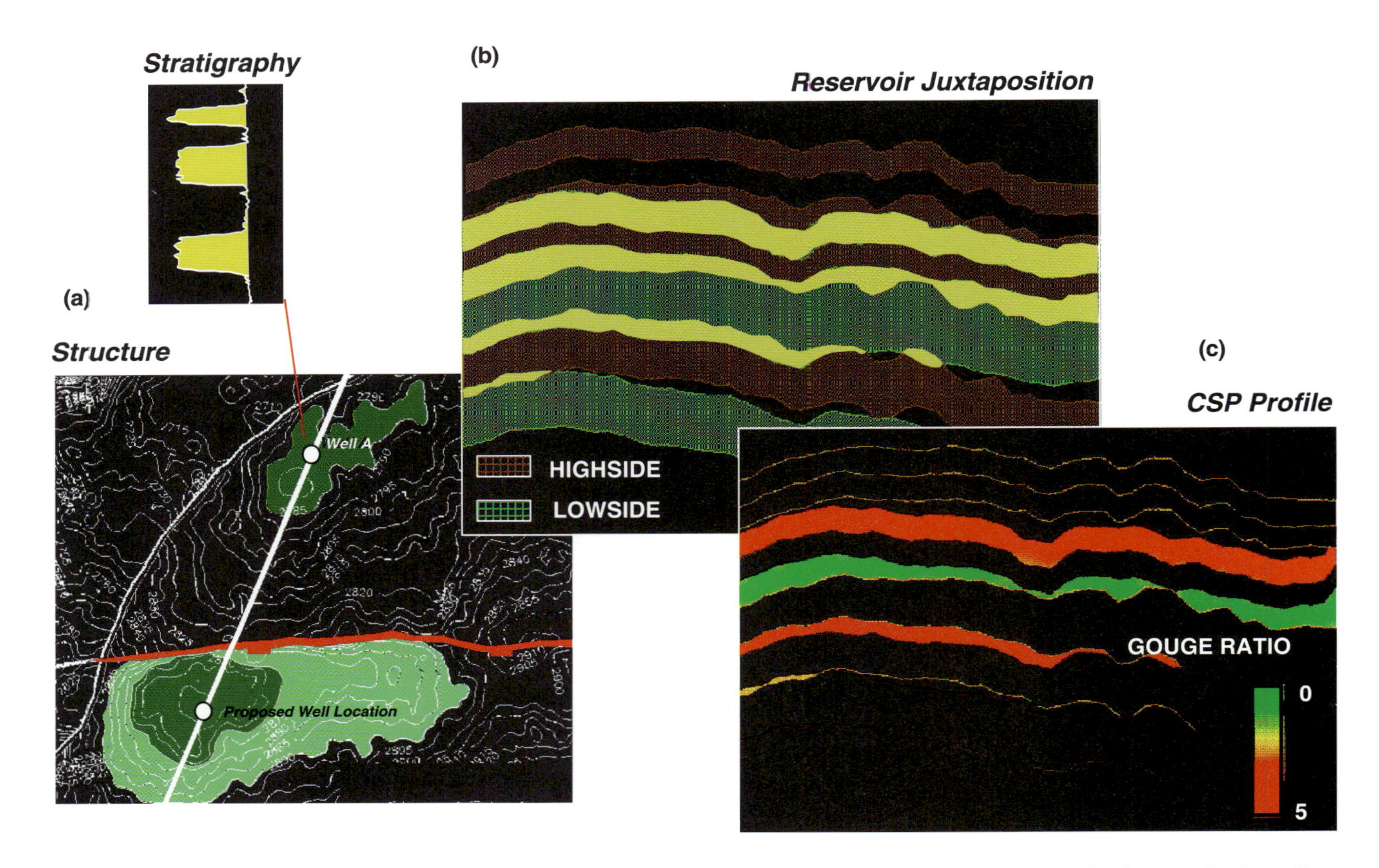

Figure 3—Overview of standard fault seal analysis techniques. (a) Construction of a fault plane profile from a single well and fault map. The well is projected southward from a known accumulation onto the trap-bounding fault (red). (b) Fault plane profile showing areas of reservoir to reservoir juxtaposition in yellow. (c) Fault plane profile from (b) with clay smear ratio overlain on areas of reservoir juxtaposition. Higher clay smear ratios (red) correlate with a higher sealing potential. Courtesy of Badley Earth Science.

(Freeman et al., 1998). If a fault zone model is incorporated, it will have the net effect of sealing some of the leakpoints identified in the juxtaposition analysis, which increases the potential for fault seal and greater column heights. These techniques, when calibrated to existing field pressure and hydrocarbon contact information, have become industry standards and are well represented in the literature of the last decade.

Limitations

The biggest limitation of current fault seal techniques is the high degree of inherent uncertainty combined with poor data resolution. Successful fault seal prediction is completely dependent on an accurate model for both the fault zone architecture as well as the juxtaposed stratigraphy. In reality, faults are complex, chaotic phenomena, and the sealing properties of a fault do not appear to scale well. Small errors in either the structural or stratigraphic data can result in a completely incorrect seal prediction.

Figures 4–7 show some examples of where unresolvable details of a fault zone could radically change its sealing properties. Examples of structural uncertainties that impact reservoir juxtaposition include errors in fault throw mapping, the presence of unresolvable slivers within the fault plane (Figures 4 and 5) (Childs et al., 1997), and the details of relay zones between two linked faults (Peacock and Sanderson, 1994; Childs et al., 1996). Stratigraphic uncertainties can be equally important, particularly where reservoirs are stacked. For example, the presence of a thin sandy bed juxtaposed against a fault-dependent closure near the crest of a structure can result in an underfilled trap (Figure 6). Finally, predictive models for the distribution of capillary seals are based on coarse empirical models. Field observations of well-exposed fault zones in the western United States (e.g., Foxford et al., 1998; J. Holl et al., unpublished data, 1997) show that gouge distribution is discontinuous and may not provide a capillary seal over the scale of a hydrocarbon trap, even when an averaged model would predict seal (Figure 7).

Management of this high level of uncertainty is a key direction for future work in fault seal. By using stochastic techniques to model the appropriate ranges in fault

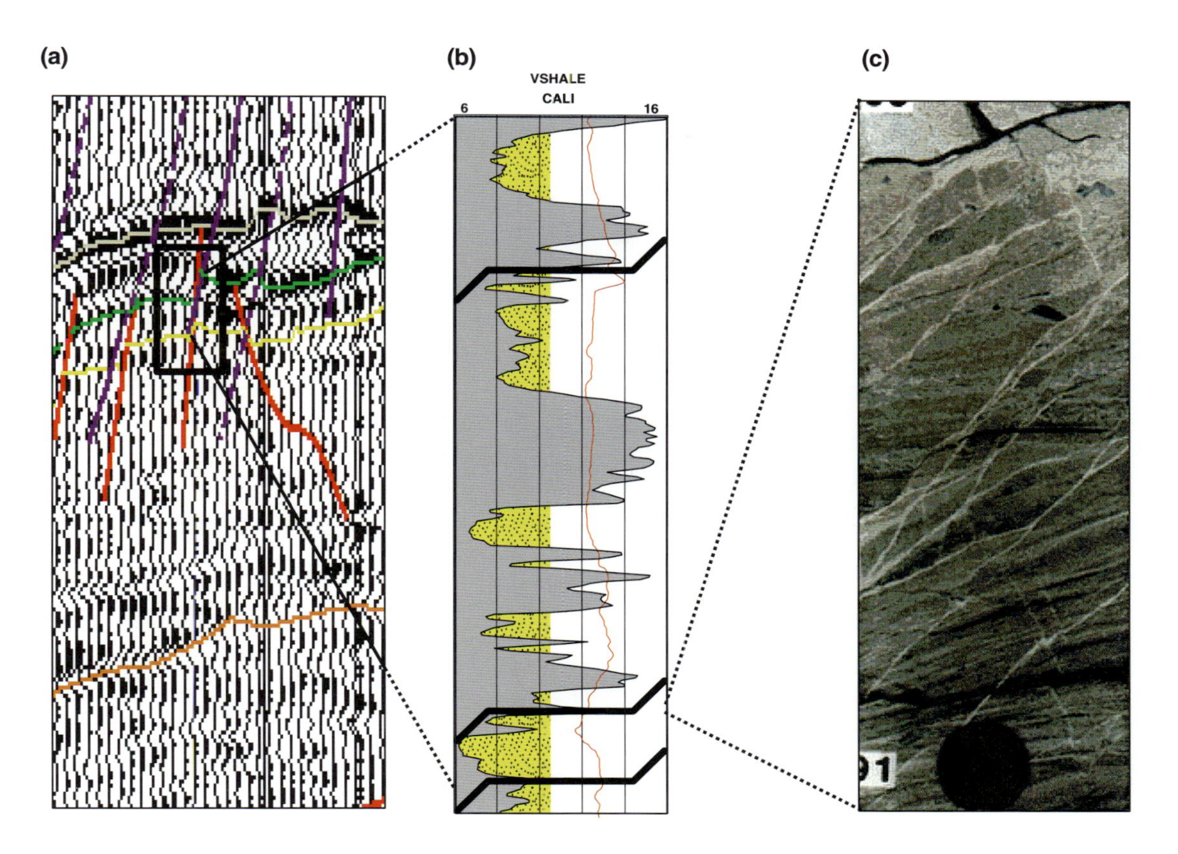

Figure 4—Fault data resolution example from the U.K. North Sea. (a) The 3-D seismic data can (poorly) resolve a single fault at the well bore location. (b) Detailed correlation of the well to surrounding wells reveals that there are at least three separate fault strands that remove the reservoir section. (c) Core reveals a complex zone of deformation bands and small offset faults associated with a single strand.

throw, fault plane stratigraphy, and fault zone composition, a probabilistic fault seal prediction can be generated that may be more applicable to prospect risking. At the same time, seismic technology is reaching a threshold where, with high-quality data, fault seal analysis becomes a secondary technology. Where seismic data can directly resolve the presence or absence of hydrocarbons on both sides of a fault, there is little need for an independent analysis of its sealing capacity.

PRODUCTION FAULT SEAL

A key problem in many producing fields is reservoir compartmentalization, both as a function of stratigraphic and fault-derived discontinuities. The significance of faults in this context is the degree to which they will impede fluid flow during production and, ultimately, the net effect they have on recovery efficiency. Fault seal prediction to a reservoir engineer may not be the same problem to the exploration geologist, and miscommunication between the two is common.

The dynamics of a producing hydrocarbon field can be very different from a trap under active hydrocarbon

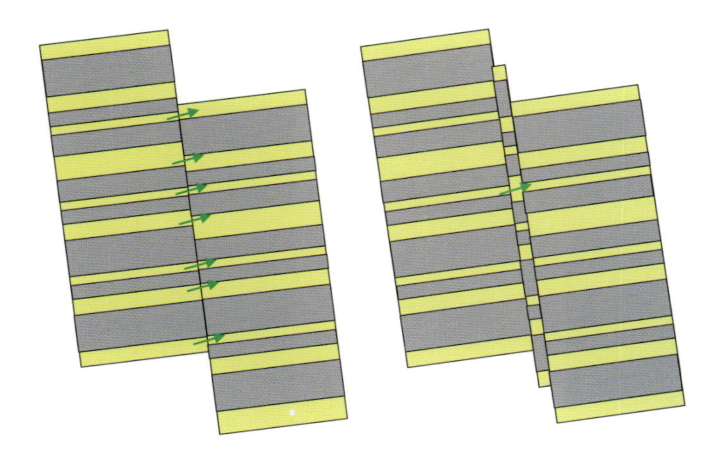

Figure 5—Schematic diagram showing the potential effect of fault slivers on juxtaposition analysis. The left example shows a single fault across which the sandstone beds (yellow) are well connected, with a total of seven leak-points. By introducing a thin sliver and accommodating the fault displacement on two strands, the number of leak-points is reduced to one, illustrating a basic limitation of juxtaposition analysis based on seismic data.

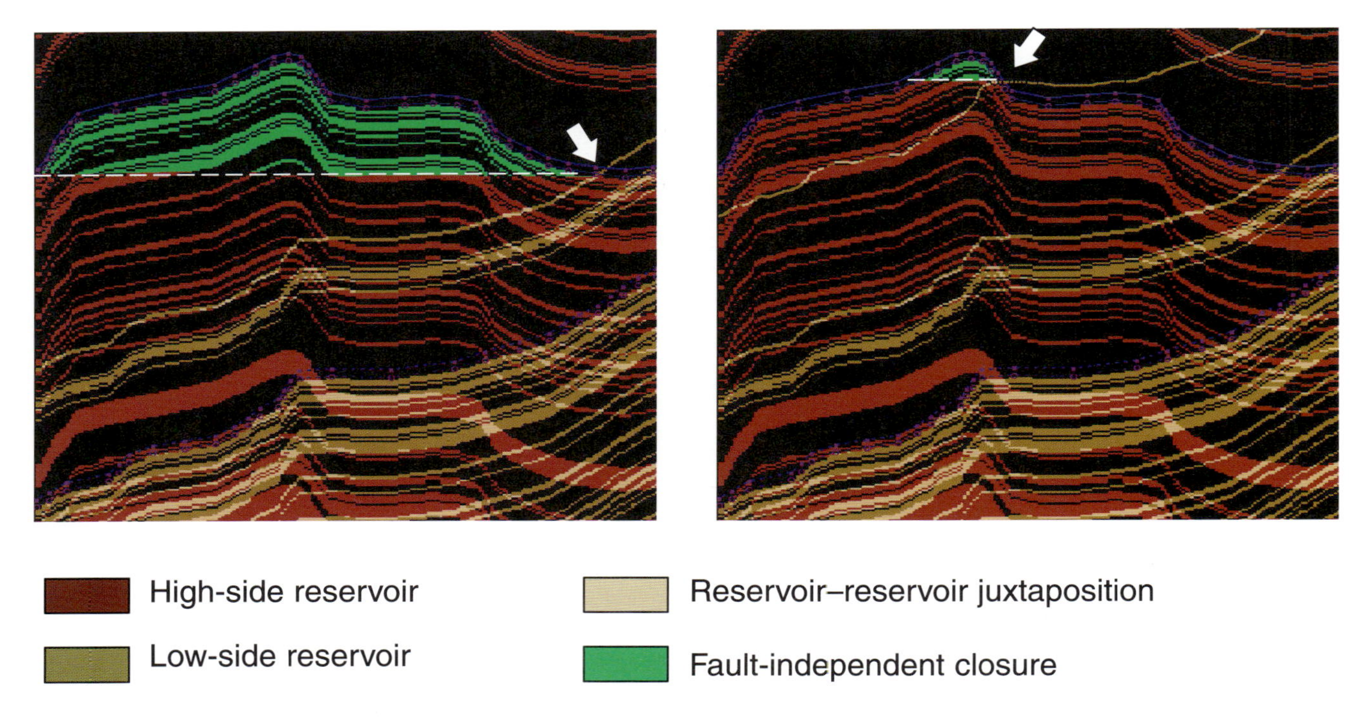

Figure 6—Schematic example on a fault plane profile of the effect of a thin "thief sand" on fault seal prediction. On the left, the high-side accumulation (green) is limited by a leakpoint well down from the crest of the structure (arrow). On the right, the accumulation is severely limited by the addition of a thin permeable bed juxtaposed near the structural crest (arrow). This model assumes that fault gouge does not contribute to the seal.

Figure 7—Schematic illustration of a possible limitation in empirical gouge models for exploration fault seal analysis. The left example shows a schematic fault plane. On the right, a mask with 90% randomly distributed "sealing gouge" is applied. Unless the gouge is continuous across the entire fault plane, the fault is likely to leak over geologic time.

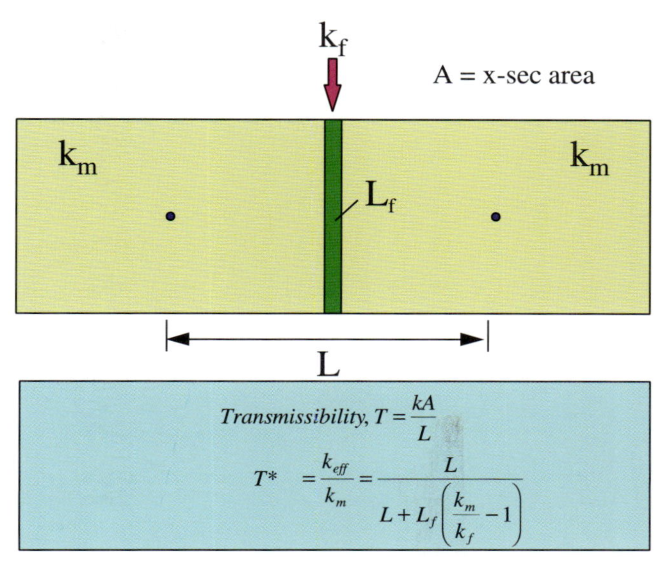

Figure 8—Calculation of fault transmissibility. Fault transmissibility is a function of both the permeability (k_f) of the fault zone and its thickness (L_f). In this simulation, T^* is the transmissibility multiplier, and k_{eff}, k_m, and k_f are the effective, matrix, and fault zone permeabilities, respectively.

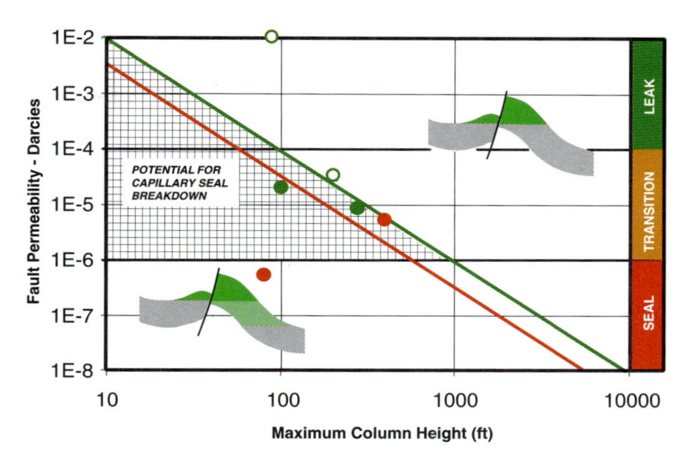

Figure 9—Graph of hydrocarbon column height versus fault permeability. See text for discussion.

charge. Pressure gradients are high (typically hundreds to thousands of psi), and production time scales are short. In this setting, the capillary seals that control exploration-scale faulted traps are unstable, and the *transmissibility* of the fault is the key parameter that governs fluid flow (Figure 8). Transmissibility is a function of the effective permeability, the cross-sectional area, and thickness of the fault zone.

In a review of examples of "anomalous" fault seals within Exxon affiliates and in the literature, we found a number of cases of fault seal breakdown: faults that provided exploration seals but were demonstrated to leak over the production time scale. The key to this apparent anomaly is the difference between capillarity and transmissibility. A capillary seal capable of maintaining a significant hydrocarbon column can be overcome by production-induced pressure differentials. Once a capillary seal is breached, the transmissibility of the fault is the property that controls the rate of fluid flow.

Figure 9 summarizes differences between fault seal to the explorationist (as inferred from contact information) and the reservoir engineer (as inferred from fluid flow rates across a fault). In this analysis, capillary entry pressure is converted to average fault zone permeability. Given an observed column height (horizontal axis), fault zone permeability can be constrained to be either above the diagonal line (red is gas, green is oil) if contacts are common or below the line if contacts are different across the fault. Because fault zone transmissibility is proportional to permeability given a constant fault geometry, the idealized production response of the fault can be represented by horizontal regions. The area of potential capillary seal breakdown is hatchured on Figure 9 and corresponds well to the field studies used in this analysis. Capillary seal breakdown is likely when a fault seals a relatively small hydrocarbon column and the reservoir is produced via pressure depletion. An interesting corollary to this analysis is that a fault that leaks over geologic time can theoretically be an effective seal during production. The most likely conditions for this behavior occur when the hydrocarbon column downdip of the fault is relatively large.

CORMORANT FIELD EXAMPLE

Mature fields with extended production histories are a natural laboratory for the study of fluid flow across faults. A number of published field studies (Goldthorpe and Chow, 1985; Bentley and Barry, 1991; Freeman et al., 1998; Ottesen Ellevset et al., 1998) have demonstrated how fault properties can be used as input to a reservoir simulation. In a recent study of Cormorant Block IV in the U.K. North Sea, we attempted to test whether the inclusion of a fault-transmissibility modifier based on geologic properties could significantly improve the match between production history and reservoir simulation.

The central objective of this study was to design a method to predict fault zone transmissibility and to incorporate it into reservoir simulation at a cell scale. Two steps were required: (1) selection of a model of fault gouge formation to predict composition at any point along the fault plane, and (2) translation of gouge composition to a numerical value for transmissibility reduction across the fault plane. Our approach was to integrate the field geologic framework into a full-field simulation, then use the simulation as a tool to calibrate models for fault transmissibility. The project involved remapping the field, revising the existing sequence stratigraphy and log analysis, building a cell-based geologic model, and then scaling

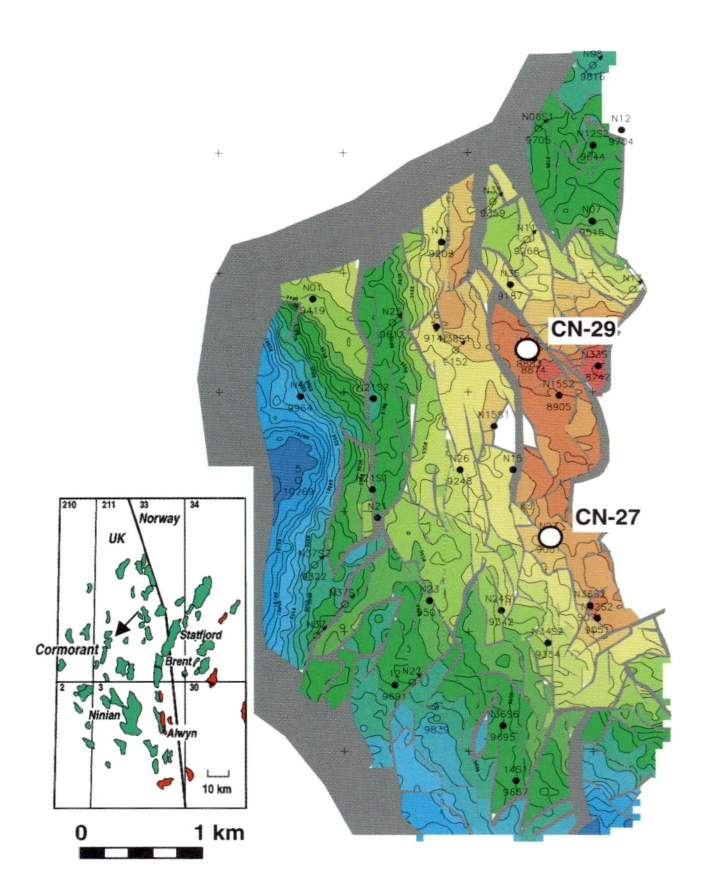

Figure 10—Depth structure map on top of the Rannoch Formation, an intrareservoir horizon at Cormorant Block IV. This surface was used as the basis for the geologic model and reservoir simulation.

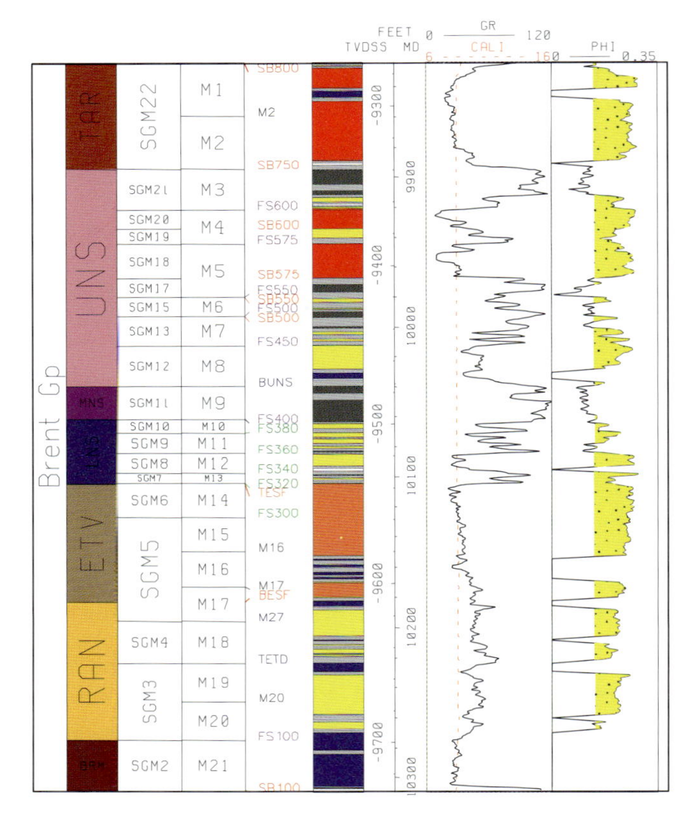

Figure 11—Type well through the Brent reservoir, Cormorant Block IV, showing (from left to right) the lithostratigraphy, geologic model zones, simulator zones, sequence stratigraphic surfaces, petrofacies, depth, gamma-ray, and porosity logs.

up to a full-field reservoir simulation. A central tenet of the study was that the transmissibility algorithm be applied uniformly to every fault in the model; adjusting individual fault properties to improve model performance was not allowed.

Geologic Setting

Cormorant is an Upper Jurassic oil field in the U.K. sector of the North Viking Graben (Taylor and Dietvorst, 1991). It consists of four fault-bounded blocks. They are westerly dipping, rotated, and eroded, and they form distinct oil accumulations. Cormorant Block IV (CBIV) is unique in that it is in the hanging wall of a major normal fault (Figure 10). The reservoir section consists of the Middle Jurassic Brent Group (Figure 11). Reservoir quality is generally good, with reservoir porosities up to 30% and permeabilities of 10 to >1000 md. There is a marked decrease in reservoir quality with depth and below the oil–water contact. The oil is highly undersaturated, with API gravity varying from 34° to 36°, bubble points from 1040 to 2970 psi, and viscosity from 0.56 to 1.18 cp. Oil production at Cormorant relies on injection of water downdip to provide pressure support and to displace oil updip toward producing wells.

Development of Cormorant Block IV proved to be more difficult than anticipated, mainly due to fault segmentation. The initial development plan of 1975, based on 2-D seismic data, required 15 wells with one row of injectors and two parallel rows of producers. As development proceeded and 3-D seismic data became available, it became clear that the field was more structurally complex than initially thought. To date, more than 40 development wells have been drilled, and nearly half have penetrated one or more faults in the reservoir section.

In part because of difficulties in producing the field, Cormorant Block IV is an exceptional resource of geologic and engineering data. The study database included a 1992 vintage 3-D seismic survey, over 40 exploration and development wells with more than 3000 ft of core and repeat formation test pressure measurements in most development wells, and 13 years of production data. A recent facies and sequence stratigraphic framework was available (Jennette and Riley, 1996), as well as complete documentation from a recent Shell U.K. 3-D interpretation and full-field reservoir simulation.

Fault Transmissibility Models

Field observations of faults indicate that, in many clastic settings, fault zone composition is related to the percentage of sandstone and shale in the faulted section and to the throw on the fault. Two models for fault gouge formation were compared in this study: gouge ratio and clay smear potential. *Gouge ratio* assumes complete mixing of sandstone and shale along the fault zone and is calculated as a moving average of the sandstone–shale ratio at any one point along the fault (Freeman et al., 1998). In contrast, *clay smear potential* treats shale beds as coherent bodies that shear into the fault zone over a limited distance (Smith, 1980; Lindsey et al., 1993). Both algorithms were used in this study for comparison. In addition, two other cases were modeled: a simple multiplier and juxtaposition only (Figure 12).

In the gouge ratio model, transmissibility of a fault connection is calculated as proportional to the moving average of the sandstone–shale ratio of the material that slides by a given point on the fault. Using a log-derived shale attribute from the geologic model, the gouge ratio was calculated for each faulted cell connection in the simulator (Figure 12). Values of gouge ratio were converted to transmissibility by iteration, adjusting the conversion parameters until an acceptable history match was achieved.

The clay smear potential method differs from gouge ratio in that intrareservoir shales are assumed to taper into the fault zone rather than mix into a uniform gouge (Figure 11). The clay smear method was implemented manually by identifying the most continuous shale layers (mainly in the Ness Formation) and by developing a matrix of transmissibility multipliers that are dependent on the simulator layer numbers on each side of the fault. This approach is similar to that described by Bentley and Barry (1991).

The simple multiplier model assumed no knowledge of the geology and was built using single, average fault transmissibility for the entire model. This method was run to evaluate the utility of a quick-pass approach to handling fault gouge effects on reservoir performance. The juxtaposition-only model was the base case, in which the effects of gouge on transmissibility were not included in the calculation.

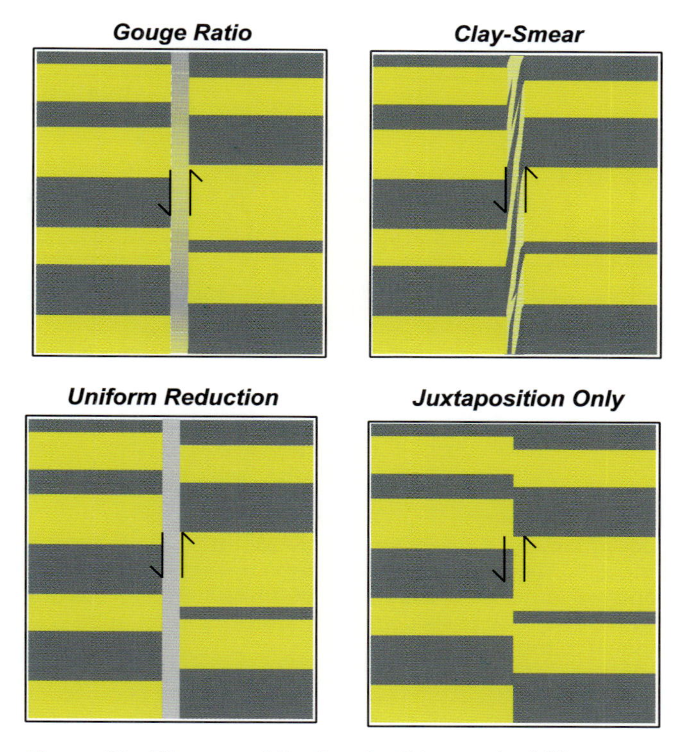

Figure 12—Diagram of the four fault transmissibility models compared in the Cormorant study. The gouge ratio model assumes uniform mixing of sandstone and shale within the fault zone. Clay smear assumes ductile shearing of the shale beds a finite distance into the fault zone. The uniform reduction model applies a single factor to fault transmissibility, and the juxtaposition-only model treats the faults as transparent to flow.

Results

Results from the four models demonstrate that accounting for gouge through explicit fault transmissibility modifications improves the water cut match at Cormorant. The net effect of decreasing fault transmissibility is to focus injected water within fault blocks, decreasing the time required for water to move within a single fault-bound compartment.

The CN-29 well is the first and most prolific oil producer drilled at Cormorant Block IV. Water injector CN-27 provides pressure support to CN29 (Figure 10). This injection–production pair dominates the early production history of the central part of the field and is used to illustrate the impact of the fault transmissibility models.

Plots of water cut versus time (Figure 13) illustrate the improvement to the water cut match of the CN-29 due to fault transmissibility modeling. The gouge ratio and clay smear models provide a close match to the observed water cut in this well. In the uniform reduction model, the overall water cut in the CN-29 is consistently low. The juxtaposition-only model also exhibits a low water cut, and first water arrives at the well about 6 months late.

Note that there is little difference between the gouge ratio and clay smear models for this particular well. Water-flood patterns are broadly similar between the two models, although local differences occur. These observations indicate that simulation results are not particularly sensitive to the details of the algorithm used for transmissibility modeling. In detail, however, the clay smear model appears to produce a slightly better match to field performance.

Figure 14 compares the juxtaposition-only and clay smear models using a simulator slice from the upper part

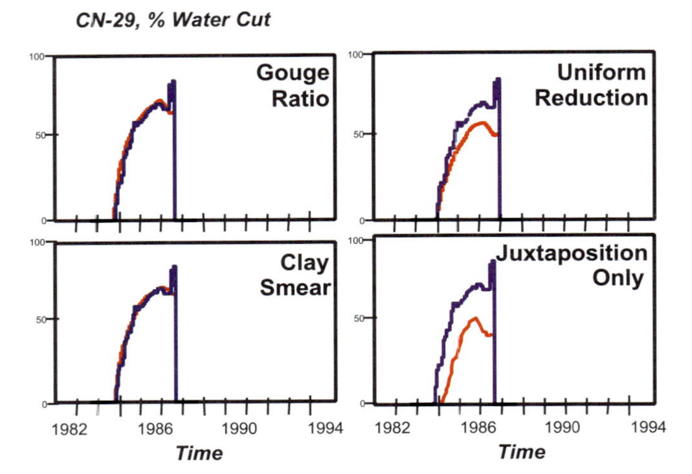

Figure 13—Comparison of percentage of water cut performance for CN-29 well from 1984–1988, Cormorant Block IV. The blue line is the observed well performance, and red is the simulation. Both of the fault transmissibility models (gouge ratio and clay smear) show a good match to the well.

of the Etive Formation reservoir. The Etive is the thickest and highest permeability reservoir in the field, and it dominated the early production performance of the field. Shortly after the onset of water injection at the CN-27 (Figure 14a), the two models are virtually identical. Over time (Figure 14b, c), the more restrictive faults in the clay smear model tend to confine the injected water to a single fault block, so that the water-flood front moves faster to the north and south. First water at the CN-29 (Figure 14c) occurs 3.0 years into the clay smear model, matching field performance. In the juxtaposition-only model, first water arrives about 6 months later and the overall water cut remains about half of that observed in the field (Figure 14d).

Discussion

This study shows that fault transmissibility modeling can improve the water cut history match for Cormorant Block IV by 20–25%. Subseismic faults were not required to match field performance, nor was it necessary to make nongeologically based changes to fault properties. These results are similar to other published studies of fault transmissibility modeling (e.g., Goldthorpe and Chow, 1985; Bentley and Barry, 1991; Freeman et al., 1998; Ottesen Ellevset et al., 1998), all of which cite some improvement in predicting field performance. In the Cormorant study, we were able to model all of the faults (over 70) in the field in a consistent manner and compare the results of different fault transmissibility models while holding all other variables constant. The results show that the improvement in water cut match was not particularly sensitive to the details of the fault transmissibility algorithm (Figures 13 and 14). This may mean that simple threshold-based models (e.g., Freeman et al., 1998) are adequate to predict fault behavior in a production setting, and it implies that "average" fault zone properties (implicit in gouge models) are useful indicators of fault transmissibility. The same does not necessarily hold true in an exploration setting, where small gaps in a capillary seal that would not be predicted in a gouge model may cause the fault to leak over geologic time.

Our experience in building and calibrating the Cormorant geologic model and reservoir simulation showed that by far the most critical parameter in controlling water-flood patterns is the fault architecture: fault distribution, linkage, and throw combined with field stratigraphy. No amount of fault transmissibility adjustment can salvage a model in which the overall fault architecture is not generally correct. Building an accurate structural model and scaling it up to simulation grid blocks proved to be the most time-consuming part of the model-building process, in part due to the necessity of verticalizing the faults.

Since this study was completed, considerable progress has been made in software development to facilitate fault modeling. All the major vendors have released fault-modeling packages designed to guide the interpreter or modeler through the process of building a network of geometrically accurate fault grids. At least one vendor has introduced a module for fault transmissibility modeling. As software tools improve and seismic data increase in resolution, it will become practical to incorporate fault properties into geologic models as a standard practice.

CONCLUSIONS

Whether or not a fault will "seal" depends on a complex set of factors, including who is asking the question. Faced with a fault-dependent prospect, the explorationist can apply a set of established tools to predict the distribution of capillary seals on a fault or set of faults. The production geologist or reservoir engineer is more likely to be concerned with the impact of faults on fluid flow during production, when fault transmissibility is the controlling factor.

In either case, there is no substitute for an accurate 3-D understanding of fault architecture and how reservoirs juxtapose across faults. Fault seal prediction at any scale requires both accurate structure and accurate stratigraphy—that is why it is such a challenge to do it consistently well.

Acknowledgments*—The permission of Esso Exploration UK Limited and Shell Expro UK to publish data from Cormorant field is gratefully acknowledged. Badley Earth Science kindly gave permission to use illustrations from their FAPS documentation for Figure 3. The comments of an anonymous reviewer and David Reynolds improved the final manuscript.*

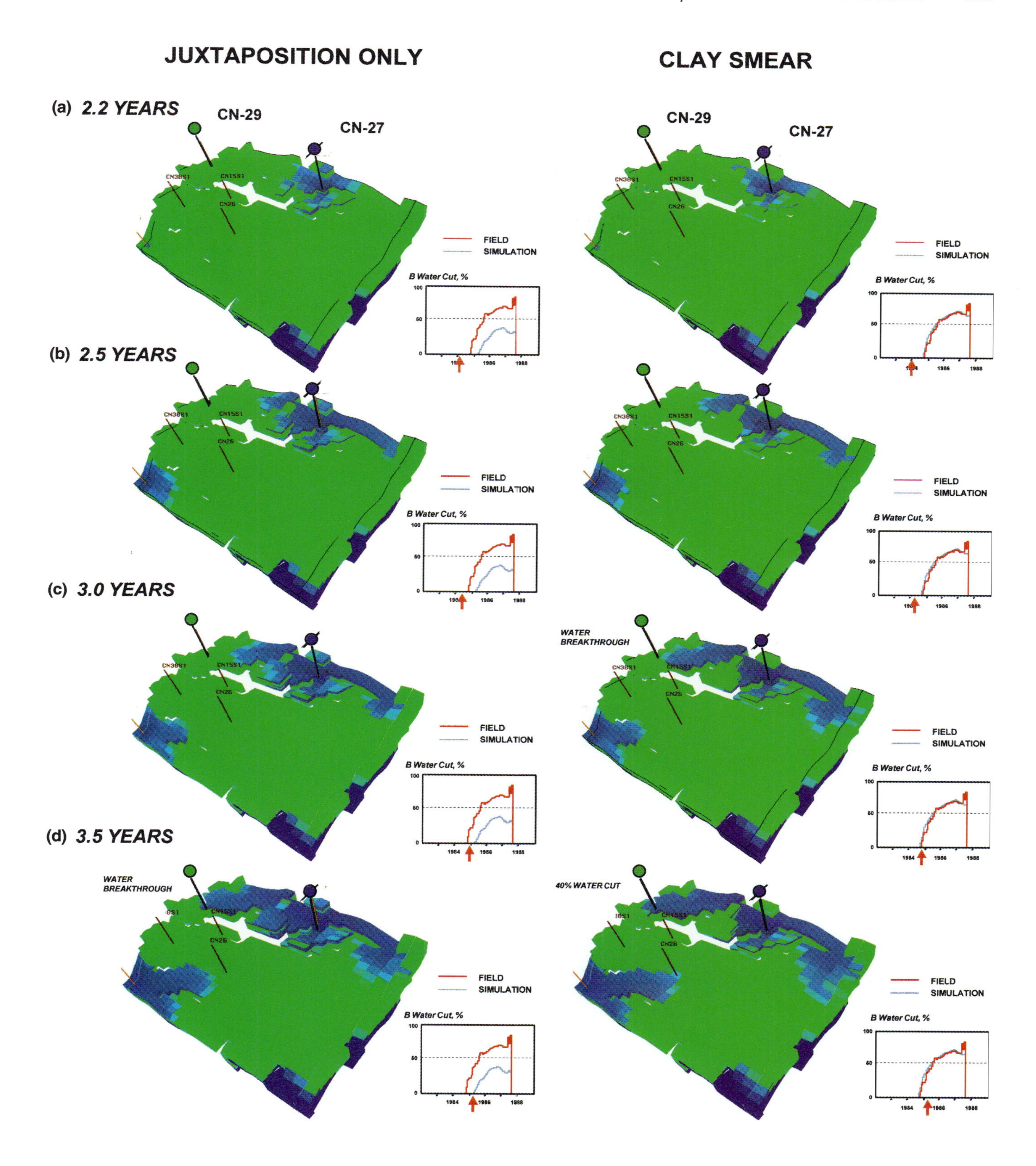

Figure 14—Comparison of water-flood performance between the juxtaposition-only and clay smear models. See text for detailed discussion.

REFERENCES CITED

Bentley, M. R., and J. J. Barry, 1991, Representation of fault sealing in a reservoir simulation: Cormorant Block IV, UK North Sea: Society of Petroleum Engineers, SPE 22667, p. 119–126.

Childs, C., J. Watterson, and J. J. Walsh, 1996, A model for the structure and development of fault zones: Journal Geological Society of London, v. 153, p. 337–340.

Childs, C., J. J. Walsh, and J. Watterson, 1997, Complexity in fault zones and its implications for fault seal prediction, *in* hydrocarbon seals—importance for exploration and production: Special Publication of the Norwegian Petroleum Society, v. 7, p. 71–72.

Downey, M. W., 1984, Evaluating seals for hydrocarbon accumulation: AAPG Bulletin, v. 68, p. 1752–1763.

Foxford, K. A., J. J. Walsh, J. Watterson, I. R. Garden, S. C. Guscott, and S. D. Burley, 1998, Structure and content of the Moab fault zone, Utah, U.S.A., and its implications for fault seal prediction, *in* G. Jones, Q. J. Fisher, and R. J. Knipe, eds., Fault sealing and fluid flow in hydrocarbon reservoirs: Geological Society of London Special Publication 147, p. 87–103.

Freeman, B., G. Yielding, D. T. Needham, and M. E. Badley, 1998, Fault seal prediction: the gouge ratio method, *in* M. P. Coward, T. S. Daltaban, and H. Johnson, eds., Structural geology in reservoir characterization: Geological Society of London Special Publication 127, p. 19–25.

Gibson, R. G., 1994, Fault zone seals in siliciclastic strata of the Columbus Basin, offshore Trinidad: AAPG Bulletin, v. 78, p. 1372–1385.

Goldthorpe, W. H., and Y. S. Chow, 1985, Unconventional modelling of faulted reservoirs: a case study: Society of Petroleum Engineers, SPE 13526, p. 279–294.

Jennette, D. C., and C. O. Riley, 1996, Influence of relative sea-level on facies and reservoir geometry of the Middle Jurassic lower Brent Group, UK North Viking Graben, *in* J. A. Howell and J. F. Aitken, eds., High resolution sequence stratigraphy: innovations and applications: Geological Society of London Special Publication, v. 104, p. 87–113.

Knott, S. D., 1993, Fault seal analysis in the North Sea: AAPG Bulletin, v. 77, p. 778–792.

Lindsey, N. G., F. C. Murphy, J. J. Walsh, and J. Watterson, 1993, Outcrop studies of shale smears on fault surfaces: Special Publication of the International Association of Sedimentologists, v. 15, p. 113–123.

Ottesen Ellevset, S., R. J. Knipe, T. Svava Olsen, Q. J. Fisher, and G. Jones, 1998, Fault controlled communication in the Sleipner Vest field, Norwegian continental shelf: detailed, quantitative input for reservoir simulation, and well planning, *in* G. Jones, Q. J. Fisher, and R. J. Knipe, eds., Fault sealing and fluid flow in hydrocarbon reservoirs: Geological Society of London Special Publication 147, p. 283–297.

Peacock, D. C. P., and D. J. Sanderson, 1984, Geometry and development of relay ramps in normal fault systems: AAPG Bulletin, v. 78, p. 147–165.

Schowalter, T. T., 1979, Mechanics of secondary hydrocarbon migration and entrapment: AAPG Bulletin, v. 63, p. 723–760.

Smith, D. A., 1966, Theoretical considerations of sealing and nonsealing faults: AAPG Bulletin, v. 50, p. 363–374.

Smith, D. A., 1980, Sealing and nonsealing faults in the Louisiana Gulf coast salt basins: AAPG Bulletin, v. 64, p. 145–172.

Taylor, D. J., and J. P. A. Dietvorst, 1991, The Cormorant field, Blocks 211/21a, 211/26a, UK North Sea, *in* I. L. Abbotts, ed., United Kingdom oil and gas fields, 25 Year Commemorative Volume: Geological Society of London Memoir 14, p. 73–81.

Watts, N., 1987, Theoretical aspects of cap-rock and fault seals for single- and two-phase hydrocarbon columns: Marine and Petroleum Geology, v. 4, p. 274–307.

Weber, K. J., G. Mandl, W. F. Pilaar, F. Lehner, and R. G. Precious, 1978, The role of faults in hydrocarbon migration and trapping in Nigerian growth fault structures: Offshore Technology Conference 10, OTC 3356, p. 2643–2653.

Converse, D. R., P. H. Nicholson, R. J. Pottorf, and T. W. Miller, 2000, Controls on overpressure in rapidly subsiding basins and implications for failure of top seal, *in* M. R. Mello and B. J. Katz, eds., Petroleum systems of South Atlantic margins: AAPG Memoir 73, p. 133–150.

Chapter 11

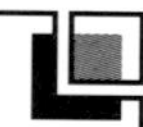

Controls on Overpressure in Rapidly Subsiding Basins and Implications for Failure of Top Seal

D. R. Converse

ExxonMobil Upstream Research Company
Houston, Texas, U.S.A.

P. H. Nicholson

ExxonMobil International Ltd.
London, U.K.

R. J. Pottorf

T. W. Miller

ExxonMobil Upstream Research Company
Houston, Texas, U.S.A.

Abstract

Understanding the principal mechanisms responsible for overpressure development in sedimentary basins is crucial for exploration and safe drilling in overpressured basins. Two mechanisms of overpressure generation, compaction–disequilibrium and stress unloading, have significant impacts on subsurface pressure distributions over geologic time. The contribution of different overpressure mechanisms can be analyzed using stress distribution and velocity data. In wells from both the North Sea and the Far East, mistaking the overpressure mechanism can lead to errors in pressure prediction. Further controls on overpressure are faults or stratigraphic changes that act as impediments to subsurface flow to create pressure cells. In the North Sea, faults are capable of supporting large pressure differences. Graben-bounding faults can also be important in transmitting deep fluids to shallower sandstones along the graben margin. In one example, the overpressure distribution away from a fault is controlled by the stratigraphic distribution of sandstones and thin interconnecting siltstones. Accurate pressure prediction from seismic velocities can help delineate pressure cell boundaries and aid in exploration. A key point is that a pressure difference across a pressure cell boundary indicates slow flow on a geologic time scale, not necessarily a lack of flow. Another level of control on subsurface pressure distribution is from leakage via top seal failure or fault leakage. Hydraulic seal failure is an important failure mechanism in many overpressured basins. Careful estimation of the total minimum stress is required to predict hydraulic top seal failure and is done using an empirical leakoff pressure versus depth trend. Combining both pressure and failure stress predictions, a simple method predicts the maximum hydrocarbon column that a top seal or fault can support. Often, a sizable hydrocarbon column can be trapped beneath a leaking top seal.

Subsurface overpressures in sedimentary basins can vary greatly over geologic time. Estimation of the stress variation is critical to evaluation of seal integrity and possible migration pathways. Evolution of overpressure can result in changes in migration pathways via the development of new leak points. We estimated the basic stress variation using a proprietary 1-D model of burial and stress evolution. Most models of pressure evolution are uncalibrated, as there are few paleobarometers in typical hydrocarbon-bearing sections. To address the question of controls on overpressure through time, we applied a useful paleobarometer derived from a comparison of coeval aqueous and hydrocarbon fluid inclusions trapped in the same cement. The difference in physical behavior of the inclusions allowed calculation of the pore pressure when the fluid inclusions were trapped. Combination of these results with a model of pressure history permitted estimation of the time of overpressure development and possibly hydrocarbon entrapment. The results of these calculations can be important, especially if overpressure development plays a major role in controlling hydrocarbon migration pathways and integrity of the seal.

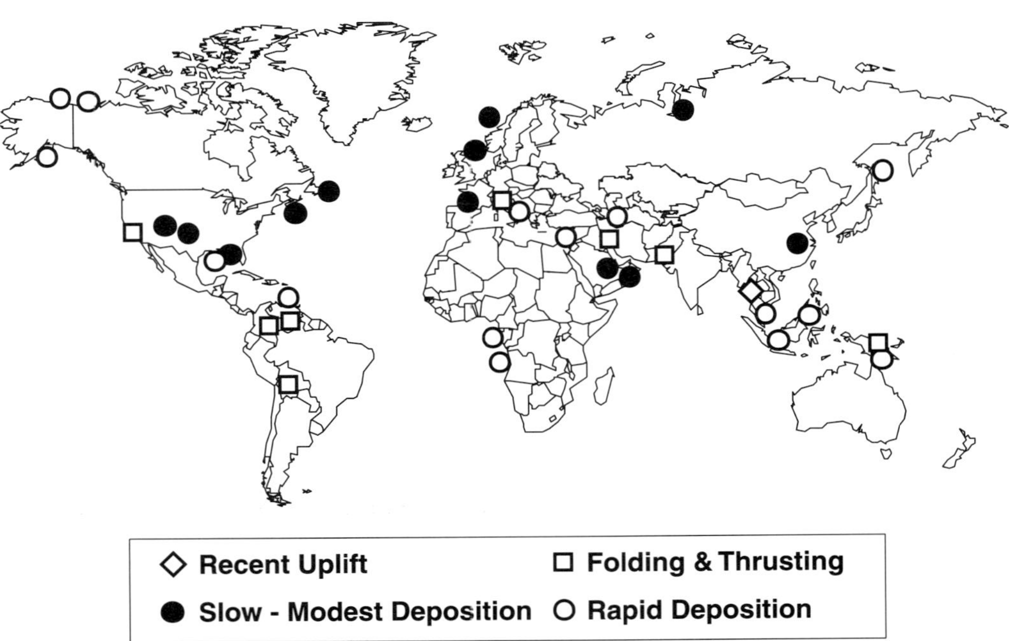

Figure 1—World map showing selected basins with hydrocarbon production that are overpressured. Our focus is on basins with rapid deposition, such as in the Gulf of Mexico, Beaufort–MacKenzie Delta, west Africa, Caspian Sea, and the Far East. Overpressures also develop in basins with slow to modest burial rates, in basins with tectonic deformation, and in some uplifted areas.

INTRODUCTION

In this chapter, we define *overpressure* as the fluid pressure above that measured in a static column of water at the same depth. Overpressured strata occur globally in a wide variety of tectonic environments (e.g., Fertl et al., 1994), from relatively quiescent cratonic basins with slow sedimentation rates, to rapidly deposited deltaic basins, thrust belts, and uplifted blocks, as shown in Figure 1. In many basins, a composite tectonic history contributes to the current overpressure configuration.

Hydrocarbon accumulations occur in reservoirs with pressures that range from hydrostatic to almost lithostatic (Figure 2). Many fields in production have high formation pressures such that the pressure gradient to surface exceeds 0.192 MPa/m (0.85 psi/ft), and in some fields the gradient approaches 0.23 MPa/m (1.0 psi/ft). A significant number of recent highly overpressured discoveries shown in Figure 2 are in the process of commercialization.

Examination of our global database does not indicate any limit for the occurrence of significant hydrocarbon accumulations. Thus, the explorationist should be optimistic about exploration targets in deep overpressured basins. Both oil and gas/gas condensate fields are likely to be overpressured, even severely overpressured. Gas fields, however, comprise more of the very deep, highly overpressured fields, a bias that probably reflects the cracking of oil reservoired at high temperatures to gas. In some basins, overpressure cutoffs have been claimed as the approximate limits for commercial hydrocarbon accumulations. For example, a limit of 0.192 MPa/m was reported as the upper bound for surface pressure gradients of commercial hydrocarbon accumulations in the Gulf of Mexico by Leach (1994) and Timko and Fertl

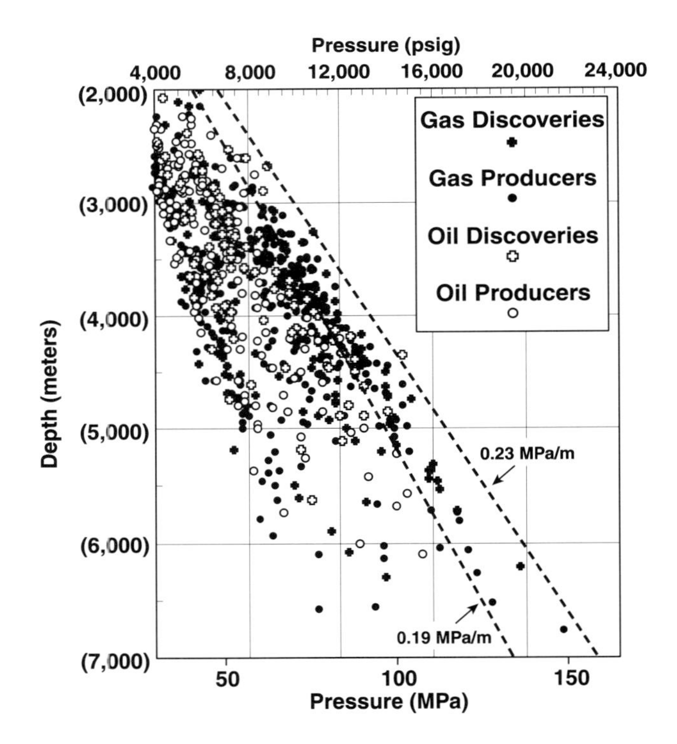

Figure 2—Graph of the worldwide distribution of oil and gas fields with respect to pressure and depth. Data are from a global compilation focusing on the characteristics of overpressured fields with significant overpressures, although some normally pressured fields are included for comparison. Note that with increasing depth and temperature, the number of gas fields relative to oil fields increases.

(1971). It is difficult to critically evaluate such a limit because the number of drilled penetrations of highly overpressured sections in these basins tends to be limited.

OVERPRESSURE MECHANISMS

Understanding the controls on fluid overpressure and the relationship between fluid pressure and hydrocarbon distribution can be important for successful hydrocarbon exploration in many basins. Two major mechanisms commonly contribute to the development of observed fluid overpressures within rapidly subsiding basins:

1. **Compaction–disequilibrium**—Overpressure results from the rapid deposition of low-permeability sediments that retard the escape of formation fluids (sometimes referred to as undercompaction; e.g., Magara 1978).
2. **Stress unloading**—Overpressure results from an increase in pore fluid volume (not structural inversion) which can be due to hydrocarbon generation, clay dehydration, diagenetic changes, and the transmission of overpressured fluids via faults (Chapman, 1994). *In our experience, this mechanism is often overlooked.*

Locally, the buoyant pressure of a large hydrocarbon column can also contribute significantly to overpressure in the reservoir section.

Determination of the principal mechanism of overpressure generation is important for correctly predicting reservoir quality and pathways of migrating and leaking hydrocarbons. Compaction–disequilibrium typically results in undercompacted sediments that may retain higher porosity and permeability than normally compacted sediments, such as in the Gulf of Mexico (Bowers, 1994, his figure 7). In an overpressured basin with significant stress unloading due to hydrocarbon generation, identifying regional leak points where overpressured fluids are transmitted along faults or through seals can play a role in exploration success. Reservoir stability during production is also very dependent on the cause of overpressure, as is successful predrill pressure prediction.

(a)

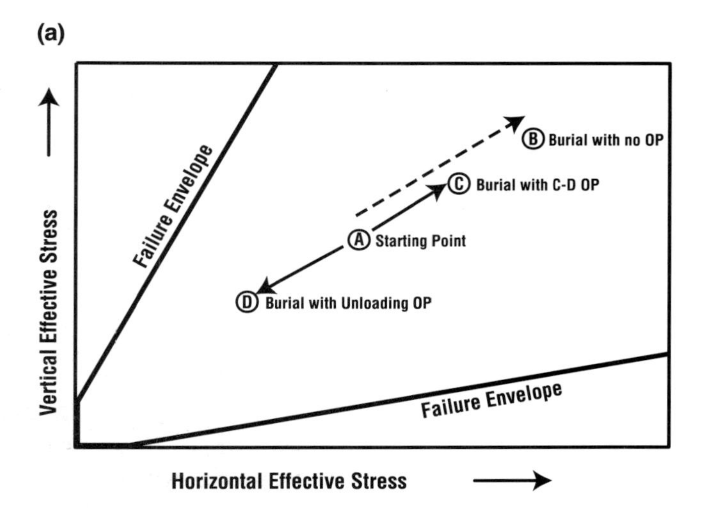

(b)

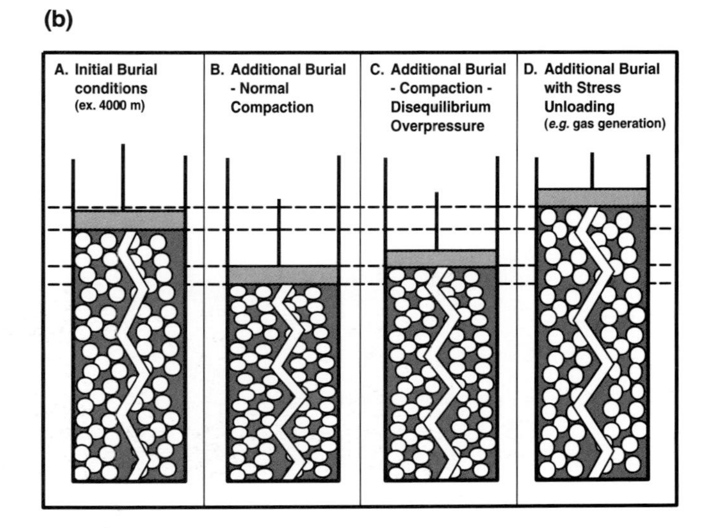

Figure 3—(a) Graph of vertical versus horizontal effective stress showing stress trajectories for strata undergoing normal compaction, compaction-disequilibrium and stress unloading. A typical failure envelope for shale is shown for reference. (b) The physical changes undergone by porous rocks following the stress trajectories in (a). See text for discussion.

Effect of Overpressure Mechanisms on Subsurface Stress Distribution and History

Different mechanisms of overpressure lead to different subsurface stress distributions and history (e.g., Miller, 1995; Miller et al., 1998), as shown in Figure 3a. The *vertical effective stress* is defined as the total vertical stress (lithostat) minus the fluid pressure, and the *horizontal effective stress* is defined as the total horizontal stress minus the fluid pressure. For a sediment undergoing normal burial with no overpressure development, the effective stresses increase with depth, as shown by the trajectory from A to B in Figure 3a. In the case of overpressures developed due to compaction–disequilibrium, the effective horizontal and vertical stresses within a stratigraphic unit continue to increase with burial (trajectory A to C), but not as rapidly as in the case of normal burial (in an extreme case the effective stresses can remain constant). In contrast, overpressures that arise from unloading mechanisms (such as gas generation) result in a decrease of effective horizontal and vertical stresses with burial (trajectory A to D). *Only stress "unloading" mechanisms lead to lower effective stresses as a stratigraphic unit is buried.*

Figure 3b depicts the physical changes undergone by a representative porous rock volume that follows the stress

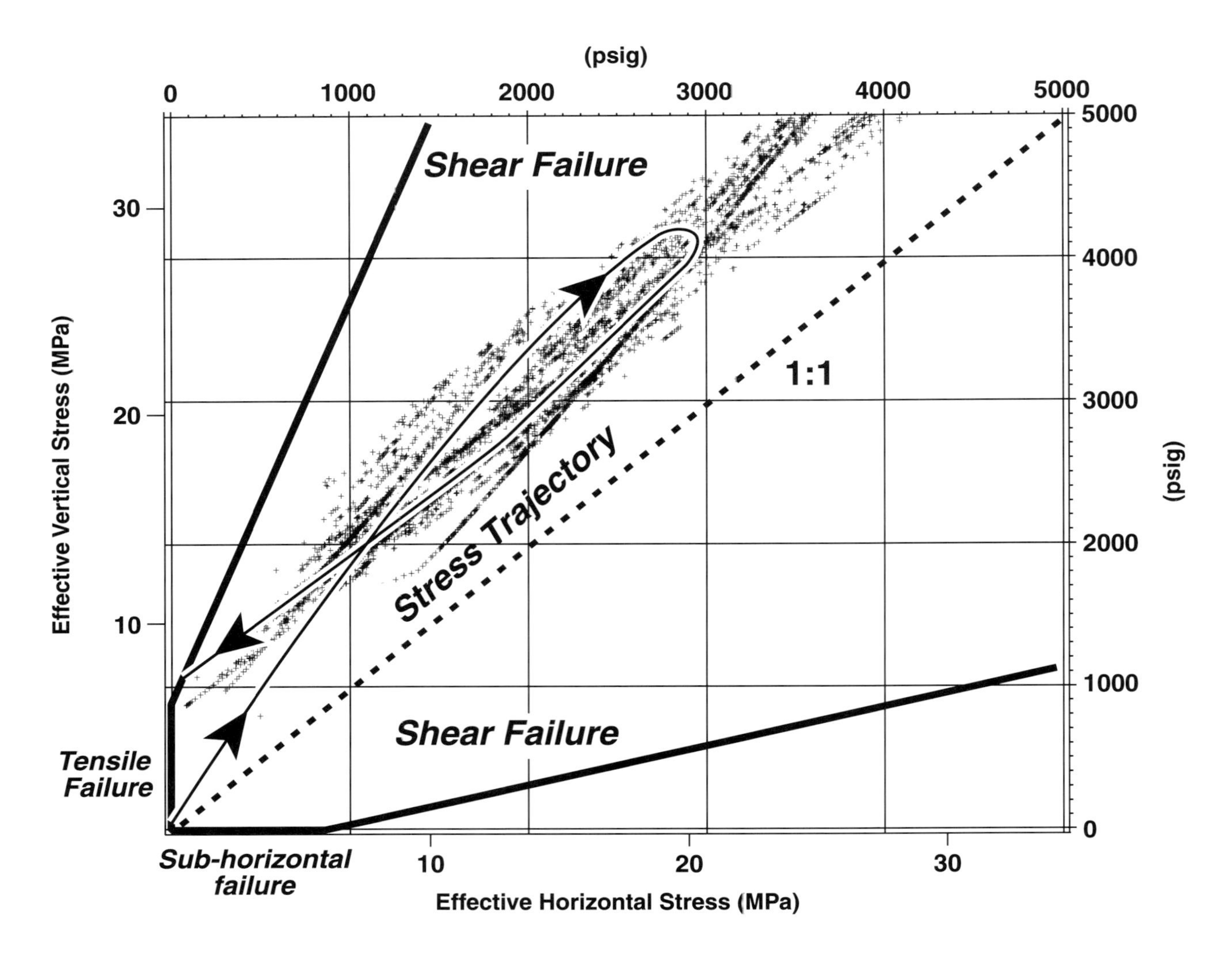

Figure 4—Effective stresses for the Central North Sea and North Viking Graben plotted relative to a generic shale failure envelope (dark lines), modified after Hareland et al. (1993). The dotted line represents where the horizontal and vertical effective stresses are equal.

trajectories shown in Figure 3a. The starting volume is represented by A (at a nominal depth of 4000 m). Additional burial with normal compaction leads to a slight decrease in volume (B). In the case of compactional disequilibrium (C), the fluid pressure increases and compaction is retarded, yielding a smaller decrease in volume than in B. In the case of stress unloading (D), the fluid volume actually increases and results in a slight increase in the sediment volume.

Subsurface Distribution: North Sea Example

A significant component of overpressure in the Mesozoic sections of the North Sea is derived from stress unloading. The evolution of stress within a stratigraphic unit is difficult to capture because most measurements reflect only the present stress configuration. Effective stresses for the Central North Sea and North Viking Graben are plotted in Figure 4 relative to a generic shale failure envelope (after Hareland et al., 1993). The points plotted represent more than 16,000 wireline pressure measurements and more than 1,400 leakoff tests.

The vertical effective stresses in Figure 4 were determined by first calculating the lithostatic load (assuming pure shale parameters) and then subtracting the fluid pressure measurements at each depth. The horizontal effective stresses were calculated using empirical leakoff trends. We assumed that the leakoff pressures were equal to the total minimum horizontal stress and then subtracted the fluid pressure at each depth. The assumption that the leakoff pressure equaled the total minimum horizontal stress caused the effective vertical stress to always exceed the effective horizontal stress. However,

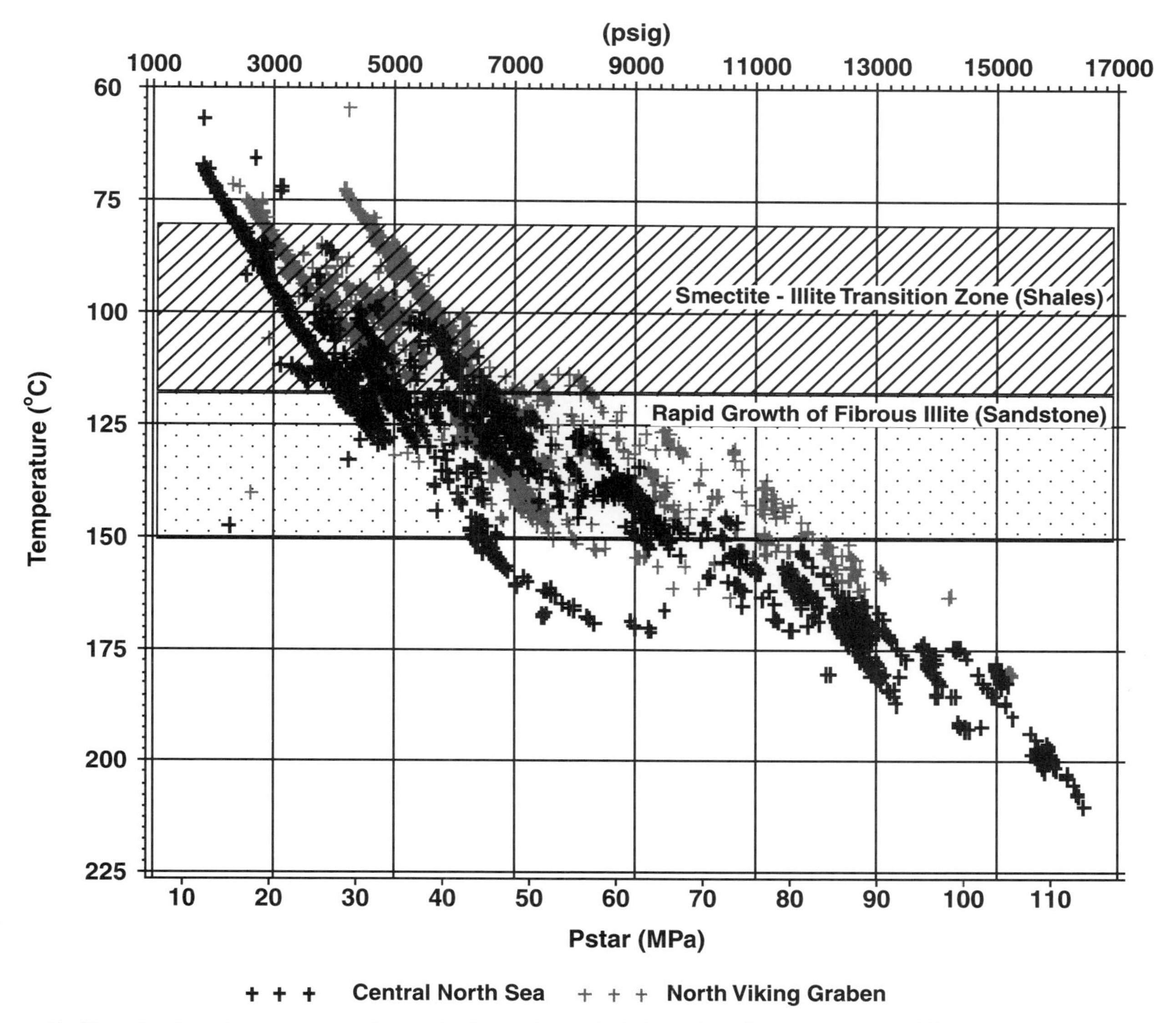

Figure 5—Plot of subsurface pressures from wireline tests against the subsurface temperature. The data from both the Central North Sea and the North Viking Graben appear to exhibit similar behavior with an abrupt increase in pressures at temperatures of 115–140°C. Some low pressures persist to greater temperatures in areas of the Central North Sea, such as where the Mesozoic section is juxtaposed against the permeable Rotliegendes Sandstone.

data from the North Sea suggest that in some deeper overpressured horizons, the total vertical stress may be the minimum stress.

A schematic representation of the stress pathway through time for the Mesozoic sedimentary rocks is overlain on Figure 4 as the line with arrowheads. This pathway was determined by 1-D pressure modeling calibrated against the present stress distribution. As indicated by the typical profile of decreasing effective stress with burial, incorporation of a stress unloading term *was required* to match a present "generic" configuration. The calculated trajectory shows an initial increase in effective stress with burial and compaction to a maximum value followed by stress unloading, which results in a significant reduction in both vertical and horizontal effective stresses. With increasing time and burial depth, the trajectory approaches the failure envelope, increasing the probability of rock failure.

There is a clear relationship between the distribution of overpressure and temperature in the North Sea, providing additional support for the stress unloading contribution. In both the North Viking Graben and the Central North Sea, there is a significant increase in overpressure between 115 and 140°C (Figure 5). This increase in pore pressure is largely due to stress unloading that may originate via hydrocarbon generation, hydrocarbon cracking, and diagenetic changes affecting permeability. As shown in Figure 5, the change in overpressure corresponds

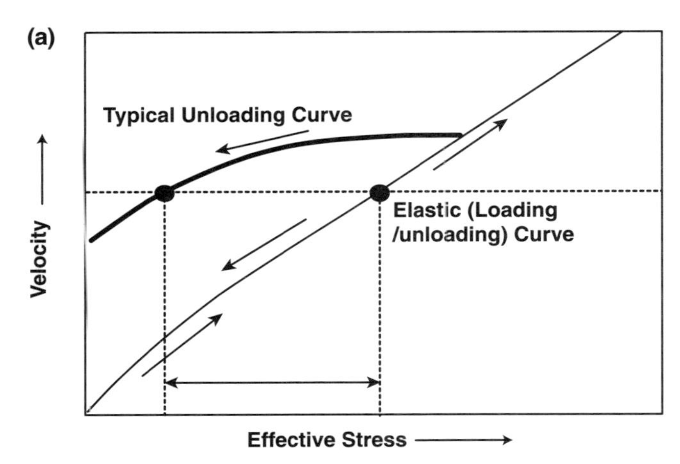

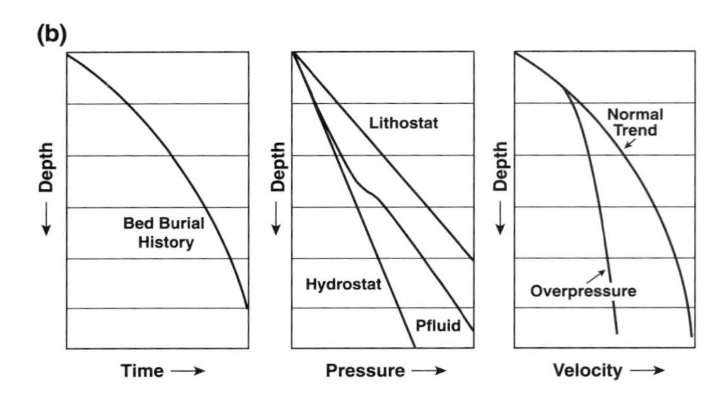

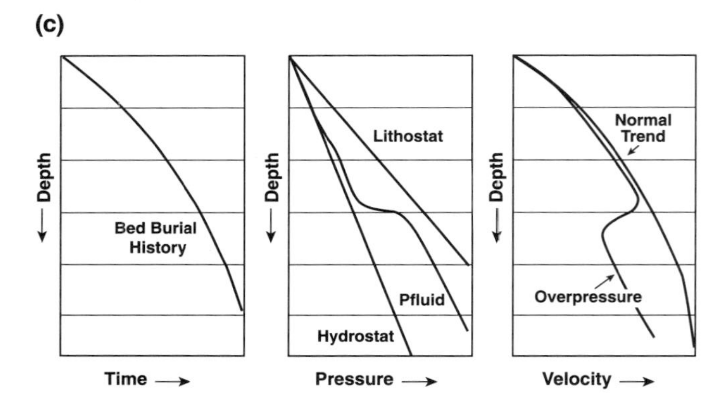

Figure 6—(a) Plot showing the variation of velocity with effective stress for both the compaction–disequilibrium (elastic) and stress unloading mechanisms. Note that for a specific velocity, the overpressure associated with the stress unloading curve is much greater than the compaction-disequilibrium curve. Modified from Bowers (1993). (b) Plots showing the evolution of pressure and velocity with burial depth for a single shale stratigraphic interval for a compaction–disequilibrium case. (c) Plots showing the evolution of pressure and velocity with burial depth for a single shale stratigraphic interval for a stress unloading case.

reasonably closely to the temperature of rapid growth of fibrous illite and silica cement (Bjorkum and Nadeau, 1998) as well as to the temperature range for main stage hydrocarbon generation (Tissot and Welte, 1984).

Velocity Expression of Overpressure

Recognition of the mechanism of overpressure generation is also very important for predicting pressures in wells while drilling. Prediction of overpressures from changes in sonic velocity curves or seismic gathers assuming a compaction–disequilibrium mechanism is very different (typically much lower) from the prediction of overpressures assuming an unloading (source) mechanism. Many empirical correlations exist for predicting fluid pressure based on sonic velocities (e.g., Hottmann and Johnson, 1966; Weakley, 1991; Kan and Sicking, 1994). These methods typically ignore the cause of overpressure. Bowers (1994) published a method for predicting overpressure that uses both the effective stress concept and sonic velocities as well as including the mechanism of overpressure generation. We use this method routinely to calculate overpressure.

The importance of distinguishing between the pressure mechanisms in accurate pressure prediction is illustrated in Figure 6. As discussed by Bowers (1994), sonic velocity varies with compaction and unloading. Velocities increase with compaction, which is generally an irreversible process; most rocks do not exhibit elastic unloading. Typically upon unloading, small decreases in velocity are associated with large changes in effective stress (Figure 6a). The key question for accurate pressure prediction is, What is the effective stress indicated by the velocity?

Figure 6b shows the evolution of pressure and velocity with burial depth for a single shale stratigraphic interval for a compaction–disequilibrium case. For the burial history shown on the left, the center plot shows the well profile of fluid pressure (effective stress = lithostat − P_{fluid}) and the plot on the right shows the shale sonic velocity profile. The sonic velocity continues to increase with depth but only very slowly as the fluid pressure increases. Both the effective stress and the overpressure continue to increase with depth. For comparison, Figure 6c shows the evolution of pressure and velocity with burial depth for a single shale stratigraphic interval for an unloading case. Here, the unloading mechanism is dominant below a key depth, as indicated by the sharp cutback in the shale velocity with depth. The effective stress decreases and the fluid pressure increases significantly at the depth of the velocity cutback. A much greater fluid pressure is predicted relative to the pressure caused by compaction–disequilibrium.

Velocity Expression: Far East Example

Using Bowers (1994) technique (described above), we evaluated the two different mechanisms for a well in a rapidly deposited basin in the Far East (Figure 7). The

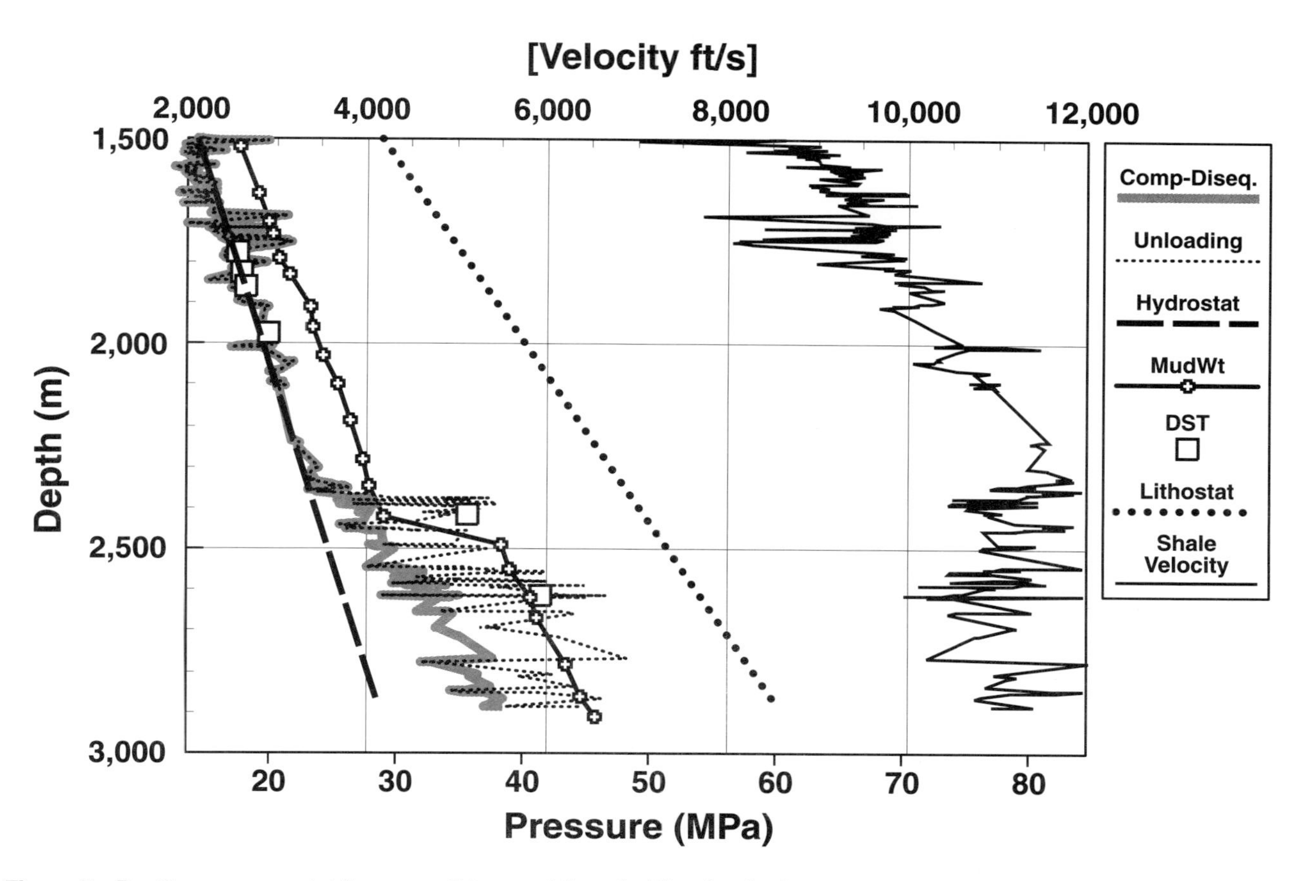

Figure 7—Depth–pressure plot from a well in a rapidly subsiding basin that compares the different predictions of subsurface pressures obtained from well log analysis using compaction–disequilibrium and unloading models. The fluid pressure predicted from a compaction–disequilibrium model underestimated the drill-stem pressures by 1200–1500 psi in the overpressured section. The pressure predicted from an unloading model more accurately matched the measured fluid pressures in the reservoirs.

shale velocity curve was derived from the sonic log by setting a shale cutoff parameter (a gamma ray cutoff) and specifying a minimum shale thickness. The shale velocity increased gradually with depth until about 2200–2300 m, below which depth the average shale velocity decreased. The mud weights used to drill the well and the results of drill-stem tests (DSTs) are superimposed on Figure 7. The upper part of the well is hydrostatic, but the deeper part is significantly overpressured.

The predictions of both the compaction–disequilibrium model and stress unloading models are shown in Figure 7. The compaction–disequilibrium model predicted pressures that were 8.3–10.3 MPa (1200–1500 psi) less than the DST pressures. The predictions of the unloading model were much more accurate. However, the unloading exponent had to be calibrated at a separate location. The likely stress unloading mechanism is the transmission of overpressured fluids from an underlying graben section. Although the average or trend predictions are sensible, both models predicted high-frequency variations in pressure relative to depth that most likely represent noise. The noise is probably due to variations in shale lithology and well bore geometry.

In the case of the North Sea examples discussed later, seismic resolution of overpressure magnitude is typically less than that from predictions based on geologic models and offset well data. However, in less mature basins, the pressure predictions from seismic data are key inputs to seal and migration models.

GEOLOGIC CONTROLS

Pressure Cells

At a more local scale, the subsurface pressure distribution is also strongly influenced by geologic factors such as faults and variations in stratigraphic facies. We define a *pressure cell* as a geographic area in which the stratigraphic interval of interest has a single water head or excess pressure:

$$P_{excess} = P_{measured} - \rho_w \bullet g \bullet z,$$

where ρ_w is the in situ water density, g is gravitational acceleration, and z is the depth (similar to Hunt, 1990).

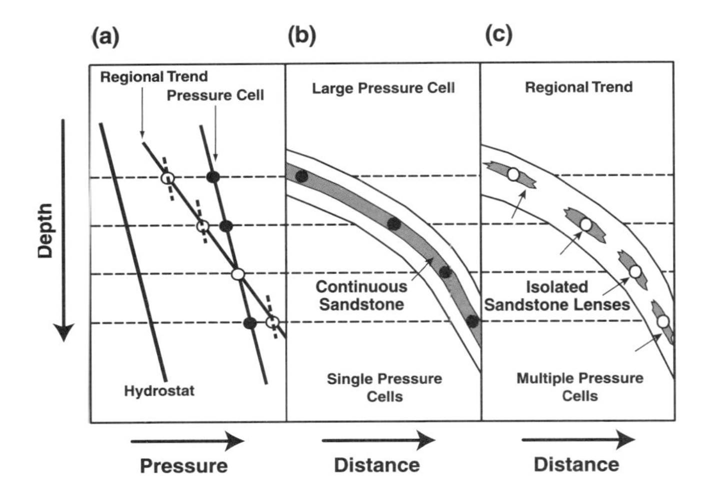

Figure 8—(a) Plots of three pressure–depth trends: a normal hydrostatic, a regional pressure trend, and a single pressure cell. (b) Schematic cross section of a single continuous sandstone with significant depth extent whose pressure–depth trend is represented by the pressure cell trend in part (a). Note that the pressure cell trend parallels the hydrostatic trend. (c) Schematic cross section of isolated sandstone lenses over a large depth range. The pressure-depth trend in each lens parallels the hydrostatic trend, but the pressure between each lens increases due to increasing stress unloading with depth. This pressure trend is represented as the regional pressure trend in part (a).

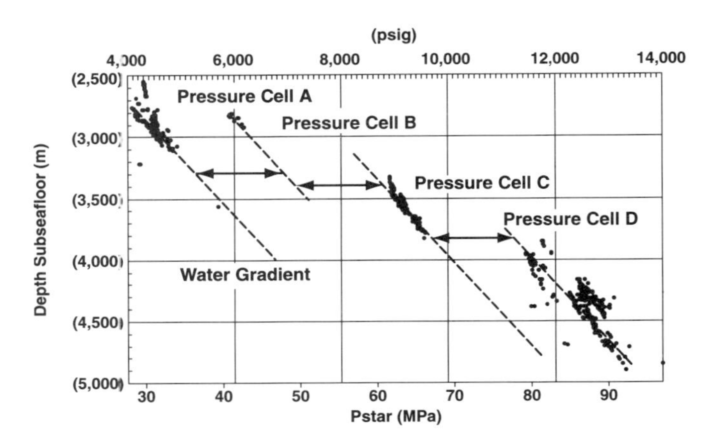

Figure 9—Plot of four separate pressure cells (A–D) developed in four adjacent fault blocks. The excess pressure increases progressively with depth between cells A, C, and D, suggesting a depth-dependent pressure mechanism. The contrast in excess pressure between cells A and B at the same depth indicates that the faults can hold significant differences in excess pressure.

Pressure cells have been identified by many authors (e.g., Hunt, 1990; Gaarenstroom et al., 1993; Nicholson and Hart, 1995).

How pressure cells can modify subsurface pressure distributions is shown in Figure 8. In this example, the mechanism responsible for overpressure continues to increase the overpressure with depth, forming a regional pressure trend. Geologically, such a system might consist of shales with isolated sandstone lenses that each closely reflect the pressures of the neighboring shales. Now consider the distribution of pressure that is well connected over a significant vertical extent. In this case (Figure 8b), the pressure in the sandstone follows either a hydrostatic gradient as shown or a hydrodynamically modified gradient if there are significant aquifer flows. The difference in pressures at the crest of the structure between the large pressure cell and the overlying seals can lead to large pressure gradients that can affect fracture development.

The geologic controls on the subsurface distribution of pressure cells can be faults, as inferred for the pressure cells shown in Figure 9. These faults appear to be capable of supporting pressure differences of thousands of pounds per square inch, although it is important to understand that this pressure difference reflects retarded flow on a geologic time scale, not a lack of flow. These boundaries may or may not retain commercial quantities of hydrocarbons. We will return to this theme later, when we consider fault leakage. See Darby et al. (1996) for additional examples of structurally controlled pressure cells in the North Sea.

Pressure cells can also be dependent on both stratigraphic and structural controls (Nicholson et al., 1995), as shown in Figure 10. In this example, the reservoir units occur as discrete bodies within shales. The overpressure is not developed in situ but is associated with pressure and fluid transmission along a graben-bounding fault (on the left) that taps a deep highly overpressured section. The fluid overpressures are dissipated via fluid movement away from the graben-bounding fault through thin, low-permeability silt conduits connecting the individual reservoir bodies. The fault serves as the major migration conduit for mature hydrocarbons exiting the deep graben. Sandstone bodies with very low excess pressures lying near the graben-bounding fault are probably not hydraulically connected to the fault and have low exploration potential. Thus, correct pressure prediction can be very important for exploration success. The required resolution may be on the order of only tens of megapascals.

Leakage Mechanisms

The ultimate control on subsurface pressures is rock failure resulting in fluid leakage. Fluid leakage from a hydrocarbon trap can occur via three principal mechanisms:

1. Capillary seal failure, when the buoyancy pressure within the hydrocarbon column exceeds the capillary entry pressure of the seal (Berg, 1975; Schowalter, 1979);

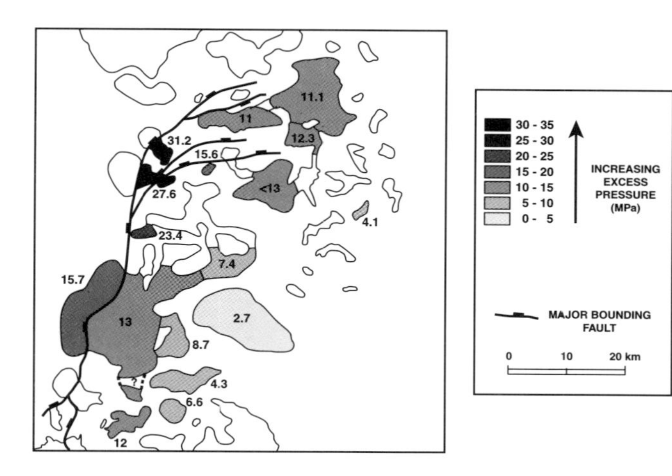

Figure 10—Map of pressure cells showing the influence of a combination of structural and stratigraphic components. Pressures were measured in irregularly shaped sandstone bodies that rest within shale. The overpressure is not developed in situ but is associated with pressure (and fluid) transmission along a graben-bounding fault on the left. The overpressure is dissipated as fluids move away from the fault along less permeable silty zones (not shown) into geographically restricted zones of sandstone accumulation.

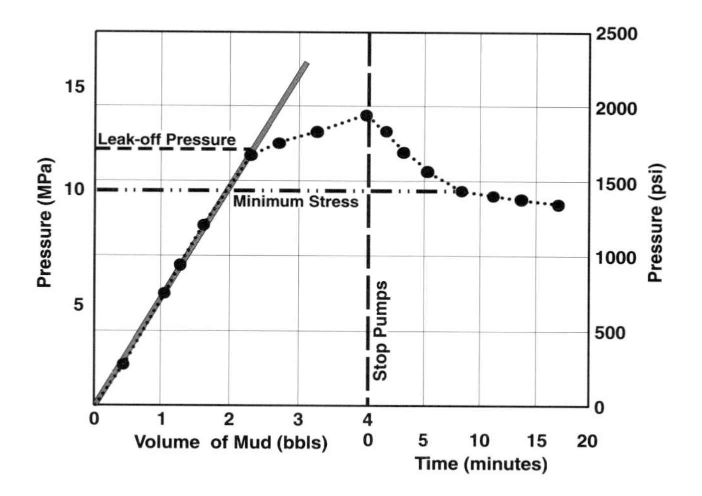

Figure 11—Plot of an extended leakoff test showing an estimate of the total minimum stress. However, the most common available data are the leakoff pressures, which are greater than the total minimum stress (see text for discussion). Modified from Kunze and Steiger (1992).

2. Mechanical failure, when the seal is either fractured or displaced due to tectonic stresses (Watts, 1983; Koch et al., 1990); and
3. Hydraulic seal failure, when the combination of normal or high aquifer pressures plus the hydrocarbon buoyancy pressure exceeds the fracture pressure of the seal (Mandl and Harkness, 1987; Watts, 1987).

The focus here is on hydraulic seal failure and its impact on both trap integrity and migration pathways.

Establishing Fracture Criteria

The first step in predicting hydraulic seal failure is to establish the criteria for fracturing. The limit of rock integrity in sedimentary basins is commonly determined from leakoff tests, which measure the total pressure required to fracture the rock in the region near the well bore (Figure 11) (Kunze and Steiger, 1992). In almost all cases, leakoff tests are made in shales or other low-permeability rock for reasons of drilling safety. The extended leakoff test will yield the fracture closing stress (minimum stress), but in practice, leakoff tests are rarely extended. In the few examples of extended leakoff tests or mini-frac tests that were available to us from the North Sea, the difference between the leakoff pressure and the minimum stress was less than 5%. Perhaps of more importance was the uncertainty in whether or not leakoff was actually attained—many tests are not carried to a clear leakoff because of fears that the well bore could be damaged. In our analysis, we assumed that the leakoff pressures are approximately equal to the minimum horizontal stress. Breckels and van Eekeln (1982) and Daines (1982) made similar approximations in their analysis of sedimentary basins.

Comparison of leakoff tests from different parts of a basin requires adjusting the data to compensate for water depth variation, as shown in Figure 12. All leakoff pressures are plotted with respect to depth below the mudline, and the pressure due to the sea water column above the mudline is subtracted from the measured leakoff pressure (P_{star}). This correction can be substantial in basins with a wide range of water depths.

Leakoff pressures within a sedimentary basin vary with depth as a function of lithology and fluid pressure. In basins in which we have worked, lithology has had a secondary effect on the leakoff pressure and will be ignored in this discussion. The effect of fluid pressure can be quite significant. Leakoff data from the North Sea are plotted against depth below mudline in Figure 13. At depths below 3200–3400 m, there is a substantial increase in the leakoff pressures that corresponds to the sharp increase in fluid pressure.

Total horizontal stresses (as measured by leakoff tests) increase with fluid pressures because of a combination of several factors. The most important is that in the subsurface, lateral strains are much more constrained than vertical strains. The physical reasons for this (Miller et al., 1998) are (1) that the earth's surface is free to move so that the overburden load is essentially independent of vertical deformation, and (2) that adjacent rocks prevent free horizontal deformations so that horizontal loads increase or decrease when lateral deformations are outward or inward, respectively. Consequently, increased pore pres-

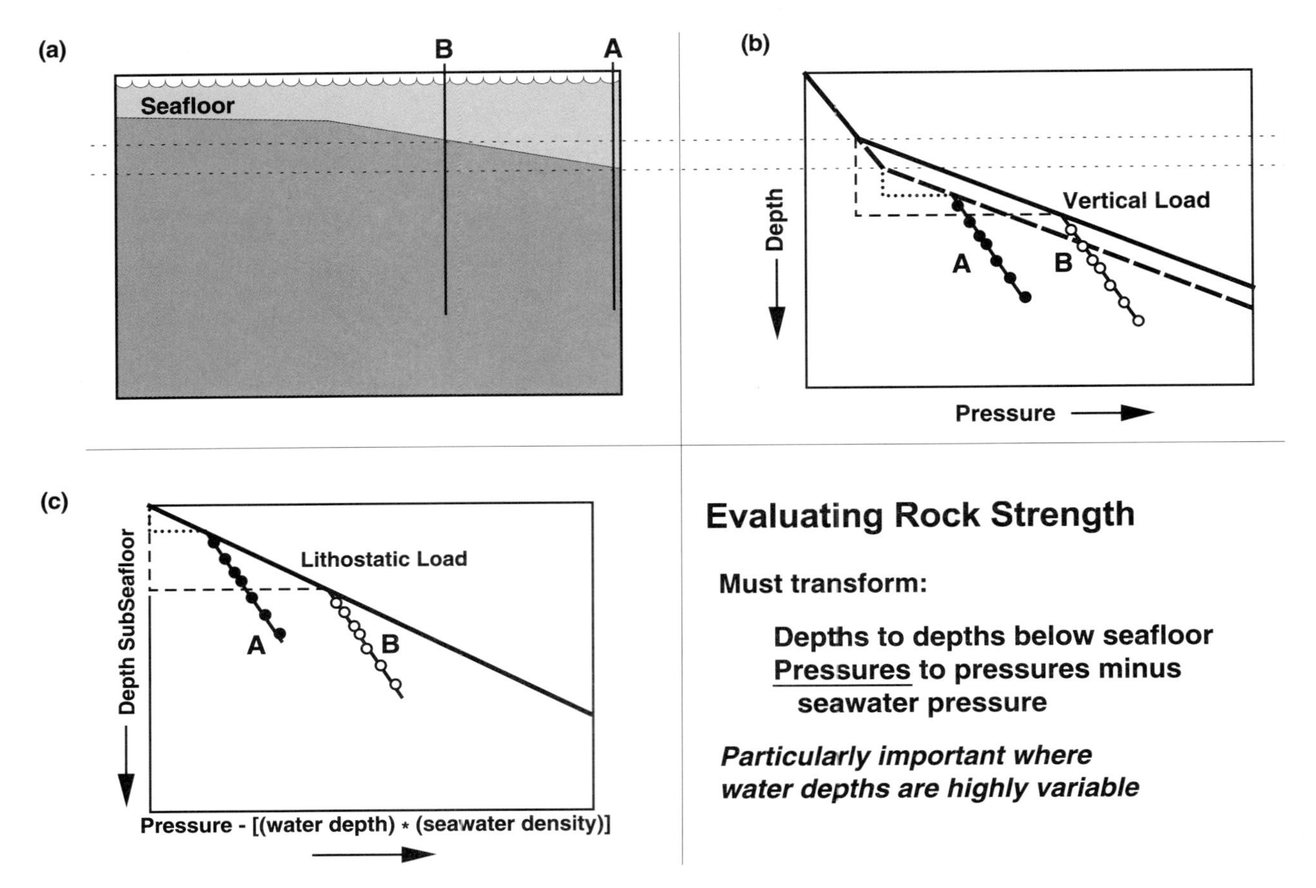

Figure 12—Comparing fracture and formation pressures from subsurface reservoirs from areas with varying water depths requires transformation of depths and stresses relative to the sea floor. (a) This transformation is shown by two wells, A and B, in different water depths. (b) Untransformed pressure–depth plot showing the difference in vertical load between the two wells. (c) Plot showing the depths transformed to depth below mudline and subtracting the pressure due to the sea water column. Here, both wells share a common vertical stress load as a function of depth below mudline.

sures, which tend to expand the rock, increase the horizontal total stresses, while the vertical total stresses are locally unchanged.

Our choice of the failure trend in Figure 13 is different from that of Breckels and van Eekeln (1982) and Gaarenstroom et al. (1993) who use the minimum envelope of the leakoff trend. Based on our global experience, we prefer to use a mean value derived from the empirical relationship between leakoff pressure and depth, and use the standard deviation about the fit to evaluate the uncertainty. There are a few commercial hydrocarbon-bearing traps in which the fluid pressure exceeds the minimum envelope of the regional or even subregional leakoff trend, but in our experience, none exceed the mean leakoff trend.

LEAKAGE MODEL

Our overall model is that hydraulic seal failure occurs when the fluid pressure in the hydrocarbon column approaches the leakoff trend (Figure 14) (Watts, 1987; Gaarenstroom et al., 1993; Converse et al., 1995; Heum, 1996). This model successfully reproduces the pressure distribution in the North Sea, as shown in Figure 13 where the formation pressures from wireline tests approach but do not exceed the mean leakoff trend. Hydrocarbon retention in a leaking trap can depend on the source of and controls on overpressure.

Three basic styles of leakage can occur within a single basin. Each has different implications for hydrocarbon exploration:

> **Case 1**—The aquifer pressure is significantly less than the fracture pressure, but the hydrocarbon buoyancy causes the fluid pressure at the crest to equal the fracture pressure.

In this case, a delicate balance exists between the hydrocarbon pressure and the fracture pressure. The addition of hydrocarbons to the trap raises the crestal pressure, which causes hydrocarbons to leak from the crest until the pressure falls below or equal to the fracture pressure (Mandl and Harkness, 1987; Engelder and

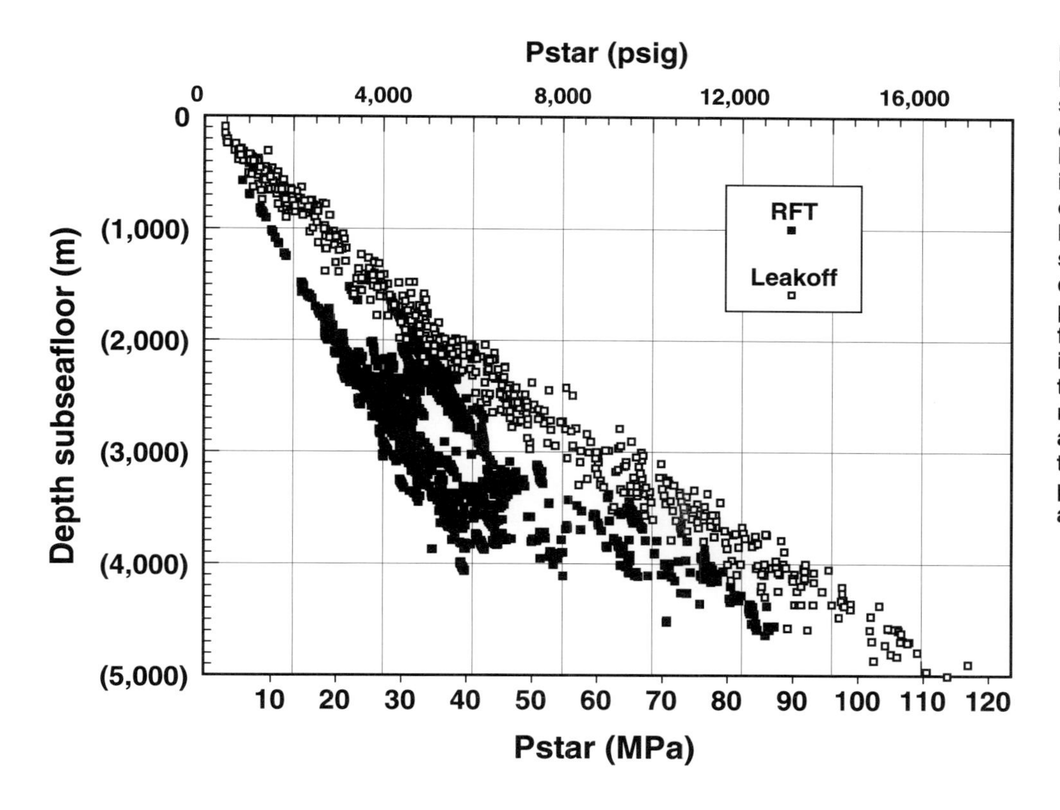

Figure 13—Plot of adjusted leakoff and formation pressures (adjusted for water depth = P_{star}) versus depth below mudline. Rock integrity is the ultimate control on overpressure. Here, both the leakoff and formation pressures are shown versus depth below the sea floor. The pressure axis has been modified to compensate for varying water depths. It is clear that formation pressures do not exceed leakoff pressures, although in some locations, the formation and leakoff pressures approach one another.

Leftwich, 1997). Some overshooting and undershooting of the equilibrium configuration can be expected, but a significant hydrocarbon column should be retained.

Case 2—The aquifer pressure fluctuates with time due to fluid losses at other locations.

Here, the same delicate pressure balance exists between the hydrocarbon pressure and the seal fracture pressure as in the previous model. However, the column height will fluctuate with time—shrinking as the aquifer pressure builds up in the pressure cell and expanding as the aquifer pressure is relieved at another location (Nicholson and Hart, 1995).

Case 3—Tectonic deformation reduces the stresses needed to open or form fractures in the seal.

In this case, because the hydrocarbon buoyancy has little effect on the opening or closing of fractures, the entire hydrocarbon column can be lost. Examples include reservoirs above remobilized salt diapirs, active thrust belts, and reactivated horst blocks (Converse et al., 1994; Koch et al., 1990).

Application of Leakage Model

This model can be useful in prospect evaluation. The following parameters are required to apply this model: (1) pressure prediction in the aquifer leg, (2) crestal elevation, (3) in situ hydrocarbon gradient, and (4) regional or local leakoff trend. With this information, the *maximum* column height for either all gas or all oil can be easily calculated, as shown graphically in Figure 14. In this example, the maximum gas column is less than the gas column that would be predicted from a structural assessment of the prospect. To determine the maximum column height graphically, start from the leakoff trend at the depth of the structural crest and assume that the hydrocarbon pressure equals the leakoff pressure. Then extend the in situ hydrocarbon gradient downward until it intersects the aquifer gradient. The maximum oil column height will always exceed the maximum gas column height due to the greater buoyancy pressure in the gas column (Watts, 1987).

The most sensitive parameter in this prospect evaluation is the prediction of the formation pressure. An error of just a few MPa (several hundred psi) in the prediction of the formation pressure can lead to a mistake in the interpretation of the seal integrity.

Example of Hydraulic Top Seal Leakage

In our example from the North Sea (Figure 15a), the formation pressures range from essentially hydrostatic to significantly overpressured (~14 MPa (2000 psi) above hydrostatic). The leakoff pressures exhibit a relatively small standard deviation about a linear least-squares fit to the data. In the shallowest oil reservoir in the field of interest, the formation pressures closely approach the mean leakoff trend and significantly exceed the minimum envelope to the regional leakoff pressures. As

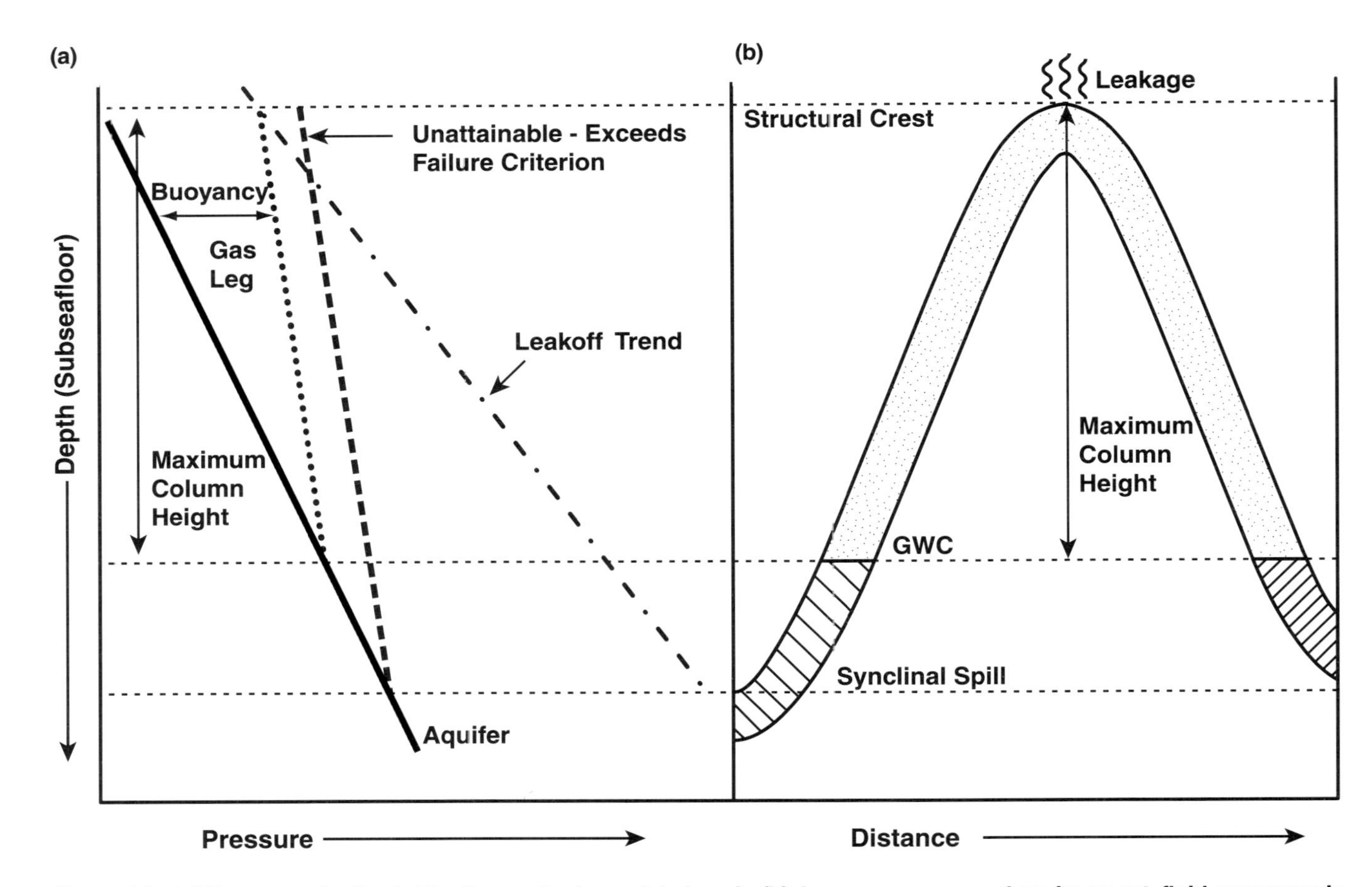

Figure 14—(a) Pressure–depth plot for the geologic model show in (b). In an overpressured environment, fluid pressures in reservoirs can approach the minimum seal stress (commonly approximated as the leakoff pressure). The maximum column height that can be trapped in a reservoir depends on the fluid pressure, the buoyancy between the hydrocarbon and the formation waters, and the leakoff pressure. Note that in this example, the trap is not filled to structural spill.

shown on the seismic line in Figure 15b, there are pronounced zones of disruption (possibly gas chimneys) above this reservoir, in addition to other leakage features such as shallow gas reservoirs. The probable source of the gas is exsolution of gas from leaking oils. Another field in this same area (not shown) exhibits a similar relationship between formation pressure and the leakoff pressure trend, but instead of overlying gas chimneys, significant hydrocarbon shows were found throughout the seal interval during drilling.

A key feature in these examples is that commercial quantities of hydrocarbons are still retained although the top seals clearly leak hydrocarbons. This relationship between leakage and retention is consistent with a model (Case 1, previously detailed) of a delicate balance between hydrocarbon pressure and fracture pressure controlling episodic losses of hydrocarbons while the seal retains a significant hydrocarbon column (Mandl and Harkness, 1987). It is also clear from this example that, if the trapped hydrocarbon were gas, then the hydrocarbon column would be much smaller.

Example of Hydraulic Leakage Affecting Migration

Two Mesozoic gas condensate accumulations in the North Sea are shown in Figure 16a. These accumulations are separated by a fault (between wells A and B). The eastern accumulation is a highly overpressured, fairly dry gas condensate. The western accumulation is less overpressured (14 MPa less), with a variable hydrocarbon composition that becomes wetter to the west.

The source rock for the reservoired hydrocarbons is absent in the vicinity of this field. Our hydrocarbon system analysis of this area determined that the hydrocarbons were derived from a source kitchen to the east and migrated westward through well B to well D and beyond. The composition of the migrating hydrocarbons probably evolved through time as the source continued to mature. This chemical evolution may be captured by the chemical variation across the larger gas condensate field (Figure 16b) in which the eastern well contains a dry gas and the western wells contain wet gases.

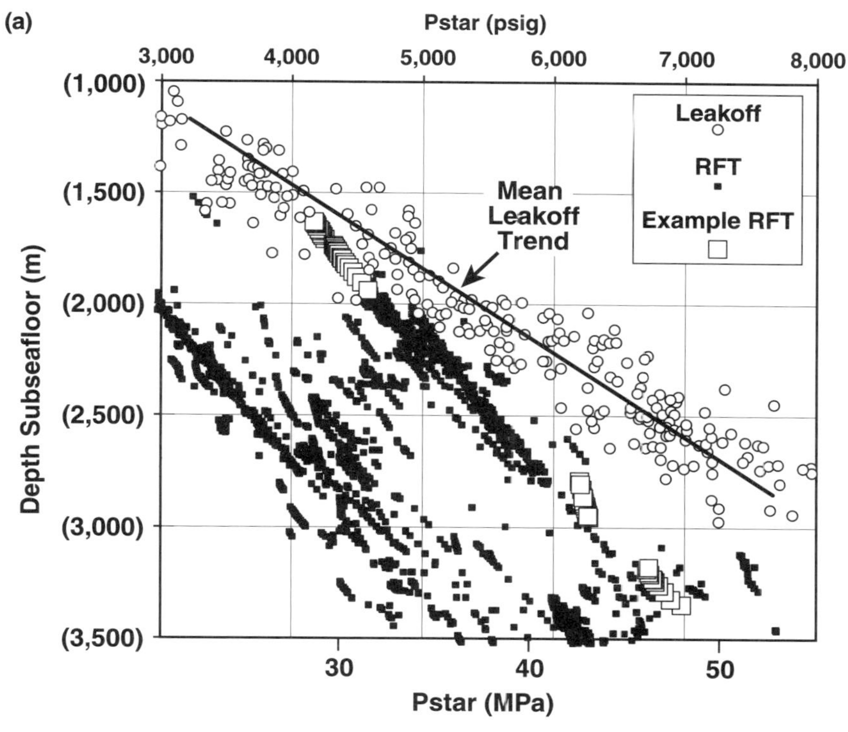

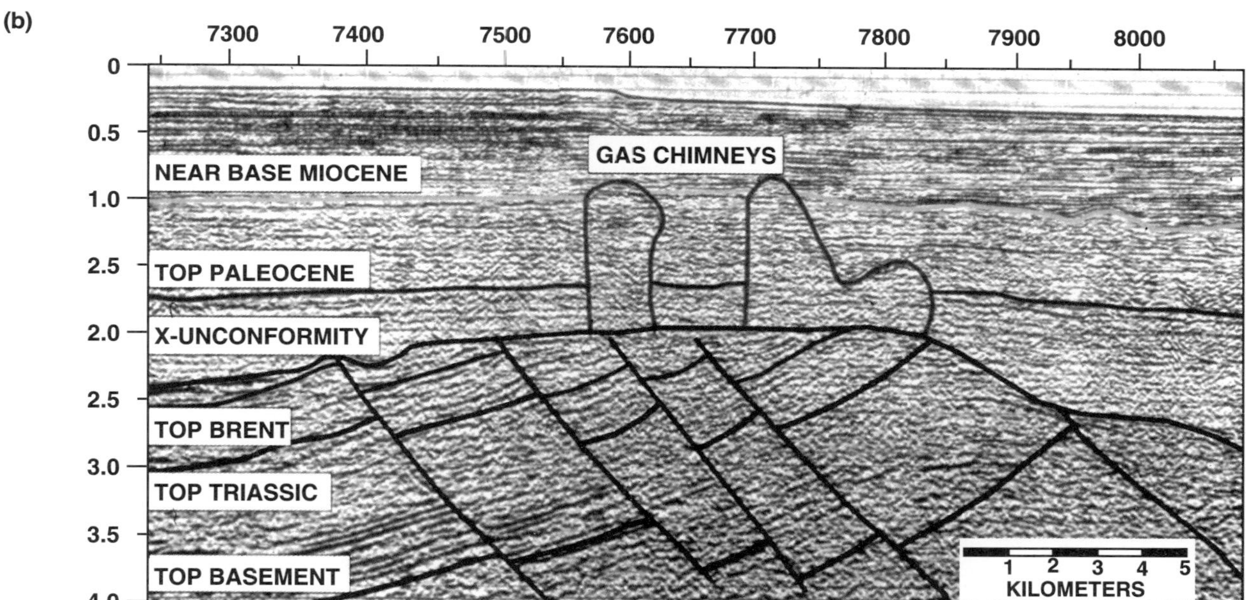

Figure 15—(a) Plot of formation and leakoff pressures for the North Viking Graben against depth below mudline. The formation pressures for the example field (large open squares) closely approach the mean leakoff trend. Leakage is expected near the crest of the column. (b) Seismic profile across example field showing well-developed gas chimneys indicating substantial fluid leakage above the main reservoirs.

This model of chemical evolution of the migrating hydrocarbons is supported by the fluid inclusion data. Hydrocarbon fluid inclusions found in reservoir samples from well B exhibit critical behavior. This behavior is interpreted to reflect the addition of gas to an oil phase at elevated pressures and temperatures when the inclusions formed. The reservoired hydrocarbons in wells A and B are very similar, but they are different from the reservoired hydrocarbons in wells C and D. The unlabeled well between wells A and C in Figure 16a contains a gas condensate with a composition that is intermediate between the two end-members (not shown).

Consider just the relationship between the hydrocarbons in wells A and B in Figure 16. Both wells contain nearly identical hydrocarbons, but the reservoir pressures are very different. The large difference in overpressure between the two wells indicates that the fault zone separating the two accumulations has low permeability. The very similar compositions of the hydrocarbons on either side of the fault are interpreted to indicate that hydro-

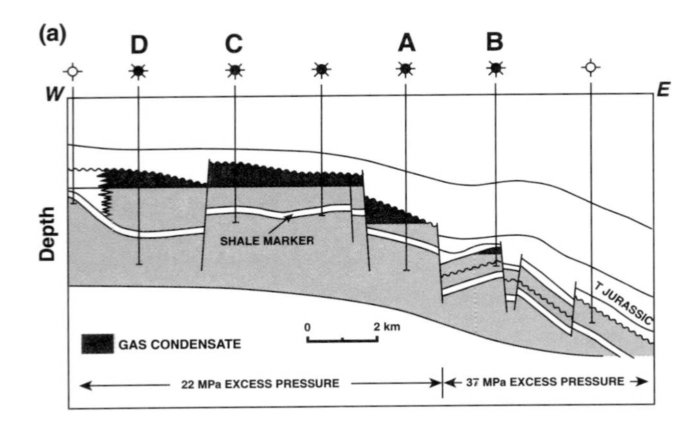

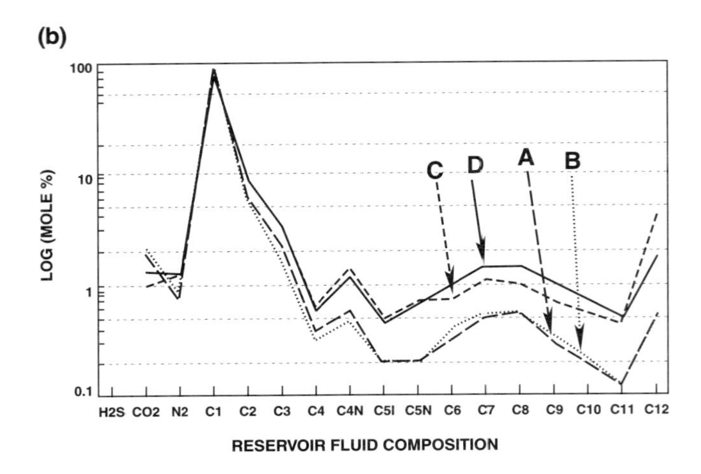

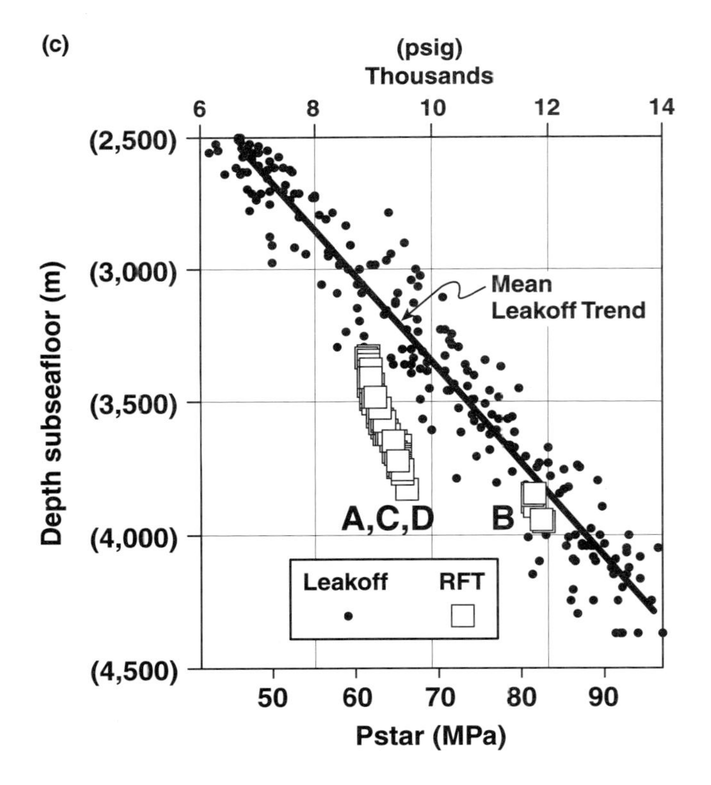

←**Figure 16—(a) Schematic cross section of two Mesozoic gas condensate accumulations showing the large difference in excess pressure between well B and the other wells. Migration analysis suggests that all reservoired hydrocarbons migrated from a deep source kitchen to the east through well B to A to C to D. (b) Plot of reservoir fluid composition from wells across the field. The variation in fluid compositions may capture the chemical evolution of the hydrocarbons along the migration pathway. (c) Pressure–depth plot showing that the trap with well B cannot support any larger hydrocarbon column currently.**

carbons have flowed across this fault but probably at slow rates. However, detailed seismic mapping suggests that the current hydrocarbon column penetrated in well B does not extend to the fault between wells A and B. Therefore, hydrocarbons are not currently migrating between the two accumulations. Migration must have occurred in the past. Why did migration cease?

The crestal hydrocarbon pressures in well B in Figure 16c are at or very close to the mean leakoff pressure trend. Hydraulically induced failure of either the top seal or the crestal bounding fault is likely to constrain the present height of the gas column in the B structure. Any additional gas reaching the structure should be lost via seal leakage. In the western accumulation, the crestal hydrocarbon pressures are also high, but pressures just attain the minimum envelope of the leakoff pressures. Top seal or fault leakage due to hydraulic failure is possible but less likely for this accumulation. The gas column height is probably controlled by either structural or stratigraphic controls within the field.

In this basin, the overpressure doubled in the Pliocene–Holocene, as shown later. Our model for hydrocarbon migration between wells A and B is that prior to the Pliocene, the formation pressures were lower and the maximum potential column height was greater. Hydrocarbons filled the B structure down to the fault zone between wells A and B and then migrated up the fault to fill the main accumulation. Migration was probably slow due to the low permeability of the fault zone unless later diagenetic changes reduced the fault zone permeability. As the pressure increased in the vicinity of well B, the maximum hydrocarbon column that could be supported decreased due to hydraulic top seal or fault leakage. Thus, the gas–water contact moved up the B structure, until the contact no longer intersected the fault separating wells A and B. Migration between the two accumulations ceased, and all subsequent hydrocarbons migrating from the east through structure B have been lost vertically via hydraulically induced leakage into either the chalk or overlying Tertiary section.

We do not fully understand the compositional variation among wells A, C, and D in Figure 16. It may represent incomplete mixing of a dry gas charge coming through wells B and A with an earlier wet gas charge residing in wells C and D. Alternatively, an unrecognized fault or stratigraphic barrier may compartmentalize the reservoir. The common magnitude of overpressure suggests good communication of formation waters. However, capillary barriers may hamper hydrocarbon communication.

MODELING OVERPRESSURE HISTORY

The previous example emphasizes the need to understand the evolution of stresses through time. The stress distribution in the past may be more important than the present-day stresses in determining trap adequacy and hydrocarbon migration pathways. Some key questions are (1) what was the stress distribution when migration was occurring and (2) how did the stress at the trap vary after hydrocarbon accumulation?

We used a 1-D ExxonMobil proprietary program to model the evolution of fluid and fracture pressures through time for a deep well in the same basin as presented in the previous example. Model reproduction of the present pressure distribution required inclusion of a stress unloading term—either a hydrocarbon volume change or reduction of permeability due to diagenesis. The burial history is shown in Figure 17a. The key features are the relative slow burial in the Jurassic and Early–middle Cretaceous followed by more rapid burial in the Late Cretaceous and Tertiary. Note the very rapid burial during the Neogene. Figure 17b shows the predicted overpressure histories for three selected beds (Mesozoic–early Tertiary). The key features of the geopressure histories are (1) essentially hydrostatic pressures until the Oligocene, (2) gradual increases in overpressure from the Oligocene until about 5 Ma, and (3) a doubling of the overpressure from 5 Ma to the present day. This increase in overpressure resulted in a change in migration pathways, as shown in the previous example.

Modeling stress evolution through time is difficult; many rock properties are not well known (e.g., permeability) so that the pore and fracture pressure predictions typically have large uncertainties. The sensitivity required to evaluate a prospect for hydraulic seal adequacy is typically on the order of a few MPa (few hundred psi) or less. Calibration with a paleobarometer is required to have a sufficient level of confidence in the pore fluid and fracture pressure predictions. As described in the next section, fluid inclusions can be used as paleobarometers.

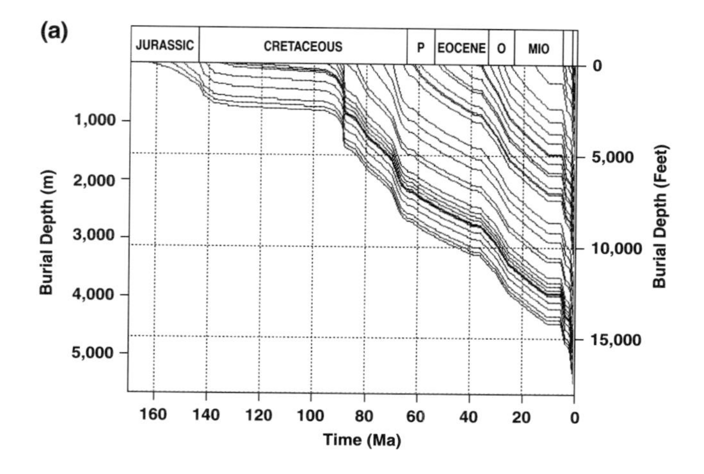

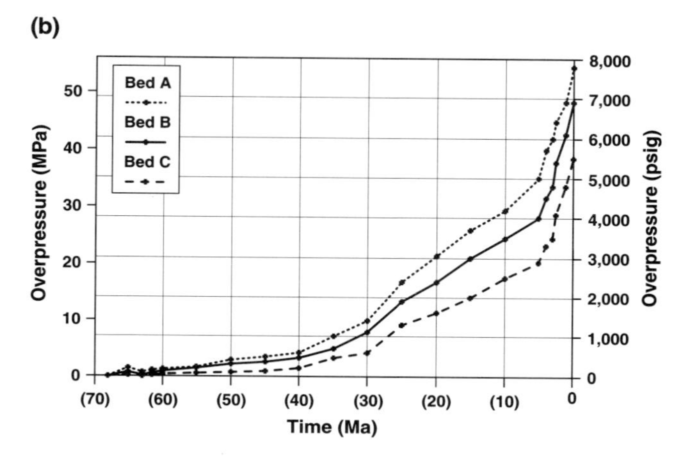

Figure 17—(a) Burial history of a well from a deeper location within the basin where much of the basin overpressure probably originated. (b) Evolution of overpressure in three Jurassic–Paleocene beds calculated using an ExxonMobil proprietary program. Although overpressure probably originated in the early–middle Tertiary, overpressures are calculated to have doubled since 5 Ma.

PALEOPRESSURES FROM FLUID INCLUSIONS

Determination of paleopore pressures from fluid inclusions requires the coexistence of co-eval aqueous and hydrocarbon inclusions in the same rock samples (Figure 18). The homogenization temperatures from the aqueous and hydrocarbon inclusions combined with an understanding of the PVT properties of the fluid determines the possible range of pressure and temperature at the time that the inclusions were trapped (Figure 19) (Narr and Burruss, 1984; Emery and Robinson, 1993).

Our first example is an evaluation of the crossing isochore technique in a case where coexisting oil and water inclusions were trapped at or very near present-day temperatures and pressures. An oil bubble point curve is shown in Figure 19 with a hydrocarbon isochore (line of constant density) extending from the bubble point curve at the homogenization temperature. A second isochore associated with the homogenization temperature of the aqueous inclusions is also shown. The slopes of the two isochores differ because of the different behavior of the water and hydrocarbon fluids. The intersection of these two isochores defines the predicted pressure and temperature at which the fluids were trapped.

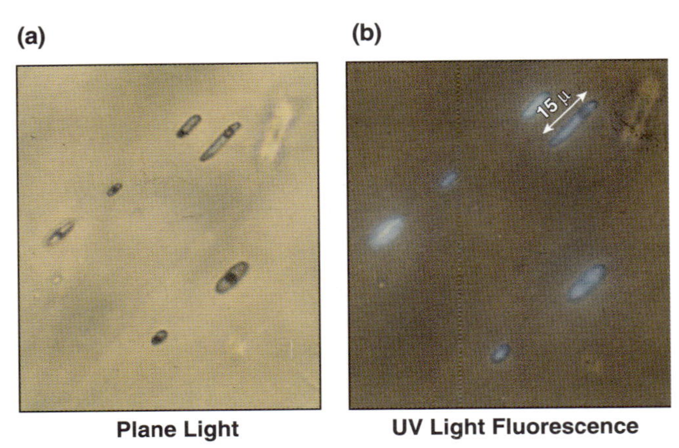

Figure 18—Photomicrographs of hydrocarbon inclusions taken in (a) plane light and (b) ultraviolet (UV) light. (c) Aqueous and hydrocarbon fluid inclusions shown in partial ultraviolet and transmitted light. The hydrocarbon inclusions fluoresce a bright blue in UV light, while water inclusions remain dark. Note typical examples of water and hydrocarbon inclusions containing liquid (L) and vapor (V) and a three-phase inclusion with water, oil, and vapor.

In Figure 19, the square indicates the predicted temperature and pressure and the asterisk records the current bottom hole conditions. The good match between the pressure–temperature prediction based on fluid inclusion analysis and the actual pressure–temperature conditions confirms the validity of this approach.

This technique was applied to a more complicated example to estimate the paleopressure and the time of fluid trapping (Figure 20). In this case, our 1-D proprietary model predicts a pressure evolution that is significantly elevated above hydrostatic pressures (shown for reference). Knowledge of the paleopressure and temperature of hydrocarbon trapping can be combined with the pressure model to predict the age of trapping.

The homogenization temperatures of the hydrocarbon and aqueous inclusions are 99°C and 135°C, respectively. These homogenization temperatures are extended to the modeled bubble point curves for oil and a methane–water mixture. The appropriate isochores for the two fluids are then projected away from the bubble point curves. The intersection of the isochores occurs at 138°C and 55.2 MPa (8000 psi). The trapping conditions determined by this technique are very similar to the conditions predicted by the overpressure model at 5 Ma (Figure 17b).

Knowledge of the age and conditions of trapping can be used to better understand hydrocarbon migration and possible leakage and thus can have an impact on hydrocarbon exploration. We have used these concepts to model the pressure, temperature, and time of trapping in a number of basins worldwide. Unfortunately, it is difficult to confirm the trapping times independently. If multiple generations of fluid inclusions are present in cements with different temporal relationships, additional constraints can be applied to the pressure model.

CONCLUSIONS

Overpressure is common and occurs in many tectonic settings. Both compaction–disequilibrium (undercompaction) and stress "unloading" mechanisms can produce large overpressures. Each mechanism has a different impact on subsurface pressure distribution and stress history. In our experience with ExxonMobil, stress unloading mechanisms are often overlooked as important causes of overpressure in rapidly subsiding basins.

Local controls on overpressure are provided by both structural and stratigraphic elements. In the examples presented here, faults that act as pressure boundaries appear to be zones of restricted flow on a geologic time scale. The combination of overpressure prediction with hydrocarbon migration analysis provides a potent tool for successful exploration in many plays.

The ultimate control on overpressures in rapidly subsiding basins is rock failure that results in leakage. In many overpressured basins, hydraulic failure of either top seals or faults also acts to limit the maximum hydrocarbon column in a trap, to define the migration pathways, or both. Evaluation of the potential for hydraulic failure can be important for prospect assessment in both frontier and mature plays.

In basins with either early hydrocarbon migration or prolonged periods of migration, evaluation of paleoseal integrity may be critical. A number of software packages exist to model geopressure history, but the problem is in the verification of the model. The fluid inclusion method discussed in this chapter is a method to calibrate the pressure model and thus more confidently evaluate prospects.

***Acknowledgments**—We would like to acknowledge the invaluable contributions of our ExxonMobil colleagues in the development of our current understanding of controls on overpressure. This paper benefited significantly from thoughtful reviews by Lori Summa, Gary Gray, and Fred Weaver. The authors wish to thank ExxonMobil Upstream Research Company, ExxonMobil International Ltd., and Shell U.K. Exploration and Production for permission to publish this paper.*

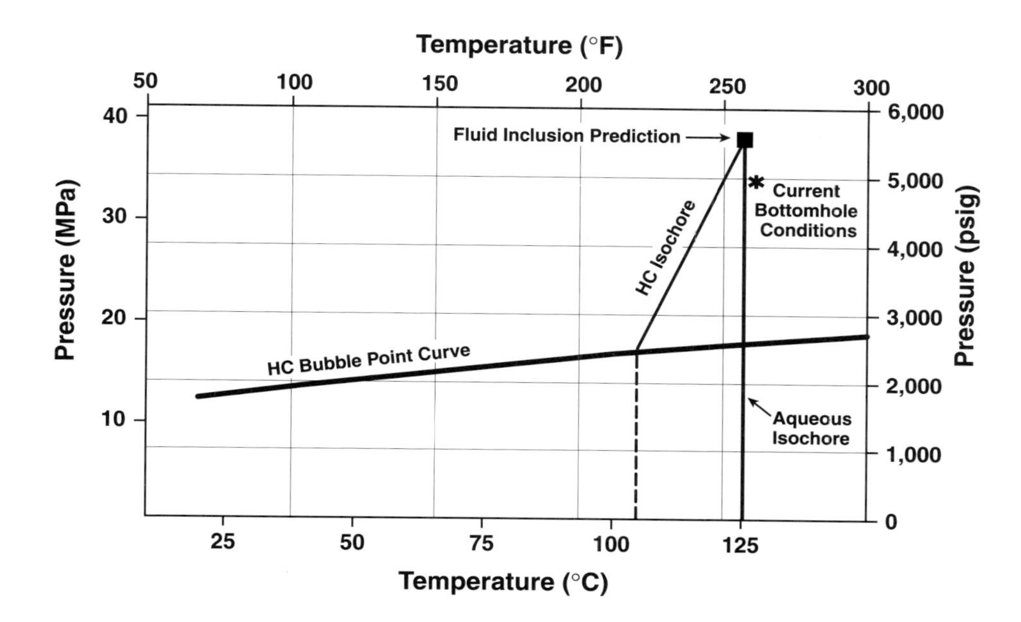

Figure 19—Pressure–temperature plot showing how paleopressures can be determined from samples with co-eval aqueous and hydrocarbon fluid inclusions. In this example, the isochore intersection predicts fluid inclusion trapping temperatures and pressures similar to current conditions.

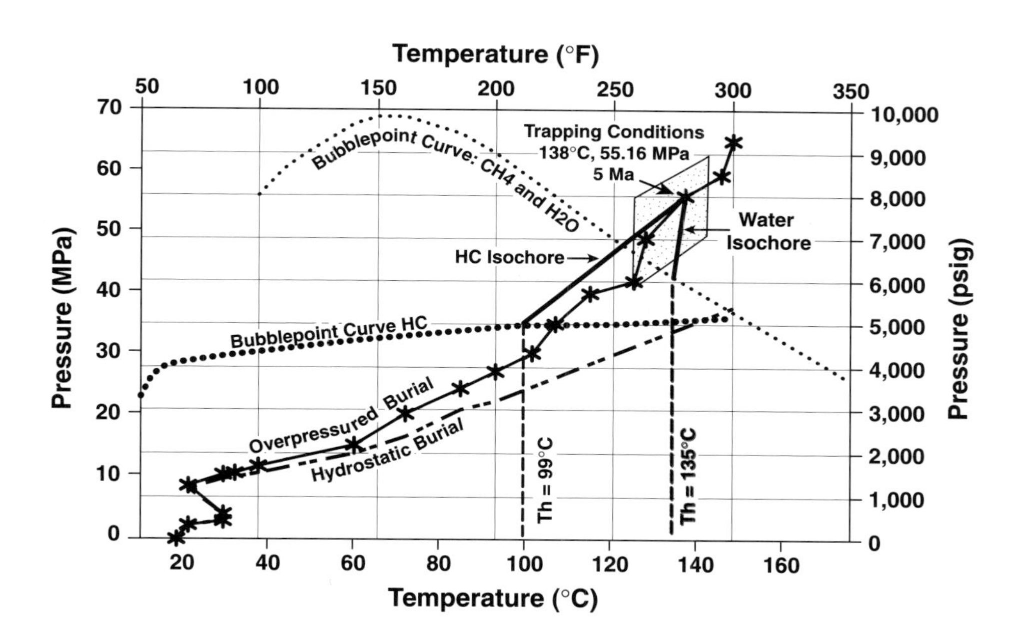

Figure 20—Pressure–temperature plot showing how the fluid inclusion data can be used to distinguish between a hydrostatic and an overpressure model for the reservoir. Interpretation of the fluid inclusion data support the overpressure model, with trapping occurring about 5 Ma at a temperature of 138°C and a pressure of 55.2 MPa (8000 psi). The rhomboid area defines the uncertainty interval in both temperature and pressure predictions.

REFERENCES CITED

Berg, R. R., 1975, Capillary pressures in stratigraphic traps: AAPG Bulletin, v. 59, p. 939–956.

Bjorkum, P. A., and P. H. Nadeau, 1998, Temperature controlled porosity/permeability reduction, fluid migration, and petroleum exploration in sedimentary basins: Australian Petroleum Production & Exploration Association Limited Journal, v. 38, p. 453–464.

Bowers, G. L., 1994, Pore pressure estimation from velocity data: accounting for overpressure mechanisms besides undercompaction: International Association of Drilling Contractors/Society of Petroleum Engineers, 27488, Drilling Conference, Dallas, Texas, p. 515–530.

Breckels, I.M., and van Eekeln, H.A.M., 1982, Relationship between horizontal stress and depth in sedimentary basins: Journal of Petroleum Technology, v. 34, p. 2191–2199.

Chapman, R. E., 1994, Abnormal pore pressure: essential theory, possible causes, and sliding, *in* W. H. Fertl, R. E. Chapman, and R. F. Hotz, eds., Studies in abnormal pressures: Developments in Petroleum Science (Elsevier), v. 38, p. 51–91.

Converse, D. R., P. H. Nicholson, G. T. Cayley, and G. G. Gray 1994, Geologic controls on subsurface pore pressure distribution (abs.): AAPG Hedberg Research Conference, Program with Abstracts, Denver, Colorado.

Converse, D. R., G. G. Gray, P. H. Nicholson, and G. T. Cayley, 1995, Geologic controls on subsurface pore distribution (abs.): AAPG Annual Convention, Program with Abstracts, Houston, Texas, p. 18A.

Daines, S. R., 1982, Prediction of fracture pressures for wild-

cat wells: Journal of Petroleum Technology, v. 34, p. 863–872.
Darby, D., R. S. Haszeldine, and G. D. Couples, 1996, Pressure cells and pressure seals in the UK Central Graben: Marine and Petroleum Geology, v. 13 p. 865–878.
Emery, D. E., and A. Robinson, 1993, Inorganic geochemistry: applications to petroleum geology: Oxford, Blackwell Scientific, 254 p.
Engelder, T., and J. T. Leftwich, Jr., 1997, A pore-pressure limit in overpressured south Texas oil and gas fields, in R. C. Surdam, ed., Seals, traps, and the petroleum system: AAPG Memoir 67, p. 255–267.
Fertl, W. H., R. E. Chapman, and R. F. Hotz, eds., 1994, Studies in abnormal pressures: Developments in Petroleum Science (Elsevier), v. 38, 454 p.
Gaarenstroom, L., R. A. J. Tromp, M. C. de Jong, and A. M. Brandenburg, 1993, Overpressures in the central North Sea: implications for trap integrity and drilling safety, *in* J. R. Parker, ed., Petroleum geology of northwest Europe: Proceedings of the 4th Conference, v. 2, p. 1305–1313.
Hareland, G., C. E. Polston, and W. E. White, 1993, Normalized rock failure envelope as a function of rock grain size (abs.): International Journal of Rock Mechanics and Mining Sciences & Geomechanics Abstracts, v. 30, n. 7, p. 715–717.
Heum, O. R., 1996, A fluid dynamic classification of hydrocarbon entrapment: Petroleum Geoscience, v. 2, p. 145–158.
Hottmann, C. E., and R. K. Johnson, 1966, Estimation of formation pressures from log-derived shale properties: Journal of Petroleum Technology, v. 17, p. 717–722.
Hunt, J. M., 1990, Generation and migration of petroleum from pressured fluid compartments: AAPG Bulletin, v. 74, p. 1–12.
Kan, T. K., and C. J. Sicking, 1994, Predrill geophysical methods for geopressure detection and evaluation, *in* W. H. Fertl, R. E. Chapman, and R. F. Hotz, eds., Studies in abnormal pressures: Developments in Petroleum Science (Elsevier), v. 38, p. 155–186.
Koch, P. S., G. M. Skerlac, and M. S. Manwaring, 1990, The effect of strain on top-seal integrity: central North Sea (abs.): Geological Society of London, Conference on Cap Rocks, London.
Kunze, K. R., and R. P. Steiger, 1992, Accurate in-situ stress measurements during drilling operations: Society of Petroleum Engineers, 67th Annual Conference, Washington, D.C., SPE 24593, p. 491–499.
Leach, W. G., 1994, Distribution of hydrocarbons in abnormal pressure in south Louisiana, U.S.A., *in* W. H. Fertl, R. E. Chapman, and R. F. Hotz, eds., Studies in abnormal pressures: Developments in Petroleum Science (Elsevier), v. 38, p. 391–428.
Magara, K., 1978, Compaction and fluid migration: practical petroleum geology: Developments in Petroleum Science (Elsevier), v. 9, 319 p.
Mandl, G., and R. M. Harkness, 1987, Hydrocarbon migration by hydraulic fracturing, *in* M. E. Jones and R. M. F. Preston, eds., Deformation of sediments and sedimentary rocks: Geological Society of London Special Publication 29, p. 39–53.
Miller, T. W., 1995, New insights on natural hydraulic fractures induced by abnormally high pore pressures: AAPG Bulletin, v. 79, p. 1005–1018.
Miller, T. W., C. H. Luk, and D. L. Olgaard, 1998, The interrelationships between overpressure mechanisms and in situ stresses: Proceedings of the American Association of Drilling Engineers Industry Forum on Pressure Regimes in Sedimentary Basins and Their Prediction, Session 2, Del Lago, Texas.
Narr, W. M., and R. C. Burruss, 1984, Origin of reservoir fractures in Little Knife field, North Dakota: AAPG Bulletin, v. 68, p. 1087–1100.
Nicholson, P. H., and S. Hart, 1995, The prediction of formation pressures, Central Graben, UK North Sea, 1995 (abs.): Offshore Europe Conference, Society of Petroleum Engineers, Aberdeen, Scotland.
Nicholson P. H., D. R. Converse, G. T. Cayley, and G. G. Gray 1995, Geologic controls on subsurface pore pressure distribution, North Sea (abs.): Petroleum Group Conference, Geological Society of London, Bath, U.K.
Schowalter, T. T., 1979, Mechanics of secondary hydrocarbon migration and entrapment: AAPG Bulletin, v. 63, p. 723–760.
Timko, D. J., and W. H. Fertl, 1971, Relationship between hydrocarbon accumulation and geopressure and its economic significance: Journal of Petroleum Technology, v. 23, p. 923–933.
Tissot, B. P., and D. H. Welte, 1984, Petroleum formation and occurrence (2nd ed.): Berlin, Springer-Verlag, 699 p.
Watts, N. L., 1983, Microfractures in chalks of Albuskjell field, Norwegian sector, North Sea: possible origin and distribution: AAPG Bulletin, v. 67, p. 201–234.
Watts, N. L., 1987, Theoretical aspects of cap-rock and fault seals for single- and two-phase hydrocarbon columns: Marine and Petroleum Geology, v. 4, p. 274–307.
Weakley, R. R., 1991, Use of surface seismic data to predict formation pore pressure worldwide: Society of Petroleum Engineers, Western Regional Meeting, Long Beach, California, SPE 21752, p. 37–42.

Souto Filho, J. D., A. C. F. Correa, E. V. Santos Neto, and L. A. F. Trindade, 2000, Alagamar–Açu petroleum system, onshore Potiguar Basin, Brazil: a numerical approach for secondary migration, *in* M. R. Mello and B. J. Katz, eds., Petroleum systems of South Atlantic margins: AAPG Memoir 73, p. 151–158.

Chapter 12

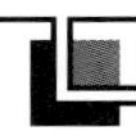

Alagamar–Açu Petroleum System, Onshore Potiguar Basin, Brazil: A Numerical Approach for Secondary Migration

J. D. Souto Filho

Petrobrás E & P–RNCE
Natal, Brazil

A. C. F. Correa

State University of Campinas
Campinas, Brazil

Permanent address: *Petrobrás Training Center*
Salvador, Brazil

E. V. Santos Neto

L. A. F. Trindade

Center of Excellence in Geochemistry
Petrobrás R & D Center
Rio de Janeiro, Brazil

Abstract

The main objective of this study was to investigate secondary migration in the Açu Formation, including reconstruction of migration pathways, estimation of velocity of the migration fronts, and identification of key variables controlling the migration process. The Potiguar Basin in northeastern Brazil is an appropriate geologic setting for studying secondary migration. Almost all of the oil accumulated in the onshore Açu sandstones was generated from offshore pods of active source rocks in the Alagamar Formation and has migrated laterally for long distances. The Açu Formation, which is the main carrier bed and reservoir of this petroleum system, dips seaward as a regional monocline structure, with gentle folds, normal faults, and facies changes. In this study, we reconstructed the history of secondary migration using basin-scale modeling of 8272 grid cells. Rock and fluid properties used in this investigation are those typically found in most of the oil accumulations in the Açu Formation.

Petroleum migration pathways and economic oil accumulations in the Alagamar–Acu petroleum system were controlled by northeast–southwest structural noses at the top of the Açu Formation. Estimated ratios of oil displacement for the earlier oil migration fronts range from 3.0 to 7.0 cm/year. Oil in the main onshore accumulations of the Potiguar Basin may have been trapped between 1 and 5 m.y. after the beginning of secondary migration. Significant loss of petroleum may have occurred along seepages from onshore outcrops of the Açu Formation.

INTRODUCTION

Significant volumes of petroleum have been found in the Potiguar Basin, making it the second largest oil producing basin in Brazil just behind the Campos Basin. Most of the oil in the basin has been discovered in the Açu Formation (Albian–Turonian) which also contains a freshwater aquifer. Geochemical correlations have shown that oils accumulated in the Açu sandstones were generated in marine–evaporitic and lacustrine intervals of the Aptian Alagamar Formation (Rodrigues et al., 1983; Mello et al., 1988; Santos Neto et al., 1990; Trindade et al., 1992). The spatial relationship between these oil accumulations and the occurrence of mature source rocks suggests long-distance lateral migration from offshore pods of effective source rocks (Santos Neto et al., 1990). Heterogeneities in oil compositions reveal trends related to the amount of source mixing, which in turn, reflect the timing of both oil generation and migration (Trindade et al., 1992). In the Areia Branca trend, the oils that migrated greater distances correlate to fresh and brackish water lacustrine source rocks, whereas the oils that migrated

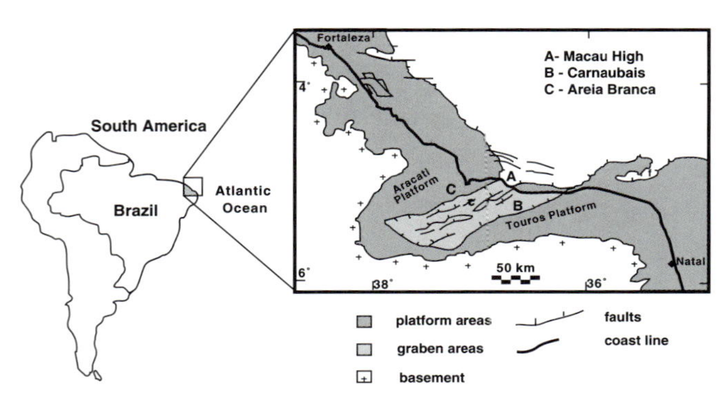

Figure 1—Location map and schematic tectonic framework of the Potiguar Basin.

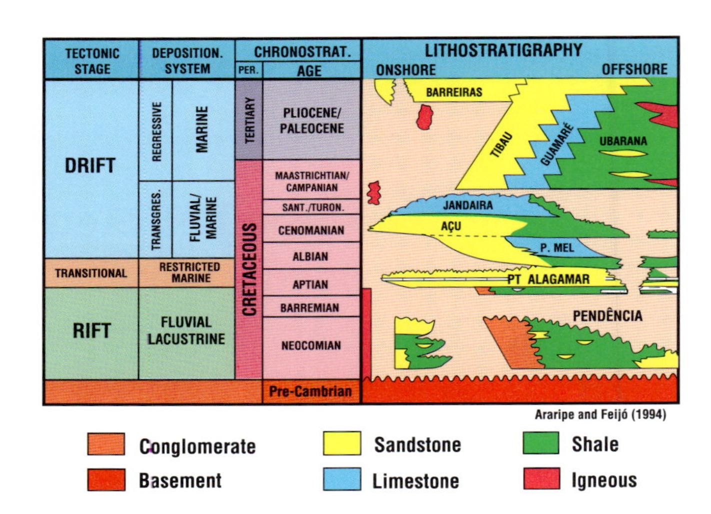

Figure 2—Stratigraphic chart of the Potiguar Basin. Modified from Araripe and Feijo (1994).

shorter distances correlate to the younger and shallower hypersaline source beds (Trindade et al., 1992). Hydrogen and carbon isotopes in the Potiguar oils show variations that reflect generation from different source rocks (Santos Neto, 1999). Maturity parameters in oils, given by isomer ratios of C_{29} steranes, show decreasing values from offshore toward onshore, suggesting the main trends of the migration pathways (Santos Neto et al., 1990).

Alternatively, Souza et al. (1983) studied the secondary migration in the Potiguar Basin using oil potential maps, an approach developed by Hubbert (1953) and Dahlberg (1982). These maps were made from a combination of water potential maps and structural maps for the top of the Açu Formation.

The main objective of this study was to investigate the processes of secondary migration in the Açu Formation using a flux simulator. The main mechanisms controlling the migration of hydrocarbons from the source to the trap are discussed. These include (1) spatial evolution of migration pathways in the Açu Formation, (2) estimates of the velocities of the migration fronts and the arrival time of oil in the areas favorable for accumulation, and (3) large-scale definition of main factors controlling secondary migration.

GEOLOGIC SETTING OF POTIGUAR BASIN

The Potiguar Basin is the most northeastern of the Brazilian marginal basins, covering an area of 48,000 km^2, with 21,000 km^2 of that area onshore (Figure 1). The basin was filled with approximately 6000 m of sediments that are divided into the following sequences : (1) continental (rift phase), (2) marine transgressive (transitional and part of the drift phase), and (3) marine regressive, or drift phase (Souza, 1982; Bertani et al., 1990) (Figure 2). The continental sequence (Neocomian) is composed of lacustrine and deltaic sedimentary rocks of the Pendência Formation. The marine transgressive sequence (Aptian–Santonian) corresponds to proximal sediments deposited under fluvial–deltaic conditions and distal sediments deposited in lagoon to restricted marine environments. The Alagamar and Açu Formations comprise this sequence. The marine regressive sequence (Campanian–Holocene) was deposited under variable water-depth conditions and is preserved only offshore.

The structural framework of the basin is defined by two major faulting systems that are oriented northeast–southwest (onshore) and northwest–southeast (offshore). These fault systems that define the tectonic features (central trough and platforms) have controlled the distribution of sediments in the basin. The central trough is bounded by faults that define horsts that separate asymmetric grabens filled with sedimentary rocks of the rift phase. The platform areas (see Figure 1) contain sedimentary sequences deposited during the transitional phase (lacustrine to restricted marine) and drift phase (open marine).

ALAGAMAR–AÇU PETROLEUM SYSTEM

Source Rock Intervals

The Alagamar Formation contains organic-rich intervals that are the major source for most of the petroleum in the Potiguar Basin (Mello et al., 1988; Santos Neto et al., 1990). These organic-rich intervals are found in three lithostratigraphic subunits of the Alagamar Formation that represent, from the base to the top, the transition from lacustrine paleoenvironments (Upanema Member) to marine–evaporitic conditions (Ponta do Tubarão beds) evolving to restricted marine conditions (Galinhos Member).

The source rocks of the lacustrine section consist mostly of dark gray to black shales with total organic

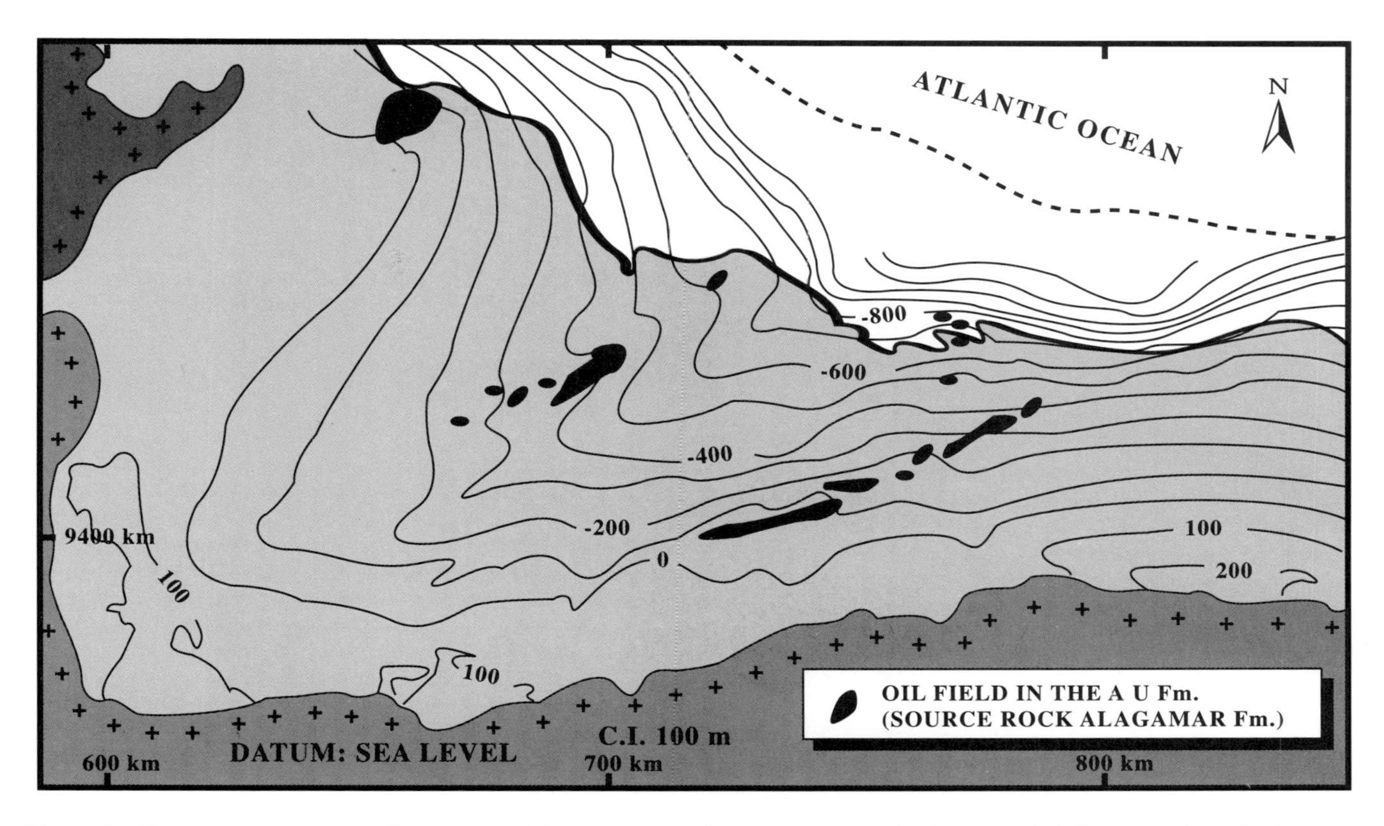

Figure 3—Structure contour map for the top of the Açu Formation showing the distribution of oil fields, onshore Potiguar Basin. Modified from Vasconcelos et al. (1990) and Lima Neto (1993).

carbon (TOC) contents up to 4% and good hydrocarbon source potential, with values of S_2 reaching 35 kg HC/ton rock. Organic matter is mostly type I and type II, with hydrogen index (HI) values ranging from 100 to 700 mg HC/g TOC and oxygen index (OI) values lower than 100 mg CO_2/g TOC (Mello et al., 1988; Cerqueira, 1995). Low HI values are due to the conversion of organic matter to petroleum with increasing depth of burial. The marine–evaporitic to restricted marine sequence is comprised of black shales and marls that have TOC values up to 6% and S_2 values up to 40 kg HC/ton rock. HI values are higher than 500 mg HC/g TOC, and OI are lower than 50 mg CO_2/g TOC (Mello et al., 1988; Cerqueira, 1995). Although the effective source rock interval is usually thinner than 200 m, the relatively high contents of hydrogen-enriched organic matter with good to excellent hydrocarbon source potential allow the genetic classification of this petroleum system as charged (Demaison and Huizinga, 1991).

Basin modeling has shown that source rock beds of the onshore Alagamar Formation are immature, whereas parts of the offshore section reached the onset of oil generation during the Campanian and the peak of oil generation in the Miocene–Holocene. There is no evidence for overmaturation of the Alagamar section (Mello, 1987). Geochemical data obtained from wells in the offshore Alagamar section support this conclusion. Carbon isotopic and compositional data indicate that associated gases were formed during primary cracking just before the peak of the oil window (Prinzhofer et al., 1998). Thus, the Alagamar–Açu petroleum system can be characterized as a typical oil-prone system.

Migration and Entrapment

Traps in the the Açu Formation have a strong structural component and are located mostly along the Carnaubais and Areia Branca trends. Oil accumulations in the Açu Formation are trapped in northeast–southwest trends that dip to the northeast (Figure 3). The main postrift structural event that affected the Açu Formation occurred during the late Santonian–middle Campanian. This occurred contemporaneously with a prominent biostratigraphic hiatus represented by an unconformity that is present regionally over the continental platform (Lima Neto, 1993). Tectonic reactivation created an important compressional component that produced folding and faulting (Cremonini, 1995). A less intense, regional east–west compression event that occurred during the Tertiary generated north–south trending folds with large wavelengths (Cremonini, 1993).

The tectonic evolution of the Potiguar Basin lead to the formation of effective migration pathways through faults, unconformities, and fractures. Breakout analysis performed in wells has shown that east–northeast oriented fractures have remained open since the Eocene–

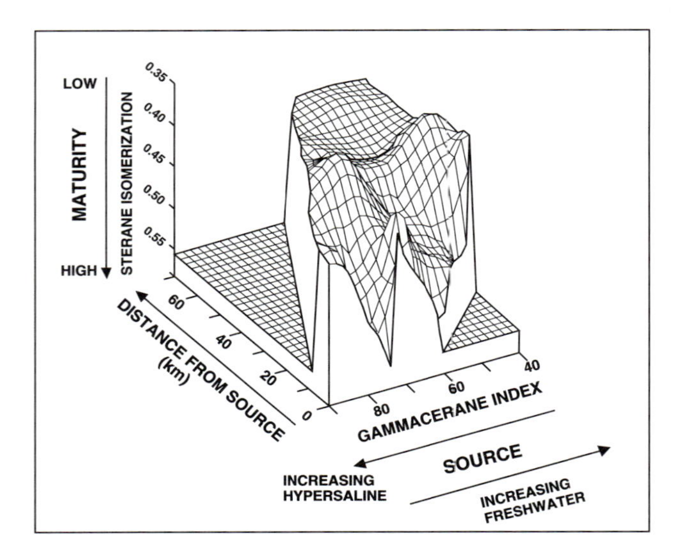

Figure 4—Compositional variability of mixed oils of the Potiguar Basin, represented by gammacerane indices (source dependent) and sterane isomerization ratios (maturity dependent), versus distance of migration from the source rock. The surface is defined by a total of 19 points. From Trindade et al. (1992).

Holocene (Lima et al., 1993). This supports the hypothesis that most of the Alagamar oil migrated via fractures, produced by the regional tension fields, to previously formed structural traps.

The regional monocline structure of the postrift section and the presence of regional seals (base of Açu unit IV and the Ponta do Tubarão beds of the Alagamar Formation) combined with a moderate degree of structural deformation allowed local breaching of seals, focusing the hydrocarbons to traps. These features are characteristic of a laterally drained, high-impedance petroleum system (Demaison and Huizinga, 1991).

Geochemical Evidence for Secondary Migration

The existence of onshore oil accumulations generated from offshore source rocks of the Alagamar Formation presents compelling evidence for long-distance lateral migration (Santos Neto et al., 1990; Trindade et al., 1992). Oils accumulated along the Estreito–Guamaré and Areia Branca trends and the Macau High have bulk geochemical, biomarker, and carbon and hydrogen isotopic compositions that reflect marine–evaporitic to mixed marine–evaporitic and lacustrine sources.

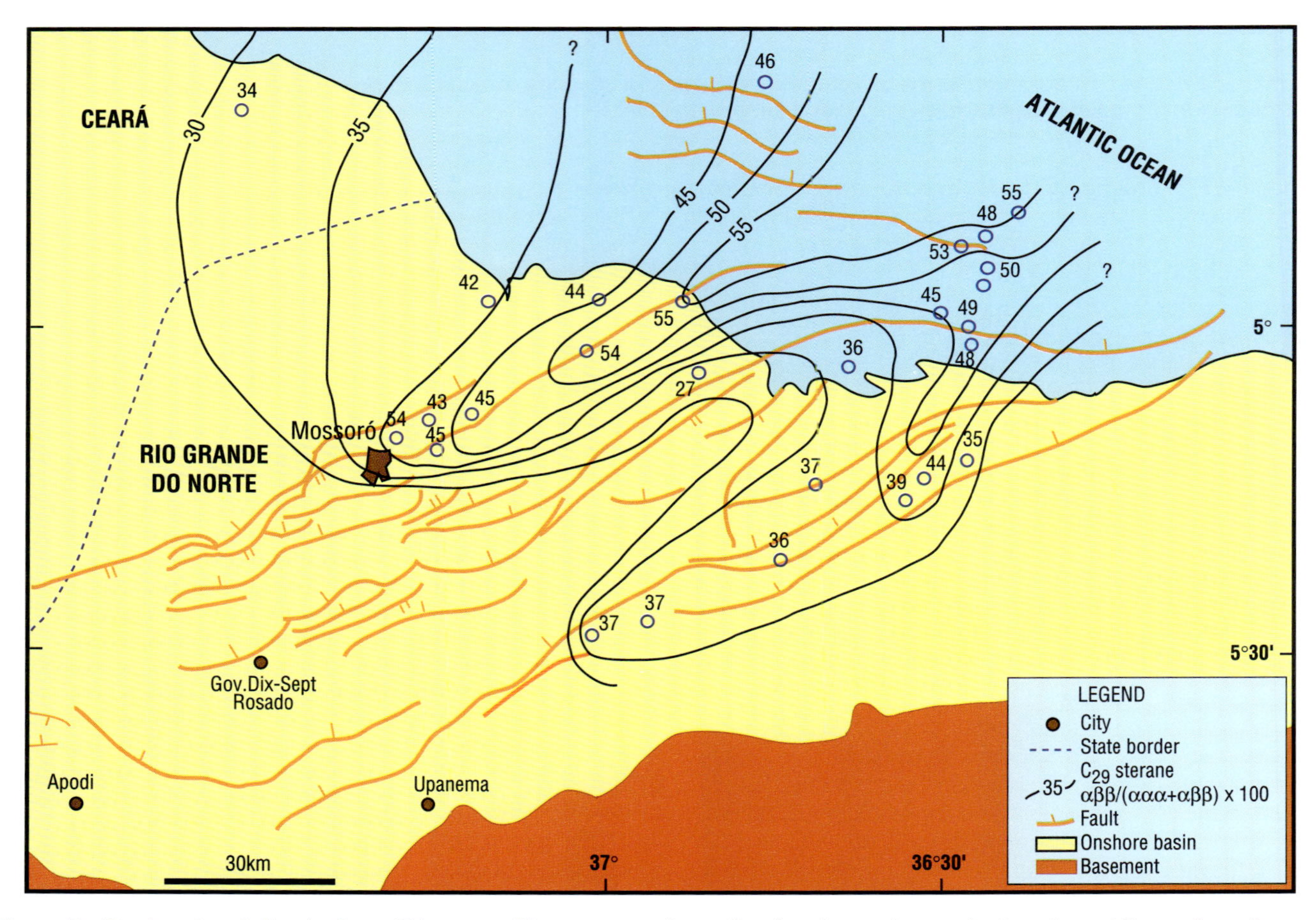

Figure 5—Regional variation in C_{29} αββ/(ααα + αββ) sterane ratios reflecting thermal maturity trends and the major migration pathways for the Alagamar oils. Modified from Santos Neto et al. (1990).

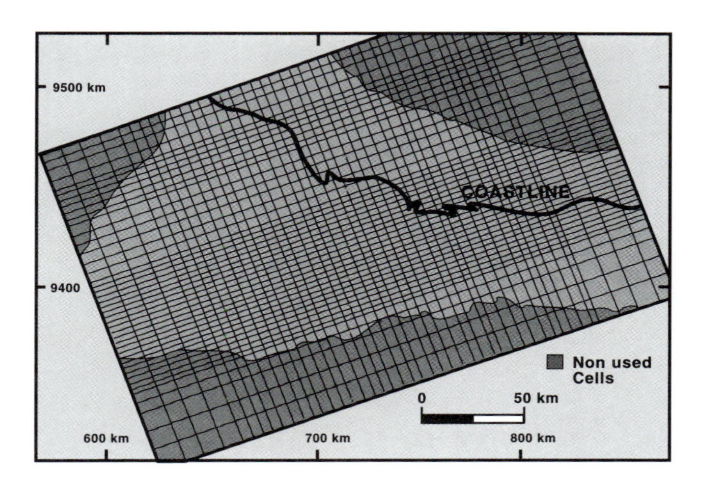

Figure 6—Grid network used in the numerical modeling of secondary migration in the Alagamar–Açu petroleum system.

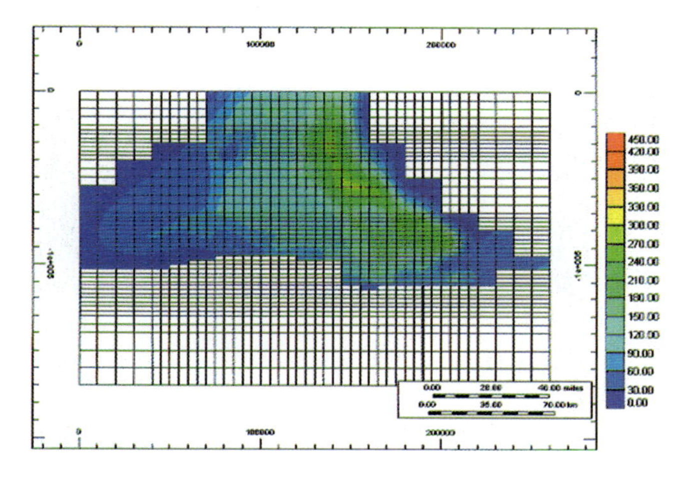

Figure 7—Isopachous map showing the thickness (in meters) of the lowermost Açu Formation subunit (Açu I).

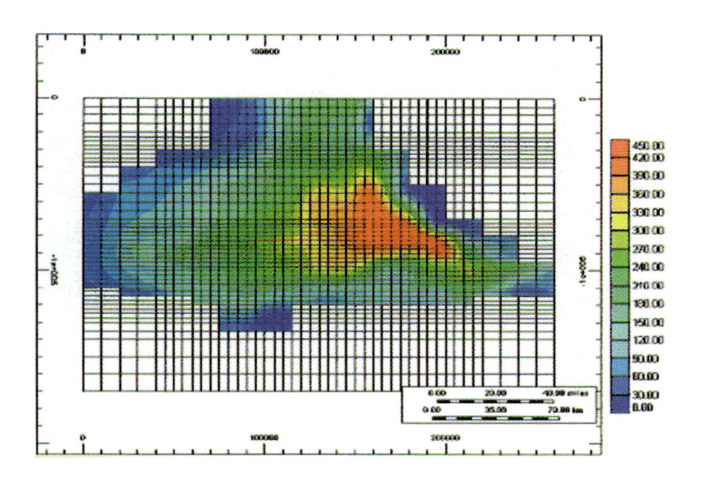

Figure 8—Isopachous map showing the thickness (in meters) of subunit Açu II.

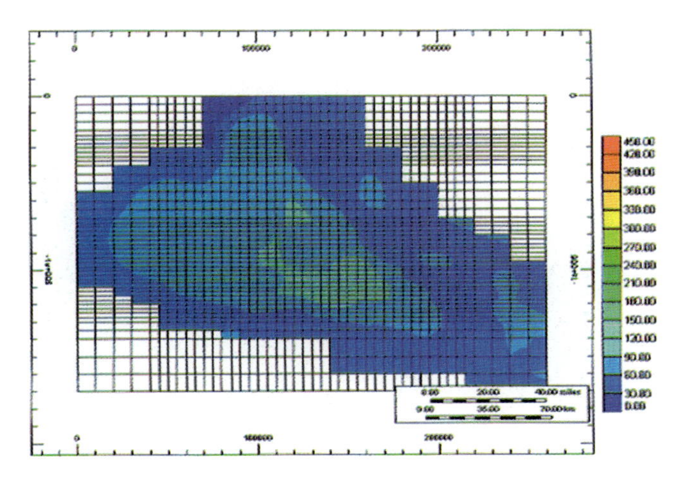

Figure 9— Isopachous map showing the thickness (in meters) of subunit Açu III.

The 3-D presentation of the heterogeneities in the oil compositions given by gammacerane index (source dependent) and sterane isomerization ratios (maturity dependent) versus distance inferred for secondary migration along the Areia Branca area shows a complex pattern (Figure 4). One general trend, however, is clear—the more migrated oils tend to have a greater lacustrine contribution and lower thermal maturity and the less migrated oils display a stronger marine–hypersaline signature and higher maturity. Such data are compatible with continuous subsidence of the offshore portion of the Potiguar Basin and increasing maturation of the source rocks through time (Mello, 1987). Thus, older lacustrine organic-enriched sedimentary rocks reached the oil window earlier than the marine–hypersaline shales and marls. Younger, more mature, lighter oils displaced older oils that had already filled traps next to the oil kitchen, with the older oils then migrating to the next trap updip following Gussow's model (Gussow, 1954). In Figure 5, the areal distribution of the C_{29} $\alpha\beta\beta/(\alpha\alpha\alpha + \alpha\beta\beta)$ sterane ratios in Alagamar oils allows the identification of the main pathways of secondary migration in the onshore part of the Potiguar Basin, which corroborates the suggested trend of secondary migration.

NUMERICAL MODELING

This study reconstructed the history of secondary migration using a three-phase numerical simulator called *IMEX®* (by Computer Modeling Group) which considers gas solubility variations in the oil as a function of pressure.

Basin-scale simulation was done utilizing variable size grid cells (Figure 6). In composing a 3-D model, the simulation was done for four rock layers, as defined by Vasconcelos et al. (1990), that correspond to the correla-

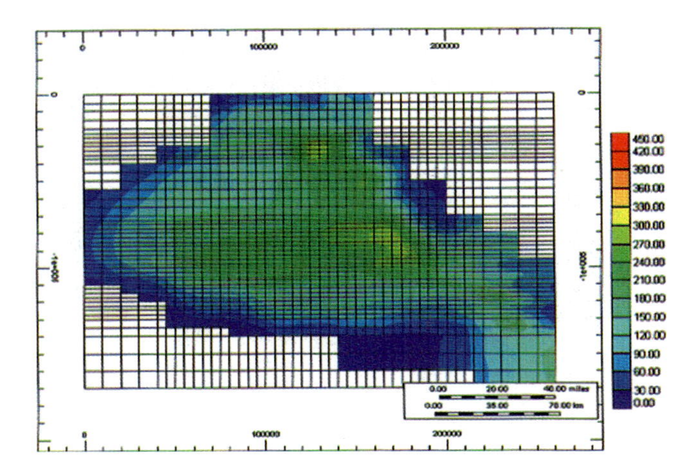

Figure 10— Isopachous map showing the thickness (in meters) of the uppermost Açu Formation subunit (Açu IV).

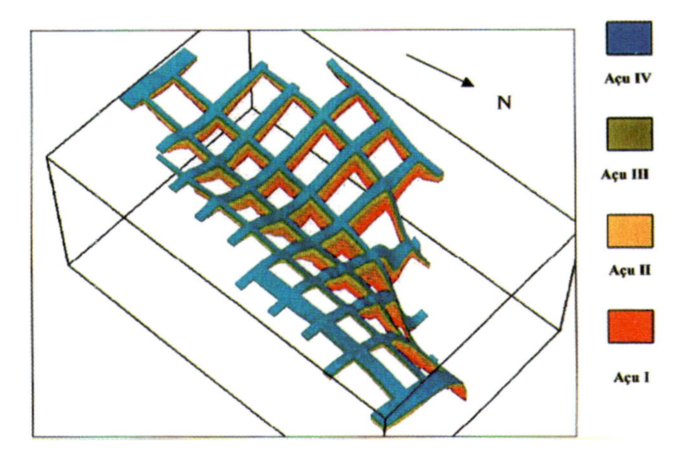

Figure 11—Schematic 3-D model for the Açu Formation showing the geometry and thicknesses of the four subunits (Açu I to IV).

tion subunits of the Açu Formation—Açu I (oldest) through Açu IV (youngest) (Figures 7–10). A structural map on the top of the Açu Formation was used as datum to compose the external geometry of the rock units.

The 3-D model contains 8272 grid cells. Figure 11 shows the external geometry of the Açu Formation. Tables 1 and 2 present rock and fluid properties corresponding to average values of the most important oil accumulations in the Açu Formation. Initially, it was assumed in the numerical modeling that oil was present at the base (Figure 12) of the Açu Formation (Açu I) which overlaps the offshore oil kitchen. The initial time was set at 20 Ma (Miocene), which was estimated to be the main phase of oil generation in the Alagamar Formation (Mello, 1987; Trindade et al., 1992).

RESULTS AND DISCUSSION

The main phases of oil displacement in the Açu Formation were identified by patterns observed on oil saturation maps for the Açu I to IV subunits at different time slices. During the first stage of migration, oil concentrated at the top of Açu III due to buoyant rise and because the base of the overlying subunit (Açu IV) was a shaly section with good lateral continuity that acted as a regional seal. Lateral migration became evident at 100,000 years after the beginning of secondary migration when the oil was displaced through Açu III at a rate of 3.0–7.0 cm/year. During this time, the migration front reached the Macau High, which is a basement high located near the present coast line in the same area where the oil migrated into Açu IV (Figure 13). Note that oil saturation maps (Figures 13–16) are shown only for Açu III because this is the most important carrier bed in the Açu Formation.

Table 1—Average Porosities and Permeabilities in Subunits of the Açu Formation

Unit	Porosity (%)	Permeability (md)
Açu I	21	1450
Açu II	23	1350
Açu III	21	750
Açu IV	29	1200

Table 2—Fluid Properties Used in the Numerical Modeling of Secondary Migration in the Alagamar–Açu Petroleum System

Property	Units
Oil density	0.904 g/cm^3
Gas density	6.500×10^{-4} g/cm^3
Water density	1.000 g/cm^3
Oil compressibility	1.657×10^{-6}
Volume formation factor of water	1.0150
Compressibility of water	5.2×10^{-7} at 42.7 kg/cm^2
Bubble pressure	1.200 kg/cm^3

At 500,000 years after the beginning of secondary migration, the oil fronts were controlled by the Areia Branca and Carnaubais structural trends (Figure 14). At 2 m.y. after the beginning of secondary migration, the oil reached the southern area of the basin (Carnaubais trend), where oil migration took place over the Aracati Platform (Figure 15). It is thought that oil seeps at this time may have been significant along outcrops of the Açu Formation at the southern and southeastern sides of the Carnaubais trend.

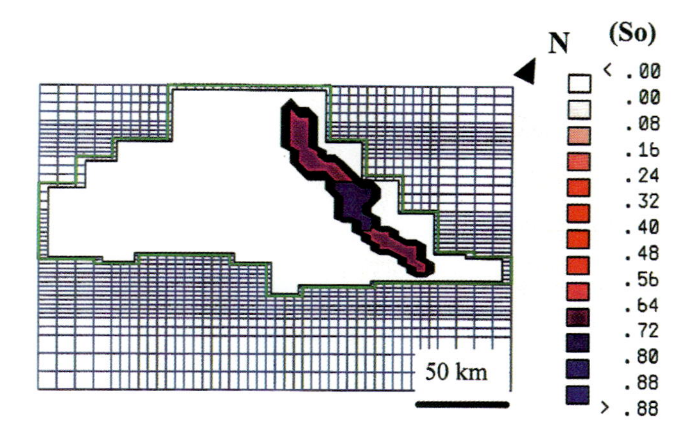

Figure 12—Oil saturation map for subunit Açu I, the base of the Açu Formation, before secondary migration is assumed to have begun (zero time).

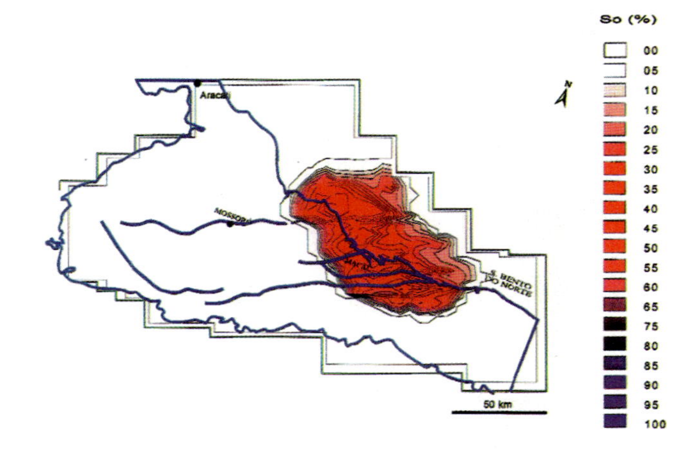

Figure 14—Oil saturation map of subunit Açu III at 500,000 years after the beginning of secondary migration.

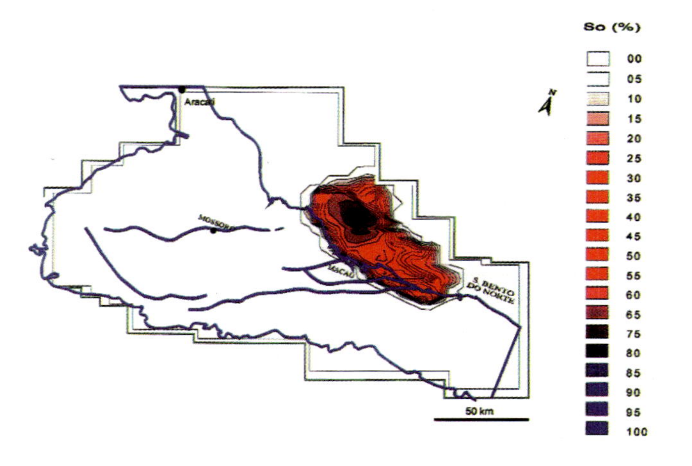

Figure 13—Oil saturation map of subunit Açu III at 100,000 years after the beginning of secondary migration.

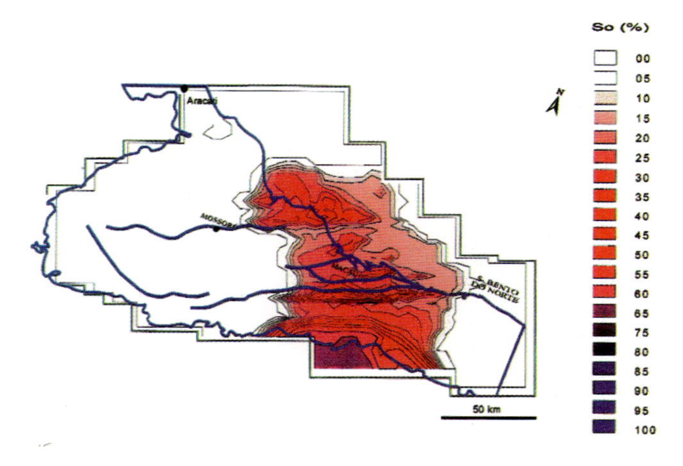

Figure 15—Oil saturation map of subunit Açu III at 2 m.y. after the beginning of secondary migration.

At 2–3 m.y. after the beginning of secondary migration, the rates of oil displacement were significantly diminished (<2 cm/year), and oil was concentrated in the areas shown in Figure 16.

CONCLUSIONS

The combination of geochemistry and basin-scale modeling of secondary oil migration in the Açu Formation indicates the following:

1. Major migration pathways and economic oil accumulations were controlled mainly by northeast–southwest structural noses at the top of the Açu Formation.
2. Estimated ratios of oil displacement for the earlier oil migration fronts range from 3.0 to 7.0 cm/year.

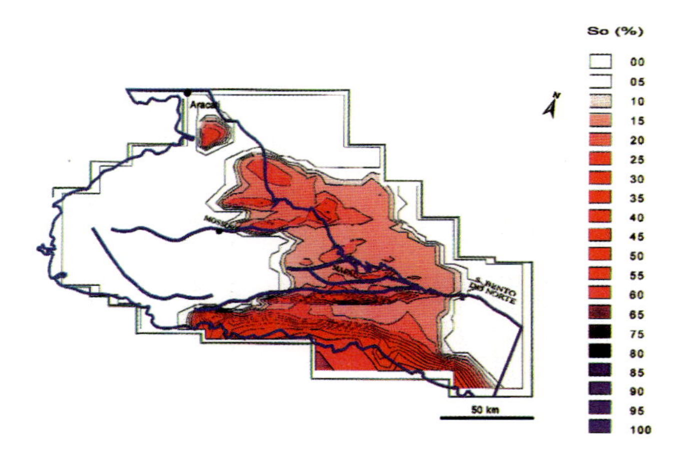

Figure 16—Oil saturation map of subunit Açu III unit at 10 m.y. after the beginning of secondary migration.

3. Oil may have reached the Macau High about 100,000 years after the beginning of secondary migration.
4. The most important onshore oil accumulations in the Potiguar Basin may have been trapped between 1 and 5 m.y. after the beginnng of secondary migration.
5. Significant amounts of petroleum may have been lost through seepage along outcrops of the Açu Formation in the northwestern part of the Aracati Platform and the southern part of the Carnaubais trend.
6. The boundaries of the Alagamar–Açu petroleum system have been defined by the water potential field produced by the geometry of the top of the Açu Formation (Açu IV).

Acknowledgments*—We would like to thank Petrobrás for the authorization to publish this paper.*

REFERENCES CITED

Araripe, P. T., and F. J. Feijó, 1994, Bacia Potiguar: Boletim de Geociências da Petrobras, v. 8, p. 127–141.

Bertani, R. T., I. G. Costa, and R. M. D. Matos, 1990, Evolução tectono-sedimentar, estilo estrutural e habitat do petróleo na Bacia Potiguar, *in* G. P. Raja Gabaglia and E. J. Milani, eds., Origem e Evolução de Bacias Sedimentares: Petrobras, Rio de Janeiro, p. 291–310.

Cerqueira, J. R., 1995, Geoquímica do Campo de Ubarana e adjacências (Bacia Potiguar): Master's thesis, Universidade Federal da Bahia, Salvador, Brasil, 192 p.

Cremonini, O. A., 1993, Caracterização estrutural e evolução tectônica da área de Ubarana, porção submersa da Bacia Potiguar, Brasil: Master's thesis, Universidade Federal de Ouro Preto, Ouro Preto, Brasil, 143 p.

Cremonini, O. A., 1995, A reativação tectônica da Bacia Potiguar no Cretáceo Superior: Fifth Simpósio Nacional de Estudos Tectônicos, Gramado, Brazil, v. 1, p. 277–280.

Dahlberg, E. C., 1982, Applied hydrodynamics in petroleum exploration: New York, Springer-Verlag, 161 p.

Demaison, G., and B. J. Huizinga, 1991, Genetic classification of petroleum systems: AAPG Bulletin, v. 75, p. 1626–1643.

Gussow, W. C., 1954, Differential entrapment of oil and gas: a fundamental principle: AAPG Bulletin, v. 38, p. 816–853.

Hubbert, M. K., 1953, Entrapment of petroleum under hydrodynamic conditions: AAPG Bulletin, v. 37, p. 1954–2026.

Lima, C. C., C. M. Bentz, L. E. N. Fonseca, F. F. Lima Neto, and G. L. N. Gusso, 1993, Correlações entre a direção do campo de tensões neotectônicas, a topografia e estruturas geológicas na Bacia Potiguar: Petrobras Internal Report, Rio de Janeiro, 179 p.

Lima Neto, F. F., 1993, Geologia da Bacia Potiguar e de suas acumulaçés de petróles: Petrobras Internal Report, Rio de Janeiro, 33 p.

Mello, M. R., P. C. Gaglianone, S. C. Brassell, and J. R. Maxwell, 1988, Geochemical and biological marker assessment of depositional environments using Brazilian offshore oils: Marine and Petroleum Geology, v. 5, p. 205–223.

Mello, U. T., 1987, Evolucão termomecânica da Bacia Potiguar–RN, Brasil. Master's thesis, Departamento de Geologia da Escola de Minas, Universidade de Ouro Preto, Ouro Preto, Brasil, 186 p.

Prinzhofer, A., E. V. Santos Neto, T. Takaki, M. R. Mello, and S. Roos, 1998, The gas system in the Potiguar Basin: relations with the petroleum genesis and migration (abs.): 6th Latin American Congress on Organic Geochemistry, Isla Margarita, Venezuela.

Rodrigues, R., J. B. L. Françolin, and H. P. Lima, 1983, Avaliação geoquímica preliminar da Bacia Potiguar terrestre: Petrobras Internal Report, Rio de Janeiro, 67 p.

Santos Neto, E. V., 1999, Use of hydrogen and carbon stable isotopes characterizing oils from the Potiguar Basin (onshore), northeastern Brazil: AAPG Bulletin, v. 83, p. 496–518.

Santos Neto, E. V., M. R. Mello, and R. Rodrigues, 1990, Caracterização geoquímica dos óleos da Bacia Potiguar: XXXVI Congresso Brasileiro de Geologia, v. 2, p. 974–985.

Souto Filho, J. D., 1994, Utilização de simulador numérico na análise do processo de migração secundária de petróleo: Master's thesis, Universidade Estadual de Campinas, São Paulo, Brazil, 164 p.

Souza, O. R., J. D. Souto Filho, and F. F. Lima Neto, 1983, Acumulações de petróleo sob condições hidrodinâmicas na Bacia Potiguar: Sociedade Brasileira de Geologia, 33rd Congresso Brasileiro de Geologia, Salvador, Brasil, v. 3, p. 1395–1409.

Souza, S. M., 1982, Atualização da litoestratigrafia da Bacia Potiguar: Sociedade Brasileira de Geologia 32nd Congresso Brasileiro de Geologia, Salvador, Brasil, v. 5, p. 2392–2406.

Trindade, L. A. F., S. C. Brassell, and E. V. Santos Neto, 1992, Petroleum migration and mixing in the Potiguar Basin, Brasil: AAPG Bulletin, v. 76, p. 1903–1924.

Vasconcelos, E. P., F. F. Lima Neto, and S. Roos, 1990, Unidades de correlação da Formação Açu, Bacia Potiguar, *in* Sociedade Brasileira de Geologia 36th Congresso Brasileiro de Geologia 36, Natal, Brasil, v. 1, p. 227–240.

Gonzaga, F. G., F. T. T. Gonçalves, and L. F. C. Coutinho, 2000, Petroleum geology of the Amazonas Basin, Brazil: modeling of hydrocarbon generation and migration, *in* M. R. Mello and B. J. Katz, eds., Petroleum systems of South Atlantic margins: AAPG Memoir 73, p. 159–178.

Chapter 13

Petroleum Geology of the Amazonas Basin, Brazil: Modeling of Hydrocarbon Generation and Migration

F. G. Gonzaga
Petrobrás E&P
Rio de Janeiro, Brazil

L. F. C. Coutinho
Petrobrás E&P
Salvador, Brazil

F. T. T. Gonçalves
Petrobrás CENPES
Rio de Janeiro, Brazil

Abstract

The Amazonas Basin is a 500,000-km^2 intracratonic basin in northern Brazil. The ~6000-m lithologic section encloses mainly Paleozoic sedimentary rocks intruded by Triassic–Jurassic diabase dikes and sills, and subsequently buried by Cretaceous–Tertiary rocks. Geochemical and geologic data point to the Upper Devonian marine black shales from Barreirinha Formation as the main hydrocarbon source rocks.

Data from 11 selected wells were used to perform thermo-mechanical modeling. Backstripping and stratigraphic analyses indicate four extensional events: Ordovician–Early Devonian, Devonian–Early Carboniferous, Middle Carboniferous–Permian, and Cretaceous–Tertiary. The tectonic subsidence curve of each well was compared to theoretical subsidence curves to define the extensional factors and determine the heat flow history. The integration of 1-D basin modeling with geologic and geochemical data suggests that the Barreirinha Formation source rock started to generate petroleum during the Late Carboniferous. Modeling of primary migration indicates that the main phase of oil expulsion began when the source rock attained a transformation ratio of ~50% and a maturation level of 0.80% R_o. The main phase of petroleum generation and expulsion occurred from Late Carboniferous to Permian time and was completed by the Early Triassic. Any later tectonic event remobilized those hydrocarbons previously trapped.

Preliminary volumetric calculations indicate that up to 1 trillion bbl of oil equivalent were expelled from the source rock. Because of the long distances of both vertical and horizontal migration, it is believed that an important amount of the expelled hydrocarbon was dispersed along migration pathways. A significant part could also have been remobilized and lost during Cretaceous uplift of the basin margins.

INTRODUCTION

The Amazonas Basin is a 500,000-km^2 intracratonic basin located in northern Brazil within the Amazon rain forest. It is separated from the Marajó Basin by the Gurupá arch on the east and from the Solimões Basin by the Purus arch on the west (Figure 1). Although geochemical data suggest a significant hydrocarbon source potential, no commercial petroleum accumulations have yet been discovered. The basin's sedimentary fill is about 6 km thick and consists mainly of Paleozoic rocks (Ordovician–Permian) intruded by Triassic–Jurassic diabase dikes and sills, and subsequently buried by Cretaceous–Tertiary rocks (Figure 2).

This chapter describes the main results of a multidisciplinary survey carried out using seismic and well data in addition to oil, gas, and rock samples. Geochemical methods included elemental and visual kerogen analysis, Rock-Eval pyrolysis, gas chromatography, and mass spectrometry. Also, numerical modeling was performed using 1-D software (BaSS/Petrobras for thermo-mechanical modeling and Genex/Institut Français du Pétrole for geochemical modeling). The integration of geologic and geochemical data with modeling results allowed charac-

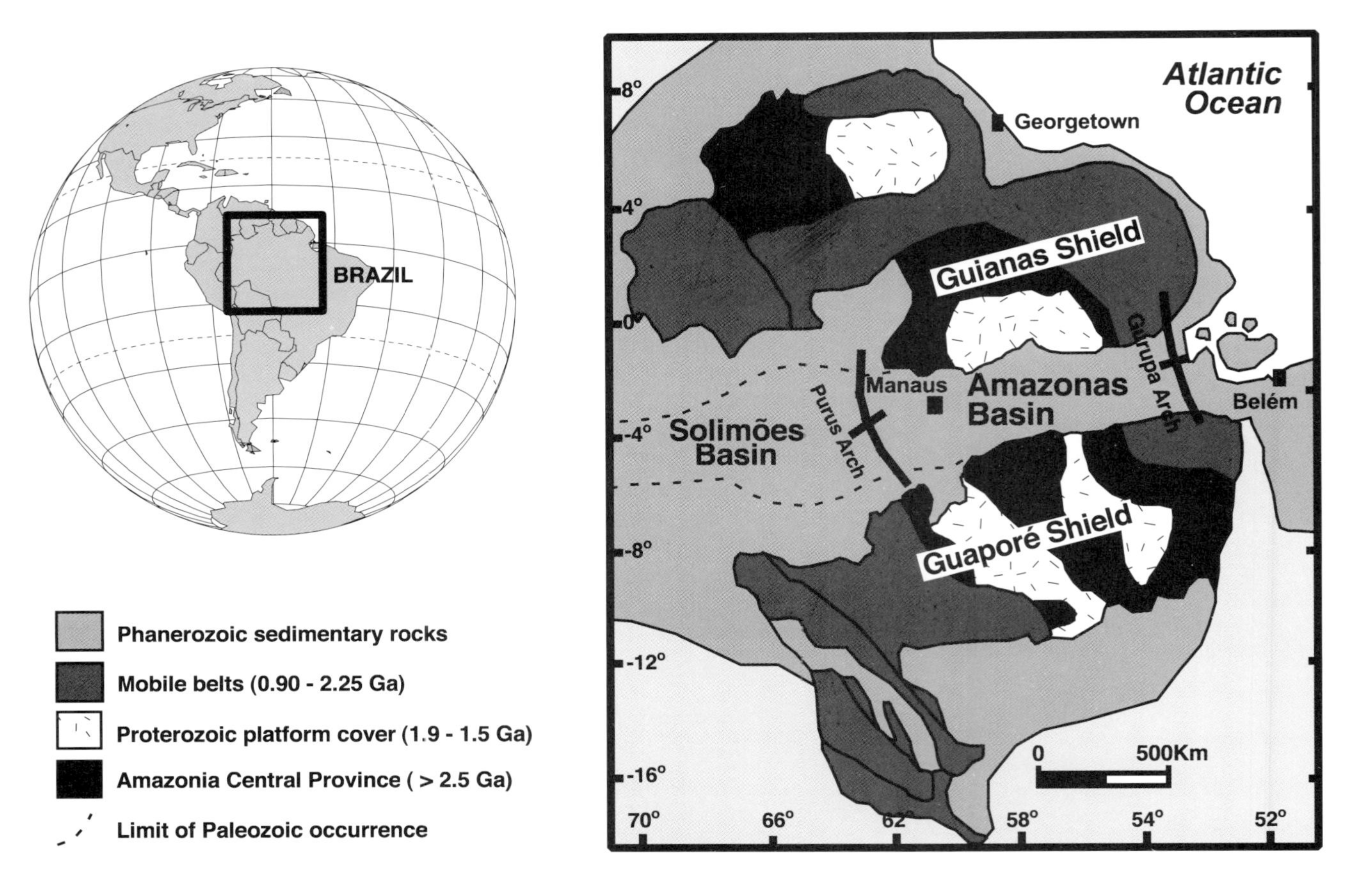

Figure 1—Location map and geologic setting of the study area. Notice the elongated form of the Amazonas and Solimões Basins in the east–west direction. Modified from Matos and Brown (1992).

terization of the petroleum system, development of a new tectonic evolution model, and assessment of the timing and volume of hydrocarbon generation and migration.

GEOLOGIC SETTING

Stratigraphy

The sedimentary record of the Amazonas Basin can be grouped into four transgressive and regressive megasequences separated by wide regional unconformities: (1) Upper Ordovician–Lower Devonian; (2) Middle Devonian–Lower Carboniferous; (3) middle Carboniferous–Permian and (4) Cretaceous–Tertiary (Cunha et al., 1994) (Figure 2). These megasequences overlie a Middle Proterozoic (?) platform cover, which comprises sedimentary rocks of the Purus Group, including reddish kaolinitic sandstones with conglomerate beds (Prosperança Formation) and brown and white dolomites (Acari Formation).

The Upper Ordovician–Lower Devonian megasequence, named the Trombetas Group, is bounded at the top by a Lower Devonian unconformity and comprises glacial and marine clastic sedimentary rocks. This group is subdivided into the following formations (from bottom to top): the Autás-Mirim, represented by Lower Ordovician fluvial–marine sandstones and neritic shales; the Nhamundá, composed of Lower Silurian coastal–neritic sandstones and glacial deposits; the Pitinga, comprising Upper Silurian marine shales and diamictites; and the Manacapuru, comprising Upper Silurian–Lower Devonian sandstones and neritic–coastal shales.

The Middle Devonian–Lower Carboniferous megasequence is bounded at the top by a Lower Carboniferous unconformity and is composed of marine clastic sedimentary rocks with glacial incursions from the Urupadi and Curuá Groups. The Urupadi Group (Middle Devonian) is made up of neritic–deltaic sandstones, siltstones, and shales of the Maecuru and Ererê Formations. The Curuá Group (Upper Devonian–Lower Carboniferous) comprises (from bottom to top) the Barreirinha, Curiri, Oriximiná, and Faro Formations. The Barreirinha Formation (upper Frasnian–lower Famenian) is represented by dark gray to black shales deposited during the Frasnian global transgressive event. The Curiri Formation (upper Famenian) comprises glacial marine diamictites, shales, and siltstones. The Oriximiná (upper

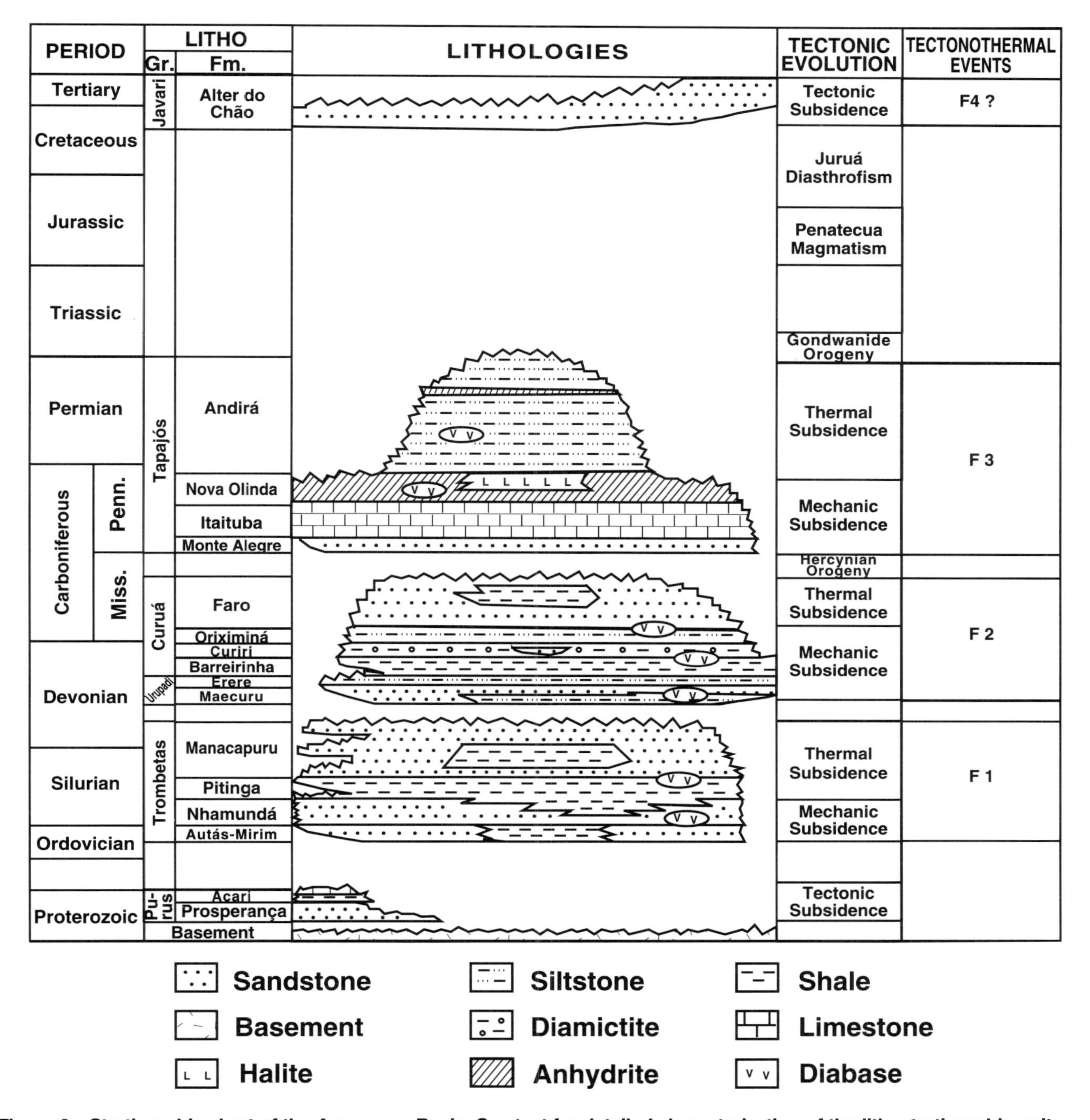

Figure 2—Stratigraphic chart of the Amazonas Basin. See text for detailed characterization of the lithostratigraphic units and their ages. Modified from Cunha et al. (1994).

Famenian–Tournasian) and Faro (Tournaisian–Visean) Formations are represented by fluvial–deltaic sandstones and shales associated with storm deposits.

The middle Carboniferous–Permian megasequence, truncated at the top by a Cretaceous unconformity, is composed of clastic and chemical sedimentary rocks of the Tapajós Group, which marks a climatic change from cold to warm and dry. It is subdivided (from bottom to top) into the Monte Alegre, Itaituba, Nova Olinda, and Andirá Formations. The Monte Alegre (lower Bashkirian) consists of eolian sandstones and lacustrine and interdune shales. The Itaituba (Bashkirian) and Nova Olinda (Moscovian) Formations comprise infratidal carbonates and shales and sabkha evaporites. The Andirá Formation (Permian) is represented by reddish terrestrial sandstones, siltstones, and shales.

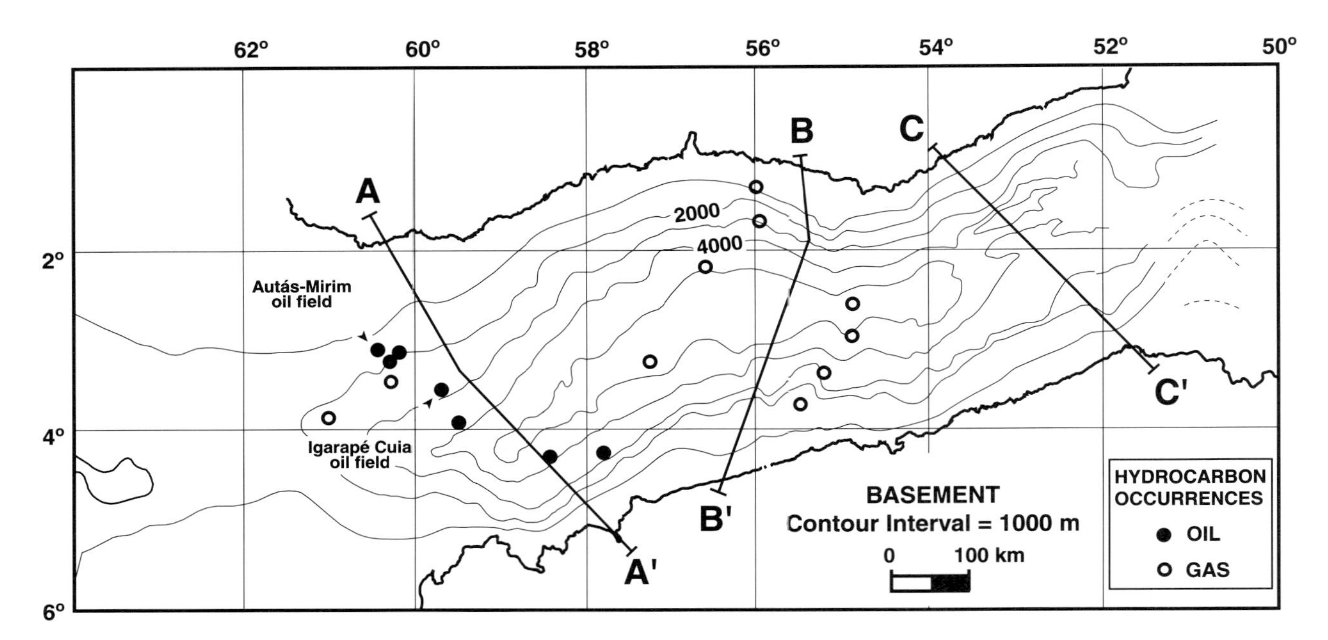

Figure 3—Structural contour map of the Precambrian basement in the Amazonas Basin based on well and seismic data. (Note locations of cross sections shown in Figure 4.)

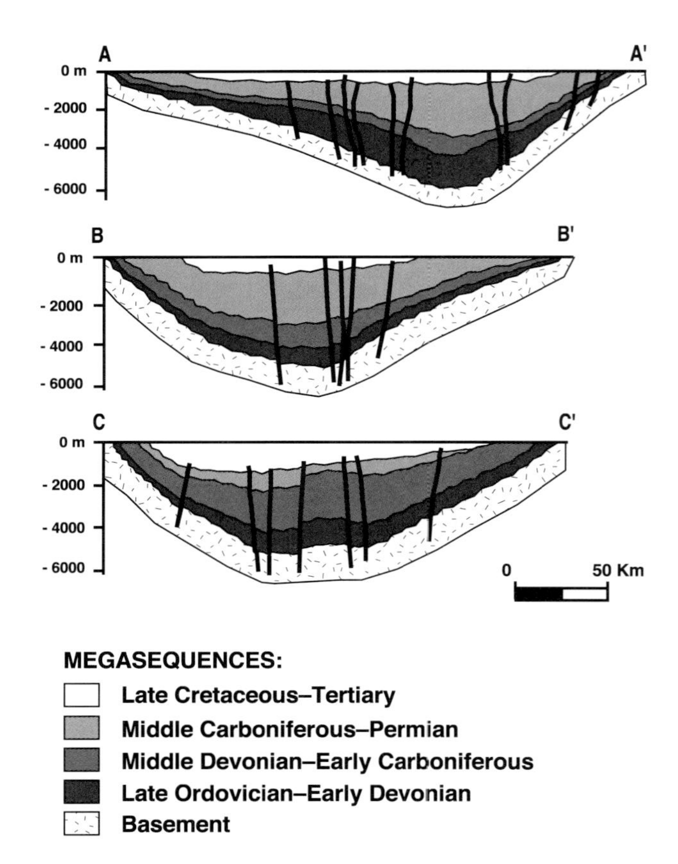

Figure 4—Geologic cross sections through the Amazonas Basin (see Figure 3 for location). Note the asymmetric character of the sedimentary section and the reversal of polarity between cross sections A–A' and B–B'. Also note the attenuation of the asymmetry toward the east (C–C').

The Cretaceous–Tertiary megasequence overlies the Paleozoic strata and is represented by the sedimentary beds of the Alter do Chão Formation (Javari Group). This formation is made up of anastamosed fluvial sandstones and argillites ranging in age from Late Cretaceous to Paleogene.

The Paleozoic megasequences were affected by an Early Jurassic (200 ± 20 Ma) magmatic event named the Penatecaua episode (Figure 2). These rocks are tholeiitic in nature and composed of mid-oceanic ridge basalt (MORB) type quartz diabase, probably formed in shallow magmatic chambers (Mizusaki et al., 1992). The intrusive bodies were emplaced in distinctive levels across the basin. In the west, the occurrences of sills are mainly in the Nova Olinda Formation, whereas in the east, they occur at all levels of the lithologic column, near or within the reservoir and source rocks.

The paleontologic and geochronologic data are based on the work of Daemon and Contreiras (1971), Savini and Altiner (1991), and Grahn (1992). The geologic time scale used in this work is from Harland et al. (1989). On the basis of these data, stratigraphic sections and chronostratigraphic isopach maps were constructed. In most of the wells drilled in the basin, the chronostratigraphic limits coincide with lithostratigraphic limits.

Tectonic and Structural Framework

The Amazonas Basin and its neighboring Solimões Basin together form a trench-shaped basin about 2600 km long and up to 500 km wide. These basins were developed over the Amazonian craton, bounded to the north by the Guyana Shield and to the south by the Guaporé Shield (Figure 1). This craton has a complex

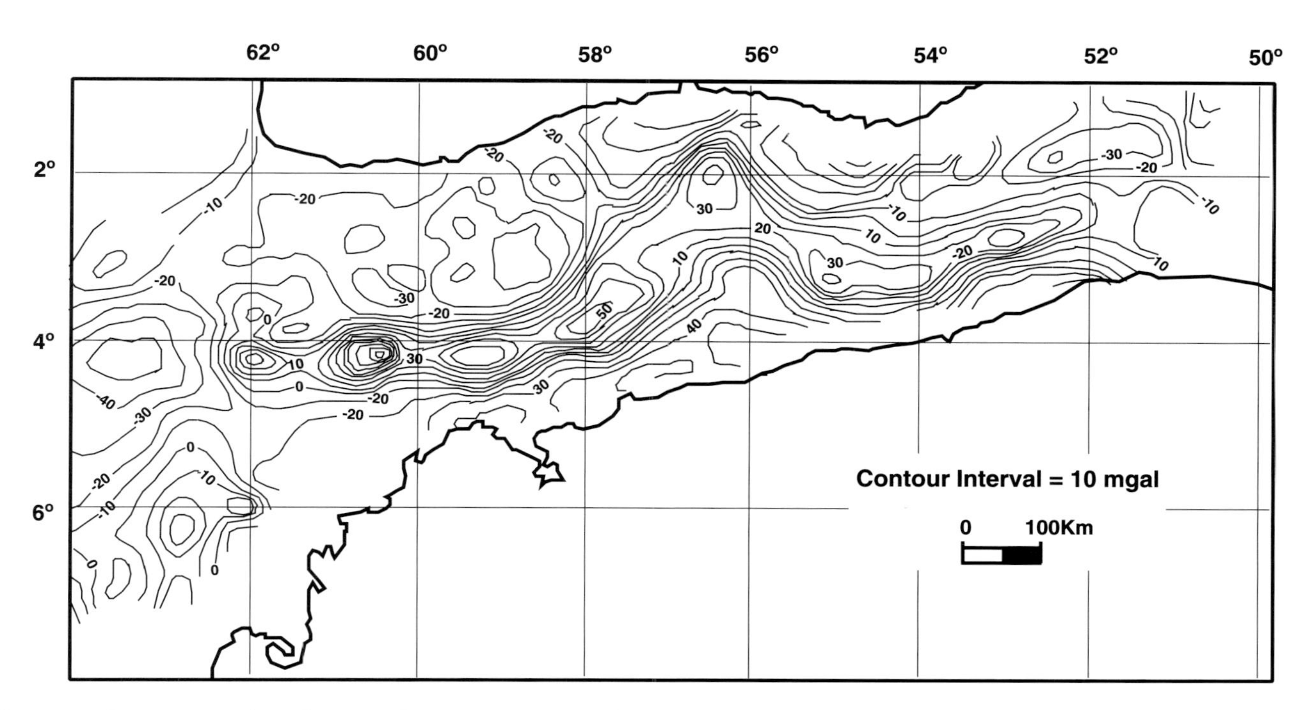

Figure 5—Bouguer gravimetric map of the Amazonas Basin (Linsser, 1958). Notice the alignment of the positive anomalies along the basin depocenter, showing a dog-leg pattern (compare with map in Figure 3).

Precambrian evolution as a consequence of three Proterozoic mobile belts (Maroni-Itacaiúnas, Rio Negro Juruena, and Rondoniano) accreted onto an Archean nuclei (Amazônia Central Province) (Cordani et al., 1984; Caputo,1984).

The basement structural map of the Amazonas Basin reveals a characteristic dog-leg pattern with depths reaching 6000 m (Figure 3). The same shape is outlined by isopach maps for all chronostratigraphic units. An asymmetric geometry can be seen in geologic cross sections throughout the basin (Figure 4). The Bouguer gravity map shows a trend of positive anomalies (up to 90 mgals) along the basin depocenter (Figures 3 and 5). Surface topography is a little higher along the basin margins (~100 m above sea level) than in the trough area (up to 50 m above sea level).

Geologic history and backstripping studies show that the basin infill was characterized by a cyclic succession of alternating high and low accumulation rate periods (Figure 6). According to Zalán (1991), this succession suggests the occurrence of episodic reactivations, which could be related to orogenies.

The opening of the North Atlantic Ocean gave rise to a period of intense magmatism during the Early Jurassic. The igneous event in the Amazonas Basin is probably related to a hot spot (Aires, 1985), although the track of its position through time remains controversial (Wilson, 1997). Our backstripping studies reveal that no significant thermal subsidence was associated with this magmatic event (Figure 6).

Two important tectonic events affected the basin: the Juruá (Early Jurassic–Early Cretaceous) and the Tertiary (late Paleogene–Holocene). The Juruá transpressional tectonic event was responsible for the development of northeast-trending reverse faults and asymmetric anticlines, which affected both the Paleozoic sedimentary rocks and the diabase sills. The Tertiary tectonics had a transcurrent nature and are represented by transpressional and transtensional structures. Such structures were amplified as a result of halokinesis in the salt depocenter areas of the basin.

Palinspastic reconstructions based on stratigraphic correlations and the thermal maturity of sedimentary sections indicate a basin extent larger than present. Our studies indicate that 1800 m of section were eroded along the basin margins. Apatite fission track studies (PECTEN/IDEMITSU, 1989) support this interpretation and suggest that the related uplift occurred during the Late Cretaceous (~110 Ma).

PETROLEUM GEOLOGY

Source Rocks

Evaluation of the petroleum source potential (S_2) was based on a large geochemical data set from surface and well samples (cores and cuttings), which includes the results of 3948 total organic carbon (TOC), 1986 Rock-Eval pyrolysis, 180 stable carbon isotope, 280 organic petrographic, and 59 gas chromatography (GC) and gas chromatography–mass spectrometry (GC-MS) analyses. TOC and Rock-Eval data in immature samples indicate that

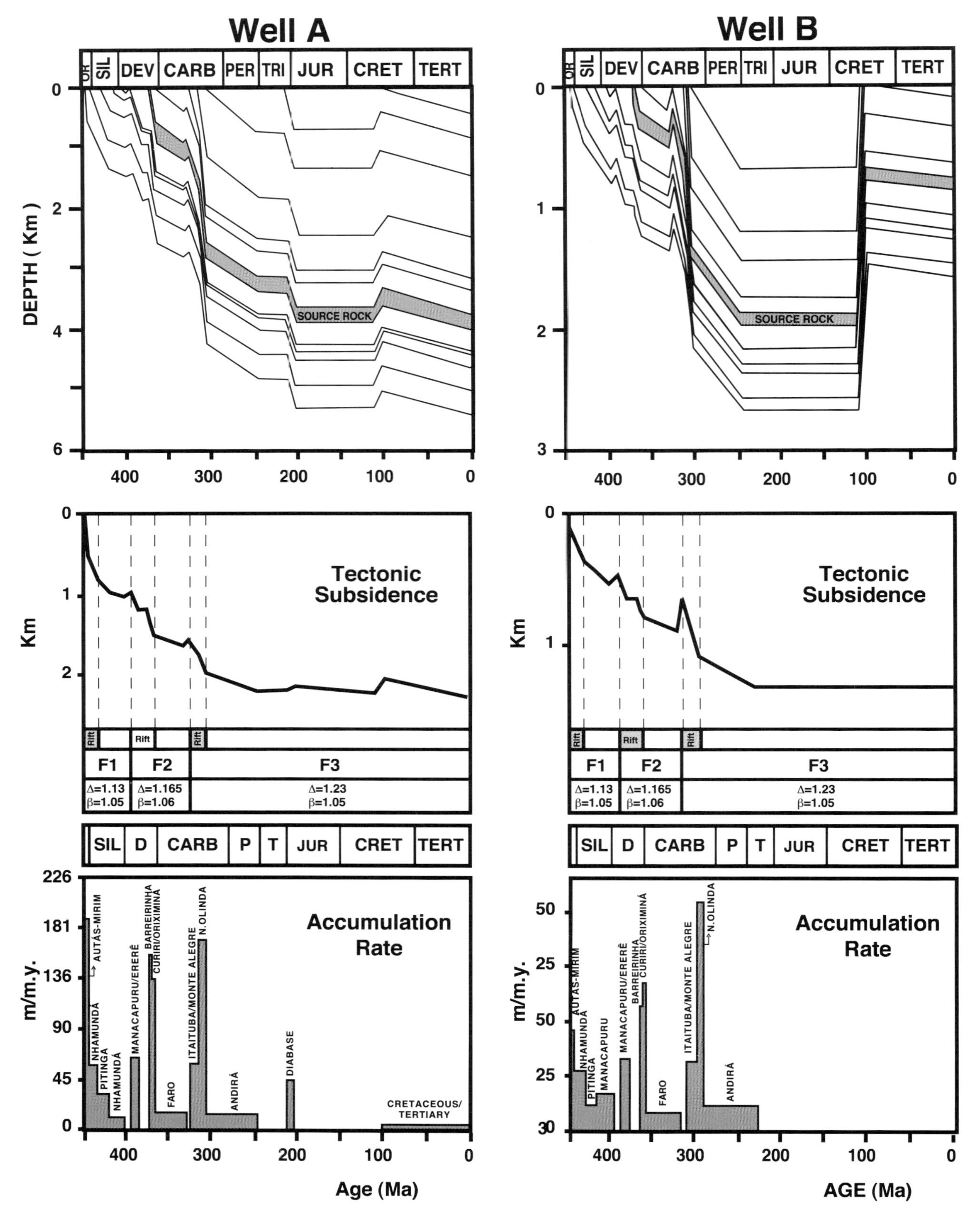

Figure 6—Geologic history, backstripping, and accumulation rate diagrams of two selected wells from the Amazonas Basin (see Figure 8 for well locations). Notice that well A (basin center) and well B (basin margin) are at different scales.

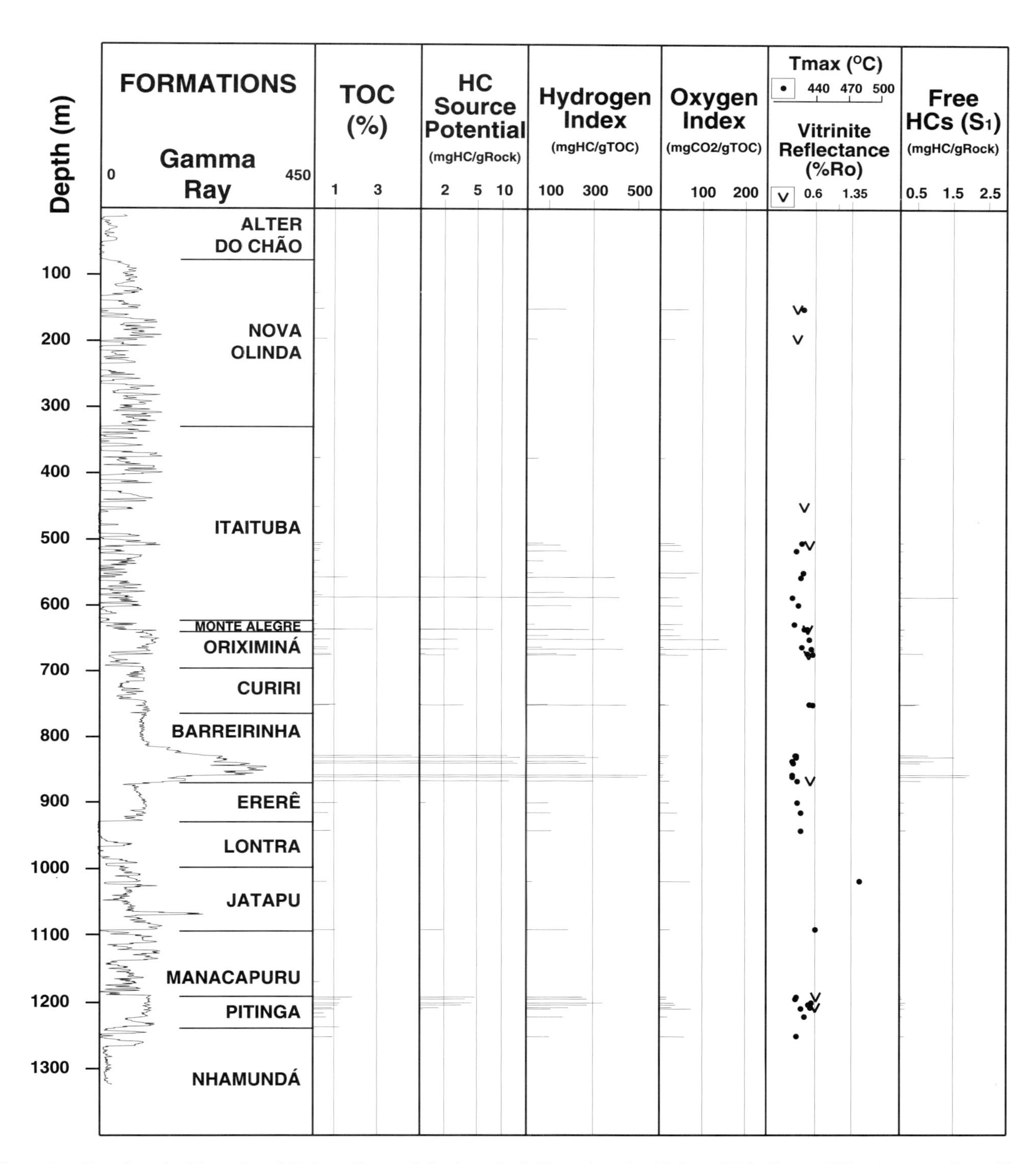

Figure 7—Geochemical log of well B (see Figure 8 for location). Note that the highest TOC, S_2, and HI values are found in the basal radioactive part of the Barreirinha Formation.

the Pitinga, Barreirinha, and Curiri Formations are the only units with significant source potentials, as shown on a typical geochemical log in Figure 7.

The thickness of the Pitinga Formation ranges from 20 to 40 m at the basin margin to 120 m in the depocenter, with TOC values generally lower than 2%, S_2 about 4 mg HC/g rock, and hydrogen and oxygen indices (HI and OI) indicating a dominance of type II kerogen. In a few samples from the basal Pitinga (upper Llandoverian–lower Wenlockian), TOC and S_2 values reach up

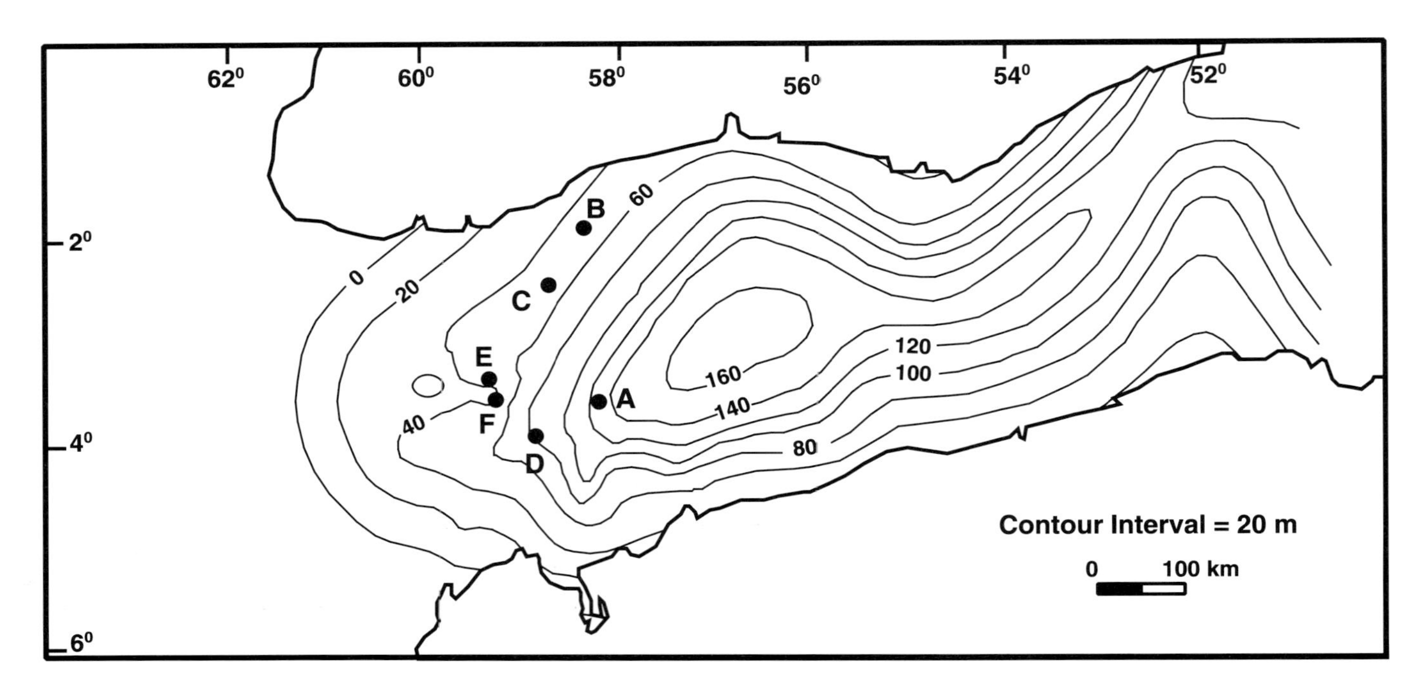

Figure 8—Isopach map of the radioactive shales of the lowermost Barreirinha Formation (see stratigraphic column and gamma ray log in Figure 7). Note the locations of some of the wells (A–F) selected for this study.

to 4% and 14 mg HC/g rock, respectively. Organic petrographic analyses show the predominance of liptinite and subordinately amorphous organic matter. The liptinite is characterized by a predominance of tasmanites and by a low relative abundance of acritarchs and sporomorphs.

The Barreirinha Formation can be subdivided in two distinctive parts (Figure 7): a basal section, informally called the "lower Barreirinha" (lower Frasnian–lower Famenian) or "radioactive Barreirinha," consisting of black shales with high values of gamma ray and resistivity; and an upper section, or "upper Barreirinha," made up of dark gray shales. The lower Barreirinha ranges in thickness from 30–40 m at the basin margin to 150–160 m in the depocenter (Figure 8). Geochemical analyses from immature to early mature samples show high TOC values (3–8 %), good HC source potential (S_2 of 5–20 mg HC/g rock), and HI varying from 100 to 400 mg HC/g TOC, thus indicating a dominance of type II kerogen. Such values, together with the predominance of amorphous organic matter over liptinite, indicate a deep bathyal anoxic environment. Liptinite is composed of tasmanites, acritarchs, spores, pollen, and vegetable cuticles. The upper Barreirinha ranges in thickness from 30 m at the basin margin to 150 m in the depocenter, with TOC values ranging from 1 to 2% and a maximum source potential of 4 mg HC/g rock. The HI is generally lower than 200 mg HC/g TOC, and the OI is about 300 mg CO_2/g TOC, indicating a predominance of type III or oxidized kerogen.

Shales of the Curiri Formation have both low TOC values (1–2%) and low S_2 values (up to 3 mg HC/g rock); petrographic data and HI and OI values point to a predominance of type III or oxidized kerogen.

The genetic classification of petroleum systems proposed by Demaison and Huizinga (1991) was applied to the potential source rocks of the Amazonas Basin. Rock-Eval data and the estimated maximum thickness of each formation were taken into consideration to perform the source potential index (SPI) calculation. Assuming that the hydrocarbon drainage was essentially lateral, we concluded that only the lower Barreirinha section has a significant SPI value (Figure 9).

Thermal Maturity

The integration of optical (spore coloration index, vitrinite reflectance, and fluorescence), chemical (T_{max}), and molecular parameters (sterane isomerization ratios) allowed an assessment of the thermal evolution. Along the northern and southern flanks and on the western platforms, where the lower Barreirinha is shallow (1500 m depth), the maturation is low (<0.65% R_o). In the central trough, source rock maturity reaches 1.0% R_o at about 4000 m depth. Gradients extrapolated from the wells suggest that the source rock may reach 1.3–1.4% R_o in the depocenter as a result of the overburden effect. A higher degree of maturation (R_o > 1.4 %) was reached only because of the heating effect of the diabase dikes and sills.

Maturity data indicate that the thermal evolution of the source rock was controlled mainly by subsidence history and that heat from igneous intrusions played an important role only in those areas where dikes and sills

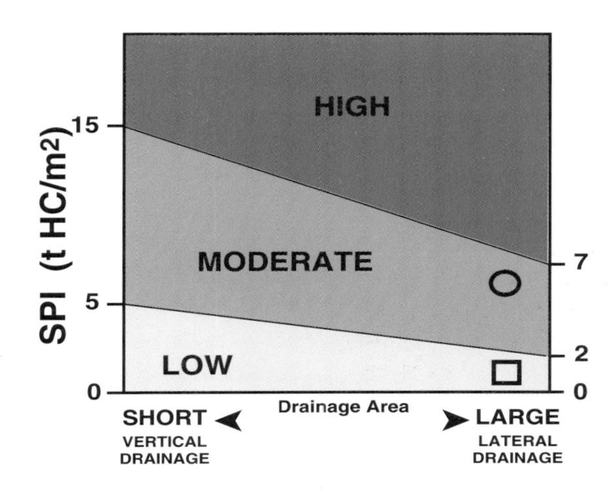

FORMATIONS	S1+S2 (mgHC/gRock)	h (m)	ρ (ton/m3)	SPI
□ Curiri Fm.	1.4	230	2.52	0.81
□ Barreirinha Fm. (upper)	1.9	160	2.52	0.71
○ Barreirinha Fm. (lower)	15.0	160	2.52	6.05
□ Pitinga Fm.	3.2	120	2.52	0.96

Figure 9—Calculated source potential index (SPI) for some of the lower Paleozoic formations of the Amazonas Basin. SPI calculation method after Demaison and Huizinga (1991).

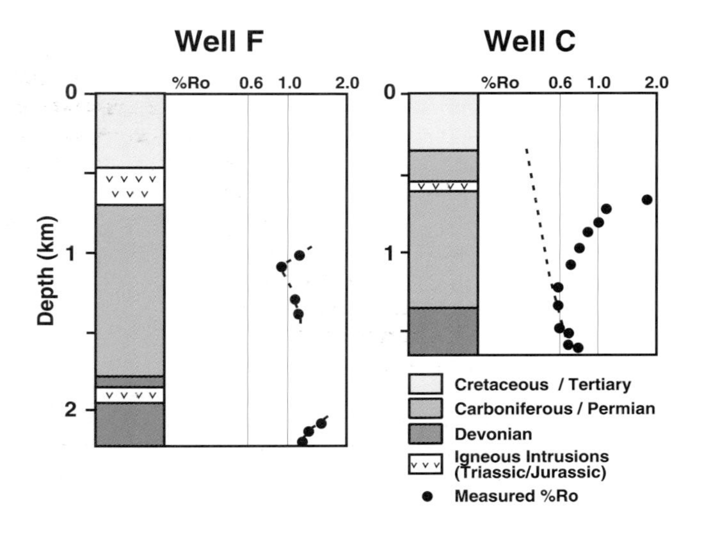

Figure 10—Vitrinite reflectance (% R_o) versus depth plots of two wells from the Amazonas Basin (see Figure 8 for well locations). Observe the thermal effect of the igneous intrusions on the sedimentary section.

were intruded into the Devonian sequence (e.g., Mullin, 1988) (Figure 10). Thus, in the eastern part of the basin, where most dikes and sills intruded the Devonian sequence, the source rock is overmature, while in the western part, where the intrusions are far from the source rock, the maturation was controlled by subsidence.

Traps, Reservoirs, and Seals

The main hydrocarbon occurrences in the Amazonas Basin are found in sandstones of the Monte Alegre, Curiri, and Ererê Formations. Well data (electric logs) and laboratory measurements indicate that Monte Alegre sandstones have the most favorable permeability and porosity characteristics for reservoiring hydrocarbons.

Monte Alegre sandstones were deposited in a desert environment, varying from terrestrial to transitional (sabkha), including fluvial, eolian, and shoal massif sandstone deposits. Such sandstone reservoirs occur throughout the basin, reaching thicknesses of 80 m. The best reservoir characteristics are associated with dune facies, which have porosities of about 20–25% and permeabilities ranging from 150 to 380 md. The seal for Monte Alegre reservoirs is composed of evaporites, carbonates (mudstones), and shales of the Itaituba Formation (Figure 11).

Curiri reservoir beds were deposited in a marine environment under glacial conditions. They are represented by sandstone lenses that were deposited in incised valleys as a response to relative sea level drops. Reservoir thicknesses range from a few meters to tens of meters, while porosity and permeability values vary widely from 6 to 20% and 1 to 400 md, respectively. The sandstone lenses are embedded within shales and diamictites which represent the seals.

The Ererê Formation is represented by barrier island sandstone bodies deposited over an erosional surface during the Frasnian transgressive event. Such reservoirs occur throughout the basin, reaching up to 10 m in thickness with maximum porosities and permeabilities of 20% and 10 md, respectively. Seals are formed by shales in the base of the Barreirinha related to the peak of the transgression event.

Petroleum occurs both in stratigraphic and structural traps. The main stratigraphic accumulation is the Autás-Mirim oil field, which contains 0.1×10^6 m^3 (0.63×10^6 bbl) of oil and 10.9×10^6 m^3 (0.39×10^9 ft^3) of gas in place trapped within Curiri sandstone lenses. The most important structural accumulation is the Igarapé Cuia oil field, with 0.03×10^6 m^3 (0.19×10^6 bbl) of oil and 5.90×10^6 m^3 (0.21×10^9 ft^3) of gas in place trapped within Monte Alegre sandstones. The structure is an anticline related to the Juruá tectonic event.

Petroleum Characterization

Oil and gas shows occur throughout the basin within Silurian, Devonian, and Carboniferous beds. Most of the gas occurrences are located in the central-eastern part of the basin (Figure 3), where the Devonian–Carboniferous sequence is strongly affected by igneous intrusions and the source rock is overmature. Otherwise, oil accumula-

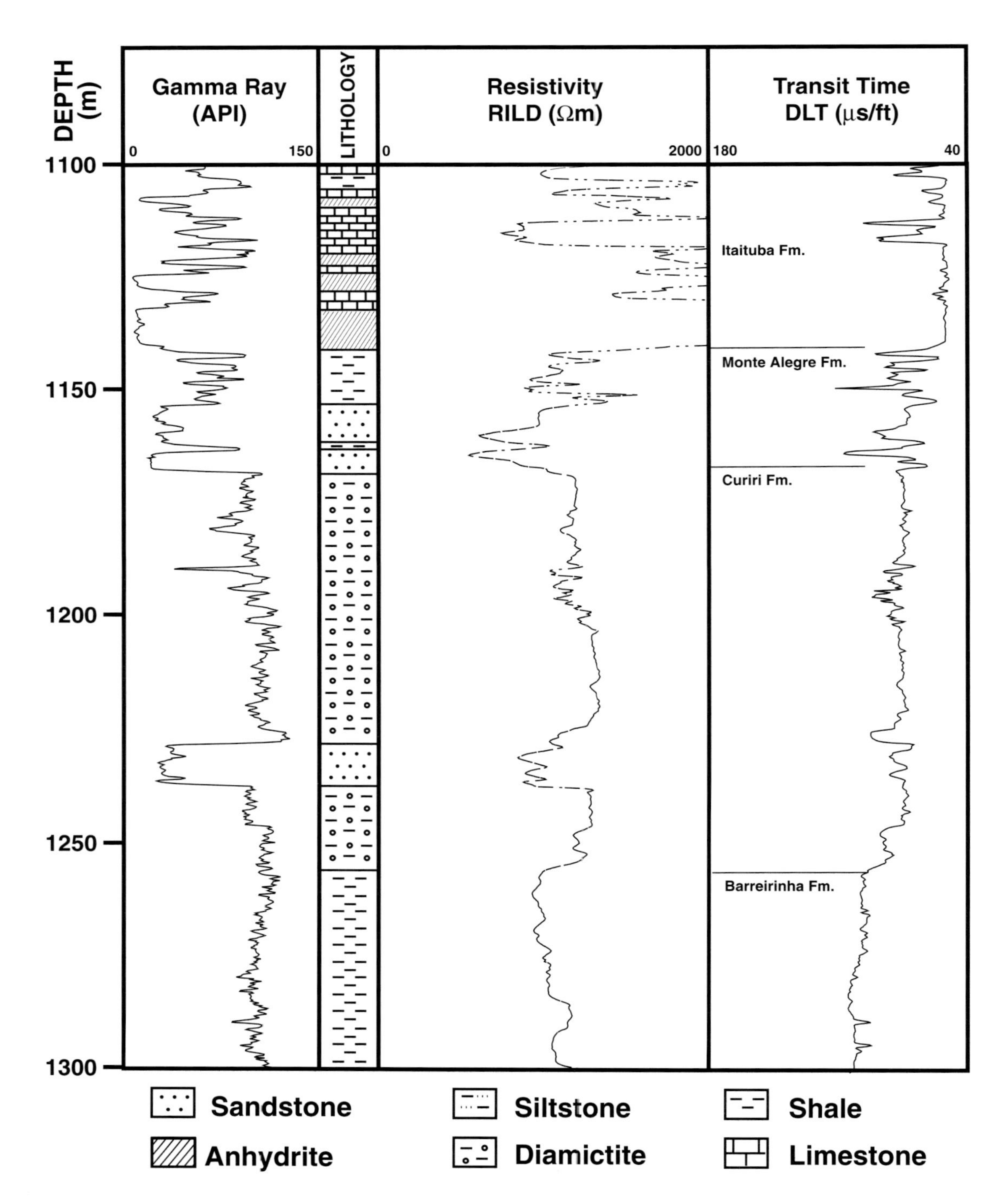

Figure 11—Sedimentary column from a selected well in the Amazonas Basin showing the lithologic composition and well log response. The main reservoirs are the sandstone layers from the Curiri and Monte Alegre Formations.

tions are restricted to the western part of the basin (Figure 3), where the intrusions are generally far from the source and reservoir rocks. The few gas occurrences in the western area are related to the local effects of diabase dikes.

A set of 12 oil samples from Monte Alegre, Curiri, and Ererê reservoirs were submitted to stable carbon isotope, GC, and GC-MS analyses. Bulk, isotopic, and molecular data revealed similar characteristics for the analyzed samples, indicating a common origin (Gonçalves et al., 1995). The most important features include the following: a high content of saturated hydrocarbons (51–77%), $\delta^{13}C$ ranging from –28.3 to –30.0‰, a predominance of low

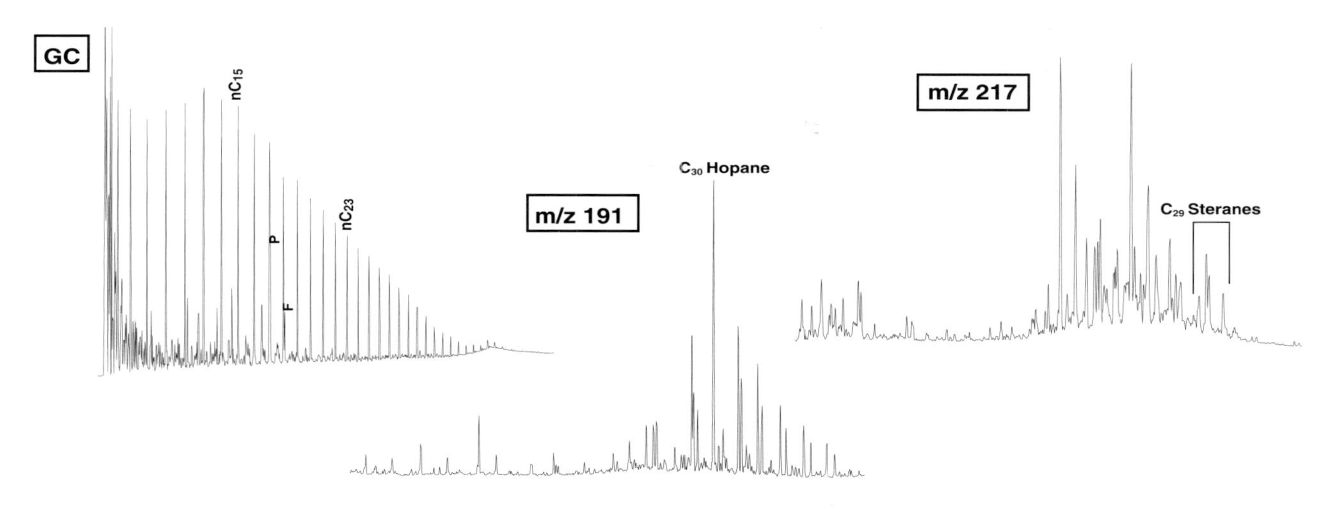

Figure 12—Gas chromatograms of total alkanes and partial *m/z* 191 and *m/z* 217 chromatograms for a typical oil sample of the Amazonas Basin, with $\delta^{13}C = -29.8‰$.

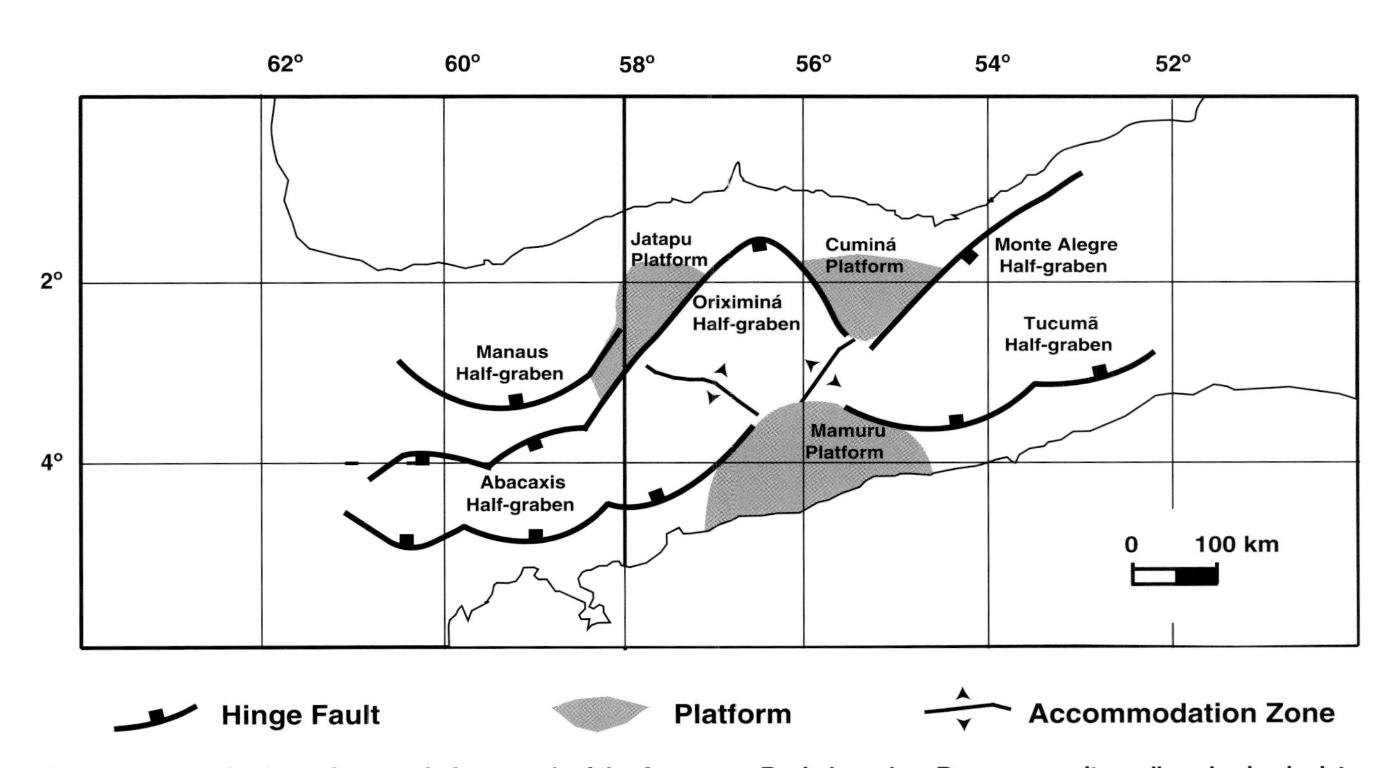

Figure 13—Map of schematic tectonic framework of the Amazonas Basin based on Bouguer gravity, well, and seismic data (Coutinho and Gonzaga, 1996). Compare with Figures 3, 4, and 5. Notice the alternation of platforms and half-grabens as well as the presence of internal highs (accomodation zones) typical of rift basins.

molecular weight *n*-alkanes (<n-C_{15}), a pristane/phytane ratio of about 2.0, hopane/sterane ratios of 2.0–7.0, a low relative abundance of gammacerane, a high proportion of C_{31} to C_{35} homohopanes, and a predominance of C_{29} regular steranes over their C_{28} and C_{27} counterparts (Figure 12). Biomarker maturity ratios (C_{29} $\alpha\alpha\alpha$ 20*S*/20*S* + 20*R* and C_{29} $\alpha\beta\beta/\alpha\alpha\alpha + \alpha\beta\beta$) indicate that the oils have reached a high degree of thermal evolution (0.8–1.0% R_o).

There is good correlation between biomarker and isotopic features for the oil samples and the organic extracts from lower Barreirinha black shales. Such features are similar to those observed for other Devonian marine oils (e.g., the Illinois and Williston Basins) (Arneth, 1984; Bethke et al., 1991).

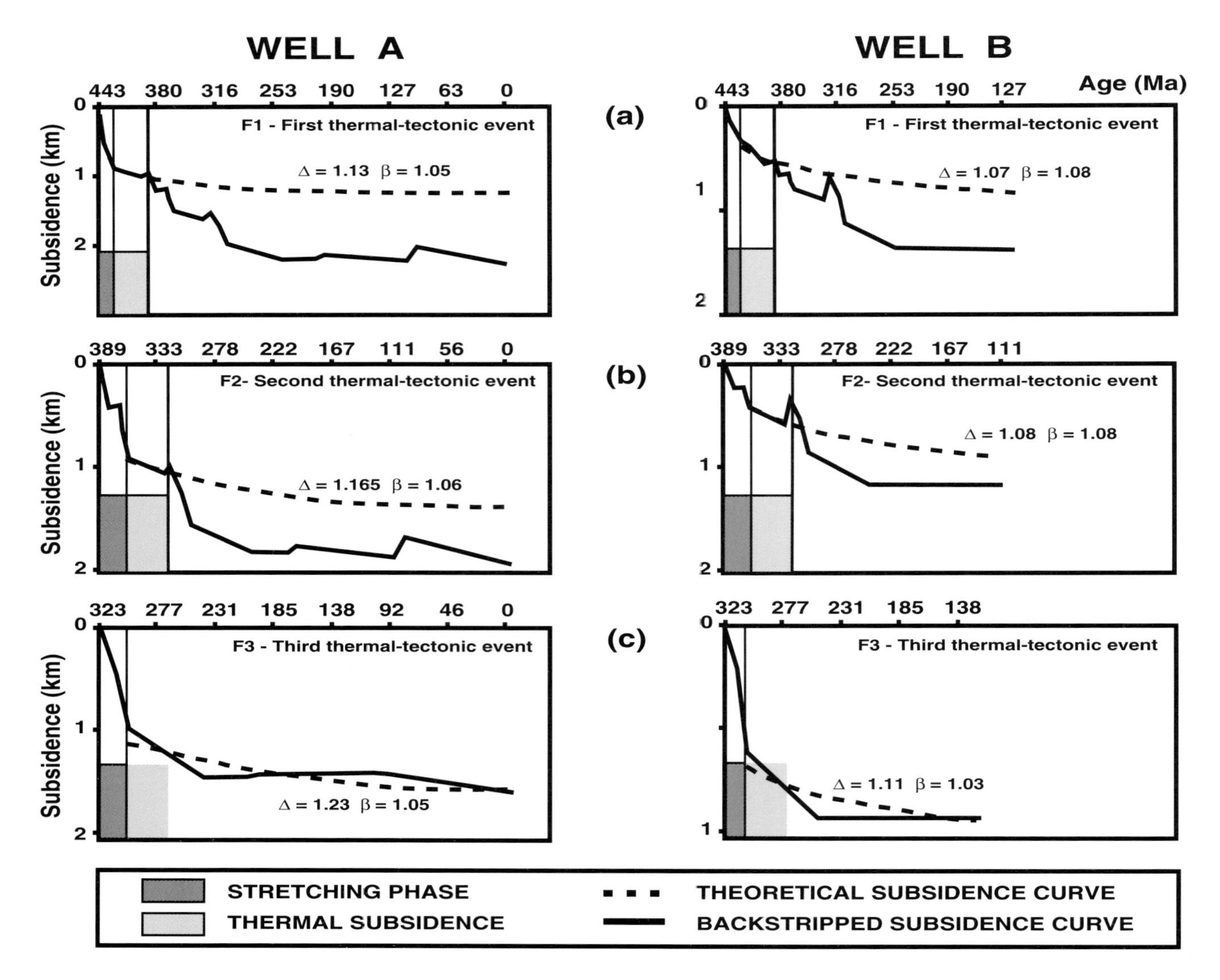

Figure 14—Subsidence plots for the (a) F1, (b) F2, and (c) F3 tectono-thermal events of two wells in the Amazonas Basin (see Figure 8 for well locations). The crustal (Δ) and subcrustal (β) stretching factors were estimated by comparing the observed backstripped subsidence with the theoretical curve calculated using the model from Royden and Keen (1980).

THERMAL AND TECTONIC EVOLUTION

Tectonics and Sedimentation

A rift model origin for the Amazonas Basin has been postulated since the first gravity survey was carried out in the 1950s (Linsser, 1958). Modeling studies performed by Nunn and Aires (1988) led to the conclusion that gravimetric highs indicate anomalous dense body intrusions at the base of the crust as a result of extensional processes since Ordovician time. Backstripping analyses indicate four extensional events: Ordovician (F1 event), Devonian (F2 event), Carboniferous (F3 event), and Cretaceous (F4 event) (Figure 6). Due to poor biostratigraphic control in the Cretaceous sequence, the last event (F4) will not be detailed here.

Each extensional event began with a period of low accumulation rates followed by high rates and ended with a period of again low accumulation rates. The first period is interpreted as the rift phase and the last period as the thermal subsidence phase. Each phase is also characterized by its typical lithofacies (Figure 2).

The lowermost part of each rift sequence is represented by deposition of high-energy sediments reflecting initial low accumulation rates (fluvial sandstones of the Autás-Mirim Formation, F1 event; sandstones of the Maecuru Formation, F2 event; and eolian sandstones of the Monte Alegre and carbonates of the Itaituba, F3 event). The high accumulation rate phases represent the main subsidence periods during which deposition of low-energy sediments prevailed (Pitinga shales chronoequivalents to the Nhamundá Formation, F1 event; shales and siltstones of the Barreirinha, Curiri, and Oriximiná Formations, F2 event; and evaporites of the Nova Olinda Formation, F3 event).

Lithofacies of the thermal sequences are represented by deposition of high-energy sediments and low accumulation rates (Manacapuru sandstones, F1 event; Faro sandstones, F2 event; and siltstones and sandstones of the Andirá Formation, F3 event). The eventual presence of a body of water or a rise in sea level allowed the deposition of low-energy sediments in these sequences (e.g., the Pitinga Formation, F1 event; and grayish carbonates at the top of the Andirá Formation, F3 event). The Paleozoic sedimentation pattern is compatible with that found in other continental rift models (e.g., McHargue et al., 1992; Lambiase and Bosworth, 1992).

Continental rifts studies have shown that the structural dog-leg patterns are the result of complex arrangements of half-grabens interconnected by accommodation zones (Rosendahl et al., 1986). The analysis of Bouguer gravity maps, geologic cross sections (Figures 4 and 5), and seismic sections allowed definition of the Amazonas Basin tectonic framework and characterization of its basic structural elements. These include the Abacaxis, Manaus, Oriximiná, Tucumã, and Monte Alegre half-grabens and the Jatapu, Mamuru, and Cuminá platforms, as well as the accommodation and hinge zones (Coutinho and Gonzaga, 1996) (Figure 13).

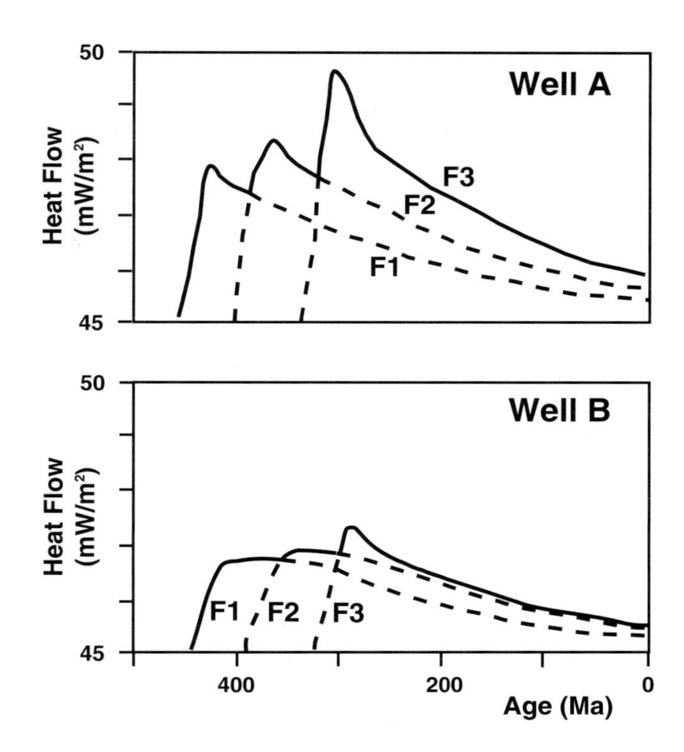

Figure 15—Heat flow history plots of two wells from the Amazonas Basin (see Figure 8 for well locations), showing the result of superposition of the three tectono-thermal events (F1, F2, and F3). Note that in well A (basin center), heat flow values are higher than in well B (basin margin).

Thermal and Mechanical Modeling

Reconstruction of the heat flow history of the Amazonas Basin was based on a unidimensional extensional model (Royden and Keen, 1980). Crustal and mantle stretching factors were estimated by comparison of tectonic subsidence curves calculated for each well with theoretical subsidence curves from the applied model. The original crustal thickness was taken from geophysical studies which estimate a value of 50 km (Aires, 1985; Matos and Brown, 1992), whereas the initial lithospheric thickness was estimated as 200 km. To simulate superposition of the successive extensional events, the original crustal thickness was progressively stretched, whereas the original lithospheric mantle thickness was partially reconstituted in response to thermal anomaly decay (Figure 14).

Because the last tectonic thermal event in the basin occurred in the Late Cretaceous (F4 event), the present-day lithosphere has attained a steady-state heat flow, with a measured average geothermal gradient of 23°C/km and a surface temperature of 27°C (Zembruscki and Kiang, 1989). Assuming an average conductivity of 2.0 W/m°C for the sedimentary section (Carvalho and Lobo, 1986), we calculated a mean heat flow of 45 mW/m^2, which is in good agreement with those found in tectonically stable areas (Polyak and Smirnov, 1968). This value comprises both the mantle heat flow due to a lithospheric thickness of 200 km (22 mW/m^2) and the crustal radiogenic heat flow (23 mW/m^2) (Jaupart, 1986). The radiogenic heat of sedimentary rocks was not considered in the model, and because the basin is in a hydrostatic condition, the convective heat flow is thought to be minimal.

The heat flow history of each well consists of the envelope of the superpositions of the heat flow curves of each extensional event (Figure 15). In response to the differences in stretching factors, the heat flow in the platform is lower than that in the trough. According to the model, the highest values of basal heat flow (up to 50 mW/m^2) were attained during the Carboniferous event, which was the main extensional phase. There is good correlation between the present temperatures calculated by simulation and the extrapolated bottom hole temperatures. Vitrinite reflectance curves calculated by the Easy™ % R_o method (Sweeney and Burnham, 1990) are also in agreement with measured data (Figure 16). Thermal modeling results, together with maturity and thermal data (vitrinite reflectance, temperature, and apatite fission track), indicate that no significant regional thermal event was associated with Penatecaua magmatism.

GENERATION AND MIGRATION MODELING

Assessment of the timing and volume of hydrocarbon generation and migration was done using an approach based on the integration of natural series, laboratory

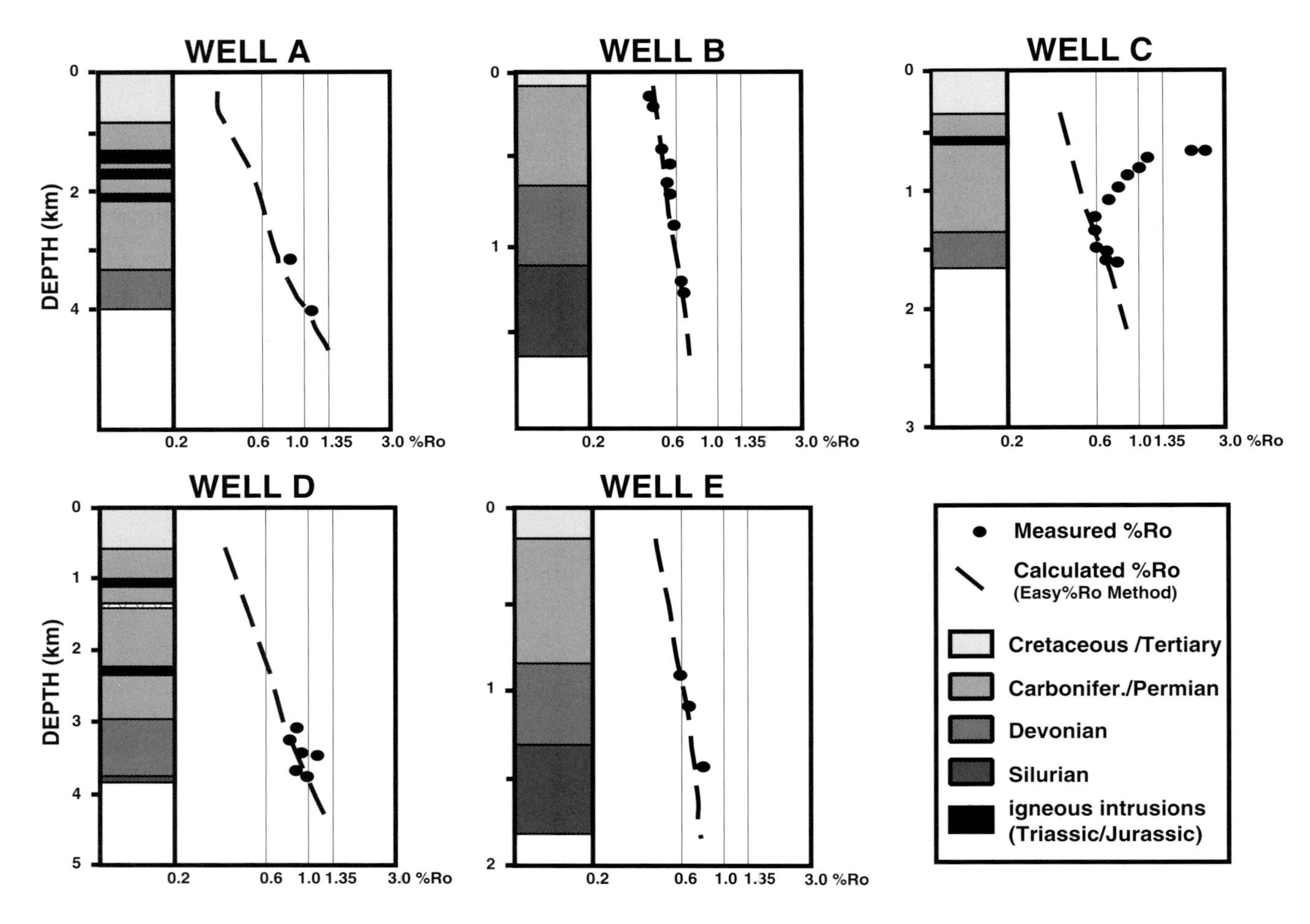

Figure 16—Vitrinite reflectance versus depth plots for five selected wells from the Amazonas Basin (see Figure 8 for well locations). Notice the good correlation between the measured vitrinite reflectance data and the calculated curve using the method from Sweeney and Burnham (1990).

simulations (hydrous pyrolysis), and kinetic modeling studies. Regression analyses of a large data base of Rock-Eval (S_2 and HI values) and maturation data (vitrinite reflectance) were used to estimate the original characteristics of the source rock: TOC of 5%, S_2 of 15 mg HC/g rock, and HI of 300 mg HC/g TOC (Figure 17). Such parameters were used to calculate present-day transformation ratios and expulsion efficiencies of source rocks in each well.

A geochemical study was carried out using a representative selection of immature samples to assess the kinetic variability of the Barreirinha Formation source rocks. Three activation energy (Ea) distribution patterns were recognized. The first set of samples presents an almost single Ea, with values from 49 to 55 kcal/mol and frequency factors (FF) ranging from 1.47×10^{12} to 1.75×10^{14}/s. The second set shows distributions of Ea mostly from 52 to 56 kcal/mol and FFs ranging from 5.02×10^{13} to 2.14×10^{14}/s. In the third set of samples, high values of Ea (56–59 kcal/mol) prevail, with FFs of about 3.00×10^{14}/s. Visual kerogen, pyrolysis, and elemental analyses suggest a strong influence of mineral matrix and kerogen composition on Ea distribution and frequency factors, whereas HI and sulfur content have a small effect. The kinetic distribution of the second set is associated with the prevailing organic facies of the lower Barreirinha source rock and was therefore selected as input for kinetic modeling.

Modeling shows that lower Barreirinha kerogen attains appropriate thermal conditions to start generation at about 1800 m depth. Such conditions were reached between Carboniferous time (in the depocenter) and Permian time (in the platform areas) (Figure 18). Most of the petroleum generation was completed by Early Triassic time. Modeled present-day transformation ratios range from 10–20% in the platforms to 95–100% in the depocenter, which is in agreement with the transformation ratio curves obtained from regression analyses of natural series geochemical data (Figure 19).

Expulsion simulation and biomarker maturity data from oil samples indicate that the main phase of oil expulsion began when the source rock attained a transformation ratio of about 50% and a maturation level of 0.80% R_o. Such conditions were reached only in the central trough area (Figure 18). Most of the petroleum was expelled between the Early Permian and the Early Triassic. Integration of palinspastic reconstructions, kinetic modeling results, and natural series studies allowed the mapping of oil kitchen evolution through time (Figure 20).

Modeling results and maturation data show that source rock thermal evolution was controlled mainly by subsidence history. The heat effect of diabase intrusions played an important role only in those areas where the dikes and sills were intruded into the Devonian sequence (central-eastern part of the basin). Due to the small thicknesses of diabase sills and Cretaceous–Tertiary sedimentary beds in most of the basin, only small amounts of petroleum have been expelled since Jurassic time. According to the modeling, in the eastern part of the basin, the thick piles of Cretaceous–Tertiary sedimentary rocks could have caused an important petroleum expulsion event if the source rocks had not been exhausted by the heat effect of Jurassic igneous intrusions.

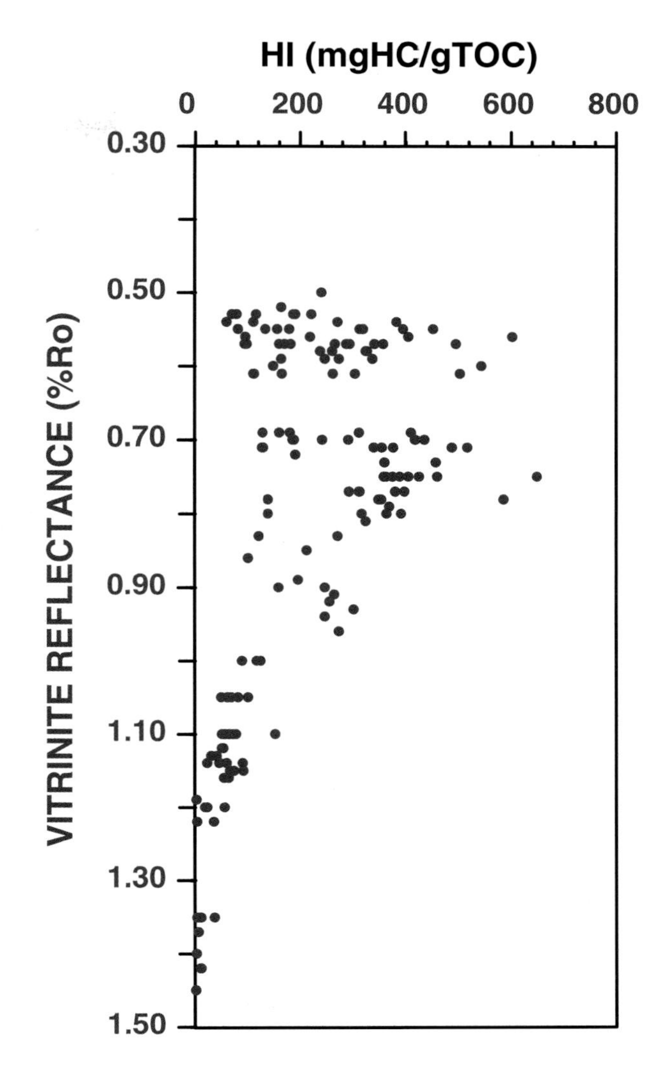

Figure 17—Crossplot showing depletion of the hydrogen index (HI) as a function of increasing maturity (vitrinite reflectance) for a representative set of samples from the radioactive shales of the Barreirinha Formation in the Amazonas Basin.

PETROLEUM SYSTEM SUMMARY

The integration of geologic and geochemical data with modeling results allowed determination of the volume of petroleum generated and expelled in the Amazonas Basin, as well as the identification of a single petroleum system in the basin: the Barreirinha–Curiri(!) petroleum system, summarized in Figure 21.

The source rock is the Upper Devonian marine black shales of the Barreirinha Formation, the main reservoir rocks are the Upper Devonian sandstones of the Curiri Formation, and the seal rocks are the shales and diamictites also of the Curiri. Petroleum is trapped mainly in stratigraphic traps of Late Devonian age. Other potential traps are also presented in Figure 21, such as structural traps formed during the Juruá tectonic event. Geochemical modeling shows that the main phase of petroleum generation and expulsion occurred from Late Carboniferous to Permian time and was completed by the Early Triassic. Any later tectonic event remobilized those hydrocarbons previously trapped.

The most important petroleum accumulations are located outside of or close to the outer limits of the generation pod, which indicates medium- to long-distance migration (up to 150 km). Oil maturity (peak of oil generation) indicates that these accumulations were not sourced by the thermal effect of igneous intrusions, which cooked the source rock to the overmature stage. The huge thickness of barren shales (up to 1000 m) that cover the Barreirinha source rock in the oil kitchen area suggests that a significant amount of petroleum was expelled downward to the Ererê sandstones and then updip to the hinge zone. The other part of the petroleum generated was expelled upward through faults to Curiri sandstone lenses or to Monte Alegre sandstones (Figure 22).

Preliminary volumetric calculations indicate that up to 1 trillion bbl of oil equivalent were expelled from the source rock (Figure 23). Because of the long distances of both vertical and horizontal migration, it is believed that an important amount of the expelled hydrocarbon was dispersed along migration pathways. A significant part could also have been remobilized and lost during Cretaceous uplift of the basin margins.

Acknowledgments*—The authors are indebted to Petrobrás–Petróleo Brasileiro S.A. for support and permission to publish this work.*

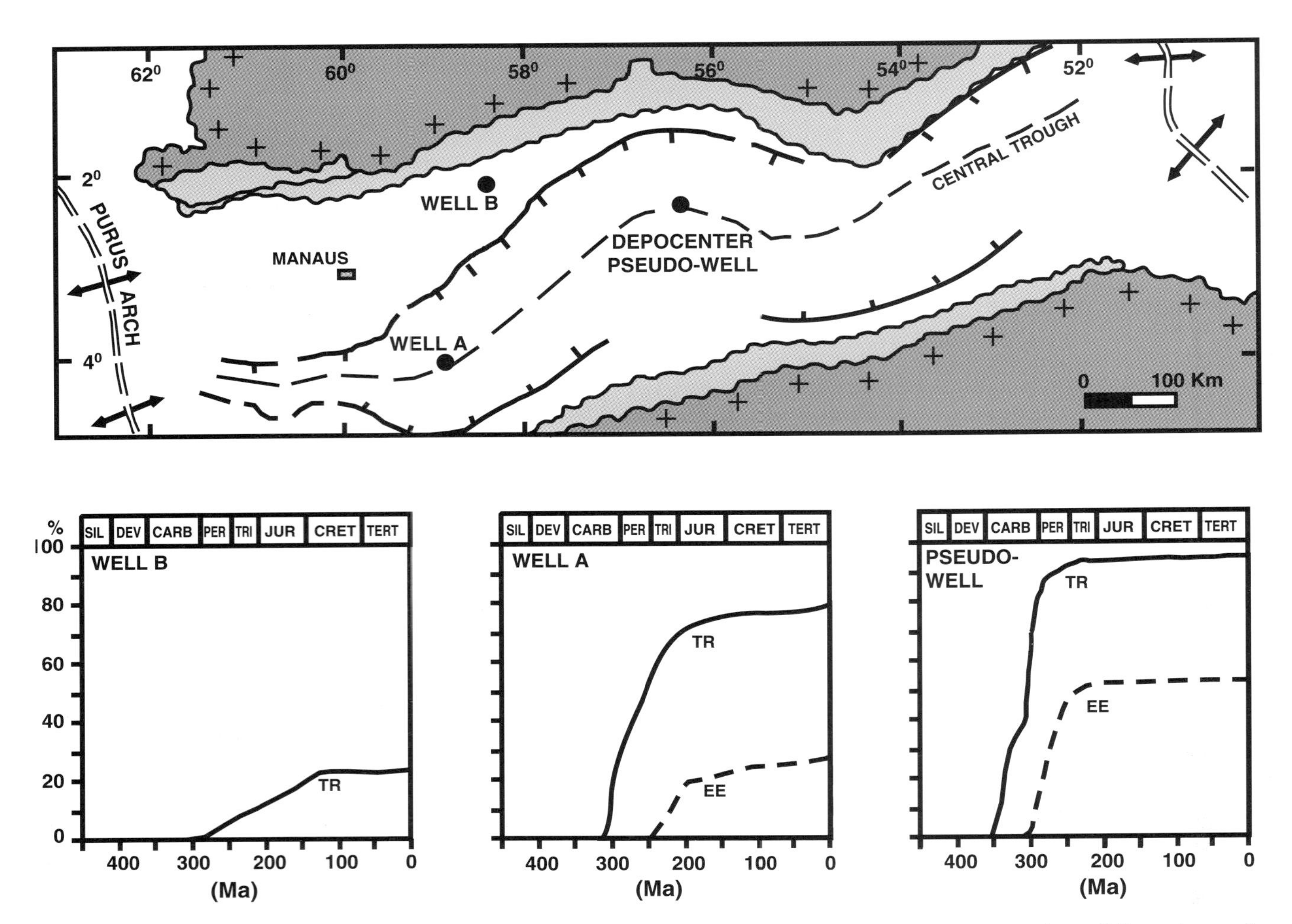

Figure 18—Evolution of the transformation ratio and expulsion efficiency in two wells and one pseudo-well (based on seismic data) in the Amazonas Basin. Notice the increasing kerogen conversion and petroleum expulsion from well B (basin margin) to the pseudo-well (depocenter).

REFERENCES CITED

Aires, J. R., 1985, Evolução tectônica da Bacia do Baixo Amazonas, Brasil: Master's Thesis, Federal University of Ouro Preto, Ouro Preto, Brazil, 116 p.

Arneth, J. D., 1984, Stable isotope and organo-geochemical studies on Phanerozoic sediments of the Williston Basin, North America: Isotope Geoscience, v. 2, p. 113–140.

Bethke, C. M., J. D. Reed, and D. F. Oltz, 1991, Long-range petroleum migration in the Illinois Basin: AAPG Bulletin, v. 75, n. 5, p. 925–945.

Caputo, M. V., 1984, Stratigraphy, tectonics, palaeoclimatology and palaogeography of northern basins of Brasil: Ph.D. thesis, California University, Santa Barbara, California, 586 p.

Carvalho, H. da S., and P. F. S. Lobo, 1986, Estudo do fluxo de calor e movimentação de fluidos na Bacia do Médio Amazonas: Internal Report, Petrobrás, Rio de Janeiro.

Cordani, U., B. B. B. Neves, and R. A. Fuck, 1984, Estudo preliminar de integração do pré-Cambriano com os eventos tectônicos das bacias sedimentares brasileiras: Série Ciência-Técnica-Petróleo, Petrobrás, Rio de Janeiro, n. 15.

Coutinho, L. F. C., and F. G. Gonzaga, 1996, Evolução tectonossedimentar e termomecânica da Bacia do Amazonas: Anais do XXXIX Congresso de Geologia, Soc. Bras. Geoc., Salvador, v. 5, p. 342–346.

Cunha, P. R. C., F. G. Gonzaga, L. F. C. Coutinho, and F. J. Feijó, 1994, Bacia do Amazonas, *in* F. J. Feijó, ed., Estratigrafia das bacias sedimentares do Brasil: Boletim de Geociências da Petrobrás, Rio de Janeiro, v. 8, n. 1, p. 47–55.

Daemon, R. F., and C. J. A. Contreiras, 1971, Zoneamento palinológico da Bacia do Amazonas: Internal Report, Petrobrás, Belém.

Demaison, G., and B. J. Huizinga, 1991, Genetic classification of petroleum systems: AAPG Bulletin, v. 75, n. 10, p. 1626–1643.

Gonçalves, F. T. T., F. G. Gonzaga, L. F. C. Coutinho, and J. A. Triguis, 1995, Petroleum geochemistry of the Amazonas Basin, Brazil: oil–source correlation and assessment of hydrocarbon generation and expulsion: Selected Papers, 17th International Meeting on Organic Geochemistry, San Sebastian, p. 444–446.

Grahn, Y., 1992, Revision of Silurian and Devonian strata of Brazil: Geobios, v. 25, n. 6, p. 703–723.

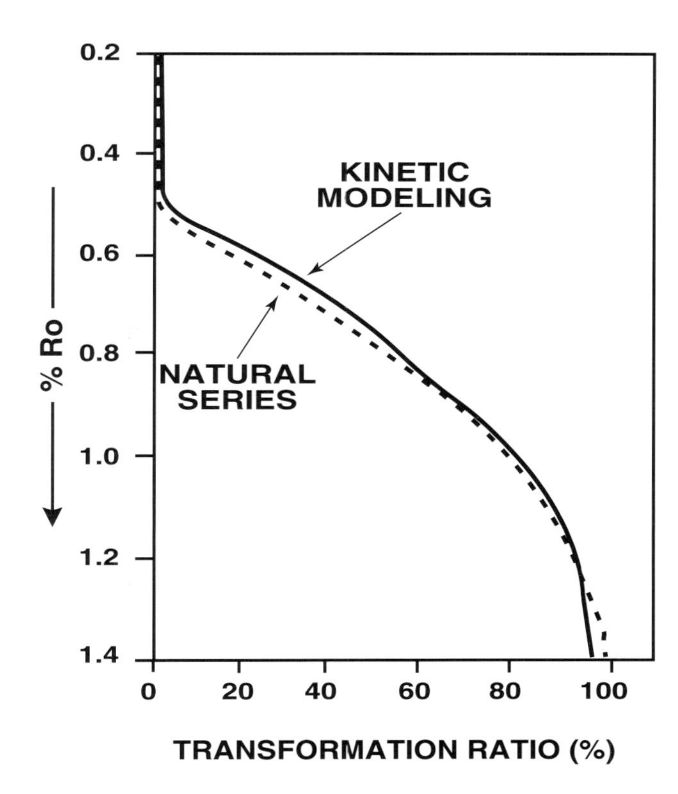

Figure 19—Graph showing the correlation of observed and calculated transformation ratio curves as a function of vitrinite reflectance for the studied wells in the Amazonas Basin. The observed curve (dashed) represents the HI depletion from the natural series (see Figure 18), and the calculated curve (solid) was obtained from kinetic modeling.

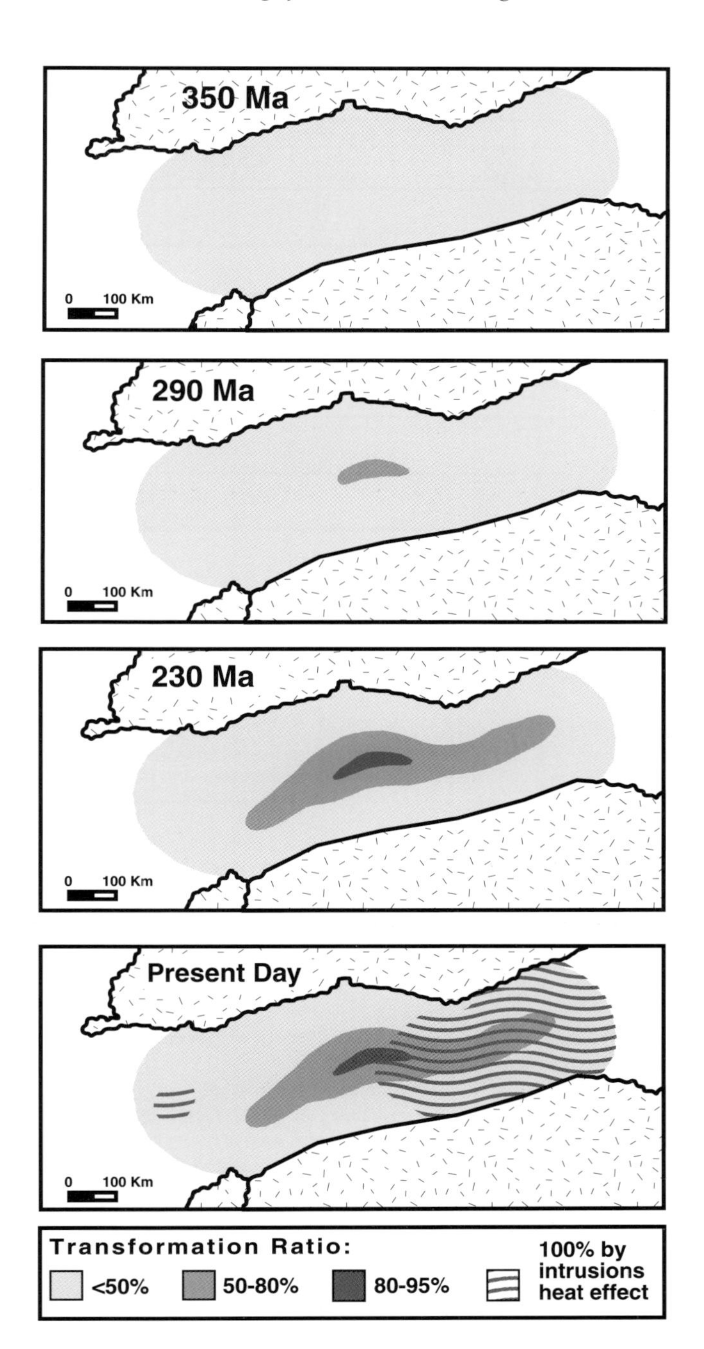

Figure 20—Location and extent of the oil kitchen in the Amazonas Basin at four different ages. Note that the present-day configuration of the oil kitchen was reached by the end of the igneous event (200 Ma) and has remained the same since then.

Harland, W. B., R. L. Armstrong, A. V. Cox, L. E. Craig, A. G. Smith, and D. G. Smith, 1989, A geologic time scale: Cambridge, U.K., Cambridge University Press, 263 p.

Jaupart, C., 1986, On the average amount and vertical distribution of radioactivity in the continental crust, *in* J. Burrus, ed., Thermal modeling in sedimentary basins: 1st IFP Exploration Research Conference, Carcans, Éditions Technip, p. 33–47.

Lambiase, J. J., and W. Bosworth, 1992, Structural controls on sedimentation in continental rifts: ELF Aquitaine Memoir 13, p. 71–77.

Linsser, H., 1958, Interpretation of the regional gravity anomalies in the Amazonas area: Internal Report, Petrobrás/ Depex, Rio de Janeiro.

Matos, R. M. D., and L. D. Brown, 1992, Deep seismic profile of the Amazonian craton (northern Brazil): Tectonics, v. 11, n. 3, p. 621–633.

McHargue, T. R., T. L. Heidrick, and J. E. Livingston, 1992, Tectonostratigraphic development of the interior Sudan rifts, Central Africa: Tectonophics, v. 213, p. 187–202.

Mizusaki, A. M. P., J. R. Wanderley Filho, and J. R. Aires, 1992, Caracterização do magmatismo básico das bacias do Solimões e do Amazonas: Internal Report, Petrobrás, Rio de Janeiro.

Mullin, R. P., 1988, Maturation model for middle Amazon Basin, Brazil: Anais do XXXV Congresso Brasileiro de Geologia, Belém, Pará, v. 6, p. 2457–2471.

Nunn, J. A., and J. R. Aires, 1988, Gravity anomalies and flexure of the lithosphere at the middle Amazon Basin, Brazil: Journal of Geophysical Research, v. 93, p. 415–428.

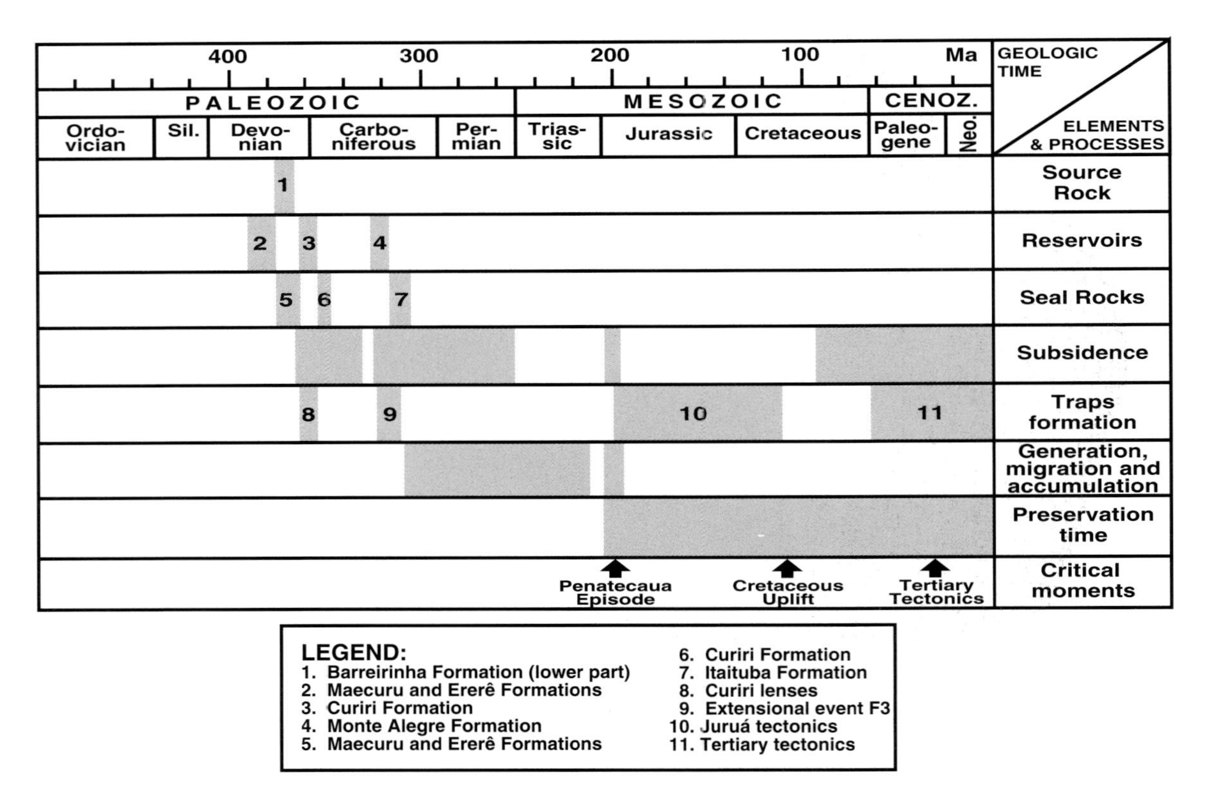

Figure 21—Petroleum system events chart for the Amazonas Basin.

PECTEN/IDEMITSU, 1989, Middle Amazon Basin: Phase II Report, p. 171–208.

Poliyak, B. G., and Y. A. Smirnov, 1968, Relationship between terrestrial heat flow and the tectonics of the continents: Geotectonics, v. 4, p. 205–213.

Rosendahl, B. R., D. J. Reynolds, P. M. Lorber, C. F. Burgues, J. McGill, D. Scott, J. J. Lambiase, and S. J. Derksen, 1986, Structural expressions of rifting: lessons from Lake Tanganyika, Africa, *in* L. E. Frostick, et al., eds., Sedimentation in the African rifts: Geological Society of London Special Publication, v. 25, p. 29–43.

Royden, L., and C. E. Keen, 1980, Rifting process and thermal evolution of the continental margin of eastern Canada determined from subsidence curves: Earth and Planetary Science Letter, v. 51, p. 343–361.

Savini, R., and D. Altiner, 1991, Pennsylvanian foraminifera and carbonate microfacies from the Amazonas and Solimões basins: biostratigraphic, paleoecologic and paleogeographic results: Internal Report, Petrobrás, Rio de Janeiro.

Sweeney, J. J., and A. K. Burnham, 1990, Evaluation of a simple model of vitrinite reflectance based on chemical kinetics: AAPG Bulletin, v. 74, n. 10, p. 1559–1570.

Wilson, M., 1997, Thermal evolution of the central Atlantic passive margins: continental break-up above a Mesozoic super-plume: Journal of the Geological Society of London, v. 154, p. 491—495.

Zalán, P. V., 1991, Influence of pre-Andean orogenies on the Paleozoic intracratonic basins of South America: Simposio Bolivariano, Bogotá, v. 1.

Zembruscki, S. G., and C. H. Kiang, 1989, Avaliação do gradiente geotérmico regional das bacias sedimentares brasileiras: Boletim de Geociências da Petrobrás, v. 3, n. 3, p. 215–227.

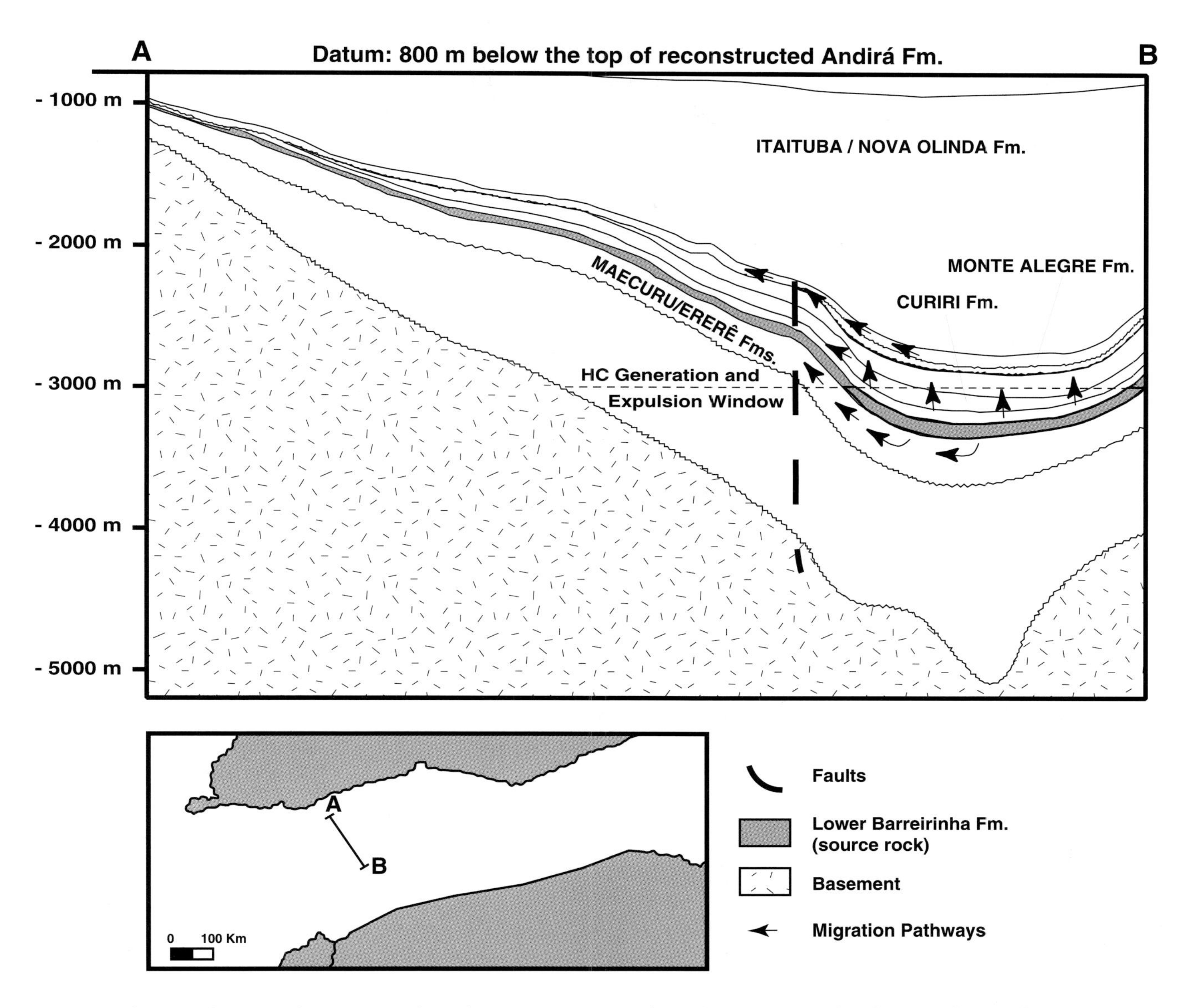

Figure 22—Schematic geologic cross section of the western part of the Amazonas Basin after a palinspastic reconstruction for the end of Permian time (main phase of petroleum expulsion). This section shows the spatial relationship among the pod of active source rock, the main carrier beds and reservoirs, and the inferred migration pathways.

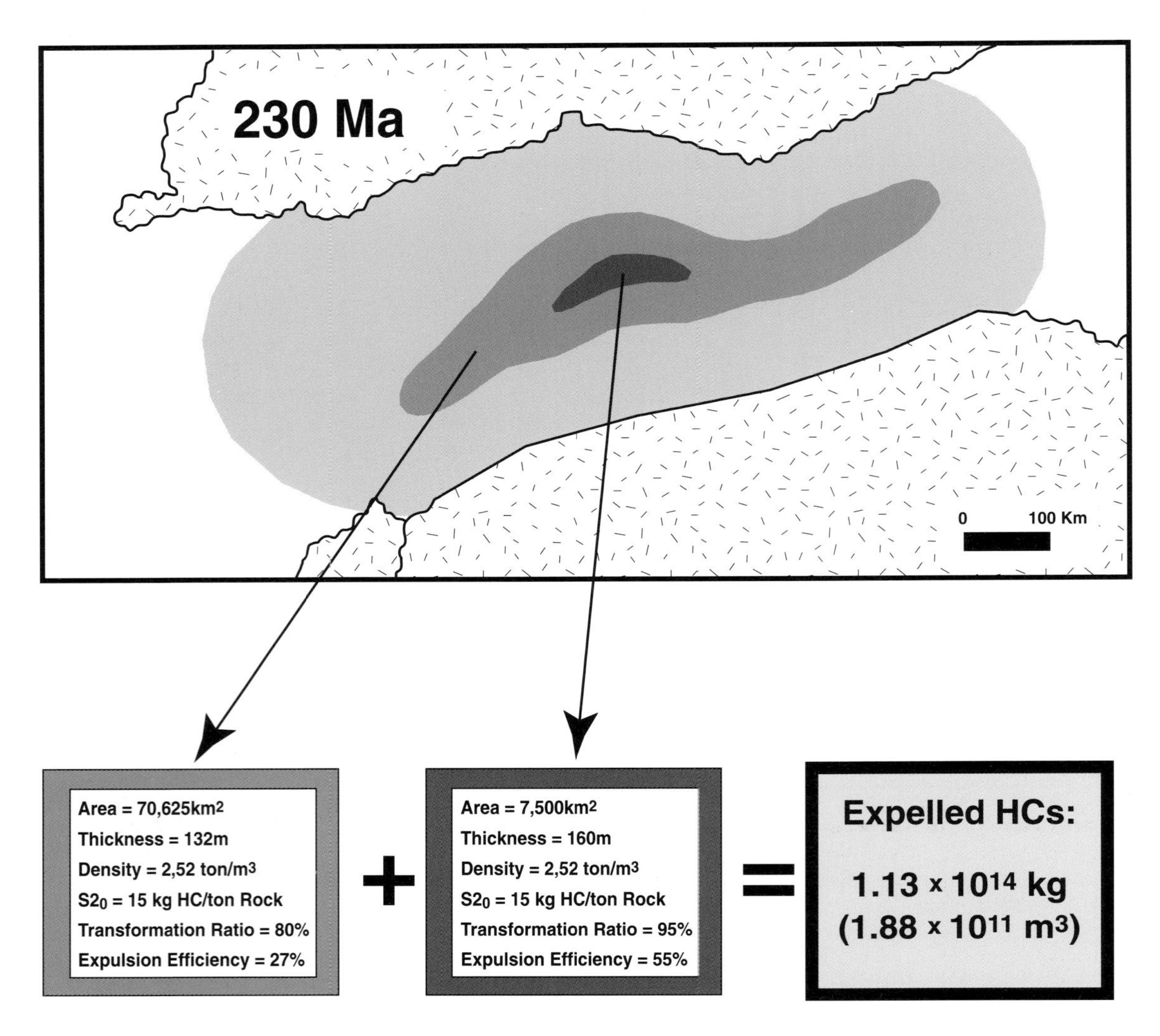

Figure 23—General scheme for the calculation of the amount of petroleum expelled from the Barreirinha Formation source rock in the Amazonas Basin.

Penteado, H. L. B., and F. Behar, 2000, Geochemical characterization and compositional evolution of the Gomo Member source rocks in the Recôncavo Basin (Brazil), *in* M. R. Mello and B. J. Katz, eds., Petroleum systems of South Atlantic margins: AAPG Memoir 73 p. 179–194.

Chapter 14

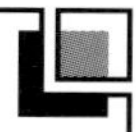

Geochemical Characterization and Compositional Evolution of the Gomo Member Source Rocks in the Recôncavo Basin, Brazil

Henrique Luiz de Barros Penteado

Petrobrás R&D Center
CENPES/SUPEP/DIVEX/CEGEQ
Rio de Janeiro, Brazil

Françoise Behar

Institut Français du Pétrole
Rueil Malmaison, France

Abstract

The lacustrine shales of the Gomo Member (Candeias Formation, Lower Cretaceous) are the main source rocks with type I kerogen in the Recôncavo Basin in northeastern Brazil. We investigated variations in source rock facies of the Gomo Member and their respective initial petroleum potentials, as well as the compositional changes of petroleum fluids generated with increasing maturity. To address these objectives, a comprehensive analytical procedure involving Rock-Eval pyrolysis of whole-rocks and kerogens, elemental analyses, and whole-rock and kerogen extracts was adopted on a series of samples covering the range from immature to highly mature levels. Whole-rock and kerogen Rock-Eval data indicated a strong mineral matrix retention effect that led to previous underestimation of the petroleum potential of the Gomo Member source rocks. The organic matter in low-TOC samples was shown to have lower petroleum potential and to be more oxidized than in high-TOC samples. Source rock facies variations were observed from one basin compartment to another, and initial HI values as a function of TOC were proposed for each area.

Although the amounts of kerogen extracts, indicative of kerogen retention capacity, do not to vary considerably with depth (20–60 mg/g TOC), the amounts of whole-rock extracts increase considerably until the peak of petroleum generation and accumulation is attained. The trend in whole-rock extracts is accompanied by a substantial increase in absolute amounts and in relative contribution of saturates in the C_{15+} fraction, whereas the amounts of NSOs and aromatics do not vary significantly. Thus, we propose a partial secondary cracking of NSOs and aromatics within the source rocks before petroleum expulsion to explain changes in extract composition. With further maturity and expulsion, whole-rock extracts and the relative contribution of saturates decrease.

INTRODUCTION

The Recôncavo Basin is a Late Jurassic–Early Cretaceous asymmetric rift system in the State of Bahia, northeastern Brazil (Figure 1). Its sedimentary column, which has thicknesses up to 8000 m, can be divided into (1) a fluvial, eolian, and lacustrine prerift sequence and (2) a synrift sequence composed of lacustrine shales at the base and deltaic and fluvial sandstones and shales in its upper part (e.g., Figueiredo et al., 1994; Caixeta et al., 1994) (Figure 2).

Previous geochemical studies (e.g., Daniel et al., 1989; Mello et al., 1994) allowed recognition of the synrift lacustrine shales of the Gomo Member of the Candeias Formation (Early Cretaceous age) (Figure 2) as the main source rocks containing type I kerogen in Recôncavo Basin. The relatively low values of total organic carbon (TOC ~1%), source rock potential (S_2 up to 5 kg HC/g rock), and hydrogen index (HI) were attributed to the high degree of maturity of these source rocks in most of the basin, especially in the structural lows. In areas where the Gomo Member source rocks are immature, Mello et al. (1994) reported TOC contents up to 10% and HI values up to 700 mg HC/g TOC. Rock extract to oil correlations using gas chromatograms, biomarkers, and carbon isotopic data (e.g., Gaglianone and Trindade, 1988; Mello et al., 1994) clearly established the relationship between the Gomo Member source rocks and the oil accumulations in various petroleum systems within Recôncavo Basin.

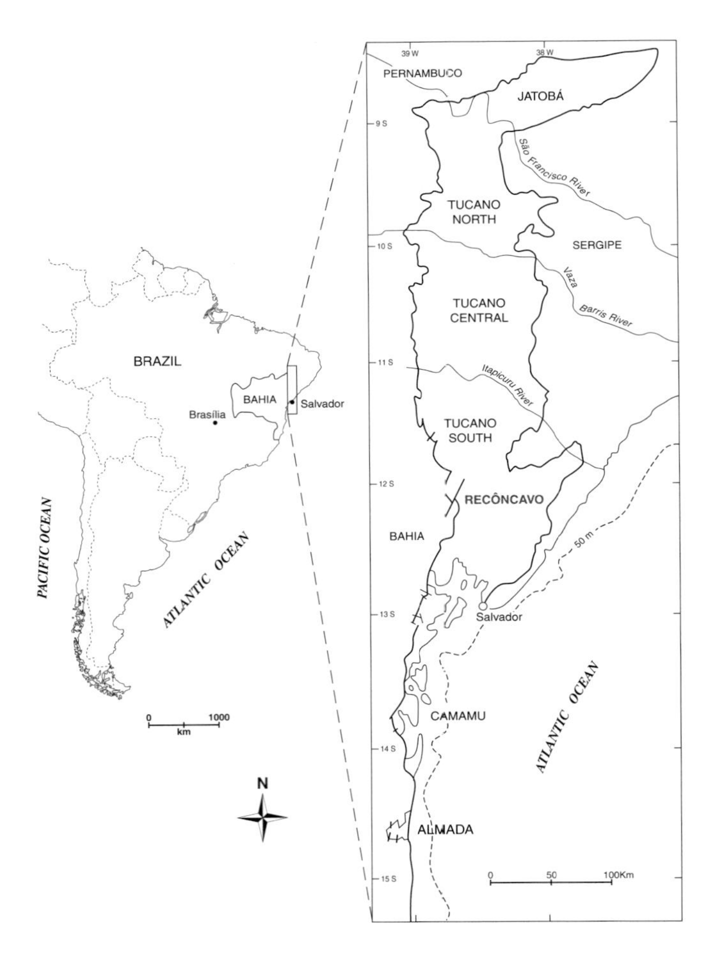

Figure 1—Location map of the Recôncavo Basin. Modified from Santos et al. (1990).

Although much knowledge has been gathered through these studies, a number of questions remain concerning the Gomo Member source rocks. Because source rocks are mature to overmature for petroleum generation in the depocenters (e.g., Mello et al., 1994), variations in source rock facies must be investigated in low-maturity areas and then extrapolated to the presently mature areas. Accordingly, the initial source rock petroleum potential and the quality of the organic matter, expressed as the S_2 peak and the HI from Rock-Eval pyrolysis, must be determined for each source rock facies.

In a recent study, Penteado (1996) observed that HI values of isolated immature kerogens of the Gomo Member are higher than those of their corresponding whole-rocks. The HI values of kerogens are much closer to those of type I organic matter reported in the literature than those of whole-rocks. Furthermore, a clear increase of HI values with TOC content in source rock samples has been observed, suggesting that a mineral matrix retention effect (Horsfield and Douglas, 1980; Espitalié et al., 1980 and 1984; Orr, 1981; Dembicki et al., 1983; Katz, 1983; Langford and Blanc-Valleron, 1990; Dembicki, 1990) and/or variations in organic matter facies may be influencing Rock-Eval data of whole-rocks. Therefore, a better understanding of variations in maturity and petroleum potential of the immature Gomo Member source rocks throughout the basin is necessary as input and calibration data for basin modeling studies that aim to provide a more comprehensive assessment of generated volumes of petroleum.

The first objective of this study was the geochemical characterization of immature and mature source rocks of the Gomo Member (Candeias Formation), addressing the problems of facies variations, initial petroleum potential, and mineral matrix retention effects. Whole-rock samples were selected using Rock-Eval data. Rock-Eval pyrolysis and elemental analyses were further performed on isolated kerogens to check the values obtained for whole-rocks and to observe variations in original composition, natural oxidation, TOC content, and maturity of the source rocks. This enabled us to establish values of initial HI for each representative immature source rock facies and to calculate the evolution of the kerogen transformation ratio with depth.

The second aim of this work was to investigate compositional changes in solvent extracts with maturity and to compare these with data from reservoired oils. Extracts were obtained from whole-rocks and isolated kerogens of Gomo Member samples at various maturity levels. The amounts and composition of whole-rock and kerogen extracts were compared to evaluate kerogen retention capacity. Quantitative data on C_{15+} saturates, aromatics, and resins and asphaltenes (NSOs) from both whole-rock and kerogen extracts were used to examine the role of secondary cracking within the source rocks before the main phase of petroleum expulsion.

This work was performed in conjunction with a 2-D compositional basin modeling study in the Recôncavo Basin that aimed to provide a better understanding of the timing of petroleum generation, expulsion, and migration, as well as related compositional changes in extracts of lacustrine source rocks and expelled petroleum (Penteado, 1999). Data provided by this work were used as input and calibration data for a 2-D compositional basin modeling study. Given the large scope of the basin modeling study, which involves basin evolution, thermal history, and modeling of petroleum generation, expulsion and migration, those results will be presented and discussed in a forthcoming paper.

METHODS

For the extraction and fractionation process, cuttings samples of source rocks were hand picked and ground to 80 mesh (0.177 mm). The first extraction was performed on whole-rocks with dichloromethane (1 g/10 mL) and magnetic stirring under reflux for 1 hour. After acid treatment and elimination of minerals from whole-rocks,

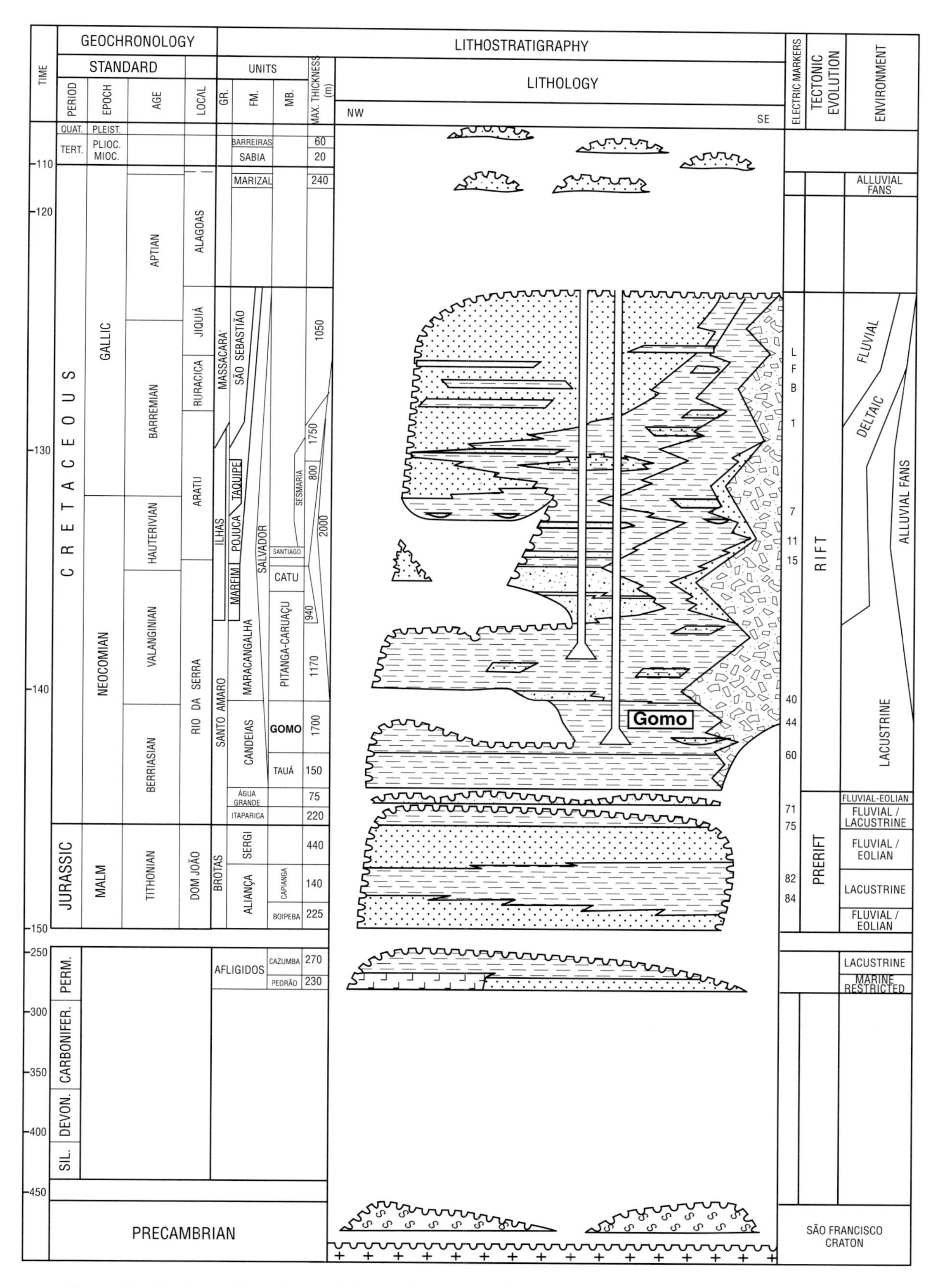

Figure 2—Stratigraphic column of the Recôncavo Basin, Brazil. Modified from Caixeta et al. (1994).

isolated kerogens were submitted to another extraction. In both whole-rock and kerogen extractions, the total dichloromethane (DCM) extract was concentrated then recovered and evaporated in order to estimate the amount of the C_{15+} extract. Extracts were fractionated by liquid chromatography (Behar et al., 1989) for recovery of C_{15+} saturates and with a mixture of n-C_5/DCM (65/35, v/v) for recovery of C_{15+} aromatics. The weight of NSOs (resins plus asphaltenes) was calculated by the difference between the total extract and the recovered hydrocarbons (saturates plus aromatics).

The first step in kerogen preparation was to isolate the organic matter by eliminating minerals with nonoxidant acids under an nitrogen atmosphere to prevent oxidation of the isolated fraction (Durand and Nicaise, 1980). Elemental analyses of keogens were performed by an outside specialized laboratory. The following elements were estimated in weight percent of the sample: carbon, hydrogen, nitrogen, oxygen, total sulfur, and iron. Ashes after combustion were also weighed to cross-check the obtained data. All determinations were run in duplicate or triplicate according to preset reproducibility criteria. Carbon, hydrogen, and nitrogen were determined on one aliquot of the kerogen using a thermal conductivity detector of gases resulting from combustion at 1000°C. Oxygen was measured on another aliquot by pyrolyzing kerogen under nitrogen flow and reducing pyrolysis gases of carbon into carbon monoxide. After further transformation into carbon dioxide, quantitation of carbon gases was performed by coulometry. Total sulfur, including pyritic and organic sulfur, was obtained by oxidation of a third aliquot of kerogen and quantitative coulometry of sulfur dioxide. The determination of iron was necessary to calculate the amount of pyrite (FeS_2) in the sample and thus, by difference and mass balance, the amounts of pyritic and organic sulfur.

Classic Rock-Eval pyrolysis procedures (e.g., Espitalié et al., 1985) were adopted for the bulk quantification of generated petroleum in both whole-rocks and kerogens.

GEOCHEMICAL CHARACTERIZATION OF SELECTED SAMPLES

Rock-Eval Pyrolysis of Whole-Rocks

Samples were selected to determine accurate values of initial HI. The selection of wells and samples was based on the TOC, Rock-Eval, and vitrinite reflectance data provided by Petrobras S.A. and compiled by Aragão et al. (1998).

As the first step, immature rocks were chosen with various TOC values in three different compartments of the basin: south, central, and northeast (Figure 3). Immature samples were collected at depths ranging from 400 to 2000 m, with vitrinite reflectance (R_o) values below 0.55 %. Next, a second set of moderately mature source rocks was selected with a maturity stage corresponding approximately to the main phase of the oil window (2200–2600 m depth, 0.55–0.65% R_o) and having various TOC contents. Finally, in each compartment, a third set of rocks was chosen with roughly the same range of TOC but at higher maturity stages than those of the previous sets (>2700 m depth, 0.65–1.0% R_o) (Figure 3).

For the immature and moderately mature series of rocks, samples shown in Table 1 are sorted by compartment and increasing TOC. For the highly mature series, they are sorted by compartment but with increasing depth. Although the separation of samples by groups of increasing maturity was based on depths and several geochemical parameters, temperatures at the maximum of the S_2 peak (T_{max}) obtained during Rock-Eval pyrolysis show a large dispersion for any given group. Contamination of samples by migrated oil and uncertainties in the determination of T_{max} might be responsible for such dispersion of values in each set of samples. A trend in increasing T_{max} values with maturity is more evident in results from analyses performed on whole-rocks. It should be noted, however, that several studies have indicated that T_{max} values do not increase considerably with maturity for type I kerogens compared to other types of kerogen (e.g., Espitalié et al., 1985).

Results shown in Table 1 and Figures 4 and 5 confirm the existence of a systematic relationship between the HI values of immature whole-rocks and the TOC content as previously indicated by Penteado (1996) for a larger data set of samples from the Gomo Member. The lowest HI values correspond to the samples with the lowest TOC values. Samples of the immature series with TOC lower than 1% have HI values below 200 mg/g TOC, which is at least three times less than those of the richest source rocks. In the graph of TOC versus S_2 (Figure 4), the function that best fits the data points seems to be of exponential or polynomial type, thus differing from the linear function proposed by Langford and Blanc-Valleron (1990). The slope of the function best-fitted to the points (which indicates the HI of the series of samples) increases steadily with TOC and then seems to stabilize at a TOC greater than 2%. This tendency for stabilization of values is more evident in the graph of TOC versus HI (Figure 5), in which HI values appear to converge to about 550 mg/g TOC for samples with TOC greater than 2%. Such behavior of data in the TOC versus HI graph has been previously observed by Orr (1981), Katz (1983), and Dembicki (1990).

This variation in HI values for rocks with TOC can be the result of either (1) a mineral matrix effect during pyrolysis or (2) a partial alteration of the organic matter in low-TOC sediments under natural depositional conditions. In fact, it is well known that a relatively large proportion of NSOs may be adsorbed by the mineral matrix (Horsfield and Douglas, 1980; Espitalié et al., 1980, 1984; Orr, 1981; Katz, 1983; Dembicki et al., 1983; Dembicki, 1990; Langford and Blanc-Valleron, 1990). Because of NSO adsorption, heavy compounds are aromatized to coke instead of being swept away by the carrier gas during Rock-Eval pyrolysis, thus leading to underes-

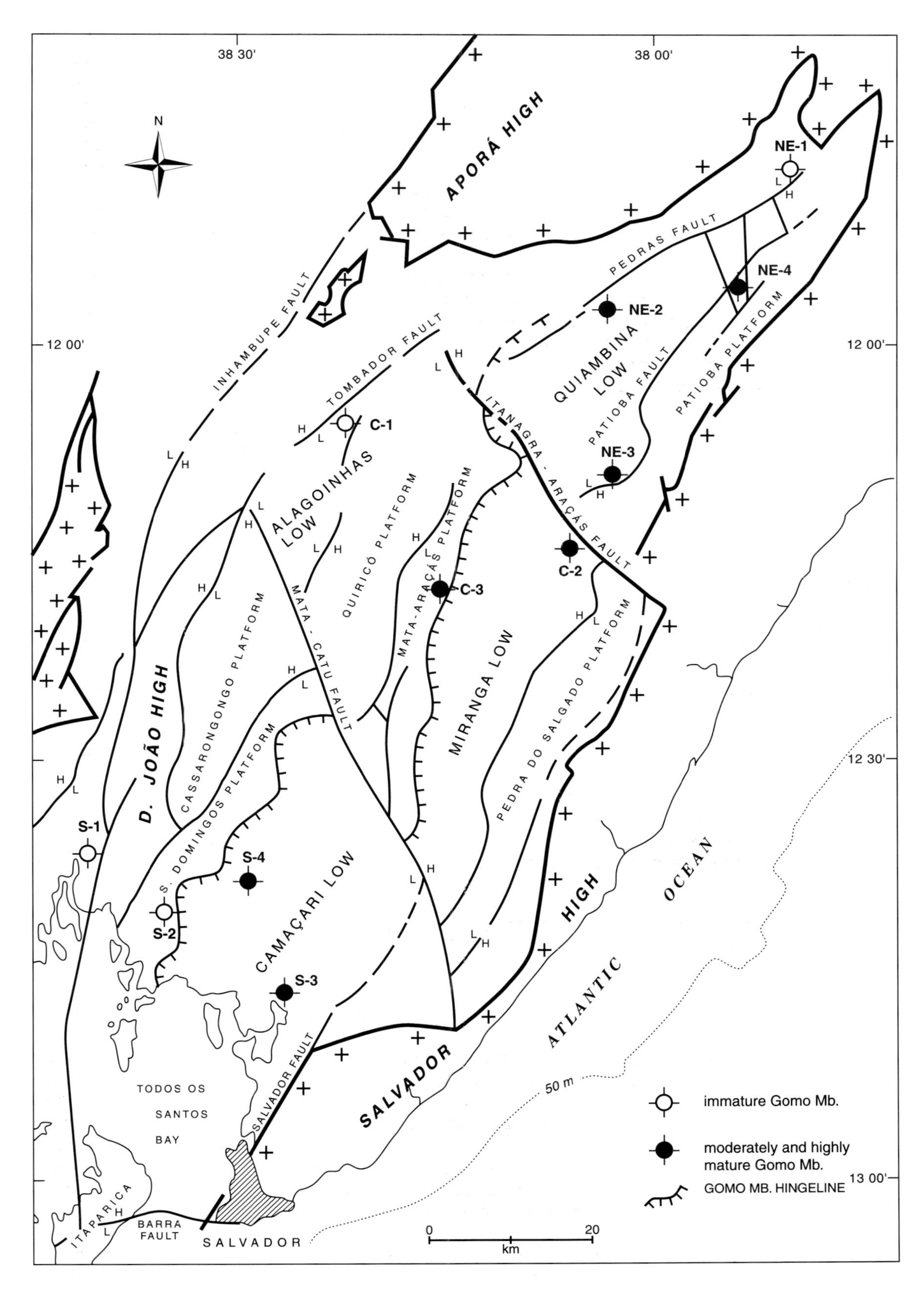

Figure 3—Structural map of Recôncavo Basin with location of wells from which samples of the Gomo Member (Candeias Formation) were studied. Modified from Santos et al. (1990).

Table 1—Geochemical Characterization of Selected Whole-Rocks and Isolated Kerogens of the Gomo Member

				Whole-Rocks					Isolated Kerogens						
Well	Depth (m)	Area	Sample No.	TOC (%)	S_2 (mg/g)	HI (mg/g C)	OI (mg/g C)	T_{max} (°C)	TOC (%)	HI (mg/g C)	OI (mg/g C)	T_{max} (°C)	H/C	O/C	FeS_2 (%)
Immature Series															
S-1	450	South	151739	2.4	11.6	490		441	74.0	663		440	1.50	0.057	8.2
S-2	1044	South	156733	0.8	1.3	164	96	445	49.0	556	11	440	1.27	0.076	38.9
S-2	990	South	156732	1.2	2.9	239	126	448	60.2	612	16	442	1.32	0.065	27.6
S-2	936	South	156731	2.9	14.8	501	63	447	64.4	853	11	447	1.48	0.047	21.9
C-1	1968	Central	156734	0.7	1.2	169	95	438	59.8	421	24	437	1.15	0.078	26.9
C-1	2022	Central	156735	2.7	13.3	492	22	443	58.2	728	6	447	1.38	0.048	29.1
C-1	2040	Central	156736	4.5	24.2	535	40	441	69.8	759	7	444	1.35	0.045	20.2
NE-1	1206	NE	156738	0.6	0.7	126	137	437	63.9	398	26	437	1.08	0.096	20.3
NE-1	1224	NE	156739	1.9	5.8	303	47	438	69.2	446	25	440	1.14	0.072	17.5
NE-1	1260	NE	156740	3.9	20.8	535	10	439	69.6	645	13	446	1.31	0.052	16.2
Moderately Mature Series															
S-3	2257	South	156747	0.6	0.9	164	109	437	70.8	158	27	430	0.75	0.103	13.1
S-4	2634	South	156743	0.6	0.9	152	97	445	50.2	205	21	440	0.84	0.087	40.6
S-4	2607	South	156742	1.6	4.4	281	19	442	64.3	242	19	443	0.90	0.052	26.6
S-4	2598	South	156741	2.7	8.3	309	22	443	76.5	249	14	437	0.89	0.046	11.3
C-2	2271	Central	156750	1.0	2.7	267	80	445	52.8	487	18	440	1.15	0.064	36.8
C-3	2286	Central	156745	1.4	3.6	247	69	443	43.0	317	23	435	0.94	0.081	50.0
C-3	2250	Central	156744	3.8	14.1	367	26	445	62.3	530	15	449	1.19	0.045	27.0
NE-2	2298	NE	156737	1.5	4.6	306	79	447	60.1	372	19	442	1.07	0.054	29.1
NE-3	2460	NE	156746	1.6	3.6	222	43	443	70.4	310	19	437	0.90	0.064	16.4
Highly Mature Series															
S-3	2865	South	156748	0.9	0.8	97	57	447	64.9	105	21	442	0.66	0.085	22.9
S-3	3471	South	156749	0.4	0.5	107	279	439	66.1	125	23	439	0.79	0.122	17.2
C-2	2745	Central	156751	0.8	1.8	212	72	446	50.9	425	21	441	1.09	0.067	40.4
C-2	3090	Central	156752	0.7	1.0	148	76	445	32.7	204	19	438	0.86	0.113	61.6
C-2	3510	Central	156753	0.9	1.9	205	54	445	49.9	545	20	437	1.15	0.074	39.9
C-2	4020	Central	156754	1.5	0.6	39	34	444	57.8	52	13	438	0.57	0.058	33.9
C-2	4035	Central	156755	1.0	0.8	83	71	446	48.9	127	15	438	0.72	0.084	41.4
NE-4	3348	NE	156756	0.5	0.3	68	300	450	61.7	92	24	444	0.71	0.115	24.4

timated HI values. Since the slope of the best-fit curve in the TOC versus S_2 graph (Figure 4) increases with TOC and then tends to stabilize, a possible mineral matrix effect would seem to be proportionately more important for low-TOC samples. If there is only a mineral matrix effect, one would expect to obtain the same HIs for isolated kerogens of immature samples with various TOC contents taken in the same compartment. The second explanation, that oxidation of organic matter occurred in its environment of deposition, can be verified by looking at the atomic ratios H/C and O/C measured on kerogens. Furthermore, if variations in organic matter facies are the only phenomenon responsible for this behavior, similar HI values should be expected for each whole-rock and its corresponding isolated kerogen, with an increase in HI directly proportional to TOC content.

Although HI values of source rocks alone do not allow a distinction between the two possible intervening effects, the oxygen index (OI) might provide a first clue of the existence of variations linked to organic matter quality. Results shown in Table 1 and Figure 6 show that organic-lean whole-rock samples have higher OI values as well as lower HI values. However, care must be taken in considering these measurements. Indeed, accurate OI values are difficult to acquire in source rocks with type I organic matter, especially if they are lean in organic matter and if carbonate minerals are present (e.g., Katz, 1983; Dembicki, 1990). In this set of samples, carbonate contents (estimated difference in mass before and after acid treatment) range from 22 to 40%.

Rock-Eval Pyrolysis and Elemental Analysis of Kerogens

Table 1 shows Rock-Eval pyrolysis data of isolated kerogens and elemental analyses in terms of TOC, H/C, and O/C atomic ratios and pyrite content. These analytical results allow us to make the following observations about immature samples of the Gomo Member (first

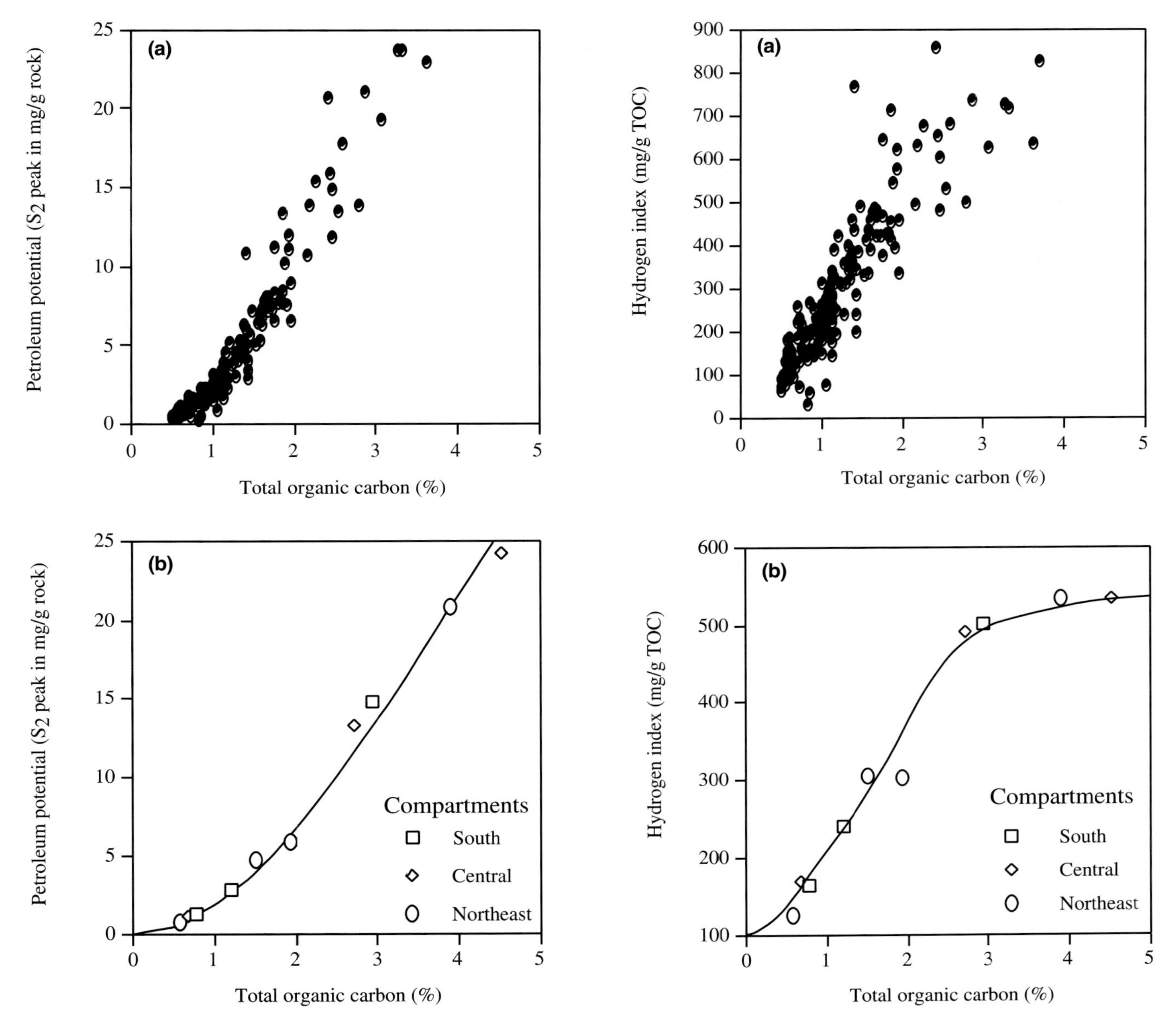

Figure 4—Graph of total organic carbon (TOC, in %) versus petroleum potential (Rock-Eval S_2 peak, in mg/g rock) data of immature whole-rocks of the Gomo Member (Candeias Formation). Data from (a) the southern basin compartment (Penteado, 1996) and (b) selected samples from the present study.

Figure 5—Graph of total organic carbon (TOC, in %) versus hydrogen index (HI, in mg/g TOC) data of immature whole-rocks of the Gomo Member (Candeias Formation). Data from (a) the southern basin compartment (Penteado, 1996) and (b) selected samples from the present study.

series of samples at the top of Table 1):

1. HI values of isolated kerogens are systematically higher than those of their corresponding whole-rocks. For TOCs lower than 1%, one has to multiply the HI of the whole-rock by a correction factor of 3 to 5 to obtain that of its respective kerogen. For TOCs from 1 to 2%, this correction factor is between 1.5 and 3, and for TOCs greater than 2%, it is very close to 1 (Figure 7). Therefore, a mineral matrix retention effect influences the HI values of whole-rock samples. In addition, this effect is more important for low-TOC samples, as evidenced by the correction factors.
2. As for the HI values of immature whole-rocks, HI values of immature kerogens show a clear increase with TOC content. The ratio between the highest and lowest HIs of kerogens in each compartment is between 1.5 and 1.8. Thus, the HI values of whole-rocks are controlled by variations in organic

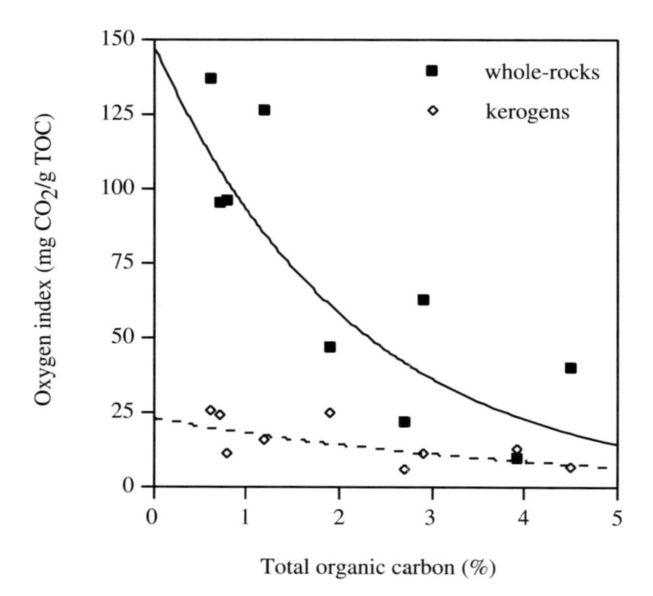

Figure 6—Graph of oxygen index values (OI, in mg CO_2/g TOC) from immature whole-rock and kerogen samples as a function of TOC (in %).

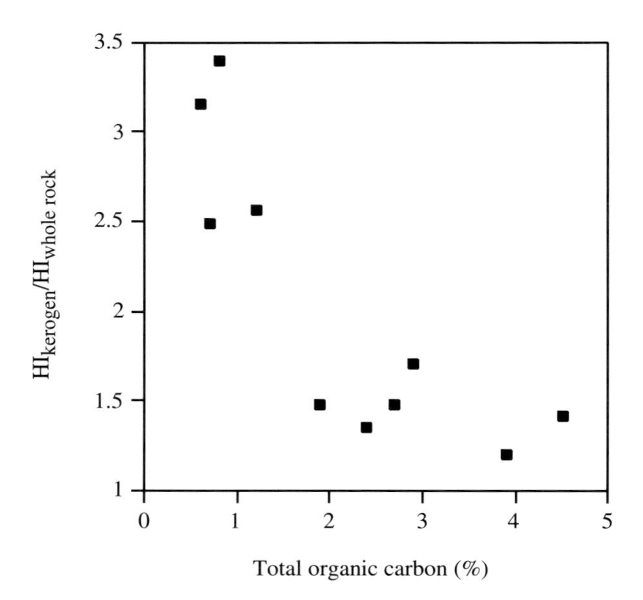

Figure 7—Variations in the $HI_{kerogen}/HI_{whole\text{-}rock}$ ratio of immature samples of the Gomo Member (Candeias Formation) as a function of TOC (in %).

matter quality resulting from depositional conditions, coupled with a mineral matrix retention effect.

3. For each one of the three basin compartments, kerogens of low-TOC samples have lower HI and higher OI values than those of high-TOC samples. This indicates that, in organic-poor sedimentary beds of the Gomo Member, the organic matter is more susceptible to partial oxidation. However, OI values for kerogens are almost always lower than those of their respective rocks, and the increase in OI with decreasing TOC is less evident for kerogens than for whole-rocks (Figure 6). Because the OI is better quantified for isolated kerogen than for organic-poor source rocks, one can guess that the values determined on whole-rocks were overestimated possibly because of a contribution of CO_2 from carbonates (e.g., Katz, 1983; Dembicki, 1990).
4. Elemental analyses of extracted kerogens show that low-TOC samples have lower H/C and higher O/C ratios (Figures 8 and 9), thus confirming that oxidation of organic matter was more intense in organic-lean rocks.
5. With regards to the geographic location of samples, it appears that samples with comparable TOC contents have higher HI and H/C and lower OI and O/C as one moves from the northeastern to the southern compartment (Figure 8). Given the greater extent of Gomo Lake in the southern compartment, organic input from terrigenous sources would probably be more dispersed and preservation conditions of the lacustrine organic matter better than those in the central and northeastern compartments.
6. Only one kerogen sample in the southern compartment has an HI higher than 800 mg/g TOC, which can be attributed to characteristic type I organic matter. The highest indices in the central and northeastern compartments are 759 and 645 mg/g TOC, respectively. These values could indicate the possibility of mixed kerogen types. Nevertheless, gas chromatograms of pyrolysates from both organic-rich (TOC >2%) and organic-poor (TOC <1%) immature samples show features characteristic of type I organic matter (Penteado, 1999). Therefore, it appears that type I kerogen predominates in all samples even though HI values are relatively low.
7. All samples are rich in pyrite.
8. The highest HI values of kerogens (>600 mg/g TOC) were obtained for O/C ratios lower than 0.05 and for H/C ratios higher than 1.3.

For the moderately and highly mature samples of the Gomo Member (middle and lower series in Table 1), we can make the following observations:

1. Once again, HI values for kerogens were found to be higher than those for whole-rocks. This clearly shows a matrix retention effect on HI values, even for mature samples.
2. As observed in immature kerogens, HI values for the moderately mature kerogens increase with TOC content. However, this is not the case for the highly mature series because TOC values are within a narrow range (from 0.4 to 1.5%), whereas the depth of these samples (hence their maturity) varies widely (from 2745 to 4035 m).

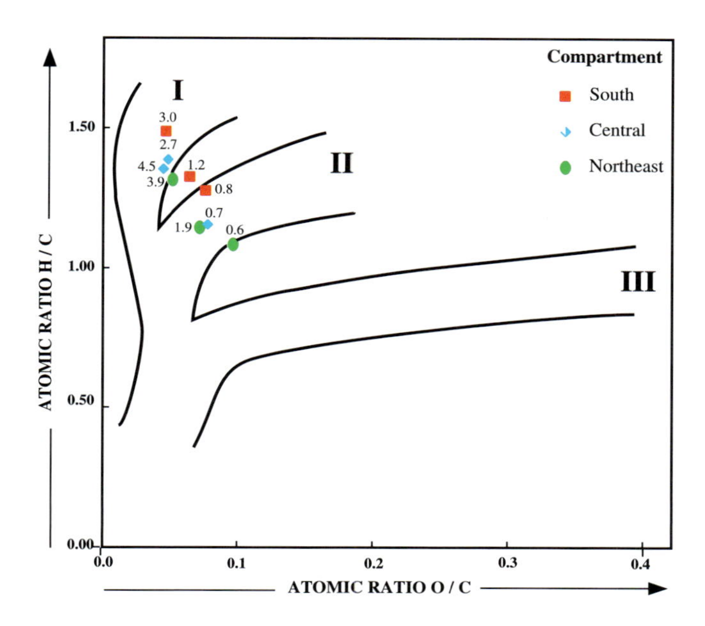

Figure 8—Van Krevelen diagram with data on immature kerogens of the Gomo Member (Candeias Formation) by compartment. Plotted values are the TOC contents (in %) of whole-rocks.

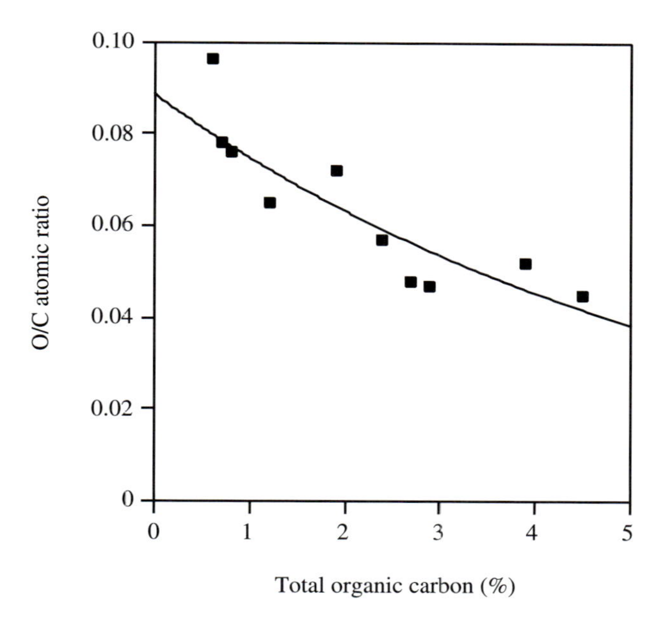

Figure 9—O/C atomic ratio of immature kerogens of the Gomo Member (Candeias Formation) as a function of TOC (in %).

3. Predictably, HI values and H/C ratios of mature kerogens are much lower than those of immature ones. At the same time, O/C ratios seem to increase slightly with maturation.
4. A special remark should be made concerning sample 156753 (C-2 well, 3510 m). Kerogen HI and H/C values are anomalously high for the depth at which this sample was collected. Indeed, these values are comparable to those obtained for samples collected 1200 m higher in the section. It is assumed that fallen material from upper sections of the well may have contaminated this particular cuttings sample.

In conclusion, this preliminary step shows the need for comparing geochemical characteristics of both whole-rocks and isolated kerogens for a correct definition of the initial HI values of immature rocks and for calculation of the transformation ratio of mature rocks. In Gomo Member source rocks, HI values are strongly dependent on both TOC and depositional conditions, particularly oxidation. The matrix retention effect generally overshadows differences in organic matter quality. Our results indicate that selected rocks in the three compartments have a very good petroleum potential (in the range of 650–850 mg/g TOC) when the TOC is higher than 2%. This potential can be about 200–250 mg/g TOC lower when the TOC is less than 2%. Moreover, we also observed that HI values and H/C ratios depend on the geographic location: for comparable TOC contents, they decrease from the south to the northeast (Figure 8).

Some authors (e.g., Espitalié et al., 1980; Dembicki, 1990) have argued that S_2 values obtained from Rock-Eval pyrolysis need not be corrected for the mineral matrix retention effect because kerogen exists within a mineral matrix in subsurface. Therefore, according to their reasoning, such a retention phenomenon would also occur under natural maturation conditions. Monin et al. (1980), however, considered that high pyrolysis temperatures induce interactions between kerogen and the mineral matrix and that data only from isolated kerogens should be taken into account. Whatever the actual processes that occur under natural maturation conditions, the mineral matrix retention effect clearly influenced the typing of the organic matter present in the sedimentary beds of the Gomo Member. Lower HI values and higher OI values from Rock-Eval pyrolysis of whole-rock samples have led previous authors to underestimate the organic matter quality and to overestimate the maturity of the Gomo Member source rocks.

As a simplified approach to the problem of HI variations with TOC, the values shown below can be proposed for the initial HI values for each compartment:

Compartment	TOC <2%	TOC >2%
Southern	HI ~575 mg/g TOC	HI ~850 mg/g TOC
Central	HI ~450 mg/g TOC	HI ~750 mg/g TOC
Northern	HI ~400 mg/g TOC	HI ~650 mg/g TOC

In the mature samples of the second and third series in Table 1, the matrix retention effect as a function of TOC is also noticeable. Widely varying TOC contents and matu-

Table 2—Absolute Amounts and Relative Distribution of Fractionations of Whole-Rock (ext 1) and Kerogen (ext 2) C_{15+} Extracts into Saturates (SAT), Aromatics (ARO), and NSO for Immature Samples from the Gomo Member

Well	Depth (m)	Area	Sample No.	TOC (%)	$HI_{kerogen}$ (mg/g C)	S_1/TOC (mg/g C)		Absolute Amounts (mg/g C) Total	SAT	ARO	NSO	Distribution (%) SAT	ARO	NSO
S-1	450	South	151739	2.4	663	8	ext 1	49	21	4	24	43	9	48
							ext 2	22	1	0	21	5	0	95
							total	**71**	**22**	**4**	**45**	**31**	**7**	**62**
S-2	1044	South	156733	0.8	556	4	ext 1	70	30	7	33	42	11	47
							ext 2	30	4	4	22	13	13	74
							total	**100**	**34**	**11**	**55**	**34**	**11**	**55**
S-2	990	South	156732	1.2	612	5	ext 1	77	35	9	33	45	11	44
							ext 2	34	4	3	28	11	7	82
							total	**111**	**38**	**12**	**61**	**35**	**10**	**55**
S-2	936	South	156731	2.9	853	10	ext 1	83	37	15	31	45	18	37
							ext 2	27	2	2	23	7	9	84
							total	**110**	**39**	**17**	**54**	**36**	**15**	**49**
C-1	1968	Central	156734	0.7	421	4	ext 1	110	43	14	53	39	13	48
							ext 2	49	6	5	38	12	11	77
							total	**159**	**49**	**19**	**91**	**31**	**12**	**57**
C-1	2022	Central	156735	2.7	728	16	ext 1	108	62	13	33	57	12	31
							ext 2	36	6	2	28	17	5	78
							total	**144**	**68**	**15**	**61**	**47**	**11**	**42**
C-1	2040	Central	156736	4.5	759	24	ext 1	101	53	14	34	53	14	33
							ext 2	38	3	1	34	8	2	90
							total	**139**	**56**	**15**	**68**	**40**	**11**	**49**
NE-1	1206	NE	156738	0.6	398	4	ext 1	113	31	10	72	28	9	63
							ext 2	37	4	4	29	9	12	79
							total	**150**	**35**	**14**	**101**	**23**	**10**	**67**
NE-1	1224	NE	156739	1.9	446	183	ext 1	418	265	56	97	63	14	23
							ext 2	40	3	3	34	9	7	84
							total	**458**	**268**	**59**	**131**	**59**	**13**	**28**
NE-1	1260	NE	156740	3.9	645	95	ext 1	195	114	34	47	59	17	24
							ext 2	32	2	2	28	7	6	87
							total	**227**	**116**	**36**	**75**	**51**	**16**	**33**

ration levels prevent us from doing any further analysis on the variation of organic matter quality by compartment for the mature series.

Characterization of Total and Whole-Rock Extracts

As described in the section on Methods, the selected source rock samples were submitted to extraction with dichloromethane prior to and after kerogen isolation (extracts 1 and 2, respectively). From this point on, the sum of whole-rock and kerogen extracts will be referred to as total extract. The C_{15+} extracts were fractionated into saturates, aromatics, and NSOs.

The results for samples of the immature series from depths of 450–2040 m are presented in Table 2 and in Figures 10 and 11. From these data, the following observations can be made:

1. Total extracts vary between 71 and 458 mg/g TOC (Figure 10).
2. In two of the three samples from the NE-1 well (1224 and 1260 m depth, nos. 156739 and 156740), the amounts of total extract (458 and 227 mg/g TOC, respectively) are very high compared to the range of 100–160 mg/g TOC for all other samples. Kerogen elemental analyses and Rock-Eval data indicate that these two samples are as immature as the others in this series. For both samples, the sum of total C_{15+} extracts plus residual hydrogen index (HI_r) exceeds 870 mg/g TOC. It appears that the larger amounts of total extracts are due to natural petroleum accumulations. The higher values for the S_1/TOC ratio (95 and 183 mg/g TOC compared to 4–24 mg/g TOC for other samples) support this interpretation.
3. The average composition of the total extracts of the immature series is 35% saturates, 11% aromatics, and 54% NSOs.
4. If the two samples containing natural petroleum accumulations are not taken into account, the amounts of whole-rock extracts range from 49 to 113 mg/g TOC (Figure 10).

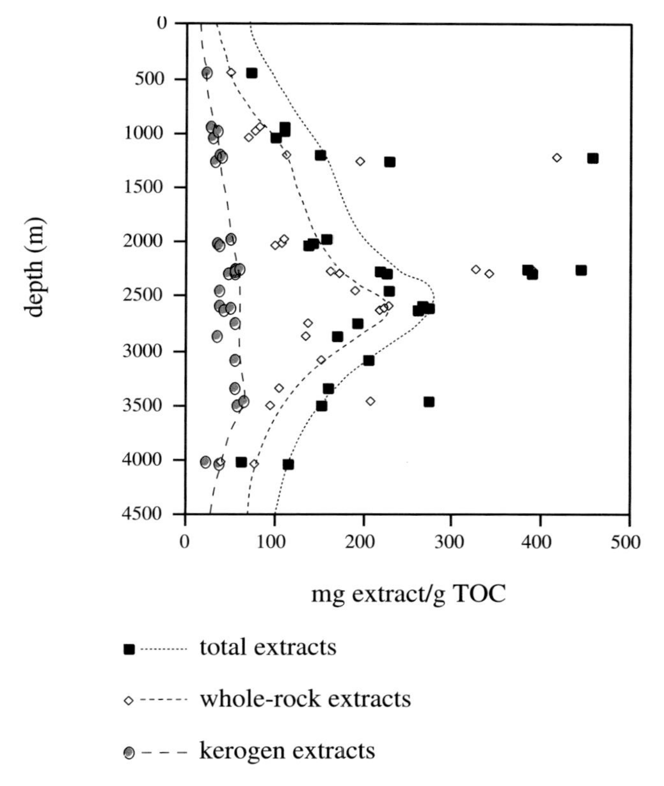

Figure 10—Variations in the absolute amounts (in mg/g TOC) of total, whole-rock, and kerogen extracts with depth.

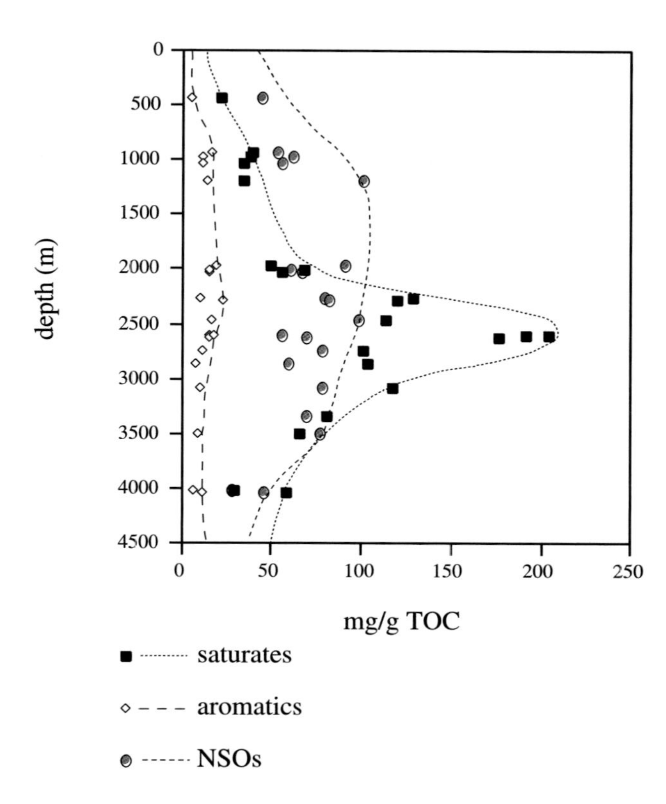

Figure 11—Variations in the absolute amounts (in mg/g TOC) of saturates, aromatics, and NSOs from total extracts with depth.

5. Whole-rock extracts are composed mostly of NSOs (31–63%) and saturates (28–59%); aromatics represent 9–18%.
6. An average of 100–150 mg/g TOC can be taken as representative of the amount of total extracts at the beginning of the oil window. This amount added to the HI values gives the true initial petroleum potential, which can be estimated as 750–950 mg/g TOC for source rocks with a TOC higher than 2% and as 500–675 mg/g TOC for those with a TOC lower than 2%. This is in good agreement with our previous data obtained for other immature samples (Penteado, 1996) and confirms the selection of average values that were defined above as input data for basin modeling studies.

Results from extracts of the moderately mature samples are shown in Table 3. Such samples lie in the depth interval of 2250–2630 m. Analytical data indicate the following:

1. The amounts of total extracts (219–446 mg/g TOC) (Figure 10) and S_1/TOC ratios are much higher than those of the immature series (Table 2) and HI values are accordingly lower (158–530 mg/g TOC), thus indicating that the selected samples have already generated petroleum compounds.
2. The sample from the C-3 well at 2250 m depth (no. 156744) has a large amount of total extract (446 mg/g TOC) which, when added to its residual HI, gives a total petroleum potential of 976 mg/g TOC. Such a value is too high when compared to those of the immature samples in the central compartment (Table 1) and might be explained by contamination from migrated petroleum. The same phenomenon may also have affected the sample at 2286 m from the same C-3 well (no. 156745), although to a lesser degree. The S_1/TOC ratios seem to corroborate such an interpretation.
3. In total extracts of moderately mature samples, saturates largely predominate (50–74%) and their absolute amounts increase in comparison to those of immature rocks: 114–255 versus 34–68 mg/g TOC (Figure 11). At the same time, the absolute amounts of NSOs from the two series are very similar: 55–115 versus 45–101 mg/g TOC for the immature rocks. The amounts of aromatics are also similar for the two series.
4. The amounts of whole-rock extracts (excluding samples from the C-3 well) are from 160 to 330 mg/g TOC, about three times the values in immature samples (Figure 10). In the moderately mature samples with no migrated petroleum, whole-rock extracts constitute 75–86% of the total extracts.

Table 3—Absolute Amounts and Relative Distribution of Fractionations of Whole-Rock (ext 1) and Kerogen (ext 2) C_{15+} Extracts into Saturates (SAT), Aromatics (ARO), and NSO for Moderately Mature Samples from the Gomo Member

Well	Depth (m)	Area	Sample No.	TOC (%)	$HI_{kerogen}$ (mg/g C)	S_1/TOC (mg/g C)		Absolute Amounts (mg/g C) Total	SAT	ARO	NSO	Distribution (%) SAT	ARO	NSO
S-3	2257	South	156747	0.6	158	71	ext 1	326	247	12	67	76	4	20
							ext 2	60	8	4	48	13	7	80
							total	**386**	**255**	**16**	**115**	**66**	**4**	**30**
S-4	2634	South	156743	0.6	205	31	ext 1	220	159	14	47	73	6	21
							ext 2	42	17	2	23	40	4	56
							total	**262**	**176**	**16**	**70**	**67**	**6**	**27**
S-4	2607	South	156742	1.6	242	63	ext 1	224	179	13	32	80	6	14
							ext 2	50	25	2	23	50	5	45
							total	**274**	**204**	**15**	**55**	**74**	**6**	**20**
S-4	2598	South	156741	2.7	249	89	ext 1	229	176	16	37	77	7	16
							ext 2	37	16	2	19	42	5	53
							total	**266**	**192**	**18**	**56**	**72**	**7**	**21**
C-2	2271	Central	156750	1.0	487	15	ext 1	163	119	9	35	73	6	21
							ext 2	56	10	1	45	18	2	80
							total	**219**	**129**	**10**	**80**	**59**	**5**	**36**
C-3	2286	Central	156745	1.4	317	105	ext 1	341	250	29	62	73	9	18
							ext 2	49	12	2	35	25	4	71
							total	**390**	**262**	**31**	**97**	**67**	**8**	**25**
C-3	2250	Central	156744	3.8	530	279	ext 1	391	305	28	58	78	7	15
							ext 2	55	4	2	49	7	3	90
							total	**446**	**309**	**30**	**107**	**69**	**7**	**24**
NE-2	2298	NE	156737	1.5	372	18	ext 1	172	108	19	45	63	11	26
							ext 2	54	12	4	38	22	8	70
							total	**226**	**120**	**23**	**83**	**53**	**10**	**37**
NE-3	2460	NE	156746	1.6	310	44	ext 1	191	109	15	67	57	8	35
							ext 2	38	5	1	32	13	3	84
							total	**229**	**114**	**16**	**99**	**50**	**7**	**43**

5. Whole-rock extracts of the moderately mature rocks are composed mostly of saturates (57–80%), with aromatics representing 4–11% and NSOs 14–35%.

Samples from the highly mature series of the Gomo Member were collected from depths between 2860 and 4030 m. From the extract data of this series, one can draw the following conclusions (Table 4):

1. The amounts of total extracts (64–207 mg/g TOC) (Figure 10) and S_1/TOC ratios are lower than those of the moderately mature series and tend to decrease with depth. The only exception seems to be the sample at 3471 m from the S-3 well (no. 156749), which contains total extract amounts comparable to those detected in samples in the 2250–2630 m interval. As in other cases (see above), this sample might represent a local natural accumulation.
2. Whole-rock extracts constitute 62–80% of total extracts, and their absolute amounts (41–153 mg/g TOC) (Figure 10) are accordingly smaller than those of the moderately mature series. The percentages of saturates in whole-rock extracts decrease with depth (from 71 to 51%).

From extract data covering such a broad depth and maturity range, one can see that the amounts of total extracts increase steadily down to depths of about 2600 m and then decrease progressively downward (Figure 10). This increase is accompanied by an enrichment in saturates both in absolute and relative amounts at the expense of NSOs and aromatics (Figure 11). Therefore, the interval between 2250 and 2600 m can be understood to be in the main phase of petroleum generation and accumulation in the source rocks. Deeper below, the decrease in absolute amounts and percentages of saturates can be interpreted as indicative of petroleum expulsion. This is because saturates are more stable and can be more easily expelled than the heavier molecules of aromatics and NSOs, as evidenced by a number of studies on natural samples and by experimental work (e.g., Vandenbroucke, 1972; Leythaeuser et al., 1988; Lafargue et al., 1990; Littke et al., 1990; Wilhelms et al., 1990; Sandvik et al., 1992; Rudkie-wicz et al., 1994). Thus, the depth of 2600 m can be considered as the point where the effects of significant petroleum expulsion overwhelm petroleum accumulation.

It has been demonstrated in previous work (Behar et al., 1997) that, through artificial maturation of kerogen in open system pyrolysis, generation of NSOs is always associated with hydrocarbon production at any maturity

Table 4—Absolute Amounts and Relative Distribution of Fractionations of Whole-Rock (ext 1) and Kerogen (ext 2) C_{15+} Extracts into Saturates (SAT), Aromatics (ARO), and NSO for Highly Mature Samples from the Gomo Member

Well	Depth (m)	Area	Sample No.	TOC (%)	$HI_{kerogen}$ (mg/g C)	S_1/TOC (mg/g C)		Absolute Amounts (mg/g C)				Distribution (%)		
								Total	SAT	ARO	NSO	SAT	ARO	NSO
S-3	2865	South	156748	0.9	105	15	ext 1	136	96	7	33	71	5	24
							ext 2	34	7	1	26	21	3	76
							total	**170**	**103**	**8**	**59**	**60**	**5**	**35**
S-3	3471	South	156749	0.4	125	14	ext 1	209	128	18	63	61	9	30
							ext 2	66	7	4	55	11	6	83
							total	**275**	**135**	**22**	**118**	**49**	**8**	**43**
C-2	2745	Central	156751	0.8	425	8	ext 1	138	91	11	36	66	8	26
							ext 2	54	10	1	43	18	2	80
							total	**192**	**101**	**12**	**79**	**53**	**6**	**41**
C-2	3090	Central	156752	0.7	204	15	ext 1	153	104	9	40	68	6	26
							ext 2	54	14	1	39	26	2	72
							total	**207**	**118**	**10**	**79**	**57**	**5**	**38**
C-2	3510	Central	156753	0.9	545	8	ext 1	95	51	7	37	54	7	39
							ext 2	57	15	2	40	26	4	70
							total	**152**	**66**	**9**	**77**	**43**	**6**	**51**
C-2	4020	Central	156754	1.5	52	4	ext 1	41	21	5	15	51	12	37
							ext 2	23	8	1	14	35	4	61
							total	**64**	**29**	**6**	**29**	**45**	**10**	**45**
C-2	4035	Central	156755	1.0	127	7	ext 1	77	46	9	22	60	12	28
							ext 2	38	12	2	24	32	5	63
							total	**115**	**58**	**11**	**46**	**50**	**10**	**40**
NE-4	3348	NE	156756	0.5	92	22	ext 1	105	60	10	35	57	10	33
							ext 2	55	21	34		38	62	
							total	**160**	**81**	**79**		**51**	**49**	

level. This means that, in mature source rocks, an increase in both aromatics and NSOs should be expected along with an observed increase in the absolute amounts of saturates. If this is not so, a portion of these classes of compounds must have been lost. This loss cannot be explained by preferential expulsion, whereby saturates would be more easily expelled (e.g., Lafargue et al., 1990; Littke et al., 1990; Wilhelms et al., 1990; Sandvik et al., 1992; Rudkiewicz et al., 1994). Therefore, the only explanation is partial secondary cracking of NSOs and aromatics within the source rocks. In fact, it has already been shown that NSOs and a portion of the aromatics are the most labile compounds and that they can start to degrade before maximum production of saturates from primary cracking is achieved, thus contributing to an increase in saturates (Behar et al., 1997). This is in good agreement with our kinetic scheme (Behar et al., 1991) in which the same thermal stability was assumed for both aromatics and NSOs. Thus, it is necessary to couple primary and secondary cracking reactions in source rocks to explain the compositional features of extracts from mature source rocks.

Characterization of Kerogen Extracts

Kerogen extracts, which can be thought of as an indication of kerogen retention capacity by absorption and/or adsorption, have been found in amounts ranging from 20 to 60 mg/g TOC which seem to increase slightly with depth (Figure 10 and Tables 2–4). The amounts of kerogen extracts usually represent less than 30% of total extracts. Compared to whole-rock extracts, kerogen extracts are always enriched in NSO compounds (45–95% of their composition).

If we consider the amounts of each class of compounds detected in total extracts of immature and moderately mature samples (Tables 2 and 3), only 3–12% of all saturates were in kerogen extracts, whereas 6–34% of all aromatics and 29–56% of all NSOs were kerogen extracts. Hence, kerogen retention capacity for saturates is lower than for aromatics and NSOs. The predominance of NSOs in kerogen extracts (generally >60%) is in good agreement with the larger size of their molecules and the consequent greater difficulty of expulsion from the kerogen network into the porous medium in the source rocks. In highly mature samples (Table 4), kerogen retention capacity appears to be slightly higher for all compound classes, with 7–28% saturates, 11–23% of aromatics, and 44–54% NSOs detected in kerogen extracts.

COMPARISON OF EXTRACT AND OIL COMPOSITIONAL DATA

One can see that the compositions of moderately mature whole-rock extracts (Table 3), here considered to come mostly from pores and thus susceptible to expulsion, are very similar to those of reservoired oils

(Gaglianone and Trindade, 1988). If one takes into account only the oils located in shallow reservoirs (<2000 m depth) to remove the possibility of secondary cracking reactions, the oil compositions in Recôncavo Basin are 50–80% saturates, 10–23% aromatics, and 6–35% NSOs. Once secondary cracking reactions and selective sorption on kerogen have caused compositional changes in petroleum in source rock pores, it seems that bulk expulsion or expulsion with partial retention of heavy compounds in the source rock pore system might reasonably account for the composition of oils in reservoirs. Moreover, retention of heavy compounds on mineral surfaces in the carrier bed pore space might also occur and lead to further petroleum compositional changes, although this effect is probably minor compared to those in the source rocks.

It is worth noting that the present composition of whole-rock extracts does not necessarily correspond to the composition at the time of significant petroleum expulsion. The extent to which secondary cracking reactions have occurred before petroleum expulsion depends on the kinetic parameters of NSOs and aromatics, as well as on the thermal history of the basin. The thermal stability and kinetic parameters of secondary cracking of NSOs and aromatics must be further investigated by laboratory experiments such as those performed by Behar et al. (1997) and McKinney et al. (1997). Another approach to these questions has been provided by a 2-D compositional modeling study in the Recôncavo Basin which tested kinetic parameters for secondary cracking that allow good calibration of observed extract compositions (Penteado, 1999).

Thus far, it is not clear how important selective retention of compounds was at the time of peak petroleum expulsion from the Gomo Member source rocks. However, its effects are noticeable in the remaining whole-rock extracts of the highly mature series, in which saturates are impoverished both in absolute and in relative amounts. Further quantitative studies on selective retention of heavy compounds in source rocks are definitely needed.

In summary, on the basis of our data, it appears that both secondary cracking reactions and selective retention of compounds in source rocks intervened in the process of compositional differentiation between source rock extracts and reservoired petroleum. The importance of such phenomena must be investigated in other series of source rocks with different kerogen types. Improvements in basin modeling software that aim at reliable predictions of petroleum composition in exploration prospects will strongly depend on the acquisition of secondary cracking kinetic parameters and retention factors of heavy compounds.

CONCLUSIONS

It has been shown that petroleum potential and HI values for the richer facies of the Gomo source rocks are characteristic of type I kerogen. Previous Rock-Eval data on immature whole-rocks led to an underestimation of these parameters due to an important mineral matrix retention effect during pyrolysis, which is stronger for organic-lean rocks. Therefore, it is recommended to cross-check HI values of whole-rocks with results obtained from isolated kerogens whenever possible. This procedure will allow a better estimation of organic matter quality and its petroleum potential.

A more intense oxidation of the lacustrine organic matter in low-TOC samples has been evidenced by HI values and H/C and O/C ratios. Furthermore, variations in organic facies from one basin compartment to another have been recognized, with better conditions for organic matter preservation in the south compartment. Thus, initial HI values as a function of initial TOC content were proposed for source rocks of each compartment. Knowledge of variations in organic matter quality and its petroleum potential has a major influence on the assessment of volumes of generated petroleum in each basin compartment.

To investigate petroleum compositional differences related to source rock maturity, we examined whole-rock and kerogen extracts of samples from the Gomo Member from a wide range of depths. Kerogen extracts, found in amounts from 20 to 60 mg/g TOC for all maturity levels, are rich in NSOs (60–90%). These amounts do not increase substantially with depth, suggesting that kerogen has a retention capacity that is attained quite early in source rock evolution. In contrast, the amounts of whole-rock extracts were shown to increase with depth until peak petroleum generation and accumulation was attained (2200–2600 m) and then to decrease gradually farther below due to significant petroleum expulsion.

For mature samples, the compositions of whole-rock extracts (57–80% saturates, 4–11% aromatics, and 14–35% NSOs), which are deemed susceptible to expulsion, closely resemble those of reservoired oils in the basin. Quantitative data on the C_{15+} fraction demonstrated that a substantial increase in saturates with maturation is not accompanied by a corresponding increase in NSOs and aromatics. Thus, partial secondary cracking of NSOs and aromatics is interpreted to have taken place within the Gomo Member source rocks before substantial petroleum expulsion occurred. Thus, bulk petroleum expulsion or expulsion with partial retention of heavy compounds might be considered as a plausible process. Therefore, secondary cracking kinetic parameters, coupled with retention factors for heavy compounds, constitute essential input data for compositional basin modeling software that aim at reliable predictions of petroleum composition in exploration prospects.

Acknowledgments—*We are greatly indebted to Petrobras for permission to publish and for funding of the first author's Ph.D. thesis work. We would like to thank Jussara M. D. Cortes and the personnel of the Sample Preparation Laboratory of the Geochemistry Section in Petrobras R&D Center for their help. Technical discussions with Marcio R. Mello, Luiz Antonio F.*

Trindade, Carla V. Araújo, and Maria Alice Aragão were greatly appreciated. François Lorant, Claudette Leblond, Thierry Lesage, Ramón Martinez, François Marquis, and Daniel Pillot are gratefully acknowledged for technical assistance and discussions. We are also very thankful to Jean-Luc Rudkiewicz, Luiz Antonio F. Trindade, Alain Prinzhofer, and Félix T. T. Gonçalves for their critical reviews. Special thanks are due to H. Dembicki, Jr., for review of a previous version of this chapter.

REFERENCES CITED

Aragão, M. A. N. F., L. A. F. Trindade, C. V. Araújo, O. B. Silva, A. A. Scartezini, F. H. Oswaldo, J. A. Canário, and A. P. Garcia (abs.), 1998, Distribution and controls of lacustrine source rocks in the Recôncavo Basin, Brazil, *in* M. R. Mello and P. O. Yilmaz, eds., Extended Abstracts Volume, AAPG International Conference and Exhibition, Rio de Janeiro, p. 306.

Behar, F., C. Saint-Paul, and C. Leblond, 1989, Analyse quantitative des effluents de pyrolyse en milieu ouvert et fermé, Revue de l'Institut Français du Pétrole, v. 44, p. 387–397.

Behar, F., S. Kressman, J.-L. Rudkiewicz, and M. Vandenbroucke, 1991, Experimental simulation in a confined system and kinetic modelling of kerogen and oil cracking: Organic Geochemistry, v. 19, p. 173–189.

Behar, F., M. Vandenbroucke, Y. Tang, F. Marquis, and J. Espitalié, 1997, Thermal cracking of kerogen in open and closed systems: determination of kinetic parameters and stoichiometric coefficients for oil and gas generation: Organic Geochemistry, v. 26, p. 321–339.

Caixeta, J. M., G. V. Bueno, L. P. Magnavita, and F. . Feijó, 1994, Bacias do Recôncavo, Tucano e Jatobá: Boletim de Geociências da Petrobrás, v. 8, p. 163–172.

Daniel, L. M. F., E. M. Souza, and L. F. Mato, 1989, Geoquímica e modelos de migração de hidrocarbonetos no Campo de Rio do Bu. Integração com o compartimento nordeste da Bacia do Recôncavo: Boletim de Geociências da Petrobrás, v. 3, p. 201-214.

Dembicki, Jr., H., 1990, Mineral matrix effects during analytical pyrolysis of source rocks: Bulletin of the Association of Petroleum Geochemical Explorationists, v. 6, p. 78–105.

Dembicki, Jr., H., B. Horsfield, and T. T. Y. Ho, 1983, Source rock evaluation by pyrolysis-gas chromatography: AAPG Bulletin, v. 67, p. 1094–1103.

Durand, B., and G. Nicaise, 1980, Procedure for kerogen isolation, *in* B. Durand, ed., Kerogen-insoluble organic matter from sedimentary rocks: Editions Technip, Paris, p. 35–53.

Espitalié, J., M. Madec, and B. Tissot, 1980, Role of mineral matrix in kerogen pyrolysis: influence on petroleum generation and migration: AAPG Bulletin, v. 64, p. 59–66.

Espitalié, J., K. Senga Makadi, and J. Trichet, 1984, Role of the mineral matrix during kerogen pyrolysis, *in* P. A. Schenck, J. W. de-Leeuw, and G. W. M. Lijmbach, eds., Advances in organic geochemistry 1983: Organic Geochemistry, v. 6, p. 365–382.

Espitalié, J., G. Deroo, and F. Marquis, 1985, La pyrolyse Rock-Eval et ses applications, Deuxième partie: Revue de l'Institut Français du Pétrole, v. 40, p. 755–784.

Figueiredo, A. M. F., J. A. E. Braga, H. M. C. Zabalaga, J. J. Oliveira, G. A. Aguiar, O. B. Silva, L. F. Mato, L. M. F. Daniel, L. P. Magnavita, and C. H. L. Bruhn, 1994, Recôncavo Basin, Brazil: a prolific intracontinental rift basin, *in* S. M. Landon, ed., Interior rift basins: AAPG Memoir 59, p. 157–203.

Gaglianone, P. C., and L. A. F. Trindade, 1988, Caracterização geoquímica dos óleos da Bacia do Recôncavo: Geochimica Brasiliensis, v. 2, p. 15–39.

Horsfield, B., and A. G. Douglas, 1980, The influence of minerals on the pyrolysis of kerogens: Geochimica et Cosmochimica Acta, v. 44, p. 1119–1131.

Katz, B. J., 1983, Limitations of "Rock-Eval" pyrolysis for typing organic matter: Organic Geochemistry, v. 4, p. 195–199.

Lafargue, E., J. Espitalié, T. Jacobsen, and S. Eggen, 1990, Experimental simulation of hydrocarbon expulsion, *in* B. Durand and F. Behar, eds., Advances in organic geochemistry 1989: Organic Geochemistry, v. 16, p. 121–131.

Langford, F. F., and M.-M. Blanc-Valleron, 1990, Interpreting Rock-Eval pyrolysis data using graphs of pyrolizable hydrocarbons vs. total organic carbon: AAPG Bulletin, v. 74, p. 799–804.

Leythaeuser, D., R. G. Schaeffer, and M. Radke, 1988, Geochemical effects of primary migration of petroleum in Kimmeridge source rocks from Brae field area, North Sea. I: Gross composition of C_{15+}-soluble organic matter and molecular composition of C_{15+}-saturated hydrocarbons: Geochimica et Cosmochimica Acta, v. 52, p. 701–713.

Littke, R., D. Leythaeuser, M. Radke, and R. G. Schaeffer, 1990, Petroleum generation and migration in coal seams of the Carboniferous Ruhr Basin, northwestern Germany, *in* B. Durand and F. Behar, eds., Advances in organic geochemistry 1989: Organic Geochemistry, v. 16, p. 247–258.

McKinney, D. E., F. Behar, and P. G. Hatcher, 1997, Experimental determination of the reaction kinetics and product distributions of ^{13}C-labeled n-C_{25} mixed with Arabian light marine oil: are mixture effects important? (abs.): Abstracts of the 18th International Meeting on Organic Geochemistry, Part I, Maastricht, The Netherlands, p. 155–156.

Mello, M. R., E. A. M. Koutsoukos, W. Mohriak, and G. Bacoccoli, 1994, Selected petroleum systems in Brazil, *in* L. B. Magoon and W. G. Dow, eds., The petroleum system—from source to trap: AAPG Memoir 60, p. 499–512.

Monin, J. C., B. Durand, M. Vandenbroucke, and A. Y. Huc, 1980, Experimental simulation of the natural transformation of kerogen, *in* A. G. Douglas and J. R. Maxwell, eds., Advances in organic geochemistry 1979: Physics and Chemistry of the Earth, v. 12, p. 517–530.

Orr, W. L., 1981, Comments on pyrolytic hydrocarbon yields in source-rock evaluation, *in* M. Bjøroy et al., eds., Advances in Petroleum Geochemistry 1981: p. 775–787.

Penteado, H. L. B., 1996, Caractérisation géochimique et modélisation de la maturation des roches mères dans la partie sud du Bassin de Recôncavo, Brésil: Mémoire de Diplôme d'Etudes Approfondies, Ecole Nationale Supérieure du Pétrole et des Moteurs–Institut Français du Pétrole, 85 p.

Penteado, H. L. B., 1999, Modélisation compositionnelle 2D de la genèse, expulsion et migration du pétrole dans le Compartiment Sud du Bassin de Recôncavo, Brésil: Ph.D. Thesis, Université Pierre et Marie Curie, Paris, 233 p.

Rudkiewicz, J.-L., O. Brévart, J. Connan, and F. Montel, 1994, Primary migration behaviour of hydrocarbons: from laboratory experiments to geological situations through fluid flow models, *in* N. Telnaes, G. van Graas, and K. Øygard, eds., Advances in organic geochemistry 1993: Organic Geochemistry, v. 22, p. 631–639.

Sandvik, E. I., W. A. Young, and D. J. Curry, 1992, Expulsion from hydrocarbon sources: the role of organic absorption, *in* C. B. Eckardt, J. R. Maxwell, S. R. Larter, and D. A. C. Manning, eds., Advances in organic geochemistry 1991: Organic Geochemistry, v. 19, p. 77–87.

Santos, C. F. dos, J. A. Cupertino, and J. A. E. Braga, 1990, Síntese sobre a geologia das bacias do Recôncavo, Tucano e Jatobá, *in* G. P. R Gabaglia and E. J. Milani, eds., Origem e evolução das bacias sedimentares brasileiras: Rio de Janeiro, Petrobrás, p. 235–266.

Vandenbroucke, M., 1972, Etude de la migration primaire: variation de composition des extraits de roche à un passage roche mère/réservoir, *in* H. R. Gaertner and H. Wehner, eds., Advances in organic geochemistry 1971: International Series of Monographs on Earth Sciences, v. 33, p. 547–565.

Wilhelms, A., S. R. Larter, D. Leythaeuser, and H. Dypvik, 1990, Recognition and quantification of the effects of primary migration in a Jurassic clastic source rock from the Norwegian continental shelf, *in* B. Durand and F. Behar, eds., Advances in organic geochemistry 1989: Organic Geochemistry, v. 16, p. 103–113.

Magnavita, L. P., 2000, Deformation mechanisms in porous sandstones: implications for development of fault seal and migration paths in the Recôncavo Basin, Brazil, *in* M. R. Mello and B. J. Katz, eds., Petroleum systems of South Atlantic margins: AAPG Memoir 73, p. 195–212.

Chapter 15

Deformation Mechanisms in Porous Sandstones: Implications for Development of Fault Seal and Migration Paths in the Recôncavo Basin, Brazil

Luciano P. Magnavita

Petrobrás/E & P–BA
Salvador, Brazil

Abstract

Fault zones in porous sandstones consist of a series of light-colored, more resistant strands of comminuted or pulverized rock, to which various terminology has been applied. Resulting fault geometry is variable but can be predicted under certain fractal distributions, thus a multifractal approach is able to discriminate different fault sets. Texturally, deformed sandstones along individual microfaults are characterized as breccia and gouge zones. Fault evolution involves a fracture processes, probably followed by diffusion mass transfer mechanisms, which make possible a balance in the silica budget. Analysis of failure criteria suggests that, initially, deformation occurs as a dilation phase through strain-softening processes, which change to strain-hardening because of reduction in grain size and porosity enhanced by healing of opening microcracks. The drastic reduction in porosity and permeability results in a membrane seal, the develpment of which depends on such factors as cataclasis, shale smear, diagenesis, and hydrocarbon migration. Along this membrane seal, displacement pressures are higher than in the undeformed sandstone, so the fault zone acts as a potential barrier capable of focusing oil migration. This is constrained by mercury injection curves indicating reductions in pore throat width of at least two orders of magnitude between the intact sandstones and the deformed rock.

These observations are exemplified by the Mata–Catu transfer zone, the most prolific trend in the Recôncavo Basin. Isotopic analyses of gases along and across the trend corroborate that the trend is in a key positon to allow long-distance oil migration. Thus, some oil accumulations in fields at great distances from the oil kitchen can be explained.

INTRODUCTION

Faults in porous sandstones do not appear as discrete, well-defined surfaces. On the contrary, they consist of a series of light-colored, more resistant strands of comminuted or pulverized rock each with a slip on the order of a few millimeters to a few centimeters (Aydin, 1978; Pittman, 1981; Jamison and Stearns, 1982; Underhill and Woodcock, 1987). The mechanisms involved in the development of these fault zones have been investigated (Blenkinsop and Rutter, 1986; Lloyd and Knipe, 1992; Hippler, 1993), and many studies have emphasized the potential seal of these faulted sandstones (Antonellini and Aydin, 1994; Fowles and Burley, 1994).

Faults can act both as barriers and seals (Smith, 1966; 1980; Downey, 1984; Hippler, 1993). Many studies have highlighted the role of faults as seals in areas such as the North Sea (Knott, 1993; Jones and Knipe, 1996; Hippler, 1997), the Gulf of Mexico (Berg and Avery, 1995; Alexander and Handschy, 1998), Trinidad (Gibson, 1994), and the Niger Delta (Weber et al., 1978; Bouvier et al., 1989). The studies are based on a variety of techniques, including logs (Berg and Avery, 1995), petrophysics (Antonellini and Aydin, 1994; Fowles and Burley, 1994), and juxtaposition diagrams (Allan, 1989; Knipe, 1997).

The fault zones studied in this chapter are affecting sandstones of the Recôncavo-Tucano-Jatobá rift (Figure 1). The main points discussed are as follows: (1) the geome-

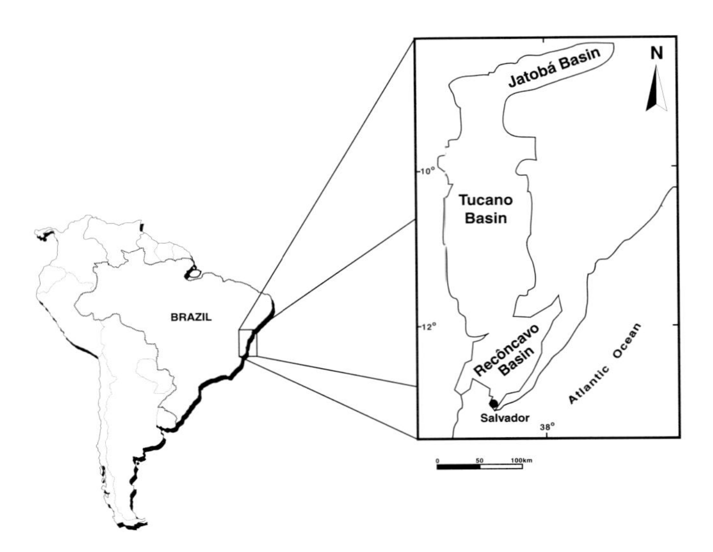

Figure 1—Location map of the Recôncavo Basin in the Recôncavo-Tucano-Jatobá rift.

tric characterization of the fault zones, (2) the mechanisms involved in their evolution, (3) the potential for the development of hydrocarbon seals, and (4) the role of the fault zones in focusing secondary migration.

The Recôncavo Basin, the oldest petroleum-producing basin in Brazil, is in a mature stage of exploration. Commercial oil production dates back to the early 1940s and eventually led to the discovery of about 85 oil and gas fields. Petroleum accumulation in the basin can be grouped into four basic systems, with the Sergi/Água Grande–Candeias(!) system being the main one, accounting for 57% of the proven oil volume in the basin, or 2.7 billion bbl (Figueiredo et al., 1994). The Sergi Formation is the main reservoir (eolian–fluvial system), averaging 18% porosity and 800 md permeability. Since the lacustrine freshwater shales of the synrift Candeias Formation (Lower Cretaceous) are above the Sergi Formation (Upper Jurassic), traps are typically structural (horsts and tilted blocks) and hydrocarbon secondary migration may rely on pathways along some fault zones. Also, the majority of Recôncavo structural traps requires that at least one fault be a sealing fault. The Mata–Catu fault zone controls the main oil trend in the basin and thus will be used as an example to illustrate fault trapping and migration in this chapter.

TERMINOLOGY

The bleached strands of granulated rock observed along the fault zones in the Recôncavo-Tucano-Jatobá rift match examples described elsewhere in porous sandstones of various ages, such as the Entrada (Upper Jurassic), Navajo (Lower Jurassic), Wingate (Triassic), Mesa Rica (Cretaceous), and Simpson Group (Ordovician) in the United States and the New Red Sandstone (Permian–Triassic) in Scotland. The terminology applied to describe these strands is extremely varied; they have been called *shear fractures* (Engelder, 1974; Jamison, 1989), *deformation bands* (Aydin, 1978; Aydin and Johnson, 1978; Antonellini and Aydin, 1994), *granulation seams* (Pittman, 1981), *microfaults* (Jamison and Stearns, 1982; Underhill and Woodcock, 1987), *gouge-filled fractures* (Nelson, 1985), *fault shears* (Blenkinsop and Rutter, 1986), *shear bands* (Berg and Avery, 1995), *tabular compaction zones* (Kulander et al., 1990), and *cataclastic slip bands* (Fowles and Burley, 1994). Where two or more of these strands come closer, the term *zone* has been commonly added to the name.

Another matter of controversy is related to the surfaces that show evidence of considerable slip as indicated by slickenside striations, generally called *slip surfaces* (Aydin and Johnson, 1978; Aydin and Johnson, 1983; Underhill and Woodcock, 1987; Antonellini and Aydin, 1994). In the present chapter, the term *fault* is retained and applied to relatively tabular or planar discontinuities that show displacements greater than 0.5 cm, that is, at least five to ten times greater than the width of the discontinuity (see Wise et al., 1984, for complete definition). Where the displacement cannot be determined macroscopically, the term *microfault* is used. When the faults appear closely spaced, the term *zone* is added to the name, thus becoming fault zone or microfault zone. The term *shear fracture* is retained to describe the initiation of movement across a previously deformed microcrack.

GEOMETRY

Typically, faults in porous sandstones do not appear as discrete, well-defined surfaces. On the contrary, they consist of a series of light-colored, more resistant strands of comminuted or pulverized rock, each with a slip on the order of a few millimeters to a few centimeters.

Along a fault zone, the minor faults wander along strike and may touch each other at a point or merge over a short distance, resulting in a braided appearance with lenses of intact rock surrounded by brittle, deformed strands of crushed rock, creating elongated crests (Figure 2a). The presence of other sets results in more complex geometric arrangements, such as lozenges, hexagons, and different type of stars (Figure 2b).

Assemblages of these elongated crests may be continuous for hundreds of meters or even a few kilometers (Figure 2c). These complex arrays have been associated with 3-D strain in which three or four sets of faults may form simultaneously (Aydin and Reches, 1982; Engelder, 1987; Underhill and Woodcock, 1987). Although some complex patterns may have locally originated from 3-D strain, on a regional basis, kinematic indicators and cross-cutting relationships in some examples indicate that the faults constitute conjugates formed as a response to plane strain shear (Magnavita, 1992).

Figure 2—Typical geometry of deformation bands in examples of sandstone outcrops from the Recôncavo-Tucano-Jatobá rift. (a) General aspect of a typical braided geometry in the Massacará Group (Aptian). Note the wandering of the fault traces, with faults touching each other at a point or for a short distance along strike. (b) Transversal section across an elongated crest in the Ilhas Group (Hauterivian). Note the extreme complexity of the microfault patterns. (c) Typical exposure of an elongated crest in the Massacará. (d) Conjugate fault pattern in the Massacará. The main fault dips to the east (to the right in the photograph).

Relative Age of Conjugate Sets

Based on the study of offsets and truncations of three sets of conjugate faults trending north–northeast, east–northeast, and northwest–southeast, Magnavita (1992) suggested that the faults must have formed contemporaneously but not simultaneously during deformation. This is corroborated by the fact that all sets mutually truncate one another. Age relationships indicate that the deformation alternated, with a few members of one set forming and then being offset by a few members of a second set. Following this, new members of the first set formed and offset some members of the second set. The same fluctuation can be extrapolated to the third set. Analogous relationships have been described by Zhao and Johnson (1991) in the Middle Jurassic Entrada Sandstone in the United States. This is also similar to the evolution of conjugate sets put forward by Bretan et al. (1996).

Apart from the extreme complexity exhibited by the fault zones, they generally show a prevailing conjugate fault plane along which most of the motion took place (Figure 2d). The crests are generally a few meters high (4–7 m) with widths varying from a few meters to several tens of meters. However, fault zones may reach hundred of meters wide and tens of meters high (Magnavita, 1997).

Fault Patterns and Fractal Distribution of Faults

In the Recôncavo-Tucano-Jatobá rift, the braided and polygonal patterns are present at the microscopic, field, and map scales (Magnavita, 1992). In all cases, the fault arrangements similarly represent zones of minor displacement that are generated before a relatively major through-going surface of discontinuity is formed. Consequently, the geometry of the fault arrangements are scale invariant and thus reflect a fractal, self-similar distribution for the faults in these sandstones.

The distribution of fracture systems has the property of scale invariance, showing the same distribution pattern at many scales. This scale invariance or fractal distribution (Mandelbrot, 1967) follows a nonlinear or power law expressed by

$$N = l^{-D}, \tag{1}$$

where N is the number of fractures with a determined attribute (e.g., length, throw, or spacing) greater than or equal to l, and D is the power law exponent, normally called the *fractal dimension*. When plotted on a log-log axis, a straight line of slope $-D$ results. The power law nature of fault size distributions is consistent with fault systems being self-similar fractals (Nicol et al., 1996).

In 2-D analysis, the method more often used is box counting. In this method, square grids of different sizes are placed over the image that depicts the fractures (e.g., maps, photographs, or line-drawings), and the number of boxes that contain fractures are counted. This number is plotted on a log-log axis as a function of the length of the side of the square in such a way that the distributions obeying Equation (1) are arranged along a straight line with slope $-D$, the fractal dimension of the set. The method just described applies to uniform fractals, which are described by D. However, the majority of fractal objects that occur in nature are not uniform, thus admitting many fractal exponents and resulting in a *multifractal.* These multifractal objects are characterized by a spectrum of fractal exponents $D(q)$, where q is a real parameter.

The multifractal spectrum $D(q)$ is obtained using an algorithm of box counting on digitalized images of fracture systems. The generation function $\chi(q, l)$, which defines the multifractal scaling in terms of exponents $\tau(q)$, is expressed as

$$\chi(q) = \ln \tau(q), \tag{2}$$

where ln is the natural logarithm,

$$\tau(q) = D(q)(q-1), \tag{3}$$

and the $(q - 1)$ term ensures that the moments are independent of box size.

Because multifractals represent how a quantity of mass of material is distributed over a geometric object or support in each box (Belfield, 1994), the box counting method does not refer to the presence or absence of just one event (a fracture in this case) included in the box under investigation. In fact, a weight μ_i is given equal to the mass (number of fractures) existing in the box under consideration. The mass μ_i is elevated to a power q, where q expresses the mass momentum of the mass distribution that defines the pattern. In this way, for $q = 0$, the fractal dimension $f(q)$ corresponds to the fractal dimension D. If $q < 0$, the boxes with lower mass are preferably weighted relative to the denser regions. If $q > 0$, the heavier cells (more mass) are preferably weighted relative to the lighter regions.

To apply the multifractal concept, Magnavita and Souza (1995) selected three pictures of outcrops of the Recôncavo-Tucano-Jatobá rift and plotted q versus $D(q)$ on a graph (Figure 3a). For each picture, a line drawing was made and different slide windows were run using a program called *MULFRA,* developed by the Institutt for Energiteknikk (IFE) of Norway. For the first image (Figure 3b), there were no good correlations because part of the image shows a regular, uniform pattern. The curves are set apart, and for $q = 0$, the distribution is close to the Euclidian dimension ($D = 2$). The second picture (Figure 3c) shows a more fractal image, that is, more self-similar. However, the last picture (Figure 3d) is the one that visually presents the most self-similar image, resulting in better clustering of each slide window, which can be related to a more fractal object. Although not conclusive to predict fault distribution, this kind of analysis may be further applied to contrast different fault sets formed under various tectonic regimes, such as extensional and strike-slip regimes.

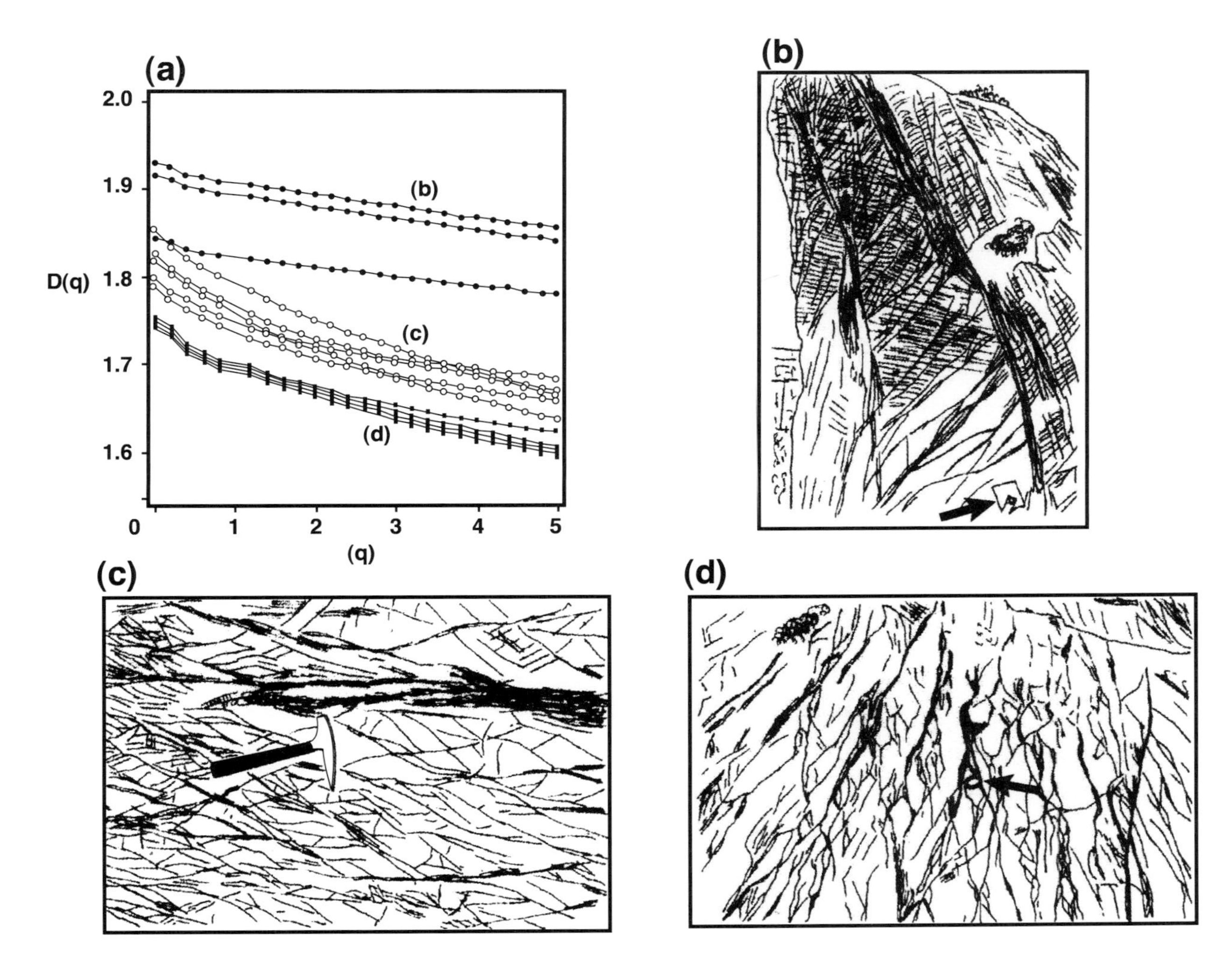

Figure 3—Application of the multifractal concept according to Magnavita and Souza (1995). (a) Multifractal spectrum of faults and microfaults picked on line drawings drawn from photographs of outcrops. Three slide windows were run for (b) and five for (c) and (d). See text for discussion. (b) Regular patterns of microfaults more typical of a Euclidian object of fractal dimension close to *D* = 2. (Arrow points to notebook for scale.) (c) Line drawing showing a more self-similar pattern of microfaults. (d) Self-similar pattern of microfaults indicative of a fractal object. (Arrow points to compass for scale.)

TEXTURAL CHARACTERIZATION

The sandstones are genetically characterized as fluvial and eolian. Petrographically, they can be classified as quartz arenites (predominant), subarkoses, and sublitharenites. Quartz, feldspar, and fragments of metamorphic rocks constitute the main components. In terms of textural maturity, the quartz grains are generally subangular to subrounded, or subrounded to rounded, according to its respective fluvial or eolian origin.

In the field, it is always difficult to positively establish grain comminution. Under the microscope, the microfaults exhibit progressive deformation from the undisturbed parent rock to the center of a 1- to 2-mm-wide zone of highly comminuted grains, which can be texturally classified as a gouge zone (Figure 4). This zone is characterized by a band of mechanically comminuted grains in a silt and clay matrix in which grain size, sorting, and porosity are substantially reduced in comparison to the original parent sandstone. This highly localized deformation decreases outward through a breccia zone, which shows a random fabric of granulated (crushed) grains sunk in a brown, fine matrix with comminuted grains. In some samples, a percolation zone along which syn- to posttectonic cementation took place is also apparent between the brecciated zone and the parent rock (Magnavita, 1992).

As pointed out by Roque (1990), the microscopic characterization of both the gouge and breccia zones (his internal and external deformed zones) is not always possible; in many cases, only the brecciated zone is present. However, it should be stressed that, if deforma-

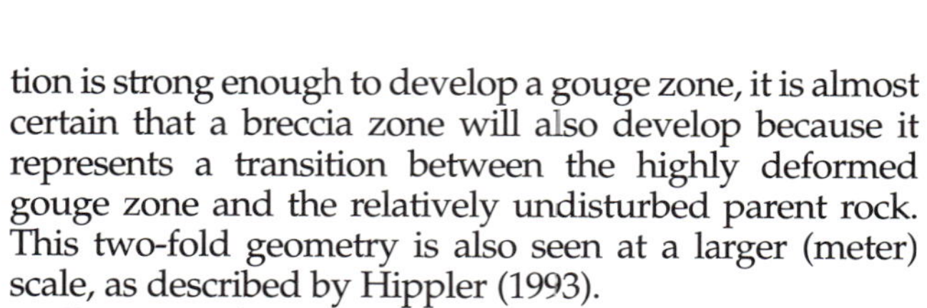

Figure 4—Internal geometry and textural characteristic of an individual fault in optical thin-section (crossed polars). A clay–silt zone of highly comminuted grains (gouge zone) passes abruptly into a zone of crushed fragments immersed in a finer ferruginous matrix (breccia zone).

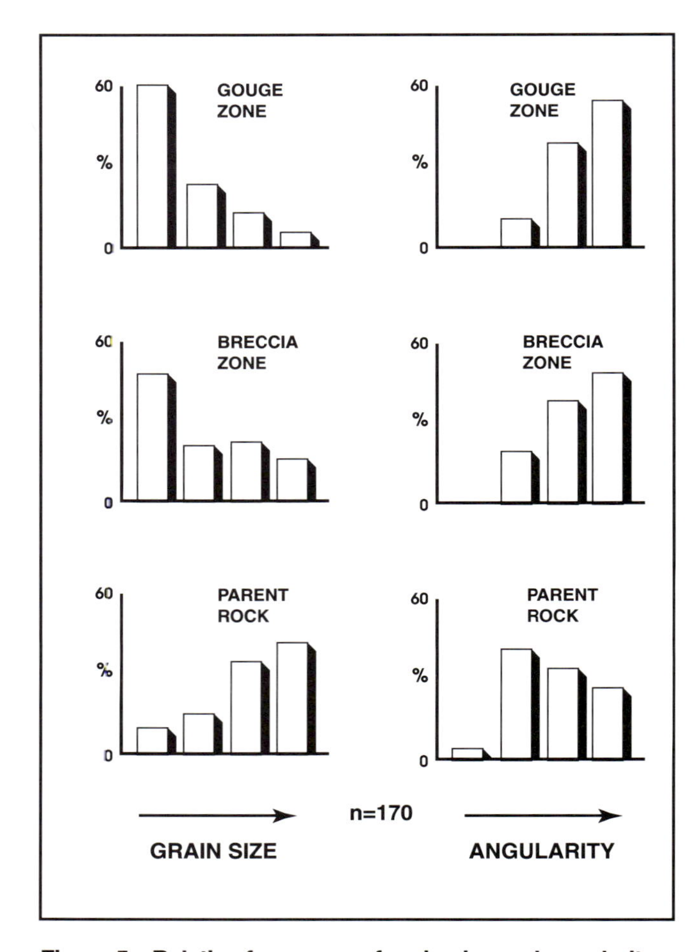

Figure 5—Relative frequency of grain size and angularity distributions across an individual fault. Grains size varies from silt to medium sand, and grain angularity varies from rounded to angular. Adapted from Roque (1990).

tion is strong enough to develop a gouge zone, it is almost certain that a breccia zone will also develop because it represents a transition between the highly deformed gouge zone and the relatively undisturbed parent rock. This two-fold geometry is also seen at a larger (meter) scale, as described by Hippler (1993).

Grain size decreases drastically within the microfaults, with the reduction as much as three size classes, from medium sand to silt (Figure 5). Relative frequency confirms a grain size reduction accompanied by a decrease in sorting (expressed by a predominance of angular grains in the breccia–gouge zone) from the undeformed parent rock to the most deformed gouge zone, where more than 75% of the grains are silt and only about 10% of the grains are subrounded. Analogous conclusions were obtained by studies carried out on similar faults in sandstones from other areas (e.g., Engelder, 1974; Jamison and Stearns, 1982; Underhill and Woodcock, 1987; Fowles and Burley, 1994).

MECHANICAL EVOLUTION OF THE FAULT ZONES

The fault zones exposed in the rift developed at shallow crustal levels, generating gouges and breccias. These noncohesive faults indicate that deformation occurred at depths of less than 5 km (Scholz, 1990). This is corroborated by the depth of the basement found in studies of buried subsidence (Chang et al., 1992). Main deformation mechanisms associated with faulting were fracture processes and diffusive mass transfer (Magnavita, 1992). Fracture and damage of the sandstones involved (1) development of isolated fractures, (2) linkage of these fractures, and (3) localization of displacement to form breccias and gouges. Further shear resulted in displacement along the through-going surfaces of discontinuity, thus forming a fault. These processes are discussed in the following sections.

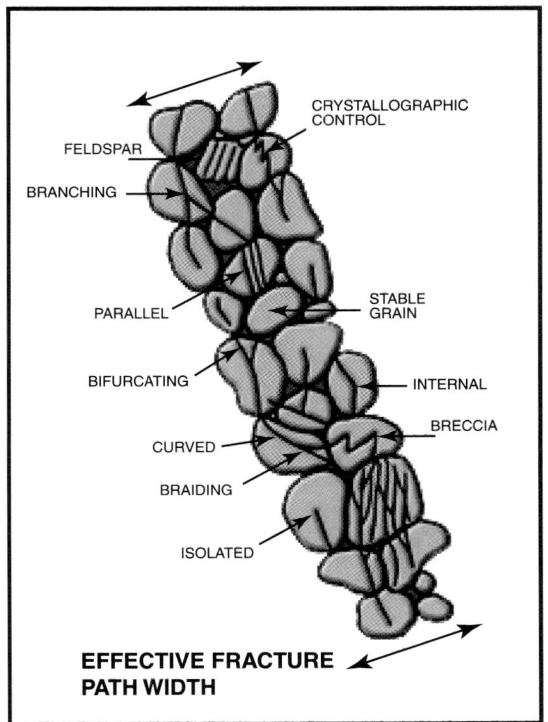

(b)

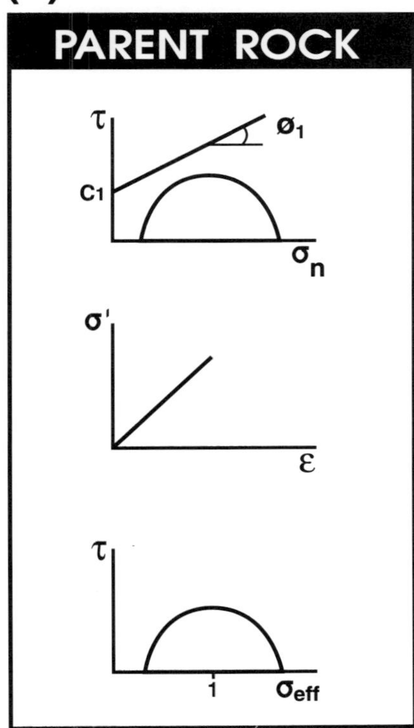

Figure 6—Initial intragranular microcracking of a fault zone. (a) Irregular paths of microcracking resulting in large effective fracturing width. Adapted from Lloyd and Knipe (1992). (b) Failure criteria during initial evolution of the microfault. Graphs for stages of deformation (from top to bottom) are shear stress versus normal stress, differential stress versus strain, and shear stress versus effective stress.

Intragranular Extensional Microcracking: The First Stage of Fault Generation

Apparently, intragranular mechanisms seem to have dominated the initial fracturing in the fault zone. The small number of contacts in the relatively well-sorted sandstones with no ductile matrix resulted in areas of high stress concentration and consequent pervasive microcracking. As shown by Lloyd and Knipe (1992), different microcrack geometries, some inherited from the parent sandstone, may have a crucial role in influencing the exact fracture processes and its subsequent linkage (Figure 6a). The microscopic heterogeneity results in an irregular distribution of fracture strengths on the grain-size scale, which in turn may result in more extensive linking fracture arrays.

The intragranular microfractures clearly represent extensional microcracks because no significant displacement is shown either parallel or normal to the trace fractures. In this way, the rock still has not entirely lost its cohesion and the process may predominate in the neighborhood of a fault zone in the parent rock. Thus, at the first buildup of stress, the parent rock may maintain the cohesion if the elastic resistance of the rock has not been surpassed (Figure 6b). Nevertheless, some extensional microcracks within the grains may be formed in this stage. The effective stress must have been low because of a probable high pore pressure during the first stages of deformation, as evidenced by many fluidization features found in the sandstones.

Linkage of Intragranular Microcracks: The Second Stage of Fault Generation

The next stage involved the development of extension microcracks at impinging grain contacts, followed by the formation of small shear faults by linking of the microcracks, as described elsewhere (Blenkinsop and Rutter, 1986; Engelder, 1987; Lloyd and Knipe, 1992). Intragranular microfractures tend to link grain contacts through stress trajectories created as a consequence of stress concentrations at grain–grain contacts (Lloyd and Knipe, 1992).

The irregular paths followed by the linkage process result in an irregular fracture trace with an effective width of at least one and usually several grain diameters (Figure 7a). Because of this, and also because fracture traces are commonly irregular within individual grains, shear displacement along these isolated fractures is likely to induce local regions of extension and/or compression, interfacial friction, and grain comminution by irregular fracturing. All these account for the initial formation of the breccia zone. There is an increase in the coefficient of internal friction (ϕ) due to the higher angularity and lower porosity observed in the zones (Figure 7b). However, decrease in strength (or cohesion, C) predominates during the crushing process as the more important factor. Furthermore, because this is basically a dilatancy phase, an increase in fluid volume occurs and the pore fluid pressure may drop. As crushing proceeds, small fragments obliterate the throats between pores, and this

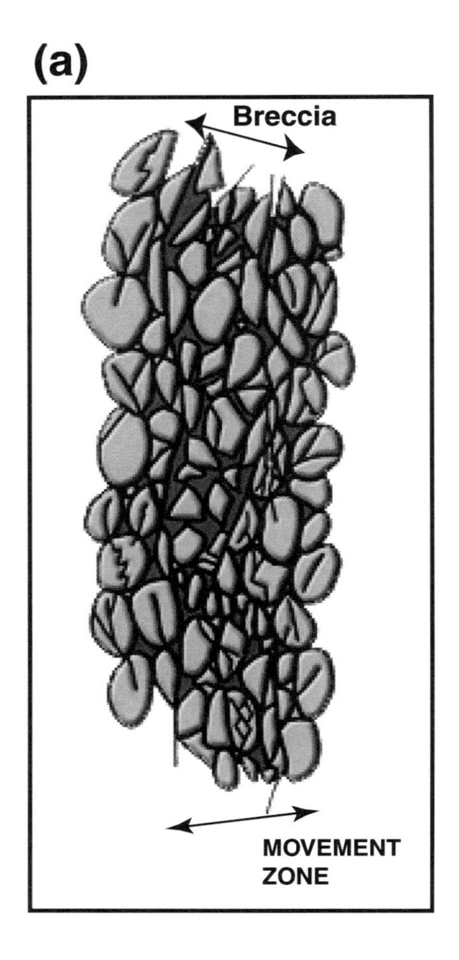

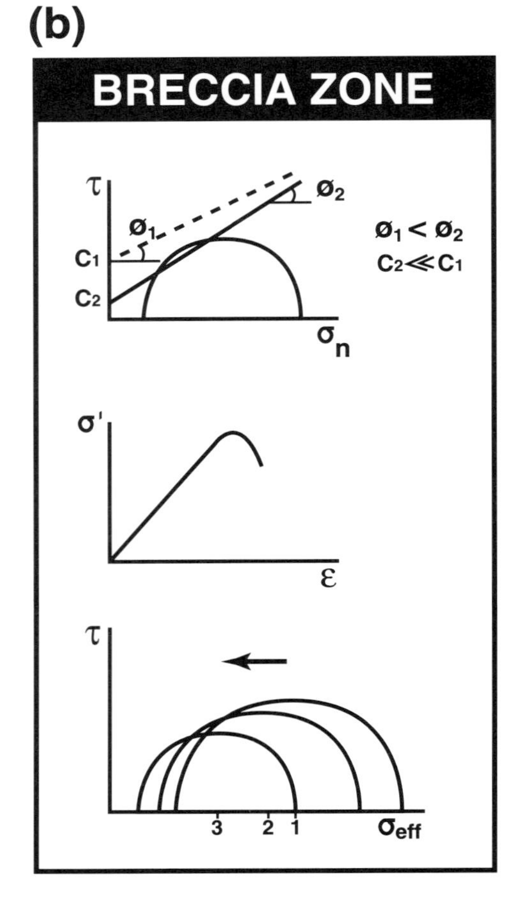

Figure 7—Development of intergranular fracturing of a fault zone. (a) Irregular paths followed by intergranular linkage resulting in large effective fracturing width. Adapted from Lloyd and Knipe (1992). (b) Failure criteria and the evolution of the breccia zone. Graphs for stages of deformation (from top to bottom) are shear stress versus normal stress, differential stress versus strain, and shear stress versus effective stress.

reduction of pore space may be accompanied by an increase in pore pressure. This causes a drop in the effective stress, shifting the Mohr circle in the direction of the tensional field (Figure 7b), thus promoting extensional microcracking. Consequently, a strain-softening phenomenon occurs during this phase.

Displacement Across Shear Fractures: The Third Stage of Fault Generation

With increasing displacement, the tip of a newly formed shear fracture propagates under mode II (in-plane shear or sliding) or mode III (anti-plane shear or tearing) propagation in which the displacements are in the plane of the crack (Atkinson, 1987). These shear fractures are subsequently exploited by shear displacement, leading to massive failure and the formation of a zone of cataclastic material having little or no cohesion. In this way, the breccia zones internally develop into fine-grained gouge zones containing fewer large clasts. The result is the typical breccia–gouge zones observed across individual faults (Figure 8a).

Although the cohesion has dropped considerably since the previous phase, there is still an increase in the coefficient of friction due to a decrease in sorting, more intimate grain interlocking, and lower porosity (Figure 8b). As the process continues, the proportion of fine-grained matrix increases and fluid flow (water plus silica) is concentrated into the dilatant zones, leading to syntectonic cementation. A further reduction in pore space forces the fluid to migrate outward from the zone of maximum deformation, resulting in a decrease in pore pressure. The outcome is an increase in the effective stress, moving the Mohr circle to the right (Figure 8b). All these factors together cause the rock to be strain-hardened during the development of the gouge zone.

Widening of the Gouge Zone

Although the width of the gouge zone appears to be roughly proportional to cycles of deformation (Scholz, 1987; Watterson et al., 1998), measurements of gouge thickness along individual faults in the field do not indicate a clear relationship between thickness and displacement (Engelder, 1974). Using a slightly different argu-

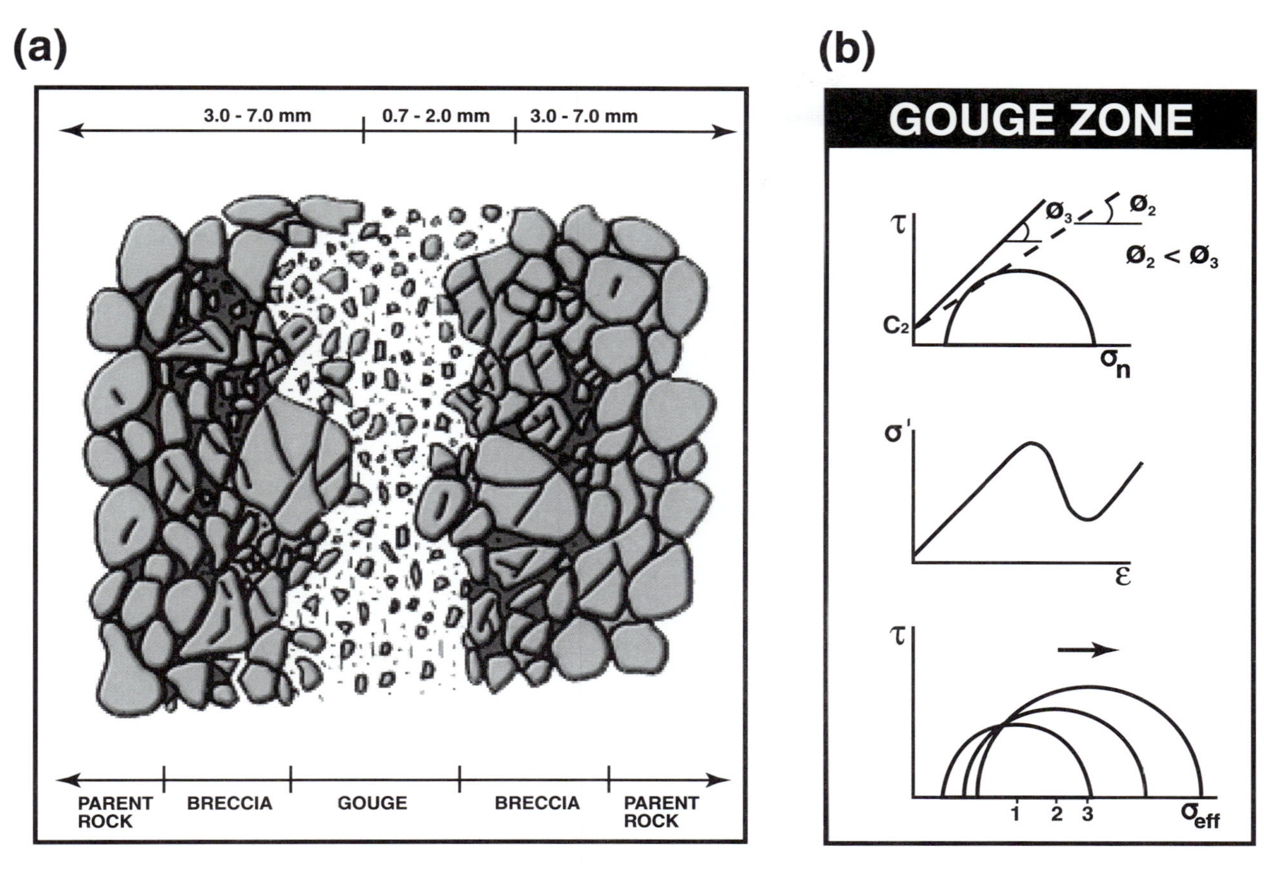

Figure 8—Final stage of evolution of a fault zone. (a) Final breccia–gouge texture of a microfault. (b) Failure criteria for the final evolution for the breccia–gouge texture. Graphs for stages of deformation (from top to bottom) are shear stress versus normal stress, differential stress versus strain, and shear stress versus effective stress.

ment, Lloyd and Knipe (1992) concluded that, during the initial fracturing phase in the developing of microbreccia–cataclasite zones in quartzites from northwest Scotland, there was no simple relationship between decreasing grain size and increasing displacement, although progressive comminution generally accompanied displacement.

One interesting aspect of the evolution of these gouge zones is the preservation of rounded grains within the comminuted zone (Figure 9). Similarly, Aydin (1978) observed that almost all large grains in faults in the Entrada and Navajo Sandstones were well-rounded quartz grains. According to Lloyd and Knipe (1992), such assimilated fragments probably represent examples where brittle intergranular fractures isolated original grains. Apparently, the preserved grain constitutes a large, hard particle in a weaker matrix. As the fault widens, these particles could rotate during the deformation because the asperities in the sliding walls would not interfere with one another or with existing grains in the gouge zone. Obviously, the rotation would favor rounded grains in such a way that the stronger grains would rotate and move along the weaker gouge zone.

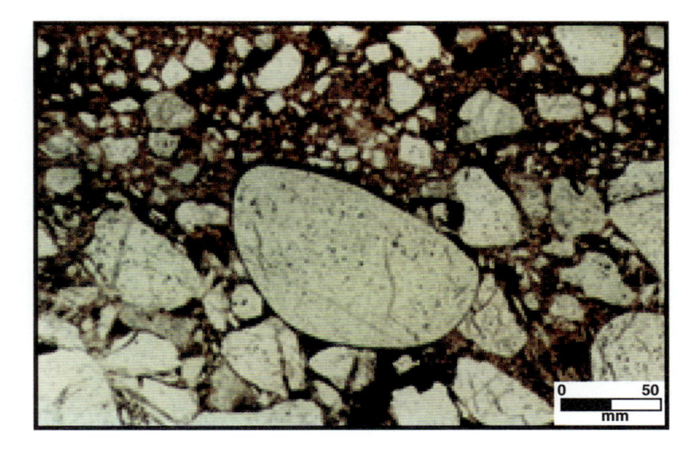

Figure 9—Optical thin-section showing different degrees of comminution in the breccia–gouge zone (uncrossed polars). The rounded grain is preserved probably because it constitutes a large, hard particle in a weaker matrix.

DIFFUSIVE MASS TRANSFER AND HEALING OF THE FAULTS

In this work, no detailed diagenetic analyses have been carried out to establish whether different phases of dissolution, diffusion, and precipitation of silica cements existed during the evolution of the sandstones. This precludes a more accurate interpretation on the healing processes acting along the faults, and thus some of the suggestions put forward at this stage are speculative. Nevertheless, preliminary inferences are possible from the standard petrographic analyses and by analogy with other regions.

General Observations

According to Roque (1990), two diagenetic phases of cementation can be recognized in the sandstones along the Recôncavo-Tucano-Jatobá rift, the first by silica and ferruginous cements and the second by a carbonate phase. Although these observations are probably valid for the general diagenetic evolution of the sandstones, in the present context, they should be regarded as an oversimplification of the diagenetic evolution of the fault zones. Here, diagenesis is expected to be more complex because of the localized tectonic stress, generating phases of dissolution and precipitation of silica.

Features of dissolution and precipitation appear to be significant across the breccia–gouge zones. Pressure solution and overgrowth features in the breccia zone and its external surroundings clearly indicate diffusive mass transfer (DMT) during the evolution of the sandstones. According to Knipe (1989), DMT is a deformation mechanism that promotes compaction and cementation through three processes: (1) solution of material at a source, (2) migration or diffusion of the material along some mass transfer pathway, and (3) precipitation of the material in sinks (Knipe, 1989; Lloyd and Knipe, 1992) or along free grain surfaces for crystal growth (James et al., 1986). In other words, the silica is transferred from a region of silica excess (exporter) to one deficient in silica (importer or sink), thus maintaining a balance in the silica budget. Because DMT commonly results in cementation, it can be considered as a microstructural strengthening process that causes strain hardening (Lloyd and Knipe, 1992).

According to Engelder (1987), in the absence of fluids, microcracks in quartz grains show no tendency to heal. However, when aqueous fluid is introduced, microcracks heal using the cement that is locally derived and transported by diffusion along the crack surface, at rates that depend on temperature, fluid pressure, chemistry, and crack dimensions. Fast healing rates (4 hours) have been found in small cracks during experiments at high temperatures (Brantley et al., 1990), whereas macrofractures transport most of the fluid volume but seal slowly. Where fractures are healed by the same material as the host grain (i.e., quartz), quasi-simultaneous DMT

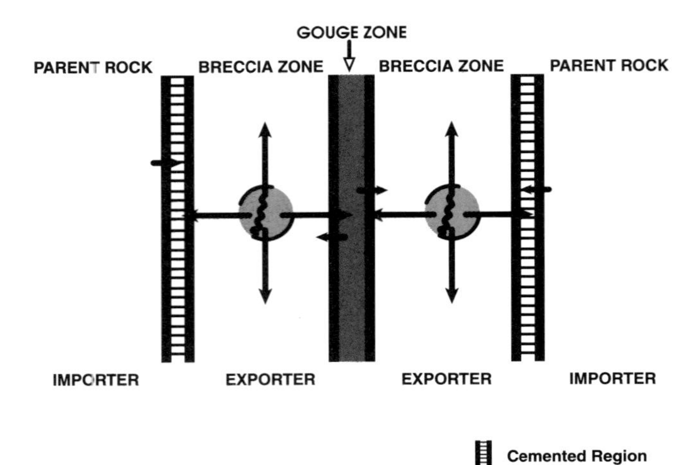

Figure 10—Silica budget across an individual fault. The breccia zone acts as a major silica exporter, whereas the gouge zone and the parent rock act as importers. Diffusive mass transfer (DMT) is expected to occur in all directions, but predominant directions are shown by longer arrows.

processes have been assumed (Blenkinsop and Rutter, 1986; Lloyd and Knipe, 1992).

Role of Breccia Zone as a Silica Exporter

Cataclasis along the breccia–gouge zone induces tectonic pressure solution due to closer and forced interaction between quartz grains, thus generating a region for silica exportation. The preferred orientation of pressure solution contacts parallel or subparallel to the most deformed zones corroborates this assumption (Magnavita, 1992). In this context, because of its relatively higher volume compared to the gouge zone, the breccia zone is considered to be the most important source of silica in the region of an individual microfault.

Size reduction also promotes a great amount of pressure solution (Houseknecht, 1988); however, fine-grained and undulose quartz inhibits silica overgrowths (James et al., 1986). Because grain size is reduced in the breccia zone, it is more likely that cementation takes place around the neighboring undeformed parent rock. Indeed, a higher degree of cementation resulting in tighter packing is sometimes seen along the zone between the breccia zone and the parent rock (Magnavita, 1992). Alternatively, it might be argued that the cementation could have occurred as a consequence of posttectonic migration of fluids along the region of contact between the breccia zone and the parent rock; this is still an open matter.

It is suggested that, during the reciprocal diffusion process, the breccia zone acted as a major silica exporter (Figure 10). Silica diffuses and precipitates as quartz over-

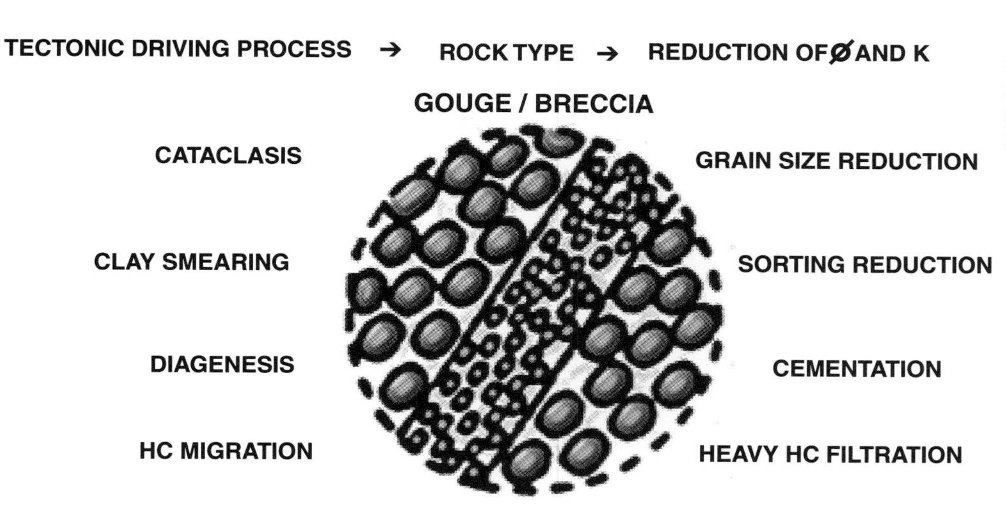

Figure 11—Main tectonic driving processes, rock types, and resulting texture in the generation of a potential membrane seal.

growths along the border of the breccia zone with the parent rock, and dilatant zones are also cemented in a quasi-simultaneous healing process. The cementation was probably important in the gouge zone, but only more detailed techniques such as cathodoluminescence and electron microscopy can clarify this aspect of faulting evolution. Also, cyclic phases of dissolution and precipitation of silica might be an important mechanism that causes strain hardening of the fault zones.

SEALING POTENTIAL

Because porosity and permeability are drastically reduced in the gouge and breccia zones, these zones act as barriers for any posttectonic fluid percolation. The faults develop a potential membrane seal with entry pressures that may be high enough to maintain a significant hydrocarbon column height. This capillarity given by small pore aperture size, a property that determines the trapping potential of a rock, is drastically reduced as a consequence of the bimodal distribution of grain sizes in the breccia–gouge zone. Because the abundance of faults and microfaults is normally associated with major faults, some fault zones may act as seals even when sandstone is placed against sandstone by faulting. In this regard, the main factors to be considered in the development of a membrane seal are cataclasis, shale smear, diagenesis, and hydrocarbon migration (Figure 11).

Membrane Potential Sealing

The potential sealing of the membrane is driven by tectonic processes (faulting), which generate a fault-related rock (gouge or breccia), resulting in reduction of controlling fluid flow parameters (porosity and permeability). The membrane acts as an entry pressure barrier that can retard or shut off oil production, or even, in some cases, trap hydrocarbons on geologic time scales. A sealing fault in the strict sense would have a membrane seal along the fault surface.

Cataclasis is the most common tectonic driving process in producing membrane seal in sandstones, in the case of sand-on-sand contacts. The resulting gouge and breccia rock types present a reduction both in grain size and sorting. The seal is not considered to be effective on geologic time scales (Hippler, 1997). This cataclasis generates microfaults with very complex and highly unpredictable geometries, creating a problem with connectivity and continuity of the sealing membrane.

Shale smear is the most effective process in forming a membrane because of the high values of displacement pressure for shale compared to that for sandstones (Schowalter, 1979). In this process, shale filaments are injected by shearing into the fault because of the high ductility of the material, resulting in a drastic reduction in sorting, creating a narrow zone of nonreservoir shale gouge within the fault zone. This process is favored by small throw because the shale can preserve its integrity and continuity.

Tectonically driven diagenesis occurs due to grain interaction during fault movement. An increase in tectonic stress localized along the contacts between quartz grains dissolves silica by pressure solution. This silica is transferred to the interstitial space, causing cementation. This is generally a heterogeneous process that is difficult to characterize and predict.

Hydrocarbon migration is another process that can increase the membrane sealing potential during secondary migration along the fault zone. Oil is driven into the fault during a dilation phase, and the heavier portions are left behind within the shear fractures. This is also a very particular and unpredictable process.

Entrapment Potential

The prediction of trap capacity for faulted traps must consider both the seal capacity of the rocks above and below the reservoir and the seal capacity of rocks laterally equivalent to the reservoir (Schowalter, 1979). However, for a seal to be truly effective, it also must be relatively homogeneous and unfractured (Downey, 1984).

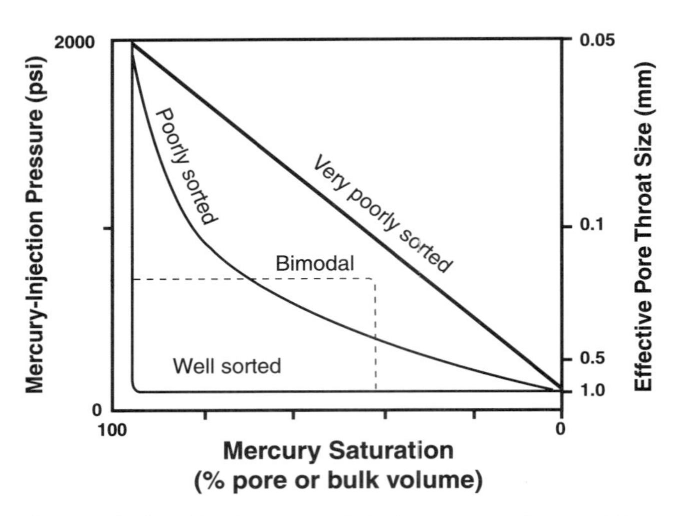

Figure 12—Idealized mercury injection curve shapes. All curves have identical displacement pressures and minimum unsaturated pore volumes. The extreme differences in saturation profiles are due to differences in pore throat size distributions. From Vavra et al. (1992).

The trapping potential of a rock is determined by the pore aperture size (or throat), a property that is drastically reduced as a consequence of the bimodal distribution of grain sizes in the breccia–gouge zone. The capillary pressures between oil and water in rock pores are responsible for trapping oil (Smith, 1966, 1980; Berg, 1975). In practice, the sealing capacity of a rock type is determined by the relative displacement pressure, which is the pressure required to force a connected hydrocarbon filament into the largest interconnected pore of a preferentially water-wet rock (Schowalter, 1979). These measurements are made through the establishment of mercury injection curves, which simulate pore size distribution and reflect the largest throat through which the pore volume can be accessed. In Figure 12, three theoretical mercury injection curves are offered. Samples dominated by throats of similar size (i.e., well sorted) have broad, flat plateaus. The plateau is steeper for more poorly sorted samples and is absent (the slope approaches 45°) for samples with very poorly sorted throats.

Comparison of curves of mercury injection (Pittman, 1981, 1992; Vavra et al., 1992) shows that crushed sandstones have significantly poorer reservoir qualities than undisturbed samples, thus creating the potential for fault-sealing traps. Data from mercury injection curves can be used to estimate the distribution of pore volume accessible by throats of a given effective size using the equation

$$r_c = 2\sigma \cos \theta / P_c \qquad (4)$$

where r_c is the radius of the capillary tube (in cm), σ is the interfacial tension of the air/mercury system, θ is the air/mercury/solid contact angle (the wettability term), and P_c is the capillary pressure (in dynes/cm^2).

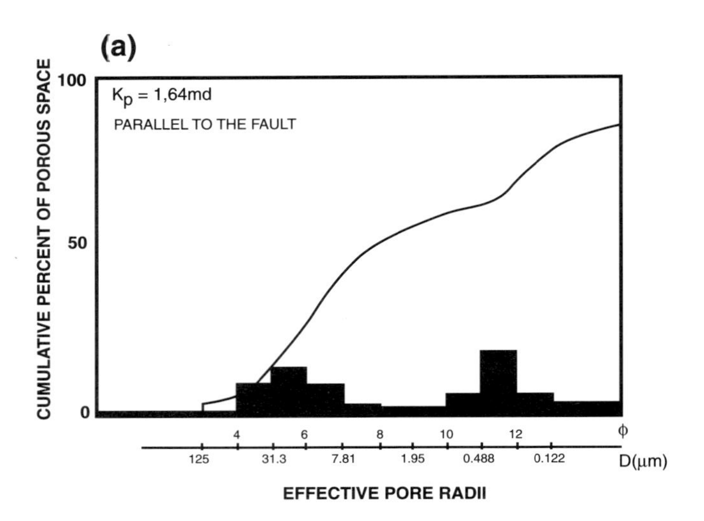

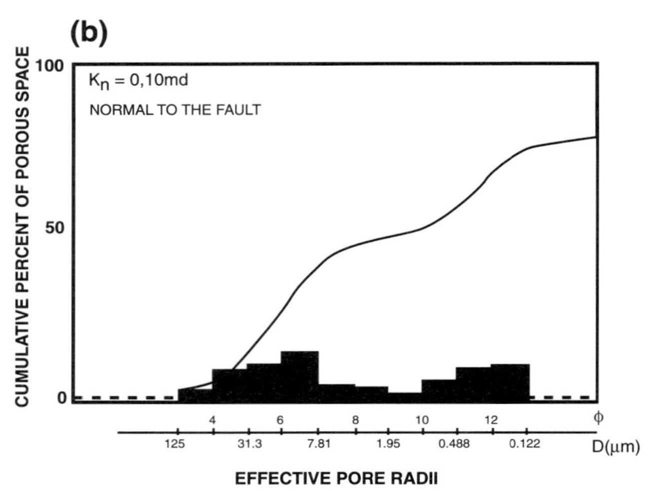

Figure 13—Capillary data taken (a) parallel and (b) normal to a microfault. See text for discussion.

The reduction in pore throat (or pore aperture) in a fault zone of the Recôncavo-Tucano-Jatobá rift is highlighted in Figure 13. Because of the core plug (2 cm in diameter) is much wider than the microfault (a few millimeters), permeability and porosity represent an average of parent rock and microfault properties. However, the resulting mercury injection curve indicates a bimodal distribution of pore apertures. This is explained by the two different pore textures sampled in both normal and parallel injection of mercury. During the beginning of the experiment, the nonwetting phase (mercury) penetrates the largest pores of the parent rock (8–125 μm, mode about 30 μm). As mercury saturation progressively increases, thus reaching higher entry pressures, smaller pore apertures are wetted by mercury, and another mode is established with many very small pores of the same size in the deformed breccia–gouge zone. These pore diameters are in the range of 0.12–0.50 μm, a reduction of at least two orders of magnitude, which results in differences in capillary pressure of the same magnitude (see equation 4), which could constitute a difference capable of trapping hydrocarbons. The similarity of pore size aperture is shown by the two plateaus in the curve (Figure 13b).

ROLE OF THE MATA–CATU TRANSFER ZONE IN FOCUSING SECONDARY MIGRATION

The Mata–Catu transfer zone has produced the most prolific trend in the Recôncavo Basin (Figure 14). The fault is confined to the Recôncavo Basin, terminating abruptly against the Aporá High to the northwest and diffusely near the Salvador High to the southeast. The fault surface flips along strike, with the northwestern segment dipping to the northeast and the southeastern part dipping to the southwest. This results in a hinge or pivotal geometry, like a pair of scissors, with a null (or piercing) point where there is virtually no vertical movement. The faults in the area show two preferential directions: one parallel to the main trend (N 30°–40° W) and another orthogonal to it (N 30°–40° E). This combination has created the faulted traps with faults anchored in the basement that typify the trend.

The Mata–Catu fault evolved during two tectonic phases, which was determined after it was recognized that the structures mapped near the top of prerift strata were not entirely reflected on maps at younger stratigraphic levels (Ghignone and Andrade, 1970). According to Magnavita (1992), some faults formed during the rifting climax (Valanginian), whereas others were generated after the deposition of the youngest synrift sequences in the Barremian–Aptian, when some of the existing faults were also reactivated.

In outcrop, the fault zone is typified by several sets of faults and microfaults, creating topographic highs as a consequence of more resistant deformed sandstones. Although direct data such as cores are not available for this work, the deformation must persist at depth. Our previous discussion indicates that the fault zone would act as a barrier for fluid migration across it and as a focus for migration along it, allowing migration up to the distant fields located to the west.

Oil Fields Along the Mata–Catu Trend

The Mata–Catu trend connects the depocenters of the Recôncavo Basin with distant areas close to the flexural margin of the half-graben. The oil kitchen coincides with the main lows (Camaçari and Miranga lows), and the trend is in a key position to allow long-distance oil migration, thus explaining some oil accumulation in fields at great distances from the oil kitchen (~40 km).

The fields along the trend produce almost only from prerift reservoirs, which are filled by oil generated from synrift source rocks. This created a situation in which the reservoirs lie stratigraphically below the source rock. In this way, the petroleum system must rely on faults to exist. The large majority of production comes from fluvial and eolian sandstones of the Sergi and Água Grande Formations.

Migration Pathways Along the Trend

To investigate hydrocarbon migration, a cross section was drawn parallel to an inferred migration pathway passing through the Remanso, Água Grande, and Buracica oil fields (Figure 15). The section crosses the null point where fault throw equals zero. All traps are characterized by tilted blocks that appear as horsts in cross sections, with faults anchored in the basement. Along the trend, some faults must have leaked and others must have sealed, and the migration path suggests vertical migration along faults and lateral migration along carrier beds.

According to Penteado et al. (1998), the prerift accumulations are characterized by downward migration of petroleum from the source rocks to the underlying carrier beds followed by updip secondary migration, with normal faults acting as conduits. In the present case, the Sergi and Água Grande Formations worked as one carrier bed system each time the oil migrated through them.

From the cross section in Figure 15 and the map in Figure 16a, the following points can be concluded:

1. The reservoirs of the Remanso field were filled directly from the source rocks in the depocenter.
2. The west fault of the Remanso field (F1 in Figure 15) acted as a sealing fault (juxtaposition seal), thus impeding migration toward the Água Grande field along this pathway.
3. The location of the Água Grande field indicates petroleum migration from the source rock into the prerift reservoirs because the oil kitchen is still in contact with the oil accumulation.
4. The fault that limits the Água Grande field to the northwest (F2 in Figure 15) is a sealing fault, trapping the oil in the field by juxtaposition and not allowing migration toward the Buracica field along this pathway.

The Buracica field is at a great distance from the oil kitchen (~40 km). However, there are no physical limits to the distance oil or gas can migrate laterally or vertically in a given geologic situation (Schowalter, 1979). They will migrate as long as the buoyant force of the hydrocarbon column is greater than the resistant force of the carrier bed.

The Buracica eastern fault (F3 in Figure 15) may have acted as a conduit and as a seal during its history. Either it was active at the time hydrocarbons reached the distant area in the basin or it leaked due to vertical migration to fill the reservoir during secondary migration. Thus,

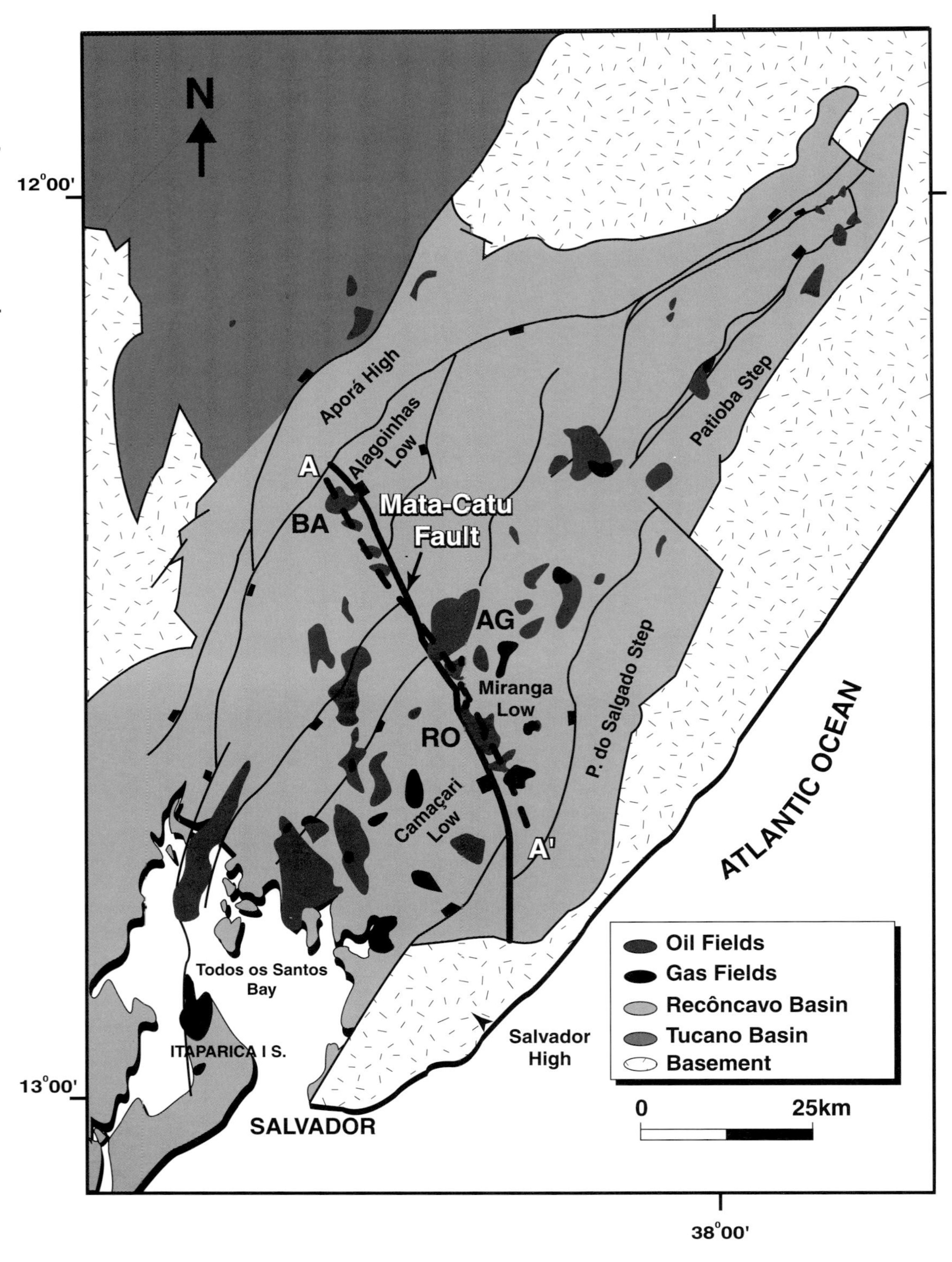

Figure 14—Map of the structural framework of the Recôncavo Basin with main oil and gas fields shown. Cross section A–A' is shown in Figure 15. Abbreviations of oil fields: RO = Remanso, AG = Água Grande, and BA = Buracica. Adapted from M. A. F. Aragão (unpublished Petrobrás data, 1995).

hydrocarbons would have entered the fault–reservoir system via the hanging wall of F3, risen up toward the fault due to buoyancy, and accumulated at the reservoir–fault interface. This would have increased the buoyant force of the growing hydrocarbon column, ultimately exceeding the capillary entry pressure of the fault zone (fault membrane), whereupon hydrocarbons would have begun to enter the fault and migrate upward.

Because of its complex movement history, amount of throw, and intensity of deformation, the Mata–Catu fault would have acted as a seal for cross migration via both juxtaposition and membrane seal mechanisms. In this

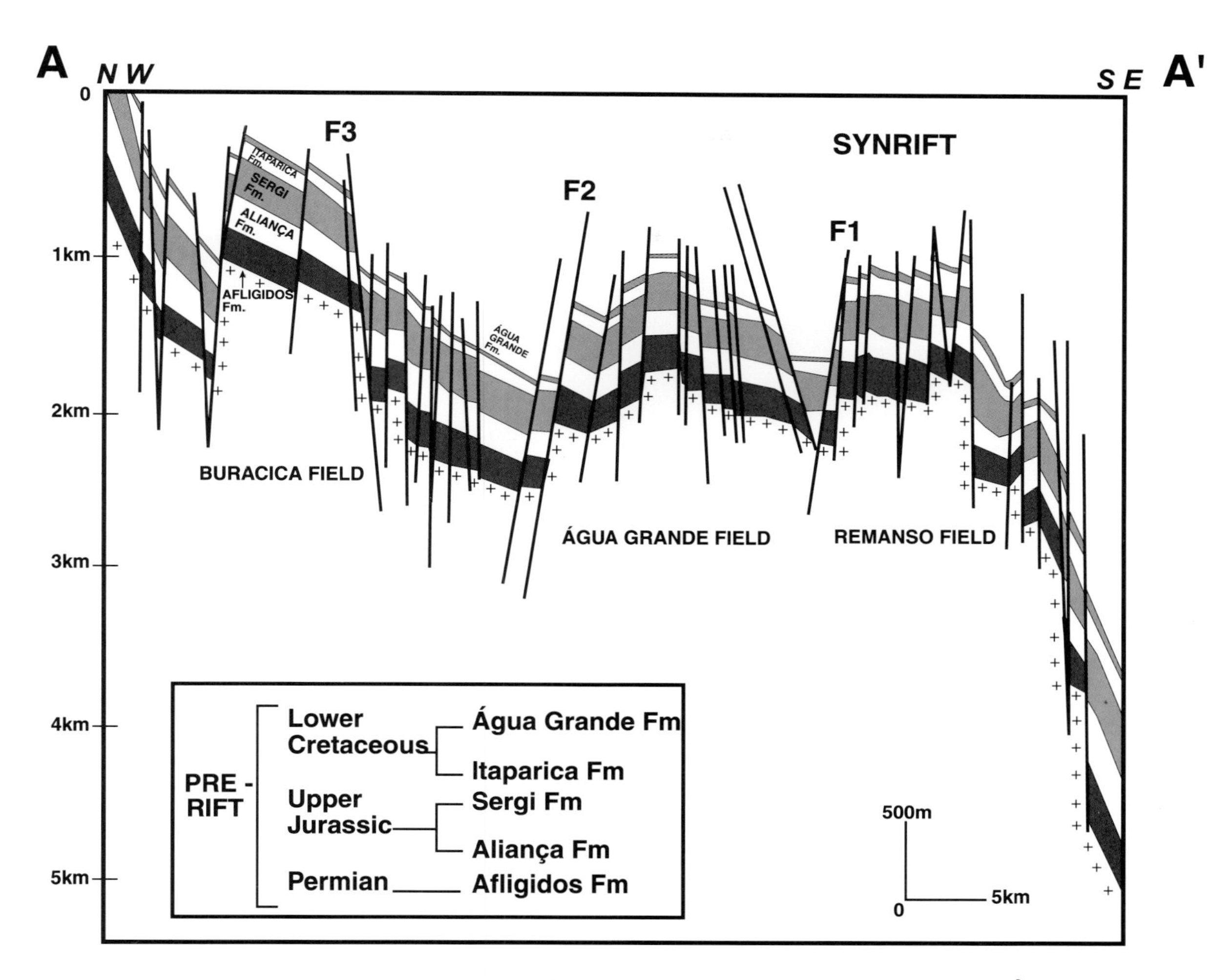

Figure 15—Cross section A–A' along the Mata–Catu trend, which crosses the null point between the Água Grande and Buracica fields. See Figure 14 for location.

way, it focused the oil to migrate parallel to its southeastern side, up to the null point, reaching the east fault of the Buracica field, which would leak during vertical migration, allowing hydrocarbon to be trapped in the reservoirs. The northeastern faults, either because of their extensional history or their lesser amount of throw, would have acted as conduits for lateral migration in cases where reservoir is juxtaposed against reservoir without full development of a membrane seal.

Maturity Versus Migration Distance Across and Along the Mata–Catu Fault

Carbon isotopic data obtained in the wells along and across the Mata–Catu fault (Tikae et al., 1997; Prinzhofer et al., 2000) indicate great isotopic variations, suggesting not only thermal evolution but also influence from isotopic fractionation processes (Prinzhofer et al., 2000). Such processes are important because they allow assumptions about migration distance, maturity, and location of the source rock pod (Prinzhofer and Huc, 1995; Prinzhofer et al., 2000). Figure 16b shows a plot of a principal component analysis (V_1/V_2 or maturity/migration diagram) (Prinzhofer and Huc, 1995; Prinzhofer et al., 1999) on which it is possible to distinguish two contrasting data trends between gases pooled in reservoirs located along and perpendicular to the Mata–Catu fault. The gases recovered from fields located along the trend clearly show great variation in maturity and migration values. This can be explained geologically because the hydrocarbon kitchen that feeds the fault in this area is located deep in the southeastern Camaçari low, corroborating a southeast–northwest secondary migration pathway. Also, the curvature of the migration line (Figure 16b), which represents the fields along the fault, changes abruptly, indicating a major contrast between the fields closer to the oil kitchen (southeast) and those more distant from it (northwest). The maturity also changes along strike.

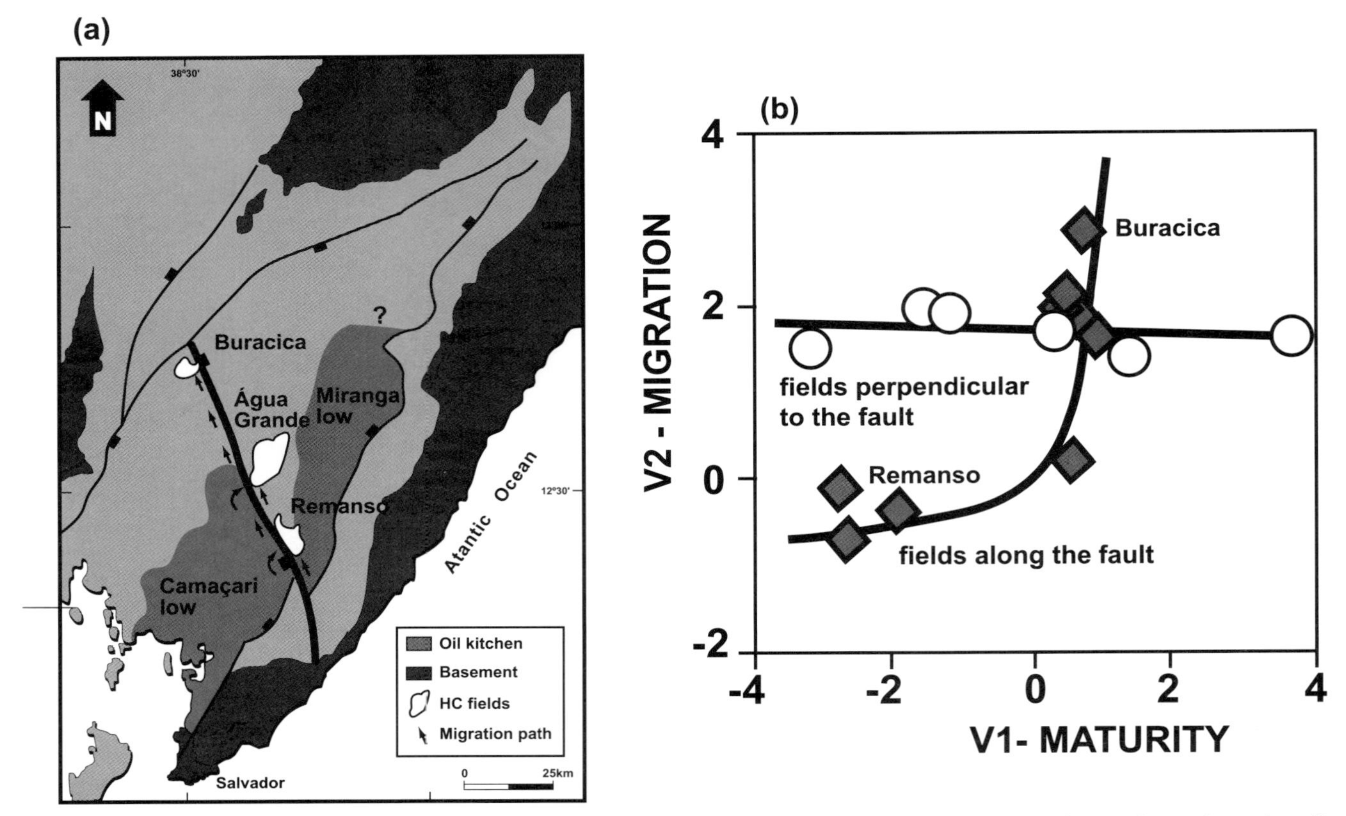

Figure 16—Migration pathway along the Mata–Catu trend. (a) Structural map showing the migration pathway from the oil kitchen in the Camaçari low to the Buracica field. (b) Graph of maturity (V_1) versus migration distance (V_2) along and across the Mata–Catu trend. Adapted from Prinzhofer and Huc (1995). See text for discussion.

In contrast, the gases recovered perpendicular to the Mata–Catu trend exhibit a great variation only in maturity, with negligible migration effects. Fields located in the Camaçari and Miranga lows (orthogonal to the trend) are along a straight line with respect to the distance of migration, although maturity increases toward the Miranga low. This is geologically validated by the fact that several hydrocarbon kitchens are present, buried at different depths in the area. In these cases, the pods are located just below the reservoirs, with very short distances of migration.

CONCLUSIONS

We can draw the following conclusions:

1. The fault zones in the Recôncavo-Tucano-Jatobá rift occur as elongated crests of grouped strands of granulated sandstone. Each strand corresponds to an individual fault or microfault.
2. A multifractal analysis was able to discriminate different fault arrangements and can be used to contrast different fault sets formed under various tectonic regimes.
3. Sandstones across individual faults are texturally characterized by an internal zone of gouge bordered by two roughly symmetric zones of breccia, leading to a characteristic breccia–gouge zone geometry. The deformation history for the evolution of the faults probably involved an earlier phase of fracture followed by diffusive mass transfer (DMT) mechanisms.
4. The fracture processes probably occurred through intragranular fracturing of the grains, controlled by previous heterogeneities in the rock. Progressive deformation made possible the linkage of these earlier formed extensional microcracks through a wide zone of deformation. Further shear resulted in displacement along the through-going surfaces of discontinuity, thus forming faults.
5. DMT of silica must have been an important mechanism in healing newly formed microfractures in a quasi-simultaneous process. During the deformation, it is likely that the breccia zone acted as a silica exporter for both the gouge zone and the parent rock, which would have acted as importers to allow a balance in the silica budget.
6. Analysis of failure criteria suggests that, initially, deformation occurred as a dilational phase that formed the breccia zones through strain-softening processes, followed by grain reduction which

enhances high pore pressure. As deformation proceeded and grain size and porosity reduced, the fault strain hardened during development of the gouge zone; this was enhanced by healing of opening microcracks.

7. The main factors to be considered in the development of a membrane seal are cataclasis, shale smear, diagenesis, and hydrocarbon migration.
8. The potential sealing of the membrane created along the deformed sandstone is driven by tectonic processes (faulting), which generate a fault-related rock (gouge or breccia), resulting in reduction of controlling fluid flow parameters (porosity and permeability).
9. Mercury injection curves indicate reductions in pore throat width of at least two orders of magnitude between the undisturbed sandstone and the deformed rock. This suggests that the membrane would be able to trap hydrocarbons because pore throat size (controlling capillary entry pressure) is the fundamental parameter in controlling whether a fault acts as a seal to hydrocarbon migration over geologic time scales.
10. The Mata–Catu trend played a key role in trapping hydrocarbons and connecting the depocenters of the Recôncavo Basin with distant areas close to the flexural margin of the half-graben, focusing long-distance secondary migration along the Sergi–Água Grande carrier beds and leaking faults.
11. Isotopic analysis of gases along and across the trend corroborates that the trend is in a key position to allow long-distance oil migration, thus explaining some oil accumulation in fields at great distances from the oil kitchen.

Acknowledgments*—I am highly indebted to Marcio R. Mello from Petrobrás for his help with the isotopic analyses and for being the stimulus throughout this work. Nivaldo Destro is acknowledged for his comments on a first version of this chapter. The manuscript has improved significantly from discussions with Ben Clennell of the University of Bahia. Reviews by P. Szatimari and P. O. Yilmaz contributed to the final version. Ricardo Marins and Miriam L. S. Mercês are thanked for the drawings. Petrobrás is acknowledged for allowing the publication of this chapter.*

REFERENCES CITED

Allan, U. S., 1989, Model for hydrocarbon migration and entrapment within faulted structures: AAPG Bulletin, v. 73, p. 803–811.

Alexander, L. L., and J. W. Handschy, 1998, Fluid flow in a faulted reservoir system: fault trap analysis for the Block 330 field in Eugene Island, South Addition, offshore Louisiana: AAPG Bulletin, v. 82, n. 3, p. 387–411.

Antonellini, M., and A. Aydin, 1994, Effect of faulting on fluid flow in porous sandstones: petrophysical properties: AAPG Bulletin, v. 78, n. 3, p. 355–377.

Atkinson, B. K., 1987, Introduction to fracture mechanics and its geophysical applications, *in* B. K. Atkinson, ed., Fracture mechanics of rock: London, Academic Press, p. 1–26.

Aydin, A., 1978, Small faults formed as deformation bands in sandstone: Pure and Applied Geophysics, v. 116, p. 913–930.

Aydin, A., and A. M. Johnson, 1978, Development of faults as zones of deformation bands and as slip surfaces in sandstone: Pure and Applied Geophysics, v. 116, p. 931–942.

Aydin, A., and A. M. Johnson, 1983, Analysis of faulting in porous sandstones: Journal of Structural Geology, v. 5, n. 1, p. 19–31.

Aydin, A., and Z. Reches, 1982, Number and orientation of fault sets in the field and in experiments: Geology, v. 10, p. 107–112.

Belfield, W. C., 1994, Multifractal characteristics of natural fracture apertures: Geophysical Research Letters, v. 21, n. 24, p. 2641–2644.

Berg, R. R., 1975, Capillary pressures in stratigraphic traps: AAPG Bulletin, v. 59, n. 6, p. 939–956.

Berg, R. R., and A. H. Avery, 1995, Sealing properties of Tertiary growth faults, Texas Gulf Coast: AAPG Bulletin, v. 70, n. 3, p. 375–393.

Blenkinsop, T. G., and E. H. Rutter, 1986, Cataclastic deformation of quartzite in the Moine thrust zone: Journal of Structural Geology, v. 8, n. 6, p. 669–681.

Bouvier, J. D., C. H. Kaars-Sijpesteijn, D. F. Kluesner, C. C. Onyejekwe, and R. C. Van der Pal, 1989, Three-dimensional seismic interpretation and fault sealing investigations, Nun River field, Nigeria: AAPG Bulletin, v. 73, n. 11, p. 1397–1414.

Brantley, S. L., B. Evans, S. H. Hickman, and D. A. Crerar, 1990, Healing of microcracks in quartz: implications for fluid flow: Geology, v. 18, p. 136–139.

Bretan, P. G., A. Nicol, J. J. Walsh, and J. Watterson, 1996, Origin of some conjugate or "X" fault structures: The Leading Edge, July, p. 812–816.

Chang, H. K., R. O. Kowsmann, A. M. F. Figueiredo, and A. A. Bender, 1992, Tectonics and stratigraphy of the East Brazil Rift: an overview: Tectonophysics, v. 213, p. 97–138.

Downey, M. W., 1984, Evaluating seals for hydrocarbon accumulations: AAPG Bulletin, v. 68, n. 11, p. 1752–1763.

Engelder, T., 1974, Cataclasis and the generation of fault gouge: GSA Bulletin, v. 85, p. 1515–1522.

Engelder, T., 1987, Joints and shear fractures in rock, *in* B. K. Atkinson, ed., Fracture mechanics of rock: London, Academic Press, p. 27–69.

Figueiredo, A. M. F., J. A. E. Braga, H. M. C. Zabalaga, J. J. Oliveira, G. A. Aguiar, O. B. Silva, L. F. Mato, L. M. F. Daniel, L. P. Magnavita, and C. H. L. Bruhn, 1994, Recôncavo Basin: a prolific intracontinental rift basin, *in* S. M. Landon, ed., Interior rift basins: AAPG Memoir 59, p. 157–203.

Fowles, J., and S. Burley, 1994, Textural and permeability characteristics of faulted, high porosity sandstones: Marine and Petroleum Geology, v. 11, n. 5, p. 608–629.

Ghignone, J. I., and G. de Andrade, 1970, General geology and major oil fields of the Recôncavo basin, Brazil, *in* M. T. Halbouty, ed., Geology of giant fields: AAPG Memoir 14, p. 337–358.

Gibson, R. G., 1994, Fault zone seals in siliciclastic strata of the Columbus basin, offshore Trinidad: AAPG Bulletin, v. 78, n. 9, p. 1372–1385.

Hippler, S. J., 1993, Deformation microstructures and diagenesis in sandstone adjacent to an extensional fault: implications for the flow and entrapment of hydrocarbons: AAPG Bulletin, v. 77, n. 4, p. 625–637.

Hippler, S. J., 1997, Microstructures and diagenesis in North Sea fault zones: implications for fault-seal potential and fault-migration rates, *in* R. C. Surdam, ed., Seals, traps, and the petroleum system: AAPG Memoir 67, p. 85–101.

Houseknecht, D. W., 1988, Intergranular pressure solution in four quartzose sandstones: Journal of Sedimentary Petrology, v. 58, n. 2, p. 228–246.

James, W.C., G. C. Wilmar, and B. G. Davidson, 1986, Role of quartz type and grain size in silica diagenesis, Nugget Sandstone, south-central Wyoming: Journal of Sedimentary Petrology, v. 56, n. 5, p. 657–662.

Jamison, W. R., 1989, Fault-fracture strain in Wingate Sandstone: Journal of Structural Geology, v. 11, n. 8, p. 959–974.

Jamison, W. R., and D. W. Stearns, 1982, Tectonic deformation of Wingate Sandstone, Colorado National Monument: AAPG Bulletin, v. 66, n. 12, p. 2584–2608.

Jones, G., and R. J. Knipe, 1996, Seismic attribute maps; application to structural interpretation and fault seal analysis in the North Sea basin: First Break, v. 14, n. 12, p. 449–461.

Knipe, R. J., 1989, Deformation mechanisms—recognition from natural tectonites: Journal of Structural Geology, v. 11, n. 1/2, p. 127–146.

Knipe, R. J., 1997, Juxtaposition and seal diagrams to help analyze fault seals in hydrocarbon reservoirs: AAPG Bulletin, v. 81, n. 2, p. 187–195.

Knott, S. D., 1993, Fault seal analysis in the North Sea: AAPG Bulletin, v. 77, n. 5, p. 778–792.

Kulander, B. R., S. L. Dean, and B. J. Ward, Jr., 1990, Fractured core analysis: interpretation, logging, and use of natural and induced fractures in core: AAPG Methods in Exploration Series n. 8, 88 p.

Lloyd, G. E., and R. J., Knipe, 1992, Deformation mechanisms accommodating faulting of quartzite under upper crustal conditions: Journal of Structural Geology, v. 14, n. 2, p. 127–143.

Magnavita, L. P., 1992, Geometry and kinematics of the Recôncavo-Tucano-Jatobá rift, NE Brazil: Ph.D. dissertation, University of Oxford, Oxford, U.K., 493 p.

Magnavita, L. P., 1997, Deformational mechanisms in porous sandstones: implications for the development of fault seal in the Recôncavo-Tucano-Jatobá rift (abs.): AAPG/ABGP Hedberg Research Symposium, Program with Abstracts, Rio de Janeiro, Brazil.

Magnavita, L. P., and M. S. Souza, 1995, Caracterização de sistemas de fraturas através de análise multifractal: V Simpósio Nacional de Estudos Tectônicos, Gramado, p. 414–415.

Mandelbrot, B. B., 1967, How long is the coast of Britain? Statistical self-similarity and fractional dimension: Science, v. 156, p. 636–638.

Nelson, R. A., 1985, Geologic analysis of naturally fractured reservoirs: Petroleum Geology Engineering, Gulf Publishing Company, v. 1, 320 p.

Nicol, A., J. J. Walsh, J. Watterson, and P. A. Gillespie, 1996, Fault size distributions: are they really power law?: Journal of Structural Geology, v. 18, n. 2/3, p. 191–197.

Penteado, H. de B., J-L. Rudkiewicz, and F. Behar, 1998, 2-D compositional modeling in a typical rift setting: the Recôncavo basin in northeastern Brazil (abs.): AAPG Annual Convention, Program with Abstracts, Salt Lake City, Utah, A520.

Pittman, E. D., 1981, Effect of fault-related granulation on porosity and permeability of quartz sandstones, Simpson Group (Ordovician), Oklahoma: AAPG Bulletin, v. 65, n. 11, p. 2381–2387.

Pittman, E. D., 1992, Relationship of porosity and permeability to various parameters derived from mercury injection–capillary pressure curves for sandstone: AAPG Bulletin, v. 76, n. 2, p. 191–198.

Prinzhofer, A., and A. Y. Huc, 1995, Genetic and post-genetic molecular and isotopic fractionations in natural gases: Chemical Geology, v. 126, n. 3/4, p. 281–290.

Prinzhofer, A., M. R. Mello, and T. Takaki, 2000, Geochemical characterization of natural gas: a physical multivariable approach and its application in maturity and migration estimates: AAPG Bulletin, v. 84, n. 8, p. 1152–1172.

Roque, N. C., 1990, Análise estrutural das falhas ocorrentes nas sub-bacias do Tucano Sul e Central–Bahia: Master's thesis, Universidade Federal de Ouro Preto, 121 p.

Scholz, C. H., 1987, Wear and gouge formation in brittle faulting: Geology, v. 15, p. 493–495.

Scholz, C. H. 1990, The mechanics of earthquakes and faulting: Cambridge, Cambridge University Press, 439 p.

Schowalter, T. T., 1979, Mechanics of secondary hydrocarbon migration and entrapment: AAPG Bulletin, v. 63, n. 5, p. 723–760.

Smith, D. A., 1966, Theoretical considerations of sealing and nonsealing faults: AAPG Bulletin, v. 50, n. 2, p. 363–374.

Smith, D. A., 1980, Sealing and nonsealing faults in Louisiana Gulf Coast salt basin: AAPG Bulletin, v. 64, n. 2, p. 145–172.

Tikae, T., M. R. Mello, E. Santos Neto, A. Prinzhofer, and M. H. Hessel, 1997, Evoluçâo térmica das rochas geradoras das bacias sedimentares brasileiras por meio de isótopos estáveis de carbono em hidrocarbonetos gasosos: Petrobras Internal Report, 65 p.

Underhill, J. R., and N. H. Woodcock, 1987, Faulting mechanisms in high-porosity sandstones; New Red Sandstone, Arran, Scotland, *in* M. E. Jones and R. M. Preston, eds., Deformation of sediments and sedimentary rocks: Geological Society of London, Special Publication 29, p. 91–105.

Vavra, C. L., J. G. Kaldi, and R. M. Sneider, 1992, Geological applications of capillary pressure: a review: AAPG Bulletin, v. 76, n. 6, p. 840–850.

Watterson, J., C. Childs, and J. J. Walsh, 1998, Widening of fault zones by erosion of asperities formed by bed-parallel slip: Geology, v. 26, n. 1, p. 71–74.

Weber, K. J., G. Mandl, W. F. Pilaar, F. Lehner, and R. G. Precious, 1978, The role of faults in hydrocarbon migration and trapping in Nigerian growth fault structures: SPE/AIME Tenth Annual Offshore Technology Conference Proceedings, v. 4, p. 2643–2653.

Wise, D. U., D. E. Dunn, J. T. Engelder, P. A. Geiser, R. D. Hatcher, S. A. Kish, A. L. Odom, and S. Schamel, 1984, Fault-related rocks: suggestions for terminology: Geology, v. 12, p. 391–394.

Zhao, G., and A. M. Johnson, 1991, Sequential and incremental formation of conjugate sets of faults: Journal of Structural Geology, v. 13, n. 8, p. 887–895.

Haack, R. C., P. Sundararaman, J. O. Diedjomahor, H. Xiao, N. J. Gant, E. D. May, and K. Kelsch, 2000, Niger Delta petroleum systems, Nigeria, *in* M. R. Mello and B. J. Katz, eds., Petroleum systems of South Atlantic margins: AAPG Memoir 73, p. 213–231.

Chapter 16

Niger Delta Petroleum Systems, Nigeria

Richard C. Haack

P. Sundararaman

Jacob O. Diedjomahor

Hongbin Xiao

Nicholas J. Gant

Chevron Petroleum Technology Co.
San Ramon, California, U.S.A.

Eric D. May

Chevron Overseas Petroleum Inc.
San Ramon, California, U.S.A.

Ken Kelsch

P.T. Caltex Pacific Indonesia
Rumbai, Indonesia

Abstract

Regional integration of results from conventional exploration geochemistry, structural analysis, and gravity–magnetic data provide a comprehensive new understanding of Niger Delta petroleum systems. Nigeria is the 12th largest producer of crude oil in the world. Daily oil production from the Niger Delta is 2.1 million bbl, and recoverable reserves are estimated to be about 22.5 billion bbl. Historically, structural play types have dominated, although large stratigraphic traps have also been discovered. The basin has matured through one cycle of successful exploration, and future success depends on linking the geology of the shelf and onshore areas to deep-water areas and exploiting new play types in older producing areas.

Three petroleum systems are present in the Niger Delta and delta frame: Lower Cretaceous (lacustrine), Upper Cretaceous–lower Paleocene (marine), and Tertiary (deltaic). One biodegraded seep oil from Nigerian tar sands along the northern flank of the Dahomey Embayment has been correlated to Neocomian source rocks in Ise-2 well. A source rock extract and pyrolyzate of the seep are similar to the Bucomazi petroleum system in the Lower Congo Basin. Oil recovered from Paleogene sandstones in Shango-1 well are inferred to be derived from Upper Cretaceous–lower Paleocene source rocks identified in Epiya-1 well, consisting of type II and II–III kerogens. The principal source for oil and gas in the Niger Delta is the Tertiary deltaic petroleum system, consisting of type II, II–III, and III kerogens. On the basis of oils and source rocks, source facies variation characteristic of this system has been regionally mapped in the northwestern part of the delta. Similar trends exist delta-wide and are responsible, along with burial, for controlling the complex distribution of gas and oil across the delta.

INTRODUCTION

The Niger Delta Basin, situated at the apex of the Gulf of Guinea on the west coast of Africa, is one of the most prolific deltaic hydrocarbon provinces in the world (Figure 1). The sedimentary basin occupies a total area of about 75,000 km^2 and is at least 11 km deep in its deepest parts. Reserves are currently estimated to be 22.5 billion bbl of oil, 770 million bbl of condensate, and 124 TCFG. Current daily oil production is 2.1 million bbl, and daily condensate production is 85,000 bbl. The present reserve to production ratio is about 30 years. Oil exploration by Shell D'Arcy, which resumed after World War II in the southern part of Nigeria, resulted in the first well being drilled in 1951. After about 15 wells, the first commercial oil discovery in the Tertiary delta was confirmed at Oloibiri-1 well in 1956.

Over the past 40 years, the Niger Delta has matured through one cycle of exploration during which a variety of different structural play types have been exploited. The 3-D bar graph in Figure 2 summarizes the relationships among different play types and recoverable reserves for most of the delta. Important relationships apparent in the graph have previously been noted by Weber (1987). Each

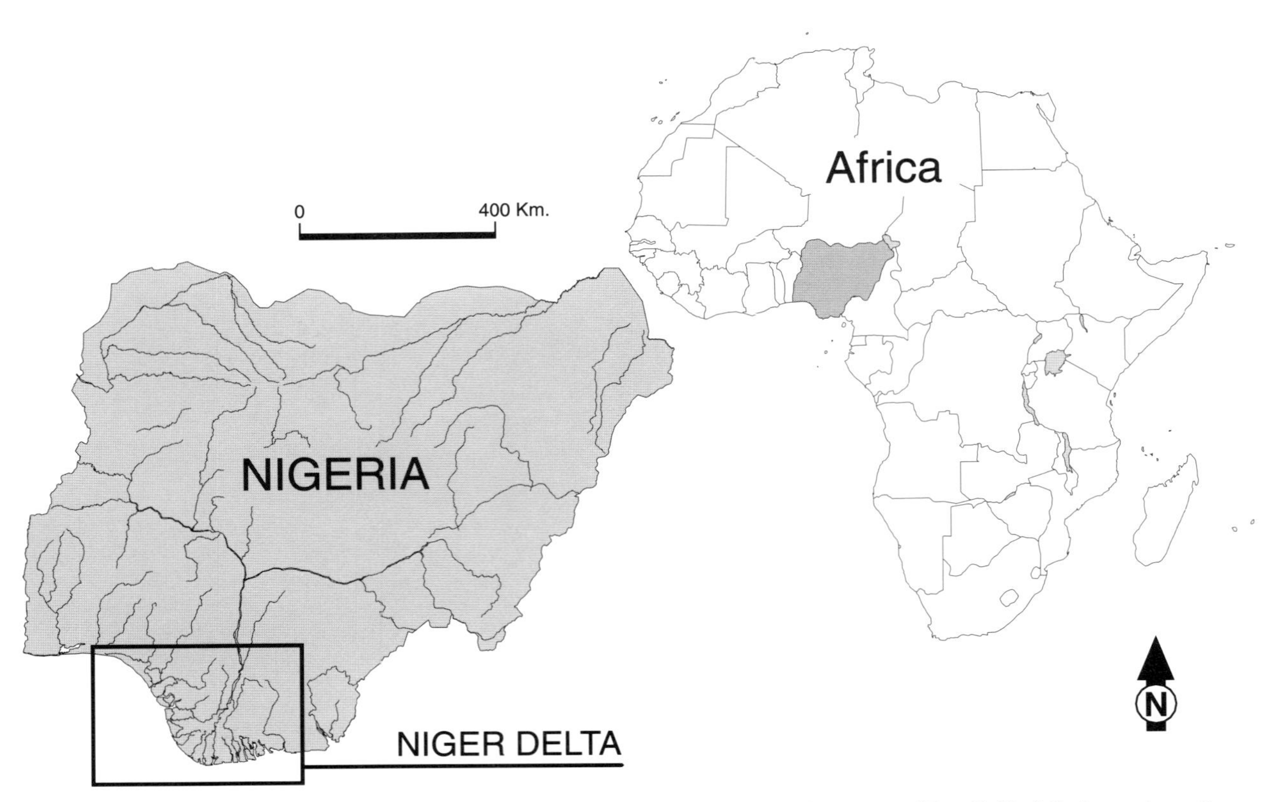

Figure 1—Index map for the study area. The Niger Delta Basin is located at the apex of the Gulf of Guinea along the west-central coast of Africa.

play type is associated with large oil fields. Most of the structural–stratigraphic traps have been discovered in the southeastern part of the delta, but they have also been noted elsewhere, particularly offshore in deep water. A variety of new combination trap styles have been drilled with encouraging results, and they represent one possible exploration future for Nigeria.

The Tertiary petroleum system of the Niger Delta has most recently been discussed in papers by Ekweozor and Daukoru (1994) and Stacher (1995). A considerable amount of new information, however, has become available since publication of those papers, including penetration of a Tertiary oil-prone source rock sequence in Aroh-2 well. As a result, three petroleum systems are now defined for the Niger Delta, new and interesting regional trends have been recognized and mapped, and a much better understanding of factors controlling the complex distribution of gas and oil across the delta has been achieved (Haack and Sundararaman, 1994; Haack et al., 1997; 1998). In this chapter, we present the following:

- a new model, called a *deformation cell*, for evolution of the delta;
- definition of three petroleum systems; and
- new models for Tertiary deltaic source rocks and episodic hydrocarbon migration for the Niger Delta.

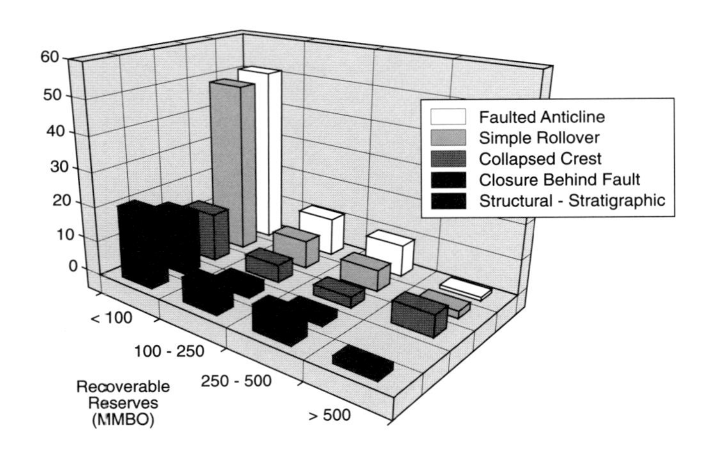

Figure 2—Frequency distribution of the size of Niger Delta oil fields, measured in millions of barrels of recoverable reserves, by play type. Five generalized play types are recognized, of which four are structural. The structural–stratigraphic category includes a variety of different play types that are being actively pursued.

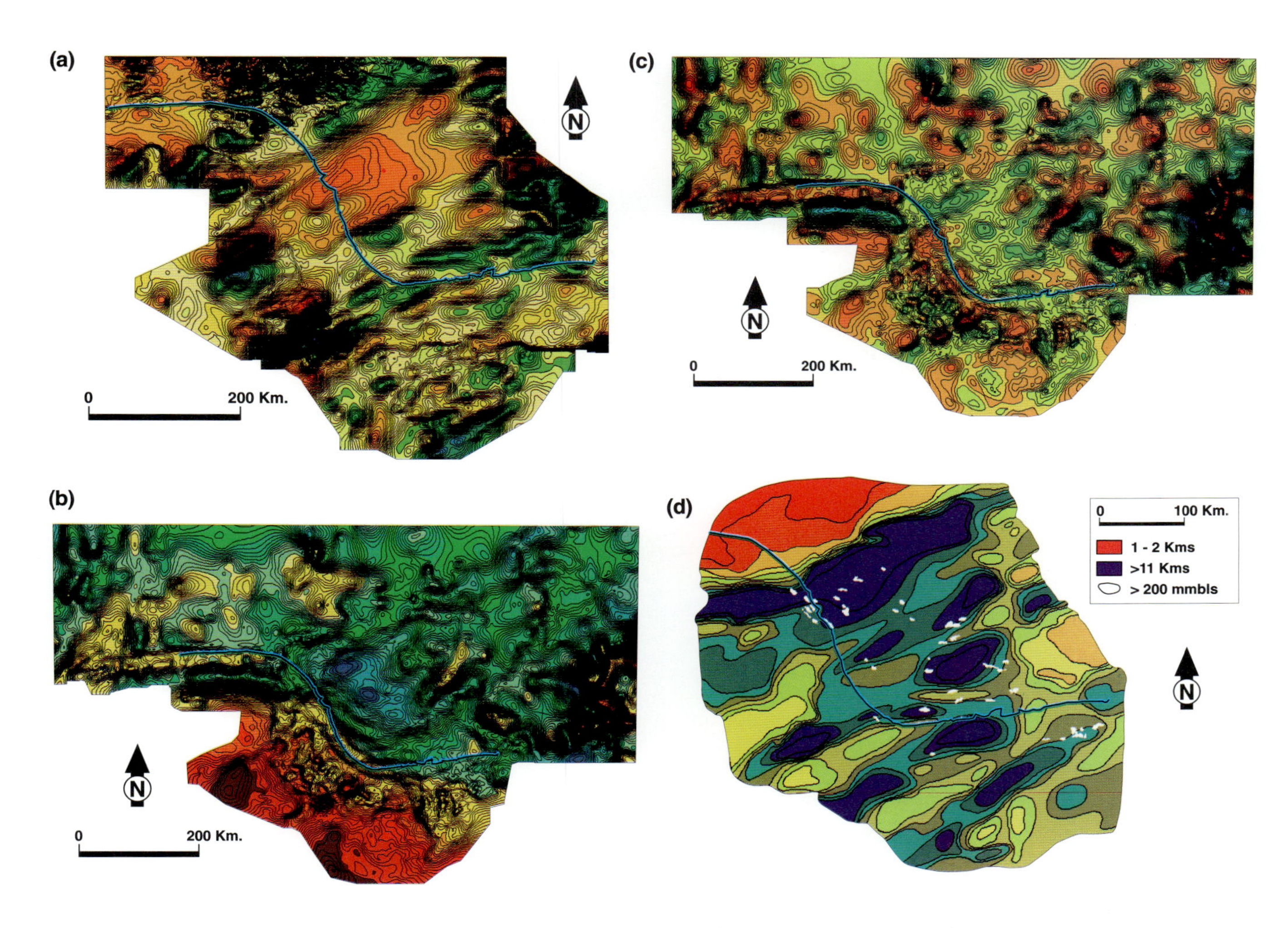

Figure 3—Regional total intensity magnetic and Bouguer gravity maps. (a) Merged air and marine total intensity magnetic map. Warm colors represent magnetic highs, cool colors are magnetic lows. Because the Niger Delta lies 16° S of the geomagnetic equator, positive basement features produce negative total intensity anomalies (lows) and basement deeps generate positive anomalies (highs), opposite of the gravity response. (b) Merged land and marine Bouguer gravity map. Warm colors (southwest) represent gravity highs; cool colors are gravity lows. (c) Residual gravity map derived from several regional gravity trend surfaces. The surface selected to subtract from the Bouguer data captures the greater bulk of the dynamic range in the gravity data and preserves anomalies of moderate wavelength. Warm colors represent positive residuals; cool colors are negative residuals. (d) Regional basement structure inferred from gravity and magnetic data. Large oil fields with >200 million bbl recoverable reserves are shown.

Other important contributions on the geology and hydrocarbon habitat of the Niger Delta include publications by Hospers (1965), Frankl and Cordry (1967), Short and Stäuble (1967), Weber and Daukoru (1975), Evamy et al. (1978), Knox and Omatsola (1989), Doust and Omatsola (1990), and Damuth (1994). In addition, key contributions on the geochemistry of oils and source rocks include papers by Ekweozor et al. (1979a, b), Ekweozor and Okoye (1980), Ekweozor and Daukoru (1984), Lambert-Aikhionbare and Ibe (1984), Udo (1985), Nwachukwu and Chukwurah (1986), Ekweozor and Udo (1988), Bustin (1988), and Ekweozor and Nwachukwu (1989).

REGIONAL GEOLOGY

Basement Configuration

All available gravity and magnetic data for the Niger Delta Basin and adjacent areas offshore were merged and analyzed for the purpose of mapping basement structures. Strong northeast to southwest trends dominate the magnetic signal across the delta (Figure 3a). These trends are parallel to fracture zone patterns in the deep ocean and are believed to have formed during opening of the Mid-Atlantic Ocean in Late Jurassic–Early Cretaceous

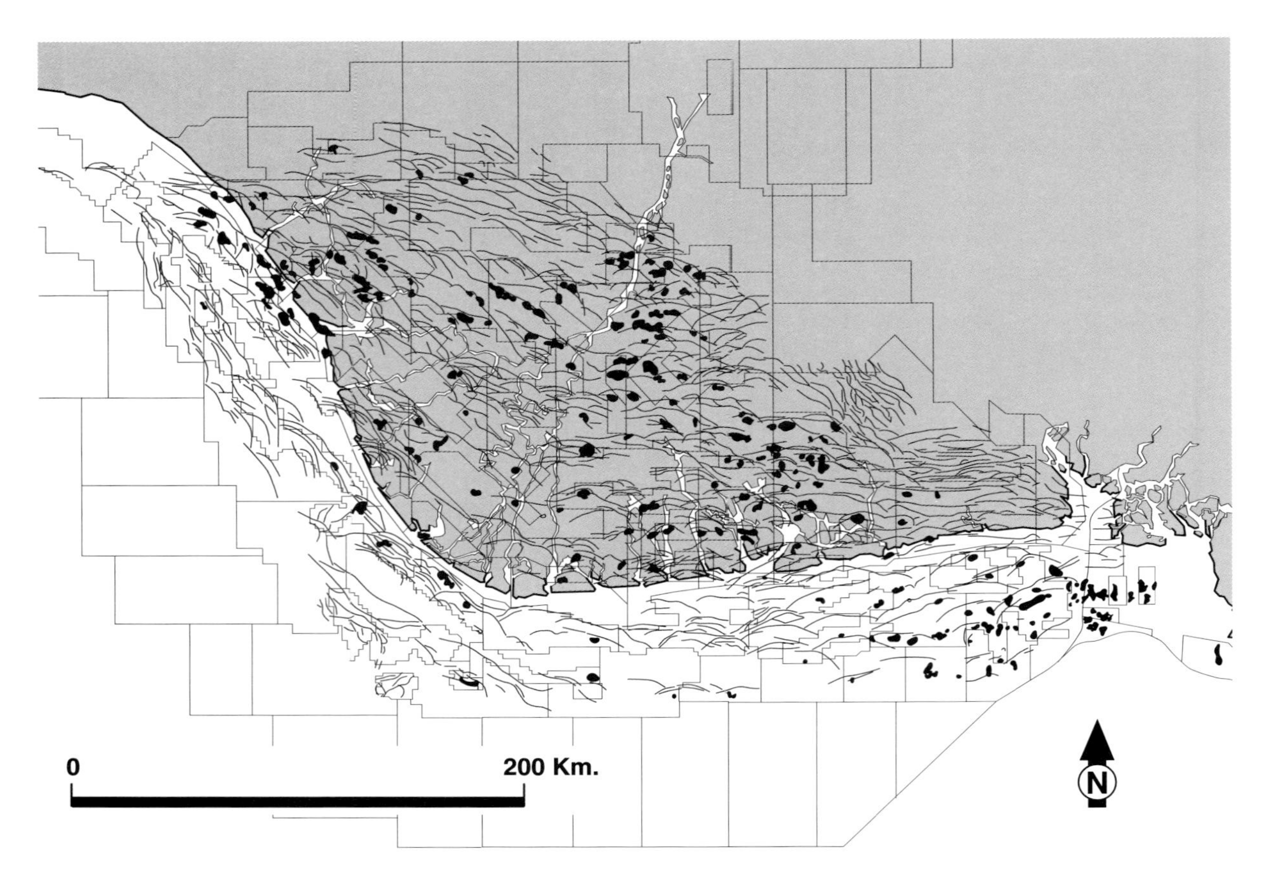

Figure 4—Tertiary structural trends of the Niger Delta. Regional normal fault framework is based on the published map of Evamy et al. (1978) and the mapping of Chevron earth scientists. Oil fields are shown in black.

time. Parallel structural trends extend through the Anambra Basin and into the Benue Trough northeast of the Niger Delta.

Regional patterns expressed in the Bouguer gravity data are different from those indicated by magnetic data. A large gravity low, due to the presence of a thick Tertiary deltaic sediment load, is present beneath the region of the delta. Conversely, there is a large gravity high associated with thinned continental crust offshore (Figure 3b). Only in the extreme northeastern part of the delta, where the basement is shallower and Cretaceous rocks crop out, can structural trends parallel to those observed in the magnetic data be observed. To resolve these apparent discrepancies and map basement structure, a regional gravity surface, capturing most of the dynamic range in the data and retaining all anomalies of moderate wavelength, was selected to subtract from the Bouguer gravity data.

The residual gravity map exhibits a significantly different but geologically acceptable anomaly pattern (Figure 3c). In the residual gravity map, effects due to thinned transitional crust and thick Tertiary basin fill are mitigated. This methodology was applied to the Bouguer data to better image structural trends in the basement. Those structural trends extending northeast to southwest, parallel to trends in the magnetic data, can now be observed. By inferring the structural fabric from the residual gravity map and combining it with depth to basement calculations from the magnetic data, a structure map of the basement was constructed (Figure 3d). This map represents an approximation of the structural configuration of the Tertiary basin at some point in time prior to very early development of the Niger Delta. In this map, a series of northeast–southwest trending structural highs and lows are present beneath the delta. Structural lows are deeper than 11 km and, in some areas, probably still influence drainage patterns today. These lows are inferred to represent subbasins and, as discussed later, possibly separate hydrocarbon kitchens. There is, in fact, a relationship between each of the subbasins onshore and the largest oil fields in the delta, those with more than 200 million bbl of recoverable oil.

Tertiary Structural Trends

Tertiary structural trends are oriented in a general northwest–southeast direction (Figure 4). These regional structure-building faults, which have been controlling deposition of the deltaic growth section since middle–late Eocene time, occur at an angle approximately 90° to the

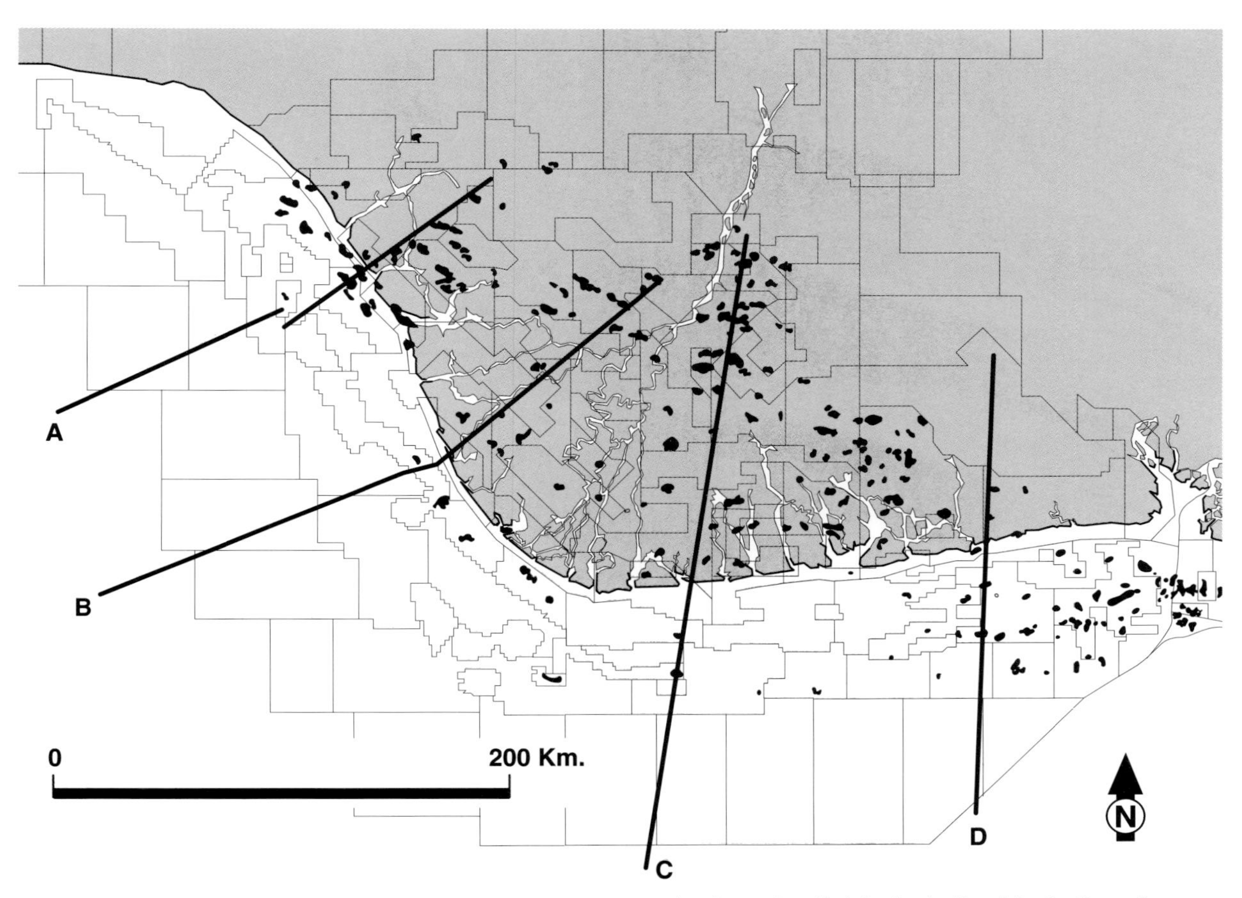

Figure 5—Location of four regional dip cross sections, based on seismic and well data, including biostratigraphy, across the Niger Delta. Oil fields are shown in black.

older pregrowth structures. The resulting Tertiary structures are regionally extensive anticlinal features developed on the downthrown side of growth faults, which dip in a southerly direction. The southern flanks of these structures generally dip moderately to steeply, and there is typically stratigraphic expansion into major counterregional growth faults as well. In fact, in many cases the amount of growth on counterregional faults greatly exceeds growth on the regional faults.

As illustrated in Figure 4, the regional Tertiary fault framework is quite complicated. With this complexity and northwest–southeast regional structural grain, long-distance lateral migration of oil and gas updip to the northeast across major faults is unlikely. More likely, hydrocarbon migration is predominantly vertical and restricted to major structural compartments that are controlled by growth faulting and deposition. Stacher (1995) has expressed a similar view.

Tertiary Structural Model

Four regional dip cross sections across the Niger Delta have been compiled from 2-D and 3-D seismic data, well control, and biostratigraphy (Figure 5). The transects, converted to depth and structurally balanced, demonstrate that regional structural and stratigraphic patterns across the delta are similar (Figure 6). They also document higher rates of progradation in the more central parts of the delta, where the amount of extension is greater. All hydrocarbon production in the Niger Delta occurs in the growth section, which is separated by regional décollements from the underlying pregrowth section (Cretaceous–Paleocene strata). The regional décollements pass through the mobile shale.

Two styles of deformation are characteristic of the Niger Delta and occur in each of the four cross sections. The domino style, characterized by a series of parallel to subparallel normal faults that pervasively break up hanging-wall strata, occurs in the most landward onshore parts of the delta. The listric style of deformation contains relatively intact hanging-wall strata with localized faulting associated with curvature of the major structure-building fault. Most faults extending offshore beyond the domino zone are listric, and listric faults in deep water are commonly associated with shale tectonism.

Throughout most of the Niger Delta, the amount of extension that has occurred through time cannot be balanced by the amount of compression that can be

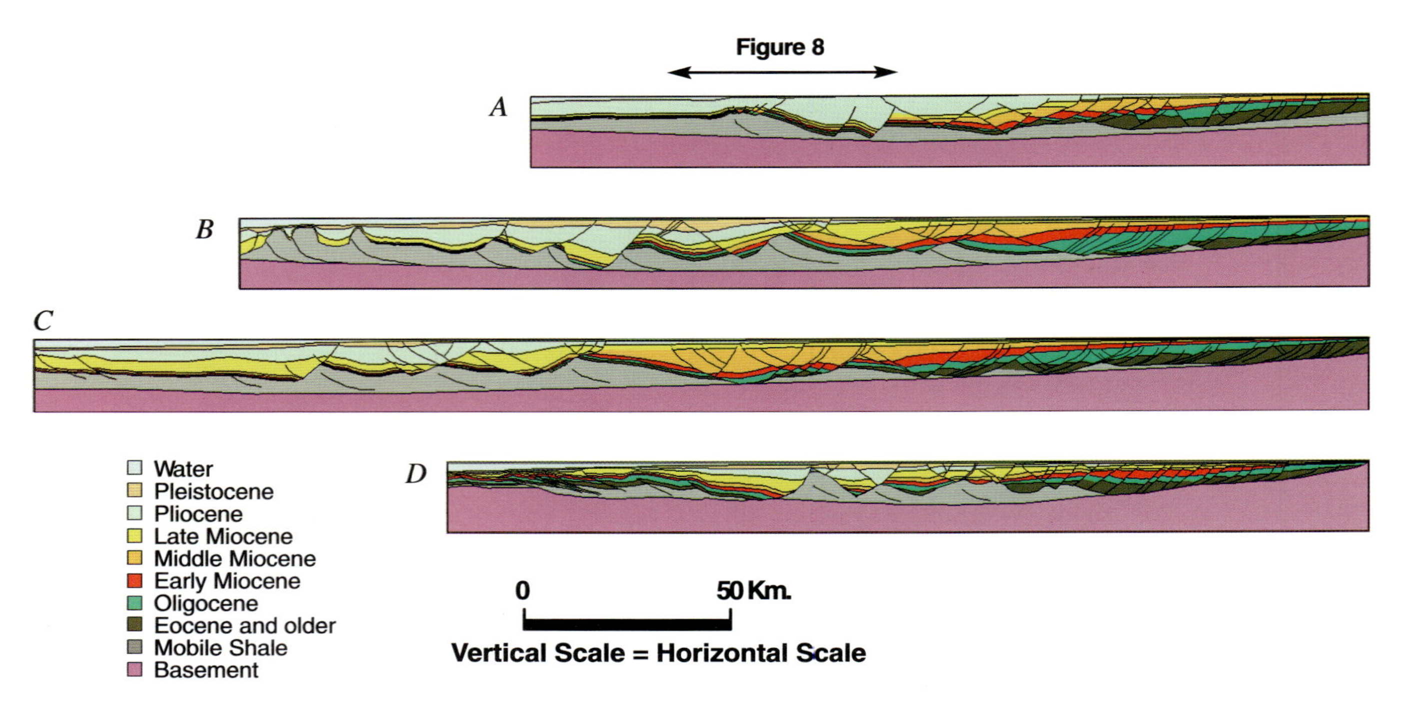

Figure 6—Regional cross sections A, B, C, and D extending across the Niger Delta into deep water. Each section is drawn at true scale (no vertical exaggeration).

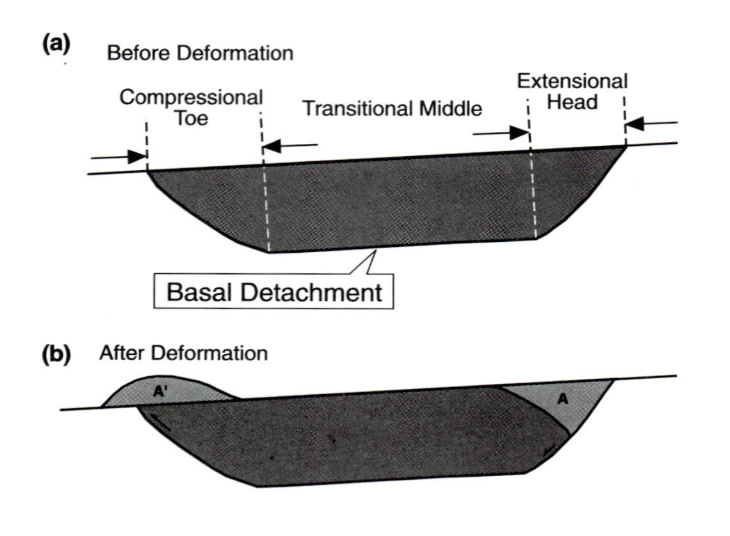

Figure 7—Idealized conceptual model for a deformation cell. (a) Idealized cell before deformation. (b) Idealized cell after deformation, where vacated area A caused by normal faulting equals excess area A' caused by compressional thrust faulting. Vacated area A represents accommodation space for deposition of sediments.

measured in the thrust belt out in front of the delta in deep water. To structurally balance rock volume when constructing cross sections, the concept of a *deformation cell* was developed. Figure 7 illustrates an idealized cross section of a cell before and after deformation. A deformation cell contains a through-going detachment with extensional faults upslope, a transitional middle detachment with a slope that is approximately parallel to the surface slope, and a compressional thrust downslope. After the cell slips for some distance, the vacated area A, caused by normal faulting, will be equal to the excess area A' caused by compressional thrusts. This model links creation of accommodation space for deposition of sediments with structural growth.

Figure 8 illustrates a seismic example of an active deformation cell from cross section A. The extensional head, transitional middle, and compressional toe are labeled. Positive bathymetry above the compressional toe creates gravitational instability, which results in slippage and downslope motion of the next younger cell. The compressional toe, offset by a new normal fault, will eventually become decapitated and wind up at the bottom of the next younger cell. Older decapitated compressional toes have been observed in 3-D seismic data in the northwestern part of the delta.

Figure 9 is a schematic diagram that illustrates progressive evolution of a deformation cell. Progradational sedimentation on an unstable slope drives large-scale formation of the cell. Figure 9a depicts an undeformed basin filled with Upper Cretaceous–Paleocene shale. Figure 9b shows that the first cycle of sedimentation produces a series of domino faults upslope and a

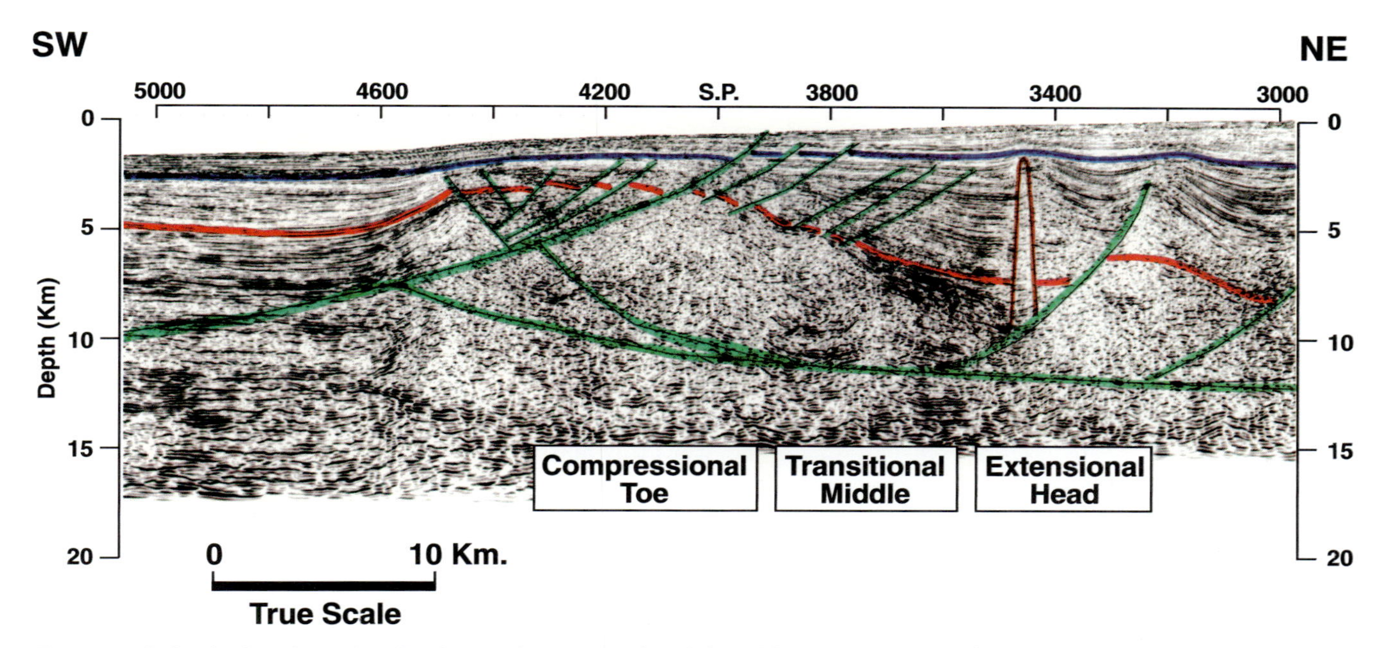

Figure 8—Seismic data from the distal part of cross section A (see Figure 6 for location), illustrating the key characteristics of an active deformation cell. The compressional toe creates a positive topographic expression on the sea floor. The resulting gravitational instability initiates outboard motion in the next younger deformation cell that develops. Considerable stratigraphic thinning occurs in the growth section between the red and blue horizons onto the compressional thrust front.

compressional toe downslope, which creates a new shelf-slope break. In Figure 9c, the next cycle of sedimentation forms a new deformation cell containing a megastructure. The counterregional fault of the megastructure probably results from the counterregional bend in the basal detachment. The compressional toe of the first cell has been decapitated and translated into the deeper part of the megastructure in the younger cell. This process is repeated through time, as younger and younger cycles of sedimentation form new cells.

PETROLEUM SYSTEMS

A *petroleum system* is the "oil and gas machine" that generates and concentrates hydrocarbons within a basin over geologic time. It is composed of two subsystems (Demaison and Huizinga, 1991): (1) a generative subsystem defined by source rocks and oil characteristics, and (2) a migration–entrapment subsystem defined by timing of generation, type of drainage (lateral or vertical), and type of trap. This regional study of the hydrocarbon habitat of the Niger Delta is focused on evaluating the oil machine at the basin scale, that is, at the scale of a petroleum system.

On the basis of available geochemical data, we recognize three petroleum systems in the Niger Delta, informally named as follows:

1. Lower Cretaceous petroleum system,

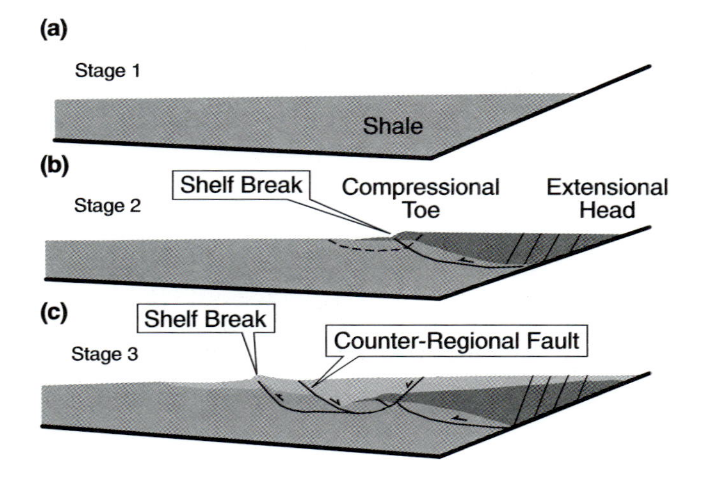

Figure 9—Conceptual model for the progressive evolution of a deformation cell in the Niger Delta. Progradational sedimentation on an unstable slope drives the large-scale development of the cell. (a) A pristine basin filled with Upper Cretaceous–Paleocene shale. (b) The first cycle of sedimentation produces a series of domino faults upslope and a compressional toe downslope. (c) The next cycle of sedimentation forms a new deformation cell containing a megastructure.

2. Upper Cretaceous–lower Paleocene petroleum system, and
3. Tertiary petroleum system.

Lower Cretaceous source rocks are identified from one of the rift grabens that occur along the northwestern margin of the delta. Upper Cretaceous–lower Paleocene source rocks are also recognized from the same area. Tertiary deltaic source rocks, the principal source for oil and gas in the Niger Delta, are time transgressive and occur in regionally extensive outer shelf to slope facies beneath the parallic sequence and in transgressive shelf facies within it. On the regional cross sections in Figure 6, source facies occur within growth strata of Eocene–Pliocene age and in older pregrowth strata.

Lower Cretaceous Petroleum System

One biodegraded seep oil from Nigerian tar sands on the northern flank of the Dahomey Embayment has been correlated to Lower Cretaceous (Neocomian) source rocks in Ise-2 well (J. Dahl, personal communication, 1990). The lacustrine alga *Botryococcus* was also identified in samples from the source rock interval in that well (A. R. Daly, personal communication, 1990). The source rock extract from Ise-2 well and pyrolyzate of the seep sample have geochemical characteristics similar to oils from the Lower Congo Basin that have been derived from Lower Cretaceous lacustrine Bucomazi source rocks (K. E. Peters, personal communication, 1993).

The approximate aerial extent of the Lower Cretaceous petroleum system is shown in Figure 10. Lower Cretaceous source potential, based on the geochemical profile from Ise-2 well, is estimated to be 5.24 tons HC/m^2 rock. This would provide a low to moderate hydrocarbon charge for a vertically drained rift basin (Demaison and Huizinga, 1991). A previous in-house estimate of source potential was 12.07 tons HC/m^2 rock. This estimate, however, is too high because obvious nonsource lithologies were mistakenly included in the calculation. As in the Dahomey Basin, a Lower Cretaceous petroleum system could also be present within the rift sequence of the Benue Trough northeast of the Niger Delta.

Upper Cretaceous–Lower Paleocene Petroleum System

The inferred regional extent of the Upper Cretaceous–lower Paleocene petroleum system is illustrated in Figure 11 and identified by the geochemical profile for Epiya-1 well. This system consists of type II and II–III oil-prone kerogen and includes source rock strata in the marine Araromi, Awgu, and Imo shales, which were deposited along the northwestern margin of the delta. Oils recovered from Paleogene sandstones in Shango-1 well are inferred to be derived from this source (K. E. Peters, personal communication, 1993). Contrary to the inferred origin of tar sand bitumen discussed above, other geochemists believe bitumen in the tar belt of the Dahomey Basin could have been derived from Upper Cretaceous marine shales. Presently, not enough geochemical data from the tar sands are available to confidently determine whether the source is lacustrine or marine. More sampling and evaluation of the Nigerian tar sands are needed.

The source potential for the Upper Cretaceous–lower Paleocene petroleum system, based on the profile for Epiya-1 well, is estimated to be 10.5 tons HC/m^2 rock. This would provide a normal hydrocarbon charge for a vertically drained basin (Demaison and Huizinga, 1991). If, however, richness increases distally, as is likely, source potential could be higher. Conversely, if thickness decreases distally, even if richness increases, then source potential might not increase and could even decrease farther offshore.

Although data are sparse, Upper Cretaceous source rocks have been identified along the eastern margin of the Niger Delta in outcrops (Inyang et al., 1995) and in the subsurface (J. A. Adakomola, personal communication, 1992). Oil has also been discovered in Cretaceous strata from the Anambra Basin. Oil produced from Seme field in the Republic of Benin, west of the Niger Delta, is also derived from Upper Cretaceous source rocks (J. Dahl, personal communication, 1991).

Upper Cretaceous–lower Tertiary marine Iabe and Landana source rocks in the Lower Congo Basin, Cabinda, consist of type II kerogen derived primarily from marine organic matter (Schöellkopf and Patterson, 1997). Although an Upper Cretaceous–lower Tertiary petroleum system is important for generating oil throughout much of west Africa, no large petroleum accumulations have been identified to date from Upper Cretaceous–lower Paleocene source rocks in the region of the Niger Delta. Nevertheless, this older petroleum system could be a major source for liquid hydrocarbons in deep-water areas of the delta. For this system to be effective in deep water, however, it would require sufficient hydrocarbon charge through migration pathways connecting pregrowth source strata to reservoirs in Tertiary traps within the growth section.

Tertiary Petroleum System

The aerial extent of the Tertiary petroleum system is shown in Figure 12a, and its source rock richness is represented by the hydrogen index (HI) versus oxygen index (OI) plot and geochemical profile for Aroh-2 well (Figures 12b and c). Variation in source richness, as illustrated stratigraphically by the profile, from dry gas prone to oil prone, can also be inferred and mapped regionally on the basis of geochemical signatures in oils. The predominance of gas-prone source rocks encountered in most wells that have been analyzed from the Niger Delta and implicit variation in source potential have been recognized by many workers and discussed thoroughly in the literature (Ekweozor and Daukoru, 1994, and references therein; Core Laboratories, 1994; Stacher, 1995; Haack and Sundararaman, 1994; Haack et al., 1997; 1998). This chap-

(a)

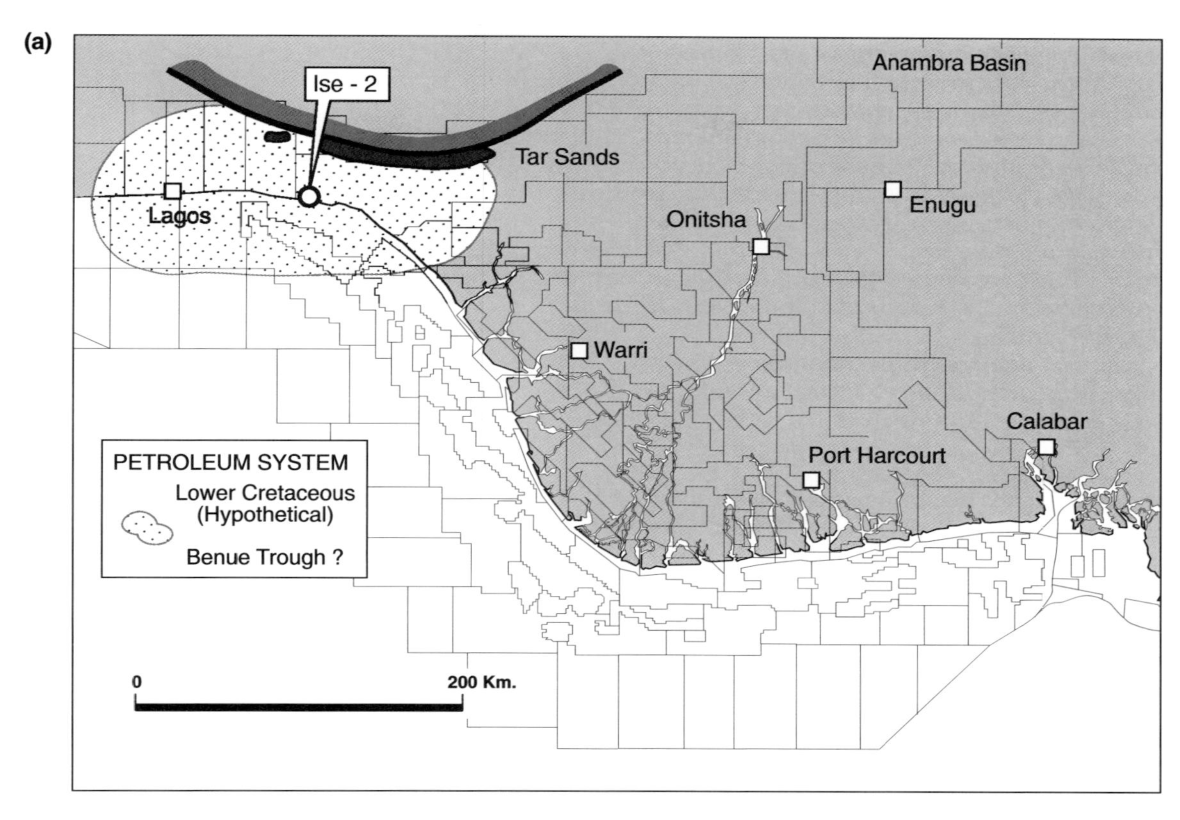

(b)

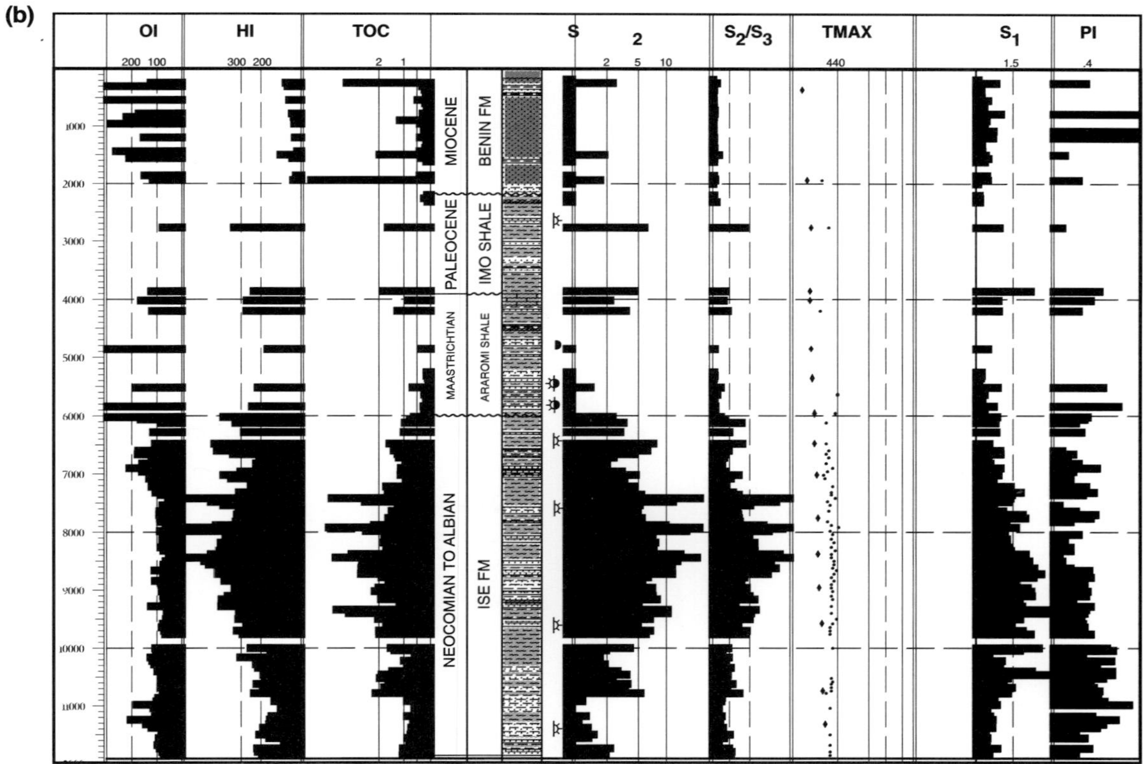

Figure 10—(a) Map showing the Lower Cretaceous petroleum system and the location of the Nigerian tar sands on the northern flank of the Dahomey Embayment. (b) Geochemical profile for Ise-2 well, representative of the Lower Cretaceous petroleum system. The source potential in this well, calculated for Neocomian source rock, is 5.24 tons HC/m² rock. The top of the oil window is at ~8500 ft (~2590 m).

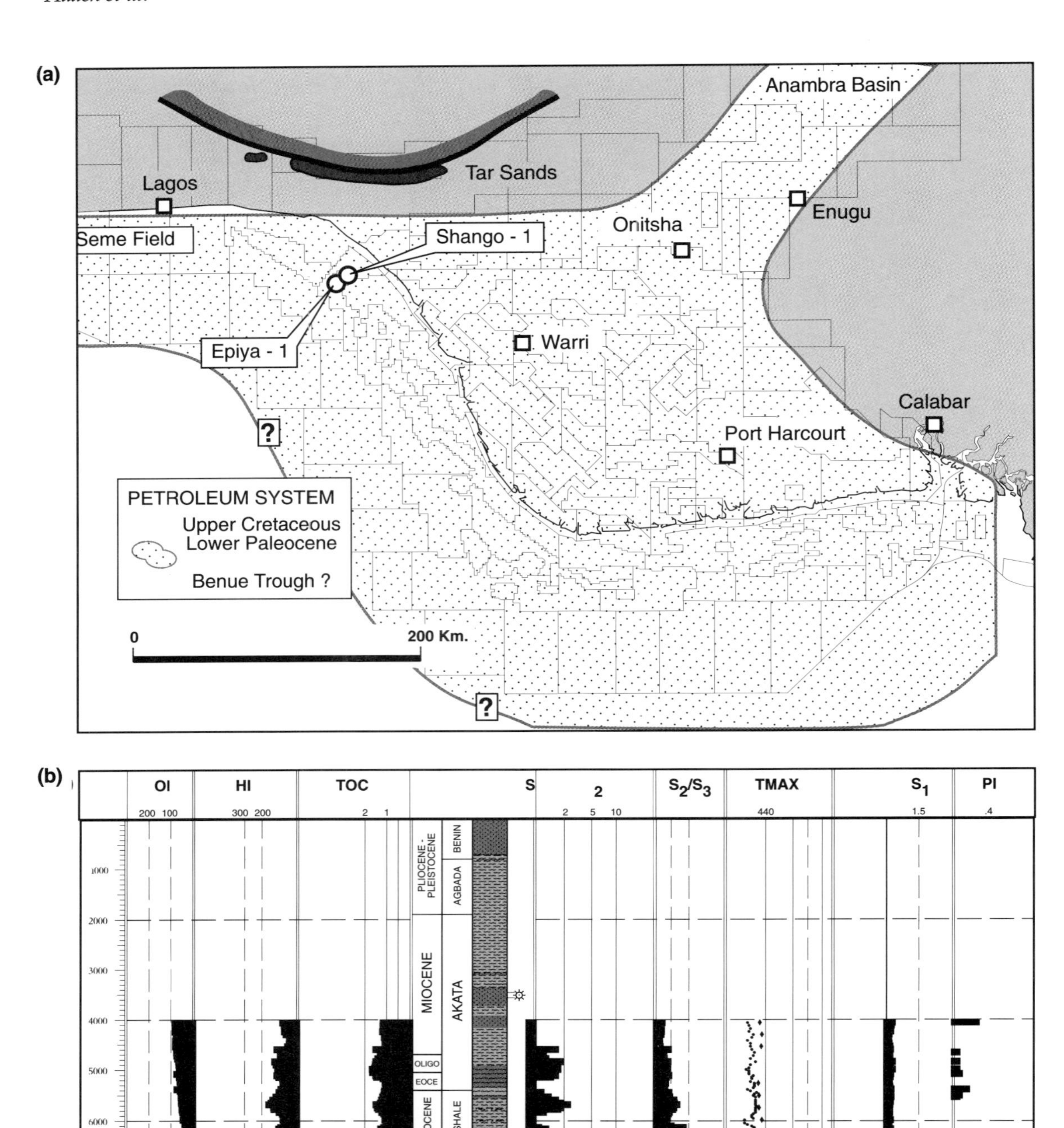

Figure 11—(a) Map showing the Upper Cretaceous–Lower Paleocene petroleum system. (b) Geochemical profile for Epiya-1 well, representative of this petroleum system. The source potential in this well is 10.5 tons HC/m² rock. The top of the oil window is at ~9000 ft (~2740 m).

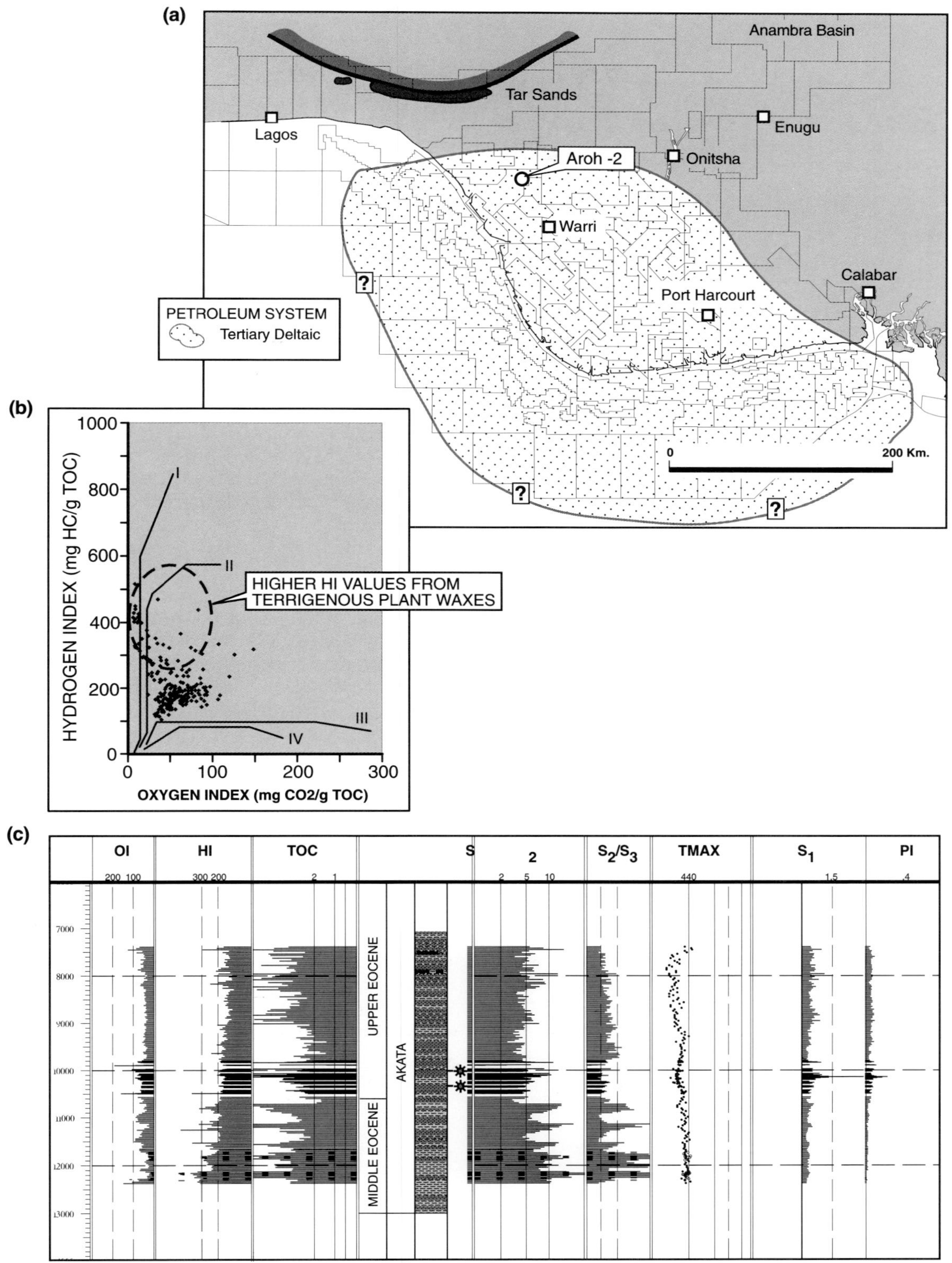

Figure 12—(a) Map showing the Tertiary (deltaic) petroleum system. (b) HI/OI plot showing the higher HI values characterizing the richer source rock section that result from the presence of terrigenous plant waxes in the kerogen. (c) Geochemical profile for Aroh-2 well, representative of this petroleum system. Source potential for the analyzed middle Eocene interval is 27 tons HC/m^2 rock, and for the more liquid-prone source rock sequence at 10,500–12,400 ft (3200–3780 m) is 15 tons HC/m^2 rock. All samples were preextracted to remove Petrofree. As a result, S_1 values are minimum values. (Preextraction does not alter or enhance other pyrolysis or TOC results.) The top of the oil window is at ~10,500 ft (~3200 m).

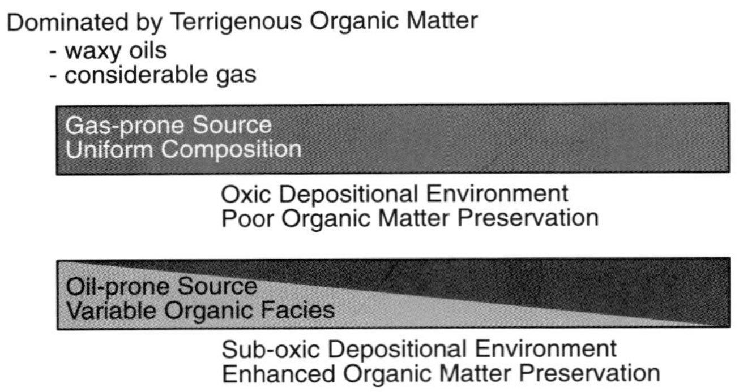

Dominated by Terrigenous Organic Matter
- waxy oils
- considerable gas

Gas-prone Source
Uniform Composition

Oxic Depositional Environment
Poor Organic Matter Preservation

Oil-prone Source
Variable Organic Facies

Sub-oxic Depositional Environment
Enhanced Organic Matter Preservation

Figure 13—Schematic diagram of the Niger Delta source rock model. Niger Delta source rocks, dominated by terrigenous organic matter, generate considerable gas and waxy oils. Oil-prone source rocks were deposited in suboxic depositional environments where hydrogenicity was preserved. Gas-prone source facies are compositionally more uniform due to degradation of organic matter in oxic environments.

ter documents the evidence for Tertiary oil-prone source rocks in the Niger Delta and provides a new understanding of the Tertiary petroleum system.

In Aroh-2 well, thermally mature oil- and gas-prone middle Eocene source strata occur upthrown to a normal fault. The richer source facies, consisting of type II oil-prone kerogen, progressively increase in abundance with depth in the lower part of the well. Extracts from these source rocks correlate to oils recovered from nearby wells. Results from pyrolysis gas chromatography of one of the richer samples from this source sequence indicate that the oil-prone deltaic source facies generate primarily light to normal oils, some wet gas, and very little dry gas. These results correspond well to the chemical composition of liquid hydrocarbons characteristic of Niger Delta oil fields. Source potential for this 4200-ft (1280-m) interval is 27 tons HC/m^2 rock, indicative of a supercharged petroleum system. Although this value is slightly inflated by the thick gas-prone section overlying the richer source rocks, more than 55% of the source potential is due to the more liquid-prone source rock section at 10,500–12,400 ft (3200–3780 m).

Based on foraminifera biostratigraphy, the deeper, richer source rocks in Aroh-2 well are inferred to have been deposited in an outer neritic to bathyal paleoenvironment. The relatively high concentration of C_{29} steranes derived from land plants in this source facies (42–57%) implies that abundant terrigenous organic matter was being deposited and preserved in deep water. Perhaps a zone of lower oxygen content began to develop in the water column impinging on the shelf–slope break during middle–late Eocene time. Although there is no evidence for anoxia in any of the biomarker data from oils and source rock extracts, something happened during the middle–late Eocene to enhance preservation of terrigenous organic matter in the vicinity of the upper slope in the northwestern part of the delta.

Figure 13 presents a conceptual model for Tertiary deltaic source rocks in the Niger Delta encompassing the complete range of source potential observed, from dry gas prone to oil prone. Depending on where you are in the delta, source rocks range in age from middle Eocene to Pliocene and they contain predominantly terrigenous organic matter. These source rocks generate light waxy oils or gas. The more gas-prone end-member of the organic facies spectrum was deposited in an oxic depositional environment in which organic matter was poorly preserved. Oil-prone organic facies, however, were deposited in suboxic depositional environments in which enhanced preservation of terrigenous organic matter was possible. Different oil-prone source facies have been recognized and mapped across the delta. The differences undoubtedly result from variation in the composition of organic matter components, which contribute to kerogen of the richer source facies.

Regional variation in oil-prone source facies, inferred on the basis of biomarker and carbon isotope data, is illustrated in Figure 14a. This map, summarizing the results from cluster analysis, illustrates the distribution of different subfamilies of oils across the delta. This discrimination of subfamilies is based on source-related geochemical parameters, which are known to be significant from analyses of about 250 oils from the Niger Delta. [Note: in Nigeria, accepted petroleum industry nomenclature and practice makes use of the red color to indicate oil and green to indicate gas (Nigerian Committee on Standardisation of Geological Nomenclature and Symbols, 1984). These conventions are followed here.]

A MORE DETAILED LOOK AT THE TERTIARY (DELTAIC) PETROLEUM SYSTEM

Results from regional geochemical investigations indicate that kerogen derived from terrigenous Tertiary organic matter deposited in marine shales is the principal source for hydrocarbons in the Niger Delta. There appears to be a minor unquantified contribution from marine organic matter to the kerogen, based on carbon isotope and biomarker data, but the primary source is organic matter derived from Tertiary land plants. Variability in preservation of terrigenous organic matter due to changes in redox potential through time caused considerable regional variation in deltaic organic facies. These laterally and temporally varying source facies, coupled with differences in maturity resulting from regional variability in geothermal gradient and burial, are the two principal factors that have controlled the complex distribution of oil and gas across the delta.

More than 35 years of drilling experience in the Niger Delta has demonstrated that oil-prone source rocks are rarely, if ever, encountered in wells. Because the richer source rocks are deeper than strata normally penetrated by wells, regional trends in oil-prone source facies must be inferred on the basis of geochemical characteristics from produced oils (Peters and Moldowan, 1993). This methodology works well as long as migration pathways are predominantly vertical and source-related geochemical signatures are not significantly altered during migration. Because of extensive normal faulting, long-distance lateral migration of hydrocarbons across major bounding faults is unlikely in the Niger Delta. Furthermore, in many areas of the delta, the common association of seismic amplitudes from 3-D data and normal faults indicates that normal faulting does play a key role in hydrocarbon migration. Such an association supports earlier assessments of the Niger Delta as a classic example of a vertically drained petroleum system (Weber and Daukoru, 1975; Evamy et. al., 1978; Demaison and Huizinga, 1991). The results and discussion that follow are based on the methodology of inferring regional source facies trends from biomarkers in oils and pertain to the richer oil-prone source facies of the delta.

Regional Source Rock Trends Inferred from Oils

Regional variability in organic matter input, determined from the contour map in Figure 14b of relative abundance of C_{29} steranes, effectively demonstrates how source facies change across the Niger Delta. Sample control for this map is shown in Figure 14a. The areas in green in Figure 14b have the highest relative abundance of C_{29} steranes (>50%), indicating the highest organic matter contribution from land plants, and represent proximity to paleodrainage systems during source rock deposition. The areas in blue have the lowest relative abundance of C_{29} steranes (about 40% or less) and hence the lowest land plant contribution, which suggests source rock deposition under more marine-influenced conditions. The easily recognized deltaic lobes on this map correspond closely to paleodrainage systems identified by Weber and Daukoru (1975). The more proximal lobate patterns occur outboard of the Anambra Basin (Cretaceous delta) in an area where the early Tertiary delta began to develop. Those lobes undoubtedly represent paleodrainage systems, which supplied sediments to the early pre-Miocene Niger Delta. The two more distal lobes in the eastern and western parts of the delta represent later drainage systems, which began to develop during the Miocene.

In the map of Figure 14b, there is close correspondence between the more marine-influenced areas (blue) and oil reserves. About 75% of Chevron Nigeria Limited joint venture reserves occur in areas where source facies contain less terrigenous organic matter. The same relationship holds true for total reserves from all operators across the entire delta (unpublished Petroconsultant data, 1994).

One difficulty with the map in Figure 14b is that it does not discriminate the oil-rich northwestern and southeasternmost parts of the delta from the more gas-rich central offshore part. This anomaly was resolved by mapping another biomarker parameter sensitive to the degree of preservation of organic matter in the depositional environment—the homohopane index (HHI). Better preservation of organic matter results in more hydrogen-rich kerogen, higher Rock-Eval HI values, and more liquid-prone source rocks. This source rock signal is detected in oils generated from those richer source rocks as a higher value for the HHI (Peters and Moldowan, 1993). This approach works well in the Niger Delta. One caveat, however, is that the HHI is affected by maturity and biodegradation. In this study, all data were screened for those effects before mapping.

The contour map in Figure 14c of the HHI, or in other words, the degree of preservation of organic matter, illustrates some interesting patterns. There is a general correspondence between deltaic lobes, inferred on the basis of C_{29} sterane content, and areas where HHI values are low. These green areas in Figure 14c represent more highly oxygenated parts of the Niger Delta during source rock deposition and consist of more gas-prone source facies than do the more oil-prone red areas. This map, in fact, now effectively discriminates a more gas-prone source facies in the central offshore part of the delta from the more oil-rich areas in the northwest and southeast. Furthermore, the dark red areas delimit the arcuate band of major oil fields so characteristic of the delta. There are other more subtle trends recognized offshore in the northwestern part of the delta that are also consistent with trends in this map.

In the map of Figure 14c, there is a close correspondence between oil reserves and the red areas, which are inferred to be more liquid prone and characterized by higher HHI values. Almost 75% of Chevron Nigeria Limited joint venture reserves are from the areas colored in red on this map, and Chevron's operated production from these areas is more than twice as liquid rich as production from green areas is. There is a similar delta-wide correlation between total reserves and inferred source rock richness (unpublished Petroconsultant data, 1994). Although not surprising, these relationships indicate that source rock facies control to some extent where oil accumulations are large, and conversely, where the gas–oil ratio (GOR) is high. The correlation, however, is not perfect, and other factors such as maturity, reservoir distribution, migration timing, and traps are also important.

Relationship Between Oil Families and Basement Structure

In Figure 14d, the distribution of different oil subfamilies is superimposed onto the structure map of the basement, revealing interesting relationships between some of the subbasins and subfamilies of oils. These subbasins, in fact, could represent separate hydrocarbon kitchens for the Niger Delta. The correspondence is not perfect, but there is enough overlap in the data to suggest possible genetic relationships among subbasins, different source facies, and oil generation.

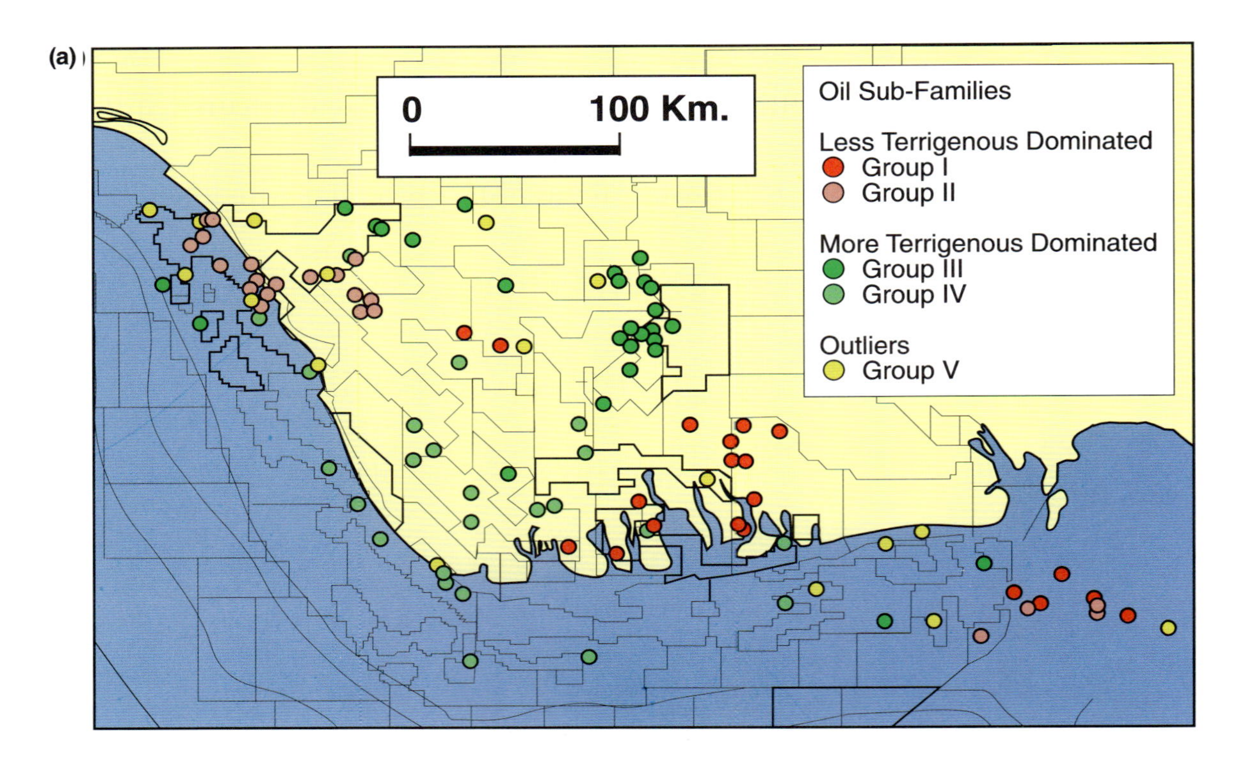

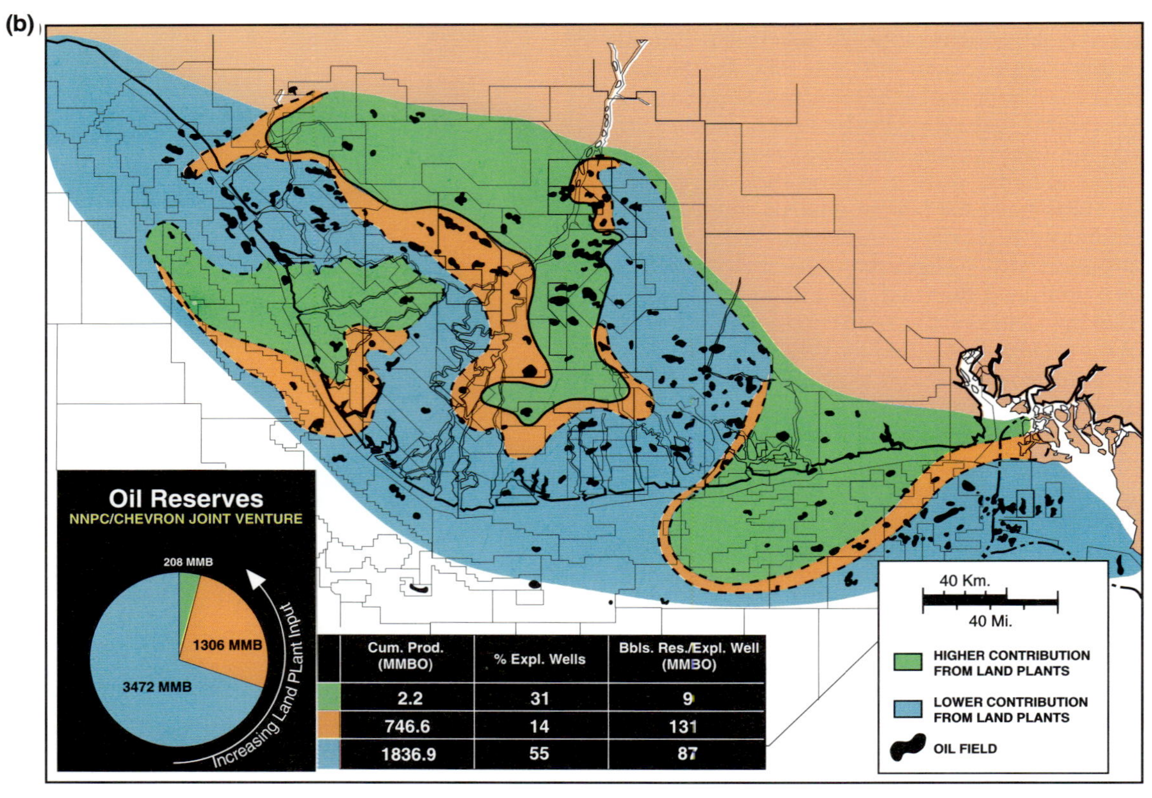

Figure 14—Regional geochemical maps of the Tertiary deltaic petroleum system based on selected biomarker parameters. (a) Four subfamilies of oils are recognized based on results from multivariate cluster analysis. (b) Contour map of relative abundance of C_{29} steranes (Peters and Moldowan, 1993, p. 182) demonstrates regional variation in abundance and distribution of terrigenous organic matter in source rocks of this petroleum system. Green areas represent highest relative abundance (C_{29} steranes > 50%), blue areas are lowest relative abundance (C_{29} steranes ≤ 40%).

(Figure 14 continues on next page.)

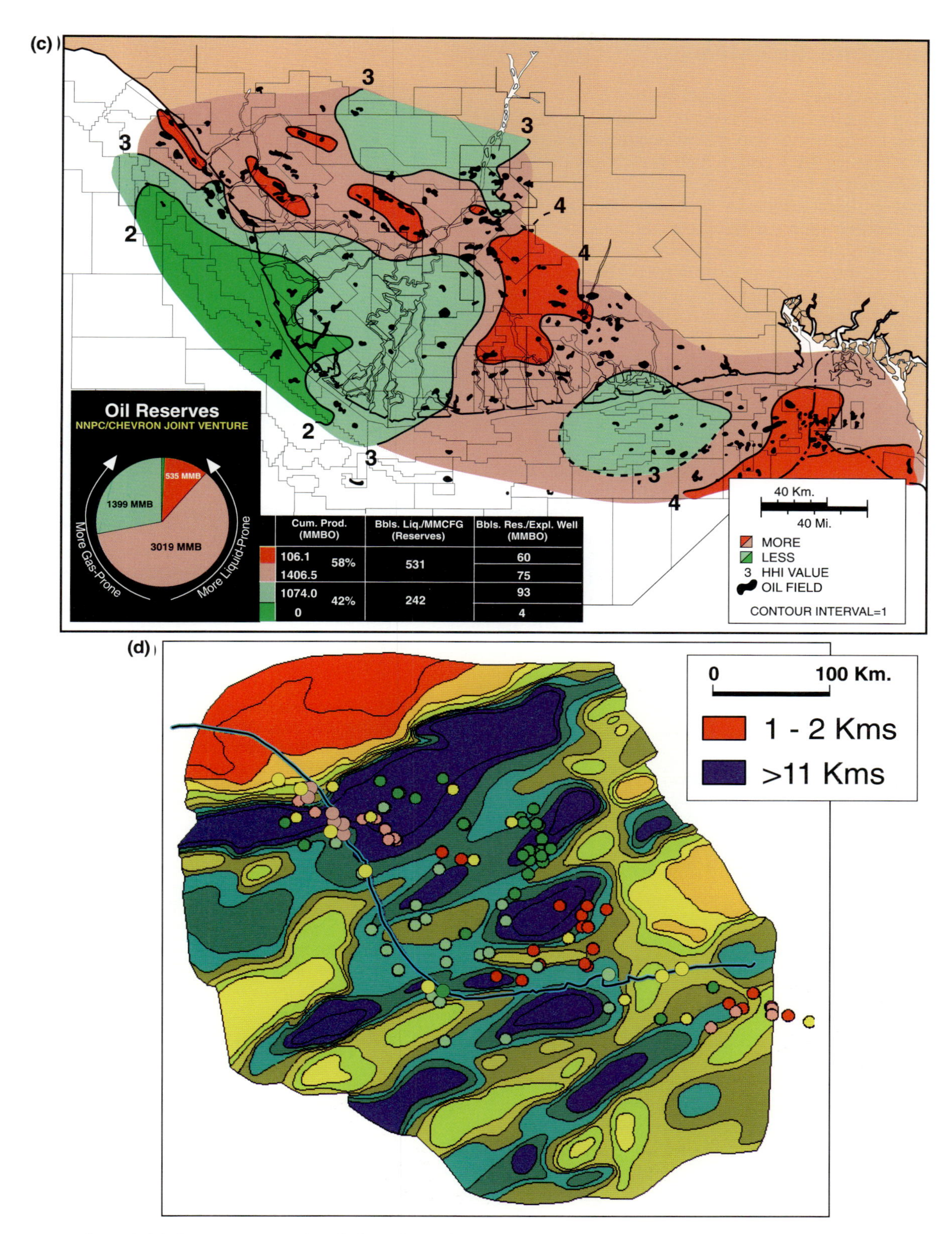

(Figure 14 continued) **(c) Contour map of variation in HHI (C_{35} HHI × 100) (Peters and Moldowan, 1993, p. 147), which indicates degree of variation in preservation of organic matter across the delta. Red areas represent better preservation of organic matter, hence more liquid prone, and green areas are poorer preservation, hence more gas prone. (d) Map showing the relationship between basement structure and oil subfamilies for the Tertiary petroleum system defined from cluster analysis. Patterns suggest that subbasins are charged by hydrocarbons from different end-members of source rock systems.**

Oil-Prone Source Rock Model

We propose a conceptual oil-prone source rock model for the Niger Delta which is based on detailed geologic and geochemical investigation in the northwestern part of the delta (Figure 15a) and which builds on understanding achieved from regional work. Oil fields within the northwest delta subbasin received hydrocarbon charge from Eocene source rocks. Large oil fields occur in the center of this subbasin, which directly overlies the deepest part of the hydrocarbon kitchen. Hydrocarbon charge southwest of the regional fault system, bounding the trend which includes Meren and Okan fields, is from a younger source rock sequence. Oil-prone source rocks from the delta contain both terrigenous and marine organic matter. The presence of marine organic matter is based on geochemical data and could not be verified microscopically because marine precursors could not be positively identified in available samples.

The relationships between organic facies and key geochemical signatures, including pristane/phytane ratio, HHI, C_{29} sterane content, and saturate carbon isotope value, are illustrated in Figure 15b. The oil-prone source rocks recovered from Aroh-2 well provide documentation for the terrigenous-dominated end-member on the left side of Figure 15b. The well occurs in the northeastern part of the northwest delta subbasin (shown in Figure 15a). Source rock samples from the more marine-influenced end-member on the right side of Figure 15b have not been recovered, but sufficient data from oils confirm its presence. For example, the change in organic facies recognized in the northwestern part of the delta, on the basis of C_{29} sterane content (Figure 14b), saturate carbon isotope values (Figure 15a), and results from cluster analysis (Figure 14a), strongly support the model of source facies variation proposed here. In addition, a corresponding change in organic facies can also be recognized on the shelf in the southeastern part of the delta (Figures 14a–d).

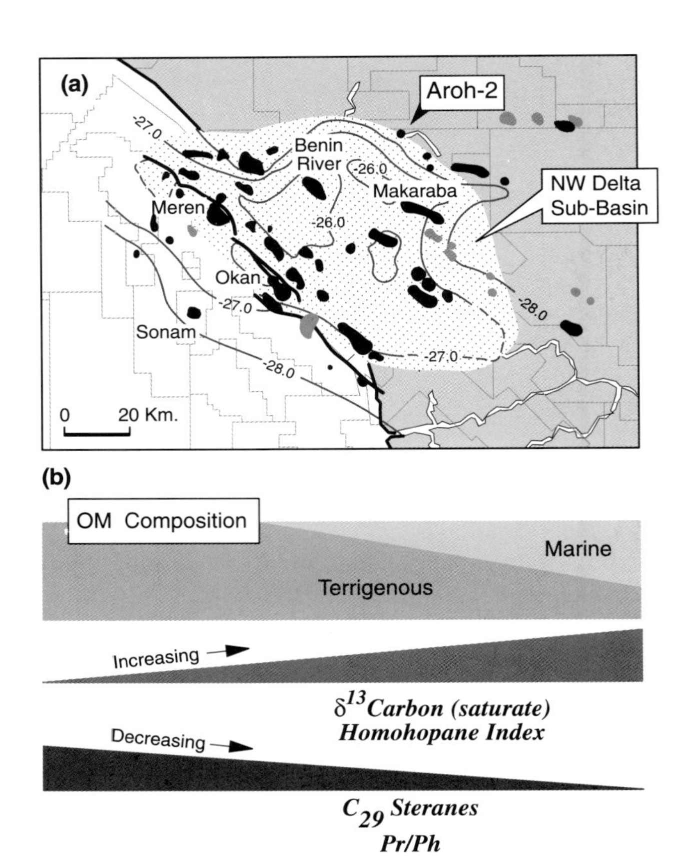

Figure 15—Oil-prone source rock model for the Niger Delta. (a) Contour map of saturate carbon isotope distribution from the northwestern part of the delta. Isotope values measured in oils from fields (sampled oil fields are dark gray; unsampled oil fields are light gray) vary from –29.3 to –24.0‰, which represents natural source facies variation characteristic of the Tertiary petroleum system. Values are heaviest (less negative) toward the center of the subbasin, where more marine organic matter is preserved. (b) Schematic diagram showing variation in source rock facies. Oil-prone source rocks from Aroh-2 well document one end-member (left side). Source rocks representing the more marine-influenced end-member (right side) have not been recovered.

Regional Maturity Trends Inferred from Oils

Biomarker data were also used to map regional trends in thermal maturity (Figure 16). The sterane isomerization parameter $20S/20S + 20R$ was selected for mapping. As expected, higher maturity is observed in the more proximal parts of the Niger Delta where geothermal gradients are higher. Furthermore, within each depobelt, defined by the regional faults in Figure 16, maturity decreases away from each major bounding fault. This pattern, repeated within all of the depobelts, implies that the major bounding faults do act as migration pathways for hydrocarbon withdrawal from deep within the source rock sequence. Because major faults have the deepest penetration into source rocks, they provide conduits for more mature hydrocarbons to migrate vertically and accumulate in structures formed by the major bounding faults.

Hydrocarbon Generation–Migration Model

We propose a simple episodic hydrocarbon migration model based on insights gained from regional study of the Niger Delta and focused on trying to understand underlying causes for the complex distribution of gas and oil regionally and within fields. As already demonstrated, variation in source facies, from gas prone to oil prone, is one of the important controls of regional trends in GOR. We now discuss additional controls implied by the proposed migration model.

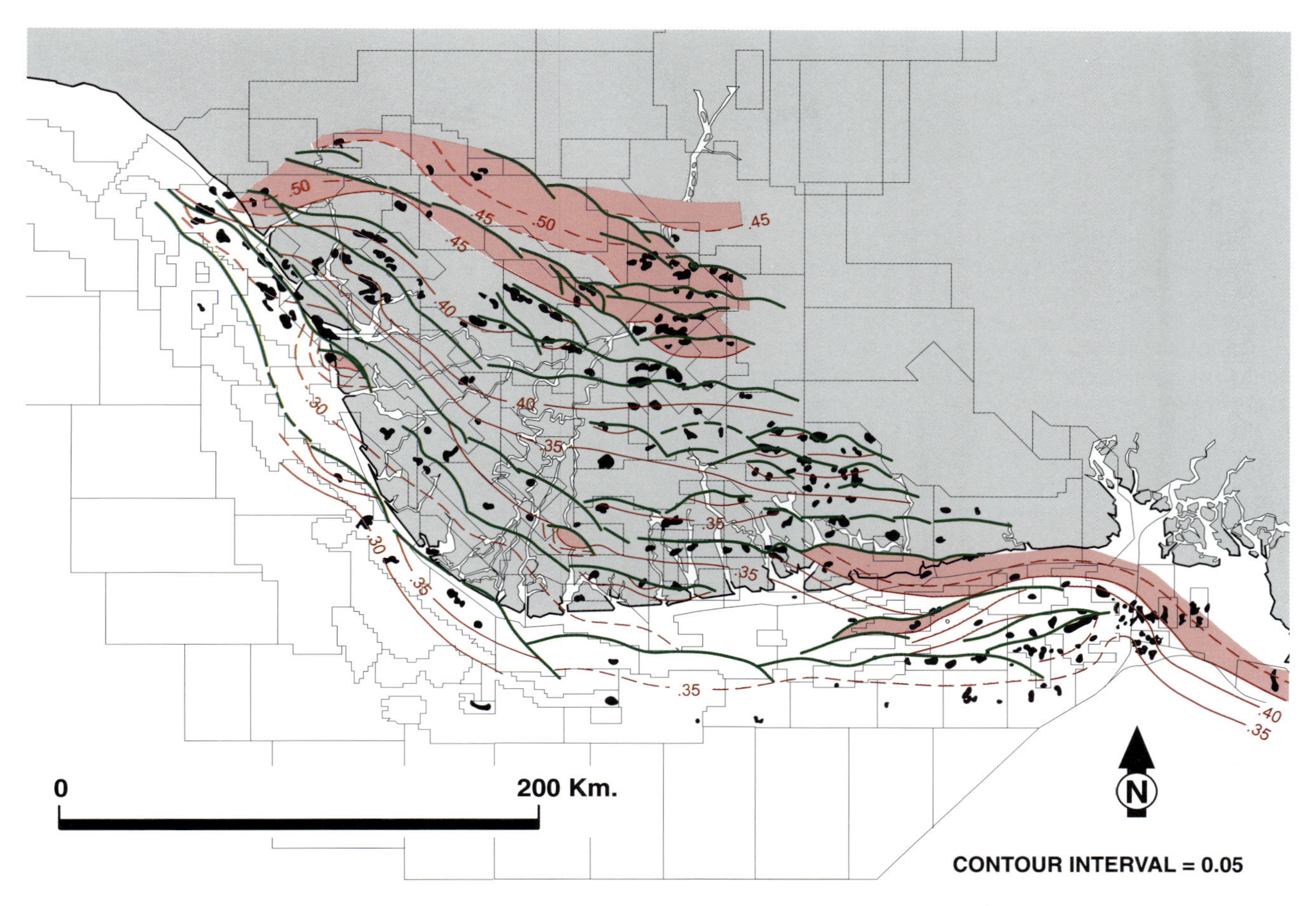

Figure 16—Contour map of regional thermal maturity across the Niger Delta inferred on the basis of sterane isomerization biomarker parameter 20*S*/20*S* + 20*R* in oils (Peters and Moldowan, 1993, p. 237). Major regional bounding faults shown in green define regional maturity compartments. Maturity contours are shown in red. Contour interval is 0.05.

The model illustrated in Figure 17 depicts burial of a typical source rock sequence in the Niger Delta: oil-prone source rocks overlain by gas-prone source rocks (e.g., Aroh-2 well). Oil is generated from oil-prone source rocks as they mature. With deeper burial, the oil-prone source, no longer capable of generating oil, begins to generate lighter hydrocarbon products. At the same time, shallower gas-prone source rocks are also likely to enter the generative window. These source rocks, which are not capable of generating oil, provide a second source for gas.

Traps, migration pathways, and connectivity of sandstone beds across faults progressively change through time concurrent with increasing burial and source rock maturation within each deformation cell. This dynamic evolution of the plumbing system, through which expelled hydrocarbons migrate, results in different traps and migration pathways being available to migrating hydrocarbons at different times. Because the composition of generated hydrocarbons also changes through time, as predicted by the model, older traps are likely to receive more liquid-rich accumulations than younger traps.

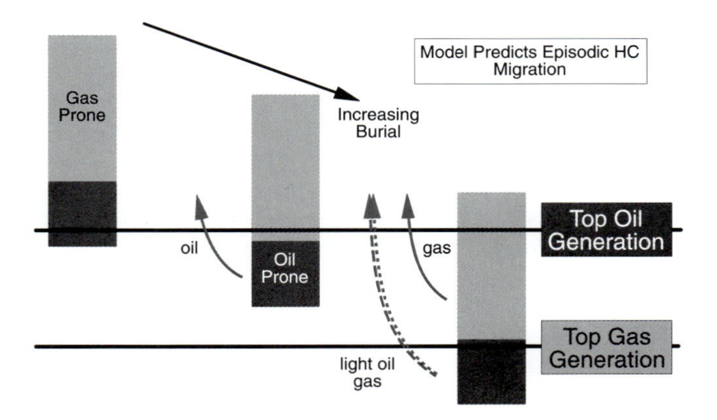

Figure 17—Conceptual generation–migration model for the Niger Delta. This model predicts that episodic hydrocarbon migration should be common for the Niger Delta. The principal assumption built into the model is that richer, oil-prone source rocks in the Akata are overlain by more gas-prone source rocks in the Agbada. All available source rock data are consistent with this assumption.

Even with this simple episodic migration model, understanding the present-day distribution of gas and oil in the Niger Delta is much more complicated because secondary processes, such as biodegradation and remigration, tend to alter the original composition of trapped hydrocarbons. Furthermore, gas solubility varies considerably within fields and across the delta due to varying hydrocarbon compositions and pressure-volume-temperature relationships. Nevertheless, the proposed model does provide a new framework for studying the complex GOR distributions across the Niger Delta within the context of an evolving and dynamic petroleum system. Within this context, the timing of generation-migration-entrapment becomes a focal point for further investigation.

CONCLUSIONS

Structural lows are inferred to be deeper than 11 km, based on gravity and magnetic data, and might provide separate hydrocarbon kitchens for hydrocarbon generation in the Niger Delta. Different subfamilies of oils, inferred to be derived from subtly different source rock facies, are associated with separate subbasins onshore. In addition, there is correspondence between subbasins and the largest oil fields in the delta (>200 million bbl).

The concept of a deformation cell is proposed to provide a better understanding of how to balance rock volume structurally in an extensional deltaic setting. The process–response model links creation of accommodation space for deposition of sediments to extensional growth faults and compressional thrust fronts.

Three petroleum systems are defined for the Niger Delta, and oils are correlated to each. The Lower Cretaceous petroleum system, characterized by lacustrine source rocks, occurs in the northwestern part of the delta and might also be present in the Benue Trough. The Upper Cretaceous–Lower Paleocene petroleum system, characterized by marine source rocks, is defined for the northwestern part of the delta. Correlative marine source rocks are undoubtedly present beneath the entire Tertiary delta and extend northeastward into the Anambra Basin. No significant oil accumulations are derived from either of these petroleum systems. For either to charge Tertiary reservoirs would require hydrocarbon migration pathways from the pregrowth source rock section through the mobile shale into the growth section.

The Tertiary deltaic petroleum system is defined by oil-prone source rocks in Aroh-2 well. This petroleum system is the principal source for hydrocarbons in the Niger Delta. Laterally and temporally varying source rock facies dominated by terrigenous organic matter, as well as regional differences in thermal maturity, are the principal factors controlling the complex regional distribution of oil and gas across the delta. There is reasonably good correspondence between oil reserves and richer source rock facies inferred on the basis of biomarker and carbon isotopic signatures from oils.

A simple episodic hydrocarbon migration model provides a new framework for studying the complex GOR distribution in the Niger Delta on a subregional scale and within fields. The model predicts changes in the composition of generated hydrocarbons through time, which occur within the geologic framework of an evolving and dynamic petroleum system, including hydrocarbon generation, migration pathways, reservoirs, and traps. Secondary processes and variable gas solubility, beyond the scope of this model, also play an important role in understanding the distribution of gas and oil in the Niger Delta.

Acknowledgments—*This work would not have been possible without support from the management of Chevron Nigeria Limited and our joint venture partner Nigerian National Petroleum Corporation. Their approval for publication of this work is appreciated. In addition, many thanks to colleagues from Shell Petroleum Development Company of Nigeria Limited, especially Daniel Trümpy, for permitting data trades. Those data provided us with a regional perspective that otherwise would be difficult to achieve. A special thanks to many Chevron colleagues and former Chevron colleagues, who over the years have inspired, supported, and critically reviewed many of the ideas presented in this paper, including Alan Nunns, Tom Heidrick, Mark Koelmel, Peter Berry, Brad Huizinga, Jeremy Dahl, and Mike Moldowan. There are others, of course, with whom we have worked over the years; we thank them too. We also thank Barry Katz, Mike Hoffman, and Ed Colling for reviewing and editing this paper. Their comments have greatly improved its quality. All of the figures were drafted by Rob Haitsma in the Chevron Drafting Department in San Ramon.*

REFERENCES CITED

Bustin, R. M., 1988, Sedimentology and characteristics of dispersed organic matter in Tertiary Niger Delta: origin of source rocks in a deltaic environment: AAPG Bulletin, v. 72, p. 277–298.

Core Laboratories, 1994, Nigeria geochemical study: Houston, Contractor report, June.

Damuth, J. E., 1994, Neogene gravity tectonics and depositional processes on the deep Niger Delta continental margin: Marine and Petroleum Geology, v. 11, n. 4, p. 320–346.

Demaison, G., and B. J. Huizinga, 1991, Genetic classification of petroleum systems: AAPG Bulletin, v. 75, p. 1626–1643.

Doust, H., and E. Omatsola, 1990, Niger Delta, *in* J. D. Edwards and P. A. Santagrossi, eds., Divergent/passive margin basins: AAPG Memoir 45, p. 201–238.

Ekweozor, C. M., and E. M. Daukoru, 1984, Petroleum source bed evaluation of the Tertiary Niger Delta: reply: AAPG Bulletin, v. 68, p. 390–394.

Ekweozor, C. M., and E. M. Daukoru, 1994, Northern delta depobelt portion of the Akata–Agbada(!) petroleum system, Niger Delta, Nigeria, *in* L. B. Magoon and W. G.

Dow, eds, The petroleum system—from source to trap: AAPG Memoir 60, p. 599–613.

Ekweozor, C. M., and J. I. Nwachukwu, 1989, The origin of tar sands of southwestern Nigeria: Nigerian Association of Petroleum Explorationists Bulletin, v. 4, p. 82–94.

Ekweozor, C. M., and N. V. Okoye, 1980, Petroleum source bed evaluation of the Tertiary Niger Delta: AAPG Bulletin, v. 64, p. 1251–1259.

Ekweozor, C. M., and O. T. Udo, 1988, The Oleananes: origin, maturation and limits of occurrence in southern Nigeria sedimentary basins: Organic Geochemistry, v. 13, p. 131–140.

Ekweozor, C. M., J. I. Okogun, D. E. U. Ekong, and J. R. Maxwell, 1979a, Preliminary organic geochemical studies of samples from the Niger Delta, Nigeria: Part 1, analysis of crude oils for triterpanes: Chemical Geology, v. 27, p. 11–28.

Ekweozor, C. M., J. I. Okogun, D. E. U. Ekong, and J. R. Maxwell, 1979b, Preliminary organic geochemical studies of samples from the Niger Delta, Nigeria: Part 2, analyses of shales: Chemical Geology, v. 27, p. 29–37.

Evamy, B. D., J. Haremboure, P. Kamerling, W. A. Knaap, F. A. Malloy, and P. H. Rowlands, 1978, Hydrocarbon habitat of Tertiary Niger Delta: AAPG Bulletin, v. 62, p. 1–39.

Frankl, E. J., and E. A. Cordry, 1967, The Niger Delta oil province—recent developments onshore and offshore: Seventh World Petroleum Congress, Proceedings, Mexico City, Mexico, v. 1B, p. 195–209.

Haack, R. C., and P. Sundararaman, 1994, Source rock variability and thermal maturity as controlling factors for business opportunities in the Niger Delta (abs.): Nigerian Association of Petroleum Explorationists Book of Abstracts, p. 17.

Haack, R. C., P. Sundararaman, and J. Dahl, 1997, Niger Delta petroleum system (abs.): AAPG/ABGP Hedberg Research Symposium, Extended Abstracts Volume, Rio de Janeiro, Brazil.

Haack, R. C., P. Sundararaman, J. O. Diedjomahor, N. J. Gant, and J. Dahl, 1998, Niger Delta petroleum systems (abs.): AAPG International Conference, Extended Abstracts Volume, Rio de Janeiro, Brazil, p. 936–937.

Hospers, J., 1965, Gravity field and structure of the Niger Delta, Nigeria, West Africa: GSA Bulletin, v. 76, p. 407–422.

Inyang, M. I., C. M. Ekweozor, and L. M. Pratt, 1995, Mid-Cretaceous anoxic events in southeastern Nigeria sedimentary basins—geochemical signatures and petroleum potential implications: Nigerian Association of Petroleum Explorationists, Official Programme, p. 34.

Knox, G. J., and E. M. Omatsola, 1989, Development of the Cenozoic Niger Delta in terms of the "escalator regression" model and impact on hydrocarbon distribution, *in* W. J. M. van der Linden et al., eds., 1987 Proceedings KNGMG Symposium on Coastal Lowlands, Geology, and Geotechnology: Dordrecht, The Netherlands, Klumer Academic Publishers, p. 181–202.

Lambert-Aikhionbare, D. O., and A. C. Ibe, 1984, Petroleum source bed evaluation of the Tertiary Niger Delta: discussion: AAPG Bulletin, v. 68, p. 387–394.

Nigerian Committee on Standardisation of Geological Nomenclature and Symbols, 1984, Standardisation of Geological Nomenclature and Symbols: Nigerian National Petroleum Corporation, 110 p.

Nwachukwu, J. I., and P. I. Chukwurah, 1986, Organic matter of Agbada Formation, Niger Delta, Nigeria: AAPG Bulletin, v. 70, p. 48–55.

Peters, K. E., and J. M. Moldowan, 1993, The biomarker guide: Englewood Cliffs, N.J., Prentice Hall, 363 p.

Schöellkopf, N. B., and B. A. Patterson, 1997, Petroleum systems of Cabinda, Angola (abs.): AAPG/ABGP Hedberg Research Symposium, Extended Abstracts Volume, Rio de Janeiro, Brazil.

Short, K. C., and A. J. Stäuble, 1967, Outline of geology of Niger Delta: AAPG Bulletin, v. 51, p. 761–779.

Stacher, P., 1995, Present understanding of the Niger Delta hydrocarbon habitat, *in* M. N. Oti and G. Postma, eds., Geology of deltas: Rotterdam, A. A. Balkema, p. 257–267.

Udo, O. T., 1985, Some aspects of the petroleum geochemistry of the Opuama clay channel complex of the Niger Delta: Ph.D. dissertation, University of Ibadan, Ibadan , Nigeria.

Weber, K. J., 1987, Hydrocarbon distribution patterns in Nigerian growth fault structures controlled by structural style and stratigraphy: Journal of Petroleum Science and Engineering, v. 1, p. 91–104.

Weber, K. J., and E. M. Daukoru, 1975, Petroleum geology of the Niger Delta: Proceedings of the Ninth World Petroleum Congress, Tokyo, Japan, v. 2, p. 209–221.

Akaegbobi, I. M., J. I. Nwachukwu, and M. Schmitt, 2000, Aromatic hydrocarbon distribution and calculation of oil and gas volumes in post-Santonian shale and coal, Anambra Basin, Nigeria, *in* M. R. Mello and B. J. Katz, eds., Petroleum systems of South Atlantic margins: AAPG Memoir 73, p. 233–245.

Chapter 17

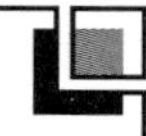

Aromatic Hydrocarbon Distribution and Calculation of Oil and Gas Volumes in Post-Santonian Shale and Coal, Anambra Basin, Nigeria

I. M. Akaegbobi

Department of Geology
University of Ibadan
Ibadan, Nigeria

J. I. Nwachukwu

Department of Geology
Obafemi Awolowo University
Ile-Ife, Nigeria

M. Schmitt

Geochemische Analysen
Sehnde, Federal Republic of Germany

Abstract

The generation and distribution of C_{11+} aromatic hydrocarbons has been investigated in 24 rock samples from the Campanian–Maastrichtian Nkporo shales and coals of the Anambra Basin in the lower Benue Trough. The study focused on dicyclic and tricyclic aromatic hydrocarbons identified on gas chromatograms. The distribution of these compounds and their alkyl homologs are strongly controlled by thermal maturation of organic matter. Compositional variation in the organic macerals as indicated by optical observation seemed to be of minor significance.

A gradual depth trend was observed for compound ratios calculated from the relative variations of the methyl aromatics in each family. The methylphenanthrene index (MPI 1) in particular, as well as other indices derived from dicyclic aromatics, demonstrated a good correlation with maturity. Theoretically calculated vitrinite reflectance values from MPI 1 range from 0.60 to 0.89% R_c. These values are consistent with measured reflectance values (0.61–0.83% R_o) and T_{max} values (430–471°C) for type III kerogen within the corresponding depth intervals. These results confirm that the Nkporo Shale and the Lower Coal Measures (Mamu Formation) are mature within the depth range of 1555–2389 m.

Volumetric estimates of the hydrocarbons generated by these shales and coals show that about 779 million bbl of oil or 4.57×10^{13} ft^3 of gas may have been generated and probably entrapped in the basin. The basin's low source potential index (4.02) indicates an undercharged system and hence low exploration potential for the basin.

INTRODUCTION

Aromatic hydrocarbons are one of the major components of crude oils and organic extracts in coals, ancient sedimentary rocks, and recent sediments. These alkylated polycyclic aromatic compounds (PAHs) are believed to be derived from degradation of nonaromatic natural precursors (Blumer and Youngblood, 1975; Wakeham et al., 1980a, b). Hase and Hites (1976) also observed that the complex aromatic mixtures contained in rock extracts can be generated through biosynthetic processes in sediments. However, living organisms may synthesize large quantities of benzol derivatives in nature. The distribution pattern of these complex aromatic hydrocarbon compounds may be controlled by postdepositional alterations due to thermal maturation of sedimentary organic matter.

Aromatic hydrocarbons present in recent and ancient strata are derived from nonaromatic biological precursors, possibly polycyclic organic compounds such as

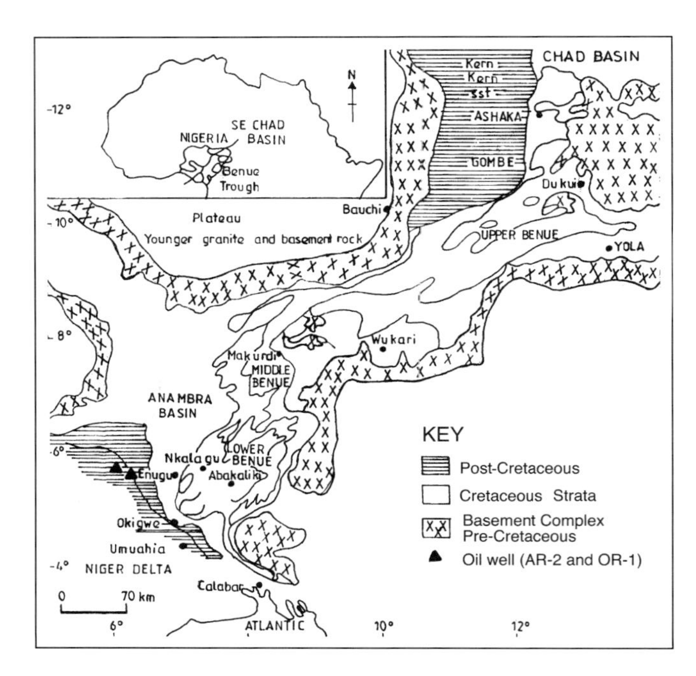

Figure 1—Geologic map of the Benue Trough showing the Anambra Basin and study well locations.

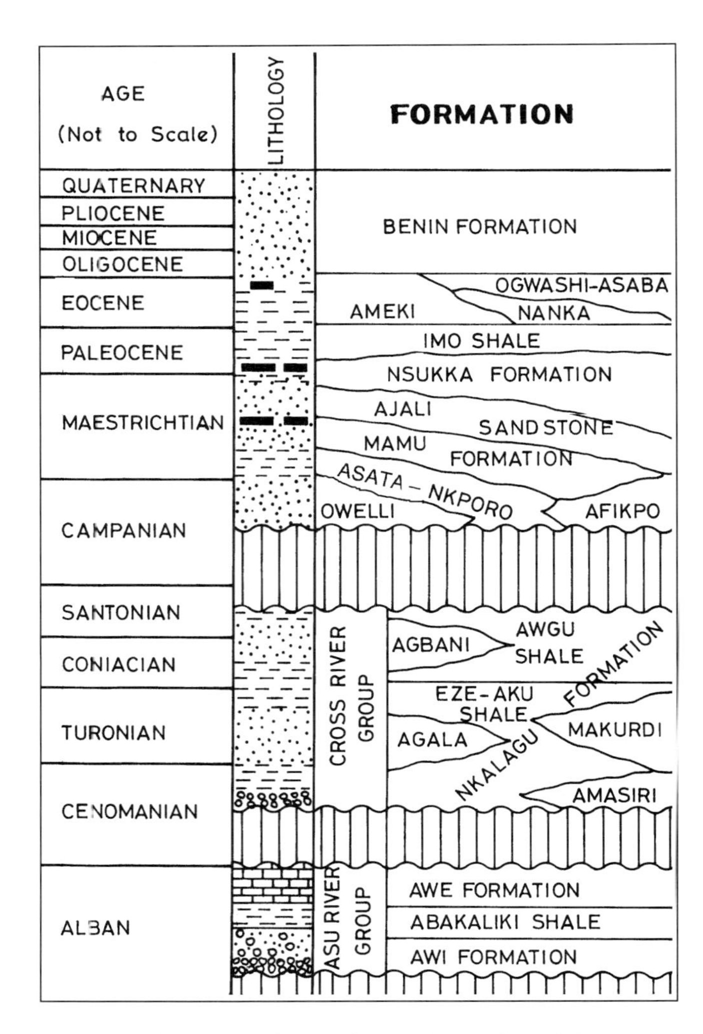

Figure 2—Stratigraphic and lithologic section of the lower Benue Trough and the Anambra Basin.

steroids and triterpenoids, which are converted into steranes and triterpanes during diagenesis (Garrigues et al., 1988; Radke, 1988; Strachen et al., 1988; Loureiro and Cardoso, 1990). It is assumed that phenanthrenes, naphthalenes, picene, chrycene, and their homologs that are present in ancient sediments are derived from these precursors through aromatization processes of their saturated rings. Therefore, it is generally accepted that the maturity of organic materials and facies are reflected in the distribution of aromatics.

This study has two main objectives. The first objective is to examine these aromatic thermal maturity indices in the Anambra Basin (lower Benue Trough) in two wells, Oda River-1 (OR-1) and Anambra River-2 (AR-2), from which ditch cuttings were available down to total depth. The second objective is to assess the petroleum potential of post-Santonian shales and coals not affected by Santonian tectonism and magmatic activity. Previous organic geochemical investigations of shales of the lower Benue Trough were largely based on conventional techniques (Rock-Eval pyrolysis, vitrinite reflectance measurements, and studies of bitumen, clay mineral, and fluid inclusions) (e.g., Ekweozor and Gormly, 1983; Unomah and Ekweozor, 1993; Akande and Erdtmann, 1998). Such an assessment is considered timely because early exploratory drilling in the Anambra Basin has confirmed the presence of gas and oil in Cretaceous formations (Avbovbo and Ayoola, 1981), but no commercial discoveries have yet been made and exploratory activity remains minimal.

GEOLOGIC SETTING

The Anambra Basin is located in the southwestern end of the Benue Trough (Figure 1) and contains up to 9 km of siliciclastic sedimentary strata (Whiteman, 1982). The basin was initiated during the Late Cretaceous and covers an area of about 40,000 km^2. It extends northward to the lower Benue River and also forms a boundary with the Tertiary Niger Delta to the south. The structural setting and general geology of the Anambra Basin have been documented by various workers (e.g., Agagu and Ekweozor, 1982; Petters and Ekweozor, 1982; Agagu et al., 1986; Nwachukwu et al., 1988; Nwajide, 1990; Nwajide and Reijers, 1997). The stratigraphy of the infilling sedimentary sequence becomes younger southward, from pre-Albian to Tertiary in the present Niger Delta Basin. The sedimentary thicknesses (>2500 m) increase to the south, attaining a maximum thickness of 12,000 m in the central Niger Delta.

Aptian–Albian pyroclastics were the earliest strata to be deposited unconformably on the basement complex in the lower Benue Trough. The basement complex is formed of undifferentiated Precambrian rocks of the west African craton, consisting mainly of migmatites, gneisses, intrusives, extrusives, and diverse metasedimentary rocks. They have been variously affected by extensive metamorphism, metasomatism, and mineralization during the course of their geotectonic evolution (McCurry, 1976; Rahaman, 1976). During the Albian–Santonian, the proto-Anambra Basin was a platform only thinly covered by older strata. This sedimentary sequence consists of the 3-km-thick transgressive Albian Asu River Group (Figure 2), which is unconformably overlain by the 2-km-thick Cenomanian–early Santonian strata of the Cross River Group. This includes shales, limestones, and sandstones of the Nkalagu Formation, which consist of the Eze-Aku and Awgu Shales and their interfingering local facies equivalents (Amasiri, Makurdi, Agala, and Agbani sandstones) (Petters and Ekweozor, 1982).

Widespread Santonian tectonism due to rifting and compressional upliftment of the Abakaliki–Benue Trough affected the sedimentary rocks of the Benue Trough, which, in turn, resulted in folding and subsidence of the Anambra Basin. Post-Santonian strata unconformably overlie the Cross River Group and include the paralic Enugu (Asata) Shale, the Nkporo Shale, the coal measures of the Mamu and Nsukka Formations, and the Ajali Sandstone.

The Paleocene Imo Shale was deposited as a transgressive facies. During the Eocene, the Ameki Group was deposited, the Anambra Basin was filled, and the Niger Delta prograded southward. The basin-fill sediments are typical of mixed continental, fluvial–deltaic, and shallow-marine sedimentation.

SAMPLES AND METHODS

We studied 24 ditch samples from the Oda River-1 (OR-1) and Anambra River-2 (AR-2) wells located in the north-central part of the Anambra Basin (Figure 1). These samples were retrieved at various depth intervals and cover all the lithologic units ranging from the Imo Shale through the coal measures to the Nkporo Shale (Figure 3). The samples are mostly shale, siltstone, and lignite which, under normal light binocular microscope and as hand specimens, contain dark substances with a characteristic odor believed to be bitumen. They are generally fine to very fine grained, dark brown to dark gray, silty and lignitic shales and claystones that are slightly bituminous and marly (Table 1). The two wells penetrated Upper Cretaceous–Paleocene sedimentary rocks and have total depths of 2390 m (OR-1) and 2180 m (AR-2). The wells were drilled with oil-base mud, and so the samples were thoroughly cleaned with hydrogen peroxide and sun dried for 48 hours. Other sample contaminants, including wall nut shells, lignite particles, and rubber chippings used for mud stabilization, were hand-picked before pulverization of samples.

Details of the analytical procedure are as follows. Pulverized coal and shale samples (20 g) derived mainly from the Imo Shale, Upper Coal Measure, Lower Coal Measure, and Asata–Nkporo Shale were extracted with dichloromethane in a Soxhlet apparatus for 48 hours at a temperature of 40°C. Elemental sulfur was removed by addition of copper powder. The coal and rock extract was first fractionated by medium-pressure liquid chromatography (MPLC) into C_{15+} saturated and C_{11+} aromatic hydrocarbon fractions (Radke et al., 1980). Further separation of the aromatic fractions into four ring classes (AF1–AF4) was accomplished by means of semi-preparative high-performance liquid chromatography (HPLC) (Radke et al., 1980, 1984). The naphthalene, phenanthrene, and sulfur-containing aromatics were subclassified as AF1B (diaromatics) and AF2 (triaromatics and pericondensed tetra-aromatics). The aromatic subfractions were dissolved in xylene and analyzed on an HP5731A gas chromatograph (GC) equipped with gerstel inlet splitter, flame ionization detector (FID), and tracor flame photometric detector (FPD). Chromatographic conditions were as follows: glass capillary column, 25-m length, 0.3-mm i.d., SE-54 silicone gum phase; temperature programmed at 100°C (2 min), 3°C/min to 250°C, hold 20 min; carrier gas helium 5 mL/min; split ratio inlet 1:10; outlet 1:1.

THERMAL MATURITY

The geology, organic facies, and maturation of the shales and carbonaceous shales of the lower Benue Trough have been discussed by a number of authors (Unomah and Ekweozor, 1993; Akaegbobi, 1995; Akaegbobi and Schmitt, 1998). These previously published works on the organic-rich shales focused primarily on the saturated hydrocarbon fractions (e.g., *n*-alkane distribution, carbon preference index, and Pr/Ph ratio) and on optical studies of isolated kerogens and coal. However, aromatic-dependent geochemical parameters have recently become of great interest, as the geochemical relevance of this hydrocarbon group in maturity studies has become obvious (Radke et al., 1984; Boreham et al., 1988; Garrigues et al., 1988; Radke, 1988). Information about the thermal maturation of sedimentary organic matter can be derived from systematic changes in the isomer distribution pattern of methylated di- and triaromatic hydrocarbons and of sulfur heterocycles contained in the rock and coal extracts and crude oil. The maturity of the Upper Coal Measures, Lower Coal Measures (Mamu Formation), and Asata–Nkporo Shale ranges from 0.65 to 0.85 $\%R_o$ and total organic carbon (TOC) ranges from 0.98 to 4.71% (Table 1). Certain factors make it possible for terrigenous type III kerogen and coals to generate and expel hydrocarbons in liquid or gaseous phase (see Hunt, 1991; Snowdon, 1991; Fleet and Scott, 1994). It is therefore important to properly understand the

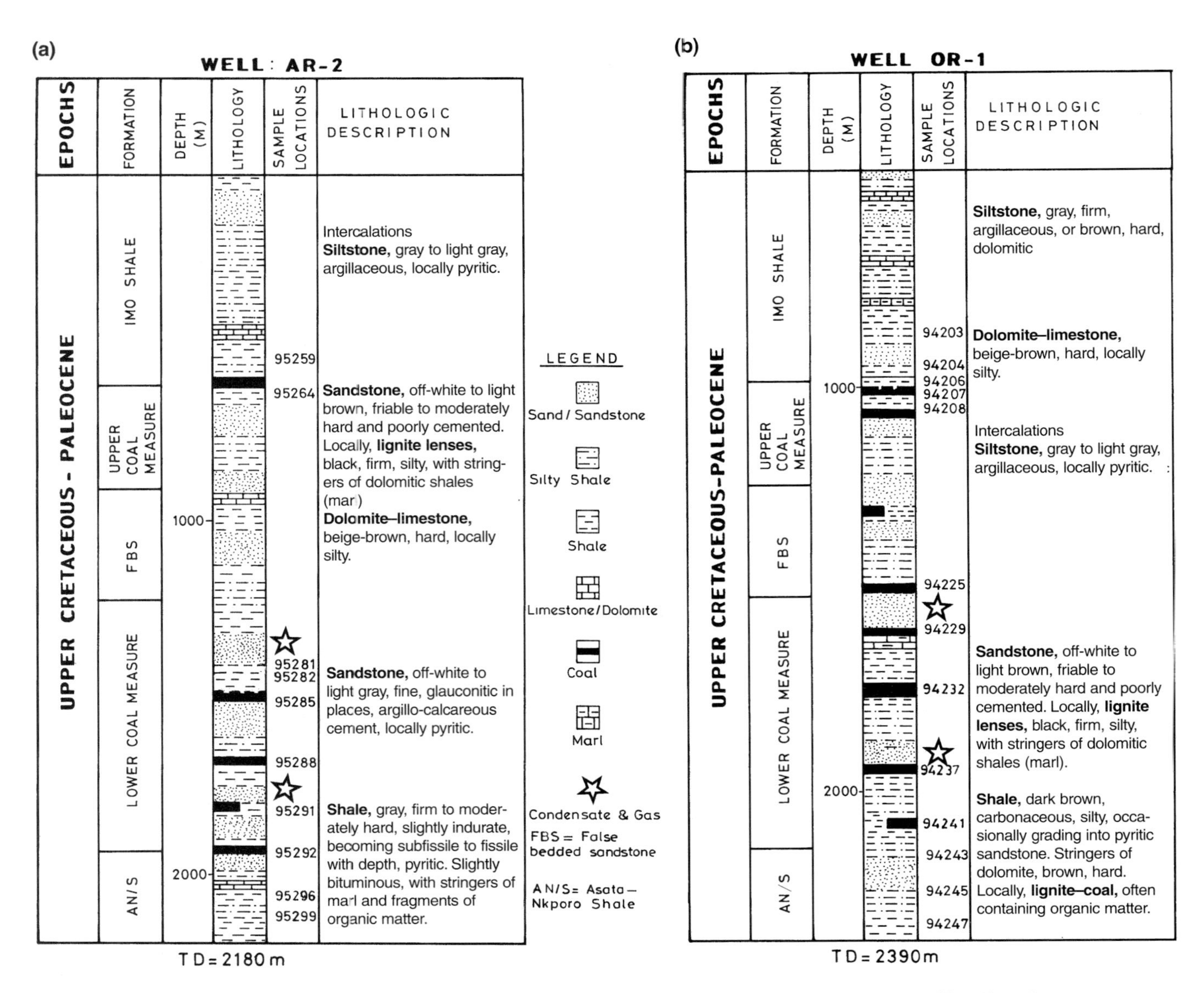

Figure 3—Lithologic columnar sections from (a) Anambra River-2 (AR-2) well and (b) Oda River-1 (OR-1) well.

relative contributions of hydrocarbons from type III kerogens and coals in mature facies. This can help petroleum explorers to plan both the charging system and the hydrocarbon phase in a prospect before drilling.

The key factor in the recognition of oil-prone and gas-prone terrigeneous organic matter and coal is the relative enrichment of hydrogen to carbon atoms in the organic matter (Fleet and Scott, 1994). Thus, the association of the microscopic components of terrigeneous organic matter and coals, with special emphasis on hydrogen content, becomes one of the criteria for using petrography in source potential evaluation. Because a quantitative understanding of the hydrocarbon-generating capability of individual macerals is lacking, it becomes neccessary to combine petrography with geochemistry in source rock studies (Powell and Boreham, 1994). Observations of organic matter types (Powell, 1978; Hunt, 1991) generally suggest that for a source rock to be effective, organic matter content must fall within the range of 10–20% for type I, 20–30% for type II, and >30% for type III kerogen. Bulk atomic H/C ratios fall in the range of 0.80–0.90, and hydrogen indices (HI) from Rock-Eval analysis are >200 mg HC/g TOC, with a liptinite content of >15%. Maceral group analysis revealed an elevated vitrinite content (Table 2); this vitrinitic material was identified as mainly hydrogen-rich desmocollinite. Liptinites account for only about 20% of the maceral content, whereas resinite and other nonidentifiable alginites (algodetrinite) are rare (Table 2). Certain hydrogen-rich vitrinites along with a few resinites have filled in the pore spaces of both the shale and coaly facies. These macerals seem to be derived from plant organs containing lignin and cellulose.

Table 1—Rock-Eval Pyrolysis and Vitrinite Reflectance Data for Analyzed Samples

Sample No.	Depth (m)	Lithology	TOC (%)	HI	OI	PI	T_{max} (°C)	R_o (%)
94203	850	Shale	1.61	9	22	0.07	430	0.56
94204	900	Shale	1.20	10	35	0.07	427	0.61
94206	960	Shale	1.16	12	45	0.06	425	0.51
94207	1000	Coal	2.68	22	58	0.06	424	0.65
94208	1020	Coal	3.28	51	78	0.09	430	0.75
94225	1500	Coal	0.98	54	91	0.10	430	—
94229	1600	Coal	1.78	57	65	0.08	430	0.60
94232	1750	Coal	3.91	99	32	0.07	431	0.65
94237	2000	Coal	1.92	50	180	0.12	440	—
94241	2200	Coaly shale	1.74	29	25	0.17	447	0.73
94243	2300	Bit. shale	2.17	23	45	0.10	434	0.81
94245	2385	Bit. shale	1.78	27	63	0.12	442	—
94247	2390	Bit. shale	2.02	29	44	0.16	441	0.79
95259	560	Shale	1.51	35	107	0.07	426	0.50
95264	620	Coaly shale	2.77	40	263	0.03	443	0.73
95281	1550	Coal	4.17	99	54	0.04	443	0.74
95282	1555	Coal	4.71	114	57	0.05	440	0.79
95285	1600	Coal	3.81	88	64	0.04	437	0.80
95288	1750	Coal	2.94	42	77	0.07	438	—
95291	1890	Coal	1.67	19	89	0.03	450	0.75
95292	1900	Coal	2.50	50	72	0.08	437	0.82
95296	2055	Bit. shale	1.83	24	78	0.08	471	0.85
95299	2130	Bit. shale	3.60	18	34	0.17	442	—
95300	2145	Bit. shale	2.02	15	76	0.06	444	0.83

Table 2—Maceral and Vitrinite Reflectance Data for Analyzed Samples

Sample No.	Depth (m)	Lithology	Vitrinite (%)	Liptinite (%)	Inertinite (%)	Min. Comp. (%)	Pyrite (%)	R_o (%)
94203	850	Shale	55	10	5	21	9	0.56
94204	900	Shale	28	15	8	32	17	0.61
94206	960	Shale	35	23	7	15	20	0.51
94207	1000	Coal	34	28	10	23	5	0.65
94208	1020	Coal	57	15	9	11	8	0.75
94225	1500	Coal	—	—	—	—	—	—
94229	1600	Coal	65	18	4	8	5	0.60
94232	1750	Coal	38	29	5	19	9	0.65
94237	2000	Coal	—	—	—	—	—	—
94241	2200	Coaly shale	45	30	16	4	5	0.73
94243	2300	Bit. shale	31	22	8	27	12	0.81
94245	2385	Bit. shale	—	—	—	—	—	—
94247	2390	Bit. shale	35	18	8	22	17	0.79
95259	560	Shale	32	12	11	37	8	0.50
95264	620	Coaly shale	43	18	7	22	10	0.73
95281	1550	Coal	48	15	12	18	7	0.74
95282	1555	Coal	42	25	6	15	9	0.79
95285	1600	Coal	55	12	9	16	8	0.80
95288	1750	Coal	—	—	—	—	—	—
95291	1890	Coal	55	15	6	14	10	0.75
95292	1900	Coal	53	18	9	12	8	0.82
95296	2055	Bit. shale	65	10	7	7	11	0.85
95299	2130	Bit. shale	—	—	—	—	—	—
95300	2145	Bit. shale	35	22	12	26	5	0.83

C_{11+} Aromatic Hydrocarbon Yield

The quantitative generation of polycyclic aromatic hydrocarbons was carried out for selected samples. This methodic approach (after Radke et al., 1986) was based on the fact that the concentrations of aromatic hydrocarbons in sedimentary extracts increase with increasing maturity of the source rock in response to thermal effects (Radke et al., 1986). The quantification of individual aromatic compound groups was accomplished by using several internal standards; the results are summarized in Tables 3 and 4. Data interpretation was limited to the aromatic groups of naphthalene, phenanthrene, and their alkylated isomers.

The rock and coal extracts contain relatively high amounts of aromatic components (Table 3). Although the distribution patterns of aromatic constituents relative to one another do not show any significant variations among the sample sets, the concentration of the aromatics relative to the aliphatic fraction may be considered high despite the low hydrogen index (HI) of the samples (Table 3). Naphthalene and its methylated homologs seem to be the largest individual components in the extracts that also contain high concentrations of phenanthrene and its methyl isomers (Table 4). The C_{11+} aromatic hydrocarbon distribution appears to be controlled by the degree of maturation. Although there is no clear depth trend, yields between 3 and 16 mg of C_{11+} aromatic HC/g TOC in the sample sets, with vitrinite reflectance ranging from 0.51 to 0.85% R_o, are considered to have attained a stage corresponding to the onset of oil and gas formation.

Gas chromatography (GC) of the total aromatic fractions from the Asata–Nkporo Shale reveal an extremely complex mixture of aromatic compounds displaying a wide molecular weight range from alkyl homologs of simple benzenes up to highly condensed polycyclic aromatic hydrocarbons such as naphthalene, phenanthrene, and anthracene. Some of the aromatic compounds contain heterocyclic structural elements of the thiophenic types. Separation of the aromatic fraction into subgroups of two- and three-ring aromatic systems followed by high-resolution GC analysis with simultaneous flame ionization and sulfur-selective flame photometric detection generally improved in the assessment of the aromatic distribution within the shale and coal samples. A typical gas chromatogram is shown in Figure 4. Maturity evaluation of the sedimentary organic matter in the Upper Coal Measures, the Lower Coal Measures, and the Asata–Nkporo Shale was focused mainly on the di- and tricyclic aromatic distribution pattern. With increasing depth in both studied wells (OR-1 and AR-2), features of the di- and triaromatics as they relate to their relative abundance are detailed in Table 4 and Figure 5 and summarized as follows:

1. The distribution of phenanthrene and its methyl isomers used in defining the triaromatic maturity indices change progressively with maturation.
2. The relative concentration of phenanthrene (μg/g TOC) increases almost constantly with depth in response to thermal stress.
3. The relative concentration of methyl- and dimethylphenanthrene increase with depth.
4. The naphthalenes and their homologs generally show clear variations in their relative concentration with burial depth.
5. Methyl- and ethylnaphthalene increase with depth, along with a corresponding increase in the relative abundance of dimethylnaphthalene.
6. The relative concentration of the sum total of the naphthalenes and its isomers is greater than that of the phenanthrenes.

The relative abundance of individual aromatic ring classes in the total aromatic fraction appears to be maturity and/or temperature controlled. The mono- and diaromatics are more abundant in samples with vitrinite reflectance of $<0.70\%$ R_o. The methylnaphthalene homologs are important only within a narrow maturity range (0.6–0.8% R_o). At higher maturity levels (0.83%R_o), the distribution is dominated by the characteristic pattern of methyl- and dimethylnaphthalenes (Figure 5). A marked influence from maturity and/or temperature is indicated by an abrupt increase in the isomer ratios TNR, MNR, and DNR at 0.80–0.83% R_o (Table 4). (For abbreviations and definitions of aromatic isomer ratios, see Table 5.) However, below 0.80% R_o maturity stage, methylnaphthalene dominates over dimethylnaphthalene. (Figure 5). The increase in the relative abundance of dimethylnaphthalene at higher rank can be attributed to a thermal rearrangement of 1-methylnaphthalene with increasing thermal stress. The rearrangement of the 1,5-dimethylnaphthalene, which belongs to the isomer of the α,α-type to yield the α,β- and β,β-type isomers explains the mechanism behind the relatively high values for the DNR.

With regards to other bulk geochemical data, such as the production index (PI), extract yield, and *n*-alkane spectrum (Tables 2 and 3), the discrepancy in the relative concentration of the various di- and triaromatics could be attributed to hydrocarbon expulsion effects. Apparently, the higher abundances of dimethyl- and trimethylnaphthalenes relative to phenanthrenes can be attributed to selective expulsion mechanisms. However, no preferential expulsion of methylphenanthrenes and dimethylphenanthrenes relative to naphthalenes was observed for the type II kerogen source rock of the Kimmeridge Formation from the Brae field during primary migration (Leythaeuser et al., 1988).

A migration effect as the possible cause of the relatively low concentrations of phenanthrenes may not be generally accepted, but it must be mentioned that the sample sets are of type III and III–II source rocks. It is therefore likely that higher plants contributed significantly to the source of compounds that serve as precursors for these isomers of the diaromatics. According to published data on primary migration (Leythaeuser et al.,

Table 3—TOC and Bitumen Extract Yield for the Two Wells

Sample No.	Depth (m)	TOC (%)	Yield (ppm)	Bit Ratio (mg Ext./g TOC)	Saturates (%)	Aromatics (%)	NSO (%)
94203	850	1.61	379	24	21.5	5.8	56.1
94204	900	1.2	337	27	22.8	2.3	39.8
94206	960	1.16	609	53	18.6	4.7	32
94207	1000	2.68	1135	42	6.8	6.6	15.3
94208	1020	3.28	1513	46	40.6	13.7	33.8
94225	1500	0.98	—	—	—	—	—
94229	1600	1.78	572	56	11.7	14.7	28.9
94232	1750	3.91	—	—	—	—	—
94237	2000	1.92	—	—	—	—	—
94241	2200	1.74	1161	66	37.5	11.6	24.5
94243	2300	2.17	1266	58	44.1	13.3	22.8
94245	2385	1.78	912	51	17.6	15.5	24.1
94247	2390	2.02	1602	79	18.4	14.3	22.6
95259	560	1.51	571	37	32.2	8.3	52.4
95264	620	2.77	564	20	26.5	15.0	38.0
95281	1550	4.17	1870	45	46.5	15.3	33.5
95282	1555	4.71	1849	39	49.2	13.0	32.4
95285	1600	3.81	—	—	—	—	—
95288	1750	2.94	564	19	26.5	15.0	38.0
95291	1890	1.67	799	48	18.7	17.50	28.3
95292	1900	2.5	394	16	32.8	15.8	40.9
95296	2055	1.83	627	34	42.2	9.5	29.0
95299	2130	3.6	1974	55	15.1	13.4	26.5
95300	2145	2.02	1049	52	37.3	16.8	35.4

Table 4—Kerogen Type and Aromatic Ratios[a] for Samples

Sample No.	Depth (m)	Lithology	Kerogen Type	T_{max} (°C)	R_o (%)	R_c (%)	MPI	MPR	MNR	DNR	TNR
94203	850	Shale	III	430	0.56	—	—	—	—	—	—
94204	900	Shale	III	427	0.61	0.60	0.42	—	1.60	3.84	0.46
94206	960	shale	III	425	0.51	—	—	—	—	—	—
94207	1000	Coal	III	424	0.65	0.64	0.44	0.86	1.59	3.33	0.58
94208	1020	Coal	III	430	0.75	—	—	—	—	—	—
94225	1500	Coal	III	430	—	—	—	—	—	—	—
94229	1600	Coal	III	430	0.60	0.67	0.51	0.85	1.56	4.20	0.72
94232	1750	Coal	III	431	0.65	0.70	0.53		1.45	3.80	0.71
94237	2000	Coal	III	440	—	—	—	—	—	—	—
94241	2200	Coaly sh	III	447	0.73	0.82	0.78	0.97	1.46	3.85	0.73
94243	2300	Bit. shale	III/II	434	0.81	0.83	0.80	1.00	1.39	4.07	0.75
94245	2385	Bit. shale	III/II	442	—	—	—	—	—	—	—
94247	2390	Bit. shale	III/II	441	0.79	0.84	0.82	1.02	1.37	3.95	0.78
95259	560	Shale	III	426	0.50	—	—	—	—	—	—
95264	620	Coaly sh	III	443	0.73	0.70	0.51	0.89	1.48	4.10	0.76
95281	1550	Coal	III	443	0.74	—	—	—	—	—	—
95282	1555	Coal	III	440	0.79	0.79	0.72	0.94	1.46	4.22	0.63
95285	1600	Coal	III	437	0.80	0.80	0.76	0.98	1.45	4.31	0.75
95288	1750	Coal	III	438	—	—	—	—	—	—	—
95291	1890	Coal	III	450	0.75	0.83	0.82	0.99	1.50	4.45	0.78
95292	1900	Coal	III	437	0.82	—	—	—	—	—	—
95296	2055	Bit. shale	III/II	471	0.85	0.88	0.86	1.10	1.66	4.84	0.83
95299	2130	Bit. shale	III/II	442	—	—	—	—	—	—	—
95300	2145	Bit. shale	III/II	444	0.83	0.89	0.88	1.15	1.38	4.86	0.82

[a]See Table 5 for definition of ratios.

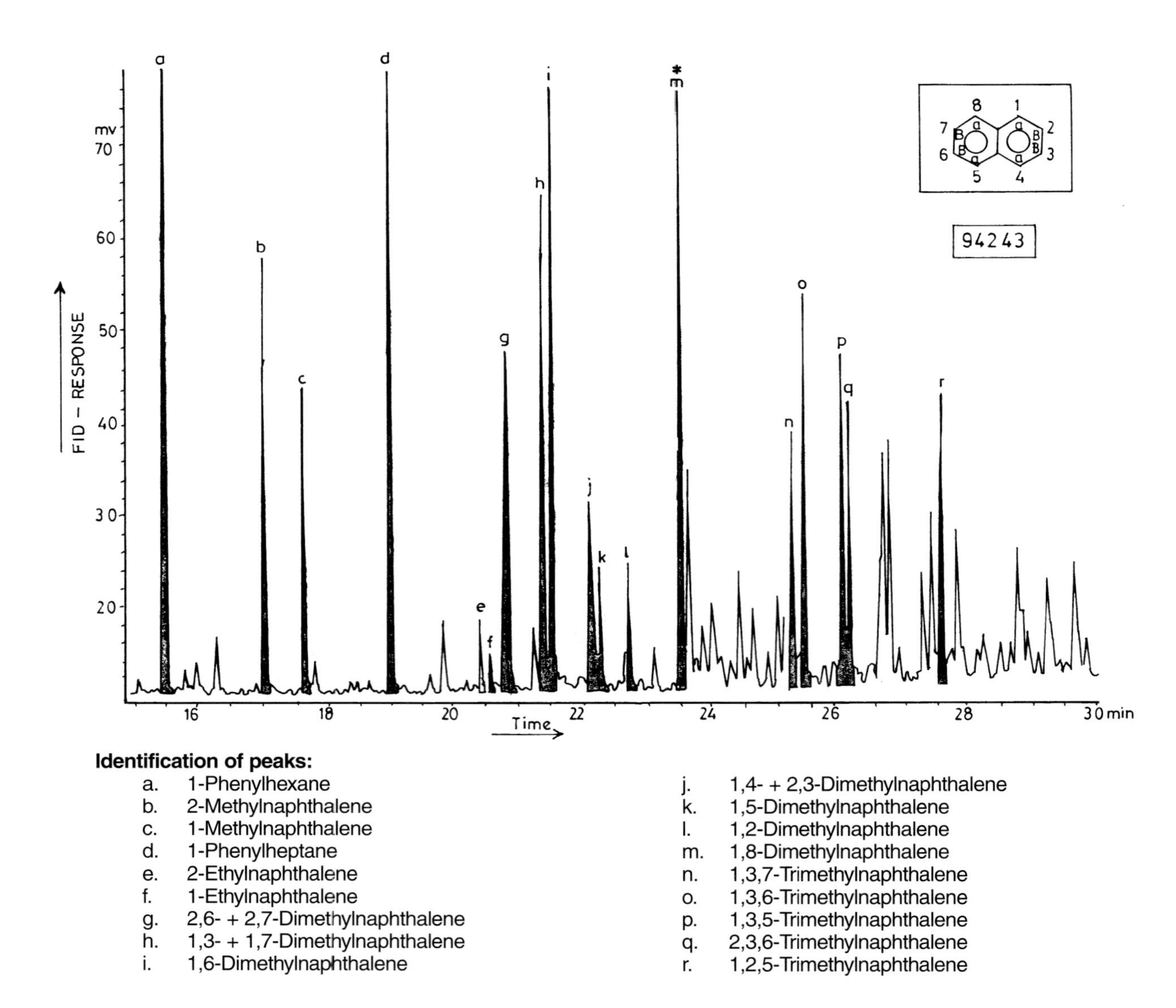

Figure 4—Gas chromatogram of the aromatic fraction of sample 94243 from Oda River-1 well. The internal standard (peak "m") is marked with a star.

Table 5—Definition of Aromatic Isomer Ratios

Name	Abbreviation	Definit on	Reference
Alkylnaphthalene ratios:			
Methylnaphthalene ratio	MNR	[2 – MN]/[1 – MN]	Radke et al. (1982b)
Dimethylnaphthalene ratio	DNR	{[2.6 – DMN]+[2.7 – DMN]}/[1.5 – DMN]	Radke et al. (1982b)
Trimethylnaphthalene ratio	TNR	[2.3.6 – TMN]/{[1.3.5 – TMN] + [1.4.6 – TMN]}	Alexander et al. (1985)
Alkylphenanthrene ratios:			
Methylphenanthrene ratio	MPR	[2 – MP]/[1 – MP]	Radke et al. (1982b)
Methylphenanthrene indices	MPI 1	1.5{[2 – MP] + [3-MP]}/{P+[1 – MP] + [9 – MP]}	Radke et al. (1982a)
	MPI 2	3[2 – MP]/{P + [1 – MP] + [9 – MP]	Radke et al. (1982a)
Calculated vitrinite reflectance	% R_c	0.60 MPI 1+0.40 (for %R_o < 1.35) –0.60 MPI 1 + 2.30 (for %R_o > 1.34)	Radke and Welte (1983)

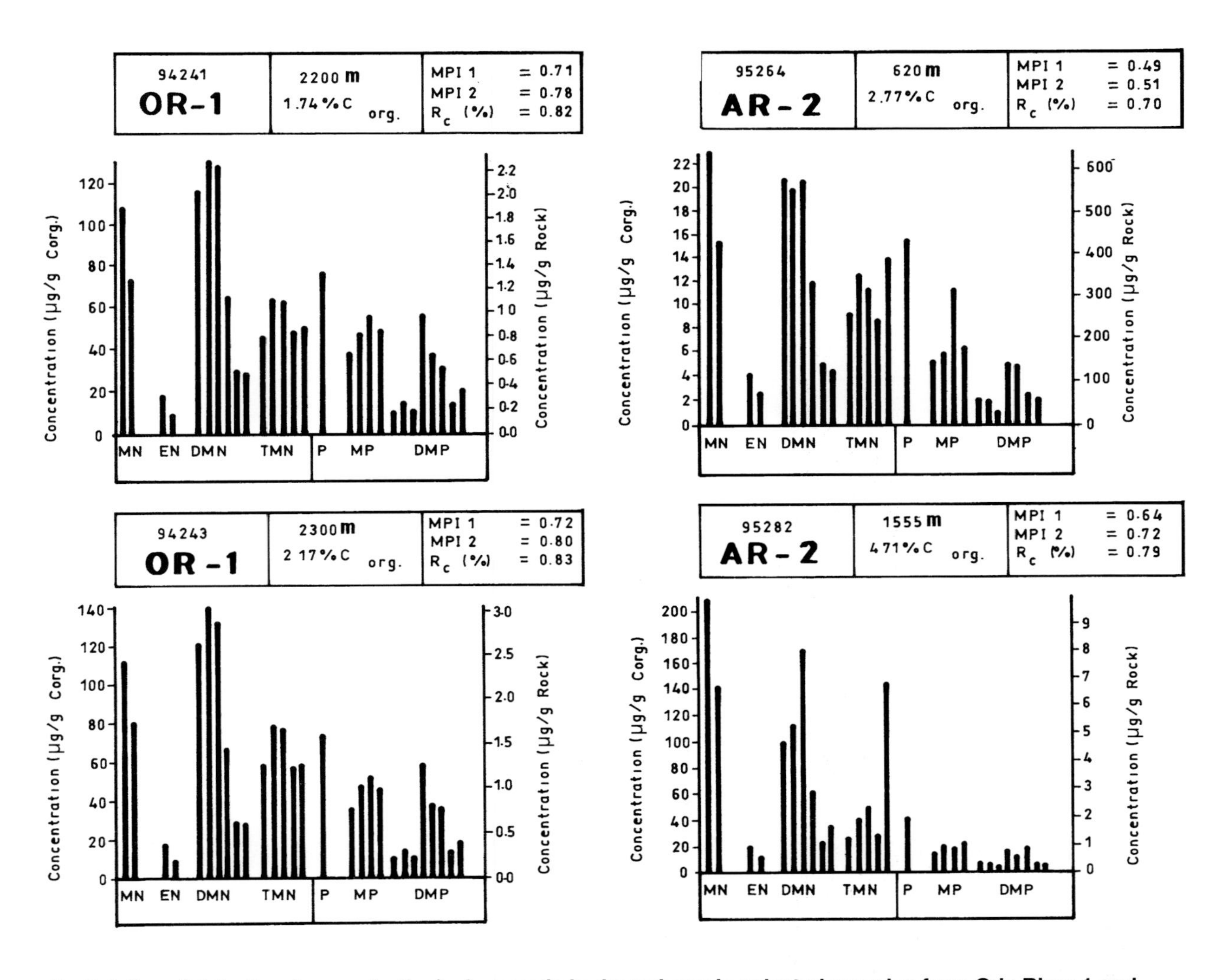

Figure 5—Relative distribution (concentration) of aromatic hydrocarbons in selected samples from Oda River-1 and Anambra River-2 wells.

1987), the expulsion mechanism of hydrocarbon depends on the type of organic matter contained in the source rock. The relatively low concentration of the phenanthrenes might have been caused by preferential or selective expulsion of hydrocarbons. This line of thought is supported by the fraction effects often observed in source rocks containing type III kerogen.

The methylphenanthrene indices (MPI 1 and MPI 2) (Tables 4 and 5) were found to generally increase linearly with depth (r = 0.85) within the zone of the oil window (Figure 6). MPI 1 and other aromatic hydrocarbon maturity ratios discussed earlier have several advantages over their sterane and hopane counterparts. The compounds upon which they are based represent a much greater portion of oils and source rock extracts and are thus less susceptible to contamination. In addition, the aromatic maturity indicators have a wider dynamic range because they evolve continually throughout the oil window (Alexander et al., 1985; Garrigues et al., 1988). Furthermore, measurement of the key compounds does not require gas chromatography–mass spectrometry but can be accomplished through conventional GC.

The observed changes in the values of MPI 1 in these wells (type III kerogens) suggest that alkylation reactions (Radke et al., 1986, 1990; Strachen et al., 1988; Leischner, 1994) probably become more prominent at higher maturity levels. Thus, the distribution of aromatic hydrocarbons (especially in type III source rocks) is most likely controlled by both dealkylation and alkylation reactions, with the former being more pronounced at the lower maturity levels and the later becoming dominant at higher maturity levels.

The relative abundances of the alkyl homologs of the aromatic hydrocarbons (phenanthrenes) were used to calculate the vitrinite reflectance (% R_c) after the method of Radke (1988) and Radke et al. (1982a, b). The data revealed MPI 1-based calculated vitrinite reflectance values ranging from 0.67 to 0.84% R_c in the depth interval

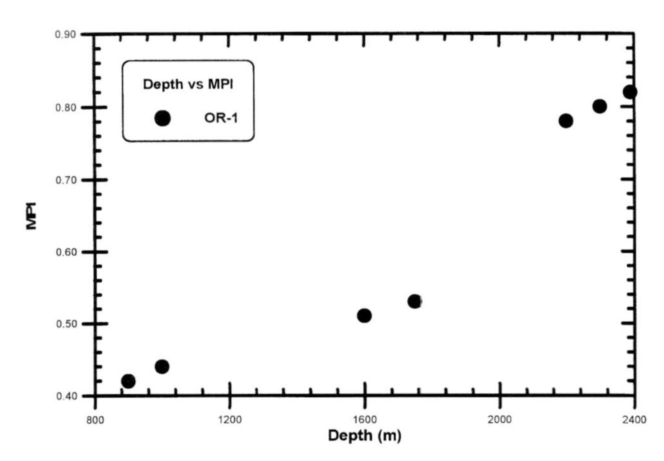

Figure 6—Crossplot of the methylphenanthrene index (MPI) versus depth for Oda River-1 well.

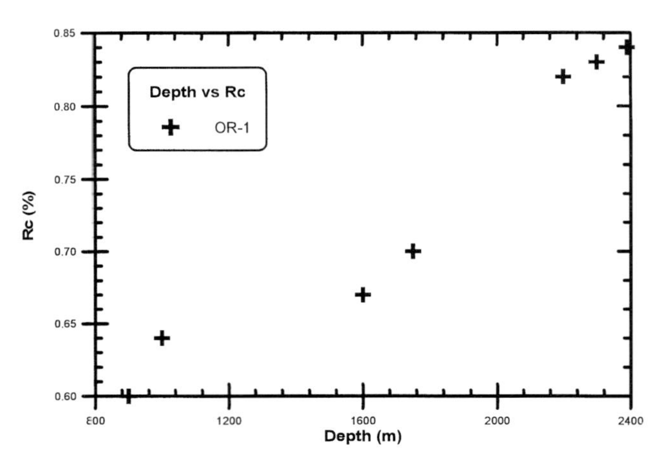

Figure 7—Crossplot of the calculated vitrinite reflectance (% R_c) versus depth for Oda River-1 well.

of 1600–2390 m in OR-1 well (Table 4) and 0.70 to 0.89% R_c at 1555–2145 m in the AR-2 well (Table 4). These values also increase linearly with depth (Figure 7), and they correlate well with measured reflectance values (0.61–0.83% R_o) and T_{max} values (430–471°C) for type III kerogen within the corresponding depth intervals for both wells. These results confirm that the type III kerogen of the Asata–Nkporo Shale and the coal measures within this depth range for both study wells are mature and attained the threshold of oil formation.

Estimation of Hydrocarbon Volumes

From the previous discussion, it is clear that the aromatic data have been used to constrain the thermally mature source intervals of the Lower Coal Measures (Mamu Formation) and the Asata–Nkporo Shale in the two study wells. Since these sections also contain adequate amounts of organic matter (see Table 1), they must have generated some hydrocarbons.

The volume of hydrocarbons generated by the Lower Coal Measures and the Nkporo Shale was estimated using the method of Schmoker (1994). The areal distribution of these formations in the lower Benue Trough are shown in Figures 8 and 9. Stratigraphic thicknesses were obtained from the OR-1 well and published sections (Petters, 1978). An average measured TOC of 2.2 wt. % and a shale density of 2.6 g/cm^3 were assumed for the Nkporo Shale. The net thickness of the mature Nkporo is 480 m over an area of 300 km^2. The effective source rock volume is 144 km^3. Thus, the total mass (M) of organic carbon (in g TOC) is given by M (g TOC) = $2.2/100 \times 2.6 \times 144 \times 10^{15}$ cm^3, which is equal to 8.24×10^{15} g.

The average present hydrogen index (HI_p) is 24 mg/g TOC at 0.81% R_o, whereas the initial hydrogen index (HI_o) for the immature type III Nkporo Shale is 126 mg/g TOC (Unomah and Ekweozor, 1993). The difference between HI_p and HI_o approximates the mass of hydrocarbon (R) generated per gram of TOC. Thus, the total mass of hydrocarbons generated is the product of M and R, or 8.4×10^{11} kg HC. This is equivalent to 6300 million bbl of 30° API oil or 4.10×10^{13} ft^3 of methane (Schmoker, 1994, figure 19.3).

A similar calculation for the Lower Coal Measures (Mamu Formation) was carried out using the following data: net source rock thickness (including coals) = 581.25 m; total area of formation = 78 km^2; net source rock volume = 45.32 km^3; average TOC of shale and coal = 2.7 wt. %; average HIp = 64 mg/g TOC; and average HI_o = 126 mg/g TOC. The total mass of hydrocarbons generated is 1.98×10^{11} kg HC, which gives a value of 1490 million bbl of oil or 9.8×10^{12} ft^3 of methane. Therefore, the total volume of oil generated by the Nkporo Shale and Lower Coal Measures is 7790 million bbl or 5.08×10^{13} ft^3 of gas. This volume of oil far exceeds the threshold value of 50 million bbl required for expulsion of oil (Macgregor, 1994).

Expulsion efficiency, however, is a function of source rock richness and varies greatly, being higher for gas-prone than oil-prone source rocks. Published values for shale source rocks are 10–15% (Hunt, 1979; Tissot and Welte, 1984), 20–30% for the Kimmeridge Clay (Goff, 1983), and up to 90% for gas-prone type III source rocks (Mackenzie et al., 1983; Leythaeuser et al., 1984). Therefore, assuming 10% expulsion efficiency for oil and 90% for gas, a total of 779 million bbl of 30° API oil or 4.57×10^{13} ft^3 of gas may have been expelled by the Nkporo Shale and Lower Coal Measures (Mamu Formation) and probably entrapped in the basin. This may account for the hydrocarbon shows in early exploratory drilling in the basin.

To further assess the petroleum potential of the Nkporo Shale and Lower Coal Measures, the source potential index (SPI) of Demaison and Huizinga (1994) was calculated for the two study wells. An average SPI of 4.02 was obtained assuming a vertically drained basin. This low SPI value indicates an undercharged system. Such petroleum systems are often associated with small reserves or are nonproductive (Demaison and Huizinga, 1994). This is in agreement with the volumes calculated

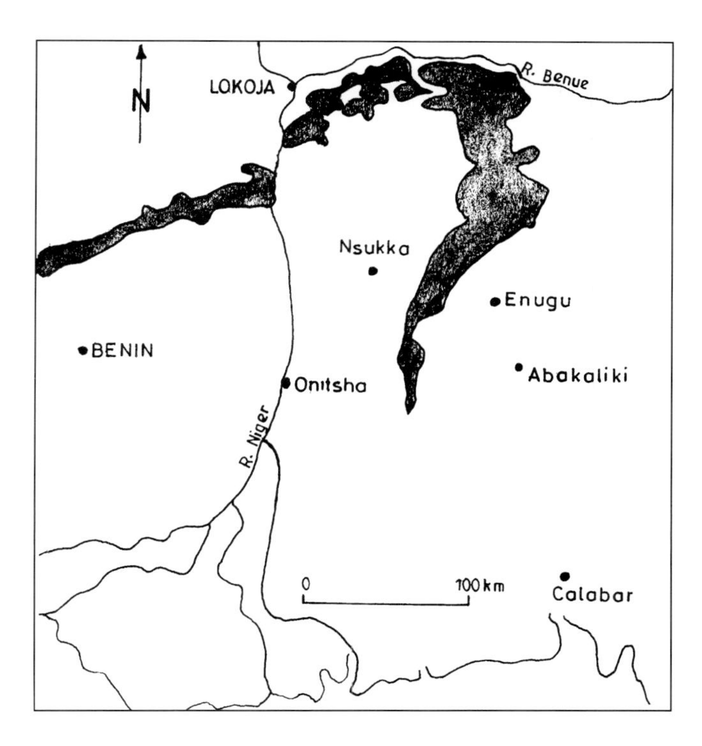

Figure 8—Areal distribution of the Asata–Nkporo Shale in the Anambra Basin.

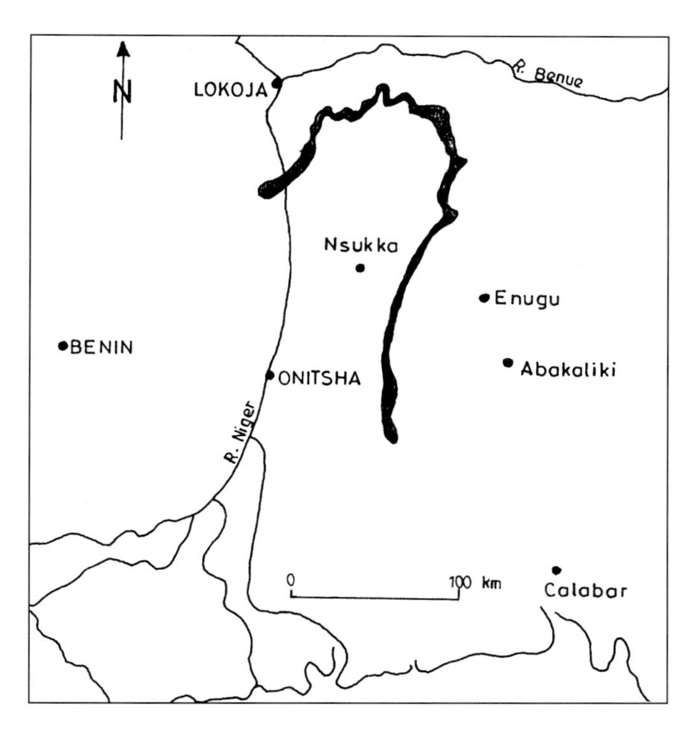

Figure 9—Areal distribution of the Lower Coal Measures (Mamu Formation) in the Anambra Basin.

(or estimated) in the present study. The low potential of the Lower Coal Measures could be due, at least in part, to oil and gas adsorption on the coals (Hunt, 1991). This is supported by geochemical analysis of Upper Cretaceous coals in the Enugu (Asata) Shale from the Anambra Basin by Hagemann and Pickel (1991). They confirmed that some hydrocarbon generation and mobilization did take place inside the coals but that it is unlikely that oil has migrated out of the coals. Thus, the undercharged nature of this petroleum system suggests limited exploration potential based on these volumetric estimates.

CONCLUSIONS

This study has shown that organic matter in post-Santonian shales and coals of the Anambra Basin has generated aromatic hydrocarbons. The variation of these hydrocarbons with depth showed that concentrations of phenanthrene increase almost constantly with increasing depth of burial. Methylphenanthrene and dimethylphenanthrene also increase in relative concentrations with depth. The napthalenes and their homologs generally show variations in relative concentrations with depth. The relative concentration of the sum total of the naphthalenes and their isomers is greater than that of the phenanthrenes.

Methylphenanthrene indices (MPI 1 and MPI 2) and their other isomer ratios were found to generally increase linearly with depth and are temperature and/or maturity sensitive, especially within the oil window. At <0.8% R_o, methylnaphthalene dominates over dimethylnapthalene, while at higher maturity, the reverse is true. This is attributed to thermal rearrangement of 1-methylnapthalene with thermal stress. Thus, the distribution of aromatic hydrocarbons in these samples appears to be controlled by both dealkylation reactions at lower maturity levels and alkylation reactions at higher maturity levels.

The Nkporo Shale and sections of the Lower Coal Measures are mature and may have generated and expelled about 779 million bbl of oil or 4.57×10^{13} ft^3 of gas in the Anambra Basin. These volumes are in agreement with a calculated source potential index of 4.02 (Demaison and Huizinga, 1994) which indicates an undercharged petroleum system. This suggests limited exploration potential for the basin, based on volumetric estimates.

Acknowledgments*—The authors wish to thank Elf Petrolum Nigeria Ltd. and their staff (Jean Bie, Dan Ndefo, C. J. Ikelionwu, and Theo Duze) for providing the analyzed samples and for permission to publish the results of this research. Our thanks also go to the German Academic Exchange Services (DAAD) for financial support of I. M. A. and to D. H. Welte, M. Radke, and E. Biermann (KFA Jülich, FRG) for analytical support. Chevron Nigeria Ltd. is also gratefully acknowledged for financial support of J. I. N. for the Hedberg Research Conference. This paper benefitted greatly from comments, suggestions, and reviews by B. J. Katz and others, for which we are grateful.*

REFERENCES CITED

Agagu, O. K., and C. M. Ekweozor, 1982, Source rock characteristics of Senonian shales in the Anambra syncline, southern Nigeria: Journal of Mining and Geology, v. 19, p. 52–61.

Agagu, O. K, E. A. Fayose, and S. W. Petters, 1986, Stratigraphy and sedimentation in the Senonian Anambra Basin of eastern Nigeria: Journal of Mineralogy and Geology, v. 22, p. 25–36.

Akaegbobi, I. M., 1995, The petroleum province of southern Nigeria—Niger delta and Anambra Basin: organic geochemical and organic petrographic approach: Ph.D. dissertation, Technical University, Berlin, Germany, 182 p.

Akaegbobi, I. M., and M. Schmitt, 1998, Organic facies, hydrocarbon source potential and the reconstruction of the depositional paleoenvironment of the Campano–Maastrichtian Nkporo Shale in the Cretaceous Anambra Basin, Nigeria: Nigerian Association of Petroleum Exploration Bulletin, v. 13, n. 1, p. 1–19.

Akande, S. O., and B. D. Erdtmann, 1998, Burial metamorphism (thermal maturation) in Cretaceous sediments of the southern Benue Trough and Anambra Basin, Nigeria: AAPG Bulletin, v. 82, n. 6, p. 1191–1206.

Alexander, R., R. I. Kagi, S. J. Rowland, P. N. Sheppard, and T. V. Chirila, 1985, The effects of thermal maturity on distributions of dimethylnaphthalenes and trimethylnaphthalenes in some ancient sediments and petroleums: Geochimica et Cosmochimica Acta, v. 49, p. 385–395.

Avbovbo, A. A., and O. Ayoola, 1981, Petroleum prospects of southern Nigeria's Anambra Basin: Oil & Gas Journal, v. 79, p. 334–347.

Blumer, M., and W. W. Youngblood, 1975, Polycyclic aromatic hydrocarbons in soils and recent sediments: Science, v. 188, p. 53–55.

Boreham, C. J., I. H Crick, and T. G. Powell, 1988, Alternative calibration of the methylphenanthrene index against vitrinite reflectance: application to maturity measurements in oils and sediments: Organic Geochemistry, v. 12, p. 289–294.

Demaison, G., and B. J. Huizinga, 1994, Genetic classification of petroleum systems using three factors: charge, migration, and entrapment, *in* L. B. Magoon and W. C. Dow, eds., The petroleum system—from source to trap: AAPG Memoir 60, p. 73–91.

Ekweozor, C. M., and J. R. Gormly, 1983, Petroleum geochemistry of Late Cretaceous and early Tertiary shales penetrated by Akukwa-2 well in the Anambra Basin, southern Nigeria: Journal of Petroleum Geology, v. 6 p. 207–216.

Fleet, A. J., and A. C. Scott, 1994, Coal and coal bearing strata as oil-prone source rocks: an overview, *in* A. C. Scott and A. J. Fleet, eds., Coal and coal-bearing strata as oil-prone source rocks?: GSA Special Publication, v. 77, p. 1–8.

Garrigues, P., R. De Sury, M. L. Angelin, J. Bellocq, J. L. Oudin, and M. Ewald, 1988, Relation of methylated aromatic hydrocarbon distribution pattern to the maturity of organic matter in ancient sediments from the Mahakam delta: Geochimica et Cosmochimica Acta, v. 52, p. 375–384.

Goff, J. C., 1983, Hydrocarbon generation and migration from Jurassic source rocks in the E. Shetland Basin and Viking Graben of the North Sea: Geological Society of London Journal, v. 140, p. 445–474.

Hagemann, H. W., and W. Pickel, 1991, Characteristics of Upper Cretaceous coals from Enugu (Nigeria) related to bitumen generation and mobilization: Organic Geochemistry, v. 17, p. 673–680.

Hase, A., and R. A Hites, 1976, On the origin of polycyclic aromatic hydrocarbons in recent sediments: biosynthesis by anaerobic bacteria: Geochimica et Cosmochimica Acta, v. 40, p. 1141–1143.

Hunt, J. M., 1979, Petroleum geochemistry and geology: San Francisco, W. H. Freeman, 617 p.

Hunt, J. M., 1991, Generation of gas and oil from coal and other terrestrial organic matter: Organic Geochemistry, v. 17, p. 673–680.

Leischner, K., 1994, Kalibration simulierter Temperaturgeschichten von Sedimentgestein: Ph.D. dissertation, Rheinisch–Westfälische Technische Hochschule, Aachen, Federal Republic of Germany, 309 p.

Leythaeuser, D., A. S. Mackenzie, R. G. Schaefer, and M. Bjoroy, 1984, A novel approach for recognition and quantification of hydrocarbon migration effects in shale sandstone sequences: AAPG Bulletin, v. 68, p. 196–219.

Leythaeuser, D., R. G. Schaefer, and M. Radke, 1987, On the primary migration of petroleum: Proceedings of the 12th World Petroleum Congress, Houston, Texas, v. 2, p. 227–236.

Leythaeuser, D., R. G. Schaefer, and M. Radke, 1988, Geochemical effects of primary migration of petroleum in Kimmeridge source rocks from Brae field area, North Sea—I. Gross composition of C_{15+}-soluble organic matter and molecular composition of C_{15+}-saturated hydrocarbons: Geochimica et Cosmochimica Acta, v. 52, p. 701–713.

Loureiro, M. R. B., and J. N. Cardoso, 1990, Aromatic hydrocarbons in the Paraiba valley oil shale: Organic Geochemistry, v. 15, p. 351–359.

Macgregor, D. S., 1994, Coal-bearing strata as source rocks—a global review, *in* A. C. Scott and A. J. Fleet, eds., Coal and coal-bearing strata as oil-prone source rocks?: GSA Special Publication, v. 77, p. 107–116.

Mackenzie, A. S., D. Leythaeuser, and R. G. Schaefer, 1983, Expulsion of petroleum hydrocarbons from shale source rocks: Nature, v. 301, p. 506–509.

McCurry, P., 1976, The geology of the Precambrian to lower Palaeozoic rocks of northern Nigeria—a review, *in* C. A. Kogbe, ed., Geology of Nigeria: Lagos, Nigeria, Elizabethan Publishing, p. 15–39.

Nwachukwu, J. I., A. A. Odulaja, and A. A. Akinokun, 1988, Source and reservoir rock potentials of Eze-Aku shale and associated limestones: Nigerian Association of Petroleum Exploration Bulletin, v. 3, p. 81–89.

Nwajide, C. S., 1990, Cretaceous sedimentation and paleogeography of the central Benue Trough, *in* C. O. Ofoegbu, ed., The Benue Trough structure and evolution: Braunschweig and Wiesbaden, Germany, Vieweg & Sohne Verlag, p. 19–38.

Nwajide, C. S., and T. J. A. Reijers, 1997, Sequence architecture of the Campanian Nkporo and the Eocene Nanka Formations of the Anambra Basin, Nigeria: Nigerian Association of Petroleum Exploration Bulletin, v. 12, n. 01, p. 75–87.

Petters, S. W., 1978, Stratigraphic evolution of the Benue Trough and its implications for Upper Cretaceous paleogeography of west Africa: Journal of Geology, v. 86, p. 311–322.

Petters, S. W., and C. M. Ekweozor, 1982, Petroleum geology of Benue Trough and southeastern Chad Basin, Nigeria: AAPG Bulletin, v. 66, p. 1141–1149.

Powell, T. G., 1978, An assessment of the hydrocarbon source rock potential of the Canadian Arctic Islands: Geological Survey of Canada, Paper 78, p. 12.

Powell, T. G., and C. J. Boreham, 1994, Terrestrially sourced oils: where do they exist and what are our limits of knowledge?—a geochemical perspective, *in* A. C. Scott and A. J. Fleet, eds., Coal and coal-bearing strata as oil-prone source rocks?: GSA Special Publication, v. 77, p. 11–29.

Radke, M., 1988, Application of aromatic compounds as maturity indicators in source rocks and crude oils: Marine and Petroleum Geology, v. 5, p. 224–236.

Radke, M., and D. H. Welte, 1983, The methylphenanthrene index (MPI): a maturity parameter based on aromatic hydrocarbons, *in* M. Bjoroy et al., eds., Advances in Organic Geochemistry: Chichester, U.K., Wiley, p. 504–512.

Radke, M., H. Willsch, and D. H. Welte, 1980, Preparative hydrocarbon group type determination by automated medium pressure liquid chromatography: Analytical Chemistry, v. 52, p. 406–411.

Radke, M., D. H. Welte, and H. Willsch, 1982a, Geochemical study on a well in the Western Canada Basin: relation of the aromatic distribution pattern to maturity of organic matter: Geochimica et Cosmochimica Acta, v. 46, p. 1–10.

Radke, M., H. Willsch, D. Leythaeuser, and M. Teichmüller, 1982b, Aromatic components of coal: relation of distribution pattern to rank: Geochimica et Cosmochimica Acta, v. 46, p. 1831–1848.

Radke, M., H. Willsch, and D. H. Welte 1984, Class separation of aromatic compounds in rock extracts and fossil fuels by liquid chromatography: Analytical Chemistry, v. 56, p. 2538–2546.

Radke, M., D. H Welte, and H. Willsch, 1986, Maturity parameters based on aromatic hydrocarbons: influence of organic matter type: Organic Geochemstry, v. 10, p. 51–63.

Radke, M., H. Willsch, and M. Teichmüller, 1990, Generation and distribution of aromatic hydrocarbons in coals of low rank: Organic Geochemistry, v. 15, n. 6, p. 539–563.

Rahaman, M. A., 1976, Review of the basement geology of southwestern Nigeria, *in* C. A. Kogbe, ed., Geology of Nigeria: Lagos, Nigeria, Elizabethan Publishing, p. 41–58.

Schmoker, J. W., 1994, Volumetric calculation of hydrocarbons generated, *in* L. B. Magoon and W. G. Dow, eds., The petroleum system—from source to trap: AAPG Memoir 60, p. 323–326.

Snowdon, L. R., 1991, Oil from type III organic matter: resinite revisited: Organic Geochemistry, v. 17, p. 743–747.

Strachan, M. G., R. Alexander, and R. I. Kagi, 1988, Trimethylnaphthalenes in crude oils and sediments: effects of source and maturity: Geochimica et Cosmochimica Acta, v. 52, p. 1255–1264.

Tissot, B. P., and D. H. Welte, 1984, Petroleum formation and occurrence, 2nd ed.: Berlin, Springer, 699 p.

Unomah, G. I., and C. M Ekweozor, 1993, Petroleum source rock assessment of the Campanian Nkporo Shale, lower Benue Trough, Nigeria: Nigerian Association of Petroleum Exploration Bulletin, v. 8, p. 172–186.

Wakeham, S. G., C. Schaffner, and W. Giger, 1980a, Polycyclic aromatic hydrocarbons in recent lake sediments—II. Compounds derived from biogenic precursors during early diagenesis: Geochimica et Cosmochimica Acta, v. 44, p. 415–429.

Wakeham, S. G., C. Schaffner, and W. Giger, 1980b, Diagenetic polycyclic aromatic hydrocarbons in recent sediments: structural information obtained by high performance liquid chromatography, *in* A. G. Douglas and J. R. Maxwell, eds., Advances in organic geochemistry: New York, Pergamon Press, p. 353–363.

Whiteman, A. J., 1982, Nigeria: its petroleum geology, resources and potential: London, Graham and Trotman, v. 1 and 2, 394 p.

Katz, B. J., W. C. Dawson, L. M. Liro, V. D. Robison, and J. D. Stonebraker, 2000, Petroleum systems of the Ogooué Delta, offshore Gabon, *in* M. R. Mello and B. J. Katz, eds., Petroleum systems of South Atlantic margins: AAPG Memoir 73, p. 247–256.

Chapter 18

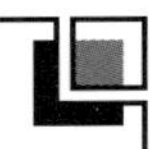

Petroleum Systems of the Ogooué Delta, Offshore Gabon

B. J. Katz

W. C. Dawson

Texaco Inc.
Upstream Technology Department
Houston, Texas, U.S.A.

L. M. Liro*

V. D. Robison

J. D. Stonebraker *(Retired)*

Texaco Inc.
International Exploration Department
Bellaire, Texas, U.S.A.

*Present address: *Veritas Exploration Services*
Houston, Texas, U.S.A

Abstract

An analysis of the petroleum geology of the Ogooué Delta region reveals the presence of two independent petroleum systems: the Madiela/Cap Lopez–Batanaga(.) and the Azile/Anguille–Anguille(!) systems. An *independent system* is defined as a suite of hydrocarbon accumulations that share a common source and reservoir couplet. The oils in the first system appear to have been generated from a pre-Cenomanian restricted (elevated salinity) marine source. These oils occur largely in Batanga Formation sandstones (Maestrichtian) and are limited mainly to the southern parts of the delta. The petroleum in the second system appears to have been generated from the Azile and Anguille Formations (Senonian) and thus developed under more normal marine conditions. This second group of oils is mainly found in Anguille reservoirs, which were deposited in submarine fans during lowstand events. This second system is geographically less restricted, being present in both the northern and southern parts of the delta.

The geographic distribution of the two systems is controlled by several factors, including the distribution of the source, the nature of the migration network, and the level of thermal maturity. Both systems share one major controlling factor: the areal limit of the two systems is in part constrained by thermal maturation. The overburden associated with the delta complex provided for appropriate levels of thermal maturation for generation to proceed. Modeling results suggest that the pre-Cenomanian marine shales and marls began generating liquid hydrocarbons during the Oligocene. The Senonian system began generating hydrocarbons during the Miocene. Both systems are still actively generating as subsidence and loading continues. The pre-Cenomanian system appears to have a more effective mechanism for vertical migration than the Azile–Anguille sourced system. This greater efficiency explains the younger age for the dominant reservoir of the pre-Cenomanian (older) system. It appears that the southward shift in the delta's depocenter may be a partial explanation for this greater efficiency.

INTRODUCTION

There have been numerous means suggested to define trends in hydrocarbon occurrence and distribution. Many of these approaches, such as play analysis, focus on a single common trait such as the reservoir unit or structural style. Although these approaches are useful, they often provide an incomplete understanding of an area's hydrocarbon prospectivity because of their narrow focus. Consequently, such approaches can often result in a number of missed opportunities. An alternative approach is to examine the hydrocarbons present and all of the elements that are necessary for their occurrence. Magoon and Dow (1994) suggested such an approach, termed *petroleum system analysis*. A *petroleum system* is defined as a "geobody" that possesses all of the necessary elements for the accumulation of commercial quantities of hydrocarbons. These elements include a trap, source rock, seal, overburden, and reservoir. In addition to the presence of these elements, it is a requirement that they share the appropriate geometric and temporal relationships for commercial quantities of hydrocarbons to accumulate and be preserved.

Convention has it that a petroleum system is named by its source and primary reservoir, that is, the reservoir unit that contains the largest volume of hydrocarbons. Included within a system's name is a symbol representing the degree of confidence that exists in the definition of the petroleum system. Magoon and Dow (1994) established three levels of confidence (with accompanying symbols): known (!), hypothetical (.), and speculative (?). The level of confidence is based on the nature of the geochemical data supporting both the presence of a source rock and the oil–source rock correlation. A known system requires a definitive oil–source rock correlation. In a hypothetical system, a source rock has been geochemically confirmed, but a definitive correlation has not been established. A petroleum system is considered speculative if no geochemical evidence exists to establish the presence of a source, but its presence is inferred through geologic and/or geophysical data.

This definition means that in any given basin it may be possible for multiple petroleum systems to be present. The presence of multiple petroleum systems requires only that more than one effective and discrete hydrocarbon source rock be present. The presence of these different effective source rocks is established through geochemical means, including a direct examination of source rock attributes (e.g., richness, quality, and thermal maturity) or indirectly through an examination of the geochemical characteristics (e.g., stable carbon isotopic and biomarker compositions) of the region's hydrocarbons. Multiple petroleum systems have been documented in a number of regions, including the Arabian platform (Alkhadhrawi, 1996), the Southern Alps of Italy (Riva et al., 1986), the northern Gulf of Mexico (Kennicutt et al., 1992), and the Ogooué Delta (Teisserenc, and Villemin, 1990).

The focus of this study is one of these regions, namely, the Ogooué Delta in offshore Gabon (Figure 1). This delta complex includes more than 30 fields, with the largest two fields, Anguille and Grondin, both containing more than 220 million bbl of recoverable reserves (Teisserenc and Villemin, 1990).

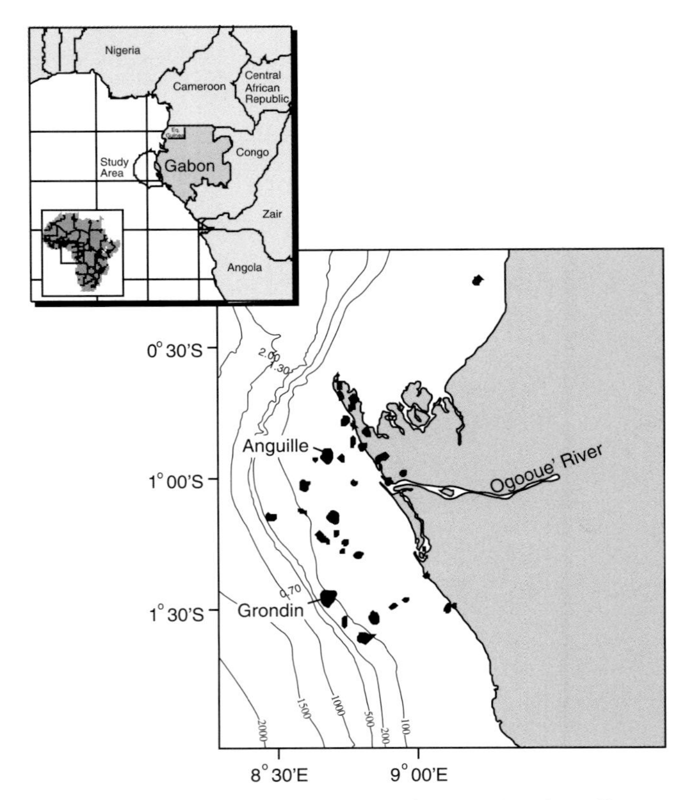

Figure 1—Index map of the Ogooué Delta showing oil fields (black patches).

BASIN DEVELOPMENT

The Ogooué Delta represents the mature and last stage of passive margin development along the west African continental margin. The west African margin was, in general, formed by a series of discrete rifting events. Karner et al. (1997) describes the margin's rift history in the following manner. Rifting along the margin began during the Berriasian, with active rifting during this initial phase being short-lived, about 3 m.y. During this period, a series of horsts and grabens developed, largely paralleling the present-day coastline. While these horsts were sites of active erosion, the grabens developed into a series of lake basins. Within these rift lakes, organic-rich sediments were often deposited which have locally provided sources of hydrocarbons (e.g., Kissenda Shale, onshore Gabon). The Eastern hinge zone demarcates the eastern limit of the broadly distributed extension, separating continental margin strata from Precambrian basement (Karner et al., 1997). The second rifting event began during the Hauterivian. The topographic relief associated with this extensional episode is thought to have been less than that associated with the initial rifting event. The eastern limit of this second rifting episode is the Atlantic hinge zone. It is located farther basinward and occurs beneath the outer shelf–upper slope transition (Karner et al., 1997). A third and final rifting episode reactivated both hinge zones, creating additional accommodation space. Rifting along the margin was terminated by Aptian time.

The onset of sea floor spreading took place during the Barremian–early Aptian and led to the development of a restricted west African salt basin. This period is often termed the *transitional phase* (Edwards and Bignall, 1988a, b). Extensive Aptian evaporites (Ezanga Formation in Gabon) developed as the combined effect of intermittent marine incursions from the south across the Walvis Ridge and a mostly arid climate in the region. The Ogooué Delta is located near the northern limits of the Aptian salt basin

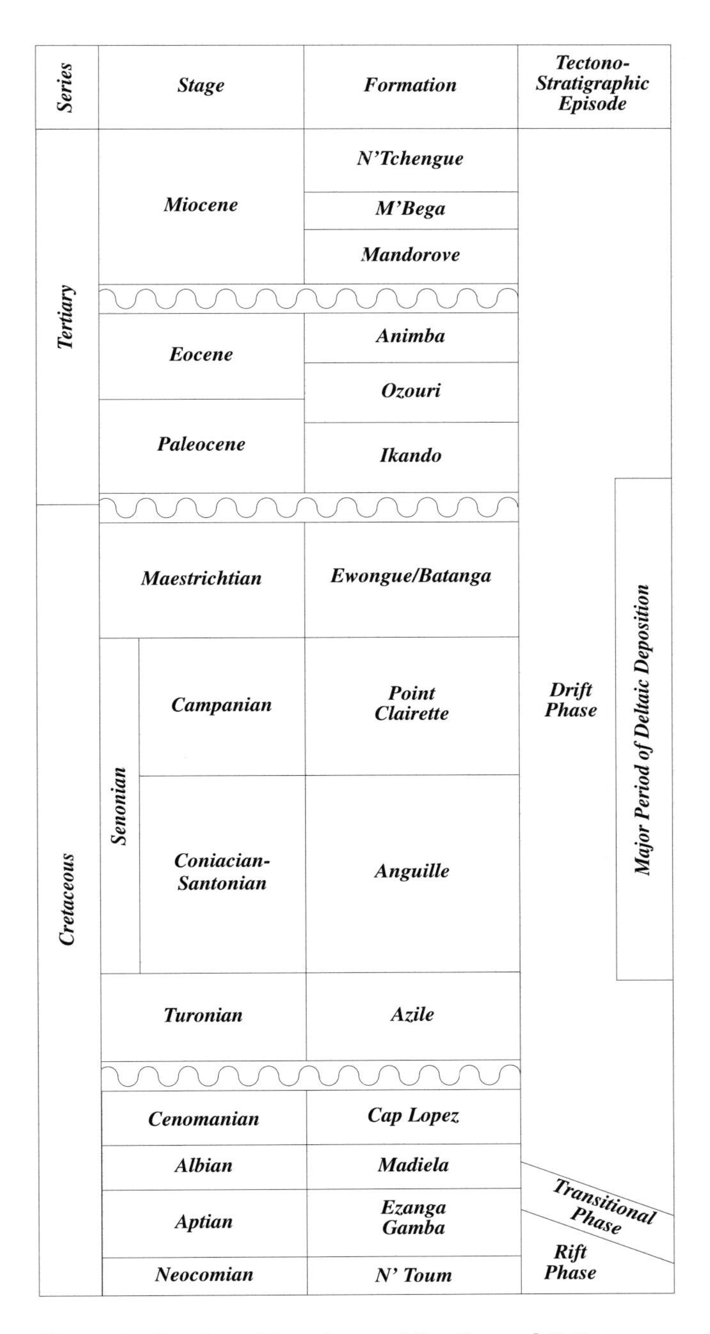

Figure 2—Stratigraphic column of the Ogooué Delta.

(Uchupi, 1992). This transitional phase was complete by Albian time, leading to open marine conditions typical of the drift stage.

Although initially during the drift phase, carbonates were locally important, the sedimentary system along the margin eventually evolved into a siliciclastic-dominated system. A series of deltas of varying sizes developed along the margin as clastic input increased. The influx of deltaic sediments resulted in both the development of growth faults and the onset of halokinesis. The final major regional event influencing the west African margin was a westward tilting caused by Neogene tectonism associated with epeirogenic movements affecting southern and western Africa (Sahagian, 1988). This tilting, plus eustatic changes and the intensification of bottom-water currents, produced a basal Miocene–Oligocene unconformity, which cut deeply into strata as old as the Albian–Cenomanian (Edwards and Bignell, 1988a, b; Karner et al., 1997).

Specifically, the Ogooué Delta is part of the northern Gabon offshore subbasin. The delta is located north of the N'Komi fracture zone. This fracture zone represents a strike-slip transfer zone related to original continental rifting (Teisserenc and Villemin, 1990). The delta region lies west of the Atlantic hinge zone. This structural position has had a significant impact on its stratigraphic development (Karner et al., 1997).

STRATIGRAPHY

As is typical of passive margins, the general stratigraphy of the Ogooué Delta region can be divided into three major tectono-stratigraphic units: (1) a basal rift sequence, (2) a transitional sequence, and (3) a drift depositional sequence (Figure 2).

Because of the thickness of the drift sequence, the rift sequence is largely unknown in the Ogooué Delta region. The oldest penetrated portion of the stratigraphic section is a lacustrine sandstone, the Barremian–Aptian N'Toum Formation. Overlying this unit are additional lacustrine facies of the lower Aptian Coniquet Formation. The transitional sequence is represented by both the highly variable middle Aptian Como Formation, which includes a series of carbonaceous shales, and the evaporitic Ezanga Formation.

The drift facies were initially deposited in two distinct regions: an eastern platform and a western basinal setting. Initially, the eastern platform was largely dominated by carbonate facies, which by Senonian time was replaced by siliciclastics. This change in facies was associated with uplift along the west African margin, which provided a major source for clastic sediment (Teisserenc and Villemin, 1990). In contrast, the western basinal setting was initially filled with fine-grained sediments. Coarse-grained clastics were introduced into this setting mainly through turbidites by the late Turonian.

A major erosional event in the Oligocene is documented across much of the region. This event generally coincided with regional uplift of central and southern Africa (Sahagian, 1988).

RESERVOIRS

The two primary producing reservoirs in the Ogooué Delta complex are the sandstones of the Anguille Formation (Coniacian–Santonian) and the Batanga Formation (Maestrichtian). In part, salt swells and early diapirs controlled the distribution of sand bodies. The Anguille reservoirs are associated with early delta development and are mostly found in the northern part of the delta complex. In contrast, the Batanga reservoirs are associated with later phases of delta development and are restricted to the southern part of the delta complex (Edwards and Bignell, 1988b).

Anguille sandstones are largely turbiditic in origin (Massonnat et al., 1992). These sandstones are fine to very fine grained, laminated, and arkosic. They exhibit maximum porosity values of ~24% and permeabilities as high as 700 md. Wireline log studies suggest that sandstones within the Anguille Formation may represent 30–50% of the gross interval (Figure 3), with net sandstone thickness ranging from 10 to 100 m. The sandstone beds tend to be limited to fairways less than 1 km wide (Teisserenc and Villemin, 1990).

Traditionally, the Anguille Formation is divided into three subunits. The upper and lower Anguille sandstones are hydrocarbon reservoirs. The middle Anguille is a nonreservoir and an effective barrier to flow, as evidenced by the different oil–water contacts for the two reservoir units (Massonnat et al., 1992).

The gross thickness of the Batanga Formation can exceed 400 m. Batanga thickness and sandstone distribution is not solely a function of depositional conditions but has been partially controlled by the extensive amount of erosion that has occurred. In the Grondin field, the net Batanga reservoir thickness ranges from 25 to 80 m (Vidal, 1980) (Figure 4). Teisserenc and Villemin (1990) noted that, elsewhere in the basin, the net Batanga reservoir may exceed 200 m. As with the Anguille, the Batanga Formation can be subdivided into three members. The upper and middle Batanga members are the primary reservoir intervals. Dolomitic cement is generally more abundant in the lower member.

There has been considerable debate as to the actual depositional setting of the Batanga sandstones. Arguments have been made for both a neritic setting, where wave and tidal action have played an important role in sand body formation (Vidal, 1980), while more recent literature has suggested that the Batanga sandstone reservoirs are stacked channelized sequences composed of thinly bedded turbidites (Smith et al., 1990). The sandstones are fine- to coarse-grained. Porosities as high as ~30% have been reported, with permeabilities up to several hundred millidarcys (Vidal, 1980).

Younger Tertiary reservoirs, including the fractured reservoirs of the Paleocene–Eocene Ozouri Group (Dunne et al., 1991, 1996) and the Miocene Mandorove Sandstone, are present but of limited volumetric importance, representing only about 6% of in-place reserves (Teisserenc and Villemin, 1990). The Mandorove sandstones are fine- to coarse-grained, with individual beds reaching thicknesses of 20–25 m and a cumulative net reservoir of 50–60 m. Further reducing the importance of these reservoirs is their crude oil quality. These younger reservoirs often contain hydrocarbons with low API gravity values resulting from biodegradation, a direct consequence of their shallow depths of burial.

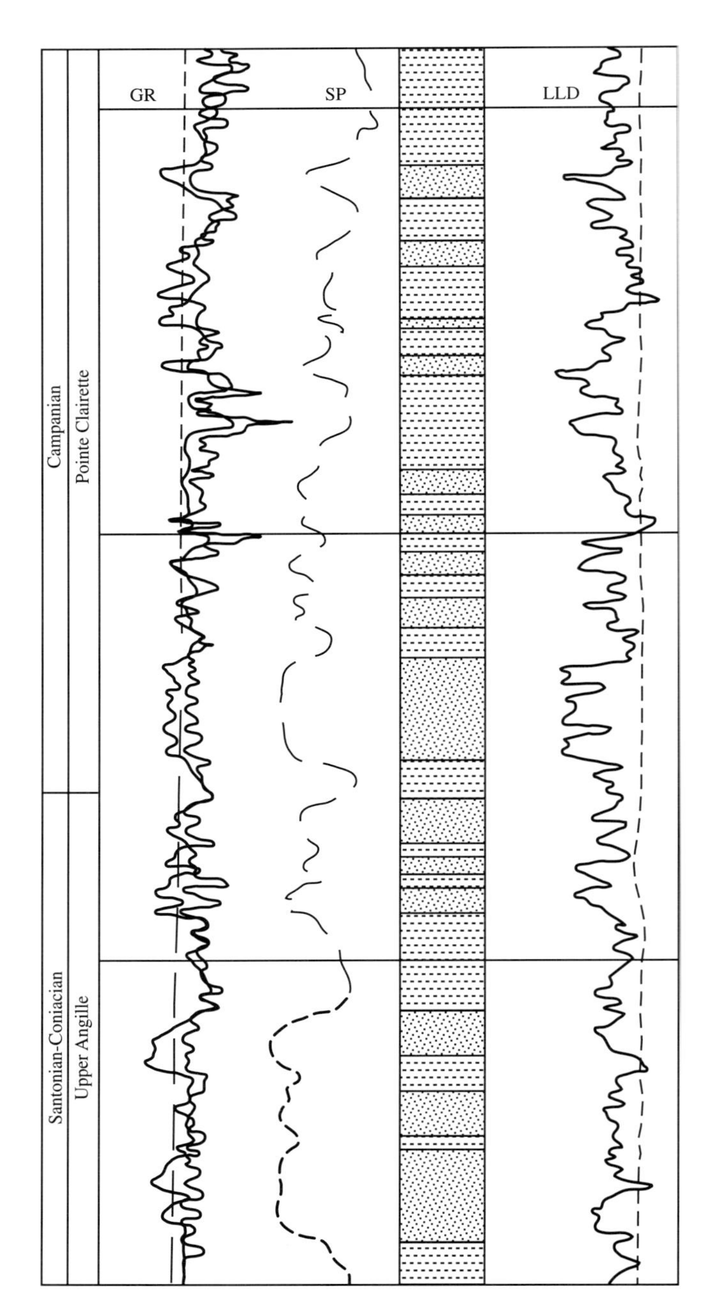

Figure 3—Representative log suite for the upper Anguille–Pointe Clairette section (shale shown by dashed pattern, sandstone by dotted pattern).

SOURCE ROCKS

Unlike many other delta sequences such as the Mississippi and Niger Deltas, in the Ogooué Delta, sufficient geochemical data are available to establish the presence of possible petroleum source rocks. Of greatest significance to this study are the Albian–Cenomanian Madiela and Cap Lopez Formations and the Turonian–Senonian Azile and Anguille Formations.

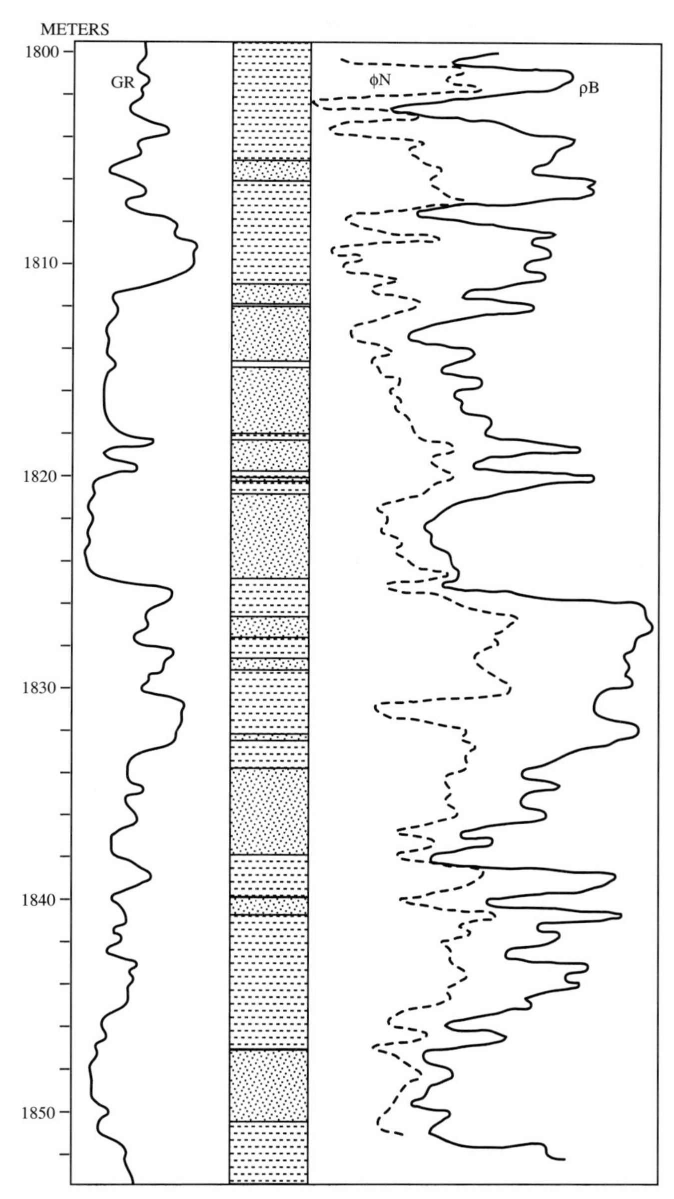

Figure 4—Representative log suite for the Batanga Sandstone reservoir section (shale shown by dashed pattern, sandstone by dotted pattern). After Teisserenc and Villemin (1990).

The Madiela and Cap Lopez Formations represent an evaporative or restricted (i.e., higher salinity) facies. Total organic carbon (TOC) contents of these units within the delta region are known to exceed 2%, with generation potentials (S_2) typically approaching 10 mg HC/g rock or greater and hydrogen indices (HI) of 400 mg HC/g TOC or greater. Examination of equivalent units elsewhere within the Aptian salt basin suggests that higher levels of organic enrichment, more elevated generation potentials, and more oil-prone tendencies than currently observed may exist in the delta (Burwood, 1999). For example, the Madiela and Cap Lopez sequence is the depositional equivalent of the Alagamar Formation, the effective source for the Potiguar Basin in Brazil (Trindade et al., 1992). The actual distribution of this organic-rich facies is poorly constrained both areally and vertically. It is not likely to be geographically widespread but may be limited to minor depressions on the Albian–Cenomanian shelf. Available well control suggests that the gross source rock interval may exceed 100 m.

The Azile and Anguille Formations represent deposition under more normal marine conditions (i.e., normal salinity levels). TOC contents of these units commonly exceed 3%, with S_2 yields typically exceeding 10 mg HC/g rock and HI often greater than 400 mg HC/g rock. Clear organic facies variations within this sequence are present. The least oil-prone (most gas-prone) material is located in closest proximity to the river mouth where terrestrial input has been greatest (Teisserenc and Villemin, 1990). As with the older source sequence, stratigraphic equivalents have also been found to be effective elsewhere along the west African margin (e.g., Iabe Formation, offshore Angola), and their attributes may also be extended throughout much of the Aptian salt basin. In contrast to the more restricted setting of the Madiela and Cap Lopez source intervals, the Azile and Anguille source intervals are geographically widespread, extending across much of the west African margin. Well control suggests that within the delta region this gross source interval may exceed 200 m.

THERMAL MATURITY

The level of thermal maturity within the delta is regionally controlled by the overburden associated with the development of the delta itself and locally by the excess heat introduced from the Loiret volcanic plug to the northwest of the delta. The level of thermal maturity throughout the delta region was assessed through a series of simulations using the *BasinMod 1-D* software package from Platte River Associates. The input was derived through a series of regional structure maps and subsurface temperature controls.

Thermal maturation modeling suggests that, at the Madiela level, the generative portion of the basin is largely restricted to the region where the Tertiary overburden exceeds 2 km (Figures 5 and 6). At the younger Azile level, the generative portion of the basin is, in general, restricted to those parts of the delta where the Tertiary overburden exceeds 2.5 km (Figures 5 and 7). The lack of an exact one-to-one correlation between thermal maturity and depth is the result of variations in geothermal gradient across the region. Variations in geothermal gradients appear to have been largely introduced through variations in salt thickness, geometry, and depth.

Modeling also provides information on the timing of hydrocarbon generation. An analysis of the burial and thermal history of the region suggests that hydrocarbon generation and expulsion from the pre-Cenomanian source rock began during the Oligocene (~30 Ma). The younger Senonian source rock sequence began generat-

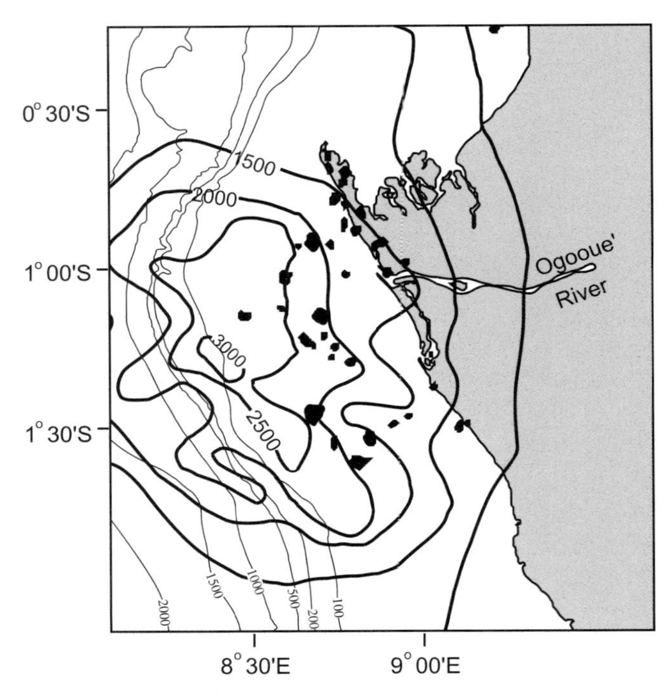

Figure 5—Simplified Tertiary strata isopach map used as input for the thermal maturation models. The distribution of oil fields (black patches) is also shown. Contour interval is 500 m.

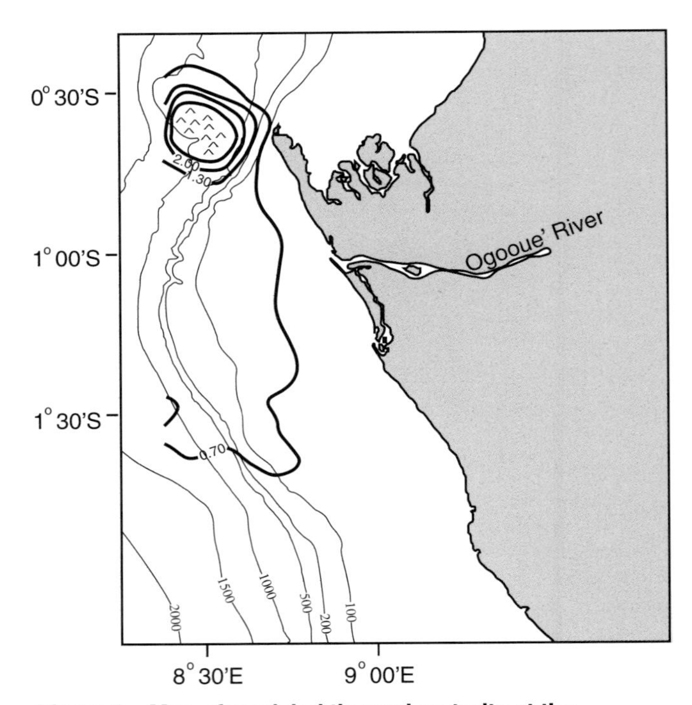

Figure 6—Map of modeled thermal maturity at the Madiella stratigraphic level showing its effective limit of hydrocarbon generation. Contour units are percent vitrinite reflectance equivalent.

ing and expelling hydrocarbons during the Miocene (~15 Ma). Generation has continued into the present as a result of continued subsidence.

In contrast, in the immediate vicinity of the Loiret volcanic plug, the hydrocarbon generation process is compressed. Entry into and exit from the oil window occurred nearly coincident with the plug's emplacement during the early Miocene (Teisserenc and Villemin, 1990).

TRAPS AND SEALS

Salt movement was responsible for most hydrocarbon trap development within the region. Fields typically formed either on the limbs of structures penetrated by salt or on top of salt anticlines or domes (Vidal, 1980). Halokinesis was initiated during the Early Cretaceous and continued into the Tertiary, resulting in deformation of both the Batanga and Anguille Formations. The relative timing of structural development was favorable relative to that of hydrocarbon generation, that is, structure development occurred prior to oil generation. Within the delta complex, the majority of the known hydrocarbon reserves are associated with nonpiercement salt domes and turtlebacks, which are anticlinal structures formed between growing salt diapirs by salt withdrawal from the margins (Figure 8).

Available reservoirs are sealed by numerous Cretaceous and Tertiary marine shales. Seal efficiency is best developed in the thin but regionally extensive

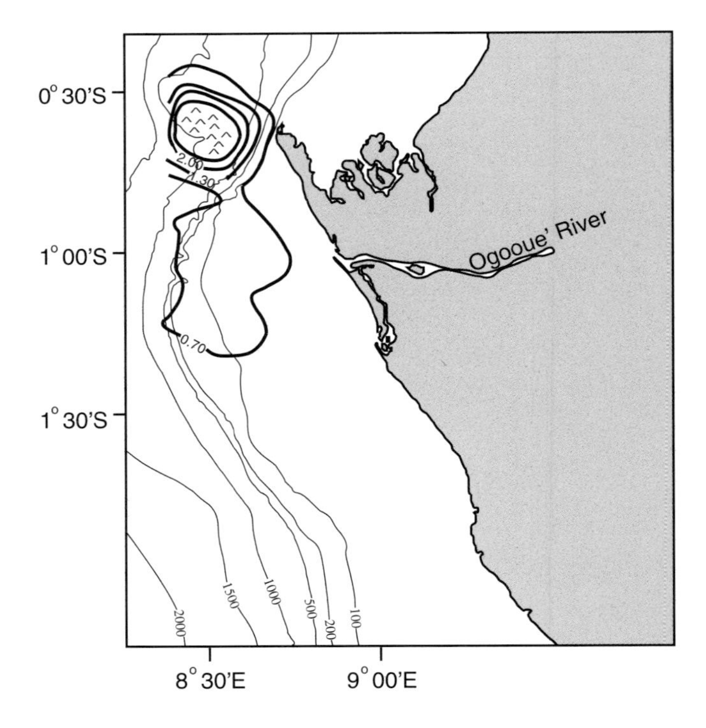

Figure 7—Map of modeled thermal maturity at the Azile stratigraphic level showing its effective limit of hydrocarbon generation. Contour units are percent vitrinite reflectance equivalent.

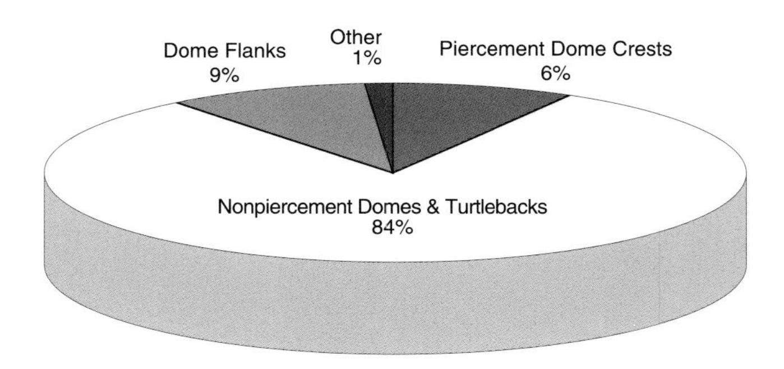

Figure 8—Pie chart of petroleum reserves based on trap type. After Teisserenc and Villemin (1990).

condensed shales associated with transgressive systems tracts. As previously noted, the pre-Miocene unconformity may erode into rocks as old as the Albian. Locally, the unconformity is responsible for the breaching of reservoirs as old as Albian–Cenomanian.

In those regions where faulting is significant, fault plane migration is likely to result in episodes of hydrocarbon remigration into shallower reservoirs. Hydrocarbon remigration via fault plane migration allows significant cumulative vertical migration but may create inefficient trapping if top seals are not well developed through lithification at the time of hydrocarbon introduction.

OIL FAMILY CHARACTERISTICS

Several oil families have been established within the Gabon petroleum province (Teisserenc and Villemin, 1990). Two of these families are present in the Ogooué Delta. The third, a lacustrine-derived oil, is known principally onshore and has been found in the Rabi-Kounga field, with its estimated reserves of 1.4 Bbbl (Boeuf et al., 1991). The first oil family displays characteristics typically associated with a more restricted marine setting. These properties include elevated sulfur contents (>1%), low pristane/phytane ratios (<1.0), a slight predominance of even *n*-alkanes, and elevated C_{35}/C_{34} homohopane and norhopane/hopane ratios (>1.0) (Figures 9a and 10a). A comparison of these geochemical attributes with the stratigraphic column suggests that the most likely source would be within the pre-Cenomanian transitional sequence, represented by the Madiela and Cap Lopez Formations.

The second oil family displays geochemical attributes typically associated with more normal marine conditions. These properties include more modest sulfur contents (<1%, except when biodegraded), moderate pristane/phytane ratios (<2.0), and low C_{35}/C_{34} homohopane and norhopane/hopane ratios (<1.0) (Figures 9b and 10b). Available stratigraphic and source rock data have established the Azile and Anguille Formations as the source for these oils.

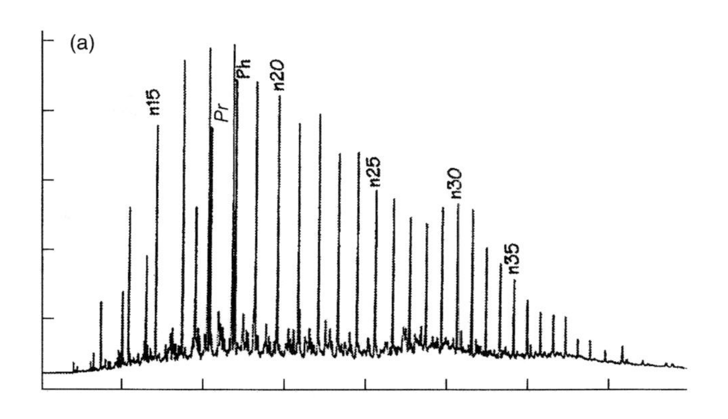

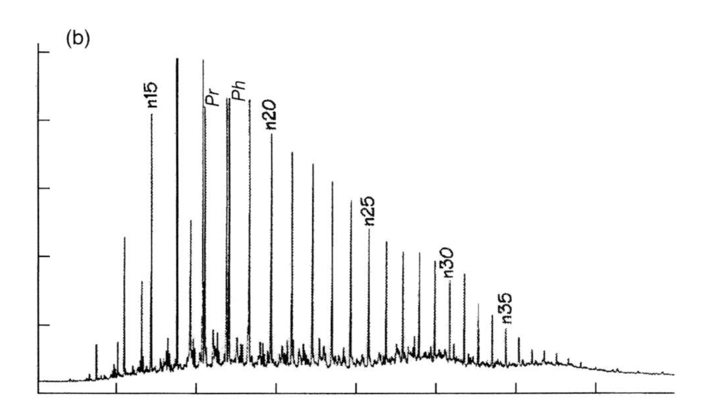

Figure 9—Whole-oil gas chromatograms of oils derived from (a) the Madiela and Cap Lopez Formations and (b) the Azile and Anguille Formations.

PETROLEUM SYSTEMS

As previously noted, the petroleum system concept requires an understanding of the temporal and spatial relationships among all geologic factors required for the development of commercial hydrocarbon accumulations. The clear presence of two distinctly different and effective source rocks define the existence of two petroleum systems within the Ogooué Delta. An examination of the stratigraphic distribution of the trapped oils derived from the two source rocks indicates that they do not share the same primary reservoirs. Oils derived from the Madiela and Cap Lopez Formations appear to occur principally in the Batanga sandstones. In contrast, the younger Azile and Anguille oils are largely reservoired within the Anguille sandstones.

Using the naming convention of Magoon (1992), the two petroleum systems identified in this study are the Madiela/Cap Lopez–Batanga(.) and the Azile/Anguille–Anguille(!) systems. Note that sufficient geochemical data are available from the Azile/Anguille–Anguille system for it to be classified as known (!), that is, a definitive oil source rock can be established. In the case of the

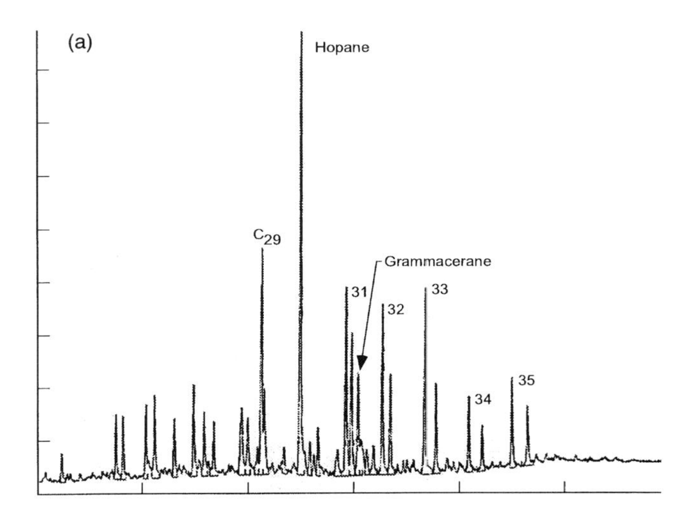

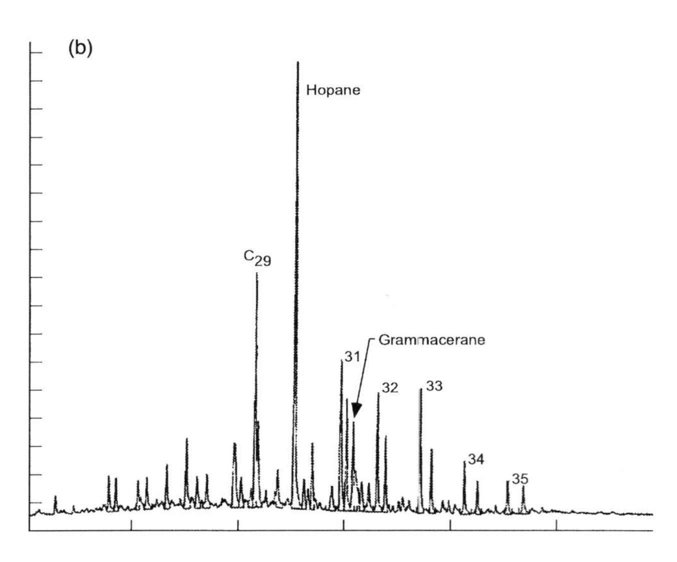

Figure 10—Ion chromatograms (*m/z* 191) of oils derived from (a) the Madiela and Cap Lopez Formations and (b) the Azile and Anguille Formations.

Madiela/Cap Lopez–Batanga system, such a correlation has not yet been established because of limited access to sample material. Therefore the system is classified as hypothetical (.).

The key temporal components of the Madiela/Cap Lopez–Batanga petroleum system are shown in Figure 11. In this system, source rock deposition occurred during the Albian. These organic-rich sedimentary rocks became effective during the Miocene and continue to expel hydrocarbons today. Hydrocarbon generation is controlled by the sedimentary overburden associated with the development of the delta itself. There are six episodes of sandstone reservoir deposition, including the Batanga Formation. Trap formation occurred initially

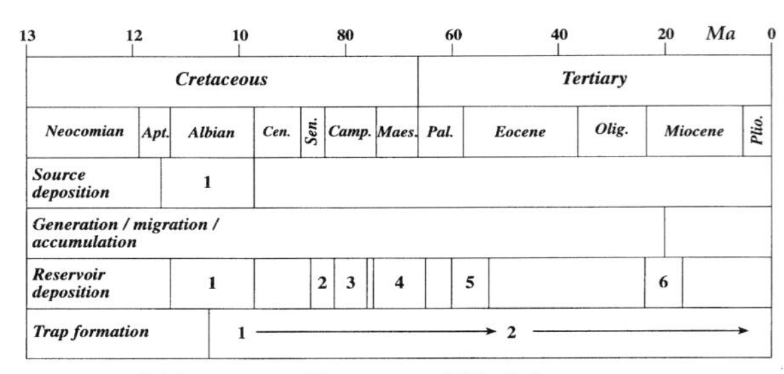

Sources: 1. Uppermost Ezanga and Madiela

Reservoirs: 1. Intra-Madiela sandstones
2. Anguille
3. Point Clairette
4. Batanga
5. Ozouri
6. Mandarove

Traps: 1. closure over salt swells
2. salt remobilization (structural and stratigraphic traps)

Figure 11—Events diagram for the Madiela/Cap Lopez–Batanga(.) petroleum system.

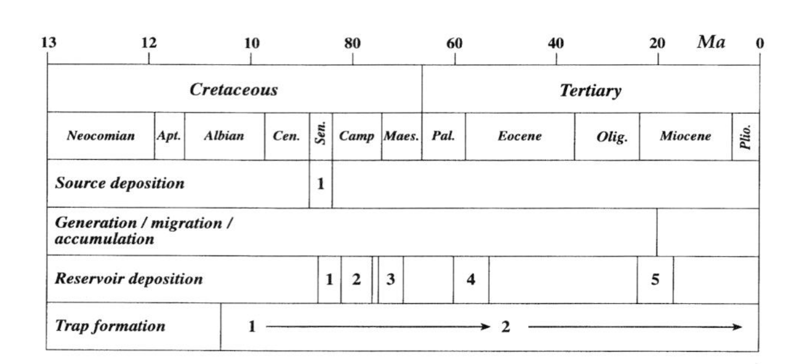

Sources: 1. Azile, Anguille

Reservoirs: 1. Anguille
2. Point Clairette
3. Batanga
4. Ozouri
5. Mandarove

Traps: 1. closure over salt swells
2. salt remobilization (structural and stratigraphic traps)

Figure 12—Events diagram for the Azile/Anguille–Anguille(!) petroleum system.

during the Albian as a result of the development of closure over salt swells. The second major episode of trap development began during the early Tertiary, probably in the Eocene, as a result of salt remobilization that formed both stratigraphic and structural traps.

Figure 12 shows the timing diagram of the Azile/Anguille–Anguille petroleum system. In this case, source rock deposition occurred during the Senonian and became effective during the Miocene. As with the older source rock, the thermal maturation of the source was controlled by the growth of the delta. There are five

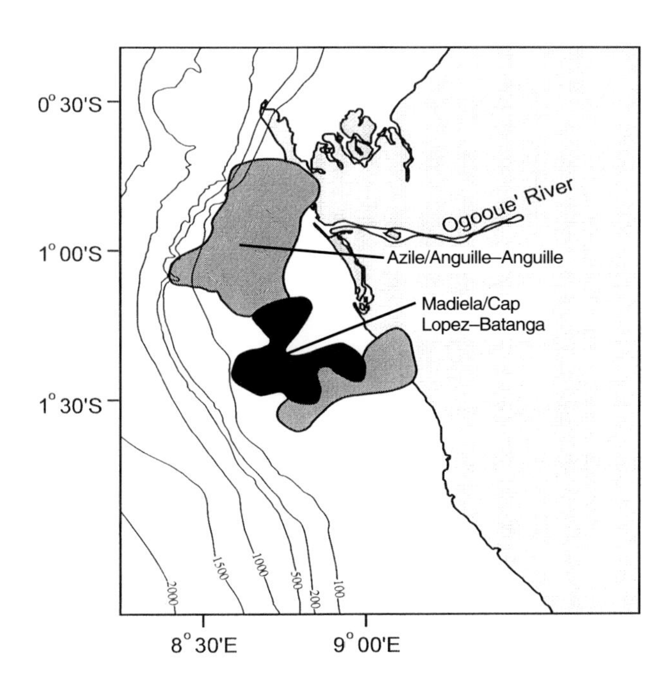

Figure 13—Map showing the distribution of the Madiela/Cap Lopez–Batanga(.) petroleum system (in black) and the Azile/Anguille-Anguille(!) system (in gray).

discrete episodes of sandstone reservoir deposition, including the main reservoir (Anguille Formation). Trap development for this system is essentially the same as for the older system.

The geographic distributions of these two petroleum systems are presented in Figure 13. A key element in the distribution of both systems is the necessary overburden for hydrocarbon generation to proceed. The systems are limited to the west by the thinning of the Tertiary overburden.

The Madiela/Cap Lopez–Batanga system appears restricted mainly to the southern area of the delta, with two oils from the northern part of the study area appearing to be part of this system. Such a distribution is consistent with the idea that the Madiela and Cap Lopez source developed within an isolated depression. Batanga penetrations within the geographic limits of the Azile/Anguille–Anguille system have revealed the presence of water-wet reservoirs with only limited hydrocarbon shows (Vidal, 1980). The Azile/Anguille–Anguille system is much less areally restricted but appears to be somewhat concentrated in the northern and central parts of the delta.

An argument can be put forward that the geographic and stratigraphic distribution of the two petroleum systems is controlled by the efficiency of vertical migration, the distribution of the source, and the nature of the reservoir. The Madiela/Cap Lopez–Batanga system appears to have a more efficient vertical migration network as well as better development of younger reservoirs. The greater areal distribution of the Azile/Anguille–Anguille system would suggest that lateral or bed-parallel migration plays an important role in the redistribution of hydrocarbons throughout the system.

CONCLUSIONS

Available data confirm the presence of two commercially significant petroleum systems within the Ogooué Delta. The first is named the Madiela/Cap Lopez–Batanga(.) system and the second, the Azile/Anguille–Anguille(!) system, following the nomenclature of Magoon (1992). Cross-formational vertical migration plays a major role in defining the Madiela/Cap Lopez–Batanga system, whereas lateral bed-parallel migration appears more instrumental in defining the limits of the Azile/Anguille–Anguille system. The presence of both systems appears to be largely influenced by sediment loading associated with the progradation of the delta. This loading played a major role both in the attainment of necessary levels of thermal maturity for hydrocarbon generation to proceed and in driving the salt tectonics within the delta, which strongly influenced in the development of traps.

Acknowledgments*—The authors thank Texaco Inc. for permission to publish this work. This paper was originally presented at the 1996 AAPG International Conference and Exhibition held in Caracas, Venezuela.*

REFERENCES CITED

Alkhadhrawi, M. R., 1996, A geochemical investigation of a suite of central Arabian crude oils: Master's thesis, King Fahd University of Petroleum and Minerals, Dhahran, Saudi Arabia, 227 p.

Boeuf, M. A. G., W. J. Cliff, and J. A. R. Hombroek, 1991, Discovery and development of the Rabi-Kounga field: a giant oil field in a rift basin onshore Gabon: Proceedings, 13th World Petroleum Congress, Buenos Aires, Argentina, v. 2, p. 33–46.

Burwood, R., 1999, Angola: source rock control for Lower Congo Coastal and Kwanza Basin petroleum systems, *in* N. R. Cameron, R. H. Bate, and V. S. Clure, eds., The oil and gas habitats of the South Atlantic: Geological Society of London Special Publication, v. 153, p. 181–194.

Dunne, L. A., P. R. Johnson, and S. B. Desantis, 1991, Geology and petrophysics of the Ozouri Group, central Gabon: Elf Aquitaine Memoir 13, p. 31–38.

Dunne, L. A., P. R. Johnson, and S. B. Desantis, 1996, Geology and petrophysics of the Ozouri Group, central Gabon (abs.): AAPG Bulletin, v. 80, p. 1287.

Edwards, A., and R. Bignell, 1988a, Hydrocarbon potential of W. Africa salt basin (1): Oil & Gas Journal, December 12, p. 71–74.

Edwards, A., and R. Bignell, 1988b, Hydrocarbon potential of W. Africa salt basin (2): Oil & Gas Journal, December 19, p. 55–58.

Karner, G. D., N. W. Driscoll, J. P. McGinnis, W. D. Brumbaugh, and N. R. Cameron, 1997, Tectonic significance of syn-rift sediment packages across the Gabon–Cabinda continental margin: Marine and Petroleum Geology, v. 14, p. 973–1000.

Kennicutt, M. C. II, T. J. McDonald, P. A. Comet, G. J. Denoux, and J. M. Brooks, 1992, The origins of petroleum in the northern Gulf of Mexico: Geochimica et Cosmochimica Acta, v. 56, p. 1259–1280.

Magoon, L. B., 1992, Identified petroleum systems within the United States—1992: USGS Bulletin, v. 2007, p. 2–11.

Magoon, L. B., and W. G. Dow, 1994, The petroleum system, *in* L.B. Magoon and W. G. Dow, eds., The petroleum system—from source to trap: AAPG Memoir 60, p. 3–24.

Massonnat, G. J., F. G. Alabert, and C. B. Giudicelli, 1992, Anguille Marine, a deep sea-fan reservoir, offshore Gabon: from geology to stochastic modelling. Proceedings, 67th Annual Society of Petroleum Engineers Technical Conference, Washington, D.C., p. 477–492.

Riva, A., T. Salvatori, R. Calcaliere, T. Ricchuto, and L. Novelli, 1986, Origin of oils in the Po basin, northern Italy: Organic Geochemistry, v. 10, p. 391–400.

Sahagian, D., 1988, Epeirogenic motions of Africa as inferred from Cretaceous shoreline deposits: Tectonics, v. 7, p. 125–138.

Smith, D. R., S. B. Desantis, L. A. Dunne, B. L. Faulker, and R. E. West, 1990, Batanga Sandstone reservoir: a turbidite stratigraphic trap (abs.): 1st Archie Conference, Abstract Volume, p. 326.

Teisserenc, P., and J. Villemin, 1990, Sedimentary basin of Gabon—geology and oil systems, *in* J. D. Edwards and P. A. Santogrossi, eds., Divergent/passive margin basins: AAPG Memoir 48, p. 117–199.

Trindade, L. A. F., S. C. Brassell, and E. V. Santos Neta, 1992, Petroleum migration and mixing in the Potiguar Basin, Brazil: AAPG Bulletin, v. 76, p. 1903–1924.

Uchupi, E., 1992, Angola Basin: geohistory and construction of the continental rise, *in* C. W. Poag and P. C. de Graciansky, eds., Geologic evolution of Atlantic continental rises: New York, Van Nostrand-Reinhold, p. 77–99.

Vidal, J., 1980, Geology of Grondin field, *in* M. T. Halbouty, ed., Giant oil and gas fields of the decade 1968–1978: AAPG Memoir 30, p. 577–589.

Gonçalves, F. T. T., R. P. Bedregal, L. F. C. Coutinho, and M. R. Mello, 2000, Petroleum system of the Camamu–Almada Basin: a quantitative modeling approach, *in* M. R. Mello and B. J. Katz, eds., Petroleum systems of South Atlantic margins: AAPG Memoir 73, p. 257–271.

Chapter 19

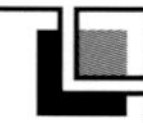

Petroleum System of the Camamu–Almada Basin: A Quantitative Modeling Approach

F. T. T. Gonçalves

R. P. Bedregal

Petrobrás R & D Center/Cenpes
Rio de Janeiro, Brazil

L. F. C. Coutinho

Petrobrás E & P Department
Salvador, Brazil

M. R. Mello

Petrobrás R & D Center/Cenpes
Rio de Janeiro, Brazil

Abstract

The Camamu–Almada Basin, located in northeastern Brazil, is part of the rift system that formed during the Early Cretaceous break-up of South America and Africa. Previous studies have characterized the occurrence of a single petroleum system in this basin, the Morro do Barro(!), which encompasses a Neocomian synrift lacustrine source rock and turbiditic reservoirs of the same age. 1-D and 2-D tectonic, thermal, and geochemical modeling was applied to better understand the evolutionary history of the petroleum system, as well as to provide a basis for better assessment of the exploration risk and petroleum potential of the Camamu–Alamada Basin.

The hydrocarbon generation modeling indicates that most of the oil was generated by the end of the rift phase (Barremian–Aptian). Present kerogen transformation ratios range from 10–20% nearby the continent (west) to 100% in the deeper areas (east). Fluid-flow modeling showed that the presence of a thick section of low-permeability shales above the source rocks favored the downward migration of petroleum to the sandstones of the Sergi Formation. Petroleum migration through normal faults, which juxtaposed source rocks of the Morro do Barro Formation to sandstones of the Sergi Formation, also played a major role in the filling of these carrier beds. Secondary migration extended considerably into the postrift phase.

INTRODUCTION

Over the past few years, application of the petroleum system concept (Magoon, 1988) has provided a new rationale to approach petroleum research, to formulate plays and prospects, and to evaluate the related exploration risk. A petroleum system includes all the essential elements (source rock, reservoir rock, seal rock, and overburden) and processes (trap formation, generation, migration, and accumulation) that are needed for a petroleum accumulation to exist (Magoon and Dow, 1994).

A petroleum system also requires the timely convergence of these elements and processes and a positive mass balance between hydrocarbon charge and losses through migration. Nevertheless, most petroleum system studies emphasize only the qualitative aspects of the system, such as describing present-day source rock maturity and distribution, assessing reservoir quality and occurrence, or performing geochemical characterization of hydrocarbons. The application of quantitative basin modeling techniques allows a better understanding of the spatial and temporal relationships between the elements and processes of petroleum systems, as well as assessing the hydrocarbon mass balance. These are critical factors when evaluating the risk associated with prospects.

The Camamu–Almada Basin, located in northeastern Brazil (Figure 1), is part of the rift system that formed during the Early Cretaceous break-up of South America and Africa. With an area of about 12,000 km^2 (10,000 km^2 of that below sea level), this basin contains some small gas accumulations in the onshore area and two oil fields in the offshore (platform) area. The basin infill comprises a thick (up to 10,000 m) succession of prerift, rift and postrift strata. A previous study (Mello et al., 1995) has characterized the occurrence of a single petroleum system in the Camamu–Almada Basin: the Morro do Barro(!)

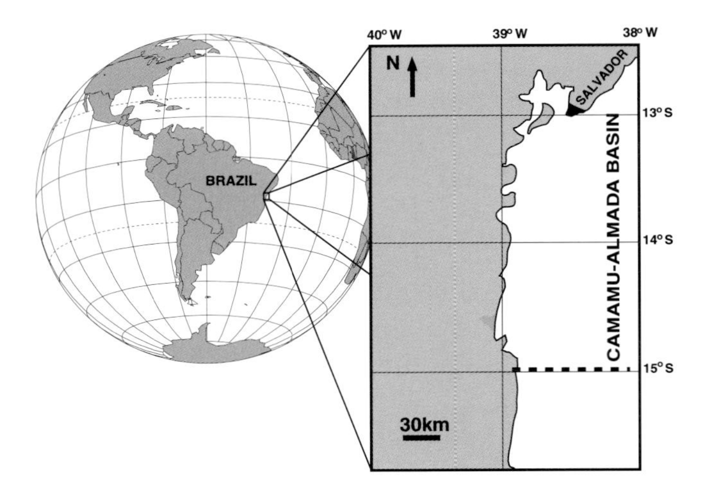

Figure 1—Location map of the Camamu–Almada Basin.

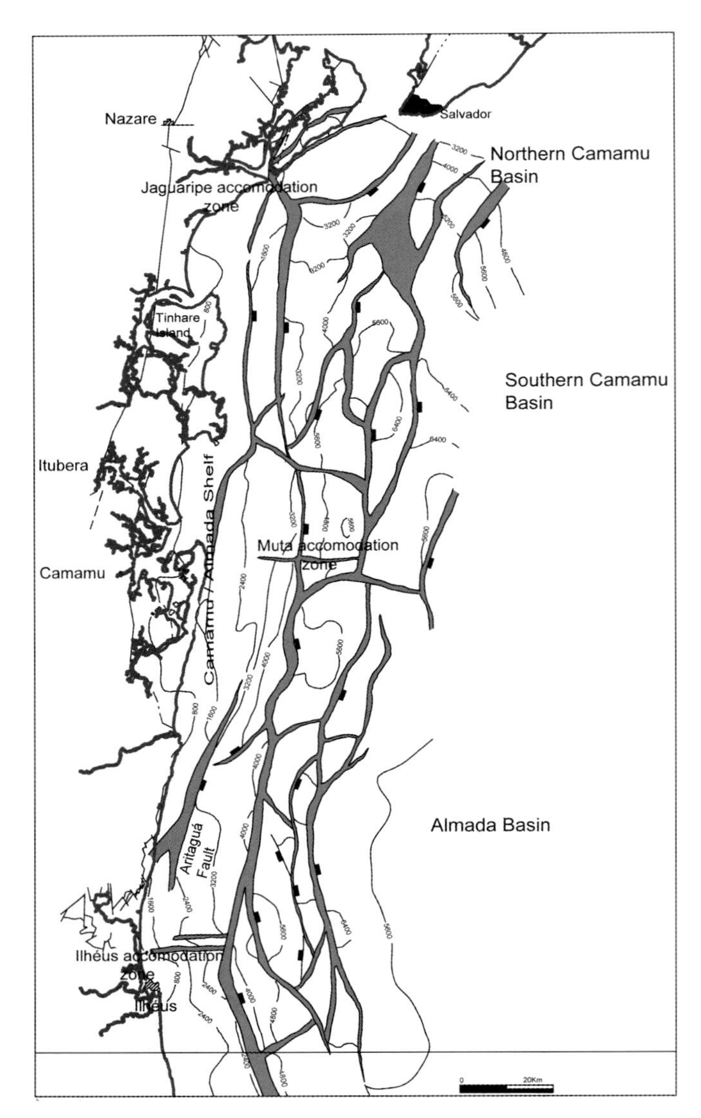

Figure 2—Simplified tectonic framework of the Camamu–Almada Basin showing the structural contour for the top of the Sergi Formation and the main faults affecting the prerift and rift section. Contour interval is 800 msec.

petroleum system. (The "!" symbol indicates a known system.) In that study, they emphasized the genetic relationship between a lower Neocomian synrift lacustrine fresh to brackish water source rock and a lacustrine turbiditic reservoir of the same age. However, some critical questions remain unsolved, such as (1) location of the pod(s) of active source rock, (2) timing of petroleum generation and expulsion, (3) hydrocarbon generation of potential source rocks in the deep offshore area, and (4) secondary migration pathways.

The objective of this study is to show how the application of an integrated 1-D and 2-D tectonic, thermal, and geochemical modeling allowed a better understanding of the evolution history of the Morro do Barro(!) petroleum system, as well as provided a basis for a better assessment of the exploration risk and the petroleum potential in the Camamu–Almada Basin.

GEOLOGIC SETTING

The origin of the Camamu–Almada Basin is associated with the rifting that resulted in the break-up of South America and Africa and the development of the Brazilian eastern marginal basins. The tectono-sedimentary evolution of the Camamu–Almada Basin, as well as other eastern Brazilian marginal basins, can be summarized as the succession of three main stages: prerift, rift, and postrift (Ponte and Asmus, 1978; Ojeda, 1982).

Located in Bahia State, northeastern Brazil, the Camamu–Almada Basin is limited to the north by an east–west accommodation zone and to the south by an east–west basement structural high. The Camamu–Almada Basin displays a complex structural framework (Figure 2) associated with the rifting of a region marked by many heterogeneities in the basement, which is mainly composed of Precambrian rocks of the Atlantic Granulitic Belt.

The tectonic framework of the basin is characterized by a series of horsts and grabens limited by NNE–SSW trending normal faults dipping to the east or west (Figure 2). Several structural features transverse to the rift axis are interpreted as accommodation zones, which may be associated with horizontal displacement or polarity inversion of the extensional faults. The interplay of these structural elements gave rise to the formation of structural platforms and restricted subbasins, which controlled rift sedimentation and hence the formation of source and reservoir rocks. In the deeper offshore area, the postrift section is affected by normal listric faults detaching on the Aptian salt.

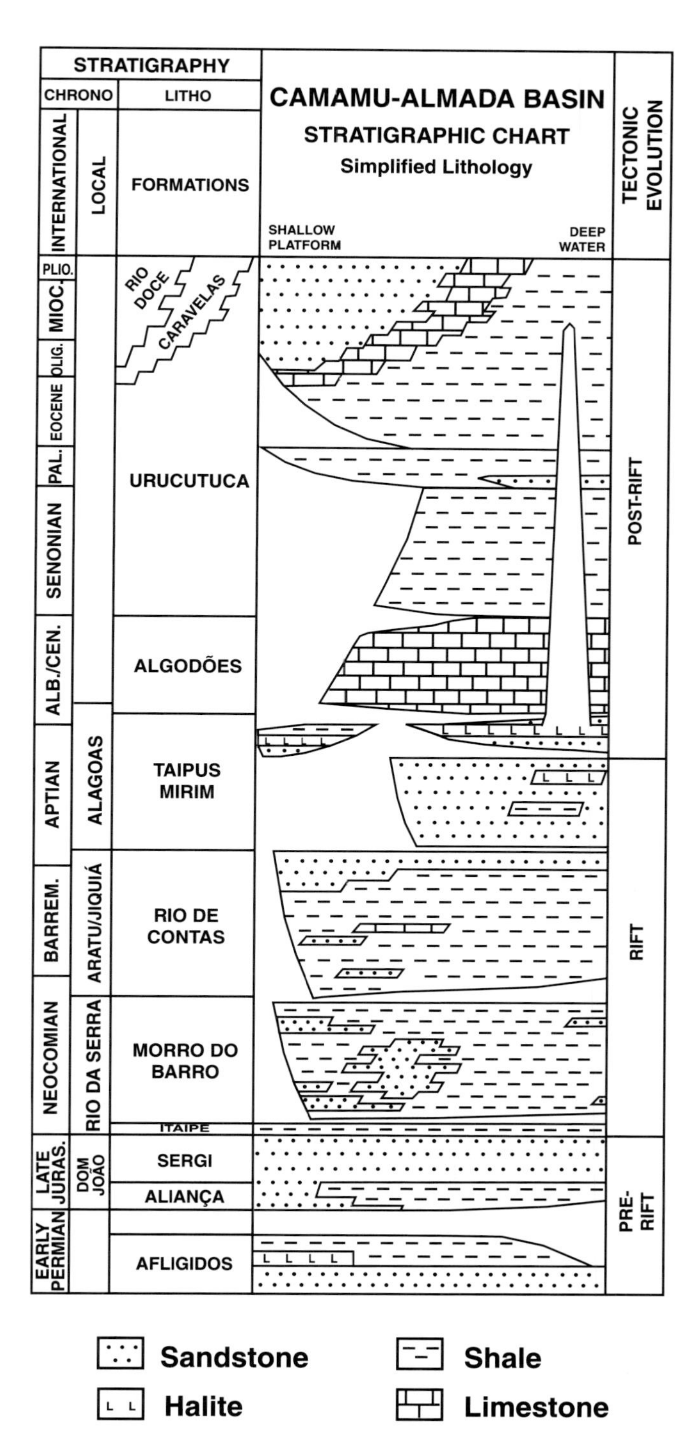

Figure 3—Stratigraphic chart for the Camamu–Almada Basin. Modified from Netto et al. (1994).

In the Camamu–Almada Basin, the prerift stage is represented by the Lower Permian Afigidos Formation, the Upper Jurassic Aliança and Sergi Formations, and the lower Neocomian Itaípe Formation (Figure 3). The Afligidos Formation consists of siliciclastic and evaporitic sedimentary rocks. The Aliança Formation comprises fluvial and lacustrine shales and sandstones, and the Sergi Formation is mainly composed of fluvial and eolian sandstones. The Itaípe Formation is represented by shales with minor sandstone layers (Netto et al., 1994). These formations have a regional distribution and were deposited in a large depression that encompassed most of the northeastern Brazilian margin.

The Early Cretaceous rift stage is characterized by intense subsidence, which resulted in the deposition of a thick succession of lacustrine strata comprising the Morro do Barro and Rio de Contas Formations (Figure 3). The Morro do Barro Formation (Neocomian) is dominated by dark gray to brown shales with interbedded sandstones, and the Rio de Contas Formation (upper Neocomian–lower Aptian) consists mainly of dark and greenish-gray to brown shales, sandstones, and minor limestone, dolomite, and marl layers (Netto et al., 1994). In this work, the lower part of Taipus–Mirim Formation (Aptian), mainly composed of interbedded shales and sandstones, is included in the rift stage due to the strong control of the rift faults on its facies distribution.

The postrift stage includes the transitional and marine phases. In the Camamu–Almada Basin, the transitional phase is represented by the Aptian siliciclastic and evaporitic sedimentary rocks of the upper part of the Taipus–Mirim Formation (Figure 3). The marine phase comprises three distinct sequences: (1) the Albian–Turonian carbonates of the Algodões Formation, which were deposited in a neritic environment in a shallow and narrow sea; (2) the Senonian open-marine shelf–slope system, characterized by siliciclastic and calcareous mudstones from the Urucutuca Formation; and (3) the Tertiary progradational sequence, which comprises the coarse-grained sandstones of the Rio Doce Formation, the platform carbonates of the Caravelas Formation, and the distal pelitic sedimentary beds of the Urucutuca Formation.

PETROLEUM SYSTEM

The petroleum system in the Camamu–Almada Basin (Mello et al., 1995) encompasses the Lower Cretaceous Morro do Barro source rocks and the resulting hydrocarbon accumulations. The system is geographically restricted to the west, north, and south by the hydrocarbon occurrences, whereas the eastern limit is unknown. The stratigraphic extent of the system is restricted to the rift and prerift sequences. Most of the petroleum is trapped in the Morro do Barro Formation reservoirs (almost 75% of original volume of oil in place), followed by Sergi Formation (about 25% of original volume) and Rio de Contas Formation reservoirs. The integration of all the geochemical and geological data allowed the definition of the Morro do Barro(!) as the unique petroleum system that occurs in the Camamu–Almada Basin (Mello et al., 1995).

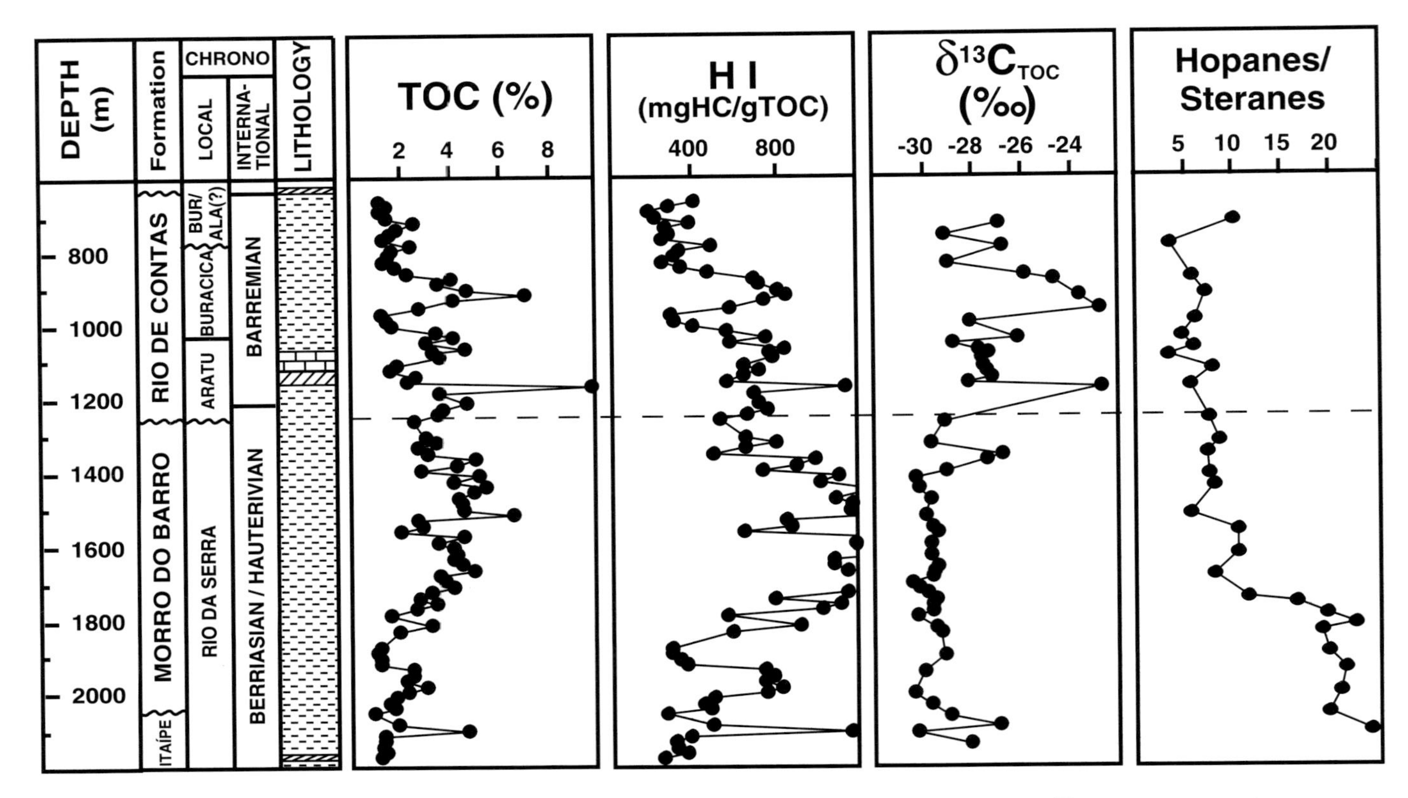

Figure 4—Variation in total organic carbon (TOC), hydrogen index (HI), carbon isotope ratio ($\delta^{13}C$), and hopanes/steranes ratio along the rift sequence in a selected well from the Camamu–Almada Basin. From Gonçalves (1997).

Petroleum Source Rock Characterization

Geochemical analyses from rock samples of 20 exploratory wells have revealed that potential source rocks occur only in the Morro do Barro and Rio de Contas Formations. Both formations comprise thick layers of shales with high total organic carbon (TOC) contents (mostly 2–5%) and high hydrogen index (HI) values (up to 1000 mg HC/g TOC), which indicate that these rocks are composed mainly of a lipid-rich type I kerogen. Organic geochemical characteristics and fossil biota indicate a fresh to brackish water lacustrine depositional environment (Mello and Maxwell, 1991; Mello et al., 1995). Stratigraphic correlations have shown a remarkable facies variation within these formations, which affects the geographic extent of the potential source rocks throughout the basin.

A detailed geochemical study of a selected offshore well in the Camamu–Almada Basin allowed assessment of the depositional paleoenvironments of the rift strata (Goncalves, 1997). In this study, 94 rock samples were analyzed using organic carbon determination, Rock-Eval pyrolysis, visual kerogen analysis, and gas chromatography and mass spectrometry (GC, GCMS, and GCMS-MS) techniques. The sedimentary column comprises a thick succession of Neocomian dark gray to brown shales of the Morro do Barro Formation overlain by the Barremian dark to greenish shales and minor carbonate and anhydrite layers of the Rio de Contas Formation (Figure 4).

In the selected well, Morro do Barro Formation samples showed high TOC (2–4%) and HI values (up to 1000 mg HC/g TOC), and fairly constant carbon isotope ratios ($\delta^{13}C$ about –29.0%). In the Rio de Contas Formation, TOC contents reach 10%, HI is lower (400–800 mg HC/g TOC), and $\delta^{13}C$ values range from –29 to –23% (Figure 4). Visual kerogen analyses point to a dominance of amorphous (>90 vol. %) organic matter, with minor amounts of woody (vitrinite and inertinite) and liptinite components (e.g., spores and *Botryococcus*). Rock-Eval T_{max}, spore coloration index, and vitrinite reflectance data indicate that the analyzed section is immature.

GC, GCMS, and GCMS-MS data of the organic extracts of rock samples from the selected well show a set of common characteristics, including the dominance of high molecular weight *n*-alkanes (>*n*-C_{23}), odd–even *n*-alkane preference, dominance of pristane over phytane, and absence of C_{30} steranes and β-carotane (Figure 5a). These features are in agreement with those reported for fresh to brackish water lacustrine source rocks from rift basins in Brazil (Mello et al., 1988; Mello and Maxwell, 1991). Hopane/sterane ratio displays an overall shift from 25 at the bottom to 5 at the top of the Morro do Barro Formation, and ranges from 1 to 7 in the Rio de Contas Formation (Figure 4).

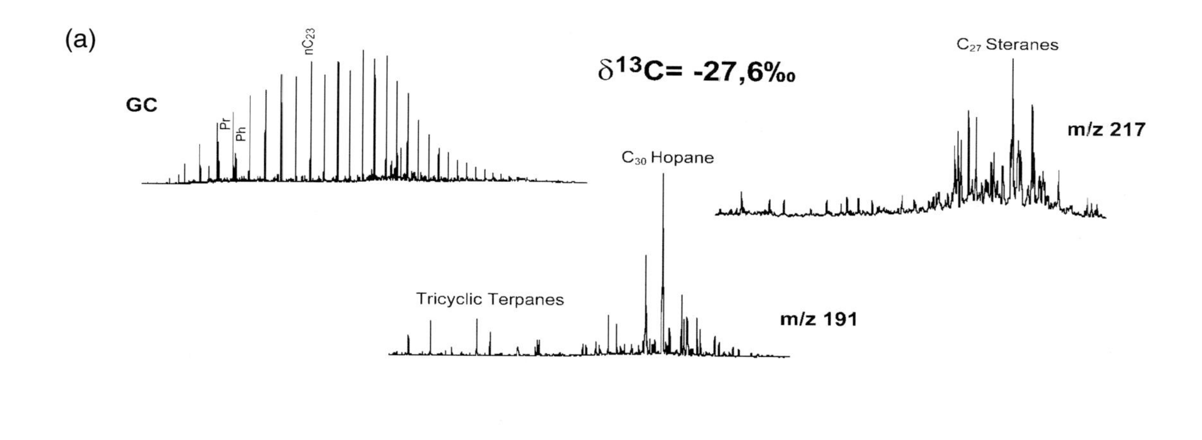

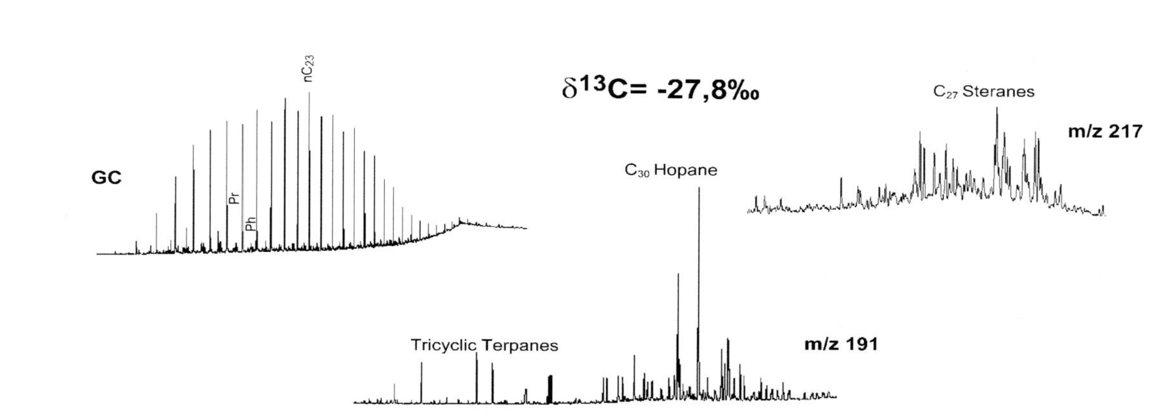

Figure 5—Gas chromatograms (GC) and mass chromatograms (*m/z* 191 and *m/z* 217) of (a) an organic extract from the Morro do Barro Formation and (b) a typical oil sample from the Camamu–Almada Basin.

Pyrolysis and organic petrography data suggest a phytoplankton and bacterial origin for the organic matter of the analyzed section. The high TOC and HI values observed in the Morro do Barro Formation, together with high hopane/sterane ratios and a strong ^{13}C isotopic depletion, reflect medium to low primary productivity and intense carbon recycling through the degradation of organic matter by anaerobic (methanogenic?) bacteria. In the Rio de Contas Formation, the positive carbon isotopic excursions coincide with the highest TOC contents and hydrogen indices and are interpreted as the result of enhanced primary productivity (Gonçalves, 1997).

Based on the integration of organic geochemical data with the $\delta^{18}O$ records from carbonates, paleontologic, and geologic information, Gonçalves (1997) identified two different phases of the Camamu–Almada rift lake evolution, which controlled the deposition of the associated source rocks:

1. Berriasian–Hauterivian (Morro do Barro Formation)—Fault-bounded deep, meromictic lake with fresh to brackish water, medium to low primary productivity, and an intense recycling of carbon through the degradation of organic matter by anaerobic bacteria. Organic preservation was greatly enhanced by the stable water column stratification and bottom anoxia.
2. Barremian (Rio de Contas Formation)—Broad and shallower (oligomictic?) lake with fresh to saline water. Enhancement of primary productivity due to increased input and recycling of nutrients favored by the lake morphology and a humid climate. More effective lake basin overturn probably resulted in conditions of anoxic–dysoxic bottom water.

Hydrocarbon Characterization and Oil–Source Correlation

Bulk and elemental analyses from oil samples of Camamu–Almada Basin petroleum accumulations show a set of common features, such as low sulfur concentration (<0.1%), high wax content (saturates >60%), and $\delta^{13}C$ values ranging from –28 to –30%. The results from GC, GCMS, and GCMS-MS analyses reveal a dominance of high molecular *n*-alkanes, pristane much higher than phytane, odd–even *n*-alkane dominance, high hopane/

sterane ratios (>15), T_s higher than T_m, low to medium relative abundance of gammacerane, and absence of β-carotane, dinosterane, and C_{30} steranes (Figure 5b).

Geochemical data are consistent with the classification of the oils from Camamu–Almada Basin as a single oil family, generated from the same petroleum source rock. Similar bulk, elemental, and biomarker features have been reported for oils sourced by fresh to brackish water lacustrine petroleum source rocks from other Brazilian rift basins (Mello et al., 1988; Mello and Maxwell, 1991).

A detailed geochemical correlation was performed among oil samples and organic extracts from potential petroleum source rocks of both the Morro do Barro and Rio de Contas Formations. Similarities in geochemical features (mainly $\delta^{13}C$ and hopane/sterane ratios) indicate that Morro do Barro shales are the source for all the petroleum accumulations in the Camamu–Almada Basin. The good oil–source correlation is illustrated in Figure 5. This conclusion is also supported by maturation data indicating that only the lower part of the Morro do Barro Formation was sufficiently buried to reach the oil window in the platform area (see explanation in the next section).

Despite similarities in the geochemical features, differences do exist among the petroleum occurrences. Slight differences in bulk and biomarker data are due to the degree of maturity, whereas some differences at the molecular level (e.g., gammacerane relative abundance and hopane/sterane ratios) point to the presence of different hydrocarbon charge areas associated with changes in the depositional environment of the source rock.

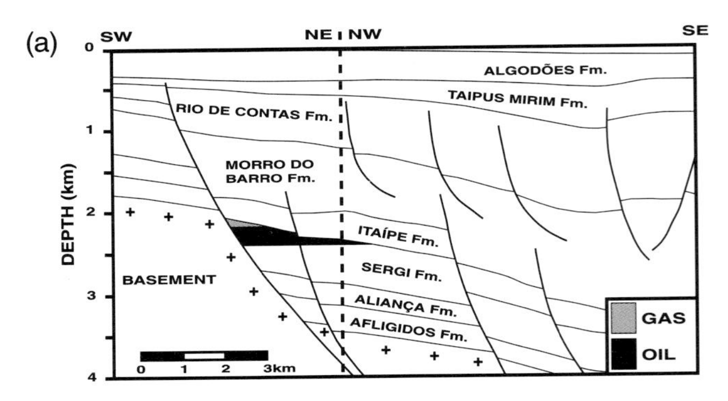

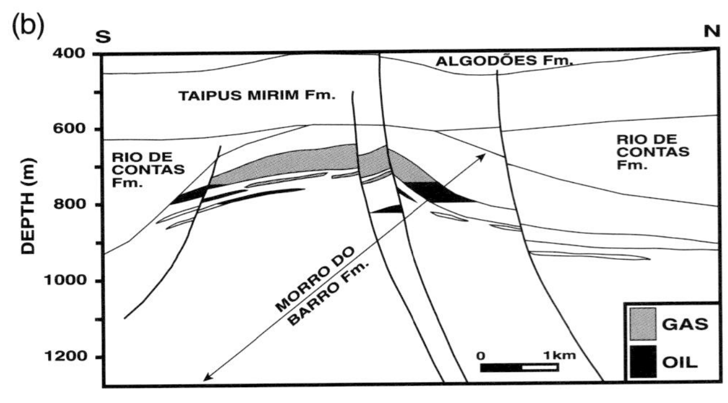

Figure 6—Schematic cross-sections showing examples of typical (a) prerift and (b) synrift traps in the Camamu–Almada Basin.

Traps, Reservoirs, and Seals

Petroleum accumulations in the Camamu–Almada Basin are associated with structural or structural–stratigraphic trapping within prerift and rift reservoirs. Most of the petroleum was found in Morro do Barro Formation reservoirs (almost 75% of original volume of oil in place), followed by the Sergi Formation (about 25% of original volume) and Rio de Contas Formation reservoirs. Two plays were selected to exemplify the petroleum system in the Camamu basin: a prerift oil field, with about 94×10^6 bbl of oil in place, and a Morro do Barro synrift oil field, with about 157×10^6 bbl of oil in place (Figure 6).

In the prerift oil field, petroleum is trapped in the Upper Jurassic fluvial and eolian sandstones of the Sergi Formation against the footwall of a major regional fault (Figure 6a). Two main reservoir facies occur: coarse-grained sandstones with porosity ranging from 14 to 23% and permeability reaching up to 500 md, interbedded with fine- to coarse-grained sandstones with 3–15% porosity and up to 7 md permeability. The sandstone reservoir is in direct lateral contact with the basement, which provides the seal together with the lacustrine shales from the Itaípe Formation.

In the synrift oil field, a residual anticline in the hanging wall of a listric fault has played an important role in the trap formation and the distribution of an oil ring and an extensive gas cap within the Berriasian sandstone reservoirs from Morro do Barro Formation (Figure 6b). Petroleum is trapped in lacustrine turbidite lobes, which comprise a succession of medium- to coarse-grained sandstones with an average porosity of 25% and a high permeability (up to 2000 md). The seal is provided by the lacustrine shales from the Morro do Barro and Rio de Contas Formations.

Source Rock Maturation and Timing of Events

The integration of TOC, Rock-Eval, and vitrinite reflectance data from the Morro do Barro and Rio de Contas Formations allowed the characterization of maturation, transformation ratio, and expulsion efficiency in the platform offshore area of the Camamu–Almada Basin. The plot of these data as a function of depth revealed a remarkable depletion of TOC and HI values, as well as an increase in production indices (PI), starting from a depth of about 2500m (Figure 7). Such features result from the thermal degradation of kerogen and indicate that most of the original hydrocarbon source potential was converted into petroleum between the depths of 2500 and 4000 m.

HI values were also used to calculate the transformation ratio (TR) evolution as a function of depth, following

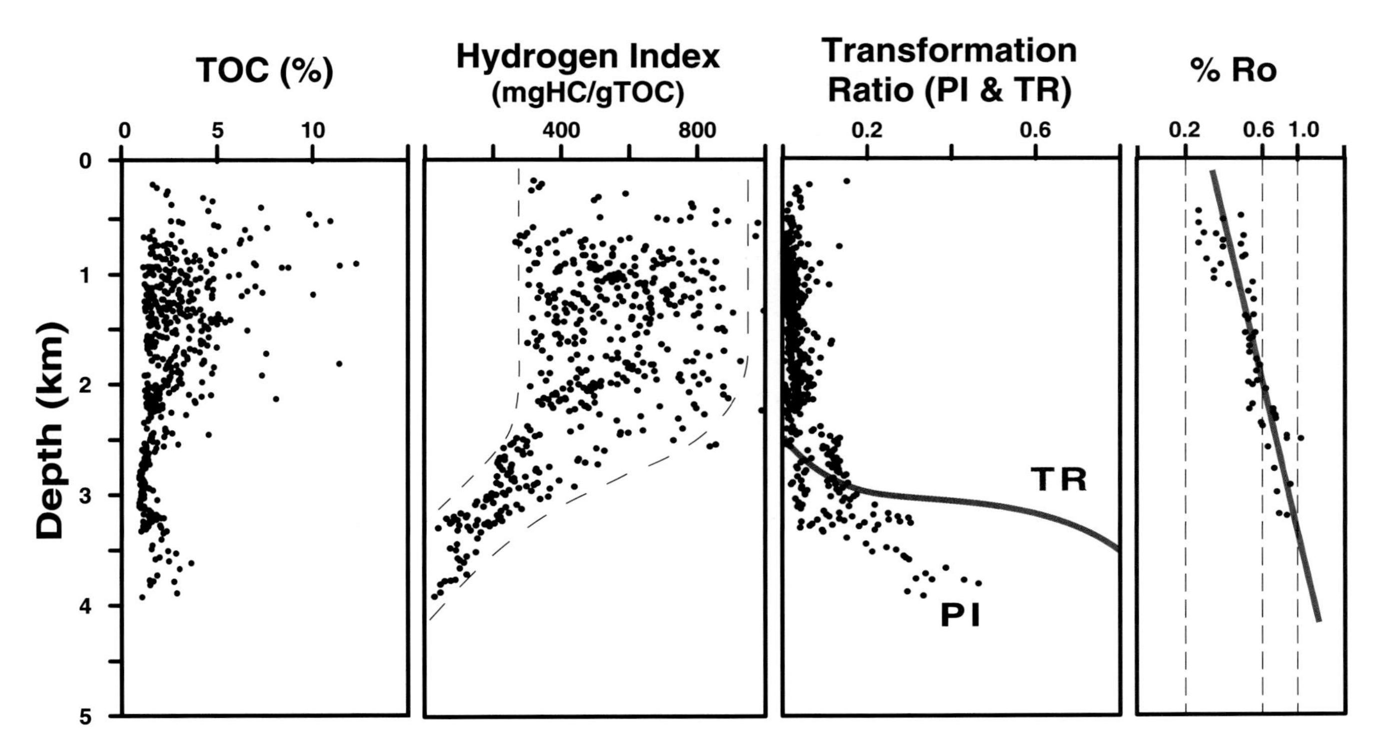

Figure 7—Variation in total organic carbon (TOC), hydrogen index (HI), production index (PI), transformation ratio (TR), and vitrinite reflectance (% R_o) as a function of depth for the Morro do Barro and Rio de Contas Formations for a set of about 500 selected samples from 12 wells throughout the Camamu–Almada Basin.

Espitalié et al. (1988). The integration of the TR and vitrinite reflectance (% R_o) curves (Figure 7) indicates that significant petroleum generation started at a depth of 2500 m, which corresponds to about 0.7% R_o, and that most of the original hydrocarbon source potential was converted (TR about 80%) at a degree of thermal evolution equivalent to 1.0% R_o. The short depth interval required to complete kerogen conversion mostly reflects the kinetic behavior of type I organic matter.

Petroleum expulsion efficiency was estimated by comparing the PI values with the calculated TR (Espitalié et al., 1988), assuming that losses of Rock-Eval S_1 values are due to primary migration. This comparison revealed that although both PI and TR increase as a function of depth, PI values are systematically lower than the calculated TR at the same depth, starting from a depth of 3000 m (Figure 7). The integration of these results with the % R_o curve suggests that petroleum expulsion started at a degree of thermal evolution equivalent to 0.80% R_o and a significant amount (up to 60–70%) of generated petroleum was expelled within the oil window. Therefore, most of the generated hydrocarbons would have been expelled as liquid petroleum.

Rock-Eval, TOC, and maturity data (vitrinite reflectance and spore coloration index) from drilled wells indicate that Morro do Barro organic-rich shales are immature in the onshore area and in most parts of the offshore platform area. Only in the outer platform does the lower part of this formation have an appropriate degree of thermal evolution to start petroleum generation and expulsion. Nevertheless, the extrapolation of maturity gradients to deep offshore areas suggests that thermal evolution of the source rock can range from mature to overmature. Such extrapolations also indicate that potential source rocks from the Rio de Contas Formation could have reached the oil window in the deeper offshore areas.

Preliminary 1-D basin modeling results from a few wells from the Camamu–Almada Basin indicate that source rocks have reached generation conditions during the rift phase (late Neocomian–Barremian) and up to the Aptian (Mello et al., 1995). However, these results are insufficient to describe the dynamics of the oil kitchen evolution across the basin, nor do they predict source rock thermal maturity in the deeper offshore area.

QUANTITATIVE MODELING

Both 1-D and 2-D tectonic, thermal, and geochemical models were applied to our data to achieve a better understanding of the hydrocarbon generation and migration processes within the Morro do Barro (!) petroleum system and to support the assessment of exploration risk and petroleum potential in the Camamu–Almada Basin.

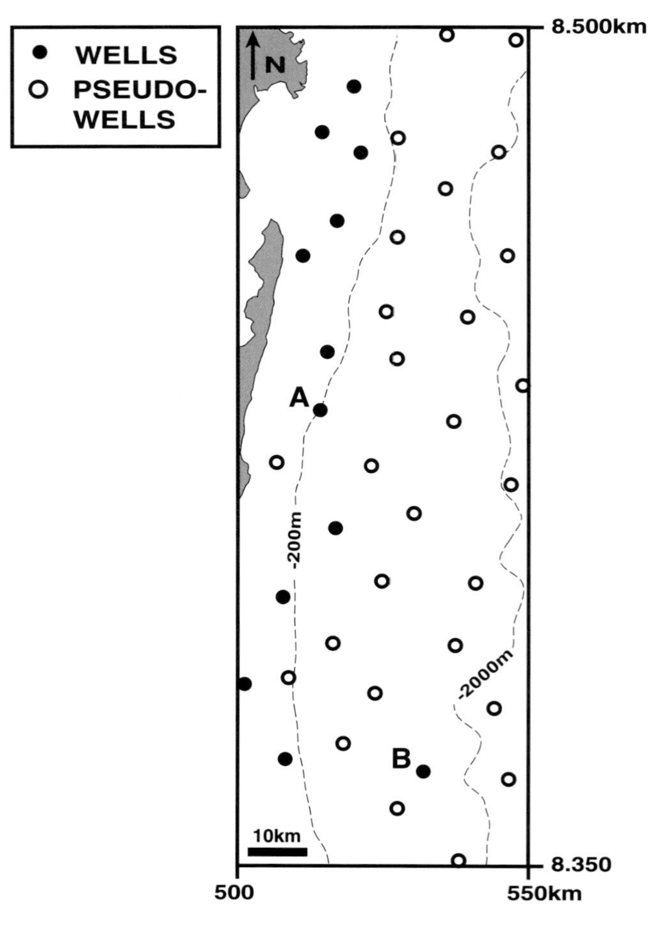

Figure 8—Location map of the selected wells and pseudowells used in the 1-D thermo-mechanical and kinetic modeling of the Camamu–Almada Basin. Contour lines represent the present bathymetry.

1-D Thermo-Mechanical Modeling

Seismic and well data were used as input data and constraints for thermo-mechanical modeling of the Camamu–Almada Basin using the *Basin Simulation System (BaSS)* program from Petrobrás. *BaSS* is a 1-D basin modeling program that reconstructs the mechanical and thermal behavior of the lithosphere during and after a rifting event based on theoretical models of rift basin formation such as those of McKenzie (1978) and Royden and Keen (1980).

Gravimetric studies from the adjacent basement area indicate crustal thickness of about 34 km. For isostatic balance, a 136-km-thick lithosphere was estimated. These values were used as initial thicknesses in our model. Compaction history of the strata were described through an exponential decay of porosity as a function of depth following Athy's law (Athy, 1930). Determination of the initial porosity (at the surface) and the porosity decay coefficient for each lithology (e.g., shale, sandstone, and carbonate) was based on exponential regression analyses of a well log data set. Paleolake and paleosea depths (during rift and drift phases, respectively) were estimated based on sedimentologic and paleontologic studies.

The amount of postrift erosion was estimated by the integration of vitrinite reflectance (% R_o) and sonic log transit time data. The assumption is that the extrapolation of both % R_o and transit time values from a single well should lead to typical values at the surface (0.2% R_o and about 180 μsec/ft transit time). When extrapolated values at the surface are lower for the transit time and higher for % R_o than the expected ones in a single well, the thickness of the eroded sedimentary section can be estimated (Dow, 1977; Heasler and Kharitonova, 1996). Although from independent sources, both % R_o and transit time methods provided consistent results, which indicate that the amount of erosion was about 100–600 m and locally as much as 2000 m.

A set of 12 exploratory wells and 27 pseudowells generated from seismic lines was chosen for the thermo-mechanical modeling of a selected area in the Camamu–Almada Basin (Figure 8). Formation lithologies, depths, thicknesses, and ages were defined for every well and pseudowell. Each well was submitted to backstripping analysis, mechanical modeling, and thermal simulation. Extrapolated borehole temperatures and vitrinite reflectance data were used to constrain the thermal model. Two wells were selected to illustrate the modeling results: well A, located in the platform area, and well B, located in the deep offshore area (Figure 8).

The burial history curves show remarkable differences between the platform and deep offshore areas (Figure 9). In the platform, most of the total subsidence is represented by the rift sequences (Morro do Barro and Rio de Contas Formations), while in the deep offshore area, the postrift strata play an important role. Postrift uplift (110 Ma) is restricted to the platform area and therefore is not observed in well B.

Stratigraphic and tectonic studies based on well and seismic data allowed the identification of three rift phases. The first rift phase started at 143Ma and is represented by the deposition of the Morro do Barro Formation and a rapid increase in basin subsidence, which are observed in both wells A and B (Figure 9). The second rift phase is associated with the beginning of Rio de Contas Formation deposition (about 125 Ma). Evidence for this phase is not clear in well B, probably due to a lack of detailed biostratigraphic data. Finally, the third rift phase is represented by a sharp increase in subsidence rate starting about 115Ma. Because the third-rift phase strata have been limited to the west by a hinge line located along the present-day shelf break to slope zone, the last phase of rift subsidence can be observed in well B but not in well A (Figure 9).

From the backstripping results, we can observe that the ratio of synrift to postrift tectonic subsidence is about 10 in well A and 1.5 in well B. The comparison of calculated and theoretical tectonic subsidence curves has

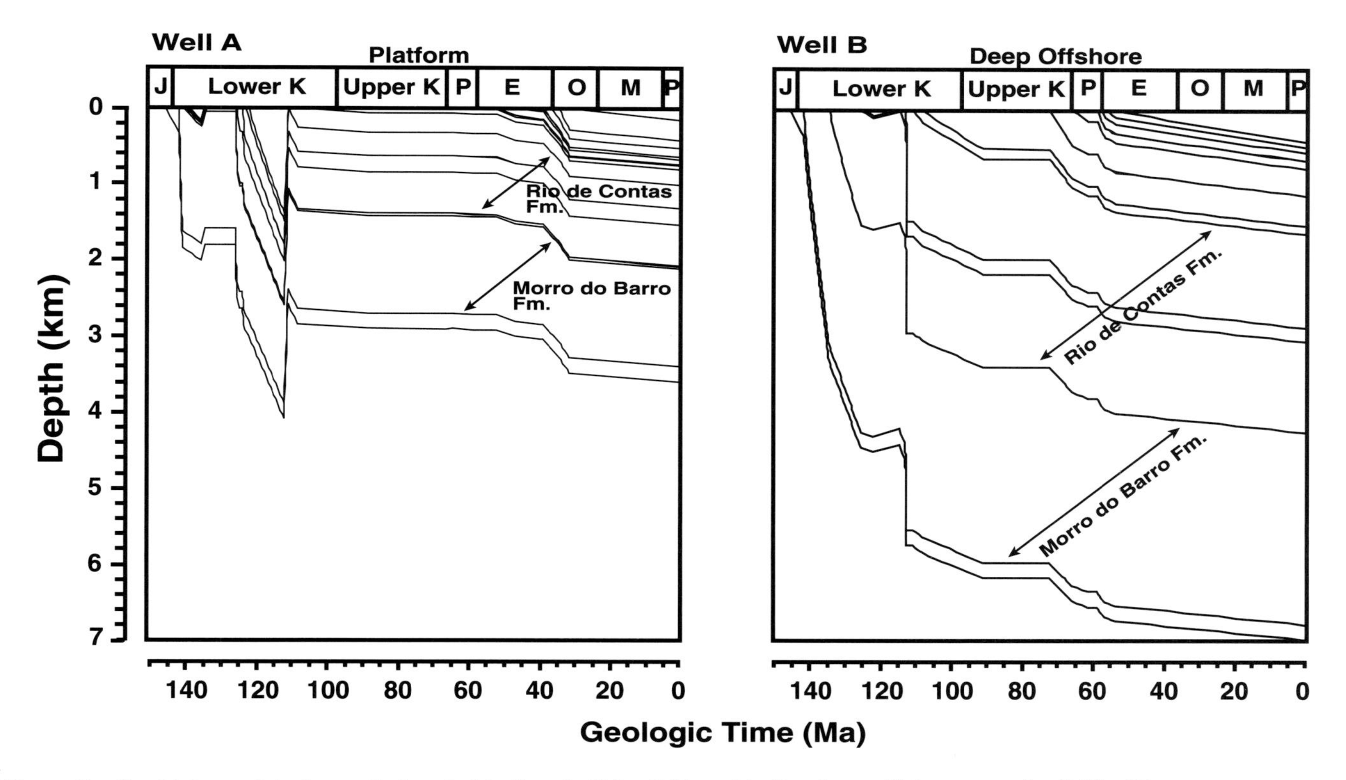

Figure 9—Geohistory plots for wells located in the shelf (well A) and in the deep offshore area (well B) of the Camamu–Almada Basin (see well locations in Figure 8). Bathymetry is not shown in the plots, but reaches about 50 m in well A and 1800 m in well B at present.

shown that a two-layer, nonuniform extension model (Royden and Keen, 1980) provides the best fit (Figure 10). The application of this model indicates that the crustal extension factor (δ) reaches up to 2.5 in the deeper areas of the basin, whereas the subcrustal extension factor (β) ranges from 1.0 in the platform to nearly 2.0 in the deeper area. Since the basal heat flow history reflects crustal and mainly subcrustal extension, in well A the maximum estimated heat flow is about 1.25 HFU, while in well B the heat flow approaches 1.95 HFU (Figure 10).

The validation of the thermo-mechanical modeling is corroborated by temperature and maturity calibration (Figure 11). Correlation of present-day calculated temperature and extrapolated borehole temperature data reveals an optimum fit, both in platform and deeper areas. Present-day calculated vitrinite reflectance using the Sweeney and Burnham (1990) model also presents a good correlation with measured data.

1-D Thermal and Kinetic Modeling

Petroleum generation history in the Camamu–Almada Basin was reconstructed using the 1-D numerical model *GENEX* from the Institut Français du Pétrole (IFP). This software reconstructs the compaction and thermal history of the sedimentary section and computes kerogen conversion into petroleum using a first-order kinetic scheme with a rate constant following Arrhenius' law. The same set of 12 exploratory wells and 27 pseudowells used in the thermo-mechanical analysis (Figure 8) was submitted to the kinetic modeling of petroleum generation.

Calculated heat flow histories obtained from the thermo-mechanical analysis were used as input for the thermal and kinetic simulations of each well. Hydrocarbon source potential was assigned both to the Morro do Barro and Rio de Contas Formations. Determination of original TOC and HI values was based on the statistical analysis of Rock-Eval and TOC data from immature samples of each formation. The kinetic parameters determined for a set of selected immature samples show a narrow activation energy distribution around 54 kcal/mol and exponential factors close to 10^{14}/sec, typical of type I kerogen. The validation of the kinetic modeling is corroborated by the correlation of the '"calculated" and "observed" transformation ratio (TR) curves (Figure 12). The calculated TR curve (dashed line in Figure 12), derived from the kinetic simulation, matches the observed TR curve (solid line in Figure 12) obtained from the HI–depth profiles (Figure 7).

Wells A and B (Figure 8) were selected to illustrate the kinetic modeling results. The TR evolution shows remarkable differences between the platform and deep offshore areas (Figure 13). In the platform area (well A),

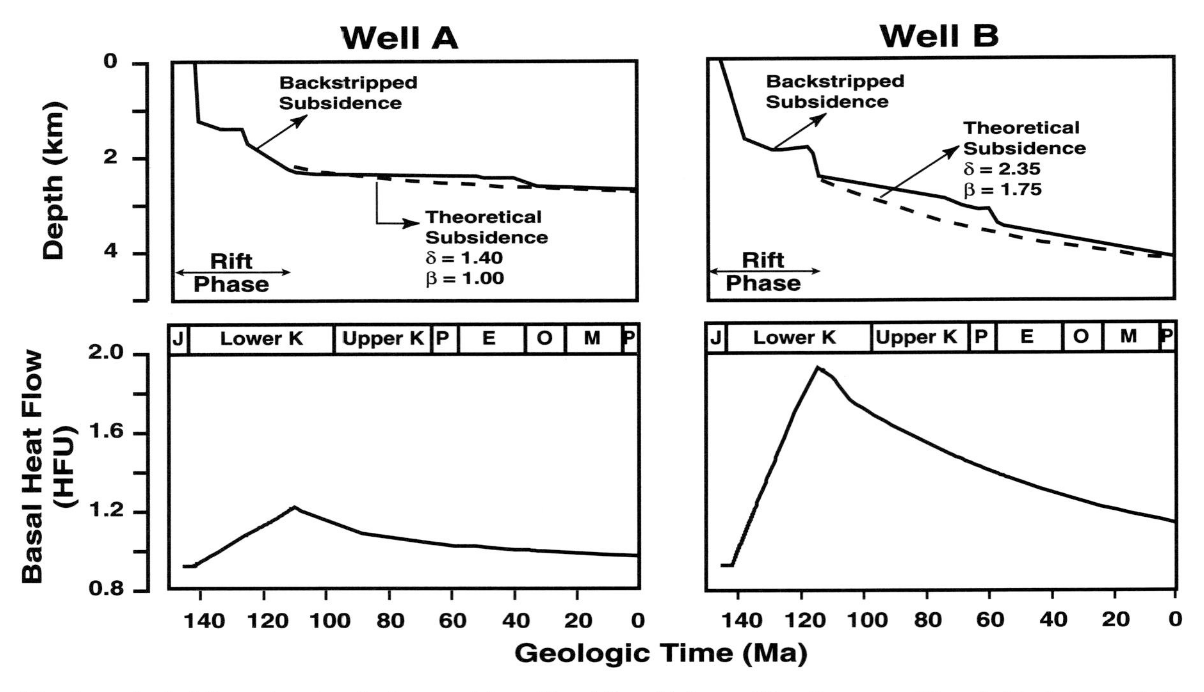

Figure 10—Appraisal of the extension factors and basal heat flow history of the Camamu–Almada Basin in wells A and B (see well locations in Figure 8). The two upper graphs show the fit between the observed tectonic subsidence curves obtained from backstripping studies and the theoretical subsidence curves calculated using the model from Royden and Keen (1980). The two lower graphs show the calculated heat flow histories for the wells.

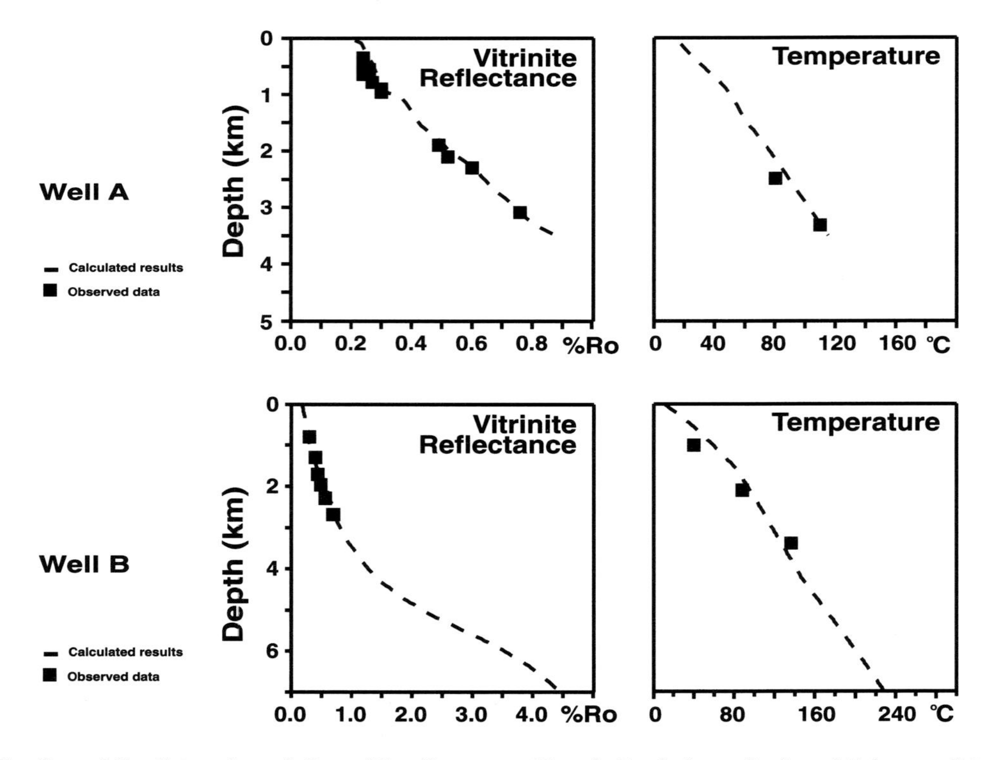

Figure 11—Calibration of the thermal modeling of the Camamu–Almada Basin in wells A and B (see well locations in Figure 8). Notice the good correlation between measured vitrinite reflectance values (squares) and calculated vitrinite reflectance curves (dashed lines) using the model from Sweeney and Burnham (1990). Also note the good fit between measured temperatures (squares) and calculated temperatures (dashed lines).

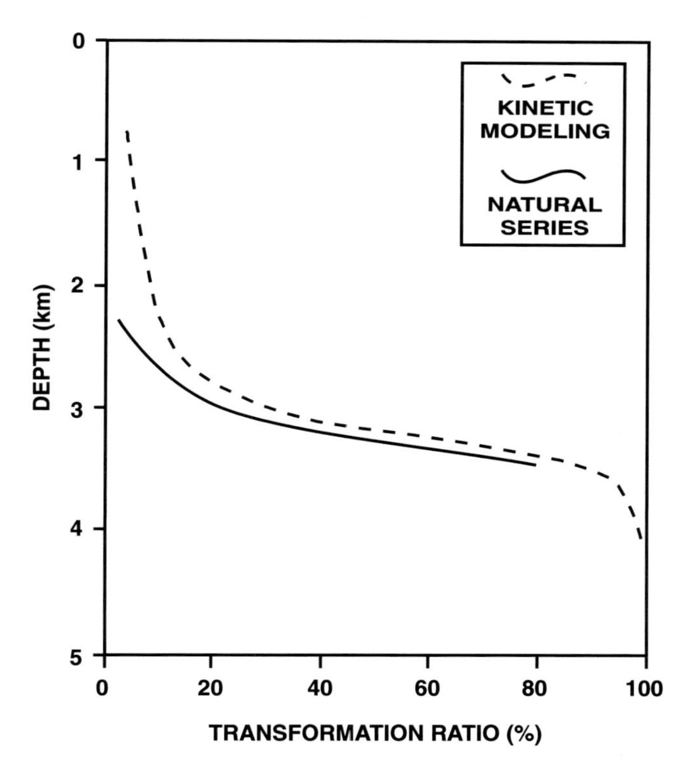

Figure 12—Graph showing the correlation of observed and calculated TR curves as a function of depth in the Camamu–Almada Basin. The observed curve (solid) represents the HI depletion from the natural series (see Figure 7), and the calculated curve (dashed) was obtained from kinetic modeling.

kerogen conversion in both the Morro do Barro and Rio de Contas source rocks started during the Early Cretaceous. Nevertheless, due to subsidence and thermal histories, TR values were very low (up to 15%) and have remained almost constant since the end of the rift phase. In the deep offshore area (well B), petroleum generation also started during the Early Cretaceous for both formations (Figure 13). The Morro do Barro source rock showed a sharp increase in the kerogen TR (up to 100%) between 140 and 120 Ma, while the Rio de Contas Formation kerogen conversion started around 130 Ma, reaching up to a TR of 95% in the Eocene.

The kinetic modeling results from the set of 12 wells and 27 pseudowells allowed the reconstruction of the kerogen conversion through time. The dynamics of petroleum generation history is illustrated by a set of maps showing the TR distribution across the study area at different ages for the Morro do Barro and Rio de Contas Formations (Figure 14). At 130 Ma, both units were immature. At 110 Ma, the Morro do Barro Formation had reached high kerogen TR values (60–100%) in most of the present-day slope and deep offshore areas, whereas high TR values for the Rio de Contas Formation were limited to some few restricted areas in the deep offshore region. From 90 Ma to the present, Morro do Barro TR values remained almost constant, reaching up to 100% in the deep offshore and ranging from 10 to 60% in the platform and slope areas (Figure 14). During the same period, Rio de Contas TR values increased in the slope and deeper offshore areas, reaching 20–100% at present, while in the platform, TR values remained almost constant, ranging from 0 to 10%. Note that, because the geographic extent

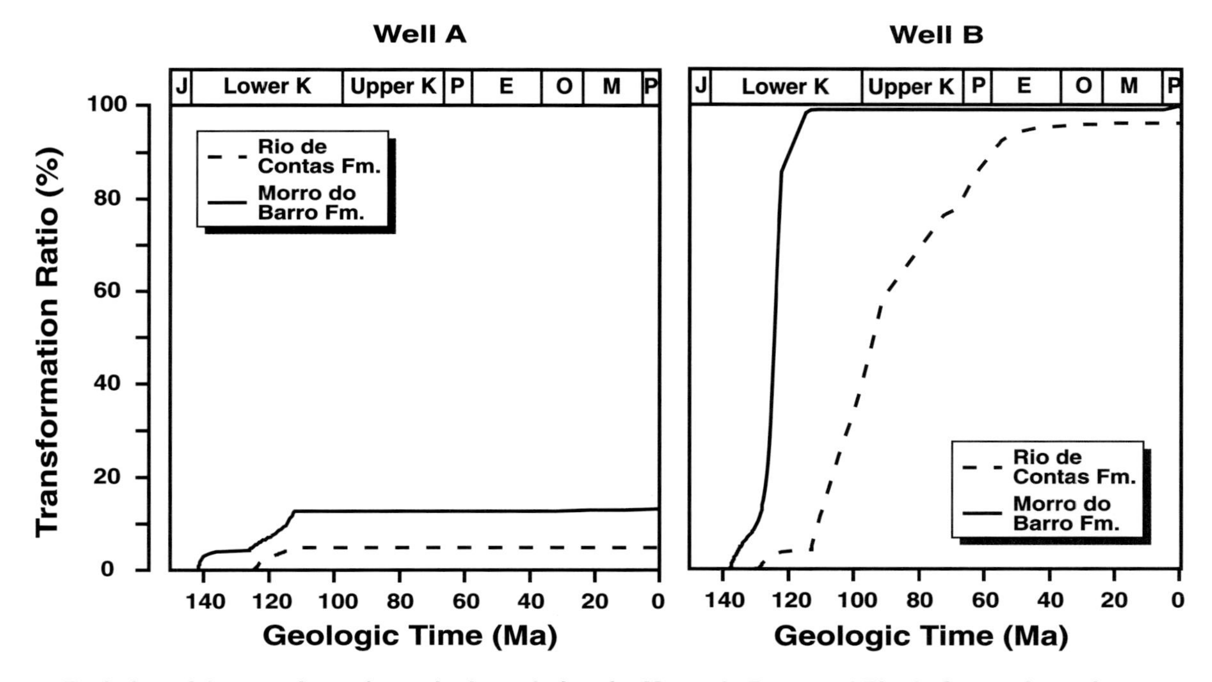

Figure 13—Evolution of the transformation ratio through time for Morro do Barro and Rio de Contas formations source rocks in the wells A and B, Camamu–Almada Basin (see location on Figure 8).

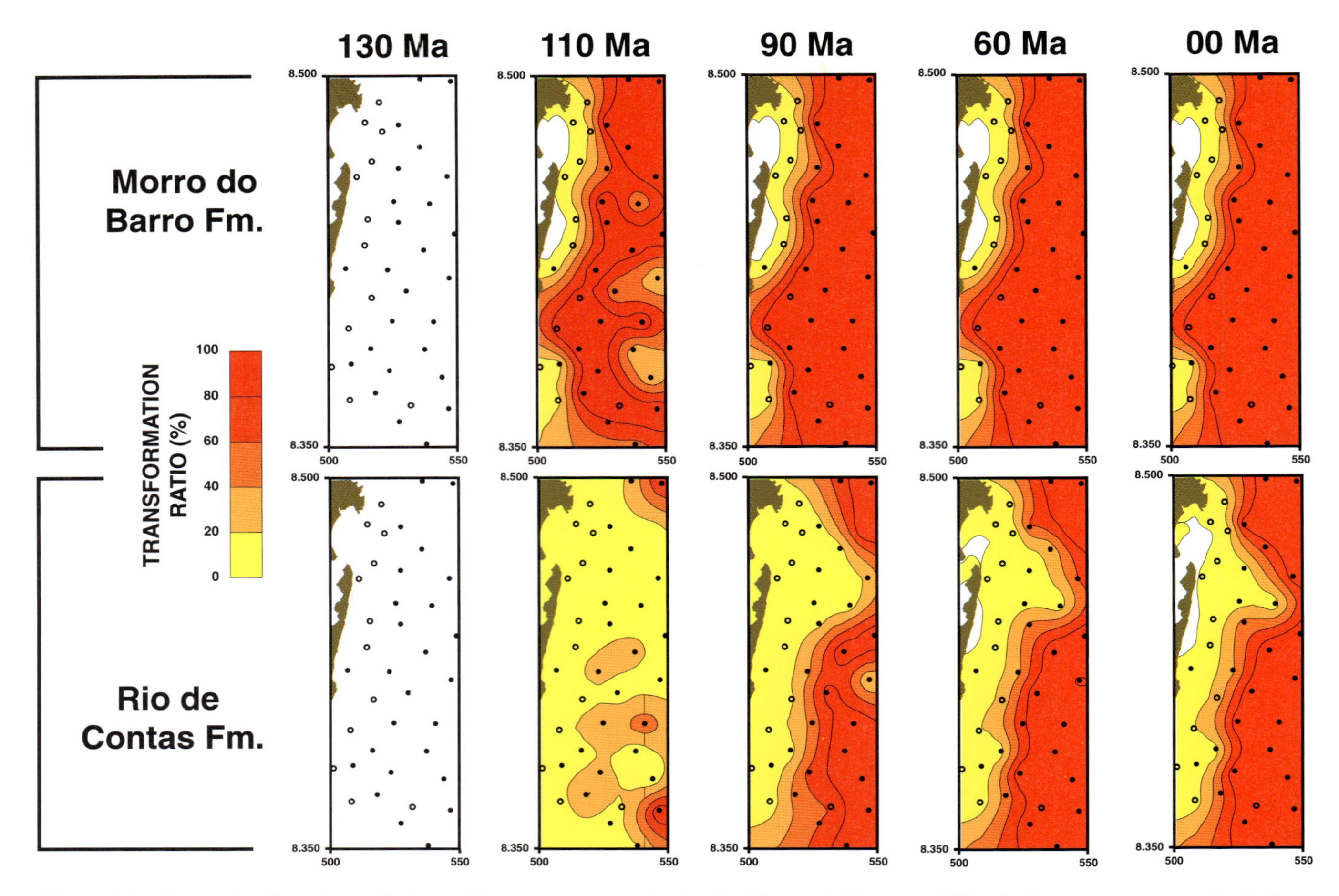

Figure 14—Maps showing the evolution of kerogen conversion in the Morro do Barro and Rio de Contas source rocks in the Camamu–Almada Basin. Each map repesents the distribution of the cumulative TR across the study area at a given age.

of the source rocks in the slope and deep offshore areas is still unknown, these maps should not be used alone for definition of the oil kitchen(s).

2-D Petroleum Generation and Migration Modeling

To improve our understanding of the interplay of the elements and processes of the Morro do Barro(!) petroleum system and to investigate the secondary migration trends on a regional scale in the Camamu–Almada Basin, IFP's *TEMISPACK* program was used along a cross section in the northern area of the basin (Figure 15). *TEMISPACK* is a 2-D integrated model that incorporates modules for the numerical simulation of such phenomena as sedimentation, compaction, erosion, heat transfer, fluid flow, source rock maturation, primary and secondary cracking, petroleum expulsion, and hydrocarbon flow (Ungerer et al., 1990).

The studied section is a 30-km-long east–west cross section with strata dipping to the east and becoming thicker toward the deep-water area, which results in a preserved sedimentary section that reaches up to 9000 m thick (Figure 15). After some simplifications to aid computation, the resulting model contained 18 geologic events (such as strata deposition and erosion) and 9 lithology types. The model was divided into 46 vertical columns, which resulted in a numerical mesh of 652 elements.

Input data for the model included strata thicknesses and ages, basement heat flow based on the thermomechanical model, petrophysical data (porosities, permeabilities, and capillary pressures), thermal properties of rocks (conductivity and specific heat), and geochemical data (source rock potential and kinetic parameters). Vitrinite reflectance, T_{max}, and borehole temperatures and pressures were used to calibrate the thermal and fluid flow model. The location and type of petroleum of the known accumulations served to constrain different scenarios, which were based on variations in the amount of erosion, heat flow histories, porosity and permeability characteristics of rocks and faults, and source rock richness.

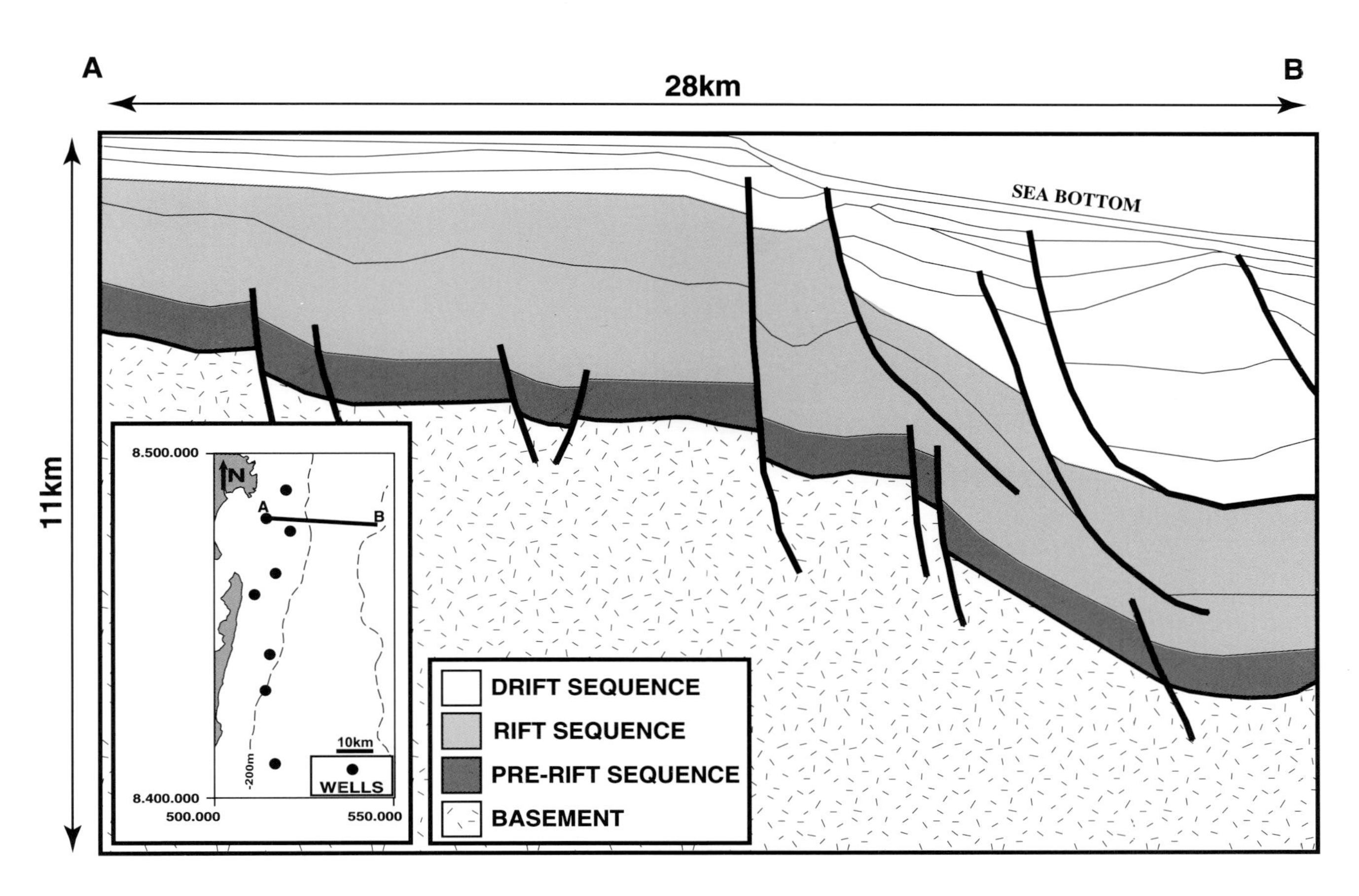

Figure 15—Geologic cross section in the northern part of the Camamu–Almada Basin used for the 2-D modeling of petroleum generation and migration.

Results of the 2-D modeling indicated that a hydrostatic pressure profile prevailed throughout basin evolution, even during periods of high sedimentation rates. Validation of the thermal and fluid flow modeling is corroborated by good correlation of calculated temperatures and pressures with measured data (not shown here). Kinetic modeling of the Morro do Barro source rocks shows that most of the oil was generated during the rift phase, which corroborates the results from the 1-D kinetic modeling. Present kerogen TR values range from 10–20% near the continent (west) to 100% in the deeper area (east).

In the studied cross section, high capillary pressures in the thick section of shales above the source rocks favored the downward migration of petroleum to the prerift sandstones of the Sergi Formation (Figure 16). Petroleum migration through normal faults, which juxtaposed source rocks of the Morro do Barro Formation onto the Sergi sandstones played a major role in the filling of these carrier beds, characterized by good lateral continuity, porosity, and permeability. Secondary migration through these carrier beds extended considerably into the postrift phase and allowed petroleum to reach the shallow prerift structural trap (actual petroleum field) located in the western part of the section (Figure 16). To test the hypothesis of an alternative petroleum system, source and reservoir characteristics were also assigned to the cells of the upper part of the Rio de Contas Formation in the deep water region, and the model predicted the occurrence of petroleum accumulations (Figure 16).

Another important pathway (not shown in the studied section) comprises migration through the lateral and upper boundary contacts of the source rocks with the intercalated sandstones. This pathway is responsible for the infilling of turbidite beds in the Morro do Barro Formation, where most of the petroleum of the Camamu–Almada Basin has been found.

CONCLUSIONS

This paper shows how the application of quantitative basin modeling techniques allow a better understanding of the spatial and temporal relationships between the elements and processes of petroleum systems, and provide a basis for a further assessment of the exploration risk and the petroleum potential in the Camamu–Almada Basin.

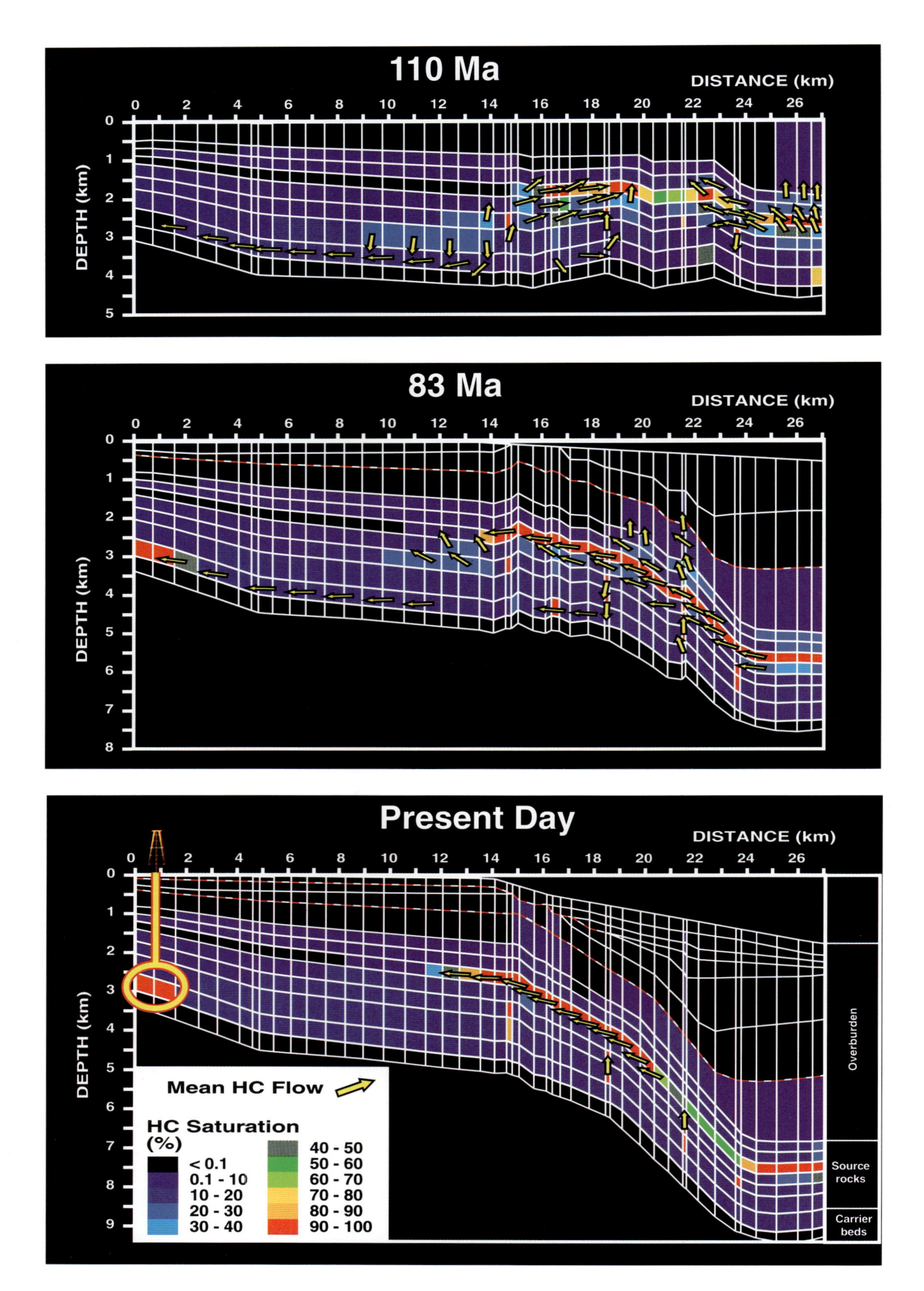

Figure 16—2-D modeling results for the studied cross section (see Figure 15) showing the evolution of petroleum migration in the Camamu–Almada Basin at three different times.

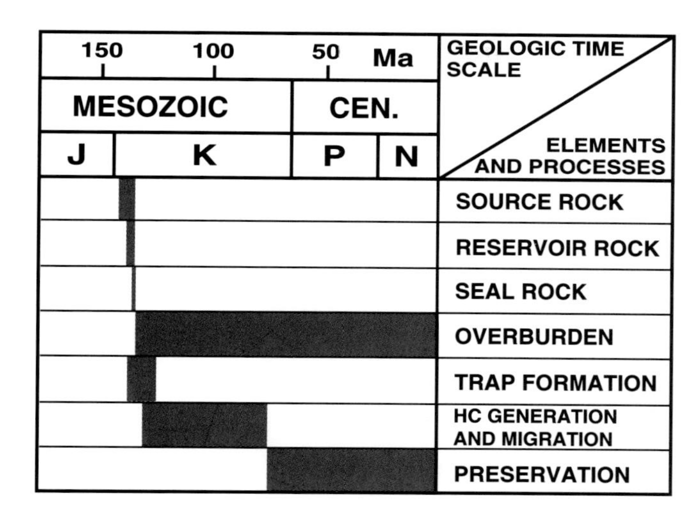

Figure 17—Events chart of the Morro do Barro(!) petroleum system in the Camamu–Almada Basin.

The integration of geologic, geochemical, and geophysical data allowed the identification of a single petroleum system in the Camamu–Almada Basin, the Morro do Barro(!) petroleum system, which encompasses the lower Neocomian lacustrine freshwater source rocks from the Morro do Barro Formation and the resulting hydrocarbon accumulations. Most of the petroleum is trapped in the Morro do Barro Formation turbiditic reservoirs, followed by the fluvial–eolian sandstone reservoirs of the Sergi Formation. The events chart in Figure 17 summarizes the stratigraphic section involved, the essential elements and processes, and the preservation time of the Morro do Barro(!) petroleum system.

Integrated 1-D thermo-mechanical and kinetic modeling revealed that, in the platform area, the Morro do Barro source rock reached low TR values (up to 30%), while in the slope and deep offshore areas, TR values showed a sharp increase (up to 80–100%) between 140 and 120 Ma. In Rio de Contas potential source rock, TR values increased in the slope and deeper offshore areas, reaching 20–100% at present, while in the platform, TR values remained almost constant, ranging from 0 to 10%. It is worth mentioning that the geographic extent of these source rocks in the slope and deep offshore areas is still unknown. 2-D modeling of a selected cross section revealed that downward migration of petroleum from the Morro do Barro Formation to the sandstones of the Sergi Formation and migration through normal faults allowed petroleum to reach the shallow and prerift traps.

Acknowledgments*—The authors would like to thank Petrobrás for permission to present and publish this paper and the Geochemistry Section of Petrobrás Research Center (Cenpes) for all the geochemical analyses.*

REFERENCES CITED

Athy, L. F., 1930, Density, porosity, and compaction of sedimentary rocks: AAPG Bulletin, v. 14, p. 1–24.

Dow, W. G., 1977, Kerogen studies and geological interpretation: Journal of Geochemical Exploration, v. 7, p. 79–99.

Espitalié, J., J. R. Maxwell, P. Y. Chénet, and F. Marquis, 1988, Aspects of hydrocarbon migration in Paris Basin as deduced from organic geochemistry survey: Organic Geochemistry, v. 13, p. 467–481.

Gonçalves, F. T. T., 1997, Caracterização geoquímica e paleoambiental do Cretáceo inferior da Bacia de Camamu, Bahia: Master's thesis, Universidade Federal do Rio de Janeiro, Rio de Janeiro, Brazil, 187 p.

Heasler, H. P., and N. A. Kharitonova, 1996, Analysis of sonic well logs applied to erosion estimates in the Bighorn Basin, Wyoming: AAPG Bulletin, v. 80, p. 630–646.

Magoon, L. B., 1988, The petroleum system—a classification scheme for research, exploration, and resource assessment, *in* L. B Magoon, ed., Petroleum systems of the United States: USGS Bulletin 1870, p. 2–15.

Magoon, L. B., and W. G. Dow, 1994, The petroleum system, *in* L. B. Magoon and W. G. Dow, eds., The petroleum system—from source to trap: AAPG Memoir 60, p. 3–24.

McKenzie, D., 1978, Some remarks on the development of sedimentary basins: Earth and Planetary Science Letters, v. 40, p. 25–32.

Mello, M. R., and J. R. Maxwell, 1991, Organic geochemical and biological marker characterization of source rocks and oils derived from lacustrine environments in the Brazilian continental margin, *in* B. J. Katz, ed., Lacustrine basin exploration—case studies and modern analogs: AAPG Memoir 50, p. 77–98.

Mello, M. R., N. Telnaes, P. C. Gaglianone, M. I. Chicarelli, S. C. Brassel, and J. R. Maxwell, 1988, Organic geochemical characterization of depositional paleoenvironments of source rocks and oils in Brazilian marginal basins, *in* L. Matavelli and L. Novelli, eds., Advances in organic geochemistry, 1987: Organic Geochemistry, v. 13, p. 31–45.

Mello, M. R., F. T. T. Gonçalves, J. L. Netto, and R. E. Witzke, 1995, Application of the petroleum system concept in the assessment of explorations risk: the Camamu Basin example, offshore Brazil: 4th Congresso Internacional da Sociedade Brasileira de Geofísica, Rio de Janeiro, Expanded Abstracts, v. 1, p. 90–92.

Netto, A. S. T., J. R. Wanderley Filho, and F. Feijó, 1994, Bacias de Jacuípe, Camamu e Almada: Boletim de Geociências da Petrobrás, v. 8, p. 173–184.

Ojeda, H. A. O., 1982, Structural framework, stratigraphy, and evolution of Brazilian marginal basins: AAPG Bulletin, v. 66, p. 732–749.

Ponte, F. C., and H. E. Asmus, 1978, Geological framework of the Brazilian continental margin: Geologische Rundschau, v. 68, p. 201–235.

Royden, L., and C. E. Keen, 1980, Rifting process and thermal evolution of the continental margin of eastern Canada determined from subsidence curves: Earth and Planetary Science Letters, v. 51, p. 343–361.

Sweeney, J. J., and A. K. Burnham, 1990, Evaluation of a simple model of vitrinite reflectance based on chemical kinetics: AAPG Bulletin, v. 74, p. 1559–1570.

Ungerer, P., J. Burrus, P. Doligez, P.-Y. Chenet, and F. Bessis, 1990, Basin evaluation by integrated two-dimensional modeling of heat transfer fluid flow, hydrocarbon generation, and migration: AAPG Bulletin, v. 74, p. 309–335.

Mohriak, W. U., M. R. Mello, M. Bassetto, I. S. Vieira, and E. A. M. Koutsoukos, 2000, Crustal architecture, sedimentation, and petroleum systems in the Sergipe–Alagoas Basin, northeastern Brazil, *in* M. R. Mello and B. J. Katz, eds., Petroleum systems of South Atlantic margins: AAPG Memoir 73, p. 273–300.

Chapter 20

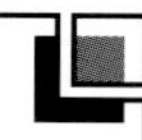

Crustal Architecture, Sedimentation, and Petroleum Systems in the Sergipe–Alagoas Basin, Northeastern Brazil

W. U. Mohriak

Petróleo Brasileiro S.A.
E&P–GEREX/GEINOF
Rio de Janeiro, Brazil

M. R. Mello

Petrobrás Research Center, DIVEX
Rio de Janeiro, Brazil

M. Bassetto

I. S. Vieira

Petróleo Brasileiro S.A.
E&P–GEREX/GEINOF
Rio de Janeiro, Brazil

E. A. M. Koutsoukos

Petrobrás Research Center, DIVEX
Rio de Janeiro, Brazil

Abstract

An integrated, multidisciplinary study of the tectonic framework, sedimentation, and petroleum systems in the Sergipe–Alagoas Basin was carried out. The methodology was based on regional integration of geologic and geophysical data, particularly seismic reflection and potential field data (gravity and magnetics), results of exploratory drilling, paleontologic and paleoenvironmental analysis of the sedimentary succession, and geochemical data from oils and source rocks. The main topics addressed were the tectono-stratigraphic evolution of the sedimentary basins in northeastern Brazil, the crustal architecture of the Sergipe–Alagoas Basin, and the petroleum systems both onshore and offshore.

Results of this study indicate that major synrift troughs are located in the proximal regions and are characterized by negative Bouguer anomalies. The proximal grabens are controlled by comparatively small synthetic and antithetic normal faults, while major rift blocks are controlled by crustal faults that dip seaward. These master faults cut through most of the crust and detach onto lower crustal horizons or even the seismic Moho. Deep-water rift blocks were affected by regional erosional episodes. The transition to pure oceanic crust is marked by wedges of seaward-dipping reflectors and igneous plugs. Some possible salt diapirs are located near the crustal limit. The petroleum systems for this basin include good source rocks in the transitional (evaporitic) and rift-phase sequences. Hydrocarbon generation and migration was effective from Late Cretaceous time onward. Exploratory plays include structural traps associated with synrift and postrift structures, as well as stratigraphic traps associated with deep-water turbidites.

INTRODUCTION

In the past few years, the exploratory interpretation of some prolific hydrocarbon provinces in the South Atlantic has advanced toward deep-water regions, such as in the Campos Basin in southeastern Brazil (Guardado et al., 1989; Mello et al., 1994) and the Gabon Basin in west Africa (e.g., Teisserenc and Villermin, 1989). It is becoming clear that a shift is underway in the petroleum evaluation of the South Atlantic margin, with regional basin analysis studies changing in scope from qualitative interpretation of seismic data in the shallow parts of the sedimentary basins, toward a more meaningful integration of geologic and geophysical data from the platform toward the boundary between continental and oceanic crust.

To date, there have been few studies analyzing the deep structure of the continental margins in the South Atlantic. Most of these investigations have been conducted by international petroleum companies and research institutions and usually refer to the African side (e.g., Rosendahl et al., 1991; Wannesson et al., 1991). A number of international scientific cruises conducted with the aim of obtaining geologic information from the ultra–deep water region of the South Atlantic (e.g.,

Gladczenko et al., 1997) lack stratigraphic control on the rift structures along the margin, which can be obtained only by exploratory boreholes drilled in the platform. Moreover, the information obtained in the ultra–deep water province is usually restricted to a few regional seismic lines that do not allow consistent mapping of regional structures.

The northeastern Brazilian margin has been the object of a multidisciplinary study aimed at (1) mapping its tectonic framework from the platform toward the oceanic crust, and (2) evaluating new exploratory frontiers in the ultra–deep water province using regional deep seismic profiles (Mohriak et al., 1993; Mohriak and Rabelo, 1994). The interpretation of this segment of the South Atlantic has brought forth many geologic complexities that, when deciphered, may increase our understanding of the tectonic evolution of passive continental margins worldwide and optimize the evaluation of their petroleum prospectivity (Mohriak et al., 1995a). Among the outstanding tectonic problems directly related to petroleum exploration, we point out geologic and geophysical evidence for the presence of wedges of seaward-dipping reflectors near the crustal limit, intraplate magmatism, oceanic fracture zones, and deep-water halokinetic structures.

This chapter discusses some of the results of the interpretation that are more related to petroleum systems, which are described and compared with analogs in other sedimentary basins worldwide. The Sergipe–Alagoas Basin can be considered as a learning school for many of the geologic concepts that have been applied in the exploration of other basins along the Brazilian margin. Several of these concepts were developed by at least two generations of geoscientists who worked in many phases of its exploration since the early 1950s. Exploratory activity in the Sergipe–Alagoas Basin resulted in the discovery of the largest oil field in onshore Brazil in the early 1960s. The first exploratory borehole ever drilled in offshore Brazil was also located in the Sergipe Basin, and it was also in this basin that Petrobrás discovered the only deep-water oil accumulation in the Brazilian margin, apart from the giant fields discovered in the Campos Basin farther to the south. This work analyzes the petroleum systems already established for the onshore and offshore regions of the Sergipe–Alagoas Basin and briefly discusses some analogies with other sedimentary basins in the South Atlantic (particularly west Africa) and the Gulf of Mexico.

BASIN ANALYSIS METHODOLOGY

The basin analysis procedures applied to the northeastern Brazilian margin include interpretation of seismic data, potential field data (mainly gravity and magnetics), and borehole data (mainly stratigraphic, paleontologic, and geochemical analyses). This approach, which involves quantitative assessment of physical properties of the sedimentary fill and also of the underlying rocks, was first applied to oceanic regions by academic institutions, and its use is now becoming more widespread along continental margins by oil industry research and exploratory groups (Pinet and Bois, 1990). This methodology results in a better analysis of petroleum systems in frontier regions by evaluation of the seismo-stratigraphic sequences within a broader framework of the crustal architecture of the basin.

The integration of potential field and regional deep seismic data has been extensively applied in many sedimentary basins worldwide as a refinement of structural and stratigraphic studies (e.g., Keen and Voogd, 1988; Keen et al., 1991, 1994; Holbrook et al., 1994). The deep seismic reflection profiles acquired in the South Atlantic, both along the Brazilian (e.g., Mohriak et al., 1990, 1993; Mohriak and Latgé, 1991) and African margins (Meyers and Rosendahl, 1991; Meyers et al., 1996) are helpful to place new constraints on the crustal architecture of the sedimentary basins, locate major depocenters, and identify master faults and transfer zones. These profiles also help to characterize the transition between continental and oceanic crust, which in the South Atlantic may play an important role in the assessment of petroleum systems.

In the 1970s and early 1980s, most seismic profiles in the offshore region of the Brazilian margin were shot only on the platform (bathymetries less than 400 m) and processed to 5–6 sec two-way traveltime, which did not allow visualization and interpretation of the deeper structures underlying the Upper Cretaceous–Tertiary stratigraphic sequences in the deep-water region. Subsequently, more modern surveys included processing to 8–10 sec traveltime. Then, in the late 1980s and early 1990s, Petrobrás acquired several experimental regional deep seismic profiles using special acquisition equipment with a very long cable (6 km) and a very powerful airgun font, with a source volume varying from 6324 in.^{3} (103.6 L) to 8832 in.^{3} (144.7 L). The large grid of regional seismic data along the Brazilian margin extends from the southern and eastern basins (Mohriak and Latgé, 1991) toward the northeastern and northern basins (Mohriak et al., 1993). This chapter discusses some of the results from the Sergipe Basin and the implications for petroleum systems in the deep-water region.

The interpretation of deep seismic data was based on integration with the regional grid of normal industry seismic data and was stratigraphically calibrated by several exploratory boreholes along the profiles (Mohriak et al., 1995a). The interpretation was conducted both on paper sections and at workstations loaded with a regional grid of seismic profiles. The extensive regional dataset of potential field data (gravity and magnetic) was used to constrain the seismic interpretation by quantitatively modeling the anomalies using standard modeling packages (Mohriak and Rabelo, 1994; Mohriak et al., 1996; Bassetto et al., 1996). The petroleum systems were analyzed on the basis of results from exploratory activity over the past few decades and from recent assessments of geochemical analysis of oils and source rocks (Araújo et al., 1994; Mello et al., 1994, 1996).

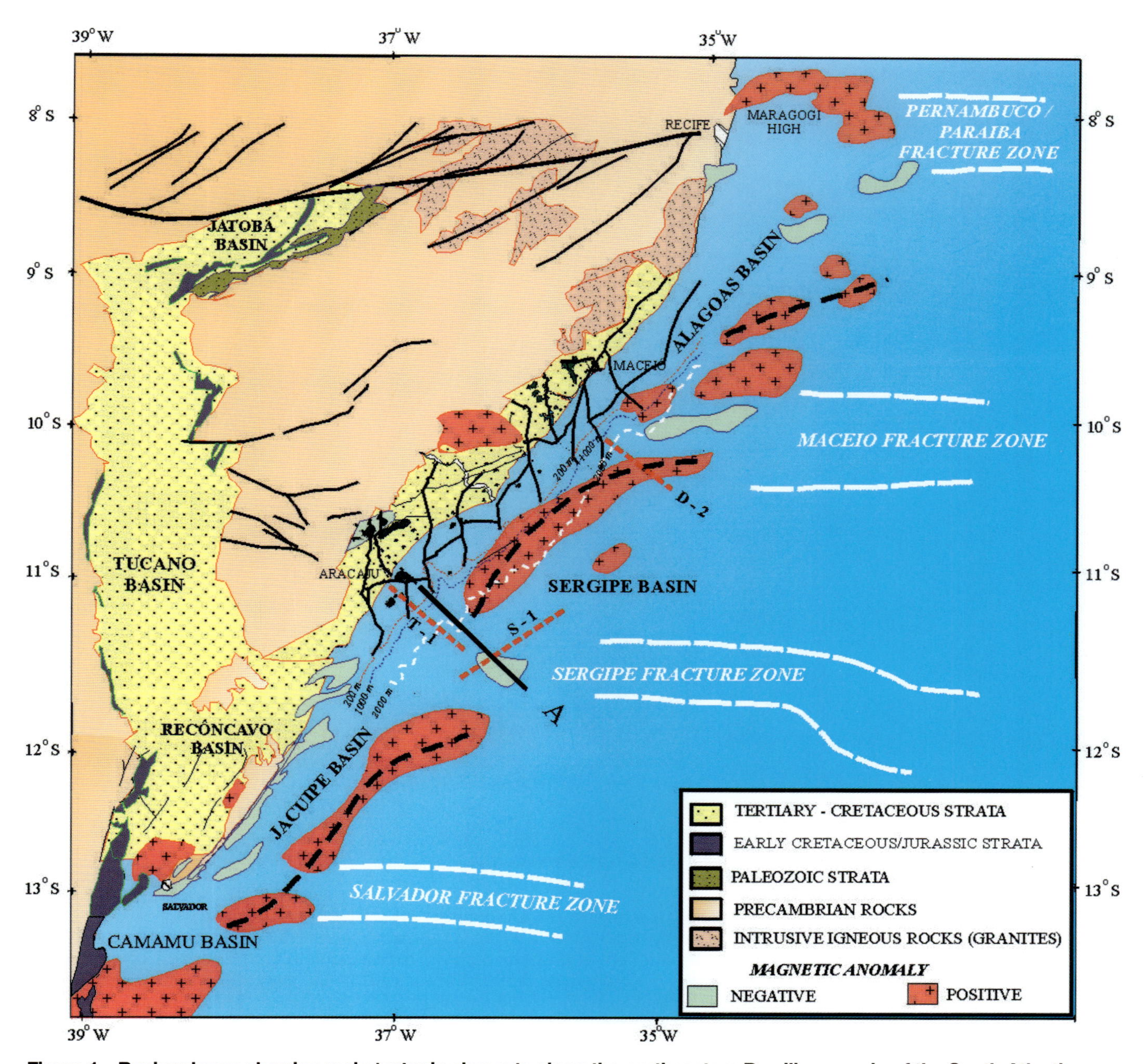

Figure 1—Regional map showing main tectonic elements along the northeastern Brazilian margin of the South Atlantic Ocean. A regional deep seismic profile extending from the platform toward the deep-water region (profile A) is shown as a thick solid line.

REGIONAL TECTONIC FRAMEWORK

The sedimentary basins in northeastern Brazil (Figure 1) form a series of asymmetric grabens, including the onshore Recôncavo-Tucano-Jatobá rift system and the offshore Jacuípe and Sergipe–Alagoas Basins. The onshore basins are characterized by depocenters separated by basement highs and transfer faults, associated with the opening of a rift system in a northwest direction oblique to the general north–south trend of the master faults (Milani and Davison, 1988). The offshore basins (particularly the Jacuípe Basin and the Sergipe subbasin) are characterized by a relatively thin sedimentary cover on land and by a series of depocenters controlled by synthetic faults on the platform and in the deep-water province (Ponte et al., 1980; Ojeda, 1982). In contrast, the Alagoas subbasin is characterized by a very thick synrift sedimentary succession onland and on the platform, with

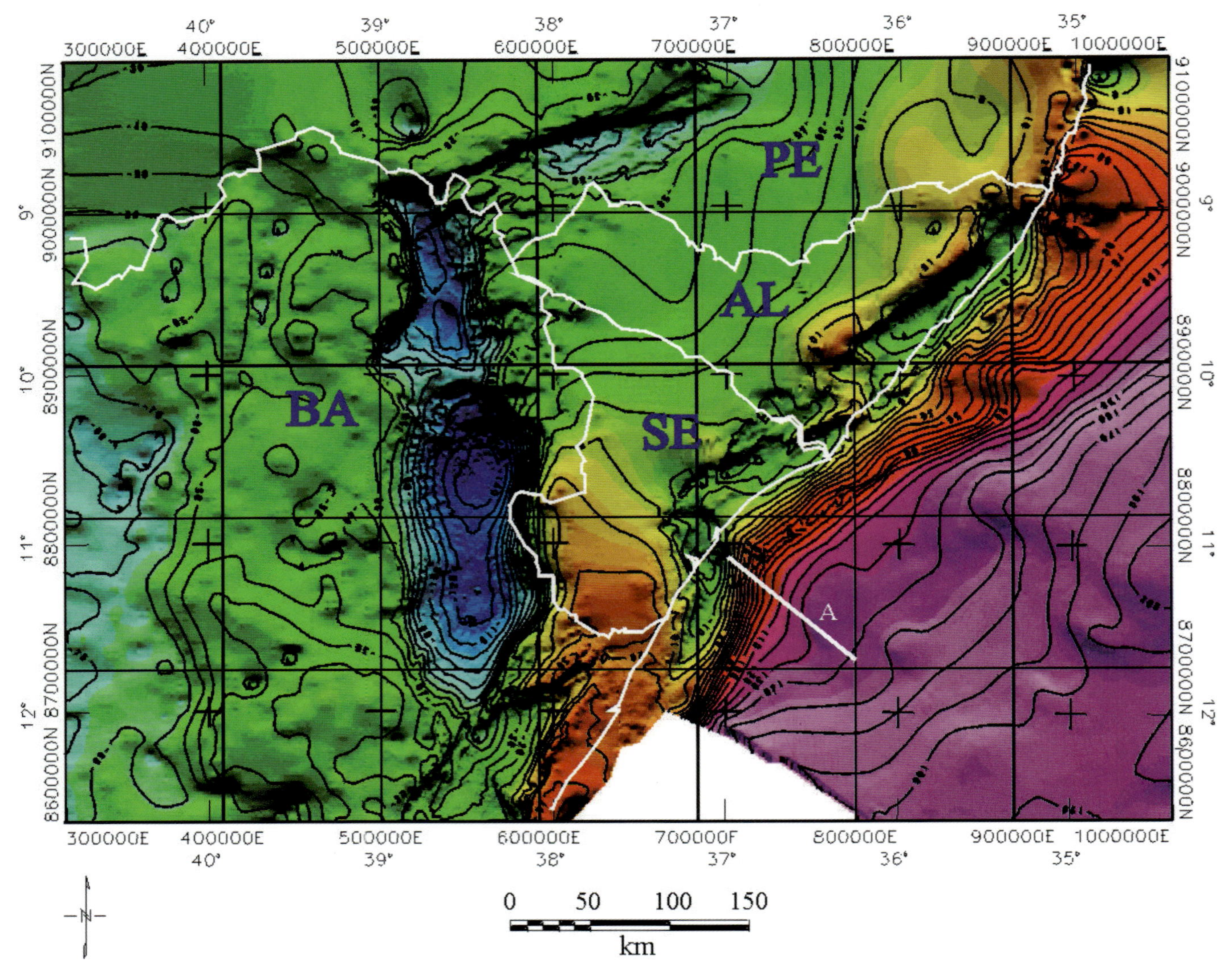

Figure 2—Regional Bouguer anomaly map for northeastern Brazil, integrating data from onshore and offshore regions. The onshore rifts (Recôncavo and Tucano) are marked by regional lows in the Bouguer anomaly (less than –100 mGal), and the continental margin is associated with gravity lows onshore and on the platform (particularly in the Mosqueiro Low in the southern Sergipe subbasin). Seaward, the Bouguer anomaly rises rapidly to over +100 mGal beyond the shelf break. Contour interval is 10 mGal.

a relatively thin sedimentary sequence deposited during thermal phase of subsidence. Toward the northeastern extremity of the subbasin, there is also evidence for inversion structures associated with transpressional tectonics (Cainelli, 1987).

Figure 2 shows the regional Bouguer anomaly map of northeastern Brazil. The aborted rifts onshore (Tucano and Recôncavo Basins) have extremely negative Bouguer anomalies (as low as –150 mGal) in contrast to the positive Bouguer anomalies in the deep-water parts of the Jacuípe and Sergipe rifts (exceeding +70 mGal beyond the shelf-edge) (Figure 2). The lack of coaxial isostatic compensation in the Tucano Basin has led several researchers (e.g., Ussami et al., 1986; Castro, 1987) to propose genetic models linking these aborted rifts to the simultaneous development of sedimentary basins along the Brazilian and African margins, following the simple shear model of Wernicke (1985). The pure shear extensional model (McKenzie, 1978) has also been invoked as the basic mechanism to explain the formation of the passive margin sedimentary basins along the northeastern margin of Brazil, based on the crustal thinning that is vertically coincident with the depocenter of the Sergipe–Alagoas Basin (e.g., Guimarães, 1988; Destro, 1994).

Figure 3 is a simplified, reduced to the pole, total intensity magnetic map of the offshore region that shows major anomalies, particularly the conspicuous positive magnetic anomalies with axes that form a megaregional system of en echelon, arcuate segments. These arches extend from the Alagoas toward the Jacuípe Basin and are

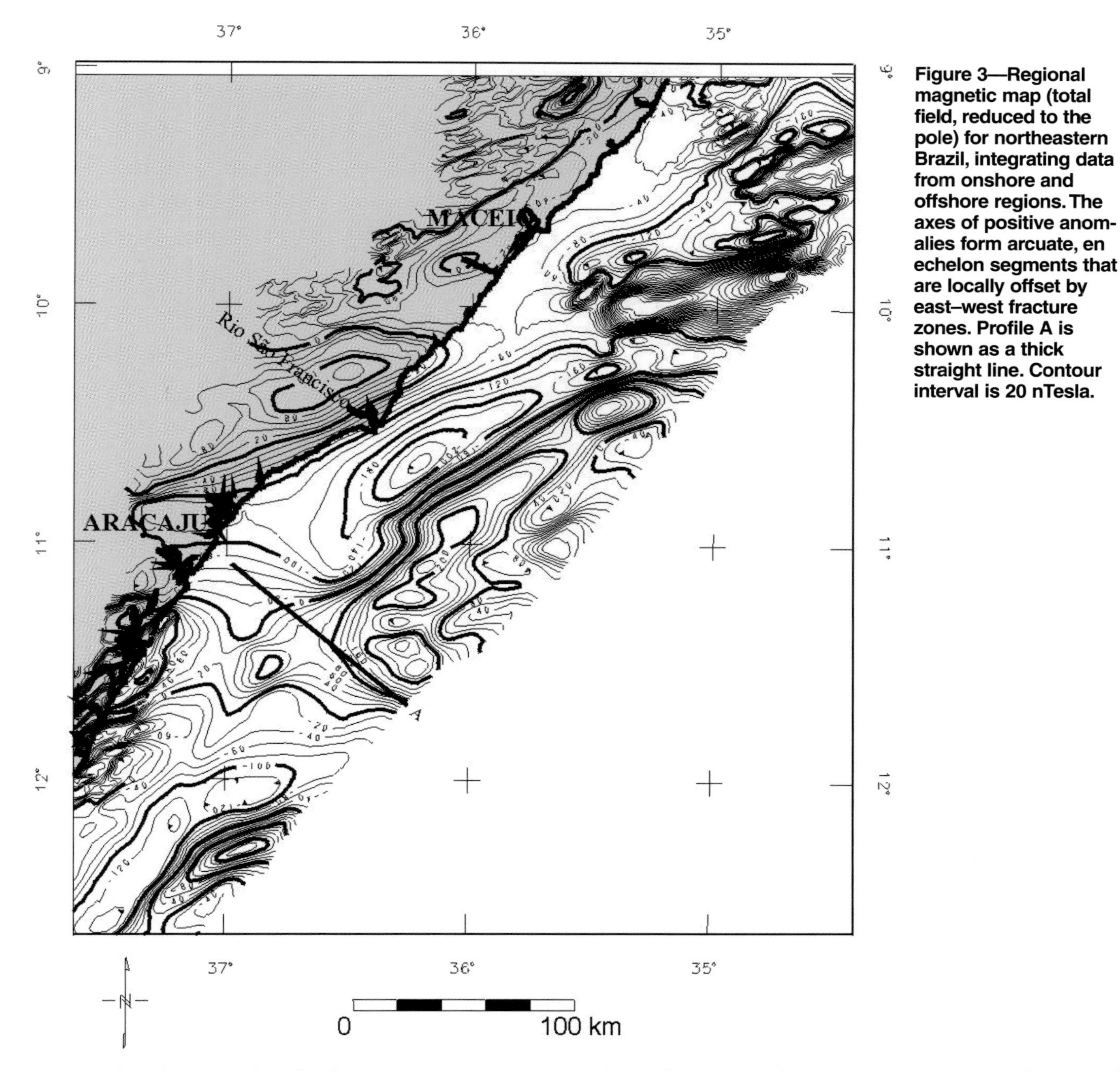

Figure 3—Regional magnetic map (total field, reduced to the pole) for northeastern Brazil, integrating data from onshore and offshore regions. The axes of positive anomalies form arcuate, en echelon segments that are locally offset by east–west fracture zones. Profile A is shown as a thick straight line. Contour interval is 20 nTesla.

intersected and apparently offset by east–west magnetic anomalies along fracture zones (Maceió fracture zone in the north, Sergipe fracture zone in the south) (Figure 1). These magnetic anomalies were first interpreted as caused by shallow basement or uplifted blocks of Precambrian granulites (Moraes and Liandrat, 1987; Liandrat et al., 1989). However, the gravity and magnetic data suggest that, rather than being caused by the shallow occurrence of Precambrian basement rocks, these anomalies have a more profound significance in the tectonic development of the margin (Mohriak et al., 1995a). In fact, the intensity of the total field magnetic anomalies bear a regional expression that is rather similar to the East Coast magnetic anomaly in North America, which has been interpreted to mark the transition from thinned continental crust to oceanic crust (Keen and Voogd, 1988; Keen et al., 1994; Holbrook et al., 1994).

Figure 4 shows the analytical signal map, which is a direct measure of the gradient of the total magnetic field and is based on derivatives of the magnetic anomaly map. It shows the main depocenters along the northeastern margin and the basement highs that limit the synrift troughs. The structural framework of the Sergipe–Alagoas Basin (Figure 1) is associated with a system of northeast–southwest trending normal faults and subsidiary east–west and northwest–southeast transverse faults. Regional seismic maps indicate that the maximum total sedimentary thickness in the basin depocenter is located near the mouth of the São Francisco River. Two synrift subbasins can be identified (Sergipe and Alagoas

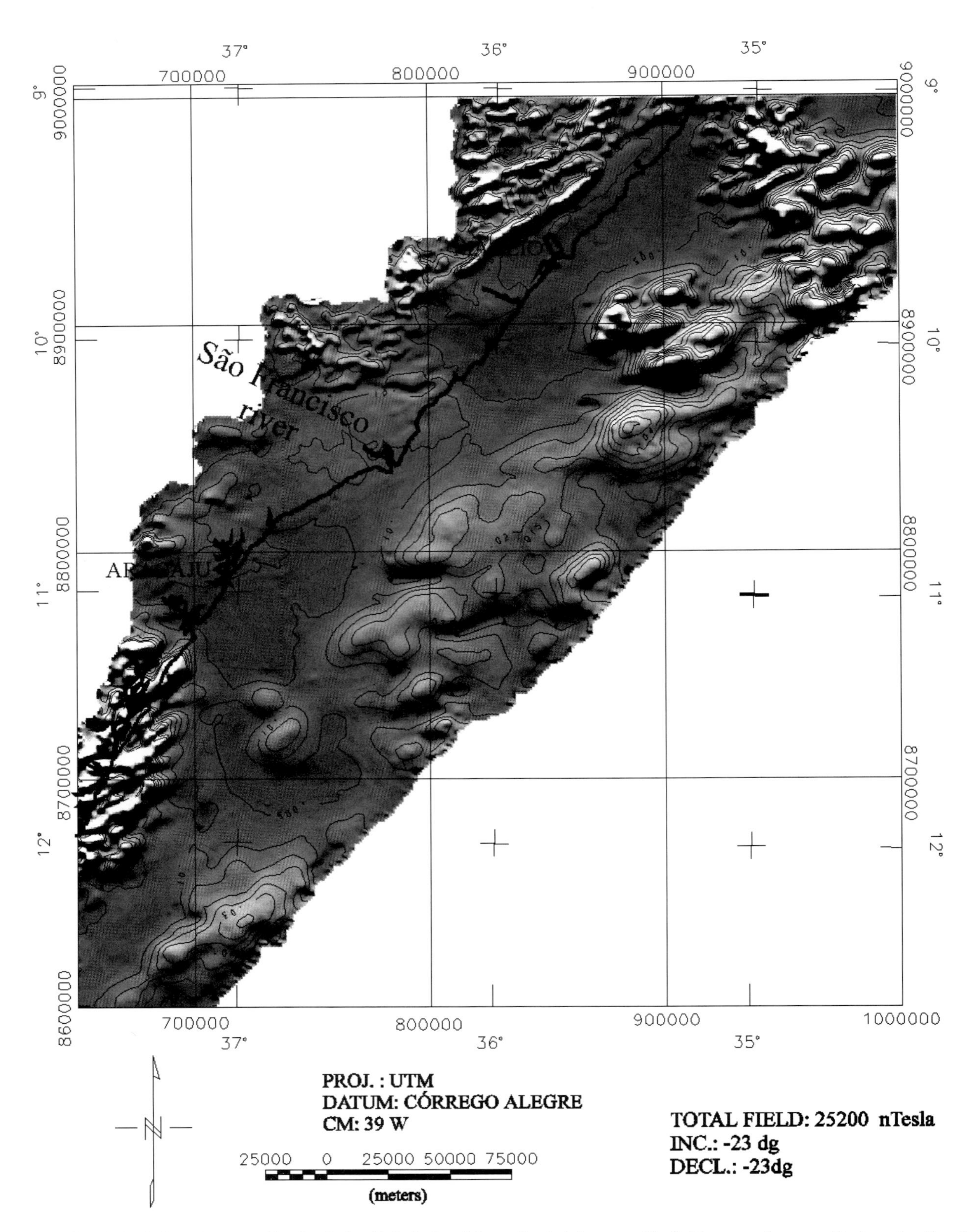

Figure 4—Analytical signal map with shaded relief, derived from the total magnetic field anomaly map. Darker shading indicate deeper basement or thicker depocenters. Northern Alagoas subbasin is characterized by anomalies indicating shallow basement and volcanics with a distinct northwest trend from oceanic crust toward continental crust. The Alagoas depocenter, near Maceió, is located both onshore and offshore, whereas the Sergipe depocenter (near Aracaju) is mainly located offshore.

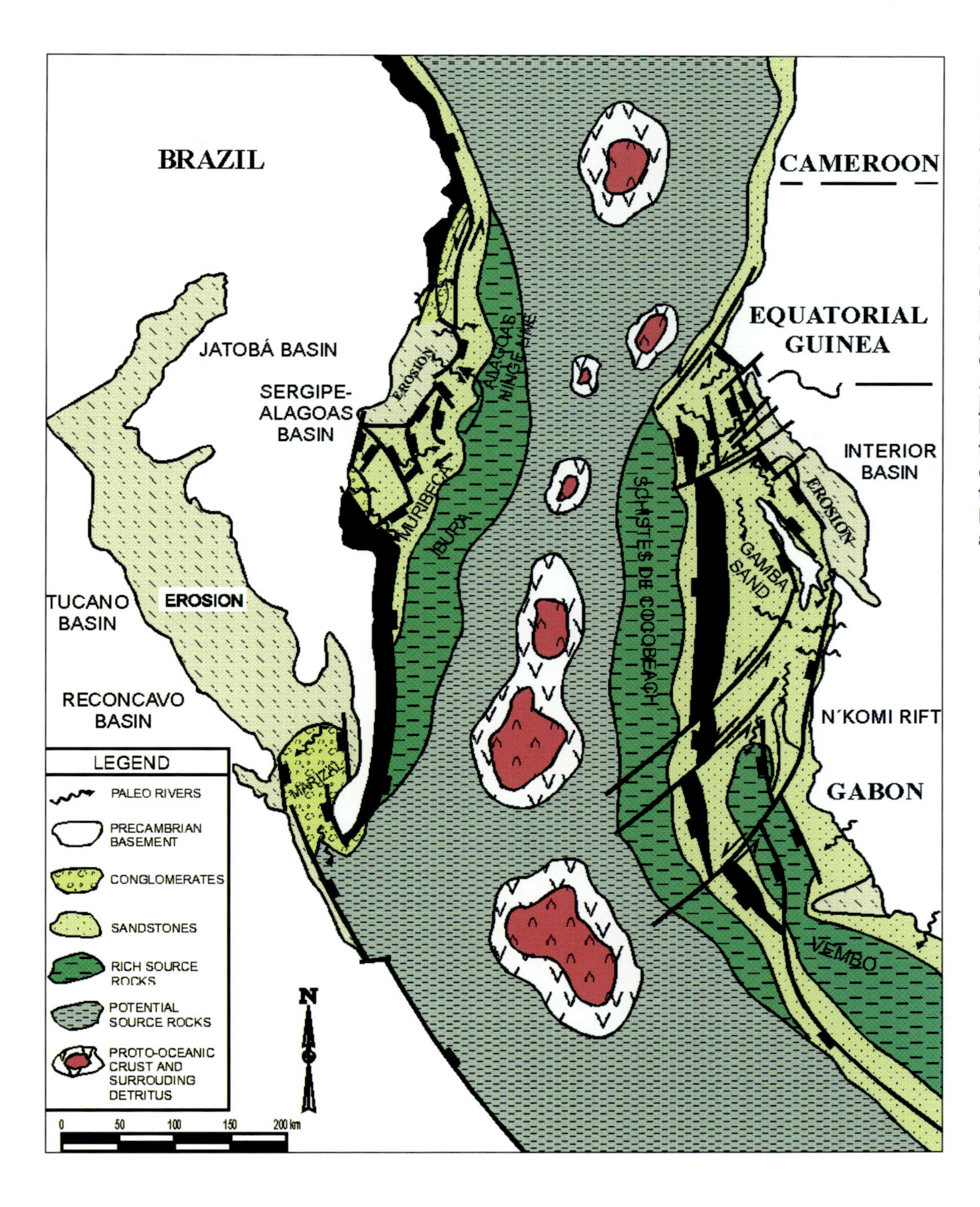

Figure 5—Paleogeographic map of the South Atlantic rift system, showing the Recôncavo-Tucano-Jatobá branch (an aborted rift system) and the Jacuípe-Sergipe-Alagoas branch (with its conjugate margin along northern Gabon and equatorial Guinea) joining the rift system developed between the eastern Brazilian and western African margin. The triple junction is characterized by a change in the direction of an incipient oceanic ridge south of the Recôncavo Basin. Modified from Bradley and Fernandez (1991).

subbasins) which are separated by a regional high located slightly north of the São Francisco River (compare Figure 4 with Figure 1).

All basins onshore and offshore of the northeastern Brazilian margin are associated with Late Jurassic–Early Cretaceous extensional processes that formed a rift system within the South American and African plates (Chang et al., 1992). The breakup of Gondwana in the Mesozoic started in the south and propagated toward the north, beginning in the Late Jurassic and probably climaxing in the Late Cretaceous, with the final separation of Africa and South America occurring along transform fractures in the equatorial rift zone (Rabinowitz and Labreque, 1979; Conceição et al., 1988; Wannesson et al., 1991; Matos, 1992). The region between the southern tip of the Jacuípe-Sergipe-Alagoas and Recôncavo-Tucano-Jatobá rift systems acted as a triple junction, forming the V-shaped branch of the South Atlantic rift-rift-rift system joining near Salvador City, south of the Recôncavo Basin (Figure 5). At this junction, these systems bifurcated and formed a propagating rift between the eastern Brazilian and western African continental rift (Bradley and Fernandez, 1991).

Volcanism was voluminous along the southern margin of Brazil (Espírito Santo, Campos, Santos, and Pelotas Basins), but almost absent during the Neocomian–Barremian rift phase of the northeastern margin (Chang

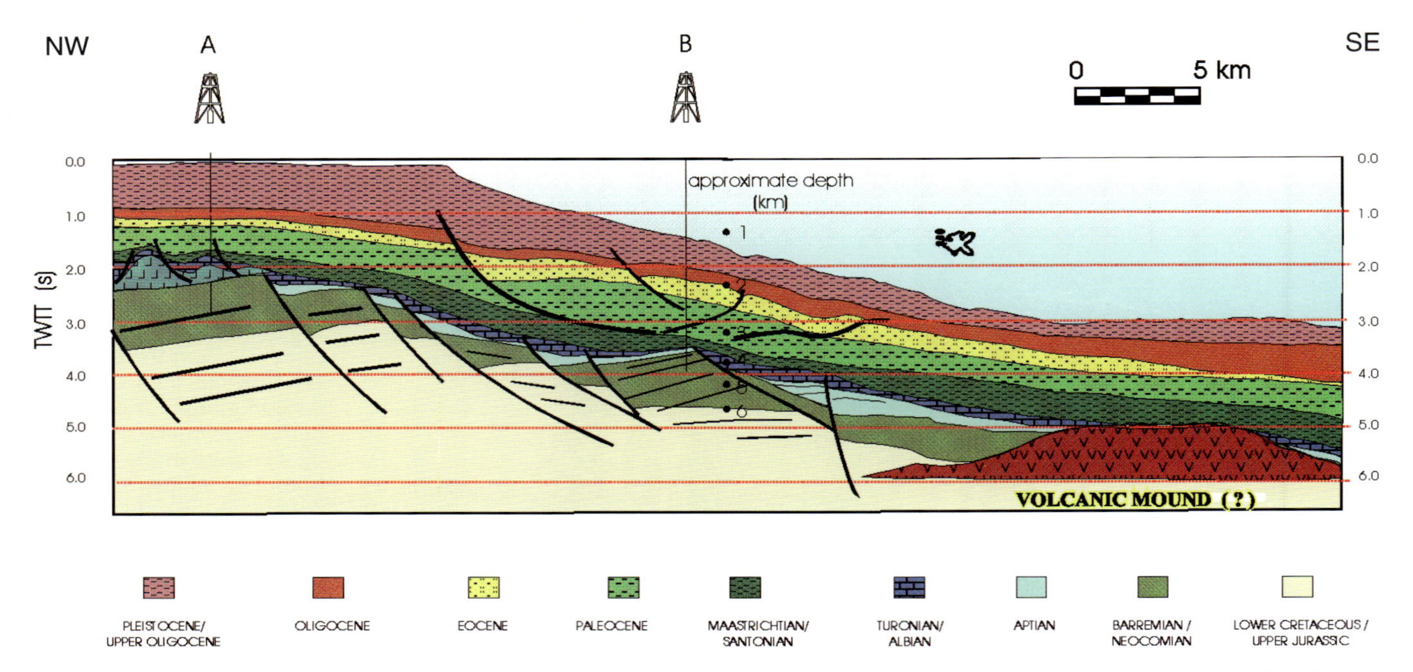

Figure 6—Regional geoseismic profile extending from the onshore region toward the deep-water province of the Sergipe subbasin. Tilted rift blocks are characterized on the platform and slope, and in deep waters, a possible volcanic mound is also identified. Modified from Bisol (1992) and Mohriak et al. (1996).

et al., 1992). Lithospheric stretching and rifting ceased in northeastern Brazil with the onset of sea floor spreading in Early–middle Cretaceous time (probably by late Aptian–early Albian) (Chang et al., 1992; Matos, 1992). There are, however, some seismic indications (e.g., offsets at the base of the Aptian salt and other younger reflectors) of localized reactivations of basement-involved normal faults in the Sergipe–Alagoas and Jacuípe Basins up to Late Cretaceous time.

A schematic section based on seismic data is shown in Figure 6, illustrating the main structural styles observed from the continent toward the deep-water region. Tilted rift blocks are characterized in the platform and in the slope, and possible volcanic mounds are characterized in the deep water region, near the transition from continental to oceanic crust (Cainelli, 1992). However, there are some indications that salt tectonics may play an important role near the crustal limit, forming salt diapirs, which would indicate that the original Aptian salt basin was not restricted to the platform but instead advanced toward the deep-water region (Mohriak, 1995a).

STRATIGRAPHIC EVOLUTION

The stratigraphic column of the Sergipe–Alagoas Basin (Figure 7) can be divided in five megasequences: the prerift, rift (or continental), transitional (or evaporitic), transgressive marine, and regressive marine. These megasequences are divided by major regional unconformities, resulting in geologic unities deposited within characteristic depositional environments (Feijó, 1994).

The prerift megasequence outcrops onland along the borders of the rift system, and maximum sedimentary thickness ranges from 100 to 300 m (Milani and Davison, 1988). It includes strata ranging from Paleozoic (mainly Carboniferous–Permian) to Mesozoic (mainly Late Jurassic–Early Cretaceous). The Jurassic unconformity divides this sequence into two units. One is the Permian–Carboniferous, which includes a siliciclastic section with glacial sediments (Batinga Formation) overlain by red beds with evaporites (Aracaré Formation). The other unit is an Upper Jurassic sequence dominated by red beds at its base (Bananeiras Formation) grading upward into fluvial–eolian sedimentary rocks (Serraria Formation).

Strata of the Early Cretaceous rift megasequence was deposited in fluvial–lacustrine environments ranging in age from Neocomian to Barremian. It is characterized by its association with active rifting and rapid variation in stratigraphic facies. The megasequence is composed of a prograding lacustrine sequence (Barra de Itiúba Formation), grading into a siliciclastic–carbonate sequence (Morro do Chaves Formation), and culminating in a predominantly siliciclastic sequence (Coqueiro Seco and Maceió Formations). Syntectonic coarse siliciclastics, associated with major border faults, are typified by conglomerates (Rio Pitanga Formation) and sandstones (Penedo Formation).

The transitional megasequence corresponds to a protooceanic stage in the South Atlantic basins. It is characterized by widespread deposition of siliciclastic, carbonates, and evaporites (Muribeca Formation) during the first marine incursions in the basin. Salt tectonics has been

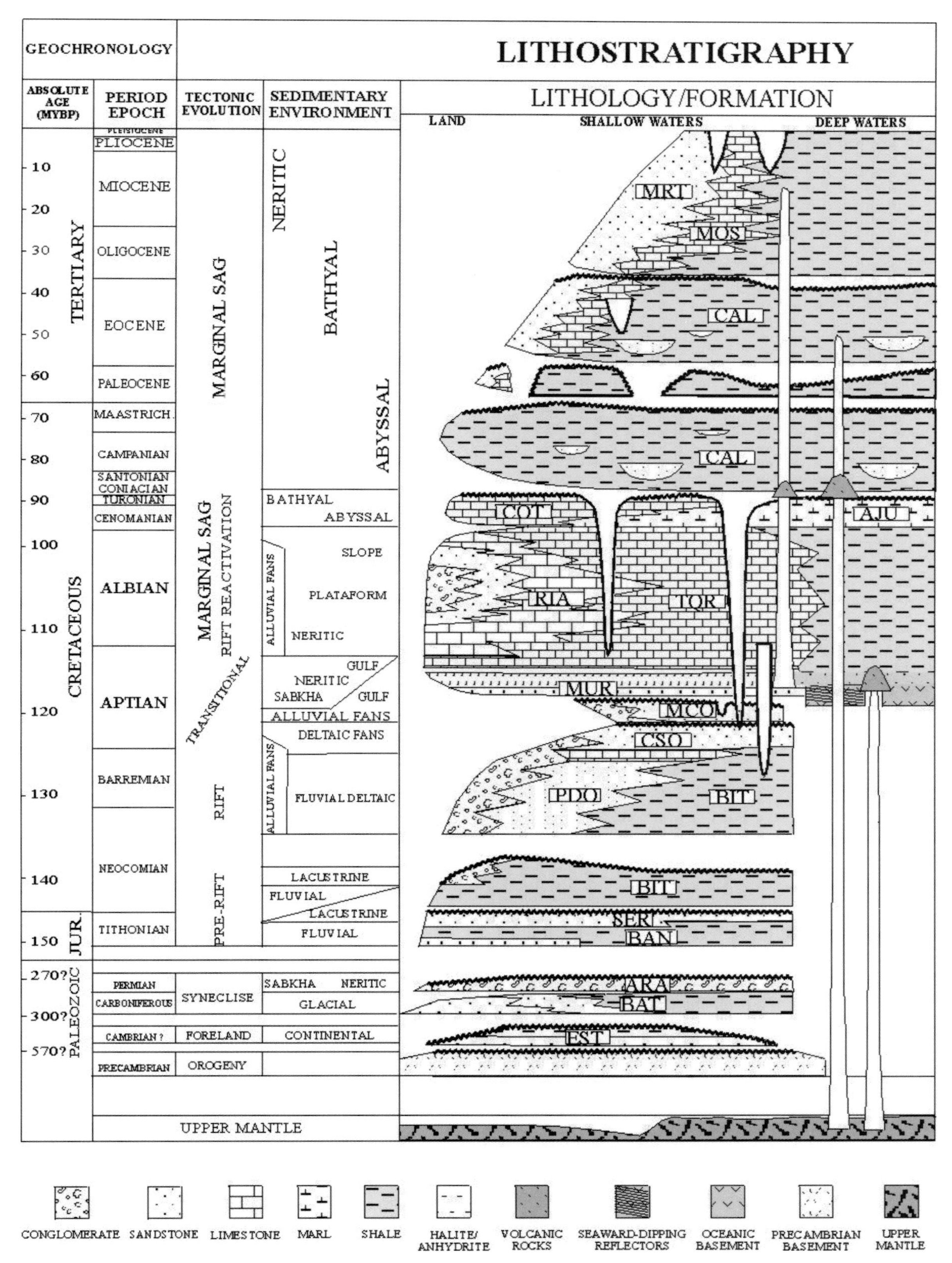

Figure 7—Tectono-stratigraphic column for the Sergipe subbasin. Abbreviations of units: EST, Estância; BAT, Batinga; ARA, Aracaré; BAN, Bananeiras; SER, Serraria; BIT, Barra de Itiúba; PDO, Penedo; CSO, Coqueiro Seco; MAC, Maceió; MUR, Muribeca; RIA, Riachuelo; TQR, Taquari; COT, Cotinguiba; AJU, Aracaju; CAL, Calumbi; MOS, Mosquero; MRT, Marituba.

considered mainly restricted to the platform, with negligible influence in deep-water environments (Cainelli, 1992; Feijó, 1994). However, elusive structures in the deep-water region may possibly be associated with salt tectonics (Mohriak, 1995a, b).

Onshore and on the continental platform, two evaporite sequences are identified, one below and the other above a regional pre-Aptian unconformity (known as the "breakup unconformity") which corresponds to a peneplain that leveled the rift blocks. The older salt deposits (Paripueira Member of the Muribeca Formation) occur mainly in local depocenters in the Alagoas subbasin, whereas the younger salt (Ibura Member) has a regional occurrence in the Sergipe subbasin (Uesugui, 1987). Above the breakup unconformity are deposits of siliciclastic rocks (Carmópolis Member) that are overlain by bituminous shale and, finally, capped by evaporites (Ibura Member). Sedimentary sections above the breakup unconformity are much less affected by basement-involved faults than the previous sequences, but in some parts of the rift system, some blocks are frequently affected by reactivations that may reach Aptian and Albian sequences.

The transgressive marine megasequence may be subdivided into an earlier phase I (Aptian–early Albian) characterized by restricted environments and predominantly carbonate deposition (Riachuelo Formation), with facies variation from proximal with siliciclastics (Angico Member) to lagoonal and distal pelitic (Taquari Formation). The subsequent phase II (late Albian–early Tertiary) was characterized by rising sea level and predominantly siliciclastic depositional environments (Koutsoukos et al., 1991). The proximal facies of phase II are characterized by sandstones (Cotinguiba Formation), the platform facies by carbonate rocks (Sapucari Member), and the distal facies by pelitic rocks (Aracaju Formation). Several levels of anoxia within maximum flooding surfaces are identified in this phase. This megasequence was affected by salt tectonics, which started to play a leading role in the development of different petroleum systems on both sides of the Atlantic by Albian time and strongly influenced Upper Cretaceous sequences (Figueiredo and Mohriak, 1984).

The regressive marine megasequence coincides with the eustatic sea level fall in the early Tertiary. This megasequence is made up of three facies: (1) the proximal facies, along the coastal region (Marituba Formation), with predominantly coarse sandstones; (2) the platform facies (Mosqueiro Formation) with mainly limestones; and (3) the distal facies (Calumbi Formation) with mostly shales and interbedded turbidites.

CRUSTAL ARCHITECTURE OF MARGIN

The offshore basins in northeastern Brazil have depocenters at depths that are beyond the reach of exploratory drilling. Some estimates suggest that basement in the deep-water region of the Sergipe Basin may be deeper than 15 km in some depocenters and that the rift thickness increases toward the continental–oceanic crustal boundary (Pontes et al., 1991; Destro, 1994). Mapping of the structural framework of the basin has been completed only recently in the deep-water region (Figure 1), and we discuss here the main results of the interpretation.

Figure 8 shows a structural map of the Sergipe–Alagoas Basin as interpreted in the late 1980s and early 1990s (Pontes et al., 1991; Lana, 1993). In this map, the tectonic framework is mainly restricted to the platform, and the transition from continental to oceanic crust is interpreted between the 2500 and 3000 m isobaths, tens of kilometers east of a regional positive magnetic anomaly. This anomaly was previously interpreted to correspond to a Precambrian block forming an outer high to the basin depocenter (Moraes and Liandrat, 1987; Liandrat et al., 1989). Some canyon channels and possible turbidite mounds are identified in the deep-water region extending beyond the 2000-m isobath (Figure 8).

Figure 9 shows part of a regional seismic profile in the deep-water region, with identification of some stratigraphic and structural features, and particularly illustrates the seismic expression of these deep-water channels. However, the focus of the present discussion is the deeper structures imaged in the profile. Some authors (Pontes et al., 1991; Destro, 1994) have interpreted the subhorizontal reflectors below a major unconformity (at 6.5 sec) as sedimentary sequences encompassing prerift and rift strata, assuming that the rift increases in thickness toward the deep-water region. This assumption was previously justified by the fact that most faults imaged in the platform are dipping basinward, and basement was not clearly imaged in the seismic profiles. The interpretation of a deep rift has profound implications for petroleum exploration because of the identification of possible amplitude anomalies, such as those observed around 6.5 sec in the middle of the profile (Figure 9), with deeper seismic reflectors imaged around 8.5–9.5 sec in the ultra-deep water region.

A detailed analysis of all geologic and geophysical data leads to other possible alternative interpretations. We discuss these possibilities by analyzing a regional deep seismic profile extending along dip direction, from the platform toward the deep-water region of the southern Sergipe subbasin. Figure 10 provides an example of such a regional seismic with no interpretation shown.

Two alternative hypotheses have been proposed for the coherent deep seismic reflectors observed in the deep-water portion of the profile (Figure 11). Noting K. Popper's assertion that the validity of a scientific hypothesis can be judged only by confronting it with a counter-hypothesis that is equally logical, we analyze two alternative geologic interpretations to assess their implications for the petroleum systems. These two end-member alternatives, each with several favorable and unfavorable arguments, have profound implications for the width of the rifts, for thermal maturation of possible source rocks, and for the petroleum exploration of these deep-water

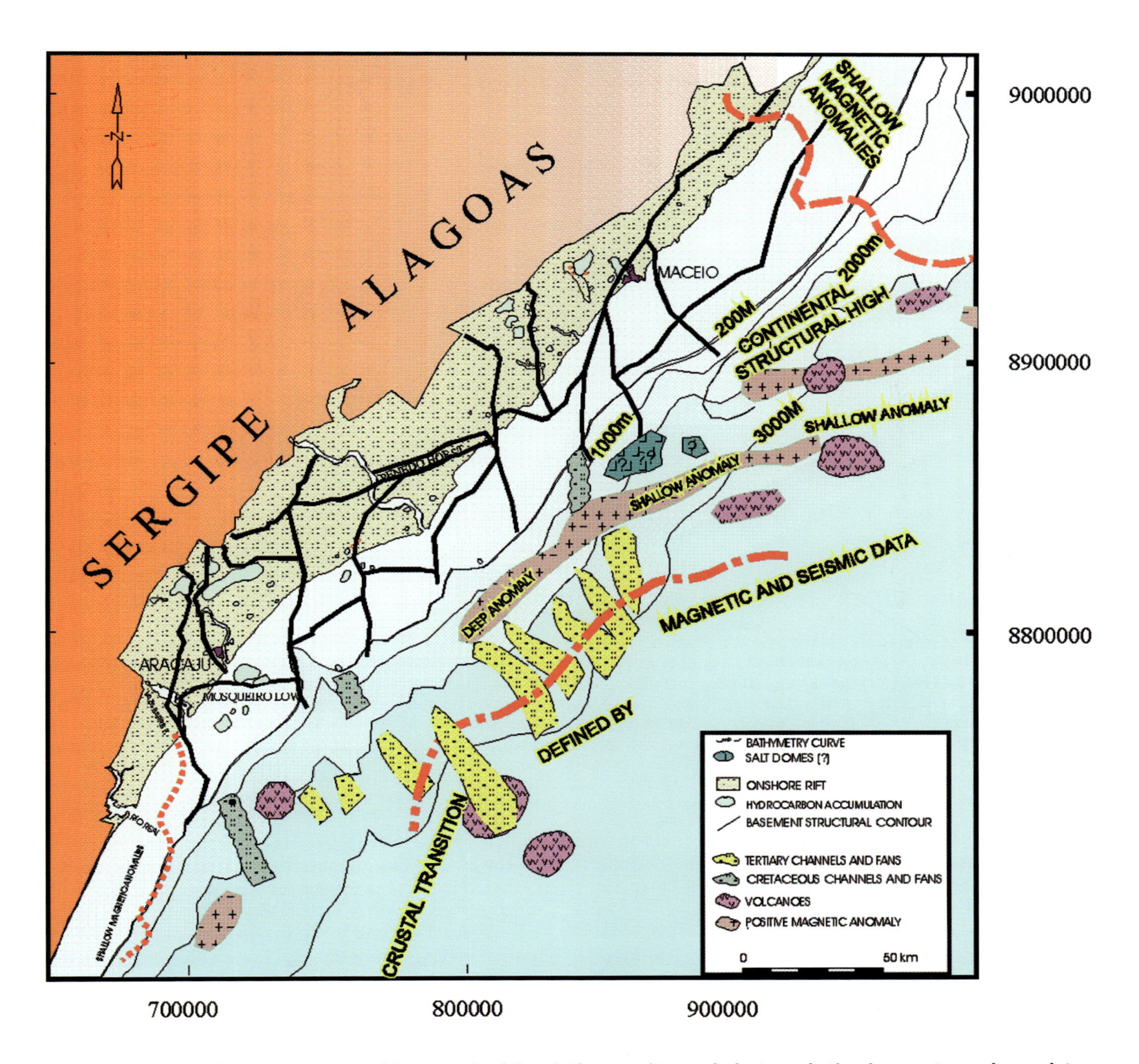

Figure 8—This map shows the structural framework of the platform and tectonic features in the deep-water region as interpreted in the early 1990s. A large number of canyon channels and fans were identified in the deep-water region, which is characterized by arcuate magnetic anomalies. Several volcanoes and possible salt structures were also recognized. Modified from Pontes et al. (1991) and Lana (1993).

provinces (Mohriak et al., 1995b). Most thermo-mechanical and geochemical modeling software requires input of tectono-physical and kinetic parameters that depend on which interpretation is adopted. These parameters are based on assumptions and interpretations of crustal and lithospheric thickness, density of mantle, crust and sedimentary layers, heat flow, thermal conductivity, rift and postrift sedimentary isopachs, and burial history. In this chapter, we discuss these two end-member possibilities by contrasting their implications for the petroleum systems. Neither of these hypotheses (Figure 11) can be proved by drilling in the near future, but the implications for risk assessment and economics of deep-water exploration are too important to be left aside. So far, we can rely only on indirect geophysical methods and tectonic analysis to assess the exploratory potential of the many diverse

Figure 9—Detail of a regional seismic profile showing multiple cut-and-fill episodes associated with erosion by canyons in the deep-water region. The channels tend to deflect to the southwest, as indicated by progradation in the canyons (Cainelli, 1992). An important horizon (base of Calumbi Formation, Santonian) is characterized by a continuous reflector at 6.0 sec two-way traveltime. Below a nearly transparent seismic sequence underlying the base of the Calumbi reflector, the profile shows strong subhorizontal discontinuous reflectors and a main unconformity at about 6.5 sec and a large-amplitude anomaly near the middle of the profile. Below a layered reflective sequence with its top at 6.5 sec is a subtle deeper seismic reflector at 8.5 sec that may occur at 9–10 sec in other profiles. Alternative geologic interpretations are proposed in the text to explain these features.

plays that might exist in these ultra–deep water regions, where an incorrect geologic interpretation could result in losses of millions of dollars.

The first hypothesis (Figure 11a) assumes that the layered reflectors of the presalt sedimentary sequence drilled in the platform continue seaward, forming an extremely thick rift basin in deep waters. The strong seismic events at 7.0–9.0 sec were attributed to impedance contrasts between sedimentary fill and Precambrian basement rocks, resulting in the interpretation of a very thick rift depocenter (>3 sec two-way traveltime) in the deep-water region. In fact, this is not abnormal because there are several basins where the synrift strata exceeds this thickness, such as in the Tucano Basin of onshore Brazil (Mohriak et al., 1997) and the Jeanne d'Arc Basin on the eastern Canadian margin (Tankard et al., 1989). The second hypothesis (Figure 11b) attributes these strong deep seismic reflectors to intracrustal horizons which, in the deep-water region of the continental margin, merge with the seismic Moho. Again, there are a few analogs for this interpretation in other passive margin sedimentary basins, such as along the Galicia margin (Boillot et al., 1989; Reston et al., 1996) and in west Africa (Meyers et al., 1996).

The latter interpretation results from the regional evaluation of several deep seismic reflection profiles in the northeastern region of Brazil (Mohriak et al., 1993, 1995a). The regional seismic profiles were systematically integrated with gravimetric and magnetic data and then comparatively analyzed using other deep seismic profiles from the South Atlantic, particularly the Campos Basin (Mohriak et al., 1990) and the Gabon Basin (Rosendahl et al., 1991; Meyers and Rosendahl, 1991; Wannesson et al., 1991; Meyers et al., 1996). Other basins that show similar features and provide valuable analogs are in the North Atlantic region, including eastern Canada (Keen and Voogd, 1988; Keen et al., 1991, 1994), the Lofoten Basin (Mjelde et al., 1993), the North Sea (e.g., Blundell, 1990), and Greenland (Larsen and Jakobsdottir, 1988). These analogs indicate that the deep seismic reflectors in the deep-water province of northeastern Brazil might correspond to intracrustal horizons or even the seismic Moho. This might be interpreted as the base of an array of strong reflectors that probably mark the transition from the lower crust to ultramafic rocks of the upper mantle (Klemperer et al., 1986).

Figure 12 shows our interpretation of the Sergipe subbasin profile based on the integration of regional

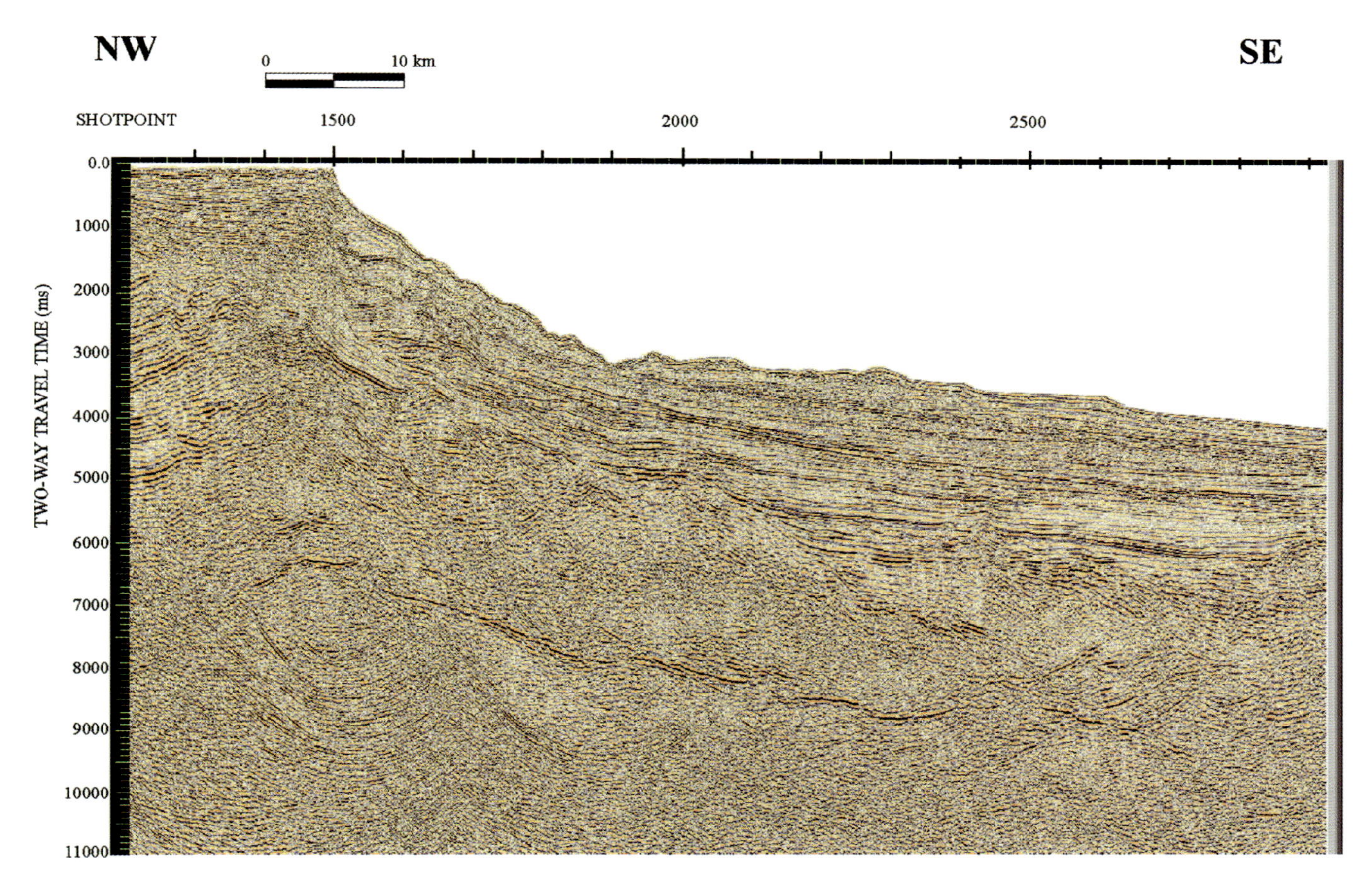

Figure 10—Regional deep seismic profile A in the Sergipe subbasin, with no interpretation shown. (See Figure 1 for location of profile A.) Basinward of the shelf edge, the seismic section shows a strong reflector from about 6. 5 to 9. 0 sec two-way traveltime.

deep seismic profiles, potential field data, and comparisons with other regional seismic profiles in the South Atlantic and elsewhere (Mohriak et al., 1998). The western portion of the seismic section is characterized by a thick sequence of layered reflectors, regionally tilted landward on the platform (low-frequency reflectors at 2.0–5.0 sec from shotpoints 1200–1400). This package has been drilled by some boreholes and is proved to correspond to Aptian–Neocomian siliclastic sedimentary rocks. Under-lying and seaward of this layered reflection package is a conspicuous array of strong reflectors in the lower part of the section which form an anticlinal structure that rises at about 8.0 sec two-way traveltime in the westernmost part of the profile. This structure has an apex that reaches about 6.5 sec near the shelf break and extends eastward as a conspicuous band of reflectors (at about 7.0–9.0 sec) throughout most of the profile, from slope to deep basin. The deep-water region is also characterized by a few landward-dipping reflectors between shotpoints 2500 and 2900, imaged at 8.0–9.0 sec (Figure 12). There is a remarkable lack of volcanic features in the platform, as confirmed by several boreholes drilled through the rift sequence. However, there are some structures that may have an igneous origin, such as the volcanic plug near shotpoint 2900, which is probably associated with the Sergipe fracture zone (~90 km from the origin, near the easternmost end of the profile). This plug is interpreted as a postrift volcanic intrusion, located close to the continental–oceanic crust boundary, as indicated by gravity and magnetic data (Mohriak et al., 1995b).

Adjacent to this volcanic feature, there are packages of reflectors with sigmoidal geometry, mostly dipping basinward but some dipping landward (at 8–9 sec, shotpoints 2500–2800, in Figure 12,). Although some interpreters postulate that these reflectors correspond to rift structures (e.g., Pontes et al., 1991; S. F. Santos, personal communication, 1994), the seismic interpretation suggests tectonic similarities to magmatic features observed in several other passive margins along the North Atlantic Ocean, particularly along the Canadian margin (e.g., Keen and Voogd, 1988; Keen et al., 1994). However, they might also represent shear zones formed during the emplacement of the volcanic plug, serving as weak zones or conduits for magmatic emplacement (McCarthy et al., 1988).

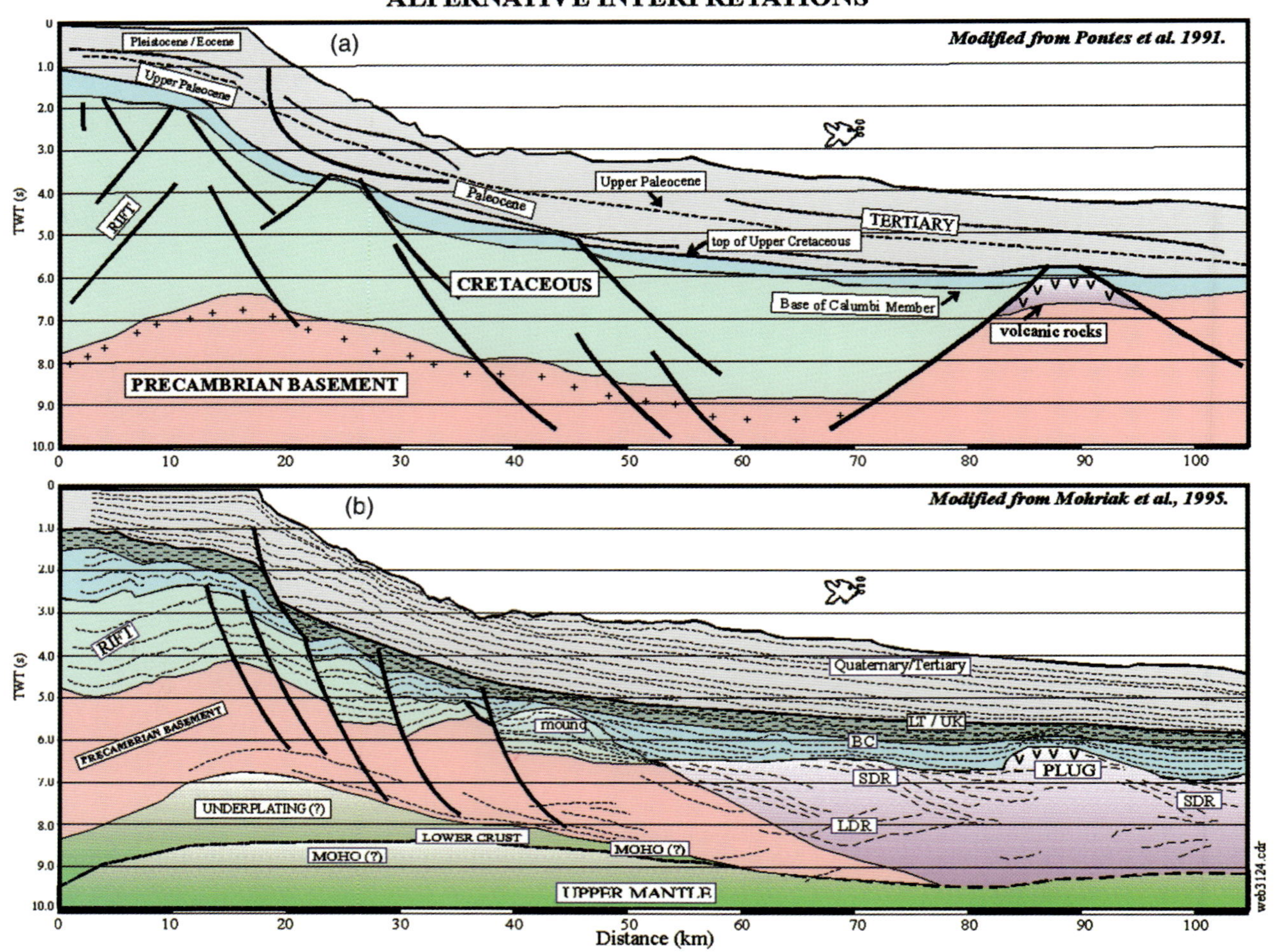

Figure 11—Alternative interpretations for regional profile A in the Sergipe subbasin. (a) This interpretation (Pontes et al. (1991) assumes that deep reflectors on regional seimic lines correspond to basement. (b) Another interpretation (Mohriak and Rabelo, 1994; Mohriak et al., 1995a; 1998) assumes that the deep seismic reflectors in the deep-water region actually correspond to lower crustal detachment horizons or the top of underplated rocks near the seismic Moho. Abbreviations: BC = unconformity at base of Calumbi Formation (Santonian); LT/UK = Upper Cretaceous–lower Tertiary sequences; SDR = seaward-dipping reflectors; LDR = landward-dipping reflectors.

The sigmoidal wedges observed on the profile (Figure 12) probably correspond to seaward-dipping reflectors formed mainly by volcanic rocks extruded during early phases of spreading at oceanic ridges (e.g., Hinz, 1981; Mutter et al., 1982; Mutter, 1985). Seaward of the volcanic plug, the basement is probably composed of plutonic and volcanic igneous rocks intercalated with volcaniclastic rocks, forming the seaward-dipping reflector wedges that ultimately grade seaward into pure oceanic crust. Thus, in the tectono-stratigraphic chart of the Sergipe subbasin, we did not include any rocks older than oceanic crust above the mantle, seaward of these volcanic plugs (see Figure 7).

The alternative interpretations of a deep rift trough (hypothesis 1) versus a seaward-dipping volcanic wedge in the distal parts of the basin (hypothesis 2) still present complex geologic problems that are not completely understood. Some of these problems are listed in Table 1. The first hypothesis assumes a very deep basement both in the platform and deep-water region, with Cretaceous strata affected by faults that penetrate Precambrian basement. The second hypothesis assumes pinch-out of the rift strata toward the crustal limit of the basin and synrift block rotation by crustal faults that detach in the lower crust reflectors. The detachment zone apparently amalgamates with the seismic Moho in the deep-water region (Mohriak et al., 1998). The magmatic underplating and massive wedges of seaward-dipping reflectors are probably contemporaneous with the emplacement of oceanic crust, postdating the rift sequence. There is very rapid crustal thinning from the

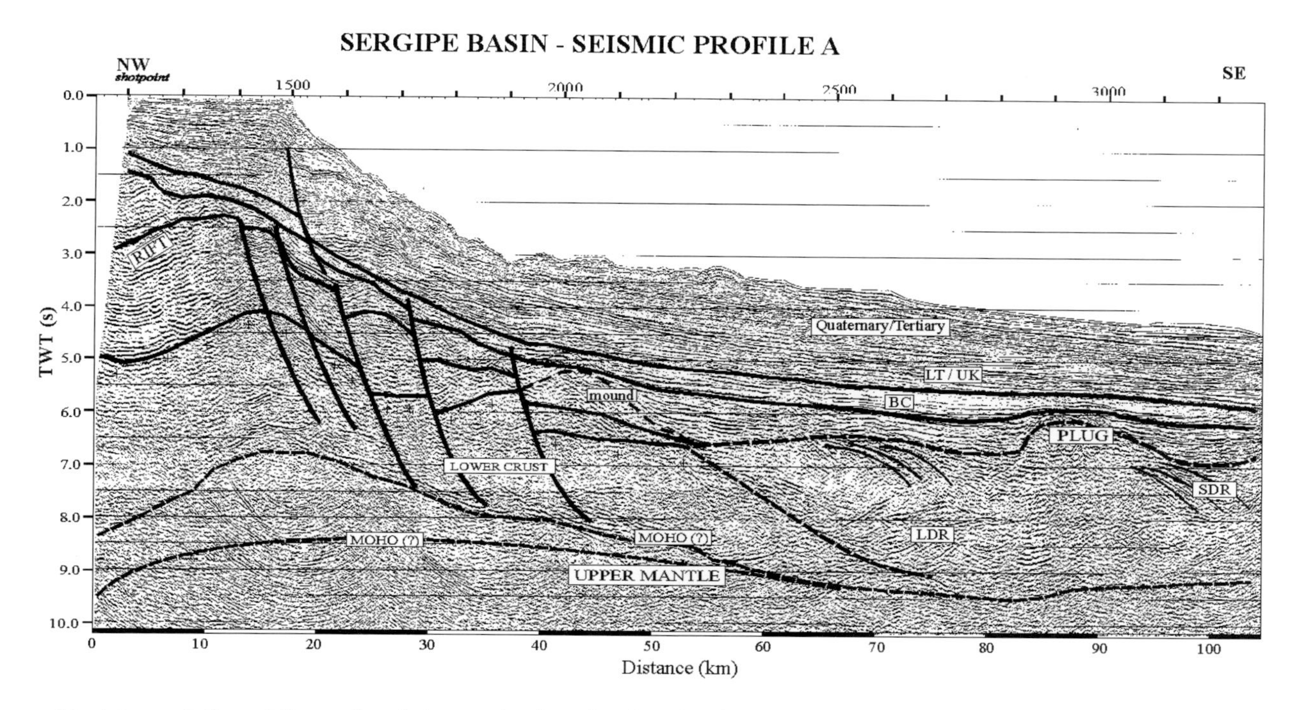

Figure 12—Interpretation of the regional deep seismic reflection line (profile A, in Figure 10) in southern Sergipe–Alagoas Basin. Important features of the deep seismic profile are discussed in the text. Abbreviations: BC = unconformity at base of Calumbi Formation (Santonian); LT/UK = Upper Cretaceous–lower Tertiary sequences; SDR = seaward-dipping reflectors; LDR = landward-dipping reflectors. Modified from Mohriak et al. (1998).

platform toward the deep-water region, where seaward-dipping reflectors onlap the Precambrian basement and, toward the deep ocean, probably merge with the volcanic rocks of the oceanic crust. Volcanic plugs in this region seem to be associated with both seaward- and landward-dipping reflectors (Figure 12).

Analysis of the gravity data corroborates the interpretation that the deep seismic reflectors correspond to horizons near the top of the upper mantle rather than the top of the basement. In the Sergipe subbasin, depocenters onshore and in shallow waters are marked by a strongly negative Bouguer anomaly (less than –20 mGal near the coastline). The gravity anomaly rapidly increases seaward of the shelf break, reaching values greater than +70 mGal, whereas the rift phase sedimentary rocks decrease in thickness toward the distal parts of the basin (Mohriak et al., 1995a).

Figure 13 shows the Bouguer anomaly plot along seismic profile A and the gravity model based on depth conversion of the seismic horizons and average densities for different layers. The Bouguer anomaly model indicates abrupt changes in Moho topography, particularly seaward of the depocenter in the Mosqueiro Low and close to the shelf break. An extremely thin continental crust extends seaward for only a few tens of kilometers beyond the shelf edge. The transition to proto-oceanic crust is interpreted to coincide with the western limit of the seaward-dipping reflectors. The transition to a pure oceanic crust occurs at the easternmost parts of the

Table 1—Problems with Alternative Geologic Interpretations of Deep Seismic Reflectors in Sergipe–Alagoas Basin

Hypothesis 1: Deep rift trough

- Incompatibility with palinspastic reconstructions between Africa and Brazil
- Incompatibility with potential field data models
- Lack of evidence for antithetic faults controlling the depocenters
- Justaposition of basement reflector with the reflector interpreted as oceanic Moho
- Apparent direct transition from rift structures to typical oceanic crust structures
- Analogies with other continental margins based on refraction and reflection seismic data
- Results of exploratory drilling in volcanic margins with similar structures

Hypothesis 2: Seaward-dipping volcanic wedge

- Very little direct evidence of volcanic rocks in stratigraphic succession
- Sigmoidal reflectors apparently indicate progradation from continent to ocean
- Apparent continuity of some seaward-dipping reflectors with sedimentary layers
- Delimitation problems of the seaward-dipping reflector wedge based only on potential field data.

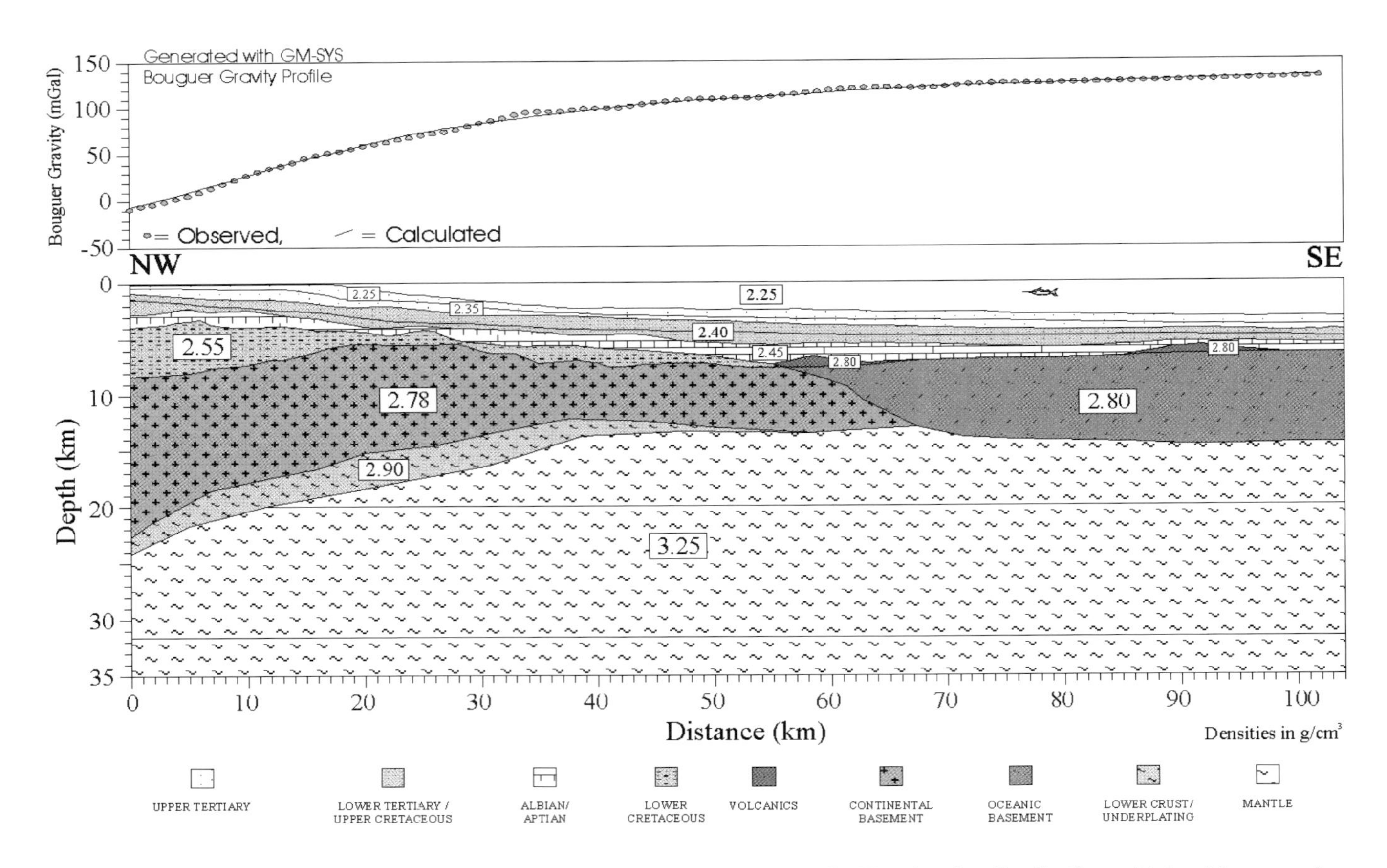

Figure 13—Gravity model for profile A along the southern Sergipe subbasin. The density distribution, obtained from sonic and density logs, is shown for each sedimentary layers (in g/cm³). The Bouguer correction for the water layer (density 1. 03 g/cm³) is 2. 25 g/cm³. The density of the crustal layers was inferred from seismic data and from worldwide analogs.

seaward-dipping reflector wedge, basinward of the volcanic plug located 90 km from the origin of the profile.

A good fit between the gravity data and the gravity model (Figure 13) was obtained by assuming a simple geologic model based on thinning of the crust (Mohriak et al., 1995a). The model indicates a mantle rising from 35 km in the onshore region, to about 25 km on the platform, and to about 20 km near the shelf edge. The Moho topography in the slope and deep basin is rather smooth, with depths of about 15–18 km across the transition to oceanic crust. However, there is growing evidence that the transition from lower crust to upper mantle is probably associated with magmatic underplating, as indicated by a combination of reflection and refraction profiling along a number of passive margin transects (e.g., LASE Study Group 1986; Sheridan et al., 1993; Holbrook et al., 1994; Keen et al., 1994). In the Sergipe subbasin, this possibility is indicated in the seismic section and in the gravity model as corresponding to an underplated layer in the lower crust (Figures 12 and 13). The strong seismic reflectors, observed in intermediate depths of the profiles, form an antiformal structure whose apex might correspond to the top of underplated rocks.

In summary, points to note on the seismic profile (Figures 12 and 13) include the following:

1. The rift layers in the Mosqueiro Low are highly rotated by a crustal-scale synthetic normal fault (beyond the westernmost limit of the profile).
2. Basinward of the shelf edge, the rift blocks are rotated by other sets of rift faults that cut most of the upper crust and extend down to the lower crust. In the deep-water region, the crustal faults apparently detach near the seismic Moho.
3. The basement reflector is rotated by these faults and depth conversion indicates that it actually rises from the depocenter in the Mosqueiro Low toward the slope and deep basin.
4. The outermost rift blocks seem to have been partially destroyed by erosion.
5. Some listric faults, detached from basement, affect the sedimentary section near the shelf break; these faults were probably controlled by mass flow triggered by salt and shale tectonics that may form mounds near the crustal limit.
6. A regional intra-Cretaceous (probably Santonian)

angular unconformity on the platform is almost flat-lying in the deep-water region and overlies horizontal sedimentary layers above the seaward-dipping reflector wedge.

7. Several arrays of reflectors with a sigmoidal geometry show seaward and landward dips in the central part of the profile.
8. Near the easternmost end of the profile are zones of landward-dipping reflectors that are probably associated with shear zones or magmatic intrusions in the crust.
9. Volcanic plugs, probably associated with leaking fracture zones, occur in the transition from the seward-dipping wedge toward the oceanic crust.
10. Near the crustal limit, there are mounds above the rift strata that may correspond to volcanic plugs or shale/salt diapirs.

SALT TECTONICS

The Sergipe–Alagoas Basin has been considered as a model for Atlantic type passive margin basins lacking a deep-water salt diapir province (Cainelli, 1992; Feijó, 1994). The conjugate margin in west Africa, north of Gabon, also seems to be largely devoid of an expressive salt diapir province, particularly the Rio Muni Basin (Turner, 1995). However, the deep-water region of northeastern Brazil is characterized by some diapiric features that have had several radically different interpretations (Mohriak, 1995a, b). These hypotheses include (1) tilted rift blocks of Neocomian age, (2) volcanic mounds and volcanic plugs, (3) leaking oceanic fracture zones, (4) transpressional features along fracture zones, (5) Precambrian basement horsts and outer highs, (6) carbonate reefs, (7) salt and shale diapirs, and (8) processing artifacts and diffractions caused by canyons in the upper stratigraphic sequences. The multiplicity of hypotheses for these features is not surprising because similar features in other sedimentary basins worldwide have also had conflicting interpretations, which sometimes can be solved only by drilling. For example, we can refer to the complex distinction among salt diapirs, tilted blocks, and volcanic intrusions in the Porcupine Basin (Tate, 1993) and along the eastern margin of North America (Sheridan, 1975; Keen et al., 1991).

Regional deep seismic profiles have also been used in conjunction with potential field data to investigate this geologic problem. Figure 14 shows two segments of regional deep seismic profiles in the southern and northern parts of the Sergipe–Alagoas Basin. The volcanic plug interpretation (Figure 14a) is constrained by gravity and magnetic data (Mohriak, 1995a, b), and this feature is aligned with the Sergipe fracture zone, which is characterized by several other volcanic plugs in the oceanic crust. The seismic line showing a possible salt diapir (Figure 14b) was shot in the northern portion of the Sergipe–Alagoas basin, along a profile extending from the slope toward the oceanic crust. Regionally, this plug-like feature is located within a low in the Bouguer anomaly map, near the boundary between the continental and oceanic crusts, which in this region also seems to be associated with seaward-dipping reflectors (Mohriak, 1995a). This structure is also located near the western (landward) prolongation of a fracture zone (Maceió fracture zone) along the dip direction. Seismic data shows strong, deep seismic reflectors at the base of the crust, rising seaward along the profile toward the crustal limit. These reflectors are imaged in other seismic profiles at about the same level (9.5 sec two-way traveltime), even in abyssal regions of typical oceanic crust. A comparison of this structure with analogs in equivalent tectonic domains, such as the oceanic crust of Cape Verde Islands (McBride et al., 1994) shows remarkable similarities in crustal architecture (Mohriak et al., 1998).

EXPLORATORY ACTIVITY IN THE SERGIPE–ALAGOAS BASIN

The first exploratory efforts in the Sergipe–Alagoas Basin began in the 1940s. Initial drilling took place in the Alagoas subbasin because early field work indicated this region as more promising in terms of source rocks. The first commercial oil discovery dates from 1957 (Tabuleiro dos Martins, in Alagoas State). The more significant oil fields, however, were discovered in the Sergipe subbasin during the 1960s (Lana, 1993). This phase included the discovery of the Carmópolis oil field, which is the largest oil field discovered in onshore Brazil up to the present, with reserves reaching 1.2 MMbbl of oil. Even today, this field accounts for about 54% of the daily production of the Sergipe–Alagoas Basin.

Exploration advanced toward the offshore region of the Sergipe subbasin in the late 1960s because it was thought that the geology of the platform would be similar to the onshore portion, which at that time was responding positively to exploratory efforts, resulting in a series of hydrocarbon accumulations. The first hydrocarbon field in offshore Brazil (Guaricema oil field) was discovered in the Sergipe subbasin by the very first wildcat drilled in 1971. The drilling activity in the basin reached a peak in the late 1970s. Since the 1990s, however, it has been declining as a result of negative results in the late 1980s and early 1990s (Bacoccoli et al., 1991). All the hydrocarbon accumulations in the offshore region are located in the southern part of the Sergipe–Alagoas Basin (Figure 15). Several hydrocarbon plays have been identified in the basin, including Cretaceous turbidites, tilted rift blocks, and channelized sandstone bodies. Two fields are used here as case histories of the exploratory plays in the Sergipe subbasin: the Carmópolis oil field (presalt play) and the Guaricema-equivalent accumulations recently discovered in the Mosqueiro Low (Late Cretaceous–early Tertiary Calumbi turbidite plays).

Figure 16 shows the type of play tested in the Carmópolis field, which is presented here as an example of a presalt (rift-phase) accumulation. The oil field is

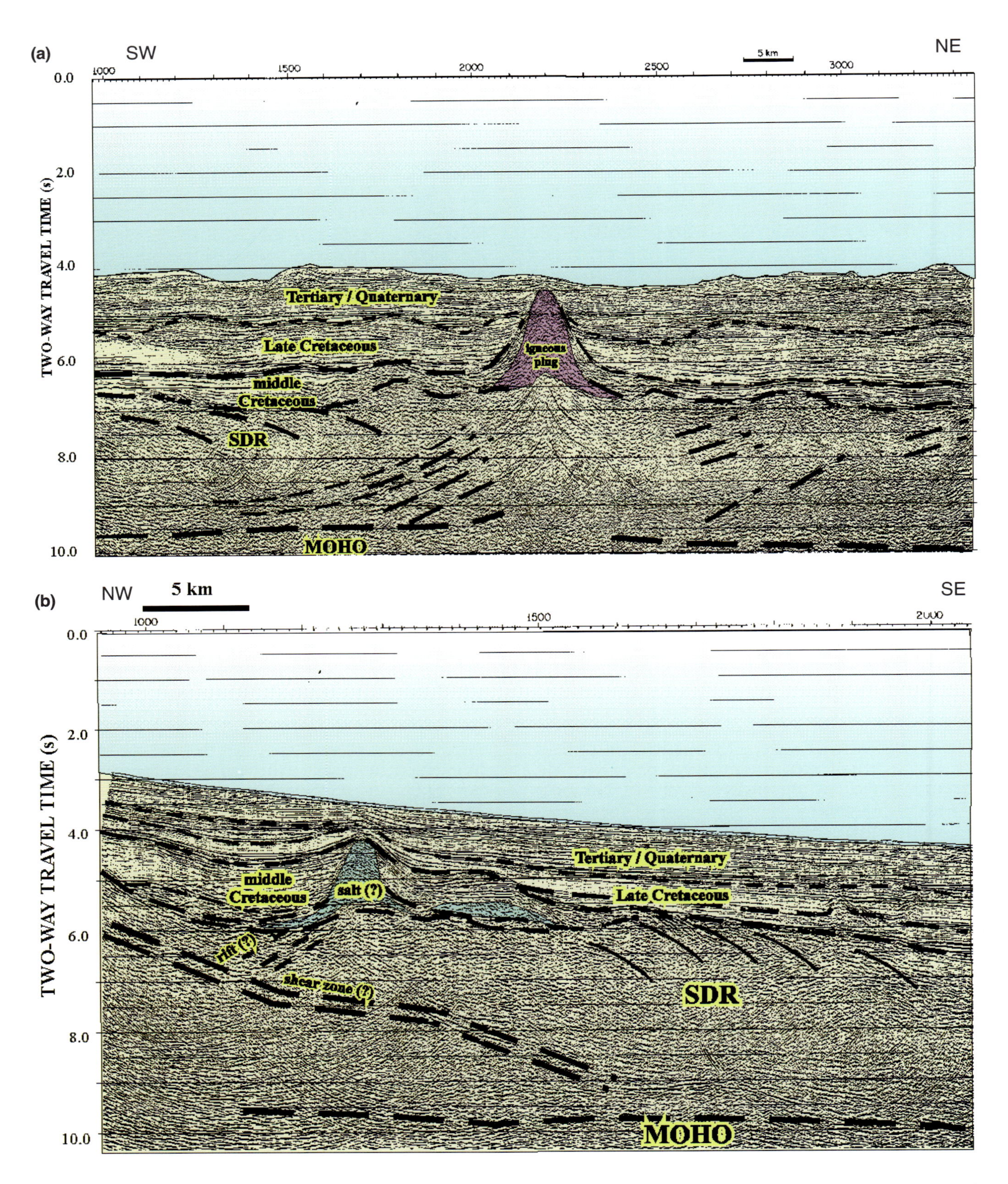

Figure 14—Segments of regional deep seismic profiles from the Sergipe–Alagoas Basin showing (a) a volcanic plug and (b) a possible salt diapir near the crustal limit. Seaward-dipping reflectors (SDR) are interpreted basinward of the salt diapir, and the seismic Moho is interpreted to be present at 9–10 sec, toward the oceanic crust. Modified from Mohriak et al. (1998).

Figure 15—Location map of oil fields in the Sergipe subbasin. Two hydrocarbon plays are discussed: the onshore presalt Carmópolis accumulation and the offshore postsalt accumulations associated with lower Tertiary turbidites in the Mosqueiro Low, south of Aracaju.

mainly associated with Barremian and Aptian reservoirs which are trapped by Aptian shales and evaporites. Stratigraphic control is provided by facies variations in the conglomerate rocks underlying the evaporites. The postsalt reservoirs are trapped by multiple seals associated with facies variations in the Albian limestones. Structural control in the western part of the field is associated with an anticline above a basement high; in the central and eastern parts, the sedimentary layers are draped over a tilted rift block. We also observe that the fractured basement is a reservoir rock with a relatively small contribution to the production. There are striking structural and stratigraphic similarities between the Carmópolis oil field and the Rabi-Kunga oil field of onshore Gabon (S. Sinclair, personal communication, 1997).

An example of a postrift phase play is shown in Figure 17. This type of play has been tested since the Guaricema oil discovery, and in the past 2 years, a number of small discoveries have been made in the Mosqueiro Low. Structural control is provided by listric normal faults that were triggered by salt tectonics. Stratigraphic control is provided by facies variations and stratigraphic expansion of the rocks from the footwall block (mainly shales) to the hanging wall block (which captured turbidite sandstones during episodes of fault activity). There are some similarities between this type of play and equivalent situations in the Gulf of Mexico, where stratigraphic expansion and capture of sandstone turbidites in downthrown blocks seems to control a number of hydrocarbon fields, such as in the Yegua sandstones (Brown, 1997).

GEOCHEMICAL ANALYSIS OF SOURCE ROCKS

Geochemical evaluation of the source rocks in the Sergipe–Alagoas Basin has been conducted over the past decades using a very large dataset of boreholes, drilled both onshore and offshore (Mello et al., 1988, 1994). Some of the main results of these analyses are discussed here.

The prerift megasequence (Figure 7) is composed of fluvial and deltaic–lacustrine siliciclastic rocks deposited under oxic conditions. These strata are composed mainly of hydrogen-poor organic matter with a total organic carbon (TOC) content of less than 0.5% (Mello et al., 1988, 1994). The rift megasequence contains excellent Neocomian–Barremian lacustrine source rocks. Geochemical analyses indicate that these rocks were deposited in environments ranging from fresh to brackish water and have a TOC content of up to 12%, mainly composed of hydrogen-rich organic matter (e.g., Barra de Itiúba Formation). In general, this source rock yields paraffinic and paraffinic-naphthenic oil with high API gravity and low sulfur content (<0.5%). Most of the hydrocarbon accumulations in the Alagoas subbasin are sourced from this prerift to synrift Neocomian lacustrine source rock.

In contrast, most hydrocarbon accumulations discovered in the Sergipe subbasin are mainly sourced by the protomarine Aptian marls and calcareous shales of the Muribeca and Maceió Formations both below and above the Aptian salt (Ibura Member). These source rocks average about 6% TOC and are composed mainly of hydrogen-rich type II kerogen (Mello et al., 1988). Basin modeling suggests that hydrocarbon generation and migration started during the early Tertiary (Paleocene), peaked by the late Oligocene, and is probably still occurring today in some parts of the basin (Araújo et al., 1994).

The source rocks of the transgressive marine megasequence are represented mainly by upper Albian–Turonian calcareous black shales of the Taquari and Aracaju Formations. These sedimentary rocks contain up to 6% TOC composed predominantly of algal and bacterial organic matter (Mello et al., 1988). In general, this type of source rock yields naphthenic oils with high API gravity and medium sulfur content (about 0.5%). However, in the Sergipe Basin, these sequences are characterized by minor sedimentary thicknesses and are immature in most parts of the basin (Mello et al., 1996).

The regressive marine megasequence is predominantly composed of marine siliciclastic oxidized sedimentary rocks. They show less than 0.1% TOC made up mainly of hydrogen-poor organic matter (Mello et al., 1988).

Figure 18 shows a schematic geoseismic section extending from the continent toward the platform of the

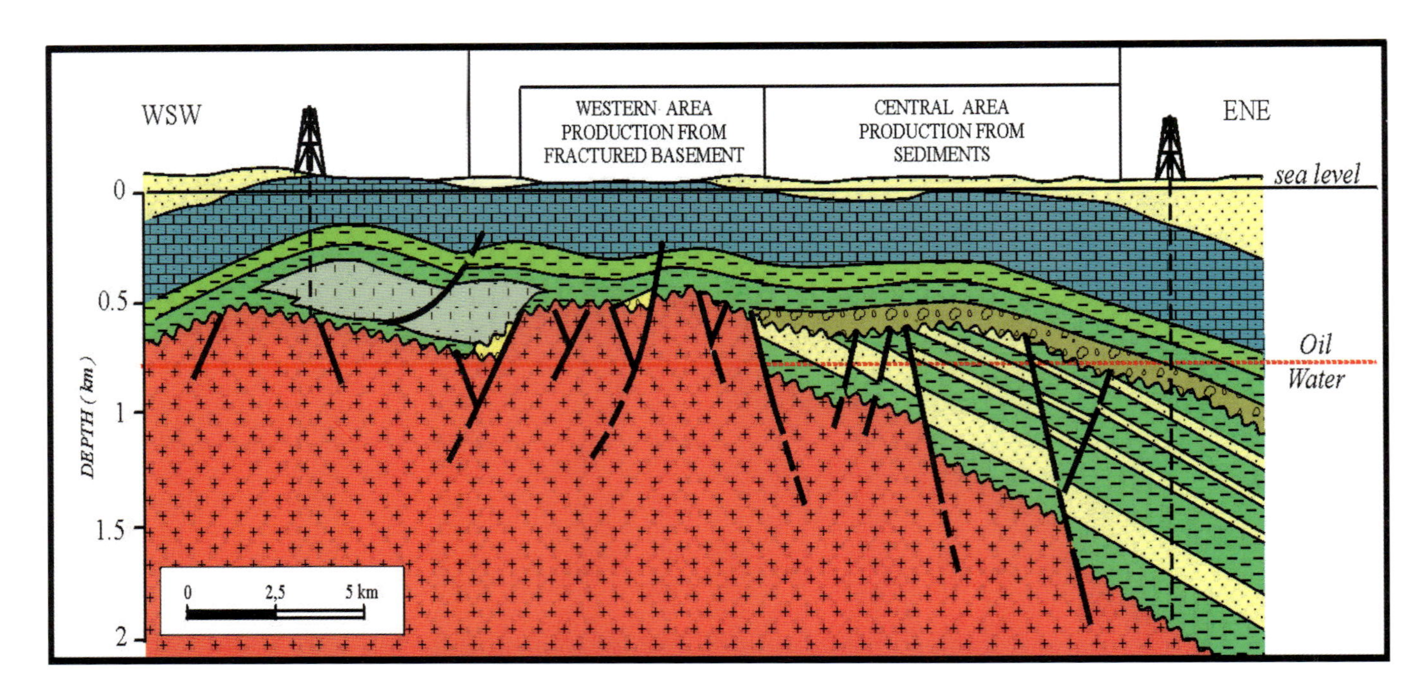

Figure 16—Cross section of the Carmópolis oil field area. Production comes from presalt (mainly Aptian–Barremian) siliciclastics, and in the western area, some production is from fractured basement rocks. Evaporite beds (mainly halite) occur in a half-graben west of the horst in the center of the section. Modified from Ponte et al. (1980).

Table 2—Tectonic and Petroleum Systems Implications for Alternative Geologic Interpretations

Deep reflector = Precambrian basement	Deep reflector = intracrustal horizon near base of crust
Very thick rift trough in deep waters affected by faults with small throws.	Tilted rift blocks are controlled by crustal-scale faults with large throws.
Pre- and synrift section may be very thick in deep waters.	Pre- and synrift section pinches out toward the crustal limit.
Faults detaching on basement surface or prerift lubricating layer.	Faults detach near the base of the crust or at the seismic moho.
Only upper part of rift is eroded in deep waters.	Extreme erosion of rift blocks occurs in deep waters.
Large anticline in platform is bounded by synthetic and antithetic faults.	Updip oil migration is favored (both landward and basinward).
Rift phase isopach much thicker than thermal phase isopach in deep waters.	Rift phase isopach thinner than thermal phase isopach in deep waters.
Rift and prerift strata extend to ultra-deep waters as prograding sequences.	Seaward- and landward-dipping reflectors have a magmatic origin.
High temperatures possible due to thick sedimentary column.	High geothermal gradient in deep waters caused by extreme crustal thinning.
Synrift and prerift source rocks are in later stage of maturation (senile).	Reduced sedimentary column and low temperatures prevent maturation of postrift source rocks.
Rift-phase exploratory plays are gas prone.	Postrift exploratory plays are oil prone.
Deep-water turbidites are deposited above rift or prerift depocenters.	Ultra–deep water turbidites are deposited on volcanic crust.

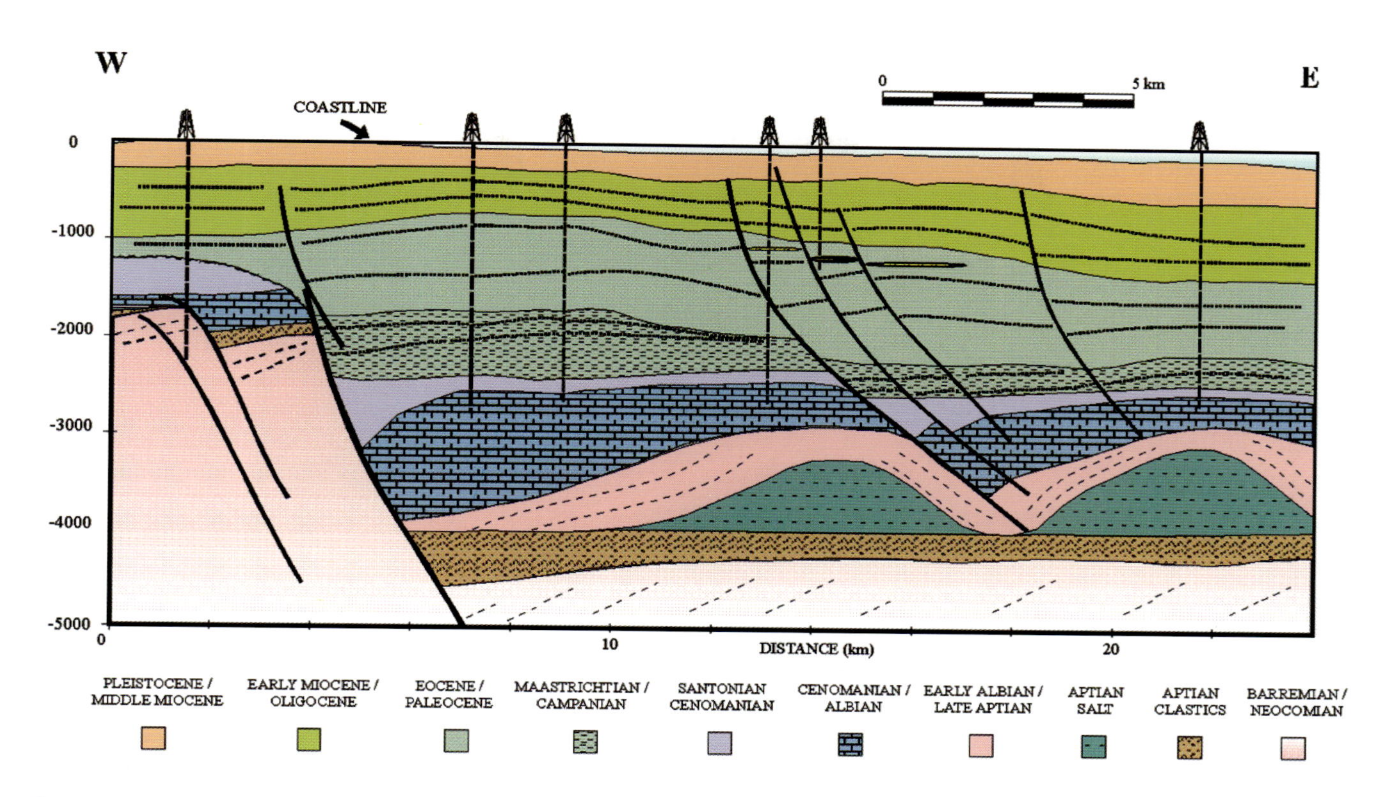

Figure 17—Regional schematic geoseismic section extending from onshore toward the platform. The Vaza Barris crustal fault near the coastline controls the Mosqueiro Low depocenter on the platform, and salt tectonics are important in creating major structures in the deeper basin. Tilted rift blocks (Aptian–Neocomian) occur deeper in the section, and the basement in the downthrown blocks may have offsets exceeding 3000 m. Lower Tertiary turbidites may occur in local lows in small roll-over structures associated with salt tectonics.

Sergipe subbasin. It illustrates the rift structures from the onshore region, characterized by shallow basement, toward the slope, where major rift faults result in large offsets for the downthrown blocks. Onshore, borehole A drilled through several stratigraphic sequences and reached the rift strata underlying the breakup unconformity (Figure 19). The geochemical analysis of the source rock intervals, drilled onshore by borehole A, is presented in Figure 20. There are rich intervals both above and below the evaporitic sequence. A small hydrocarbon accumulation was discovered in the offshore region by borehole B. This accumulation was probably generated by Aptian source rock pods located basinward of the shelf edge.

PETROLEUM SYSTEMS

Characterization of the major petroleum systems in the Sergipe–Alagoas Basin in northeastern Brazil was undertaken using a multidisciplinary approach that integrated data from several decades of exploratory activity.

In the Alagoas subbasin, most of the hydrocarbon accumulations occur in the synrift sequence, and this petroleum system is designated the Barra de Itiúba–Barra de Itiúba/Maceió(!). In the Sergipe subbasin, source rocks from the evaporite shales of the Ibura Member may source reservoirs in the Muribeca Formation, and this petroleum system is called the Ibura–Muribeca(!). The evaporitic shales may also source reservoirs in the Maceió and Calumbi Formations, and this system is designated the Ibura–Maceió/Calumbi(!). Source rocks from the Ibura Member and Maceio Formation (both below and above the salt) may feed turbidite reservoirs of the Calumbir Formation, and this is designated the Ibura/Maceió–Calumbi(!) petroleum system.

Figures 21 and 22 show the petroleum system events charts for two systems. The protomarine hypersaline source rocks of the Ibura Member of the Muribeca Formation (Figure 7) charged the Carmópolis oil field located onshore. The marine megasequence includes several offshore discoveries in Upper Cretaceous–lower Tertiary turbidites, which are associated with the Ibura/Maceió–Calumbi(!) petroleum system.

Hydrocarbons in the Carmópolis field (Figure 16) occur predominantly in coarse clastics of Lower Cretaceous alluvial fan and fan deltas and in fractured Precambrian basement. Aptian evaporites and Albian marine shales, marls, and limestones provide seals. Figure 21 shows the events chart for presalt plays in the Ibura–Muribeca(!) petroleum system. The importance of this type of play in the South Atlantic is demon-

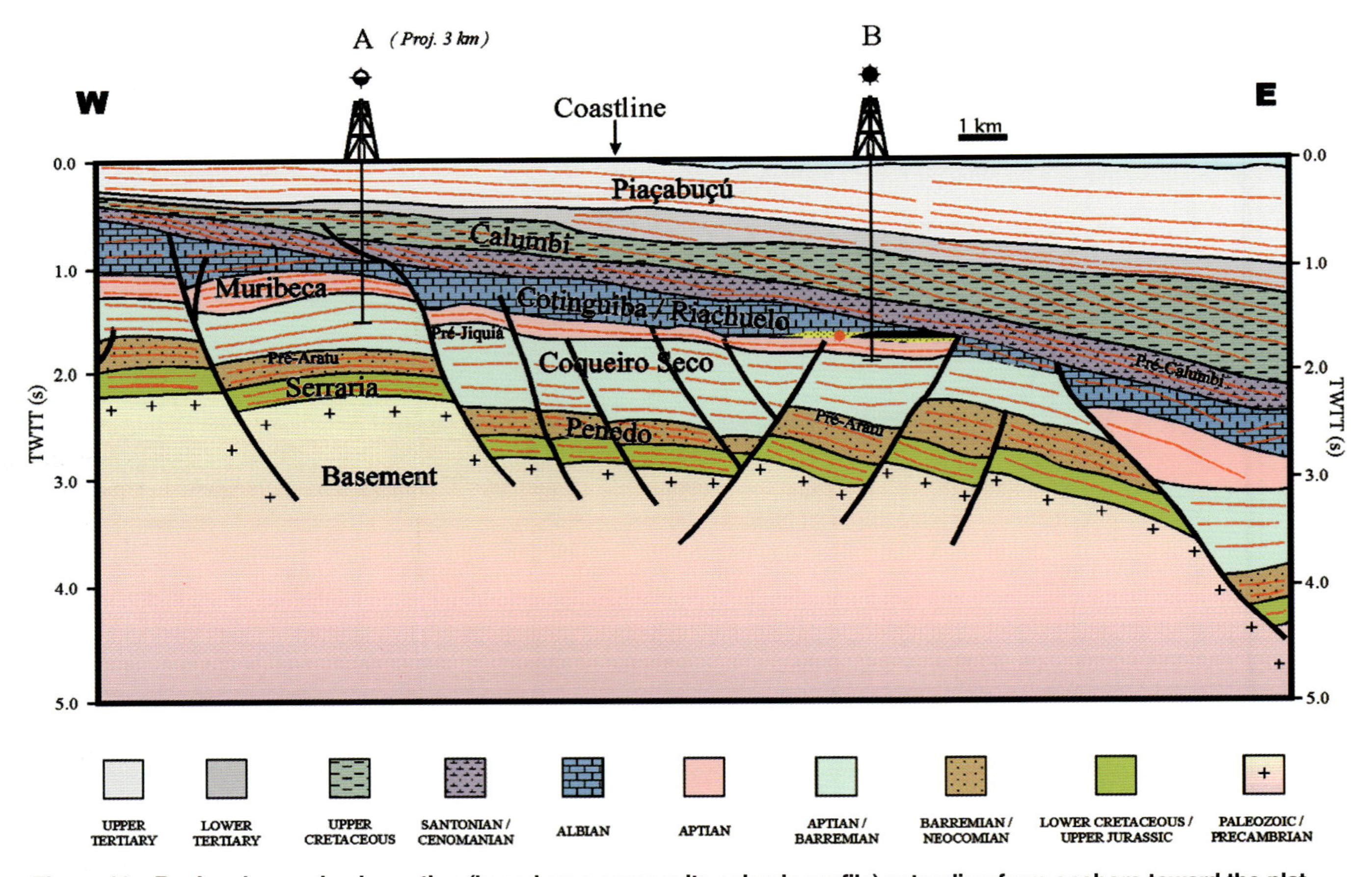

Figure 18—Regional geoseismic section (based on a composite seismic profile) extending from onshore toward the platform of the Sergipe Basin, showing the rift architecture controlled by a number of synthetic and antithetic faults in shallow waters and by a major synthetic (seaward-dipping) fault near the eastern end of the profile. Borehole A, located onshore, was drilled through the presalt formations. A small hydrocarbon accumulation is located in a rift block near borehole B. Modified from Chagas et al. (1993).

strated by the fact that more than 50% of the production of the Sergipe subbasin comes from this field alone. Its analog in west Africa (Rabi–Kunga oil field) is responsible for two-thirds of the oil production in Gabon (World Oil, 1997).

The hydrocarbons in the marine megasequence, such as the Guaricema oil field and other recent discoveries in the Mosqueiro Low (Figure 17) were mainly accumulated in Upper Cretaceous–lower Tertiary deep-water turbidites. Seals are deep-water shales and marls deposited during transgressive and regressive episodes of the marine megasequence. Figure 22 shows the events chart for this postsalt hydrocarbon play in the Ibura/Maceió–Calumbi(!) petroleum system, with a critical moment in the early Tertiary.

CONCLUSIONS

This work addresses the interpretation of the crustal architecture, sedimentation, and petroleum systems of the Sergipe–Alagoas Basin. The integration of regional seismic profiles with gravity and magnetic data provides evidence that deep seismic reflectors are related to the crustal architecture of the margin rather than to sedimentary features. This has profound consequences for deep-water petroleum systems.

The stratigraphic evolution of the Sergipe–Alagoas Basin is characterized by several phases of subsidence, from Paleozoic to the present day. Prerift sedimentary layers (Paleozoic–Upper Jurassic sequences) crop out on land and form a thin veneer above the Precambrian basement in the subsurface. Synrift strata (Neocomian–Barremian sequences) are mainly composed of siliciclastic rocks deposited in fluvial–lacustrine paleoenvironments. First, marine incursions in the basin resulted in local deposition of Aptian evaporites, which are separated from rift-phase sequences by a major erosional event corresponding to the breakup unconformity. Thermal phase sedimentary rocks include Albian–Cenomanian limestones and Turonian–Recent siliciclastics. Mohriak et al. (1998) have pointed out that the Sergipe Basin stratigraphic chart (Figure 7) should indicate that the oceanic crust region is characterized by

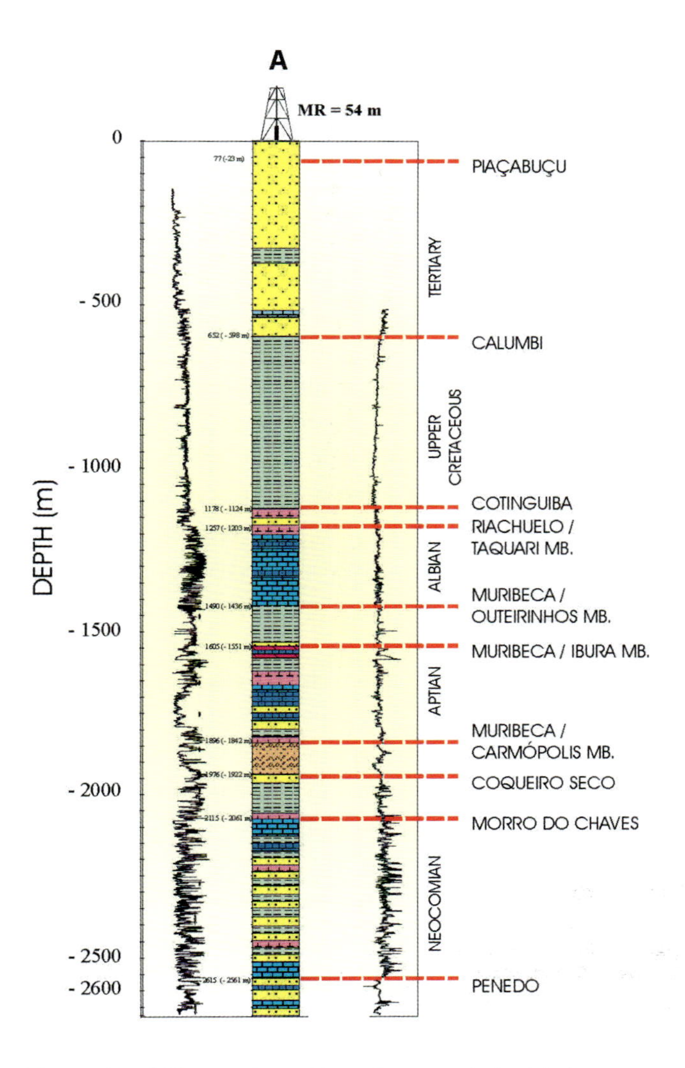

Figure 19—Lithostratigraphic column and electric log response of the stratigraphic sequences drilled by borehole A (Figure 18). The evaporite layers (Ibura Member of the Muribeca Formation) occur sandwiched between shale sequences of Aptian age. The Neocomian sequence is characterized by continental sand-rich siliciclastics below the carbonates (including coquinas) of the Morro do Chaves Formation.

several volcanic intrusions that affect the sedimentary layers up to the Tertiary.

Previous interpretations of the Brazilian northeastern margin assumed that this segment of the South Atlantic corresponds to a nonvolcanic margin. This was based on the absence of basalts at the base of the rift sequence, which are common in the volcanic margins of southeastern Brazil. However, deep seismic data suggest that the deep-water extensions of these rifts are characterized by divergent and sigmoidal reflectors. Rather than corresponding to sediment progradation or tilted rift blocks controlled by antithetic faults, these features may actually correspond to thick wedges of magmatic seaward-dipping reflectors, which are typical of volcanic margin basins.

The seismic profiles imaged several characteristic structures along the transition from continental to oceanic crust, including multiple wedges of seaward-dipping reflectors, intracrustal landward-dipping reflectors, and volcanic plugs along fracture zones. These magmatic features are associated with extensional processes and oceanic crust formation, and they apparently postdate the rift-phase lithospheric extension associated with the breakup of Gondwana in the Early Cretaceous. The extensional processes active during rifting resulted in crustal normal faults that apparently detach near the seismic Moho in the deep-water region. These faults control and rotate synrift sedimentary units along the margin. The seaward portion of the rift seems to have been uplifted and highly eroded during a postbreakup tectonic event. It is inferred that in the Sergipe subbasin, the seaward-dipping reflector wedges onlap both Precambrian crust and remnants of eroded rift blocks along the boundary with the proto-oceanic crust. The wedges mark the transition to pure oceanic crust and are locally associated with volcanic intrusions. Several volcanic plugs in deep-water region are aligned with oceanic fracture zones that apparently penetrate down through the whole crust and reach the upper mantle.

However, there are some diapiric structures located near the boundary between continental and oceanic crust that suggest deep-water salt tectonics. This interpretation is also based on a megaregional approach, involving integration of geologic and geophysical data (gravity, magnetic, and seismic). One of these structures is located within a regional Bouguer anomaly low, and magnetic data does not show any large anomaly. The small size of the structure and the regional spacing of the magnetic grid may preclude a definitive interpretation. Structural and stratigraphic features associated with salt diapirism conform with analogs from the Gulf of Mexico, North Sea, and eastern North American margin (Mohriak, 1995a, b). This has important consequences for petroleum exploration because salt tectonics plays a very important role in controlling sedimentation and creating stratigraphic and structural traps.

The Sergipe–Alagoas Basin is characterized by a series of oil accumulations related to different source rocks, including shales of the rift and transitional phases, and shales and marls of the transgressive marine megasequence. The source rocks of the rift sequence were deposited in Neocomian–Barremian time and are responsible for most of the hydrocarbon accumulations in the Alagoas subbasin. This petroleum system is the Barra de Itiúba–Barra de Itiúba/Maceió(!).

The Ibura/Muribeca(!) petroleum system is exemplified by the Carmópolis field. This play is characterized by presalt and postsalt reservoirs located on local highs formed during the rift phase. The reservoirs are mainly sourced by the black shales of the Ibura Member (Muribeca Formation), which were deposited during the transitional phase. Geochemical data indicate that most

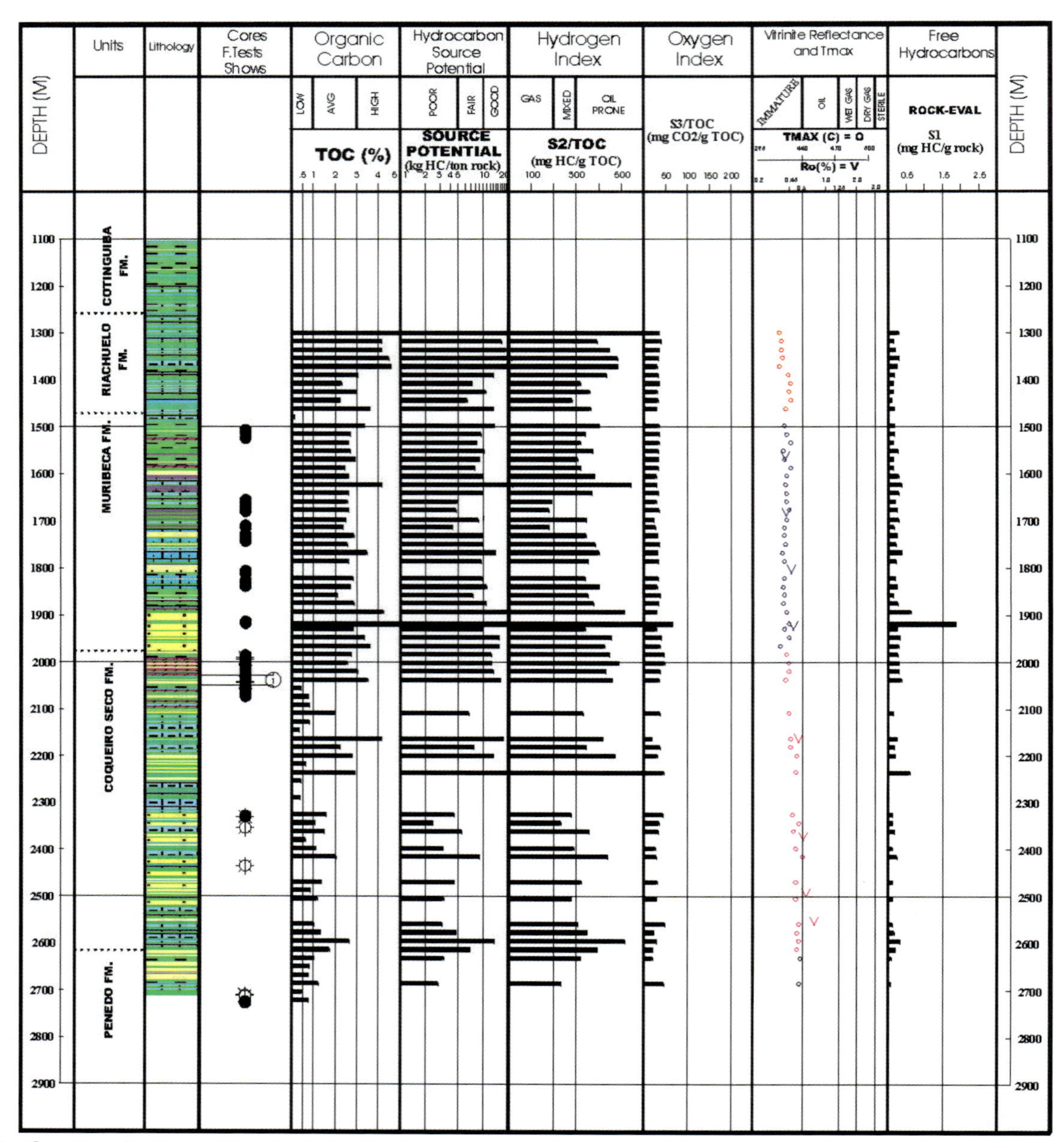

Figure 20—Geochemical log of borehole A, showing organic-rich layers of Albian–Neocomian age. Mature source rocks occur below 2400 m as indicated by vitrinite reflectance indices above 0. 6% R_o. The Aptian source rocks (immature in the borehole) have TOC contents of about 4%.

of the oil in the Sergipe subbasin was sourced by the Aptian transitional (evaporitic) shales.

The Ibura/Maceió–Calumbi(!) petroleum system is exemplified by the Guaricema oil field and other similar discoveries in lower Tertiary Calumbi Formation turbidite sandstones. The structure of this system is characterized by salt tectonic control of listric normal faults, which were efficient in capturing sands and in creating anticlinal structures involving Upper Cretaceous–lower Tertiary rocks.

Mapping the geographic extent of the petroleum systems in the Sergipe subbasin emphasizes the association of the oil fields with offshore pods of active Aptian source rocks in regional lows of the basin. The integration of these data with geochemical modeling allowed the prediction and characterization, in time and space, of the petroleum pathways from source to trap in the Sergipe–Alagoas Basin. Several plays were identified in deep-water frontier regions, including tilted rift blocks, deep-water turbidite mounds, and possibly some structures related to salt tectonics. The regional mapping and petroleum systems interpretation indicates that there is still potential for future discoveries in the platform and deep-water regions of the basin.

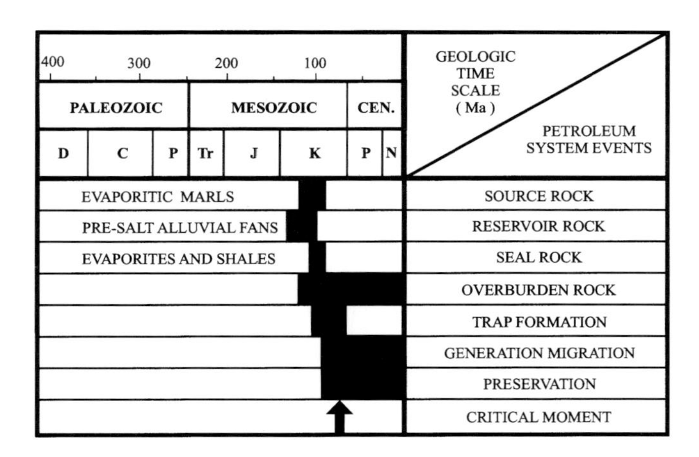

Figure 21—The Ibura–Muribeca(!) petroleum system events chart for the presalt play, Sergipe subbasin.

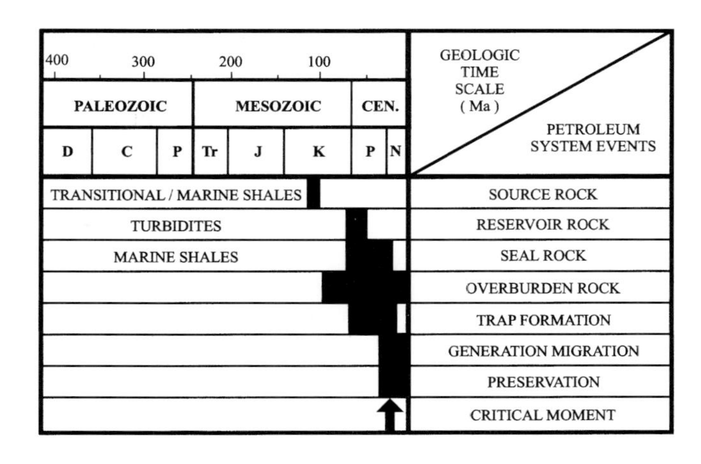

Figure 22—The Ibura/Maceió–Calumbi(!) petroleum system events chart for the postsalt play, Sergipe subbasin.

However, several large turbidite mounds that might constitute exploratory targets are possibly located on oceanic crust, reducing their exploratory potential. This does not mean that these regions cannot contain hydrocarbon accumulations, particularly because there is evidence that, in some South Atlantic basins, the Upper Cretaceous–lower Tertiary marine sequence may be a viable source rock for the deep-water accumulations recently discovered in the west African margin (World Oil, 1997). It simply means that the explorationist must cautiously analyze all alternative interpretations (as done in Table 2). To properly assess petroleum systems, it is vital to analyze the sedimentary fill of a basin, as well as its deeper crustal architecture, by multidisciplinary integration of geologic, geophysical, and geochemical tools. This approach will optimize the petroleum exploration process by reducing the risk involved in the evaluation of ultra–deep water prospects, which risk very large investments and therefore necessitate a significant premium to result in an economic discovery.

Acknowledgments*—This study has greatly benefited from many discussions with several explorationists from the Petrobrás E & P Head Office and Petrobrás operational unities in Sergipe and Bahia. Pedro V. Zalán critically read the first draft of this manuscript and contributed several suggestions. We gratefully acknowledge Petrobrás management at E & P for constant support throughout this work. We thank S. R. Michelucci, Manager of the New Frontiers Interpretation, for his constant cooperation and incentive. We also thank L. R. Guardado, Petrobrás Exploration Manager, and C. F. Lucchesi, E & P Chief Executive Officer, for permission to publish. Finally, we thank B. Katz for his cooperation and helpfulness, and the referees for critically reviewing the manuscript.*

REFERENCES CITED

Araújo, C. V., E. S. T. Frota, and G. P. Hamsi, Jr., 1994, Modelagem cinética dos intervalos potencialmente geradores da seção albo-turoniana da Bacia de Sergipe–Alagoas: Relatório Interno Petrobrás/Cenpes, 41 p.

Bacoccoli, G., coord., 1991, O futuro da exploração de petróleo na Bacia de Sergipe–Alagoas: Relatório Interno Petrobrás, Depex, Rio de Janeiro, 2 vols.

Bassetto, M., W. U. Mohriak, and I. S. Vieira, 1996, Modelagens gravimétrica e magnética utilizadas como ferramentas de apoio à análise regional de bacias sedimentares: Anais do 39° Congresso Brasileiro de Geologia, v. 5, p. 222–225.

Bisol, D. L., 1992, Atlas de seções sísmica das bacias sedimentares brasileiras. Vol. 1—Bacias maritimas da costa leste: Internal Report, Petrobrás, DEPEX/Cenpes, Rio de Janeiro.

Blundell, D. J., 1990, Relationships between deep crustal structure and sedimentary basins around Britain, *in* B. Pinet and C. Bois, eds., The potential of deep seismic profiling for hydrocarbon exploration: Paris, Editions Technip, p. 317–333.

Boillot, G., D. Mougenot, J. Girardeau, and E. L. Winterer, 1989, Rifting processes on the West Galicia margin, Spain, *in* A. J. Tankard and H. R. Balkwill, eds., Extensional tectonics and stratigraphy of the North Atlantic margins: AAPG Memoir 46, p. 363–377.

Bradley, C. H., and M. N. Fernandez, 1991. Early Cretaceous paleogeography of Gabon/northeastern Brazil—a tectonostratigraphic model based on propagating rifts, *in* R. Curnelle, ed., Géologie Africaine Mémoire 13, Elf Aquitaine, p. 17–30.

Brown, D., 1997, 3-D targets elusive Yegua sands: AAPG Explorer, May, p. 6–7.

Cainelli, C., 1987, Histórico e evidências da presença de regime transtensional/ transpressional na Bacia de Sergipe–Alagoas: Tectos I—Primeiro Seminário de tectonica da Petrobrás, Depex/Cenpes, Rio de Janeiro, p. 311–331.

Cainelli, C., 1992, Sequence stratigraphy, canyons, and gravity mass flow deposits in the Piaçabuçu Formation, Sergipe–Alagoas Basin, Brazil: Ph.D. dissertation, The University of Texas, Austin, Texas.
Castro, Jr., A. C. M., 1987, The northeastern Brazil and Gabon basins: a double rifting system associated with multiple crustal detachment surfaces: Tectonics, v. 6, p. 727–738.
Chagas, L. S., D. S. N. Chiossi, R. M. Cerqueira, G. P. Hamsi, Jr., G. Marques, and G. Lisboa, 1993, Evolução tectono-sedimentar do rift da Bacia de Sergipe–Alagoas e novas perspectivas explóratorias: Internal Report, Petrobrás, DEPEX/DENEST/SERINT, Rio de Janeiro, 76 p.
Chang, H. K., R. O. Kowsmann, A. M. F. Figueiredo, and A. Bender, 1992, Tectonics and stratigraphy of the East Brazil Rift System: an overview: Tectonophysics, v. 213, p. 97–138.
Conceição, J. C. J., P. V. Zalán, and S. Wolff, 1988, Mecanismo, evolução e cronologia do rift sul-atlântico: Boletim de Geociências da Petrobrás, v. 2, n. 4, p. 255–265.
Destro, N., 1994, Tectonism, stratigraphy, and sedimentation in the Sergipe and Alagoas Basins, NE Brazil: an overview: 14th International Sedimentological Congress, Recife, Brazil.
Feijó, F. J., 1994, Bacias de Sergipe e Alagoas: Boletim de Geociências da Petrobrás, Rio de Janeiro, v. 8, n. 1, p. 149–161.
Figueiredo, A. M. F., and W. U. Mohriak, 1984, A tectônica salífera e as acumulações de petróleo na Bacia de Campos: 33rd Congresso Brasileiro de Geologia, Rio de Janeiro, Sociedade Brasileira de Geologia, v. 3., p. 1380–1384.
Gladczenko, T. P., K. Hinz, O. Eldholm, H. Meyer, S. Neben, and J. Skogseid, 1997, South Atlantic volcanic margins: Journal of the Geological Society of London, v. 154, p. 465–470.
Guardado, L. R., L. A. P. Gamboa, and C. F. Luchesi, 1989, Petroleum geology of the Campos Basin, a model for producing an Atlantic-type basin, *in* J. D. Edwards and P. A. Santogrossi, eds., Divergent/passive margin basins: AAPG Memoir 48, p. 3–79.
Guimarães, P. T. M., 1988, Basin analysis and structural development of the Sergipe–Alagoas Basin, Brazil: Ph.D. dissertation, The University of Texas, Austin, Texas, 171 p.
Hinz, K., 1981, A hypothesis on terrestrial catastrophes: wedges of very thick oceanward dipping layers beneath passive continental margins: Geologisches Jahrbuch, v. E-22, p. 3–28.
Holbrook, W. S., G. M. Purdy, R. E. Sheridan, L. Glover III, M. Talwani, J. Ewing, and D. Hutchinson, 1994, Seismic structure of the U.S. Mid-Atlantic continental margin: Journal of Geophysical Research, v. 99, n. B9, p. 17,871–17,891.
Keen, C. E., and B. Voogd, 1988, The continental–ocean boundary at the rifted margin off eastern Canada: new results from deep seismic reflection studies: Tectonics, v. 7, p. 107–124.
Keen, C. E., B. C. MacLean, and W. A. Kay, 1991, A deep seismic reflection profile across the Nova Scotia continental margin, offshore eastern Canada: Canadian Journal of Earth Sciences, v. 28, p. 1112–1120.
Keen, C. E., P. Potter, and S. P. Srivastva, 1994, Deep seismic reflection data across the conjugate margins of the Labrador Sea: Canadian Journal of Earth Sciences, v. 31. p. 192–205.
Klemperer, S. L., T. A. Hauge, E. C. Hauser, J. E. Oliver, and C. J. Potter, 1986, The Moho in the northern Basin and Range province, Nevada, along COCORP 40°N seismic-reflection transect: GSA Bulletin, v. 97, p. 603–618.
Koutsoukos, E. A. M., Mello, M. R., Azambuja Fillho, N. C., Hart, M. B. and J. R., Maxwell, 1991, The upper Aptian–Albian succession of the Sergipe Basin, Brazil: an integrated paleoenvironmental assessment: AAPG Bulletin, v. 75, n. 3, 479–498.
Lana, M. C., 1993, Potencial petrolífero e exploração em águas profundas na Bacia de Sergipe–Alagoas: Relatório Interno Petrobrás, DEPEX/DINORD/SESEA, 36 p.
Larsen, H. C., and S. Jakobsdottir, 1988, Distribution, crustal properties and significance of seawards-dipping sub-basement reflectors off E Greenland, *in* A. C. Morton and L. M. Parson, eds., Early Tertiary volcanism and the opening of the NE Atlantic: Geological Society of London Special Publication 39, p. 95–114.
LASE Study Group, 1986, Deep structure of the US East Coast passive margin from large aperture seismic experiments (LASE): Marine and Petroleum Geology, v. 3, p. 234–242.
Lindrat, E., M. P. P. B. Hora, and R. A. V. Moraes, 1989, Resultados dos levantamentos aeromagnetométricos "offshore" no litoral brasileiro entre Salvador e João Pessoa: Anais do 1 º Congresso da Sociedade Brasileira de Geofísica, Rio de Janeiro, p. 626–631.
Matos, R. M. D., 1992, The northeastern Brazilian rift system: Tectonics, v. 11, n. 4, p. 766–791.
McBride, J. H., R. S. White, T. J. Henstock, and R. W. Hobbs, 1994, Complex structure along a Mesozoic sea-floor spreading ridge: BIRPS deep seismic reflection, Cape Verde abyssal plain: Geophysical Journal International, v. 119, p. 453–478.
McCarthy, J., J. C. Mutter, J. L. Morton, N. H. Sleep, and G. A. Thompson, 1988, Relic magma chamber structures preserved within the Mesozoic North Atlantic crust?: GSA Bulletin, v. 100, p. 1423–1436.
McKenzie, D., 1978, Some remarks on the development of sedimentary basins: Earth and Planetary Science Letters, v. 40, p. 25–32.
Mello, M. R., P. C. Gaglianone, S. C. Brassel, and J. R. Maxwell, 1988, Geochemical and biological marker assessment of depositional environment using Brazilian "offshore" oils: Marine and Petroleum Geology, v. 5, p. 205–223.
Mello, M. R., W. U. Mohriak, E. A. M. Koutsoukos, and G. Bacoccoli, 1994, Selected petroleum systems in Brazil, *in* L. B. Magoon and W. G. Dow, eds., The petroleum system–from source to trap: AAPG Memoir 60, p. 499–512.
Mello, M. R., E. A. M. Koutsoukos, W. U. Mohriak, G. Bacoccoli, A. S. T. Netto, and F. T. T. Goncalves, 1996, The petroleum system concept in the Sergipe Basin, northeastern Brazil (abs.): AAPG International Conference and Exhibition, Caracas, Venezuela, AAPG Bulletin, v. 80, n. 8, p. 1314.
Meyers, J. B., and B. R. Rosendahl, 1991, Seismic reflection character of the Cameroon volcanic line: evidence for uplifted oceanic crust: Geology, v. 19, p. 1072–1076.
Meyers, J. B., B. R. Rosendahl, and J. A. Austin, Jr., 1996, Deep-penetrating MCS images of the South Gabon Basin: implications for rift tectonics and post-breakup salt remobilization: Basin Research, v. 8, p. 65–84.
Milani, E. J., and I. Davison, 1988, Basement control and transfer tectonics in the Recôncavo-Tucano-Jatobá rift, northeast Brazil: Tectonophysics, v. 154, p. 47–70.

Mjelde, R., M. A. Sellevoll, H. Shimamura, T. Iwasaki, and T. Kanazawa, 1993, Crustal structure beneath Lofoten, N. Norway, from vertical incidence and wide-angle seismic data: Geophysical Journal International, v. 114, p. 116–126.

Mohriak, W. U., 1995a, Elusive salt tectonics in the deep-water region of the Sergipe–Alagoas Basin: evidence from deep seismic reflection profiles: 4th International Congress of the Brazilian Geophysical Society, Rio de Janeiro, Expanded Abstracts, p. 51–54.

Mohriak, W. U., 1995b, Salt tectonics structural styles: contrasts and similarities between the South Atlantic and the Gulf of Mexico, *in* C. J. Travis, H. Harrison, M. R. Hudec, B. C. Vendeville, F. J. Peel, and B. E. Perkins, eds., Salt, sediment, and hydrocarbons: GCS-SEPM Foundation 16th Annual Research Conference, Houston, Texas, p. 177–191.

Mohriak, W. U., and M. A. L. Latgé, 1991, Deep seismic survey of Brazilian passive margin basins: the southeastern region: 2nd Congresso Brasileiro de Geofísica, Salvador, Boletim de Resumos Expandidos, p. 621–626.

Mohriak, W. U., and J. H. L. Rabelo, 1994, Sísmica profunda nas bacias marginais brasileiras: integração megaregional e resultados preliminares em Sergipe–Alagoas e Jacuípe: II Seminário de Interpretação Exploratória, Petrobrás–Departamento de Exploração, Rio de Janeiro, p. 246–251.

Mohriak, W. U., R. Hobbs, and J. F. Dewey, 1990, Basin-forming processes and the deep structure of the Campos Basin, offshore Brazil: Marine and Petroleum Geology, v. 7, n. 2, p. 94–122.

Mohriak, W. U., M. C. Barros, J. H. L. Rabelo, and R. D. Matos, 1993, Deep seismic survey of Brazilian passive basins: the northern and northeastern regions: Third International Congress of the Brazilian Geophysical Society, Rio de Janeiro, Expanded Abstracts, p. 1134–1139.

Mohriak, W. U., J. H. L. Rabelo, R. D. Matos, and M. C. Barros, 1995a, Deep seismic reflection profiling of sedimentary basins offshore Brazil: geological objectives and preliminary results in the Sergipe Basin: Journal of Geodynamics, v. 20, p. 515–539.

Mohriak, W. U., M. Bassetto, and I. S. Vieira, 1995b, Deep seismic constraints on the crustal architecture of sedimentary basins in the Brazilian margin: tectonic and exploratory implications: Boletim de Resumos Expandidos, V Simpósio Nacional de Estudos Tectônicos–SNET-95, Gramado, p. 246–248.

Mohriak, W. U., M. Bassetto, and I. S. Vieira, 1996, Estrutura crustal, tectônica e sedimentação na Bacia de Sergipe–Alagoas: Anais do 39th Congresso Brasileiro de Geologia, Sociedade Brasileira de Geologia, Salvador, v. 5, p. 435–439.

Mohriak, W. U., M. Bassetto, and I. S. Vieira, 1997, Tectonic evolution of the rift basins in the northeastern Brazilian region: 5th International Congress of the Brazilian Geophysical Society, Expanded Abstracts, v. 1, p. 4–7.

Mohriak, W. U., M. Bassetto, and I. S. Vieira, 1998, Crustal architecture and tectonic evolution of the Sergipe–Alagoas and Jacuípe Basins, offshore northeastern Brazil: Tectonophysics, v. 288, p. 199–220.

Moraes, R. A. V., and E. Liandrat, 1987, Interpretação do embasamento magnético–Projeto Maragoji/Canavieiras: Tectos–I, Primeiro Seminário de Tectônica da Petrobrás, Depex, Rio de Janeiro, p. 344–356.

Mutter, J. C., 1985, Seaward dipping reflectors and the continent-ocean boundary at passive continental margins: Tectonophysics, v. 114, p. 117–131.

Mutter, J. C., M. Talwani, and P. L. Stoffa, 1982, Origin of seaward-dipping reflectors in oceanic crust off the Norwegian margin by "subaerial sea-floor spreading": Geology, v. 10, p. 353–357.

Ojeda, H. A. O., 1982, Structural framework, stratigraphy, and evolution of Brazilian marginal basins: AAPG Bulletin, v. 66, p. 732–749.

Pinet, B., and C. Bois, 1990, The potential of deep seismic profiling for hydrocarbon exploration: IFP Exploration and Production Research Conference, Paris, Editions Technip.

Ponte, F. C., J. R. Fonseca, and A. V. Carozzi, 1980. Petroleum habitats in the Mesozoic–Cenozoic of the continental margin of Brazil, *in* D. A. Miall, ed., Facts and principles of world petroleum occurrence: Canadian Society of Petroleum Geologists Memoir 6, p. 857–886.

Pontes, C. E. S., F. C. C. Castro, J. J. G. Rodrigues, R. R. P. Alves, R. T. Castellani, S. F. Santos, and M. B. Monis, 1991, Reconhecimento tectônico e estratigráfico da Bacia Sergipe–Alagoas em águas profundas: Congresso Brasileiro de Geofísica, Salvador, Boletim de Resumos Expandidos, p. 638–643.

Rabinowitz, P. D., and J. LaBreque, 1979, The Mesozoic South Atlantic Ocean and evolution of its continental margins: Journal of Geophysical Research, v. 84, n. B11, p. 5973–6002.

Reston, T. J., C. M. Krawczyk, and D. Klaeschen, 1996, The S reflector west of Galicia (Spain): evidence from prestack depth migration for detachment faulting during continental breakup: Journal of Geophysical Research, v. 101, n. B4, p. 8075–8091.

Rosendahl, B. R., H. Groschel-Becker, J. Meyers, and K. Kaczmarick, 1991, Deep seismic reflection study of a passive margin, southeastern Gulf of Guinea: Geology, v. 19, p. 291–295.

Sheridan, R. E., 1975, Dome structure, Atlantic outer continental shelf east of Delaware: preliminary geophysical report: AAPG Bulletin, v. 59, p. 1203–1211.

Sheridan, R. E., D. L. Musser, L. Glover, M. Talwani, J. I. Ewing, W. S. Holbrook, G. M. Purdy, R. Hawman, and S. Smithson, 1993, Deep seismic reflection data of EDGE U.S. mid-Atlantic continental margin experiment: implications for Appalachian sutures and Mesozoic rifting and magmatic underplating: Geology, v. 21, p. 563–567.

Tankard, A. J., H. J. Welsink, and W. A. M. Jenkins, 1989, Structural styles and stratigraphy of the Jeanne d'Arc Basin, Grand Banks of Newfoundland, *in* A. J. Tankard and H. R. Balkwill, eds., Extensional tectonics and stratigraphy of the North Atlantic margins: AAPG Memoir 46, p. 65–282

Tate, M. P., 1993, Structural framework and tectono-stratigraphic evolution of the Porcupine Seabight Basin, offshore western Ireland: Marine and Petroleum Geology, v. 10, n. 2, p. 95–123.

Teisserenc, P., and J. Villermin, 1989, Sedimentary basin of Gabon: geology and oil systems, *in* J. D. Edwards and P. A. Santogrossi, eds., Divergent/passive margin basins: AAPG Memoir 48, p. 117–199.

Turner, J. P., 1995, Gravity-driven structures and rift basin evolution: Rio Muni Basin, offshore equatorial west Africa: AAPG Bulletin, v. 79, n. 8, p. 1138–1158.

Uesugui, N., 1987, Posição estratigráfica dos evaporitos da Bacia de Sergipe–Alagoas: Revista Brasileira de Geociências, v. 17, n. 2, p. 131–134.

Ussami, N., G. D. Karner, and M. H. P. Bott, 1986, Crustal detachment during South Atlantic rifting and formation of Tucano-Gabon basin system: Nature, v. 322, p. 629–632.

Wannesson, J., J. C. Icart, and J. Ravart, 1991, Structure and evolution of adjoining segments of the west African margin determined from deep seismic profiling, *in* R. Meissner et al., eds., Continental lithosphere: deep seismic reflections: American Geophysical Union, Washington, D.C., p. 275–289.

Wernicke, B., 1985, Uniform-sense normal simple shear of the continental lithosphere: Canadian Journal of Earth Sciences, v. 22, p. 108–125.

World Oil, 1997, World Oil, August, p. 123–125.

Karner, G. D., 2000, Rifts of the Campos and Santos Basins, southeastern Brazil: distribution and timing, *in* M. R. Mello and B. J. Katz, eds., Petroleum systems of South Atlantic margins: AAPG Memoir 73, p. 301–315.

Chapter 21

Rifts of the Campos and Santos Basins, Southeastern Brazil: Distribution and Timing

Garry D. Karner

Lamont-Doherty Earth Observatory of Columbia University
Palisades, New York, U.S.A.

Abstract

Satellite-derived gravity anomalies were used to map the location and distribution of rift subbasins comprising the Campos and Santos Basins of the southeastern Brazilian continental margin. Free-air and crustal Bouguer gravity anomalies define several features. A negative–positive gravity gradient along the southeastern Brazilian margin correlates generally with termination of oceanic fracture zones, the boundary of synrift evaporites, and an abrupt change in anomaly trends from east–west to margin-parallel. The gravity gradient thus defines the location of the ocean–continent boundary and suggests that much of the São Paulo Plateau is underlain by thinned continental crust. Second, a major offshore tectonic hinge zone, consisting of a series of short-segment, en echelon, high-standing blocks subparallel to the Brazilian margin demarcates the western limit of significant continental extension. The Badejo High of the Campos Basin is part of this hinge zone trend. The Serra do Mar and Serra da Mantiqueira mountains represent an onshore hinge zone. Third, a series of major rift subbasins exist seaward of both the Campos and Santos hinge zones. These have limited along-strike continuity, implying that synrift lake communication, water chemistry, and possibly source quality and preservation were restricted to each subbasin.

Extension between west Africa and Brazil occurred during the Early Cretaceous as a series of rift pulses that culminated in the initiation of sea floor spreading. The Congo–Cabinda margin of west Africa and the Camamu–Almada margin of eastern Brazil are characterized by a common tripartite rift history: Berriasian–Hauterivian, Hauterivian–middle Barremian, and late Barremian–early Aptian. Early depth-independent, broadly distributed, and increasingly focused brittle deformation (rift phases I and II) was replaced by depth-dependent deformation dominated by plastic thinning of the lower crust and lithospheric mantle (rift phase III). The nonmarine part of the Lagoa Feia Formation correlates with rift phase II while the "transitional" part is associated with rift phase III. The intervening pre-Alagoas unconformity is equivalent to the pre-Chela unconformity on the Congo margin. The possibility of a mid-crustal detachment beneath the Brazilian margin active during rift phase III has profound implications for the hydrocarbon maturation history of the margin. Seismic reflection data from the São João da Barra Low (Campos Basin) have helped define an early synrift depositional package that likely equates with rift phase I.

INTRODUCTION

The fundamental elements of petroleum systems—the distribution and timing of source, reservoir, seal, and trap formation—are consequences of the amplitude, depth-partitioning, style, and interaction of the tectonic events responsible for basin formation. During active continental extension, the resulting depositional packages and bounding surfaces are a complex function of the spatial and temporal distribution of rifting and the subsequent structural and sedimentologic interactions within and between rift systems. In particular, the structural interplay between multiple rift events can result in complex stratigraphic stacking patterns in response to changing sediment source regions (basement, prerift, or earlier synrift), physiography, drainage modification and exploitation of accommodation zones, and eustasy (e.g., Driscoll and Karner, 1998; Karner and Driscoll, 1999a, b). Successful application of the petroleum system concept requires that the underlying geologic and tectonic controls on basin architecture, source rock distribution, and reservoir quality be well understood and quantifiable.

Extension between west Africa and Brazil occurred during the Early Cretaceous as a series of rift pulses with varying duration that culminated in the initiation of sea floor spreading. Studies of the west African margin have revealed that rifting occurred in three discrete phases, each phase having its own particular deformation style: Berriasian–Hauterivian, Hauterivian–middle Barremian, and late Barremian–early Aptian (Karner et al., 1997). Each rift phase is recorded by an onlap surface followed by an overall regressive package representing the subsequent infilling of the basin; each phase resulted in the formation of deep, anoxic, lacustrine systems. Extension style changed dramatically as a function of space and time (Karner and Driscoll, 1999a). Early, depth-independent, broadly distributed brittle deformation was progressively replaced by depth-dependent deformation in which extension tended to focus in the lower crust and lithospheric mantle as breakup was approached. This latter form of extension was characterized by a general paucity of basement-involved normal faulting and regional subsidence across the entire region.

Depth-dependent extension late in the rifting history of a margin and the accompanying postrift "thermal-type" subsidence has been explained by Driscoll and Karner (1998) and Karner and Driscoll (1999a, b) as a consequence of extension partitioning across an intracrustal detachment that effectively thinned the lower crust and lithospheric mantle. The detachment geometry and how it merges with the Moho can be mapped using the style, distribution, and timing of accommodation generated across the margin. The fact that this style of deformation appears to be most dominant during the late stages of the rifting process suggests that it may be related to the upwelling of asthenospheric heat and the associated modification of lower crustal plasticity. The counterbalancing brittle deformation is focused in a narrow region that will eventually form the ocean–continent boundary.

Within the resolution of intra-Atlantic paleontologic correlations, the eastern Brazilian rift basins conjugate to the west African margin (e.g., Camamu–Almada) show the same tripartite rift history as the Congo and Cabinda margins (Bate, 1999). In contrast, the rift sequencing of the Campos and Santos Basins is not so well defined, at least based on published studies of these basins (e.g., Mohriak et al., 1989; Guardado et al., 1989; Abrahão and Warme, 1990). Most descriptions of the rift phase Lagoa Feia megasequence of the Campos Basin and the Guaratiba and Ariri megasequences of the Santos Basin give little insight into the timing and sequencing of rift events responsible for the formation of these basins. The problem has been exacerbated by not knowing the location of the ocean–continent boundary along the southeast Brazilian margin and the oceanic or continental nature of the São Paulo Plateau. The failure to unambiguously locate the ocean–continent boundary has resulted from two factors. First, the existence of a magnetic quiet zone limits the use of correlatable marine magnetic anomalies to define either the timing or location of the ocean–continent boundary and thus the nature of the São Paulo Plateau. Second, thick accumulations of postrift sediments, often disrupted by salt diapirism, tend to obscure the synrift structures of the margin.

The purpose of this chapter is threefold: (1) to outline the general rift architecture of the Campos and Santos Basin systems using free-air and crustal Bouguer gravity anomalies across the region; (2) to help define the location of the ocean–continent boundary and nature of the São Paulo Plateau; and (3) to investigate the stratigraphic development and rift structuring of the Campos Basin by analogy with the Congo and Cabinda margins of west Africa.

RIFT ARCHITECTURE OF THE CAMPOS AND SANTOS BASINS AND NATURE OF THE SÃO PAULO PLATEAU

Because it is dominated by near-field density contrasts, the satellite-derived free-air gravity has been particularly useful in defining the general bathymetry and fabric of both thinned continental and oceanic crust (e.g., Cande et al., 1988). Early opening trends of oceanic crust in magnetic quiet zones are well-imaged, allowing for improved plate tectonic reconstructions for the South Atlantic. These images also provide information on changing stress orientations at the time of breakup and on the oceanic affinity of deeply subsided crusts adjacent to the continental margin (e.g., Matos, 1992). Six thermal plumes have likely influenced oceanic crust chemistry and thickness during the Early Cretaceous. In addition to Tristan da Cunha, these hot spots are Trindade and Martin Vas, Saint Helena, Ascension, and Fernando de Noronha (Figure 1).

Figure 2 shows a free-air gravity anomaly image for the southeastern Brazilian margin region. Clearly identified are fracture zone trends, the "edge-effect" anomaly that characterizes the shelf break, and hot spot traces (e.g., Trindade and Tristan da Cunha). Of particular importance in Figure 2 is the location and termination of the fracture zones as they approach the Brazilian margin. North of about lat. 22°S, fracture zone definition is dramatically compromised over a broad region of presumably oceanic crust. Sediment burial is unlikely the explanation for this loss of gravity signature because regions with similar sedimentary thicknesses to the south of lat. 22°S show well-defined fracture zone trends (Kumar et al., 1979; Hinz et al., 1999). The crust north of lat. 22°S, characterized by broad positive free-air gravity anomalies, has been described by Hinz et al. (1999) (on the basis of wide-angle seismic reflection data) as Albian oceanic crust with a strong reflective irregular or smooth basement relief with crustal thicknesses of 8–9 km. Hinz et al. (1999) postulate that this thickened oceanic crust was influenced by periods of high magma production during oceanic crustal accretion within the Albian–Turonian spreading center related to the Tristan da

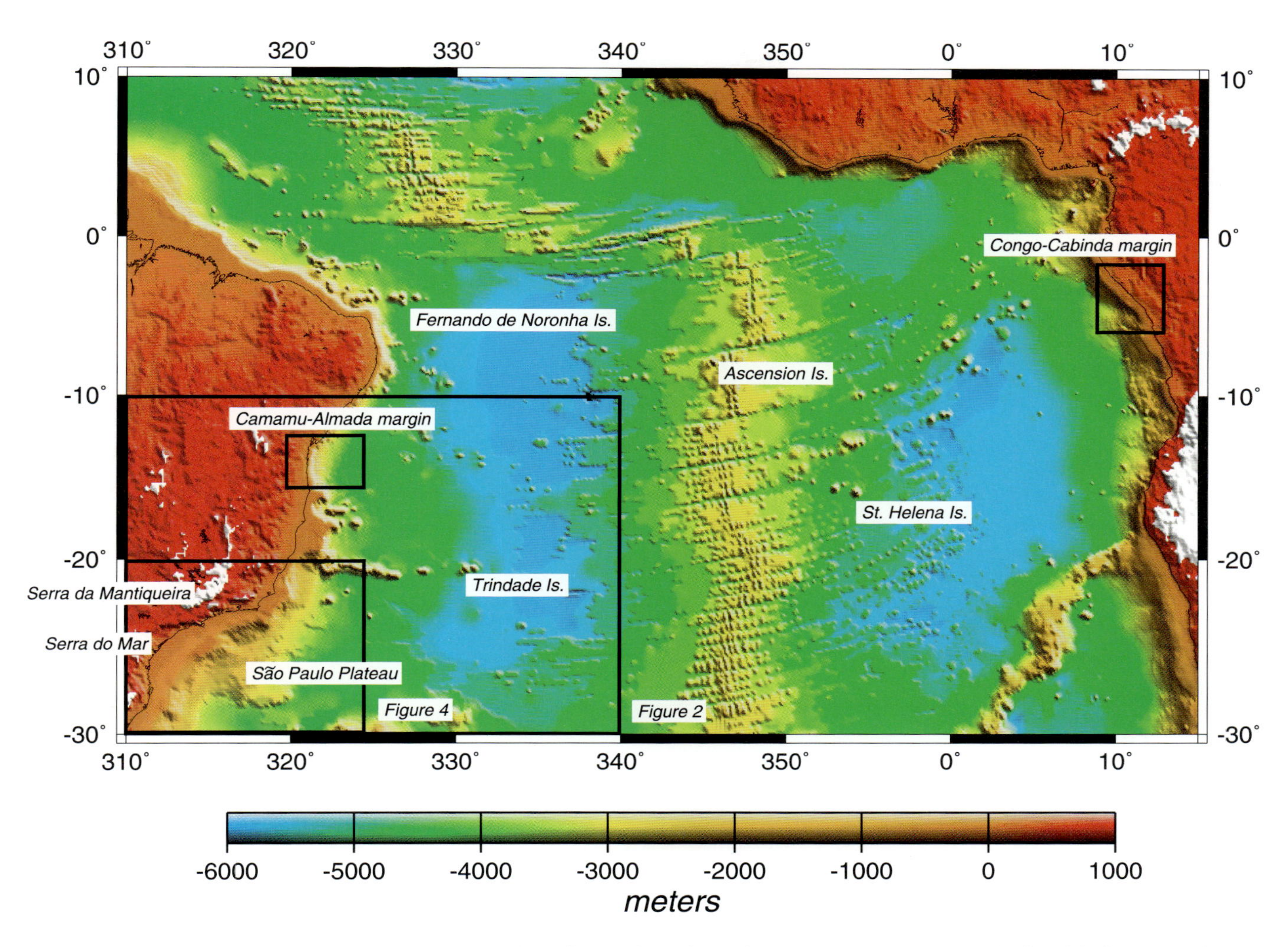

Figure 1—General bathymetry and topography of the South Atlantic region. Note the conjugate relationship between the Congo–Cabinda and the Camamu–Almada margins. The regions spanned by Figures 2 and 4 are shown, as are the locations of the onshore Serra do Mar and Serra da Mantiqueira mountains, the offshore São Paulo Plateau, and the major thermal plumes in the South Atlantic (Fernando de Noronha, Ascension, St. Helena, and Trindade).

Cunha plume. Irrespective of the details of crustal accretion, the fracture zone trends to the south of lat. 22°S can be traced a significant distance to the west, especially in the region immediately to the south of the São Paulo Plateau (Figure 3).

The existence of a magnetic quiet zone within the South Atlantic and thick accumulations of postrift sedimentary rocks across much of the offshore continental margin has made it difficult to unambiguously define the location of the ocean–continent boundary (e.g., Chang et al., 1992; Karner and Driscoll, 1999a). Offshore southeastern Brazil is no exception. The termination of oceanic fracture zones identified in the satellite gravity offers an important proxy to define the western limit of oceanic crust and thus the location of the ocean–continent boundary adjacent to the Santos and Campos Basins. Filtering the satellite gravity may help to highlight geologic features such as fracture zone and major trend changes. In similar tectonic positions on the west African margin (Karner et al., 1997) and northwest Australian margin (Driscoll and Karner, 1998), the crustal Bouguer gravity has proved useful in highlighting both the rift structure of the margin and the location of the ocean–continent boundary.

The crustal Bouguer gravity image in Figure 4 was calculated from satellite-derived free-air gravity anomalies. For the offshore southeastern Brazilian region, the crustal Bouguer gravity map was constructed by determining the Bouguer gravity anomaly of the region from the $2' \times 2'$ global free-air gravity grid of the world's oceans (Geosat) generated by Sandwell and Smith (1992) and the ETOP05 topographic grid using a sediment/water density contrast of 1170 kg/m^3. From the Bouguer gravity anomaly, a least-squares bicubic trend surface was

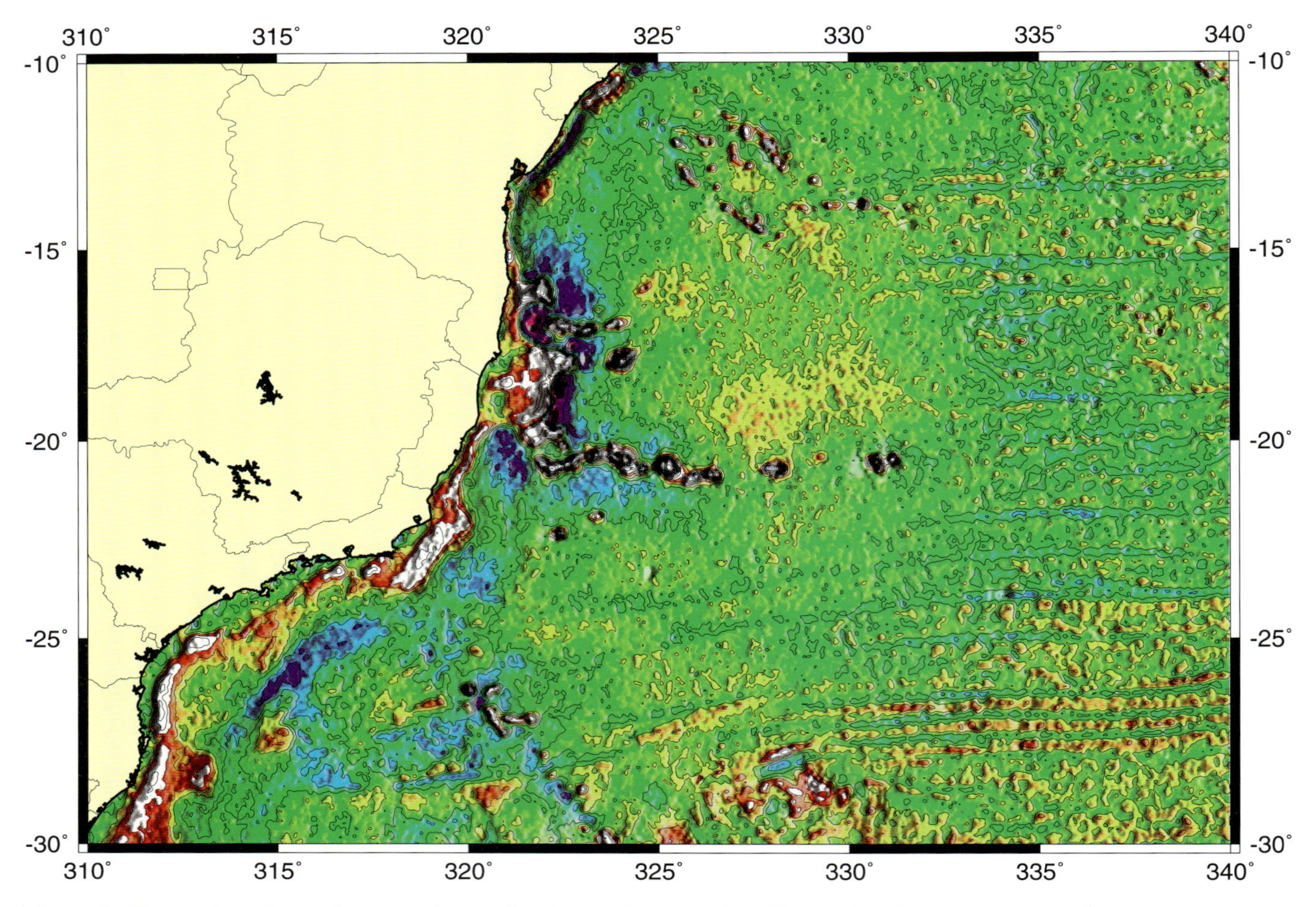

Figure 2—Geosat free-air gravity anomaly map for the southeastern Brazilian region. Clearly shown are fracture zone trends, the "edge-effect" anomaly that characterizes the shelf edge, and hot spot traces. Note the termination of the fracture zones as they approach the Brazilian margin, corresponding to the ocean–continent boundary. Much of the deep-water São Paulo Plateau is part of the oceanic crust. Contour interval is 10 mgal.

subtracted to produce the crustal Bouguer gravity map shown in Figure 4. Removing a long-wavelength gravity trend is, in effect, a form of isostatic correction but one that is based on the actual data distribution rather than on unconstrained density and flexural rigidity assumptions.

To assess the accuracy of the satellite gravity and thus its derived products, we have constructed a crustal Bouguer gravity contour map using all available Lamont-Doherty shipboard data across the study region (shown as white dots in Figure 5). The contouring and imaging were obtained using the *GMT* software package of Wessel and Smith (1995), a minimum curvature surface gridding, and the method outlined above to calculate the crustal Bouguer. As can be seen from Figure 5, the spatial resolution of the shipboard data rapidly decreases away from the main port of call, Rio de Janeiro. A comparison of the two gravity images, one derived from satellite data (Figure 4) and the other from shipboard gravity measurements (Figure 5), reveals that the shape, trend, and amplitude of the satellite gravity data are in good agreement with the shipboard gravity, giving us confidence in interpreting the satellite-derived crustal Bouguer gravity in terms of basin and crustal structure. More to the point, the satellite data, being a regionally consistent data set, contains significantly more reliable regional information than the shipboard data (compare Figures 4 and 5).

The broad regions of positive gravity anomalies seen on the crustal Bouguer image (Figure 4) that are associated with thickened oceanic crust tend to obscure the location of the fracture zones that are so obvious on the free-air gravity images (compare Figures 2 and 4). In an effort to highlight the short-wavelength components of the crustal Bouguer gravity, the gravity has been filtered further by using a high-pass filter that suppresses wavelengths of >400 km (i.e., anomalies wider than >200 km) and by illuminating the image from the north. In the resulting image of Figure 6, the shorter wavelengths associated with the oceanic fracture zones have been traced and the western terminations marked. The ocean–continent boundary (solid black line in Figure 6) should coin-

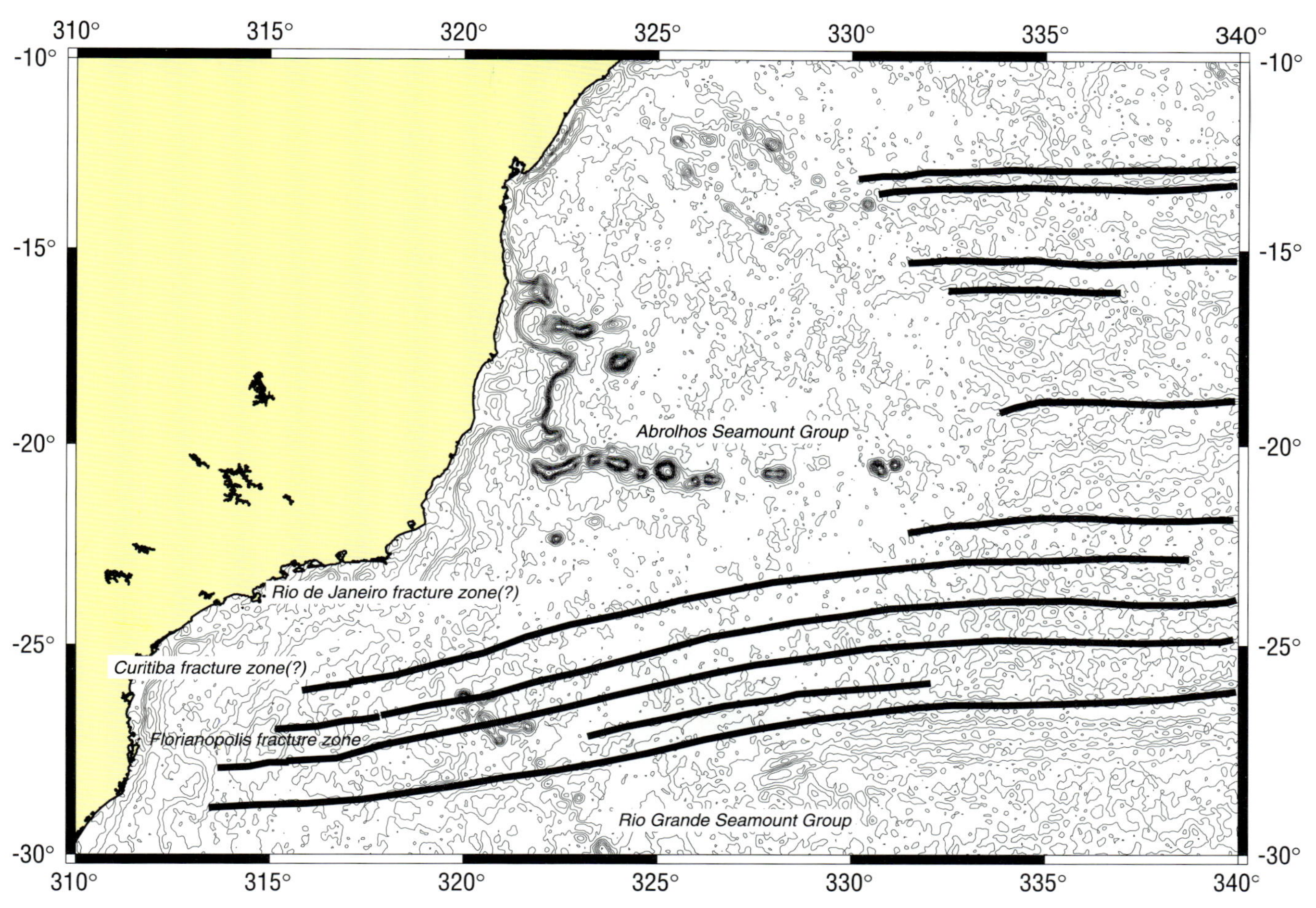

Figure 3—Fracture zone trends for the southeastern Brazilian region. Base map shows Geosat free-air gravity contours. The broad positive free-air gravity anomalies north of lat. 22°S correspond to thickened Albian oceanic crust. The Rio de Janeiro and Curitiba fracture zones are also shown. Fracture zone termination is used as a proxy to delineate the ocean–continent boundary.

cide with the westerly termination of oceanic fracture zone trends (e.g., Figure 3) and with areas characterized by an abrupt change from east–west to north–south (or margin-parallel) trending features. The location and locus of the predicted ocean–continent boundary is broadly aligned with a negative–positive gravity gradient in the eastern Campos Basin and along the eastern margin of the São Paulo Plateau.

In addition, the ocean–continent boundary approximately delineates the eastern limit of the synrift evaporites responsible for the diapiric structures observed in seismic reflection profiles across the São Paulo Plateau (e.g., Chang et al., 1992; Demercian et al., 1993). The gravity amplitude across the boundary ranges from 20 to 30 mgal (Figure 4). The source of the gravity dipole is in part a consequence of the juxtaposition of relatively thick, old oceanic crust adjacent to thinned continental crust. Thickened oceanic crust is a function of the decompressive melting of asthenospheric material and is enhanced by relatively hot mantle, as might be induced by the presence of plumes during the early stages of sea floor spreading (e.g., Gladczenko et al., 1997; Hinz et al., 1999). We conclude that the crust underlying the São Paulo Plateau is extended continental crust, consistent with conclusions from Kowsmann et al. (1982), and that the southern boundary is essentially a transform margin being controlled by the Curitiba and Florianopolis fracture zones (Macedo, 1990).

Because it highlights density contrasts within the crust, the crustal Bouguer gravity tends to accentuate the location of sedimentary basins and flexurally compensated features such as rift flanks (e.g., Campos and Santos hinge zones). The most prominent features in Figure 4 are as follows: (1) a negative–positive gravity gradient defining the ocean–continent boundary (discussed earlier), (2) positive anomalies corresponding to Aptian–Albian thickened oceanic crust and offshore crustal hinge zones, and (3) negative anomalies that correlate with the location of rift depocenters or subbasins within the Campos and Santos Basins and oceanic fracture zones. The gravity

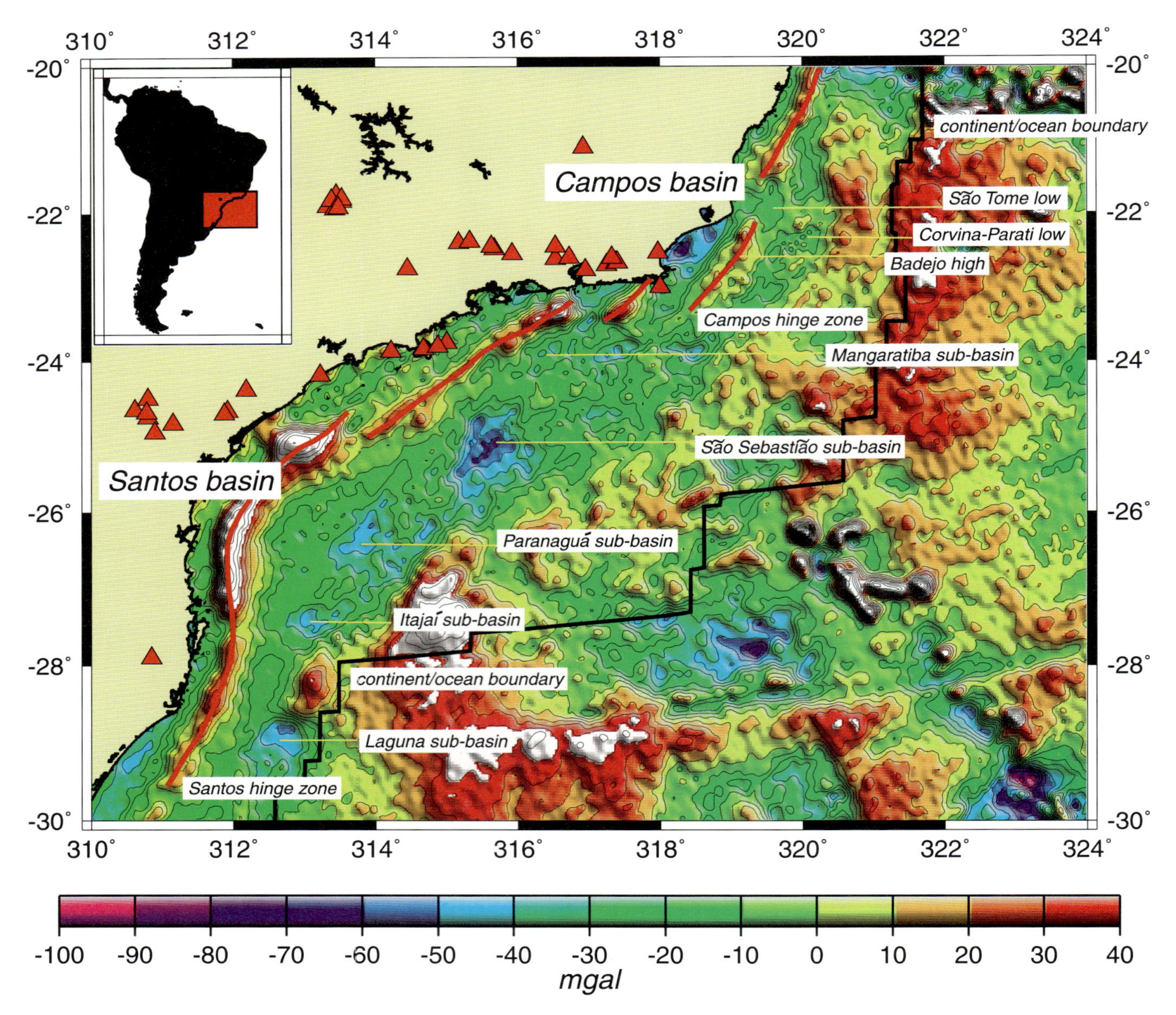

Figure 4—Satellite-derived crustal Bouguer gravity anomaly map of the southeastern Brazilian region. This anomaly accentuates sedimentary basins and flexurally supported features such as rift flanks. Prominent anomalies include a negative–positive gravity gradient generally across the ocean–continent boundary (black line); positive anomalies associated with Aptian–Albian thickened oceanic crust and the Santos and Campos hinge zones (red lines); and negative anomalies correlated with rift subbasins in the Campos and Santos Basins and oceanic fracture zones. The Laguna, Itajaí, Paranagua, São Sebastião, and Mangaratiba subbasins are predicted to have limited along-strike continuity. Neocomian–Eocene alkalic volcanic centers are shown (red triangles). Contour interval is 10 mgal.

anomalies indicate the approximate along-strike continuity of the various rift subbasins (named on Figure 4) and associated hinge zones that comprise the southeastern Brazilian margin (red lines in Figure 4).

A major tectonic hinge zone, the Campos and Santos hinge zones, trends subparallel to the Brazilian margin (Figure 4) and demarcates the western limit of significant continental extension. Hinge zones represent both a transition in major extension and a region of significant relief, the topography being a consequence of the mechanical unloading and flexural rebound of the lithosphere in response to extension (Weissel and Karner, 1989). The predicted rift flank topographies, asymmetric with their steepest face to the east, helped to create a time lag between the formation of the topography, the development of fluvial drainage networks, and the input of clastics into the evolving rift basins. This likely resulted in the formation of condensed sections in the basins away from

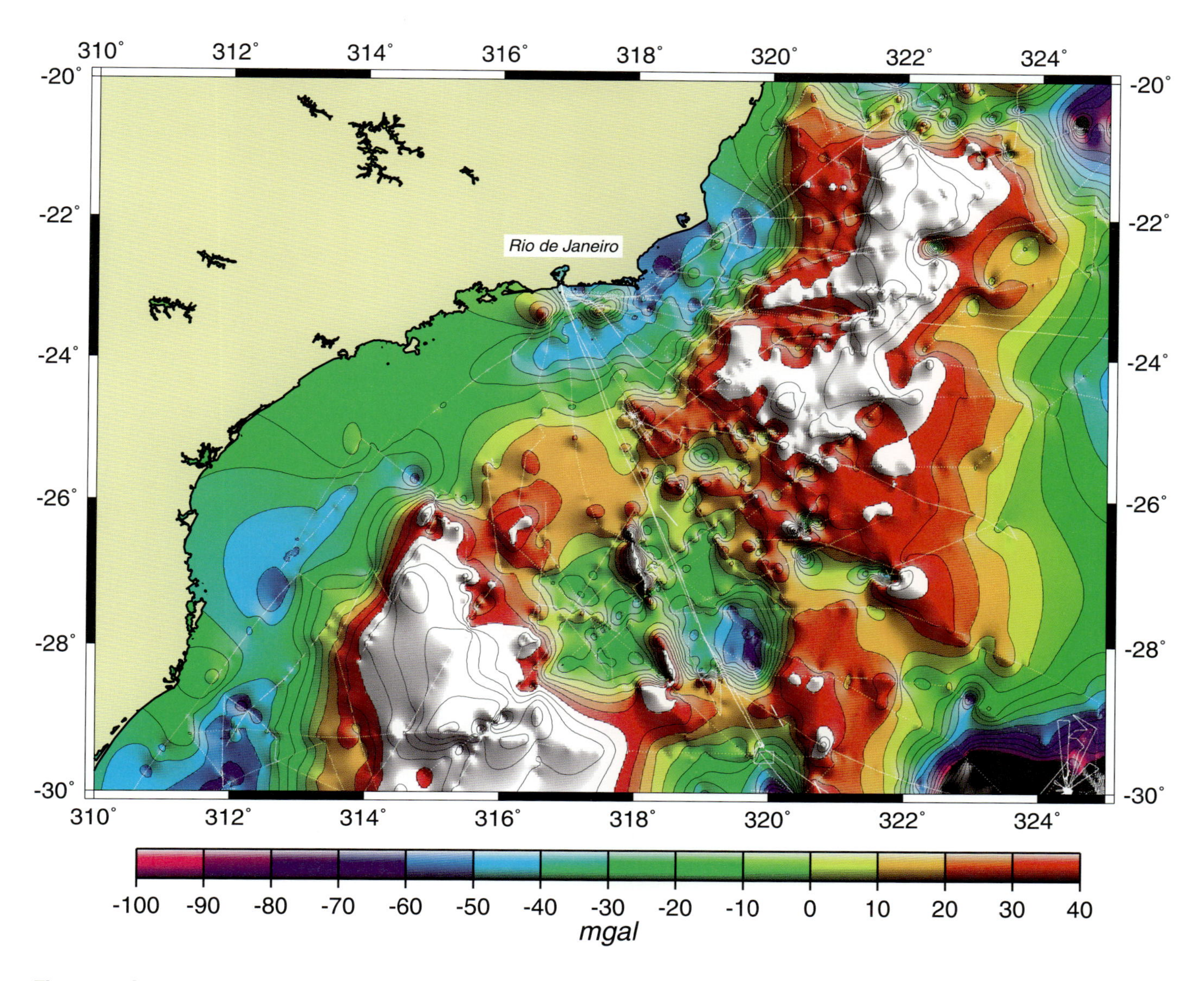

Figure 5—Shipboard-derived crustal Bouguer gravity anomaly map of the southeastern Brazilian region. These data allow an evaluation of the accuracy of the Geosat gravity data (Figure 2). Note the good agreement in anomaly amplitudes, wavelengths, and trends between the two maps in areas of adequate data coverage. This comparison gives confidence in interpreting the satellite-derived crustal Bouguer gravity in terms of basin and crustal structure. Contour interval is 10 mgal.

the accommodation zones, as was the case for west African basins (e.g., Karner et al., 1997). The Serra do Mar and Serra da Mantiqueira mountain systems along and adjacent to the coast (Figure 1) represent an onshore hinge zone, with minor extensional basins occurring between the mountain systems and the Campos and Santos hinge zones. A second offshore hinge zone may exist to the east of the Campos and Santos hinge zones, separating moderately extended crust within the Santos and western Campos Basins from highly extended continental crust across the São Paulo Plateau and eastern Campos Basin, respectively.

In detail, the offshore Campos and Santos hinge zones consist of a series of short-segment, en echelon, high-standing blocks. Sections of significant hinge zone disruption correlate approximately with the location of the Curitiba, Florianopolis, and Rio de Janeiro fracture zones and represent possible examples of where rift basin segmentation influenced subsequent spreading axis offset (Figure 4) (Macedo, 1990). Maximum disruption of the Campos and Santos hinge zones occurs across the Rio de Janeiro fracture zone, which is also the transition from the Santos to the Campos Basin. Here, the Campos offshore hinge zone jumps significantly to the east in a

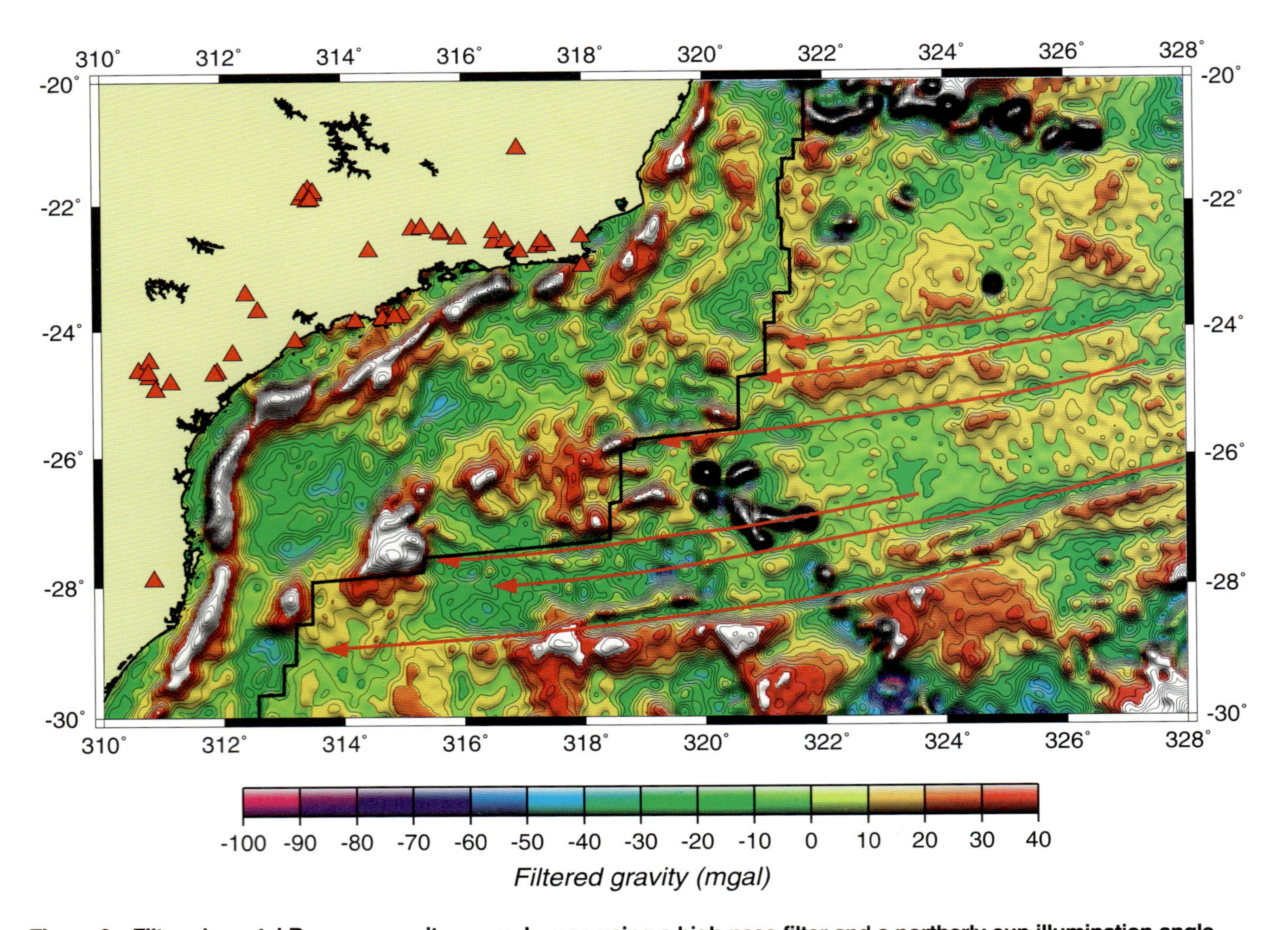

Figure 6—Filtered crustal Bouguer gravity anomaly map using a high-pass filter and a northerly sun illumination angle showing the location of major oceanic fracture zones and their approximate westerly terminations. The location of the ocean–continent boundary (black line) is based on the termination of the fracture zone trends and on abrupt changes from east–west to north–south trending features. The boundary coincides with a negative–positive gravity gradient and with the eastern limit of synrift evaporites as seen on seismic profiles across the São Paulo Plateau. Contour interval is 5 mgal.

series of steps, the easternmost of which represents a major basement feature within the Campos Basin called the Badejo High. The Badejo High was a local source of conglomerates and facilitated the development of coquinas and turbidites within the lacustrine phase of development in the Campos Basin rift system. However, the Badejo High ceased to be a controlling feature prior to deposition of the late synrift evaporites.

Seaward of the Campos and Santos hinge zones, individual gravity lows, interpreted to be rift subbasins, also show an en echelon arrangement but with limited along-strike continuity. This implies that along-strike basin connectivity, and hence lake communication, water chemistry, and possibly source quality and preservation, is spatially distinct to each subbasin system. The segmented character of the hinge zone likely controlled the spatial access of river drainage systems to the newly formed rift basins, which in turn would have influenced both source and reservoir rock quality. Source rock development in these rift subbasins should be time equivalent and correlate with the Marnes Noires, Bucomazi, Melania Shale, and Falcão source rocks of the west African margin (Henry et al., 1995; Karner et al., 1997; Bate, 1999).

The distribution and architecture of the Santos rift subbasins are poorly known from reflection seismics and well data, primarily due to a thick postrift sedimentary succession that has been deformed by salt tectonics. Based on gravity data, however, there would appear to be five main subbasins. For convenience, we have named these subbasins, from south to north, the Laguna, Itajaí, Paranagua, São Sebastião, and Mangaratiba subbasins (Figure 4). The largest subbasins with the thickest synrift sections are predicted to be the São Sebastião and Laguna. Note again that all the subbasins have limited along-strike continuity, although the size of any one of these subbasins is comparable to that of the

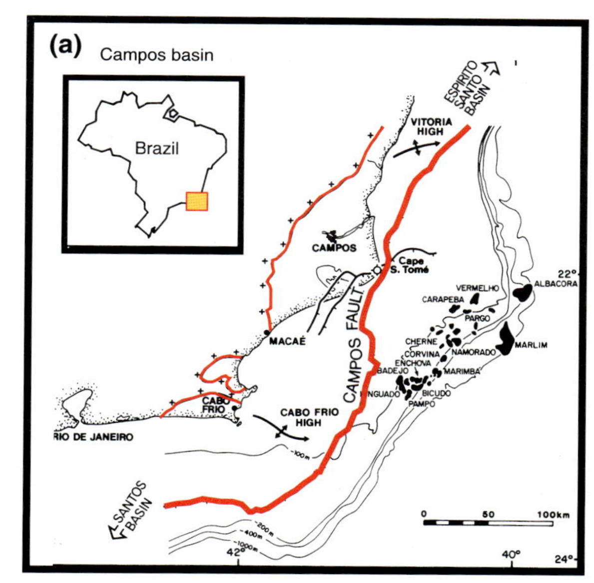

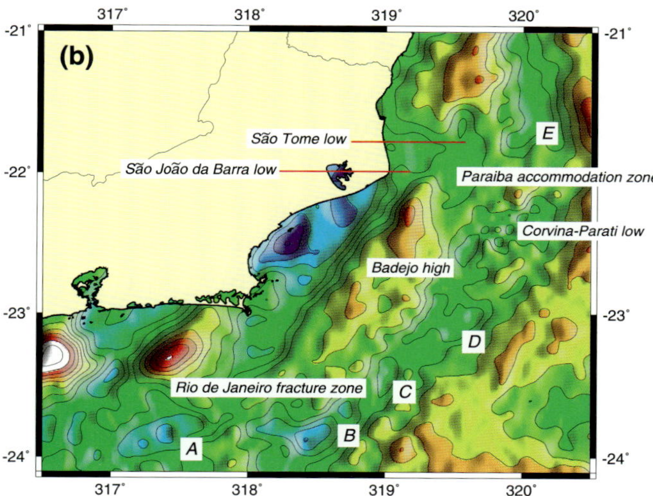

Figure 7—Comparison of (a) general structural map and (b) crustal Bouguer gravity anomaly map of the Campos Basin. Note the good correspondence of a gravity high with the Badejo High (part of the Campos-Santos hinge zone system). A series of subbasins (A through D) are juxtaposed against the southern and eastern margins of the Badejo High. Note that the major oil discoveries in the Campos Basin are associated either with synrift reservoirs east of the Badejo High or with postrift turbiditic sandstones overlying the Corvina-Parati Low, implying equivalent plays along and east of the entire Badejo High.

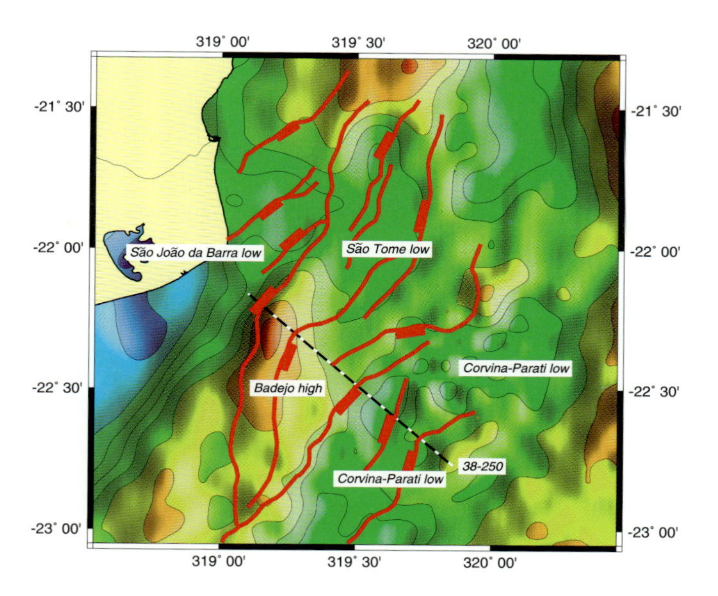

Figure 8—Major basin framework faults in the Campos Basin superposed onto the crustal Bouguer gravity anomaly map. Positive gravity anomalies are associated with footwalls of the major extensional faults, the most important of which is the Badejo High. Dashed line shows location of seismic profile 38-250 (in Figure 9).

entire presently explored Campos Basin. By early Aptian time, any local accommodation within the subbasins was sufficiently filled to be blanketed by evaporites of the Ariri Formation.

With respect to the Campos Basin, there is good correspondence between gravity highs and the Badejo High (part of the Campos and Santos hinge zones) and between gravity lows and the São Tome and Corvina–Parati Lows (Figures 4 and 7). The segmentation of the Badejo High by the Paraiba accommodation zone bisects the Campos Basin into two structural provinces. To date, all of the major oil discoveries in the Campos Basin have been found between 100 and 1000 m water depth and have been associated with either synrift reservoirs on the eastern side of the Badejo High or with postrift turbiditic sandstones that overly the Corvina–Parati Low (Figure 7) (Guardado et al., 1990). Similar hydrocarbon plays presumably exist in the northern structural province. The Campos fault, as shown in Figure 7, seems to be an amalgamation of a number of fault systems rather than a single basin-bounding fault.

Figure 8 shows the superposition of the main rift framework faults presented by Guardado et al. (1990) onto the crustal Bouguer gravity image. It can be seen that many of the positive gravity features are associated with footwalls of the major extensional faults. The crustal Bouguer image shows significantly more detail than has been presented in the literature from seismic data about the structure and arrangement of the subbasins within the Campos Basin (e.g., Guardado et al., 1990; Mohriak et al., 1989; Abrahão and Warme, 1990). For example, the Badejo High is only one of a number of high-standing en echelon blocks that form the Campos hinge zone. The hinge zone is segmented by a number of accommodation zones, the Rio de Janeiro fracture zone, and the Paraiba accommodation zone (Figure 7). The northern continuation of the Campos hinge zone is interrupted by the São Tome Low, and given its location straddling the Paraiba accommodation zone, it is likely an extensional system with a significant shear component. The São Tome Low appears to be structurally linked to a series of rift subbasins to the west of the Badejo High, collectively called the São João da Barra Low (Figure 8). Although the main basin development is immediately to the east of the Badejo High, the São João da Barra Low does appear to continue to the south toward the Rio de Janeiro fracture zone (Figure 7).

A disconcerting observation is that some of the basin-controlling faults shown in Figure 8 cross the crustal Bouguer gravity trends. This problem is in part due to limited spatial resolution of the satellite gravity. However, another problem is likely related to aligning complex or poorly imaged fault systems across relatively widely spaced 2-D seismic grids. In fact, crustal Bouguer gravity imaging might prove useful in defining the dominant trends in basins as a guide to mapping structure from seismics in the absence of 3-D seismic surveys.

The São Tome Low is structurally separated from a broad subbasin called the Corvina–Parati Low (Figures 7 and 8). In fact, a series of subbasins appear to be juxtaposed against the southern and eastern margins of the Badejo High (subbasins A-D; Figure 7). The large gravity minima associated with these southern subbasins suggests that they may represent greater crustal extension than their northern counterparts. The asymmetry of the gravity anomaly associated with the Badejo High suggests greater extension to the west relative to the eastern subbasins.

The exact opposite situation occurs for the northern continuation of the Badejo High. Here, the asymmetry of the gravity positive suggests major extension to the east of the high responsible for subbasin E (Figure 7), with the bounding fault located on the eastern side of the subbasin. Based on gravity data, structural highs are well developed between the São Tome Low, Corvina-Parati Low, and subbasin E. Although the various subbasins along the eastern margin of the Badejo High are distinct, the intervening structural highs seem to be less pronounced than between the northern subbasins. It is interesting to note the distribution of major hydrocarbon fields in the Campos Basin (Figure 7) (Guardado et al., 1990, figure 54). If our gravity interpretation is correct concerning the segmentation of the various Campos subbasins, then we would predict limited along-strike basin connectivity between subbasins A through D and the Corvina–Parati Low. Further, the interplay of Paraiba River delta lobe switching along the margin and eustasy suggests the possibility of Tertiary turbidite hydrocarbon plays along the entire Badejo High (Figure 7).

SYNRIFT STRATIGRAPHIC PACKAGES OF THE CONGO AND CABINDA MARGINS

To place into context the rift sequencing of the Campos and Santos Basins relative to the development of the South Atlantic margin, it is important to review the rift development of the Congo and Cabinda margins and the ensuing implications for the conjugate Brazilian margin. The Congo and Cabinda margins formed during the rifting of Africa from South America in the Early Cretaceous, the style and distribution of which varied both across and along the margin (Karner et al., 1997; Karner and Driscoll, 1999a). Rifting occurred in three main phases: Berriasian–Hauterivian, Hauterivian–middle Barremian, and late Barremian–early Aptian. Each rift phase resulted in the formation of deep, anoxic lacustrine systems, with the third and final rift phase creating the necessary paleoenvironments and accommodation for the major lacustrine source rocks on the west African margin (specifically, the Marnes Noires, Falcão, Bucomazi, and Melania Shale source rocks). The spatial partitioning of extension was responsible for the development of at least two major tectonic hinge zones: an inner (onshore) zone and an outer (offshore) zone subparallel to the margin.

The depositional packages associated with the final rift phase (late Barremian–early Aptian) are bounded by two regional and tectonically significant unconformities: a sequence boundary named the pre-Chela unconformity, which is associated with an early Aptian relative sea-level fall, and the slightly younger Chela unconformity produced during the ensuing flooding of the margin. The transgressive sequences overlying this unconformity comprise fluvial sandstones at the base grading upward into lagoonal facies and eventually into the evaporites of the regionally extensive Ezanga and Loeme Formations. This last rift phase, however, was accompanied by minor brittle deformation of the crust, although accommodation was rapidly being generated. This is the so-called sag phase of Teisserenc and Villemin (1990) and Henry et al. (1995). The generically equivalent depositional packages on the Brazilian margin comprise the transitional megasequence of Chang et al. (1988, 1992). Despite the lack of evidence for significant extension in general and for late rift-stage deformation in particular, this same region nevertheless underwent significant postrift subsidence.

The form of the subsidence that occurred during and after rifting implies that extension was partitioned as a function of depth. Plastic deformation thinned the lower crust, while brittle deformation of the upper crust occurred to the west near the future ocean–continent boundary. Lower crustal extension beneath an intracrustal detachment maintained the region near sea level during the final stages of the rifting process (as breakup was approached). These shallow-water regions were extremely sensitive to small-scale eustatic fluctuations. Repeated marine incursion and desiccation cycles could feasibly explain the thick salts of the Ezanga and Loeme Formations (Karner and Driscoll, 1999a). The evaporites are thus late synrift deposits, placing the timing of continental breakup as early Aptian. It is thus important that the known distribution of evaporites within the Santos and Campos Basins (e.g., Demercian et al., 1993) reside principally to the west of the ocean–continent boundary determined from the filtered crustal Bouguer gravity and the oceanic fracture zone terminations. The existence of a mid-crustal detachment beneath the west African margin, active during the late stages of continental rifting, has profound implications for the hydrocarbon maturation history of the margin. This detachment is envisioned as a zone of brittle–ductile transition that migrates vertically through the crust in response to the input of heat during rifting (Driscoll and Karner, 1998). It is this extension partitioning with depth that allows the rifting process to continue in the absence of brittle deformation.

Brazilian basins conjugate to the Congo and Cabinda margins, for example, the Camamu–Almada margin, show similar unconformity-bounded rift packages (Karner and Driscoll, 1999a). The ages of these unconformities (late Berriasian, Valanginian–Hauterivian, and late Barremian–early Aptian) and the timing of source rock development are similar to those on the west African margin, at least within the resolution of intra-Atlantic ostracod paleontology (Bate, 1999). Although there is sufficient evidence for a time transgressive onset of sea floor spreading in the South Atlantic, the same case cannot be made for the onset of rifting. Recent studies of rift initiation in the Gulf of Suez (Omar and Steckler, 1995) and the Woodlark Basin of Papua–New Guinea (Taylor et al., 1999) indicate that rift onset is synchronous over thousands of kilometers within the resolution of available data. Given this and the age of the unconformities on the Camamu–Almada margin, it would appear that the onset of rifting is essentially synchronous between eastern Brazil and west Africa. If correct, then the Campos and Santos Basins should exhibit a tripartite rift history similar to that of the Congo, Cabinda, and Camamu–Almada Basins.

CAMPOS AND SANTOS BASINS: STRUCTURAL SETTING

As with the Congo and Cabinda margins, the Campos and Santos Basins were formed in the Early Cretaceous during the breakup of Africa and South America. Rifting was superposed onto and during the emplacement of the Serra Geral tholeiites, the same basalts that were subaerially extruded across the entire Paraná Basin. The Serra Geral volcanics have a maximum thickness of 1500–2000 m and cover an area of greater than 1.1 million km^2. According to Renne et al. (1992), the Paraná volcanic province and the Etendeka volcanics of Namibia were erupted very rapidly at 133 ± 1 Ma and generally predate the opening (or sea floor spreading) of the South Atlantic. Voluminous tholeiitic basalts are diagnostic of significant partial melting of asthenospheric material at elevated temperatures. These conditions are likely the consequence of plume activity.

The Campos and Santos rift basins were developed across a major, broad intracontinental plateau of which the present southeastern Brazilian highlands are a remnant (Figure 1) (Karner et al., 1993). This plateau was likely a result of magmatic underplating associated with the emplacement of the Serra Geral basalts and/or residual lithospheric heat introduced by the same mantle plume responsible for the tholeiitic basalts (Karner et al., 1993). Today, the southeastern Brazilian highlands consist of two components (Figure 1): (1) a broad plateau with a width of about 1500–2000 km and relief of 500–1500 m (relative to sea level), and (2) a region of coastal topography represented by the Serra do Mar and Serra da Mantiqueira mountains with a width of 100–200 km and local relief of 500–2800 m. This marginal topography is most simply explained as part of the rift flank produced in response to the extension that formed the Campos and Santos Basins (Weissel and Karner, 1989; Gallagher et al., 1994). The present Serra do Mar is the erosional coastal remnant of a scarp retreat process initiated during Neocomian rifting. Petrobrás boreholes in the onshore part of the Campos Basin have drilled through 600 m of tholeiitic basalts and penetrated acid Precambrian igneous rocks similar to the granitoid rocks that outcrop along the coast (Guardado et al., 1990). These same Precambrian rocks form the coastal mountains of the Serra do Mar and Serra da Mantiqueira.

From Neocomian to Eocene time, the areas bordering both the Paraná Basin and the Campos and Santos Basins were intruded by alkaline igneous rocks (Figure 4) (Ulbrich and Gomes, 1981; Mizusaki et al., 1988). While many researchers have linked these alkaline basalts to a hot spot track (Herz, 1977; Ulbrich and Gomes, 1981; Almeida, 1983), both the pattern of radiometric dates and the distribution of alkaline rocks around the borders of the Paraná Basin has failed to confirm this hypothesis. The larger intrusive bodies have an age range from 48 to 133 Ma.

The occurrence of volumetrically insignificant alkaline igneous rocks poses a special problem in terms of their genesis. Alkaline volcanic rocks are considered to be the result of small amounts of partial melting of lithospheric material; hence their emplacement must be rapid otherwise they would solidify during ascent. Their emplacement during the Early Cretaceous–Eocene in southeastern Brazil implies that they were generated at this time. However, an obvious relationship between melt genesis and emplacement with a causative tectonic event does not exist. Thus, the simplest interpretation for the timing and distribution of the alkali basalts rimming the Paraná Basin (e.g., Figure 4) is that they represent partial melting of fertile lithosphere due to the dissipation and lateral propagation of heat out away from the region that was thermally damaged by the original Early Cretaceous plume. More importantly, because of the low degree of partial melting required to generate alkali basalts, the amount of heat advected into the lithosphere with the migrating melts is negligible. Thus, it plays little, if any, regional role in modifying the Tertiary temperature structure of the lithosphere.

SYNRIFT STRATIGRAPHIC PACKAGES OF THE CAMPOS AND SANTOS BASINS

The Neocomian synrift sequences of the Campos and Santos Basins (including the salt formations) are collectively called the Lagoa Feia (Campos) and Guaratiba and Ariri (Santos) megasequences and are characterized by fluvial–lacustrine and deltaic sedimentary rocks. These strata show extreme facies variations in response to syndepositional tectonism (Guardado et al., 1990; Abrahão and Warme, 1990) and the development of significant variations in lake bathymetry (e.g. Chang et al., 1992). Based on seismic data, the maximum synrift sedimentary thickness in the Campos Basin is 4000 m (Guardado et al., 1990). The fossil record in the synrift sequences includes ostracods, pelecypods, gastropods, and fish (Abrahão and Warme, 1990). The ostracods, in particular, have been used to define local stages (Aratu = late Hauterivian–early Barremian; Buracica = middle Barremian; Jiquia = late Barremian–early Aptian; Alagoas = early Aptian) (Braccini et al., 1997; Bate, 1990). In the Campos Basin, the oldest strata drilled are dated as latest Aratu (early Barremian). Reinterpretation of published reflection seismic data suggests that older synrift sedimentary rocks may exist. The lower sedimentary layers are frequently associated with volcanic debris and flows produced by widespread volcanism and intercalated clastics (Bertani and Carozzi, 1985; Mizusaki et al., 1988).

Alluvial fan deposits are associated with rift-generated topography, either as part of the main flanks of the Serra do Mar or local topography developed along the Campos and Santos and other hinge zones (e.g., Badejo High). However, because this "internal" hinge zone topography was eventually destroyed, the crests of these structural highs became the sites of coquina development and deposition. Lacustrine coquinas were also deposited in areas isolated from zones of high terrigeneous sediment input. Toward the depocenter of the Campos Basin, the lower clastic sequences (lower Barremian) tend to be sandy to silty fine-grained lacustrine facies (Guardado et al., 1990), while the upper units (middle–upper Barremian) consist of lacustrine carbonates and organic-rich black shales (Dias et al., 1988). It is in these structural lows (i.e., relatively deep lake systems) that the principal source rocks were deposited (e.g., Mello et al., 1993), paralleling the timing of source rock development on the west African margin. Lake waters were already relatively saline at this time (Chang et al., 1992).

The nonmarine pre-Alagoas sedimentary units are separated from the overlying units by a regional lower Aptian unconformity termed the *pre-Alagoas unconformity.* This unconformity was produced during a phase of uplift and regional truncation. Brittle deformation of the region ceased after the formation of this unconformity, leading to the development of a gentle sag phase of subsidence known as the *transitional megasequence* (e.g., Guardado et al., 1990). The Aptian–Alagoas transitional megasequence can be divided into a lower terrigeneous sequence and an upper evaporite sequence. The lower sequence represents a major influx of alluvial sediments into the basin from both sources to the north and northwest and from intrabasinal highs. Sedimentary environments ranged from alluvial fans and fan deltas to sabkhas. Maximum sedimentary thickness is about 600 m. Low ostracod diversity, indicative of stressed and restricted environments, appears to be heralding increasing salinity conditions conducive to evaporite deposition. Massive halites, reaching thicknesses of 2000 m, accumulated in the most rapidly subsiding sections of the basin. This unconformity and the transition from lacustrine strata to evaporites is surprisingly similar to the pre-Chela and Chela unconformities of the west African margin, which immediately preceded deposition of the carnalites and halites of the Loeme and Ezanga evaporites.

From this description of the synrift stratigraphy, it is not clear whether the tripartite rift history of the west African and Camamu–Almada margins is applicable to the Campos Basin. Comparison of ostracod zonations between west Africa and Brazil suggests that the age range of synrift units in the Campos Basin is early Barremian–late Aptian (ostracod zones AS7–AS11) (Bate, 1990). This implies that the rift phases of the Campos Basin correlate with rift phases II and III of the west African margin. However, these zonations are based on well data that are necessarily restricted to high blocks, such that the wells do not sample the complete stratigraphic section of the basin. Resolution of this problem is found in the reinterpretation of published seismic reflection profiles from the Campos Basin (Guardado et al., 1990).

Figure 8 shows seismic reflection profile 38-250, which traverses the São João da Barra Low, the Badejo High, and the updip portion of the Corvina–Parati Low. The interpretation presented here, modified from Guardado et al. (1990), identifies two synrift packages within the São João da Barra Low, a major rift subbasin characterized by a half-graben geometry and a westward-dipping border fault (Figure 9). The synrift packages are unconformity bounded. Rift package I comprises diverging seismic reflectors that onlap either basement or basalt. The diverging sedimentary wedge attests to differential subsidence and thus suggests an early episode of block rotation—rift phase I. Continued border fault reactivation in rift phase II induced further block rotation and thus the development of a second onlap surface and a divergent sedimentary wedge (Figure 9). It is this second wedge that possibly corresponds to the nonmarine component of the Lagoa Feia Formation. Seismic data quality east of the Badejo High makes it difficult to recognize the onlap surfaces diagnostic of rift phases I and II. The divergent wedge does consists of two packages, a lower chaotic reflector package overlain by a package with well-defined subparallel reflectors. It is unclear from present data whether these two packages represent renewed accommodation generated by discrete rift events or a change in depositional style (or both).

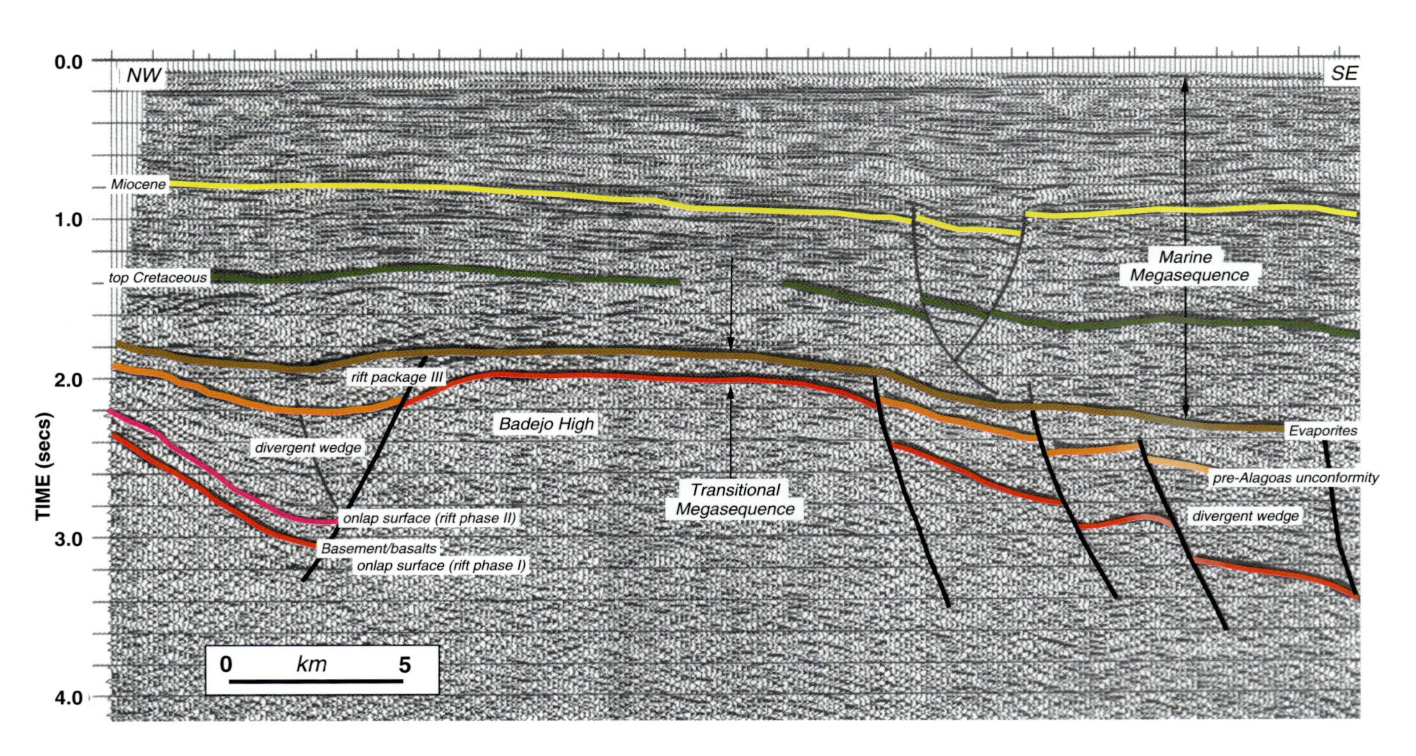

Figure 9—Regional seismic profile 38-250 showing the rift sequencing in the Campos Basin. (See Figure 8 for profile location.) See text for discussion of rift phases. Modified from Guardado et al. (1990). Note the dual structural role of the Badejo High: footwall to the Campos fault system and hanging wall to the eastern border fault controlling subbasin D.

Subsequent regional truncation across the region responsible for the formation of the pre-Alagoas unconformity was followed by gentle regional sagging in the absence of significant brittle deformation. By analogy with the west African margin, we suggest that this sag or transitional component of the Lagoa Feia Formation was in response to a third phase of rifting (rift phase III) that preferentially thinned the lower crust and lithospheric mantle across the Campos Basin margin as breakup was approached. The intervening pre-Alagoas unconformity is time and tectonically equivalent to the pre-Chela unconformity on the Congo margin (Figure 9). As in the west African margin, this uplift was responsible for the development of restricted environments conducive to the formation of evaporites, such as the upper Lagoa Feia Formation (Campos) and the Ariri Formation (Santos).

CONCLUSIONS

Free-air and crustal Bouguer gravity anomalies were used to map the distribution of rift subbasins of the Campos and Santos Basins along the southeastern Brazilian continental margin. In addition, the synrift stratigraphic development of the Congo–Cabinda and Camamu–Almada margins was used as a proxy for the Campos Basin. From these data, the following conclusions can be drawn.

First, fracture zone trends are well-expressed in the satellite free-air, crustal Bouguer, and filtered crustal Bouguer gravity images of the South Atlantic. Their westerly termination and coregistration with a negative–positive gradient in the filtered crustal Bouguer gravity anomaly identifies the location of the ocean–continent boundary. This same boundary marks an abrupt change in the gravity anomaly trend from east–west to margin-parallel features. It also delineates the eastern limit of the synrift evaporites responsible for the diapiric structures seen in seismic reflection profiles of the São Paulo Plateau. The source of the gravity gradient is in part a result of the relatively thick, old oceanic crust juxtaposed to thinned continental crust. In turn, crustal thickening was a function of early sea floor spreading in close proximity to mantle plumes, such as Tristan da Cunha.

Second, oceanic fracture zone trends and the estimated location of the ocean–continent boundary suggests that most of the crust underlying the São Paulo Plateau is extended continental crust. The southern boundary of the plateau, a transform margin, is controlled by the Curitiba and Florianopolis fracture zones.

Third, crustal Bouguer gravity anomalies provide evidence for a major offshore tectonic hinge zone that consists of a series of short-segment, en echelon, high-standing blocks located subparallel to the Brazilian margin. The Badejo High of the Campos Basin is part of this hinge zone trend. Maximum disruption of the hinge zone occurs across the Rio de Janeiro fracture zone, which is also the transition from the Santos to the Campos Basin. Seaward of the hinge zone, a series of distinct negative gravity anomalies suggests the presence of major rift subbasins. These subbasins appear to have limited along-strike continuity, implying that lake communication, water chemistry, and possibly source quality and preservation, was spatially distinct to each subbasin. The Serra do Mar and Serra da Mantiqueira mountain systems represent an onshore hinge zone, with minor extensional basins occurring between the mountains and the Campos and Santos hinge zones. A second offshore hinge zone may exist to the east of the Campos and Santos hinge zones, separating moderately extended crust in the Santos and western Campos Basins from highly extended continental crust across the São Paulo Plateau and eastern Campos Basin, respectively.

Fourth, crustal Bouguer gravity highs and lows correspond well with the Badejo High (part of the Campos and Santos hinge zones) and the São Tome and Corvina–Parati Lows. In particular, major positive gravity features are associated with the footwalls of large basin-bounding border faults. The asymmetry of the gravity anomaly associated with the Badejo High suggests greater extension to the west relative to the eastern subbasins, requiring that the Badejo High is a footwall block to the São João da Barra Low but a hanging-wall block to the easterly subbasins.

Fifth, the Congo and Cabinda margins of west Africa and the Camamu–Almada margin of east Brazil are characterized by a common tripartite rift history: Berriasian–Hauterivian, Hauterivian–middle Barremian, and late Barremian–early Aptian. Extension style changed dramatically as a function of space and time. Early, depth-independent, broadly distributed brittle deformation (rift phases I and II) was progressively replaced by depth-dependent deformation in which rifting was dominated by plastic thinning of the lower crust and lithospheric mantle (rift phase III). Interpretation of published seismic reflection data shows two synrift packages in the São João da Barra Low. Rift package I comprises diverging seismic reflectors that onlap either basement or basalt. The diverging sedimentary wedge attests to differential subsidence and thus an episode of block rotation. Continued border fault reactivation induced further block rotation and development of a second onlap surface. This fault reactivation resulted in a second stratal wedge, possibly corresponding to the nonmarine component of the Lagoa Feia Formation. Although it is difficult to recognize onlap surfaces diagnostic of rift phases I and II in the eastern subbasins, the divergent wedge does consist of two packages: a lower chaotic reflector package overlain by well-defined subparallel reflectors. It is unclear if these two packages represent renewed accommodation generated by discrete rift events or a change in depositional style (or both). Subsequent regional truncation formed the pre-Alagoas unconformity followed by gentle regional sagging without significant brittle deformation. By analogy with the west African margin, this sag or transitional component of the Lagoa Feia was in response to a third phase of rifting that preferentially thinned the lower crust and lithospheric mantle across the Campos Basin margin as breakup was approached.

Finally, the pre-Alagoas unconformity and the transition from lacustrine sediments to evaporites parallels that of the pre-Chela and Chela unconformities on the west African margin. These unconformities were formed immediately prior to the deposition of evaporites and heralded the onset of an extension phase characterized by a paucity of basement-involved normal faulting and regional subsidence. The pre-Alagoas unconformity is thus time and tectonically equivalent to the pre-Chela unconformity on the Congo margin.

Acknowledgments—*Discussions with Ricardo Bedregal, Neal Driscoll, Juliano Macedo, and Jeffrey Weissel concerning Campos Basin seismic interpretations and South Atlantic tectonics are gratefully acknowledged, as are reviews by Ricardo Bedregal, Randy Hunt, and Barry Katz. The author would also like to express his thanks for the support and patience of the editors of this volume: Barry J. Katz and Marcio R. Mello. The Wessel and Smith GMT software was used extensively in the construction of figures in this chapter. (This is Lamont-Doherty Earth Observatory publication #6008.)*

REFERENCES CITED

Abrahão, D., and J. E. Warme, 1990, Lacustrine and associated deposits in a rifted continental margin—Lower Cretaceous Lagoa Feia Formation, Campos Basin, offshore Brazil, *in* B. J. Katz, ed., Lacustrine basin exploration: Case studies and modern analogs: AAPG Memoir 50, p. 307–326.

Almeida, F. F. M., 1983, Relações tectônicas das rochas alcalinas mesozôicas da região meridional da plataforma sulamericana: Revista Brasieira de Geociências, v. 13, p. 139–158.

Bate, R. H., 1999, Non-marine ostracod assemblages of the pre-salt rift basins of west Africa and their role in sequence stratigraphy, *in* N. R. Cameron, R. H. Bate, and V. S. Clure, eds., The oil and gas habitats of the South Atlantic: Geologic Society of London Special Publication 153, p. 283–292.

Bertani, R. T., and A. V. Carozzi, 1985, Lagoa Feia Formation (Lower Cretaceous) Campos Basin, offshore Brazil—rift valley stage lacustrine carbonate reservoirs I: Journal of Petroleum Geology, v. 8, p. 37–58.

Braccini, E., C. N. Denison, J. R. Scheevel, P. Jeronimo, P. Orsolini, and V. Barletta, 1997, A revised chronostratigraphic framework for the pre-salt (Lower Cretaceous) in Cabinda, Angola: Bulletin des Centre de Recherches Exploration–Production Elf Aquitaine, v. 21, p. 125–151.

Cande, S. C., J. L. LaBrecque, and W. F. Haxby, 1988, Plate kinematics of the South Atlantic: Chron 34 to Present: Journal of Geophysical Research, v. 93, p. 13,479–13,492.

Chang, K. H., R. O. Kowsmann, and A. M. F. Figueiredo, 1988, New concepts of the development of east Brazilian marginal basins: Episodes, v. 11, p. 194–202.

Chang, K. H., R. O. Kowsmann, A. M. F. Figueiredo, and A. Bender, 1992, Tectonics and stratigraphy of the east Brazil rift system: an overview: Tectonophysics, v. 213, p. 97–138.

Demercian, S., P. Szatmari, and P. R. Cobbold, 1993, Style and pattern of salt diapirs due to thin-skinned gravitational gliding, Campos and Santos Basins, offshore Brazil: Tectonophysics, v. 143, p. 156–161.

Dias, J. L., J. Q. Oliveira, and J. C. Vieira, 1988, Sedimentological and stratigraphic analysis of the Lago Feia Formation, rift phase of the Campos basin, offshore Brazil: Revista Brasileira de Geociências, v. 18, p. 252–260.

Driscoll, N. W., and G. D. Karner, 1998, Lower crustal extension across the northern Carnarvon Basin, Australia: evidence for an eastward dipping detachment: Journal of Geophysical Research, v. 103, p. 4975–4992.

Gallagher, K., C. J. Hawkesworth, and M. S. Mantovani, 1994, The denudation history of the onshore continental margin of SE Brazil inferred from apatite fission track data: Journal of Geophysical Research, v. 99, p. 18,117–18,146.

Gladczenko, T. P., K. Hinz, O. Eldholm, H. Meyer, S. Neben, and J. Skogseid, 1997, South Atlantic volcanic margins: Journal of the Geological Society of London, v. 154, p. 465–470.

Guardado, L. R., L. A. P. Gamboa, and C. F. Lucchesi, 1990, Petroleum geology of the Campos Basin, Brazil: a model for a producing Atlantic type basin, *in* J. D. Edwards and P.A. Santogrossi, eds., Divergent/passive margin basins: AAPG Memoir 48, p. 3–80.

Henry, S., W. D. Brumbaugh, and N. R. Cameron, 1995, Pre-salt source rock development on Brazil's conjugate margin: west African examples (ext. abs.): Fourth International Congress, Brazilian Society of Geophysics, Rio de Janeiro, p. 3.

Herz, N., 1977, Timing and spreading in the South Atlantic: information from Brazilian alkaline rocks: GSA Bulletin, v. 88, p. 101–112.

Hinz, K., S. Neben, B. Schrenkenberger, H. A. Roeser, M. Block, K. Goncalves de Souza, and H. Meyer, 1999, The Argentine continental margin north of 48°S: sedimentary successions, volcanic activity during breakup: Marine and Petroleum Geology, v. 16, p. 1–12.

Karner, G. D., and N. W. Driscoll, 1999a, Tectonic and stratigraphic development of the West African and eastern Brazilian margins: insights from quantitative basin modeling, *in* N. R. Cameron, R. H. Bate, and V. S. Clure, eds., The oil and gas habitats of the South Atlantic: Special Publication of the Geological Society of London, v. 153, p. 11–40.

Karner, G. D., and N. W. Driscoll, 1999b, Style, timing, and distribution of tectonic deformation across the Exmouth Plateau, northwest Australia, determined from stratal architecture and quantitative basin modelling, *in* C. MacNiocaill and P. D. Ryan, eds., Continental tectonics: Special Publication of the Geological Society of London, v. 164, p. 271–311.

Karner, G. D., N. Ussami, and F. F. Alkmim, 1993, Tectonic significance of the topography of southeastern Brazil: speculations and implications (ext. abs.): 10th Anniversary Conference UFOP/Petrobrás, Universidade Federal de Ouro Preto, Ouro Preto, Brazil, p. 18–19.

Karner, G. D., N. W. Driscoll, J. P. McGinnis, W. D. Brumbaugh, and N. Cameron, 1997, Tectonic significance of syn-rift sedimentary packages across the Gabon–Cabinda continental margin: Marine and Petroleum Geology, v. 14, p. 973–1000.

Kowsmann, R. O., M. P. A. Costa, M. P. Boa Hora, H. P. Almeida, and P. P. Guimares, 1982, Geologia estrutural do Plato de São Paulo: Congresso Brasileiro de Geologia 32, Salvador, Anais, v. 4, p. 1558–1569.

Kumar, N., R. Leyden, J. Carvalho, and O. Francisconi, 1979, Sediment isopach of continental margins of Brazil: AAPG Map Series. Catalog 831.

Macedo, J. M., 1990, Evolução tectônica da bacia de Santos e reas continentais adjacentes, *in* G. P. de Raja Gabaglia and E. J. Milani, eds., Origem e evolução de bacias sedimentares: Petrobrás, 415 p.

Matos, R. D., 1992, The northeastern Brazilian rift system: Tectonics, v. 11, p. 766–791.

Mello, M. R., W. U. Mohriak, E. A. M. Koutsoukos, and J. C. A. Figueira, 1993, Brazilian and west African oils: generation, migration, accumulation, and correlation: Proceedings, 13th World Petroleum Congress, Buenos Aires, v. 2, p. 153–164.

Mizusaki, A. M. P., A. Thomaz Filho, and J. Valença, 1988, Volcano-sedimentary sequence of Neocomian age in Campos Basin (Brazil): Revista Brasileira de Geociências, v. 18, p. 247–251.

Mohriak, W. P., M. R. Mello, G. D. Karner, J. F. Dewey, and J. R. Maxwell, 1989, Geological and stratigraphic development of the Campos Basin, offshore Brazil: AAPG Memoir 46, p. 577–598.

Omar, G. I., and M. S. Steckler, 1995, Fission-track evidence on the initial rifting of the Red Sea: two pulses, no propagation: Science, v. 270, p. 1341–1344.

Renne, P. R., M. Ernesto, I. G. Pacca, R. S. Coe, J. M. Glen, M. Prévot, and M. Perrin, 1992, Age of Paraná flood volcanism, rifting of Gondwanaland, and the Jurassic–Cretaceous boundary: Science, v. 258, p. 975–979.

Sandwell, D. T., and W. H. F. Smith, 1992, Global marine gravity from ERS-1, Geosat and Seasat reveals new tectonic fabric: EOS Transactions, American Geophysical Union, v. 73, p. 133.

Taylor, B., A. Goodliffe, and F. Martinez, 1999, How continents break up: insights from Papua New Guinea: Journal of Geophysical Research, v. 104, p. 7497–7512.

Teisserenc, P., and J. Villemin, 1990, Sedimentary basin of Gabon—geology and oil systems, *in* J. D. Edwards and P.A. Santogrossi, eds., Divergent/passive margin basins: AAPG Memoir 48, p. 117–199.

Ulbrick, H., and C. B. Gomes, 1981, Alkaline rocks from continental Brazil: Earth and Science Reviews, v. 17, p. 135–154.

Weissel, J. K., and G. D. Karner, 1989, Flexural uplift of rift flanks due to mechanical unloading of the lithosphere during extension: Journal of Geophysical Research, v. 94, p. 13,919–13,950.

Wessel, P., and W. H. F. Smith, 1995, New version of the generic mapping tools released: EOS Transactions, American Geophysical Union, v. 76, p. 329.

Guardado, L. R., A. R. Spadini, J. S. L. Brandão, and M. R. Mello, 2000, Petroleum system of the Campos Basin, *in* M. R. Mello and B. J. Katz, eds., Petroleum systems of South Atlantic margins: AAPG Memoir 73, p. 317–324.

Chapter 22

Petroleum System of the Campos Basin, Brazil

L. R. Guardado

A. R. Spadini

J. S. L. Brandão

Petrobrás E&P
Rio de Janeiro, Brazil

M. R. Mello

Petrobrás–Cenpes
Rio de Janeiro, Brazil

Abstract

The Campos Basin, located offshore from Rio de Janeiro in southeastern Brazil, is the most prolific petroleum-bearing basin in Brazil. To date, Petrobrás has discovered more than 70 hydrocarbon accumulations in the basin, including seven giant oil fields in deep water. The Campos Basin accounts for more than 80% of the total exploitable Brazilian reserves and 75% of total oil production. The Barremian calcareous shales from the Lagoa Feia Formation are the source rocks of all the oil in the basin. Accumulations occur in a variety of siliciclastic and carbonate reservoirs, ranging from Barremian to Miocene in age. Upper Cretaceous–Tertiary turbidities contain most of the oil, but Neocomian basalts also contain commercial accumulations. Peak oil generation occurred during late Miocene time, and the most important migration pathways are through salt windows and along listric faults.

INTRODUCTION

The Campos Basin covers an area of about 100,000 km^2 (up to the 3400-m isobath). The oil fields discovered occur in water depths ranging from 80 m to more than 2600 m (Figure 1).

Because of its importance as a petroleum province, many papers have been published identifying and characterizing the petroleum system of the Campos Basin (e.g., Meister, 1984; Guardado et al., 1989; Mohriak et al., 1990; Mello et al., 1994). The abundance of petroleum in this basin is the result of the existence in time and space of all the elements and processes necessary for a world-class petroleum system. These include (1) rich, oil-prone source rocks, (2) the presence of effective migration pathways, (3) excellent reservoirs, (4) effective traps and seals, and (5) the appropriate timing for trap formation and oil migration. This study is a geologic, geophysical, and geochemical investigation of the elements and processes making up the petroleum system of the Campos Basin. In addition, geochemical results from a large number of sediment cores, cuttings, and oil samples recovered from the Lagoa Feia Formation are discussed. The objective is to identify and characterize the key elements that contributed to the establishment of the most prolific petroleum system in Brazil.

TECTONO-STRATIGRAPHIC SETTING

The Campos Basin was formed as a result of the breakup of Gondwana in the Early Cretaceous. This was followed by subsequent infilling of the rift basin with as much as 9000 m of Early Cretaceous–Holocene sediments, as shown in the stratigraphic chart of Figure 2. The basin has several main structural elements (Figure 3): (1) northeast- and northwest-trending horsts and grabens mapped at the Neocomian basalt reflector; (2) pre-Aptian structures related to the Campos fault; and (3) a salt dome province, which is an extension of the São Paulo Plateau in ultra-deep water.

The sedimentary section of the Campos Basin can be subdivided into three megasequences. (Figure 2). First,

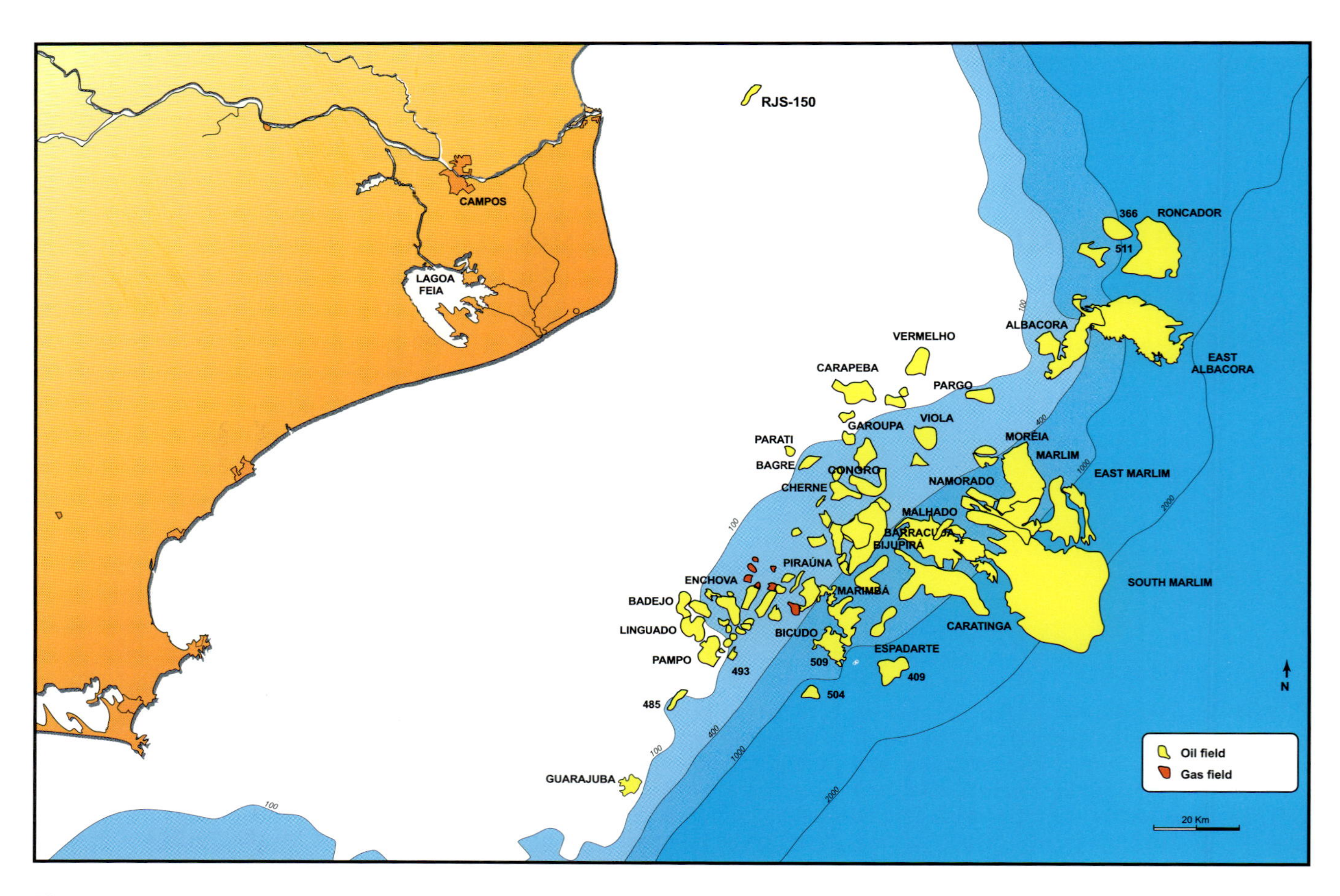

Figure 1—Map of the Campos Basin showing the distribution of oil fields discovered up to 1999.

the *nonmarine rift megasequence* is composed of lacustrine Barremian sedimentary rocks overlying Neocomian basalts. The strata were deposited in a variety of paleoenvironments that were strongly influenced by rift tectonics. These included alluvial fans, fan deltas, carbonate banks, and lacustrine environments ranging from brackish to hypersaline. This sequence contains the calcareous shales of the Lagoa Feia Formation, the most important source rock in the basin. Second, the *transitional megasequence* was deposited in Aptian time during a period of tectonic quiescence. It represents the beginning of the drift phase and contains a lower sequence, mostly composed of conglomerates and carbonates, and an upper sequence consisting of halite and anhydrite. Third, the *marine megasequence* is made up of Albian shallow-water carbonates, mudstones, and marls (Macaé Formation) at its base. This basal sequence grades upward into an Upper Cretaceous–Paleocene bathyal sequence consisting of shales, marls, and sandstone turbidites. A progradational siliciclastic sequence characterizes the remaining Neogene section. The deposition of this megasequence was strongly affected by salt tectonics.

SOURCE ROCKS AND HYDROCARBON CHARACTERIZATION

Figure 4 shows a geochemical log from a selected Campos Basin well that indicates the excellent hydrocarbon source potential of the Lagoa Feia Formation. The Barremian Lagoa Feia section is composed of well-laminated shales interbedded with carbonates. It ranges in thickness from 100 to 300 m, with typical total organic carbon (TOC) values averaging 2–6% and locally as high as 9%. The hydrogen indices (HI) are up to 900 mg HC/mg TOC, consistent with type I kerogen (Figure 4). Organic petrography shows the predominance of lipid-rich material mainly of algal and bacterial origin. On average, the organic-rich beds contain about 90% amorphous organic matter (Mello et al., 1994).

Gas chromatography (GC) and gas chromatography–mass spectrometry (GCMS) data were obtained from the oils and organic extracts from the organic-rich sections of the Lagoa Feia Formation (Figure 5). The biological marker distribution (Figure 6) suggests lacustrine brackish to saline depositional environments.

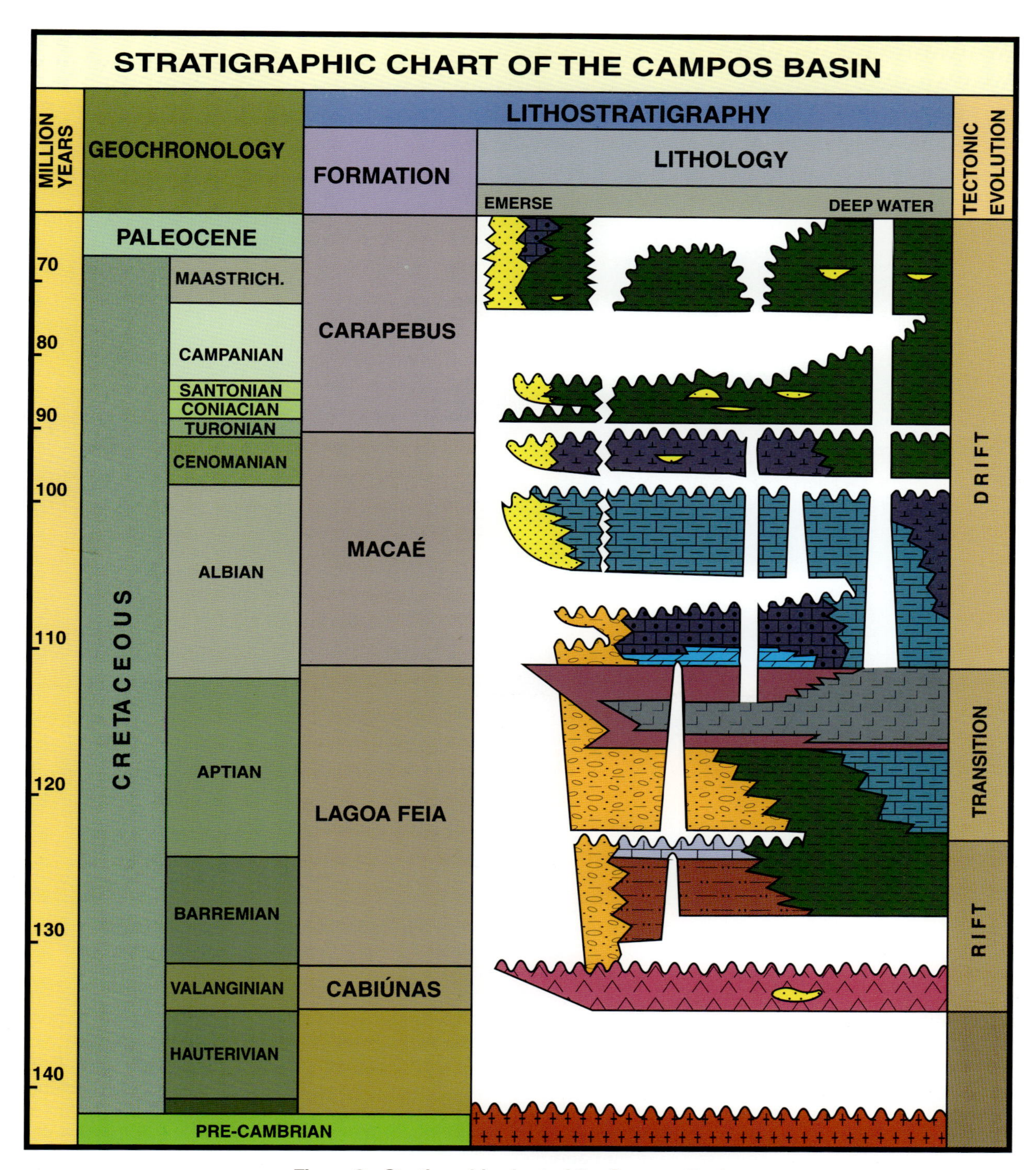

Figure 2—Stratigraphic chart of the Campos Basin.

Characteristics include the dominance of low molecular weight *n*-alkanes with a slight odd-carbon predominance; pristane greater than phytane; heavy $\delta^{13}C$ values (between –22 and –25‰); high concentrations of high molecular weight tricyclic terpanes (up to C_{40}), C_{30} dinosteranes, β-carotane, 28,30-bisnorhopane, and 4-methyl steranes; a low hopane/sterane index; and $T_s/T_m < 1$ (Figures 5 and 6) (Mello, 1988; Mello and Maxwell, 1990; Mello et al., 1994). The diagnostic molecular features that characterize the lacustrine marine-influenced systems in the Lagoa Feia Formation are the presence and abundance of 24-*n*-propylcholestanes. Such compounds are

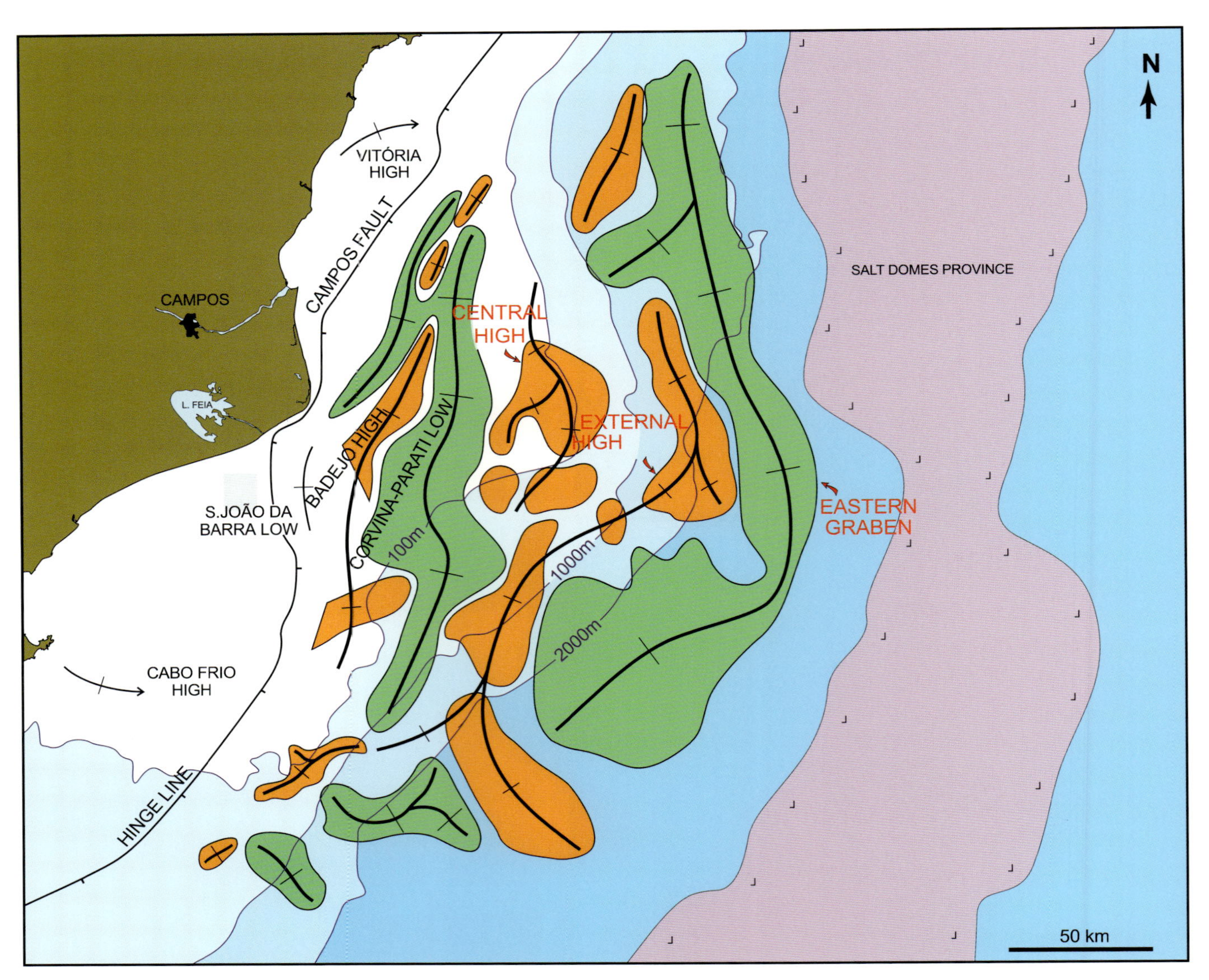

Figure 3—Structural framework map of the rift section in the Campos Basin. From Rangel and Martins (1998).

considered to be marine source indicators, derived from marine *Chrysophytae* algae (Moldowan et al., 1990; Mello and Hessel, 1998).

Most of the Campos Basin oils have API gravities ranging from 17–37° API and consist of a mixture of biodegraded and nonbiodegraded oils. This mixture resulted from more than one generation and migration pulse, coupled with biodegradation episodes that occurred during successive stages of reservoir filling (Soldan et al., 1995).

Reconstruction of the paleoenvironment of the Lagoa Feia Formation, based on bivalve and sedimentologic data (Carvalho et al., 1996), suggest the occurrence during the early Neocomian of small, interior brackish to saline alkaline lakes (without marine influence) and a large epicontinental water body (possibly a restricted gulf or lagoon). Intermittent sea water incursions into the water body brought about the onset of marine depositional conditions, eventually resulting in the emergence of typical marine bivalve species *(Agelasina* and *Remondia)* (Mello and Hessel, 1998). Blooms of cyanobacteria coincided with increased water salinity. This may have led to mass mortality of the benthic bivalves as a result of the release of cell toxins into the water (Brongersma-Sanders, 1957) and/or the reduction of water column oxygenation resulting from algae senescence and decay. These bloom cycles with a high biomass of cyanobacteria led to the deposition of the organic-rich shales that form the source rocks of the Lagoa Feia Formation (Mello and Hessel, 1998).

Paleozoologic data from coquinas indicate that during the late Barremian, the euryhaline bivalve *Arcopagella* dominated the epicontinental waters of the Campos Basin. In the early Aptian, a mixohaline species of

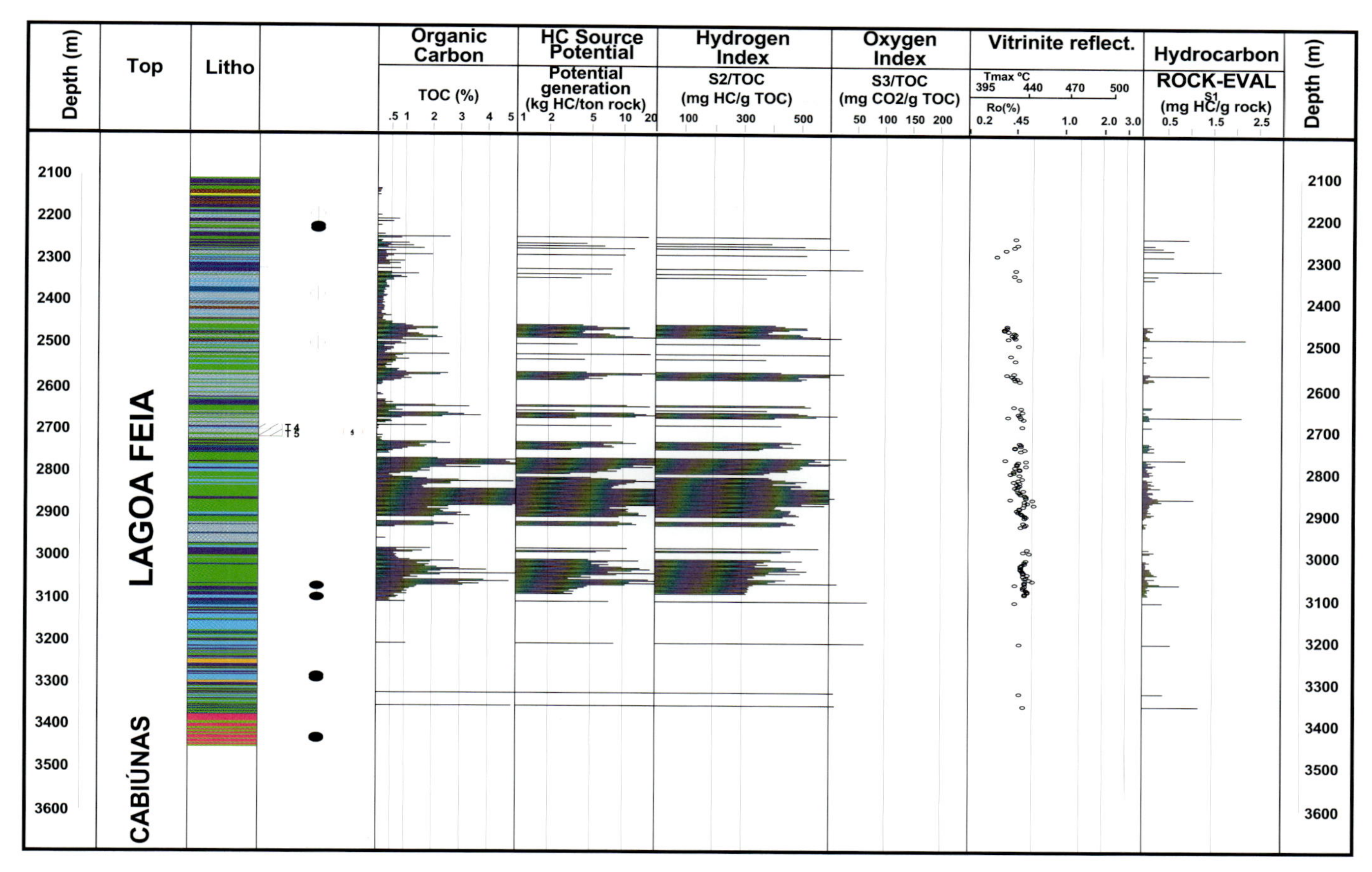

Figure 4—Selected geochemical well log from the Campos Basin showing the hydrocarbon source potential of calcareous shales of the Lagoa Feia Formation.

Camposella dominated the bottom waters in extremely high abundance. During the middle Aptian, these meridional waters extended to the northeast away from the continent, and the marine bivalves *Agelasina* and *Remondia* appeared, reflecting increased salinity (Mello and Hessel, 1998). These observations are consistent with isotopic ratios ($\delta^{13}C$) of organic matter from the Lagoa Feia shales, which show a tendency toward more positive values higher up in the sequence, also suggesting progressively increasing water salinity (Carvalho et al., 1996).

RESERVOIRS

Reservoir rocks with good to excellent permeability and porosity are widespread in the Campos Basin both in time and space. The Barremian carbonates (coquinas) of the rift sequence have porosities reaching 15–20% and permeabilities as high as 1 darcy (d). The best reservoirs consist of high-energy bivalve grainstones having intergranular porosity. Grainstones and packstones with secondary porosity (moldic–vugular) also constitute reservoirs, as do Neocomian basalts in which vesicles and fractures have created high porosity and permeability.

The Albian shallow-water shelf carbonates comprise a broad spectrum of reservoirs. The highest quality units, with porosities up to 28% and permeabilities of >1 d, are related to oolitic facies deposited in a high-energy environment. There has been little secondary cementation, and therefore much of the original intergranular porosity has been preserved. Porous (up to 30%) but less permeable (up to 100 md) reservoirs are related to oncolitic–peloidal packstones and grainstones deposited in a shallow marine environment with moderate water agitation. Fine-grained limestones deposited in a deeper and lower-energy setting also constitute reservoirs, but despite their high porosities (20–30%), they have relatively low permeabilities (maximum of a few millidarcys).

The Upper Cretaceous turbidites were deposited mainly in gentle slope troughs formed as a result of halokinesis. On the basis of foraminifera, paleobathymetry ranged from upper to lower bathyal. Porosities range from 20 to 25% and permeabilities from 100 md up to 5 d.

The main Tertiary siliciclastic reservoirs (Oligocene–Miocene) are predominantly medium- to fine-grained

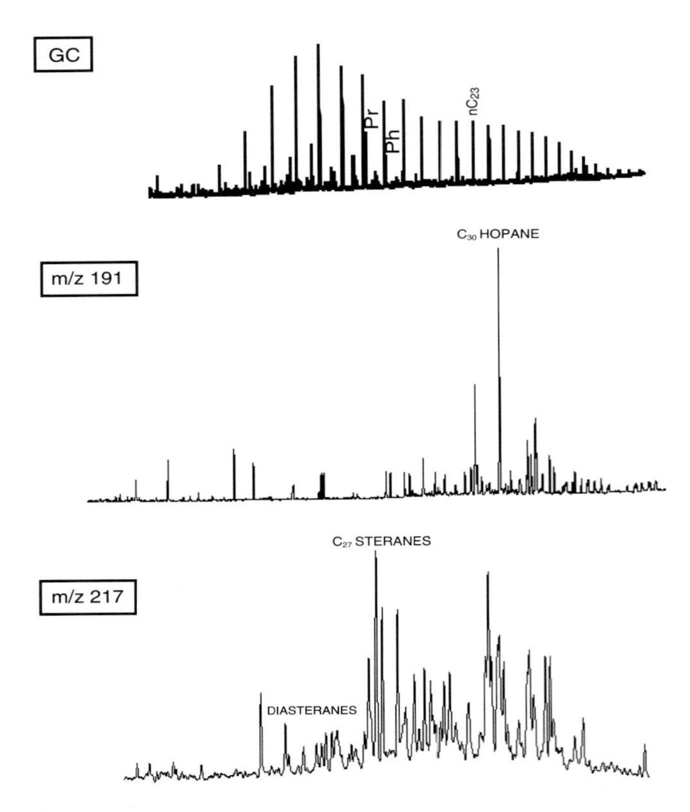

Figure 5—GC and GCMS data for a typical oil from the Campos Basin sourced from the calcareous organic-rich section of the Lagoa Feia Formation.

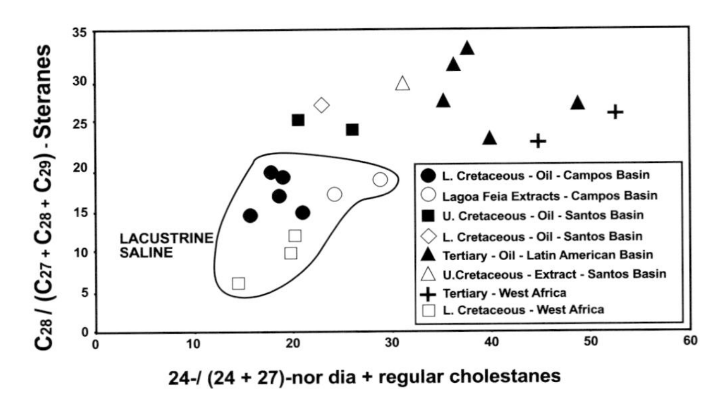

Figure 6—Plot of GCMS-MS data for South Atlantic margin oils and organic-rich extracts. Note the distribution of the lacustrine saline samples from the Campos Basin.

sandstones with porosities of about 30% and permeabilities of several darcys. These turbidite complexes developed as extensive basin-floor fans. Locally, there are also mass-wasting deposits and cross-bedded sandstones formed from tidal reworking.

PETROLEUM SYSTEM

The vast oil reserves of the Campos Basin are directly related to the widespread occurrence of the prolific Barremian lacustrine source rocks of the Lagoa Feia Formation. Petroleum generation began in Santonian–Coniacian time, reached its peak during the late Miocene, and continues until the present day (Mello et al., 1994). The oil fields located within the rift section are associated with intrabasinal highs, where fault movements created monoclinal structures that plunge basinward (Guardado et al., 1989). The oil migrated laterally from adjacent organic-rich calcareous shales and accumulated in porous coquinas that were deposited on the flanks of regional highs.

Most of the oil discovered in the Campos Basin (40 billion bbl of oil in place, with reserves estimated at 10.7 billion bbl) (Lucchesi and Gontijo, 1998) is related to the Lagoa Feia–Carapebus(!) petroleum system. The events chart for this petroleum system is shown in Figure 7. The main fields, including Marlim and Barracuda (Tertiary reservoirs) and Roncador (Cretaceous reservoir), are located at water depths between 250 and 2600 m. The giant Tertiary oil fields are predominantly stratigraphic traps (Lucchesi et al., 1995) controlled by the lateral pinch-out of turbidite reservoirs to the west and by regional dips toward the east. Seismic amplitude anomalies have delineated oil-saturated Oligocene turbidite reservoirs, as seen in Figure 8. Identification of these anomalies has substantially lowered exploration risk in the Campos Basin (Guardado et al., 1997). Although halokinesis has not played a major role in defining reservoir geometry and facies distribution, it has played an important role in determining trap geometry (Rangel and Martins, 1998).

Discovered in 1996, the Roncador field is the largest field reservoired within Cretaceous (Maastrichtian) turbidites in the basin. Its geometry is complex and is predominantly controlled by accommodation space related to halokinesis and preservation from subsequent erosion (Santos et al., 1998). It is important to note that only the uppermost reservoir zone displays a seismic amplitude anomaly (Figure 9). This is not observed in lower parts of the reservoir (Rangel et al., 1998).

Synchronous generation and entrapment provided the favorable conditions for establishment of the prolific Lagoa Feia–Carapebus(!) petroleum system. The hydrocarbon migration model for the postsalt sequence (Figure 10) includes migration from Barremian rift source rocks through salt windows and upward along listric faults (e.g., Meister, 1984; Mohriak et al., 1990; Mello et al., 1994).

CONCLUSIONS

All essential petroleum system elements and processes have resulted in one of the most prolific hydrocarbon provinces within the South Atlantic sedimentary basin system.

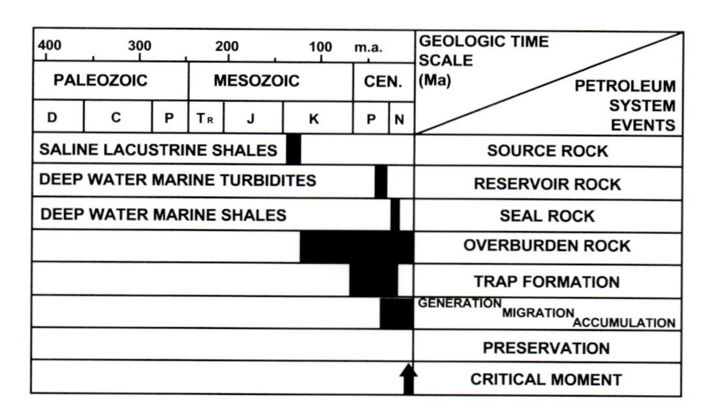

Figure 7—Events chart for the petroleum system of the Campos Basin.

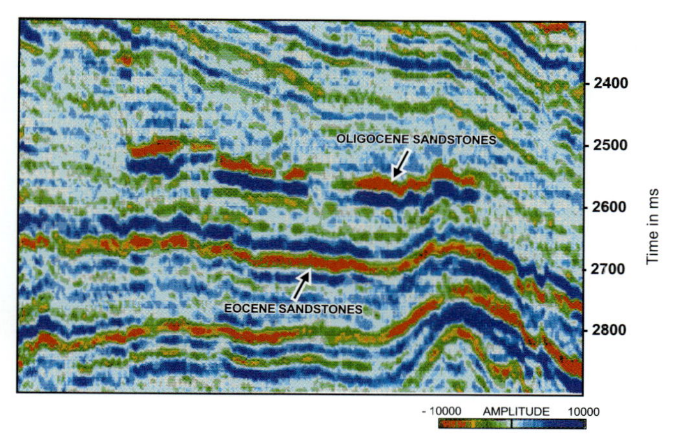

Figure 8—Part of a seismic section from the Campos Basin showing amplitude anomalies related to the oil-saturated Oligocene and Eocene turbidite reservoirs.

These include the following:

- Excellent lacustrine saline source rocks (Lagoa Feia Formation)
- High-quality Cretaceous–Tertiary turbidite sandstone reservoirs (Carapebus Formation)
- Effective traps created by halokinesis associated with excellent seals
- Salt windows and related listric fault systems associated with good carrier beds allowing migration from source rock to trap
- Synchronous generation and trap formation associated with an adequately focused drainage system.

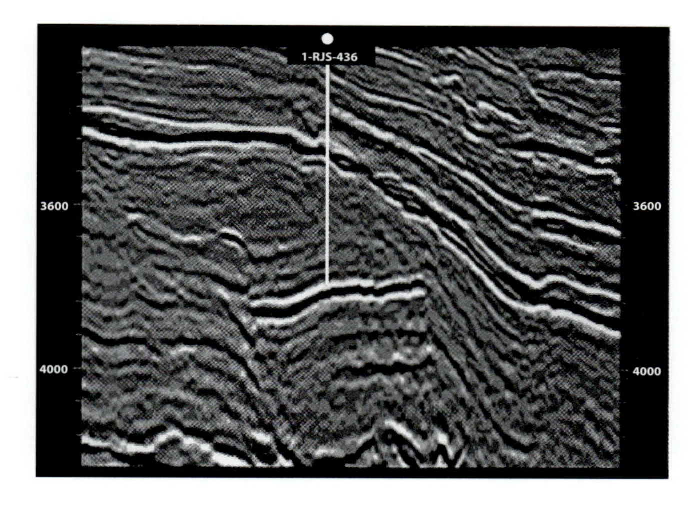

Figure 9—Part of a seismic section showing the amplitude signature of the Roncador oil field in the Campos Basin.

REFERENCES CITED

Brongersma-Sanders, M., 1957, Mass mortality in the sea: GSA Memoir 67, n. 1, p. 941–1010.

Carvalho, M. D., U. M. Praca, J. L. Dias, A. C. Silva-Telles, T. Horschutz, M. H. R. Hessel, M. S. Scuta, A. S. C. Barbosa, L. C. S. Freitas, and A. D. Sayad, 1996, Coquinas da formação Lagoa Feia da bacia de Campos estudo sedimentológico na caracterização da qualidade de reservatório: Petrobrás Internal Report, Rio de Janeiro, 188 p.

Guardado, L. R., L. A. P. Gamboa, and C. F. Lucchesi, 1989, Petroleum geology of the Campos Basin, Brazil: a model for producing Atlantic type basins, *in* J. D. Edwards and P. A. Santogrossi, eds., Divergent/passive margin basins: AAPG Memoir 48, p. 3–80.

Guardado, L. R., B. Wolff, and J. S. L. Brandao, 1997, Campos Basin, Brazil, a model for a producing Atlantic basin: OTC Proceedings, 1997 Offshore Technology Conference, Houston, Texas, p. 457–462.

Lucchesi, C. F., and J. E. Gontijo, 1998, Deep water reservoir management: the Brazilian experience: OTC Proceedings, 1998 Offshore Technology Conference, Houston, Texas, p. 625–628.

Lucchesi, C. F., C. C. Martins, C. A. Costa, and L. R. Guardado, 1995, 3-D seismic as an exploration and production tool: the Campos Basin experience: OTC Proceedings, 1995 Offshore Technology Conference, Houston, Texas, p. 507–512.

Meister, E. M., 1984, Geology of petroleum in Campos Basin, Brazil (abs.): AAPG Bulletin, v. 68, n. 4, p. 506.

Mello, M. R., 1988, Geochemical and molecular studies of the depositional environments of source rocks and their derived oils from the Brazilian marginal basins: Ph.D. dissertation, Bristol University, Bristol, U.K., 240 p.

Mello, M. R., and M. H. Hessel, 1998, Biological marker and paleozoological characterization of the early marine incursion in the lacustrine sequences of the Campos Basin, Brazil (abs.): Extended Abstracts Volume, AAPG Annual Convention, Salt Lake City, Utah, v. 2, A455.

Mello, M. R., and J. R. Maxwell, 1990, Organic geochemical and biological marker characterization of source rocks and

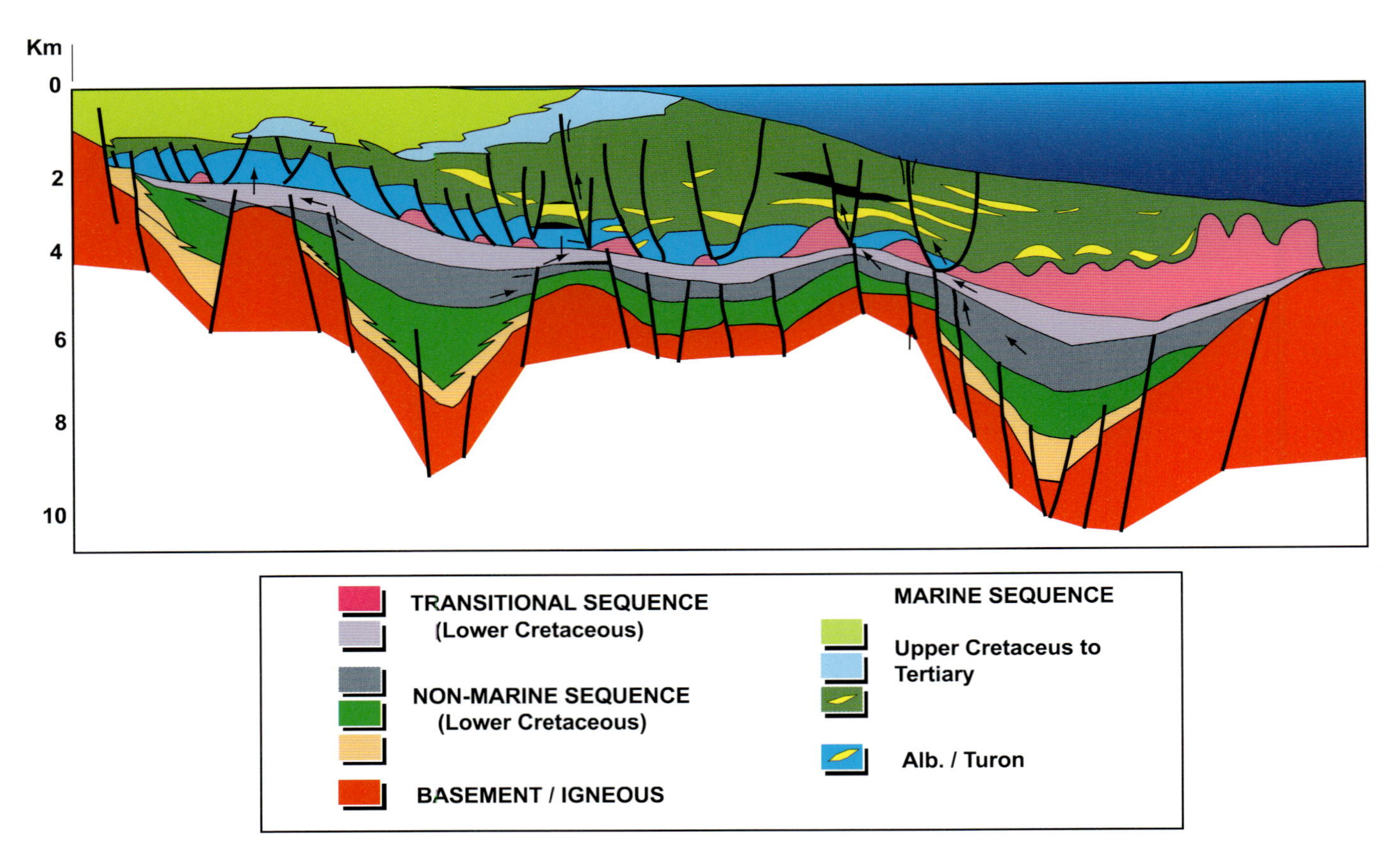

Figure 10—Schematic geologic section of the Campos Basin showing the structural and stratigraphic framework along compartments. From Rangel and Martins (1998).

oils derived from lacustrine environments in the Brazilian continental margin, *in* B.J. Katz, ed., Lacustrine basin exploration—case studies and modern analogs: AAPG Memoir 50, p. 77–99.

Mello, M. R., W. U. Mohriak, E. A. M. Koutsoukos, and G. Bacoccoli, 1994, Selected petroleum systems in Brazil, *in* L. B. Magoon and W. G. Dow, eds., The petroleum system—from source to trap: AAPG Memoir 60, p. 499–512.

Mohriak, W., M. R. Mello, J. F. Dewey, and J. R. Maxwell, 1990, Petroleum geology of the Campos Basin, offshore Brazil, *in* J. Brooks, ed., Classic petroleum provinces: Geological Society of London Special Publication, p. 119–142.

Moldowan, J. M., F. J. Fago, C. Y. Lee, S. R. Jacobson, D. S. Watt, N. E. Slougui, A. Jeganathan, and D. C. Young, 1990, Sedimentary 24-*n*-propylcholestanes, molecular fossils diagnostic of marine algae: Science, v. 247, p. 309–312.

Rangel, H. D., and C. C. Martins, 1998, Main exploratory compartments, Campos Basin, *in* Searching for oil and gas in the land of giants: Search, Rio de Janeiro, Schlumberger, p. 32–40.

Rangel, H. D., P. R. Santos, and C. M. S. P. Quintaes, 1998, Roncador field, a new giant, in Campos Basin, Brazil: OTC Proceedings, 1998 Offshore Technology Conference, Houston, Texas, p. 579–587.

Santos, P. R., H. D., Rangel, C. M. S. P. Quintaes, and J. M. Caixeta, 1998, Turbidite reservoir distribution in Roncador field, Campos Basin, Brazil (abs.): Extended Abstracts Volume, AAPG International Conference & Exhibition, Rio de Janeiro, p. 290.

Soldan, A. L., J. R. Cerqueira, J. C. Ferreira, L. A. F. Trindade, J. C. Scarton, and C. A. G. Corá, 1995, Giant deep water oil fields in Campos Basin, Brazil: a geochemical approach: Revista Latino-Americana de Geoquímica Orgânica, v. 1, n. 1, p. 14–27.

Cole, G. A., A. G. Requejo, D. Ormerod, Z. Yu and A. Clifford, 2000, Petroleum geochemical assessment of the Lower Congo Basin, *in* M. R. Mello and B. J. Katz, eds., Petroleum systems of South Atlantic margins: AAPG Memoir 73, p. 325–339.

Chapter 23

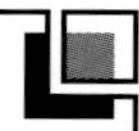

Petroleum Geochemical Assessment of the Lower Congo Basin

G. A. Cole

A. G. Requejo*

D. Ormerod

Z. Yu

A. Clifford

BHP Petroleum
Houston, Texas, U.S.A.

**Consultant to BHP*
Present address: *Geochemical Solutions International*
The Woodlands, Texas, U.S.A.

Abstract

Petroleum geochemical data and burial history, thermal, and fluid flow models show that the southern Lower Congo Basin contains dual petroleum systems. The older petroleum system includes a presalt type I lacustrine source (Barremian Bucomazi Formation time equivalent) that is present but not well understood in the deeper offshore Lower Congo Basin. The lacustrine synrift grabens are generally obscured by salt on seismic profiles, and there have been few well penetrations except on the shelf. However, based on models of burial history, thermal evolution, and fluid flow, we have found that the lacustrine sequence matures offshore and that migration is focused updip toward the shelf where the reservoired oils correlate to this source. A younger petroleum system consists of type II marine clastic shales of the Cenomanian–Turonian (Upper Cretaceous) Iabe Formation that have charged the overlying Tertiary sandstones in deep-water areas. These turbidite reservoir sandstones, although charged primarily by the Cretaceous Iabe marine clastic source, show some evidence of mixing with early oils from the Tertiary Landana Formation. Mixing of oils between the two petroleum systems is limited to those areas where the Loeme Salt has evacuated sufficiently to allow migration through an assumed permeable salt weld.

INTRODUCTION

Angola is one of the major oil producing countries of Africa. After Nigeria, it is the most significant oil producer in the sub-Saharan region, with total production approaching 650,000 bbl per day and recoverable reserve estimates of about 5.4 billion bbl (International Petroleum Encyclopedia, 1996). Angola's known oil fields are located primarily in the Cabinda enclave, both onshore and offshore. More recent deep-water discoveries in the southern Lower Congo Basin (e.g., the Girassol and Dalia fields, which are estimated to contain several billion barrels of oil each) have enhanced exploitable reserves.

The evaluation of the petroleum potential of a sedimentary basin requires a thorough understanding of the petroleum system elements that control the accumulation of oil and gas. The primary controlling elements of the petroleum system are an actively generating and expelling source pod (or a source pod that has expelled after trap formation), a reservoir unit (typically sandstone) that can trap hydrocarbons, a seal for the reservoir unit, and overburden rock (Magoon and Dow, 1994). The main process that controls the accumulation of oil and gas within a basin, however, is the migration of hydrocarbons from the actively expelling source pod to the trap. One way to understand these mechanisms, both as a process and in terms of timing, is through the use of 2-D basin modeling methods.

This chapter describes the major elements of the petroleum systems in the southern Lower Congo Basin, specifically: (1) the basic petroleum geology, (2) the principal source rocks and their generative hydrocarbon potential, (3) the relationship between reservoired hydrocarbons and effective source rocks, and (4) the migration pathways that have been modeled to determine the timing of expulsion and filling of traps. The discussion focuses on how these elements exhibit broad regional trends resulting in "compartmentalization" of the basin through salt raft tectonics.

PETROLEUM GEOLOGY

Rift phase strata deposited during the Early Cretaceous extensional opening of the South Atlantic have economic importance as both reservoir facies and highly prolific hydrocarbon source rocks. Tectonic regimes (Figure 1) associated with continental separation can be divided into four stages (Brice et al., 1982), from oldest to youngest:

- prerift,
- synrift,
- synrift II, and
- postrift (with regional subsidence).

Prerift sediments were deposited on faulted metamorphic basement prior to major continental rifting. This section consists of a sandy fluvial–lacustrine sequence that is up to 1000 m in thickness (Brice et al., 1982). In the synrift I stage, a range of graben and half-graben troughs along the early rifted margins of western Africa and Brazil formed the depocenters for organic-rich lacustrine shales. The lake systems that developed within the grabens were progressively filled by lacustrine turbidites that graded laterally and upward into organic-rich shales. In the coastal basins of Angola (Lower Congo and Kwanza), this rifting stage produced the bituminous shales of the Bucomazi Formation and its time equivalents, the primary synrift source rocks for major oil accumulations along the west African margin (Burwood et al., 1995; Burwood, 1999). These organic-rich shales can attain a maximum thickness of 1800 m (McHargue, 1990). Beyond Angola, time-equivalent organic-rich lacustrine shale deposits along the South Atlantic margin include the Marnes Noires and Melania Formations of Congo and Gabon, respectively, as well as the Lagoa Feia, Guaratiba, and Mariricu Formations of Brazil.

The synrift I stage was followed by a period of regional subsidence (synrift II stage), resulting in marine incursions. This transitional sequence consists of lacustrine carbonates and sandstones and alluvial clastics and is marked by a transition from nonmarine lacustrine conditions to marine conditions. In the Lower Congo Basin of Angola, saline waters initially entered from the south. By the end of this stage, active marine transgression culminated in deposition of carbonates and a desiccation event that led to deposition of the Loeme Salt, a regional to subregional seal between the synrift lacustrine petroleum system and the overlying marine petroleum system.

Following evaporite deposition, permanent marine conditions and carbonate-dominated sedimentation accompanied the postrift phase and passive margin opening of the South Atlantic Ocean. The Albian–earliest Cenomanian Pinda and Cabo Ledo (Moita Seca) Formations are the first postsalt units documented through well control. These units fully span continental clastics to marine shaly marls (700–800 m) and are thought to contain secondary hydrocarbon potential. Postsalt source rock deposition could have been influenced by either of two mechanisms: (1) upwelling similar to that observed in the present-day Walvis Bay area, or (2) oceanic anoxic events (OAE) resulting from an expanded oxygen minimum layer.

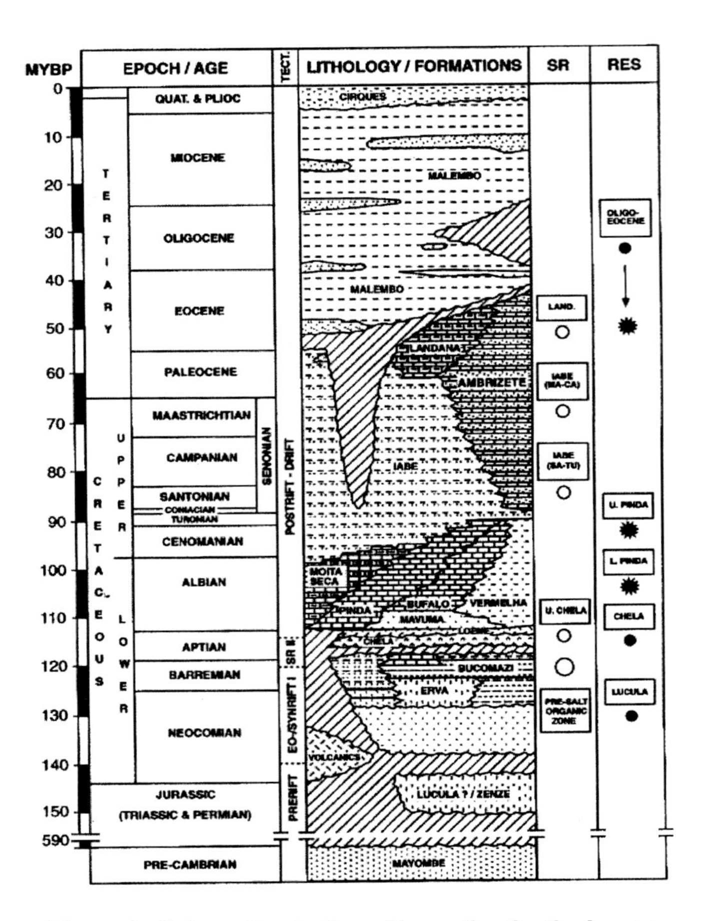

Figure 1—Schematic stratigraphic section for the Lower Congo Basin showing the primary and secondary source rock and reservoir units. After Burwood et al. (1990).

The Cenomanian–early Eocene section contains several groups and constituent formations deposited under conditions conducive to petroleum source rock formation. These include the Upper Cretaceous Iabe and the Paleogene Landana Formations of the Lower Congo Basin. The Iabe Formation encompasses the time period during which OAE source rocks would have been deposited, although the main Iabe source rock development occurred in Turonian–early Tertiary time. Salt movement during this time resulted in growth faults and the formation of salt domes and ridges. At the end of the Paleogene, the seaward edge of the continent foundered, producing a strong westward tilt and regional subsidence (postrift stage). During the Oligocene–Miocene, a thick regressive sequence consisting of turbiditic sands and silty clays was unconformably deposited across the older shelf sequence. This clastic sequence built the continental shelf to its present position. A generalized stratigraphic section highlighting reservoir and potential source units is shown in Figure 1.

During Oligocene–Miocene deposition, the resultant gravity sliding, or *salt raft tectonics* , produced detached sections referred to as "Pinda bumps," which are Upper Cretaceous shelf strata that are displaced seaward. These bumps may represent source rock "mini-kitchens" that could charge local structures depending on overburden and heat flow considerations (sufficient to mature the source rock pods). Another important feature is the extensive faulting throughout much of the Tertiary section; these faults likely acted as conduits for hydrocarbon migration.

Significant salt diapirs and ridges are also evident in the Lower Congo Basin and are most likely the result of periodic halokinesis associated with tectonic activity along numerous transfer fault zones. These diapirs increase in size and frequency westward into the ultra-deep-water areas of the Lower Congo Basin (Blocks 31 and 32). The salt, which has been active since the Albian, acts to concentrate and encapsulate potential reservoir sandstones in their vicinity. The Albian section may also contain porous carbonates that could act as reservoirs.

PETROLEUM GEOCHEMISTRY

The geochemical aspects of a petroleum system are identification of the primary source rocks responsible for any commercial accumulation of oil and gas within a basin and the correlation of reservoired oil and gas to specific source units. Major source rock intervals within the Lower Congo Basin are developed within both the pre- and postsalt sections. Primary effective source units are likely associated with Late Cretaceous flooding events in the deep-water areas and with the synrift lacustrine section both nearshore and onshore.

Synrift Presalt Lacustrine Source Rocks

Previous geochemical studies in Angola have focused largely on the presalt succession in the coastal Lower Congo Basin. Burwood et al. (1990, 1992, 1995), Mycke and Burwood (1995), and Burwood (1999) describe the source rock potential and molecular characteristics of the Bucomazi and Chela Formations. The wells studied (CABGOC 86-1 and 123-4) are in the Cabinda enclave north of our study area. To summarize these papers, the presalt Bucomazi and Chela sequence is comprised of late Barremian–Aptian sedimentary rocks that were initially deposited in a freshwater environment which evolved into a brackish water to saline depositional setting. The Bucomazi section has been subdivided into four zones, representing geochemically discrete organofacies and depositional settings.

The oldest zone, zone D, is the basal Bucomazi interval which consists of fissile, dark brown argillaceous siltstones containing predominantly amorphous organic matter. These were deposited in a deep, anoxic, freshwater lake setting with periodic slumping of shoreline sediments. Total organic carbon (TOC) contents range from 5 to 10% and hydrogen indices (HI) from 650 to 800 mg HC/g TOC, suggesting type I kerogens. Kerogen kinetics of the richest intervals show the predominance of a single activation energy of about 56–58 kcal/mol, characteristic of type I kerogens.

Zone C consists of more calcareous, microlaminated carbonaceous strata containing predominantly amorphous organic matter deposited in shallower water, anoxic, alkaline lake settings. TOC ranges from 2 to 5% and HI from 550 and 800 mg HC/g TOC, corresponding to more typical type II kerogens. Kerogen kinetics also show a narrow distribution at a lower activation energy (54 kcal/mol) than zone D, suggesting an organofacies change from that zone.

Strata of zone B were also deposited under lacustrine conditions but in shallower, brackish waters under less pervasive anoxia, marking the onset of "sheet drape" sedimentation (synrift II) of Bouroullec et al. (1991). Organic-rich claystones have TOC contents of 7–13%; otherwise TOC averages about 1%. Organofacies of the richer intervals contain type II kerogens with intermediate HI values of 400 mg HC/g TOC. In contrast to the underlying zones, kerogen kinetics show variable activation energies of broad distribution averaging 52 kcal/mol, suggesting mixed assemblages of organic matter types.

Zone A marks the conclusion of Bucomazi sedimentation. These strata represent deposition in a shallow, low-energy lacustrine environment, possibly indicating a progressive contraction and drying out of the lake. A predominantly oxic water column with dysoxic periods is indicated by relatively low TOC contents (about 1%), except at the basal interval where values exceed 5.0%. HI values up to 500 mg HC/g TOC are recorded in the organic-rich interval but are generally less than 400 mg HC/g TOC where TOC is lower. Kerogen kinetic data show a broad activation energy distribution (maximum 52 kcal/mol) typical of type II organic matter assemblages.

The focus of this chapter is the deep-water (>500 m) to ultra-deep-water (>1500 m) portion of the southern Lower Congo Basin, where the presalt section has not been penetrated (the nearest penetrations are on the shelf to the east). Figure 2 shows a crossplot of TOC versus the S_2 yield from Rock-Eval pyrolysis for potential source rocks in the southern Lower Congo Basin. On the basis of this data and the limited presalt penetrations, we believe development of good, thick organic-rich shales is questionable within the lacustrine section. However, organic-rich potential source rocks exceeding 4–5% TOC are evident within some wells. Burwood (in press) has recently summarized source rock data for this area, including both pre- and postsalt candidates.

Of greater importance is the kinetic breakdown of the lacustrine organic-rich shales. To address generation and expulsion, a 1-D section was modeled using burial history and thermal modeling software developed by Institut Français du Pétrole (IFP) and Beicip GENEX. Results are shown in Figure 3 for the lacustrine section as determined in a pseudowell constructed from a regional

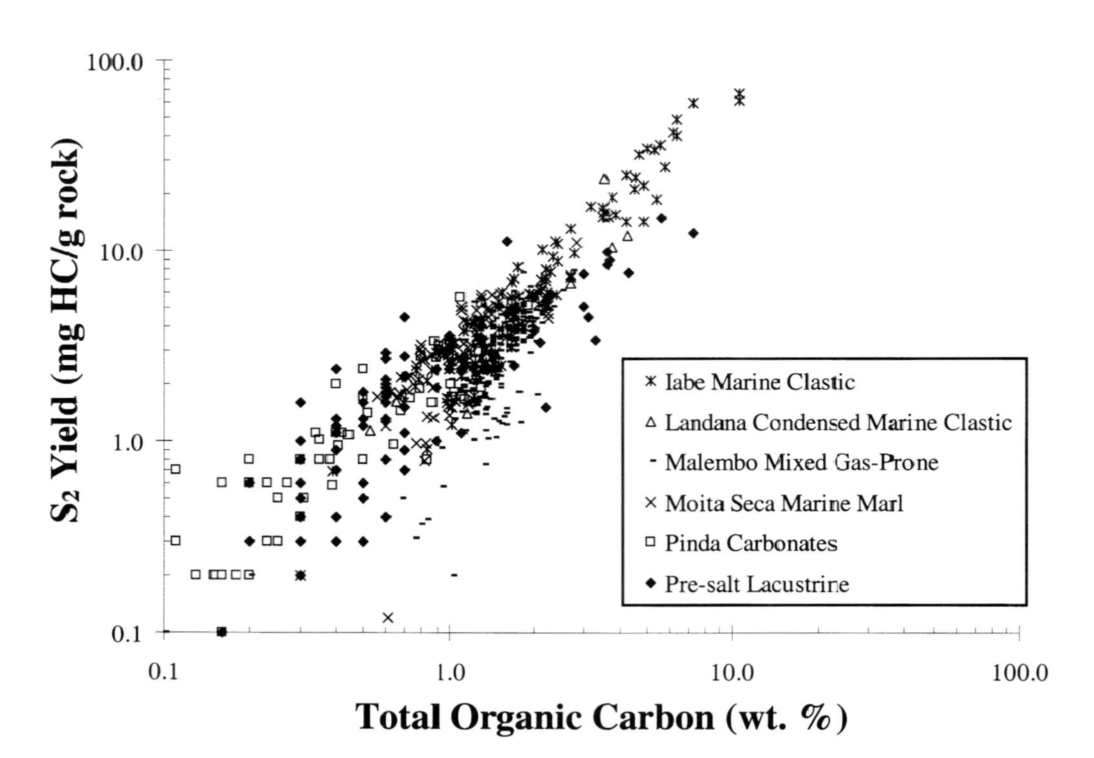

Figure 2—Crossplot between TOC (in wt. %) and the S_2 yield from Rock-Eval pyrolysis showing that the two primary source rocks are the postsalt marine Iabe Formation and presalt lacustrine units (Bucomazi Formation time equivalents).

seismic dip line across the southern Lower Congo Basin that was calibrated to the nearest well data. The presalt section was not imaged at this site but was assumed to be present; it has a thickness of 20 m, an average TOC of 6%, and an HI of 600 mg HC/g TOC. In this and other burial history and thermal models presented (both 1-D and 2-D), we used default kinetic parameters as described in the literature (Tissot and Espitalié, 1975; Ungerer et al., 1986; Ungerer and Pelet, 1987; Sweeney and Burnham, 1990). This simulation showed that the lacustrine source rock, if present at this location, would have expelled its hydrocarbons during the early Miocene. These results indicate that this source rock could have contributed significant hydrocarbons into any presalt traps (assuming the presence of a Loeme Salt seal) or could have contributed hydrocarbons into the postsalt section (Tertiary sandstones or Pinda carbonate reservoirs) through salt withdrawal windows (welded zones).

Postrift Marine Source Rocks

The earliest postrift section is the extensive Loeme Salt, which formed during the early phase of separation of the African and South American continents. The postrift phase comprises the regressive Albian–Cenomanian carbonates that encompass shallow-water to platform facies with minor shelfal sandstones. A major source rock sequence, the marine clastic Iabe Formation (Cenomanian–Turonian), occurs in the Upper Cretaceous postsalt section. Other contributing source rocks may be the uppermost Albian–Cenomanian Moita Seca and the lower Tertiary Landana. Overlying these are a series of localized deltaic systems that built into local accommodation zones within an overall transgressive regime. The overlying major regressive cycle is associated with Eocene–Holocene deltaic clastics.

The uplift in onshore Angola, in combination with major lowstands of sea level, generated large submarine fan systems within the offshore Lower Congo Basin. Deeply incised canyon systems on the shelf to slope region can be seen within the shelfal area. These acted both as bypass mechanisms for the submarine fan sediments and as a means of focusing the sediment load. The sands were shed from the major delta system that emptied into the north–south trending mini-basins formed by salt and raft tectonics. The sands filled from the north to the south, with the paths between the mini-basins varying with salt halokinesis. As particular diapirs, salt walls, and turtleback structures switched the rate of accommodation within a mini-basin, the likely overall path of the turbidite flows varied. It is this combination of postsalt source rock "pod" development and reservoir deposition/trap development that is important in the deep-water and ultra-deep-water petroleum systems of the southern Lower Congo Basin. Of even more critical importance is where these source rocks matured and how they charged the Tertiary turbidite sandstones.

The most dominant postsalt source rock is the marine clastic Cenomanian–Turonian Iabe Formation, which includes intervals having in excess of 10% TOC with excellent oil-prone quality (Figure 2) (Burwood, 1999). Secondary source rocks are the Albian–lower Cenomanian Moita Seca marine shales and marls, which have

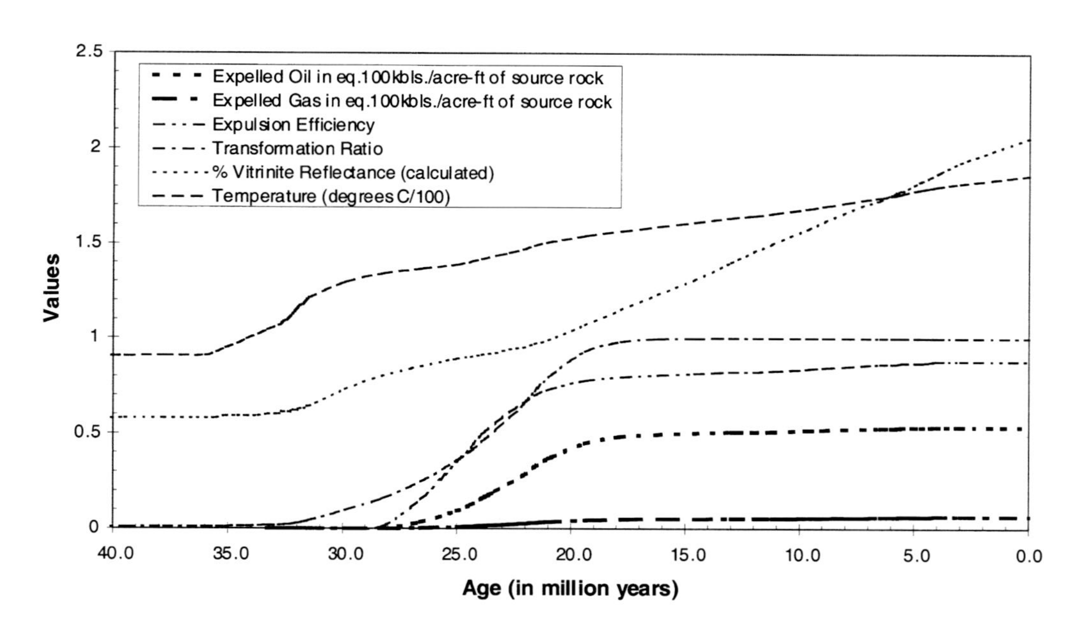

Figure 3—Calculated curves for oil and gas expulsion, expulsion efficiency, temperature, transformation ratio, and vitrinite reflectance for a presalt Bucomazi time equivalent (lacustrine) source unit in a pseudo-well in the deep-water part of the southern Lower Congo Basin. (Units used for the temperature plot are in °C/100 in order to display at this scale.)

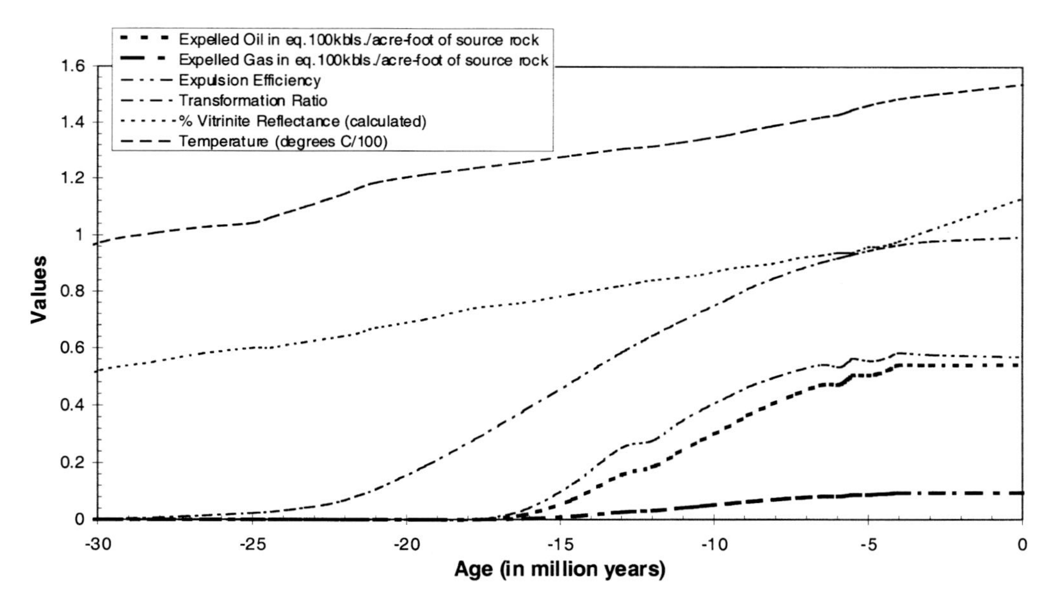

Figure 4—Calculated curves for oil and gas expulsion, expulsion efficiency, temperature, transformation ratio, and vitrinite reflectance for the postsalt marine clastic Iabe source unit in a pseudo-well in the deep-water part of the southern Lower Congo Basin. (Units used for the temperature plot are in °C/100 in order to display at this scale.)

TOC values of up to 3% with oil-prone quality, and the lower Tertiary Landana, with 4% TOC and oil-prone quality (Figure 2). The Tertiary Landana remains only a potential source rock across most of the deep-water region because it may not be sufficiently mature to expel hydrocarbons. In some localized areas, however, it has apparently contributed to accumulations, as indicated by oil–source correlations (see below).

As before, the kinetic breakdown of the postsalt marine type II Iabe organic-rich shales is significant. To address generation and expulsion for the Cretaceous Iabe source unit, the same 1-D section discussed above was used to determine the efficiency of expulsion, the amounts of oil and gas expelled at calculated (and calibrated) temperatures, and the vitrinite reflectance values. The estimates are based on a 75-m-thick Iabe source rock, with a TOC of 4% and an HI of 550 mg HC/g TOC, which was observed in one of the salt-rafted source pods in the deep-water part of the Lower Congo Basin. As illustrated in Figure 4, this unit can attain sufficient maturity to expel significant amounts of oil and associated gas within the deep-water parts of the basin. Expulsion has occurred quite recently, mainly within the last 15 m.y. Since trap formation occurred during Oligocene–Miocene time, expulsion and migration timing is very favorable for the trapping of oil and gas.

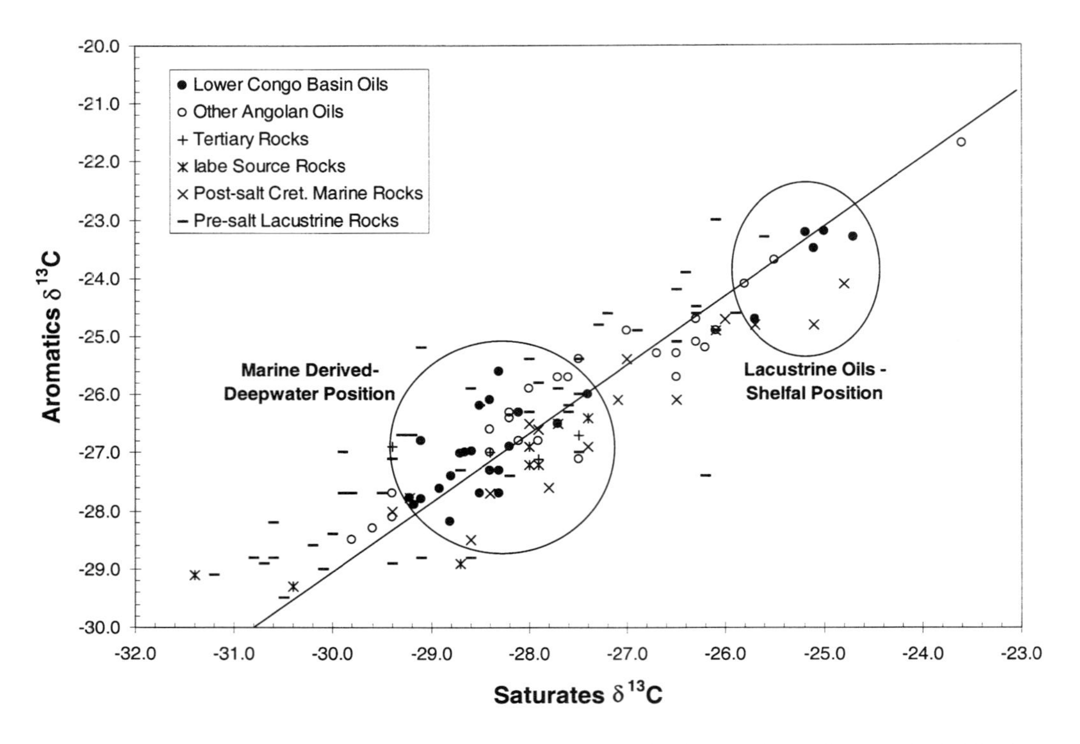

Figure 5—Crossplot of the carbon isotopic compositions of saturate and aromatic hydrocarbons isolated from soils and source rock bitumens.

Distribution of Oil Families

One way to assess the working petroleum systems within a basin is to define the number of oil families present and their respective extents, both regionally and stratigraphically. Oil families (or oil types) can be defined by comparing and contrasting geochemical characteristics such as biomarker and stable carbon isotopes of oils and source rock extracts (Curiale, 1993; Peters and Moldowan, 1993). Four geochemical characteristics were deemed most diagnostic for discriminating oil families in the Lower Congo Basin:

1. The hopane/sterane ratio, an indicator of source facies and depositional environment (Isaksen, 1991). Low values (<3) are primarily characteristic of source rocks containing type II kerogens that were deposited in distal marine and some carbonate environments, whereas higher values (>5) are derived from type I or type II/III source kerogens deposited in lacustrine and marine deltaic environments.
2. C_{26}/C_{25} tricyclic terpane ratios, which are higher for lacustrine-derived oils (values >1.0–1.1) relative to marine-derived oils (Zumberge, 1987; Burwood et al., 1992).
3. Saturate versus aromatic stable carbon isotope values and the isotopic canonical variable of Sofer (1984), both of which have been used to distinguish oil types. A more positive value for the isotopic canonical variable is characteristic of lacustrine or type III derived oils, whereas more negative values are associated with marine-derived oils.
4. Other empirical relationships observed within the biomarker data set, such as the relationship among C_{21}, C_{23}, and C_{24} tricyclic terpanes. In most cases, lacustrine-derived oils showed greater C_{23}/C_{21} and C_{24}/C_{21} tricyclic terpane ratios than marine-derived oils.

Using selected crossplots of the above properties, we were able to assess the origins and establish the different oil families that are present within the deep-water Lower Congo Basin. Figure 5 shows the relationship between the saturate and aromatic hydrocarbon isotopic compositions and illustrates how presalt lacustrine oils are isotopically heavier (more positive values) than marine postsalt oils. Source rock bitumen data are also included in Figure 5, and although some overlap is observed, distinct groupings are still evident. Figure 6 illustrates the variation in the C_{26}/C_{25} tricyclic terpane ratio relative to the isotopic canonical variable of Sofer (1984). This figure clearly shows the division between postsalt derived oils and their respective source rocks (Cretaceous Iabe, undifferentiated Tertiary, and other postsalt Cretaceous units) and presalt lacustrine derived oils and their source rocks (Bucomazi and Chela Formation time equivalents). Figure 7 illustrates differences in the C_{26}/C_{25} tricyclic terpane ratio and the hopane/sterane ratio between the postsalt marine section and the predominantly lacustrine presalt section. Postsalt marine source units and their respective oils have lower hopane/sterane ratios in relation to the lacustrine derived oils. Figure 8 shows a crossplot between the C_{26}/C_{25} tricyclic terpane ratio and the C_{24}/C_{21} tricyclic terpane ratio. Again, the postsalt marine oils and their respective source rocks can be readily differentiated from the presalt lacustrine oils and their respec-

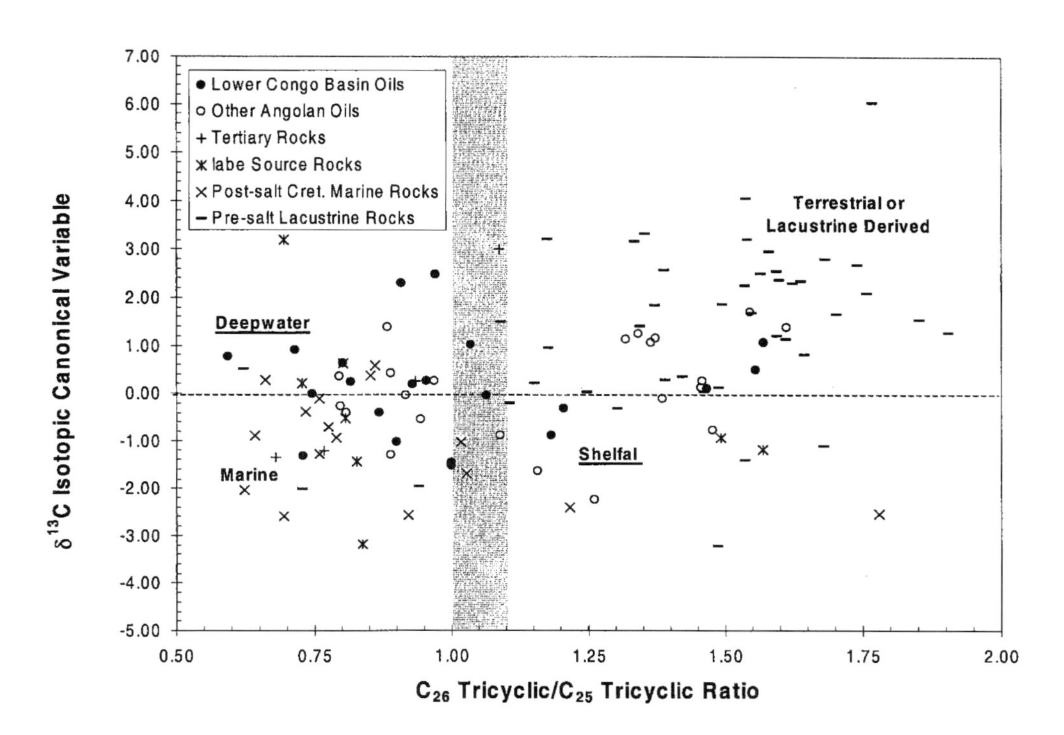

Figure 6—Crossplot of the C_{26}/C_{25} tricyclic terpane ratio and the isotopic canonical variable of Sofer (1984) for oils and source rock bitumens.

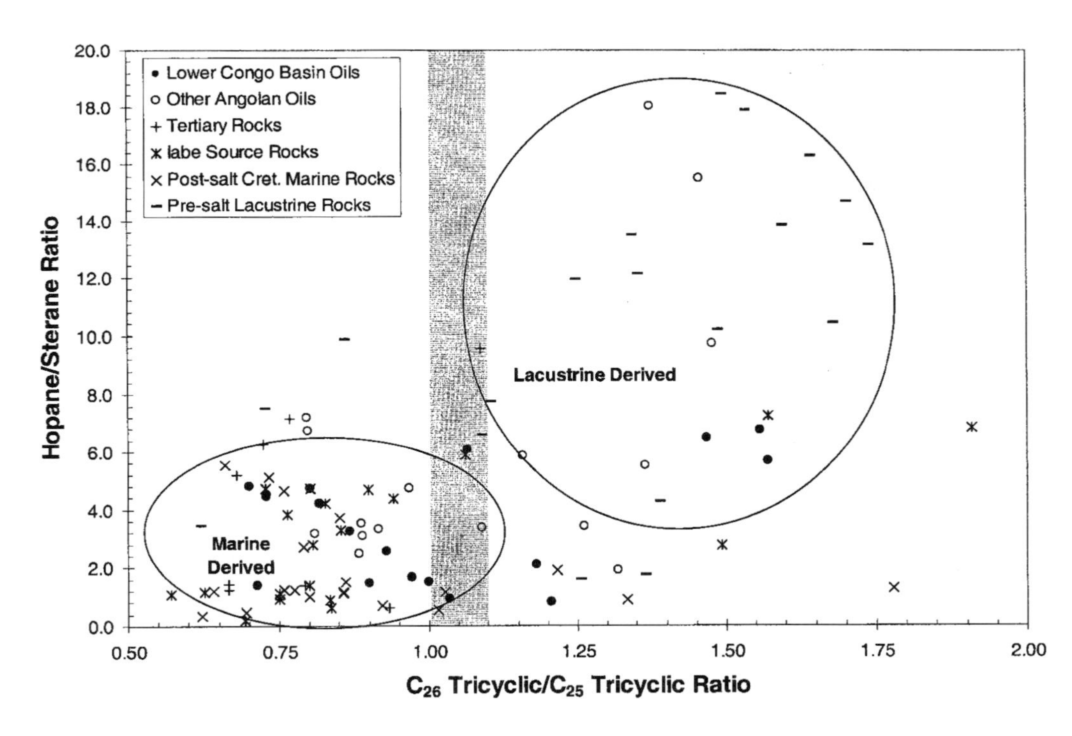

Figure 7—Crossplot of the C_{26}/C_{25} tricyclic terpane ratio and the hopane/sterane ratio for oils and source rock bitumens.

tive source units. The C_{24}/C_{21} tricyclic terpane ratio is generally lower for the postsalt marine oils and source rocks in comparison to presalt oils and lacustrine source rocks.

As summarized above, the two primary source rocks for the deep-water Lower Congo Basin oils are the lacustrine presalt section and the postsalt marine clastic Iabe Formation. A possible secondary source (where sufficiently mature) could be the Tertiary Landana Formation (formerly the Tertiary Iabe). Figure 9 shows a plot of the biomarker ratio oleanane/C_{30} hopane versus the C_{26}/C_{25} tricyclic terpane ratio. Oleanane is an indicator of organic contribution from angiosperm land plants, and its occurrence is commonly observed in post-Cretaceous source rocks (Ekweozor et al., 1979; Ekweozor and Telnaes, 1990). Based on this plot, some oils and shows in the

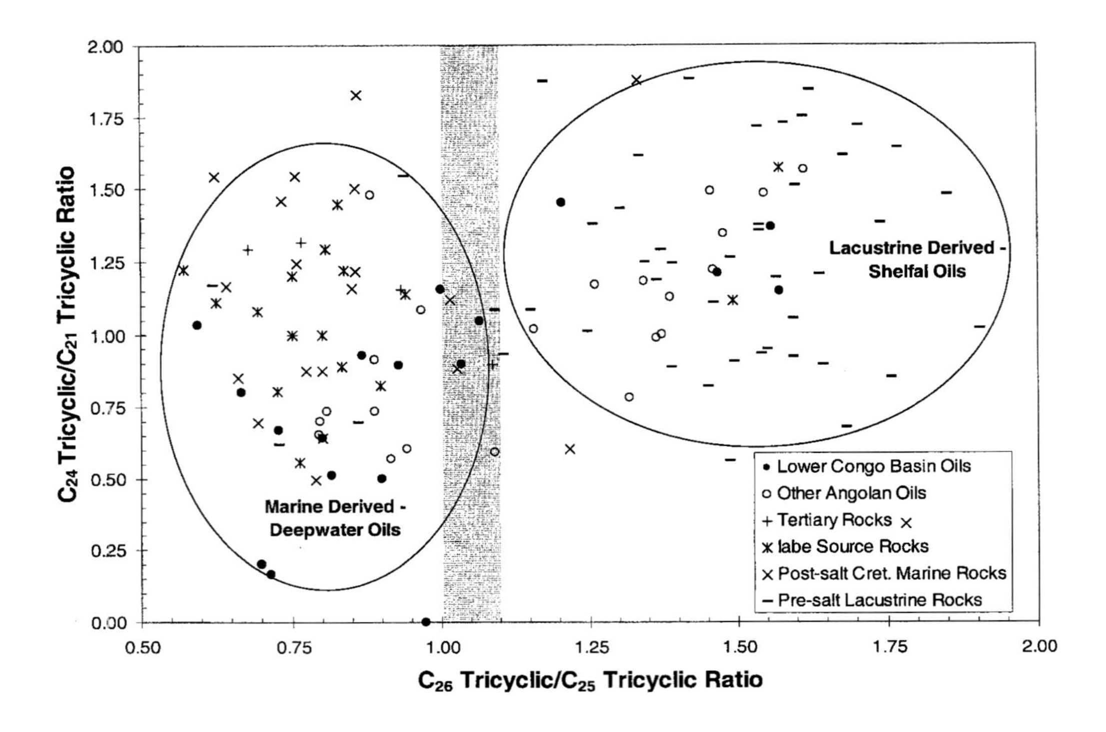

Figure 8—Crossplot of the C_{26}/C_{25} tricyclic terpane ratio and the C_{24}/C_{21} tricyclic terpane ratio for oils and source rock bitumens.

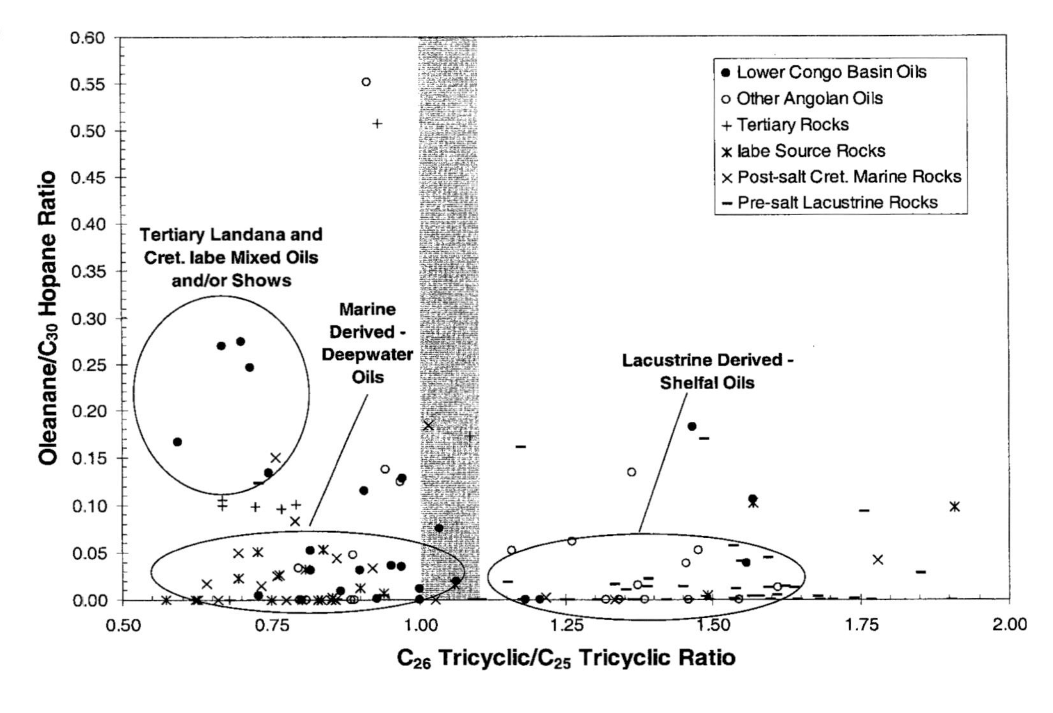

Figure 9—Crossplot of the C_{26}/C_{25} tricyclic terpane ratio and the oleanane/C_{30} hopane ratio for oils and source rock bitumens.

southern Lower Congo Basin contain elevated oleanane levels that strongly suggest some contributions from a Tertiary source rock. The only source unit with adequate source richness and potential that also contains olenane is the Tertiary Landana Formation (Figure 9).

On the basis of these data, we recognize two broad families or types of oils in the southern Lower Congo Basin. Figure 10 shows the regional distribution of these families. The shelfal oils (Pinda carbonate plays) were derived mostly from the lacustrine presalt section, whereas the deeper water areas are dominated by Iabe-derived oils or mixtures. This distribution is controlled by two factors: (1) the stratigraphic section, primarily where the Loeme Salt acts as a seal and isolates the two petro-

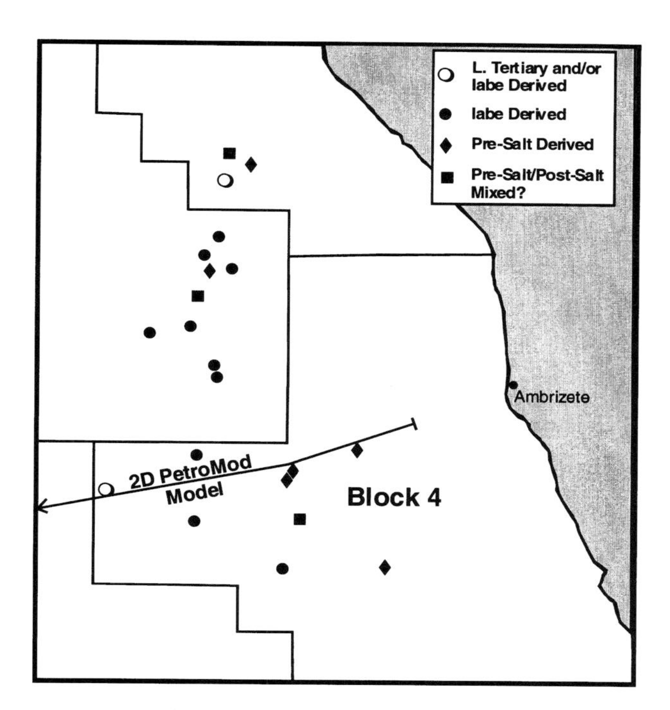

Figure 10—Map showing the distribution of oil types in the southern Lower Congo Basin. The location of the schematic section (Figure 11) used for the 2-D PetroMod model is also shown.

leum systems, and (2) the regional distribution of thermally mature, effective source rocks. Farther into the deep-water part of the basin, the characteristics of shows and oils from the westernmost wells also suggest derivation from the Iabe source unit and may even contain some contributions from oleanane-enriched facies of the Tertiary Landana Formation. No data are available from the Girassol or Dalia discoveries in western Block 17, but scout information supports an Iabe derivation with a possible contribution from the presalt lacustrine section. This suggests possible windows (weld zones) in the Loeme Salt that would allow the presalt source rocks to charge postsalt traps.

2-D Modeling of the Lower Congo Basin

Burial history, thermal, and fluid flow 2-D models constructed from seismic data can be used to assess the petroleum potential within a basin. A basic model, if calibrated correctly for geology, heat flow, and pressure considerations, can provide valuable information about the thermal maturity of the source rock sequences, the timing of source rock events such as generation and expulsion–migration, the trapping of oil and gas, and the quality of the trapped oil. For this chapter, the primary objectives of the 2-D model were to address the maturity of the source and the general migration directions. The calibration data for the model consisted of deep-water southern Lower Congo Basin lithologies, heat flow, temperature, and well pressures. In addition, the oil families described in the previous section provided information on which petroleum system charged the respective reservoir units. The 2-D modeling software called *PetroMod* (IES) was used to model the charge history and oil distributions of a selected seismic line (Figure 10). Figure 11 shows a representative schematic cross-section for this line drawn from a regional seismic dip line.

The 2-D fluid flow model shows that the distribution of oils is controlled by the Aptian salt seal and the location of possible windows through the salt. In the basic model, we assumed the most likely scenario—that the two petroleum systems were separated by the 3–10 m Loeme Salt sheet, and thus, the two source systems (or petroleum systems) were isolated from each other. Some small weld zones in the salt sheet were assumed so that we could evaluate whether these would allow any significant charge from the presalt to migrate into the postsalt section. However, since no mixed or presalt oils have been observed in the deep-water areas along this transect, we assumed that the Loeme Salt acted as an excellent seal between the two petroleum systems.

Maturity of the Source Rock Sequences

The maturity of the presalt lacustrine and postsalt Iabe marine clastic source rocks was estimated by calibration to the temperature data from the Margarida-1 and Block 4 deep-water wells, in addition to the vitrinite reflectance profile from the Margarida-1 well. This calibration study indicates that both source rocks attained sufficient maturity in the deep-water areas to contribute oil and gas to their respective petroleum systems. Figure 12 shows the present-day transformation ratios for both source rocks. At present, the presalt lacustrine source is overmature in the deep-water area, expulsion mature along the shelf–slope, and immature on the shelf. In the deep water, this source matured and expelled most of its oil during the early Miocene and would currently be expelling minor wet/dry gases. The postsalt Iabe source is post-peak expulsion mature to overmature in the deep-water areas and immature to early mature for expulsion along the shelf–slope to shelfal areas. This source rock expelled its primary oil charge in the deep-water area during the late Miocene–early Pliocene and has attained present-day maturities in the range of 0.9–1.1% R_o according to the most likely heat flow scenario.

The main questions, however, are how maturation is related to the migration of oil from these source rocks and how well the model correlates with known accumulations and shows. Figures 13 through 16 show the total petroleum saturation and oil flow vectors across the dip line at 12, 8.2, and 4 Ma and at the present day. The trends are summarized as follows.

At 12 Ma (Figure 13), the Cretaceous postsalt Iabe source was immature and had not generated sufficient oil and gas to saturate the surrounding rock. The presalt lacustrine source unit had been generating and expelling oil and associated gas since early Miocene time. At 12 Ma,

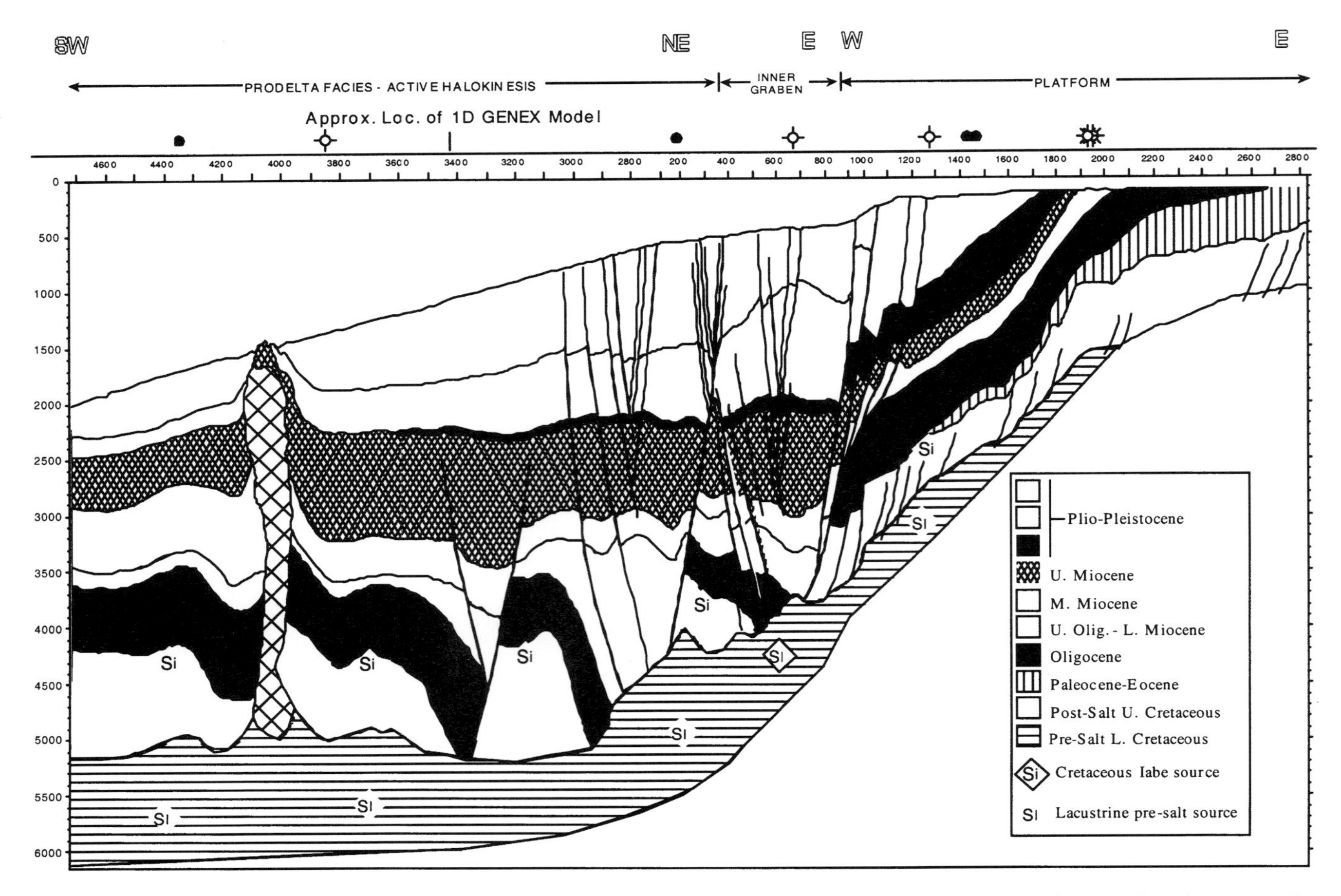

Figure 11—Schematic drawing from a regional seismic line across the southern Lower Congo Basin showing the generalized stratigraphic section, wells, and source rock intervals. This line is the basis for the 2-D *PetroMod* model. The only major interval not shown is the thin (3–10 m) Loeme Salt "sheet," which may act as a seal between the presalt and postsalt petroleum systems.

flow continued updip toward the shelf and some flow migrated through the salt welds. The oil and gas was saturating the Cretaceous section and some gas had reached the Tertiary sandstones at the 20–50 km section of the line.

At 8.2 Ma (Figure 14), the Cretaceous postsalt Iabe source had attained early maturity and was generating and expelling some oil and gas to saturate the surrounding rock. The presalt lacustrine source units continued expelling volatile oil and associated gas. At that time, flow continued updip toward the shelf with some charge migrating through the salt welds. This oil and gas was saturating the Cretaceous section, and some gas had reached the Tertiary sandstones at the 20–50 km part of the line.

At 4 Ma (Figure 15), the Cretaceous Iabe source was at peak expulsion and expelling sufficient oil and associated gas to oversaturate the surrounding shales and siltstones. The hydrocarbons began to actively migrate from the source pods formed by the salt rafts. Migration was predominantly vertical at this time. Of particular interest is the morphology of the pods and how these controlled migration. The salt rafted Iabe source pods formed in such a way that no migration occurred toward the shelf. All migration was vertically above the pods and/or enhanced by the associated faulting, and this occurred only in the deep-water areas, not the shelf. Thus, the shelf would not be expected to contain postsalt derived oils, as the Iabe source was too immature to effectively charge any Tertiary reservoirs, which is confirmed by the oil family distributions. At this time, oil was still migrating toward the shelf from the presalt lacustrine source and also through the salt windows at the 20–50 km part of the line. If the lacustrine presalt oil was actively migrating through the windows and into the Cretaceous and Tertiary reservoirs, some mixed oil types would be expected. However, geochemical characteristics of oil shows and accumulations in the deep-water southern Lower Congo Basin suggest generation from either the Iabe postsalt source or a mixture with the Tertiary Landana, but no discernible contribution from any presalt sources. This suggests that salt windows are most likely rare or nonexistent, limiting migration through the Loeme Salt sheet. The salt welds probably consist of resid-

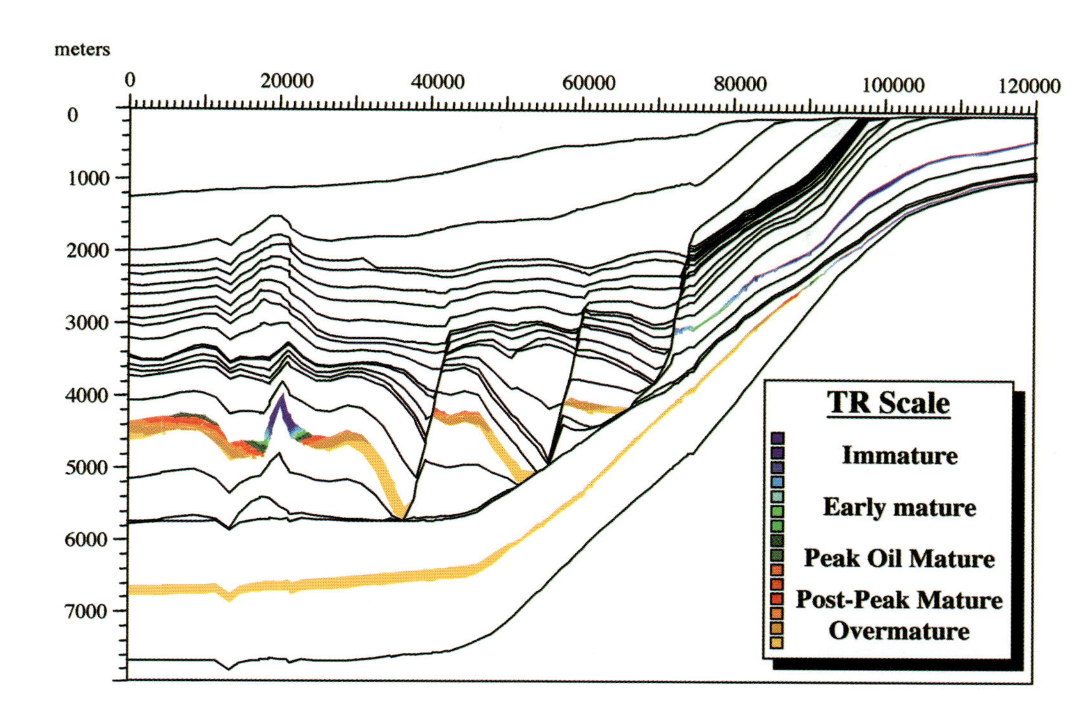

Figure 12—Transformation ratio for pre- and postsalt source rocks along the dip line in the southern Lower Congo Basin. The presalt lacustrine source is overmature in the deep-water area, expulsion mature along the shelf–slope, and immature on the shelf. The postsalt Iabe source is postpeak expulsion mature to overmature in the deep-water areas and immature to early mature for expulsion along the shelf–slope to shelfal areas.

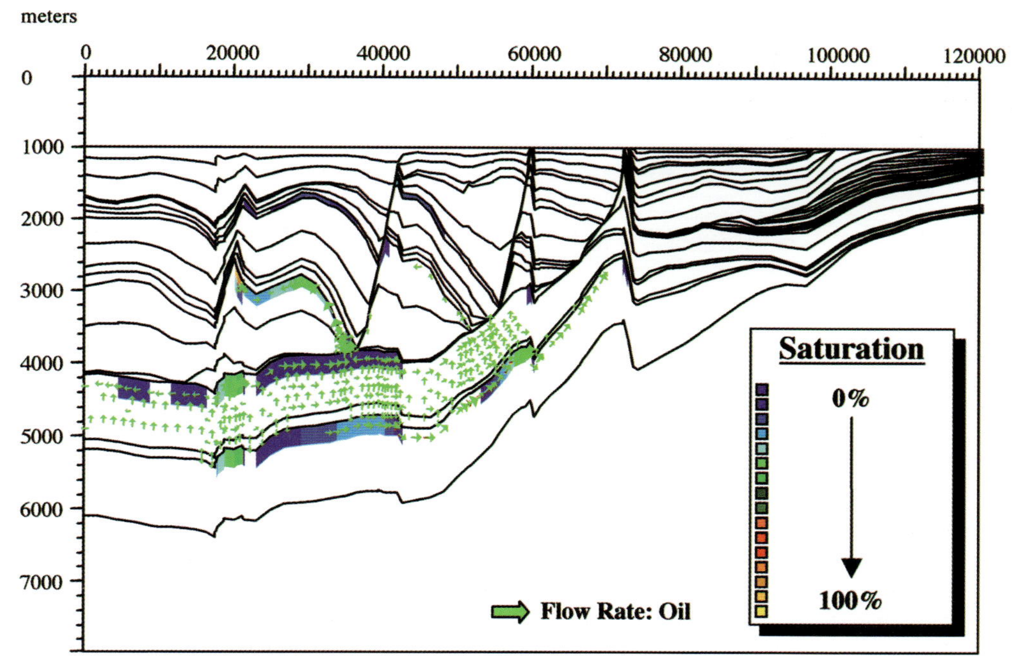

Figure 13—Petroleum saturation and flow directions (green arrows) at 12 Ma, Lower Congo Basin.

ual salt and anhydrites that still form a sufficient seal to preclude any presalt lacustrine oil migration.

Present-day total petroleum saturation (Figure 16) and petroleum saturation derived exclusively from either the Cretaceous Iabe (Figure 17) or the presalt lacustrine source rocks (Figure 18) indicate that sufficient hydrocarbons have been expelled from the postsalt Iabe source to charge traps above the mature source pods. Presalt lacustrine contributions would therefore not be required to justify a Tertiary play in the deep-water area of the southern Lower Congo Basin.

CONCLUSIONS

Two primary source rocks can be identified within the southern Lower Congo Basin. The first is a presalt lacustrine sequence time equivalent to the Bucomazi

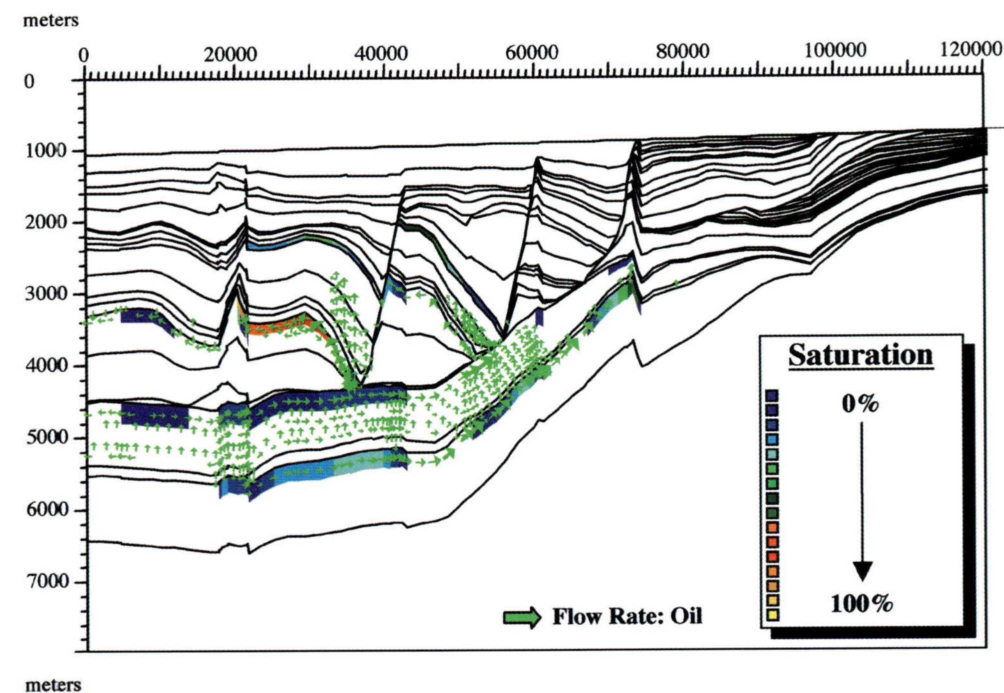

Figure 14—Petroleum saturation and flow directions (green arrows) at 8.2 Ma, Lower Congo Basin.

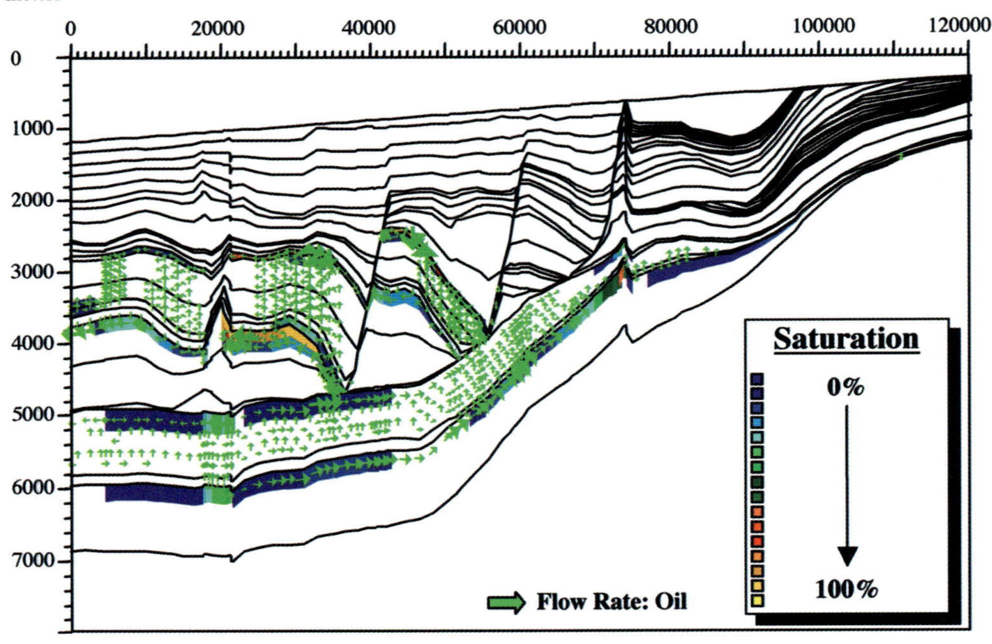

Figure 15—Petroleum saturation and flow directions (green arrows) at 4 Ma, Lower Congo Basin.

Formation, which consists of clastics with intervals of high TOC and HI that contain mostly oil-prone type I kerogen assemblages. Source rock data for this sequence are sparse in the southern Lower Congo Basin, but some thin intervals with high measured TOCs (>5%) and good residual HI values are evident. The second source rock is the postsalt (Cenomanian–Turonian) marine clastic Iabe Formation. TOC values in this unit exceed 5% in some wells, and pyrolytic yields suggest a marine type II kerogen with initial HI values of 350–550 mg HC/g TOC. This source rock is syn- to postdepositionally thickened due to salt tectonics (rafting). Also, salt rafting has removed this source from some areas of the southern Lower Congo Basin, increasing the source risk in localized areas.

Secondary source rocks include the clastics and marls of the Albian–Cenomanian Moita Seca, with TOCs reaching 2–3% and mixed oil/gas proneness, and the Tertiary

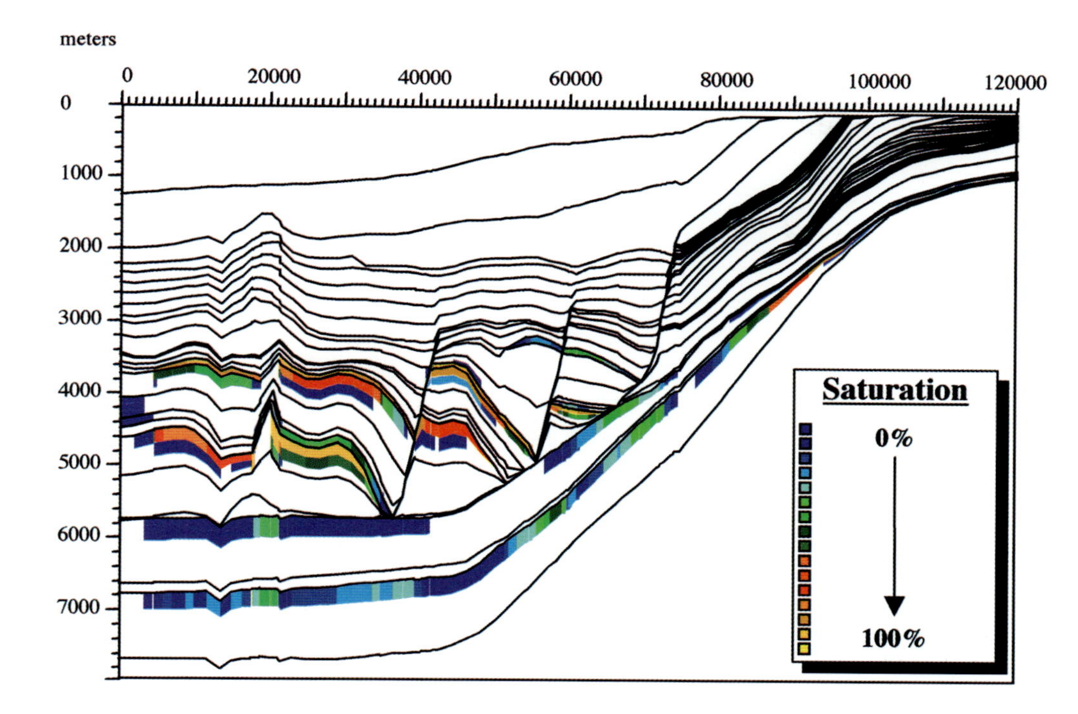

Figure 16—Present-day total petroleum saturation. Flow directions are not shown, but are similar to those in Figure 15.

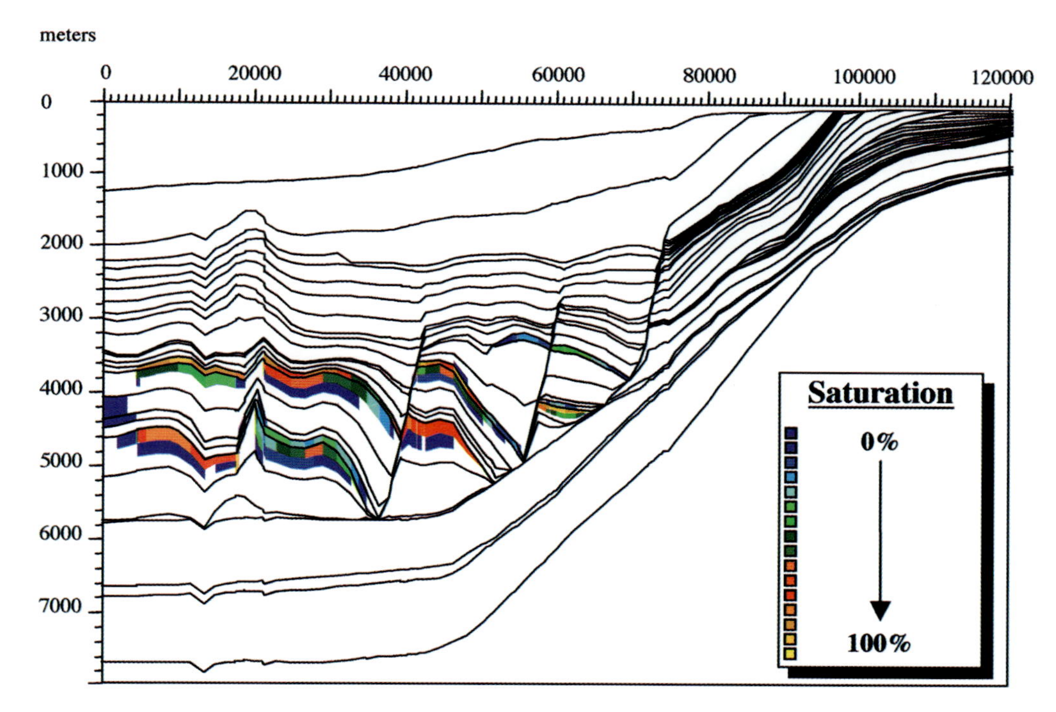

Figure 17—Present-day petroleum saturation derived from the postsalt Cretaceous labe source rocks.

(Oligocene and younger) Malembo, which contains up to 2% TOC with gas/minor oil proneness. The Malembo is not mature over most of the basin, whereas the Moita Seca could have contributed hydrocarbons into the system. The lower Tertiary (Paleocene–Eocene) Landana Formation also contains moderate to excellent source richness and potential. This unit is mostly marine in character, with some contribution from land-derived organic mater (as evidenced by biomarkers). In the southern Lower Congo Basin, the Landana is mainly immature for expulsion. However, in deep-water areas where burial was maximized from the Tertiary fill, the Landana could be sufficiently mature to expel hydrocarbons and has apparently contributed to accumulations, as indicated by oil–source correlations

Key risks regarding source aspects of the petroleum systems in the southern Lower Congo Basin are (1) source presence, that is, whether organic-rich lacustrine shales

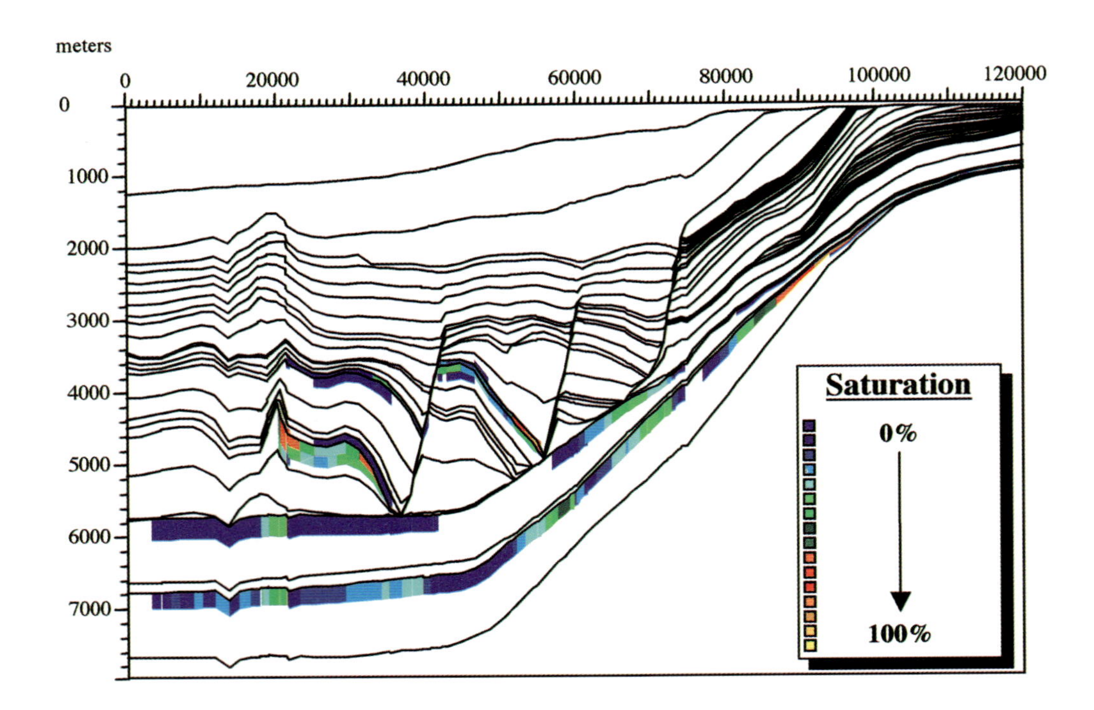

Figure 18—Present-day petroleum saturation derived from the presalt lacustrine source rocks.

are present, and (2) the thickness and lateral extent of mature postsalt Iabe marine clastic shales.

Stable carbon isotopes and biomarkers readily illustrate the relationships between Lower Congo Basin oils and potential source rocks (presalt marine Iabe, undifferentiated Tertiary, and presalt lacustrine), as well as identifying oil families. Available data indicate that the most diagnostic biomarker parameters include distributions of tricyclic terpanes, most notably the C_{26} and C_{25} homologs, and the hopane/sterane ratio. Another key geochemical property is the canonical variable derived from stable carbon isotopic compositions (Sofer, 1984). Using these relationships, oils can be correlated to either the postsalt Iabe marine clastic source or the presalt lacustrine source. Limited mixing of these principal oil types, or between the Iabe and the Tertiary Landana sources, is also indicated in some oils.

A regional dip line was modeled (using 2-D *PetroMod*) to address expulsion timing and fluid flow (migration) from the two primary source rock intervals. We drew several principal conclusions from the basic model:

1. Both source rock systems are mature in the deep water parts of the basin, given the thermal regime for this region. On the shelf, the postsalt system is immature due to minimal burial, whereas the presalt system is immature to oil mature.
2. Oil shows and discoveries of postsalt origin should predominate in Tertiary deep-water sandstones, whereas the shelf contains primarily lacustrine-sourced presalt oil accumulations. This is consistent with observed oil occurrences in this part of the southern Lower Congo Basin.
3. The Aptian Loeme Salt forms a regional seal between the postsalt marine clastic sourced system and the presalt lacustrine system. The salt acts as a barrier that prevents presalt charge from reaching Tertiary or Cretaceous reservoir intervals in the deep water and focuses the charge laterally to updip, shelfal areas. Salt seal therefore controls fluid flow and migration of the presalt lacustrine-derived hydrocarbons. This is consistent with observed oil occurrences of presalt derived oils.
4. Charge in the deep water is controlled by the salt-rafted source pods whereby the source has been thickened by the salt movement. It is primarily vertical above the slightly overpressured Tertiary strata, then modified laterally by the turbidite sandstone packages.

On the basis of the data presented, we conclude that the southern Lower Congo Basin contains dual petroleum systems. The deep-water Tertiary plays are charged primarily by the Cretaceous Iabe marine clastic type II source unit, whereas the shelfal areas are dominated by oils derived from the presalt lacustrine sequence (Bucomazi time equivalent). Mixing is limited to those areas where the Loeme Salt has evacuated sufficiently to allow migration through an assumed permeable salt weld.

Acknowledgments—*We thank Sonangol and BHP Petroleum for permission to use and publish these data. We also acknowledge constructive comments by Ralph Burwood and Barry Katz on an earlier version of this manuscript. Discussions with Bill Brumbaugh were helpful in developing some of the petroleum geology concepts.*

REFERENCES CITED

Bouroullec, J. L., J. P. Rehault, J. Rolet, J. Tiercelin, and A. Modeguer, 1991, Quaternary sedimentary processes and dynamics in the northern part of the lake Tanganyika Trough, East African rift system. Evidence of lake eustatism?: Bulletin Cent. Recherche Exploration Production Elf-Aquitaine, v. 15, p. 343–368.

Brice, S. E., M. D. Cochran, G. Pardo, and A. D. Edwards, 1982, Tectonics and sedimentation of the South Atlantic rift sequence: Cabinda, Angola, *in* J. S. Watkins and C. L. Drake, eds., Studies in continental margin geology: AAPG Memoir 34, p. 5–18.

Burwood, R., 1999, Angola: source rock control for Lower Congo coastal and Kwanza Basin petroleum systems, *in* N. Cameron, R. H. Bate, and V. Clure, eds., The oil and gas habitat of the South Atlantic: Geological Society of London Special Publication No. 153, p. 181–194.

Burwood, R., P. J. Cornet, L. Jacobs, and J. Paulet, 1990, Organofacies variation control on hydrocarbon generation: a Lower Congo coastal basin (Angola) case history: Organic Geochemistry, v. 16, p. 325–338.

Burwood, R., P. Leplat, B. Mycke, and J. Paulet, 1992, Rifted margin source rock deposition: a carbon isotope and biomarker study of a west African Lower Cretaceous "lacustrine" section: Organic Geochemistry, v. 19, p. 41–52.

Burwood, R., S. M. De Witte, B. Mycke, and J. Paulet, 1995, Petroleum geochemical characterization of the Lower Congo coastal basin Bucomazi Formation, *in* B. J. Katz, ed., Lacustrine source rocks: Berlin, Springer-Verlag, p. 235–263.

Curiale, J. A., 1993, Oil to source rock correlation: concepts and case studies, *in* M. H. Engel and S. A. Macko, eds., Organic geochemistry principles and applications: New York, Plenum Press, p. 473–490.

Ekweozor, C. M., and N. Telnæs, 1990, Oleanane parameter: verification by quantitative study of the biomarker occurrence in sediments of the Niger delta: Organic Geochemistry, v. 16, p. 401–413.

Ekweozor, C. M., J. I. Okogun, D. E. U. Ekong, and J. R. Maxwell, 1979, Preliminary organic geochemical studies of samples from the Niger delta (Nigeria), II: analysis of shale for triterpenoid derivatives: Chemical Geology, v. 27, p. 29–37.

International Petroleum Encyclopedia, 1996, Tulsa, PennWell.

Isaksen, G. H., 1991, Molecular geochemistry assists exploration: Oil & Gas Journal, March 18, p. 127–131.

Magoon, L. B., and W. G. Dow, 1994, The petroleum system, *in* L. B. Magoon and W. G. Dow, eds., The petroleum system—from source to trap: AAPG Memoir 60, p. 3–24.

McHargue, T. R., 1990, Stratigraphic development of proto-South Atlantic rifting in Cabinda, Angola—a petroliferous lake basin, *in* B. J. Katz, ed., Lacustrine basin exploration—case studies and modern analogs: AAPG Memoir 50, p. 307–326.

Mycke, B., and R. Burwood, 1995, Biomarker based stratigraphic correlation of Lower Congo coastal basin wells, *in* J. O. Grimalt and C. Dorronsoro, eds., Organic geochemistry: developments and applications to energy, climate, environment and human history: ALGOA, San Sebastian, Spain, p. 58–61.

Peters, K. E., and J. M. Moldowan, 1993, The biomarker guide: interpreting molecular fossils in petroleum and ancient sediments: Englewood Cliffs, N.J., Prentice-Hall.

Sofer, Z., 1984, Stable carbon isotope compositions of crude oils: application to source depositional environments and petroleum alteration: AAPG Bulletin, v. 68, p. 31–49.

Sweeney, J. J., and A. K. Burnham, 1990, Evaluation of a simple model of vitrinite reflectance based on chemical kinetics: AAPG Bulletin, v. 74, p. 1559–1570.

Tissot, B., and J. Espitalié, 1975, L'evolution thermique de la matiere organique des sediments: applications d'une simulation mathematique: Revue Institut Français du Pétrole, v. 30, p. 743–777.

Ungerer, P., J. Espitalié, F. Marquis, and B. Durand, 1986, Use of kinetic models of organic matter evolution for the reconstruction of paleotemperatures. Application to the case of the Gironville well (France), *in* J. Burrus, ed., Thermal modeling in sedimentary basins: Paris, Edition Technip, p. 531–546.

Ungerer, P., and R. Pelet, 1987, Extrapolation of the kinetics of oil and gas formation from laboratory experiments to sedimentary basins: Nature, v. 327, p. 52–54.

Zumberge, J. E., 1987, Prediction of source rock characteristics based on terpane biomarkers in crude oils: a multivariate statistical approach: Geochimica et Cosmochimica Acta, v. 51, p. 1625–1637.

Harris, N. B., 2000, Toca Carbonate, Congo Basin: response to an evolving rift lake, *in* M. R. Mello and B. J. Katz, eds., Petroleum systems of South Atlantic margins: AAPG Memoir 73, p.341–360.

Chapter 24

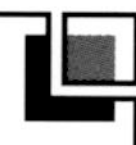

Toca Carbonate, Congo Basin: Response to an Evolving Rift Lake

Nicholas B. Harris

Department of Geosciences
The Pennsylvania State University
University Park, Pennsylvania, U.S.A.

Abstract

The Toca Formation is a Barremian lacustrine carbonate in the late rift section of the Congo Basin, equatorial west Africa. It is an important oil reservoir in Cabinda, Angola; noncommercial discoveries exist elsewhere in the basin. However, the Toca is a difficult target to explore and develop due to major sedimentary facies changes over short distances. The Toca occurs at multiple stratigraphic levels, here designated Toca 1, Toca 2, and Toca 3 in ascending order. Toca 1 and Toca 2 are coeval with the Marnes Noires Formation (Middle Bucomazi Formation), an organic carbon-rich marl. Toca 3 is probably coeval with the lowermost Argilles Vertes Formation (Upper Bucomazi), an organic carbon-poor siltstone.

The Toca shows systematic stratigraphic variations that reflect decreasing lake salinity and probably climate change. Toca 1 consists primarily of allochthonous algal-derived carbonate clasts, including oncolites, algal grainstones, and gastropods. Toca 2 contains both in situ and allochthonous deposits of algal-derived clasts and pelecypod coquina formed by filter-feeding organisms. Toca 3 consists almost entirely of pelecypod shells and lime mud (no algal-derived bioclasts) and forms in situ carbonate banks. Diagenesis and reservoir quality are related to depositional facies. Porosity and permeability were enhanced by early dolomitization and early dissolution related to subaerial exposure. Porosity and permeability were adversely affected by early compaction (particularly in allochthonous facies), calcite cementation, and late dolomitization.

INTRODUCTION

The Toca Carbonate is a Barremian lacustrine carbonate in the upper part of the synrift section of the Congo Basin, equatorial west Africa (Figure 1). It represents an important but problematic exploration target in the synrift section. It is productive in several oil fields in Cabinda (Figure 2), including the Malongo West field (240 MMBO from the Toca) and the Takula field (39 MMBO from the Toca through 1996; Lomando, 1996). The Toca has produced oil at the onshore Pointe Indienne field (Figure 2) in Republic of Congo. Similar carbonates of the Lagoa Feia Formation, Campos Basin, Brazil, have been productive, including Linguado field, where recoverable reserves will total 108 million bbl of oil (Horschutz, 1992), and Pampo field, which has produced 170 million bbl of oil (Lomando, 1996).

However, the Toca has proven to be a difficult target both to explore and to develop, exemplified by the Viodo discovery in the Republic of Congo. Conoco's VIM-1 well discovered oil in the Toca and tested 1135 billion bbl of oil per day (Harris et al., 1994). The VIM-2 well, a step-out 2 km to the north of VIM-1, found a section consisting entirely of marl. The VIM-3 well, a step-out 1 km to the east of VIM-1, found thick Toca carbonate, but the Toca here was relatively impermeable due to facies changes. Finally, Elf Congo drilled the MDJM-1 well 600 m west of VIM-1; here, the carbonate was porous and permeable but thinner than in VIM-1.

This chapter describes results of a sedimentologic study of cores from 16 wells in the Popular Republic of Congo and Cabinda, Angola, and analysis of sidewall cores and cuttings samples from additional intervals. Locations of the wells are shown in Figures 2.

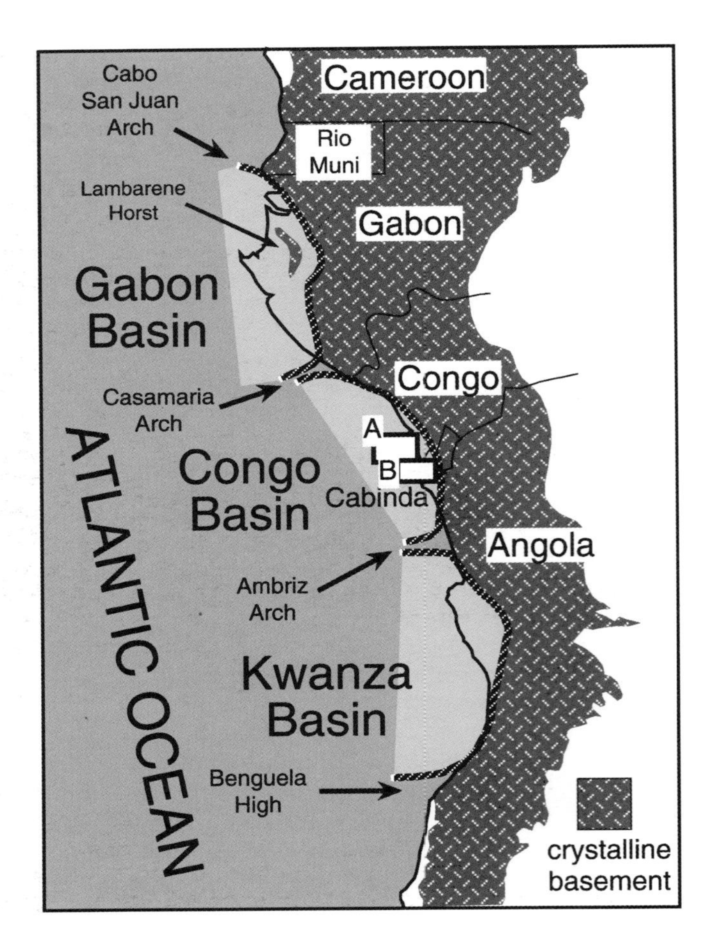

Figure 1—Location map showing principal sedimentary basins on the equatorial west African margin. The Congo Basin, the focus of this study, is the central of these basins. The Republic of Congo wells are located in the small box labeled "A" (see Figure 2a), and the Cabinda wells are in the box labeled "B" (Figure 2b). Modified from Standlee et al. (1992).

Tectonic and Structural Setting

The Mesozoic rift basins of the Atlantic margins of west Africa and Brazil developed in the latest Jurassic–Early Cretaceous as Africa separated from South America (Rabinowitz and LeBreque, 1979; Brice et al., 1982). The basins formed along mobile belts created by the late Precambrian collision of the west African, Congo, and São Francisco cratons (Maurin and Guiraud, 1993). The mobile belts constituted zones of weakness later exploited during rifting. In the Congo Basin, the mobile belt is defined by a zone of overthrusted metamorphic rocks on the eastern margin of the basin.

The Congo Basin (Figure 1) is the central of the three prominent basins on the equatorial west African margin. It is separated from the Gabon or Atlantic Basin to the north by the Casamaria Arch and from the Kwanza Basin of Angola to the south by the Ambriz Arch. Complementary basins on the Brazilian margin include the Campos, Recôncavo, Sergipe–Alagoas, and Potiguar Basins. During the Early Cretaceous, the Kwanza Basin was separated from the South Atlantic Ocean by the Walvis Ridge (Rabinowitz and LaBreque, 1979; Burke and Sengor, 1988). This feature, anchored by the long-lived Walvis hotspot, was critical to the stratigraphic development of all the Aptian salt basins. While it remained topographically high, it acted as a dam that held back the South Atlantic Ocean and permitted the rift basins to remain the site of continental sedimentation for an extended period of time. When the Walvis Ridge subsided below sea level during the Aptian, ocean water invaded all of the west African rift basins almost simultaneously.

Stratigraphic Setting

The stratigraphy of the Congo Basin is divisible into two major intervals (Figure 3): the *presalt* section, which represents continental rift sedimentation, and the *postsalt* section, which represents largely marine sedimentation on the subsiding west Africa passive margin (McHargue, 1990). The two intervals are separated by the Aptian Loeme Salt, which marks the major incursion of the South Atlantic Ocean into the rift. Figure 3 depicts the stratigraphy from the lowermost presalt unit, the Vandji Formation, up through the lowermost postsalt unit, the Senji Formation.

The presalt rift sequence can in turn be divided into *early rift* and *late rift* stages. These are equivalent to the "fault" and "sag" phases of McHargue (1990); the two intervals comprise the synrift I phase of Brice et al. (1982). The following Cabinda nomenclature and age dates are from Braccini et al. (1997), and Republic of Congo nomenclature is from Harris et al. (1994) and Karner et al. (1997) with age dates from Grosdidier et al. (1996). The early rift stage includes two sedimentary units (in ascending order):

1. Neocomian coarse basal sandstone (Vandji Formation in Congo and Lucula Formation in Cabinda), deposited in an alluvial–fluvial setting (Bracken, 1994).
2. Neocomian–Barremian lacustrine shales and sandstones deposited as sediment gravity flows (Sialivakou and Djeno Formations in Congo; Erva and Lower Bucomazi Formations in Cabinda).

The late rift stage comprises three units (in ascending order):

1. Barremian organic-rich lacustrine marls (Marnes Noires or Pointe Noires Formation in Congo and Middle Bucomazi Formation in Cabinda). The Toca Carbonate is laterally equivalent to the Marnes Noires succession and the lowermost Argilles Vertes (Grosdidier et al., 1996; Braccini et al., 1997).

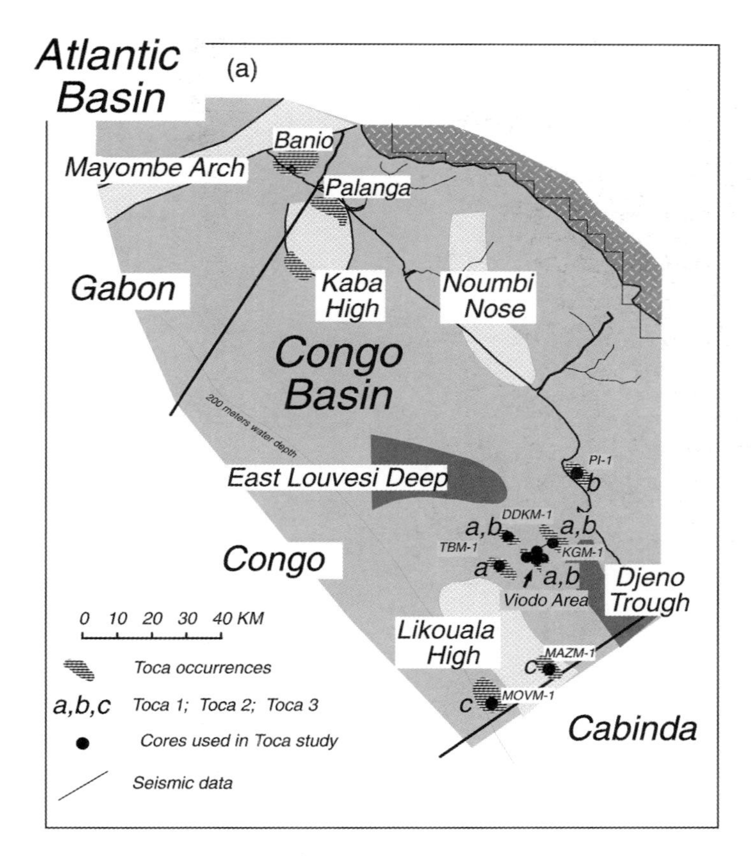

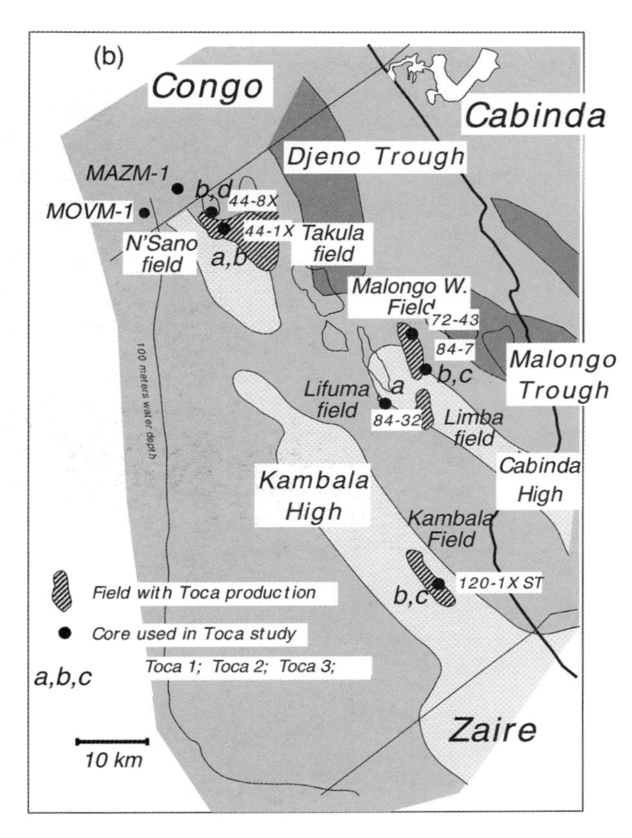

Figure 2—Detailed location maps showing Toca occurrences and seismic lines in (a) the Popular Republic of Congo and (b) Cabinda, Angola. Three stratigraphic levels of the Toca have been identified in this study and are located on these maps: Toca 1 (labeled "a"), Toca 2 ("b"), and Toca 3 ("c").

2. Barremian lacustrine siliciclastic shales with minor sandstone (Argilles Vertes or Pointe Indienne Formations in Congo and Upper Bucomazi in Cabinda).
3. Aptian sandstones and shales (Chela Formation), complexly interbedded and representing a variety of environments ranging from marine to possibly lacustrine and fluvial.

A regional unconformity, representing a period of 2 m.y. separates the Argilles Vertes or Upper Bucomazi Formation from the overlying Chela. The origin of this unconformity is unclear. The Chela, which in places has clear marine affinities (Braccini et al., 1997), is more closely related to the overlying Loeme Salt than to underlying synrift units.

TOCA STRATIGRAPHY AND DISTRIBUTION

The Toca Carbonate occurs at multiple stratigraphic levels at many locations in the Congo Basin (Figures 4 and 5), separated by shale intervals that range in thickness from 1 to >100 m. Where multiple levels are present, Toca occurrences commonly show a systematic evolution in the type of clasts that comprise the carbonate and their distribution relative to paleotopography. I suggest that this evolution is a response to changing hydrologic conditions in the lake(s) during deposition, in particular salinity. Carbonates deposited during any one particular phase of the lake share common characteristics of lithoclast type, areal distribution, and early diagenesis, determined largely by lake chemistry and physical hydrology.

The Toca occurrences are subdivided here into three levels, which in ascending stratigraphic order are Toca 1, Toca 2, and Toca 3 (Figures 4 and 5). Assignment of specific occurrences to particular Toca levels is based on lithoclast type, specifically the proportion of oncolites and gastropods to pelecypods, and the mode of occurrence, specifically the proportion of in-place versus allochthonous carbonates. The exact timing of Toca deposition at a specific location depends on lake level. Therefore two different occurrences of Toca 1 carbonate may not have been deposited synchronously if their relative topographic positions were different.

Toca 1 and Toca 2 are coeval with the Marnes Noires Formation (Middle Bucomazi), based on clear interfin-

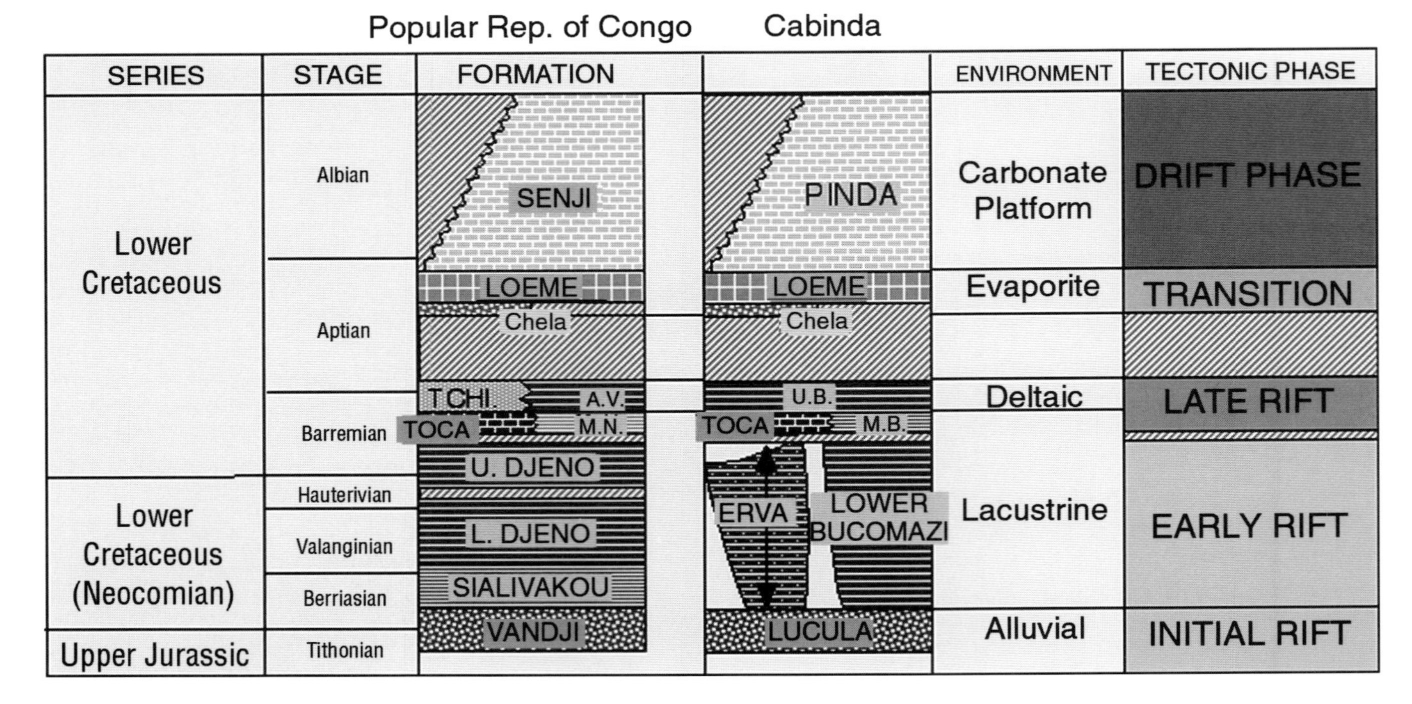

Figure 3—Jurassic–Lower Cretaceous stratigraphic chart for the Congo Basin, Popular Republic of Congo, and Cabinda, Angola. The Toca is laterally equivalent to the Marnes Noires (M.N.) and Argilles Vertes (A.V.) in the Republic of Congo and to the Middle Bucomazi (M.B.) and Upper Bucomazi (U.B.) in Cabinda. Republic of Congo stratigraphy modified from Karner et al. (1997) and Grosdidier et al. (1996); Cabinda stratigraphy from Braccini et al. (1997).

gering relationships between the units in the Viodo area, Republic of Congo. I tentatively correlate Toca 3 with the basal Argilles Vertes Formation (Upper Bucomazi) on the basis of the character of the underlying shale in the MAZM-1 well. These interpretations are consistent with the stratigraphy defined by Braccini et al. (1997) for Cabinda, although supporting palynologic and ostracode data were not available for this study.

The Toca Carbonate occurs on or near paleotopographic highs, from the first major horst block inboard of the eastern rift margin (Pointe Indienne field) westward to the oceanward side of the Outer Basin High (MOVM-1 well) on the western side of the Likouala High, Congo, and the Kambala field, Cabinda (Figure 2). The Toca does not occur along the eastern margin of the basin (with the possible exception of the Banio Limestone in southernmost Gabon), probably because that margin was the site of major siliciclastic deposition.

MARNES NOIRES FORMATION

The Marnes Noires Formation (or Middle Bucomazi Formation) generally underlies Toca 1, is the basinal equivalent to Toca 1 and 2, and overlies Toca 2 where the section has not been truncated by the basal Chela unconformity. It is a fine-grained, finely laminated, organic carbon-rich marl (Figure 6A) containing pelagic and detrital carbonate (ostracodes, micrite, and microspar), clay, silt-size quartz and feldspar, and organic matter. Couplets of dark organic carbon- and siliciclastic-rich laminae and light carbonate-rich laminae average 1 mm in thickness. Based on a typical formation thickness of 200 m and an age span of about 3 m.y. for the formation (Braccini et al., 1997), these couplets may represent sunspot (11-year) cycles. The absence of bioturbation, high hydrogen and low oxygen indices (Harris et al., 1994), and the local presence of ostracodes (E. Braccini, personal communication, 1999) indicate that bottom water conditions were anoxic to dysoxic.

The Marnes Noires Formation was deposited on a major erosional surface developed at the top of the Djeno Formation (Figure 7). Topographic relief on this surface was about 100 m in the Viodo area. In cores from the Viodo area, the transition between the Marnes Noires and Toca 1 is gradational over a thickness from 5.5 to 22 m and is characterized by numerous thin beds of muddy allochthonous carbonate (Figure 6B). Soft sediment deformation is common in some wells, suggesting rapid deposition on a topographic slope. Alternatively, it may reflect lowering lake level as the decreasing weight of the water column destabilized the shallow unlithified sediment (A. J. Lomando, personal communication, 1999).

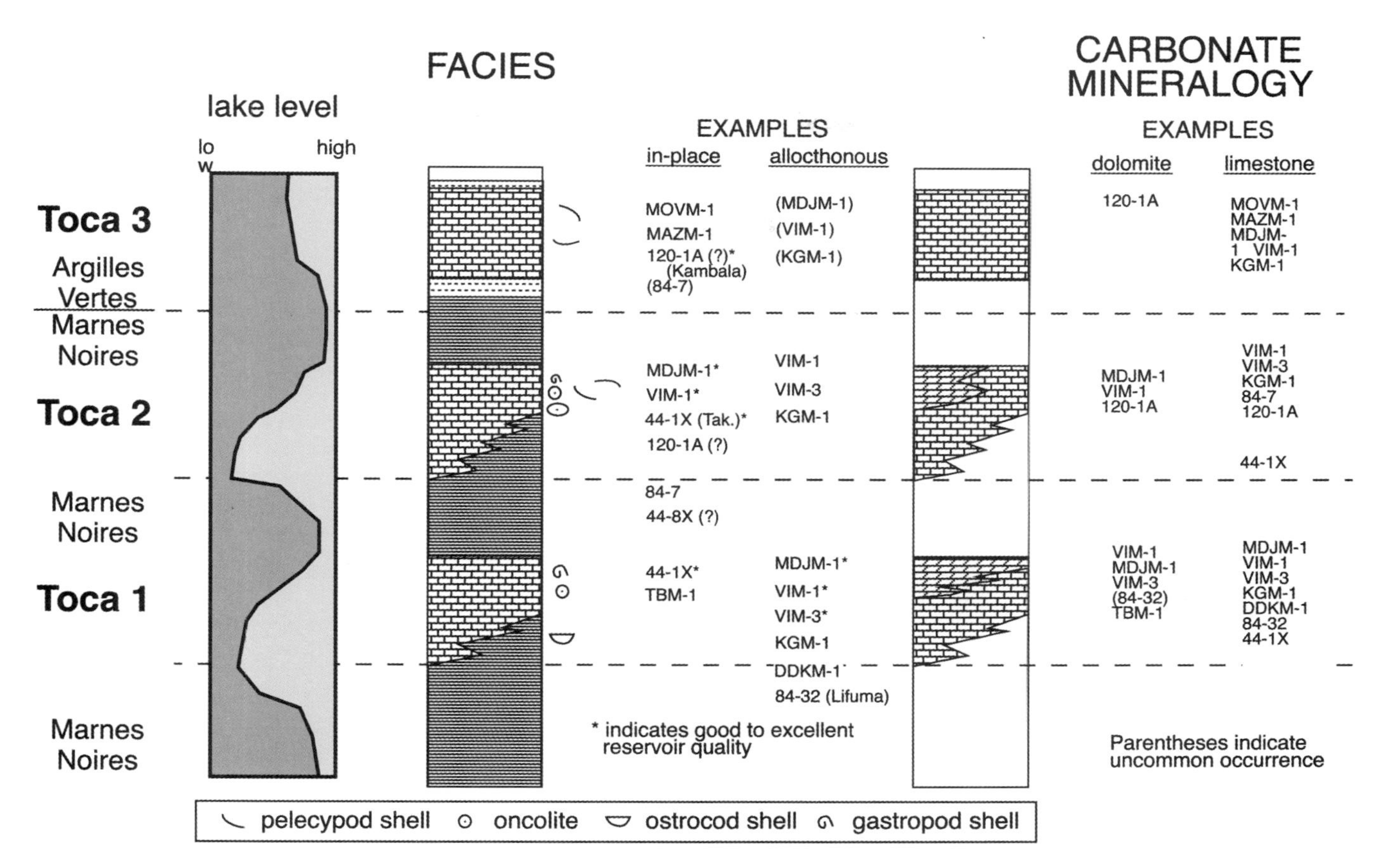

Figure 4—Stratigraphy of Toca units in the Congo Basin and inferred lake level fluctuations. Toca 1 (lowermost) consists of algal-based lithoclasts (gastropods and oncolites); Toca 2 (middle) consists of algal- and filter feeder-based lithoclasts (pelecypod shells); Toca 3 (uppermost) consists almost entirely of filter feeder-based lithoclasts.

TOCA 1

Toca 1 occurs in a north–northwest oriented belt through the center of the basin. Allochthonous Toca 1 beds (carbonate turbidites) occur on and adjacent to north–northwest trending horsts in the central part of the Congo Basin, including the Viodo and Lifuma field structures. It also occurs on a west–southwest dipping ramp in the KGM-1 well. In situ Toca 1 (algal grainstone) occurs in a horst system that includes Takula field and the TBM-1 well, west of the Viodo structure. I have not recognized Toca 1 in wells from southern Cabinda.

Basic Facies Types

The most common Toca 1 facies, found in the Viodo area and at Lifuma field, consists entirely of allochthonous material. The carbonate originally formed in shallow water but was subsequently transported to relatively deep water by sediment gravity flows (Figure 7). The carbonate now forms grainstone and packstone beds ranging in thickness from a few centimeters to 2 m.

No data provide absolute constraints on water depth. However, the fact that the carbonate is interbedded with the Marnes Noires shale (see next subsection) suggests that water depths were significant. Depths to the base of the oxygenated water column in modern East African rift lakes range from 200 to 250 m in Lake Malawi (Halfman, 1996) and from 150 to 250 m in Lake Tanganyika (Cohen, 1989).

The Toca 1 in the 44-1X well (Takula field) and probably in the TBM-1 well is an aggregate of coarse algal clasts. The general lack of carbonate mud in these samples suggests that they are shallow-water deposits, unlike the Viodo area samples.

Allochthonous Carbonate Facies

Lithologies and Sedimentology

The Toca 1 in the Viodo and Lifuma wells largely consists of stacked carbonate grainstone and packstone beds (Figures 6C and D). They are composed of abundant gastropod shells and oncolites (Figure 6E and F), with subordinate ostracodes, dolomite clasts, Marnes Noires

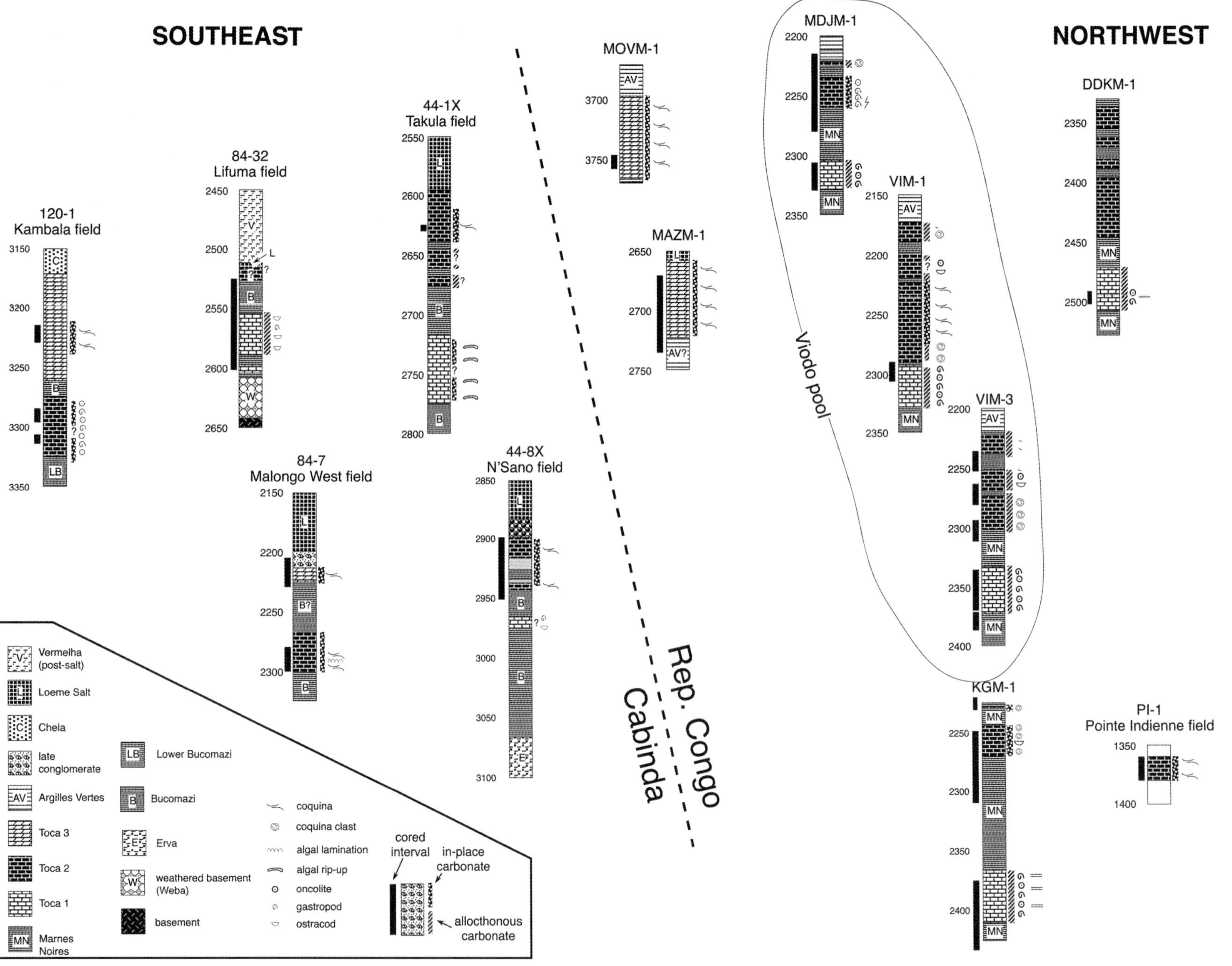

Figure 5—Results of core and cuttings analysis from Republic of Congo and Cabinda, Angola, wells.

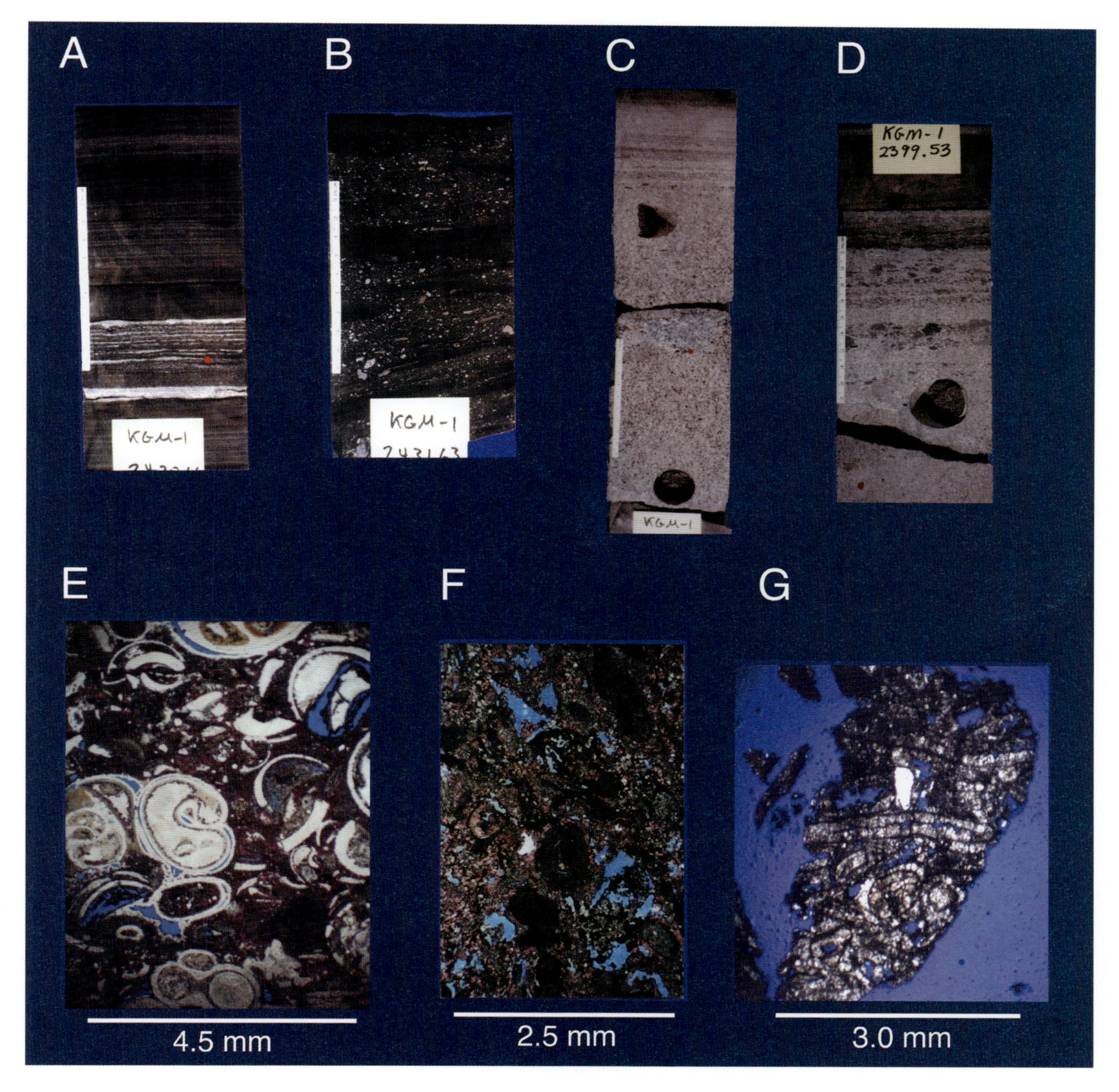

Figure 6—Core and thin-section photographs of Toca 1 carbonates and the underlying Marnes Noires shale. Scale bars in core photos are 10 cm. (A) Core photo, KGM-1 well, from transition zone underlying Toca 1. Planar laminated Marnes Noire shale and thin distal carbonate turbidites. (B) Core photo, KGM-1 well, from transition zone underlying Toca 1. Muddy carbonate turbidite, with large Marnes Noires shale rip-up clasts. (C) Core photo, KGM-1 well, from Toca 1. Massive carbonate grainstone bed (turbidite) sharply overlies Marnes Noires type shale bed. The grainstone bed coarsens upward slightly in the bottom 4 cm. It is mostly massive and coarse grained, except for the upper 8 cm which fines upward and exhibits parallel lamination. The grainstone bed grades upward into Marnes Noires type shale. (D) Core photo, KGM-1 well, from Toca 1. Top of carbonate grainstone bed (turbidite) exhibits planar lamination and abundant shale rip-up clasts. (E) Thin-section photo, KGM-1 well, from carbonate grainstone bed in Toca 1; calcite is stained dark pink. Gastropod shells are obvious throughout the photo. White shells are now composed of opaline silica, an early diagenetic replacement, in a calcite matrix. (F) Thin-section photo, VIM-3 well, from carbonate grainstone bed in Toca 1; calcite is stained dark pink. Oncolites are evident in center and top right of photo. (G) Thin-section photo, 44-1X well side-wall core (Takula field), from in situ Toca 1; calcite is not stained. Algal grainstone is probably composed of recrystallized algal rip-up clasts. Large pores between algal grains result in high permeability and good flow rates.

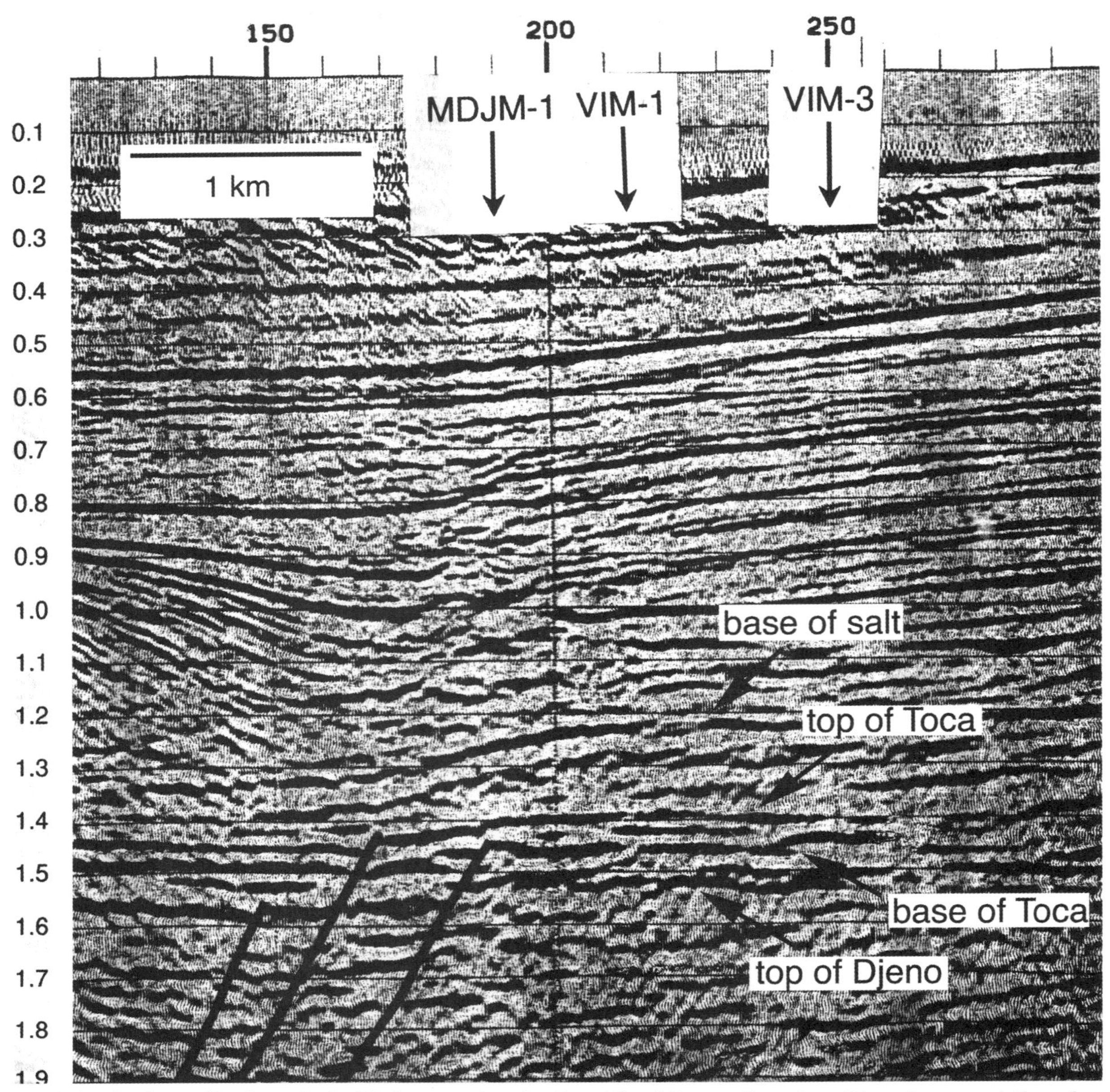

Figure 7—Seismic dip profile through the Viodo area, Popular Republic of Congo. Toca 1 and Toca 2 accumulations formed on and adjacent to a small north–northwest trending horst. The Marnes Noires Formation, generally characterized by high-amplitude parallel reflectors, overlies the Djeno Formation at a pronounced angular unconformity. The Toca forms a lensoid body characterized by bright discontinuous reflectors within the Marnes Noires.

shale clasts (Figure 6D), and siliciclastic grains. These are set in a fine-grained matrix of carbonate mud and carbonate cement. The lithoclasts suggest that their source was a shallow-water gastropod–oncolite shoal. The abundance of gastropods suggests that these gastropods were algal grazers rather than predators or scavengers. The common presence of oncolites also suggests that the shallow-water ecosystem was based on benthic algae.

In the Viodo area, the grainstone beds typically have sharp bases (Figure 6C). Beds thicker than a few centimeters are massive and very coarse grained. Near their tops, the beds commonly become finer grained, develop planar lamination, and may grade upward into Marnes Noires marls. These beds are interpreted as turbidites, containing the T_a and T_b of the Bouma sequence, deposited from high-density turbidity currents (Lowe, 1982). Thin carbonate beds commonly display parallel lamination or ripple cross-lamination, corresponding to T_b and T_c of the Bouma sequence. The turbidites were apparently derived from shallow-water carbonate

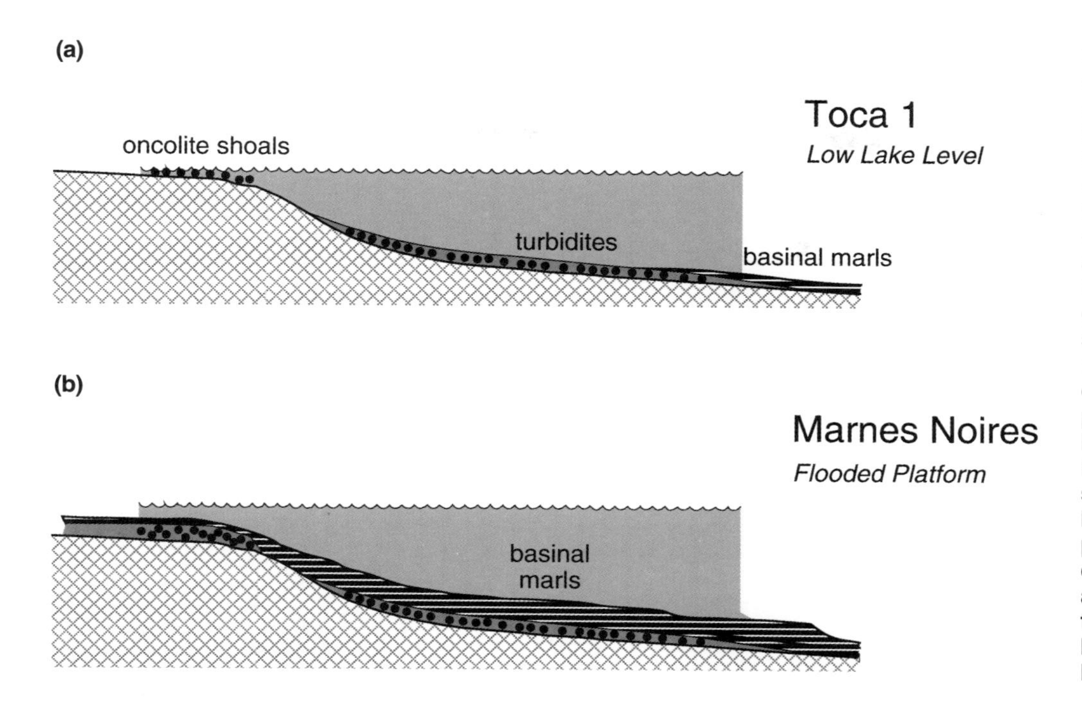

Figure 8—Schematic cross sections showing model for deposition of carbonate turbidites in Toca 1. (a) A horst block became nearly emergent during a major fall in lake level, leading to the development of gastropod and oncolite shoals. Because the carbonate grains never lithified in the shallow water, much of the material avalanched into deeper water flanking the platform, interfingering with Marnes Noires shale. (b) Platform drowned during subsequent rapid rise in lake level, shutting down production of shallow-water carbonate. Horst block and adjacent deep basin were then covered by Marnes Noires shale. Modified from Harris et al. (1994).

shoals, perhaps triggered by slumping of unconsolidated banks of oncolites and gastropod shells (Figure 8). A rapid rise in lake level drowned the shallow-water platforms and terminated deposition of Toca 1; organic carbon-rich shale was deposited across platforms and basinal areas (Figure 8).

The carbonate in intertonguing Marnes Noires shale beds is purely dolomite, demonstrated by whole-rock chemical analyses in which the molar ratio (MgO/MgO + CaO) of carbonate averaged 0.49. This suggests that the lake during Toca 1 deposition was relatively saline (Last, 1990), perhaps reflecting low precipitation and runoff from drainage basins surrounding the rift lake.

In none of these examples can I identify the shallow-water platform that sourced these turbidites. Because wells on the crest of the Viodo structure contain turbidites (MDJM-1 and VIM-1), clearly the carbonates did not come from the crest of this structure unless they were transported from a structurally higher position along strike. They may have been derived from the Tchibuela structure, several kilometers to the west. In that case, substantial thicknesses may be mounded up along the western flank of the Viodo structure.

Thickness

The allochthonous Toca 1 occurrences have carbonate thicknesses of less than 40 m. Among the wells in the Viodo area, the thinnest carbonate (20 m) occurs in the well highest on structure. Net thickness increases in an off-structure direction, suggesting that the carbonate formed mounds at the base of the slope. The proportion of shale interbeds simultaneously increases downdip.

Diagenesis and Reservoir Qualitys

Two diagenetic facies characterize most of the allochthonous Toca 1 carbonates: one consists of subequal dolomite and calcite, while the other consists entirely of dolomite. In the calcite–dolomite facies, both minerals are present in the matrix. Oncolites are most commonly preserved as dolomite (Figure 6F), ostracodes as calcite, and gastropods as dolomite, calcite, or opaline silica (Figure 6E). Some dissolution porosity is present, but intercrystalline porosity is minor. Dolomite in the matrix of the mixed calcite–dolomite facies was probably syndepositional in origin, forming in saline water at the bottom of the Marnes Noires lake.

In the pure dolomite facies, clasts are commonly dissolved, particularly gastropod shells. Much more intercrystalline porosity is present in the matrix, which enhances the permeability. Minor baroque dolomite is present in the calcite–dolomite facies, suggesting a second late phase of recrystallization at high temperature (Radke and Mathis, 1980).

The distribution of the two dominant diagenetic facies is strongly controlled by structural position. Toca 1 on the crest of the Viodo structure consists entirely of pure dolomite, whereas in off-structure wells, it is composed of the calcite–dolomite mixture. In an intermediate position, both diagenetic facies are present, with the pure dolomite facies sharply overlying the calcite–dolomite facies.

At least two alternative models may explain the distribution of the pure dolomite facies (Figure 9). One possibility is that meteoric waters entered the hydrologic system during lowstands while the overlying Toca 2 was deposited. Chemical components dissolved from Toca 2

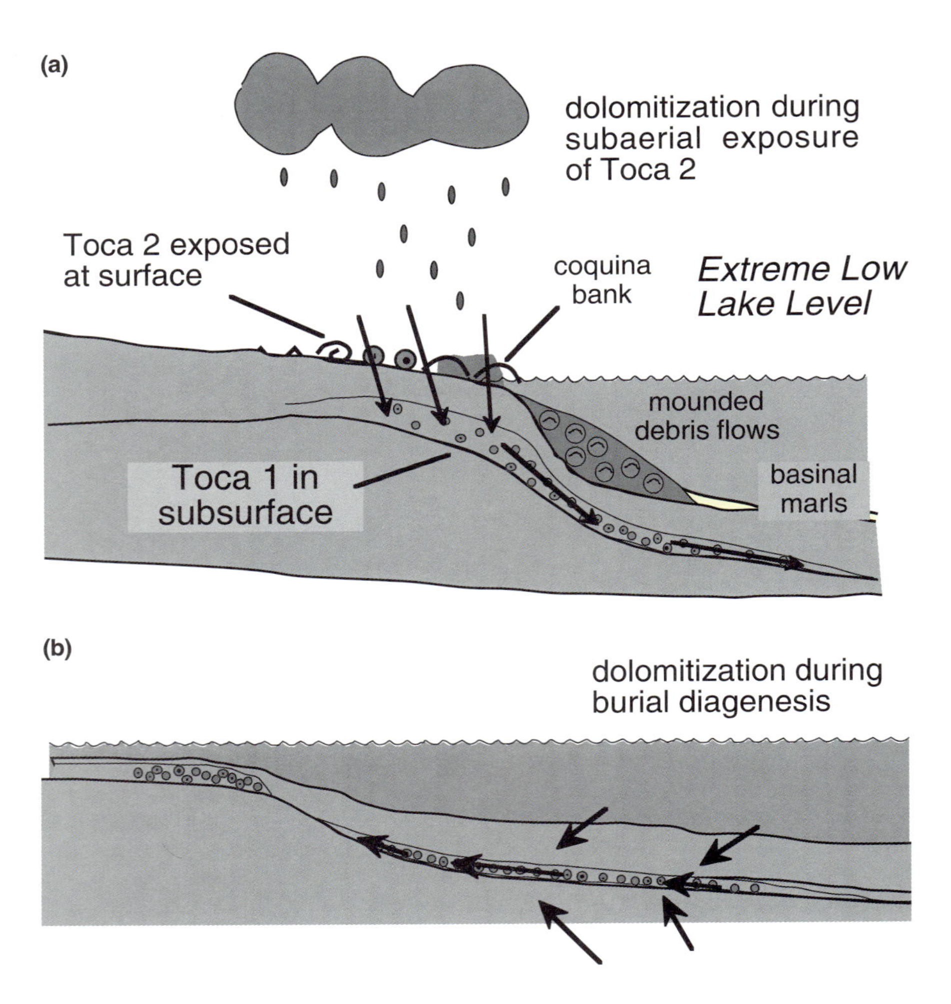

Figure 9—Schematic cross sections showing two alternative models for conversion of the mixed calcite–dolomite facies to pure dolomite in Toca 1. Arrows indicate direction of water flow. (a) Dissolution of Toca 2 during periods of extreme low lake level and subaerial exposure. Components dissolved from Toca 2 were responsible for dolomitizing the underlying Toca 1. (b) Dolomitization during burial diagenesis by formation waters that flowed into Toca 1 and migrated updip.

dolomitized Toca 1 at slightly greater depths. Alternatively, formation waters expelled from compacting marls may have entered Toca 1 carbonates during burial. These would have migrated updip, altering the carbonates along the flow path. The updip carbonates (such as in the MDJM-1 well) were most intensely converted to dolomite because flow of formation water was most concentrated in these rocks. The baroque dolomite formed during a much later event.

Reservoir quality in Toca 1 allochthonous facies is moderately good where the grainstones consist of pure dolomite; it is poorer where they consist of a mixture of calcite and dolomite. The pure dolomite facies in the Toca 1 turbidites from the Viodo area averages 15.8% porosity and 14.6 md permeability (Figure 10A and B). The calcite–dolomite facies averages 10.6% porosity and 2.8 md permeability. The fivefold enhancement of permeability in the pure dolomite results from the higher intercrystalline porosity in the turbidite matrix.

In-Place Shallow Platform Carbonates

Occurrence and Thickness

Algal grainstones, interpreted to be in situ, are present at Takula field; similar facies may also be present in the TBM-1 well. Rounded to highly elongate clasts (probably disk-shaped in three dimensions) have formed a mud-free framework with good porosity (Figure 6G). This interpretation is based on sidewall cores from the 44-1X well and on cuttings from the TBM-1 well. The lack of carbonate mud suggests that these clasts were deposited in a relatively high-energy environment.

Toca 1 in-place carbonates are significantly thicker than the allochthonous facies: 58 m in 44-1X well and 79 m in TBM-1 well. The facies appears to consist of nearly 100% net carbonate.

Diagenesis and Reservoir Quality

The Toca 1 algal grainstone at Takula field is a relatively unaltered limestone, in contrast to the allochthonous facies which generally contain dolomite. This difference suggests that the lake during deposition of Toca 1 was saline stratified. The higher salinities in deeper parts of the lake promoted dolomite formation, while fresher water conditions in shallow water resulted in deposition of calcite-dominated facies.

Similar lithologies were probably present in the TBM-1 well. However much of the section now consists of baroque dolomite, indicating later alteration at high temperatures (Radke and Mathis, 1980). The greater

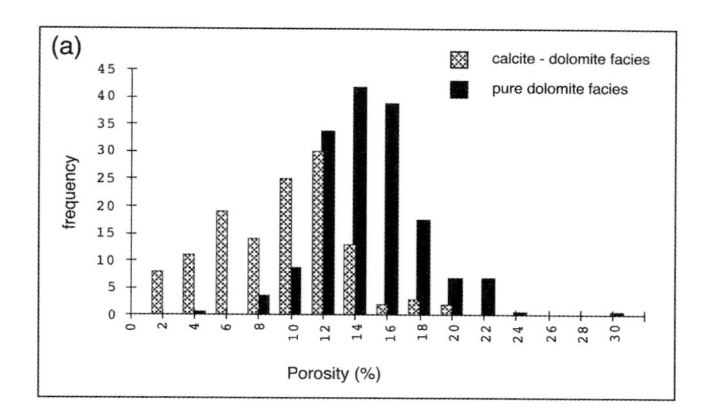

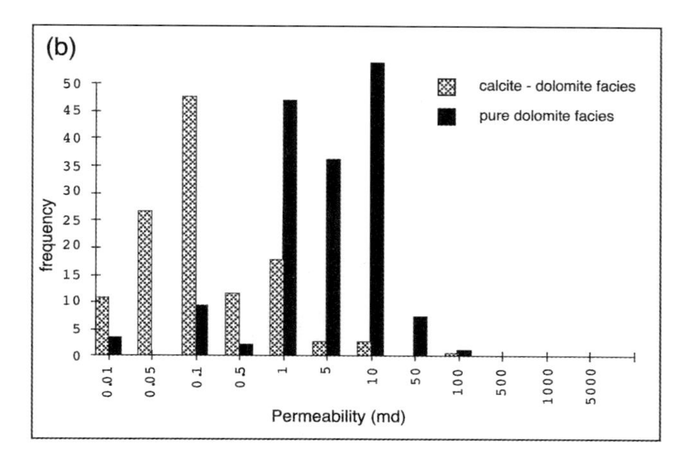

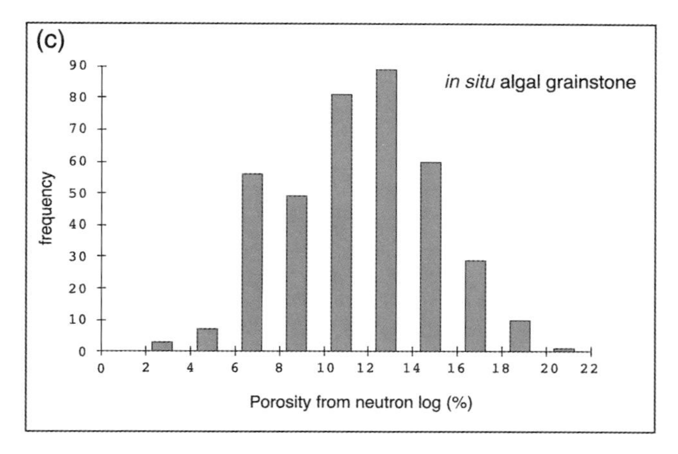

Figure 10—Graphs of porosity and permeability in Toca 1 facies: (a) Core porosity in allochthonous Toca 1 from Viodo area wells and Lifuma field. (b) Core permeability in allochthonous Toca 1 from Viodo area wells and Lifuma field. (c) Porosity from neutron logs from in situ Toca 1 at Takula field.

degree of alteration may be due to the presence of a fault, penetrated by the well, which would have acted as a conduit for late burial diagenetic formation water.

The Toca 1 interval at Takula field averages 11.8% porosity, based on a neutron porosity log (Figure 10c). Given the relatively thick nature of the section, this constitutes a fairly good reservoir. Poroperm data were not available from the TBM-1 core, but the overall poor reservoir quality in that well was evident in core and cuttings. The association of late dolomitization with poor reservoir quality is also evident at Kambala field.

TOCA 2

Toca 2 is present in several wells from Cabinda and Congo (Figures 2, 4, and 5), and it is an important reservoir in the Takula and Malongo West fields in Cabinda and in the Viodo pool in the Congo. It is on average the thickest of the Toca units, ranging from as thin as 10 m net carbonate in the Kambala field to as thick as 102 m in the VIM-1 well (Figure 5). It is the most widespread of three principal Toca levels (Figure 2), occurring as far east as the easternmost horst block (the Pointe Indienne occurrence) and as far west as the Outer Basin High (Kambala field). It is best developed along a central belt of fields that includes Viodo, Takula, and Malongo West, where the thickest reservoir occurs in coquinas at platform margins.

Basic Facies Types

The Toca 2 comprises a complex set of shallow-water platform carbonate facies, deep-water allochthonous carbonate facies, and coeval deep-water marl of the Marnes Noires Formation (Figure 11a). Shallow-water platform facies are present in wells on the crest of horst blocks in the Viodo and Takula areas. These locally include an inner belt of gastropods and oncolite grainstones and generally include platform margin deposits such as pelecypod coquinas and skeletal grainstones. Slope deposits form an extensive apron around the platform and include deep-water debris flows composed of pelecypod coquinas and skeletal grainstones. These different facies have a strong influence on reservoir quality. In general, the shallow-water platform facies have the best reservoir quality, in part because they were subaerially exposed periodically, forming secondary porosity.

The depositional model presented in Figure 11 is largely based on the Viodo area, where cored Toca 2 wells are most closely spaced. Such relationships have not been recognized in Cabinda, where wells have been much less extensively cored.

Platform Carbonate Facies

Lithologies, Sedimentology, and Thickness

Two shallow-water carbonate facies are present on the crest of the Viodo structure—a lower gastropod-rich

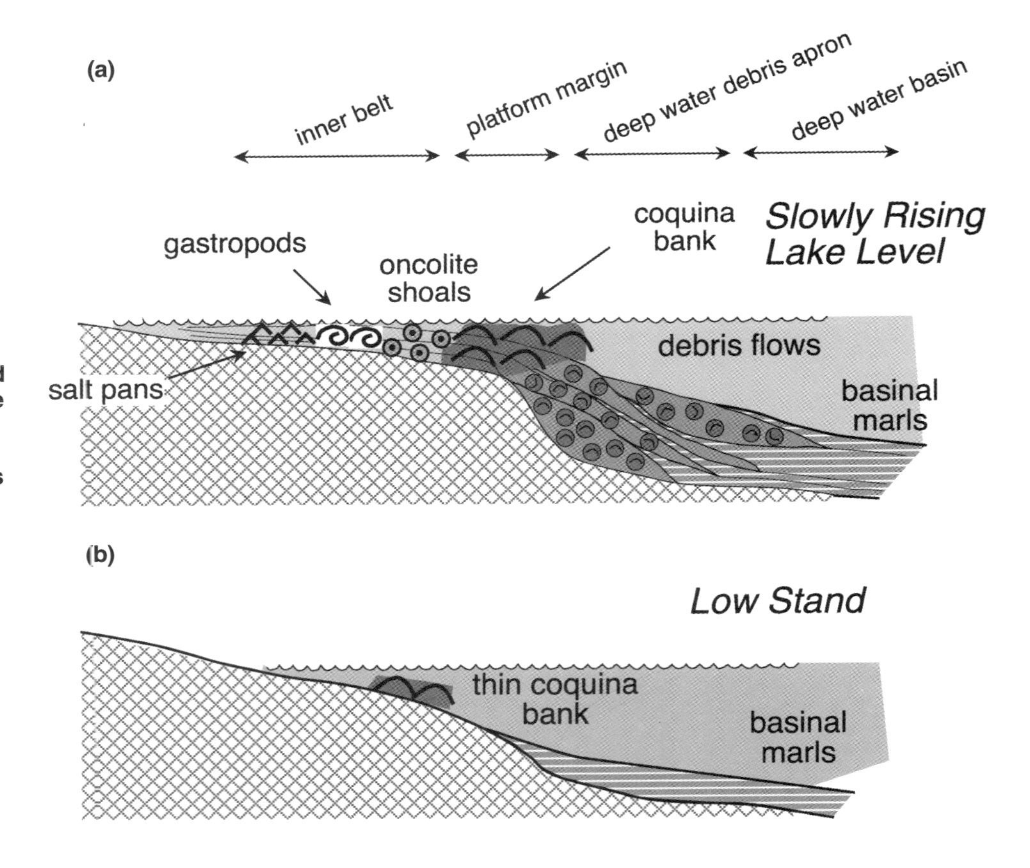

Figure 11—Schematic cross sections showing facies models for Toca 2. (a) Distribution of Toca 2 basic facies types in the Viodo area. The inner belt facies include salt pans, gastropod shoals, and oncolite shoals. The platform margin facies is a coquina and skeletal grainstone bank. Slope and deep-water facies are dominantly carbonate debris flows, with minor grain flow deposits and basinal marls. Note that at the platform margin, the coquina bank apparently prograded over its own carbonate debris flow deposits. (b) Model for thin Toca 2 carbonates, such as in the 84-7 well. During lowstands, coquina banks formed in a downdip position on a ramp. During highstands, the coquina bank drowned and mud was deposited at the site. Because the site was never subaerially exposed, no significant leaching of the coquina occurred.

grainstone and an upper coarse oncolitic zone; together they comprise 27 m of net carbonate. These were deposited in the interior of the shallow-water platform (Figure 11a). They are separated from the underlying Marnes Noires by a thin transition zone of light colored, ostracode-rich, bioturbated (?) siltstone. Coquinas and carbonate grainstones formed at the margins of shallow-water platforms in the Viodo and Takula field areas. In favorable settings, coquinas formed thick, high-quality reservoirs, locally 50 m thick.

Gastropod grainstones are typically massive, very coarse grained rocks (Figure 12A) with local bioturbation, a feature that distinguishes them from the Toca 1 grainstones. Minor oncolites and rounded shale clasts are present. Interbedded finely laminated micritic dolomite beds (Figure 12A) are evaporite deposits, formed on the fringes of the paleolake during extreme falls in lake level. The gastropod grainstone is overlain by coarse-grained, poorly sorted, and highly leached oncolite grainstone (Figure 12B and F). The leaching suggests that this interval was subaerially exposed. This is capped by a bed rich in siliciclastic grains and ostracodes, in turn overlain by deep-water deposits, including Marnes Noires shale and carbonate debris flow beds. The quartz and ostracode-rich intervals are common in transgressive facies, deposited during rises in lake level.

The dominant lithologies on the platform margins are pelecypod coquinas (Figure 12C and G) and carbonate grainstones, up to 50 m thick in the Viodo area. The carbonate grains in the latter were probably derived from broken pelecypod shells, suggesting moderately high wave energy. At Pointe Indienne field near the basin margin, beds contain pelecypod shells in a matrix of siliciclastic silt. This occurrence is on a horst block, separated from the faulted basin margin by a graben about 15–20 km wide (P. Willette, personal communication, 1997). The high abundance of interbedded siliciclastics in the section suggests that they were derived from the basin margin, implying that the graben separating the Pointe Indienne horst from the basin margin was nearly filled.

Thin coquinas and grainstones formed on carbonate ramps (Figure 11B), such as in the cored part of Toca 2 in the 84-7X well (downdip from Malongo West field). This well contains thin pelecypod coquinas, typically 5–80 cm thick (Figure 12E), and algal carbonate beds interbedded with gray calcareous shale or marl. Algal beds are fine grained with planar to wavy or mounded lamination and consist of interlaminated sparry calcite and dark brown organic (?) material. Unlike platform margin occurrences, net carbonate is about 40% of the section, compared to 75–90% in wells from the Viodo area and Takula field.

Diagenesis of Platform Deposits

All platform facies have undergone major dissolution. Significant moldic porosity has been generated as a result, particularly in the oncolite and pelecypod coquina facies

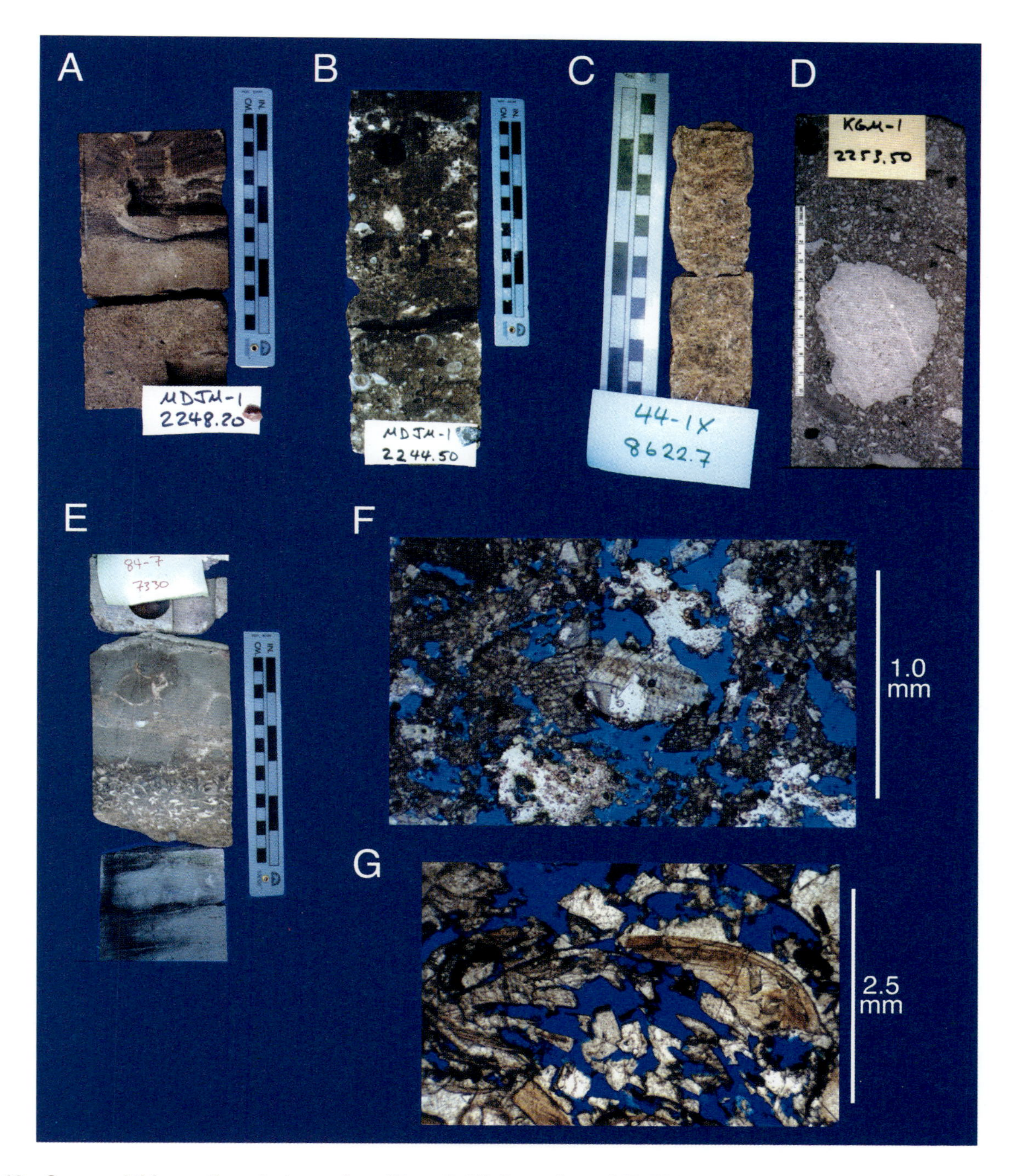

Figure 12—Core and thin-section photographs of Toca 2. (A) Core photo, MDJM-1 well, from gastropod facies in shallow-water Toca 2. Laminated micritic dolomite bed at the top is brecciated and filled with gastropod grains. The brecciation and a small amount of halite cement in the dolomite suggest that these beds were deposited in evaporative conditions on the lake edge. (B) Core photo, MDJM-1 well, from coarse oncolite facies in shallow-water Toca 2. Intense leaching and silica cement (white patches) suggest that this zone was subaerially exposed. (C) Core photo, 44-1X well (Takula field), from coquina in platform margin facies in shallow-water Toca 2. (D) Core photo, KGM-1 well, from deep-water Toca 2. Scale bar is 10 cm. Debris flow facies is from relatively deep water adjacent to platform margin. White clasts are coquina, including the 8-cm clast in the center. Dark clasts are Marnes Noires shale. The outer centimeter of the large clast has been tightly cemented, leaving a more porous inner core. (E) Core photo, 84-7 well (Malongo West field), from thin coquina in Toca 2, overlying gray shale. Coquina is overlain by brecciated calcareous mudstone. The coquina is interpreted as deposited in a ramp setting during lake level lowstand. (F) Thin-section photo, MDJM-1 well (Viodo discovery), from gastropod zone in Toca 2 (see gastropod in center). Note baroque dolomite. Small amounts of silica cement appear as white patches in lower right. (G) Thin-section photo, PI-1 well (Pointe Indienne field), from coquina in Toca 2. Growth lamellae in shell are evident; the shell has been extensively dissolved.

and where the gastropod facies has been completely dolomitized (Figure 12F). In the coarse oncolite zone, dissolution has been selective, completely removing some oncolites and leaving others relatively intact. Dissolution is significant in platform margin coquinas at Takula and Pointe Indienne fields (Figure 12G) and in the Viodo pool. In some cases, the pelecypod shells appear to be relatively unrecrystallized despite the dissolution; elsewhere, shells are substantially recrystallized.

Thin coquinas, located in downdip positions (such as the 84-7 well) and deposited during lowstands, do not have significant dissolution porosity, probably because they were not subaerially exposed. The matrix of these coquinas shows little indication of recrystallization. Shells have been dissolved and replaced by coarse sparry calcite, and minor remaining secondary porosity was later filled by coarse baroque dolomite. As a result, reservoir quality in these limestones is poor.

Dolomitization is also important, particularly affecting the oncolite and gastropod facies. The matrix of the oncolite facies has been dolomitized and partially replaced by opaline silica. The gastropod facies is variably dolomitized, in part consisting of relatively fine-grained calcite and dolomite. Where completely dolomitized, some baroque dolomite is present, indicating recrystallization at high temperatures. Dissolution is more intense in this interval, and porosity and permeability are correspondingly higher. Baroque dolomite generally does not dominate in the platform coquinas and grainstones (Figure 12F). However, the Toca 2 occurrence in Kambala field (120-1 well) is intensely dolomitized. Little porosity has been generated by the dolomitization, and moldic porosity is minor. The overlying Toca 3 in this well has been replaced by baroque dolomite. It is likely that the dolomitization of Toca 2 is related to this late high-temperature process rather than to an early diagenetic event.

Allochthonous Deep-Water Carbonate Facies

Lithologies, Sedimentology, and Thickness

A complex suite of carbonates was deposited in relatively deep water adjacent to the shallow-water platforms (Figure 11a). These deposits were cored extensively in the Viodo area. They consist of various types of sediment gravity flow deposits interbedded with Marnes Noires shale facies. The carbonate fraction of the Marnes Noires interbeds contains significantly more calcite than Marnes Noires interbeds in Toca 1, with the mole fraction (MgO/MgO + CaO) in carbonate averaging 0.19. This indicates that the lake water had become significantly fresher.

The most abundant sediment gravity flow deposits are coarse, massive carbonate conglomerates. Clasts are matrix supported, and there is no apparent grading by clast size, suggesting deposition as debris flows. These include abundant coquina clasts that range in size from small shell fragments to boulders 3.5 m in diameter (Figure 12D) derived from the platform margin coquina bank. Marnes Noires clasts and siltstone clasts are also common, typically displaying soft sediment deformation that indicates they were entrained and redeposited before lithification. Dolomite clasts, similar to the laminated dolomite from the lake margin, are present, suggesting that marginal clasts could make their way through the coquina bank into deep water.

At the base of the Toca 2 in the VIM-1 well, a thin carbonate debris flow interval is overlain by the in-place coquina bank (Figure 5), demonstrating that a coquina bank prograded basinward, building up and out over its own debris pile (Figure 11a). Debris flow deposits in the VIM-3 well are thick (Figure 5), amounting to about 35 m of net carbonate. Clearly the margin of the coquina bank was very active, regularly shedding large amounts of debris into deep water.

Other sediment gravity flow deposits include coarse oncolites (similar in appearance to the oncolites in MDJM-1, ostracode-rich packstones and fine-grained oncolites, and brown clay ooids in a micritic dolomite matrix. Sandstone is a minor facies, occurring mostly below the debris flow deposits and ranging from 3 cm to almost 1 m thick. The sandstone is generally massive, but locally shows soft sediment deformation or flame structures, suggesting rapid emplacement. The sandstone at the base of Toca 2 appears to mark the lowest lake level, with the overlying carbonates deposited during the subsequent slow rise of lake level. The sandstones are not present in the platform sequence.

Diagenesis of Allochthonous Deposits

The carbonate debris flows have undergone intense pressure solution. This reduced porosity by intense compaction within the debris flow matrix and by carbonate cementation as the carbonate that was dissolved during pressure solution was reprecipitated nearby. Small clasts are completely cemented by calcite (Figure 12D); large clasts developed a tightly cemented outer rind and a more porous core. Very large clasts may be quite porous, but this porosity is isolated by the tight debris flow matrix and does not result in overall good permeability.

Toca 2 Reservoir Quality

Reservoir quality in the Toca 2 is highly dependent on depositional facies and diagenesis. Thick shallow-water platform coquinas, gastropod grainstones, and oncolite facies all have excellent reservoir quality. Porosities of the platform facies from the Viodo wells and Takula field generally exceed 15%, averaging 18% in the limestone coquinas, 19% in the gastropod grainstones, and 23% in the oncolite zone (Figure 13). Permeabilities are correspondingly high. Mean permeabilities in these lithofacies exceed 50 md, and the coquinas average 466 md.

The excellent reservoir quality in the platform facies appears to be largely due to meteoric leaching during

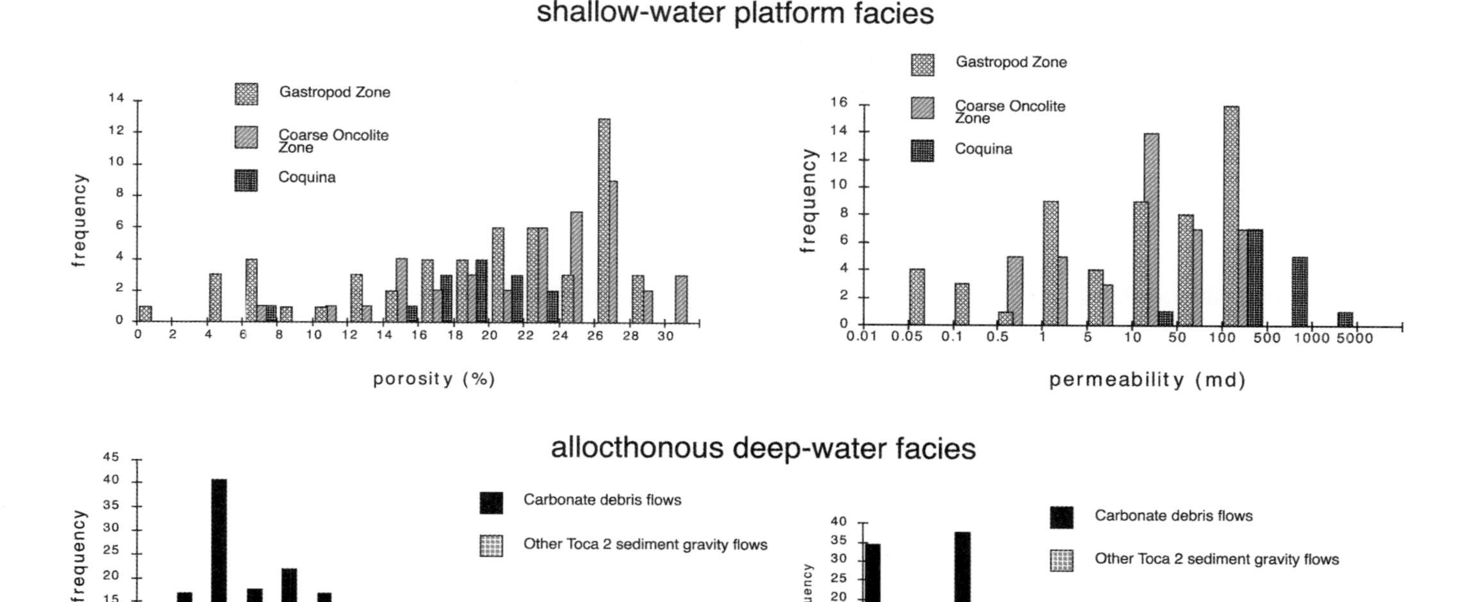

Figure 13—Toca 2 porosity and permeability data from (a) shallow-water platform facies and (b) allochthonous deep-water facies. Reservoir quality in the shallow-water platform facies of Toca 2 is typically the best of the three Toca units.

extreme lake lowstands. In contrast, thin in-place coquina and grainstone beds (such as in 84-7 and 44-8X wells) experienced little or no subaerial exposure and therefore no meteoric water dissolution. Reservoir quality in these coquinas is relatively low, averaging 5.2% porosity and 2.1 md permeability.

Porosities and permeabilities in the carbonate debris flows, the most abundant of the allochthonous deep-water facies, are very low (Figure 13), averaging 8.3% porosity and 18 md permeability. These values are actually deceptively high because they include high porosities and permeabilities in the centers of some large coquina boulders. The median permeability is 0.23 md, more accurately reflecting the low permeability of the debris flow matrix.

Other allochthonous Toca 2 facies have better reservoir quality (Figure 13). The ostracode packstone averages 12% porosity and 2.9 md permeability where it is relatively clean. In a downdip depositional site, the unit is considerably siltier and has poorer reservoir quality. The fine-grained oncolite zone is dolomitized, which may have created or enhanced matrix permeability, and has good reservoir quality, averaging 23% porosity and 24 md permeability. Sandstones have poor reservoir quality (average 8.7% porosity and 0.7 md permeability) due to poor depositional sorting and probably to authigenic clay in the pores.

Depositional Models

Deposition of the Toca 2 carbonate was triggered by a fall in lake level, which probably reached its lowest level near the base of the unit. The lowest lake level may be indicated by the thin sandstone at the base of Toca 2, with siliciclastic grains derived from exposed basement highs. During and immediately following the lake level fall, carbonate facies became established on shallow-water platforms, with oncolites and gastropods forming nearest the shoreline and pelecypod shoals in slightly deeper water at the break in slope (Figure 11a). Lake level probably fluctuated dramatically on a shorter time scale with an amplitude of 5–20 m, periodically exposing the entire shallow-water platform. During these times, the pelecypod shoal occasionally collapsed, forming subaqueous debris flows that piled up at the base of the slope. Deposition of carbonate was terminated by a rapid rise in lake level that drowned the platform.

Thin pelecypod coquinas, such as in the 84-7 and 44-8X wells, probably represent deposition on a carbonate ramp during a period of fluctuating lake level (Figure 11b). During lowstands, pelecypod coquinas and algal limestones were deposited; these were covered by shales or marls during subsequent highstands. The thinner accumulation of carbonates may be due to its depositional position in slightly deeper water; it would have come within the carbonate production zone only during extreme lake level falls. This model is consistent with the lack of carbonate dissolution in the thin coquinas, as compared to the larger accumulations in the VIM-1 and 44-1X wells.

TOCA 3

Toca 3 is best developed in wells along the Outer Basin High (Figures 2, 4, and 5), including the Likouala High in Congo and the Kambala High in Cabinda. In these wells, it ranges from 26 to 74 m in total thickness and is nearly pure carbonate. The MOVM-1 occurrence on the west side of the Likouala High is the most basinward Toca occurrence.

Stratigraphic Position

Toca 3 is interpreted to occur either within the uppermost part of the Marnes Noires or in the Lower Argilles Vertes, based on three observations:

1. The Toca in MAZM-1 overlies a shale that is relatively light in color and geochemically very different from the Marnes Noires. A shale sample recorded 0.41% total organic carbon (TOC), a hydrogen index (HI) of 271, and an oxygen index (OI) of 137; this compares to 3–16% TOC, HI of 360–860, and OI of 7–70 in the Marnes Noires Formation (Harris et al., 1994). The shale laminae below the Toca 3 also appear to be disrupted by very small burrows, a feature not seen in Marnes Noires shales.
2. The presence of coquina debris flows near the Marnes Noires - Argilles Vertes contact in the MDJM-1 and KGM-1 wells (Viodo area) indicates that coquina was being deposited at the time of this transition.
3. Two carbonate intervals are present in the 120-1X well (Kambala field). The lower carbonate contains abundant clay spherules identical to those in Toca 2 from the VIM-1 and VIM-3 wells, establishing a correlation between this unit and Toca 2 in the Viodo area. This establishes the upper Toca in the 120-1X well as a stratigraphically higher unit. Furthermore, Braccini et al. (1997) correlated the upper of two Toca units at Kambala field (120-9 and 121-2 wells) with the strata above the Middle Bucomazi Formation, and thus it is equivalent to the Upper Bucomazi or Argilles Vertes.

The Toca 3 is overlain by several different formations (Figure 5). In the MAZM-1 well, it is directly overlain by the Loeme Salt, and in the 120-1 well, it is overlain by the Chela, in both cases in unconformable relationships. In MOVM-1, it is overlain by the Argilles Vertes in apparent conformable contact. In the 84-7 well, it is overlain by a late conglomerate.

Depositional Facies, Lithologies, and Sedimentology

Toca 3 in the MAZM-1 and MOVM-1 wells consists almost entirely of 30-cm- to 2.2-m-thick cycles that grade upward from pelecypod coquina to lime mudstone (Figure 14A). At the base of the cycles, the limestone consists of tightly packed pelecypod shells (Figure 14B). The fraction of carbonate mud increases systematically upward, such that the upper part of each cycle consists of peloidal carbonate mud and ostracodes, with discontinuous blue-green clay laminae (Figure 14C). Near the base of the Toca 3 in both the MAZM-1 and MOVM-1 wells is a thin zone of coarse oncolites and stromatolites.

The "muddying-upward" cycles represent repeated abrupt small-scale deepening events in the lake, followed by gradual progradation of a muddy shoreline (Figure 15). Pelecypod coquinas were deposited just after the lake deepened. Nearshore muds then slowly swamped the pelecypods as the shoreline built outward. The origin of these cycles is not clear. It may have been tectonic, related to abrupt subsidence of the basin on the order of 1–4 m per event (estimate based on approximate decompacted thickness of the cycles). Alternatively, it may have been climatic, related to periods of high precipitation that rapidly raised lake level.

Diagenesis and Reservoir Quality

The Toca 3 exhibits two diagenetic trends. In the MOVM-1 and MAZM-1 wells, the carbonate is dominantly limestone. The original carbonate in the shells is replaced by calcite with little net gain in porosity (Figure 14B). Calcite cement partially fills interparticle pore space, probably derived from carbonate dissolved along stylolites. Some remaining porosity has been filled by small amounts of baroque dolomite.

The lack of carbonate dissolution in the Toca 3 in the MOVM-1 and MAZM-1 wells distinguishes these coquinas from the Toca 2. This difference may be due to a moderately stable lake level during deposition of the Toca 3 (lake level rises of no more than 4 m and no evidence for exposure) and very dynamic lake level fluctuations during deposition of Toca 2 (evidence for significant exposure, indicating significant lake level falls). Because there were no lake level falls of comparable magnitude during deposition of the Toca 3, carbonate dissolution occurred later and simply involved dissolution and reprecipitation of calcite with no net gain of porosity.

The reservoir quality of Toca 3 limestones (in MOVM-1 and MAZM-1 wells) is generally poor. Average porosity in the wells is 6.4–6.8% and average permeability is 1.1–2.78 md (Figure 16). Maximum permeability in the Toca 3 limestones is 37 md.

The Toca 3 in the 120-1 well at Kambala field has been profoundly dolomitized (Figure 14E), and no calcite now remains. Much of the carbonate now has a striped

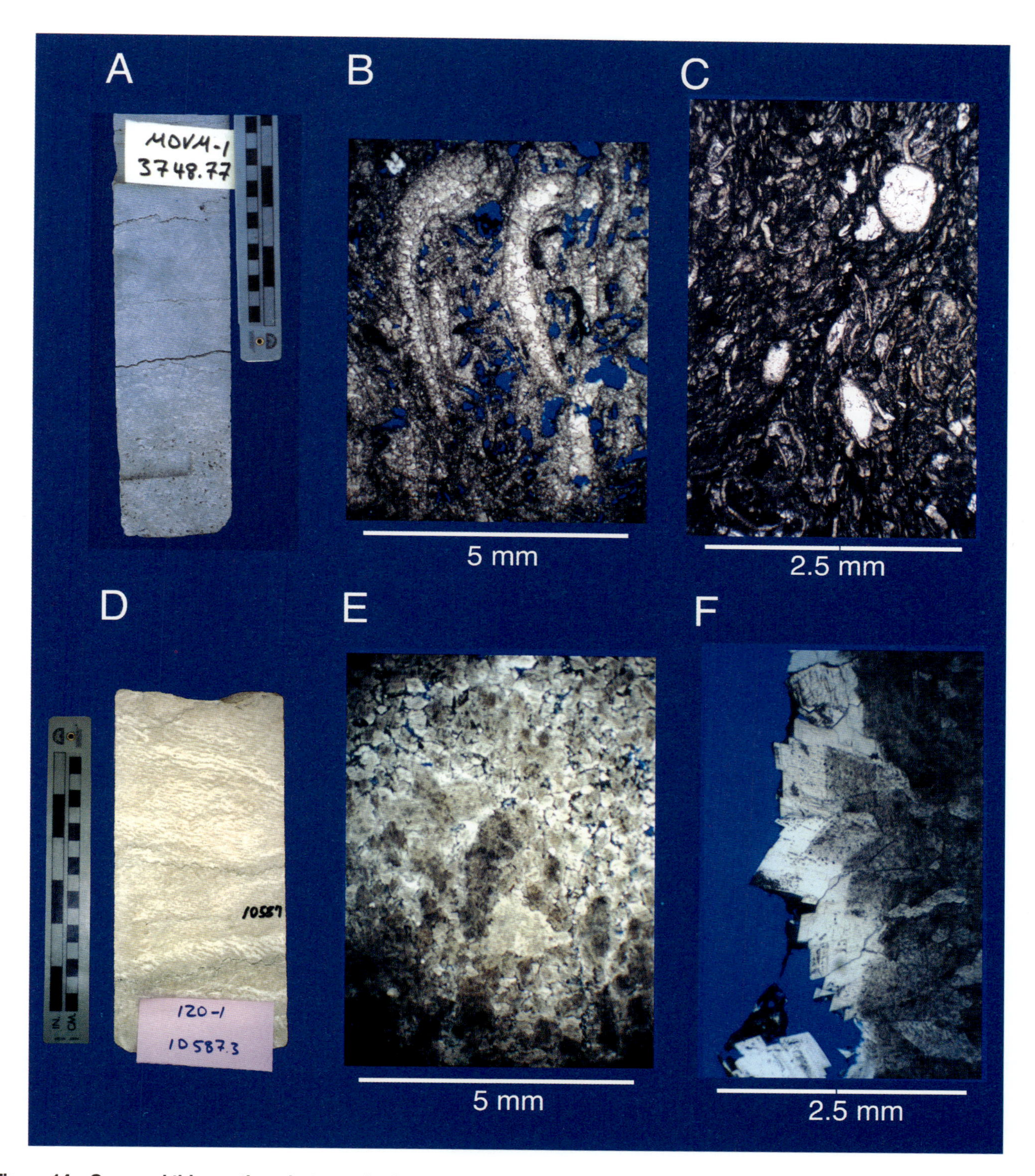

Figure 14—Core and thin-section photographs from Toca 3. (A) Core photo, MOVM-1 well, from "muddying-upward" cycle in Toca 3. Pelecypod coquina at base grades upward to lime mudstone at top. Dark wiggly lines are stylolites, common in lime mudstone. (B) Thin-section photo, MOVM-1 well, from coquina at base of muddying-upward cycle in Toca 3, equivalent to bottom of core piece in (A). Porosity is derived mostly from dissolution of small grains. Sample not stained for calcite. (C) Thin-section photo, MOVM-1 well, from lime mudstone near top of a muddy-upward cycle, equivalent to top of core piece in (A). Rock is ostracode rich and peloidal. Sample not stained for calcite. (D) Core photo, 120-1 well (Kambala field), from completely dolomitized section in Toca 3 similar to zebra dolomite of Devonian carbonates in Alberta, Canada (Dravis and Muir, 1992). (E) Thin-section photo, 120-1 well, from massive dolomite. "Ghosts" of pelecypod shells are evident in center of photo. (F) Thin-section photo, 120-1 well, from clear baroque dolomite growing into vug.

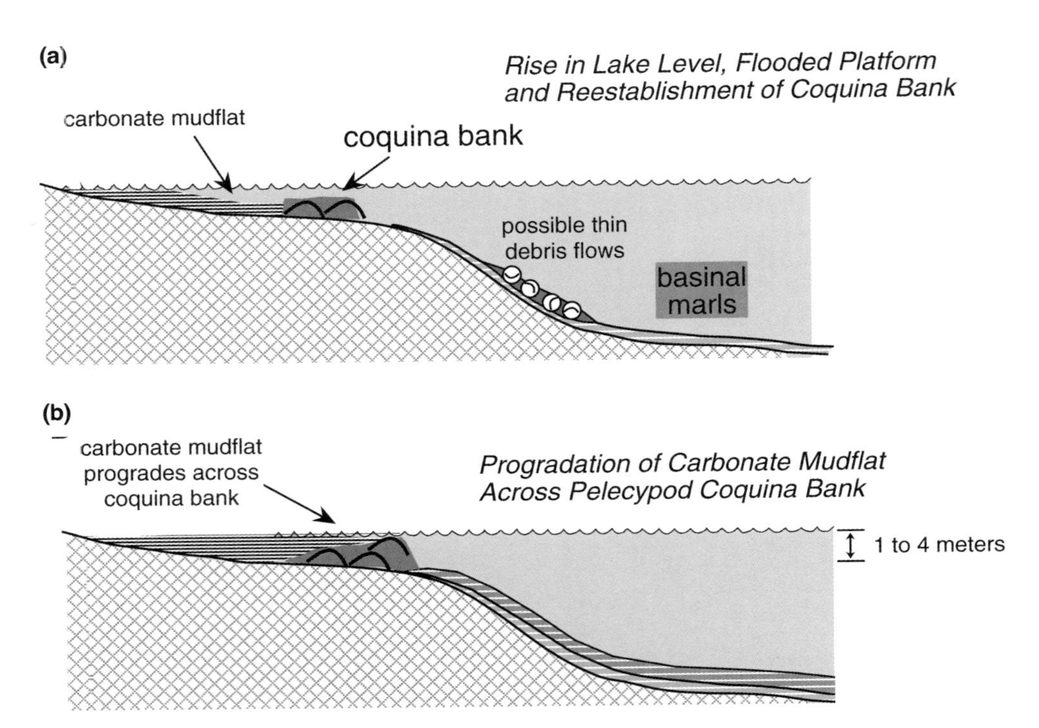

Figure 15—Schematic cross sections showing the origin of a shallowing- or muddying-upward cycle in Toca 3 at MOVM-1 and MAZM-1 wells. (a) With a 1–4 m rise in lake level, coquinas formed on the shelf. A small volume of debris may have been shed into deeper water near the platform margin. (b) A nearshore carbonate mud flat built out and prograded across the coquina bank. This was followed by a flooding event that drove the carbonate mudflat backward and permitted reestablishment of the coquina bank on the platform margin.

appearance (Figure 14D, similar to the "zebra dolomite" in the Middle Devonian Elk Point pools on the Comet Platform in Alberta, Canada (Dravis and Muir, 1992). Large vugs, partially to completely filled with baroque dolomite, are also common (Figure 14F). Zebra dolomite is interpreted as a late high-temperature diagenetic feature associated with fracturing, suggesting that the carbonate in the 120-1X well was subjected to the passage of high-temperature fluids, perhaps moving up from considerable depth along a major fault. The effect of late high-temperature dolomitization has been to destroy porosity. Porosities are generally low in 120-1X, averaging 4% (Figure 16). Permeabilities were substantially affected by fracture and vug development and are highly anisotropic. The average permeability parallel to fractures is 43 md, whereas the average permeability perpendicular to the fractures is 9.6 md, which is lower by a factor of 4.5.

CONCLUSIONS

The Toca Carbonate in the Congo Basin underwent a systematic stratigraphic evolution that is reflected in the type of bioclasts, mode of deposition, diagenesis, and reservoir quality. The oldest Toca level, Toca 1, is almost completely dominated by algal-based carbonate lithoclasts, oncolites, and grazing gastropods. The intermediate Toca 2, deposited during a period of dramatically fluctuating lake level, contains these algal-based components as well as a significant proportion of pelecypod coquinas (pelecypods are filter-feeding organisms). The uppermost unit, the Toca 3, is almost exclusively pelecypod coquinas, with the algal-based components virtually absent.

The sequence from Toca 1 to Toca 3 thus represents a major paleoecologic transition from an algal-based carbonate system to a filter feeder-based system. This transition is also marked by a systematic decrease in lake salinity, indicated by decreasing dolomite relative to calcite in interbedded shales. It may be that the high salinity during early stages of Marnes Noires deposition reduced calcium availability to organisms. This would have mostly affected larger or more heavily shelled organisms such as pelecypods. A similar phenomenon occurs in modern Lake Turkana (A. Cohen, personal communication, 1999), where bivalves do not survive at high alkalinity (~15–20 mEq/L); gastropods are less sensitive. As salinity decreases, bivalves form an increasingly larger part of the ecosystem.

The carbonates undergo a parallel evolution in their mode of deposition. Toca 1 is predominantly found as sediment gravity flow deposits that form aprons around the flanks of intrabasin platforms; occurrences of significant thickness on the platforms are rare. This may have been due to high wave energy that winnowed a cementing mud. Toca 2 contains both in situ shallow-water platform deposits and deep-water sediment gravity flow deposits. The platform margin coquinas underwent lithification, then collapsed into deep water during extreme lowstands. Toca 3 deposition was essentially restricted to shallow-water margins; no significant allochthonous deposits were noted.

Finally, reservoir quality is intimately tied to the mode of deposition. Shallow platform facies in Toca 1 and particularly Toca 2 make excellent reservoirs. Allochthonous deposits, in contrast, generally have relatively low porosity and permeability due to incorporation of carbonate mud or intense compaction.

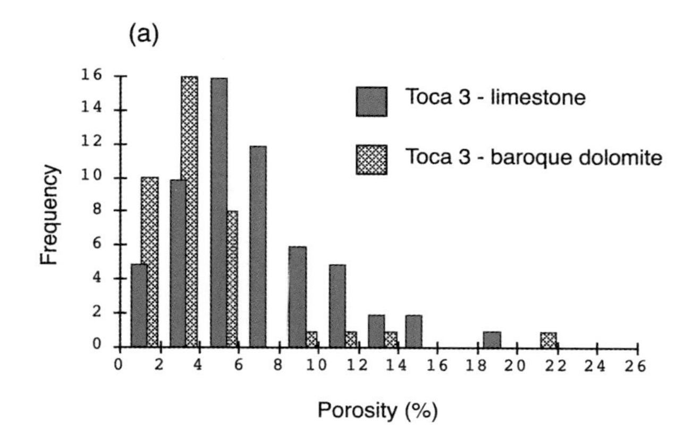

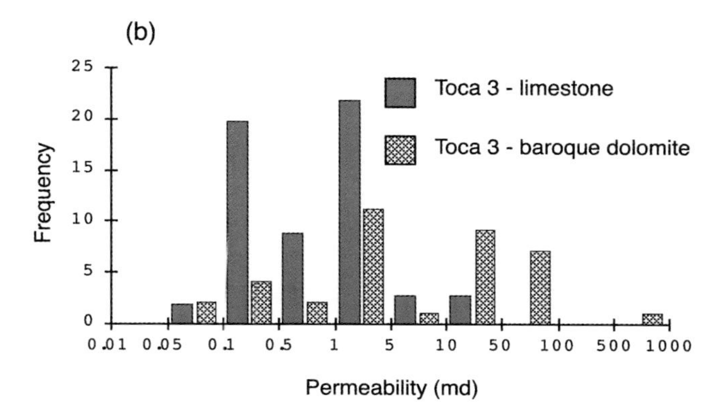

Figure 16—Graphs of (a) porosity and (b) permeability showing reservoir quality of Toca 3 limestones. Reservoir quality is relatively poor compared to other Toca units, but in spite of this, Toca 3 in the MAZM-1 well tested relatively well.

Acknowledgments*—I thank Elf Congo, Occidental International Oil and Gas Inc., and Chevron Overseas Petroleum Inc. on behalf of Cabinda–Gulf Oil Company for providing cores and other data. Occidental provided access to wells drilled by Conoco Congo Ltd. I also thank A. J. Lomando, A. S. Cohen, and B. J. Katz for their thoughtful editing that greatly improved the manuscript.*

REFERENCES CITED

Braccini, E., C. N. Denison, J. R. Scheevel, P. Jeronimo, P. Orsolini, and V. Barletta, 1997, A revised chrono-litho-stratigraphic framework for the pre-salt (Lower Cretaceous) in Cabinda, Angola: Bulletin des Centres de Recherches Exploration–Production Elf-Aquitaine., v. 21, p. 125–151.

Bracken, B. R., 1994, Syn-rift lacustrine beach and deltaic sandstone reservoirs—pre-salt (Lower Cretaceous) of Cabinda, Angola, west Africa, *in* A. J. Lomando, B. C. Schreiber, and P. M. Harris, eds., Lacustrine reservoirs and depositional systems: SEPM Core Workshop No. 19, p. 173–200.

Brice, S. E., M. D. Cochran, G. Pardo, and A. D. Edwards, 1982, Tectonics and sedimentation of the South Atlantic rift sequence; Cabinda, Angola, *in* J. S. Watkins and C. L. Drake, eds., Studies in continental margin geology: AAPG Memoir 34, p. 5–18.

Burke, K., and A. M. C. Sengor, 1988, Ten metre global sea-level change associated with South Atlantic Aptian salt deposition: Marine Geology, v. 83, p. 309–312.

Cohen, A. S., 1989, Facies relationships and sedimentation in large rift lakes and implications for hydrocarbon exploration; examples from lakes Turkana and Tanganyika: Palaeogeography, Palaeoclimatology, Palaeoecology, v. 70, p. 65–80.

Dravis, J. J., and I. D. Muir, 1992, Origin of zebra dolomites and associated secondary porosity in Devonian sequences of western Canada (abs.): Program with Abstracts, AAPG Annual Convention, Calgary, p. 32.

Grosdidier, E., E. Braccini, G. DuPont, and J.-M. Moron, 1996, Biozonation du Crétacé inférieur non marin des bassins du Gabon et du Congo: Géologie de l'Afrique et de l'Atlantique Sud: Actes Colloques Angers 1994, p. 67–82.

Halfman, J. D., 1996, CTD-transmissometer profiles from Lakes Malawi and Turkana, *in* T. C. Johnson and E. O. Odada, eds., The limnology, climatology and paleoclimatology of the East African lakes: Amsterdam, Gordon and Breach Publishers, p. 169–182.

Harris, N. B., P. Sorriaux, and D. F. Toomey, 1994, Geology of the Lower Cretaceous Viodo Carbonate, Congo Basin: a lacustrine carbonate in the South Atlantic rift, *in* A. J. Lomando, B. C. Schreiber, and P. M. Harris, eds., Lacustrine reservoirs and depositional systems: SEPM Core Workshop No. 19, p. 143–172.

Horschutz, P. M. C., L. C. S. de Freitas, C. V. Stank, A. da S. Barroso, and W. M. Cruz, 1992, The Linguado, Carapeba, Vermelho, and Marimbá giant oil fields, Campos Basin, offshore Brazil, *in* M. T. Halbouty, ed., Giant oil and gas fields of the decade 1978–1988: AAPG Memoir 54, p. 137–153.

Karner, G. D., N. W. Driscoll, J. P. McGinnis, W. D. Brumbaugh, and N. R. Cameron, 1997, Tectonic significance of syn-rift sediment packages across the Gabon–Cabinda continental margin: Marine and Petroleum Geology, v. 14, p. 973–1000.

Last, W. M., 1990, Lacustrine dolomite; an overview of modern, Holocene, and Pleistocene occurrences: Earth Science Reviews, v. 27, p. 221–263.

Lomando, A. J., 1996, Lacustrine carbonate reservoirs of Cabinda, Angola; the challenge of a unique reservoir type (ext. abst.): Extended Abstract Volume, AAPG International Hedberg Research Conference, Pau, France.

Lowe, D. R., 1982, Sediment gravity flows; II, Depositional models with special reference to the deposits of high-density turbidity currents: Journal of Sedimentary Petrology, v. 52, p. 279–297.

Maurin, J.-C., and R. Guiraud, 1993, Basement control in the development of the Early Cretaceous west and central rift system: Tectonophysics, v. 228, p. 81–95.

McHargue, T. R., 1990, Stratigraphic development of proto-South Atlantic rifting in Cabinda, Angola—a petroliferous lake basin, *in* B. J. Katz, ed., Lacustrine basin exploration: AAPG Memoir 50, p. 307–326.

Rabinowitz, P. D., and J. LaBreque, 1979, The Mesozoic South Atlantic Ocean and evolution of its continental margins: Journal of Geophysical Research, v. 84, p. 5973–6002.

Radke, B. M., and R. L. Mathis, 1980, On the formation and occurrence of saddle dolomites: Journal of Sedimentary Petrology, v. 50, p. 1149–1168.

Standlee, L. A., W. D. Brumbaugh, and N. R. Cameron, 1992, Controlling factors in the initiation of the South Atlantic rift system, *in* R. Curnelle, ed., Géologie Africaine; 1er colloque de Stratigraphie et de paleogeographie des bassins sedimentaires ouest-africains; 2e colloque africain de Micropaleontologie: Bulletin des Centres de Recherches Exploration–Production Elf-Aquitaine Memoire 13, p. 141–152.

Schoellkopf, N. B., and B. A. Patterson, 2000, Petroleum systems of offshore, Cabinda, Angola, *in* M. R. Mello and B. J. Katz, eds., Petroleum systems of South Atlantic margins: AAPG Memoir 73, p. 361–376.

Chapter 25

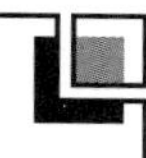

Petroleum Systems of Offshore Cabinda, Angola

Noelle B. Schoellkopf

Brooks A. Patterson

Chevron Overseas Petroleum Inc.
San Ramon, California, U.S.A.

Abstract

Multiple petroleum systems and numerous source–reservoir pairs are recognized in offshore Cabinda. In Block 0, the principal petroleum systems are the Bucomazi–Vermelha(!), Iabe/Landana–Pinda(!), and Malembo–Pinda(!). These systems arise from three prolific source rock units: the lacustrine Lower Cretaceous Bucomazi Group, the marine Upper Cretaceous–lower Tertiary Iabe and Landana Formations, and the marine Tertiary Malembo Formation. Cabinda oils can be separated into source families on the basis of composition using various analytical techniques. Quantitative biomarker analysis is useful in differentiating the genetic origin of these oils, especially when they have multiple sources.

The Lower Cretaceous Bucomazi source rock, the major contributor to the oil produced in offshore Cabinda, contains four units rich in type I kerogens: the lower Bucomazi and time-equivalent Erva Formation, the middle ("organic") Bucomazi, and the upper Bucomazi. These units together average a high SPI of 33 tons HC/m^2 rock. Oils generated from the Bucomazi are trapped in pre-rift Lucula sandstones, synrift Erva sandstones, and lacustrine Toca carbonates, and also have migrated through salt windows into the marine postrift Pinda, Vermelha, and Iabe. The marine Upper Cretaceous Iabe and Paleocene–Eocene Landana shales are also excellent source rocks in the outer shelf to slope depositional environments. They consist of type II kerogen with high average SPIs (>15 tons HC/m^2 rock). Oils from these source rocks have migrated primarily to reservoirs in the Pinda and Malembo. The Malembo has fair source potential in western offshore Cabinda (Blocks B and C), with type II and II–III kerogen and moderate SPIs (5–15 tons HC/m^2 rock). Oils generated from the Malembo have been found in Pinda and Malembo reservoirs. Effective seals are plentiful and include Bucomazi and Toca shales, the Loeme Salt, anhydritic and argillaceous beds in the Pinda Group, marine shales atop a transgressive Cenomanian–Eocene megasequence (Vermelha, Iabe, and Landana Formations), and Malembo shales.

Just as the locations of the source facies have shifted westward through time, the generative depressions have also moved westward following the evolution of the west African margin. Oil generation in Block 0 has been active from the Late Cretaceous to the present, with the most recent generation occurring in the west. The petroleum systems have evolved from east to west with differences in source origins, reservoirs, and timing.

INTRODUCTION

Block 0 is located offshore of the province of Cabinda, Angola. The block extends from the coastline to about the 200-m isobath, as shown in Figure 1. It is divided into areas A, B, and C and presently excludes the Vanza field area. The leaseholder of Block 0 is the Cabinda Association, which is a partnership of four companies: Sonangol (41%), Chevron (39.2%), Elf (10%), and Agip (9.8%). Cabinda Gulf Oil (CABGOC), a subsidiary of Chevron, has been the operator of this block since 1967 and of the original concession since 1957.

Block 0 is of special interest to the study of petroleum systems because of the large volumes of oil present and its active generation of hydrocarbons from multiple sources. The reserve figures given here can be found in press releases and other sources in the public domain. The U.S. Energy Information Administration (website *www.eia.doe.gov*) cites 1.2 billion bbl as proven reserves for Block 0, which is a conservative estimate. Cumulative oil

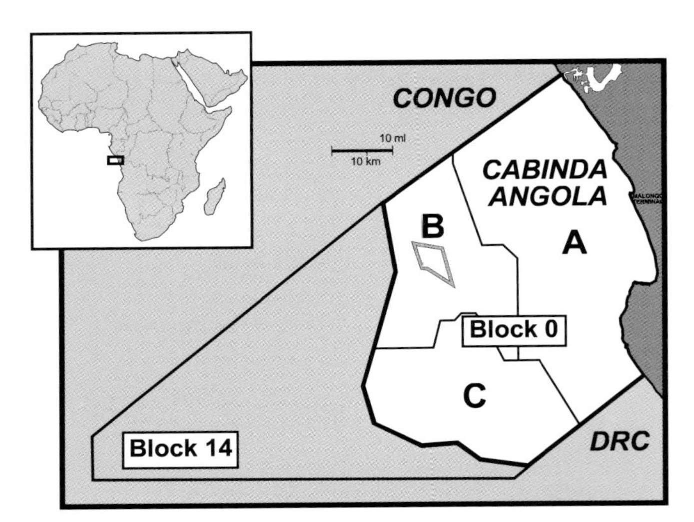

Figure 1—Location map of offshore Cabinda, west Africa, showing areas A, B, and C of Block 0, the main producing block.

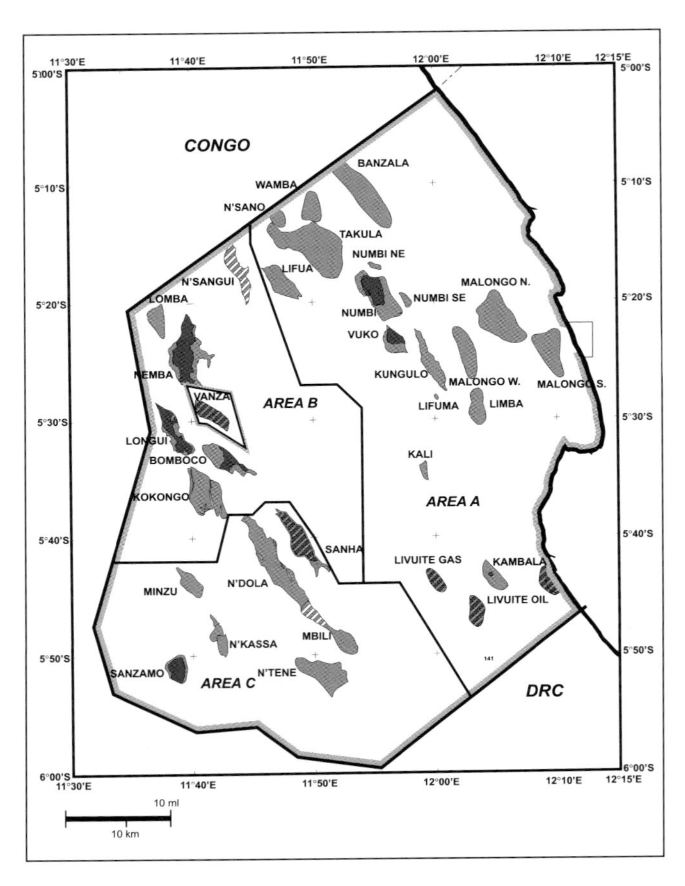

Figure 2—Map of oil and gas fields in Block 0, offshore Cabinda. Key to patterns: dark gray, gas or condensate; light gray, oil; dark diagonals, oil and gas; light diagonals, under delineation.

production by the end of 1998 was 2.1 billion bbl. While specific values for reserves of original oil in place are confidential, a reasonable estimate for Block 0 is at least an order of magnitude greater than the proven reserves.

At the end of 1999, the average daily production from Block 0 was 510,000 bbl. Early field discoveries and development were primarily in area A. Over the last two to three decades, numerous new fields have been discovered in areas B and C (Figure 2). In 1994, Kokongo was the first field in areas B and C to be brought on production. Other area B and C fields are being developed (e.g., Lomba and Nemba). The combined daily production from areas B and C exceeds 160,000 bbl/day (1998 data) and is expected to increase as more fields are developed.

Three prolific source rock "megasystems" have been recognized and correlated with the oils from these reservoirs. The relative importance of each petroleum system varies across Block 0, as described in the next section.

STRATIGRAPHY

The stratigraphic section of offshore Cabinda can be divided into three distinct geologic packages (Figure 3): (1) the presalt nonmarine *rift and sag section,* (2) the salt and postsalt marine Cretaceous–lower Tertiary *drift and passive margin subsidence section,* and (3) the *Tertiary* (Oligocene–Holocene) *sedimentary wedge.* This is a fairly standard view of the South Atlantic rift and passive margin evolution (Lehner and DeRuiter, 1977; Reyre, 1984; Meyers et al., 1996). The presalt stratigraphy has been discussed extensively by Brice et al. (1982) and McHargue (1990). Like McHargue and Meyers and co-workers we consider the pre-Chela unconformity to be the *break-up unconformity* separating the nonmarine rift section below from the marine passive margin section above.

These three geologic packages include three excellent source rock "megasystems." These are the lacustrine shales of the Lower Cretaceous Bucomazi, the marine Upper Cretaceous Iabe and lower Tertiary Landana shales and marls, and the Tertiary Malembo shales. These source megasystems charge numerous reservoirs in Block 0.

We define a *source rock megasystem* as a sequence of units containing geochemically similar source rocks which can be grouped together (but not differentiated with confidence) by their oil signature alone. For example, the Erva Formation and the lower, middle, and upper Bucomazi Group are all part of the Bucomazi source rock megasystem. Oils from these units have a Bucomazi signature and can show significant heterogeneity (Burwood et al., 1990, 1995), but at present, we cannot reliably differentiate lower Bucomazi, Erva, or middle Bucomazi provenance based on their oil characteristics. Similarly, we cannot separate Iabe and Landana oil signatures with any confidence. The source megasystem terminology arose as a matter of convenience based on oil signatures. It allowed us to sidestep stratigraphic nomenclature controversies about the definition of the upper,

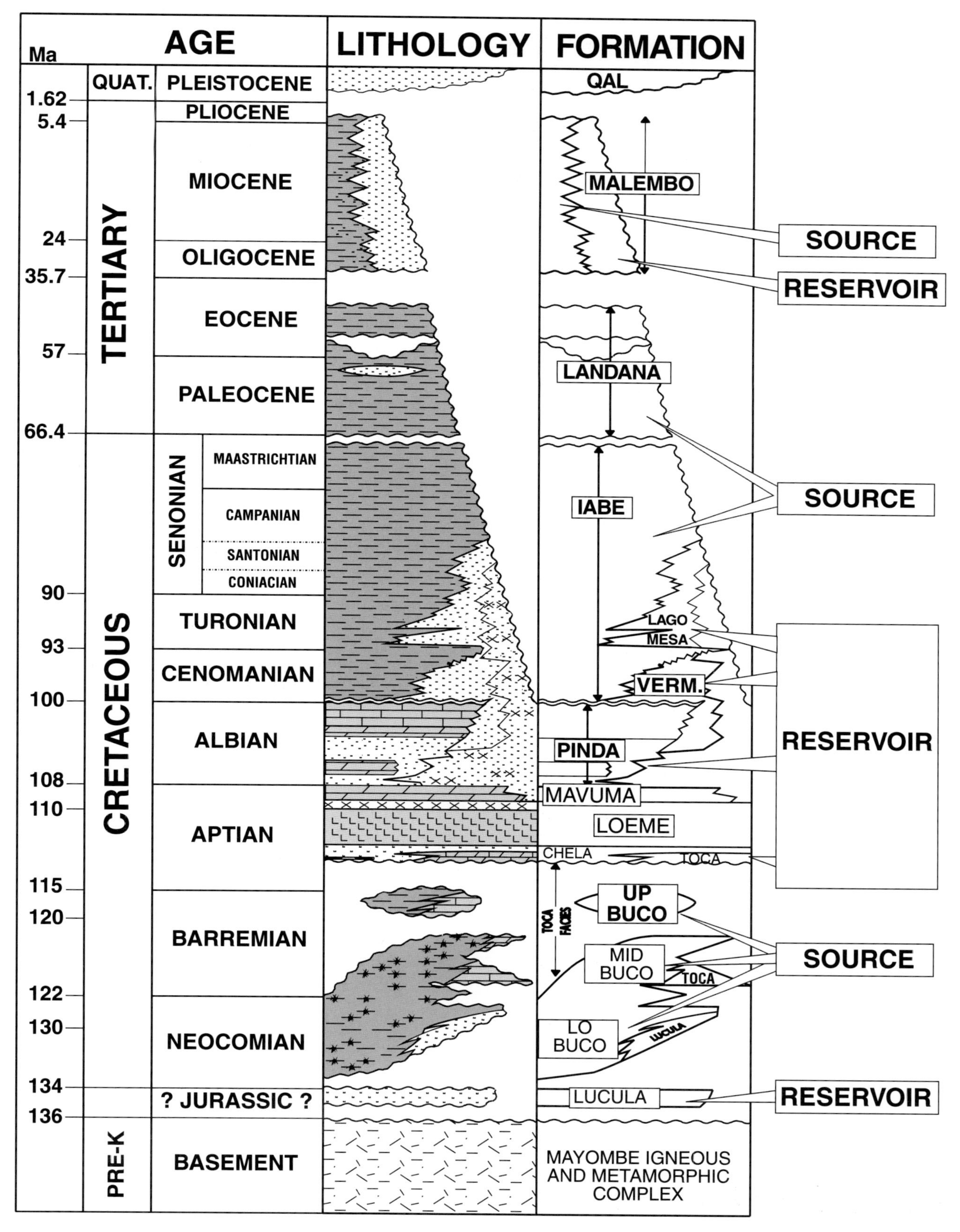

Figure 3—Generalized stratigraphic column for offshore Cabinda.

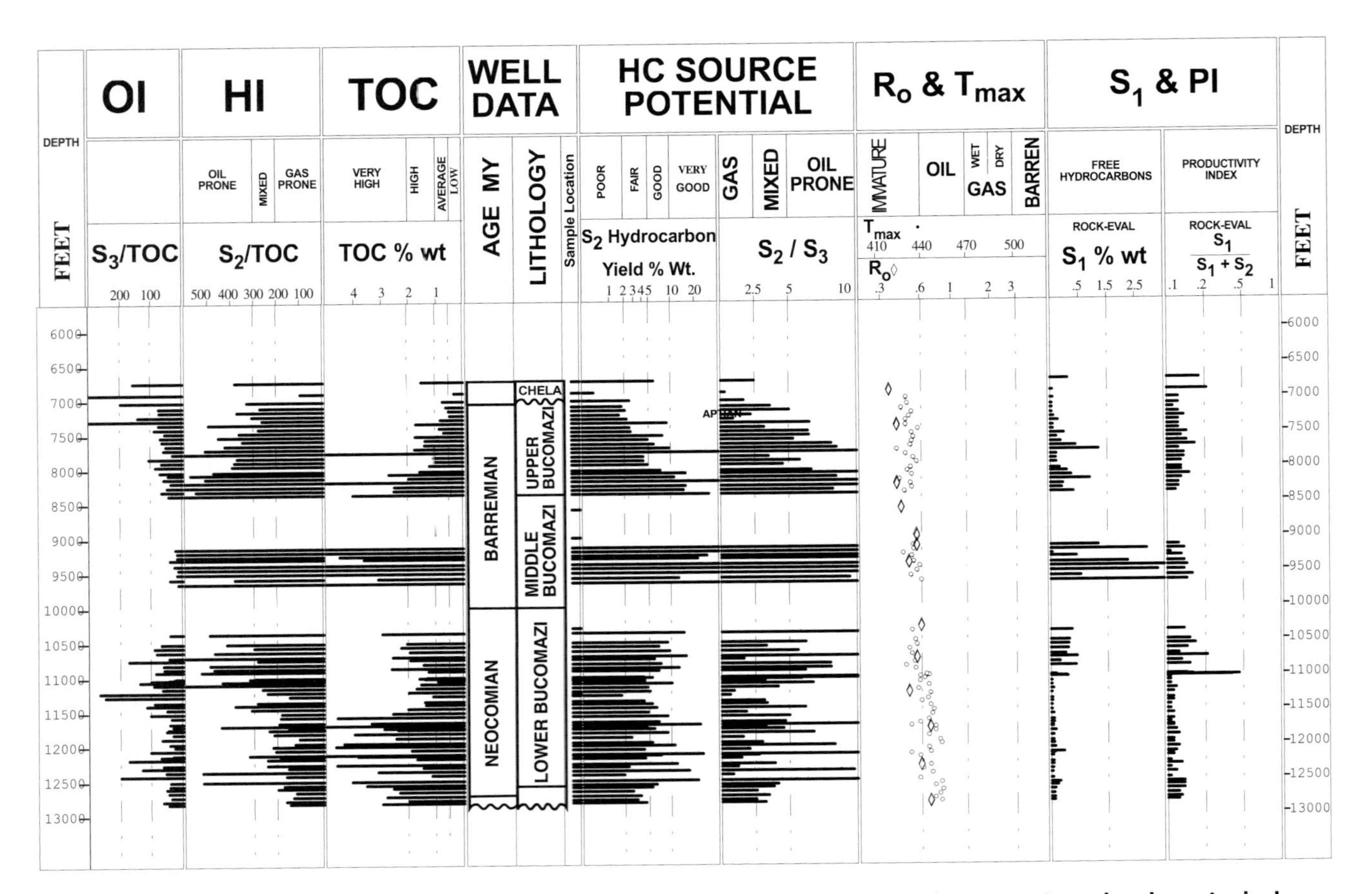

Figure 4—Geochemical log from a selected well in the Lower Cretaceous Bucomazi megasystem showing a typical signature for the presalt interval.

middle, and lower Bucomazi and Erva as "formations" versus "members" and the unresolved detailed vertical and lateral relationships of these units in the deeper section (Franks and Nairn, 1973; Brice et al., 1982; McHargue, 1990).

Lower Cretaceous Bucomazi Megasystem

The Lower Cretaceous Bucomazi source rock megasystem is responsible for most of the oil found in Block 0 in offshore Cabinda. Also, it is the sole contributor to the oil produced in area A of this block. It includes several units: the lower Bucomazi, the Erva Formation (considered time-equivalent to the lower Bucomazi), the middle ("organic") Bucomazi, and the upper Bucomazi. We have followed the informal nomenclature of McHargue (1990) for the presalt stratigraphy.

The geochemical log (Figure 4) is our standard method of presenting large amounts of source rock analyses from a well. Analytical results from ditch cuttings, sidewall cores, and even conventional core samples are conveniently displayed and interpreted in this manner. Preferred analytical spacing for ditch samples is every 30–60 ft for total organic carbon (TOC) and pyrolysis and every 300–500 ft for vitrinite reflectance. The geochemical log allows us to display the source richness and quality alongside the age, formation, and sampling and to evaluate the maturity and contaminants. The analytical results are easily interpreted when the data are presented in this fashion. The data presented include the oxygen index (OI), hydrogen index (HI), TOC, age, unit names, sampling frequency, source potential (pyrolysis S_2 and S_2/S_3), vitrinite reflectance (R_o), pyrolysis maturity (T_{max}), and contamination indicators (pyrolysis S_1 and production index, PI). We refer the reader to earlier papers by Espitalié et al. (1985, 1986) and Peters (1986) for commonly used interpretation guidelines.

Average source rock properties for Block 0 cited in this chapter are based on analyses of more than 50 wells from Block 0. Geochemical logs from three wells are shown to illustrate the general character of the three source rock megasystems. A typical geochemical log for the presalt interval (Figure 4) shows that the lower, middle, and upper Bucomazi all contain very good source rocks (TOC > 2%) with good to excellent hydrocarbon source potential ($S_2 > 5$). The kerogen type is classified using standard HI/OI crossplots (Peters, 1986). The Bucomazi consists mostly of type I and some type I–II kerogen, as evidenced by the high HI values (>500) and low oxygen indices (<100). This may be similar to the "type II_{sup}/I and

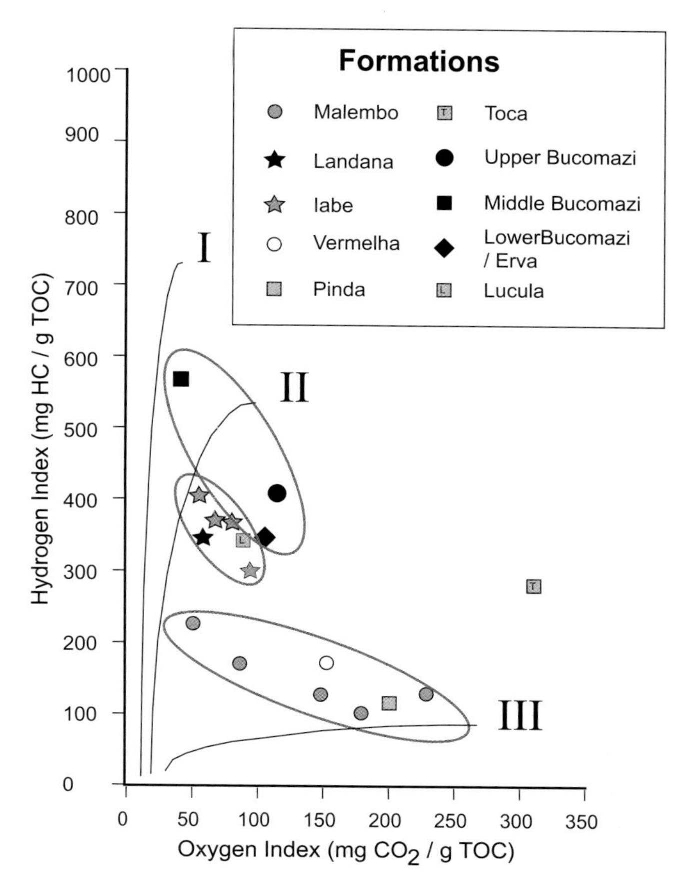

Figure 5—HI/OI crossplot of average source quality of Cabinda stratigraphic units. Note that averages are not fully corrected for maturity effects.

II/II_{sup}" organofacies described by Burwood et al. (1990). We interpret some samples with lower HI values (300–500) and low OI values (<100), particularly in the upper Bucomazi, as slightly degraded type I kerogen because in our experience these kerogens exhibit a monoenergetic type I kinetic distribution.

Note that HI values become depressed by maturation, so that pyrolysis results from the lower Bucomazi underestimate the initial HI. The lower Bucomazi shown on the geochemical log in Figure 4 has reached oil window maturity, as shown by the maturity indicators (R_o > 0.6% and T_{max} > 440°C). The empty gaps in the data at about 8400 and 9600 ft are due to lack of samples. Although the middle Bucomazi is the richest of the three units, it is evident that the lower and upper Bucomazi also have excellent source potential. The bulk of the presalt section (for wells in basinal positions) contains rich type I and type I–II kerogens.

Although the average TOC content of the middle Bucomazi across Block 0 exceeds 5%, average TOC values for the upper and lower Bucomazi are generally lower, around 2–3%. This high organic content in the middle Bucomazi is the reason why this unit was originally referred to as the "organic" Bucomazi. Similarly, average HI values are highest for the middle Bucomazi but remain consistently high for the upper and lower Bucomazi as well, as shown in the HI/OI diagram in Figure 5. Note that the average HI values for the Bucomazi are depressed on this plot because they have not been corrected for maturity (which decreases the HI). Plots of immature samples alone show higher HI values and are distributed more clearly as type I and I–II kerogens on the HI/OI diagrams.

Across Cabinda, the combined presalt source units average a very high source potential index (SPI) of 33 tons HC/m^2 rock, following the scale of Demaison and Huizinga (1991). Previously cited SPI values for the Bucomazi section of Cabinda were higher (SPI = 46), perhaps due to a more vigorous maturity screening applied by the previous authors. In any case, the values are extremely high, indicative of the excellent charging capacity of these units.

The presence, thickness, and areal distribution of these lacustrine source rocks are tied to the evolution of rift systems preceding the separation of Africa and South America. As described previously by other authors (Brice et al., 1982; McHargue, 1990; Burwood et al., 1995), the Bucomazi was deposited during the Early Cretaceous in lacustrine and associated lake margin settings controlled by the evolving rift.

Initially, the lakes in the newly formed grabens and half-grabens were probably discontinuous, isolated from one another by emergent transfer zones such as the Malongo and Kambala transfer zones (see McHargue, 1990, figure 6, a map of presalt structural features). Thus, the lower Bucomazi lake system may have had little if any communication among the Takula, Malongo, and Kambala subbasins. Active fault zones and emergent horsts shed clastic sediments into the grabens, resulting in synchronous deposition of sandy lithofacies along the basin margins and more shaly lower and middle Bucomazi lithofacies in the basin centers (as shown in McHargue, 1990, figure 6).

Later, during deposition of the remaining middle Bucomazi, there was less topographic relief (less sand and less fault movement), and the lake system apparently went through a period of increased anoxia common to most of the Lower Congo subbasins. Although we do not know whether the lake systems were connected at this time, we observe that similar, extremely rich middle Bucomazi type source rocks were deposited in numerous subbasins during the Barremian. As seen in Figure 4, the TOC, HI, and SPI values are commonly "off the scale" in this interval.

By late Bucomazi time, the lake system became increasingly less restricted and water depths were probably shallower. As the rate of subsidence slowed and the topographic relief decreased through the late Bucomazi, the shales became more calcareous and less anoxic, resulting in lower TOC and HI values and higher OI values, as seen in the upper part of the geochemical log in Figure 4.

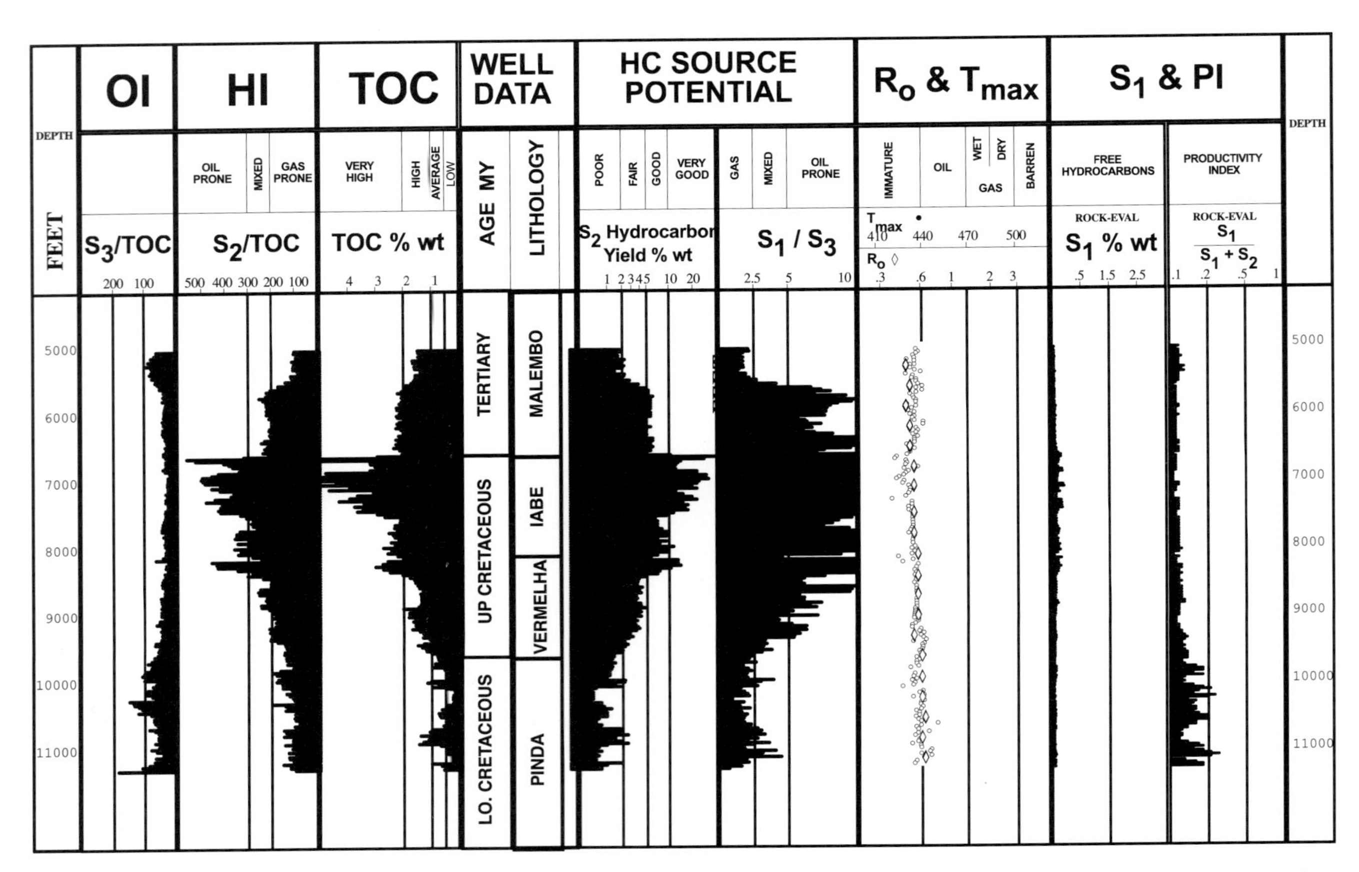

Figure 6—Geochemical log from a selected well in the Upper Cretaceous Iabe Formation. Note its excellent source rock characteristics.

Upper Cretaceous Iabe and Paleocene–Eocene Landana Megasystem

The base of the Iabe Formation is defined at the top of the Vermelha Formation, where the Vermelha can be recognized either lithologically or as a log marker. The lowermost Iabe is mostly early Turonian in age. The top of the Iabe Formation generally occurs at an unconformity below the Landana; often the upper part of the Iabe is missing. Where present, the uppermost Iabe is Maastrichtian in age. Iabe lithologies grade from shales and marls in western Block 0 to sandy reservoirs with interbedded carbonates and shales in eastern area A.

The Landana Formation is Paleocene–Eocene in age. Its lower limit occurs at the top of the Iabe Formation, near the Cretaceous–Tertiary boundary. Its upper limit is placed at the Oligocene unconformity. In eastern Block 0, the top of the Landana is easily recognized on logs by characteristic "hot" shales with extremely elevated gamma ray signatures. The top of the Iabe is also picked on the basis of one of these characteristic "hot" gamma ray markers (Alto log pick), where present.

In western Block 0, where the Landana is thin (condensed section) and often eroded by the overlying Oligocene unconformity, it consists primarily of deep-water shales. To the east it passes into mixed shales and carbonates and eventually into mixed carbonates and sandstones. In areas B and C, the Landana and uppermost Iabe are cut by multiple unconformities and are frequently eroded. The Landana in this area is generally extremely thin, less than 200 m (600 ft) thick and can be considered as a "condensed" section.

As shown in Figure 5, the Iabe and Landana source rocks consist primarily of type II kerogen. Some rich intervals from the upper parts of the two units plot in the type I field, with HI values in excess of 700. Total average SPI for the two formations combined is in the high range (>15 tons HC/m^2 rock). The geochemical log in Figure 6 shows that the marine Upper Cretaceous Iabe Formation is an excellent source rock when sampled from outer shelf to slope depositional settings. Although not shown on this log, the Landana Formation shows similar source rock characteristics in the same depositional environments.

Where these source rocks are well developed, TOC content commonly exceeds 2%, HI is greater than 500, and OI values are extremely low. Hydrocarbon yield and S_2/S_3 values indicate that these are rich oil-prone source rocks. As can be seen in Figure 6, Iabe source quality is poorer at its base and grades upward into source rocks

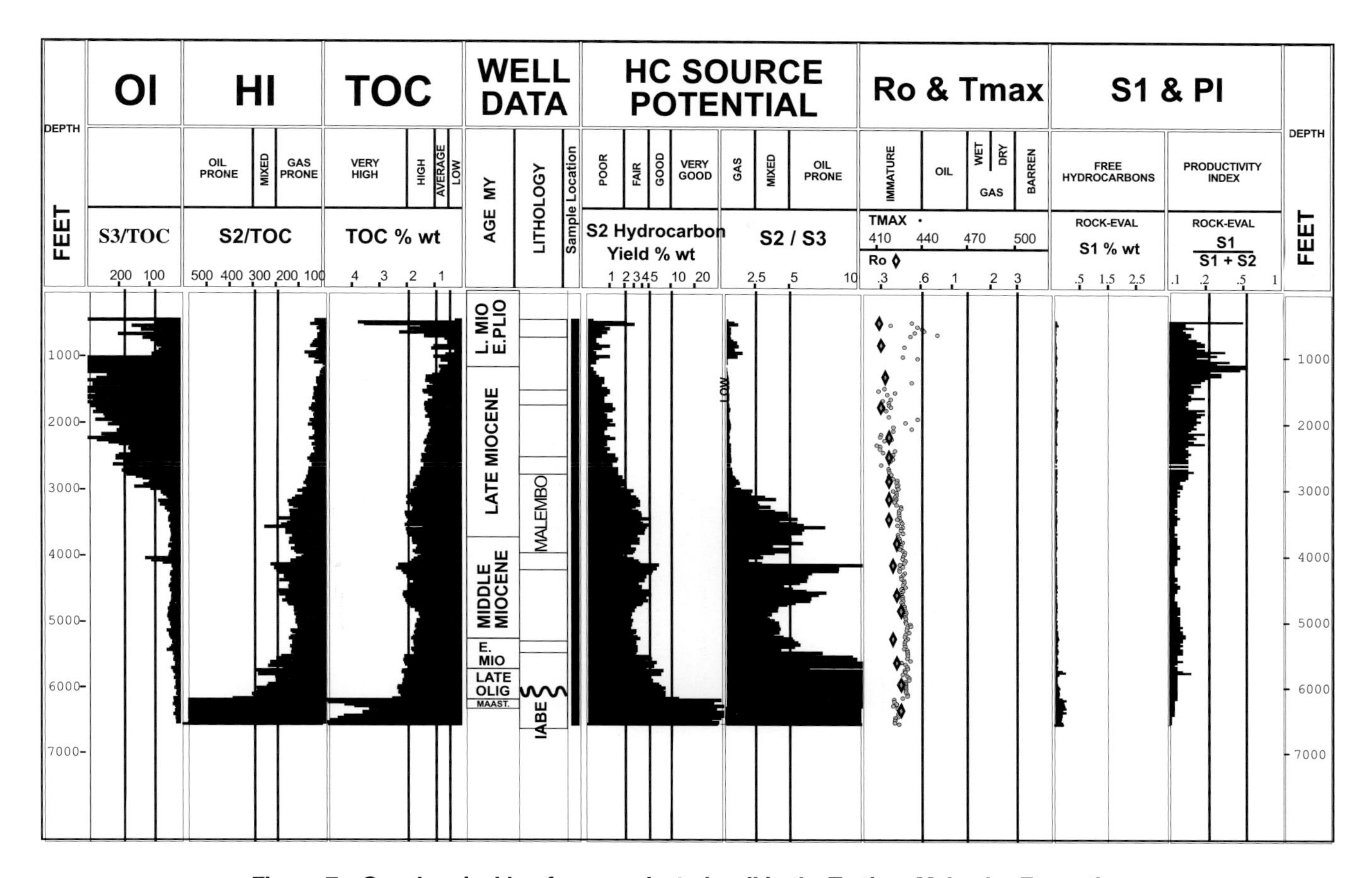

Figure 7—Geochemical log from a selected well in the Tertiary Malembo Formation.

that are richer and more oil-prone in the younger section. This is not an effect of maturity but of rising sea level from Turonian to Maastrichtian time. The source rocks are richest and best developed around the paleoshelf margin and slope, especially in the latest Cretaceous. There is evidence of Maastrichtian–Eocene upwelling associated with the richest source intervals: abundant fish bones, phosphatic zones, and mixed foraminiferal populations are indicative of upwelling.

The vertical and areal distributions of these source rocks appear to be controlled by the intersection of the Late Cretaceous–early Tertiary oceanic oxygen-minimum zone with the continental margin outer shelf and slope, a source rock depositional model described by Demaison and Moore (1980). This resulted in a ribbon-like geometry along the trace of the paleoshelf margin.

Oligocene–Miocene Malembo Megasystem

The base of the Malembo Formation occurs at the middle Oligocene unconformity, which truncates the underlying Landana and Iabe Formations. The top of the Malembo traditionally coincides with the top of the Miocene, below the Pliocene–Pleistocene surface sands. In western Block 0, the Malembo consists of thick sequences of shales and sandstones, deposited under outer shelf to bathyal conditions. In area A, the Malembo is generally thin to absent.

The Malembo Formation has moderate to good source potential in western offshore Cabinda (Blocks B and C). As illustrated by the geochemical log in Figure 7, it contains fairly to moderately rich type II and type II–III kerogen. Its average SPI values are in the moderate range (5–15 tons HC/m^2 rock). Locally, parts of the Malembo have very good source potential, although it is generally poorer in organic carbon and hydrogen than the Iabe and Landana megasystem.

In general, the source quality decreases upsection through the Malembo. This is evident from the geochemical log in Figure 7, which shows decreasing HI and increasing OI upward through the upper Miocene section. Three source rock cycles with improved TOC and S_2 values can be seen on this log: in the Oligocene, middle Miocene (doublet), and lower upper Miocene. The sections with improved source quality generally coincide with periods of relative sea level highstands, whereas the poorer sections are generally associated with regressive intervals and periods of sediment progradation on the margin.

The best source rocks in the Malembo are in the Oligocene section, when it is present. We believe that this is due to multiple factors: (1) the effects of high sea level, (2) waning phases of the intense coastal upwelling and strong anoxia seen in the Maastrichtian–Eocene, and (3) possibly the enhanced development of Oligocene shales in regional "troughs" associated with listric growth faulting and slippage on the salt.

It is not clear whether the preferred distribution of improved Oligocene source rock in the Tertiary troughs is due to syndepositional paleobathymetric effects or to later structural movement. We believe structural movements and preferential erosion from the structural highs were the primary control on the enhanced distribution of source rock in the troughs, because we see no evidence that paleobathymetric variation in shallower foraminiferal data coincides with the trough distribution. However, because of the poor seismic imaging and stratigraphic control in the deepest parts of the troughs, and the intense structural movement and rotation along the nearby listric growth faults, it is difficult to prove this argument conclusively.

RESERVOIRS

The sedimentary section of offshore Cabinda includes many reservoirs and effective traps. The presalt reservoirs include the prerift Lucula Sandstone and synrift Neocomian sandstones and Toca carbonates. These strata were deposited as eolian, alluvial, and lacustrine turbidite sands and as lacustrine carbonate shoals (see detailed descriptions by McHargue, 1990). Traps are formed in the Lucula and Toca on upthrown tilted basement blocks, with Bucomazi shales and Loeme Salt acting as seals. In some cases, the Lucula reservoir may connect with an overlying Toca reservoir, as in Malongo West. The reservoir properties of the Toca have been enhanced by diagenesis associated with periods of uplift and exposure.

Postsalt reservoirs include marine lagoonal, shelf, bar, and beach sandstones and sandy carbonates in the Mavuma, Pinda, Vermelha, and Iabe Formations. In most of these units, trap geometry is controlled by salt-cored rollover anticlines and associated faults related to syndepositional extensional gliding of the postsalt section on the salt. Primary reservoir quality is generally higher in strata of the high-energy shallow depositional environments of the rollovers and worsens toward the downthrown side of the growth faults (Dale et al., 1992). The seals for most of these reservoirs are generally transgressive marine shales or, more rarely, thin supratidal dolomites and anhydrites.

The Aptian Mavuma Formation produces oil in the Kungulo field where it consists of dolomitic sandstones and sandy dolomites. Its reservoir properties vary from poor to fair.

As shown in Figure 8, the bulk of Block 0 reserves reside in the Pinda and Vermelha Formations, accounting for nearly 40% and more than 25%, respectively, of the original oil in place. The Albian Pinda Formation is the principal reservoir in areas B and C, accounting for about

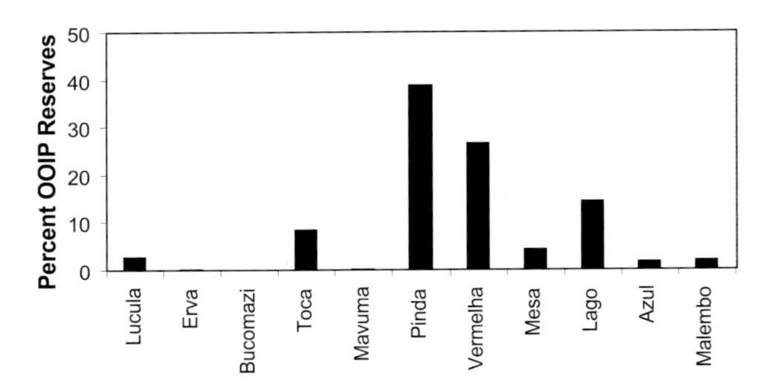

Figure 8—Distribution within the various stratigraphic units of reserves of original oil in place in Block 0, offshore Cabinda.

90% of the reserves from these areas (Figure 9). It produces from bar, shoal, shoreface, and channel sandstones and sandy carbonates, as in Takula field in area A. As described by Dale et al. (1992), reservoir porosity often exceeds 20% in the sandstones but is less in the limestones. Permeability is highly variable, averaging about 150 md for sandstones and 10–20 md for limestones in the Takula area. Traps occur in salt-cored rollover anticlines and in isolated "rafts" of Cretaceous strata sealed by transgressive marine shales above and often sealed laterally by Malembo or Iabe shales across growth faults.

The Cenomanian Vermelha Formation has the most significant production in Block 0 and is the second most important unit in terms of oil in place reserves (Figure 8). It is by far the most significant reservoir in area A (Figure 9). The Vermelha is best developed in the Greater Takula trend of Wamba, Takula, Numbi, and Vuko (described by Dale et al., 1992). The Vermelha reservoirs consist of littoral sandstones with excellent reservoir properties, averaging about 25% porosity and 1000 md permeability in the better intervals (Dale et al., 1992). Published original oil in place estimates for the Vermelha in the Greater Takula area are 2.8 billion bbl. This value does not include more recently discovered Vermelha reserves from other fields, such as Limba, Lifua, N'Sano, and N'Sangui.

The Turonian Mesa and Coniacian Lago Members of the Iabe Formation are the principal reservoirs in the Banzala and Malongo North fields, respectively. Together, they add up to the third most significant reserves in Block 0 (Figure 8). These reservoirs consist of very fine grained sandstone and siltstone, with interbedded sandy limestone and dolomite, that was deposited in coastal and shelf paleoenvironments. Porosity and permeability are highly variable. Reservoir quality is generally better in the high-energy shallow depositional environments in the syndepositional structural highs and worsens toward the growth faults. The seals are generally provided by marine-flooding shales. Trap types are primarily anticlinal rollover structures associated with growth faults.

Reservoirs of the Oligocene–Miocene Malembo Formation consist of channelized or ponded sandstones from deep-water slope environments. The reservoirs

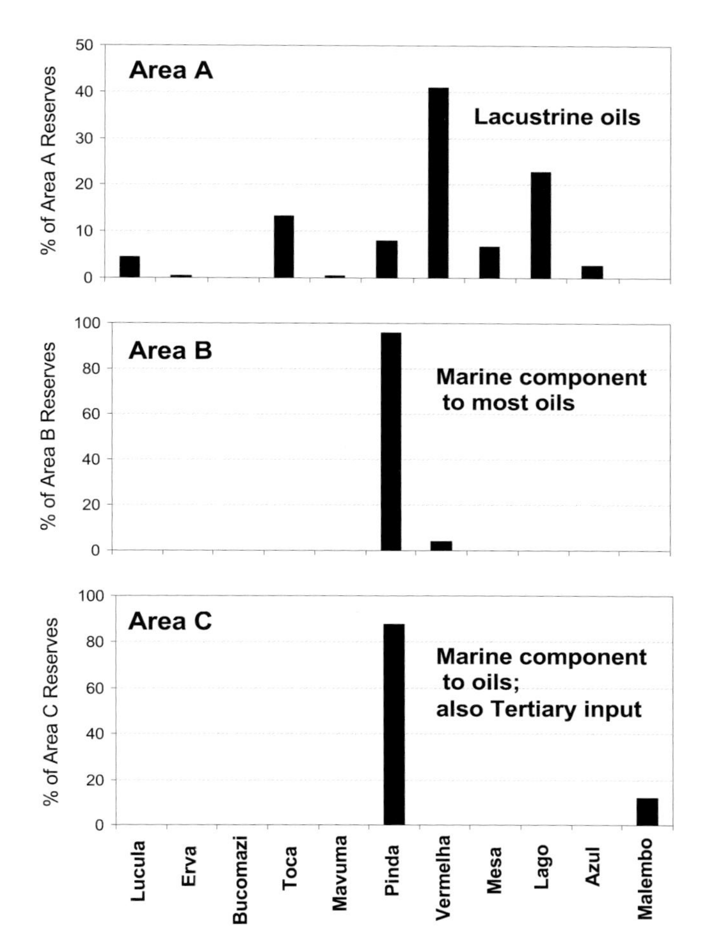

Figure 9— Distribution within the stratigraphic units of reserves of original oil in place in areas A, B, and C of Block 0, offshore Cabinda. Note the different oil types that distinguish the three areas.

discovered to date have been of Miocene age and typically combine both structural and stratigraphic trapping components. Malembo reservoirs also feature prominently in recent discoveries in nearby Block 14. Although sometimes discontinuous, these reservoirs can have good porosity and permeability and excellent formation tests.

GENETIC ORIGIN OF OILS AS IDENTIFIED BY GEOCHEMISTRY

Marine and lacustrine oils show variability in a number of geochemical parameters that can be used to identify the provenance of the oils. These parameters include carbon and hydrogen isotopes, sulfur and wax content, and various biological marker compound ratios such as the gammacerane index, oleanane index, regular sterane and monoaromatic steroid ratios, C_{26}/C_{25} tricyclic terpane ratios, and the presence or absence of C_{30} sterane. The most diagnostic differences occur between nonmarine- and marine-sourced oils. More subtle oil characteristics also exist that can be used to separate Malembo from Iabe/Landana oil provenance.

In their studies of west African oils, Burwood et al. (1990) have shown that carbon isotopic signatures can be used to separate marine and nonmarine oils. Bucomazi-sourced oils in offshore Cabinda usually have carbon isotopic signatures that fall along the waxy trend line, as shown in Figure 10. All of these oils from area A can be shown by other methods, such as biomarker analyses, stratigraphic position, and basin modeling, to have originated solely from the presalt interval.

Bucomazi-sourced oils from areas B and C fall between the waxy and nonwaxy trend lines. Some of this isotopic variation is because some of these oils are of mixed marine and lacustrine origin. Some of the variation may also be due to compositional variations in Bucomazi-sourced oils, such as those noted by Burwood et al. (1990, 1995). However, the fact that area B and C oils are typically of mixed origin (based on reliable biomarker indicators), and that the various source units have overlapping isotopic ranges, makes the use of isotopic signatures alone not diagnostic.

Offshore Cabinda oils also vary in sulfur content and other parameters, such as wax content. The oils with a marine component typically have slightly higher sulfur contents (>0.5%) than the lacustrine oils (<0.3%), and the lacustrine oils typically are waxier than the marine oils. Again, however, these general compositional characteristics of the oils are not diagnostic. In most cases, the marine oils are oil mixtures, in which dilution of sulfur and wax content may make the provenance unclear. Furthermore, results not presented here indicate that the Bucomazi generates oils of variable wax content. Where the oils are biodegraded (causing an increase in sulfur content and a decrease in wax content), these parameters may be altered to the point of becoming useless.

The most reliable method of identifying the genetic origin of offshore Cabinda oils is through the use of biological markers. These biomarker techniques and an extensive supporting west African data base were developed at Chevron over the years by co-workers, including W. Seifert, J. M. Moldowan, K. E. Peters, and B. J. Huizinga, among others. As shown in Figure 11, C_{27-29} regular sterane and especially monoaromatic steroid ratios are very diagnostic in distinguishing marine from lacustrine oils (Moldowan et al., 1985; Peters and Moldowan, 1993). Note that both source rock extracts and oils were analyzed. The distribution of oils analyzed using biomarkers can be inferred from the oil component crosses on the maturity maps shown later (Figures 14–16).

A very diagnostic marine indicator is the presence of C_{30} steranes (4-desmethyl), a compound believed to be genetically related to marine algae (Moldowan et al., 1984, 1985). Except in cases of severe biodegradation and alteration, these compounds are always present in oils originating from the Malembo and Iabe Formations but are absent in Bucomazi oils. Conversely, high gammacer-

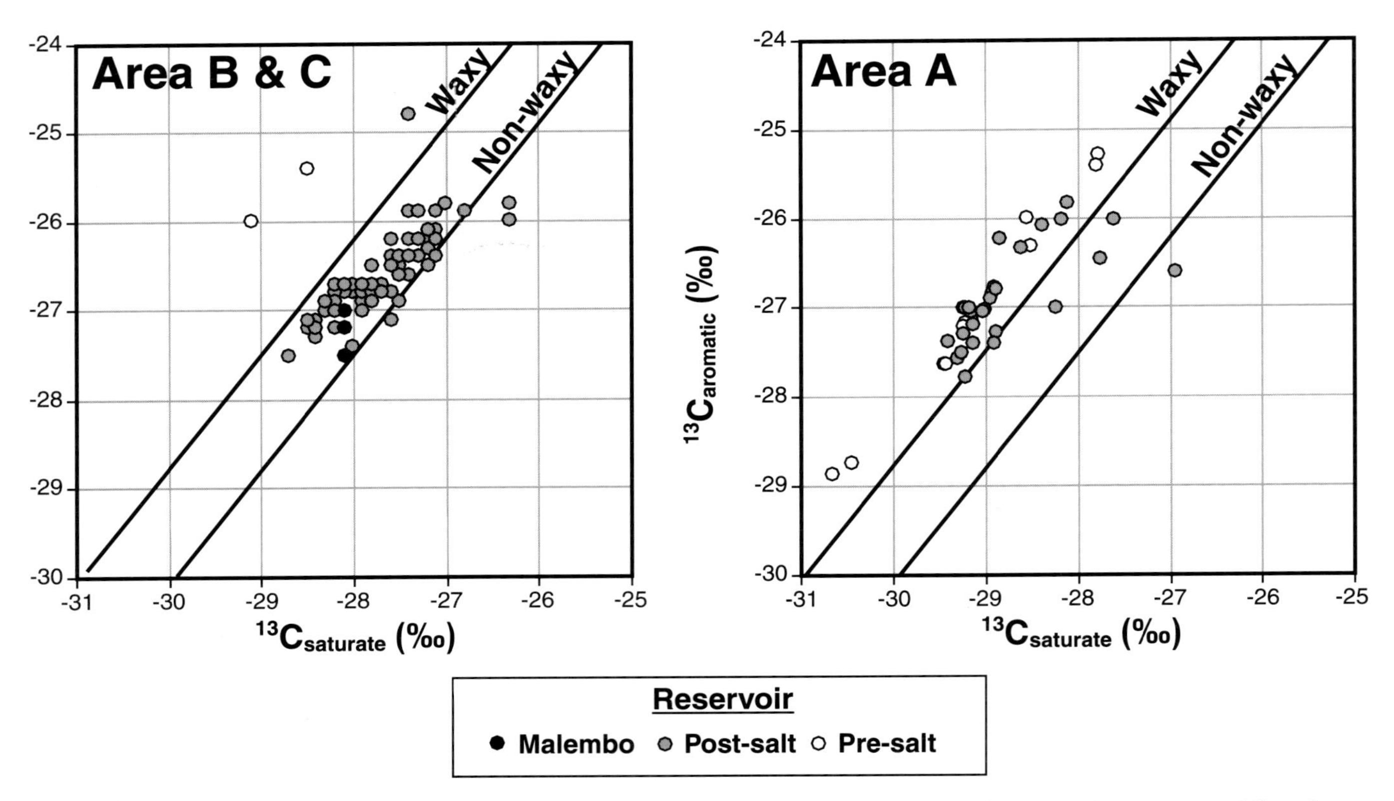

Figure 10—Plots of carbon isotopic signatures for Cabinda oils. Bucomazi-sourced oils fall along the waxy trend line. Area B and C oils are mostly of mixed origin.

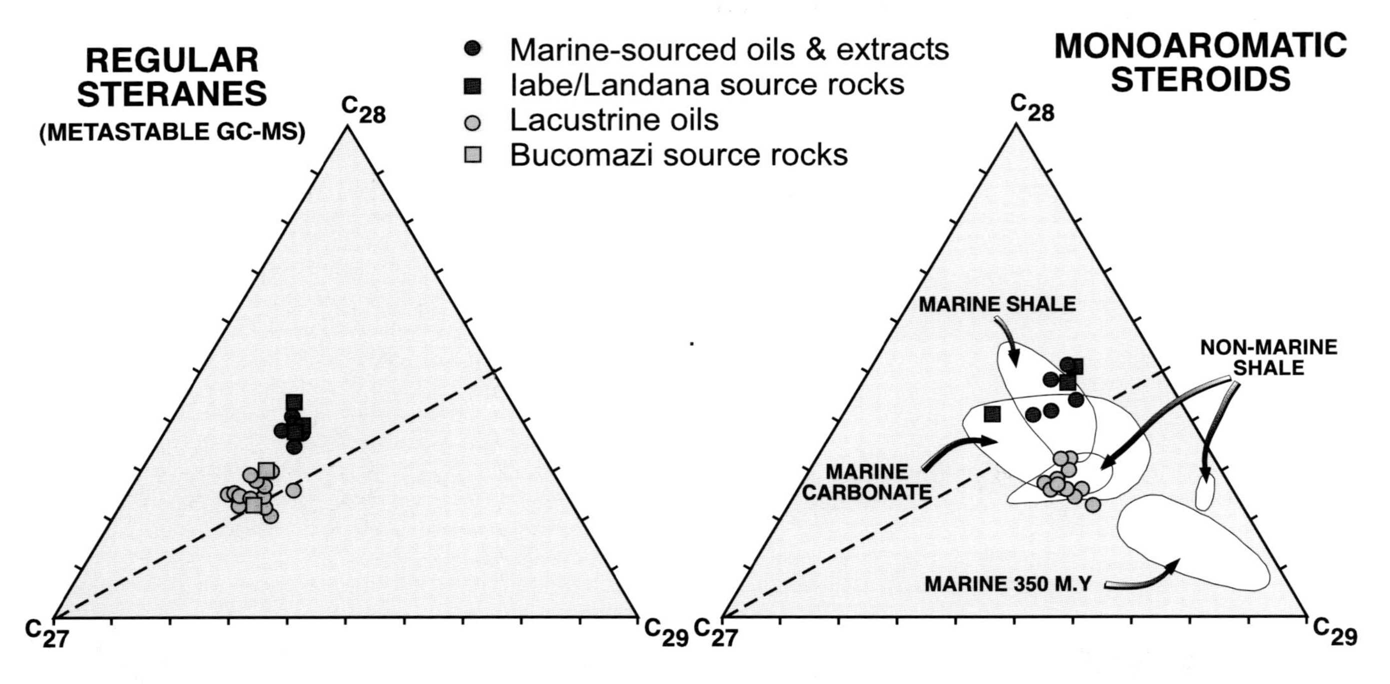

Figure 11—Ternary plots of C_{27-29} regular sterane and monoaromatic steroid ratios showing the distinction between marine oils and lacustrine oils. Note the correlation with source rock extracts. After Moldowan et al. (1985).

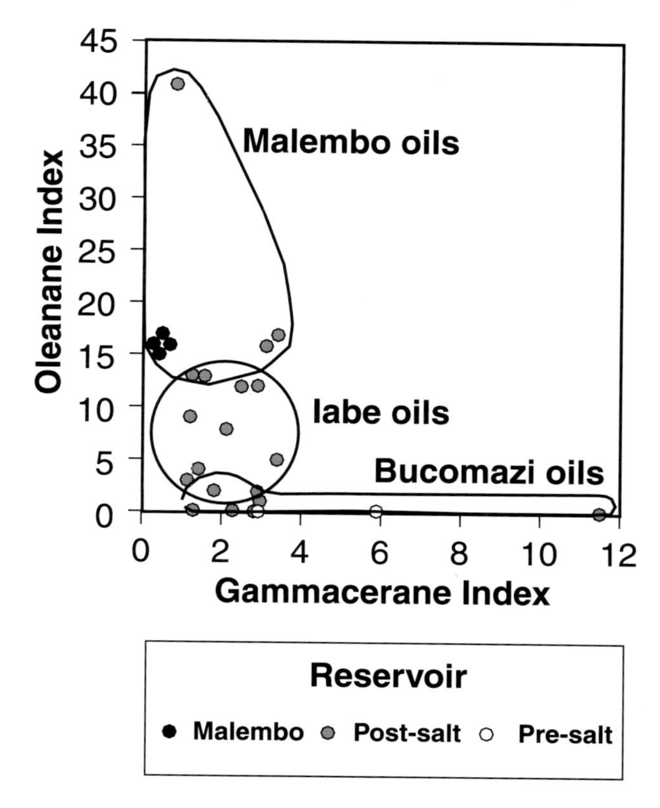

Figure 12—Crossplot of the gammacerane index and the oleanane index. Note the variability of the gammacerane index in the Bucomazi oils.

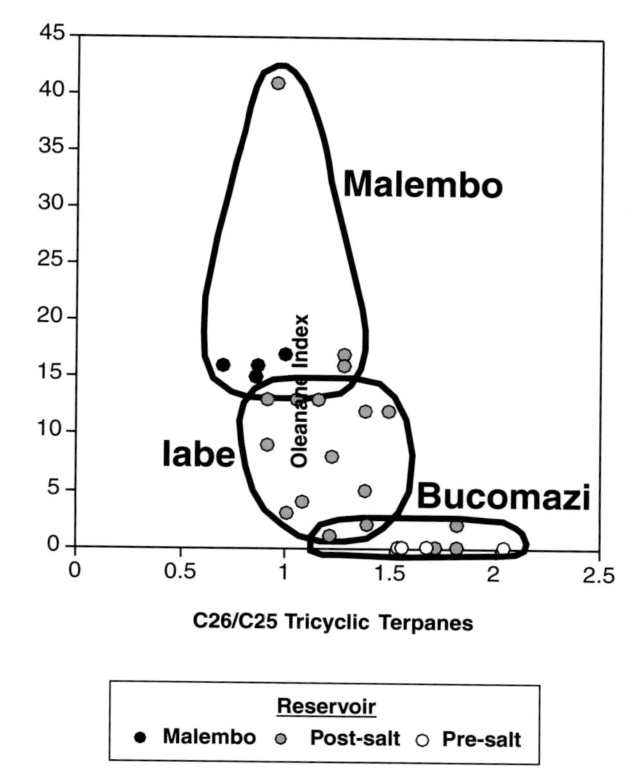

Figure 13—Crossplot of the C_{26}/C_{25} tricyclic terpane ratio and the oleanane index. The C_{26}/C_{25} ratios are highly variable in Bucomazi oils.

ane indices ([gammacerane/17α,21β(H)-hopane] × 100) have been found exclusively in oils known or thought to have originated in the presalt. Note, however, that not all Bucomazi-sourced oils have high gammacerane indices, as shown in Figure 12. As mentioned in Moldowan et al. (1985), very high gammacerane indices probably indicate hypersaline source rock environments (either marine or nonmarine).

Another compound ratio useful for identifying the genetic origins of offshore Cabinda oils is the oleanane index ([oleanane/oleanane + hopane] × 100). The increased abundance of 18α-(H)-oleanane in the Tertiary section is commonly ascribed to the proliferation of angiosperms and higher land plants during this time (Riva et al., 1988; Moldowan et al., 1994). As Figures 12 and 13 illustrate, high oleanane indices are characteristic of oils believed to have come from the Malembo.

Evidence pointing to a Malembo provenance includes

1. Oils with high oleanane indices consistently have structural positions adjacent to mature Tertiary troughs (see the maturity map in Figure 16).
2. Organic matter descriptions from Cabinda wells indicate that higher land plant matter is abundant in the Malembo but rare in the deep-water Iabe and Landana section.
3. Analyses of west African source rocks by Moldowan et al. (1994) showed low amounts of oleanane present in upper Campanian–lower Oligocene samples in contrast to high or variable amounts in upper Oligocene–upper Miocene samples. The highest oleanane index was measured in an upper Oligocene sample from the Malembo Formation of Angola.

Elevated C_{26}/C_{25} tricyclic terpane ratios (low C_{25}/C_{26}) are characteristic of some but not all Bucomazi oils (Figure 13). Tricyclic terpane ratios (C_{25}/C_{26}) have been used as marine versus lacustrine indicators, and some authors have suggested that high amounts of C_{25} (low C_{26}) in the Bucomazi are related to early marine incursions (Burwood, 1999). This ratio is highly variable in the Bucomazi (Figure 13). While we agree that some type I–II kerogens are present, we see no paleontologic evidence of marine character in the Bucomazi. The tricyclic terpane ratios may reflect some other poorly understood environmental changes in the Bucomazi (salinity?) that are not necessarily related to marine incursion. However, in postsalt oils interpreted as originating from Iabe and Malembo source rocks, low amounts of C_{26} relative to C_{25} is consistent with a marine source.

Table 1—Source–Reservoir Couplets Based on Biomarker Correlation, Block 0

Source	Reservoir	Source–Reservoir Couplet
Malembo	Malembo	Malembo–Malembo
	Pinda	Malembo–Pinda
Iabe/Landana	Malembo	Iabe/Landana–Malembo
	Pinda	Iabe/Landana–Pinda
Bucomazi	Iabe	Bucomazi–Iabe
	Vermelha	Bucomazi–Vermelha
	Pinda	Bucomazi–Pinda
	Toca	Bucomazi–Toca
	Bucomazi–Erva	Bucomazi–Bucomazi/Erva
	Lucula	Bucomazi–Lucula

Note: Dominant reservoirs used to name the petroleum systems are in boldface.

MULTIPLE SOURCE–RESERVOIR PAIRS

As seen from the previous discussion, offshore Cabinda oils can be separated into source families on the basis of composition by using various geochemical analytical techniques. Quantitative biomarker analysis is particularly useful in differentiating the origin of these oils, especially when they are mixed and originate from multiple source systems.

As shown in Table 1, multiple source–reservoir pairs have been documented in Cabinda through the use of geochemistry. Bucomazi oils are trapped in prerift Lucula sandstones, synrift Erva sandstones, and Toca lacustrine carbonates. In addition, Bucomazi oils have migrated through salt windows into the marine postrift Mavuma, Pinda, Vermelha, and Iabe Formations. In some cases, these oils may have migrated considerable distances both vertically and laterally between the generative depression and their ultimate reservoir destination, such as from the Takula subbasin to the Takula field (about 20–30 km).

Oils from Iabe and Landana source rocks have migrated primarily to reservoirs in the Pinda and Malembo Formations. Generally, mature Iabe and Landana source rocks are found to be in close proximity to or in direct contact with these reservoirs. Malembo-sourced oils have been found in the Pinda and Malembo Formations. These oils also appear to have migrated relatively short distances to the reservoir.

PETROLEUM SYSTEMS

The petroleum systems are named according to a convention established by Magoon (1988) and Magoon and Dow (1994). The system is named for the source rock and the most important reservoir rock, separated by a dash. The level of certainty that a particular source rock has generated the hydrocarbons trapped in that reservoir is indicated by a symbol, as follows: known (!), hypothetical (.), or speculative (?).

Unlike Burwood (1999), we did not divide the Bucomazi into multiple petroleum systems because we believe that the geochemical signatures of the oils do not correlate reliably to individual stratigraphic units in the presalt section. The Vermelha Formation was selected as the reservoir component of the name rather than the Pinda Formation because we are quantitatively uncertain about how much of the Pinda reserves (Figure 8) originated from presalt versus postsalt sources. Ideally, the amount of oil in place in the Pinda should be divided among the component source origins, but this cannot be done reliably at this time. However, we suspect that if the multiple sources of the Pinda-reservoired oils were taken into account, the Vermelha would be the dominant reservoir in Block 0, as it is in area A (Figure 9). In terms of cumulative production, the Vermelha is clearly the winner because the Pinda fields are just beginning to be brought into production. Thus, we named this the *Bucomazi–Vermelha(!) petroleum system.*

The *Iabe/Landana–Pinda(!) petroleum system* is named on the basis of the presence of Pinda-reservoired oils with marine signatures but with low amounts of oleanane, implying a lack of contribution from the Malembo. Some of these fields are of mixed provenance, with oils co-sourced from the Bucomazi or the Malembo. The marine signature is dominant in the case of a Bucomazi–Iabe/Landana mixture, and it is difficult to quantify the amount of oil contributed by each source. Iabe/Landana–Malembo mixtures are difficult to recognize geochemically, but can be inferred from maps of maturity and hydrocarbon charge.

There was little if any controversy over naming the *Malembo–Pinda (!) petroleum system* because the Malembo is the most obvious source candidate and the Pinda clearly the dominant reservoir in areas B and C, where this petroleum system is active.

Hydrocarbon Charge and Petroleum System Distribution

Bucomazi Generative Areas

A regional map of Bucomazi generative areas is shown in Figure 14. This map was prepared as a summary of several other maps, which showed source rock maturity at several different stratigraphic levels (basement, and lower, middle, and upper Bucomazi). Projecting 1-D modeling results onto the smoothed depth structure maps created the initial maturity maps, which were then combined to create the summary map in Figure 14. Oils identified through biomarker analysis as purely Bucomazi in origin are shown as black dots. Some of these are located in presalt and others in postsalt reservoirs. Oils of mixed origin with a Bucomazi (lacustrine) component are shown as X's.

Iabe/Landana Generative Areas

The Iabe/Landana section is mature in western areas B and C, as summarized in the simplified map of Figure

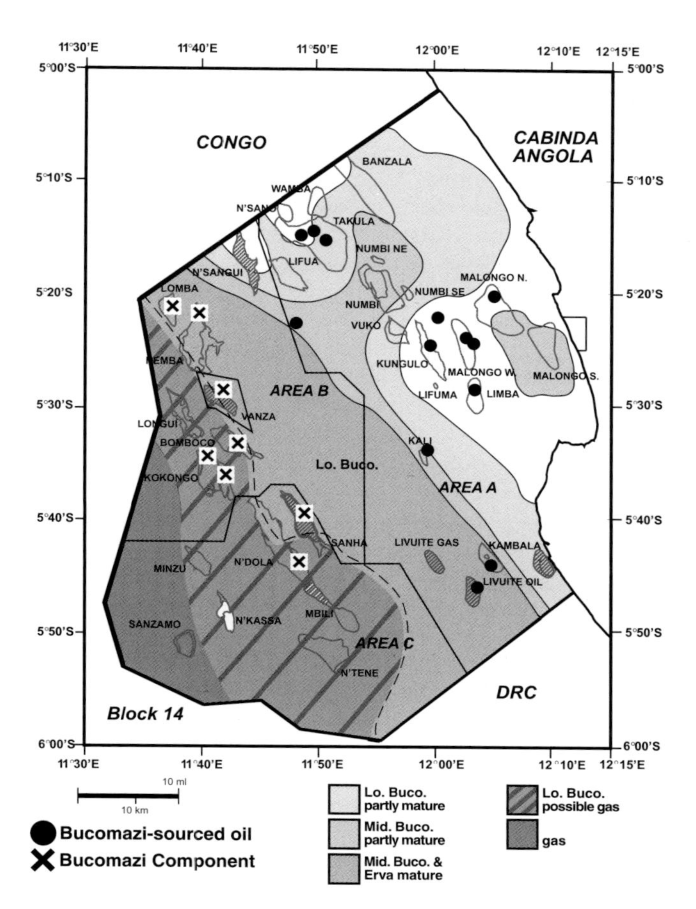

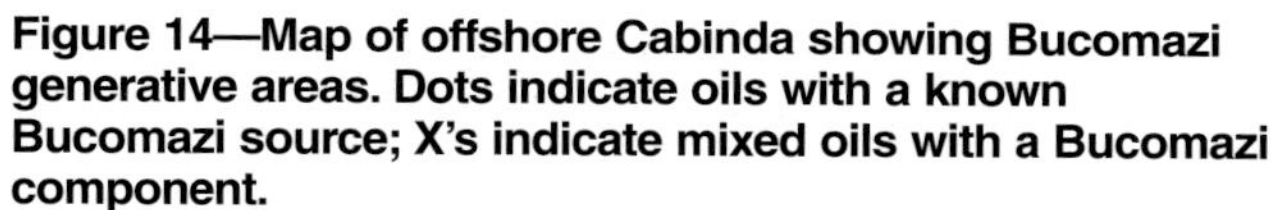

Figure 14—Map of offshore Cabinda showing Bucomazi generative areas. Dots indicate oils with a known Bucomazi source; X's indicate mixed oils with a Bucomazi component.

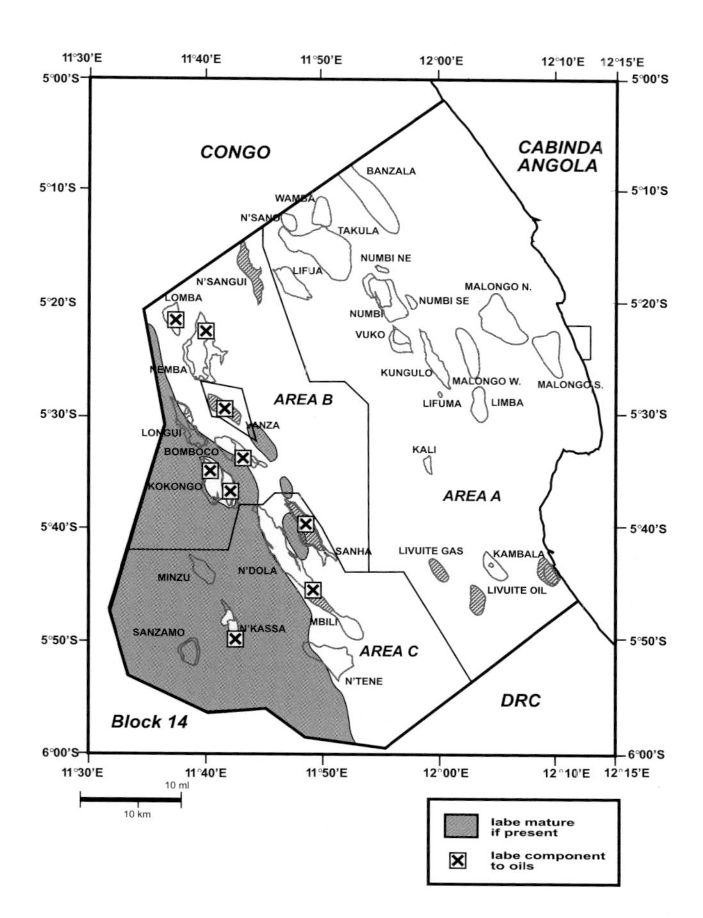

Figure 15—Map of offshore Cabinda showing Iabe/Landana generative areas. Dots indicate oils with a known Iabe/Landana source; X's indicate mixed oils with an Iabe/Landana component.

15. On a more detailed scale, some of the Tertiary troughs should be excluded from the generative areas because the Iabe is probably not present in these areas. Similarly, erosional unconformities locally remove portions of the upper Iabe, in which case no hydrocarbons can be generated. This map should be understood to indicate areas in which the Iabe is likely to be mature *if it is present.*

Oils of mixed origin with an Iabe/Landana component are indicated with X's in Figure 15. Most of the oils analyzed to date that have an Iabe/Landana component have had either small amounts of gammacerane (interpreted to indicate a Bucomazi contribution) or greater or lesser amounts of oleanane (of Malembo or Iabe origin, depending on the amount). Only one oil, from the Pinda in the Nkassa area, is interpreted as coming solely from an Iabe/Landana source. Note that we interpret another oil from the same well, but reservoired in the Malembo, to be sourced from the Malembo. The maturity maps and biomarker signatures confirm the effectiveness of the Iabe/Landana petroleum megasystem in areas B and C in the western part of Block 0.

Malembo Generative Areas

As shown in Figure 16, the Malembo Formation in Block 0 is mature only in the areas where troughs filled with Tertiary strata have formed due to rafting of the underlying marine section on the salt. Only in the vicinity of these troughs is the Malembo deep enough to be mature in this block.

Oils with a Malembo source component are shown with X's in Figure 16. All of these oils are located in reservoirs above or adjacent to downthrown mature Malembo source rocks in the Tertiary troughs. The juxtaposition of Malembo source rocks with Pinda reservoir rocks is shown schematically on the cross section in Figure 17.

Effective Regional Seals and Migration Pathways

Effective sealing units are plentiful in Block 0. These include shales in the Bucomazi and Toca Formations, the Loeme Salt, anhydritic and argillaceous layers within and

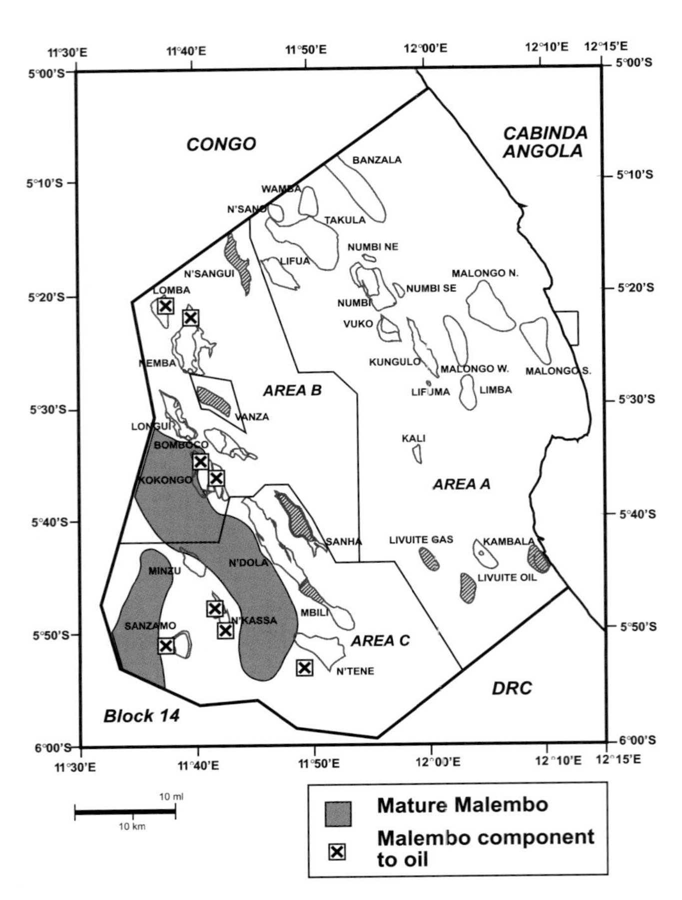

Figure 16—Map of offshore Cabinda showing Malembo generative areas. X's indicate mixed oils with a Malembo component.

at the top of the Pinda Formation, transgressive shales of Cenomanian–Eocene age (Vermelha, Iabe, and Landana Formations), and shales within the Malembo Formation.

Several particularly effective sealing units have acted as regional constraints on migration. These include the middle Bucomazi, the salt, and the deep-water Iabe shales. Although vertical (fault-related) migration pathways predominate in offshore Cabinda, significant lateral migration has occurred within the presalt and also along the Chela Formation. Frequent oil staining in the Chela indicates that it has acted as a regional conduit for presalt oils.

The availability of salt windows is a key component in the migration history of Bucomazi-sourced oils. These salt windows provide the only migration routes into the postsalt interval. The salt windows occur in association with listric normal faults, which have rotated the salt and postsalt section along the fault in such a way to create a gap (salt window), as shown schematically in Figure 17.

Oils from Iabe/Landana source rocks generally migrate into Pinda and Malembo reservoirs either across or along some of the major growth faults. Pinda reservoirs containing oils with an Iabe/Landana component are usually in direct lateral fault contact with mature Iabe source rocks.

Malembo-sourced oils occur in the vicinity of the Tertiary troughs, either in Pinda reservoirs juxtaposed against the Malembo by growth faults or in Malembo reservoirs located directly above or on fault "feeders" above the mature source. Malembo to Malembo migration may occur both along faults and through sandstone-dominated parts of the section.

Timing of Generation

Oil generation has been active in Block 0 from the Late Cretaceous to the present. As the sedimentary depocenters and source facies have shifted through time, the hydrocarbon charge systems have also evolved westward with the Tertiary progradation of the west African margin. The timing of events for the three petroleum systems in Block 0 is summarized in Figure 18.

Although the timing of oil generation in the Bucomazi–Vermelha petroleum system varies slightly among subbasins, most of the oil was generated from Cenomanian to Paleocene time. In areas B and C, the generation of oil from the Bucomazi was essentially completed prior to the deposition of the Tertiary strata. Later burial beneath the Tertiary generated primarily gas and condensate. In area A, the rate of generation slowed considerably after the Cretaceous due to lack of additional overburden. Thus, for all three areas, the "critical moment" for the Bucomazi–Vermelha system is approximately at the end of the Cretaceous.

The source rocks of the two younger petroleum systems, the Iabe/Landana–Pinda and the Malembo–Pinda, are immature in area A. In areas B and C, the timing of generation is essentially synchronous for these two petroleum systems because of the adjacent structural positions and similar burial histories of the two source rocks. The deeper portions of the two source rocks matured from middle–late Miocene time to the present (~15–0 Ma), whereas the shallower source units of both formations matured only recently (5–0 Ma). Thus, in areas B and C, the critical moment for both the Iabe/Landana–Pinda and the Malembo–Pinda petroleum systems is the present day.

Summary

As summarized in Figure 17, oils from the three petroleum systems discussed here have migrated into multiple stratigraphic levels across Block 0 and have frequently mixed together in areas B and C. Geochemical methods have allowed us to correctly correlate these oils to their respective sources. At least 10 source–reservoir pairs have been documented in Block 0 (Table 1). We have been able to test these concepts through basin modeling and thus map the extent of these three petroleum systems.

We hope that improved understanding of petroleum systems will result in better prediction of hydrocarbon

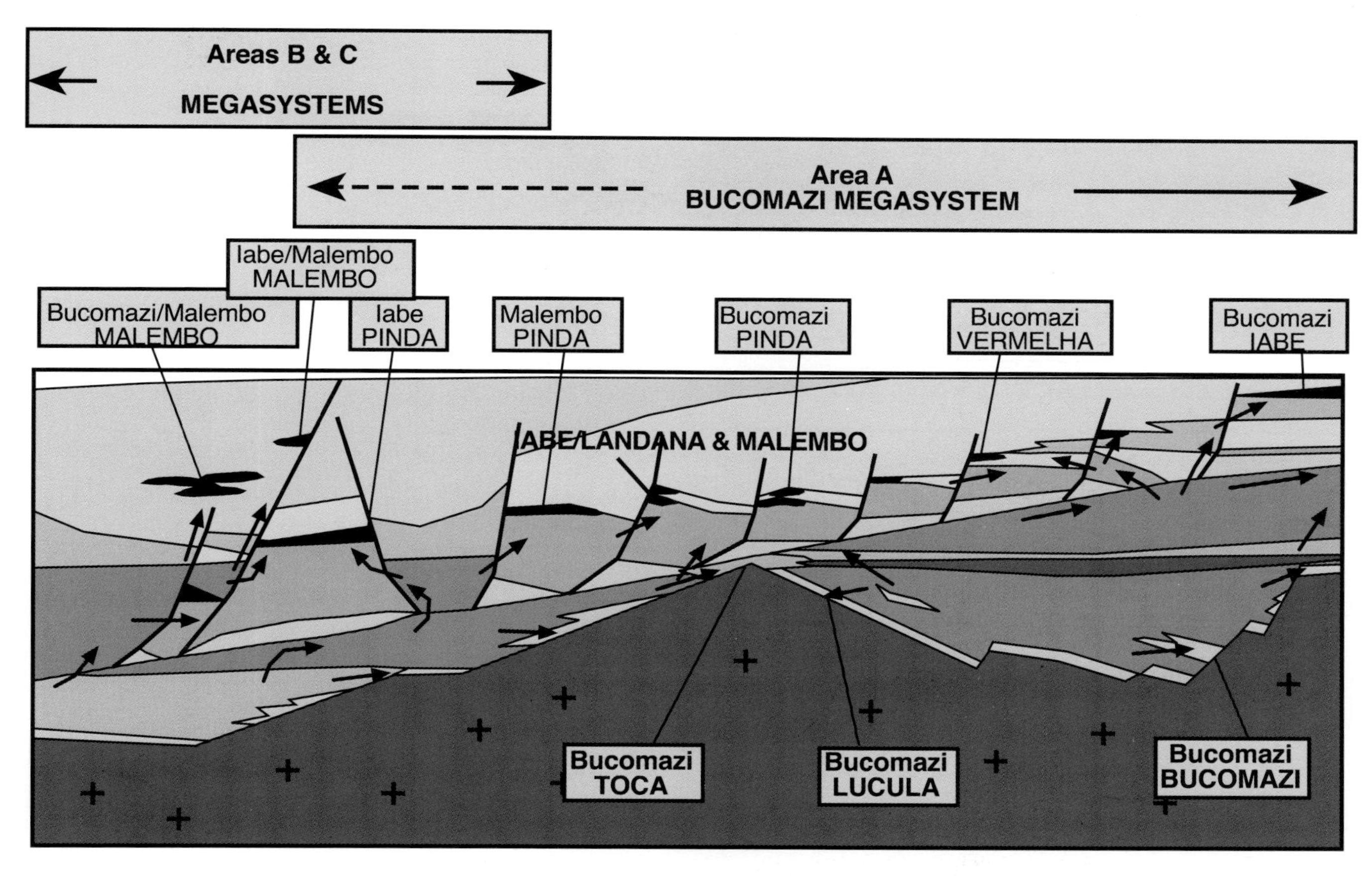

Figure 17—Schematic WSW-ENE cross section of offshore Cabinda sedimentary section showing how oils from the three source rock systems form multiple source–reservoir pairs. Salt windows provide the only migration routes from the Bucomazi into the postsalt section.

charge and effective migration routes and will eventually provide predictive capabilities for oil and gas compositions.

Acknowledgments*—We would like to thank the Cabinda Association, which includes Sonangol, Agip, Elf, and Chevron, for their support of this and other related geochemical and basin modeling studies in Block 0. Special acknowledgment goes to the Area A Post-Salt Study Team, who provided numerous maps and geological support. This team included Henry Bretthauer (CABGOC), Patrick Orsolini (Elf), Lucio Casimiro and Paulino Jeronimo (Sonangol), and Alberto de Carvalho (AGIP). Assistance from Jean-Michel Gaulier of the Institut Français du Pétrole is also gratefully acknowledged. Other Angola staff in the United States provided assistance with reserves estimates. We would also like to thank the contributions of former Chevron geochemists whose excellent work over the years contributed to the knowledge base that made this chapter possible: Gerard Demaison, Mike Moldowan, Brad Huizinga, Ken Peters, and Mary Rose Cassa.*

REFERENCES CITED

Brice, S. E., M. D. Cochran, G. Pardo, A. D. Edwards, 1982, Tectonics and sedimentation of the South Atlantic rift sequence: Cabinda, Angola, *in* J. S. Watkins and C. L. Drake, eds.), Studies in continental margin geology: AAPG Memoir 34, p. 5–18.

Burwood, R., 1999, Angola: source rock control for Lower Congo Coastal and Kwanza Basin petroleum systems, *in* N. Cameron, R. H. Bate, and V. Clure, eds., The oil and gas habitats of the South Atlantic: Geological Society of London Special Publication, v. 153.

Burwood, R., P. J. Cornet, L. Jacobs, and J. Paulet, 1990, Organofacies variation control on hydrocarbon generation: a Lower Congo Coastal Basin (Angola) case history: Advances in Organic Geochemistry 1989, v. 16, n. 1–3, p. 325–338.

Burwood, R., S. M. De Witte, B. Mycke, and J. Paulet, 1995, Petroleum geochemical characterisation of the Lower Congo Coastal Basin Bucomazi Formation, *in* B. J. Katz, ed., Petroleum source rocks: Berlin, Springer-Verlag, p. 235–263.

Dale, C. T., J. R. Lopes, and S. Abilio, 1992, Takula oil field and the greater Takula area, Cabinda, Angola, *in* M. T.

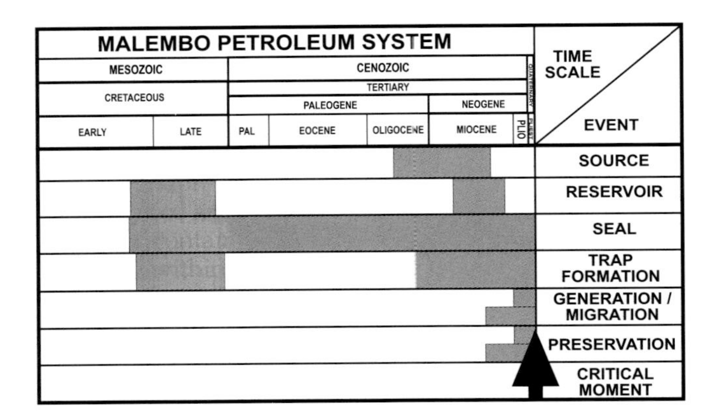

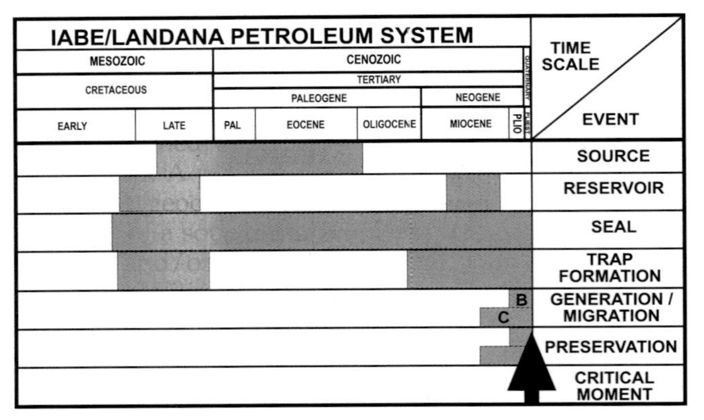

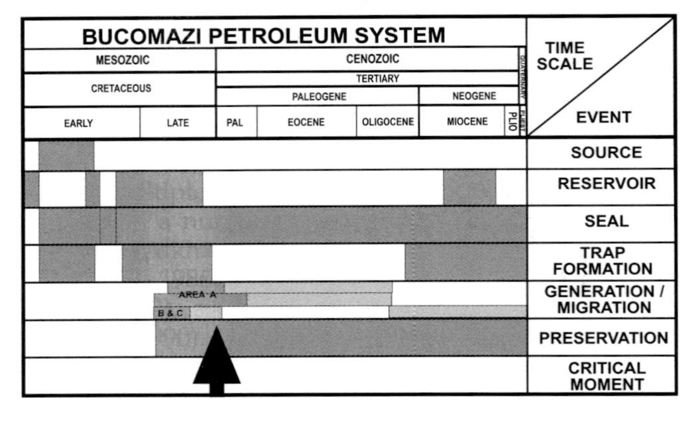

Figure 18—Events charts for the three identified petroleum systems in Block 0, offshore Cabinda. Note the difference in timing of oil generation among the petroleum systems. (Timing of events have been sketched in approximately and are not intended to be precise.)

Halbouty, ed., Giant oil and gas fields of the decade, 1978–1988: AAPG Memoir 54, p. 197–215.

Demaison, G., and B. J. Huizinga, 1991, Genetic classification of petroleum systems: AAPG Bulletin, v. 75, n. 10, p. 1626–1643.

Demaison, G. J., and G. T. Moore, 1980, Anoxic environments and oil source bed genesis: AAPG Bulletin, v. 64, n. 8, p. 1179–1209.

Espitalié, J., G. Deroo, and F. Marquis, 1985, La pyrolyse Rock-Eval et ses applications, Deuxième partie: Revue de l'Institut Français du Pétrole, v. 40, n. 6, p. 755–784.

Espitalié, J., G. Deroo, and F. Marquis, 1986, La pyrolyse Rock-Eval et ses applications, Troisième partie: Revue de l'Institut Français du Pétrole, v. 41, n. 1, p. 73–89.

Franks, S., and A. E. M. Nairn, 1973, The equatorial marginal basins of west Africa, *in* A. E. M. Nairn and F. G. Stehli, eds., The ocean basins and margins: Vol. 1, the South Atlantic: New York, Plenum Press, p. 301–350.

Lehner, P., and P. A. C. DeRuiter, 1977, The structural history of the Atlantic Margin of Africa: AAPG Bulletin, v. 61, n. 7, p. 961–981.

Magoon, L. B., 1988, The petroleum system—a classification scheme for research, exploration, and resource assessment, *in*, Petroleum systems of the United States: USGS Bulletin, v. 1870, p. 2–15.

Magoon, L. B., and W. G. Dow, 1994, The petroleum system, *in* L. B. Magoon and W. G. Dow, eds., The petroleum system—from source to trap: AAPG Memoir 60, p. 3–24.

McHargue, T. R., 1990, Stratigraphic development of proto-South Atlantic rifting in Cabinda, Angola—a petroliferous lake basin, *in* B. J. Katz, ed., Lacustrine basin exploration case studies and modern analogs: AAPG Memoir 50, p. 307–326.

Meyers, J. B., B. R. Rosendahl, H. Groschel-Becker, J. A. Austin, Jr., and P. A. Rona, 1996, Deep penetrating MCS imaging of the rift-to-drift transition, offshore Douala and North Gabon Basins, west Africa: Marine and Petroleum Geology, v. 13, n. 7, p. 791–835.

Moldowan, J. M., 1984, C_{30}-steranes, novel markers for marine petroleums and sedimentary rocks: Geochimica et Cosmochimica Acta, v. 48, n. 12, p. 2767–2768.

Moldowan, J. M., W. K. Seifert, E. J. Gallegos, 1985, Relationship between petroleum composition and depositional environment of petroleum source rocks: AAPG Bulletin, v. 69, n. 8, p. 1255–1268.

Moldowan, J. M., J. Dahl, B. J. Huizinga, F. J. Fago, L. J. Hickey, T. M. Peakman, and D. W. Taylor, 1994, The molecular fossil record of oleanane and its relation to angiosperms: Science, v. 265, p. 768–771.

Peters, K. E., 1986, Guidelines for evaluating petroleum source rock using programmed pyrolysis: AAPG Bulletin, v. 70, n. 3, p. 318–329.

Peters, K. E., and J. M. Moldowan, 1993, The biomarker guide: interpreting molecular fossils in petroleum and ancient sediments: Englewood Cliffs, New Jersey, Prentice-Hall, 363 p.

Reyre, D., 1984, Remarques sur l'origine et l'évolution des bassins sédimentaires africains de la côte atlantique: Bulletin de la Société Géologique de France (7), t. XXVI, n. 6, p. 1041–1059.

Riva, A., P. G. Caccialanza, and F. Quagliaroli, 1988, Recognition of 18β(H)oleanane in several crudes and Tertiary–Upper Cretaceous sediments: definition of a new maturity parameter: Advances in Organic Geochemistry 1987, v. 13, n. 4–6, p. 671–675.

Araújo, L. M., J. A. Trigüis, J. R. Cerqueira, and L. C. da S. Freitas, 2000, The atypical Permian petroleum system of the Paraná Basin, Brazil, *in* M. R. Mello and B. J. Katz, eds., Petroleum systems of South Atlantic margins: AAPG Memoir 73, p. 377–402.

Chapter 26

The Atypical Permian Petroleum System of the Paraná Basin, Brazil

L. M. Araújo

Petrobrás E & P/GEREX/GESIP
Rio de Janeiro, Brazil

J. A. Trigüis

UENF
Macaé, Brazil

J. R. Cerqueira

L. C. da S. Freitas

Petrobrás/Cenpes/DIVEX/CEGEQ
Rio de Janeiro, Brazil

Abstract

In the Permian petroleum system of the Paraná Basin, organic-rich source rocks (Assistência Member of the Irati Formation) were thermally matured by the heat of Cretaceous basic igneous intrusions (diabase sills) of varying thicknesses (few centimeters up to 240 m). This constitutes an atypical petroleum system, characterized by the synchronism of generation and migration processes with the magmatism (138–127 Ma). During this period, countless generation and migration pulses occurred from multiple generation kitchens that were not related to the basin depocenter, in sharp contrast to basins in which thermal maturation is controlled by increasing burial.

The total volume of hydrocarbons generated by the thermal effect of the magmatism (petroliferous charge) was calculated, mapped, and interrelated with other processes (trapping, migration, and accumulation) and essential elements (reservoirs and seal) to unravel the temporal and spatial relationships of this atypical Permian petroleum system.

INTRODUCTION

Magoon and Dow (1994) defined petroleum systems as *atypical* when the source rocks are thermally matured by igneous intrusions. In this context, the petroleum system related to the Permian source rocks (Assistência Member of the Irati Formation) of the intracratonic Paraná Basin constitutes an atypical petroleum system in about two-thirds of the basin (700,000 km^2). This is because the Kazanian source rocks, which are thermally immature due to burial alone (<3000 m), reached thermal maturity essentially due to the heat effect of diabase sills up to 240 m thick.

Our assessment of the effect of igneous thermal alteration on source rocks is based on organic carbon content, Rock-Eval pyrolysis, Fisher elemental pyrolysis, petrographic kerogen analysis, biomarkers, fluid inclusion geothermometry, and nuclear magnetic resonance analyses. The aim of this paper is to present the results of quantification of the hydrocarbons generated by the thermal effect of the igneous intrusions and to discuss the implications for patterns of primary, secondary, and tertiary migration, the volumetric mass balance of the available charge of the atypical Permian petroleum system, and the efficiency of its essential elements.

SYSTEMATICS FOR EVALUATION OF THE PETROLEUM SYSTEM

Initially, an area was selected on the eastern border of the basin (Figure 1). To study the behavior of geochemical and optical parameters in 15 shallow wells (<100 m) with sections thermally altered by igneous intrusions and in two sections with original parameters, 700 samples were analyzed at intervals of 0.4 m, where diabase sills vary from zero up to 28.4 m. By following the behavior of the geochemical and optical parameters in areas affected by the heat of igneous intrusions in these control wells and by comparing the data with those from the immature sec-

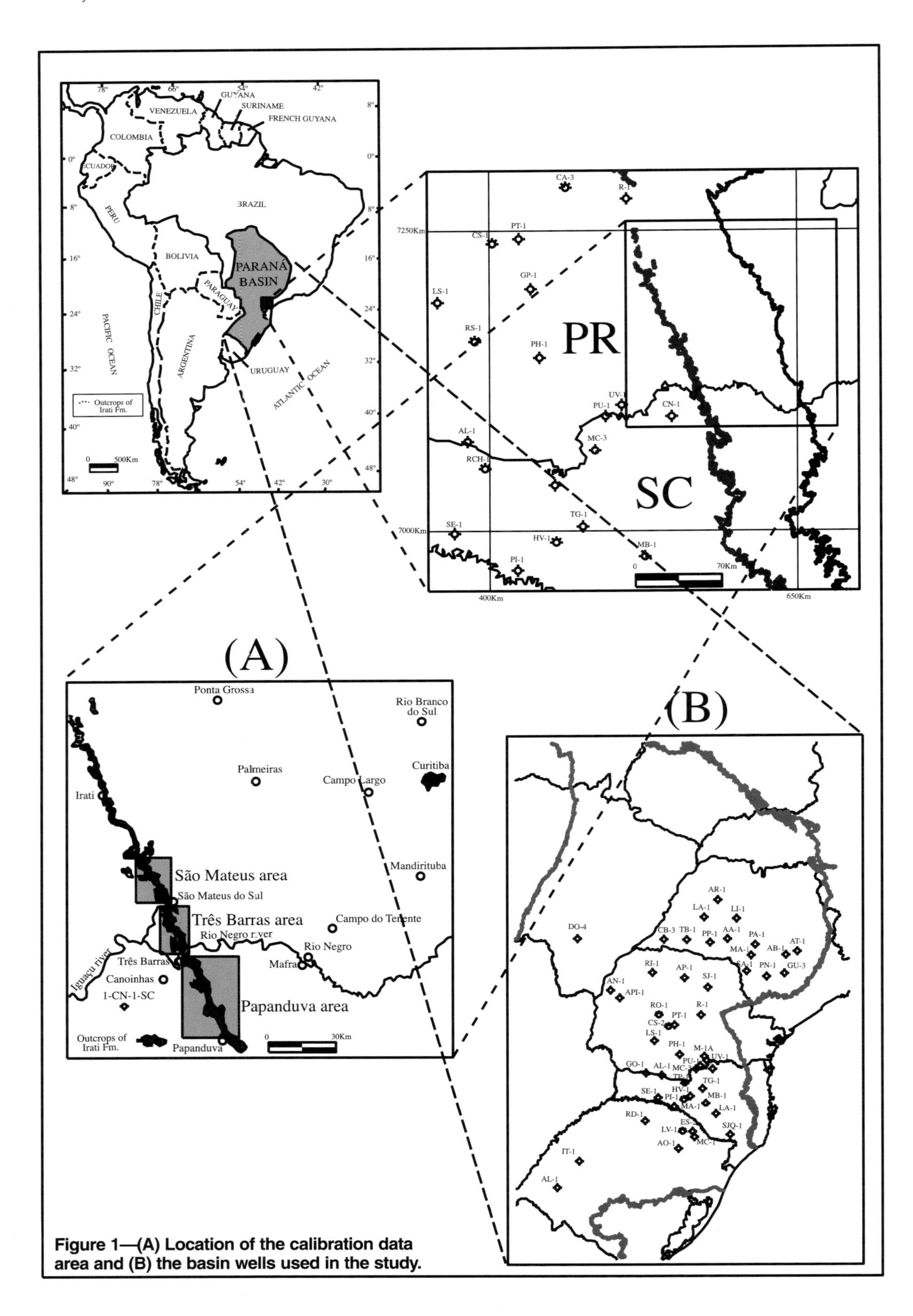

Figure 1—(A) Location of the calibration data area and (B) the basin wells used in the study.

tions, the pattern of variation in these parameters was determined in relation to different thicknesses of diabase sills and the average depletion of the total organic carbon (TOC) content was calculated.

The behavior of geochemical parameters along the eastern border was extrapolated for other areas of the basin, where 48 wells were studied (Figure 1). On this basis, it was possible to relate the results of the control area with the rest of the basin, where the Assistência Member of the Irati Formation extends across an area of approximately 700,000 km^2.

Once we defined the area of calibration of the geochemical and optical parameters and determined the percent average depletion of TOC caused by igneous intrusive thermal effects, it was possible to reconstruct the original TOC based on the medium residual TOC of the 48 selected wells and to estimate the original generation potential (S1 and S2) of the sections affected by the heat of the igneous intrusions. After calculating the original generation potential of the sections altered by the diabase sills, it was possible to measure the residual generation potential in the laboratory ($S1_r$ and $S2_r$) to determine the mass of petroleum produced by the heat of igneous intrusion, the migrated mass of hydrocarbons, and the volume of hydrocarbons migrated per area. This petroleum charge was then related to the elements and processes that constitute the atypical Permian petroleum system of the Paraná Basin.

ASSESSMENT OF HYDROCARBON GENERATION CAUSED BY THE HEAT OF IGNEOUS INTRUSIONS

The thermal effect of igneous intrusions was observed in data derived from Rock-Eval, Fisher pyrolysis, optical parameters (Spore Color Index, or SCI), and vitrinite reflectance (% R_o) taken above and below intrusions in a thickness equivalent to that of the intrusive body, with an error of ±10%, also observed by Sad and Saraiva (1982). The transition zone between altered and unaltered source rocks is abrupt when monitored by Fisher pyrolysis data and transitional when based on SCI and % R_o data.

The original hydrogen index (HI) varies from 319 to 924 mg HC/g TOC, but the residual values of HI in the thermally affected zone are about 10% of the original value and decrease to almost zero where the thickness of the igneous intrusion is greater than the source rock thickness. Such results were also observed by Cerqueira and Santos Neto (1990) and Ferreira et al. (1994). Maximum TOC depletion (50%) was observed near the contacts with the igneous intrusions; TOC values gradually increase away from the intrusive contacts. Average TOC depletion in the transition zone was 30%. Similar results have been observed by Clayton and Bostick (1986).

Nuclear magnetic resonance (NMR) indicates that the aromatization of the kerogen decreases away from the intrusive contacts, varying from 100% at the contacts to 20% in the transition zones. Similar results have been observed by Simoneit et al. (1978, 1981), Dennis et al. (1982), Peters et al. (1983), and Saxby and Stephenson (1987).

In the thermally altered zone, saturated to aromatic hydrocarbon ratios increase and the isoprenoid/*n*-alkane ratios decrease. Furthermore, a shift toward the *n*-alkane distribution (higher concentration of *n*-alkanes with low molecular weight) was observed, similar to the results of Clayton and Bostick (1986).

Biomarker maturity parameters derived from C_{32} hopane ($22S/22S + 22R$), C_{30} hopane and moretane ($\alpha\beta/\alpha\beta + \beta\alpha$), T_m/T_s (trisnorhopane/trisnorneohopane), C_{29} $\alpha\alpha\alpha$ sterane ($20S/20S + 20R$), and C_{20} sterane ($\alpha\beta\beta/\alpha\beta\beta + \alpha\alpha\alpha$) measured in samples with vitrinite reflectance values lower than 1% R_o indicated progressive transformation toward isomerization equilibrium. However, when vitrinite reflectance was greater than 1% R_o, a reversal trend of molecular maturity indices was observed, caused by low-maturity biomarker signatures that occur in samples exposed to rapid heating rates or to high absolute temperatures (as monitored by Raymond and Murchison, 1992).

The most frequent activation energies of the Irati kerogen range between 43 and 48 kcal/mol, with the preexponential factor between 1.23×10^{10} and 2.13×10^{12}, showing a transformation rate greater than 80% in the thermally altered zone. In some cases, a drastic decay in all activation energy levels was observed.

The basic premise for geochemical calibration of the data was based on the observation of a bulk depletion in the organic carbon mass of the kerogen in the thermally affected zone. This depletion is well defined by an almost complete transformation of the source rock generation potential (S2). The mass of organic carbon consumed in the kerogen thermal degradation is proportional to the volume of hydrocarbons generated by the system in liquid and/or gaseous form.

The first step in reconstituting the original generation potential (S1 and S2) of the thermally affected zone was to determine the distribution and thicknesses of the diabase sills (Figure 2) and to correlate these with the depletion percentages of the organic carbon caused by igneous heat input. For diabase sills with thicknesses equal to or greater than the thickness of the source rocks, a maximum decrease of 30% was adopted for the TOC value, which corresponds to the average value observed in wells monitored along the border of the basin.

To make the TOC reconstitution possible and systematic and to decrease the error margins, the method considers that, despite facies variations, the system will be depleted in organic carbon mass by an average of 30%. This value of 30%, based on the calibration of optical and geochemical parameters (vitrinite reflectance, SCI, and Rock-Eval pyrolysis data), supplied the correction factors that were applied to the residual TOC for each interval variation in the extent of the thermally altered zone, whose thickness is equivalent to the igneous intrusive thickness.

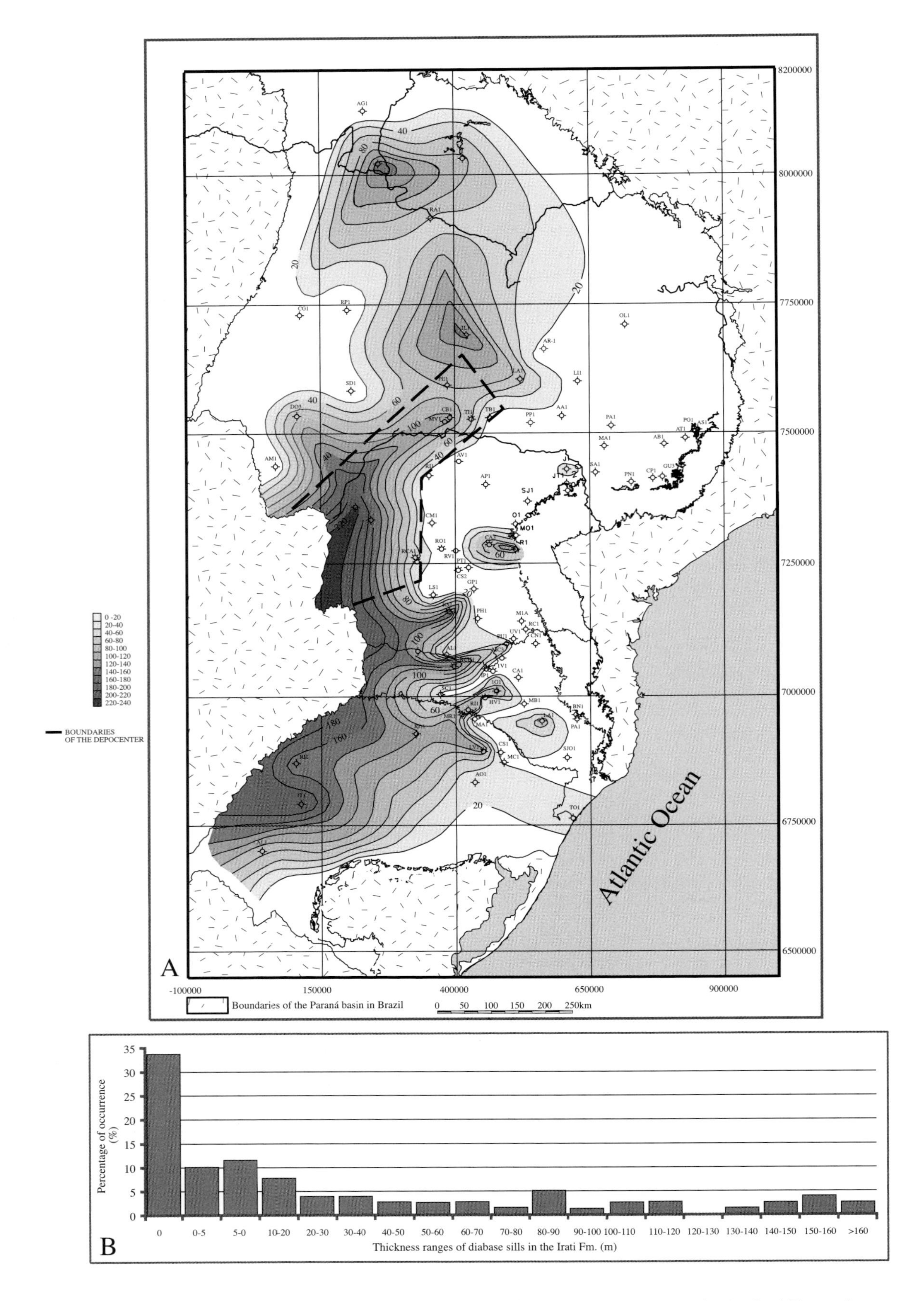

Figure 2—(A) Distribution of diabase isoliths and (B) isopach percentages in the Irati Formation.

After correction of the residual TOC in the original TOC, one section of the Assistência Member of the Irati Formation was selected that had good correlation between the TOC and the generation potential for the evaluation of the original generation potential (S1 and S2) of the thermally affected section. In the selected section, the relationship between the TOC content and the generation potential of the insoluble fraction (S2) presented a degree of statistical correlation of 94%, defined by the minimum square method. The relationship between the TOC and the soluble fraction (S1) gave a correlation of 87%. The generation potential and the soluble fraction calculated by the correlation equations of the standard well is given by

$$S2_c = 7.432 \times TOC - 5.195 \;\; (R_2 = 0.94)$$

and

$$S1_c = 0.4965 \times TOC + 0.1275 \;\; (R_2 = 0.87).$$

The area where the optical and geochemical parameters were thermally affected by igneous intrusions was designated as the thermal halo. To determine the Assistência Member thermal halo, the direct relationship was used, and the extent of the thermally altered zone was considered to be equivalent to the thickness of the intrusive.

The results plotted on a map provided the identification of two different areas (Figure 3). The first, corresponding to two-thirds of the basin area, includes Assistência Member sections with diabase sills having thicknesses equivalent to or greater than the source rocks (thermal halo ≥100%). This area roughly coincides with the area where the diabase intruded into the Irati Formation is more than 20 m thick (Figure 3). The second area, equivalent to the remaining third of the basin area, corresponds to the area where the Irati section has a thermal halo varying from 0% (absence of intrusions) to 100%.

Only wells that revealed a section with a sufficient number of reliable geochemical analyses (more than 10) in the Irati Formation were used for the reconstitution of TOC and original S1 and S2. Because of this restriction, only 48 wells were selected. The scarcity of geochemical data in the northern part of the basin made geochemical reconstitution unfeasible in this area.

After characterizing the percent thermal halo for each investigated section of the Irati Formation, relative corrections were applied to calculate the depletion of TOC from the sections where the residual mean TOC was determined (Figure 4). After reconstitution of the residual TOC in each section affected by diabase sills (Figure 5), the original TOC was applied to the correlation equations (TOC versus S1 and TOC versus S2) obtained for the original section. This was done to calculate the mass of original free hydrocarbons ($S1_o$) (Figure 6) and the original generation potential ($S2_o$) (Figure 7). The reconstituted genetic potentials compared with the residual mass of free hydrocarbons ($S1_r$) (Figure 8) and the residual potential ($S2_r$) (Figure 9) were used to quantify the volume of hydrocarbons removed from the source rocks in response to the atypical generation caused by the heat of igneous intrusive bodies.

MASS BALANCE CALCULATIONS

The systematic calculations applied by Espitalié et al. (1987) in Paris basin were adapted to Paraná basin. Pyrolysis parameters were used to calculate the amount of produced hydrocarbons and to monitor its distribution in the Paleozoic charge system. The applied method can be summarized in three stages:

1. Calculation of the mass of produced hydrocarbons (ΔSp), obtained by the difference between the original genetic potential ($S1_o + S2_o$, which was calculated through reconstituted TOC using the correlation equations between TOC versus S1 and TOC versus S2 in the standard well) and the residual generation potential ($S2_r$, which was measured in the laboratory from the intervals altered by the heat of intrusion) (Figure 10):

$$\Delta Sp = (S1_o + S2_o) - S2_r \text{ (kg HC/ton rock)}.$$

2. Calculation of the mass of hydrocarbons produced per unit of area ($\Delta Sp/1\ km^2$), in a section of 1 km^2, with thickness (h) equivalent to the thickness of the Assistência Member (for each analyzed section, with TOC > 1% and organic matter of type I + II) and constituted by rocks with an average density of 2.5 ton/m^3:

$$(\Delta Sp/km^2) = \Delta Sp \times h \times 2.5 \times 10^3 \text{ (ton HC/km}^2).$$

3. Calculation of the mass of hydrocarbons migrated per unit of area (Sm/km^2), determined by the difference between the residual hydrocarbons ($S1_r$) and the generated hydrocarbons per unit of area ($\Delta Sp/km^2$):

$$(\Delta Sm/km^2 = S1_r - (\Delta Sp/km^2) \text{ (ton HC/km}^2).$$

To make use of the standard volumetric unit in the petroleum industry (m^3), a conversion factor of 1.66 was introduced that represents the average density of hydrocarbons (600 kg/m^3) generated under a pressure of 20 MPa and a temperature higher than 200°C (Figure 11):

$$\text{Volume of hydrocarbons } (m^3/km^2) = 2.5 \times \Delta Sm \times h \times 10^3 \times 1.66 \text{ (m}^3 \text{ HC/km}^2).$$

According to Espitalié et al. (1987), the distribution of the hydrocarbon mass along the perimeter of the source rocks defines the compartmentalization of three different areas: (1) the depleted area, where $\Delta Sm/km^2$ is negative,

(Text continues on p. 391.)

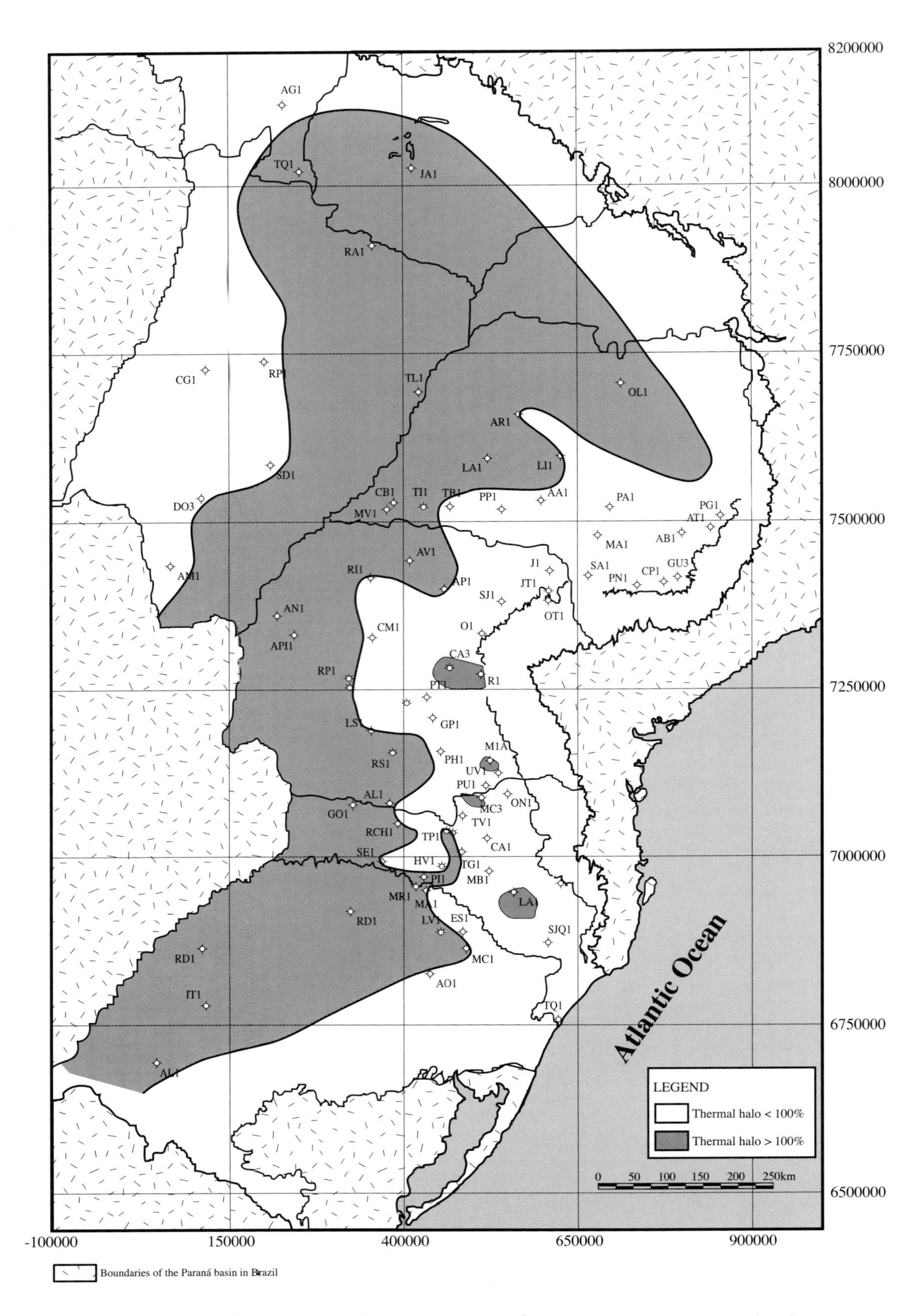

Figure 3—Thermal halo map showing the effect of igneous intrusions on the Irati Formation.

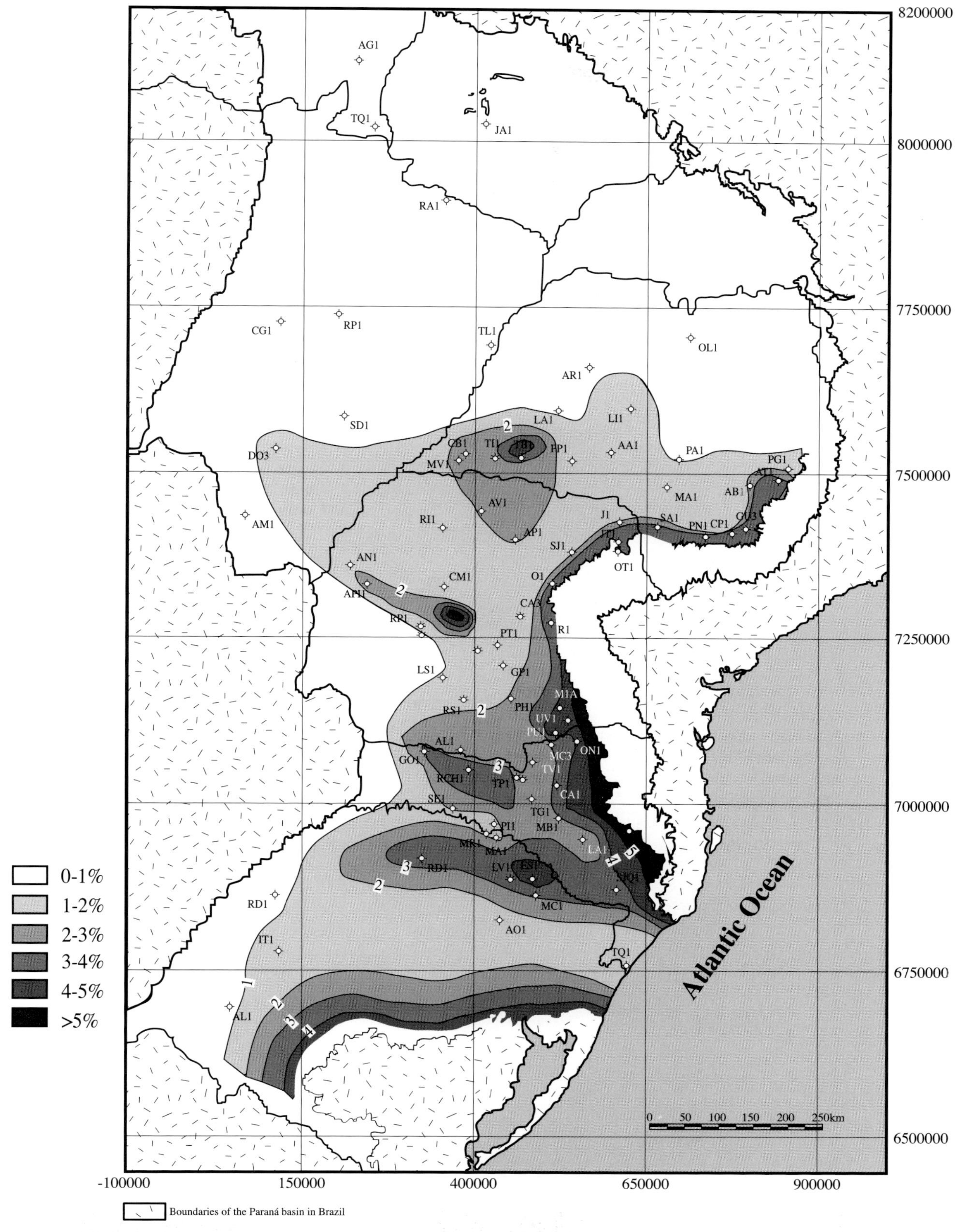

Figure 4—Average residual organic carbon distribution in the Assistência Member of the Irati Formation, Paraná Basin (CI = 1).

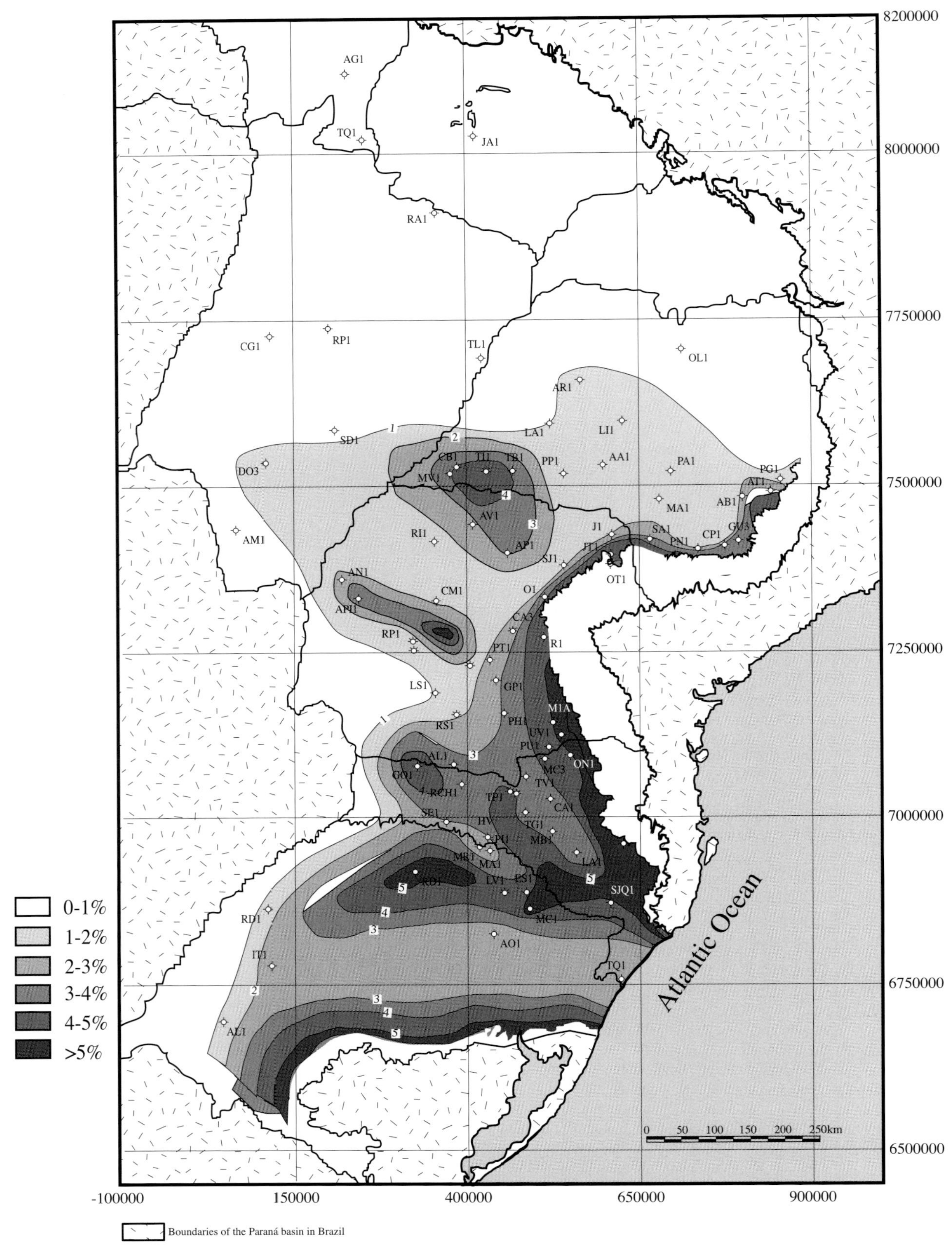

Figure 5—Reconstituted average organic carbon distribution in the Assistência Member of the Irati Formation, Paraná Basin (CI = 1).

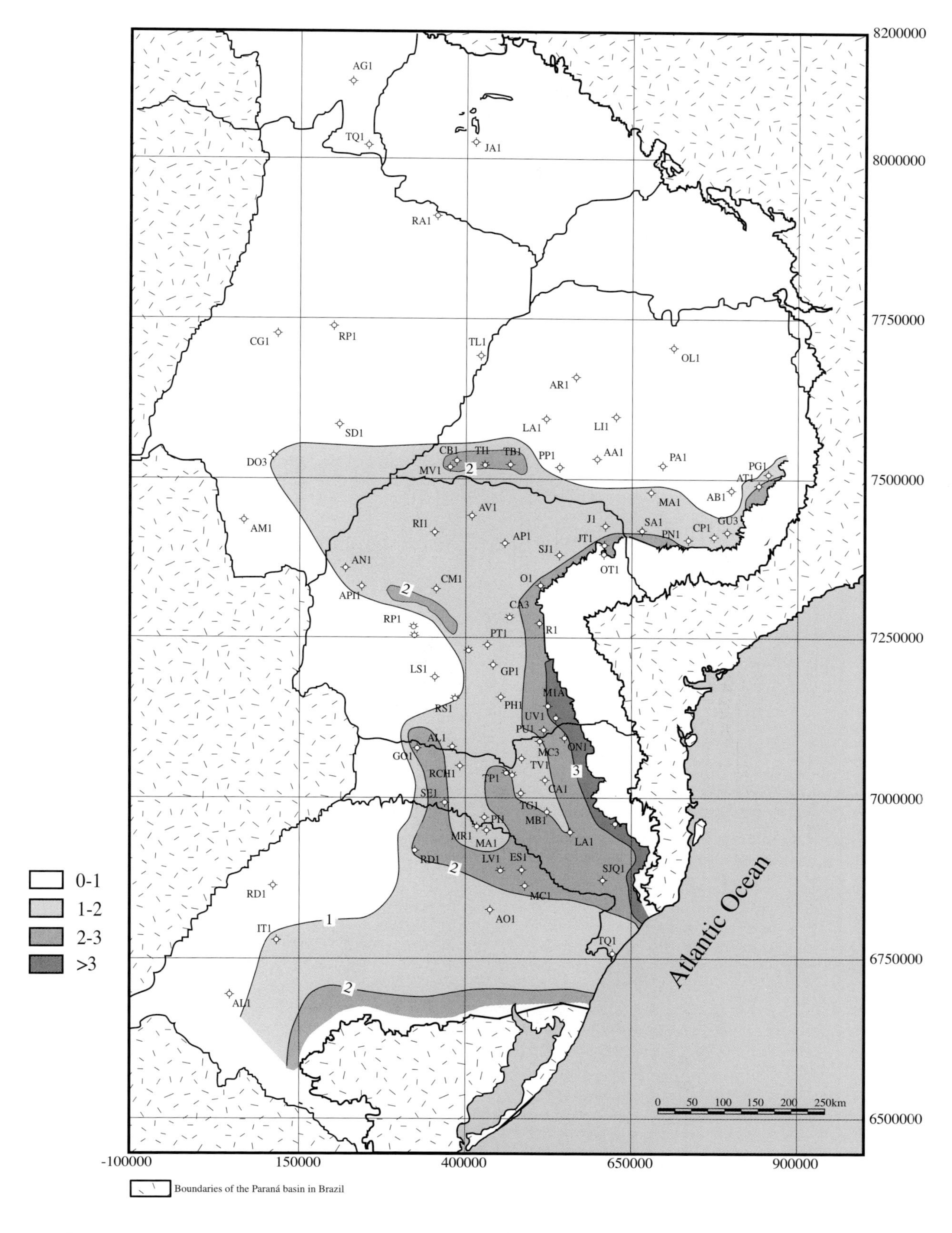

Figure 6—Original mass of free hydrocarbons in the Assistência Member of the Irati Formation, reconstituted through an equation of correlation ($S1_o$ = kg HC/ ton rock; CI = 1).

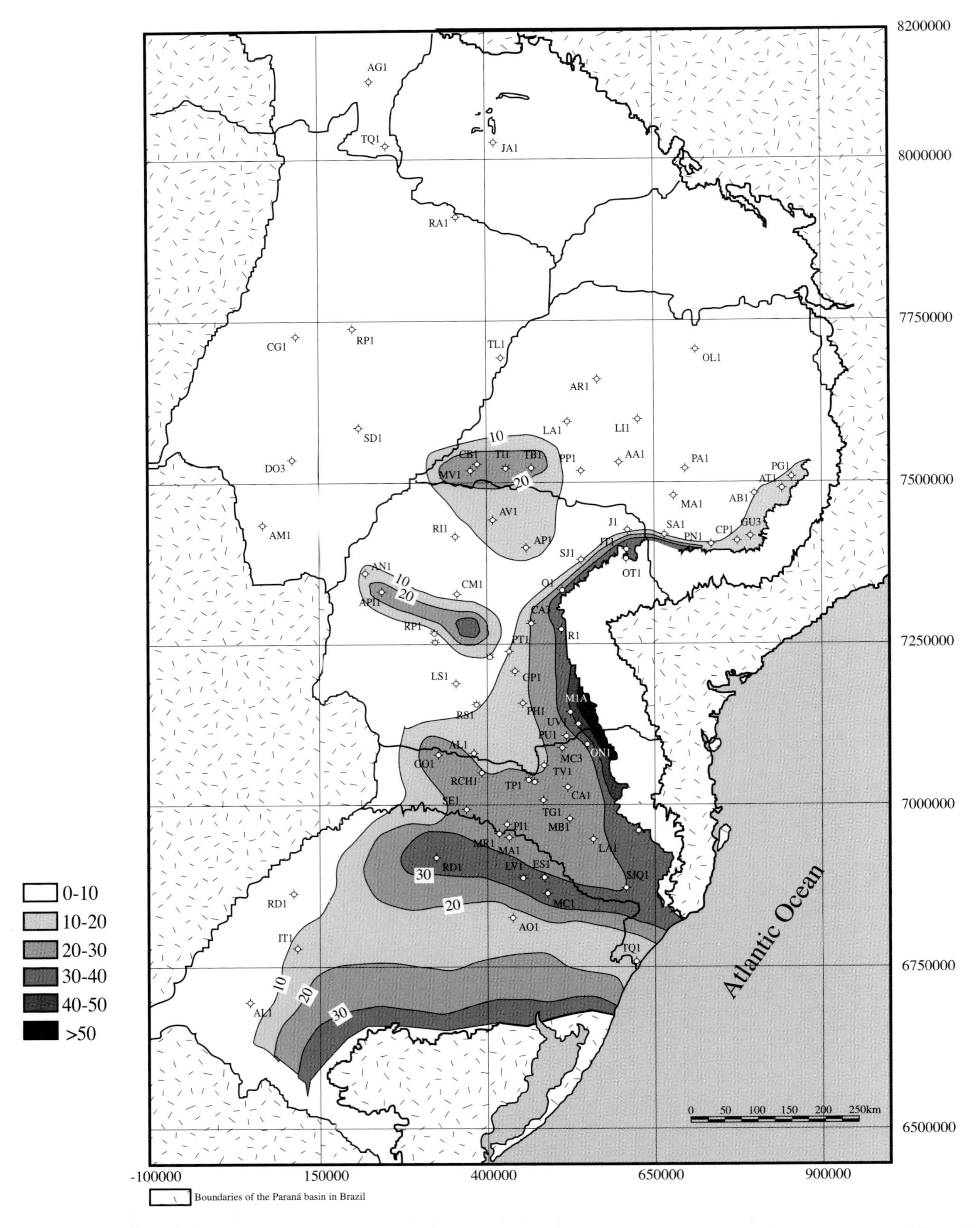

Figure 7—Average original source rock potential in the Assistência Member of the Irati Formation, reconstituted through an equation of correlation (S2 = kg HC/ ton rock; CI = 10).

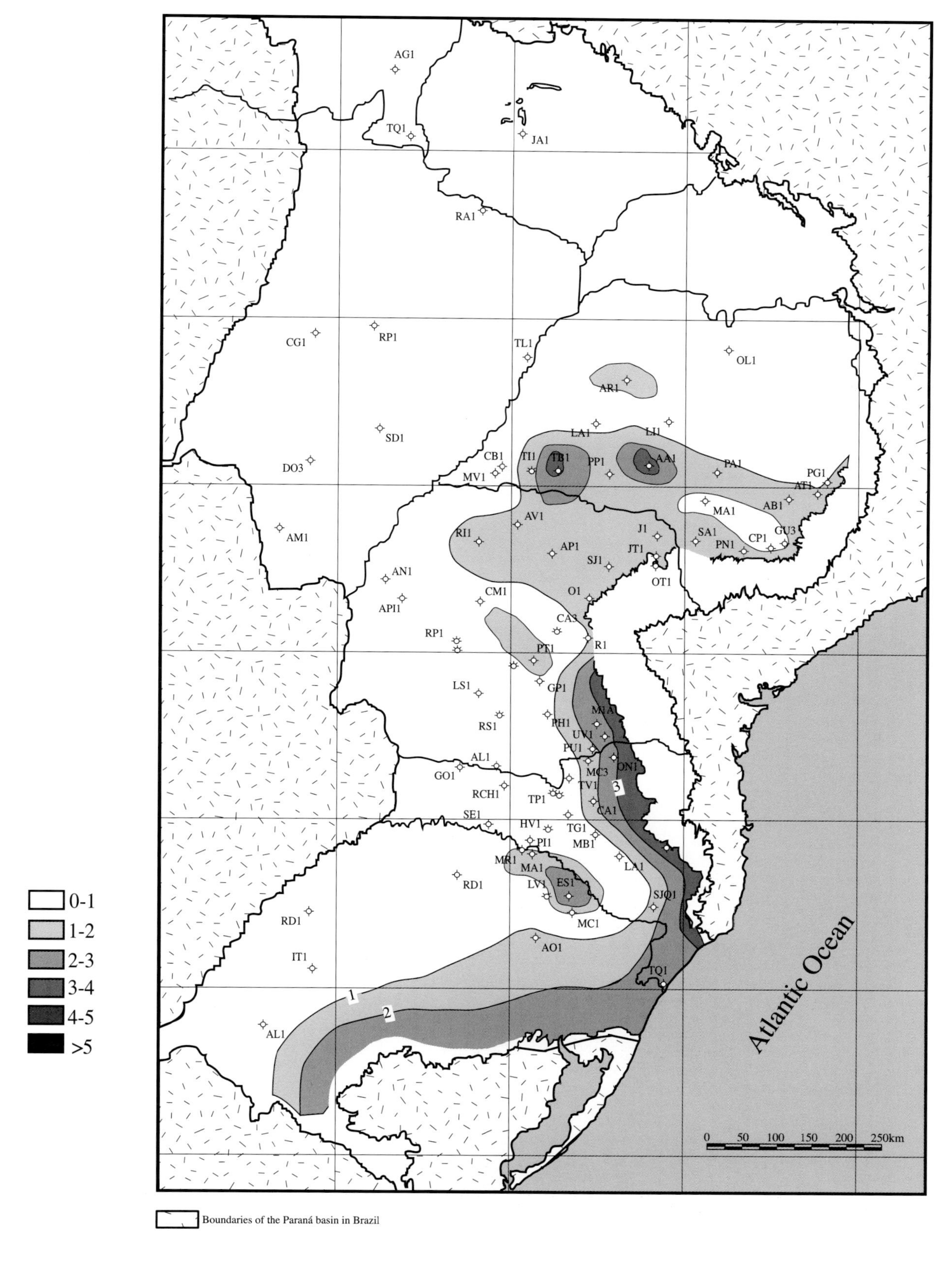

Figure 8—Residual mass of free hydrocarbons in the Assistência Member of the Irati Formation ($S1_r$ = kg HC/ ton rock; CI = 1).

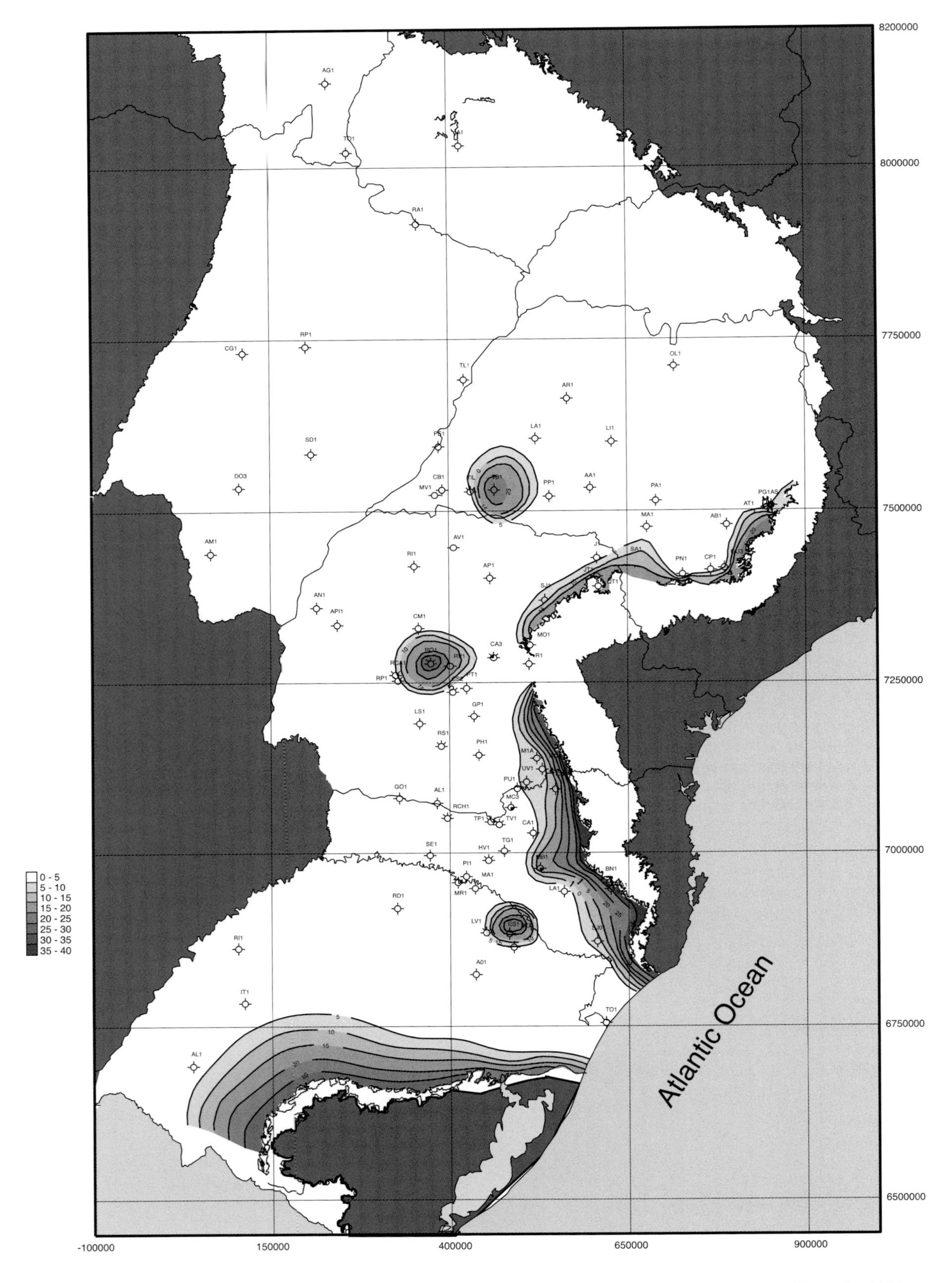

Figure 9—Average residual source rock potential in the Assistência Member of the Irati Formation ($S2_r$ = kg HC/ ton rock; CI = 5).

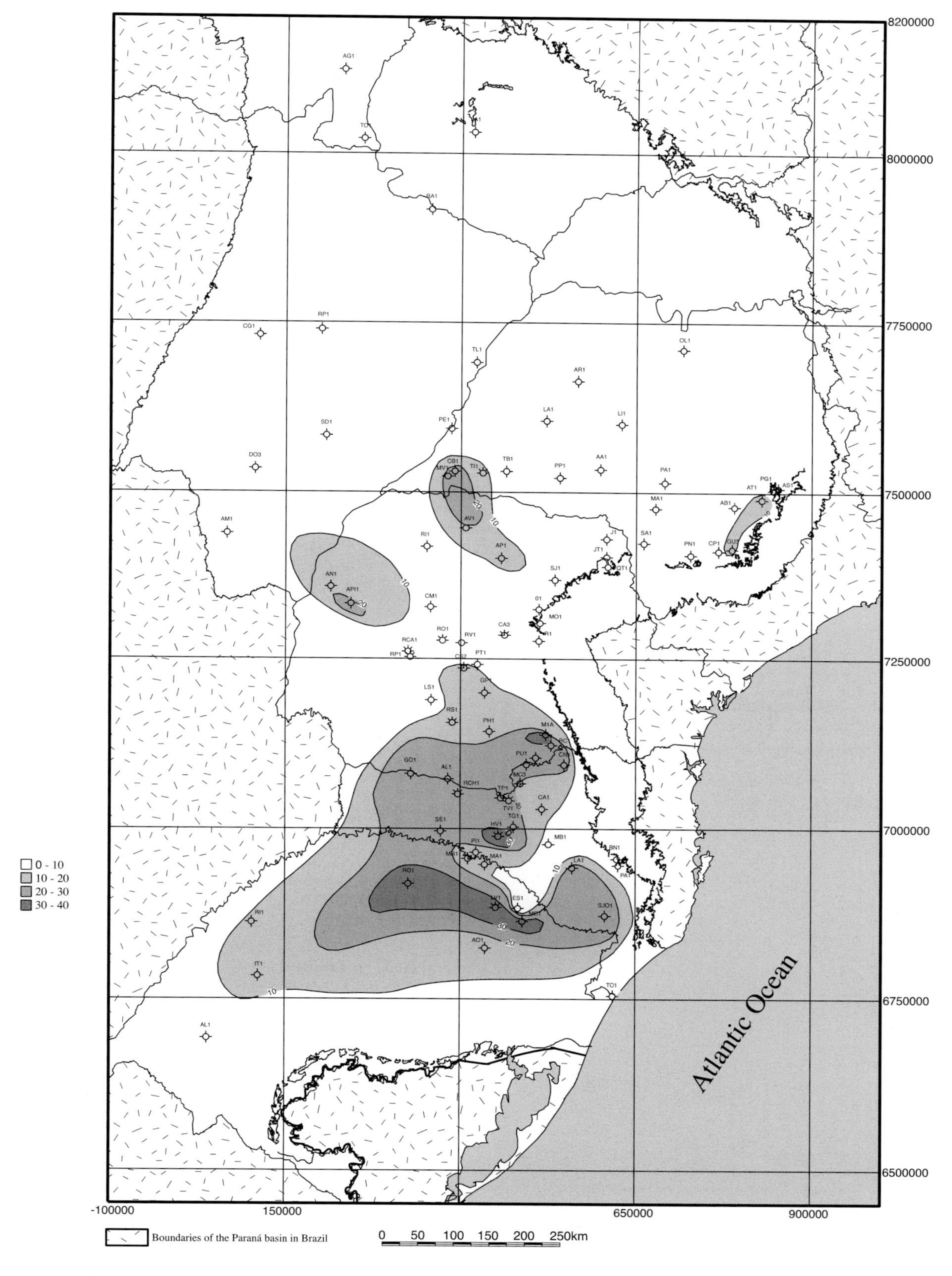

Figure 10—Distribution of the hydrocarbon mass produced by the thermal effect of igneous intrusives on the Assistência Member of the Irati Formation ($\Delta Sp = S2_o + S1_o - S2_r$ kg HC/ ton rock; CI = 10).

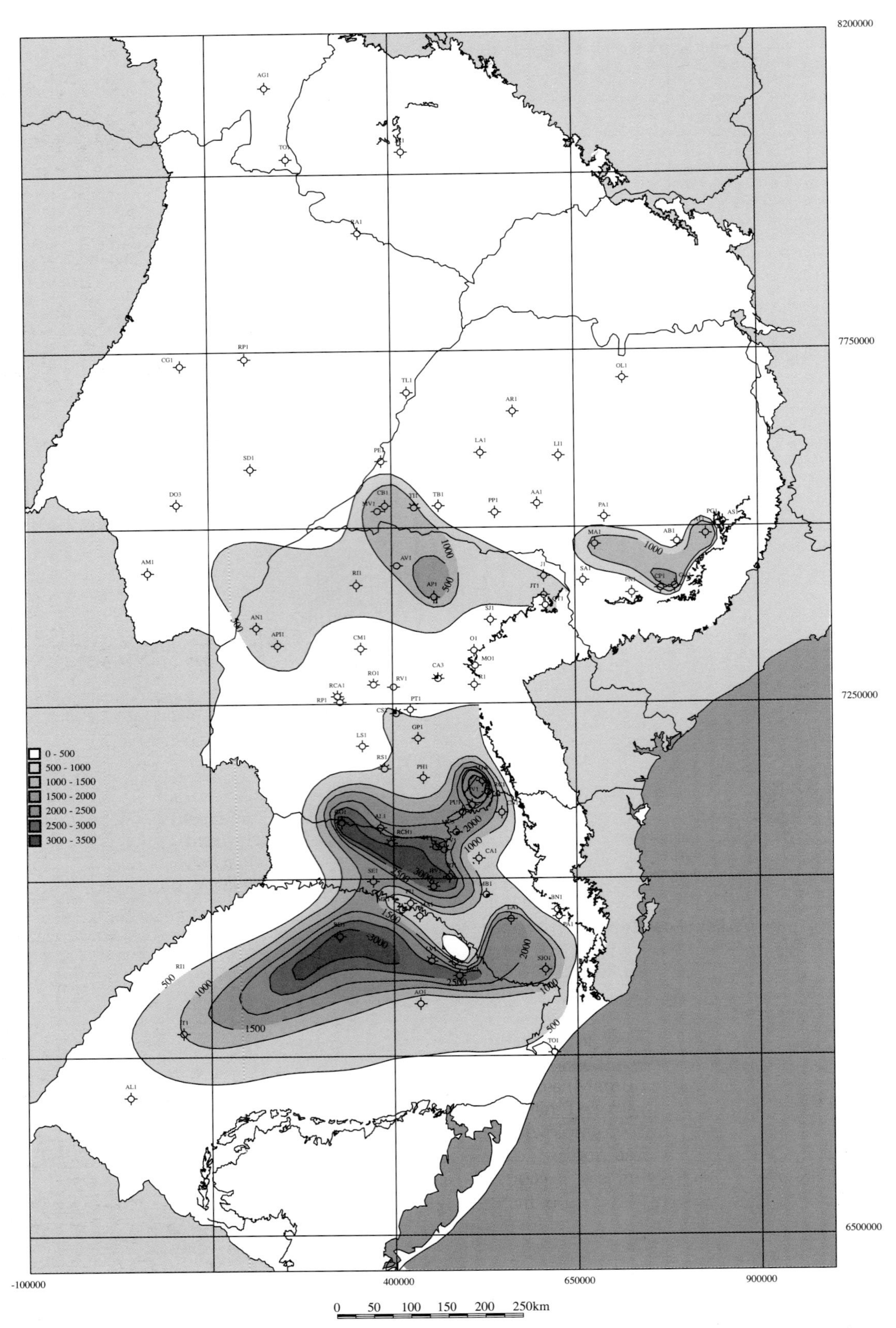

Figure 11—Areal distribution of the volume of hydrocarbons migrated due to the thermal effect of igneous intrusives on the Assistência Member of the Irati Formation (V m^3/km^2 = 2.5 × 1.66 × ΔSm × h m^3 HC/km^2; CI = 500 × 10^3).

indicating lateral or vertical migration outside of the source rocks; (2) the accumulation area, where $\Delta Sm/km^2$ is positive, indicating lateral migration to the bordering areas of the generating system; and (3) the nonmigration area, where $\Delta Sm/km^2$ is equal to zero, indicating that the generated mass remained in the source rocks.

The charge calculated by the equations, that is, the volume of available hydrocarbons to be trapped (Sluijk and Nederlof, 1984) varied from zero to 3500×10^3 m^3/km^2 (Figure 11). The volume of migrated hydrocarbons (m^3) per area (km^2) was delineated in the geochemical map by curves of isocharge of 500×10^3, with maximum isopletes of 3500×10^3.

The calculated charge is more expressive south of the Ponta Grossa arch (fault zone of the Rio Alonzo), it being the maximum isocharge distributed in Santa Catarina and Rio Grande do Sul states (Figure 11). Here, the source rock is richer and the thickness of the intrusives is greater than the source rock thickness. In São Paulo State, the maximum isocharge of 1500×10^3 was mapped in a restricted area. This decrease in the charge was due to a decrease in the organic content and to the small thickness of the diabase sills, mainly in the eastern part of the state. Basically, the configuration and location of the Irati Formation kitchen reflects the spatial distribution of the diabase sills that affected the organic content contained in the bituminous shale and marls. In comparative terms, the migrated maximum amount of the Irati section (3500×10^3 m^3/km^2) converted to the source potential index (SPI = 2.1 ton HC/m^2), established by Demaison and Huizinga (1991) to classify the available charge in vertical drainage, would be considered abnormally low to supply the trapped subsystems effectively.

The discrimination of the type of hydrocarbons generated (liquid or gaseous) by intrusive heating constitutes an unsolved problem in the evaluation of the petroleum charge. The maturity level based on geochemical and optical parameters suggests that the measured thermal stress is compatible with the liquid hydrocarbons window. However, there is no parameter that can be used to infer the speed of the intrusive emplacement or the speed of primary migration in order to assess the relative amount of liquid hydrocarbons cracking to gaseous hydrocarbons in the source rocks. The drastic hydrogen loss (>90%) in a thermal maturity level equivalent of 0.7 and 2% R_o can be attributed to the direct generation of high amounts of gaseous hydrocarbons. The quantitative approach used in this paper considered the petroleum charge generated only as liquid hydrocarbons.

GEOPRESSURING AND PRIMARY MIGRATION

Experiments conducted by Lafargue et al. (1994) on the lower Toarcian shales of Paris Basin have highlighted the importance of geopressurization of the source rocks as an essential factor in primary migration (expulsion of the hydrocarbons from the source rocks toward the carrier beds). This demonstrates that the maximum expulsion occurs when the pressure reaches values between 300 and 500 kgf/cm^2. As a consequence of internal pressurization of the source rocks, geopressured compartments originate in connection with the development of diagenetic seals above the generation kitchens. In most of the studies published by Ortoleva (1994), the geopressured compartments were related to the phase of kerogen thermal cracking.

Some evidence of geopressurization in the Paraná Basin, attributed both to kerogen thermal cracking and to cracking of oil to gas, were preserved, in spite of further intense tectonic compartmentalization. Geopressured zones were detected in two wells drilled in the basinal depocenter. In one well, the existence of a geopressured compartment was confirmed (1.22 $kgf/cm^2/m$, 1.74 psi/m) owing to a diagenetic seal immediately above the Irati Formation. The pressure shift between the compartments normally and abnormally pressurized above and below the seal is 85 kgf/cm^2 (Figure 12). The geopressurized compartment with a pressure gradient of 1.22 $kgf/cm^2/m$ was characterized from the Irati Formation (Kazanian) to the Ponta Grossa Formation (Devonian).

The diagenetic seal above the Permian source rocks provides evidence that the thermobaric reactions in the source rocks supplied dissolved silicate and carbonate for cementation, similar to what has been observed in other basins (Hunt, 1990; Al-Shaieb et al., 1994; Bradley and Powley, 1994; Shepherd et al., 1994; Surdan et al., 1994). The seal was preserved for 127 million years, probably because of the basin center location of the area, where the effects of uplift common along the borders were attenuated. The occurrence of Rio Bonito Formation sandstones in the depocenter with porosities of 20% at 4000 m of depth would also indicate that acidic fluids coming from the Devonian source rocks dissolved cement and created secondary porosity.

The great thickness of the geopressurized compartment (~2500 m from Permian to Devonian source rocks) is possibly due to the pressure equalization that occurred between the two generating systems. The Devonian system pressurization was caused by the thermal cracking of oil to gas during magmatism. In a sealed system, the cracking of 1% of oil can be enough to increase the pore pressure from the hydrostatic to the lithostatic limit (Gaarenstroom et al., 1993).

The relationship between the residual mass of free hydrocarbons ($S1_r$) present in the shale and marls of the Irati Formation (Figure 8) and the mass of produced hydrocarbons ($S2_o + S1_o - S2_r$) by the thermal igneous effect (Figure 10) indicates that the efficiency of expulsion of hydrocarbons from the source rocks was almost 100%. This high efficiency, greater than that verified by Cooles et al. (1986), Mackenzie et al. (1987), Espitalié et al. (1987), Larter (1988), and Pepper (1991) (75–90%), can be attributed to the igneous promotion of secondary cracking (oil to gas) of the remaining liquid hydrocarbons that are normally retained in the organic matter or in the rock miner-

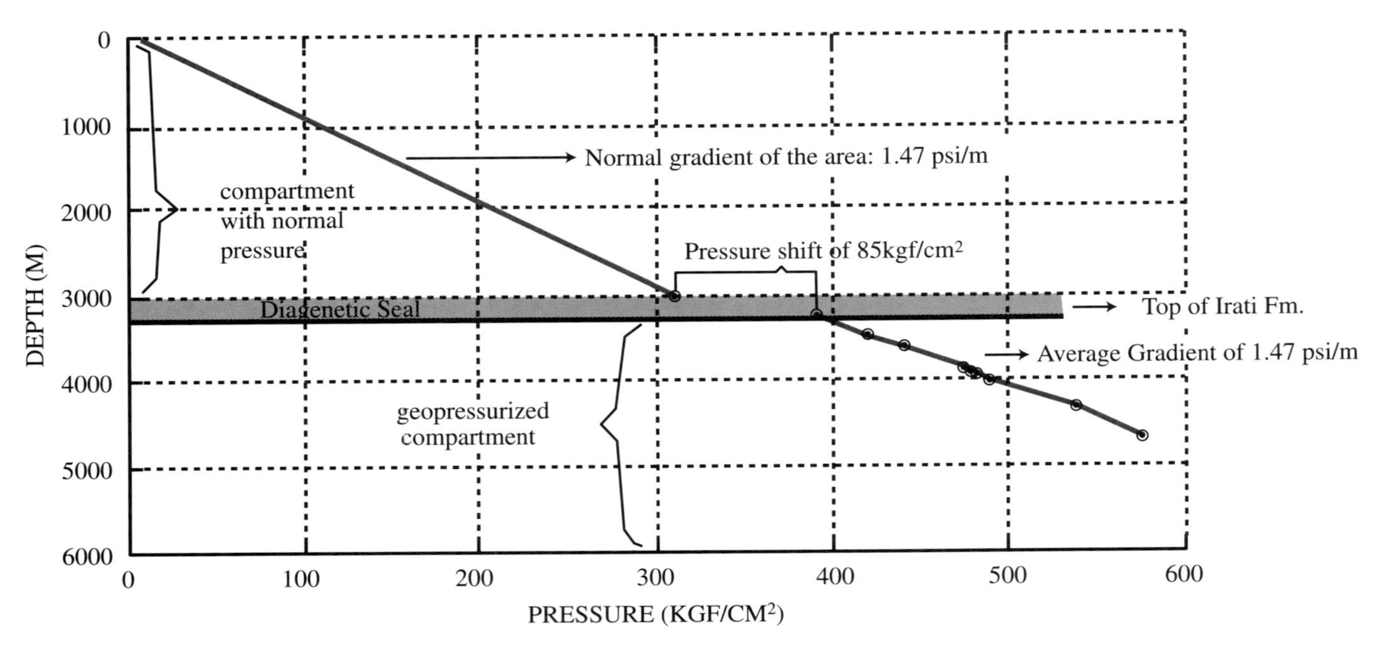

Figure 12—Geopressurized compartment observed at well 1-API-1-PR (central Paraná Basin) with a diagenetic seal above the Permian source rock (Irati Formation). Pressure data obtained from cable tests (RFT).

al matrix. However, gaseous hydrocarbons cannot be correctly measured during drilling. Sokolov et al. (1971) concluded that the sampling of source rocks with preservation of original temperature and pressure conditions contain 98% more gas than the samples collected during conventional drilling.

Qualitative analysis of the hydrocarbon occurrences sampled during drilling of the Irati Formation indicates that, in the area with a thermal halo equal to or greater than the source rock thickness (thickness of intrusions ≥ source rocks), no occurrence of liquid hydrocarbons was reported. This corroborates the hypothesis of complete cracking of the oil remaining in the source rocks to gaseous hydrocarbons. In wells more recently drilled in the basin, which used special muds and more accurate detection equipment, gas was detected during penetration of the Irati Formation. These observations reinforce the hypothesis that the expulsion efficiency from the Irati Formation has been similar to the level observed in other basins (75–90%).

A petrographic investigation of microfractures in thin sections from a well affected by a 27.2-m-thick diabase sill revealed that 90% of the microfractures were distributed parallel to bedding and were partially filled with calcite. Quantitative analysis of the relative abundance of microfractures indicates a high concentration of horizontal fractures at about 9 m from the contact with the igneous body and another increase at about 18.5 m from the basal contact. This reinforces the hypothesis that the microfractures are more likely related to thermobaric causes due to kerogen cracking than to mechanical shear caused by intrusion of the sill. The open microfractures observed near the contact between the intrusion and source rock are interpreted to reflect the overlap of fracturing caused both by hydrocarbon generation and insertion of the sill.

To evaluate the primary migration efficiency, relative quantification of residual bitumen was done on thin sections from two Irati sections intruded by sills of 28.4 m and 2 m thickness, respectively. In the well with the thicker sill, rare occurrences of residual bitumen were observed, while in the well with the thinner sill, larger amounts were observed. The reverse relationship between relative abundance of residual bitumen in source rocks and thickness of the sills indicates that the efficiency of the primary migration is directly related to heat dissipation from the sills. Thinner sills would allow faster reestablishment of the thermal balance, reflected in decreased efficiency of the primary migration.

Analysis of the homogeneization temperatures (Th) measured in aqueous and oil fluid inclusions (Fuzikawa, 1994) within calcite and quartz veins cutting two Irati sections intruded by diabase sills 27.2 m and 20.6 m thick, respectively, allowed the following inferences:

1. In the well affected by the thinner sill, water inclusions with Th ranging from 40° to 240°C indicate that the filling of the fractures began during the generation pulses and continued during uplift of the basin borders started in the Cretaceous. The temperatures of the oil inclusions, varying from 60° to 260°C, indicate that during primary migration, hydrocarbon pulses were displaced under different temperatures. This results from the differ-

ence between the maximum level of heat in the source rocks and the cooling that occurred during migration. The higher homogeneization temperatures were recorded in samples closer to the sill, both for aqueous and oil inclusions, which is consistent with the temperature gradient expected from contact with the intrusion.

2. In the well intruded by the thicker sill, the aqueous inclusions showed homogeneization temperatures ranging from 40° to 150°C. The sample closest to the igneous body showed mainly aqueous inclusions and some oil inclusions with homogeneization temperatures from 50° to 70°C. A possible explanation for this is the intense shearing and thermal expansion that occur during intrusion. During the cooling of the sill, the pressure relief caused by the open fractures caused a reversion in the direction of the flow, carrying hydrocarbons toward the contact zone which had maturity geochemical parameters (biomarkers) incompatible with the optical parameters (% R_0 and SCI), as verified by Trigüis (1986) and Mendonça Filho (1994). The immature parameters of the biomarkers indicating thermal maturity ($T_S/T_S + T_m$), tricyclics/pentacyclics, C_{32} homohopanes ($22S/22S + 22R$), C_{29} steranes ($20S/20S + 20R$), and C_{29} steranes ($\alpha\beta\beta/\alpha\beta\beta + \alpha\alpha\alpha$) were determined along the sections affected by the sills in both wells. They register a reverse primary migration of less thermally evolved hydrocarbons toward the contact, as well as migration of thermally evolved hydrocarbons away from the contact.
3. The trapping of innumerous oil inclusions with homogeneization temperatures from 50° to 130°C in the Palermo Formation, 15.5 m below the base of the Irati source rocks, indicates that the vertical propagation of the fractures extended beyond the limits of oil generation. This reinforces the theory that petroleum can migrate vertically both upward or downward from the source rocks when the differential pressure is great enough. The capture of oil inclusions under a high-temperature spectrum, in the section underlying the Irati Formation indicates that, during the multiple pulses of migration, the fractures were kept opened by the geopressurization of the generation system.

The efficiency of the primary migration process in the Irati Formation can be attributed to the existence of a network of predominantly horizontal fractures and microfractures and to the extensive tectonic compartmentalization of regional and local scale. This facilitated the contemporaneity between primary and secondary migration through activation of the faults that bordered the compartments. The structural framework, reinterpreted by Marques et al. (1993), matched with the geochemical map of the petroleum charge, shows that the Permian kitchen is segmented by 20 structural megafeatures (rupture lines and faults zones) (Figure 13). These form a mosaic whose limits correspond to zones of weakness and pressure relief, where the primary migration converged away from intrusion sites during magmatism.

Most of these megafeatures were reactivated during the Cretaceous magmatic event due to forces mainly along the northwest faults zones, where an intricate network of dikes was intruded. This occurred in connection with the transtensional movements caused by the initial stages of Gondwana's breakup and the vertical uplift of the sedimentary section caused by the insertion of horizontal and subhorizontal sills that form about 12% of the total column of the basin (O. A. Zanotto, personal communication, 1998). According to Ar/Ar dating (Turner et al., 1994), the Serra Geral magmatism occurred between 137 and 127 Ma. Milani (1997) inferred from Ar/Ar ages from of a well drilled in the basin depocenter that both intrusive and extrusive magmatism occurred simultaneously between 138 and 128 Ma.

The large vertical movements of blocks, creating a tectonic megacompartmentalization, generated zones of pressure relief where the flows converged. The framework of this megacompartmentalization was broken into smaller blocks, creating multiple areas of pressure relief of a few square kilometers. Investigation of four areas in Guiabá Paulista (SP), Taquara Verde (SC), Três Pinheiros (SC), and Matos Costa (SC) which had available seismic control and closely spaced wells, confirmed the local compartmentalization at the level of the Permian source rocks, as shown in Figure 14, which reveals bordering faults at about every 2 km. The sills mapped into the source rocks vary in thickness and stratigraphic level, indicating that the vertical buoyancy of the blocks activated the bordering faults of the compartments, creating zones of pressure relief to which primary migration converged.

The activation of each compartment probably occurred in a sequential way, considering the time and spatial differentiation in the insertion of the igneous intrusions. The resulting sequential mechanism of local pressure relief transformed the generation kitchen in a mosaic composed of several separate and unsynchronized kitchens. From this, it can be concluded that primary migration occurred mostly laterally along short distances (1–10 km) to the bordering faults of the compartments, which were the conduits for vertical migration to the sandstone reservoirs.

Because of the contemporaneity of the generation and primary migration, the countless intrusions dated between 127 and 138 Ma (Turner et al., 1994; Milani, 1997) caused multiple generation–migration pulses from different kitchens. These had a high expulsion efficiency due to the creation of a network of horizontal microfractures that interconnected the geopressurized kitchens to the vertical pressure relief faults (Figure 15). Due to this sequence of events, it can be imagined that great losses of petroleum charge occurred due to an excess of vertically focused flow and to fragmentation of the petroleum charge that reached the rock reservoir in intermittent pulses.

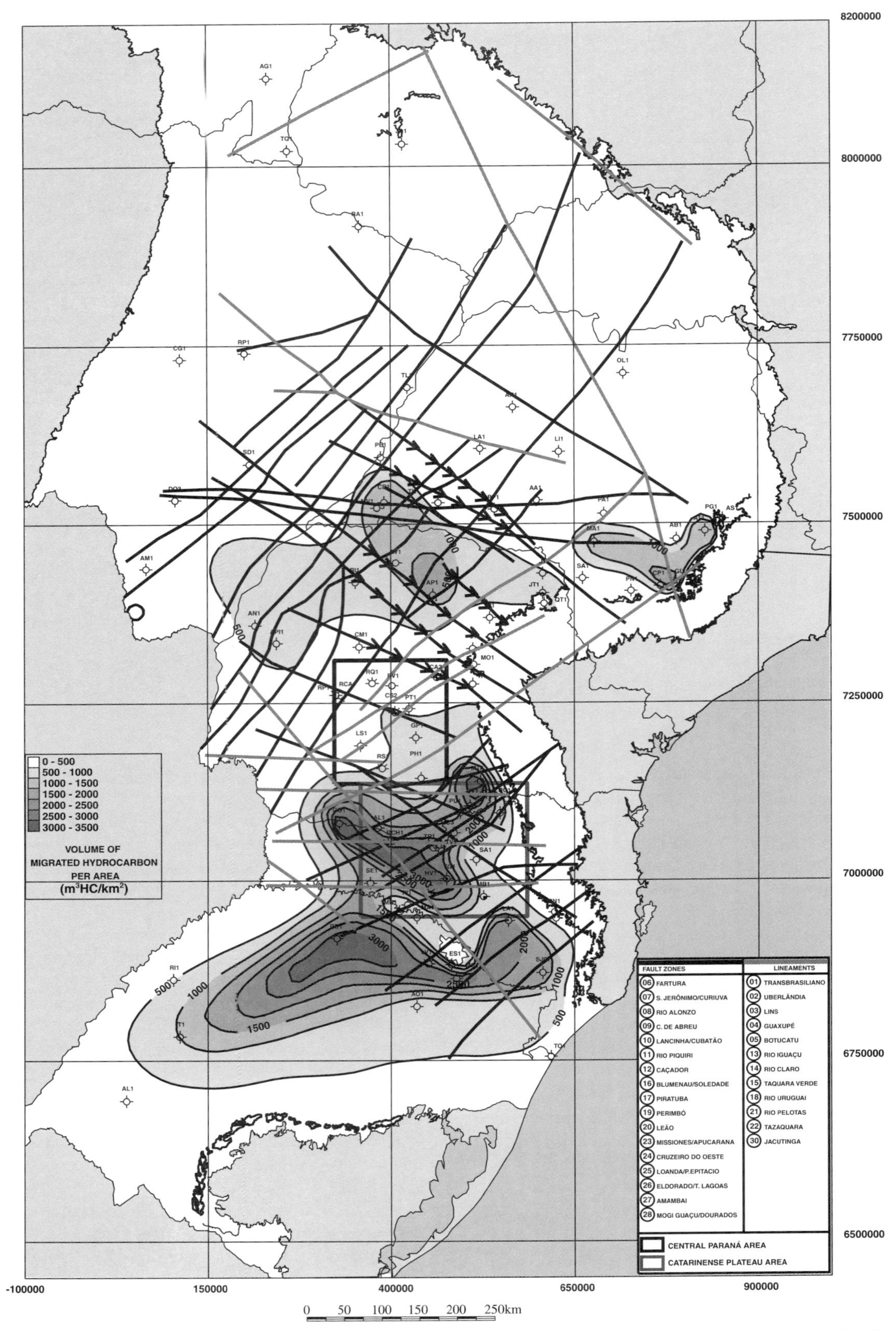

Figure 13—Structural framework of the Paraná Basin and the Permian hydrocarbon kitchen resulting from the thermal effect of diabase sills on the Irati Formation (CI = 500 × 10³).

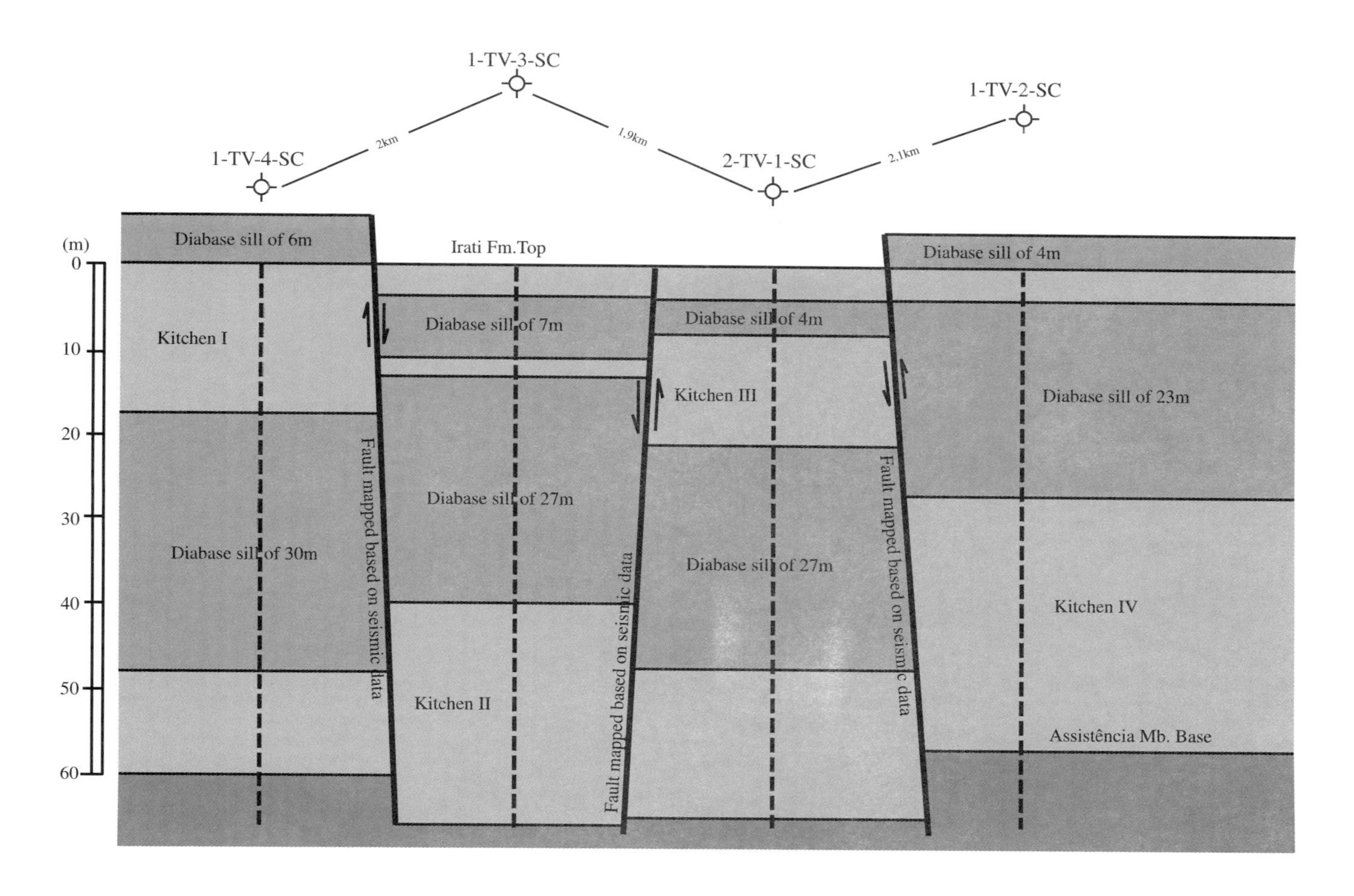

Figure 14—Spatial positioning of the diabase sills of the Irati Formation in the Taquara Verde area (SC). The thickness range and the insertion of different stratigraphic levels in restricted areas are indicative of the complex distribution of igneous intrusives and the activation of compartmented kitchens of generation.

SECONDARY MIGRATION

In the secondary migration process, vertical movement of the petroleum charge through conductive faults usually happens due to the differential pressure between abnormally and normally pressurized compartments. This generally involves short geologic time intervals and short migration courses on the order of 1500 m (England, 1994). However, in Campos Basin, it can exceed distances of 3000 m. The second stage of secondary migration into the carrier beds (reservoirs) is controlled under hydrostatic conditions by the buoyancy pressure generated by a less dense hydrocarbon column in a water-saturated system. When the permeability of the carrier beds is greater than 1 md, after reaching the base of the seal via buoyancy, the flow tends to migrate laterally (England et al., 1987).

The secondary migration process began in the Paraná Basin by the focused flow of hydrocarbons through the bordering faults that limit the generation compartments. Via this vertical migration through fault conduits, the petroleum charge reached the carrier beds situated 100–200 m above the source rocks. After migrating through the shaly section of the Serra Alta Formation, the petroleum charge gradually migrated laterally into thin bodies of fine sandstones and carbonates of the Teresina Formation and into thick bodies of fluvial-eolian sandstones of the Rio do Rasto, Pirambóia, and Botucatu Formations.

The temporary and spatial interaction of migration and accumulation processes of this atypical Permian petroleum system occurred during two phases of trap formation:

1. The first phase occurred during the Permian–Triassic Cape–La Ventana orogenic event close to the Gondwana southern margin (De Wit and Ransome, 1992). This was related to the cratonward propagation of compressional forces inside the Paraná Basin which created structural inversions prior to deposition of the Pirambóia and Botucatu Formations (Milani, 1992).

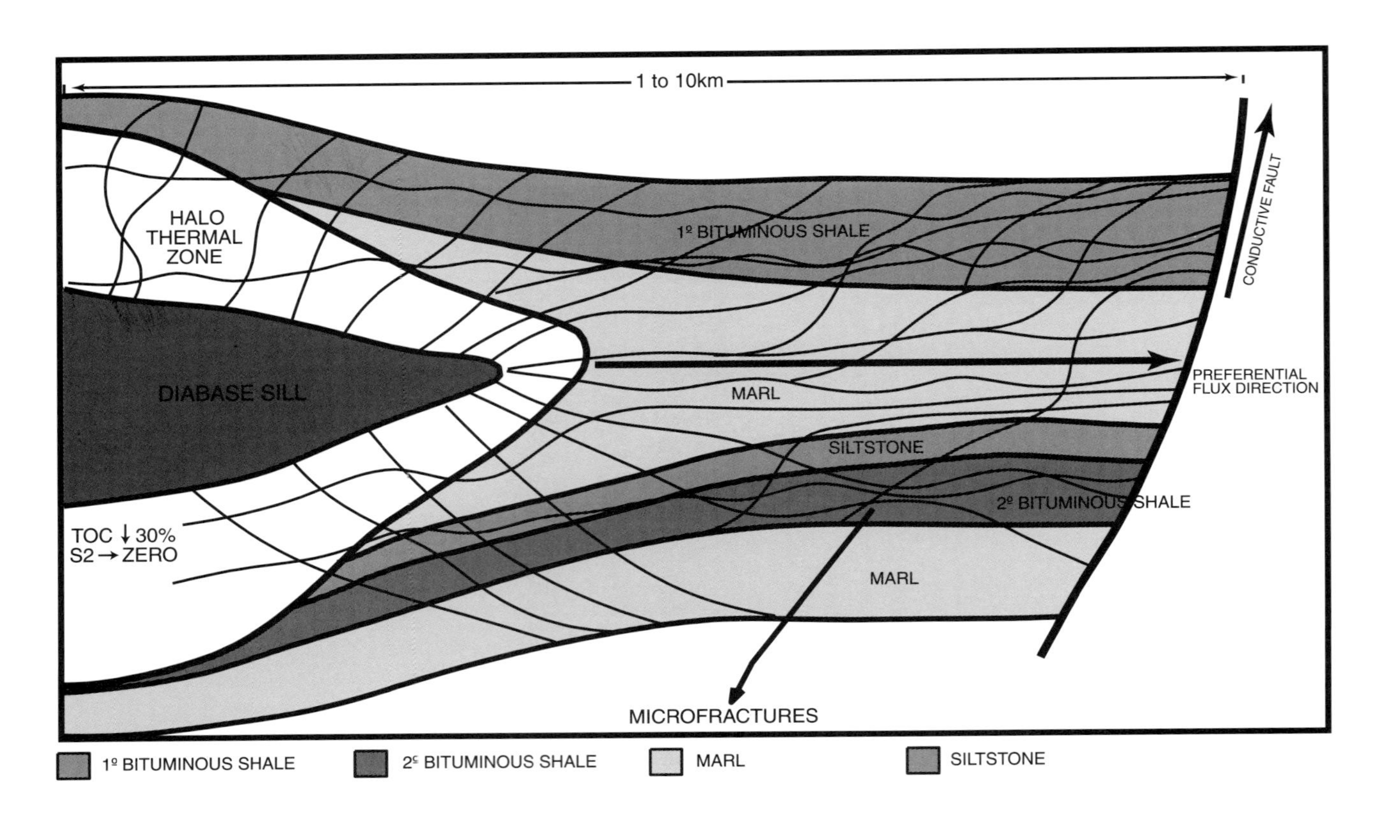

Figure 15—Generation and primary migration model of the Permian petroleum system of Paraná Basin fractures and microfractures resulting from thermal cracking of the Irati Formation kerogen due to the effects of magmatic intrusion. The preferential pattern of horizontal fractures was observed in thin sections.

2. The second phase, concomitant to the generation–migration process (127–138 Ma), occurred during the active rifting of Gondwana. It was related to transtensional forces overlapped on the previous structural framework imprinted in the Jurassic–Triassic section (Pirambóia and Botucatu Formations) creating complex structural inversions difficult to characterize in seismic mapping.

The contemporaneity of the generation–migration process with the second phase of trap formation decreases the potential for the Pirambóia and Botucatu Formations to trap petroleum because of uncertainties about the synchronism between events. Based on this premise, it is reasonable to suppose that the greatest chance of storage of petroleum charge occurred in the structures originated in the Cape–La Ventana event.

Lateral migration along long distances demands special geologic conditions, such as the presence of reservoir rocks of regional extent or migration routes through regional unconformities. This has been verified in the Alberta Basin of Canada, where migration of about 100 km has been suggested for the Athabasca tar sand (Deroo et al., 1977). Due to their continental extent (839,000 km^2) (Araújo et al., 1995), fluvial-eolian sandstones of the Botucatu and Pirambóia Formations (Jurassic–Triassic) that overlay the Permian source rocks in the Paraná Basin constitute carrier beds favorable to long-distance migration. However, the fragmentation of the petroleum charge supplied to secondary migration made displacements along long distances unfeasible. Larger continuous phases of hydrocarbons were needed to compensate for losses during migration. In addition to this restriction, the extensive tectonic compartmentalization tended to focus the flow vertically.

The only known hydrocarbon occurrences in the Jurassic–Triassic sequence are found in an area of 7000 km^2 in the eastern part of the basin (São Paulo state), where 19 important shows of residual biodegraded hydrocarbons in Pirambóia fluvial-eolian sandstones were mapped (Figure 16). The geochemical signature of these hydrocarbons indicates a Permian origin (Cerqueira and Santos Neto, 1986). In this area, the orogenic Cape–La Ventana event produced positive movements (Milani, 1992) and an erosive episode that created a unconformity surface on which rests the Pirambóia reservoir a distance of 50–150 m from the Permian source rocks (Figure 16). As the basal Teresina Formation is mainly shaly in this area, the first available carrier beds are the Pirambóia sandstones. The petroleum charge of 1000×10^3 m^3

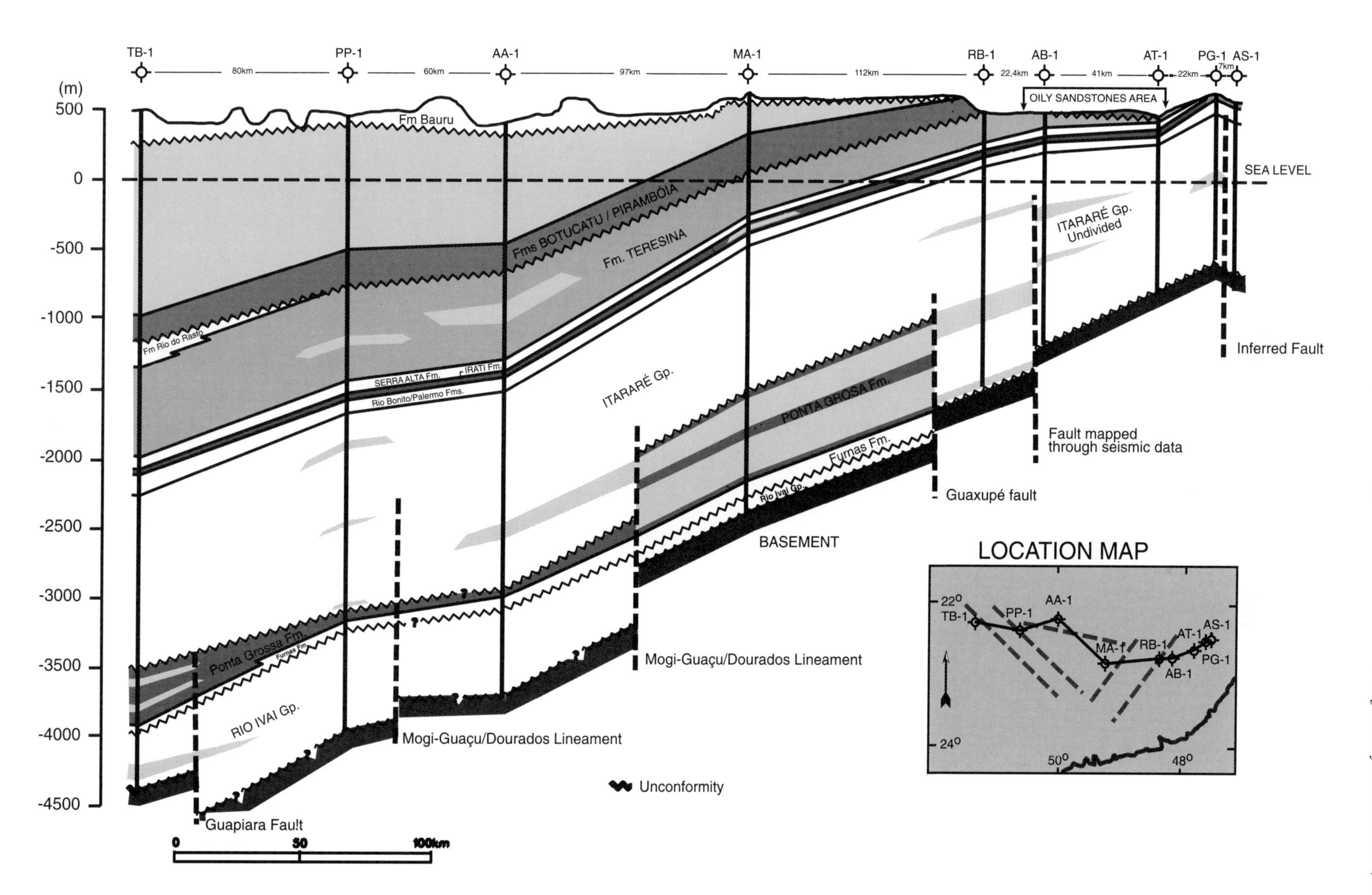

Figure 16—Geologic section showing the Permian–Triassic unconformity (Cabo–La Ventana orogeny) which created the close proximity of Permian source rocks with the Pirambóia Formation sandstones in the area of tar sand occurrences.

HC/km^2 calculated for this area was most likely focused directly toward this carrier bed instead of migrating laterally. The inferred residual volume of 60,000 × 10^3 m^3 HC (Hettich, 1981) represents approximately 1% of the produced volume, considering the area of 7000 km^2. An amount of 0.25 m^3 HC/m^3 of sandstone, measured in the main occurrences of the tar sands, indicates that these areas correspond to residual petroleum accumulation biodegraded by meteoric gravitational flow during the marginal uplift that began in the middle Cretaceous.

Calculation of a volumetric balance demands a determination of the residual hydrocarbon saturation after secondary migration. In most carrier beds, the residual hydrocarbon saturation after secondary migration varies from 1 to 10% (England et al., 1987). Macgregor and Mackenzie (1987) determined the residual hydrocarbon saturation in the Sumatra Basin to be about 2%. Larter and Hortad (1992), studying the reservoirs of the East Shetland Basin, concluded that when the residual hydrocarbon saturation is about 3%, the balance between the petroleum charge and the lost volume is positive and petroleum accumulations exist along the migration route. When the residual hydrocarbon saturation is about 6%, the mass balance is negative and no hydrocarbon accumulations exist along the migration route.

Calculation of the volume of petroleum lost (V_p) along the routes of migration is given by the equation $V_p = \phi f V_r$, where ϕ is porosity, f is the residual hydrocarbon saturation, and V_r is the rock volume crossed by the flow (Mackenzie and Quigley, 1988). The difference between the petroleum charge and the lost volume corresponds to the volume trapped. As there is no measurement of residual hydrocarbon saturation in the Paraná Basin, the cutoffs (3% and 6%) determined by Larter and Hortad (1992) were adopted to proceed calculation of the volumetric balance of the petroleum charge (m^3 HC/km^2) using the above equation.

Two areas with significant petroleum charge were selected to calculate the volumetric balance: the Catarinense Plateau and north-central Paraná Basin close to the basin depocenter (Figure 13). In the Catarinense Plateau area (SC), where the average petroleum charge is 1500 × 10^3 m^3 HC/km^2 and ϕ =14% (Teresina and Rio Rasto Formations), it was calculated (assuming f = 6%) that the hydrocarbon charge would be dispersed (lost) in a thickness of 178 m. Assuming f = 3%, ϕ = 20%, and So = 70% (oil saturation), the storage thickness would be 6.4 m. The supply through the faults started from the Pirambóia Formation base and migrated vertically 100 m up to the base of the Serra Geral Formation, a sealing package of volcanic rocks as thick as 2000 m.

For the north-central Paraná Basin, where the average petroleum charge is 500 × 10^3 m^3 HC/km^2, it was calculated that this charge would be dispersed in a thickness of 59 m. For a vertical buoyancy of 150 m up to the base of the volcanic seal, the volumetric balance would be negative due to the lost volume of 900 × 10^3 m^3 HC/km^2, which is larger than the calculated petroleum charge of 500 × 10^3 m^3 HC/km^2.

Two characteristics of this atypical Permian petroleum system support the hypothesis that lateral migrations took place over a short distance: (1) the excessive number of vertical faults that focused flow vertically, and (2) fragmentation of the petroleum charge caused by multiple generation–migration pulses.

TERTIARY MIGRATION

The uplift of the borders of the Paraná Basin related to the South Atlantic rifting added to the propagation of compressional stresses from the western Andean orogeny. This created a structural asymmetry that resulted in the regional basculation of the eastern border toward the western, northern, and southern borders of the basin. Apatite fission trace data indicate that the first uplift pulses of the eastern border began 110 Ma (Lelarge, 1993), reaching a climax between 100 and 90 Ma and causing erosion of 3000 m of section. Zanotto (1993) also suggests 3000 m of erosion using the vitrinite reflectance method. The erosion intensified during the Cenozoic, causing the erosional scarp to retreat more than 80 km from the original border, which provided the onset of the gravitational hydraulic head contemporaneous with the tertiary migration.

Due to gravitational drive, the speed of water displacement into the Botucatu and Pirambóia sandstones (with 17% average porosity) can reach values of about 1000 km/0.22 m.y. (based on δ^{13}C data from Silva, 1983). This gravitational water movement is 4.5 times faster than liquid hydrocarbon displacement via buoyancy (assuming a carrier bed with permeability of 1000 md; England et al., 1991) and resulted in the onset of the Paraná Basin hydrodynamic traps.

In the Catarinense Plateau area along a 150-km extent of the Peixe River, dozens of oil occurrences have been detected in the basalt from seeps along the railroad and in countless shallow wells. Oils of up to 33° API have been collected here and geochemically correlated to the extracts of Permian source rocks. In these shallow wells, oil occurs in the fractured and vesicular zones at the tops of the lava flows, usually near the topographic surface (~100 m depth). The incipient degree of bacterial degradation provides evidence of recent tertiary migration vertically from the Permian–Jurassic section. Intense renewal of meteoric water is active in the basaltic aquifer.

Long-distance lateral tertiary migration (<80 km) in the Paraná Basin is quite unlikely because of three factors:

1. The existence of an intricate network of dikes, sills, and fault zones indicates short-distance migrations with a tendency for vertical migration.
2. The small thickness of possible accumulations in the Jurassic–Triassic sections calculated from volumetric balance (<10 m) would be retained in the pores as residual fraction, only a few hundreds of meters from the trap.

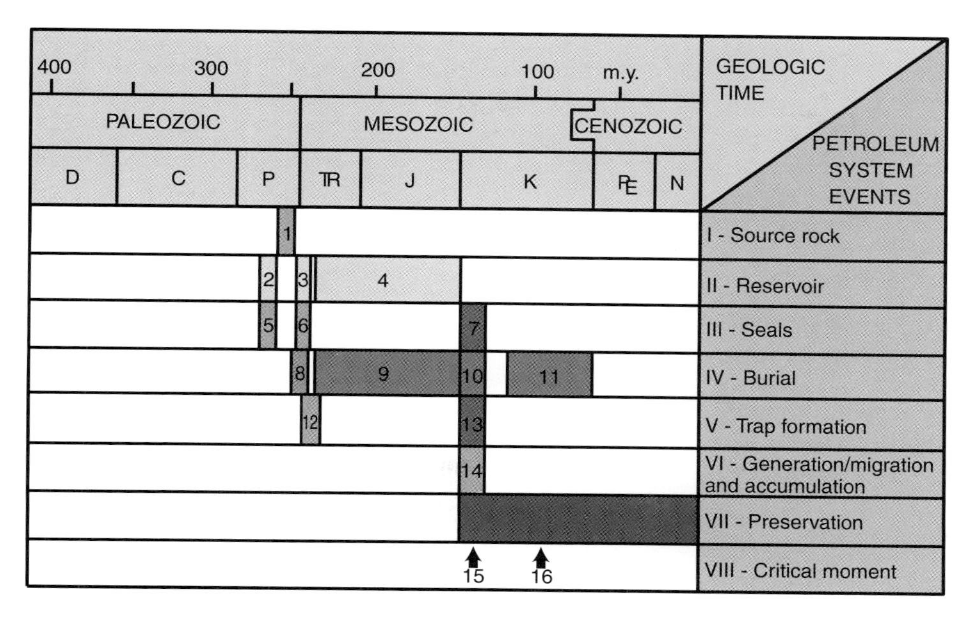

Events Key

I. Source rock
   1. Assistencia Mmbr of Irati Fm., Kazanian age

II. Reservoirs
   2. Palermo and Rio Bonito ss
   3. Teresina and Rio do Rasto ss
   4. Pirambóia and Botucatu ss

III. Seals
   5. Palermo and Rio Bonito sh
   6. Teresina and Rio do Rasto sh
   7. Serra Geral igneous rocks

IV. Burial
   8. Late Permian–Early Triassic
   9. Late Triassic–Jurassic
   10. Early Cretaceous
   11. Late Cretaceous

V. Trap formation
   12. La Ventana orogeny
   13. Rifting phase

VI. Generation/migration and accumulation
   14. Igneous intrusives thermal effect

VII. Preservation

VIII. Critical moments
   15. Rift–drift phase
   16. Late Cretaceous–Cenozoic uplift

Figure 17—Events chart showing the relationship of the essential elements and processes, as well as the time of preservation and critical moments of the Permian petroleum system of Paraná Basin.

3. The intense meteoric gravitational flux, which provided a renewal of water in the Botucatu and Pirambóia aquifer approximately 500 times over a period of 100 m.y. (Araújo et al., 1995), would biodegrade liquid hydrocarbons moving toward the gravitational flow, decreasing API gravity and increasing oil viscosity. Basically, all reservoirs that overlie the Permian source rocks belong to an open hydrodynamic environment (Araújo, 1989), with an intense flux of meteoric water that has been flushing the porous sandstones since the uplift of the basin margin.

EVENTS CHART

The events chart (Figure 17) summarizes the temporal relationships of the essential elements and processes of this atypical Permian petroleum system. Due to the intrusion of diabase sills from a few centimeters up to 240 m in thickness during the Cretaceous magmatic event, high levels of thermal alteration affected the Permian source rocks (average thickness 19 m) that reached the liquid and gaseous hydrocarbon generation windows (VI in Figure 17).

The magmatic event that occurred between 127 and 138 Ma (Turner et al., 1994; Milani, 1997) produced the reactivation of megastructures and a complex secondary network of faults that compartmentalized the petroleum system into small blocks (1–10 km^2 in area). The activation of each compartment probably occurred in a sequential way in response to time and spatial differences in the insertion of the igneous intrusions. This sequential mechanism of local pressure relief transformed the generation kitchen into a mosaic composed of several different separate and unsynchronized kitchens. It can be inferred that primary migration was lateral, traveling short distances (1–10 km) to the bordering faults of the compartments, which were the conduits for upward movement toward the reservoirs.

The contemporaneity of the events of generation, primary migration, and secondary migration promoted multiple generation–migration pulses from different kitchens. Expulsion efficiency was high due to the creation of a network of horizontal fractures that interconnected the abnormally pressurized kitchens with the faults that limited each block.

Through this vertical fault-focused migration, the petroleum charge reached the carrier beds between 100 and 200 m above the source rock. After passing the Serra Alta Formation shaly section, the fragmented petroleum charge gradually migrated laterally into thin bodies of fine sandstone and grainstone of the Teresina Formation and into thick bodies of fluvial-eolian sandstone of the Rio do Rasto, Pirambóia, and Botucatu Formations (II in Figure 17).

Temporary and spatial interaction of the migration and accumulation processes of this atypical Permian petroleum system occurred during two phases of trap creation. The first was during the Permian–Triassic Cape–La Ventana orogeny, close to the southern Gondwana margin (III, 5 and 6, in Figure 17). The second phase occurred during Gondwana rifting, concomitant to the generation–migration process (magmatic event), when transtensional stresses affecting the Jurassic–Triassic section (Pirambóia and Botucatu Formations) overlapped a structural framework previously imprinted during the Permian–Triassic (III, 7, in Figure 17).

The critical moments happened during the magmatic event most likely due to diachronism between the creation of structural inversion (trap formation) and the secondary migration pulses (VIII, 15, in Figure 17). A second phase of critical moments occurred during the uplift of the Paraná Basin margins related to South Atlantic rifting and the Andean orogeny, which climaxed about 100 and 90 Ma (Lelarge, 1993). The uplift caused the erosion of the volcanic Serra Geral sequence, reaching the basal formations of the sedimentary column. Consequently, a gravitational hydraulic head was developed concomitant to the beginning of tertiary migration (VIII, 16, in Figure 17). The resulting intense meteoric gravitational flux, with a renewal of water about 500 times over a period of 100 m.y. (Araújo et al., 1995), constitutes a critical moment for the integrity of the petroleum system.

CONCLUSIONS

The Irati source rocks, having a large range of thicknesses (from a few centimeters up to 240 m), were intruded by diabase sills related to Serra Geral magmatism (138–127 Ma). The thermal effect extends above and below the intrusions in a thickness equivalent to that of the intrusive bodies. In the thermally affected zones, hydrogen indices decay to 10% and TOC is depleted to about 30% of its original value.

Mass–balance calculations indicate that, as a consequence of this intrusion-driven generation, $500–3500 \times 10^3$ m^3 HC/km^2 (equivalent oil) of hydrocarbons migrated out of the source rock in 50% of the Paraná Basin. The homogeneization of the temperatures measured from oil fluid inclusions in calcite and quartz veins in the intrusion-affected zone range from 60° to 260°C, indicating a large range of capture temperature. This probably resulted from the difference between the maximum level of heat in the source rocks and the cooling that occurred during migration, as well as from the reverse migrations of less thermally evolved hydrocarbons toward the intrusions.

The high efficiency of primary migration verified in the Irati thermally altered zones can be attributed to two factors: (1) the presence of a network of predominantly horizontal microfractures and fractures detected in the examination of thin sections, and (2) the extensive regional and local tectonic compartmentalization that allowed the simultaneous occurrence of both primary and secondary migration.

Large-scale block buoyancy produced by the large volume of intrusions generated complex differential relief causing a convergent flow path. The produced charge was fragmented due the presence of several oil kitchens created by the diachronism of several intrusive pulses with a large range in thickness. The presence of thousands of compartments recorded in the seismic sections indicates short distances of primary lateral migration from the intrusive bodies, estimated to be from 1 to 10 km in distance.

Secondary migration occurred mostly vertically through the compartment boundary faults up to the carrier beds located 100–200 m above the Irati source rocks. The migration of the charge within the carrier beds is interpreted to have occurred along short distances due to the irregular relief created by the differential intrusion thicknesses, the fragmentation of the charge in thousands of primary pulses, and the intense tectonic compartmentalization caused by the igneous intrusions.

The uplift of the margins of the basin, which occurred about 100 Ma, caused the remigration of hydrocarbons from possible traps with consequent loss of a substantial portion of the original charge.

Acknowledgments—*The authors would like to thank Petrobrás, Rio de Janeiro, for permission to publish this paper, Márcio Rocha Mello for the invitation to contribute to this Memoir, and Luiz Fernando De Ros for an early review of the manuscript. Additional thanks to Eugênio V. dos Santos Neto for reviews and suggestions on the text and figures and to Paulo R. Veloso for transforming all colored figures into black and white.*

REFERENCES CITED

Al-Shaieb, Z., J. O. Puckette, A. A. Abdalla, and P. B. Ely, 1994. Three levels of compartmentation within the overpressured interval of the Anadarko Basin, *in* P. J. Ortoleva, ed., Basin compartments and seals: AAPG Memoir 61, p. 69–85.

Araújo, L. M., 1989. Ambientes hidrodinâmicos da Bacia do Paraná—análise preliminar: Internal report, Petrobrás/DEPEX/NEXPAR, Curitiba, 23 p.

Araújo, L. M., A. B. França, and P. E. Potter, 1995, Aqüífero gigante do Mercosul no Brasil, Argentina, Paraguai e Uruguai: mapas hidrogeológicos das Formações Botucatu, Pirambóia, Rosário do Sul, Buena Vista, Misiones e Tacuarembó: UFPR/Petrobrás, Curitiba, 16 p.

Bradley, J. S., and D. E. Powley, 1994, Pressure compartments in sedimentary basins: a review, *in* P. J. Ortoleva, ed., Basin compartments and seals: AAPG Memoir 61, p. 3–27.

Cerqueira, J. R., and E. V. Santos Neto, 1986, Projeto Análise da Bacia do Paraná (Geoquímica Orgânica): Internal report, Petrobrás/CENPES/SINTEP, Rio de Janeiro, 3 vol.

Cerqueira, J. R., and E. V. Santos Neto, 1990, Caracterização geoquímica das rochas geradoras de petróleo da Formação Irati e dos óleos a ela relacionados, Bacia do Paraná, Brasil: II Congresso Latinoamericano de Geoquímica Orgânica, Caracas, Venezuela, p. 26.

Clayton, J. L., and N. H. Bostick, 1986, Temperature effects on kerogen and on molecular and isotopic composition of the organic matter in Pierre Shale near an igneous dike: Organic Geochemistry, v. 10, p. 135–143.

Cooles, G. P., A. S. Mackenzie, and T. M. Quigley, 1986, Calculation of petroleum masses generated and expelled from source rocks: Organic Geochemistry, v. 10, p. 235–245.

Demaison, G., and B. J. Huizinga, 1991, Genetic classification of petroleum systems: AAPG Bulletin, v. 75, p. 1626–1643.

Dennis, L. W., G. E. Maciel, P. G. Hatcher, and B. R. T. Simoneit, 1982, ^{13}C nuclear magnetic resonance studies of kerogen from Cretaceous black shales thermally altered by basaltic intrusion and laboratory simulations: Geochimica et Cosmochimica Acta, v. 43, p. 901–907.

Deroo, G., T. G, Powell, B. Tissot, and R. G. McCrossan, 1977, The origin and migration of petroleum in the Western Canadian sedimentary basin, Alberta; a geochemical and thermal maturation study: Geological Survey of Canada Bulletin, v. 262, 136 p.

De Wit, M. J., and I. D. Ransome, 1992, Regional inversion tectonics along the Southern margin of Gondwana, *in* M. J. De Wit and I. D. Ransome, eds., Inversion tectonics of the cape fold belt, Karoo and Cretaceous basins of Southern Africa. Rotterdan, Balkema, p. 15–21.

England, W. A., 1994, Secondary migration and accumulation of hydrocarbons, *in* L. B. Magoon and W. G. Dow, eds., The Petroleum system—from source to trap: AAPG Memoir 60, p. 211–217.

England, W. A., A. S. Mackenzie, D. M. Mann, and T. M. Quigley, 1987, The movement and entrapment of petroleum fluids in the subsurface: Journal of the Geological Society, v. 144, p. 327–347.

England, W. A., A. L. Mann, and D. M. Mann, 1991, Migration from source to trap, *in* R. K. Merrill, ed., Source and migration processes and evaluation techniques: AAPG Treatise of Petroleum Geology, p. 23–46.

Espitalié, J., F. Marquis, and L. Sage, 1987, Organic geochemistry of the Paris Basin, *in* J. Brooks and K. W. Glennie, eds., Petroleum geology of north west Europe: London, Graham and Trotman, v. 1, p. 71–86.

Ferreira J. C., J. R. Cerqueira, and E. S. T. Frota, 1994, Processo não convencional de geração e migração de petróleo e suas implicações exploratórias: Internal Report, Petrobrás/CENPES/DIVEX/SEGEQ, Rio de Janeiro, 66 p.

Fuzikawa, K., 1994, Microscopia e medidas das temperaturas de homogeneização das inclusões fluidas em vênulas de carbonatos das amostras de sedimentos da Bacia do Paraná, CNEM, Belo Horizonte, 12 p.

Gaarenstroom, L., R. A. J. Tromp, M. C. de Jong, and A. M. Brandenburg, 1993, Overpressure in the Central North Sea: implications for trap integrity and drilling, *in* J. R. Parker, ed., Petroleum geology of northwest Europe: Proceedings of the 4th Conference, Petroleum Geology 86, The Geological Society of London, p. 1305–1313.

Hettich, M., 1981, Arenitos oleígenos da Formação Pirambóia, São Paulo—Segunda Parte: cadastramento regional de ocorrências: Internal Report, GEOSOL/SIX, São Mateus do Sul, 133 p.

Hunt, J. M., 1990, Generation and migration of petroleum from abnormally pressured fluid compartments: AAPG Bulletin, v. 74, p. 1—12.

Lafargue, E., J. Espitalié, T. M. Broks, and B. Nyland, 1994, Experimental simulation of primary migration: Organic Geochemistry, v. 22, p. 575–686.

Larter, S., 1988, Some pragmatic perspectives in source rock geochemistry: Marine and Petroleum Geology, v. 5, p. 194–204.

Larter, S., and I. Horstad, 1992, Migration of hydrocarbon into Brent Group reservoirs: some observations from the Gullfaks Field, Tampen Spur area, North Sea, *in* A. C. Morton, R. S. Haszeldine, M. R. Giles, and S. Brown, eds., Geology of the Brent Group: Geological Society of London Special Publication, v. 61, p. 441–452.

Lelarge, N. L. M. V., 1993, Thermochronologie par la method des traces de fission d'une marg passive (Dome de Ponta Grossa, SE, Brasil): Thesis de docteur de l'Universit Joseph Fourier, Grenoble, France, 244 p.

Macgregor, D. S., and A. S. Mackenzie, 1987, Quantification of oil generation and migration in the Malaca Strait region: Proceedings of 15th Annual Convention of Indonesian Petroleum Association, Jakarta, p. 305–309.

Mackenzie, A. S., and T. M. Quigley, 1988, Principles of geochemical prospect appraisal: AAPG Bulletin, v. 72, p. 399–415.

Mackenzie, A. S., I. Price, D. Leythaeuser, P. Muller, M. Radke, and R. G. Schaefer, 1987, The expulsion of petroleum from Kimmeridge clay source-rocks in the area of Brae field, UK continental shelf, *in* J. Brooks and K. W. Glennie, eds., Petroleum geology of north western Europe: London, Graham and Trotman, p. 865–877.

Magoon, L. B., and W. G. Dow, 1994, The petroleum system, *in* L. B. Magoon and W. G. Dow, eds., The petroleum system—from source to trap: AAPG Memoir 60, p. 3–24.

Marques, A., O. A. Zanotto, O. B. Paula, M. A. M. Astolfi, A. B. França, and E. A. Barbosa, 1993, Compartimentação tectônica da Bacia do Paraná: Internal Report, Petrobrás/DEPEX/NEXPAR, Curitiba, 26 p.

Mendonça Filho, J. G.,1994, Estudo petrográfico e organogeoquímico de amostras de folhelhos da Formação Irati, Permiano Superior da Bacia do Paraná, no Estado do Rio Grande do Sul: Tese de Mestrado, IG-UFRGS, Porto Alegre, 212 p.

Milani, E. J., 1992, Intraplate tectonics and the evolution of the Paraná Basin, SE Brazil, *in* M. J. De Wit and I. D. Ransome, eds., Inversion tectonics of the cape fold belt, Karoo and Cretaceous basins of southern Africa: Rotterdam, Balkema, p. 101–108.

Milani, E. J., 1997, Evolução Tecto-Estratigráfica da Bacia do Paraná e seu Relacionamento com a Geodinâmica Fanerozóica do Gondwana Sul-Ocidental: Tese de doutorado em Geociências, Instituto de Geociências, Universidade Federal do Rio Grande do Sul, RS, Porto Alegre, 255 p.

Ortoleva, P. J, 1994, Basin compartments and seals: AAPG Memoir 61, 477 p.

Pepper, A. S., 1991, Estimating the petroleum expulsion behaviour of source rocks: a novel quantitative approach, *in* W. A. England and A. J. Fleet, eds., Petroleum migration: Geological Society Special Publication 59, p. 9–31.

Peters, K .E., J. K. Whelan, J. M. Hunt, and M. E. Tarafa, 1983, Programmed pyrolysis of organic matter from thermally altered Cretaceous black shales: AAPG Bulletin, v. 67, p. 2137–2146.

Raymond, A. C., and D. G. Murchison, 1992, Effect of igneous activity on molecular-maturation indices in different types of organic matter: Organic Geochemistry, v. 18, p. 725–735.

Sad, J. H. G., and N. T. A. Saraiva, 1982, Rochas oleígenas da Formação Irati: Internal Report, Geosol-Geologia e Sondagens Ltda e Superintendência da Industrialização do Xisto.

Saxby, J. D., and L. C. Stephenson, 1987, Effect of an igneous intrusion on oil shale at Rundle (Australia): Chemical Geology, v. 63, p. 1–16.

Shepherd, L. D., P. A. Drzewieckie, J. M. Bahr, and J. A.Simo, 1994, Silica budget for a diagenetic seal, *in* P. J. Ortoleva, ed., Basin compartments and seals: AAPG Memoir 61, p. 369–385.

Silva, R. B. G., 1983, Estudo hidroquímico e isotópico das águas subterrâneas do aqüífero Botucatu no Estado de São Paulo: Tese de Doutorado, USP, São Paulo, 133 p.

Simoneit, B. R. T., S. Brenner, K. E. Peters, and I. R. Kaplan, 1978, Thermal alteration of Cretaceous black shale by basaltic intrusions in the eastern Atlantic: Nature, v. 273, p. 501–504.

Simoneit, B. R. T., S. Brenner, K. E. Peters, and I. R. Kaplan, 1981, Thermal alteration of Cretaceous black shale by diabase intrusions in the Eastern Atlantic—II. Effects on bitumen and kerogen: Geochimical et Cosmochimica Acta, v. 45, p. 1581–1602.

Sluijk, D., and M. H. Nederlof, 1984. Worldwide geological experience as a systematic basis for prospect appraisal, *in* G. Demaison and R. J. Murris, eds., Petroleum geochemistry and basin evaluation: AAPG Memoir 35, p. 15–26.

Sokolov, V. A., A. A. Geodekyan, C. G. Grigoryev, A. Ya. Krems, V. A. Stroganov, L. M. Zorkin, M. I. Zeidelson, and S. J. Vainbaum, 1971, The new methods of gas surveys, gas investigations of wells and some practical results, *in* R. W. Boyle, ed., Geochemical exploration: Canadian Institute of Mining and Metallurgy Special Volume 11, p. 538–544.

Surdan, R. C., Z. S. Jiao, and R. S. Martinsen, 1994, The regional pressure regime in Cretaceous sandstones and shales in the Powder River Basin, *in* P. J. Ortoleva, ed., Basin compartments and seals: AAPG Memoir 61, p. 213–235.

Trigüis, J. A., 1986, An organic geochemistry investigation of heat-affected sediments in the Paraná Basin (Brasil): Ph.D. Thesis, University of Newcastle, Newcastle, U.K., 203 p.

Turner, S., M. Regelous, S. Kelley, C. Hawkesworth, and M. Mantovani, 1994, Magmatism and continental break-up in the South Atlantic: high precision Ar/Ar geochronology: Earth and Planetary Science Letters, Amsterdam, v. 121, p. 333–348.

Zanotto, O. A., 1993, Erosão pós-Cretáceo na Bacia do Paraná, com base em dados de reflectância da vitrinita: Simpósio Sul-brasileiro de Geologia, 5, Sociedade Brasileira de Geologia, Resumos, Curitiba, p. 58.

Bushnell, D. C., J. E. Baldi, F. H. Bettini, H. Franzin, E. Kovas, R. Marinelli, and G. J. Wartenburg, 2000, Petroleum systems analysis of the eastern Colorado Basin, offshore northern Argentina, *in* M. R. Mello and B. J. Katz, eds., Petroleum systems of South Atlantic margins: AAPG Memoir 73, p. 403–415.

Chapter 27

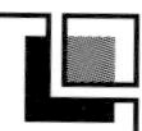

Petroleum Systems Analysis of the Eastern Colorado Basin, Offshore Northern Argentina

David C. Bushnell

YPF/Maxus
Dallas, Texas, U.S.A.
Currently: Consultant
Dallas, Texas, U.S.A.

Jorge E. Baldi

Repsol YPF
Buenos Aries, Argentina
Currently: Repsol YPF
Rio de Janeiro, Brazil

Fernando H. Bettini

Humberto Franzin *(Retired)*

Repsol YPF
Buenos Aries, Argentina

Edward (Ned) Kovas

Repsol YPF
Dallas, Texas, U.S.A.
Currently: Repsol YPF
Buenos Aires, Argentina

Raúl Marinelli *(Retired)*

Repsol YPF
Buenos Aries, Argentina

Grant J. Wartenburg

Maxus (subsidiary of Repsol YPF)
Dallas, Texas, U.S.A.
Currently: Repsol YPF
The Woodlands, Texas, U.S.A.

Abstract

The Colorado Basin is a large (100,000-km^2), deep (15,000-m) Cretaceous and Tertiary clastic sedimentary basin located offshore from southern Buenos Aires Province, Argentina. Static components of the petroleum systems—source, reservoir, seal, and overburden—are either present or can be reasonably predicted from well and seismic data. However, the best structural traps are buried beneath 7000 m of sedimentary rocks. Basin-edge stratigraphic traps are suggested in the geometry of the seismic data but are difficult to delineate stratigraphically. Exploratory drilling has so far been unsuccessful.

It is the timing of the dynamic petroleum system factors that has negatively impacted the success of the potential petroleum systems in the basin. A 4000-m Cretaceous sandstone section has pushed the most likely source rock candidate completely through the oil window before deposition of the regional seal. Any hydrocarbons that might have been expelled would have migrated vertically to the surface and been lost. Migration paths, although simple and easily delineated, tend to be dispersive near the central basin but focus on broad areas of the basin flanks.

INTRODUCTION

This chapter represents part of a large, multidisciplinary study of the offshore northwestern Argentina basins performed jointly by the exploration staff of Yacimientos Petrolíferos Fiscales (YPF) in Buenos Aires and YPF/Maxus in Dallas during 1995 and 1996 (YPF/Maxus staff, 1996). The Colorado Basin is one of a string of extensional basins lining the Argentine and Brazilian margins of the southwestern Atlantic Ocean (Uliana et al., 1989; Marinelli et al., 1996) (Figure 1). It is located off the coast to the south of Buenos Aires Province, having dimensions of 200 × 500 km and containing more than 15,000 m of Cretaceous and Tertiary clastic strata (Figures 2 and 3). An additional several thousand meters of Paleozoic clastics subcrop beneath the basin as an extension of the Claromecó Basin located onshore to the north of the Colorado Basin (Juan et al., 1996).

It is not certain whether any petroleum systems exist in the Colorado Basin that could yield commercial hydrocarbon deposits, as exploration to date has yielded largely negative results. Rather than describing known petroleum systems, this study examines the potential petroleum systems that might exist or may have existed in the eastern offshore part of the basin. As a conceptual guide, we use a series of related papers from AAPG Memoir 60: Blanc and Connan (1994), Demaison and Huizinga (1994), Deming (1994), Downey (1994), Magoon and Dow (1994), and Smith (1994).

STATIC PETROLEUM SYSTEM COMPONENTS

All of the static components of a family of potential petroleum systems in the Colorado Basin have been either proved or reasonably hypothesized from available seismic and well data: abundant reservoirs, regional seals, source rocks, overburden, and basin margin structural and stratigraphic traps (Keeley and Light, 1993; YPF/Maxus staff, 1996). Components of these potential petroleum systems are shown diagrammatically in Figure 4 and are reviewed individually in the following subsections. The regional stratigraphy of the source and reservoir rocks is thoroughly covered in a well-referenced paper by Fryklund et al. (1996).

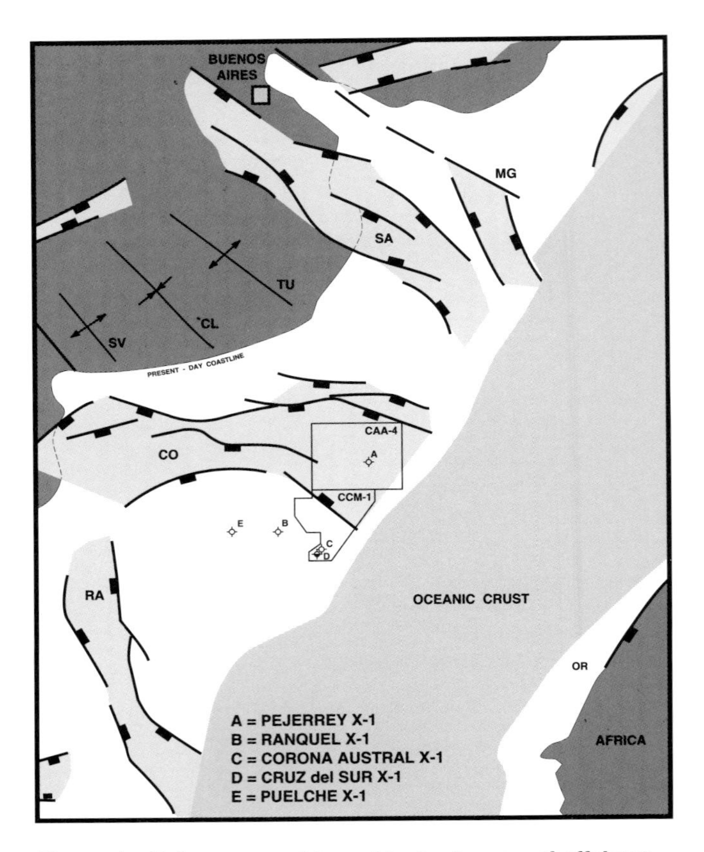

Figure 1—Paleogeographic and tectonic map of offshore northern Argentina at the end of Jurassic time. Tectonic features (in light gray) influenced deposition during the Cretaceous. Exploration blocks are outlined and wells are listed for the study area in the eastern Colorado Basin (CO) in the center of the figure. Other abbreviations: MG, Martin Garcia High; SA, Salado Basin; TU, Tandil Uplift; CL, Claromeco Basin; SV, Sierra de la Ventana; Rawson Basin; OR, Orange Basin. After Uliana (1989).

Source Rocks

Proving the existence of an adequate source rock with optimal timing of generation and expulsion is one of the most difficult challenges in frontier basin exploration, and the Colorado Basin proves no exception. However, a few meters of oil saturation in Cretaceous reservoirs below 3000 m in the Cruz del Sur x-1 well indicate that at least some source rocks exist and that they have generated and expelled petroleum. In addition, oil-prone source rocks were described from the same well, and gas-prone source rocks were seen in the Puelche x-1 well. It is these limited oil shows and source rock occurrences that have provided much of the impetus for our search for a petroleum system in the basin that might be capable of generating and trapping major hydrocarbon reserves. Some of the most likely source rock candidates are discussed below.

Paleozoic Source Rocks

The possibility of a Paleozoic oil or gas source in the Colorado Basin has been recognized since the penetration of kerogen-rich Permian shales by the Puelche x-1 well (west of the study area) in 1977 (Yacimientos Petrolíferos Fiscales, 1977) (see Figure 1 for well locations). In addition, some gas-prone Permian shales were seen in the Cruz del Sur x-1 well (Fryklund, 1994–1995). This subcropping Paleozoic section appears to be an offshore extension of the Claromecó Basin of southern Buenos Aires Province, where organic lacustrine shales have been reported in the Permian–Carboniferous Pillahuinco Group (Keeley and Light, 1993). Unfortunately, the upper Paleozoic section cannot be readily mapped in the study area. It occurs mainly as isolated fault blocks subcropping the Colorado Basin rift and sag sequences with highly angular unconformities. Seismic expression of these subcropping fault blocks is well illustrated in Juan et al. (1996, their figures 4, 6, 7, and 8). None of the remnants appear large enough to charge a major petroleum accumulation, and their structural complexity suggests that any hydrocarbons would have likely been generated and expelled in a previous tectonic cycle.

Synrift Lacustrine Shales

Rich lacustrine source shales and marls were deposited during the Early Cretaceous in several of the southern Atlantic margin basins (Tissot et al., 1980). The Jiquia Formation of the Espírito Santo Basin (Estrella et al., 1984)

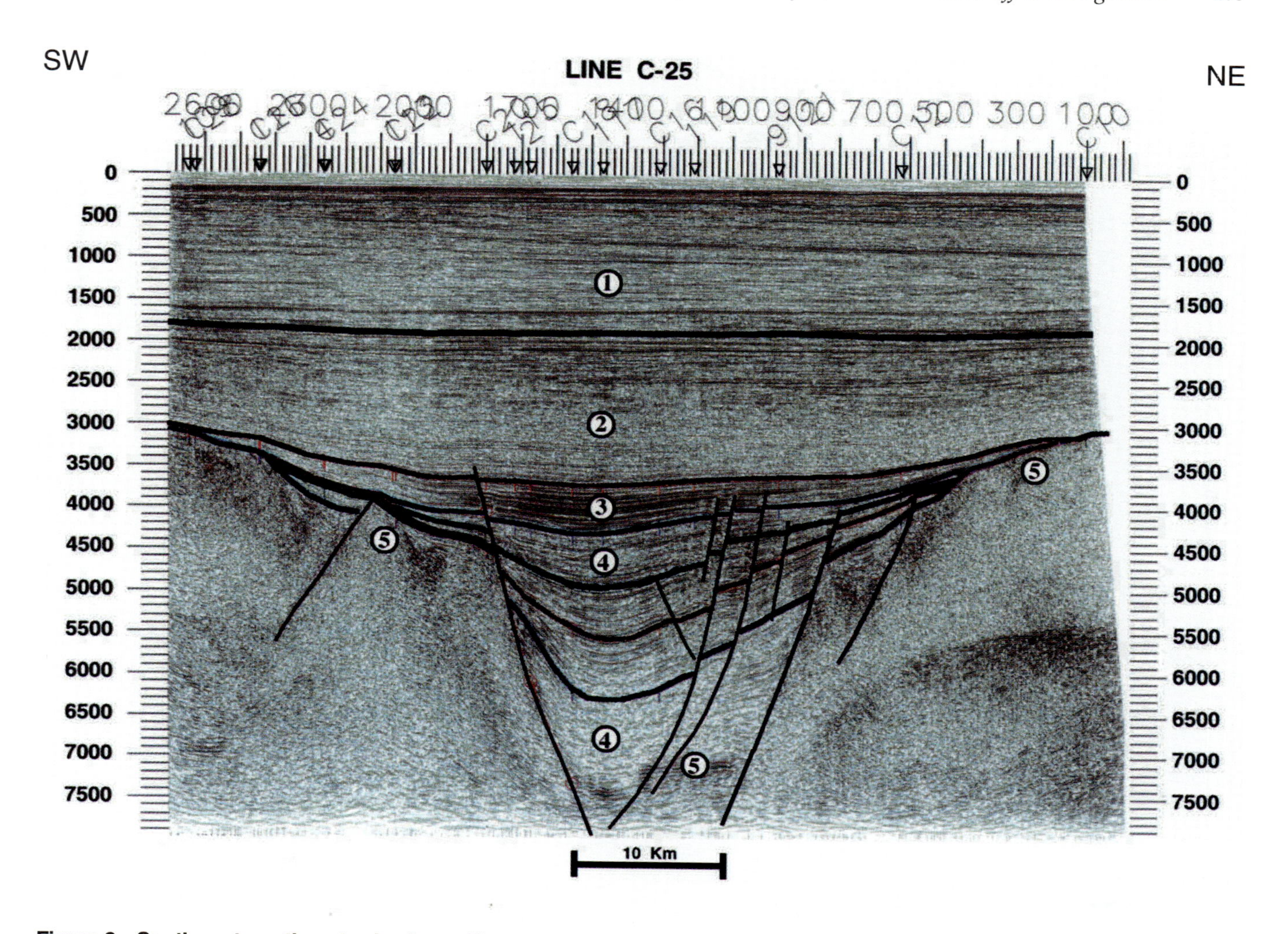

Figure 2—Southwest–northeast seismic profile across the central graben of the eastern Colorado Basin (see Figure 3 for location). Vertical scale is in milliseconds. Key: (1) Tertiary passive margin shale seal; (2) Upper Cretaceous Colorado and Fortin Formations; (3) Aptian(?) hypothetical source rocks; (4) synrift deposits; (5) prerift section.

and the Lagoa Feia Formation of the Campos Basin (Guardado et al., 1989) serve as a type examples from farther north in Brazil. It is possible that coeval lacustrine facies may have developed in the Early Cretaceous central trough of the Colorado Basin, but because of its great depth, they have never been penetrated. Seismic resolution is poor, but some strong continuous reflectors can be seen in Figure 2 below 4.5 sec (about 9000 m) which may represent a deep lacustrine facies within the graben. However, the reflectors also may indicate evaporites or other lithologies. Fryklund et al. (1996) described a bed of mature source rocks penetrated by the Cruz del Sur x-1 well that may be of synrift origin, although in a smaller marginal basin. The rocks average 2.4% TOC and have an hydrogen index (HI) of 500.

Basal Sag Sequence

By far the most promising source rock candidate is the basal "sag" sequence which occurs in the central part of the basin (Tissot et al., 1980). It neither crops out nor is penetrated by wells, but it is well represented on 2-D seismic profiles. It can be seen in Figure 2 at about 4.0 sec two-way traveltime, or 7000 m depth. It is characterized by a thick package of very strong, laterally continuous reflectors that show little change in character or thickness over tens of kilometers, suggesting a widespread, low-energy depositional environment. This unit onlaps the breakup unconformity before reaching the area of the exploratory wells. However, because the unconformably overlying Albian Fortin Formation is of marginal marine origin at the basin margin, the basal sag sequence is believed to be a marine deposit, possibly of Aptian age. Whether the sequence is of deep- or shallow-marine origin cannot be definitely determined, but there is no seismic stratigraphic evidence of a deep marine basin. The strong reflectors may be organic shales or anhydrites. Both lithologies are known from the Aptian sequence of the southern Atlantic (Foresman, 1978).

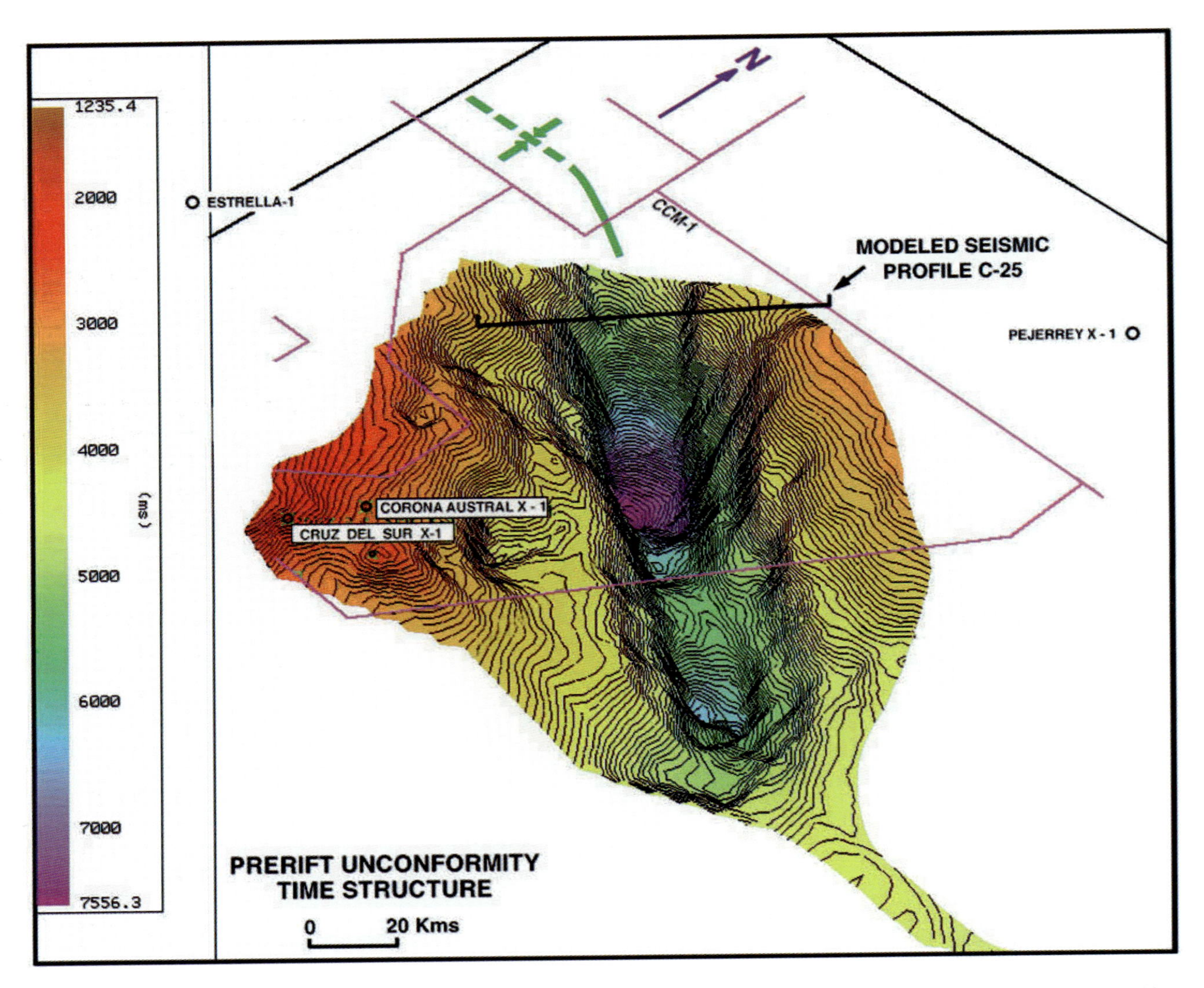

Figure 3—Structural relief map of the eastern Colorado Basin. Time structural contours are drawn on the prerift (pre-Upper Jurassic) unconformity. Note the location of the seismic profile in Figure 2.

Fortin Formation

A second widespread stratigraphic unit that contains potential source rocks is in the basinal extension of the Fortin Formation. This formation is dominated by sandstone where it has been penetrated on the south flank of the basin (Fryklund et al., 1996), but it includes some paralic to neritic shales. It thickens from 100–250 m along the southern basin margin to more than 1000 m in the basin depocenter (Figure 2), where some limited source facies could have developed. Although there is little direct seismic stratigraphic support for a source rock in the basin center, the Fortin Formation includes the Turonian–Cenomanian time–rock sequence that comprises rich marine source rocks along much of the southern Atlantic margin (Tissot et al., 1980). Consequently, this hypothetical source was examined from a petroleum systems standpoint to determine if it could be the source for the oil shows in the Cruz del Sur x-1 well.

Pedro Luro Formation

Although the Pedro Luro and Barranca Final Formations are predominantly bathyal shales, only the Pedro Luro has any chance of being thermally mature. In the Corona Austral x-1 well, TOC measurements of up to 1.35% indicate source potential (Fryklund, 1994–1995), and this facies could become even richer in kerogen in the depocenter.

Reservoir Rocks

Sandstones, many of which are of reservoir quality, occur in most formations in the eastern Colorado Basin. Reviews of the regional stratigraphy of these units are provided by Fryklund et al. (1996) and Juan et al. (1996). The summaries below are derived mainly from well reports (Yacimientos Petroliferos Fiscales, 1977; Fryklund, 1994–1995) and focus on the specific roles of the sand-

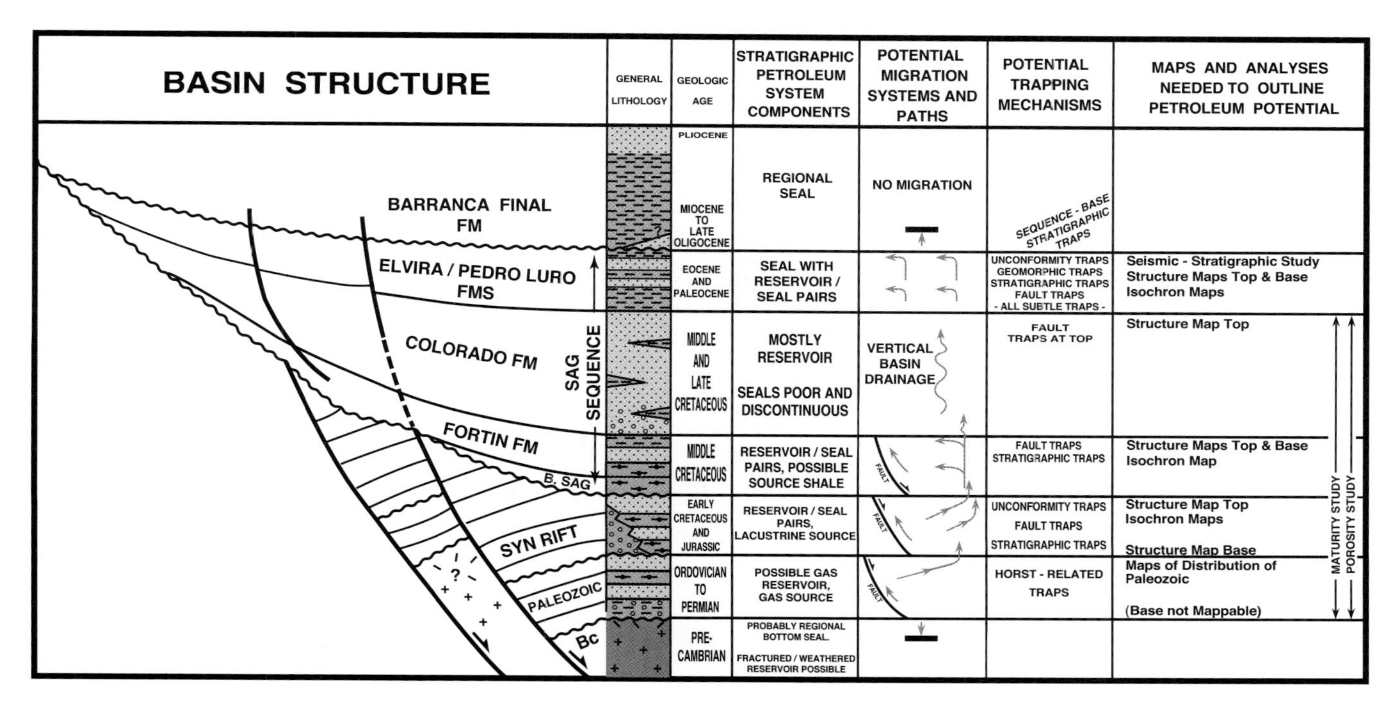

Figure 4—Chart showing the basin structure, stratigraphy, petroleum system components, and potential exploration plays of the Colorado Basin.

stones as reservoirs and migration systems in the petroleum systems of the eastern Colorado Basin. They are presented here in order of deposition.

Paleozoic and Basement Subcrop

The thick Paleozoic section of the Claromecó Basin, discussed above as a potential source rock, is well exposed in the Sierra de la Ventana Mountains just to the north of the Colorado Basin (Juan et al., 1996). In those extensive outcrops, the authors observed little porosity in the highly indurated quartzitic sandstones and diamictites. Pre-Cretaceous sandstones were penetrated in all of the offshore wells in the eastern Colorado Basin except Ranquel x-1 and have been variously described as hard, quartzitic, quartz cemented, and having low porosity and permeability (Yacimientos Petrolíferos Fiscales, 1977; Fryklund, 1994–1995). While the pre-Cretaceous sandstones may provide a poor-quality gas reservoir, it is doubtful that they would yield commercial oil. However, in areas where the subcrop unconformity is overlain by shale, sufficient permeability may exist for short-range petroleum migration.

Synrift Section

No wells penetrate the rift sedimentary rocks of the central graben of the eastern Colorado Basin, but sandstones of probably equivalent age were drilled in smaller flanking fault blocks in both the Corona Austral x-1 and Cruz del Sur wells. In the latter, a 280-m-thick section of sandstone was penetrated that had as much as 15-20% log porosity and measurable oil saturations (Fryklund, 1994–1995). Similar sandstones can be expected in other Cretaceous fault blocks in the area and constitute reservoir objectives. In the Cruz del Sur well, sandstones with little observed porosity were found interbedded with andesites and tuffs.

A similar mix of clastic and volcanic lithologies is likely to be present in the central graben buried to depths greater than 7000 m with temperatures in excess of 250°C. Modeled sandstone porosity is less than 2% (YPF/Maxus staff, 1996). These rocks may have been effective reservoirs and migration systems during the Late Cretaceous before deep burial.

Fortin and Colorado Formations

Coarse clastics of mostly nonmarine origin are the dominant lithology in the Colorado and Fortin Formations of the sag sequence. In the Estrella x-1, Cruz del Sur x-1, and Corona Austral x-1 wells, these Upper Cretaceous units consist mainly of continental to shallow marine sandstones and conglomerates, but minor marine shale intercalations are described in well reports (Yacimientos Petrolíferos Fiscales; 1977 Fryklund, 1994–1995). Although the sandstones are generally described as texturally and mineralogically immature, porosities of up to 32% are reported. The thickness of the sequence ranges from 700 to 1100 m in the wells and thickens to nearly 4000 m in the basin center (Figure 2). Because of the general paucity of seals in the Upper Cretaceous section, effective reservoirs and migration systems are likely to be found only at the top of the Colorado Formation, beneath the regional Pedro Luro seal, and in the Fortin Formation, where local reservoir–seal pairs may occur.

Elvira Formation

Fine-grained, glauconitic, marine sandstones are found in the lower part of the Eocene Elvira Formation in the wells (Figure 4). They lie on the unconformity that separates the Elvira and Pedro Luro Formations, but in the eastern Colorado Basin, they appear to represent a low-energy, deeper water facies. Unfortunately, these sandstones appear to be isolated from the deeper petroleum migration pathways by the underlying Pedro Luro shales.

Seal Rocks

Above the massive reservoirs of the Upper Cretaceous, the section becomes totally marine and consists mainly of bathyal siltstones and claystones more than 2000 m thick. This unit includes the Pedro Luro, Elvira, and Barranca Final Formations and forms an unbroken regional seal over the entire eastern Colorado Basin.

Nonmarine to paralic, intraformational shale seals are also observed in the Colorado and Fortin Formations, but they are thin and less likely to be continuous. On a regional scale, they may only slightly impede vertical petroleum migration. However, the basal sag sequence appears to be a regional seal over the basin axis (Figures 2 and 4) whether it is a shale or an evaporite.

The igneous basement and Paleozoic clastics of the prebasinal terrane (Juan et al., 1996) serve as an effective bottom seal in a regional sense. There may, however, be local areas of good permeability related to weathering or fracturing immediately beneath the unconformity. Where sealed by overlying shales, these may provide reservoirs or migration systems of limited extent.

Overburden Rocks

Deming (1994, p. 165) defines overburden rock as "... that series of mostly sedimentary rock that overlies the source rock, seal rock, and reservoir rock." In the Colorado Basin, it is the reservoir and seal rocks that actually constitute the overburden rock, and this displacement from the classic position contributes to the problem in maturity and timing discussed later. The entire Upper Cretaceous and Tertiary sequence provides an overburden that is more than 6000 m thick over the hypothetical Lower Creta-ceous source rock. Because of this great thickness of overburden, it becomes necessary to examine the thermal history of the hypothetical source rocks at several stages of burial.

Traps

The Lower Cretaceous extensional tectonism that created the deep central graben had ceased by Aptian time and was never rejuvenated in the eastern Colorado Basin (Figure 2). Seismic data show that the extremely thick sag and passive margin sequences are almost bereft of structural traps in the central basin near the hypothetical source rocks. A few localized, poorly defined structural traps on the basin flanks have been drilled with very limited success. The oil shows that were logged and wireline tested in Cruz del Sur x-1 (Fryklund, 1994–1995) are the only positive indication of a working petroleum system. On the northern flank, a regional nose and migration focus was tested unsuccessfully by the Pejerrey x-1 well, which found abundant reservoirs but no hydrocarbon shows. Suggestions of stratigraphic trap geometry can be seen in the seismic data, especially on the basin flanks, but the lithologic components of these potential traps are as yet unknown. Although potential stratigraphic onlap trap geometry can be mapped seismically on the basin flanks, there is little suggestion in the wells or seismic data of the necessary lateral seals to make these traps effective.

Potential fault trap geometry is evident in seismic data deep in the central graben beneath the hypothetical source rock (Figures 2 and 4). Although the section is nowhere exposed and has never been penetrated, strong reflectors suggest the possibility of reservoir–seal pairs and the potential for fault traps on both sides of the graben. However, that section is now buried to depths in excess of 7000 m and is considered no longer prospective.

DYNAMIC PETROLEUM SYSTEM COMPONENTS

Maturity and Timing

Maturity modeling was done using the Platte River 1-D (PC platform) and 2-D (UNIX platform) basin modeling programs. Modern heat flow was estimated using the stratigraphy and temperature logs from two wells on the south flank of the basin. One well, the Cruz del Sur x-1, supplied abundant maturity data, mainly in the form of vitrinite reflectance measurements. These data were supplemented, and the results corroborated, by apatite fission track analyses performed on a few key samples. Figure 5 shows the burial history diagram of this well. It also illustrates an interpretation of the bathymetric history based on stratigraphic facies, biostratigraphic data, and the sea level curves of Haq et al. (1988). During the Tertiary, deposition did not keep pace with subsidence, resulting in bathyal conditions and consequently a cold sedimentary surface. Only during the Pliocene–Pleistocene was the basin filled to its present neritic depths. It can be seen in Figure 5 that the shallowing of the sea floor from 1000 m to its present depth was caused by rapid sediment fill, not tectonic uplift. The bathyal sediment surface temperature for maturity modeling was estimated using water temperature versus latitude and depth diagrams for the Atlantic Ocean (Williams, 1962).

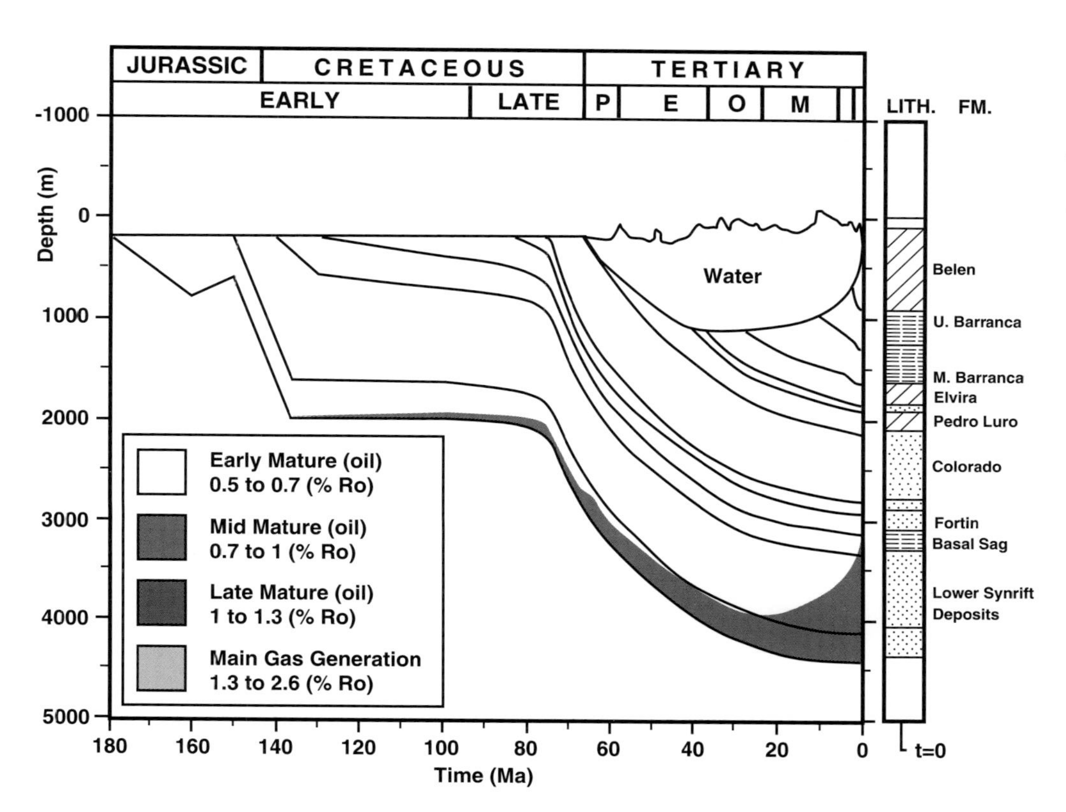

Figure 5—Schematic diagram showing the burial history of the Colorado Basin and the thermal maturity windows based on data from the Cruz del Sur x-1 well.

An excellent match was found between predicted and observed maturity in this well, so the calculated heat flow of 56.2 mW/m^2 was used as a reference for the basin. We recognize that using a single control point to represent the thermal history of a large extensional basin is less than ideal. Because the well is in a basin flank position 60 km from the edge of the rift, it may record a temperature history that is cooler than that closer to the rifting activity. Maturity estimates may therefore err significantly; generation and expulsion may have initiated earlier than portrayed because of higher heat flow.

The 2-D maturity modeling reveals a critical timing problem in the petroleum system dynamics. Maturity models were created for three critical times in the geologic history of the basin to illustrate this problem. First, Figure 6 shows a north–south profile modeling the basin at 70 Ma, in Maastrictian time. During the Maastrictian, the Aptian source rock candidate was apparently at peak maturity, but the regionally sealing Pedro Luro shales had not yet been deposited. Any oil generated and expelled is likely to have migrated vertically through the sandstones of the Fortin and Colorado Formations to the surface and lost. Although there is a possibility of local seals within the Fortin Formation, it is doubtful that these would have been sufficiently continuous to channel hydrocarbons 30–60 km to the basin margins where trapping may have been possible. The basal sag sequence, itself a possible seal in the central basin graben, does not reach the basin edges.

In Figure 7, the maturity model reflects conditions at 43 Ma, during the middle Eocene. This display was generated to reflect conditions immediately after deposition and partial compaction of the regional shale seal of the Pedro Luro Formation. At this time, the Fortin Formation was apparently in the peak generation window. As previously mentioned, the Fortin has at least hypothetical source potential because of its Albian age, but there is little suggestion in the seismic data of the presence of extensive strata deposited in low-energy environments. An important observation here is that at the time when the petroleum system gained a seal atop the main migration system, the principal source rock candidate in the basal sag sequence had already passed through the oil window. It is unlikely, therefore, that much of the oil that might have been generated in that hypothetical source rock was preserved. Gas generation and expulsion, however, may be continuing to the present day.

The final model (Figure 8) was generated for the basin at the present time. A regional seal is apparently present at the base of the Pedro Luro and younger shales, and any petroleum reaching that seal is most likely diverted toward the basin margins. The modern oil generation window is thought to be centered in the upper Colorado Formation, which is seen in the wells mostly as a thick continental sandstone with no source potential. However, the possibility of an Upper Cretaceous marine invasion is shown.

Although the most prospective potential petroleum systems of the Colorado Basin appear to fail for oil accumulations because of timing factors, continued gas generation and migration is likely to be occurring over longer time spans.

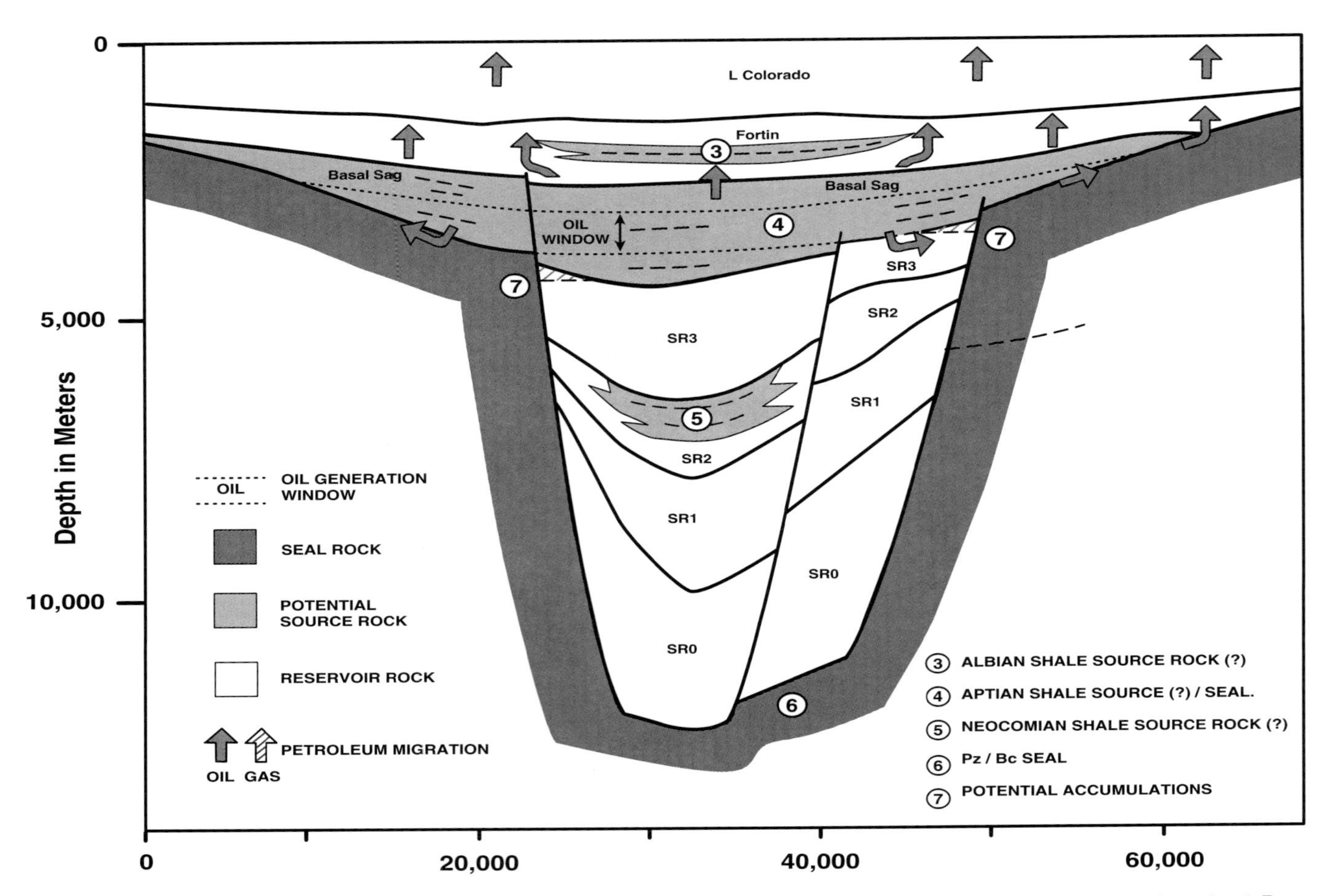

Figure 6—North–south thermal maturity model of the Colorado Basin during Maastrichtian time (70 Ma) based on the 2-D seismic profile in Figure 2.

Lateral Migration

The base of the regional Pedro Luro shale seal has provided the major drainage control in the Colorado Basin since its deposition in the Paleocene. There appear to be few impediments to long-distance migration at the top of the Colorado Formation. Figure 9 shows the present-day structure on this surface along with the drainage pattern that should result based on a seismic time map at the top of the overlying Pedro Luro Formation. Dashed arrows in Figure 9 show predicted lateral migration directions and originate at 10-km intervals along the basin axis. These lateral migration patterns are largely dispersive in the central basin, but they show a broad focus over a 40-km stretch along the southern edge of the mapped area. Both the Corona Austral x-1 and Cruz del Sur X-1 wells are located well away from this migration focus. On the north flank, a large, low-relief structural nose is visible near the Pejerrey x-1 well which focuses drainage from most of the north half of the basin.

In Figure 10, the present-day migration pattern is shown for the Fortin Formation based on a seismic time map on the top Fortin reflector. The pattern is similar to that at the base of the Pedro Luro except that it is more pronounced. Migration foci are more sharply defined along the southern edge of the study area. On the northern flank, virtually all of the potential migration paths converge on the subtle structural nose at the Pejerrey well. However, the regional continuity of seals within the Fortin Formation is uncertain at best, and vertical "leakage" toward the base of the Pedro Luro Formation can be expected. Long-distance hydrocarbon migration to the southern edge of the basin and to the Pejerrey structural nose is considered unlikely to prevail in the Fortin Formation.

A Paleocene–Eocene migration map for the Fortin Formation was constructed using an isochron map of the interval between the top of the Eocene Pedro Luro Formation and the top of the Fortin Formation (Figure 11). This map simulates Fortin structure at the end of Paleocene time, with some inaccuracies due to seismic velocity variations. Because of the regular and symmetric nature of Tertiary tectonic subsidence, predicted patterns of migration appear little changed from the present-day pattern at that level.

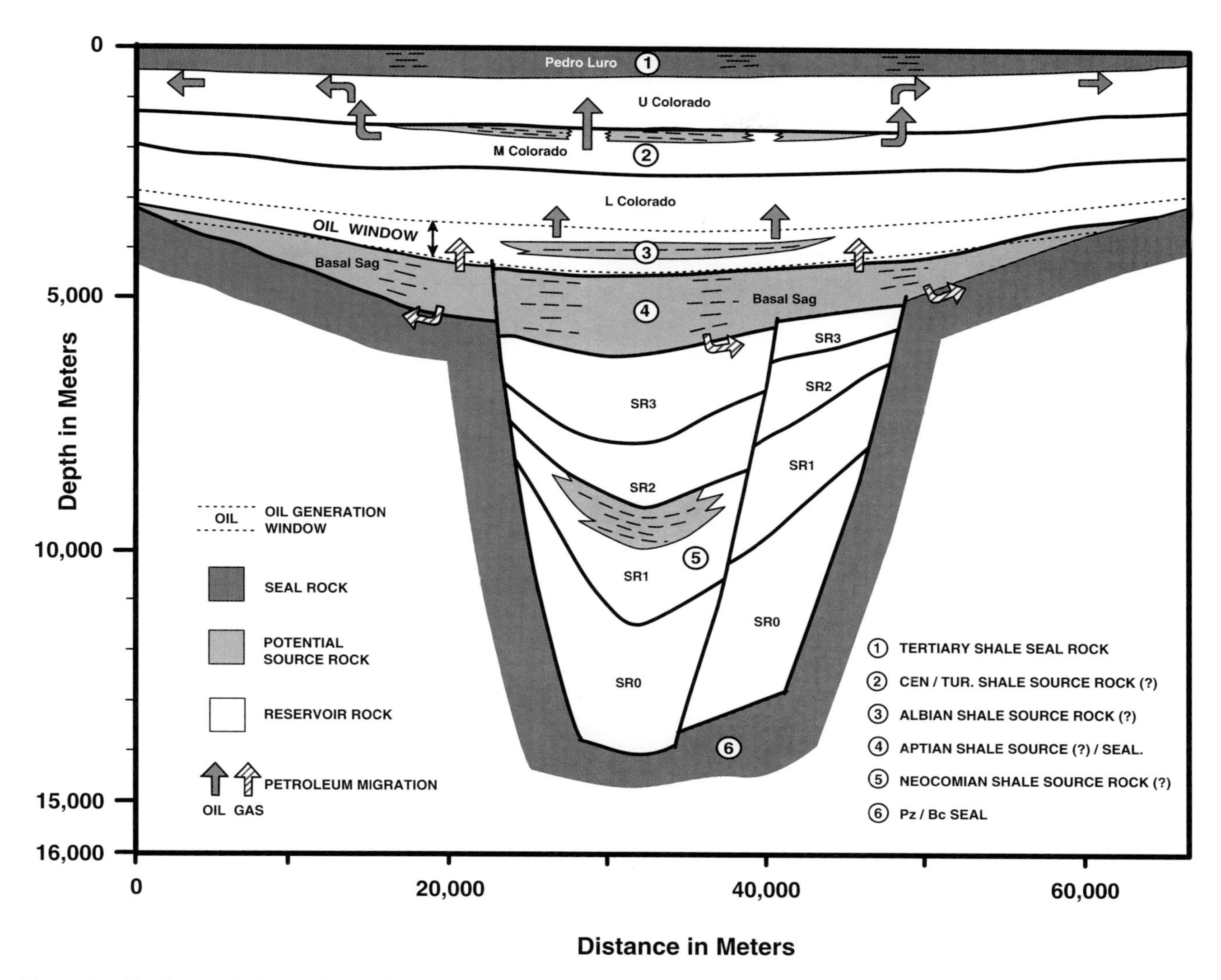

Figure 7—North–south thermal maturity model of the Colorado Basin during middle Eocene time (43 Ma) based on the 2-D seismic profile in Figure 2.

Preservation

During the Late Cretaceous and Tertiary passive margin phase, there were no tectonic disturbances and no significant erosional events to impact the preservation of any hydrocarbon accumulations that may have formed. However, immediately after deposition of the Colorado Formation reservoir sequence in the Late Cretaceous, the western part of the study area was blanketed by tuffs and basalts (Yacimientos Petrolíferos Fiscales, 1977; YPF/Maxus staff, 1996). The volcanic event may have affected the preservation of any petroleum that had been trapped in the area at that time. Preservation is probably poor in the structural traps in the central rift of the basin because that section is presently at temperatures in excess of 300°C. These problems are diagrammed in the petroleum system events chart of Figure 12.

CONCLUSIONS

The exploration prospectivity of the eastern Colorado Basin can be analyzed in terms of static and dynamic components of the petroleum system. Previous drilling showed the existence of static components that attracted modern hydrocarbon exploration. Sandstone reservoirs are abundant in at least three formations. The basin is blanketed by a shale seal, and some intraformational seals are also present. However, the geometry and stratigraphy of hydrocarbon traps are poorly developed except at great depths. Although minor oil and gas shows have been reported, the most promising high-volume source rock has only been inferred on the basis of seismic data. Even if geochemical evidence existed for a source, the dynamic factors prove to be the most serious weaknesses of the petroleum system. Maturity modeling shows the

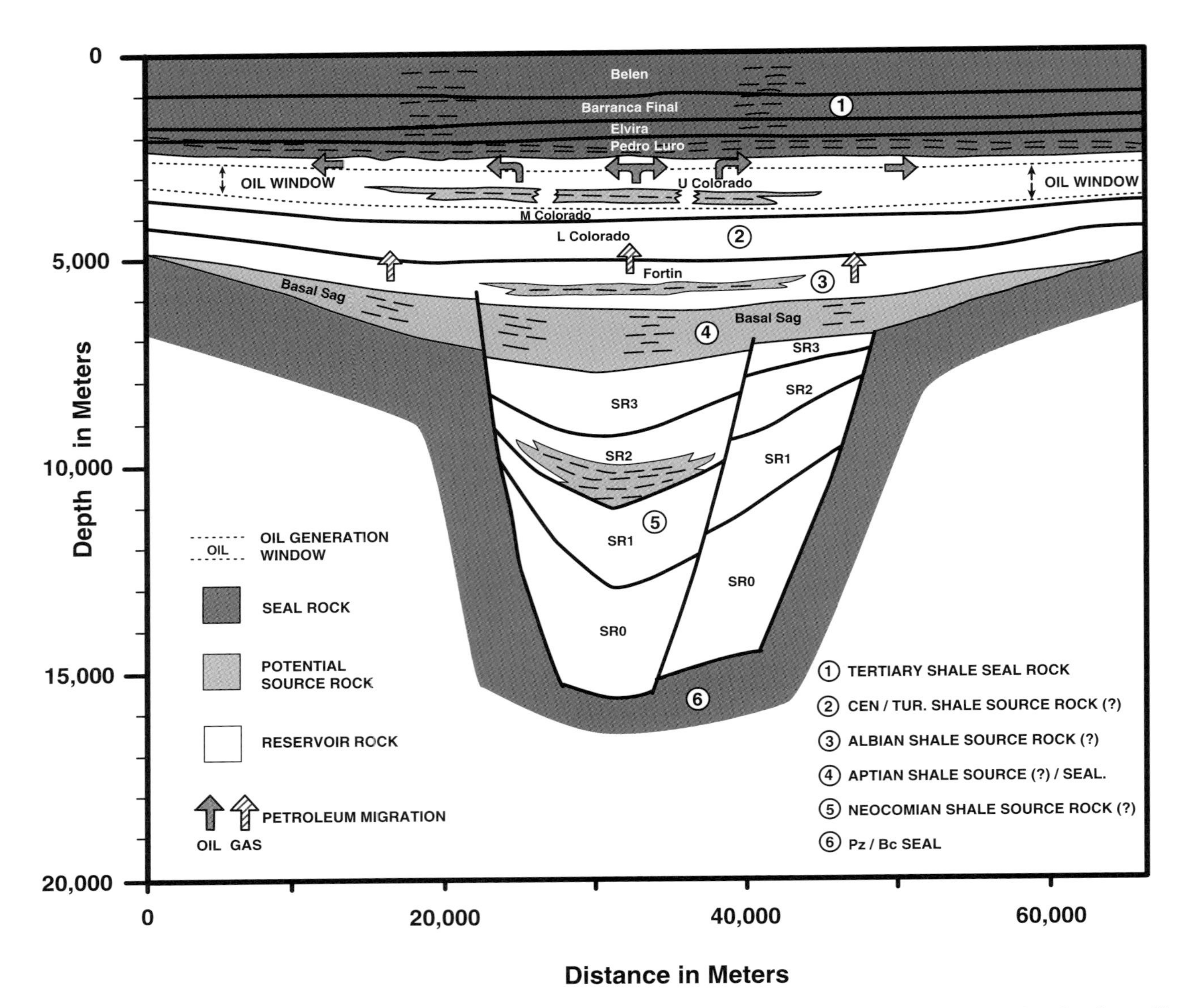

Figure 8— North–south thermal maturity model of the Colorado Basin for the present day based on the 2-D seismic profile in Figure 2.

prospective source rock to have passed through the oil window before deposition of the regional seal, so that any expelled oil would have been lost to the surface. Although clear migration patterns can be drawn, most migration is dispersive to the basin flanks, where structural traps are almost lacking and stratigraphic traps poorly defined. A possible exploration direction might involve stratigraphic gas prospects along the basin margins, but the lack of convincing direct hydrocarbon indicators has so far discouraged this approach.

REFERENCES CITED

Blanc, P., and J. Connan, 1994, Preservation, degradation, and destruction of trapped oil, *in* L. G. Magoon and W. G. Dow, eds., The petroleum system—from source to trap: AAPG Memoir 60, p. 237–247.

Demaison, G., and B. J. Huizinga, 1994, Genetic classification of petroleum systems using three factors: charge, migration, and entrapment, *in* L. G. Magoon and W. G. Dow, eds., The petroleum system—from source to trap: AAPG Memoir 60, p. 73–89.

Deming, D., 1994, Overburden rock, temperature, and heat flow, *in* L. G. Magoon and W. G. Dow, eds., The petroleum system—from source to trap: AAPG Memoir 60, p. 165–186.

Downey, M. W., 1994, Hydrocarbon seal rocks, *in* L. G. Magoon and W. G. Dow, eds., The petroleum system—from source to trap: AAPG Memoir 60, p. 159–164.

Estrella, G., M. R. Mello, P. C. Gaglionone, R. L. M. Asevedo, K. Tsubone, E. Rossetti, J. Concha, and I. M. R. A. Brüning, 1984, The Espírito Santo Basin (Brazil) source rock characterization and petroleum habitat, *in* G. Demaison and R. J.

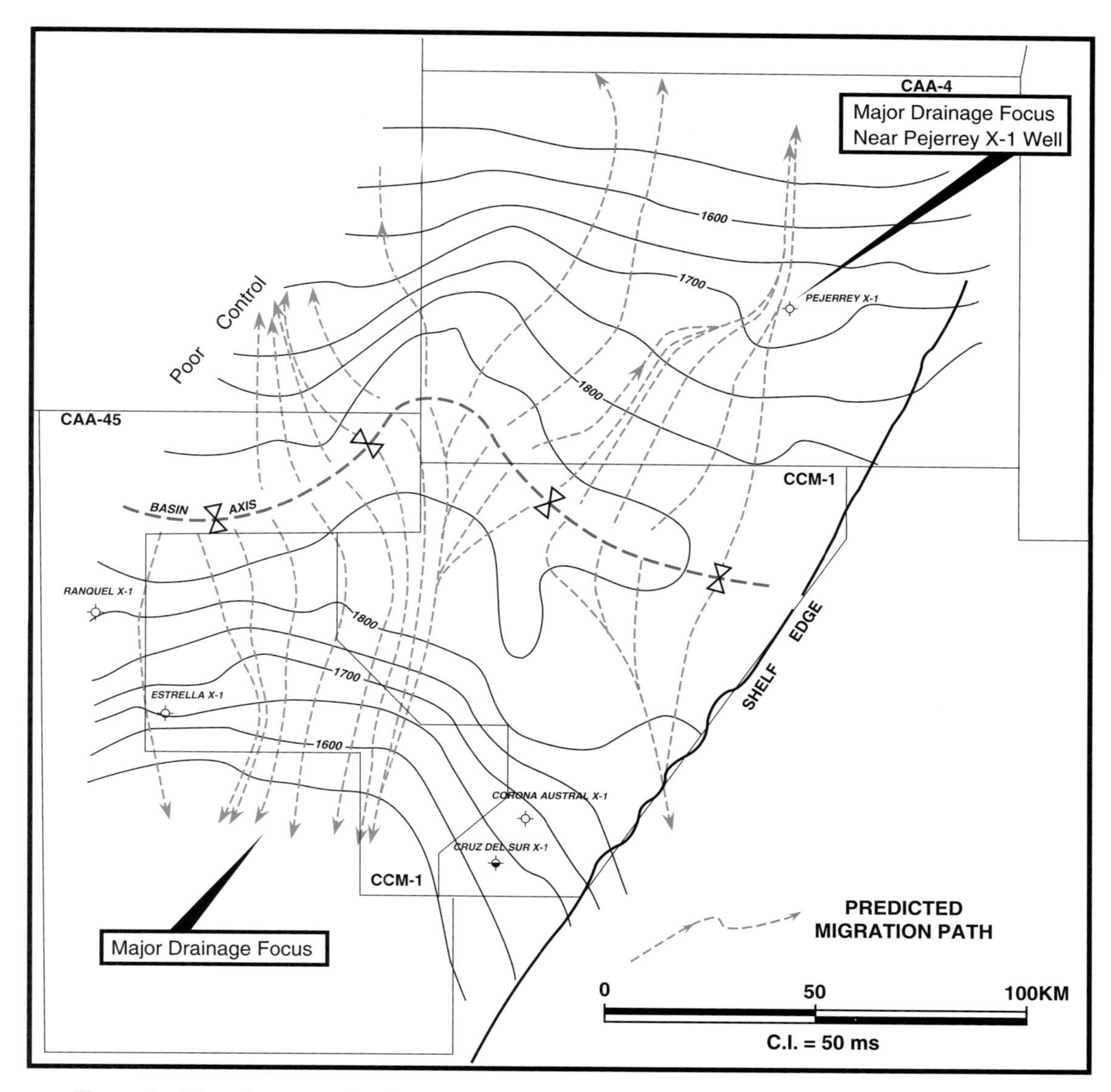

Figure 9—Migration map showing present-day migration in the top of the Colorado Formation.

Murris, eds., Petroleum geochemistry and basin evaluation: AAPG Memoir 35, p. 253–272.

Foresman, J. B., 1978, Organic geochemistry of DSDP Leg 40, continental rise of southwest Africa, *in* J. H. Natland, ed., Initial Reports of the Deep Sea Drilling Project, Washington, D.C., U.S. Government Printing Office, v. 40, p. 557–567.

Fryklund, R. E., 1994–1995, Well summary reports: Cruz del Sur x-1, Estrella w-1, and Corona Austral x-1, Block Colorado Marine Basin (CCM-1). Post-drill interpretation: UTAL unpublished report, Buenos Aires, Argentina.

Fryklund, R. E., A. Marshal, and J. Stevens, 1996, La cuenca del Colorado, *in* V. A. Ramos and M. A. Turic, eds., Geologiá y recursos naturales de sa plataforma continental Argentina: XIII° Congreso Geológicó Argentino—III° Congreso de Exploración de Hydrocarburos, p. 135–158.

Guardado, L. R., L. A. P. Gamboa, and C. F. Lucchesi, 1989, Petroleum geology of the Campos Basin, Brazil, a model for a producing Atlantic type basin, *in* J. D. Edwards and P. A. Santogrossi, eds., Divergent/passive margin basins: AAPG Memoir 48, p. 3–80.

Haq, B. U., J. Hardenbol, and P. R. Vail, 1988, Mesozoic and Cenozoic chronostratigraphy and cycles of sea-level change, *in* C. K. Wilgus, B. S. Hastings, C. G. Kendall, H. W. Posamentier, C. A. Ross, and J. C. Van Wagoner, eds., Sea-level changes: an integrated approach: SEPM Special Publication 42, p. 71–108.

Juan, R. del C., J. de Jager, J. Russell, and I. Gebhard, 1996, Flanco norte de la cuenca del Colorado, *in* V. A. Ramos and M. A. Turic, eds., Geologiá y recursos naturales de la plataforma continental Argentina: XIII° Congreso Geológicó Argentino—III° Congreso de Exploración de Hydrocarburos, p. 117–134.

Keeley, M. L., and M. P. R. Light, 1993, Basin evolution and prospectivity of the Argentine continental margin: Journal of Petroleum Geology, v. 16, n. 4, p. 451–464.

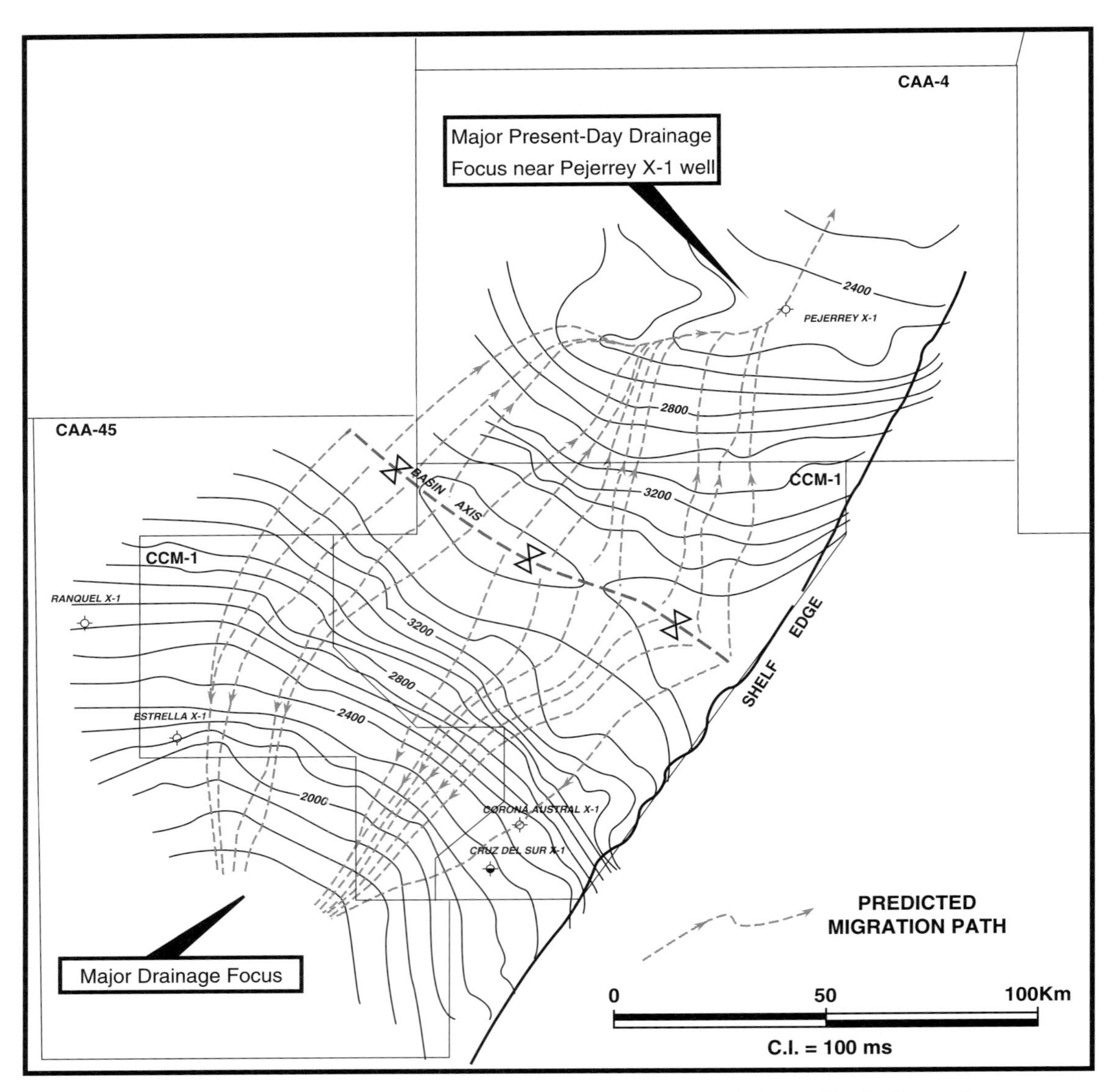

Figure 10—Migration map showing present-day migration within the Fortin Formation.

Magoon, L. B., and W. G. Dow, 1994, The petroleum system, *in* L. G. Magoon and W. G. Dow, eds., The petroleum system—from source to trap: AAPG Memoir 60, p. 3–24.

Marinelli, R. V., G, A. Rebay, and H. J. Franzin, 1996, Cuencas del Talud Continental, *in* V. A. Ramos and M. A. Turic, eds., Geologiá y recursos naturales de la plataforma continental Argentina: XIII° Congreso Geológicó Argentino—III° Congreso de Exploración de Hydrocarburos, p. 343–358.

Smith, J. T., 1994, The petroleum system logic as an exploration tool in a frontier setting, *in* L. G. Magoon and W. G. Dow, eds., The petroleum system—from source to trap: AAPG Memoir 60, p. 25–50.

Tissot, B. P., J. R. Delteil, G. Demaison, A. Combaz, and P. Masson, 1980, Paleoenvironment and petroleum potential of middle Cretaceous black shales in Atlantic basins: AAPG Bulletin, v. 64, p. 2051–2063.

Uliana, M. A., K. T. Biddle, and J. Cerdan, 1989, Mesozoic extension and the formation of Argentine sedimentary basins, *in* A. J. Tankard and H. R. Balkwill, eds., Extensional tectonics and stratigraphy of the North Atlantic margins: AAPG Memoir 46, p. 559–614.

Williams, J., 1962, Oceanography: Boston, Little Brown, 242 p.

Yacimientos Petrolíferos Fiscales, 1977, Well summary reports: Ranquel x-1 and Puelches x-1, Colorado Basin, Argentina. Post-drill interpretation: YPF unpublished report, Buenos Aires.

YPF/Maxus staff, 1996, Geology and petroleum potential of the offshore Colorado Basin: YPF/Maxus unpublished internal report, Buenos Aires and Dallas.

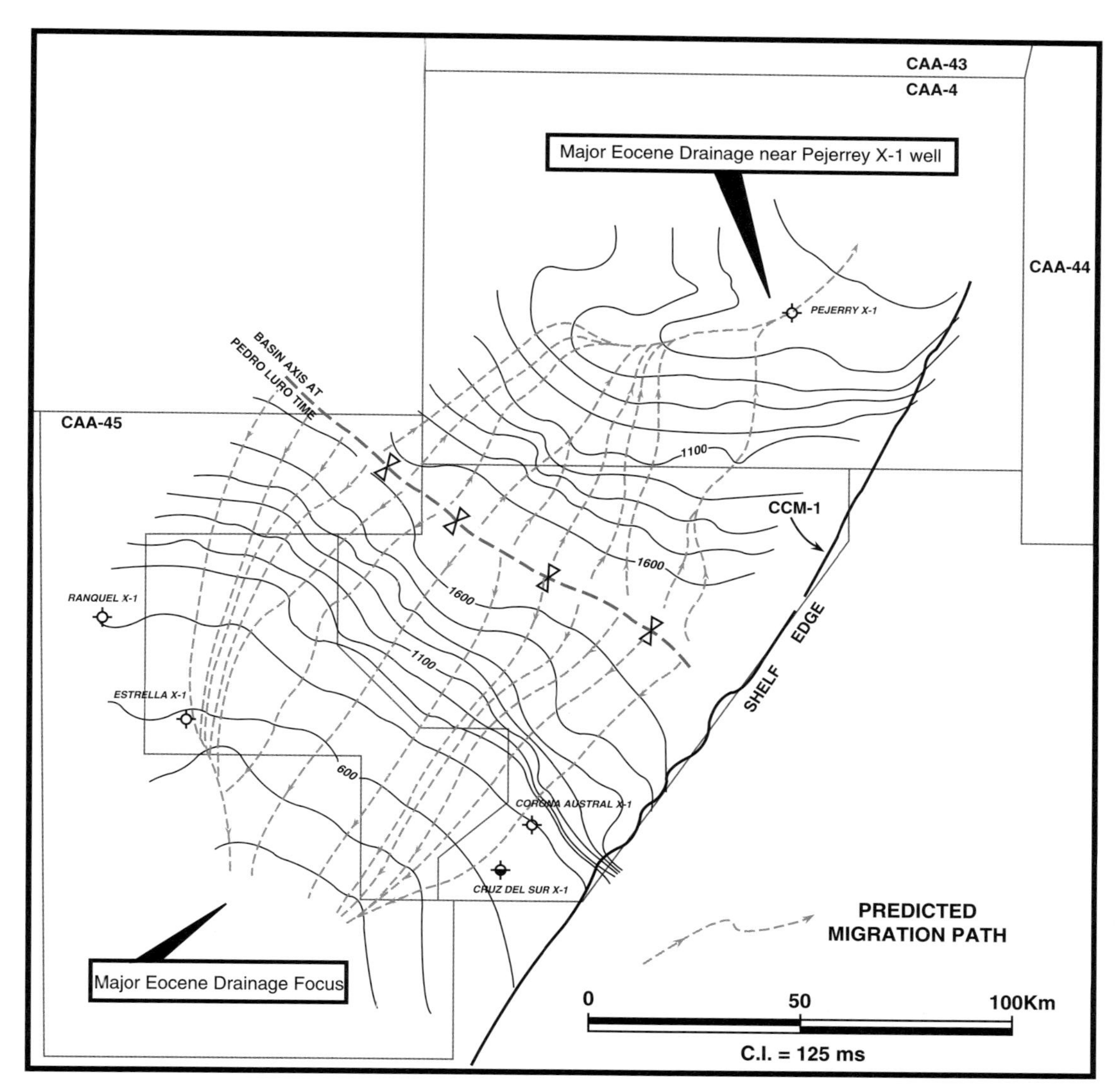

Figure 11—Migration map showing Eocene migration within the Fortin Formation.

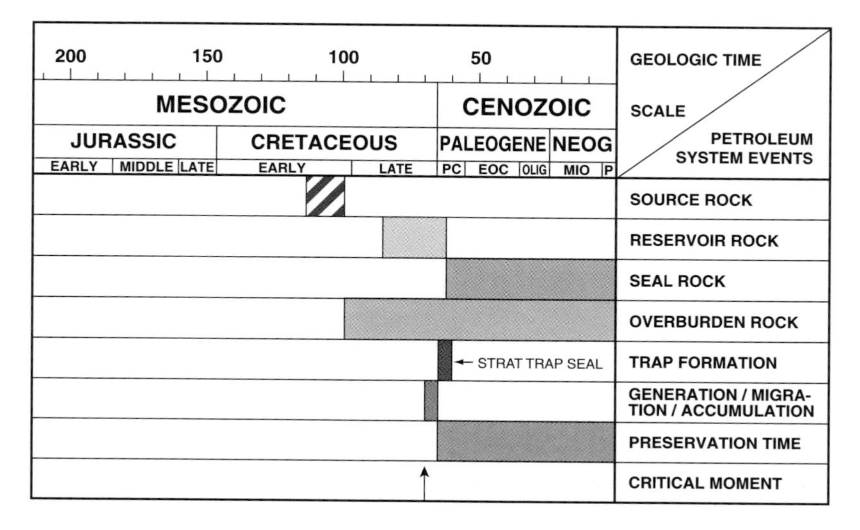

Figure 12—Petroleum system events chart for the eastern Colorado Basin.

Otis, R. M., and N. Schneidermann, 2000, A failed hydrocarbon system—Rawson Basins, Argentina, *in* M. R. Mello and B. J. Katz, eds., Petroleum systems of South Atlantic margins: AAPG Memoir 73, p. 417–427.

Chapter 28

A Failed Hydrocarbon System—Rawson Basins, Argentina

Robert M. Otis

Nahum Schneidermann

Chevron Overseas Petroleum, Inc.
San Ramon California, U.S.A.

Abstract

The business environment of the petroleum industry over the past 10 years has exerted increasing pressure on companies to improve exploration success with a corresponding reduction in costs and cycle time. Companies have responded by establishing processes for geologic risk assessment, volumetric estimation, and economic analysis. The "hydrocarbon system" concept is integral to all of these processes because it focuses evaluation drilling programs to address fundamental uncertainties of the petroleum geology. In this manner, companies can significantly cut the cycle time to key decisions, move quickly into appraisal or to other opportunities, and thereby reduce costs and allow evaluation of additional prospective areas.

The hydrocarbon system concept was used in a frontier exploration program in the Rawson Basins, offshore Argentina. The focus of the postulated play was a test designed to prove or disprove the presence of a hydrocarbon system in any of the basins, rather than the more classic approach of testing the most economically viable (i.e., largest) prospect. The exploration program commenced with acquisition and interpretation of seismic and potential field data. A geologic model was developed assuming a hydrocarbon system composed of a lacustrine source rock associated with early rifting of the Atlantic Ocean, interbedded shales and fluvial–deltaic sandstones for seal and reservoir rocks, and structural closures for trap. An evaluation drilling program was developed to test the presence of an active hydrocarbon, and a single obligation well was positioned to test for a lacustrine source rock in the basin center, migration pathways, and reservoir and seal. Drilling encountered a sequence of red beds that lacked discrete reservoir sandstones or organic shales. The well penetrated the "basement" Paleozoic strata and was abandoned without any drill-stem testing.

The absence of a hydrocarbon system was clear, and the decision to abandon the area was validated. The appraisal of the Rawson Basins was done quickly, had clear objectives for evaluation (to prove or disprove the hydrocarbon system), and resulted in a sound and timely decision, based on clear, definitive data, to pursue other opportunities elsewhere.

INTRODUCTION

The Rawson Basins are a series of three sedimentary basins located about 250 km east of the Valdez Peninsula, offshore Argentina (Figure 1). In 1988, a partnership consisting of Esso Sociedad Anonima Petrolera Argentina, Esso Exploration Argentina, S.A., Chevron International (Argentina) LTD., and Companias Asociadas Petroleras, S.A. (CAPSA), obtained the right to explore and produce hydrocarbons in three offshore blocks: CRM-1, CRM-2, and CRM-3. The joint venture was obligated to acquire seismic data and drill one well. The three basins were largely unexplored, having only 2-D seismic data on a 20–50 km grid with the nearest well being in the Valdez Basin, about 250 km to the west (Figure 1).

The primary objective was to acquire sufficient seismic and well data under the obligations of the joint venture and be able to determine a cost-effective program to evaluate these basins. The strategy to meet this objective was to use a systematic approach to quickly identify the hydrocarbon systems in these basins and to locate the obligation well in order to evaluate the potential for generation and entrapment of hydrocarbons for the dominant hydrocarbon system. Further investment would depend on the outcome of the first well.

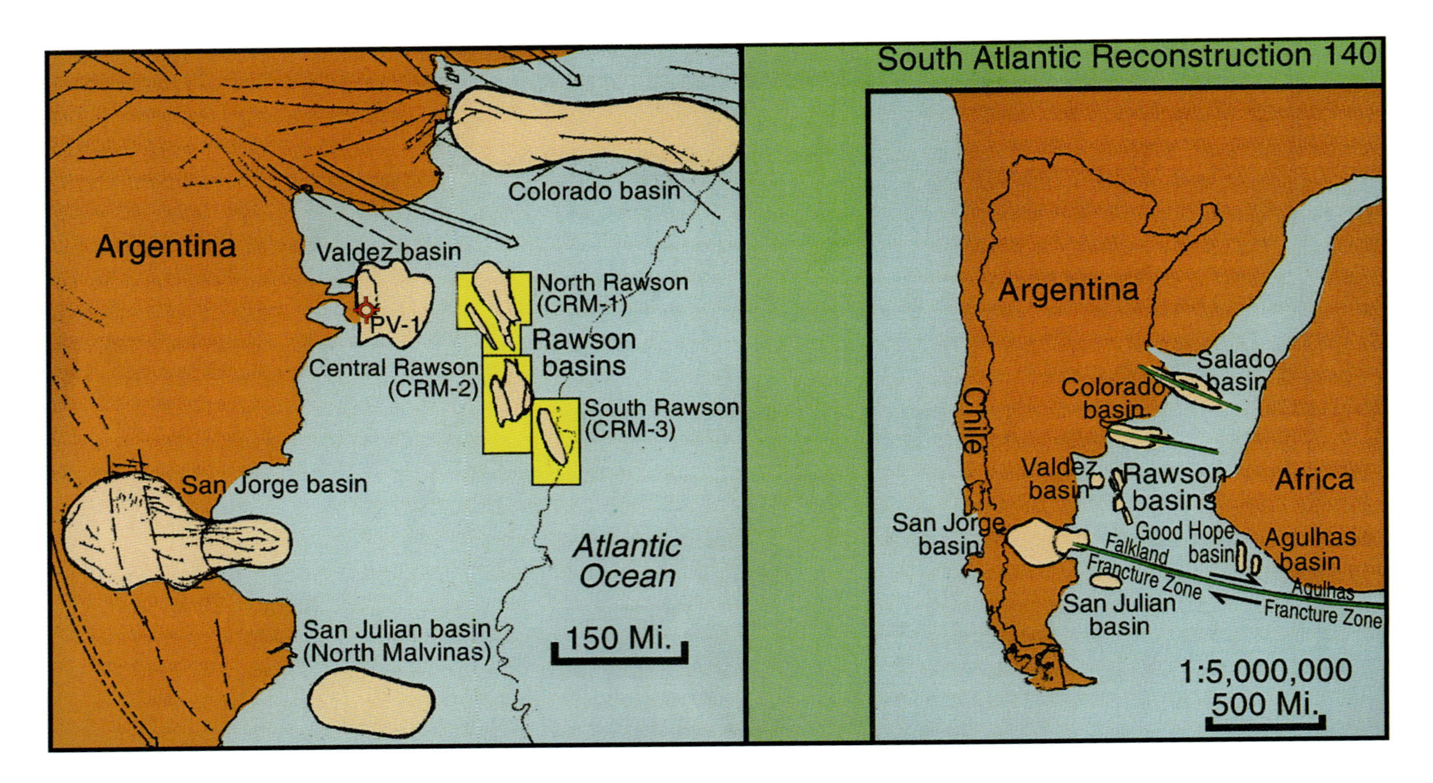

Figure 1—Location map showing the Rawson Basins, a system of three basins located about 250 km offshore from the Valdez Peninsula, Argentina. Blocks CRM-1, CRM-2, and CRM-3 each encompass one of the three basins.

Implementation of the strategy utilized the exploration evaluation process (Otis and Schneidermann, 1997). This process began with a review of data from the Rawson Basins and the development of a model for an expected hydrocarbon system. Next, under the umbrella of the hydrocarbon system concept, plays were developed and prospects evaluated. The overall potential of the hydrocarbon system was assessed economically using risk and volumetric estimates of the prospects based on the techniques discussed by Otis and Schneidermann (1997). Finally, the exploration well was located to prove or disprove the presence of a hydrocarbon system, rather than the more classic approach of testing the most economically viable (i.e., largest) prospect. Although the test was negative in that the hydrocarbon system was absent, the project partners gained a large degree of confidence about their decision to abandon these three offshore blocks.

EXPLORATION EVALUATION PROCESS

The process of evaluating a prospect is fully described in Otis and Schneidermann (1997). The foundation of this process is knowledge of petroleum geology, in particular, the concepts of the hydrocarbon system, play, and prospect evaluation. These ideas have been developed over the past 25 years with contributions from Dow (1972, 1974), Nederlof (1979), Perrodon (1980, 1983, 1992), Demaison (1984), Ulmishek (1986), White (1988, 1993), Demaison and Huizinga (1991), Magoon (1987, 1988, 1989), and Magoon and Dow (1994).

The entire evaluation process is illustrated in Figure 2 and begins with a basin review. The basin review provides the geologic setting and potential or known source rocks are identified. A basin model is developed for hydrocarbon generation, and finally, play concepts are constructed for the generation of prospects where reservoir and trap can preserve hydrocarbons. Once the basic geologic framework is in place, geologic risk is assessed by assigning a probability of success of finding producible hydrocarbons. In parallel, volumetrics are estimated as a probability distribution of recoverable hydrocarbons. Petroleum and facilities engineering support provides estimates of production profiles and costs for facilities and transportation. Based on this information, economics are calculated corresponding to pessimistic, mean, and optimistic reserve estimates, so that a go/no-go decision can be made. A decision to go ahead is usually followed by the drilling of a well, and results are documented to provide feedback for the knowledge base and process improvement.

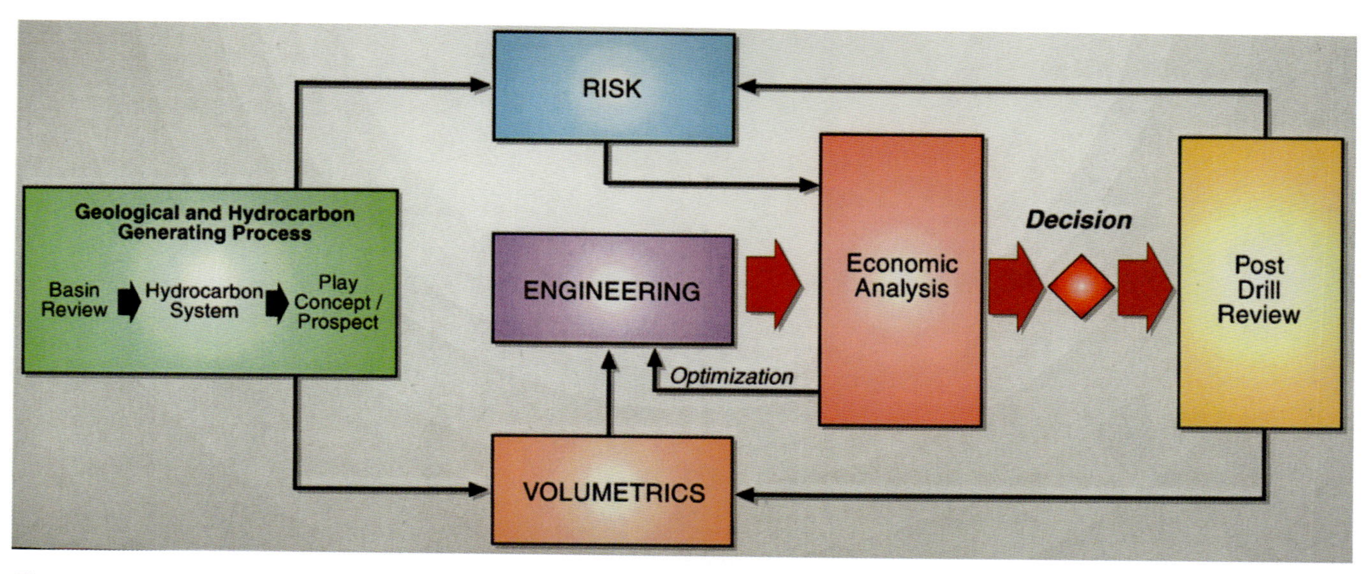

Figure 2—Flow chart for the exploration evaluation process. From Schneidermann and Otis (1997).

HYDROCARBON SYSTEM

The term *hydrocarbon system* is used by us to mean a volume of sedimentary rock containing genetically related hydrocarbons and charged by a single pod of active source rock. This definition requires that manifestations of hydrocarbons are present (seeps, shows, or a producing well), but it can also be applied by analogy to frontier areas such as the Rawson Basins. (As used here, a hydrocarbon system is similar to a petroleum system except that it may include gas as well as oil.)

The Rawson Basins were first observed as three potential sedimentary basins on a large-grid 2-D seismic program acquired in 1975. These three basins are part of a larger pattern of rift basins extending north–south along the South Atlantic coastline which formed in Jurassic–Cretaceous time during the opening of the South Atlantic Ocean (Ludwig et al., 1979). Some of these basins have proven highly productive on both sides of the Atlantic, including the Lower Congo Basin of west Africa, the Recôncavo Basin of Brazil, and the San Jorge Basin of Argentina. Others have been largely unproductive, including the neighboring Valdez and Colorado Basins (Figure 1). At the time of acquiring the contracts on the Rawson concessions, the only known fact was that basins with sufficient sedimentary thickness to support a hydrocarbon system were present.

The hydrocarbon systems of productive rift basins along the South Atlantic margins generally have lacustrine source rocks. The Bucomazi and Sengi shales in the Lower Congo Basin, the Gomo and Marfim shales in the Recôncavo Basin (Mello et al., 1994), and the Pozo D-29 shale in San Jorge are examples of lacustrine source rocks that support multi-billion barrel hydrocarbon systems. Reservoirs are typically fluvial sandstones capped by interbedded shales. However, many other basins are filled with terrigenous red beds and show little evidence of source or reservoir rocks.

As part of the contract, 4000 km of new 2-D seismic and gravity–magnetic data at a 4 km × 10 km grid over the North Rawson Basin (Block CRM-1) and at an 8 km × 20 km grid over the Central and South Rawson Basins (Blocks CRM-2 and CRM-3) were acquired. The Bouguer residual gravity map (Figure 3) supported the assumption of three deep basins having fluvial or fluvial–deltaic sedimentary fill. The seismic data likewise confirmed the presence of three strata-filled basins (Figure 4). Densities typical of sedimentary rocks were used to estimate a depth to basement and tie it to a seismic marker. Other prominent reflections were selected and tied to events associated with the South Atlantic opening using sequence stratigraphic analogies to other South Atlantic rift basins.

Maps were constructed on the main horizons. The structural setting was confirmed to be rift basins with main bounding faults and a series of synthetic and antithetic normal faults. The North Rawson Basin in Block CRM-1 is the deepest basin, characterized by a major east-bounding fault that is detached in the deep crust and an antithetic fault on the west. A central "deep" occupied the eastern part of the North Rawson Basin, with the western portion being significantly shallower. The entire rift sequence in this basin is covered by an Upper Cretaceous–lower Tertiary unconformity overlain by Tertiary shales. This unconformity and the overlying Tertiary shales are seismically tied to a Valdez Basin well located about 250 km west of the North Rawson Basin (Figure 1). Evidence of angular truncation is seen on seismic profiles below this unconformity. Multiple structural highs were found along the western antithetic fault, but

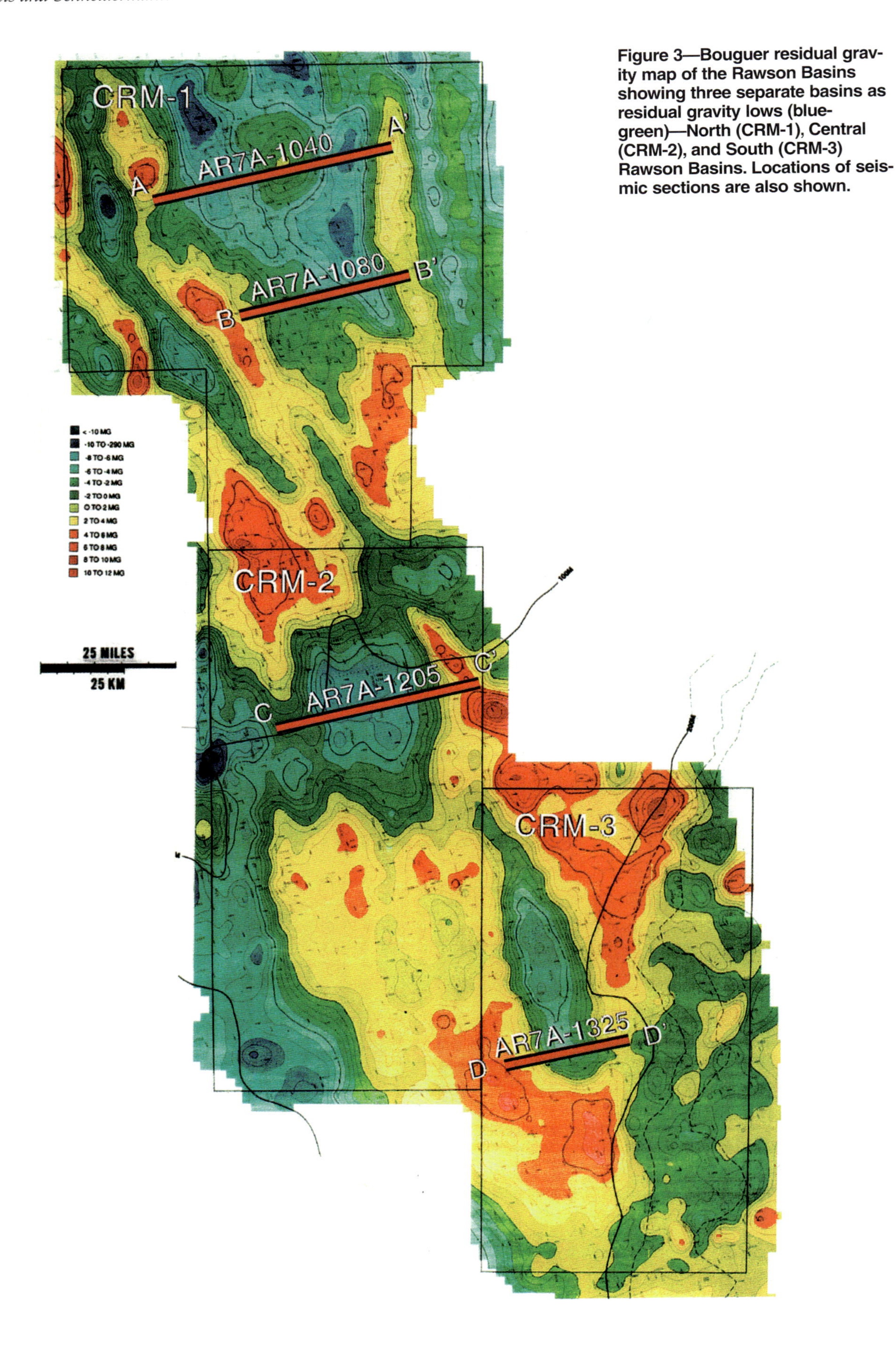

Figure 3—Bouguer residual gravity map of the Rawson Basins showing three separate basins as residual gravity lows (blue-green)—North (CRM-1), Central (CRM-2), and South (CRM-3) Rawson Basins. Locations of seismic sections are also shown.

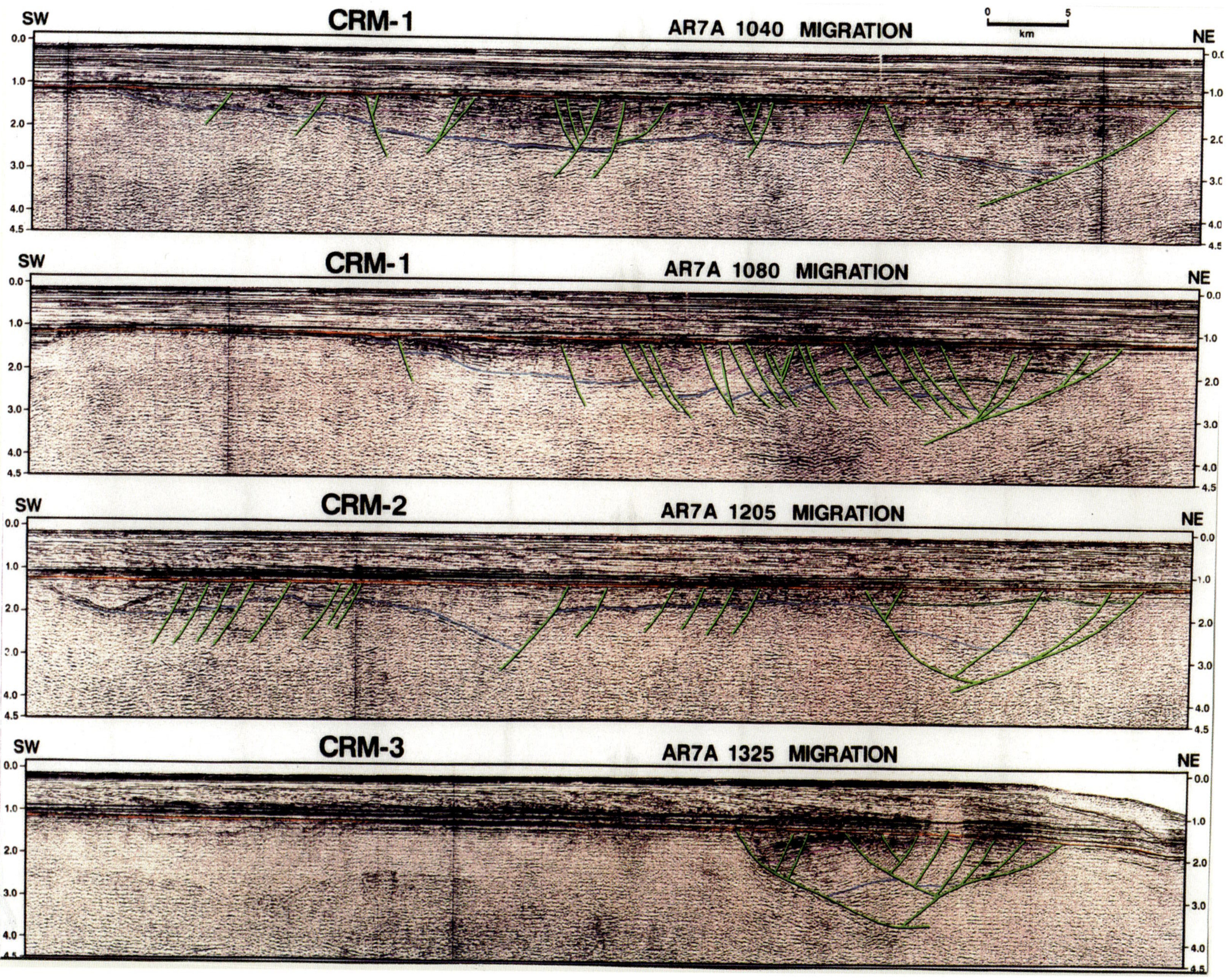

Figure 4—Representative seismic profiles from the North (CRM-1), Central (CRM-2), and South (CRM-3) Rawson Basins showing interpreted rift basins with characteristic major bounding faults and rotated normal fault blocks. See Figure 3 for locations of seismic lines.

structures are lacking in the deep basin center of the North Rawson. Two other rift basins—Central and South Rawson—were confirmed in Blocks CRM-2 and CRM-3, and structures were mapped throughout all the basins.

By analogy with other known South Atlantic margin basins, these rift basins were likely occupied either by a terrigenous fluvial–deltaic system (typically, red beds) or a lake system. Lacustrine sedimentation would provide a source rock for the hydrocarbon system. If present, this source rock could have been deposited in the deep centers of any of the basins in early stages of development or in the shallower parts of the basins during later stages. By analogy, either or both of these scenarios would have provided a rich source rock to charge any reservoir–trap configuration that was properly plumbed to the active source rock. Note that the amount of erosion at the main unconformity is unknown and thus introduces increased risk concerning the actual maturity of the postulated source rock.

Once the structural setting of the three basins was established, a depositional model for reservoir and trap relationships was postulated and a stratigraphic column constructed (Figure 5). Our hypothesis of the basic stratigraphic setting is a slowly subsiding basin with deposition of sand in a fluvial–deltaic setting sourced by rivers from the west and the presence of shallow lakes at the basin depocenter. Late in the depositional history of the basins, some uplift and/or erosion occurred as evidenced by the angular truncation at the main unconformity.

A basin model incorporating these postulated structural and deposition histories was used to estimate the timing of source rock maturity and expulsion of hydrocarbons. Typical parameters for thermal gradient, conductivity, and rock type were used in the model. Modeling indicated that a lacustrine source rock typical of the South Atlantic margin would be in the late oil to early gas window in the deepest part of the North Rawson Basin and that the bulk of all three basins would be in the early to middle oil window. Migration pathways would be upward along faults from the basin center and along sandstone layers updip away from the centers of the basins.

Thus, characteristics of the postulated hydrocarbon system for the Rawson Basins can be summarized as follows:

1. Lacustrine source rocks were deposited in the three basins created by an opening rift system.
2. Sediments were buried to depths sufficient to generate hydrocarbons by Late Cretaceous time in the deepest part of the North Rawson Basin with possible generation still continuing today in all three basins.
3. Migration of hydrocarbons occurred upward along faults to sandstone conduits and/or to structural–stratigraphic traps.
4. Migration of hydrocarbons also occurred laterally along sandstone conduits to structural–stratigraphic traps updip and away from the centers of the basins.
5. Tertiary shales overlying the main unconformity provide the regional top seal to the hydrocarbon system.

PLAYS AND PROSPECTS

A *play*, by our definition, is one or more accumulations of hydrocarbons identified by having in common certain geologic features, including the reservoir, trap, seal, charge, preservation, engineering character of its location, or any combination of these. They have unique geologic and engineering features and thus have general risk, volumetric, engineering, and economic profiles that can be used for prioritization. In contrast, a *prospect* represents an individual potential accumulation that is perceived to belong to an individual play. Each prospect has a probability of geologic success and a range of potential hydrocarbon volumes within its trap confines (Otis and Schneidermann, 1997).

Within the overall geologic framework of the Rawson Basins, the following plays were developed (Figure 6):

1. **Basal sandstone**—A basal sandstone, deposited at the beginning of rifting, is trapped in tilted fault blocks and sealed by overlying lacustrine shales. Migration pathways are vertical along the western bounding fault and lateral along sandstones that extend into the centers of the basins.
2. **Basin margin**—Fluvial–deltaic sandstone reservoirs deposited along the western deep basins margin (Blocks CRM-1 and -2) are trapped in tilted fault blocks and sealed by overlying lacustrine shales. Migration pathways are vertical along the western bounding fault and lateral along sandstones that extend into the centers of the basins.
3. **Fluvial**—Fluvial to fluvial–deltaic sandstone reservoirs deposited in the western parts of the basins (Blocks CRM-1 and -2) are trapped in tilted fault blocks and sealed by overlying lacustrine shales. Migration pathways are vertical along the west-bounding fault, then lateral along sandstones that extend long distances from the centers of the basins.
4. **Alluvial fan**—Alluvial fan and fan delta sandstone reservoirs deposited along major faults bounding the deep basin margins (Blocks CRM-1, -2, and -3) are trapped stratigraphically against the bounding faults and sealed vertically by overlying lacustrine shales. Migration pathways are direct from interbedded lacustrine shales and lateral along the sandstones that extend into the centers of the basins.
5. **Lacustrine deltaic**—Lacustrine deltaic sandstone reservoirs deposited within the deep basin depocenters are trapped in tilted fault blocks and sealed by overlying lacustrine shales. Migration pathways are direct from interbedded lacustrine shales and lateral along the sandstones that extend into the centers of the basins.

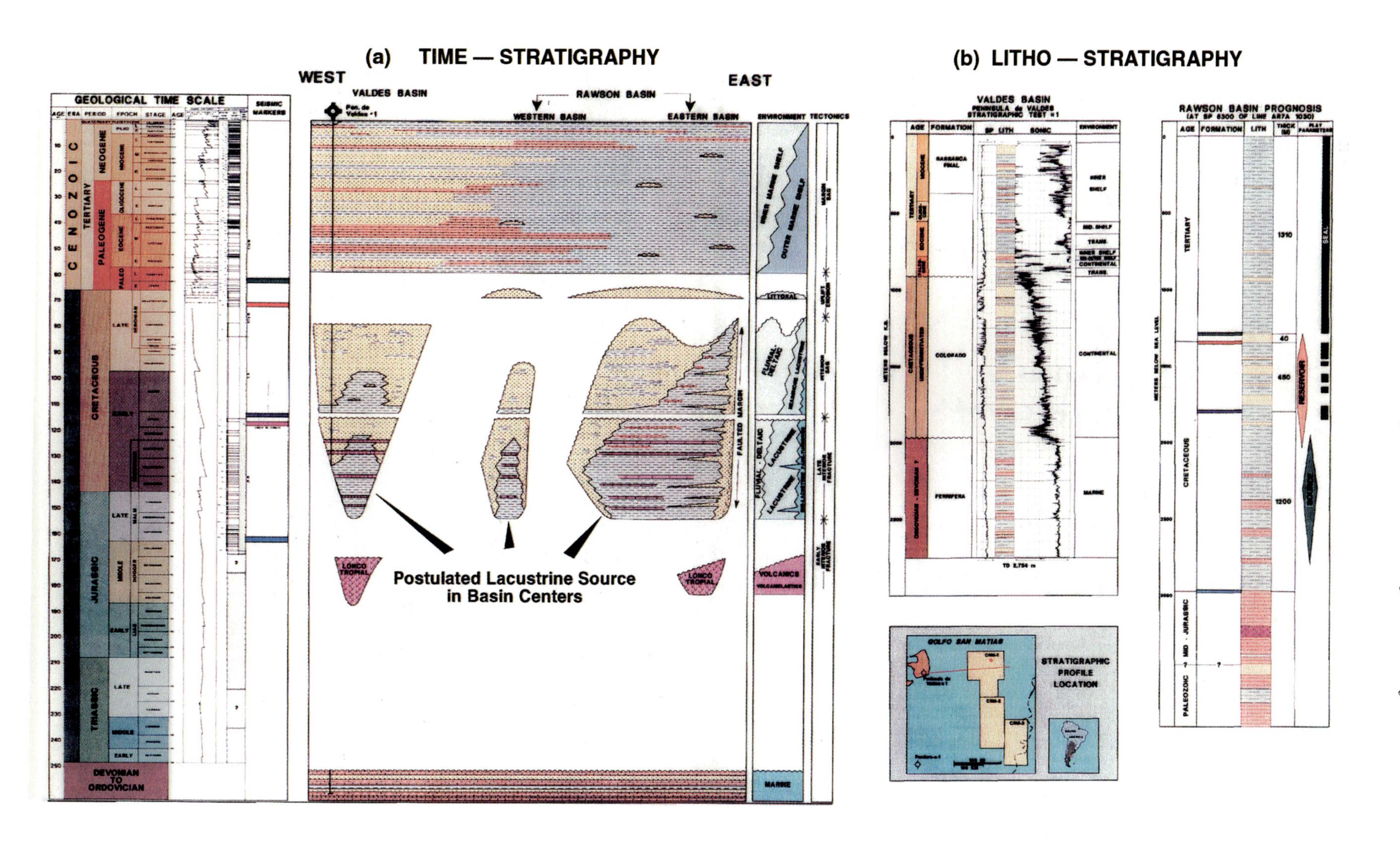

Figure 5—(a) Geologic time scale (left) alongside an interpreted west–east time stratigraphic profile from the Pozo de Valdez #1 well in the Valdez Basin to the North Rawson Basin, Block CRM-1. (b) Lithologic stratigraphic columns from the Valdez Basin well (left) and from a hypothetical well located in the center of the North Rawson Basin (right). Postulated source, reservoir, and seal rocks are shown.

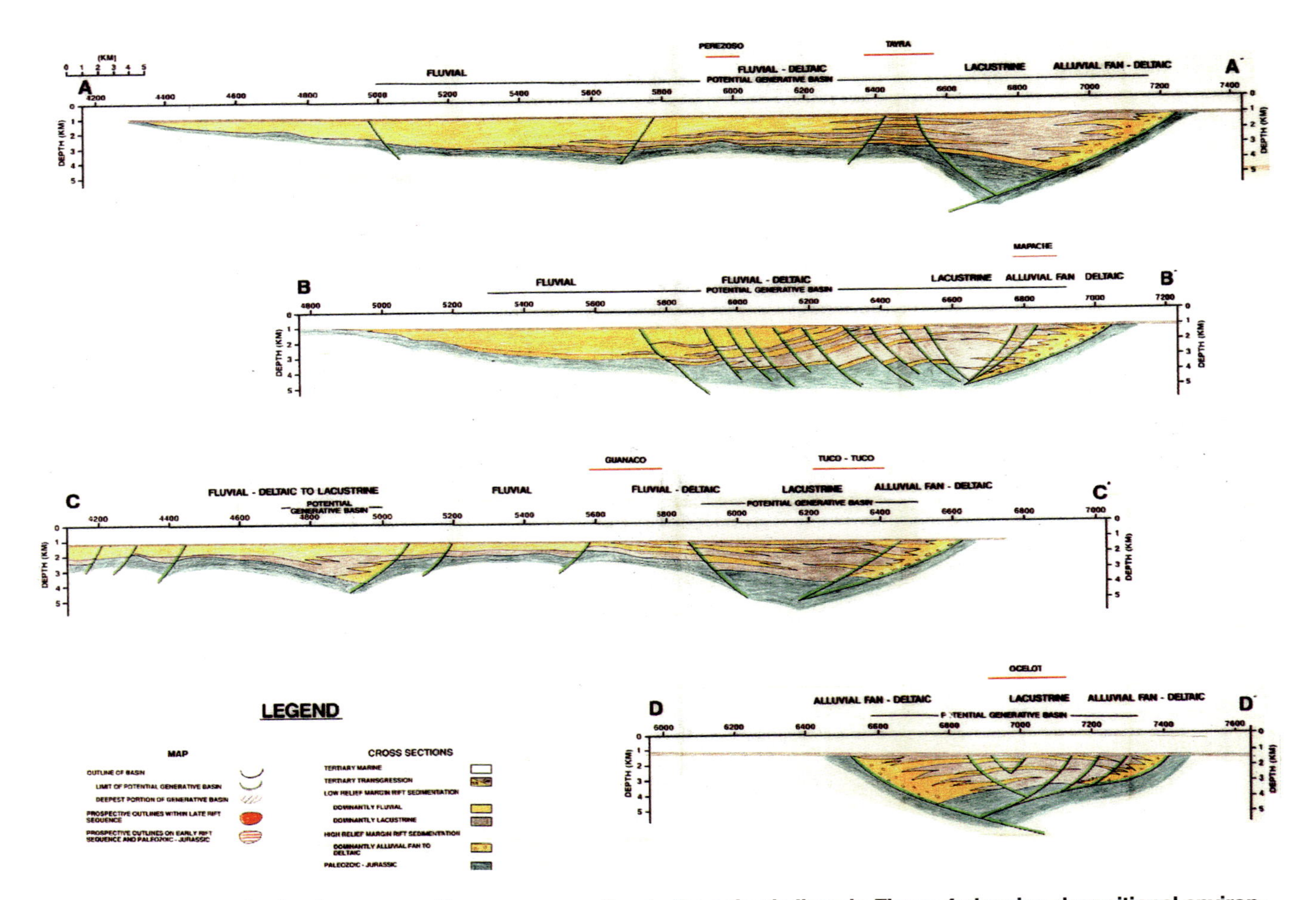

Figure 6—Interpreted seismic cross sections corresponding to the seismic lines in Figure 4 showing depositional environments and plays identified within the Rawson Basins. Plays include basal sandstone, basin margin, fluvial, alluvial fan, and lacustrine deltaic (see text for details).

Because the Rawson Basins were largely unknown entities, plays were prioritized on the basis of experience. Several general considerations were used to prioritize the plays. First, stratigraphic traps are riskier than fault traps, which are riskier than unfaulted anticlines. Second, fluvial–deltaic reservoirs have better permeabilities than alluvial fans or fan deltas. Third, traps far from the pod of active source rock are riskier than those in close proximity due to long-distance migration. Finally, larger depocenters have a better chance for lake development and thus generate more oil than small depocenters.

On the basis of these considerations, we focused on structural traps along the margin of the deep center of the North Rawson Basin in Block CRM-1. However, prospects were developed for all plays, and volumetric estimates and risk assessments were made for all plays and prospects. The plays were prioritized as follows:

1. The basal sandstone and basin margin plays in the North Rawson Basin of Block CRM-1 had the most potential for both hydrocarbon charge and reservoir. These plays also had the greatest number of prospects, giving them the highest upside potential.
2. Although lower in overall risk, a lacustrine deltaic play in the Central Rawson Basin of Block CRM-2 had little follow-up potential (low upside).
3. The fluvial play had limited charge potential, lower reservoir quality, and long-distance lateral migration risk.
4. The alluvial fan play had high trap risk and lower reservoir quality.

Once the plays had been prioritized, the main focus was to test prospects from the basal sandstone and basin margin plays in the North Rawson Basin. Multiple prospects were mapped for these plays. Two prospects were shown to be of great interest—Pudu, a large faulted four-way dip-enclosed structure located west of the basin margin, and Tayra, a smaller faulted structure located adjacent to the basin margin (Figures 7 and 8). Our intentions were to test the North Rawson Basin with a mini-

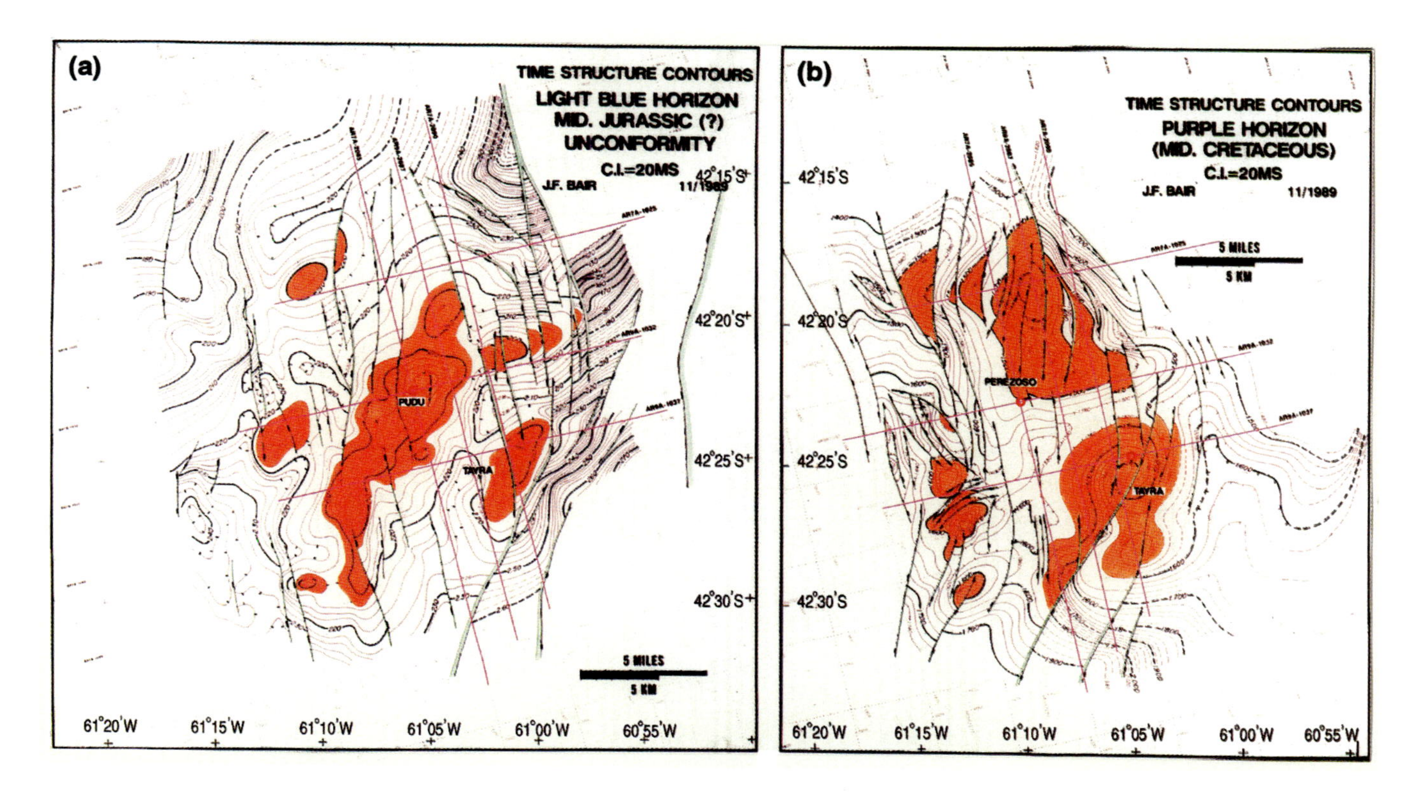

Figure 7—Time structure contour maps of (a) the Pudu and Tayra prospects drawn on the Middle Jurassic unconformity and (b) the Perezoso and Tayra prospects drawn on the middle Cretaceous unconformity. Based on size and multiple objectives, the Pudu and Tayra were identified to test the postulated hydrocarbon system in the Rawson Basins.

mum number of wells (one), and the choice was between Pudu (the largest structure) and Tayra (the best test of hydrocarbon systems).

The decision was made to drill the Tayra prospect because this well would provide the best test of the hydrocarbon system. If the well was unsuccessful in finding significant hydrocarbons, we would conclude that either the lacustrine source or the migration pathways were absent. Without one or both of these factors, the absence of a hydrocarbon system would be confirmed and the area could be abandoned without further expenditures. Likewise, because of the limited potential of the other two basins, these could also be relinquished with little or no remorse.

RESULTS AND CONCLUSIONS

The Tayra well was drilled in the North Rawson Basin, and the results are shown in Figure 9. Briefly, the well encountered a sequence of Tertiary shales and claystones above the main unconformity and undifferentiated Cretaceous red beds between the unconformity and Paleozoic basement. The well confirmed the absence of a reservoir, migration pathways (shows were lacking), and an active source rock in the basin center. Thus, the absence of a hydrocarbon system was confirmed, and the concession was relinquished.

In conclusion, this effort required only a 3-year cycle time to acquire and interpret the necessary data, to evaluate and test the potential of a large area, and to move ahead to a decision on future activity. In this case, the decision was to relinquish the holding, but because this decision was predicated on a process based on the identification and evaluation of the hydrocarbon system, it was considered sound and carried no regrets. This approach allowed us to focus limited resources (both human and financial) on other higher potential areas throughout the world.

Our report was designed to provide the reader with a "real life" example of the impact the search for a hydrocarbon system can have on the process of exploration. Active indications of a "live" hydrocarbon system, such as seeps, oil flowing tests, and oil shows, are required for approval to continue exploration. By themselves, these active indications are inadequate to demonstrate the presence or absence of an economically viable accumulation. However, the absence of these indicators in a properly positioned test of a viable hydrocarbon system is a sufficient reason to exit the exploration venture.

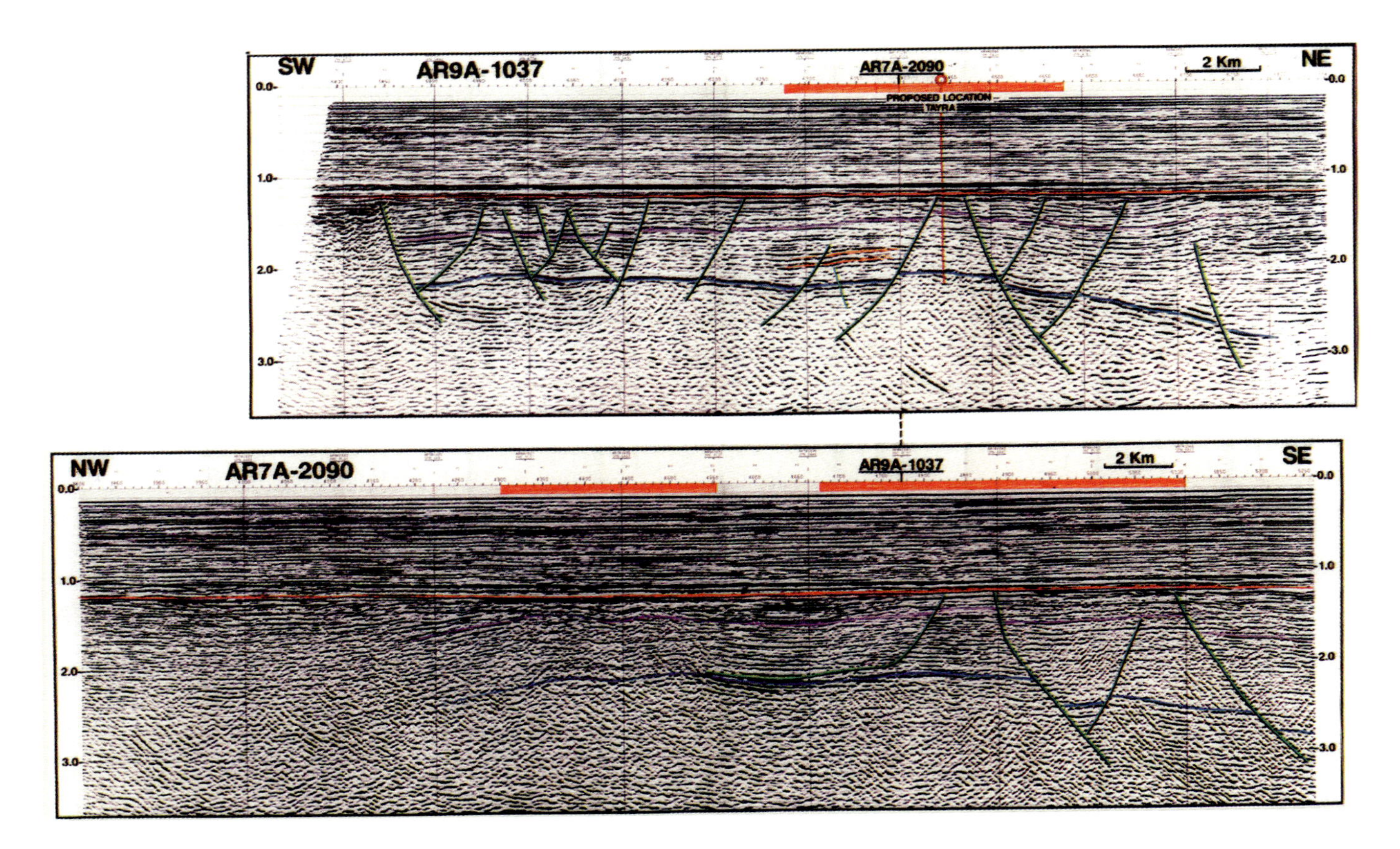

Figure 8—Two seismic profiles from the North Rawson Basin showing the structure and location relative to the postulated source kitchen in the basin center. The Tayra play is located directly adjacent to the basin center and was thus considered the best test of the hydrocarbon system. See Figure 7 for locations of seismic lines.

Acknowledgments—*We thank the many dedicated, creative, and highly professional earth scientists at Esso (operator), Chevron, and CAPSA who worked over the years on the concession and contributed data and interpretations. We also thank M. D. Spafford for his help in preparation of the displays and the convenors for inviting us to contribute to the conference. Finally, we thank Chevron Overseas Petroleum for permission to publish this paper. The interpretations and conclusions presented here are not necessarily those of the management and staff of the partnership, and we alone take responsibility for them.*

REFERENCES CITED

Demaison, G., 1984, The generative basin concept, *in* G. Demaison and R. J. Murris, eds., Petroleum geochemistry and basin evaluation: AAPG Memoir 35, p. 1–14.

Demaison, G., and B. J. Huizinga, 1991, Genetic classification of petroleum systems: AAPG Bulletin, v. 75, p. 1626–1643.

Dow, W. G., 1972, Application of oil correlation and source rock data to exploration in Williston basin (abs.): AAPG Bulletin, v. 56, p. 615.

Dow, W. G., 1974, Application of oil correlation and source rock data to exploration in Williston basin (abs.): AAPG Bulletin, v. 58, n. 7, p. 1253–1262.

Ludwig, W. J., J. I. Ewing, C. C. Windisch, A. G. Lonardi, and F. F. Rios, 1979, Structure of Colorado Basin and continent-ocean crust boundary off Bahia Blanca, Argentina, *in* J. S. Watkins, L. Montadert, and P. W. Dickerson, eds., Geological and geophysical investigations of continental margins: AAPG Memoir 29, p. 113–124.

Magoon, L. B., 1987, The petroleum system—a classification scheme for research, resource assessment, and exploration (abs.): AAPG Bulletin, v. 71, p. 587.

Magoon, L. B., 1988, The petroleum system—a classification scheme for research, exploration and resource assessment, *in* L. B. Magoon, ed., Petroleum systems of the United States: USGS Bulletin 1870, p. 2–15.

Magoon, L. B., 1989, The petroleum system—status of research and methods, *in* L. B. Magoon, ed., The petroleum system—the status of research and methods, 1990, USGS Bulletin 1912, p. 1–9.

Magoon, L. B., and W. G. Dow, eds., 1994, The petroleum system—from source to trap: AAPG Memoir 60, 655 p.

Mello, M. R., E. A. M. Koustsoukos, W. U. Mohriak, and G. Bacoccoli, 1994, Selected petroleum systems in Brazil, *in* L. B. Magoon and W. G. Dow, eds., The petroleum

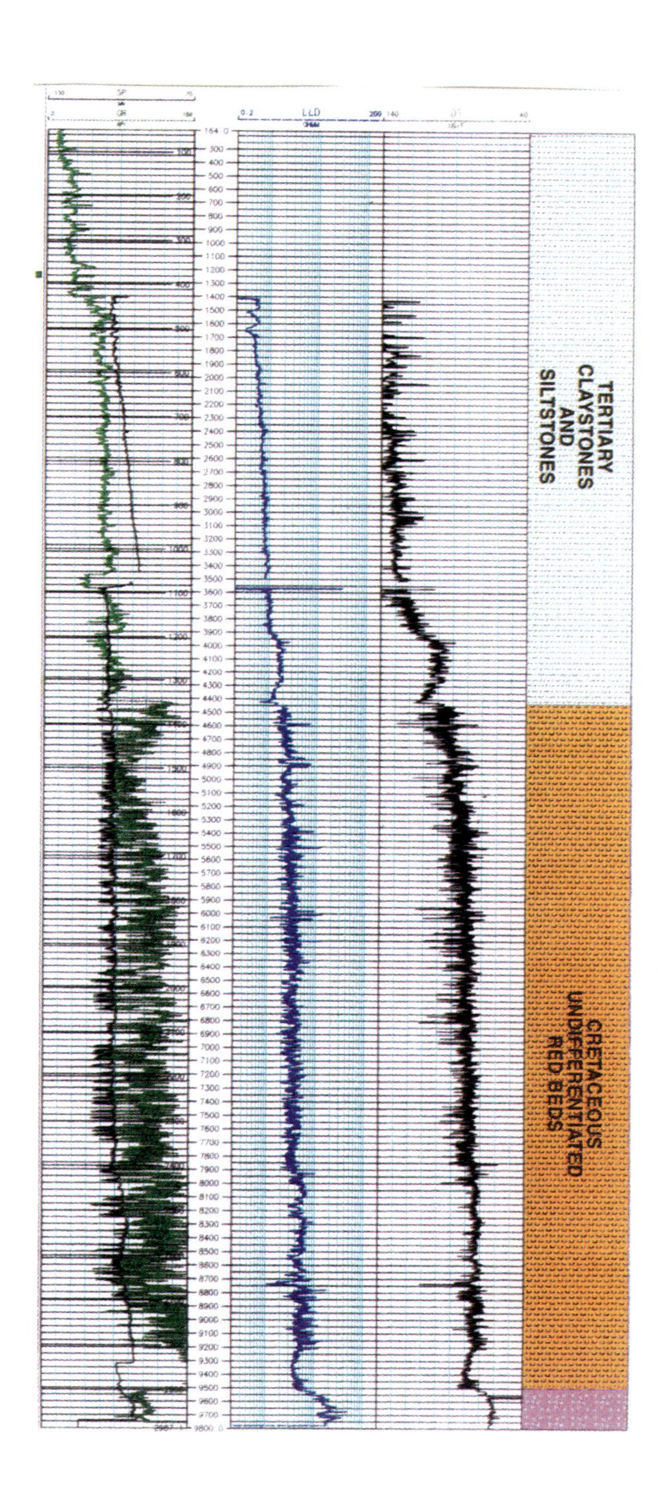

Figure 9—Well log from Tayra X-1 well. Results showed a basin filled with undifferentiated red beds. These results indicate no active source rocks and no evidence of hydrocarbon migration (shows) which confirmed the absence of a hydrocarbon system.

system—from source to trap: AAPG Memoir 60, p. 499–512.

Nederlof, M. H., 1979, The use of habitat of oil models in exploration prospect appraisal: Proceedings of the 10th World Petroleum Congress, Bucharest, Rumania, p. 13–21.

Otis, R. M., and N. Schneidermann, 1997, A process for evaluating exploration prospects: AAPG Bulletin, v. 81, n. 7, p. 1087–1109.

Perrodon, A., 1980, Geodynamique petroliere. Genese et repartition des gisements d'hydrocarbures: Paris, Masson–Elf Acquitaine, 381 p.

Perrodon, A., 1983, Dynamics of oil and gas accumulations: Pau, Elf Acquitaine, p. 187–210.

Perrodon, A., 1992, Petroleum systems: models and applications: Journal of Petroleum Geology, v. 15, n. 3, p. 319–326.

Schneidermann, N., and R. M. Otis, 1997, Use of petroleum systems in risk assessment of plays and prospects (abs.), *in* J. V. C. Howes and R. A. Noble, eds., Proceedings of International Conference on Petroleum systems of SE Asia and Australasia: Indonesian Petroleum Association, Jakarta, Indonesia, p. 1020.

Ulmishek, G., 1986, Stratigraphic aspects of petroleum resource assessment, *in* D. D. Rice, ed., Oil and gas assessment—methods and applications: AAPG Studies in Geology 21, p. 59–68.

White, D. A., 1988, Oil and gas play maps in exploration and assessment: AAPG Bulletin, v. 72, n. 8, p. 944–949.

White, D. A., 1993, Geologic risking guide for prospects and plays: AAPG Bulletin, v. 77, p. 2048–2061.

Holtar, E., and A. W. Forsberg, 2000, Postrift development of the Walvis Basin, Namibia: results from the exploration campaign in Quadrant 1911, *in* M. R. Mello and B. J. Katz, eds., Petroleum systems of South Atlantic margins: AAPG Memoir 73, p. 429–446.

Chapter 29

Postrift Development of the Walvis Basin, Namibia: Results from the Exploration Campaign in Quadrant 1911

Erik Holtar

Norsk Hydro Exploration and Production International
Oslo, Norway

Arne Willy Forsberg

Norsk Hydro Canada Oil & Gas Inc.
Calgary, Alberta, Canada

Abstract

The exploration campaign in offshore Namibia gained new momentum in 1992 when new licenses were awarded in the independent Namibia. Seismic and well data from Quadrant 1911 has brought new insight to the stratigraphy of the vast Namibian shelf.

A stratigraphic breakdown of the northern Namibian offshore area is proposed. The post-breakup succession is divided into seven major stratigraphic units or groups, from W1 (oldest) to W7 (where W denotes the Walvis Basin). Each group is described in terms of geometry, with examples from seismic expression and structural maps, and lithofacies as seen in well data.

The W1 Group consists of volcanic and volcanoclastic rocks, succeeded by the W2 Group shallow-marine carbonates. After a tectonic event causing block faulting, the W3 Group consisting of siltstones and claystones with limestone stringers was deposited. The W4 Group consists of claystones and siltstone–sandstones, including claystones with very good hydrocarbon source rock potential. In Quadrant 1911, the W4 Group also includes a large volcanic center. The W5 Group consists of mudstones with interbeds of sandstones and thin limestones. Following the Cretaceous–Tertiary boundary, the W6 Group mudstones, claystones, and marls were deposited, followed by claystones of the W7 Group.

The sedimentary succession of Quadrant 1911 postdates the Neocomian Etendeka plateau basalts found in onshore Namibia. After onset of the drift phase in late Hauterivian time, the Walvis Basin subsided and a marine transgression eventually took place. Shallow-marine platform carbonates prevailed until an Albian tectonic event resulted in complex block faulting and the formation of several subbasins. Subsequent volcanic activity created a series of volcanic centers localized near the Walvis Ridge bathymetric feature. The southern African craton was uplifted, leading to the formation of large-scale westward-prograding wedges. Later sedimentation largely followed the evolution of a passive continental margin, responding to relative sea level changes and paleoclimate.

INTRODUCTION

In 1992, Quadrant 1911 was awarded as exploration license 001 to a group consisting of Norsk Hydro (operator), Saga Petroleum, and Statoil. Since then, a total of 8000 km of seismic lines has been acquired by the NH9206 and NH9404 surveys, covering the 11,619-km^2 quadrant situated in the Walvis Basin, Namibia. This basin was undrilled until the 1911/15-1 well was finished in early 1994. Prior to this well, the nearest offshore exploration wells to the north were the Block 9 wells of the Angolan Kwanza Basin and, to the south, the Kudu wells of the Orange Basin. In mid-1995, the 1911/10-1 well was drilled, and a total of four wells have now been drilled in the Walvis Basin.

The seismic grid varies from 2 × 2 km on the basin flank areas to 2 × 4 km in the basin areas and on the eastern platform. A high-resolution aeromagnetic survey covers all of Quadrant 1911. In addition, regional lines of good quality were used to establish a framework for the area. These include the ECL-89 and ECL-91 surveys and the Nopec/Namcor regional survey N2R-93 which covers areas to the north, south, and east of the quadrant. In the latter survey, regional line N2R-93-229, recorded to 14 sec two-way traveltime, has been particularly useful.

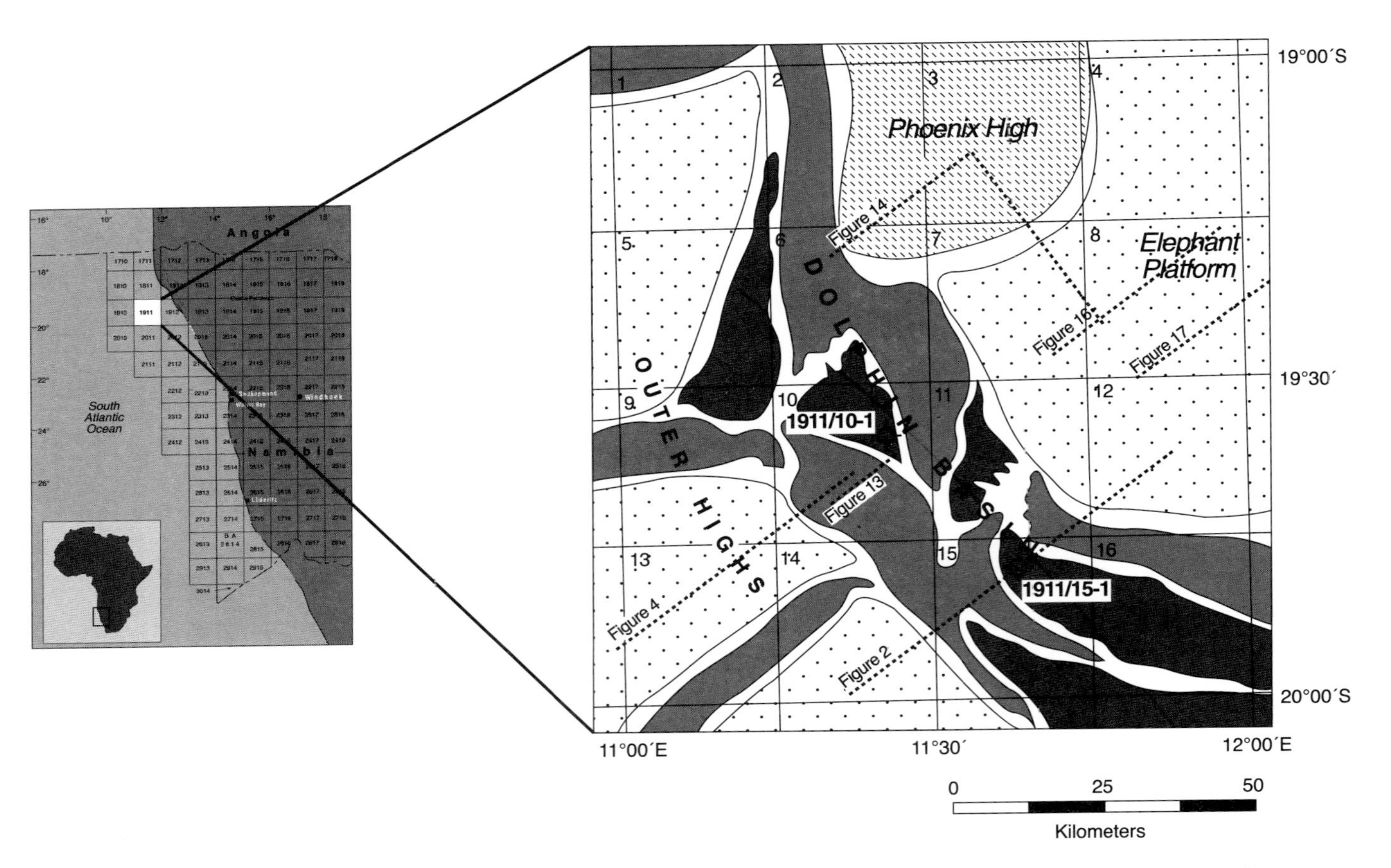

Figure 1—Location map showing the structural elements of Quadrant 1911 in Walvis Basin, offshore Namibia, southern Africa.

There are several structural elements in Quadrant 1911 (Figure 1). To the east lies the relatively stable Elephant Platform, where the main passive marine shelf developed. The central part of the quadrant is dominated by a complex rift graben system called the Dolphin Basin. To the west are a few highs that have a thinner cover of postrift sedimentary rocks; these are collectively called the Outer Highs. To the north, the outstanding feature is a large volcanic center named the Phoenix High.

The stratigraphic breakdown is based on seismic mapping of geometrically confined depositional units (Figure 2). Data from wells 1911/15-1 and 1911/10-1 were used to further describe and interpret these units. We propose that the post-breakup succession be divided into seven major units, suggested here to be ranked as stratigraphic groups, ranging from W1 (oldest) to W7 (youngest), where W denotes the Walvis Basin. The described part of the stratigraphic section broadly corresponds to the transitional and thermal sag tectono-stratigraphic sequences of Light et al. (1993).

The transitions between the individual groups represent significant changes in gross depositional patterns reflected in lithologic breaks or major changes in geometry (thickness distribution). Each of the seven major units consists of one to six subunits that are ranked as formations (Figure 3). Most of these are regarded as classic transgressive, highstand to lowstand tract sequence cycles related to relative sea level changes. Some units, especially in the Early–middle Cretaceous succession, are confined by lithologic changes that are interpreted to be related more to local tectonic events.

"BASEMENT" ROCKS

Onshore Namibia, the youngest predrift rocks are the Etendeka Group, as described by Milner et al. (1995), which are interbedded basalts, quartz latites, and minor latites. This group overlies the Paleozoic Karoo Sequence, and at its base, the volcanics are overlying and interbedded with eolian sandstones of the Etjo Formation. Ages ($^{40}Ar/^{39}Ar$) from stratigraphically equivalent units in the South American Paraná Basin (Renne et al., 1992) date the volcanics at 133 ± 1 Ma (Valanginian), which is supported by similar studies of the Etendeka Group in Namibia.

Light et al. (1993) proposed a subdivision of the Namibian shelf into five main tectono-stratigraphic sequences: basin and range, synrift I, synrift II, transitional, and thermal sag. In the Walvis Basin, the synrift II sequence is a very thick northward- and westward-thickening trough believed to be contemporaneous with the Etendeka volcanics. The transitional sequence represents

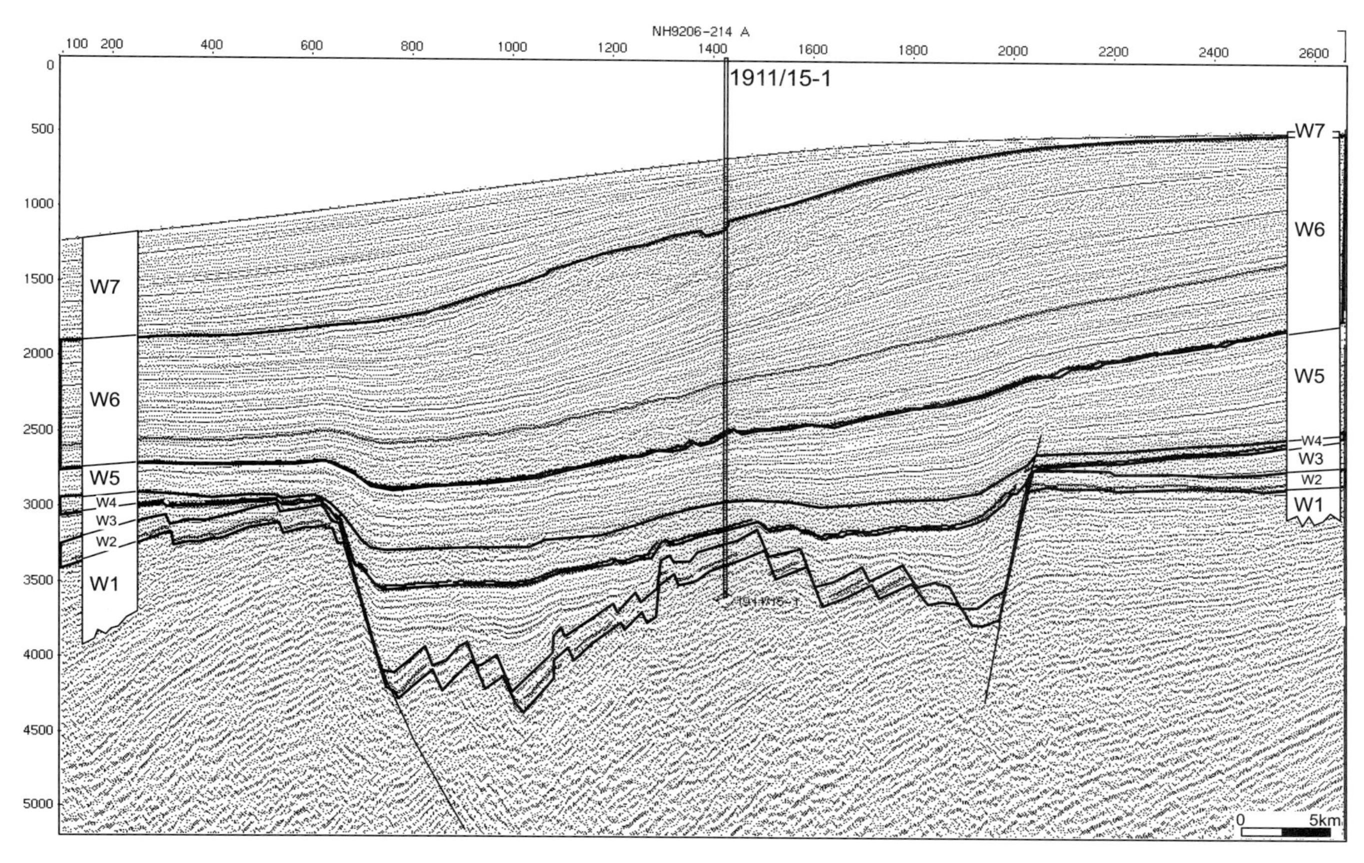

Figure 2—Seismic section across Quadrant 1911 (west–east) in the area of the 1911/15-1 well showing the suggested stratigraphic subdivision into the W1 through W7 Groups.

the Hauterivian–middle Aptian interval, and eolian sandstones and volcanics in wells on the Kudu gas discovery far to the south on the Namibian shelf mark the onset of thermal sag following the end of rifting. The thermal sag sequence comprises the middle Aptian–Holocene section which mainly represents a passive margin.

A strong seismic event in the eastern part of Quadrant 1911 is interpreted to be the top of Paleozoic or older basement. The event horizon dips steeply to the west, and disappears below 7 sec two-way traveltime before reaching the well locations. Eastward, the event approaches the surface before reaching the coast. The event is intersected by a few north–south trending normal faults with throw down to the east. A reliable correlation from this platform area into the basin and further onto the outer highs has not been established. Generally, the reflectors describe wedge-shaped packages that thicken to the southwest. The uppermost of these events has been labeled w100.

W1 GROUP

We have used the w100 reflector to define the base of the lithocolumn as described in this chapter. In the 1911/15-1 well location, the reflector ties in just below the total depth of the well and represents the shallowest possible base of the basaltic beds that constitute the W1 Group. Vertical seismic profile data from the well suggests that there is a marked increase in velocity (i.e., density) at this depth. The interval between reflectors w100 and w201 thins toward the northeast. The latter reflector ties into the well at 3947 m at the top of the youngest basalt. To the southwest, the interval between w100 and w201 thickens abruptly and displays a pattern of strongly divergent seaward-dipping wedges (Figure 4). Regional mapping based on a relatively open seismic grid shows a decrease in thickness from 3 sec two-way traveltime in Quadrant 2010 to 0 sec along a NNW–SSE trend that crosses the Elephant Platform and the northeastern corner of Quadrant 1911. The inferred continent–ocean boundary is situated just southwest of Quadrant 1911 (Figure 5).

Interpretation and 2-D modeling of aeromagnetic data indicate a gradual eastward decrease in thickness over the area of Quadrant 1911 from >1000 m in the west to a few hundred meters in the east. In the extreme northeastern comer of the quadrant, a strong linear magnetic anomaly indicates the termination of the entire lava wedge. The w100 and w201 reflectors accordingly converge in this area. The thickness and distribution of

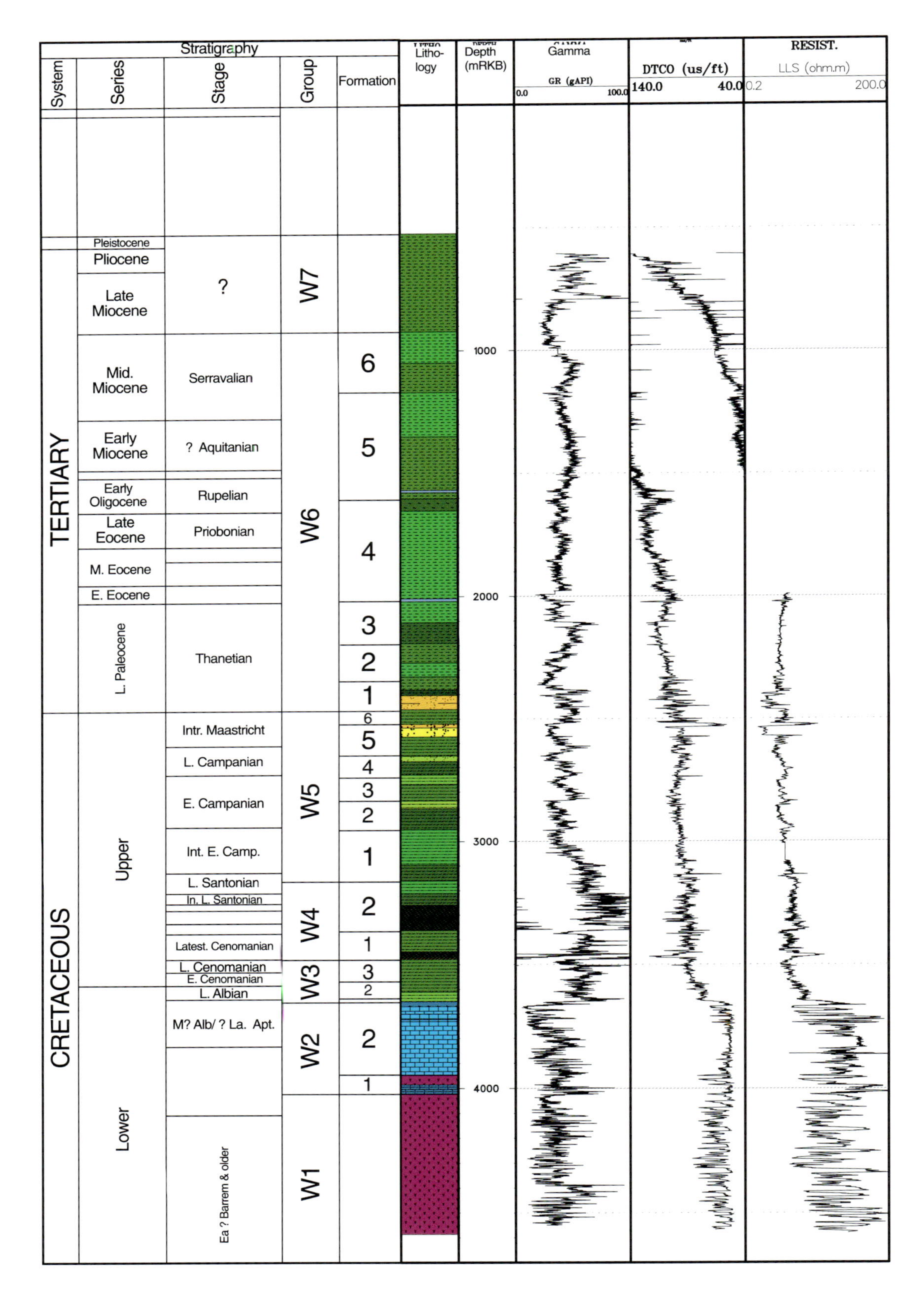

Figure 3—Summary electric log from well 1911/15-1 shown with the stratigraphic column for this part of the Walvis Basin and the suggested stratigraphic division into the W1 through W7 Groups. DTCO is the sonic velocity log.

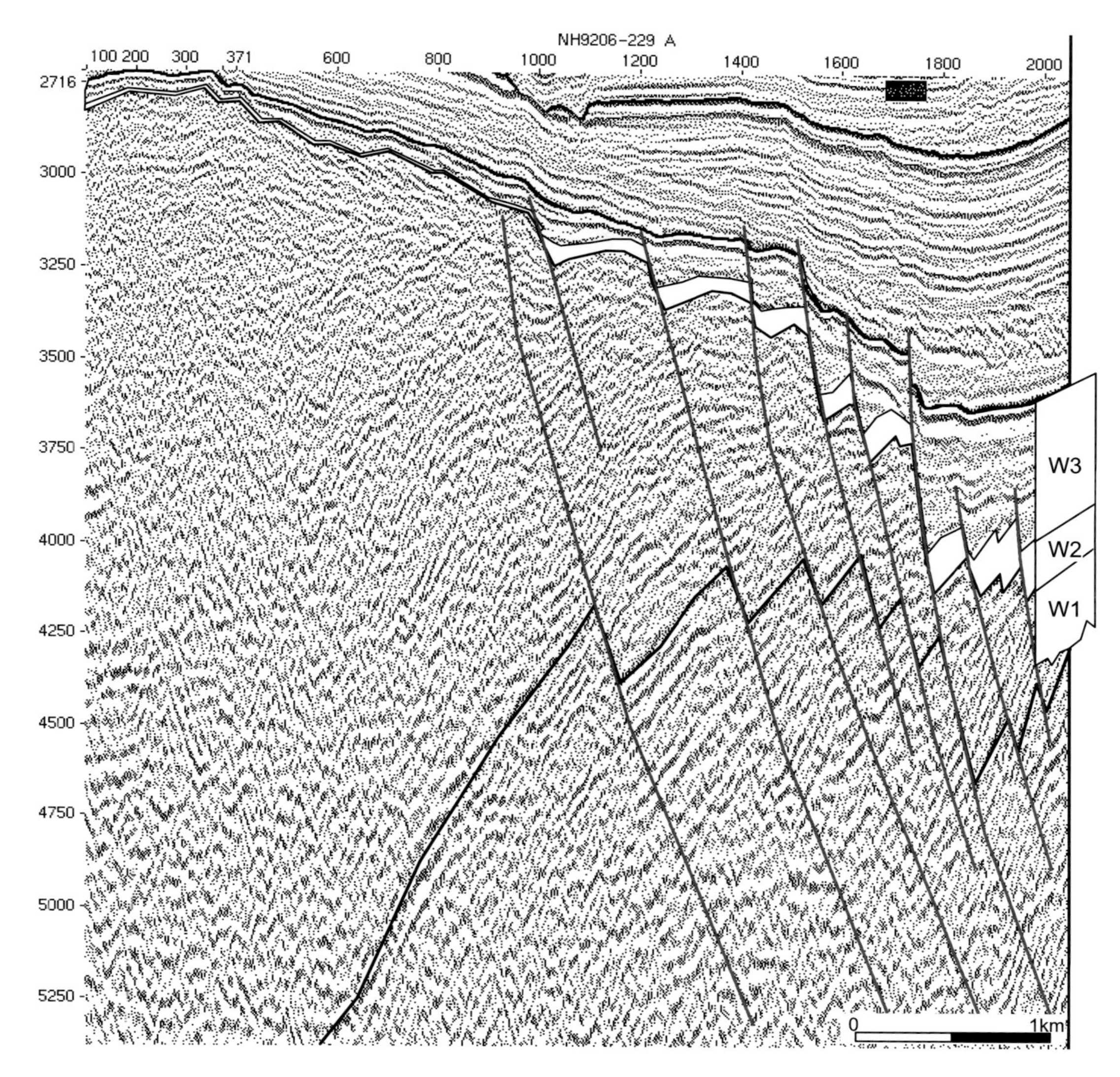

Figure 4—Seismic section through the Outer Highs (west–east) showing seaward-dipping reflectors within the W1 Group.

the basalt flows indicate that their origins are connected to the breakup of the Atlantic Ocean. The westward-dipping reflectors represent stacked subaerial flood basalts that flowed to the east from the spreading axis. The South Atlantic volcanic margin is described in more detail by Gladczenko et al. (1997).

In the 1911/15-1 well, Group W1 consists of a series of basaltic lava flows separated by tuffaceous layers and occasionally very thin beds of siliciclastic strata. The base of this unit was not reached when the total depth of the well was set at 4586 m. The top of the >563-m-thick continuous flow series is at 4023 m. The electric logs show that individual lava flows range in thickness from less than 1 m to about 20 m (Figure 6). One core (at 4340–4352 m) incorporates the base and top of two thick flows and the interstitial sequence. On the basis of different lithologic and textural characteristics, the cored section can be divided into four intervals:

- Interval 1 (4340–4343.1 m) comprises the lower part of one lava flow. Above 4342.25 m, the flow unit is massive with rare amygdules. Below this depth, the rock is slightly reddened and amygdules are abundant. Also, feldspar crystals change to a more acicular morphology. This interval probably represents the base of the flow.
- Interval 2 (4343.1–4345.4 m) comprises a composite flow unit produced by repeated surging of lava. Thin, nonamygdaloidal layers are glassy and probably represent freezing at the very base of each surge, all of which are thin enough to be amygdaloidal throughout.
- Interval 3 (4345.4–4347.25 m) is a unit containing several interbeds of siliciclastic sedimentary rocks. The base is marked by a distinctive reddened and silicified argillaceous siltstone. This is interpreted as a pedogenically modified lacustrine horizon

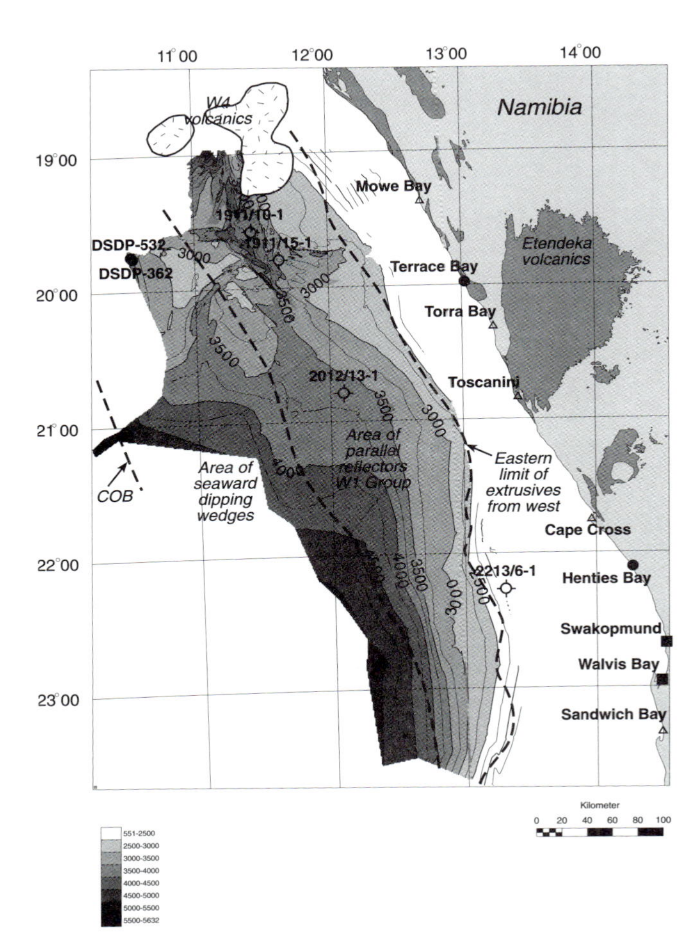

Figure 5—Structural map (in time units) of the top of the W2 Group, Walvis Basin, Namibia. Note the onshore outline of the Etendeka Group volcanics. Also note the dashed line marking the eastern limit of basalts from the spreading center. The dashed line on the west marks the onset of the seaward-dipping wedges. Contour interval is 500 msec two-way traveltime.

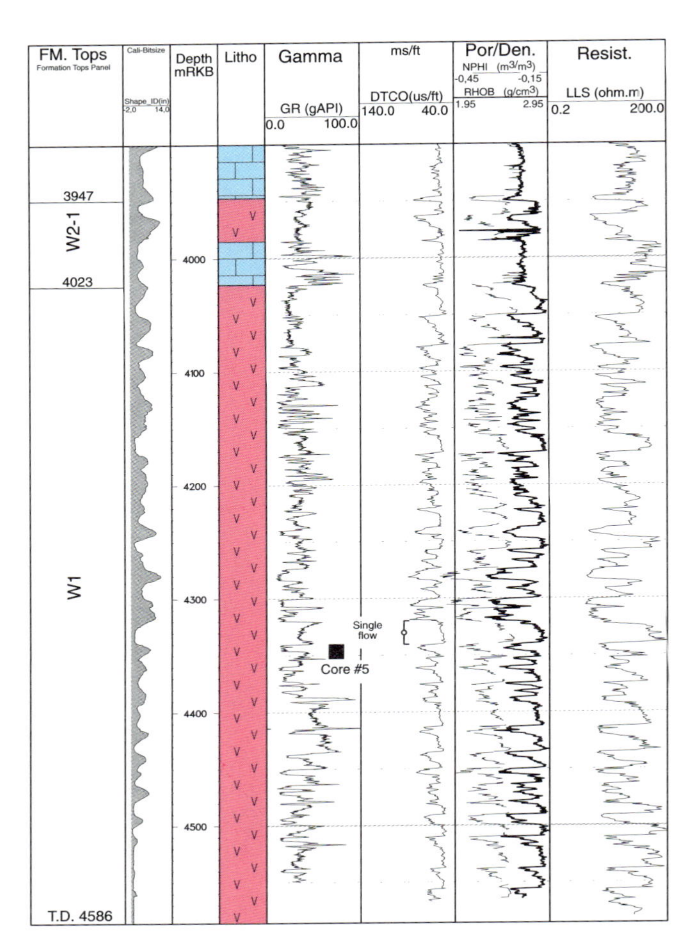

Figure 6—Wireline logs of the W1 Group volcanics from well 1911/15-1.

truncating the underlying lava flow. It is overlain by a thin, reddened amygdaloidal lava flow. Throughout this flow unit are thin, irregular siltstone layers that are thought to have been washed into fractures and open hollows within the flow. Thin-section investigations of the siltstones reveal that the mineralogic composition differs from what would be expected if they were derived solely from erosion of the surrounding lavas. They do contain plagioclase feldspar of volcanic origin, but they also contain common monocrystalline quartz and micas. This indicates a major contribution from a nonvolcanic basement source. The common presence of highly unstable aegerine augite or hornblende grains also suggests that the source was very proximal to the site of deposition.

- Interval 4 (4347.25–4352 m) comprises the top of a single thick lava flow unit. It is characterized by an upward increase in modal size, size range, and abundance of amygdules. Local reddening and significant mineralized fractures mark the top of the flow. Flow banding structures become visible as amygdule abundance decreases downward.

Attempts to date the lavas radiometrically have thus far not yielded a conclusive age. The occurrence of two lava surges interbedded with the lower part of the overlying carbonate sequence suggests that there is no significant gap in time between the W1 and W2 Groups. Barremian marine microfossils recorded in cuttings from the WI Group are probably caved from the W2-1 Formation, the lower unit of the W2 Group (see Figure 3).

In the Kudu gas field, southwest of Luedertiz, continental deposits consisting of a mixture of clastic and volcanoclastic strata of Barremian and older age are encountered in the reservoir sequence. These deposits include the main reservoir in the Kudu 9A-3 well and possibly also in the 9A-2 well. This reservoir, named the "lower gas sand" by Wickens and McLachlan (1990), comprises a medium-grained anhydritic sandstone interbedded with subaerial basalts and volcanoclastics. The depositional environment of the sandstones is interpreted as eolian, possibly a coastal dune complex (Wickens and McLachlan, 1990). Based on these observations, it seems reasonable to assign an age not younger than Barremian to the lavas.

W2 GROUP

Seismic and Stratigraphic Characterization

The pattern of seaward-dipping reflectors typical of the W1 Group terminates at the w201 reflector. The depositional geometry between the w201 and w202 reflectors is represented by a parallel band of strong reflectors. The w202 reflector ties in near the top of the W2 Group carbonates in wells 1911/15-1 and 1911/10-1.

The main depositional trend of the w201–w202 interval seems to be parallel to the present-day coastline (Figure 7). To the south of Quadrant 1911, a westward thinning of the interval is observed. This may reflect a depositional thinning onto an outer paleohigh, or it may be the effect of more compactable argillaceous facies. No trace of erosion is observed in this area. On the crests of the Outer Highs, significant postdepositional erosion has reduced the thickness of the w201–w202 interval and partially removed it (see Figure 4). To the northeast, this interval also thins, probably by onlap onto a landward rise. Generally, the w201–w202 interval thickens considerably in the northwestern corner of Quadrant 1911, where it shows apparent onlap from the west. It is unclear whether the interval in this area represents the W2 Group as seen in wells, a basinward equivalent of the W2 Group, or a series of late-stage lava flows possibly belonging to the W1 Group.

In the 1911/15-1 well, the carbonate deposits of the W2 Group rest directly on volcanics at 4023 m. The boundary between W1 Group volcanics and W2 Group carbonates is represented by a distinct increase in logged gamma-ray response (see Figure 6) which coincides with a decrease in neutron density. The top of the carbonates is at 3652 m, giving a total thickness of 371 m for the carbonates.

Subdivision of the W2 Group

We suggest a division into two stratigraphic units (or formations, in this case) for the W2 Group. The lower part of the lower W2-1 Formation is comprised of 41 m of muddy limestones, marls, and glauconitic claystones. High, spiky gamma-ray readings (up to 100° API) in this part of the formation are due to uranium enrichment and could indicate stagnant environments during deposition. Analysis of sidewall cores from this interval does not show elevated total organic carbon (TOC) contents (<0.15%). These deposits are separated from the W2-2 Formation by a 35-m-thick basaltic package consisting of two major flows (see Figure 6). The w201 reflector ties in

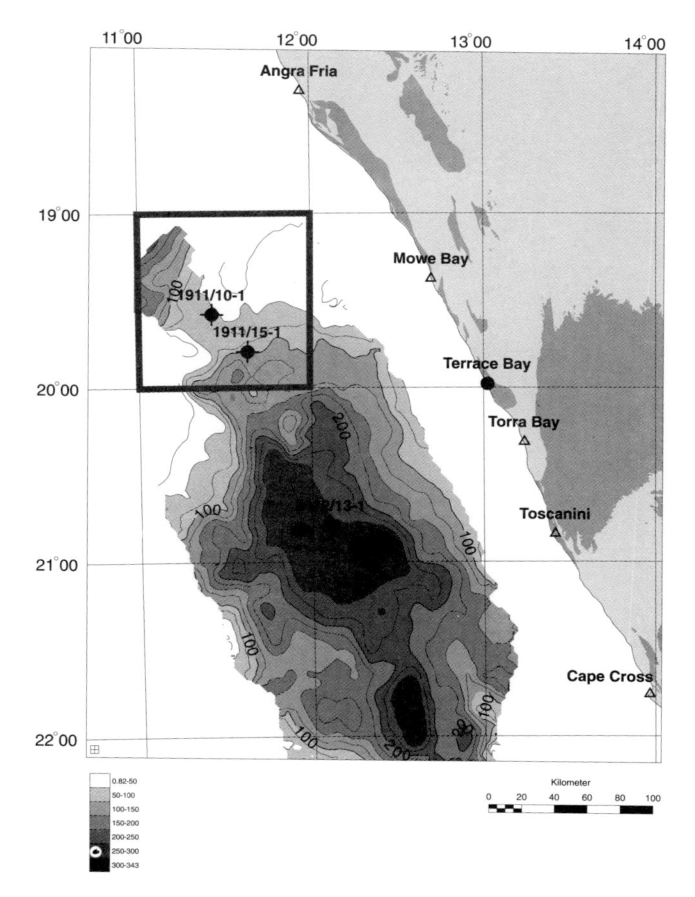

Figure 7—Subregional interval thickness (time) map of the W2 Group, Walvis Basin, offshore Namibia. Contour interval is 50 msec two-way traveltime.

with the top of the youngest flow at 3947 m in the well. Including the basalts, the thickness of the W2-1 Formation is 76 m.

There is a sharp upward decrease on the neutron density log at the transition between the upper basaltic bed of the W2-1 Formation and the basal limestones of the W2-2 Formation. The 295-m-thick W2-2 Formation is mainly composed of micritic mudstones with a substantial amount of algally bound pelloidal grainstones and packstones. Oolites are frequently described in cuttings from the uppermost 50–60 m.

One core was cut in the upper part of the W2-2 Formation (3724–3742.5 m). It reveals algal–pelletal grainstone and packstone facies that are interbedded with subordinate pelletal–foraminiferal packstones and grainstones, as well as lime mudstones. These deposits contain very little siliciclastics and no evaporites. The few quartz grains that are seen in thin-sections are very fine sand to silt size and are all well rounded.

Bulk cuttings from this group invariably yielded TOC contents less than 1%. Picked shale lithologies typically have TOC contents in the 1–2% range with hydrogen

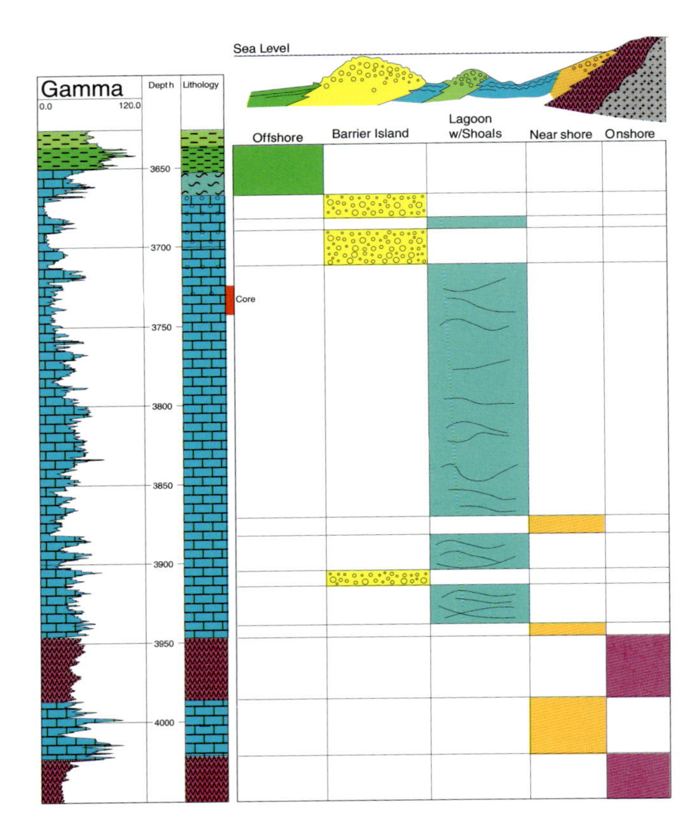

Figure 8—Schematic stratigraphic column showing depositional transgressive facies development over time and space in the W2 Group carbonates. Also shown (on the left) are the wireline logs from well 1911/15-1, Walvis Basin.

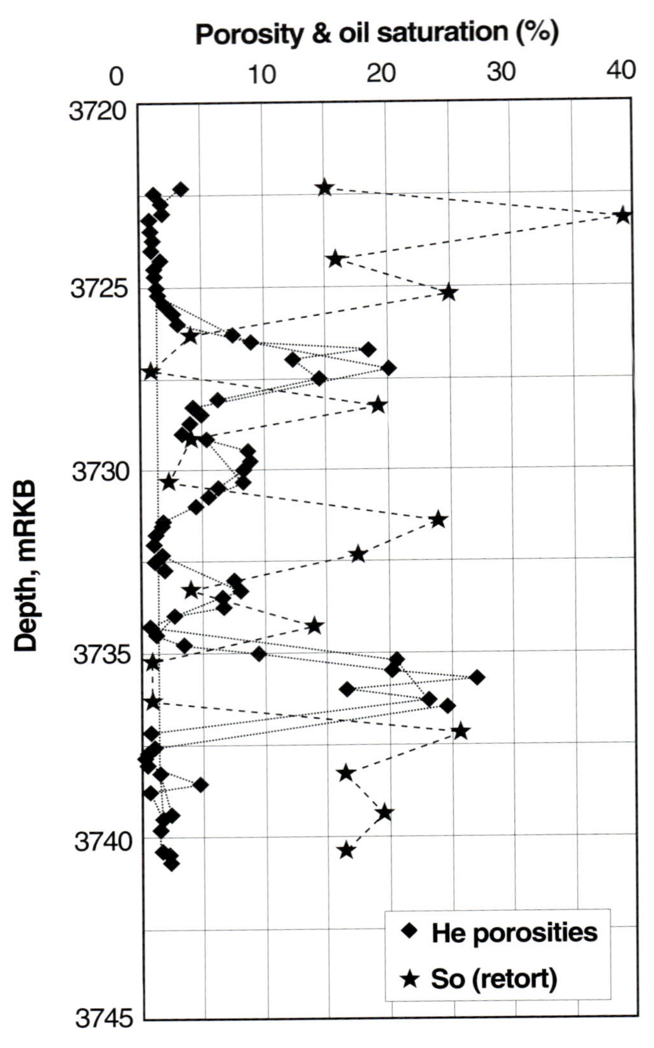

Figure 9—Plot of porosity and oil saturation versus depth for the W2 Group carbonates in well 1911/15-1. (RKB is rig kelly bushing, the reference for wireline log depths.)

indices (HI) in the range of 100–200 mg HC/g TOC (type III kerogen). Based on these screening data, the carbonates can be regarded only as potential source rocks if there are significant lateral facies changes.

Well-preserved diagnostic microflora and microfauna are scarce in the lower interval of the W2 Group. Barremian marine microfossils recorded in cuttings from the volcanics of the W1 Group are probably caved from the W2-1 Formation. The maximum age of the W2-1 Formation is thus most probably Barremian. The most probable age range for the W2-2 Formation is late Aptian–middle Albian. The W2 Group is thus assigned a Barremian–middle Albian age. This implies that the basal carbonate deposits are somewhat older than the massive salt deposits and overlying Pinda Group carbonates and evaporites of the Kwanza Basin in Angola (Abilio, 1986).

The succession of various depositional facies represented in the W2 Group is summarized in Figure 8. The return of lava flows in the lower part of the group indicates a reemergence of the area after the initial marine transgression. The coarser grained zones encountered in the core are interpreted as middle platform shoal deposits, while the thicker oolithic zones in the uppermost part of the W2-2 Formation are considered to be platform margin grainstones. The development from restricted lagoonal environments to more high-energy (barrier island) environments, as evidenced by the upward increasing proportion of grainstones containing ooides, indicates a landward shift of facies during a general marine transgression.

W2 Group Reservoir

The main reservoir target in well 1911/15-1, drilled into the top of rotated prerift blocks, turned out to be shallow-marine carbonate deposits of Aptian–Albian age. Prior to drilling, this interval was regarded as the most prospective within the license area, mainly because of its high number of large structural closures as defined by the w202 reflector. The lateral distribution of the different carbonate facies is difficult to predict, especially the distribution of its reservoir properties.

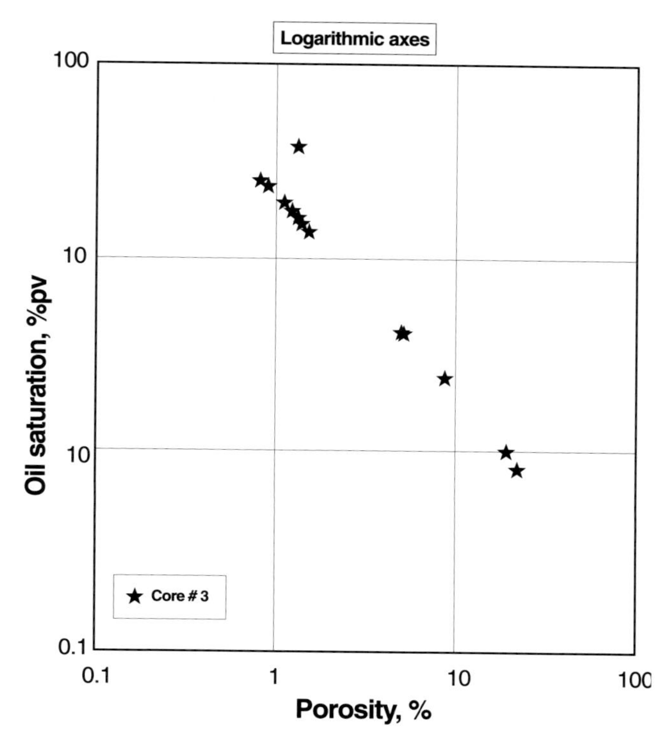

Figure 10—Plot of residual oil saturation versus porosity from retort analysis for the W2 Group carbonates in well 1911/15-1.

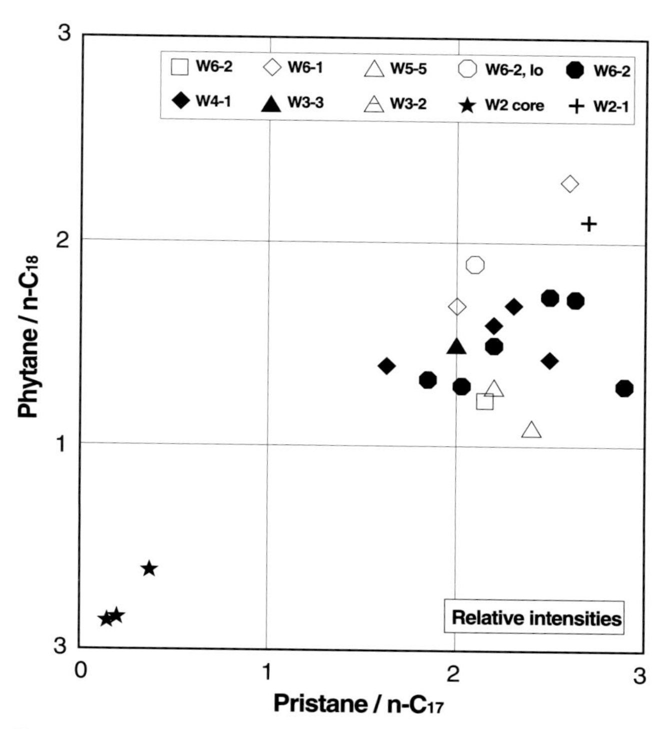

Figure 11—Plot of phytane/n-C_{18} ratios versus pristane/ n-C_{17} ratios for rock extracts from various formations in well 1911/15-1.

The cored part of the carbonates revealed zones with very good reservoir properties. Porosities of 15–25% were measured within the algal–pelletal grainstone and packstone facies (Figure 9). A high proportion (70–95%) of this porosity occurs in macropores. These intervals have permeabilities ranging from 10 to 500 md and seem to correlate well with low-velocity intervals distinguished on the sonic velocity log (see the DTCO log in Figure 3). According to the logs (sonic velocity and gamma), such high-porosity zones are especially found in the upper one-third (about 100 m) of the W2 Group limestone. The net to gross ratio in this interval (3650–3750 m) is estimated to about 35%, with an average porosity of 20%.

Enhanced oil saturation (15–40%) was recorded in samples from low-porosity zones of the carbonates in the core (Figure 9). The porosity versus oil saturation plot in Figure 10 shows an inverse relationship between the two parameters. Such a relationship is generally found in depleted reservoirs. Saturated fraction gas chromatograms (C_{15+} GC) from core extracts have a lighter end-biased appearance. They also show low isoprenoid/n-alkane ratios and carbon preference indexes near 1.0, indicating a mature source (Tissot and Welte, 1984).

Figure 11 shows the phytane/n-C_{18} versus pristane/n-C_{17} ratios for the saturated fraction of the core extracts plotted together with those derived from other parts of the W2 Group. The difference in isoprenoid/n-alkane ratios between the core samples and the samples from other parts of the sequence is striking. While ratios from all other samples indicate that the host rocks have not yet reached the main hydrocarbon-generating zone, the ratios for the core extracts indicate a very mature source rock. The ratios from the core extracts are in fact about the same as or even lower than ratios found in many condensates.

The source of these hydrocarbons has not yet been established. However, based on our observations, we believe that the traces of residual oil encountered in the W2-2 Formation core extracts may represent remnants of hydrocarbons that have migrated from a mature source kitchen, either in fluid phase as a light oil or in a gaseous phase as a condensate.

W3 GROUP

Seismic and Stratigraphic Characterization

In the 1911/15-1 well, the base of the W3 Group is characterized by a significant downward drop in gamma-ray (Figure 8) and transit-time log responses at 3652 m, coinciding with a marked neutron density increase. The top is at 3479 m, where the w401 reflector ties into the well. Its total thickness is 173 m in the well. On structural highs, the W3 Group is thin to absent.

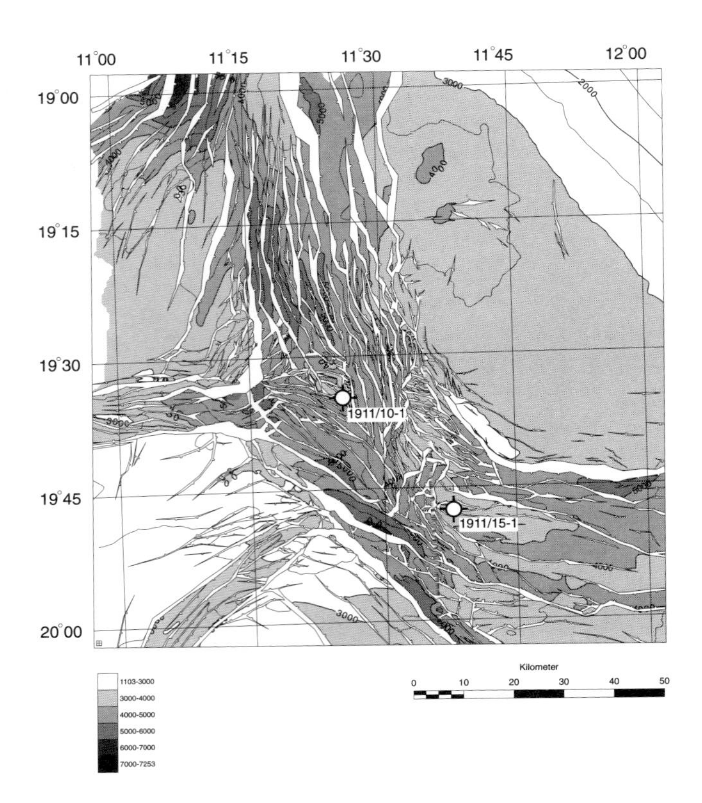

Figure 12—Structural depth map for the base of the W3 Group, Quadrant 1911, Walvis Basin, Namibia. Contour interval is 1000 m (below mean sea level).

In contrast to the parallel band of strong reflectors seen below, the seismic data exhibit a transparent and divergent pattern above reflector w202, which defines the base of the W3 Group (Figure 2). The divergent pattern shows that the area experienced a period of differential subsidence and tilting of individual fault blocks. The organization of faults on the structural depth map in Figure 12 reveals that a three-pronged graben feature was formed as a consequence of this major tectonic restructuring of Quadrant 1911. Farther south, the relief dies out, and the surface becomes a more smooth westward-dipping slope (see Figure 5).

The top of the W3 Group is defined at the prominent w401 reflector, which is very continuous but exhibits some degree of deterioration to the east. Within the W3 Group, the w301 reflector defines the top of the major sedimentary wedges that thin onto the crest of most rotated fault blocks, including the drilled 1911/15 structure (Figure 2). Onlap onto the w202 reflector occurs in most areas. The transparent pattern below the w301 reflector may in certain areas be continuous over the highs and cover several fault blocks. In other areas, this pattern appears to be limited to the structurally lower part of the individual blocks.

The interval above the w301 reflector shows rather strong, continuous, and generally more widely distributed parallel reflectors. At its lower boundary, the interval seems to exhibit some degree of onlap or downlap. Above underlying faults or flexures, the parallel internal pattern may change abruptly and become chaotic and open with a mounded character. These features terminate upward at the w302 reflector, where the pattern becomes significantly more concordant over the entire basin area.

Although the thickness of the interval between the w202 and w401 reflectors may reach up to 800 msec two-way traveltime in some of the half-grabens in Quadrant 1911, it is still rather thin on a regional scale. In a depocenter in Quadrant 2112 to the southeast, the interval reaches thicknesses up to 1200 msec.

Within Quadrant 1911, the W3 Group has significantly higher interval velocities derived from seismic stacking velocities than seen in the wells. The interval velocities increase with increasing thickness of the group, reaching values of more than 5000 m/sec. The high interval velocities are clearly not compatible with shales or porous sandstones and instead may suggest that the lithologies of the W3 Group are more calcareous in the rift basins compared to what is observed in the well. The transition from the carbonate environments of Group W2 to the siliciclastics of Group W3 may be regarded as overall transgressive and is apparently related to a tectonic phase that occurred in the middle Albian.

Bulk cuttings from the W3 group yielded TOC contents from 0.3 to 2% with HI values generally in the range of 50–250 mg HC/g TOC. The richer samples are found scattered throughout the group. As the group thickens dramatically into local graben areas, the organic facies may improve downflank into more basinal settings. Vitrinite reflectance values of about 0.5% R_o and high pristane/n-C_{17} and phytane/n-C_{18} ratios (Figure 11) in the saturated hydrocarbon fraction of the extracts indicate low or moderate maturity.

Subdivision of W3 Group

Based on seismic mapping and log response, a threefold division of the W3 Group is proposed. The W3-1 Formation is easily distinguishable on well logs by its high, spiky gamma-ray signature (3652–3637 m) (Figure 8) that contrasts with both the underlying carbonates of the W2 Group and the overlying claystones of the W3-2 Formation. Lithologically, this unit is dominated by glauconitic siltstones grading upward to claystones with limestone stringers. The top of the W3-1 Formation is at 3637 m, giving it a thickness of only 15 m. Although this is actually below seismic resolution at the well, by defining the w301 reflector as the top of the formation, its thickness increases to more than 1000 m in graben areas.

The W3-1 Formation was deposited during or just after the main period of differential subsidence. This unit is of middle Albian age and it is very condensed in the well. As the W3-1 Formation thickens dramatically into graben areas, the lithologies probably become more varied. The recorded high gamma log and low sonic velocity log signatures indicate that the unit may be a potential source rock for hydrocarbons, especially downflank from highs, such as the one drilled by the 1911/15-1 well. Maturation modeling (using the *PetroMod 2D* software package) suggest that the thick, downflank deposits of the W3-1 Formation reached maturities well into the condensate–gas zone as early as Turonian–Santonian time. Although a direct correlation between these potential petroleum source rocks and the residual hydrocarbons encountered in one of the cores has not been established, they may very well have been parts of the same paleo–petroleum system.

On well logs, the W3-2 Formation can be distinguished from the W3-1 Formation by a sharp upward decrease in gamma log response, coinciding with a moderate increase in neutron density. The W3-2 Formation consists of 68 m of light gray claystones, occasionally grading upward to silty claystones. The top is at 3569 m where the w302 reflector ties in. Although the thickness is not as great as for some parts of the underlying W3-1 Formation, it reaches more than 800 m in the central parts of graben areas. Lithologic facies changes within this unit can be expected in the chaotic mounded sections which may be composed of carbonate buildups or sandy deposits.

The W3-2 Formation was deposited during a mainly passive, intermediate stage of sedimentary fill into middle Albian age grabens. Faulting seems to have been restricted to the main basin-bounding faults in the west and east and occurred with little or no tilting of the minor fault blocks. The observed change from silty lithologies in the W3-1 to more argillaceous lithologies in the W3-2 suggests deposition in deeper and/or calmer waters. The chaotic mounded sections seen on the crests of tilted fault blocks are interpreted either as buildups of biogenic carbonates or as siliciclastic sand ridges created by wave and current action. Low-angle accretionary clinoforms observed on the eastern platform suggest that sediment input from the Namibian mainland caused a westward progradation of a coastline which may have reached into the Quadrant 1911 area at the time. The age of the unit is late Albian, possibly stretching into the early Cenomanian.

The upper part of the W3 Group is assigned to the W3-3 Formation. In well 1911/15-1, it consists of silty claystones grading upward into siltstones. The main criteria for identifying the transition between units W3-3 and W3-2 on well logs are moderate upward decreasing gamma-ray responses coinciding with decreasing sonic velocity and neutron density responses. The w401 reflector ties into the top of the W3-3 Formation in the well at 3479 m (3122 msec). W3-3 thickness is 90 m, but it increases to close to 400 m in a depocenter southwest of the well.

The W3-3 Formation is interpreted to represent a late-stage passive infill into the Albian topography. Thickness variations in the basin can mostly be attributed to accommodation space created by differential compaction of underlying W3 Group strata. Its history of deposition spans most of the Cenomanian.

W4 GROUP

Seismic and Stratigraphic Characterization

The base of the W4 Group corresponds to the w401 reflector, which ties in at 3479 m to the 1911/15-1 well. This reflection is attributed to a positive impedance contrast between the higher velocity silty deposits of the W3-3 Formation and the low-velocity, high gamma-ray shales at the base of the W4-1 Formation. The reflector is very strong and continuous, only deteriorating to the east where a reduction in the impedance contrast probably reflects a facies change. A shallower reflector, the w501, seismically defines the top of the W4 Group. This ties in at 3163 m to the well. The total thickness of the W4 group is thus 316 m.

Age dating based on palynology indicates that the depositional history of this group started in the latest Cenomanian, continued through the Turonian and Coniacian, and ended sometime in the late Santonian. This gives the W4 Group a depositional time span of 6–8 m.y. and makes it roughly time equivalent to the clastic Cabo Ledo, Itombe, and N'Golome Formations in the Kwanza Basin of Angola (Abilio, 1986).

The age of the W4-2 Formation ranges from early–middle Turonian to late Santonian. Based on the seismic observations described above, this is the most probable time of growth of the Phoenix High volcano and some of the other bathymetric features that have been linked to the Walvis Ridge in this area.

In large parts of the graben area, the internal seismic pattern of the W4 Group changes upward from parallel with an onlapping lower boundary to oblique reflectors that generally dip eastward (Figure 13). The bounding surface between these two different internal patterns coincides with the base of a second high gamma-ray shale package at 3366 m in the 1911/15-1 well (top of the W2-1 Formation; see Figure 3). Based on these log and seismic features, we have divided the W4 Group into two formations.

Although the W4-1 Formation is only 72 msec (113 m) thick at the well location, its sedimentary sequence attains a thickness of as much as 150 msec in the deeper part of the basin. The entire W4 Group thins eastward onto the Elephant Platform, where the two formations become seismically inseparable. This eastward thinning is in contrast to the drastic thickening of the overlying W5-1 Formation observed toward the east.

In the northern part of Quadrant 1911, a volcanic center, the Phoenix High, is distinguished by seismic data (Figure 14) and high-resolution magnetic data. By following the various reflectors that define the external and internal geometry of this feature to where they converge and interfinger with the interpreted sedimentary packages, its relationship to the W4 Group becomes evident. The growth of the Phoenix High volcano is thus interpreted to have been contemporaneous with the deposition of sediments in the W4 Group (probably the W4-2

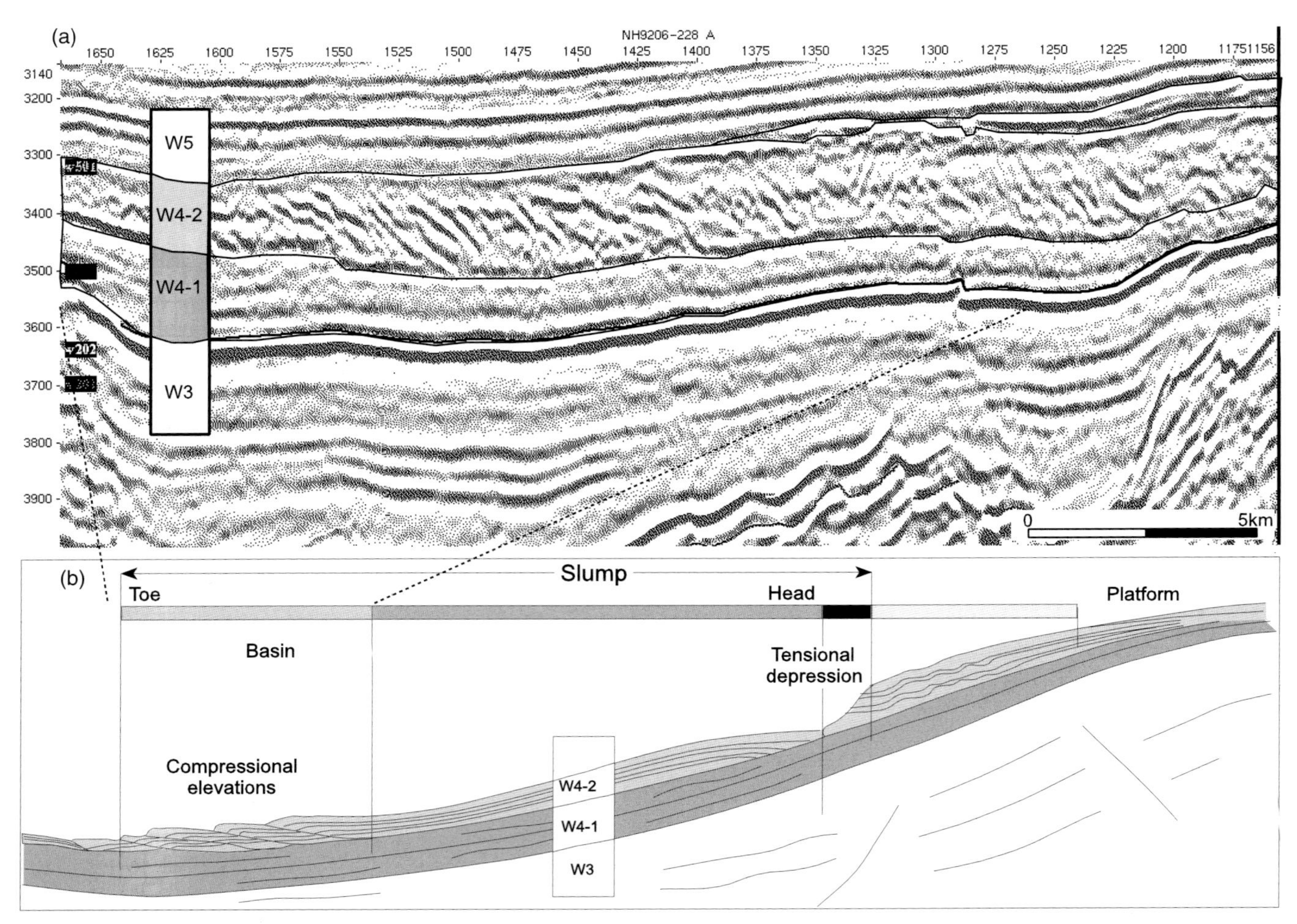

Figure 13—(a) Seismic section (west–east) showing the internal seismic pattern of the W4 Group. (b) Slump model for the W4 Group interpreted from the above seismic section.

Formation). Internal seismic patterns also indicate that the growth of the volcanic complex may have triggered soft sediment rearrangement during deposition of the W4-2 Formation (Figure 13b).

The seismic profile (Figure 14) beneath this volcanic sequence is difficult to interpret. It is probable that the main fault separating the platform to the east from the graben to the west continues approximately north–south and that the volcanics now straddle these two tectonic elements. At least two more volcanic centers are identified, one as a northward continuation of the Phoenix High and one just to the northwest of Quadrant 1911 (Figure 5). Based on mapped reflectors that lap onto these other volcanic features, they seem to have been contemporaneous with the Phoenix High, or possibly slightly younger. The outlines of the Phoenix High and the other volcanic features are also detectable on detailed bathymetric maps of the area and have been related to the Walvis Ridge bathymetric feature by previous authors (Light et al., 1993).

W4 Group Source Rock

In the 1911/15-1 well, the most characteristic log features of the W4 Group are the two distinct high gamma-ray carbonaceous claystone intervals that occur at the base of each subunit. The main lithology of the W4-1 Formation is greenish gray claystone with minor interbeds of siltstone. The formation totals 113 m in thickness, and the lower 22 m is comprised of dark yellow-brown argillaceous and calcareous siltstones with fairly large amounts of carbonaceous material. Log signatures and weak shows on cuttings and sidewall cores gave early indications of source rock type deposits. Further analysis proved that, within this interval, the TOC content varies from 5% to >10%. Rock-Eval pyrolysis (Figure 15a) indicates very good type II kerogen, with HI values reaching 600 mg HC/g TOC. The rest of the W4-1 claystones also exhibit source rock quality, with typical TOC values in the range of 1–6% and HI values generally ranging from 200 to 400 mg HC/g TOC.

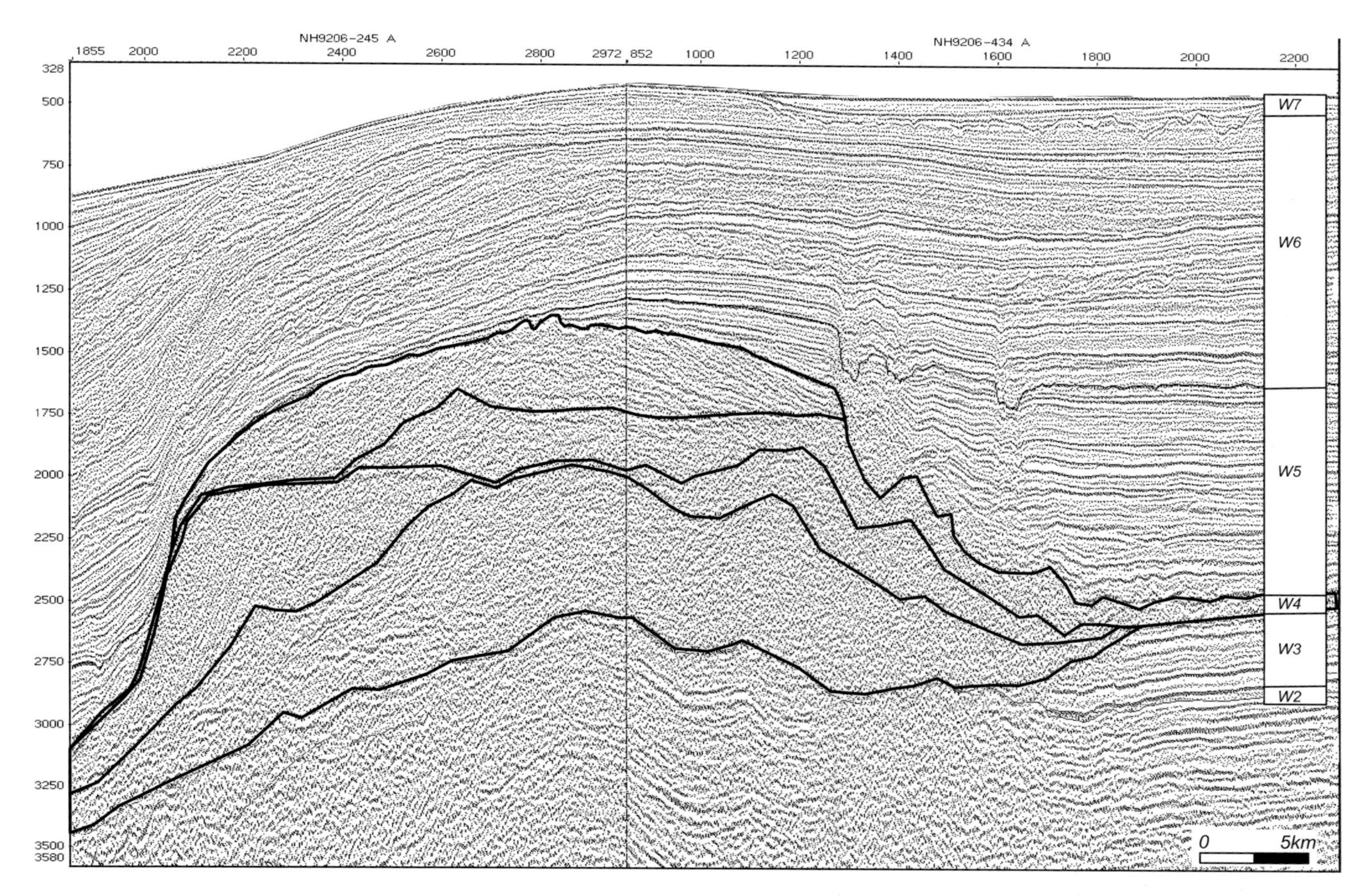

Figure 14—Seismic section across the Phoenix High, Walvis Basin. This is a composite section, running WSW-ENE (on the left) and NNW-SSE (on the right).

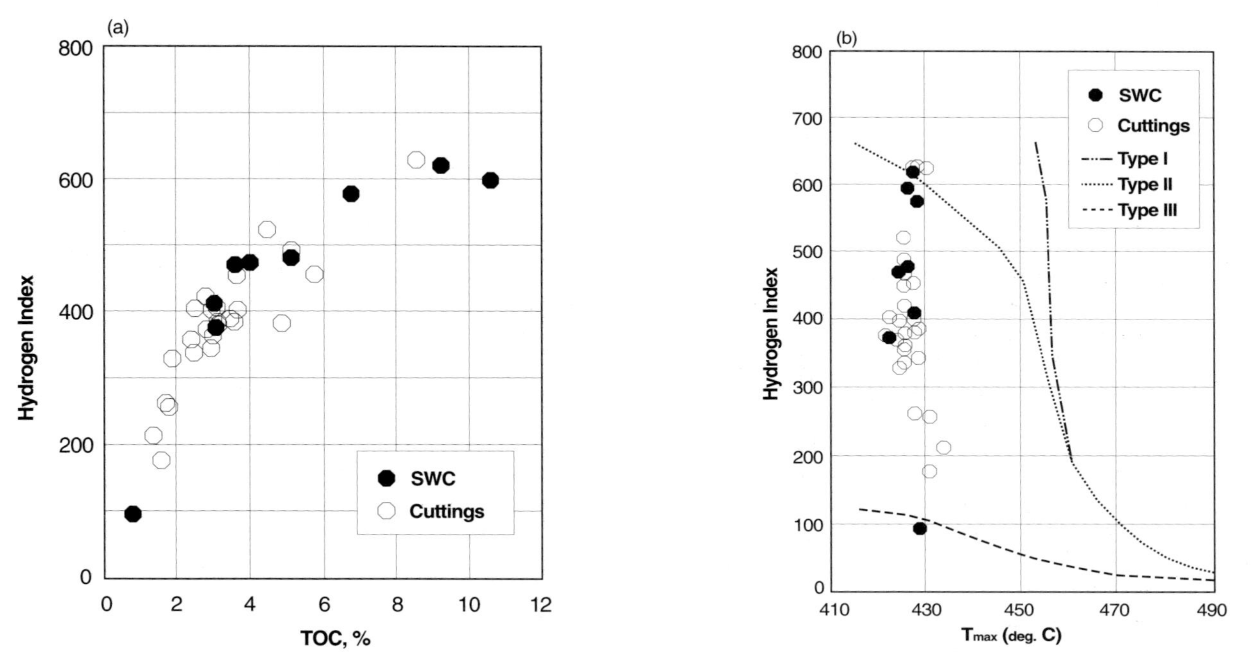

Figure 15—Geochemical data from sidewall cores (SWC) and cuttings of the W4 Group, well 1911/15-1. (a) Plot of HI versus TOC. (b) Plot of HI versus T_{max}.

The W4-2 Formation is comprised of dusky yellow-brown claystones and siltstones. The latter sometimes grade into very fine sandstones. Carbonaceous material is common throughout, and the TOC content is generally 2–5% with HI values typically at 350–500 mg HC/g TOC. As indicated by T_{max} data (Figure 15b) and by the weak shows recorded while drilling, these sedimentary rocks have reached maturities only in the very upper part of the oil-generating zone in the location of the 1911/15-1 well.

The high content and quality of the organic matter in these sediments and the high uranium content in the lower part indicates an initially anoxic paleoenvironment followed by dysoxic conditions. Bottom water circulation was probably poor or absent.

W5 GROUP

Seismic and Stratigraphic Characterization

In the 1911/15-1 well, the W5 Group consists of dark gray to olive gray mudstones with interbeds of sandstones and thin limestones. Thin beds of very fine grained sandstone are found in the upper part of some of the formations (Figure 3). The base of the group is at 3163 m and the top at 2463 m. This total thickness of 700 m increases to more than 1200 m on the Elephant Platform to the east. Palynologic age dating indicates that the depositional history of the W5 Group lasted from late Santonian to late Maastrichtian time.

The base of the W5 Group marks the end of a tectonically active period that involved volcanism and erosion of footwall blocks. As a response to significant differential subsidence between the exposed African continent and the offshore area, large sedimentary clinoforms began to reach the Quadrant 1911 area at this time and a typical passive margin shelf was established. The depositional geometry of the W5 Group is hence dominated by large prograding wedges that accumulated in the northeastern part of Quadrant 1911, whereas parallel laminated beds of moderate thicknesses are typical for the western area.

Subdivision of W5 Group

We suggest subdividing the W5 Group into six stratigraphic units. Each of these is interpreted to represent a genetic sequence with a transgressive system tract at the base followed by a highstand and eventually a lowstand system tract. Basin floor fans are occasionally recognized basinward of highstand or lowstand clinoforms.

The W5-1 Formation is well developed on the Elephant Platform, with thicknesses reaching more than 1000 m. On seismic lines, four systems tracts can be recognized within this formation (Helland-Hansen, 1995). The initial transgressive systems tract is generally thin in this area and not readily recognized. As can be seen in Figure 16, a major highstand event created a clear shelf edge clinoform followed by a forced regression wedge and a lowstand wedge. The forced regression wedge shows steep foreset beds that accumulated entirely in front of the previous shelf. The lowstand wedge has less steep foreset beds, and its strata partially accumulated on the shelf. During highstand, the shelf developed along a northwest–southeast trend. Deposition shifted basinward during the subsequent lowstand, mainly on the previous slope. On the seismic profile (Figure 16), we have interpreted high-density submarine fan and turbidite deposits in the Dolphin Basin that are correlated to the highstand part of the W5-1 Formation. In well 1911/15-1, the 210-m-thick W5-1 Formation is defined between 3163–2953 m. The age of this formation is late Santonian–early Campanian.

Along the southern edge of the Phoenix High, a large erosional channel cuts deeply into the W4 Group. Seismic tie to the well indicates that this feature was created during the lowstand period of W5-1 deposition and was filled during the subsequent transgression, as evidenced by the base of the W5-2 Formation. In well 1911/15-1, the total thickness of the W5-2 Formation is 117 m (2953–2836 m); its age is early Campanian. The overlying W5-3 Formation is defined at 2836–2738 m, with a thickness of 98 m, and its age is Campanian.

The w504 reflector near the top of the W5-4 Formation defines the top of a massive mounded fan system in the southeastern part of Quadrant 1911. It corresponds to the upper lowstand part of this formation. The fan system, which we would expect to contain coarse clastics, extends farther south, but it thins completely out downflank before reaching the site of well 1911/15-1. In the well, the W5-4 Formation is defined at 2738–2652 m. This unit was deposited mainly during the late Campanian.

The W5-5 Formation is defined at 2652–2522 m with a thickness of 130 m. It is dated as Maastrichtian. At the top of the formation, there is a sandstone interval (2522–2572.5 m) consisting of fine- to medium-grained (but occasionally coarse-grained) quartz with kaolinite as a secondary mineral. In the upper 12–15 m, it is tightly cemented with silica, but judging from wireline logs, the porosity in the rest of the sandstone varies between 25 and 30%. No hydrocarbon shows have been recorded in these sandstones. They are not readily detectable on seismic lines as a geometric feature, but they are assumed to be present where the seismic character appears discontinuous. The tightly silica-cemented upper part of the sandstone interval should give a pronounced reflector, but the event is not clear and continuous on the seismic. This may be caused by slumps or channeling at the well location. The sandstones are thought to represent a basin floor sheet sand shed into the basin during the lowstand part of the W5-5 Formation.

In the eastern part of the Elephant Platform, the internal reflectors of the W5-6 Formation dip to the west as the thickness of the unit increases (Figure 17). In map view, the feature is elongated northwest–southeast along the shelf edge and has a seemingly hard and irregular upper surface. Combined with observations from the overlying W6 Group, it is interpreted to be a set of carbonate mounds or reefs. This interval is not penetrated by wells in offshore Namibia.

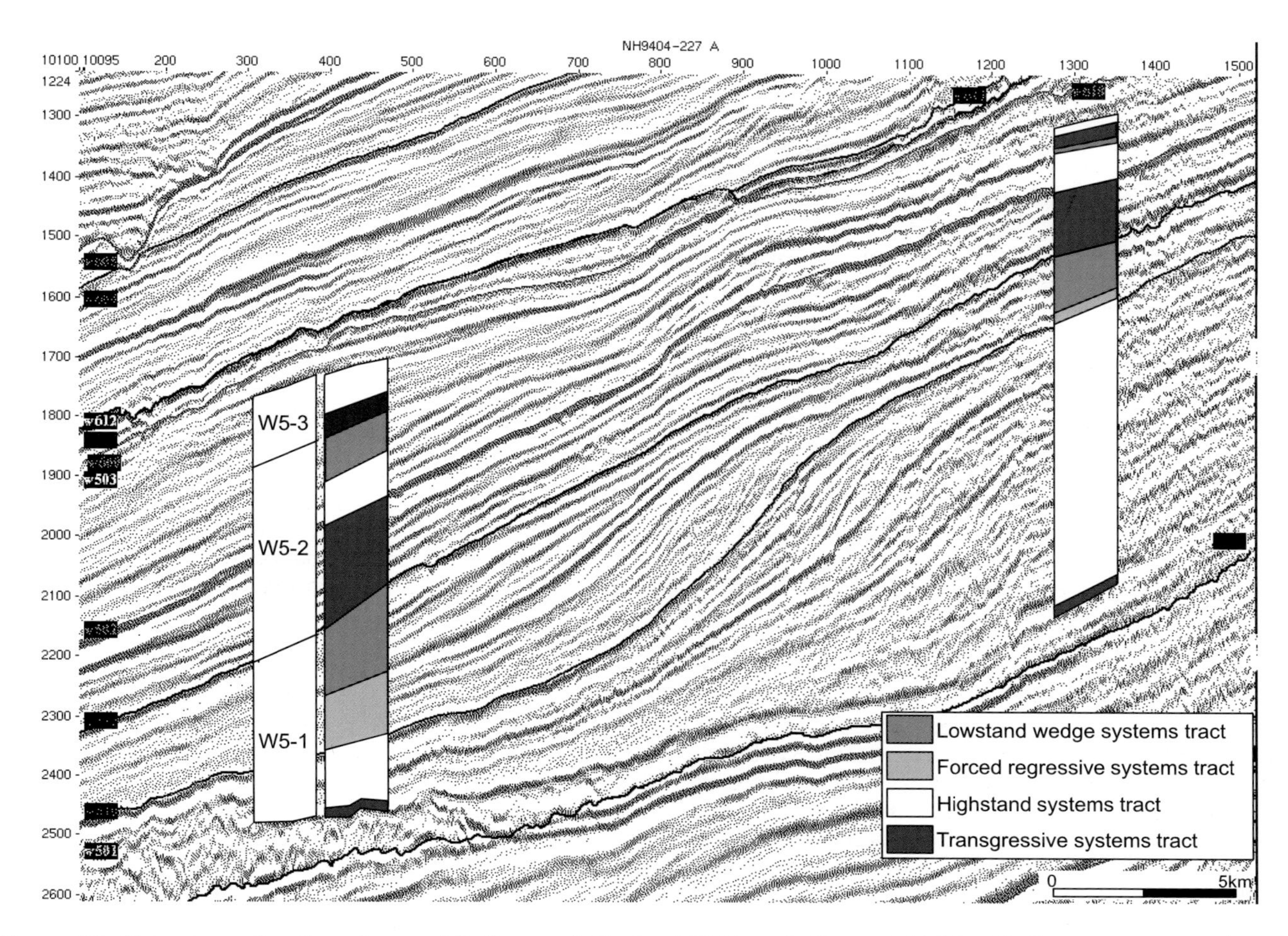

Figure 16—Seismic section from the west (left) toward Elephant Platform to the east (right) with the sequence stratigraphic nomenclature superimposed on formations of the W5 Group.

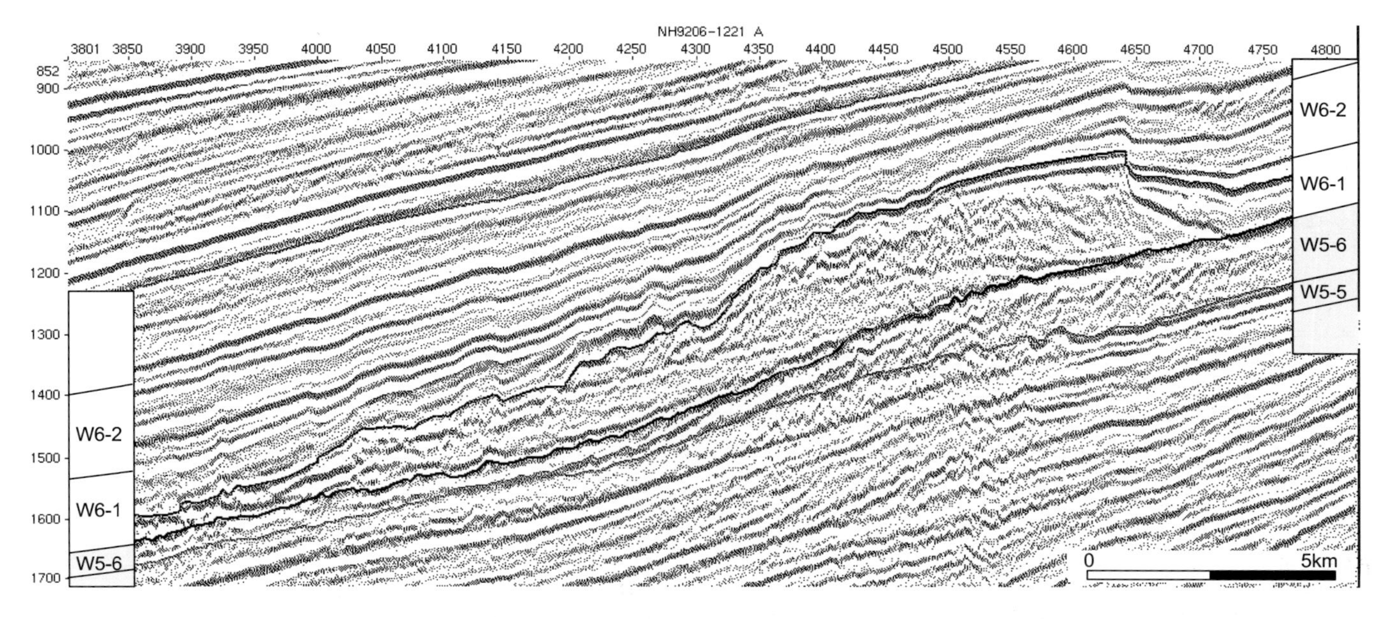

Figure 17—Seismic section from the west (left) toward Elephant Platform to the east (right) showing the reef-like features in the upper part of the W5 Group and lower part of the W6 Group.

A marly claystone interval in the W5-6 Formation in well 1911/15-1 is interpreted to represent a basinal restricted onlapping wedge associated with the mounded features near the shelf edge. In well 1911/15-1, the W5-6 Formation is defined between 2522–2463 m (59 m thick). The unit yielded Maastrichtian age microfossils.

W6 GROUP

Seismic and Stratigraphic Characterization

In well 1911/15-1, the base of the W6 Group is at 2463 m and the top at 922 m, with a resulting thickness of 1541 m. The recorded age of the W6 Group ranges from the late early Paleocene to middle Miocene. Maastrichtian foraminifera and dinocysts recorded in the basal part of the group in the well are believed to reflect Paleocene erosion and resedimentation of latest Cretaceous strata from the shelf. The base of the W6 Group is an nonconformity because the early Paleocene record is missing.

At the time of deposition of the basal W6 Group, a pronounced shelf edge was already established on the Elephant Platform. This edge reached out to, but did not cover, the Phoenix High. The main part of the W6 Group in the 1911/15-1 well is dominated by pelagic mudstones, claystones, and marls. In its lower part however, one distinct sandstone interval is recognized in the well, unconformably overlying the W5 Group. The presence of kaolinite as a major secondary mineral in these sandstones also indicates substantial weathering of micas and feldspars onshore.

On the paleoshelf edge, northwest–southeast trending large reef-like features were established in the lowermost part of the group, superimposing on the mounded features of the uppermost part of the W5 Group (Figure 17). These features remain undrilled to date, but can be regarded as leads for future hydrocarbon exploration in the area.

Subdivision of the W6 Group

The W6 Group is subdivided in six formations (Figure 3); this division is based on the same principles as for the W5 Group. The sequence stratigraphic relationships, however, are not as clear for the W6 Group.

In well 1911/15-1, the W6-1 Formation is defined at 2463–2353 m, giving a thickness of 110 m. Biostratigraphic studies indicate a maximum age of late early Paleocene for the interval. The formation comprises one major coarsening-upward unit underlain by the distal heterolithic part of a basin floor fan.

The basal W6-1 sandstone interval (2397–2463 m) consists of irregularly interbedded sandstones, siltstones, and claystones. The grain size of the quartz sand varies from fine to very coarse. Measured helium porosities from two cores in the sandstones were 26–31%, with permeabilities in the range of 150–590 md. The irregular topography of this sandstone interval makes it readily identifiable on seismic sections. The sandstones are interpreted to represent a channelized basin floor fan complex and can be traced updip to a point source east of the southeastern corner of Quadrant 1911. Deposition was restricted to the eastern slope and the southern Dolphin Basin, not reaching the Outer Highs.

In Quadrant 1911, two laterally distinct reef-like features are identified at the base of the W6-1 Formation. Both are characterized by a gentle talus slope to the east and a very steep slope to the east (Figure 17). Interpretation of seismic lines along strike strongly suggests that the southern feature predates the northern. If these features are reefs or large bioherms, their presence indicates a period of stable depositional conditions and sediment starvation in basinal areas in earliest Paleocene time.

In the W6-2 Formation, the interpreted carbonate features shift updip relative to the W6-1 Formation (Figure 17). These map out as long, continuous barrier reefs and are the youngest of such features observed in this area. The main part of the W6-2 Formation in well 1911/15-1 is interpreted to represent the basinal restricted onlapping wedge associated with the reefal features. The formation is defined from 2353 to 2197 m (156 m thick) and is late Paleocene in age.

In well 1911/15-1, the W6-3 Formation is defined at 2197–2025 m. The 172-m-thick formation is also assigned a late Paleocene age. The lower half of the formation represents a pronounced transgression, presumed to have drowned the shelf edge reef structures. During the following highstand, major sedimentary slumps apparently moved downslope from the Phoenix High and about 30 km along the shelf to the south.

Before the deposition of the W6-4 Formation, the main depocenter shifted northward. In the northern part of Quadrant 1911, the thick highstand and lowstand tracts of this formation prograded across the top of the Phoenix High, thus moving the shelf edge significantly westward. In well 1911/15-1, the 422-m-thick (2025–1603 m) marls, siltstones, and claystones of the W6-4 Formation are Eocene–earliest Oligocene in age. In the upper part of the W6-4 Formation is a large undrilled basin floor fan located centrally in the Dolphin Basin.

In well 1911/15-1, the W6-5 Formation is defined from 1603 to 1173 m. The 430-m-thick claystones of this formation were deposited in early Oligocene–middle Miocene time. The northward shift in deposition along the shelf seems to have continued, and the W6-5 Formation has its main depocenter on the Phoenix High itself. Several channels appear at the base of the formation, running normal to and down the shelf slope.

We define the W6-6 Formation from 1173 to 922 m, giving a thickness of 251 m. This unit is inferred to have been deposited during the middle Miocene, but only the lower part of the interval is represented by cuttings from wells in Quadrant 1911. The deposition is characterized by an increased thickness basinward of the previous shelf edge, bringing the shelf edge westward. The thickest deposits are to the immediate west of the Phoenix High.

W7 GROUP

In well 1911/15-1, the W7 Group is defined from 922 to 522 m, for a total thickness of 400 m. The top of this unit is the present sea floor. The group is interpreted to consist primarily of claystones and is believed to be middle Miocene–Holocene in age.

The transition between the W7 Group and the W6 Group is thought to be related to a dramatic eustatic sea level drop that occurred in middle–late Miocene time. Based on evidence of a paleocoastline on the present-day sea floor, we estimate that sea level then was about 150 m lower than it is today. The reflector w701 ties into the base of the W7 Group in the 1911/15-1 well, above the level where well cuttings were obtained. The reflector can be traced westward to the site of the DSDP 362 well (Figure 5), where it ties to this well at about 1776 m (below mean sea level), corresponding to the middle–late Miocene transition according to Bolli et al. (1978).

Near the base of the W7 Group, the shelf totally covered the Phoenix High. Reflector w701 displays elongated channels normal to and down the slope, as well as very large north–south trending channels on the Outer Highs. The seismic data shows no lithologic contrast between the channel fill and the surrounding strata consisting of claystones. The channels are interpreted to be of a nonerosional character. They may have served as feeder channels running down the shelf slope, accumulating sediments at approximately the same rate as the surroundings. Individual channels are seen to be migrating northward. Features like these seemed to have been common along a large part the west African offshore area in Miocene–Pliocene time. Their sequence stratigraphic significance in offshore Gabon has been discussed in detail by Rasmussen (1994).

Toward the top of the W7 Group, deposition shifted farther basinward. The shelf edge was established about where it is today (Figure 1).

CONCLUSIONS

Exploration activity in offshore Namibia in the northern Walvis Basin has revealed important information about the sedimentary succession that postdates the Neocomian Etendeka plateau basalts found in onshore Nantibia. The proposed stratigraphic subdivision divides the post-breakup succession into seven major lithologically and geometrically confined depositional groups—W1 to W7.

After the onset of the drift phase in late Hauterivian time, the Walvis Basin received subaerial basalt flows from a spreading center to the west. Following subsidence as the spreading center moved farther westward, a marine transgression eventually took place as the embryonic South Atlantic Ocean advanced northward. Shallow marine platform carbonates then prevailed in the area of Quadrant 1911 from the latter Barremian to middle Albian time. In Albian time, a local tectonic event resulted in complex block faulting and the formation of several subbasins and highs. This marked the termination of the carbonate platform and the transition into a clastic-dominated depositional system. In Cenomanian–Turonian time, a series of volcanic centers localized near the Walvis Ridge bathymetric feature emerged. This tectonic episode was followed by the formation of large-scale westward-prograding wedges sourced from the elevated southern African craton. Later sedimentation largely followed the evolution of a passive continental margin, responding to relative sea level changes and paleoclimate. The basal Tertiary unconformity marks a period of very limited deposition, possibly with the formation of reefal structures on the shelf.

The two wildcat wells drilled in Quadrant 1911 found good porous reservoir carbonates of Early Cretaceous age in well-defined structural closures. Compositional analysis of residual hydrocarbons, found in relatively high concentrations in less porous zones, indicate that these are not indigenous and hence point to a paleo–petroleum system. The source rock for these hydrocarbons has not been identified. Most likely it is to be found within the strata that were deposited in local grabens created as the carbonate platform broke up in middle Albian time. The time of breaching (failure of seal) of this reservoir is not known, and play types in younger potential reservoirs updip have not been tested. A rich source rock containing marine type II kerogen was deposited in the late Cenomanian–Santonian succession, contemporaneous with the growth of a large volcanic center in the northern part of Quadrant 1911. Within this part of the Walvis Basin, this source rock has barely reached the top of the oil window maturity where it was penetrated.

Acknowledgments*—We would like to thank our partners in license 001, Statoil and Saga Petroleum, for giving us the permission to publish these data. The statements in this chapter represent the opinions of the authors and not necessarily the conclusions of the whole license group. We would also like to thank our colleagues in Norsk Hydro, Jan Robert Eide and Arne Rasmussen, for constructive cooperation and fruitful discussions. Bob Martin has had a good hand with the figures.*

REFERENCES CITED

Abilio, M. S., 1986, The geology and hydrocarbon potential of Angola, *in* SADC Energy Sector, Technical and Administrative Unit, eds., Oil and gas exploration in the SADC region: Proceedings from SADC Energy Sector Seminar, Arusha, Tanzania.

Bolli, H. M., W. B. F. Ryan, et al., 1978, Initial Reports of the Deep Sea Drilling Project: Washington, D.C., Government Printing Office, v. 40.

Gladczenko, T. P., K. Hinz, O. Eldholm, H. Meyer, S. Neben, and J. Skogseid, 1997, South Atlantic volcanic margins: Journal of Geological Society of London, v. 154, p. 465–470.

Helland-Hansen, W., 1995, Sequence stratigraphy theory: remarks and recommendations, *in* R. J. Steel, V. L. Felt, E. P. Johannessen, and C. Mathieu, eds., Sequence stratigraphy on the north west European margin: Amsterdam, Elsevier, Norwegian Petroleum Society Special Publication 5, p. 13–21.

Light, M. P. R., M. P. Maslanyj, R. J. Greenwood, and N. L. Banks, 1993, Seismic sequence stratigraphy and tectonics offshore Namibia, *in* G. D. Williams and A. Dobb, eds., Tectonics and sequence stratigraphy: Geological Society of London Special Publication 71, p. 163–191.

Milner, S. C., A. R. Duncan, A. Ewart, and J. S. Marsh, 1995, Promotion of the Etendeka Formation to group status: a new integrated stratigraphy: Communs Geological Survey, Namibia, v. 9, p. 5–12.

Rasmussen, E. S., 1994, The relationship between submarine canyon fill and sea-level change: an example from middle Miocene offshore Gabon, west Africa: Sedimentary Geology, v. 90, p. 61–75.

Renne, P. R., M. Ernesto, I. G. Pacca, R. S. Coe, J. M. Glen, M. Prévot, and M. Perrin, 1992, The age of Paraná flood volcanism, rifting of Gondwanaland, and the Jurassic–Cretaceous boundary: Science, v. 258, p. 975–979.

Tissot, B. P., and D. H. Welte, 1984, Petroleum formation and occurrence (2nd ed.): Berlin, Springer-Verlag, 699 p.

Wickens, H. de V., and I. R. McLachland, 1990, The stratigraphy and sedimentology of the reservoir interval of the Kudu 9A-1 and Kudu 9A-3 boreholes, *in* The Kudu offshore drilling project: Communications of the Geological Survey of Namibia, Ministry of Mines and Energy, v. 6.

Index